These safety symbols are used in laboratory and field investigations in this book to indicate possible hazards. ing of each symbol and refer to this page often. *Remember to wash your hands thoroughly afte*

PROTECTIVE EQUIPMENT

Do not begin any lab without the proper protection equip

GOGGLES

Proper eye protection must be worn when performing or observing science activities which involve items or conditions as listed below.

APRON

Wear an approved apron when using substances that could stain, wet, or destroy cloth.

SOAP

Wash hands with soap and water before removing goggles and after all lab activities.

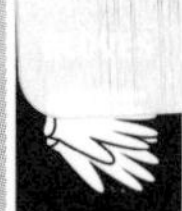

materials, chemicals, animals, or materials that can stain or irritate hands.

LABORATORY HAZARDS

Symbols	Potential Hazards	Precaution	Response
DISPOSAL	contamination of classroom or environment due to improper disposal of materials such as chemicals and live specimens	• DO NOT dispose of hazardous materials in the sink or trash can. • Dispose of wastes as directed by your teacher.	• If hazardous materials are disposed of improperly, notify your teacher immediately.
EXTREME TEMPERATURE	skin burns due to extremely hot or cold materials such as hot glass, liquids, or metals; liquid nitrogen; dry ice	• Use proper protective equipment, such as hot mitts and/or tongs, when handling objects with extreme temperatures.	• If injury occurs, notify your teacher immediately.
SHARP OBJECTS	punctures or cuts from sharp objects such as razor blades, pins, scalpels, and broken glass	• Handle glassware carefully to avoid breakage. • Walk with sharp objects pointed downward, away from you and others.	• If broken glass or injury occurs, notify your teacher immediately.
ELECTRICAL	electric shock or skin burn due to improper grounding, short circuits, liquid spills, or exposed wires	• Check condition of wires and apparatus for fraying or uninsulated wires, and broken or cracked equipment. • Use only GFCI-protected outlets	• DO NOT attempt to fix electrical problems. Notify your teacher immediately.
CHEMICAL	skin irritation or burns, breathing difficulty, and/or poisoning due to touching, swallowing, or inhalation of chemicals such as acids, bases, bleach, metal compounds, iodine, poinsettias, pollen, ammonia, acetone, nail polish remover, heated chemicals, mothballs, and any other chemicals labeled or known to be dangerous	• Wear proper protective equipment such as goggles, apron, and gloves when using chemicals. • Ensure proper room ventilation or use a fume hood when using materials that produce fumes. • NEVER smell fumes directly. • NEVER taste or eat any material in the laboratory.	• If contact occurs, immediately flush affected area with water and notify your teacher. • If a spill occurs, leave the area immediately and notify your teacher.
FLAMMABLE	unexpected fire due to liquids or gases that ignite easily such as rubbing alcohol	• Avoid open flames, sparks, or heat when flammable liquids are present.	• If a fire occurs, leave the area immediately and notify your teacher.
OPEN FLAME	burns or fire due to open flame from matches, Bunsen burners, or burning materials	• Tie back loose hair and clothing. • Keep flame away from all materials. • Follow teacher instructions when lighting and extinguishing flames. • Use proper protection, such as hot mitts or tongs, when handling hot objects.	• If a fire occurs, leave the area immediately and notify your teacher.
ANIMAL SAFETY	injury to or from laboratory animals	• Wear proper protective equipment such as gloves, apron, and goggles when working with animals. • Wash hands after handling animals.	• If injury occurs, notify your teacher immediately.
BIOLOGICAL	infection or adverse reaction due to contact with organisms such as bacteria, fungi, and biological materials such as blood, animal or plant materials	• Wear proper protective equipment such as gloves, goggles, and apron when working with biological materials. • Avoid skin contact with an organism or any part of the organism. • Wash hands after handling organisms.	• If contact occurs, wash the affected area and notify your teacher immediately.
FUME	breathing difficulties from inhalation of fumes from substances such as ammonia, acetone, nail polish remover, heated chemicals, and mothballs	• Wear goggles, apron, and gloves. • Ensure proper room ventilation or use a fume hood when using substances that produce fumes. • NEVER smell fumes directly.	• If a spill occurs, leave area and notify your teacher immediately.
IRRITANT	irritation of skin, mucous membranes, or respiratory tract due to materials such as acids, bases, bleach, pollen, mothballs, steel wool, and potassium permanganate	• Wear goggles, apron, and gloves. • Wear a dust mask to protect against fine particles.	• If skin contact occurs, immediately flush the affected area with water and notify your teacher.
RADIOACTIVE	excessive exposure from alpha, beta, and gamma particles	• Remove gloves and wash hands with soap and water before removing remainder of protective equipment.	• If cracks or holes are found in the container, notify your teacher immediately.

Your online portal to everything you need

connectED.mcgraw-hill.com

Look for these icons to access exciting digital resources

WebQuest

Assessment

Concepts in Motion

Mc Graw Hill Education

Glencoe

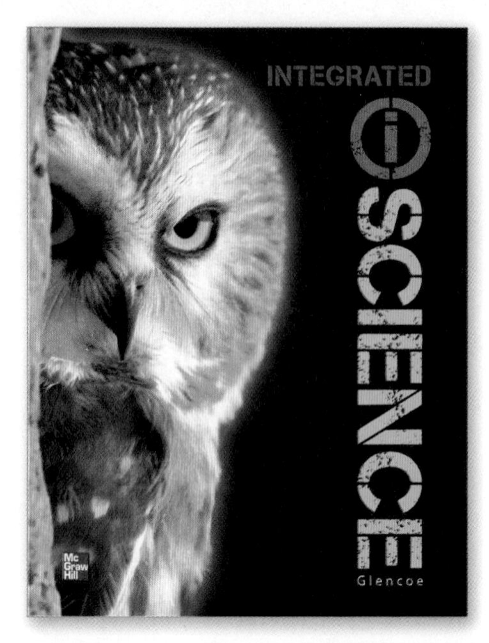

Northern Saw-Whet Owl, *Aegolius acadicus*
This small owl is nocturnal and, therefore, seldom is seen. It is only about 17 cm–22 cm in length and has a wingspan of about 50 cm–56 cm. Its habitat includes short conifers and dense thickets across most of the United States and southern Canada.

The McGraw·Hill Companies

Education

Send all inquiries to:
McGraw-Hill Education
8787 Orion Place
Columbus, OH 43240-4027

ISBN: 978-0-07-888007-0
MHID: 0-07-888007-6

Printed in the United States of America.

5 6 7 8 9 10 QVR 16 15 14 13

Contents in Brief

Authors and Contributors

Authors

American Museum of Natural History
New York, NY

Michelle Anderson, MS
Lecturer
The Ohio State University
Columbus, OH

Juli Berwald, PhD
Science Writer
Austin, TX

John F. Bolzan, PhD
Science Writer
Columbus, OH

Rachel Clark, MS
Science Writer
Moscow, ID

Patricia Craig, MS
Science Writer
Bozeman, MT

Randall Frost, PhD
Science Writer
Pleasanton, CA

Lisa S. Gardiner, PhD
Science Writer
Denver, CO

Jennifer Gonya, PhD
The Ohio State University
Columbus, OH

Mary Ann Grobbel, MD
Science Writer
Grand Rapids, MI

Whitney Crispen Hagins, MA, MAT
Biology Teacher
Lexington High School
Lexington, MA

Carole Holmberg, BS
Planetarium Director
Calusa Nature Center and Planetarium, Inc.
Fort Myers, FL

Tina C. Hopper
Science Writer
Rockwall, TX

Jonathan D. W. Kahl, PhD
Professor of Atmospheric Science
University of Wisconsin-Milwaukee
Milwaukee, WI

Nanette Kalis
Science Writer
Athens, OH

S. Page Keeley, MEd
Maine Mathematics and Science Alliance
Augusta, ME

Cindy Klevickis, PhD
Professor of Integrated Science and Technology
James Madison University
Harrisonburg, VA

Kimberly Fekany Lee, PhD
Science Writer
La Grange, IL

Michael Manga, PhD
Professor
University of California, Berkeley
Berkeley, CA

Devi Ried Mathieu
Science Writer
Sebastopol, CA

Elizabeth A. Nagy-Shadman, PhD
Geology Professor
Pasadena City College
Pasadena, CA

William D. Rogers, DA
Professor of Biology
Ball State University
Muncie, IN

Donna L. Ross, PhD
Associate Professor
San Diego State University
San Diego, CA

Marion B. Sewer, PhD
Assistant Professor
School of Biology
Georgia Institute of Technology
Atlanta, GA

Julia Meyer Sheets, PhD
Lecturer
School of Earth Sciences
The Ohio State University
Columbus, OH

Michael J. Singer, PhD
Professor of Soil Science
Department of Land, Air and Water Resources
University of California
Davis, CA

Karen S. Sottosanti, MA
Science Writer
Pickerington, Ohio

Paul K. Strode, PhD
I.B. Biology Teacher
Fairview High School
Boulder, CO

Jan M. Vermilye, PhD
Research Geologist
Seismo-Tectonic Reservoir Monitoring (STRM)
Boulder, CO

Judith A. Yero, MA
Director
Teacher's Mind Resources
Hamilton, MT

Dinah Zike, MEd
Author, Consultant, Inventor of Foldables
Dinah Zike Academy; Dinah-Might Adventures, LP
San Antonio, TX

Margaret Zorn, MS
Science Writer
Yorktown, VA

Consulting Authors

Alton L. Biggs
Biggs Educational Consulting
Commerce, TX

Ralph M. Feather, Jr., PhD
Assistant Professor
Department of Educational Studies and Secondary Education
Bloomsburg University
Bloomsburg, PA

Douglas Fisher, PhD
Professor of Teacher Education
San Diego State University
San Diego, CA

Edward P. Ortleb
Science/Safety Consultant
St. Louis, MO

Series Consultants

Science

Solomon Bililign, PhD
Professor
Department of Physics
North Carolina Agricultural and Technical State University
Greensboro, NC

John Choinski
Professor
Department of Biology
University of Central Arkansas
Conway, AR

Anastasia Chopelas, PhD
Research Professor
Department of Earth and Space Sciences
UCLA
Los Angeles, CA

David T. Crowther, PhD
Professor of Science Education
University of Nevada, Reno
Reno, NV

A. John Gatz
Professor of Zoology
Ohio Wesleyan University
Delaware, OH

Sarah Gille, PhD
Professor
University of California San Diego
La Jolla, CA

David G. Haase, PhD
Professor of Physics
North Carolina State University
Raleigh, NC

Janet S. Herman, PhD
Professor
Department of Environmental Sciences
University of Virginia
Charlottesville, VA

David T. Ho, PhD
Associate Professor
Department of Oceanography
University of Hawaii
Honolulu, HI

Ruth Howes, PhD
Professor of Physics
Marquette University
Milwaukee, WI

Jose Miguel Hurtado, Jr., PhD
Associate Professor
Department of Geological Sciences
University of Texas at El Paso
El Paso, TX

Monika Kress, PhD
Assistant Professor
San Jose State University
San Jose, CA

Mark E. Lee, PhD
Associate Chair & Assistant Professor
Department of Biology
Spelman College
Atlanta, GA

Linda Lundgren
Science writer
Lakewood, CO

Keith O. Mann, PhD
Ohio Wesleyan University
Delaware, OH

Charles W. McLaughlin, PhD
Adjunct Professor of Chemistry
Montana State University
Bozeman, MT

Katharina Pahnke, PhD
Research Professor
Department of Geology and Geophysics
University of Hawaii
Honolulu, HI

Jesús Pando, PhD
Associate Professor
DePaul University
Chicago, IL

Hay-Oak Park, PhD
Associate Professor
Department of Molecular Genetics
Ohio State University
Columbus, OH

David A. Rubin, PhD
Associate Professor of Physiology
School of Biological Sciences
Illinois State University
Normal, IL

Toni D. Sauncy
Assistant Professor of Physics
Department of Physics
Angelo State University
San Angelo, TX

Series Consultants, continued

Malathi Srivatsan, PhD
Associate Professor of Neurobiology
College of Sciences and Mathematics
Arkansas State University
Jonesboro, AR

Cheryl Wistrom, PhD
Associate Professor of Chemistry
Saint Joseph's College
Rensselaer, IN

Reading

ReLeah Cossett Lent
Author/Educational Consultant
Blue Ridge, GA

Math

Vik Hovsepian
Professor of Mathematics
Rio Hondo College
Whittier, CA

Series Reviewers

Erin Darichuk
West Frederick Middle School
Frederick, MD

Joanne Hedrick Davis
Murphy High School
Murphy, NC

Anthony J. DiSipio, Jr.
Octorara Middle School
Atglen, PA

Adrienne Elder
Tulsa Public Schools
Tulsa, OK

Carolyn Elliott
Iredell-Statesville Schools
Statesville, NC

Christine M. Jacobs
Ranger Middle School
Murphy, NC

Jason O. L. Johnson
Thurmont Middle School
Thurmont, MD

Brian McClain
Amos P. Godby High School
Tallahassee, FL

Von W. Mosser
Thurmont Middle School
Thurmont, MD

Ashlea Peterson
Heritage Intermediate Grade Center
Coweta, OK

Nicole Lenihan Rhoades
Walkersville Middle School
Walkersvillle, MD

Maria A. Rozenberg
Indian Ridge Middle School
Davie, FL

Barb Seymour
Westridge Middle School
Overland Park, KS

Ginger Shirley
Our Lady of Providence Junior-Senior High School
Clarksville, IN

Curtis Smith
Elmwood Middle School
Rogers, AR

Sheila Smith
Jackson Public School
Jackson, MS

Sabra Soileau
Moss Bluff Middle School
Lake Charles, LA

Tony Spoores
Switzerland County Middle School
Vevay, IN

Nancy A. Stearns
Switzerland County Middle School
Vevay, IN

Kari Vogel
Princeton Middle School
Princeton, MN

Alison Welch
Wm. D. Slider Middle School
El Paso, TX

Linda Workman
Parkway Northeast Middle School
Creve Coeur, MO

Teacher Advisory Board

The Teacher Advisory Board gave the authors, editorial staff, and design team feedback on the content and design of the Student Edition. They provided valuable input in the development of *Glencoe Integrated iScience*.

Frances J. Baldridge
Department Chair
Ferguson Middle School
Beavercreek, OH

Jane E. M. Buckingham
Teacher
Crispus Attucks Medical Magnet High School
Indianapolis, IN

Elizabeth Falls
Teacher
Blalack Middle School
Carrollton, TX

Nelson Farrier
Teacher
Hamlin Middle School
Springfield, OR

Michelle R. Foster
Department Chair
Wayland Union Middle School
Wayland, MI

Rebecca Goodell
Teacher
Reedy Creek Middle School
Cary, NC

Mary Gromko
Science Supervisor K–12
Colorado Springs District 11
Colorado Springs, CO

Randy Mousley
Department Chair
Dean Ray Stucky Middle School
Wichita, KS

David Rodriguez
Teacher
Swift Creek Middle School
Tallahassee, FL

Derek Shook
Teacher
Floyd Middle Magnet School
Montgomery, AL

Karen Stratton
Science Coordinator
Lexington School District One
Lexington, SC

Stephanie Wood
Science Curriculum Specialist, K–12
Granite School District
Salt Lake City, UT

Online Guide

connectED.mcgraw-hill.com

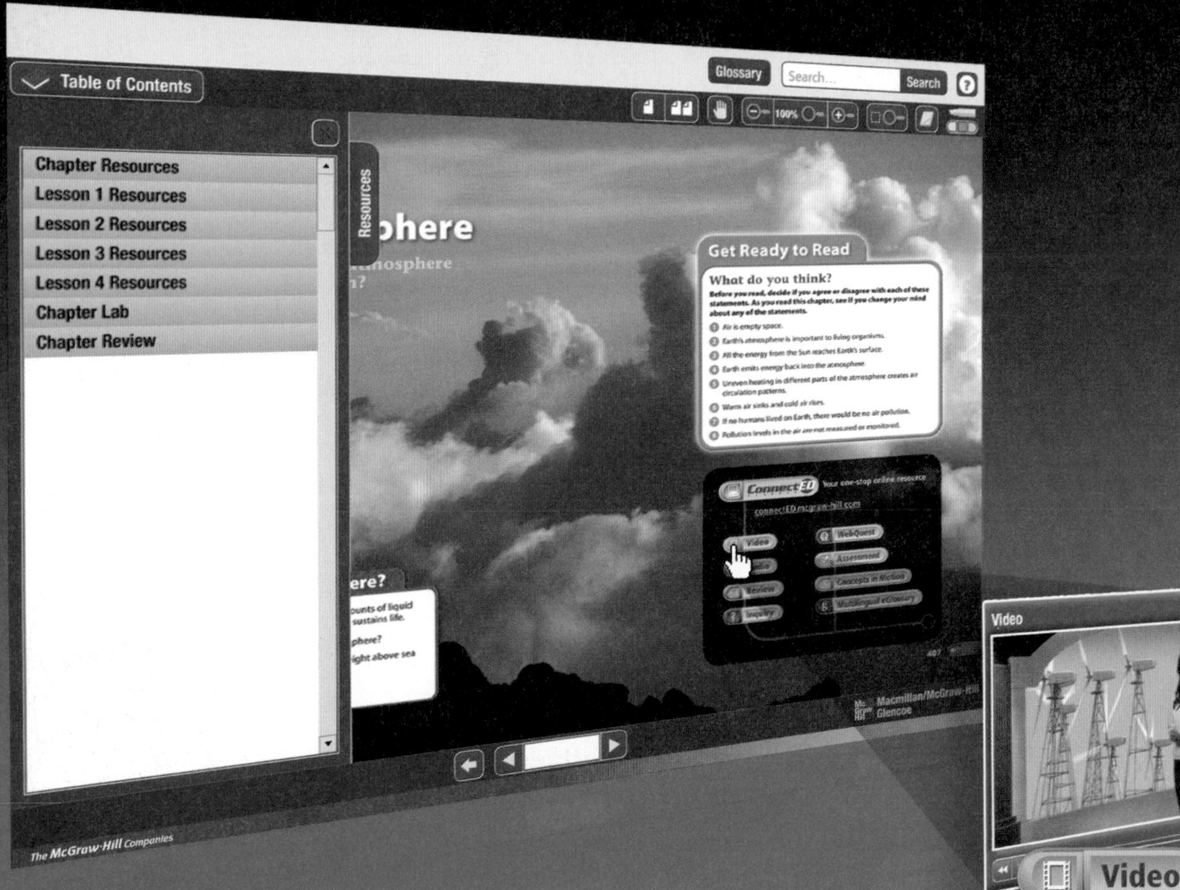

Your Digital Science Portal

See the science in real life through these exciting videos.

Click the link and you can listen to the text while you follow along.

Try these interactive tools to help you review the lesson concepts.

Explore concepts through hands–on and virtual labs.

These web-based challenges relate the concepts you're learning about to the latest news and research.

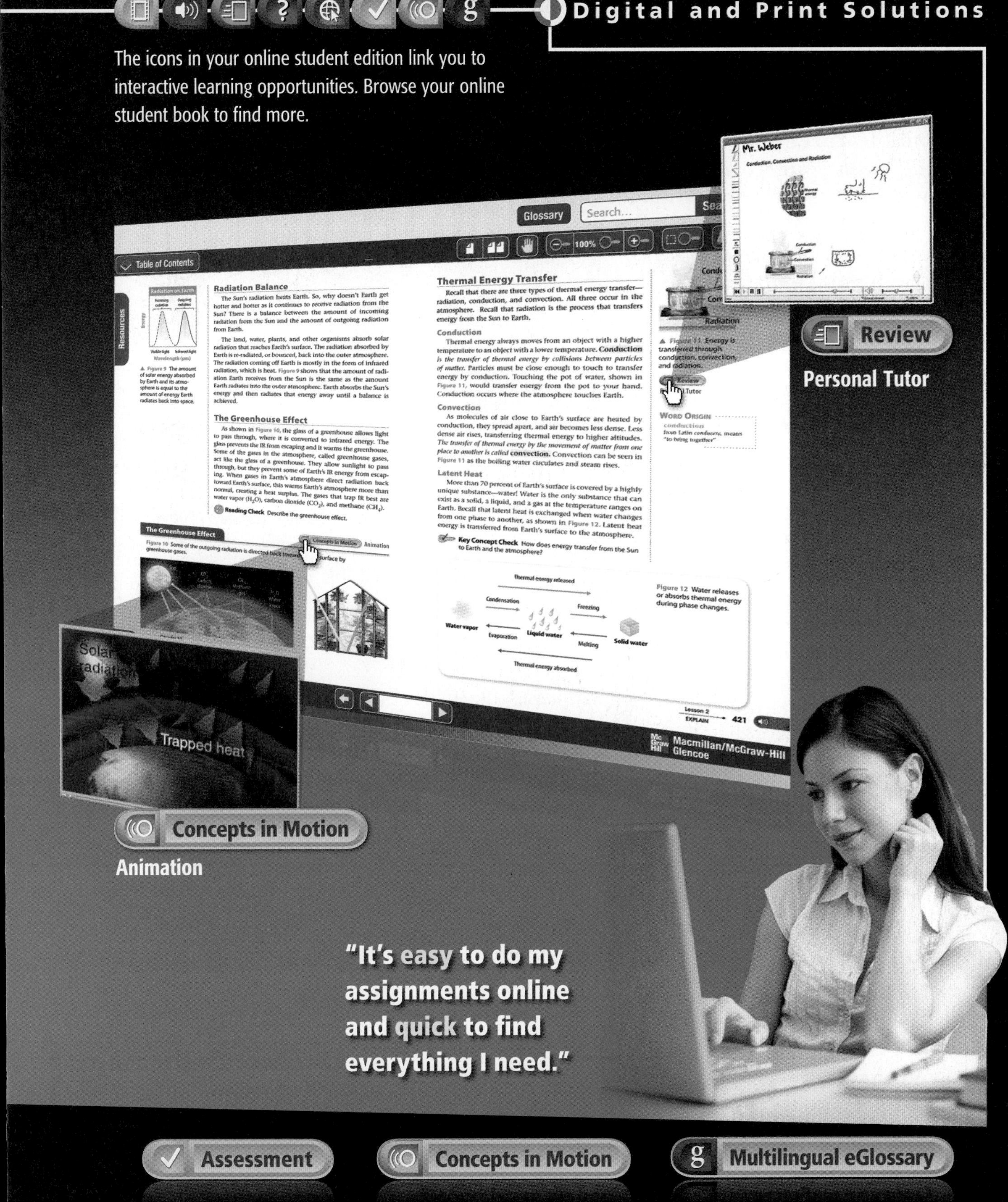

Digital and Print Solutions
The icons in your online student edition link you to interactive learning opportunities. Browse your online student book to find more.
Mr. Weber
Conduction, Convection and Radiation
Glossary
Search...
Table of Contents
Resources
Radiation Balance
The Greenhouse Effect
Thermal Energy Transfer
Conduction
Convection
Latent Heat
Word Origin
Reading Check Describe the greenhouse effect.
Key Concept Check How does energy transfer from the Sun to Earth and the atmosphere?
Figure 12 Water releases or absorbs thermal energy during phase changes.
Thermal energy released
Condensation
Freezing
Water vapor
Evaporation
Liquid water
Melting
Solid water
Thermal energy absorbed
Lesson 2
EXPLAIN
421
Macmillan/McGraw-Hill
Glencoe
Solar radiation
Trapped heat
Review
Personal Tutor
Concepts in Motion
Animation
"It's easy to do my assignments online and quick to find everything I need."
Assessment
Concepts in Motion
Multilingual eGlossary

Treasure Hunt

Your science book has many features that will aid you in your learning. Some of these features are listed below. You can use the activity at the right to help you find these and other special features in the book.

- THE BIG IDEA can be found at the start of each chapter.
- The Reading Guide at the start of each lesson lists **Key Concepts,** vocabulary terms, and online supplements to the content.
- ConnectED icons direct you to online resources such as animations, personal tutors, math practices, and quizzes.
- Inquiry Labs and Skill Practices are in each chapter.
- Your FOLDABLES help organize your notes.

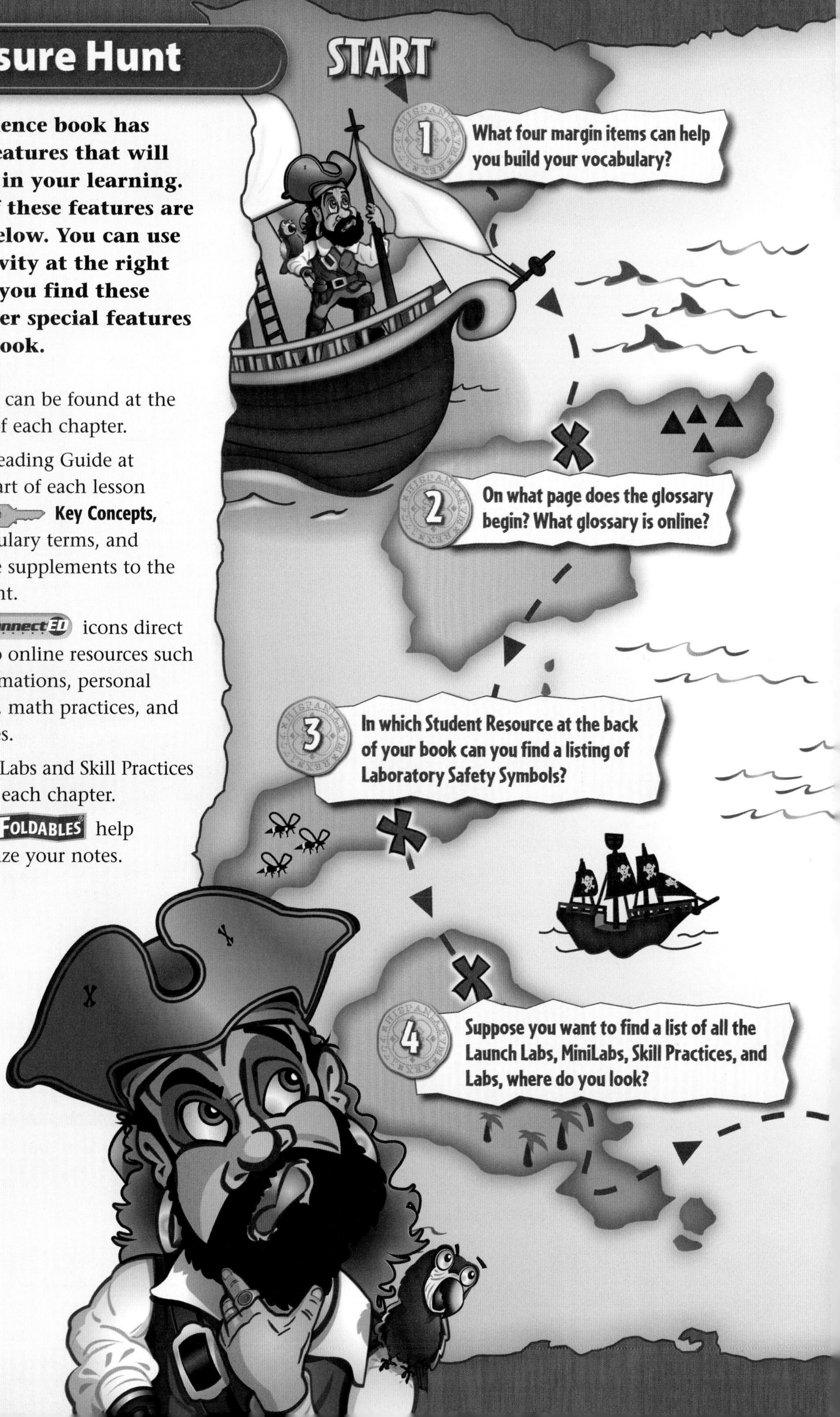

START

1. What four margin items can help you build your vocabulary?
2. On what page does the glossary begin? What glossary is online?
3. In which Student Resource at the back of your book can you find a listing of Laboratory Safety Symbols?
4. Suppose you want to find a list of all the Launch Labs, MiniLabs, Skill Practices, and Labs, where do you look?

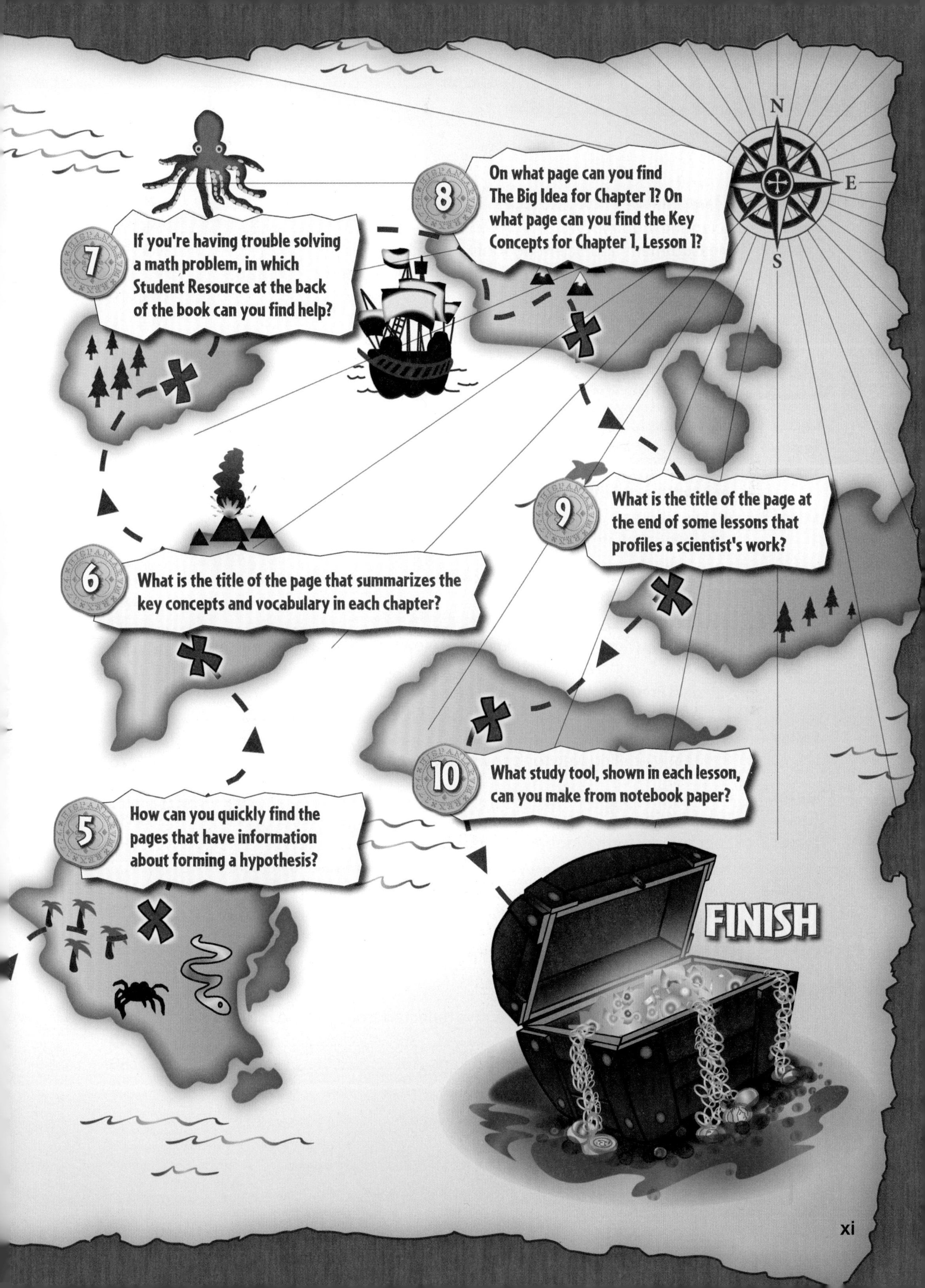

N
E
S
8
On what page can you find The Big Idea for Chapter 1? On what page can you find the Key Concepts for Chapter 1, Lesson 1?
7
If you're having trouble solving a math problem, in which Student Resource at the back of the book can you find help?
9
What is the title of the page at the end of some lessons that profiles a scientist's work?
6
What is the title of the page that summarizes the key concepts and vocabulary in each chapter?
10
What study tool, shown in each lesson, can you make from notebook paper?
5
How can you quickly find the pages that have information about forming a hypothesis?
FINISH

Table of Contents

TABLE OF CONTENTS

Student Resources

Inquiry Launch Labs

Inquiry MiniLabs

Inquiry Skill Practice

Inquiry Labs

Features

Green Science

How It Works

How Nature Works

Science & Society

Careers in Science

Nature of Science

Scientific Problem Solving

THE BIG IDEA **What is scientific inquiry?**

Sci-Fi Movie Scene?

This might look like a weird spaceship docking in a science-fiction movie. However, it is actually the back of an airplane engine being tested in a huge wind tunnel. An experiment is an important part of scientific investigations.

- Why is an experiment important?
- Does experimentation occur in all branches of science?
- What is scientific inquiry?

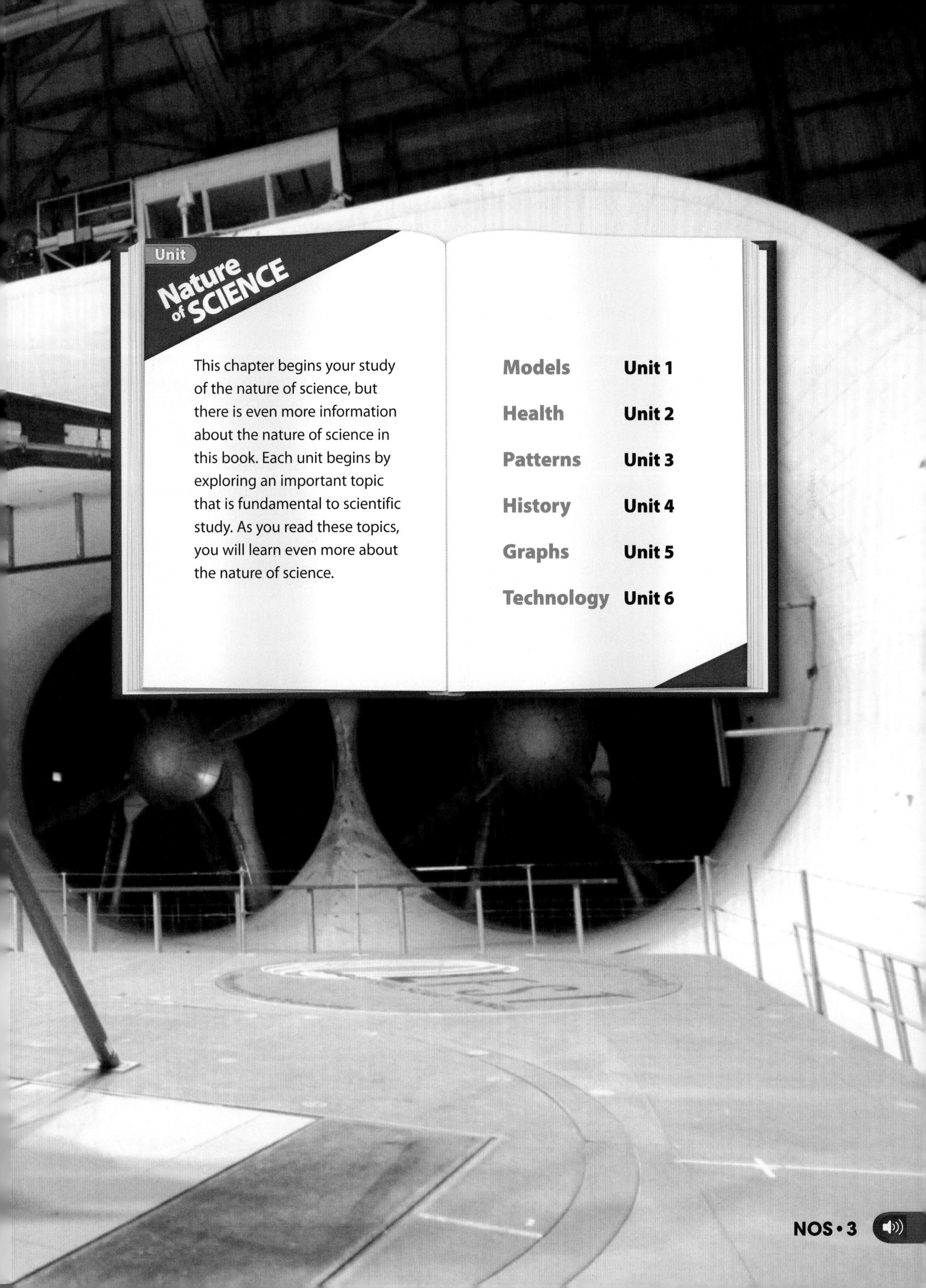
Unit
Nature of SCIENCE
This chapter begins your study of the nature of science, but there is even more information about the nature of science in this book. Each unit begins by exploring an important topic that is fundamental to scientific study. As you read these topics, you will learn even more about the nature of science.
Models Unit 1
Health Unit 2
Patterns Unit 3
History Unit 4
Graphs Unit 5
Technology Unit 6

Lesson 1 Scientific Inquiry

Reading Guide

Key Concepts
ESSENTIAL QUESTIONS

- What are some steps used during scientific inquiry?
- What are the results of scientific inquiry?
- What is critical thinking?

Vocabulary

science p. NOS 4
observation p. NOS 6
inference p. NOS 6
hypothesis p. NOS 6
prediction p. NOS 6
scientific theory p. NOS 8
scientific law p. NOS 8
technology p. NOS 9
critical thinking p. NOS 10

g Multilingual eGlossary

Video

- BrainPOP®
- Science Video

Understanding Science

A clear night sky is one of the most beautiful sights on Earth. The stars seem to shine like a handful of diamonds scattered on black velvet. Why do stars seem to shine more brightly some nights than others?

Did you know that when you ask questions, such as the one above, you are practicing science? **Science** *is the investigation and exploration of natural events and of the new information that results from those investigations.* Like a scientist, you can help shape the future by accumulating knowledge, developing new technologies, and sharing ideas with others.

Throughout history, people of many different backgrounds, interests, and talents have made scientific contributions. Sometimes they overcame a limited educational background and excelled in science. One example is Marie Curie, shown in **Figure 1.** She was a scientist who won two Nobel prizes in the early 1900s for her work with radioactivity. As a young student, Marie was not allowed to study at the University of Warsaw in Poland because she was a woman. Despite this obstacle, she made significant contributions to science.

Figure 1 Modern medical procedures such as X-rays, radioactive cancer treatments, and nuclear-power generation are some of the technologies made possible because of the pioneering work of Marie Curie and her associates.

Branches of Science

Scientific study is organized into several branches, or parts. The three branches that you will study in middle school are physical science, Earth science, and life science. Each branch focuses on a different part of the natural world.

WORD ORIGIN

science
from Latin *scientia,* means "knowledge" or "to know"

Physical Science

Physical science, or physics and chemistry, is the study of matter and energy. The physicist shown here is adjusting an instrument that measures radiation from the Big Bang. Physical scientists ask questions such as

- What happens to energy during chemical reactions?
- How does gravity affect roller coasters?
- What makes up protons, neutrons, and electrons?

Earth Science

Earth scientists study the many processes that occur on Earth and deep within Earth. This scientist is collecting a water sample in southern Mexico.

Earth scientists ask questions such as

- What are the properties of minerals?
- How is energy transferred on Earth?
- How do volcanoes form?

Life Science

Life scientists study all organisms and the many processes that occur in them. The life scientist shown is studying the avian flu virus.

Life scientists ask questions such as

- How do plant cells and animal cells differ?
- How do animals survive in the desert?
- How do organisms in a community interact?

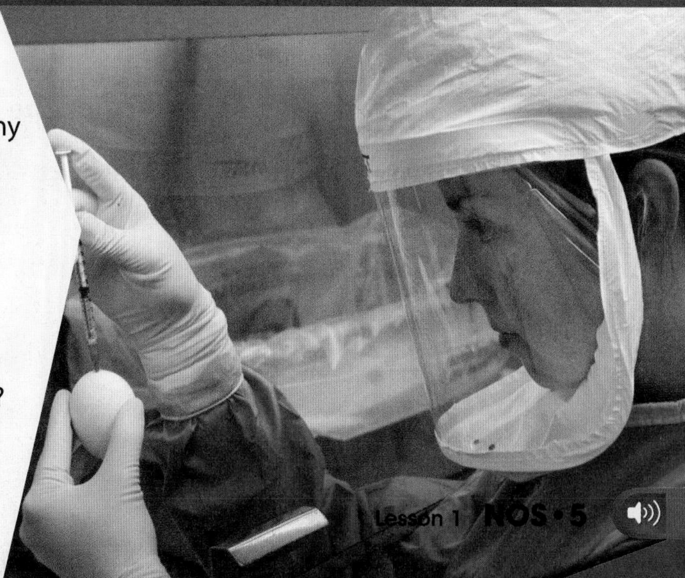

What is Scientific Inquiry?

When scientists conduct investigations, they often want to answer questions about the natural world. To do this, they use scientific inquiry—a process that uses a variety of skills and tools to answer questions or to test ideas. You might have heard these steps called "the scientific method." However, there is no one scientific method. In fact, the skills that scientists use to conduct an investigation can be used in any order. One possible sequence is shown in **Figure 2.** Like a scientist, you perform scientific investigations every day, and you will do investigations throughout this course.

 **Reading Check** What is scientific inquiry?

Ask Questions

Imagine warming yourself near a fireplace or a campfire? As you throw twigs and logs onto the fire, you see that fire releases smoke and light. You also feel the warmth of the thermal energy being released. These are **observations**—*the results of using one or more of your senses to gather information and taking note of what occurs.* Observations often lead to questions. You ask yourself, "When logs burn, what happens to the wood? Do the logs disappear? Do they change in some way?"

When observing the fire, you might recall from an earlier science course that matter can change form, but it cannot be created or destroyed. Therefore, you infer that the logs do not just disappear. They must undergo some type of change. An **inference** *is a logical explanation of an observation that is drawn from prior knowledge or experience.*

Hypothesize and Predict

After making observations and inferences, you decide to investigate further. Like a scientist, you might develop a **hypothesis**—*a possible explanation for an observation that can be tested by scientific investigations.* Your hypothesis about what happens to the logs might be: When logs burn, new substances form because matter cannot be destroyed.

When scientists state a hypothesis, they often use it to make predictions to help test their hypothesis. *A* **prediction** *is a statement of what will happen next in a sequence of events.* Scientists make predictions based on what information they think they will find when testing their hypothesis. For instance, based on your hypothesis, you might predict that if logs burn, then the substances that make up the logs change into other substances.

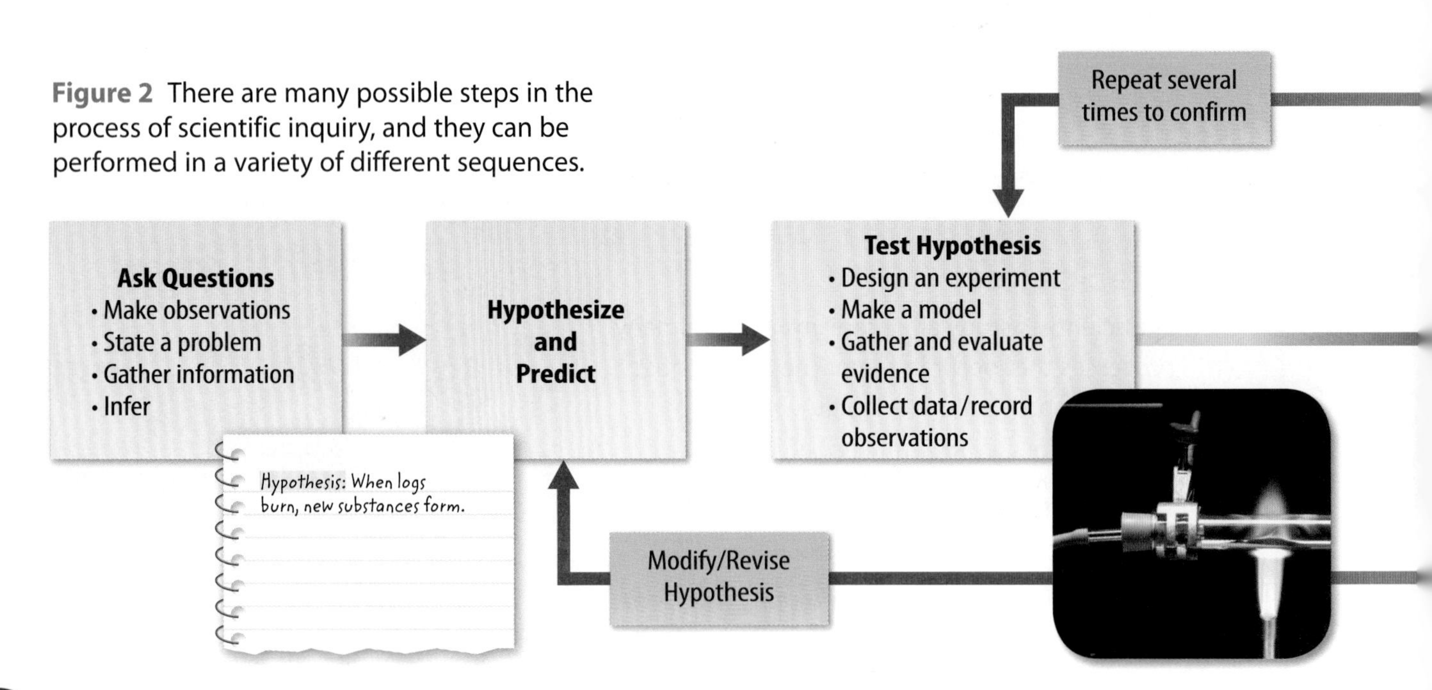

Figure 2 There are many possible steps in the process of scientific inquiry, and they can be performed in a variety of different sequences.

Test Hypothesis and Analyze Results

How could you test your hypothesis? When you test a hypothesis, you often test your predictions. If a prediction is confirmed, then it supports your hypothesis. If your prediction is not confirmed, you might modify your hypothesis and retest it.

To test your predictions and hypothesis, you could design an experiment to find out what substances make up wood. Then you could determine what makes up the ash, the smoke, and other products that formed after the burning process. You also could research this topic and possibly find answers on reliable science Web sites or in science books.

After doing an experiment or research, you need to analyze your results and findings. You might make additional inferences after reviewing your data. If you find that new substances actually do form when wood burns, your hypothesis is supported. If new products do not form, your hypothesis is not supported. Some methods of testing a hypothesis and analyzing results are shown in **Figure 2.**

Draw Conclusions

After analyzing your results, you can begin to draw conclusions about your investigation. A conclusion is a summary of the information gained from testing a hypothesis. Like a scientist does, you should test and retest your hypothesis several times to make sure the results are consistent.

Communicate Results

Sharing the results of a scientific inquiry is an important part of science. By exchanging information, scientists can evaluate and test others' work and make faster progress in their own research. Exchanging information is one way of making scientific advances as quickly as possible and keeping scientific information accurate. During your investigation, if you do research on the Internet or in science books, you use information that someone else communicated. Scientists exchange information in many ways, as shown below in **Figure 2.**

Key Concept Check What are some steps used during scientific inquiry?

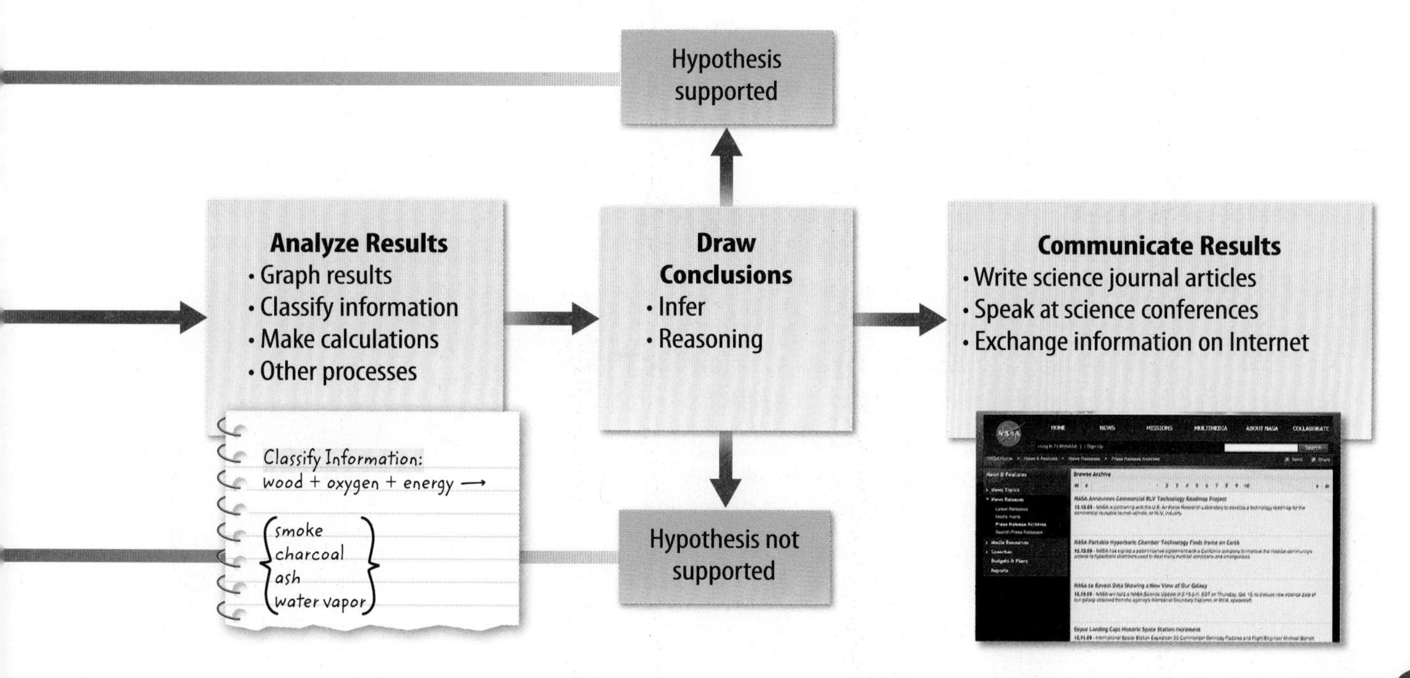

Unsupported or Supported Hypotheses

What happens if a hypothesis is not supported by an investigation? Was the scientific investigation a failure and a waste of time? Absolutely not! Even when a hypothesis is not supported, you gain valuable information. You can revise your hypothesis and test it again. Each time you test a hypothesis, you learn more about the topic you are studying.

Scientific Theory

When a hypothesis (or a group of closely related hypotheses) is supported through many tests over many years, a scientific theory can develop. A **scientific theory** *is an explanation of observations or events that is based on knowledge gained from many observations and investigations.*

A scientific theory does not develop from just one hypothesis, but from many hypotheses that are connected by a common idea. The kinetic molecular theory described below is the result of the investigations of many scientists.

Scientific Law

A scientific law is different from a societal law, which is an agreement on a set of behaviors. A **scientific law** *is a rule that describes a repeatable pattern in nature.* A scientific law does not explain why or how the pattern happens, it only states that it will happen. For example, if you drop a ball, it will fall towards the ground every time. This is a repeated pattern that relates to the law of universal gravitation. The law of conservation of energy, described below, is also a scientific law.

Kinetic Molecular Theory

The kinetic molecular theory explains how particles that make up a gas move in constant, random motions. A particle moves in a straight line until it collides with another particle or with the wall of its container.

The kinetic molecular theory also assumes that the collisions of particles in a gas are elastic collisions. An elastic collision is a collision in which no kinetic energy is lost. Therefore, kinetic energy among gas particles is conserved.

Law of Conservation of Energy

The law of conservation of energy states that in any chemical reaction or physical change, energy is neither created nor destroyed. The total energy of particles before and after collisions is the same.

However, this scientific law, like all scientific laws, does not explain *why* energy is conserved. It simply states that energy *is* conserved.

Scientific Law v. Scientific Theory

Both are based on repeated observations and can be rejected or modified.

A scientific law states that an event *will* occur. For example, energy will be conserved when particles collide. It does not explain why an event will occur or how it will occur. Scientific laws work under specific conditions in nature. A law stands true until an observation is made that does not follow the law.

A scientific theory is an explanation of *why* or *how* an event occurred. For example, collisions of particles of a gas are elastic collisions. Therefore, no kinetic energy is lost. A theory can be rejected or modified if someone observes an event that disproves the theory. A theory will never become a law.

Results of Scientific Inquiry

Why do you and others ask questions and investigate the natural world? Most often, the purpose of a scientific investigation is to develop new materials and technology, discover new objects, or find answers to questions, as shown below.

Key Concept Check What are the results of scientific inquiry?

New Materials and Technology

Every year, corporations and governments spend millions of dollars on research and design of new materials and technologies. **Technology** *is the practical use of scientific knowledge, especially for industrial or commercial use.* For example, scientists hypothesize and predict how new materials will make bicycles and cycling gear lighter, more durable, and more aerodynamic. Using wind tunnels, scientists test these new materials to see whether they improve the cyclist's performance. If the cyclist's performance improves, their hypotheses are supported. If the performance does not improve or it doesn't improve enough, scientists will revise their hypotheses and conduct more tests.

New Objects or Events

Scientific investigations also lead to newly discovered objects or events. For example, NASA's *Hubble Space Telescope* captured this image of two colliding galaxies. They have been nicknamed the mice, because of their long tails. The tails are composed of gases and young, hot blue stars. If computer models are correct, these galaxies will combine in the future and form one large galaxy.

Answers to Questions

Often scientific investigations are launched to answer *who, what, when, where,* or *how* questions. For example, research chemists investigate new substances, such as substances found in mushrooms and bacteria, as shown on the right. New drug treatments for cancer, HIV, and other diseases might be found using new substances. Other scientists look for clues about what causes diseases, whether they can be passed from person to person, and when the disease first appeared.

FOLDABLES®

Create a two-tab book and label it as shown. Use it to discuss the importance of evaluating scientific information.

Why is it important to...

...be scientifically literate? | ...use critical thinking?

Evaluating Scientific Information

Do you ever you read advertisements, articles, or books that claim to contain scientifically proven information? Are you able to determine if the information is actually true and scientific instead of pseudoscientific (information incorrectly represented as scientific)? Whether you are reading printed media or watching commercials on TV, it is important that you are skeptical, identify facts and opinions, and think critically about the information. **Critical thinking** *is comparing what you already know with the information you are given in order to decide whether you agree with it.*

 **Key Concept Check** What is critical thinking?

Be A Rock Star!

Do you dream of being a rock star?

Sing, dance, and play guitar like a rock star with the new Rocker-rific Spotlight. A new scientific process developed by Rising Star Laboratories allows you to overcome your lack of musical talent and enables you to perform like a real rock star.

This amazing new light actually changes your voice quality and enhances your brain chemistry so that you can sing, dance, and play a guitar like a professional rock star. Now, there is no need to practice or pay for expensive lessons. The Rocker-rific Spotlight does the work for you.

Dr. Sammy Truelove says, "Before lack of talent might have stopped someone from achieving his or her dreams of being a rock star. This scientific breakthrough transforms people with absolutely no talent into amazing rock stars in just minutes. Of the many patients that I have tested with this product, no one has failed to achieve his or her dreams."

Disclaimer: This product was tested on laboratory rats and might not work for everyone.

Skepticism
Have you heard the saying, if it sounds too good to be true, it probably is? To be skeptical is to doubt the truthfulness of something. A scientifically literate person can read information and know that it misrepresents the facts. Science often is self-correcting because someone usually challenges inaccurate information and tests scientific results for accuracy.

Identifying Facts and Misleading Information
Misleading information often is worded to sound like scientific facts. A scientifically literate person can recognize fake claims and quickly determine when information is false.

Critical Thinking
Use critical thinking skills to compare what you know with the new information given to you. If the information does not sound reliable, either research and find more information about the topic or dismiss the information as unreliable.

Identify Opinions
An opinion is a personal view, feeling, or claim about a topic. Opinions cannot be proven true or false. And, an opinion might contain inaccurate information.

Science cannot answer all questions.

It might seem that scientific inquiry is the best way to answer all questions. But there are some questions that science cannot answer. Questions that deal with beliefs, values, personal opinions, and feelings cannot be answered scientifically. This is because it is impossible to collect scientific data on these topics.

Science cannot answer questions such as

- Which video game is the most fun to play?
- Are people kind to others most of the time?
- Is there such a thing as good luck?

Safety in Science

Scientists know that using safe procedures is important in any scientific investigation. When you begin a scientific investigation, you should always wear protective equipment, as shown in **Figure 3.** You also should learn the meaning of safety symbols, listen to your teacher's instructions, and learn to recognize potential hazards. For more information, consult the Science Skills Handbook at the back of this book.

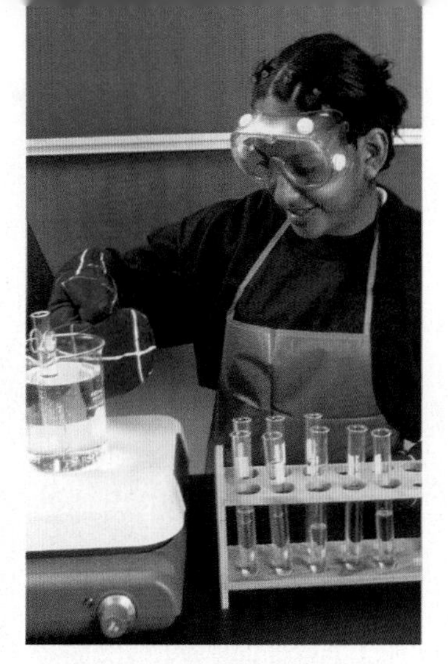

Figure 3 Always follow safety procedures when doing scientific investigations. If you have questions, ask your teacher.

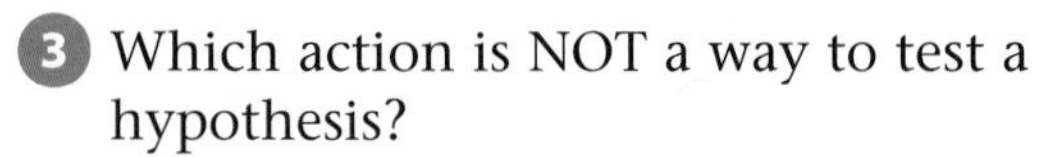

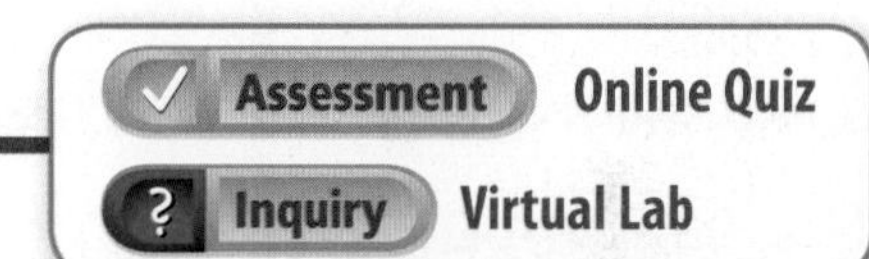

Use Vocabulary

1. **Define** *technology* in your own words.
2. **Use the term** *observation* in a sentence.

Understand Key Concepts

3. Which action is NOT a way to test a hypothesis?
   - **A.** analyze results
   - **B.** design an experiment
   - **C.** make a model
   - **D.** gather and evaluate evidence
4. **Describe** three examples of the results of scientific inquiry.
5. **Give an example** of a time when you practiced critical thinking.

Interpret Graphics

6. **Compare** Copy and fill in the graphic organizer below. List some examples of how to communicate the results of scientific inquiry.

Critical Thinking

7. **Summarize** Your classmate writes the following as a hypothesis:

   Red is a beautiful color.

   Write a brief explanation to your classmate explaining why this is not a hypothesis.

Lesson 2

Reading Guide

Key Concepts
ESSENTIAL QUESTIONS

- Why did scientists create the International System of Units (SI)?
- Why is scientific notation a useful tool for scientists?
- How can tools, such as graduated cylinders and triple-beam balances, assist physical scientists?

Vocabulary
description p. NOS 12
explanation p. NOS 12
International System of Units (SI) p. NOS 12
scientific notation p. NOS 15
percent error p. NOS 15

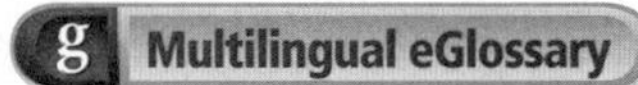

Measurement and Scientific Tools

Description and Explanation

Suppose you work for a company that tests cars to see how they perform during accidents, as shown in **Figure 4.** You might use various scientific tools to measure the acceleration of cars as they crash into other objects.

A **description** *is a spoken or written summary of observations.* The measurements you record are descriptions of the results of the crash tests. Later, your supervisor asks you to write a report that interprets the measurements you took during the crash tests. An **explanation** *is an interpretation of observations.* As you write your explanation, you make inferences about why the crashes damaged the vehicles in specific ways.

Notice that there is a difference between a description and an explanation. When you describe something, you report your observations. When you explain something, you interpret your observations.

The International System of Units

Different parts of the world use different systems of measurements. This can cause confusion when people who use different systems communicate their measurements. This confusion was eliminated in 1960 when a new system of measurement was adopted. *The internationally accepted system of measurement is the* **International System of Units (SI).**

Key Concept Check Why did scientists create the International System of Units?

Figure 4 A description of an event details what you observed. An explanation explains why or how the event occurred.

SI Base Units

When you take measurements during scientific investigations and labs in this course, you will use the SI system. The SI system uses standards of measurement, called base units, as shown in **Table 1.** Other units used in the SI system that are not base units are derived from the base units. For example, the liter, used to measure volume, was derived from the base unit for length.

SI Unit Prefixes

In older systems of measurement, there usually was no common factor that related one unit to another. The SI system eliminated this problem.

The SI system is based on multiples of ten. Any SI unit can be converted to another by multiplying by a power of ten. Factors of ten are represented by prefixes, as shown in **Table 2.** For example, the prefix *milli-* means 0.001 or 10^{-3}. So, a milliliter is 0.001 L, or 1/1,000 L. Another way to say this is: 1 L is 1,000 times greater than 1 mL.

Converting Among SI Units

It is easy to convert from one SI unit to another. You either multiply or divide by a factor of ten. You also can use proportion calculations to make conversions. An example of how to convert between SI units is shown in **Figure 5.**

Table 1 SI Base Units

Quantity Measured	Unit (symbol)
Length	meter (m)
Mass	kilogram (kg)
Time	second (s)
Electric current	ampere (A)
Temperature	kelvin (K)
Substance amount	mole (mol)
Light intensity	candela (cd)

Table 2 Prefixes

Prefix	Meaning
Mega- (M)	1,000,000 or (10^6)
Kilo- (k)	1,000 or (10^3)
Hecto- (h)	100 or (10^2)
Deka- (da)	10 or (10^1)
Deci- (d)	0.1 or $\left(\frac{1}{10}\right)$ or (10^{-1})
Centi- (c)	0.01 or $\left(\frac{1}{100}\right)$ or (10^{-2})
Milli- (m)	0.001 or $\left(\frac{1}{1,000}\right)$ or (10^{-3})
Micro- (μ)	0.000001 or $\left(\frac{1}{1,000,000}\right)$ or (10^{-6})

Figure 5 The rock in the photograph has a mass of 17.5 grams. Convert that measurement to kilograms. ▼

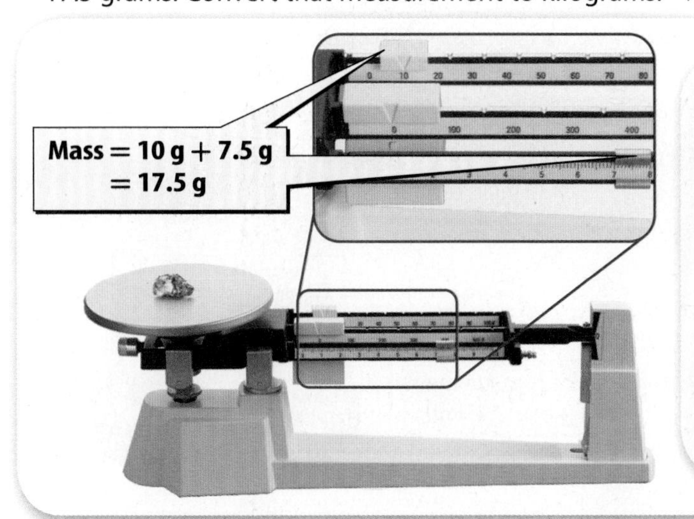

1. Determine the correct relationship between grams and kilograms. There are 1,000 g in 1 kg.

$$\frac{1\text{ kg}}{1{,}000\text{ g}}$$

$$\frac{x}{17.5\text{ g}} = \frac{1\text{ kg}}{1{,}000\text{ g}}$$

$$x = \frac{(17.5\text{ g})(1\text{ kg})}{1{,}000\text{ g}};\ x = 0.0175\text{ kg}$$

2. Check your units. The unit *grams* is cancelled out in the equation, so the answer is 0.0175 kg.

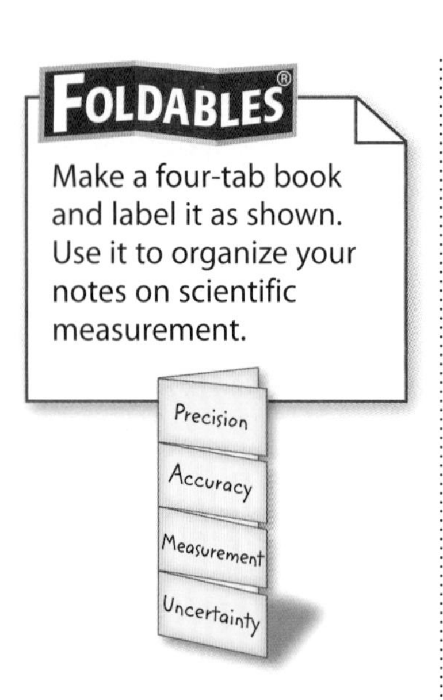

Table 3 Student Density and Error Data
(Accepted value: Density of sodium chloride, 21.7 g/cm³)

	Student A	Student B	Student C
	Density	Density	Density
Trial 1	23.4 g/cm³	18.9 g/cm³	21.9 g/cm³
Trial 2	23.5 g/cm³	27.2 g/cm³	21.4 g/cm³
Trial 3	23.4 g/cm³	29.1 g/cm³	21.3 g/cm³
Mean	23.4 g/cm³	25.1 g/cm³	21.5 g/cm³

Measurement and Uncertainty

You might be familiar with the terms *precision* and *accuracy*. In science, these terms have different meanings. Precision is a description of how similar or close repeated measurements are to each other. Accuracy is a description of how close a measurement is to an accepted value.

The difference between precision and accuracy is illustrated in **Table 3.** Students were asked to find the density of sodium chloride (NaCl). In three trials, each student measured the volume and the mass of NaCl. Then, they calculated the density for each trial and calculated the mean, or average. Student A's measurements are the most precise because they are closest to each other. Student C's measurements are the most accurate because they are closest to the scientifically accepted value. Student B's measurements are neither precise nor accurate. They are not close to each other or to the accepted value.

WORD ORIGIN
notation
from Latin *notationem,* means "a marking or explanation"

Tools and Accuracy

No measuring tool provides a perfect measurement. All measurements have some degree of uncertainty. Some tools or instruments produce more accurate measurements, as shown in **Figure 6.**

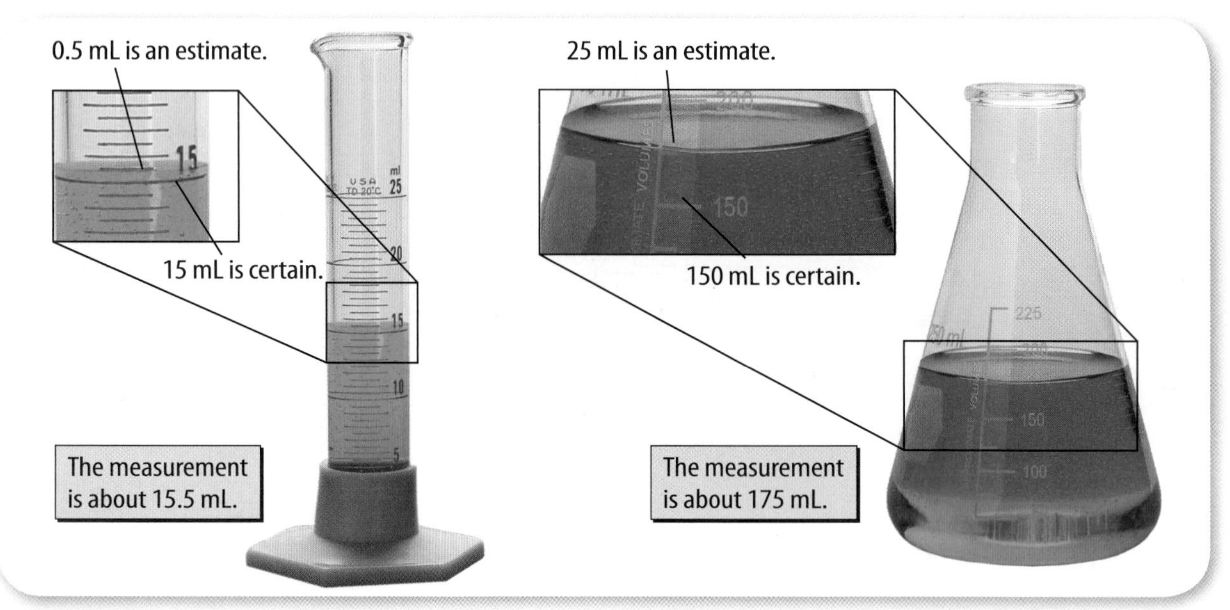

Figure 6 The graduated cylinder is graduated in 0.5-mL increments. The beaker is graduated in, or divided into, 25-mL increments. Liquid measurements taken with the graduated cylinder have greater accuracy.

Scientific Notation

Suppose you are writing a report that includes Earth's distance from the Sun—149,600,000 km—and the density of the Sun's lower atmosphere—0.000000028 g/cm^3. These numerals take up too much space and might be difficult to read, so you use **scientific notation**—*a method of writing or displaying very small or very large values in a short form.* To write the numerals in scientific notation, use the steps shown to the right.

Key Concept Check Why is scientific notation a useful tool for scientists?

Percent Error

The densities recorded in **Table 3** are experimental values because they were calculated during an experiment. Each of these values has some error because the accepted value for table salt density is 21.65 g/cm^3. Percent error can help you determine the size of your experimental error. **Percent error** *is the expression of error as a percentage of the accepted value.*

How to Write in Scientific Notation

1. Write the original number.
   - **A.** **149,600,000**
   - **B.** **0.000000028**
2. Move the decimal point to the right or the left to make the number between 1 and 10. Count the number of decimal places moved and note the direction.
   - **A.** **1.49600000** = 8 places to the left
   - **B.** **00000002.8** = 8 places to the right
3. Rewrite the number deleting all extra zeros to the right or to the left of the decimal point.
   - **A.** **1.496**
   - **B.** **2.8**
4. Write a multiplication symbol and the number *10* with an exponent. The exponent should equal the number of places that you moved the decimal point in step 2. If you moved the decimal point to the left, the exponent is positive. If you moved the decimal point to the right, the exponent is negative.
   - **A.** **1.496×10^8**
   - **B.** **2.8×10^{-8}**

Math Skills — Percent Error

Solve for Percent Error A student in the laboratory measures the boiling point of water at 97.5°C. If the accepted value for the boiling point of water is 100.0°C, what is the percent error?

1. This is what you know: **experimental value = 97.5°C**
   accepted value = 100.0°C
2. This is what you need to find: percent error
3. Use this formula:

$$\text{percent error} = \frac{|\text{experimental value} - \text{accepted value}|}{\text{accepted value}} \times 100\%$$

4. Substitute the known values into the equation and perform the calculations

$$\text{percent error} = \frac{|97.5° - 100.0°|}{100.0°} \times 100\% = 2.50\%$$

Practice

Calculate the percent error if the experimental value of the density of gold is 18.7 g/cm^3 and the accepted value is 19.3 g/cm^3.

Review
- Math Practice
- Personal Tutor

Scientific Tools

As you conduct scientific investigations, you will use tools to make measurements. The tools listed here are some of the tools commonly used in science. For more information about the correct use and safety procedures for these tools, see the Science Skills Handbook at the back of this book.

◀ Science Journal

Use a science journal to record observations, write questions and hypotheses, collect data, and analyze the results of scientific inquiry. All scientists record the information they learn while conducting investigations. Your journal can be a spiral-bound notebook, a loose-leaf binder, or even just a pad of paper.

Balances ▶

A balance is used to measure the mass of an object. Units often used for mass are kilograms (kg), grams (g), and milligrams (mg). Two common types of balances are the electronic balance and the triple-beam balance. In order to get the most accurate measurements when using a balance, it is important to calibrate the balance often.

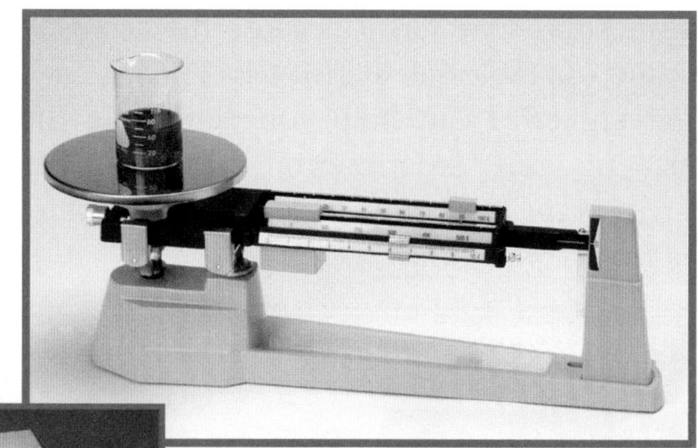

◀ Glassware

Laboratory glassware is used to hold or measure the volume of liquids. Flasks, beakers, test tubes, and graduated cylinders are just some of the different types of glassware available. Volume usually is measured in liters (L) and milliliters (mL).

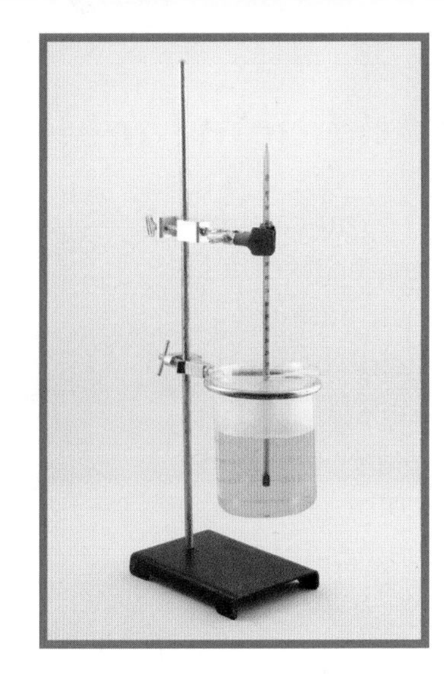

Thermometers ▶

A thermometer is used to measure the temperature of substances. Although Kelvin is the SI unit of measurement for temperature, in the science classroom, you often measure temperature in degrees Celsius (°C). Never stir a substance with a thermometer because it might break. If a thermometer does break, tell your teacher immediately. Do not touch the broken glass or the liquid inside the thermometer.

◀ Calculators

A hand-held calculator is a scientific tool that you might use in math class. But you also can use it in the lab and in the field (real situation outside the lab) to make quick calculations using your data.

Computers ▼

For today's students, it is difficult to think of a time when scientists—or anyone—did not use computers in their work. Scientists can collect, compile, and analyze data more quickly using a computer. Scientists use computers to prepare research reports and to share their data and ideas with investigators worldwide.

Hardware refers to the physical components of a computer, such as the monitor and the mouse. Computer software refers to the programs that are run on computers, such as word processing, spreadsheet, and presentation programs.

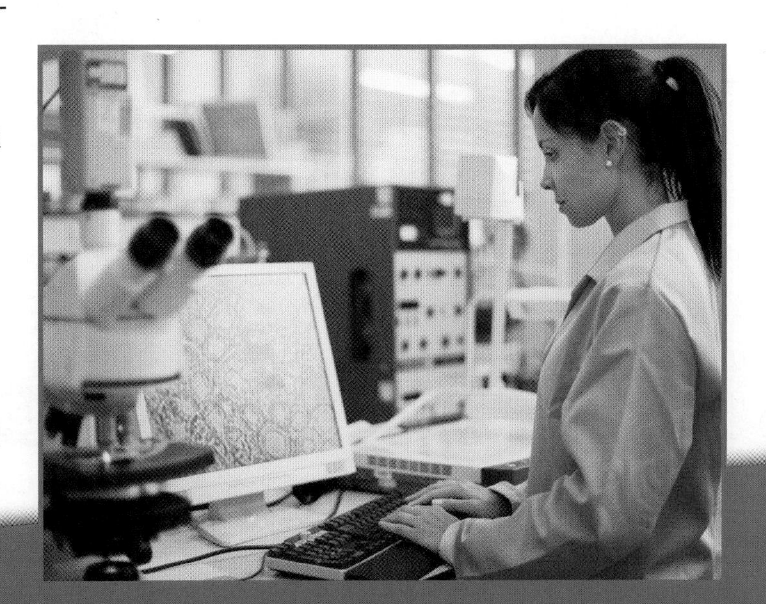

Electronic probes can be attached to computers and handheld calculators to record measurements. There are probes for collecting different kinds of information, such as temperature and the speed of objects.

Key Concept Check How can scientific tools, such as graduated cylinders and triple-beam balances, assist scientists?

Additional Tools Used by Physical Scientists

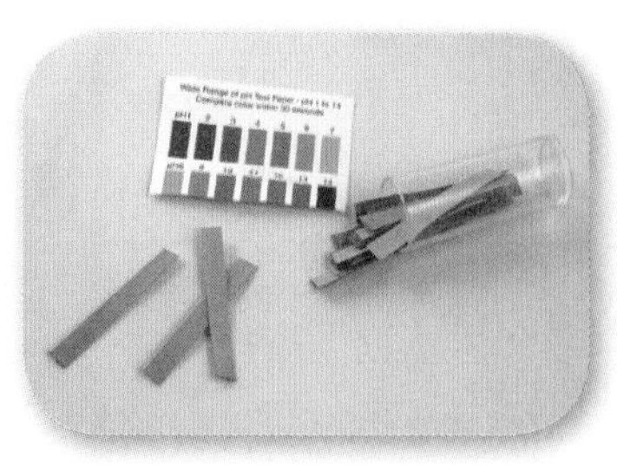

You can use pH paper to quickly estimate the acidity of a liquid substance. The paper changes color when it comes into contact with an acid or a base.

Scientists use stopwatches to measure the time it takes for an event to occur. The SI unit for time is seconds (s). However, for longer events, the units *minutes (min)* and *hours (h)* can be used.

A hot plate is a small heating device that can be placed on a table or desk. Hot plates are used to heat substances in the laboratory.

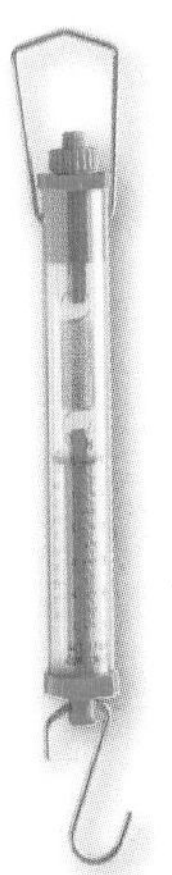

You use a spring scale to measure the weight or the amount of force applied to an object. The SI unit for weight is the newton (N).

Lesson 2 Review

Online Quiz

Use Vocabulary

1. A spoken or written summary of observations is a(n) ________, while a(n) ________ is an interpretation of observations.

Understand Key Concepts

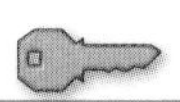

2. Which type of glassware would you use to measure the volume of a liquid?
   A. beaker
   B. flask
   C. graduated cylinder
   D. test tube

3. **Summarize** why a scientist measuring the diameter of an atom or the distance to the Moon would use scientific notation.

4. **Explain** why scientists use the International System of Units (SI).

Interpret Graphics

5. **Identify** Copy and fill in the graphic organizer below listing some scientific tools that you could use to collect data.

Critical Thinking

6. **Explain** why precision and accuracy should be reported in a scientific investigation.

Math Skills

Math Practice

7. Calculate the percent error if the experimental value for the density of zinc is 9.95 g/cm^3. The accepted value is 7.13 g/cm^3.

Skill Practice **Predict**

45 minutes

How do geometric shapes differ in strength?

If you look at a bridge, a building crane, or the framework of a tall building, you will notice that various geometric shapes make up the structure. In this activity, you will observe the strength of several geometric shapes in terms of their rigidity, or resistance to changing their shape.

Materials

plastic straws

scissors

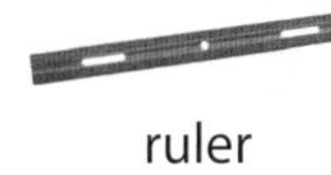

ruler

string

Safety

Learn It

When scientists make a hypothesis, they often then **predict** that an event will occur based on their hypothesis.

Try It

1. Read and complete a lab safety form.
2. You are going to construct a triangle and a square using straws. Predict which shape will be more resistant to changing shape and write your prediction in your Science Journal.
3. Measure and cut the straws into seven segments, each 6 cm long.
4. Measure and cut one 20-cm and one 30-cm length of string.
5. Thread the 30-cm length of string through four straw segments. Bend the corners to form a square. Tie the ends of the string together in a double knot to complete the square.
6. Thread the 20-cm string through three of the straw segments. Bend to form a triangle. Tie the ends of the string together to complete the triangle.
7. Test the strength of the square by gently trying to change its shape. Repeat with the triangle. Record your observations.
8. Propose several ways to make the weaker shape stronger. Draw diagrams showing how to modify the shape to make it more rigid.
9. Test your hypothesis. If necessary, refine your hypothesis and retest it. Repeat this step until you make the shape stronger.

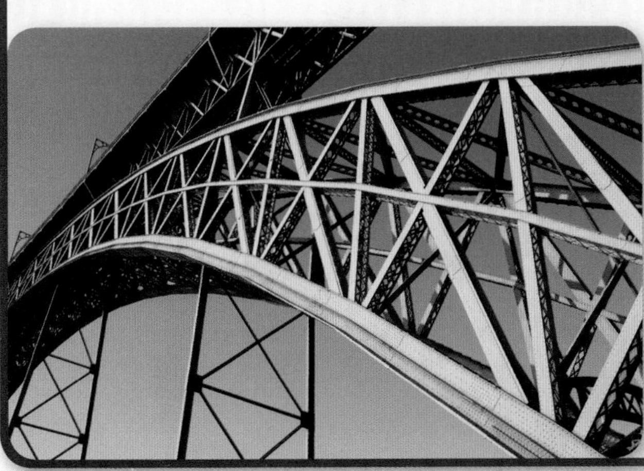

Apply It

10. Look at the photograph at to the left. Which of your tested shapes is used the most? Based on your observations, why is this shape used?
11. What modifications made your shape stronger? Why?
12. **Key Concept** How might a scientist use a model to test a hypothesis?

Lesson 3 Case Study

Reading Guide

Key Concepts
ESSENTIAL QUESTIONS

- Why are evaluation and testing important in the design process?
- How is scientific inquiry used in a real-life scientific investigation?

Vocabulary

variable p. NOS 21
constant p. NOS 21
independent variable p. NOS 21
dependent variable p. NOS 21
experimental group p. NOS 21
control group p. NOS 21
qualitative data p. NOS 24
quantitative data p. NOS 24

g Multilingual eGlossary

Video Science Video

The Minneapolis Bridge Failure

On August 1, 2007, the center section of the Interstate-35W (I-35W) bridge in Minneapolis, Minnesota, suddenly collapsed. A major portion of the bridge fell more than 30 m into the Mississippi River, as shown in **Figure 7.** There were more than 100 cars and trucks on the bridge at the time, including a school bus carrying over 50 students.

The failure of this 8-lane, 581-m long interstate bridge came as a surprise to almost everyone. Drivers do not expect a bridge to drop out from underneath them. The design and engineering processes that bridges undergo are supposed to ensure that bridge failures do not happen.

Controlled Experiments

After the 2007 bridge collapse, investigators had to determine why the bridge failed. To do this, they needed to use scientific inquiry, which you read about in Lesson 1. The investigators designed controlled experiments to help them answer questions and test their hypotheses. A controlled experiment is a scientific investigation that tests how one factor affects another. You might conduct controlled experiments to help discover answers to questions, to test a hypotheses, or to collect data.

Figure 7 A portion of the Interstate-35W bridge in Minneapolis, Minnesota, collapsed in August 2007. Several people were killed, and many more were injured.

Identifying Variables and Constants

When conducting an experiment, you must identify factors that can affect the experiment's outcome. A **variable** *is any factor that can have more than one value.* In controlled experiments, there are two kinds of variables. The **independent variable** *is the factor that you want to test. It is changed by the investigator to observe how it affects a dependent variable.* The **dependent variable** *is the factor you observe or measure during an experiment.* **Constants** *are the factors in an experiment that do not change.*

You can change the independent variable to observe how it affects the dependent variable. Without constants, two independent variables could change at the same time, and you would not know which variable affected the dependent variable.

Experimental Groups

A controlled experiment usually has at least two groups. The **experimental group** *is used to study how a change in the independent variable changes the dependent variable.* The **control group** *contains the same factors as the experimental group, but the independent variable is not changed.* Without a control, it is impossible to know whether your experimental observations result from the variable you are testing or some other factor.

This case study will explore how the investigators used scientific inquiry to determine why the bridge collapsed. Notebooks in the margin identify what a scientist might write in a science journal. The blue boxes contain additional helpful information that you might use.

Simple Beam Bridges

Before you read about the bridge-collapse investigation, think about the structure of bridges. The simplest type of bridge is a beam bridge, as shown in **Figure 8.** This type of bridge has one horizontal beam across two supports. A beam bridge often is constructed across small creeks. A disadvantage of beam bridges is that they tend to sag in the middle if they are too long.

Figure 8 Simple beam bridges span short distances, such as small creeks.

Figure 9 Truss bridges can span long distances and are strengthened by a series of interconnecting triangles called trusses. ▶

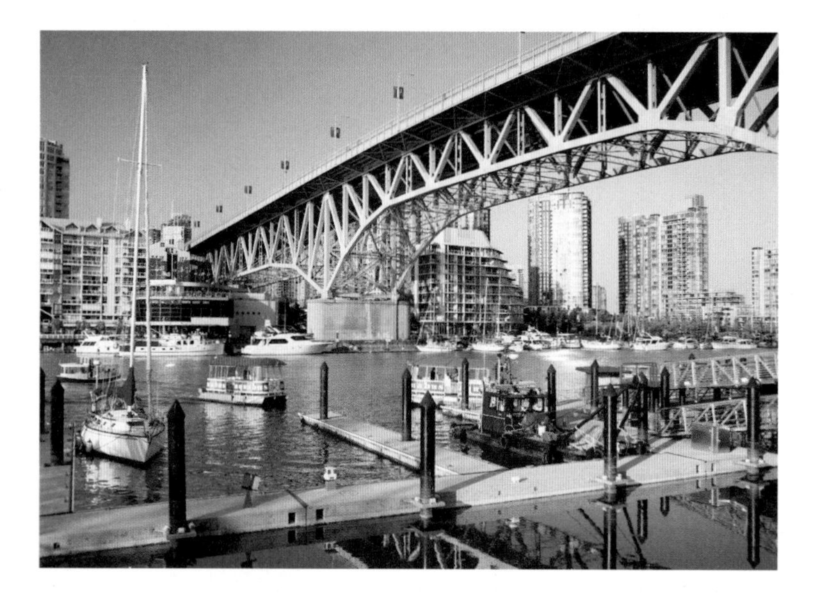

Truss Bridges

A truss bridge, shown in **Figure 9,** often spans long distances. This type of bridge is supported only at its two ends, but an assembly of interconnected triangles, or trusses, strengthens it. The I-35W bridge, shown in **Figure 10,** was a truss bridge designed in the early 1960s. The I-35W bridge was designed with straight beams connected to triangular and vertical supports. These supports held up the deck of the bridge, or the roadway. The beams in the bridge's deck and the supports came together at structures known as gusset plates, shown below on the right. These steel plates joined the triangular and vertical truss elements to the overhead roadway beams. These beams ran along the deck of the bridge. This area, where the truss structure connects to the roadway portion of the bridge at a gusset plate, also is called a node.

Reading Check What are the gusset plates of a bridge?

Figure 10 Trusses were a major structural element of the I-35W bridge. The gusset plates at each node in the bridge, shown on the right, are critical pieces that hold the bridge together. ▼

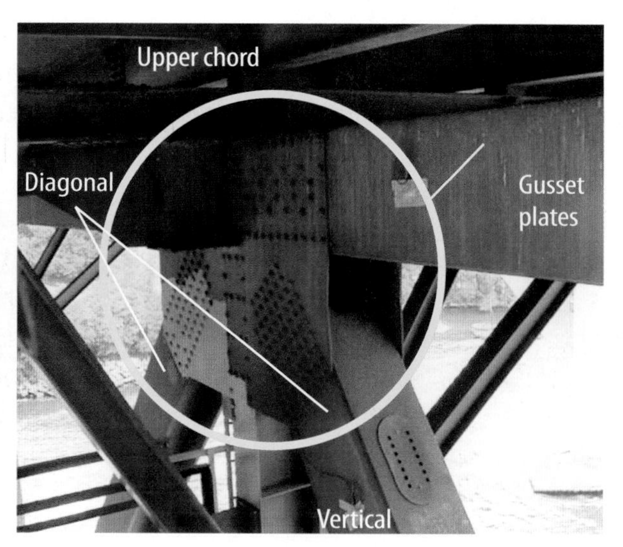

Figure 11 The collapsed bridge was further damaged by rescue workers trying to recover victims of the accident.

Bridge Failure Observations

After the I-35W bridge collapsed, shown in **Figure 11,** the local sheriff's department handled the initial recovery of the collapsed bridge structure. Finding, freeing, and identifying victims was a high priority, and unintentional damage to the collapsed bridge occurred in the process. However, investigators eventually recovered the entire structure.

The investigators labeled each part with the location where it was found. They also noted the date when they removed each piece. Investigators then moved the pieces to a nearby park. There, they placed the pieces in their relative original positions. Examining the reassembled structure, investigators found physical evidence they needed to determine where the breaks in each section occurred.

The investigators found more clues in a video. A motion-activated security camera recorded the bridge collapse. The video showed about 10 seconds of the collapse, which revealed the sequence of events that destroyed the bridge. Investigators used this video to help pinpoint where the collapse began.

Scientists often observe and gather information about an object or an event before proposing a hypothesis. This information is recorded or filed for the investigation.

Observations:
- Recovered parts of the collapsed bridge
- A video showing the sequence of events as the bridge fails and falls into the river

Asking Questions

One or more factors could have caused the bridge to fail. Was the original bridge design faulty? Were bridge maintenance and repair poor or lacking? Was there too much weight on the bridge at the time of the collapse? Each of these questions was studied to determine why the bridge collapsed. Did one or a combination of these factors cause the bridge to fail?

Asking questions and seeking answers to those questions is a way that scientists formulate hypotheses.

When gathering information or collecting data, scientists might perform an experiment, create a model, gather and evaluate evidence, or make calculations.

Qualitative data:
A thicker layer of concrete was added to the bridge to protect rods.

Quantitative data:
- The concrete increased the load on the bridge by 13.4 percent.
- The modifications in 1998 increased the load on the bridge by 6.1 percent.
- At the time of the collapse in 2007, the load on the bridge increased by another 20 percent.

Gathering Information and Data

Investigators reviewed the modifications made to the bridge since it opened in 1967. In 1977, engineers noticed that salt used to deice the bridge during winter weather was causing the reinforcement rods in the roadway to weaken. To protect the rods, engineers applied a thicker layer of concrete to the surface of the bridge roadway. Analysis after the collapse revealed that this extra concrete increased the dead load on the bridge by about 13.4 percent. A load can be a force applied to the structure from the structure itself (dead load) or from temporary loads such as traffic, wind gusts, or earthquakes (live load). Investigators recorded this qualitative and quantitative data. **Qualitative data** *uses words to describe what is observed.* **Quantitative data** *uses numbers to describe what is observed.*

In 1998, additional modifications were made to the bridge. The bridge that was built in the 1960s did not meet current safety standards. Analysis showed that the changes made to the bridge during this renovation further increased the dead load on the bridge by about 6.1 percent.

An Early Hypothesis

At the time of the collapse in 2007, the bridge was undergoing additional renovations. Four piles of sand, four piles of gravel, a water tanker filled with over 11,000 L of water, a cement tanker, a concrete mixer, and other equipment, supplies, and workers were assembled on the bridge. This caused the load on the bridge to increase by about 20 percent. In addition to these renovation materials, normal vehicle traffic was on the bridge. Did the renovation equipment and traffic overload the bridge, causing the center section to collapse as shown in **Figure 12**? Only a thorough analysis could answer this question.

Figure 12 The center section of the bridge broke away and fell into the river.

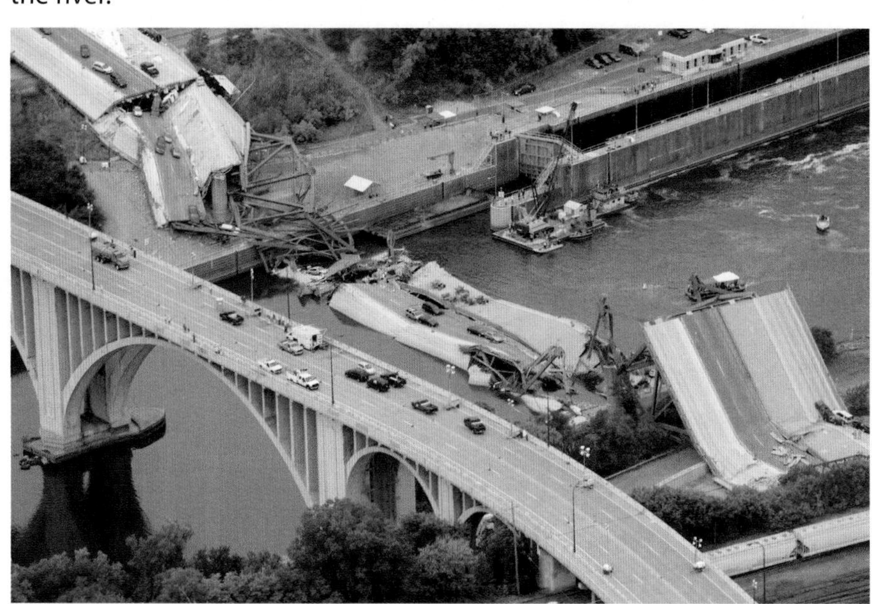

A hypothesis is a possible explanation for an observation that can be tested by scientific investigations.

Hypothesis:
The bridge failed because it was overloaded.

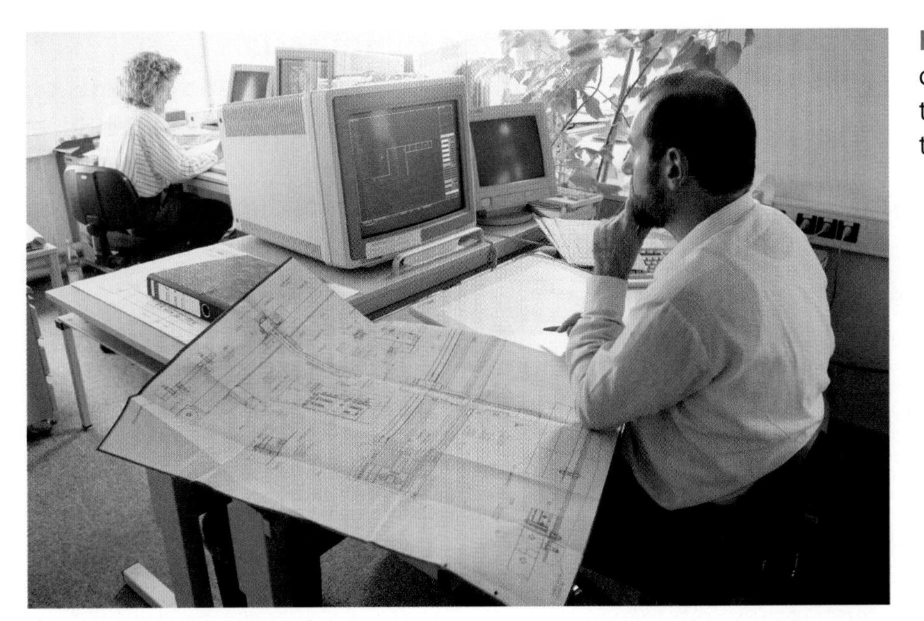

Figure 13 Engineers used computer models to study the structure and loads on the bridge.

Computer Modeling

The analysis of the bridge was conducted using computer-modeling software, as shown in **Figure 13.** Using computer software, investigators entered data from the Minnesota bridge into a computer. The computer performed numerous mathematical calculations. After thorough modeling and analysis, it was determined that the bridge was not overloaded.

Revising the Hypothesis

Evaluations conducted in 1999 and 2003 provided additional clues as to why the bridge failed. As part of the study, investigators took numerous pictures of the bridge structure. The photos revealed bowing of the gusset plates at the eleventh node from the south end of the bridge. Investigators labeled this node *U10.* Gusset plates are designed to be stronger than the structural parts they connect. It is possible that the bowing of the plates indicated a problem with the gusset plate design. Previous inspectors and engineers missed this warning sign.

The accident investigators found that some recovered gusset plates were fractured, while others were not damaged. If the bridge had been properly constructed, none of the plates should have failed. But inspection showed that some of the plates failed very early in the collapse.

After evaluating the evidence, the accident investigators formulated the hypothesis that the gusset plates failed, which lead to the bridge collapse. Now investigators had to test this hypothesis.

Hypothesis:
~~1. The bridge failed because it was overloaded.~~
2. The bridge collapsed because the gusset plates failed.
Prediction:
If a gusset plate is not properly designed, then a heavy load on a bridge will cause a gusset plate to fail.

Test the Hypothesis:
- Compare the load on the bridge when it collapsed with the load limits of the bridge at each of the main gusset plates.
- Determine the demand-to-capacity ratios for the main gusset plates.
- Calculate the appropriate thicknesses of the U10 gusset plates.

Independent Variables: actual load on bridge and load bridge was designed to handle
Dependent Variable: demand-to-capacity ratio

Testing the Hypothesis

The investigators knew the load limits of the bridge. To calculate the load on the bridge when it collapsed, they estimated the combined weight of the bridge and the traffic on the bridge. The investigators divided the load on the bridge when it collapsed by the load limits of the bridge to find the demand-to-capacity ratio. The demand-to-capacity ratio provides a measure of a structure's safety.

Analyzing Results

As investigators calculated the demand-to-capacity ratios for each of the main gusset plates, they found that the ratios were particularly high for the U10 node. The U10 plate, shown in **Figure 14,** failed earliest in the bridge collapse. **Table 4** shows the demand-to-capacity ratios for a few of the gusset plates at some nodes. A value greater than 1 means the structure is unsafe. Notice how high the ratios are for the U10 gusset plate compared to the other plates.

Further calculations showed that the U10 plates were not thick enough to support the loads they were supposed to handle. They were about half the thickness they should have been.

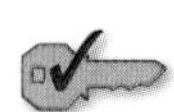

Key Concept Check Why are evaluation and testing important in the design process?

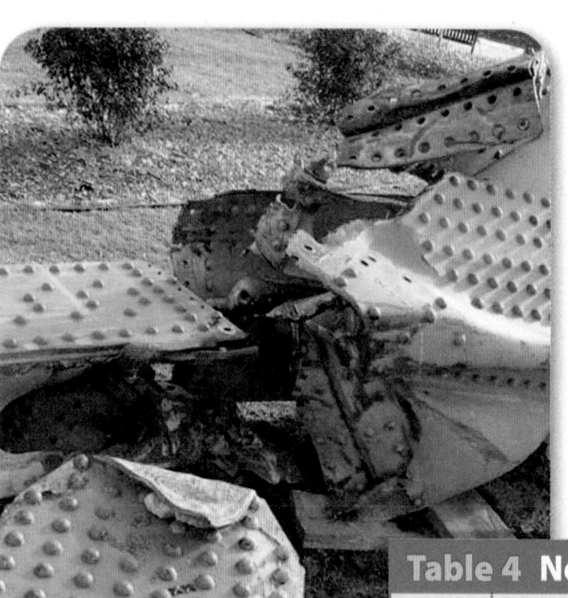

Figure 14 The steel plates, or gusset plates, at the U10 node were too thin for the loads the bridge carried.

Table 4 Node-Gusset Plate Analysis

Gusset Plate	Thickness (cm)	Demand-to-Capacity Ratios for the Upper-Node Gusset Plates					
		Horizontal loads			Vertical loads		
U8	3.5	0.05	0.03	0.07	0.31	0.46	0.20
U10	1.3	1.81	1.54	1.83	1.70	1.46	1.69
U12	2.5	0.11	0.11	0.10	0.71	0.37	1.15

Drawing Conclusions

Over the years, modifications to the I-35W bridge added more load to the bridge. On the day of the accident, traffic and the concentration of construction vehicles and materials added still more load. Investigators concluded that if the U10 gusset plates were properly designed, they would have supported the added load. When the investigators examined the original records for the bridge, they were unable to find any detailed gusset plate specifications. They could not determine whether undersized plates were used because of a mistaken calculation or some other error in the design process. The only thing that they could conclude with certainty was that undersized gusset plates could not reliably hold up the bridge.

The Federal Highway Administration and the National Transportation Safety Board published the results of their investigations. These published reports now provide scientists and engineers with valuable information they can use in future bridge designs. These reports are good examples of why it is important for scientists and engineers to publish their results and to share information.

Analyzing Results: The U10 gusset plates should have been twice as thick as they were to support the bridge.

Conclusions: The bridge failed because the gusset plates were not properly designed and they could not carry the load that they were supposed to carry.

Key Concept Check Give three examples of the scientific inquiry process that was used in this investigation.

Lesson 3 Review

Assessment Online Quiz

Use Vocabulary

1. **Distinguish** between qualitative data and quantitative data.
2. **Contrast** *variable, independent variable,* and *dependent variable.*

Understand Key Concepts 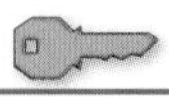

3. Constants are necessary in a controlled experiment because, without constants, you would not know which variable affected the
   A. control group.
   B. experimental group.
   C. dependent variable.
   D. independent variable.
4. **Give an example** of a situation in your life in which you depend on adequate testing and evaluation in a product design to keep you safe.

Interpret Graphics

5. **Summarize** Copy and fill in the flow chart below and summarize the sequence of scientific inquiry steps that was used in one part of the case study.

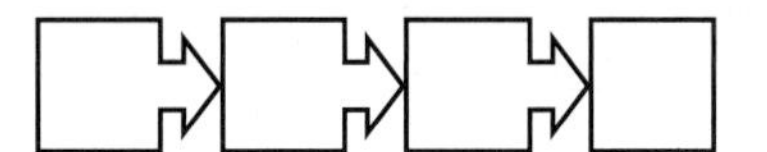

Critical Thinking

6. **Analyze** how the scientific inquiry process differs when engineers design a product, such as a bridge, and when they investigate design failure.
7. **Evaluate** why the gusset plates were such a critical piece in the bridge design.
8. **Recommend** ways that bridge designers and inspectors can prevent future bridge collapses.

Build and Test a Bridge

Materials

plastic straws

ruler

scissors

cotton string

cardboard

Also needed: notebook paper, books or other masses, balance (with a capacity of at least 2 kg)

Safety

In the Skill lab, you observed the relative strengths of two different geometric shapes. In the case study about the bridge collapse, you learned how scientists used scientific inquiry to determine the cause of the bridge collapse. In this investigation, you will combine geometric shapes to build model bridge supports. Then you will use scientific inquiry to determine the maximum load that your bridge will hold.

Ask a Question

What placement of supports produces the strongest bridge?

Make Observations

1. Read and complete a lab safety form.
2. Cut the straws into 24 6-cm segments.
3. Thread three straw segments onto a 1-m piece of string. Slide the segments toward one end of the string. Double knot the string to form a triangle. There should be very little string showing between the segments.
4. Thread the long end of the remaining string through two more straw segments. Double knot the string to one unattached corner to form another triangle. Cut off the remaining string, leaving at least a 1 cm after the knot. Use the string and one more straw segment to form a tetrahedron, as shown below.
5. Use the remaining string and straw segments to build three more tetrahedrons.
6. Set the four tetrahedrons on a piece of paper. They will serve as supports for your bridge deck, a 20-cm x 30-cm piece of cardboard.
7. With your teammates, decide where you will place the tetrahedrons on the paper to best support a load placed on the bridge deck.

Form a Hypothesis

8. Form a hypothesis about where you will place your tetrahedrons and why that placement will support the most weight. Recall that a hypothesis is an explanation of an observation.

Test Your Hypothesis

9. Test your hypothesis by placing the tetrahedrons in your chosen locations on the paper. Lay the cardboard "bridge deck" over the top.
10. Use a balance to find the mass of a textbook. Record the mass in your Science Journal.
11. Gently place the textbook on the bridge deck. Continue to add massed objects until your bridge collapses. Record the total mass that collapsed the bridge.
12. Examine the deck and supports. Look for possible causes of bridge failure.

Analyze and Conclude

13. **Analyze** Was your hypothesis supported? How do you know?
14. **Compare and Contrast** Study the pictures of bridges in Lesson 3. How does the failure of your bridge compare to the failure of the I-35W bridge?
15. **The Big Idea** What steps of scientific inquiry did you use in this activity? What would you do next to figure out how to make a stronger bridge?

Communicate Your Results

Compare your results with those of several other teams. Discuss the placement of your supports and any other factors that may cause your bridge to fail.

Extension

Try building your supports with straw segments that are shorter (4 cm long) and longer (8 cm long). Test your bridges in the same way with each size of support.

Lab Tips

- ☑ When building your tetrahedrons, make sure to double knot all connections and pull them tight. When you are finished, test each tetrahedron by pressing lightly on the top point.
- ☑ When adding the books to the bridge deck, place the books gently on top of the pile. Do not drop them.

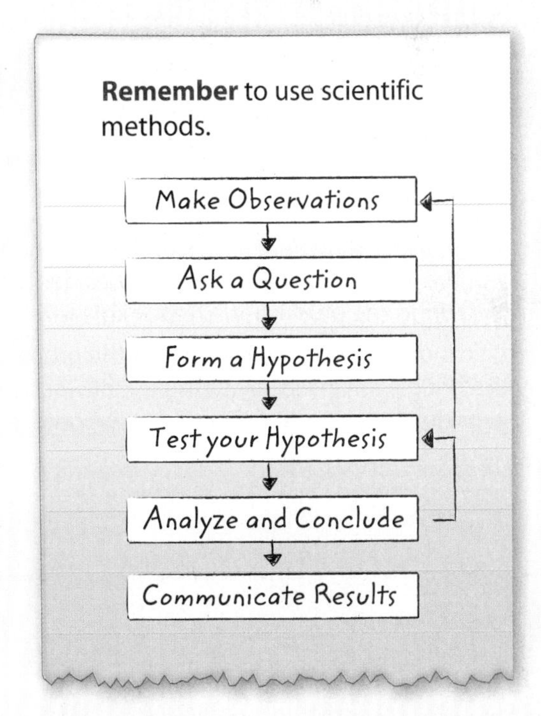

Study Guide and Review

Scientific inquiry is a collection of methods that scientists use in different combinations to perform scientific investigations.

Key Concepts Summary

Lesson 1: Scientific Inquiry

- Some steps used during scientific inquiry are making **observations** and **inferences,** developing a **hypothesis,** analyzing results, and drawing conclusions. These steps, among others, can be performed in any order.
- There are many results of scientific inquiry, and a few possible outcomes are the development of new materials and new technology, the discovery of new objects and events, and answers to basic questions.
- **Critical thinking** is comparing what you already know about something to new information and deciding whether or not you agree with the new information.

Vocabulary

science p. NOS 4
observation p. NOS 6
inference p. NOS 6
hypothesis p. NOS 6
prediction p. NOS 6
scientific theory p. NOS 8
scientific law p. NOS 8
technology p. NOS 9
critical thinking p. NOS 10

Lesson 2: Measurement and Scientific Tools

- Scientists developed one universal system of units, the **International System of Units (SI),** to improve communication among scientists.
- **Scientific notation** is a useful tool for writing large and small numbers in a shorter form.
- Tools such as graduated cylinders and triple-beam balances make scientific investigation easier, more accurate, and repeatable.

Vocabulary

description p. NOS 12
explanation p. NOS 12
International System of Units (SI) p. NOS 12
scientific notation p. NOS 15
percent error p. NOS 15

Lesson 3: Case Study—The Minneapolis Bridge Failure

- Evaluation and testing are important in the design process for the safety of the consumer and to keep costs of building or manufacturing the product at a reasonable level.
- Scientific inquiry was used throughout the process of determining why the bridge collapsed, including hypothesizing potential reasons for the bridge failure and testing those hypotheses.

Vocabulary

variable p. NOS 21
independent variable p. NOS 21
dependent variable p. NOS 21
constants p. NOS 21
qualitative data p. NOS 21
quantitative data p. NOS 21
experimental group p. NOS 24
control group p. NOS 24

Use Vocabulary

1. The ________ contains the same factors as the experimental group, but the independent variable is not changed.
2. The expression of error as a percentage of the accepted value is ________.
3. The process of studying nature at all levels and the collection of information that is accumulated is ________.
4. The ________ are the factors in the experiment that stay the same.

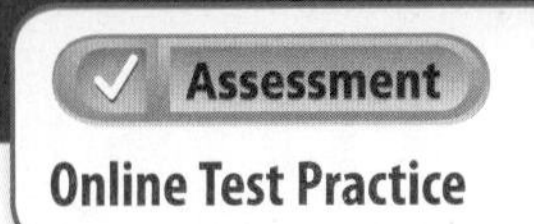

Understand Key Concepts

5 Which is NOT an SI base unit?

A. kilogram
B. liter
C. meter
D. second

6 While analyzing results from an investigation, a scientist calculates a very small number that he or she wants to make easier to use. Which does the scientist use to record the number?

A. explanation
B. inference
C. scientific notation
D. scientific theory

7 Which is NOT true of a scientific law?

A. It can be modified or rejected.
B. It states that an event will occur.
C. It explains why an event will occur.
D. It is based on repeated observations.

8 Which tool would a scientist use to find the mass of a small steel rod?

A. balance
B. computer
C. hot plate
D. thermometer

Critical Thinking

9 **Write** a brief description of the activity shown in the photo.

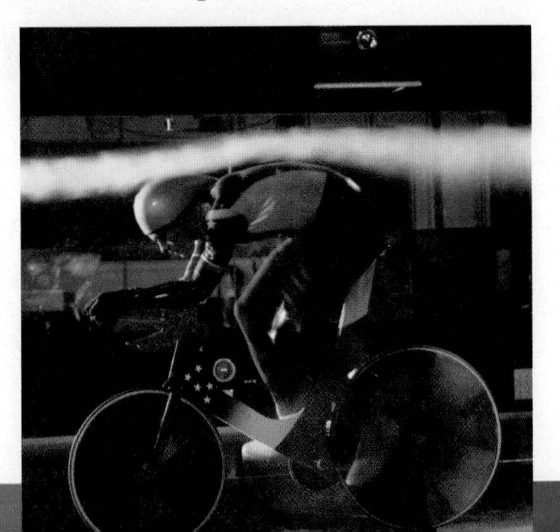

Writing in Science

10 **Write** a five-sentence paragraph that gives examples of how critical thinking, skepticism, and identifying facts and opinions can help you in your everyday life. Be sure to include a topic sentence and concluding sentence in your paragraph.

REVIEW THE BIG IDEA

11 What is scientific inquiry? Explain why it is a constantly changing process.

12 Which part of scientific inquiry does this photo demonstrate?

Math Skills

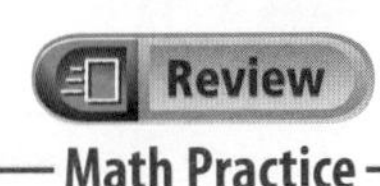

Math Practice

13 The accepted scientific value for the density of sucrose is 1.59 g/cm^3. You perform three trials to measure the density of sucrose, and your data is shown in the table below. Calculate the percent error for each trial.

Trial	Density	Percent Error
Trial 1	1.55 g/cm^3	
Trial 2	1.60 g/cm^3	
Trial 3	1.58 g/cm^3	

Unit 1

Motion & Energy

3500 B.C.
The oldest wheeled vehicle is depicted in Mesopotamia, near the Black Sea.

400 B.C.
The Greeks invent the stone-hurling catapult.

1698
English military engineer Thomas Savery invents the first crude steam engine while trying to solve the problem of pumping water out of coal mines.

1760–1850
The Industrial Revolution results in massive advances in technology and social structure in England.

1769
The first vehicle to move under its own power is designed by Nicholas Joseph Cugnot and constructed by M. Breszin. A second replica is built that weighs 3,629 kg and has a top speed of 3.2 km per hour.

1794
Eli Whitney receives a patent for the mechanical cotton gin.

1800 — 1900

1817
Baron von Drais invents a machine to help him quickly wander the grounds of his estate. The machine is made of two wheels on a frame with a seat and a pair of pedals. This machine is the beginning design of the modern bicycle.

1903
Wilbur and Orville Wright build their airplane, called the Flyer, and take the first successful, powered, piloted flight.

1976
The first computer for home use is invented by college dropouts Steve Wozniak and Steve Jobs, who go on to found Apple Computer, Inc.

? Inquiry
Visit ConnectED for this unit's STEM activity.

Models

Have you ridden on an amusement park roller coaster such as the one in **Figure 1**? As you were going down the steepest hill or hanging upside down in a loop, did you think to yourself, "I hope I don't fly off this thing"? Before construction begins on a roller coaster, engineers build different models of the thrill ride to ensure proper construction and safety. A **model** is a representation of an object, an idea, or a system that is similar to the physical object or idea being studied.

Figure 1 Engineers use various models to design roller coasters.

Using Models in Physical Science

Models are used to study things that are too big or too small, happen too quickly or too slowly, or are too dangerous or too expensive to study directly. Different types of models serve different purposes. Roller-coaster engineers might build a physical model of their idea for a new, daring coaster. Using mathematical and computer models, the engineers can calculate the measurements of hills, angles, and loops to ensure a safe ride. Finally, the engineers might create another model called a blueprint, or drawing, that details the construction of the ride. Studying the various models allows engineers to predict how the actual roller coaster will behave when it travels through a loop or down a giant hill.

Types of Models

Physical Model

A physical model is a model that you can see and touch. It shows how parts relate to one another, how something is built, or how complex objects work. Physical models often are built to scale. A limitation of a physical model is that it might not reflect the physical behavior of the full-size object. For example, this model will not accurately show how wind will affect the ride.

Mathematical Model

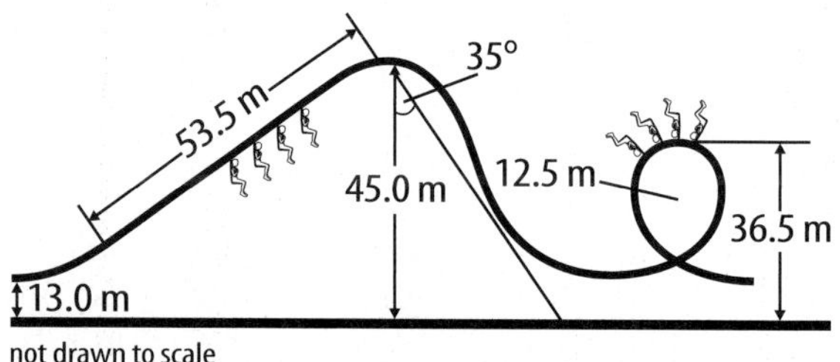

A mathematical model uses numerical data and equations to model an event or idea. Mathematical models often include input data, constants, and output data. When designing a thrill ride, engineers use mathematical models to calculate the heights, the angles of loops and turns, and the forces that affect the ride. One limitation of a mathematical model is that you cannot use it to model how different parts are assembled.

Making Models

An important factor in making a model is determining its purpose. You might need a model that physically represents an object. Or, you might need a model that includes only important elements of an object or a process. When you build a model, first determine the function of the model. What variables need to change? What materials should you use? What do you need to communicate to others? **Figure 2** shows two models of a glucose molecule, each with a different purpose.

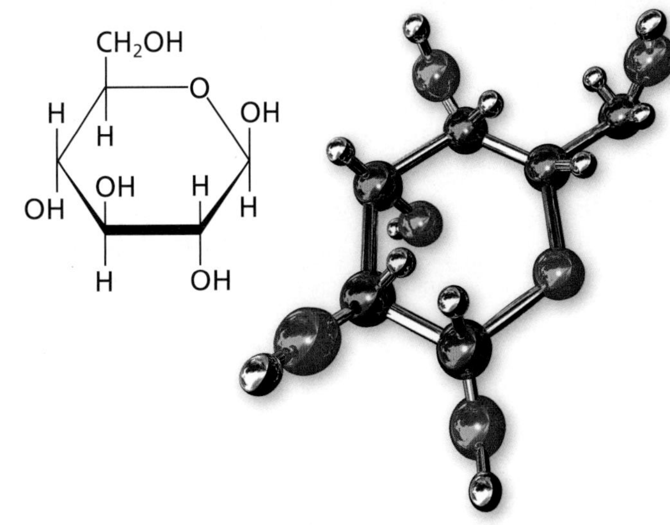

Figure 2 The model on the left is used to represent how the atoms in a glucose molecule bond together. The model on the right is a 3-D representation of the molecule, which shows how atoms might interact.

Limitations of Models

It is impossible to include all the details about an object or an idea into one model. All models have limitations. When using models to design a structure, an engineer must be aware of the information each model does and does not provide. For example, a blueprint of a roller coaster does not show the maximum weight that a car can support. However, a mathematical model would include this information. Scientists and engineers consider the purpose and the limitations of the model they use to ensure they draw accurate conclusions from models.

Computer Simulation

A computer simulation is a model that combines large amounts of data and mathematical models with computer graphic and animation programs. Simulations can contain thousands of complex mathematical models. When roller coaster engineers change variables in mathematical models, they use computer simulation to view the effects of the change.

Inquiry MiniLab — 30 minutes

Can you model a roller coaster?

You are an engineer with an awesome idea for a new roller coaster—the car on your roller coaster makes a jump and then lands back on the track. You model your idea to show it to managers at a theme park in hopes that you can build it.

1. Read and complete a lab safety form.
2. Create a blueprint of your roller coaster. Include a scale and measurements.
3. Follow your blueprint to build a scaled physical model of your roller coaster. Use **foam hose insulation, tape,** and other **craft supplies.**
4. Use a **marble** as a model for a roller-coaster car. Test your model. Record your observations in your Science Journal.

Analyze and Conclude

1. **Compare** your blueprint and physical model.
2. **Evaluate** After you test your physical model, list the design changes you would make to your blueprint.
3. **Identify** What are the limitations of each of your models?

Chapter 1

Describing Motion

What are some ways to describe motion?

Inquiry How is their motion changing?

Have you ever seen a group of planes zoom through the sky at an air show? When one plane speeds up, all the planes speed up. When one plane turns, all the other planes turn in the same direction.

- What might happen if all the planes did not move in the same way?
- How could you describe the positions of the planes in the photo?
- What are some ways you could describe the motion of the planes?

Get Ready to Read

What do you think?

Before you read, decide if you agree or disagree with each of these statements. As you read this chapter, see if you change your mind about any of the statements.

1. Displacement is the distance an object moves along a path.
2. The description of an object's position depends on the reference point.
3. Constant speed is the same thing as average speed.
4. Velocity is another name for speed.
5. You can calculate average acceleration by dividing the change in velocity by the change in distance.
6. An object accelerates when either its speed or its direction changes.

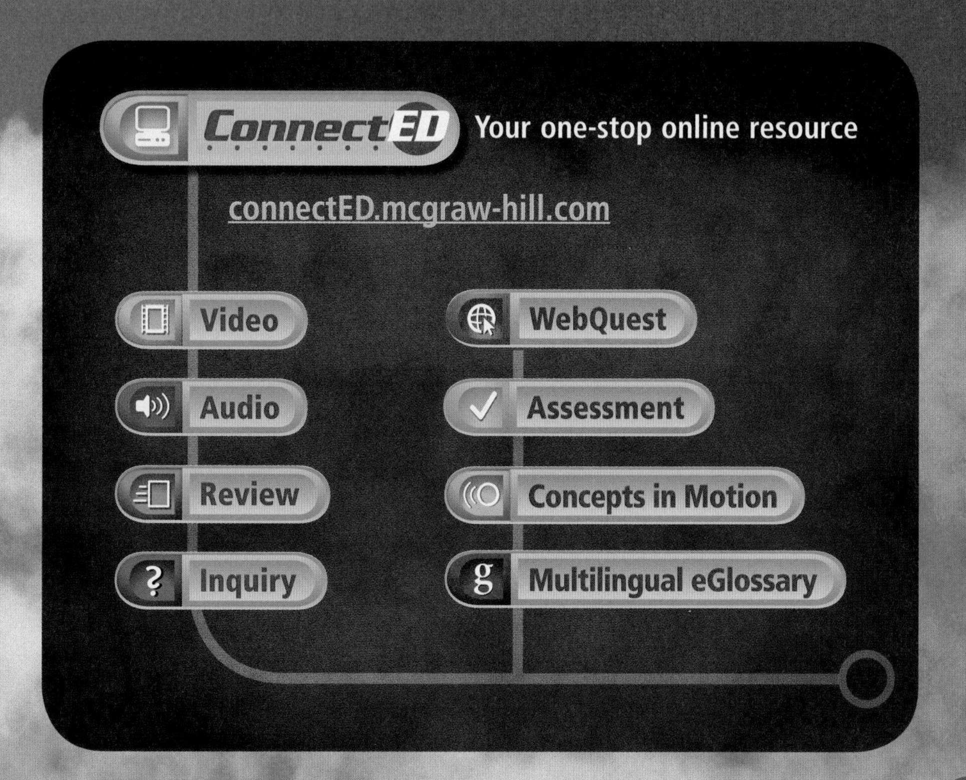

Lesson 1

Reading Guide

Key Concepts
ESSENTIAL QUESTIONS

- How does the description of an object's position depend on a reference point?
- How can you describe the position of an object in two dimensions?
- What is the difference between distance and displacement?

Vocabulary

reference point p. 9
position p. 9
motion p. 13
displacement p. 13

g Multilingual eGlossary

Position and Motion

Inquiry Where are you?

A short time ago, people on this ship probably saw only open ocean. They knew where they were only by looking at the instruments on the ship. But the situation has changed. How can the lighthouse help the ship's crew guide the ship safely to shore?

Inquiry Launch Lab

10 minutes

How do you get there from here?

How would you give instructions to a friend who was trying to walk from one place to another in your classroom?

1. Read and complete a lab safety form.
2. Place a sheet of **paper** labeled *North, East, South,* and *West* on the floor.
3. Walk from the paper to one of the three locations your teacher has labeled in the classroom. Have a partner record the number of steps and the directions of movement in his or her Science Journal.
4. Using these measurements, write instructions other students could follow to move from the paper to the location.
5. Repeat steps 3 and 4 for the other locations.

Think About This

1. How did your instructions to each location compare to those written by other groups?
2. **Key Concept** How did the description of your movement depend on the point at which you started?

Describing Position

How would you describe where you are right now? You might say you are sitting one meter to the left of your friend. Perhaps you would explain that you are at home, which is two houses north of your school. You might instead say that your house is ten blocks east of the center of town, or even 150 million kilometers from the Sun.

What do all these descriptions have in common? Each description states your location **relative** to a certain point. *A* **reference point** *is the starting point you choose to describe the location, or position, of an object.* The reference points in the first paragraph are your friend, your school, the center of town, and the Sun.

Each description of your location also includes your distance and direction from the reference point. Describing your location in this way defines your position. *A* **position** *is an object's distance and direction from a reference point.* A complete description of your position includes a distance, a direction, and a reference point.

Reading Check What are two ways you could describe your position right now?

SCIENCE USE V. COMMON USE
relative
Science Use compared (to)
Common Use a member of your family

Describing Position

Figure 1 The arrows indicate the distances and directions from different reference points.

Visual Check How do you know which reference point is farther from the table?

Using a Reference Point to Describe Position

Why do you need a reference point to describe position? Suppose you are planning a family picnic. You want your cousin to arrive at the park early to save your favorite picnic table. The park is shown in **Figure 1.** How would you describe the position of your favorite table to your cousin? First, choose a reference point that a person can easily find. In this park, the statue is a good choice. Next, describe the direction that the table is from the reference point—toward the slide. Finally, say how far the table is from the statue—about 10 m. You would tell your cousin that the position of the table is about 10 m from the statue, toward the slide.

FOLDABLES®

Fold a sheet of paper to make a half book. Use it to organize your notes about how position and motion are related.

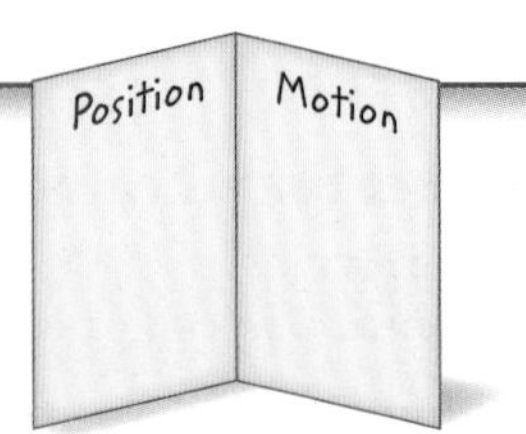

Reading Check How could you describe the position of a different table using the statue as a reference point?

Changing the Reference Point

The description of an object's position depends on the reference point. Suppose you choose the drinking fountain in **Figure 1** as the reference point instead of the statue. You could say that the direction of the table is toward the dead tree. Now the distance is measured from the drinking fountain to the table. You could tell your cousin that the table is about 12 m from the drinking fountain, toward the dead tree. The description of the table's position changed because the reference point is different. Its actual position did not change at all.

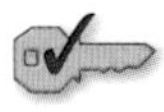

Key Concept Check How does the description of an object's position depend on a reference point?

Figure 2 If east is the reference direction, then the museum is in the positive direction from the bus stop. The library is in the negative direction.

The Reference Direction

When you describe an object's position, you compare its location to a reference direction. In **Figure 1,** the reference direction for the first reference point is toward the slide. Sometimes the words *positive* and *negative* are used to describe direction. The reference direction is the positive (+) direction. The opposite direction is the negative (–) direction. Suppose you **specify** east as the reference direction in **Figure 2.** You could say the museum's entrance is +80 m from the bus stop. The library's entrance is –40 m from the bus stop. To a friend, you would probably just say the museum is two buildings east of the bus stop and the library is the building west of the bus stop. Sometimes, however, using the words *positive* and *negative* to describe direction is useful for explaining changes in an object's position.

ACADEMIC VOCABULARY
specify
(verb) to indicate or identify

Inquiry MiniLab

10 minutes

Why is a reference point useful?

To find an object's position, you need to know its distance and direction from a reference point.

1. Read and complete a lab safety form.
2. Put a **sticky note** at the 50-cm mark of a **meterstick.** This is your reference point.
3. Place a **small object** at the 40-cm mark. It is 10 cm in the negative direction from the reference point.
4. Copy the table in your Science Journal. Continue moving the object and recording its distance, its reference direction, and its position to complete the table.

Position of Object

Distance (cm)	Reference Direction	Position (cm)
10 cm	negative	40 cm
40 cm	positive	
15 cm	positive	
	positive	75 cm
		30 cm

Analyze and Conclude

1. **Recognize Cause and Effect** How would the data in the table change if the positions were the same but the reference point was at the 40-cm mark?
2. **Key Concept** Why is a reference point useful in describing positions of an object?

Describing Position in Two Dimensions

Figure 3 You need two reference directions to describe the position of a building in the city.

Visual Check If the library is the reference point, how would you describe the position of your house in two dimensions?

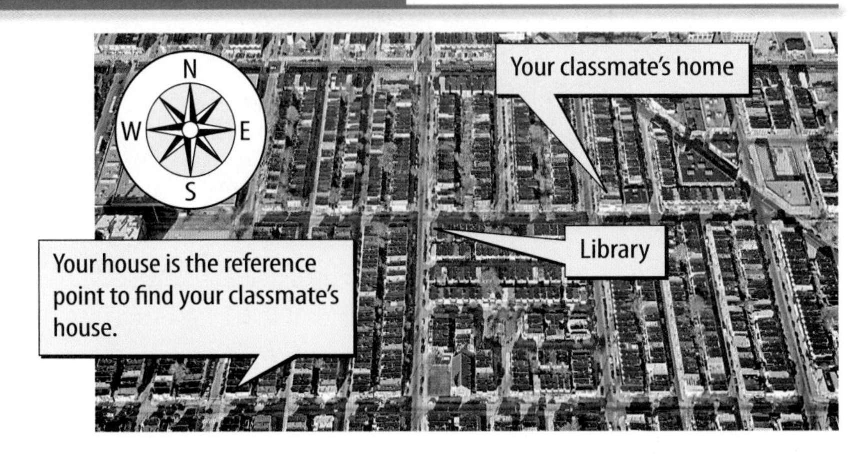

REVIEW VOCABULARY

dimension
distance or length measured in one direction

Describing Position in Two Dimensions

You were able to describe the position of the picnic table in the park in one **dimension.** Your cousin had to walk in only one direction to reach the table. But sometimes you need to describe an object's position using more than one reference direction. The city shown in **Figure 3** is an example. To describe the position of a house in a city might require two reference directions. When you describe a position using two directions, you are using two dimensions.

Reference Directions in Two Dimensions

To describe a position on the map in **Figure 3,** you might choose north and east or south and west as reference directions. Sometimes north, south, east, and west are not the most useful reference directions. If you are playing checkers and want to describe the position of a certain checker, you might use "right" and "forward" as reference directions. If you are describing the position of a certain window on a skyscraper, you might choose "left" and "up" as reference directions.

Reading Check What are two other reference directions you might use to describe the position of a building in a city?

Locating a Position in Two Dimensions

Finding a position in two dimensions is similar to finding a position in one dimension. First, choose a reference point. To locate your classmate's home on the map in **Figure 3,** you could use your home as a reference point. Next, specify reference directions—north and east. Then, determine the distance along each reference direction. In the figure, your classmate's house is one block north and two blocks east of your house.

Key Concept Check How can you describe the position of an object in two dimensions?

Describing Changes in Position

Sometimes you need to describe how an object's position changes. You can tell that the boat in **Figure 4** moved because its position changed relative, or compared, to the buoy. **Motion** *is the process of changing position.*

Word Origin

motion
from Latin *motere;* means "to move"

Motion Relative to a Reference Point

Is the man in the boat in **Figure 4** in motion? Suppose the fishing pole is the reference point. Because the positions of the man and the pole do not change relative to each other, the man does not move relative to the pole. Now suppose the buoy is the reference point. Because the man's distance from the buoy changes, he is in motion relative to the buoy.

◀ **Figure 4** The man in the boat is not in motion compared to his fishing pole. He is in motion compared to the buoy.

Distance and Displacement

Suppose a baseball player runs the bases, as shown in **Figure 5.** Distance is the length of the path the player runs, as shown by the red arrows. **Displacement** *is the difference between the initial (first) position and the final position of an object.* It is shown in the figure by the blue arrows. Notice that distance and displacement are equal only if the motion is in one direction.

Key Concept Check What is the difference between distance and displacement?

Figure 5 Distance depends on the path taken. Displacement depends only on the initial and final positions. ▼

Distance and Displacement

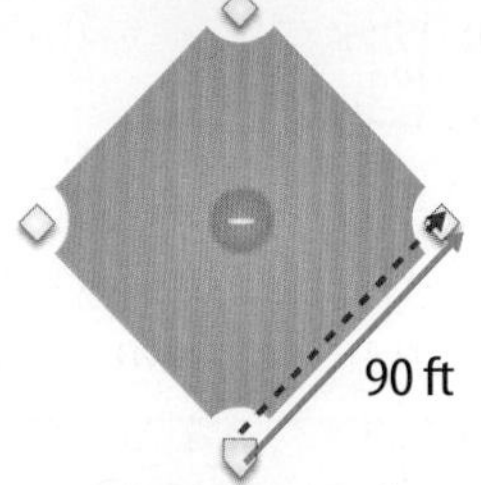

When a player runs to first base, the distance is 90 ft and the displacement is 90 ft toward first base.

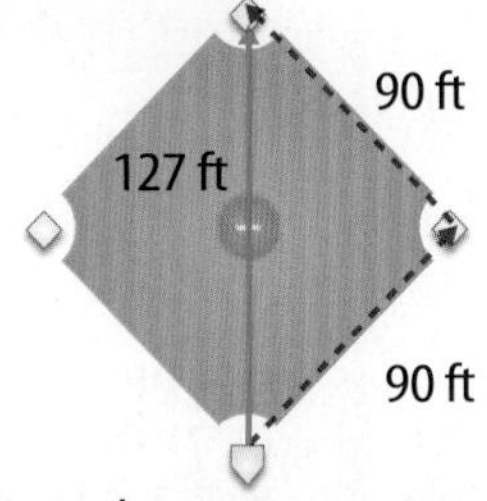

When a player runs to second base, the distance is 180 ft, but the displacement is 127 ft toward second base.

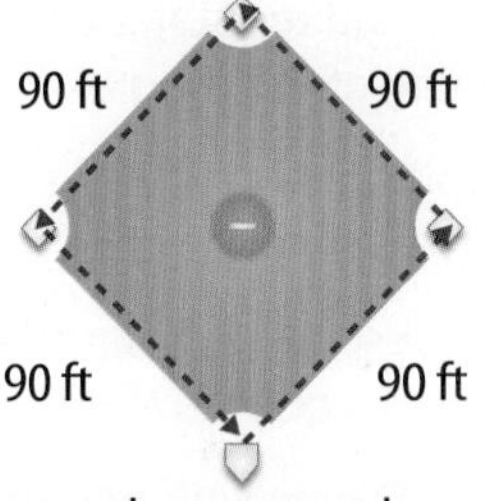

When a player runs to home base, the distance is 360 ft, but the displacement is 0 ft.

Lesson 1 Review

Online Quiz

Visual Summary

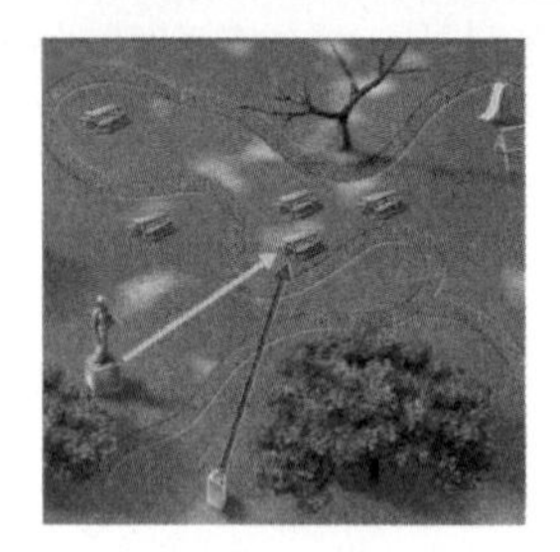

A reference point, a reference direction, and distance are needed to describe the position of an object.

An object is in motion if its position changes relative to a reference point.

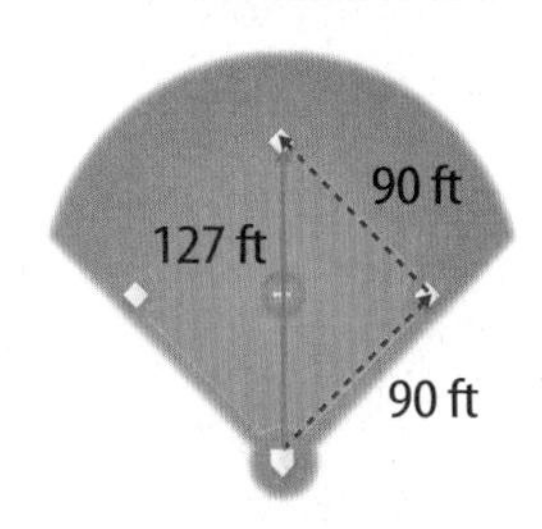

The distance an object moves and the object's displacement are not always the same.

FOLDABLES®

Use your lesson Foldable to review the lesson. Save your Foldable for the project at the end of the chapter.

What do you think NOW?

You first read the statements below at the beginning of the chapter.

1. Displacement is the distance an object moves along a path.
2. The description of an object's position depends on the reference point.

Did you change your mind about whether you agree or disagree with the statements? Rewrite any false statements to make them true.

Use Vocabulary

1. **Define** *motion* in your own words.
2. The difference between the initial position and the final position of an object is its ________.

Understand Key Concepts 

3. **Explain** why a description of position depends on a reference point.
4. To describe a position in more than one dimension, you must use more than one
   A. displacement.
   B. reference direction.
   C. reference point.
   D. type of motion.
5. **Apply** If you walk 2 km from your house to a store and then back home, what is your displacement?

Interpret Graphics

6. **Interpret** Using 12 as the reference point, how you can tell that the hands of the clock on the right have moved from their previous position, shown on the left?

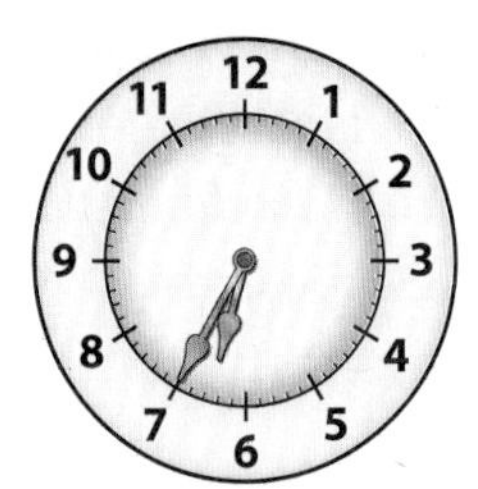

7. **Summarize** Copy and fill in the graphic organizer below to identify the three things that must be included in the description of position.

Critical Thinking

8. **Compare** Relative to some reference points, your nose is in motion when you run. Relative to others, it is not in motion. Give one example of each.

HOW IT WORKS

GPS to the Rescue!

How Technology Helps Bring Home Family Pets

You've seen the signs tacked to streetlights and telephone poles: *LOST! Golden retriever. Reward. Please Call!* Losing a pet can be heartbreaking. Fortunately, there's an alternative to posting fliers—a pet collar with a Global Positioning System (GPS) chip that helps locate the pet. Here is how GPS can help you track or locate your pet:

Satellite

Cell phone tracking display ▶

Cell phone tower

GPS Collar

1 GPS is a network of at least 24 satellites in orbit around Earth. Each satellite circles Earth twice a day and sends information to ground receivers.

2 GPS satellites act as reference points. Ground-based GPS receivers compare the time a signal is transmitted by a satellite to the time it is received on Earth. The difference indicates the satellite's distance. Signals from as many as four satellites are used to pinpoint a user's exact position.

3 GPS uses computer technology to calculate location, speed, direction, and time. The same GPS technology used to locate or guide airplanes, cars, and campers can help find a lost pet anywhere on Earth!

4 A GPS pet collar works much the same as any other GPS receiver. Once it is activated, the collar can transmit a message to a Web site or to the owner's cell phone.

It's Your Turn

DESIGN GPS technology has revolutionized the way people track and locate almost everything. Can you think of a new application for GPS technology? Write an advertisement or a TV commercial for a new idea that puts GPS technology to work!

Lesson 2

Speed and Velocity

Reading Guide

Key Concepts
ESSENTIAL QUESTIONS

- What is speed?
- How can you use a distance-time graph to calculate average speed?
- What are ways velocity can change?

Vocabulary

speed p. 17
constant speed p. 18
instantaneous speed p. 18
average speed p. 19
velocity p. 23

g Multilingual eGlossary

Inquiry How Fast?

When you hear the word *cheetah,* you might think of how fast a cheetah can run. As the fastest land animal on Earth, it can reach a speed of 30 m/s for a short period of time. Other than how fast it runs, how might you describe the motion of a cheetah?

Launch Lab

10 minutes

How can motion change?

Have you ever used a tube slide at a playground or at a water park? You can build a marble tube slide from foam tubes. You can then use the slide to observe how the motion of a marble changes as it rolls down the slide.

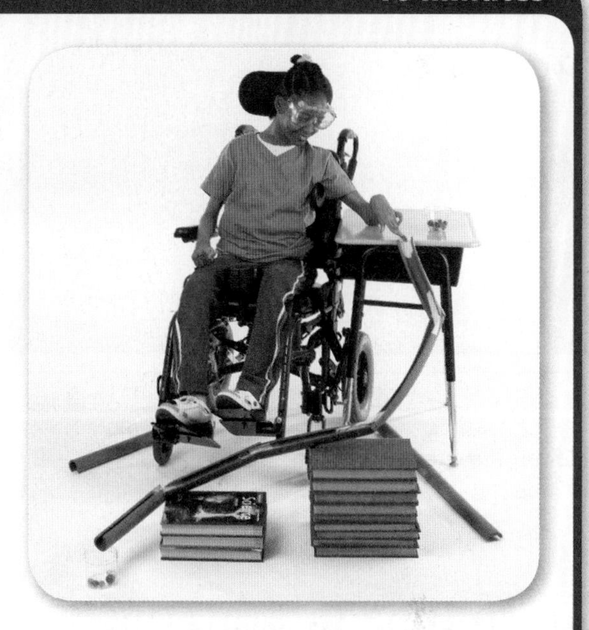

1. Read and complete a lab safety form.
2. Join **foam tubes** into a tube slide, using **masking tape** to connect the tubes. Use a desk and other objects to support the slide so that it changes direction and slopes toward the floor.
3. Drop a **marble** into the top of the slide. Observe the changes in its motion as it rolls through the tubes. Have the marble roll into a **container** at the bottom of the slide.

Think About This

1. In what ways did the motion of the marble change as it moved down the slide?
2. **Key Concept** At what parts of the tube slide did the marble move fastest? At what parts did it move slowest?

What is speed?

How fast do you walk when you are hungry and there is good food on the table? How fast do you move when you have a chore to do? Sometimes you move quickly, and sometimes you move slowly. One way you can describe how fast you move is to determine your speed. **Speed** *is a measure of the distance an object travels per unit of time.*

Key Concept Check What is speed?

Units of Speed

You can calculate speed by dividing the distance traveled by the time it takes to go that distance. The units of speed are units of distance divided by units of time. The SI unit for speed is meters per second (m/s). Other units are shown in **Table 1.** What units of distance and time are used in each example?

Table 1 Typical Speeds

Airplane 245 m/s 882 km/h 548 mph	
Car on a Highway 27 m/s 97 km/h 60 mph	
Person Walking 1.3 m/s 4.7 km/h 2.9 mph	

Table 1 Different units of distance and time can be used to determine units of speed.

Constant and Changing Speed

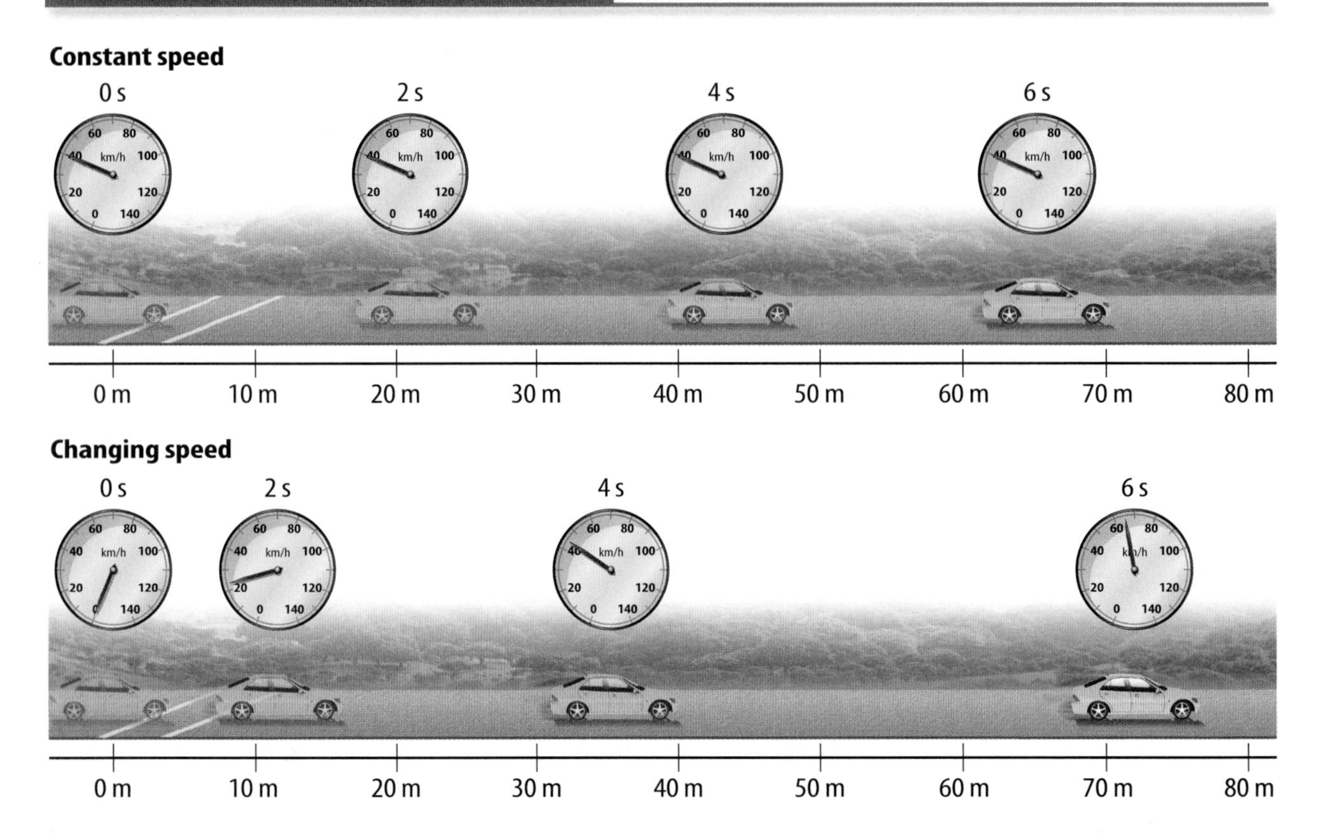

Figure 6 When the car moves with constant speed, it moves the same distance each period of time. When the car's speed changes, it moves a different distance each period of time.

Visual Check What is the bottom car's instantaneous speed at 6 s?

Constant Speed

What happens to your speed when you ride in a car? Sometimes you ride at a steady speed. If you move away from a stop sign, your speed increases. You slow down when you pull into a parking space.

Think about a time that a car's speed does not change. As the car at the top of **Figure 6** travels along the road, the speedometer above each position shows that the car is moving at the same speed at each location and time. Each second, the car moves 11 m. Because it moves the same distance each second, its speed is not changing. **Constant speed** *is the rate of change of position in which the same distance is traveled each second.* The car is moving at a constant speed of 11 m/s.

Changing Speed

How is the motion of the car at the bottom of **Figure 6** different from the motion of the car at the top? Between 0 s and 2 s, the car at the bottom travels about 10 m. Between 4 s and 6 s, however, the car travels more than 20 m. Because the car travels a different distance each second, its speed is changing.

If the speed of an object is not constant, you might want to know its speed at a certain moment. **Instantaneous speed** *is speed at a specific instant in time.* You can see a car's instantaneous speed on its speedometer.

Reading Check How would the distance the car travels each second change if it were slowing down?

Average Speed

Describing an object's speed is easy if the speed is constant. But how can you describe the speed of an object when it is speeding up or slowing down? One way is to calculate its average speed. **Average speed** *is the total distance traveled divided by the total time taken to travel that distance.* You can calculate average speed using the equation below.

Average Speed Equation

$$\textbf{average speed} \text{ (in m/s)} = \frac{\textbf{total distance} \text{ (in m)}}{\textbf{total time} \text{ (in s)}}$$

$$\bar{v} = \frac{d}{t}$$

The symbol $\bar{v}$ represents the term "average velocity." You will read more about velocity, and how it relates to speed, later in this lesson. However, at this point, $\bar{v}$ is simply used as the symbol for "average speed." The SI unit for speed, meters per second (m/s), is used in the above equation. You could instead use other units of distance and time in the average speed equation, such as kilometers and hours.

Math Skills — Average Speed Equation

Solve for Average Speed Melissa shot a model rocket 360 m into the air. It took the rocket 4 s to fly that far. What was the average speed of the rocket?

1. **This is what you know:** distance: $d = 360$ m; time: $t = 4$ s
2. **This is what you need to find:** average speed: $\bar{v}$
3. **Use this formula:** $\bar{v} = \frac{d}{t}$
4. **Substitute:** the values for d and t into the formula and divide. $\bar{v} = \frac{360 \text{ m}}{4 \text{ s}} = 90 \text{ m/s}$

Answer: The average speed was **90 m/s.**

Practice

1. It takes Ahmed 50 s on his bicycle to reach his friend's house 250 m away. What is his average speed?
2. A truck driver makes a trip that covers 2,380 km in 28 hours. What is the driver's average speed?

Review
- Math Practice
- Personal Tutor

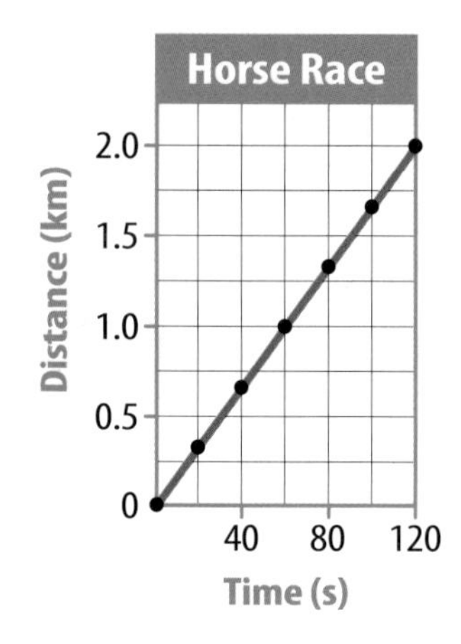

Figure 7 According to this distance-time graph, it took 2 min for the horse to run 2 km.

Visual Check How would this distance-time graph be different if the horse's speed changed over time?

Distance-Time Graphs

The Kentucky Derby is often described as the most exciting two minutes in sports. The thoroughbred horses in this race run for a distance of 2 km. The speeds of the horses change many times during a race, but **Figure 7** describes what a horse's motion might be if its speed did not change. The graph shows the distance a horse might travel when distance measurements are made every 20 seconds. Follow the height of the line from the left side of the graph to the right side. You can see how the distance the horse ran changed over time.

Graphs like the one in **Figure 7** can show how one measurement compares to another. When you study motion, two measurements frequently compared to each other are distance and time. The graphs that show these comparisons are called distance-time graphs. Notice that the change in the distance the horse ran around the track is the same each second on the graph. This means the horse was moving with a constant speed. Constant speed is shown as a straight line on a distance-time graph.

Reading Check How is constant speed shown on a distance-time graph?

Inquiry MiniLab

15 minutes

How can you graph motion?

You can represent motion with a distance-time graph.

1. Read and complete a lab safety form.
2. Use **masking tape** to mark a starting point on the floor.
3. As you cross the starting point, start a **stopwatch.** Stop walking after 2 s. Measure the distance with a **meterstick.** Record the time and distance in your Science Journal.
4. Repeat step 3 by walking at about the same speed for 4 s and then for 6 s.
5. Use the graph in **Figure 7** as an example to create a distance-time graph of your data. The line on the graph should be as close to the points as possible.

Analyze and Conclude

1. **Predict** Based on the graph, how far would you probably walk at the same speed in 8 s?
2. **Key Concept** Look back at the average speed equation. Explain how you could use your graph to find your average walking speed.

Comparing Speeds on a Distance-Time Graph

You can use distance-time graphs to compare the motion of two different objects. **Figure 8** is a distance-time graph that compares the motion of two horses that ran the Kentucky Derby. The motion of horse A is shown by the blue line. The motion of horse B is shown by the orange line. Look at the far right side of the graph. When horse A reached the finish line, horse B was only 1.5 km from the starting point of the race.

Recall that average speed is distance traveled divided by time. Horse A traveled a greater distance than horse B in the same amount of time. Horse A had greater average speed. Compare how steep the lines are on the graph. The measure of steepness is the slope. The steeper the line, the greater the slope. The blue line is steeper than the orange line. Steeper lines on distance-time graphs indicate faster speeds.

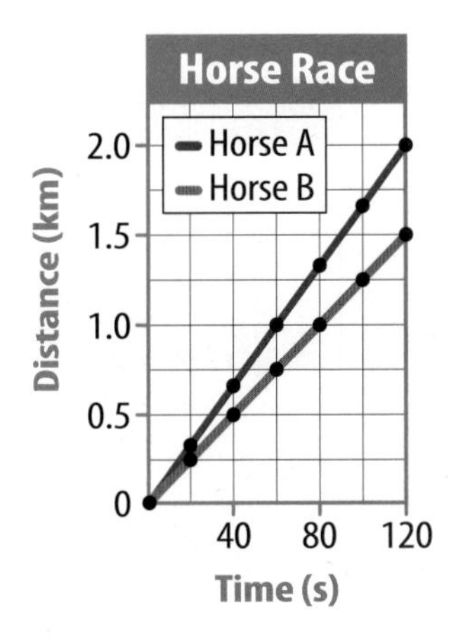

Figure 8 You can tell that horse A ran a faster race than horse B because the blue line is steeper than the orange line.

Using a Distance-Time Graph to Calculate Speed

You can use distance-time graphs to calculate the average speed of an object. The motion of a trail horse traveling at a constant speed is shown on the graph in **Figure 9.** The steps needed to calculate the average speed from a distance-time graph also are shown in the figure.

Key Concept Check How can you use a distance-time graph to calculate average speed?

Average Speed

Review Personal Tutor

Figure 9 The average speed of the horse from 60 s to 120 s can be calculated from this distance-time graph.

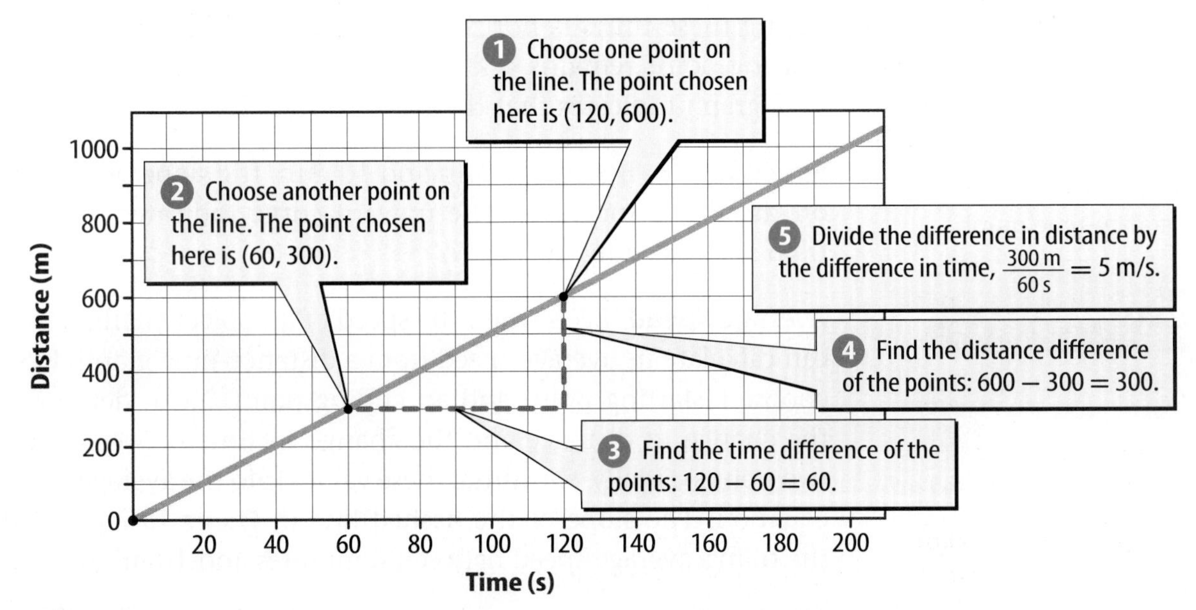

Visual Check How does the average speed of the horse from 60 s to 120 s compare to its average speed from 120 s to 180 s?

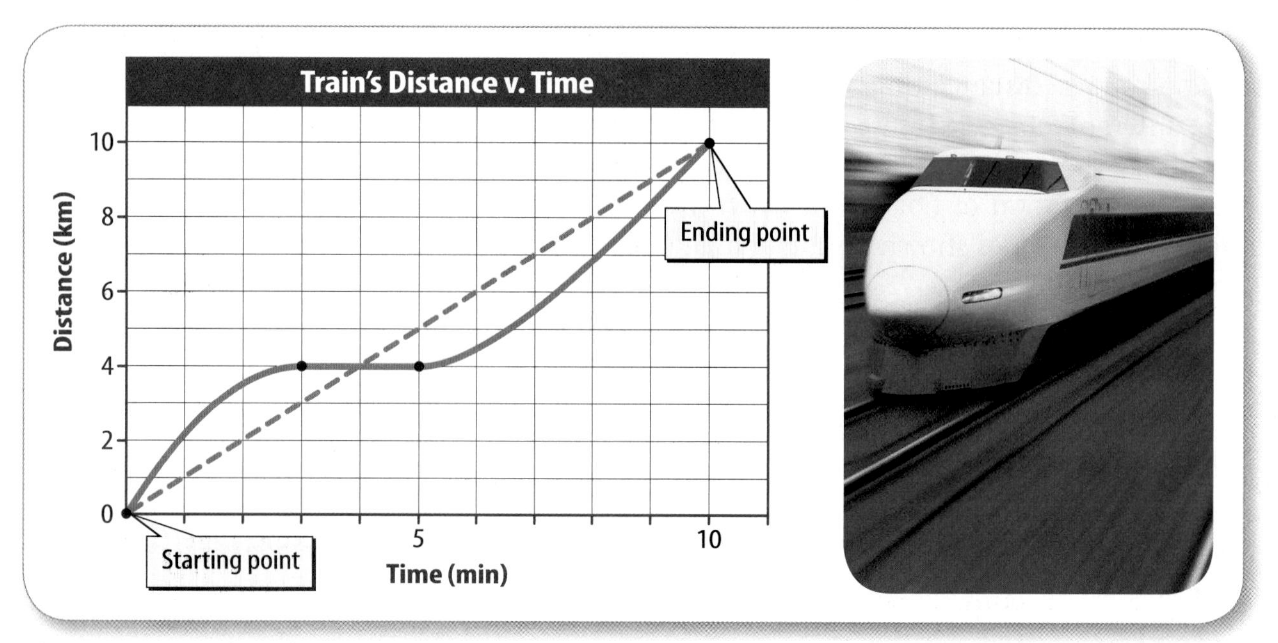

Figure 10 Even though the train's speed is not constant, you can calculate its average speed from a distance-time graph.

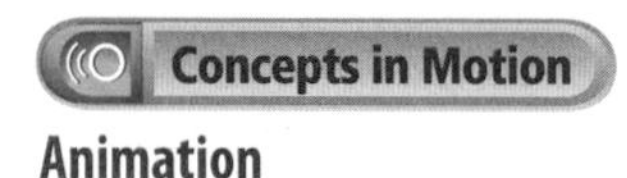

Animation

Distance-Time Graph and Changing Speed

So far, the distance-time graphs in this lesson have included straight lines. Distance-time graphs have straight lines only for objects that move at a constant speed. The graph in **Figure 10** for the motion of a train is different. Because the speed of the train changes instead of being constant, its motion on a distance-time graph is a curved line.

Slowing Down Notice how the shape of the line in **Figure 10** changes. Between 0 min and 3 min, its slope decreases. The downward curve indicates that the train slowed down.

Stopping What happened between 3 min and 5 min? The line during these times is horizontal. The train's distance from the starting point remains 4 km. A horizontal line on a distance-time graph indicates that there is no motion.

Speeding Up Between 5 min and 10 min, the slope of the line on the graph increases. The upward curve indicates that the train was speeding up.

Average Speed Even when the speed of an object changes, you can calculate its average speed from a distance-time graph. First, choose a starting point and an ending point. Next, determine the change in distance and the change in time between these two points. Finally, substitute these values into the average speed equation. The slope of the dashed line in **Figure 10** represents the train's average speed between 0 minutes and 10 minutes.

FOLDABLES®

Make a horizontal three-tab book and label it as shown. Use it to summarize how changes in speed are represented on a distance-time graph.

Reading Check What is the average speed of the train for the trip shown in **Figure 10?**

Velocity

Often, describing just the speed of a moving object does not completely describe its motion. If you describe the motion of a bouncing ball, for example, you would also describe the direction of the ball's movement. Both speed and direction are part of motion. **Velocity** *is the speed and the direction of a moving object.*

Representing Velocity

In Lesson 1, an arrow represented the displacement of an object from a reference point. The velocity of an object also can be represented by an arrow, as shown in **Figure 11.** The length of the arrow indicates the speed. A greater speed is shown by a longer arrow. The arrow points in the direction of the object's motion.

In **Figure 11,** both students are walking at 1.5 m/s. Because the speeds are equal, both arrows are the same length. But the girl is walking to the left and the boy is walking to the right. The arrows point in different directions. The students have different velocities because each student has a different direction of motion.

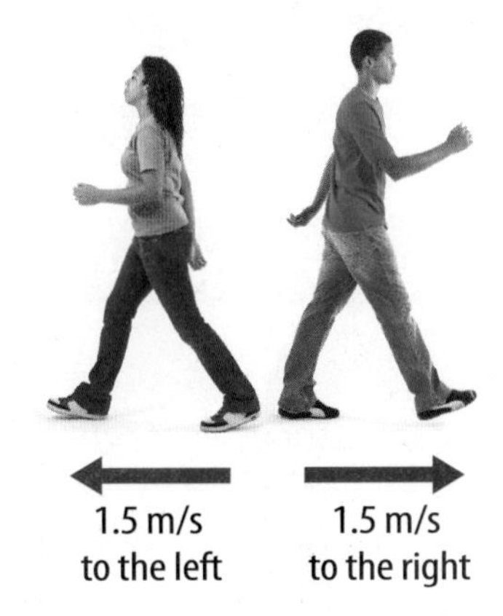

▲ **Figure 11** The students are walking with the same speed but different velocities.

WORD ORIGIN
velocity
from Latin *velocitas;* means "swiftness, speed"

Changes in Velocity

Look at the bouncing ball in **Figure 12.** Notice how from one position to the next, the arrows showing the velocity of the ball change direction and length. The changes in the arrows mean that the velocity is constantly changing. Velocity changes when the speed of an object changes, when the direction that the object moves changes, or when both the speed and the direction change. You will read about changes in velocity in Lesson 3.

 **Key Concept Check** How can velocity change?

Figure 12 The velocity of the ball changes continually because both the speed and the direction of the ball change as the ball bounces. ▼

Changing Speed and Direction

Visual Check Are there two positions of the bouncing ball in which the velocity is the same? Explain.

Lesson 2 Review

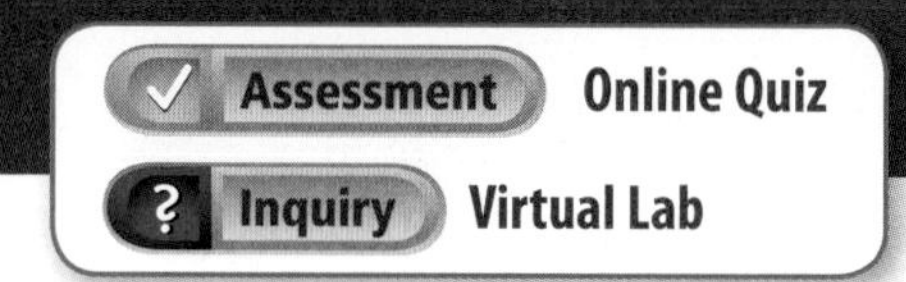

Visual Summary

Speed is a measure of the distance an object travels per unit of time. You can describe an object's constant speed, instantaneous speed, or average speed.

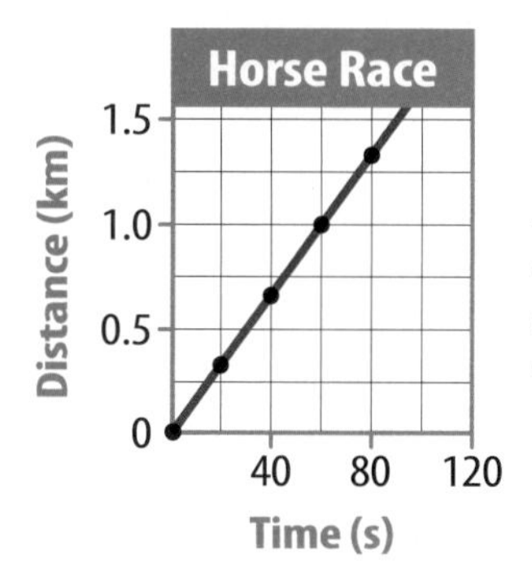

A distance-time graph shows the speed of an object.

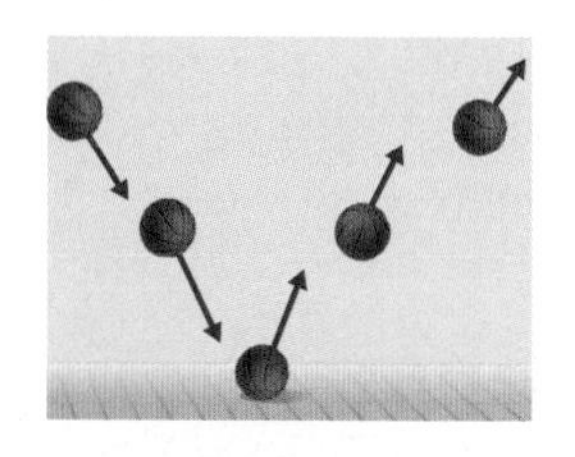

Velocity includes both the speed and the direction of motion.

FOLDABLES®

Use your lesson Foldable to review the lesson. Save your Foldable for the project at the end of the chapter.

What do you think NOW?

You first read the statements below at the beginning of the chapter.

3. Constant speed is the same thing as average speed.

4. Velocity is another name for speed.

Did you change your mind about whether you agree or disagree with the statements? Rewrite any false statements to make them true.

Use Vocabulary

1 **Distinguish** between speed and velocity.

2 **Define** *constant speed* in your own words.

Understand Key Concepts 

3 **Recall** How can you calculate average speed from a distance-time graph?

4 **Analyze** Describe three ways a bicyclist can change velocity.

5 Which choice is a unit of speed?

A. h/mi
B. km/h
C. m^2/s
D. $N{\cdot}m^2$

Interpret Graphics

6 **Organize Information** Copy and fill in the graphic organizer below to show possible steps for making a distance-time graph.

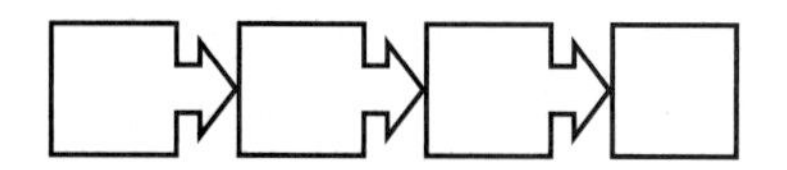

7 **Interpret** What does the shape of each line indicate about the object's speed?

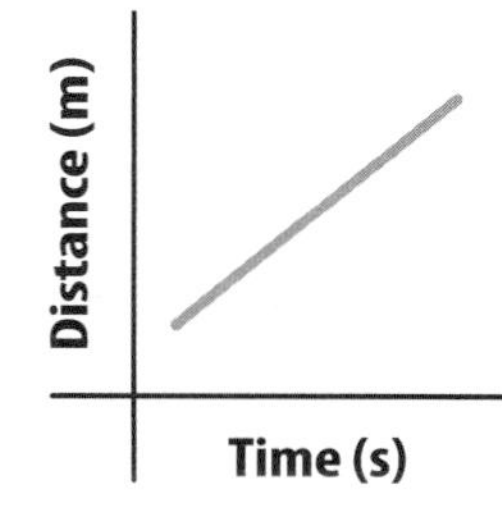

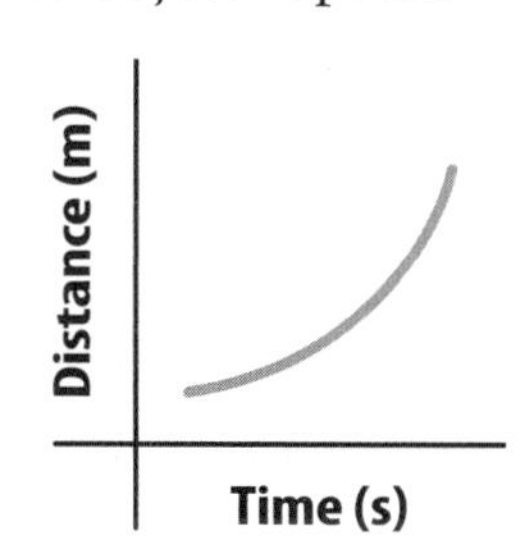

Critical Thinking

8 **Decide** Aaron leaves one city at noon. He has to be at another city 186 km away at 3:00 P.M. The speed limit the entire way is 65 km/h. Can he arrive at the second city on time? Explain.

Math Skills

Math Practice

9 A train traveled 350 km in 2.5 h. What was the average speed of the train?

25 minutes

What do you measure to calculate speed?

Materials

meterstick

stopwatch

wind-up toys (4)

calculator

graph paper

Safety

You turn on the television and see a news report. It shows trees that are bent almost to the ground because of a strong wind. Is it a hurricane or a tropical storm? The type of storm depends on the speed of the wind. A meteorologist must measure both distance and time before calculating the wind's speed.

Learn It

When you **measure,** you use a tool to find a quantity. To find the average speed of an object, you measure the distance it travels and the time it travels. You can then calculate speed using the average speed equation. In this lab, you use distance and time measurements to calculate speeds of moving toys.

Try It

1. Read and complete a lab safety form.
2. Copy the data table on this page into your Science Journal. Add more lines as you need them.
3. Choose appropriate starting and ending points on the floor. Use a meterstick to measure the distance between these points. Record this distance to the nearest centimeter.
4. Wind one toy. Measure in tenths of a second the time the toy takes to travel from start to finish. Record the time in the data table.
5. Repeat steps 3 and 4 for three more toys. Vary the distance from start to finish for each toy.

Apply It

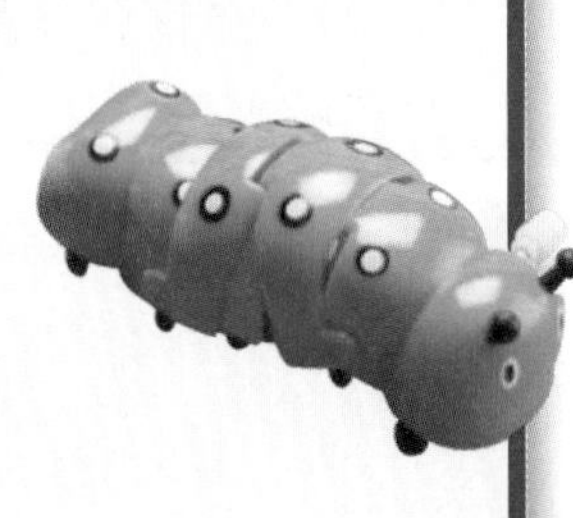

6. **Calculate** the average speed of each toy. Record the speeds in your data table.
7. **Create** a bar graph of your data. Place the name of each toy on the *x*-axis and the average speed on the *y*-axis.
8. **Key Concept** Use the definition of *speed* to explain why the average speeds of the toys can be compared, even though the toys traveled different distances.

Toy Speeds			
Toy	Distance (m)	Time (s)	Average Speed (m/s)

Lesson 3

Reading Guide

Key Concepts
ESSENTIAL QUESTIONS

- What are three ways an object can accelerate?
- What does a speed-time graph indicate about an object's motion?

Vocabulary

acceleration p. 27

Multilingual eGlossary

Video BrainPOP®

Acceleration

Inquiry Is velocity changing?

How does the velocity of this motorcycle racer change as he speeds along the track? As he enters a curve, he slows down, leans to the side, and changes direction. On a straightaway, he speeds up and moves in a straight line. How can the velocity of a moving object change?

10 minutes

In what ways can velocity change?

As you walk, your motion changes in many ways. You probably slow down when the ground is uneven. You might speed up when you realize that you are late for dinner. You change direction many times. What would these changes in velocity look like on a distance-time graph?

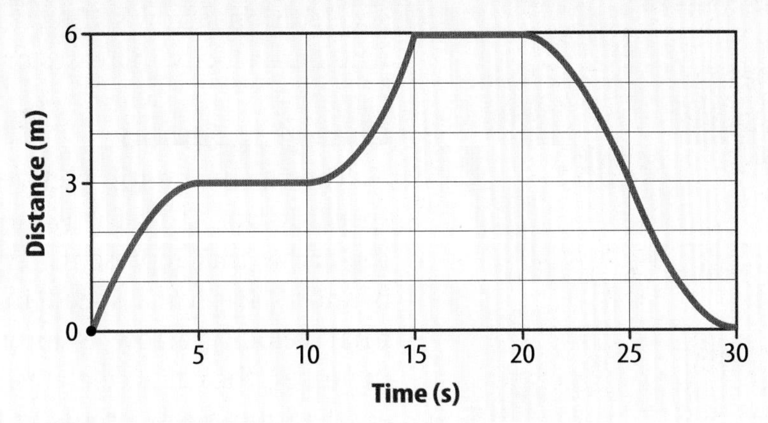

1. Read and complete a lab safety form.
2. Use a **meterstick** to measure a 6-m straight path along the floor. Place a mark with **masking tape** at 0 m, 3 m, and 6 m.
3. Look at the graph above. Decide what type of motion occurs during each 5-second period.
4. Try to walk along your path according to the motion shown on the graph. Have your partner time your walk with a **stopwatch.** Switch roles, and repeat this step.

Think About This

1. What does a horizontal line segment on a distance-time graph indicate?
2. **Key Concept** According to the graph, at what times do the following motions take place? **a.** You change direction. **b.** Your speed increases. **c.** Your speed decreases.

Acceleration—Changes in Velocity

Imagine riding in a car. The driver steps on the gas pedal, and the car moves faster. Moving faster means the car's velocity increases. The driver then takes her foot off the pedal, and the car's velocity decreases. Next the driver turns the steering wheel. The car's velocity changes because its direction changes. The car's velocity changes if either the speed or the direction of the car changes.

When a car's velocity changes, the car is accelerating. **Acceleration** *is a measure of the change in velocity during a period of time.* An object accelerates when its velocity changes as a result of increasing speed, decreasing speed, or changing direction.

You might have experienced a large acceleration if you have ever ridden a roller coaster. Think about all the changes in speed and direction you experience on a roller coaster ride. When you drop down a hill of a roller coaster, you reach a faster speed quickly. The roller coaster is accelerating because its speed is increasing. The roller coaster also accelerates any time it changes direction. It accelerates again when it slows down and stops at the end of the ride. Each time the velocity of the roller coaster changes, it accelerates.

Reading Check What is acceleration?

Ways an Object Can Accelerate

Figure 13 Acceleration occurs when an object speeds up, slows down, or changes its direction of motion.

Speeding Up

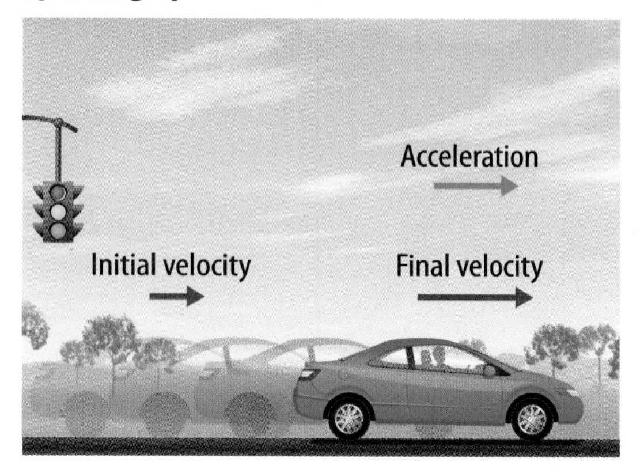

Slowing Down

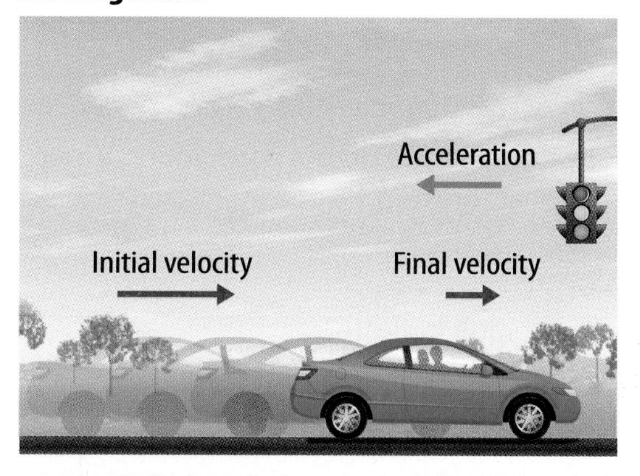

Changing Direction

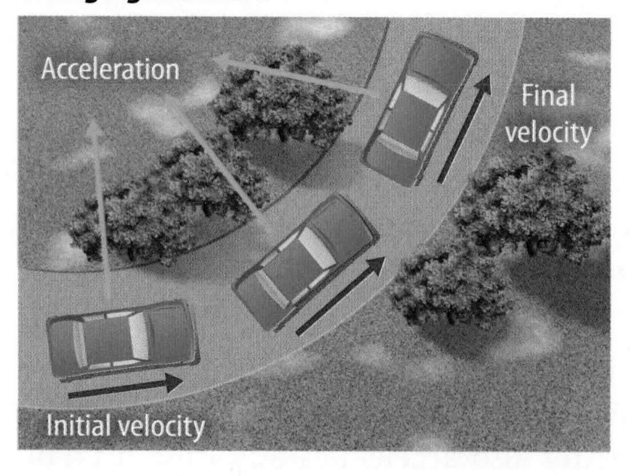

Visual Check If the car in the top picture moved faster, how would the acceleration arrow change?

Representing Acceleration

Like velocity, acceleration has a direction and can be represented by an arrow. Ways an object can accelerate are shown in **Figure 13.** The length of each blue acceleration arrow indicates the amount of acceleration. An acceleration arrow's direction depends on whether velocity increases or decreases.

Changing Speed

The car in the top picture of **Figure 13** is speeding up. At first it is moving slowly, so the arrow that represents its initial velocity is short. The car's speed increases, so the final velocity arrow is longer. As velocity increases, the car accelerates. Notice that the acceleration arrow points in the same direction as the velocity arrows.

The car in the middle picture of **Figure 13** is slowing down. At first it moves fast, so the arrow showing its velocity is long. After the car slows down, the arrow showing its final velocity is shorter. When velocity decreases, acceleration and velocity are in opposite directions. The arrow that represents acceleration is pointing in the direction opposite to the direction the car is moving.

Reading Check In what direction is acceleration if an object is slowing down?

Changing Direction

The car in the bottom picture of **Figure 13** has a constant speed, so the velocity arrow is the same length at each point in the turn. But the car's velocity changes because its direction changes. Because velocity changes, the car is accelerating. Notice the direction of the blue acceleration arrows. It might surprise you that the car is accelerating toward the inside of the curve. Recall that acceleration is the change in velocity. If you compare one velocity arrow with the next, you can see that the change is always toward the inside of the curve.

Key Concept Check What are three ways an object can accelerate?

Calculating Acceleration

Acceleration is a change in velocity divided by the time interval during which the velocity changes. Recall that "velocity" is the speed of an object in a given direction. However, if an object moves along a straight line, you can calculate its acceleration without considering the object's direction. In this lesson, "velocity" refers to only an object's speed. Positive acceleration can be thought of as speeding up in the forward direction. Negative acceleration is slowing down in the forward direction as well as speeding up in the reverse direction.

Acceleration Equation

$$\textbf{acceleration} \text{ (in m/s}^2\text{)} = \frac{\textbf{final speed} \text{ (in m/s)} - \textbf{initial speed} \text{ (in m/s)}}{\textbf{total time} \text{ (in s)}}$$

$$a = \frac{v_f - v_i}{t}$$

Acceleration has SI units of meters per second per second (m/s/s). This can also be written as meters per second squared (m/s^2).

FOLDABLES®

Make a horizontal two-tab Foldable and label it as shown. Use it to summarize information about the changes in velocity that can occur when an object is accelerating.

Changing speed | Changing direction

Acceleration

Math Skills — Acceleration Equation

Solve for Acceleration A bicyclist started from rest along a straight path. After 2.0 s, his speed was 2.0 m/s. After 5.0 s, his speed was 8.0 m/s. What was his acceleration during the time 2.0 s to 5.0 s?

1. **This is what you know:**
   initial speed: $v_i = 2.0$ m/s
   final speed: $v_f = 8.0$ m/s
   total time: $t = 5.0\text{ s} - 2.0\text{ s} = 3.0\text{ s}$

2. **This is what you need to find:** acceleration: a

3. **Use this formula:** $a = \frac{v_f - v_i}{t}$

4. **Substitute:** the values for v_i, v_f, and t into the formula; subtract; then divide.
   $a = \frac{8.0\text{ m/s} - 2.0\text{ m/s}}{3.0\text{ s}} = \frac{6.0\text{ m/s}}{3.0\text{ s}} = 2.0\text{ m/s}^2$

Answer: The acceleration of the bicyclist was **2.0 m/s^2**.

Practice

Aidan drops a rock from a cliff. After 4.0 s, the rock is moving at 39.2 m/s. What is the acceleration of the rock?

- **Math Practice**
- **Personal Tutor**

Inquiry MiniLab

10 minutes

How is a change in speed related to acceleration?

What happens if the distance you walk each second increases? Follow these steps to demonstrate acceleration.

1. Read and complete a lab safety form.
2. Use **masking tape** to mark a course on the floor. Mark start, and place marks along a straight path at 10 cm, 40 cm, 90 cm, 160 cm, and 250 cm from the start.
3. Clap a steady beat. On the first beat, the person walking the course is at start. On the second beat, the walker should be at the 10-cm mark, and so on.

Analyze and Conclude

1. **Explain** what happened to your speed as you moved along the course.
2. **Key Concept** Suppose your speed at the final mark was 0.95 m/s. Calculate your average acceleration from start through the final segment of the course.

WORD ORIGIN

horizontal
from Greek *horizein,* means "limit, divide, separate"

vertical
from Latin *verticalis,* means "overhead"

Speed-Time Graphs

Recall that you can show an object's speed using a distance-time graph. You also can use a speed-time graph to show how speed changes over time. Just like a distance-time graph, a speed-time graph has time on the **horizontal** axis—the *x*-axis. But speed is on the **vertical** axis—the *y*-axis. The figures on the next few pages compare distance-time graphs and speed-time graphs for different types of motion.

Object at Rest

An object at rest is not moving, so its speed is always zero. As a result, the speed-time graph for an object at rest is a horizontal line at $y = 0$, as shown in **Figure 14.**

Figure 14 Both the distance-time graph and the speed-time graph are horizontal lines for an object at rest.

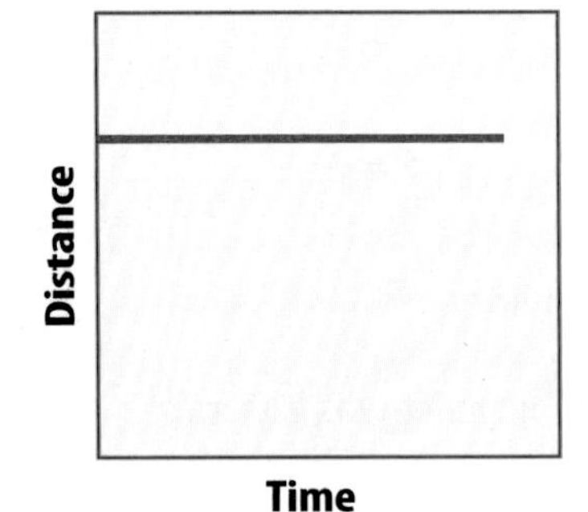

The object's distance from the reference point does not change.

Speed

Time

The speed is zero and does not change.

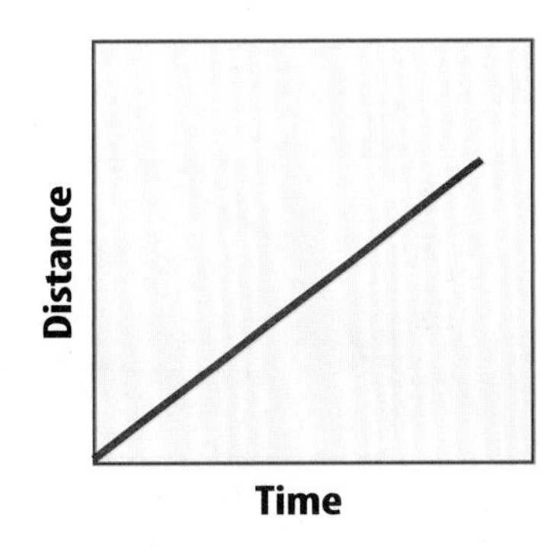

The distance increases at a steady rate over time.

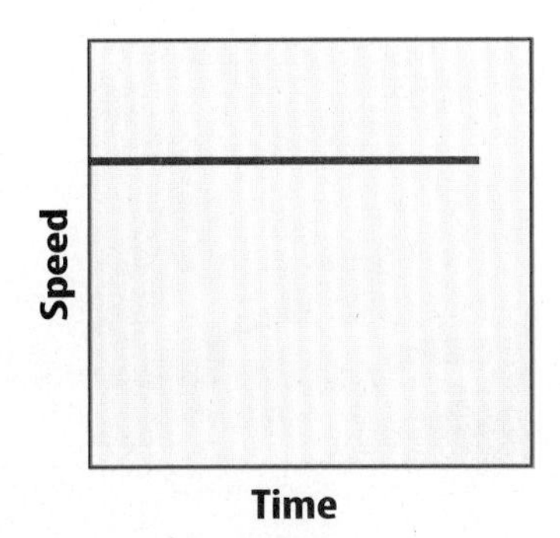

The object's speed does not change.

▲ **Figure 15** For an object moving at constant speed, the speed-time graph is a horizontal line.

Constant Speed

Think about a farm machine moving through a field at a constant speed. At every point in time, its speed is the same. If you plot its speed on a speed-time graph, the plotted line is horizontal, as shown in **Figure 15.** The speed of the object is represented by the distance the horizontal line is from the *x*-axis. If the line is farther from the *x*-axis, the object is moving at a faster speed.

Speeding Up

A plane speeds up as it moves down a runway and takes off. Suppose the speed of the plane increases at a steady rate. If you plot the speed of the plane on a speed-time graph, the line might look like the one in **Figure 16.** The line on the speed-time graph is closer to the *x*-axis at the beginning of the time period when the plane has a lower speed. It slants upward toward the right side of the graph as the speed increases.

Reading Check Why does the speed-time graph of an object that is speeding up slope upward from left to right?

Figure 16 The line on the speed-time graph for an object that is speeding up has an upward slope. ▼

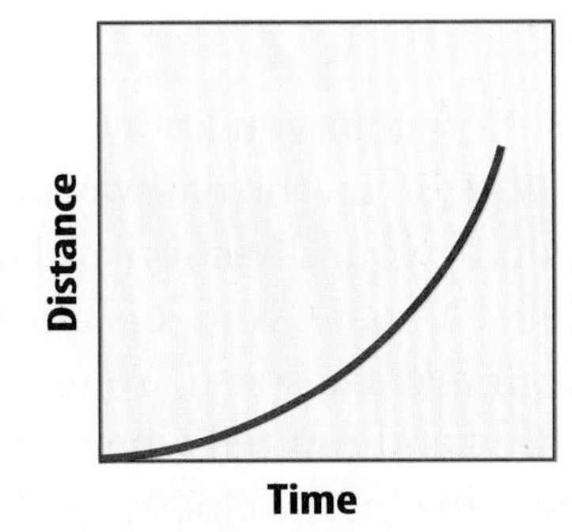

As the distance increases, the rate of increase gets larger over time.

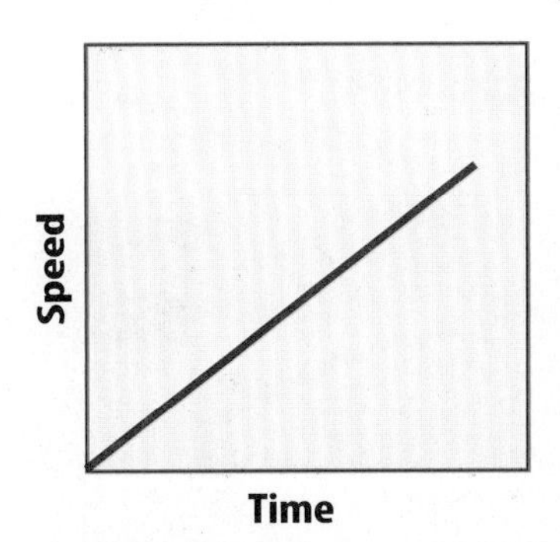

The speed of the object increases at a steady rate over time.

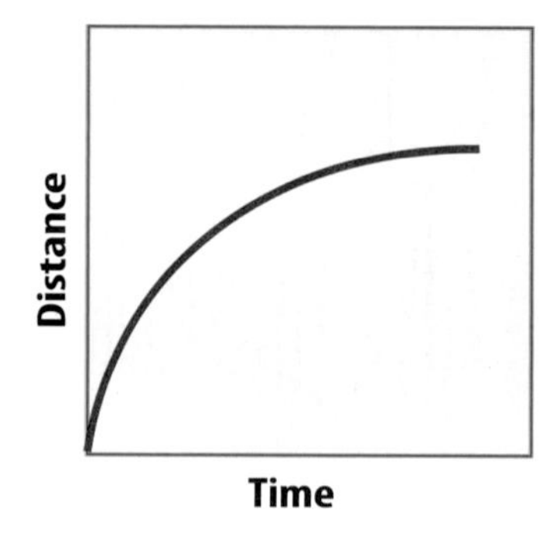

As the distance increases, the rate of increase gets smaller over time.

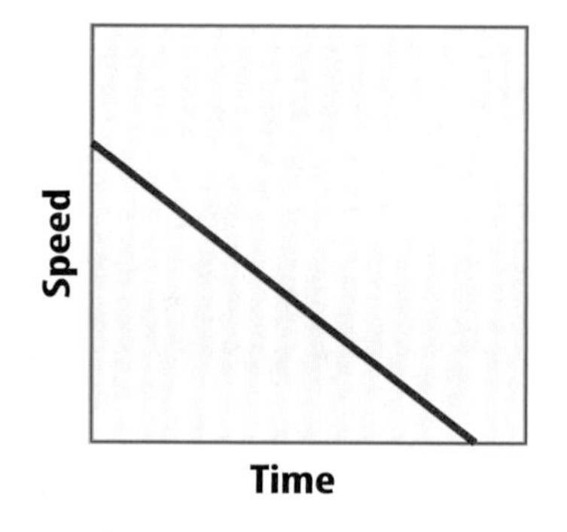

The speed of the object decreases at a steady rate over time.

▲ **Figure 17** The line on the speed-time graph for an object that is slowing down has a downward slope.

Slowing Down

The speed-time graph in **Figure 17** shows the motion of a space shuttle just after it lands. It slows down at a steady rate and then stops. Initially, the shuttle is moving at a high speed. The point representing this speed is far from the *x*-axis. As the shuttle's speed decreases, the points representing its speed are closer to the *x*-axis. The line on the speed-time graph slopes downward to the right. When the line touches the *x*-axis, the speed is zero and the shuttle is stopped.

Key Concept Check What does a speed-time graph show about the motion of an object?

Limits of Speed-Time Graphs

You have read that distance-time graphs show the speed of an object. However, they do not describe the direction in which an object is moving. In the same way, speed-time graphs show only the relationship between speed and time. A speed-time graph of the skier in **Figure 18** would show changes in his speed. It would not show what happens when the skier's velocity changes as the result of a change in his direction.

Figure 18 A speed-time graph of the motion of this skier would show changes in speed but not changes in direction. ▼

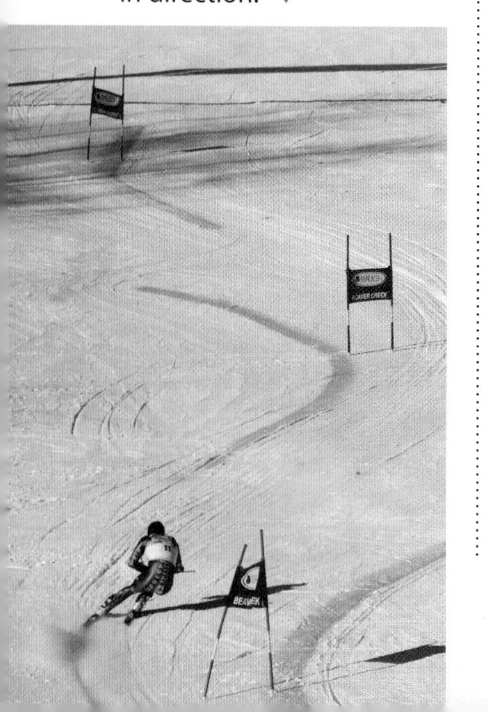

Summarizing Motion

Now that you know about motion, how might you describe a walk down the hallway at school? You can describe your position by your direction and distance from a reference point. You can compare your distance and your displacement and find your average speed. You know that you have an instantaneous speed and can tell when you walk at a constant speed. You can describe your velocity by your speed and your direction. You know you are accelerating if your velocity is changing.

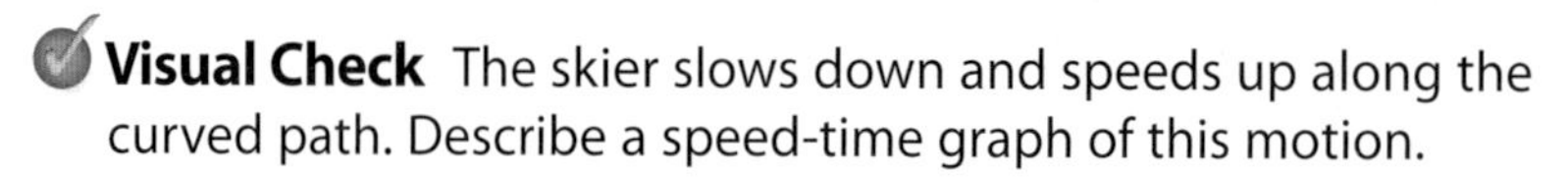

Visual Check The skier slows down and speeds up along the curved path. Describe a speed-time graph of this motion.

Lesson 3 Review

Visual Summary

An object accelerates if it speeds up, slows down, or changes direction.

Acceleration in a straight line can be calculated by dividing the change in speed by the change in time.

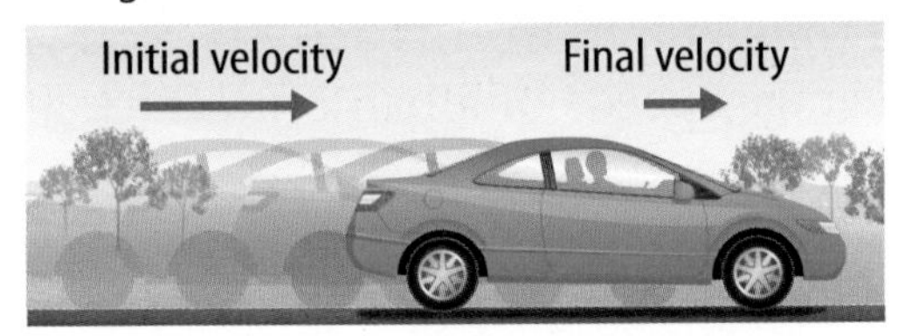

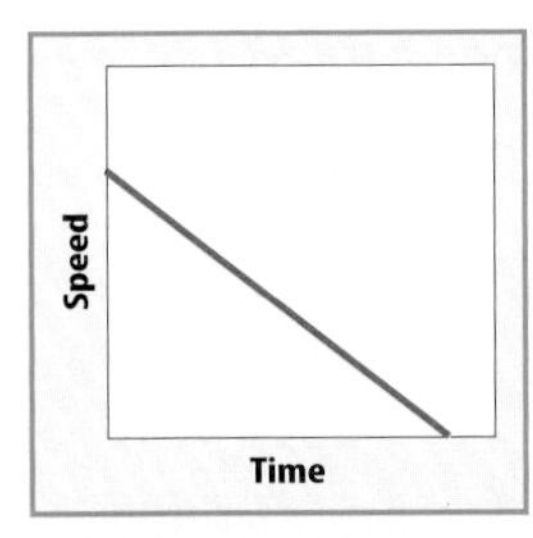

A speed-time graph shows how an object's speed changes over time.

FOLDABLES®

Use your lesson Foldable to review the lesson. Save your Foldable for the project at the end of the chapter.

What do you think NOW?

You first read the statements below at the beginning of the chapter.

5. You can calculate acceleration by dividing the change in velocity by the change in distance.

6. An object accelerates when either its speed or its direction changes.

Did you change your mind about whether you agree or disagree with the statements? Rewrite any false statements to make them true.

Use Vocabulary

1. **Define** *acceleration* in your own words.
2. **Use the term** *acceleration* in a complete sentence.

Understand Key Concepts 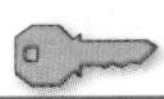

3. **Recall** how a roller coaster can accelerate, even when it is moving at a constant speed.
4. A speed-time graph is a horizontal line with a *y*-value of 4. Which describes the object's motion?
   - **A.** at rest
   - **B.** constant speed
   - **C.** slowing down
   - **D.** speeding up

Interpret Graphics

5. **Organize Information** Copy and fill in the graphic organizer below for the four types of speed-time graphs. For each, describe the motion of the object.

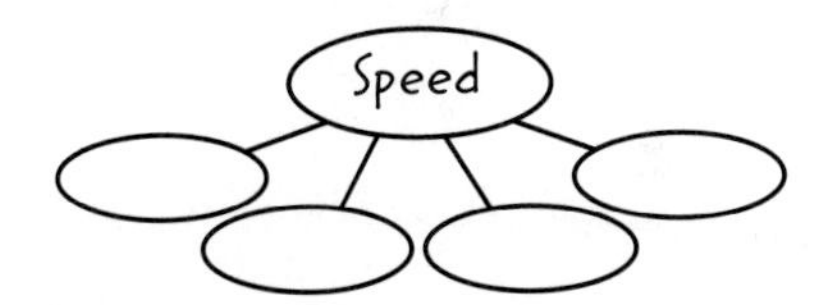

Critical Thinking

6. **Evaluate** A race car accelerates on a straight track from 0 to 100 km/h in 6 s. Another race car accelerates from 0 to 100 km/h in 5 s. Compare the velocities and accelerations of the cars during their races.

Math Skills

Math Practice

7. After 2.0 s, Isabela was riding her bicycle at 3.0 m/s on a straight path. After 5.0 s, she was moving at 5.4 m/s. What was her acceleration?
8. After 3.0 s, Mohammed was running at 1.2 m/s on a straight path. After 7.0 s, he was running at 2.0 m/s. What was his acceleration?

Materials

meltersticks (6)

stopwatches (6)

masking tape

tennis ball

Safety

Calculate Average Speed from a Graph

You probably do not walk the same speed uphill and downhill, or when you are just starting out and when you are tired. If you are walking and you measure and record the distance you walk every minute, the distances will vary. How might you use these measurements to calculate the average speed you walked? One way is to organize the data on a distance-time graph. In this activity, you will use such a graph to compare average speeds of a ball on a track using different heights of a ramp.

Ask a Question

How does the height of a ramp affect the speed of a ball along a track?

Make Observations

1. Read and complete a lab safety form.
2. Make a 3-m track. Place three metersticks end-to-end. Place three other metersticks end-to-end about 6 cm from the first set of metersticks. Use tape to hold the metersticks in place. Mark each half-meter with tape. Use books to make a ramp leading to the track.

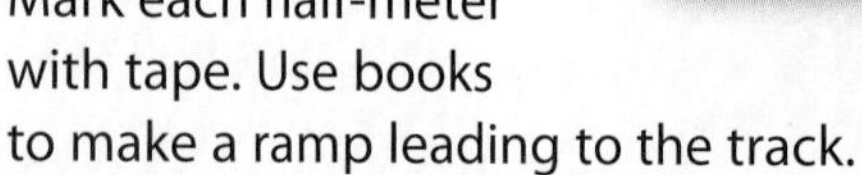

3. A student should be at each half-meter mark with a stopwatch. Another student should be by the ramp to roll a ball along the track.
4. When the ball passes start, all group members should start their stopwatches. Each student should stop his or her stopwatch when the ball crosses the mark where the student is stationed.
5. Practice several times to get consistent rolls and times.

Form a Hypothesis

6. Create a hypothesis about how the number of books used as a ramp affects the speed of the ball rolling along the track.

Test Your Hypothesis

7. Write a plan for varying the number of books and making distance and time measurements.
8. Create a data table in your Science Journal that matches your plan. A sample is shown to the right.
9. Use your plan to make the measurements. Record them in the data table.
10. Plot the data for each height of the ramp on a graph that shows the distance the ball traveled on the *x*-axis and time on the *y*-axis. For each ramp height, draw a straight line that goes through the most points.
11. Choose two points on each line. Calculate the average speed between these points by dividing the difference in the distances for the two points by the difference in the times.

Analyze and Conclude

12. **Compare** the average speeds for each ramp height. Use this comparison to decide whether your results support your hypothesis.
13. **The Big Idea** How was the distance-time graph useful for describing the motion of the ball?

Communicate Your Results

Prepare a poster that shows your graph and describes how it can be used to calculate average speed.

Design and conduct an experiment comparing the average speed of different types of balls along the track.

8

Distance (m)	Time(s)		
	2 books	3 books	4 books
0.50			
1.00			
1.50			
2.00			
2.50			
3.00			

Lab TIPS

- If the ball doesn't roll far enough, reduce the track length to 2 m.
- Practice using the stopwatches several times to gain experience in making accurate readings.

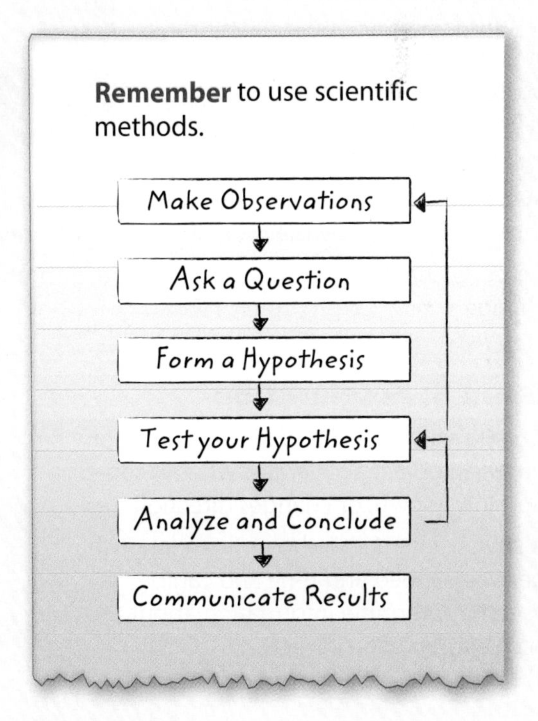

Chapter 1 Study Guide

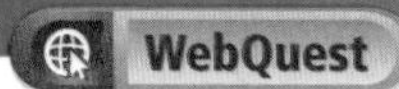

The motion of an object can be described by the object's position, velocity, and acceleration.

Key Concepts Summary

Lesson 1: Position and Motion

- An object's **position** is its distance and direction from a **reference point.**
- The position of an object in two dimensions can be described by choosing a reference point and two reference directions, and then stating the distance along each reference direction.
- The distance an object moves is the actual length of its path. Its **displacement** is the difference between its initial position and its final position.

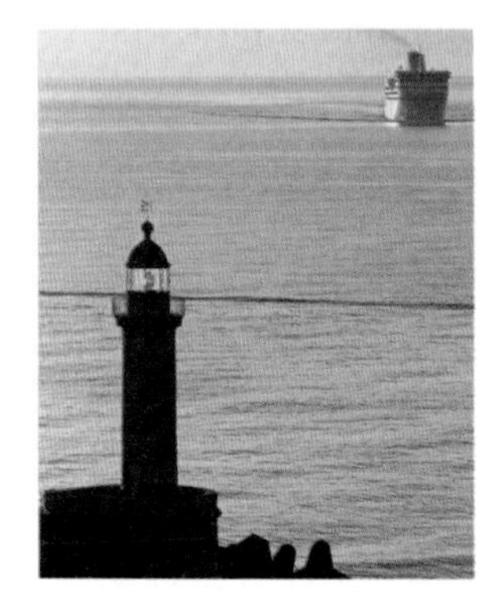

Vocabulary

reference point p. 9
position p. 9
motion p. 13
displacement p. 13

Lesson 2: Speed and Velocity

- **Speed** is the distance an object moves per unit of time.
- An object moving the same distance each second is moving at a **constant speed.** The speed of an object at a certain moment is its **instantaneous speed.**
- You can calculate an object's **average speed** from a distance-time graph by dividing the distance the object travels by the total time it takes to travel that distance.
- **Velocity** changes when speed, direction, or both speed and direction change.

Vocabulary

speed p. 17
constant speed p. 18
instantaneous speed p. 18
average speed p. 19
velocity p. 23

Lesson 3: Acceleration

- **Acceleration** is a change in velocity over time. An object accelerates when it speeds up, slows down, or changes direction.
- A speed-time graph shows the relationship between speed and time and can be used to determine information about the acceleration of an object.

Vocabulary

acceleration p. 27

- Personal Tutor
- Vocabulary eGames
- Vocabulary eFlashcards

FOLDABLES® Chapter Project

Assemble your lesson Foldables as shown to make a Chapter Project. Use the project to review what you have learned in this chapter.

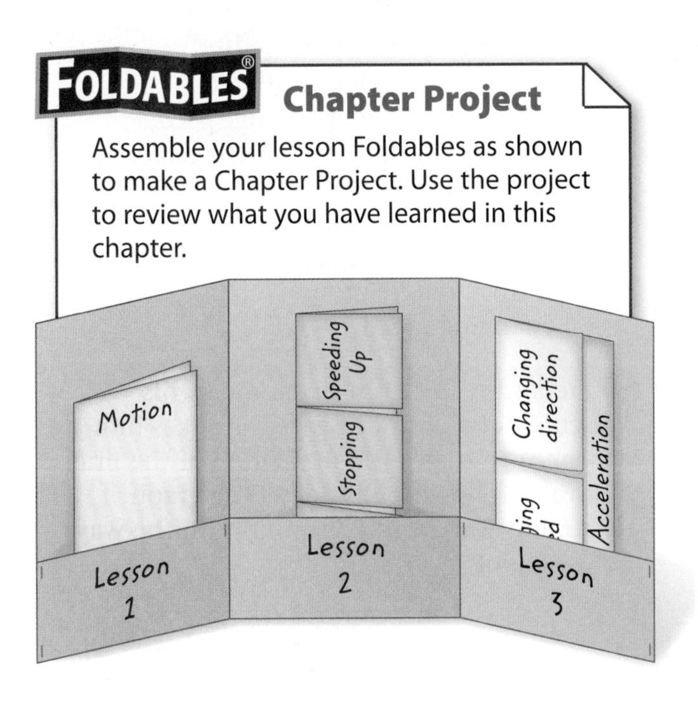

Use Vocabulary

1. A pencil's _______ might be described as 3 cm to the left of the stapler.
2. An object that changes position is in _______.
3. If an object is traveling at a _______, it does not speed up or slow down.
4. An object's _______ includes both its speed and the direction it moves.
5. An object's change in velocity during a time interval, divided by the time interval during which the velocity changed, is its _______.
6. A truck driver stepped on the brakes to make a quick stop. The truck's _______ is in the opposite direction as its velocity.

Link Vocabulary and Key Concepts

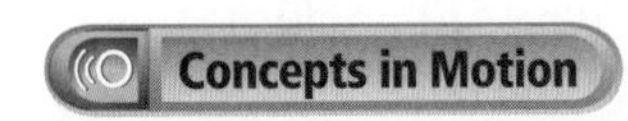

Interactive Concept Map

Copy this concept map, and then use vocabulary terms from the previous page to complete the concept map.

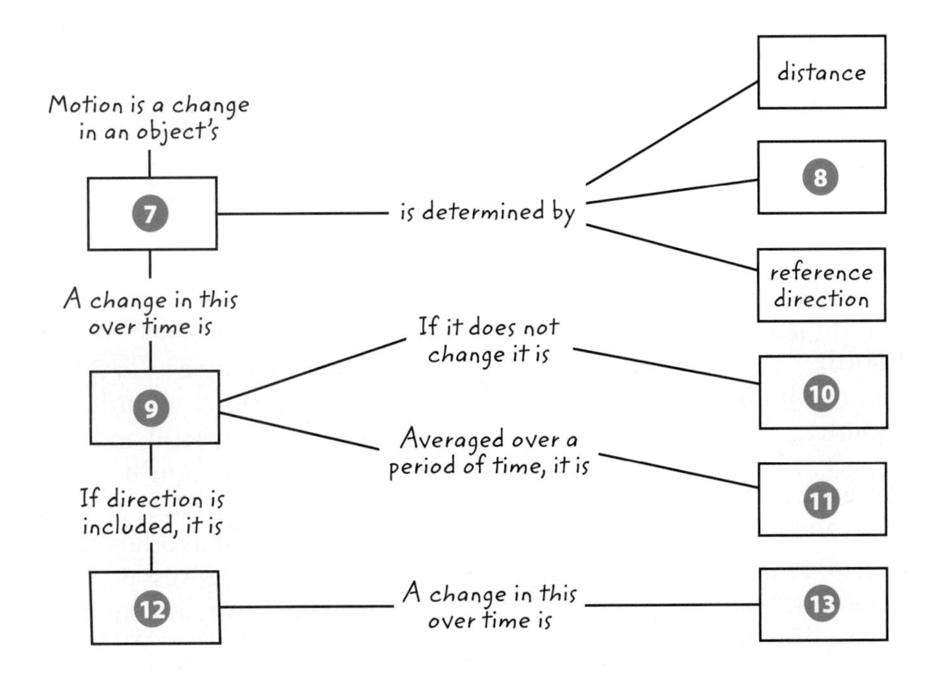

Chapter 1 Review

Understand Key Concepts 

1. An airplane rolls down the runway. Compared to which reference point is the airplane in motion?
   A. the cargo the plane carries
   B. the control tower
   C. the pilot flying the plane
   D. the plane's wing

2. Which describes motion in two dimensions?
   A. a car driving through a city
   B. a rock dropping off a cliff
   C. a sprinter on a 100-m track
   D. a train on a straight track

3. Which line represents the greatest average speed during the 30-s time period?

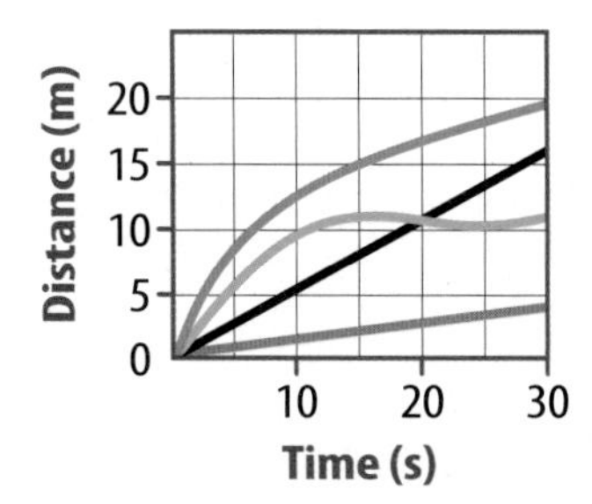

   A. the blue line
   B. the black line
   C. the green line
   D. the orange line

4. Which describes the greatest displacement?
   A. walking 3 m east, then 3 m north, then 3 m west
   B. walking 3 m east, then 3 m south, then 3 m east
   C. walking 3 m north, then 3 m south, then 3 m north
   D. walking 3 m north, then 3 m west, then 3 m south

5. Which has the greatest average speed?
   A. a boat sailing 80 km in 2 hours
   B. a car driving 90 km in 3 hours
   C. a train traveling 120 km in 3 hours
   D. a truck moving 50 km in 1 hour

6. Which describes motion in which the person or object is accelerating?
   A. A bird flies straight from a tree to the ground without changing speed.
   B. A dog walks at a constant speed along a straight sidewalk.
   C. A girl runs along a straight path the same distance each second.
   D. A truck moves around a curve without changing speed.

7. Richard walks from his home to his school at a constant speed. It takes him 4 min to travel 100 m. Which of the lines in the following distance-time graph could show Richard's motion on the way to school?

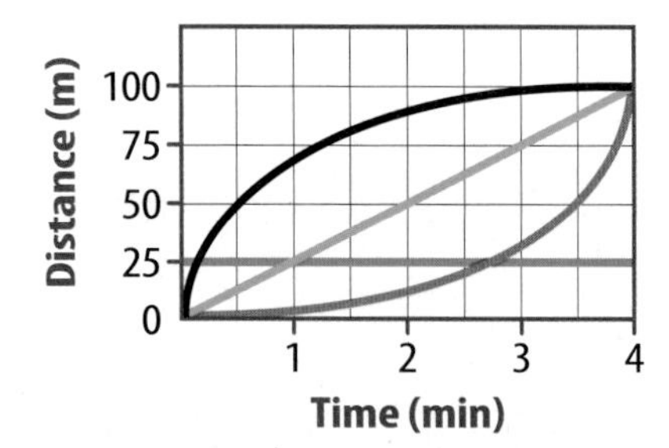

   A. the black line
   B. the blue line
   C. the green line
   D. the orange line

8. Which is a unit of acceleration?
   A. kg/m
   B. kg•m/s^2
   C. m/s
   D. m/s^2

9. Which have the same velocity?
   A. a boy walking east at 2 km/h and a man walking east at 4 km/h
   B. a car standing still and a truck driving in a circle at 4 km/h
   C. a dog walking west at 3 km/h and a cat walking west at 3 km/h
   D. a girl walking west at 3 km/h and a boy walking south at 3 km/h

Assessment
Online Test Practice

Critical Thinking

10. **Describe** A ruler is on the table with the higher numbers to the right. An ant crawls along the ruler from 6 cm to 2 cm in 2 seconds. Describe the ant's distance, displacement, speed, and velocity.

11. **Describe** a theme-park ride that has constant speed but changing velocity.

12. **Construct** a distance-time graph that shows the following motion: A person leaves a starting point at a constant speed of 4 m/s and walks for 4 s. The person then stops for 2 s. The person then continues walking at a constant speed of 2 m/s for 4 s.

13. **Calculate** A truck driver travels 55 km in 1 hour. He then drives a speed of 35 km/h for 2 hours. Next, he drives 175 km in 3 hours. What was his average speed?

14. **Interpret** Keisha measured the distance her friend Morgan ran on a straight track every 2 s. Her measurements are recorded in the table below. What was Morgan's average speed? What was her acceleration for the entire trip?

Time (s)	Distance (m)
0	0
2	2
4	6
6	8
8	14
10	20

Writing in Science

15. **Write** A friend tells you he is 30 m from the fountain in the middle of the city. Write a short paragraph explaining why you cannot identify your friend's position from this description.

REVIEW THE BIG IDEA

16. Nora rides a bicycle for 5 min on a curvy road at a constant speed of 10 m/s. Describe Nora's ride in terms of position, velocity, and acceleration. Compare the distance she rides and her displacement.

17. What are some ways to describe the motion of the jets in the photograph below?

Math Skills

Math Practice

Solve One-Step Equations

18. A model train moves 18.3 m in 122 s. What is the train's average speed?

19. A car travels 45 km in an hour. In each of the next two hours, it travels 78 km. What is the average speed of the car?

20. The speed of a car traveling on a straight road increases from 63 m/s to 75 m/s in 4.2 s. What is the car's acceleration?

21. A girl starts from rest and reaches a walking speed of 1.4 m/s in 3.0 s. She walks at this speed for 6.0 s. The girl then slows down and comes to a stop during a 10.0-s period. What was the girl's acceleration during each of the three time periods?

Standardized Test Practice

Record your answers on the answer sheet provided by your teacher or on a sheet of paper.

Multiple Choice

1 Radar tells an air traffic controller that a jet is slowing as it nears the airport. Which might represent the jet's speed?

A 700 h

B 700 h/km

C 700 km

D 700 km/h

Use the diagram below to answer question 2.

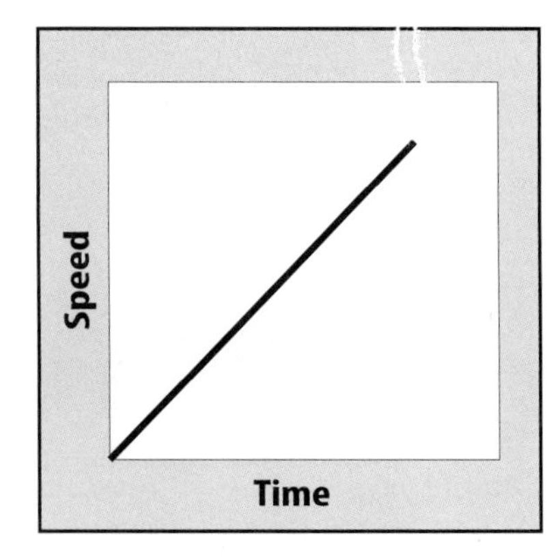

2 What does the graph above illustrate?

A average speed

B constant speed

C decreasing speed

D increasing speed

3 Why is a car accelerating when it is circling at a constant speed?

A It is changing its destination.

B It is changing its direction.

C It is changing its distance.

D It is changing its total mass.

4 Which is defined as the process of changing position?

A displacement

B distance

C motion

D relativity

5 Each diagram below shows two sliding boxes. Which boxes have the same velocity?

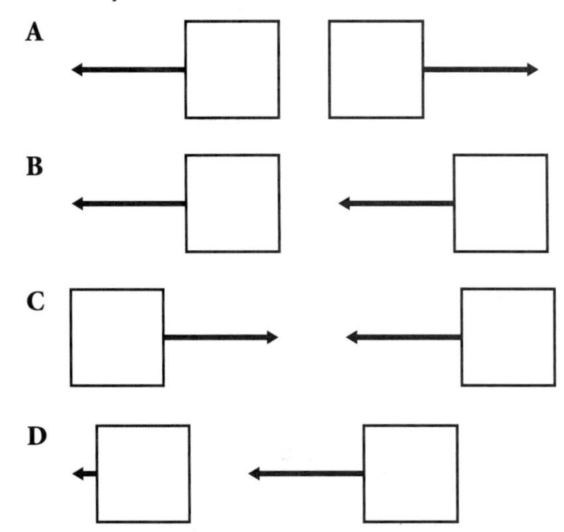

6 In the phrase "two miles southeast of the mall," what is the mall?

A a dimension

B a final destination

C a position

D a reference point

7 The initial speed of a dropped ball is 0 m/s. After 2 seconds, the ball travels at a speed of 20 m/s. What is the acceleration of the ball?

A 5 m/s^2

B 10 m/s^2

C 20 m/s^2

D 40 m/s^2

8 Which could be described by the expression "100 m/s northwest"?

A acceleration

B distance

C speed

D velocity

Use the diagram below to answer question 9.

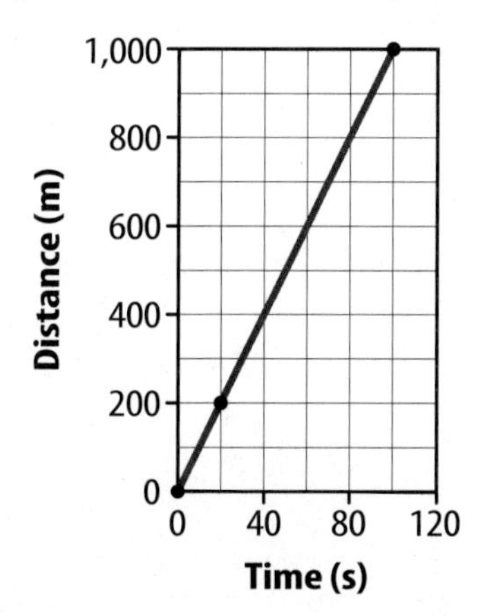

9 In the above graph, what is the average speed of the moving object between 20 and 60 seconds?

A 5 m/s

B 10 m/s

C 20 m/s

D 40 m/s

10 A car travels 250 km and stops twice along the way. The entire trip takes 5 hours. What is the average speed of the car?

A 25 km/h

B 40 km/h

C 50 km/h

D 250 km/h

Constructed Response

Use the diagram below to answer questions 11 and 12.

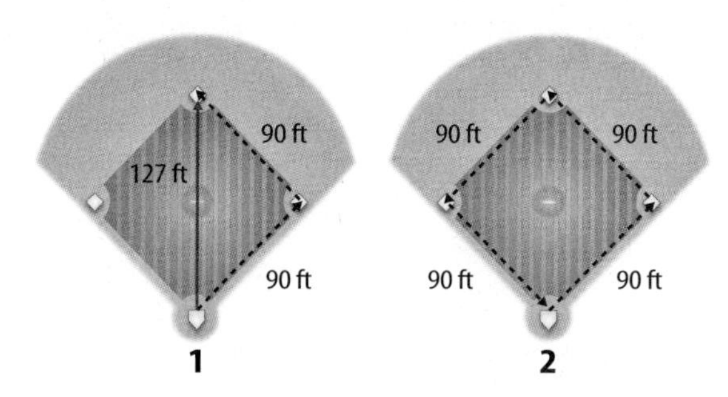

11 The dashed lines show the paths two players run on baseball diamonds. What distance does the player travel on diamond 1? How does it compare to the distance the player runs on diamond 2?

12 Calculate the displacement of the runners on diamonds 1 and 2. Explain your answers.

Use the diagram below to answer question 13.

North

Key: Each square is one city block

Soccer field

School

West

Ice cream shop

East

Home

South

13 A student walks from home to school, on to a soccer field, then to an ice cream shop, and finally home. Use grid distances and directions to describe each leg of his trip. What is the distance between the student's home and the ice cream shop?

NEED EXTRA HELP?													
If You Missed Question...	1	2	3	4	5	6	7	8	9	10	11	12	13
Go to Lesson...	2	3	3	1	2	1	3	2	2	2	1	1	1

Chapter 2

The Laws of Motion

THE BIG IDEA **How do forces change the motion of objects?**

Inquiry Why move around?

Imagine the sensations these riders experience as they swing around. The force of gravity pulls the riders downward. Instead of falling, however, they move around in circles.

- What causes the riders to move around?
- What prevents the riders from falling?
- How do forces change the motion of the riders?

Get Ready to Read

What do you think?

Before you read, decide if you agree or disagree with each of these statements. As you read this chapter, see if you change your mind about any of the statements.

1. You pull on objects around you with the force of gravity.
2. Friction can act between two unmoving, touching surfaces.
3. Forces acting on an object cannot be added.
4. A moving object will stop if no forces act on it.
5. When an object's speed increases, the object accelerates.
6. If an object's mass increases, its acceleration also increases if the net force acting on the object stays the same.
7. If objects collide, the object with more mass applies more force.
8. Momentum is a measure of how hard it is to stop a moving object.

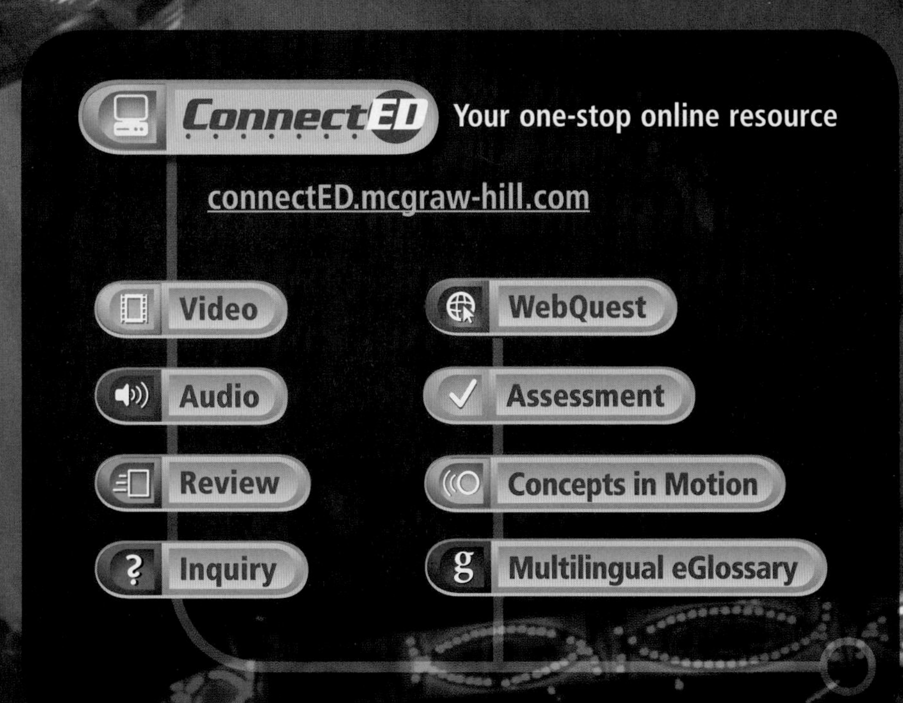

Lesson 1

Gravity and Friction

Reading Guide

Key Concepts
ESSENTIAL QUESTIONS

- What are some contact forces and some noncontact forces?
- What is the law of universal gravitation?
- How does friction affect the motion of two objects sliding past each other?

Vocabulary

force p. 45
contact force p. 45
noncontact force p. 46
gravity p. 47
mass p. 47
weight p. 48
friction p. 49

 Multilingual eGlossary

 Video

What's Science Got to do With It?

Inquiry Why doesn't he fall?

This astronaut is on an aircraft that flies at steep angles and provides a sense of weightlessness. Why doesn't he fall? He does! Earth's gravity pulls the astronaut down, but the aircraft moves downward at the same speed.

5 minutes

Can you make a ball move without touching it?

You can make a ball move by kicking it or throwing it. Is it possible to make the ball move even when nothing is touching the ball?

1. Read and complete a lab safety form.
2. Roll a **tennis ball** across the floor. Think about what makes the ball move.
3. Toss the ball into the air. Watch as it moves up and then falls back to your hand.
4. Drop the ball onto the floor. Let it bounce once, and then catch it.

Think About This

1. What made the ball move when you rolled, tossed, and dropped it? What made it stop?
2. **Key Concept** Did something that was touching the ball or not touching the ball cause it to move in each case?

Types of Forces

Think about all the things you pushed or pulled today. You might have pushed toothpaste out of a tube. Maybe you pulled out a chair to sit down. *A push or a pull on an object is called a* **force.** An object or a person can apply a force to another object or person. Some forces are applied only when objects touch. Other forces are applied even when objects do not touch.

Word Origin
force
from Latin *fortis,* means "strong"

Contact Forces

The hand of the karate expert in **Figure 1** applied a force to the stack of boards and broke them. You have probably also seen a musician strike the keys of a piano and an athlete hit a ball with a bat. In each case, a person or an object applied a force to an object that it touched. *A* **contact force** *is a push or a pull on one object by another that is touching it.*

Contact forces can be weak, like when you press the keys on a computer keyboard. They also can be strong, such as when large sections of underground rock suddenly move, resulting in an earthquake. The large sections of Earth's crust called plates also apply strong contact forces against each other. Over long periods of time, these forces can create mountain ranges if one plate pushes another plate upward.

Figure 1 The man's hand applies a contact force to the boards.

Make a vertical two-tab book from a sheet of paper. Label it as shown. Use it to organize your notes on gravity and friction.

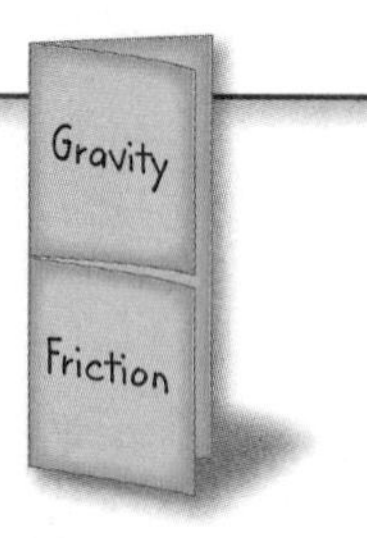

▲ **Figure 2** A noncontact force causes the girl's hair to stand on end.

Noncontact Forces

Lift a pencil and then release it. What happens? The pencil falls toward the floor. A parachutist falls toward Earth even though nothing is touching him. *A force that one object can apply to another object without touching it is a* **noncontact force.** Gravity, which pulled on your pencil and the parachutist, is a noncontact force. The magnetic force, which attracts certain metals to magnets, is also a noncontact force. In **Figure 2,** another noncontact force, called the electric force, causes the girl's hair to stand on end.

Key Concept Check What are some contact forces and some noncontact forces?

Figure 3 Arrows can indicate the strength and direction of a force. ▼

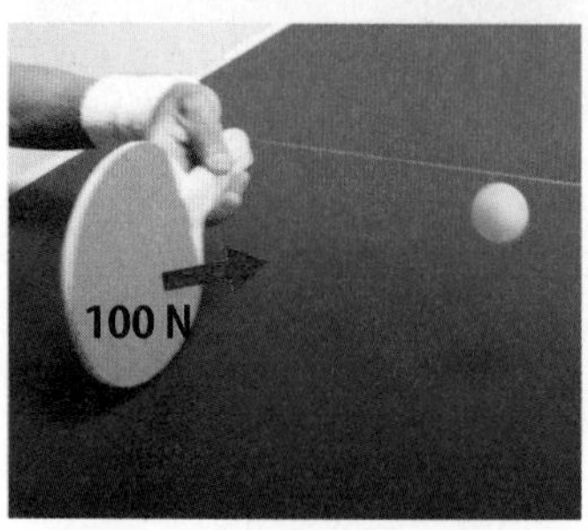

Visual Check How are the lengths of the arrows related to the different forces on the two balls?

Strength and Direction of Forces

Forces have both strength and direction. If you push your textbook away from you, it probably slides across your desk. What happens if you push down on your book? It probably does not move. You can use the same strength of force in both cases. Different things happen because the direction of the applied force is different.

As shown in **Figure 3,** arrows can be used to show forces. The length of an arrow shows the strength of the force. Notice in the figure that the force applied by the tennis racquet is stronger than the force applied by the table-tennis paddle. As a result, the arrow showing the force of the tennis racquet is longer. The direction that an arrow points shows the direction in which force was applied.

The SI unit for force is the newton (N). You apply a force of about 1 N when lifting a stick of butter. You use a force of about 20 N when lifting a 2-L bottle of water. If you use arrows to show these forces, the water's arrow would be 20 times longer.

Change in Mass

Gravitational force increases if the mass of at least one of the objects increases.

Change in Distance

The gravitational force between objects decreases as the objects move apart.

▲ **Figure 4** The gravitational force between objects depends on the mass of the objects and the distance between them.

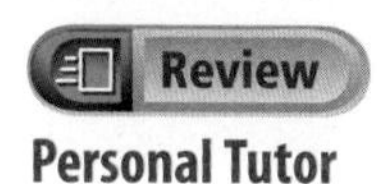

Personal Tutor

What is gravity?

Objects fall to the ground because Earth exerts an attractive force on them. Did you know that you also exert an attractive force on objects? **Gravity** *is an attractive force that exists between all objects that have mass.* **Mass** *is the amount of matter in an object.* Mass is often measured in kilograms (kg).

The Law of Universal Gravitation

In the late 1600s, an English scientist and mathematician, Sir Isaac Newton, developed the law of universal gravitation. This law states that all objects are attracted to each other by a gravitational force. The strength of the force depends on the mass of each object and the distance between them.

Key Concept Check What is the law of universal gravitation?

Gravitational Force and Mass The way in which the mass of objects affects gravity is shown in **Figure 4.** When the mass of one or both objects increases, the gravitational force between them also increases. Notice that the force arrows for each pair of marbles are the same size even when one object has less mass. Each object exerts the same attraction on the other object.

Gravitational Force and Distance The effect that distance has on gravity is also shown in **Figure 4.** The attraction between objects decreases as the distance between the objects increases. For example, if your mass is 45 kg, the gravitational force between you and Earth is about 440 N. On the Moon, about 384,000 km away, the gravitational force between you and Earth would only be about 0.12 N. The relationship between gravitational force and distance is shown in the graph in **Figure 5.**

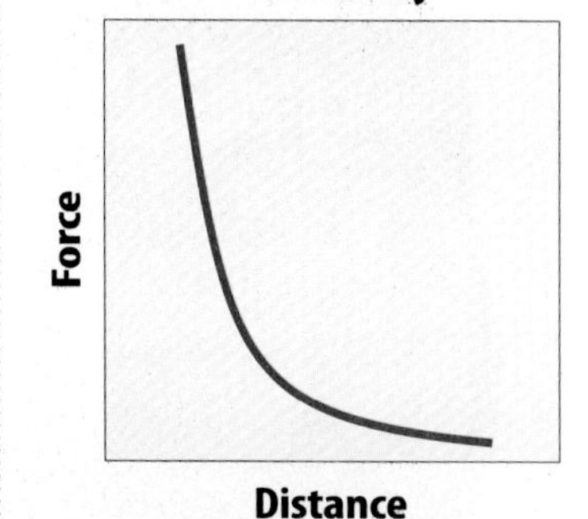

▲ **Figure 5** The gravitational force between objects decreases as the distance between the objects increases.

Reading Check What effect does distance have on gravity?

Weight—A Gravitational Force

Earth has more mass than any object near you. As a result, the gravitational force Earth exerts on you is greater than the force exerted by any other object. **Weight** *is the gravitational force exerted on an object.* Near Earth's surface, an object's weight is the gravitational force exerted on the object by Earth. Because weight is a force, it is measured in newtons.

The Relationship Between Weight and Mass An object's weight is proportional to its mass. For example, if one object has twice the mass of another object, it also has twice the weight. Near Earth's surface, the weight of an object in newtons is about ten times its mass in kilograms.

Reading Check What is the relationship between mass and weight?

ACADEMIC VOCABULARY
significant
(adjective) important, momentous

Weight and Mass High Above Earth You might think that astronauts in orbit around Earth are weightless. Their weight is about 90 percent of what it is on Earth. The mass of the astronaut in **Figure 6** is about 55 kg. Her weight is about 540 N on Earth and about 500 N on the space station 350 km above Earth's surface. Why is there no **significant** change in weight when the distance increases so much? Earth is so large that an astronaut must be much farther away for the gravitational force to change much. The distance between the astronaut and Earth is small compared to the size of Earth.

Reading Check Why is the gravitational force that a friend exerts on you less than the gravitational force exerted on you by Earth?

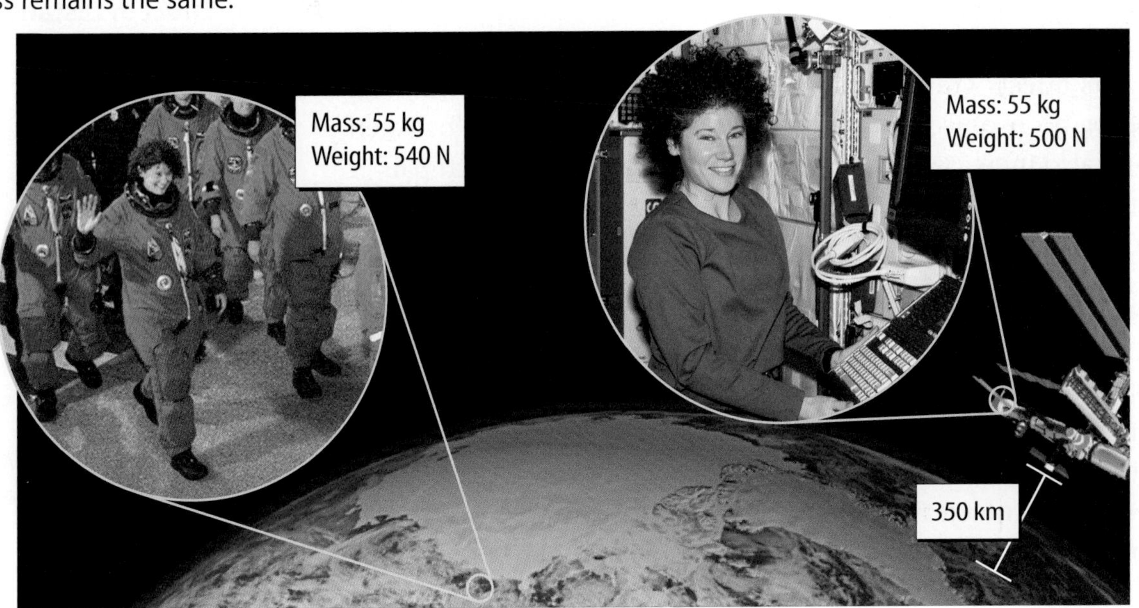

Figure 6 As she travels from Earth to the space station, the astronaut's weight changes, but her mass remains the same.

Visual Check What would be the weight of a 110-kg object on Earth? On the space station?

Static and Sliding Friction

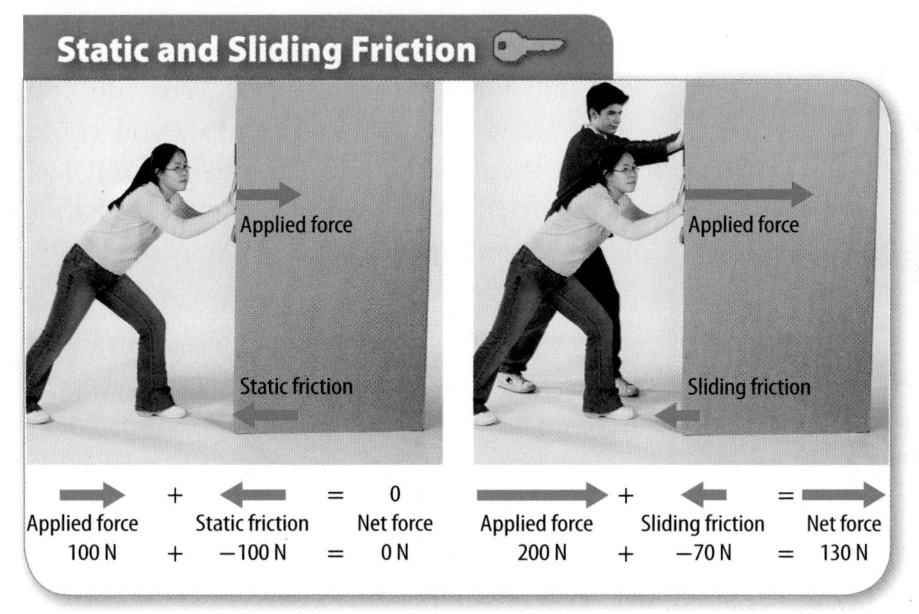

◀ **Figure 7** Static friction prevents the box on the left from moving. Sliding friction slows the motion of the box on the right.

Friction

If you slide across a smooth floor in your socks, you move quickly at first and then stop. The force that slows you is friction. **Friction** *is a force that resists the motion of two surface that are touching.* There are several types of friction.

Static Friction

The box on the left in **Figure 7** does not move because the girl's applied force is balanced by **static** friction. Static friction prevents surfaces from sliding past each other. Up to a limit, the strength of static friction changes to match the applied force. If the girl increases the applied force, the box still will not move because the static friction also increases.

SCIENCE USE V. COMMON USE
static
Science Use at rest or having no motion
Common Use noise produced in a radio or a television

Sliding Friction

When static friction reaches its limit between surfaces, the box will move. As shown in **Figure 7,** the force of two students pushing is greater than the static friction between the box and the floor. Sliding friction opposes the motion of surfaces sliding past each other. As long as the box is sliding, the sliding friction does not change. Increasing the applied force makes the box slide faster. If the students stop pushing, the box will slow and stop because of sliding friction.

Fluid Friction

Friction between a surface and a fluid—any material, such as water or air, that flows—is fluid friction. Fluid friction between a surface and air is air resistance. Suppose an object is moving through a fluid. Decreasing the surface area toward the oncoming fluid decreases the air resistance against the object. The crumpled paper in **Figure 8** falls faster than the flat paper because it has less surface area and less air resistance.

Figure 8 Air resistance is greater on the flat paper. ▼

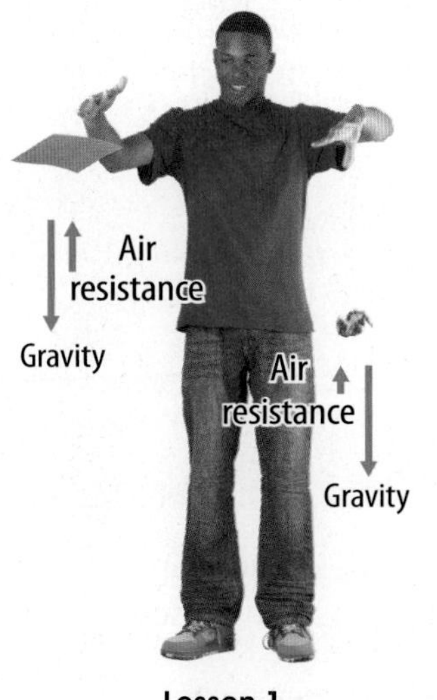

Inquiry MiniLab 10 minutes

How does friction affect motion?

Friction affects the motion of an object sliding across a surface.

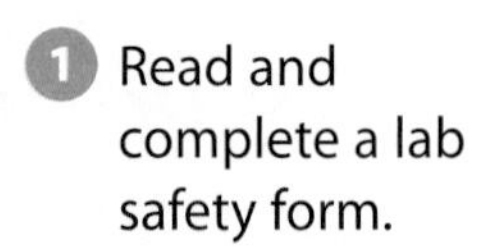

1. Read and complete a lab safety form.
2. Use **tape** to fasten **sandpaper** to a table. Attach a **spring scale** to a **wooden block** with an **eyehook** in it.
3. Record in your Science Journal the force required to gently pull the block at a constant speed on the sandpaper and then on the table.

Analyze and Conclude

1. **Compare** the forces required to pull the block across the two surfaces.
2. **Key Concept** How did reducing friction affect the motion of the block?

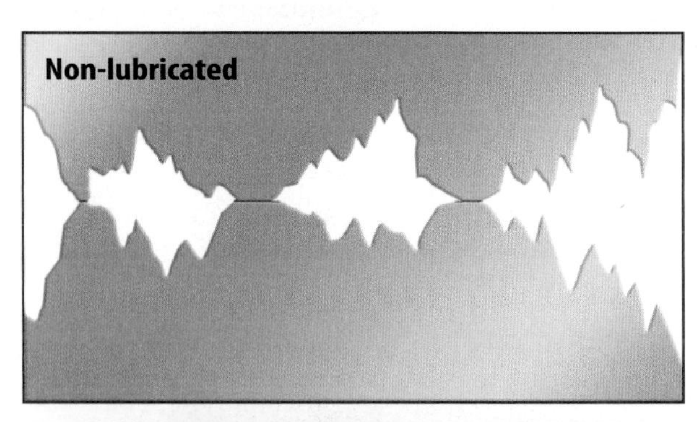

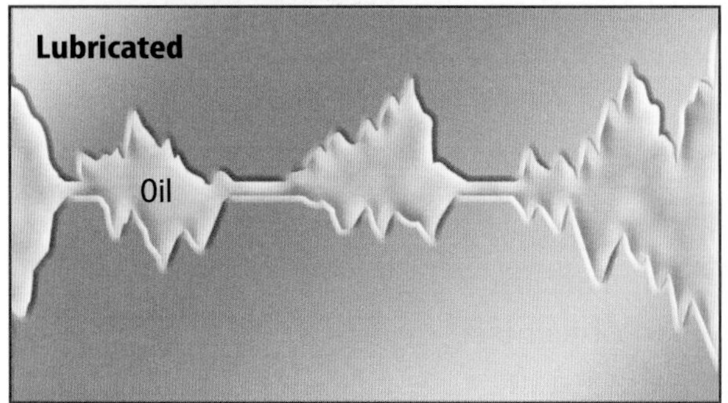

Figure 9 Lubricants such as oil decrease friction caused by microscopic roughness.

What causes friction?

Rub your hands together. What do you feel? If your hands were soapy, you could slide them past each other easily. You feel more friction when you rub your dry hands together than when you rub your soapy hands together.

What causes friction between surfaces? Look at the close-up view of surfaces in **Figure 9.** Microscopic dips and bumps cover all surfaces. When surfaces slide past each other, the dips and bumps on one surface catch on the dips and bumps on the other surface. This microscopic roughness slows sliding and is a source of friction.

Key Concept Check How does friction affect the motion of two objects sliding past each other?

In addition, small particles—atoms and molecules—make up all surfaces. These particles contain weak electrical charges. When a positive charge on one surface slides by a negative charge on the other surface, an attraction occurs between the particles. This attraction slows sliding and is another source of friction between the surfaces.

Reading Check What are two causes of friction?

Reducing Friction

When you rub soapy hands together, the soapy water slightly separates the surfaces of your hands. There is less contact between the microscopic dips and bumps and between the electrical charges of your hands. Soap acts as a lubricant and decreases friction. With less friction, it is easier for surfaces to slide past each other, as shown in **Figure 9.** Motor oil is a lubricant that reduces friction between moving parts of a car's engine.

Look again at the effect of air resistance on the falling paper in **Figure 8.** Reducing the paper's surface area reduces the fluid friction between it and the air.

Lesson 1 Review

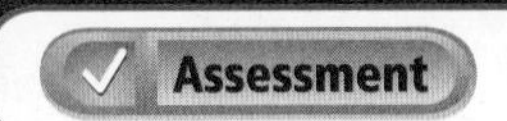

Visual Summary

Forces can be either contact, such as a karate chop, or noncontact, such as gravity. Each type is described by its strength and direction.

Gravity is an attractive force that acts between any two objects that have mass. The attraction is stronger for objects with greater mass.

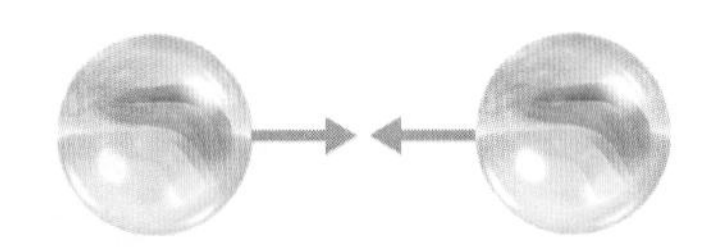

Friction can reduce the speed of objects sliding past each other. Air resistance is a type of fluid friction that slows the speed of a falling object.

FOLDABLES®

Use your lesson Foldable to review the lesson. Save your Foldable for the project at the end of the chapter.

What do you think NOW?

You first read the statements below at the beginning of the chapter.

1. You pull on objects around you with the force of gravity.
2. Friction can act between two unmoving, touching surfaces.

Did you change your mind about whether you agree or disagree with the statements? Rewrite any false statements to make them true.

Use Vocabulary

1. **Define** *friction* in your own words.
2. **Distinguish** between weight and mass.

Understand Key Concepts 

3. **Explain** the difference between a contact force and a noncontact force.
4. You push a book sitting on a desk with a force of 5 N, but the book does not move. What is the static friction?
   A. 0 N
   B. 5 N
   C. between 0 N and 5 N
   D. greater than 5 N
5. **Apply** According to the law of universal gravitation, is there a stronger gravitational force between you and Earth or an elephant and Earth? Why?

Interpret Graphics

6. **Interpret** Look at the forces on the feather.

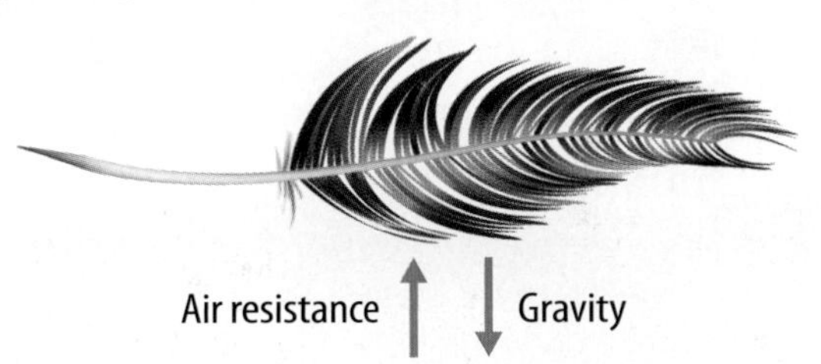

In terms of these forces, explain why the feather falls slowly rather than fast.

7. **Organize Information** Copy and fill in the table below to describe forces mentioned in this lesson. Add as many rows as you need.

Force	Description

Critical Thinking

8. **Decide** Is it possible for the gravitational force between two 50-kg objects to be less than the gravitational force between a 50-kg object and a 5-kg object? Explain.

SCIENCE & SOCIETY

Avoiding an Asteroid Collision

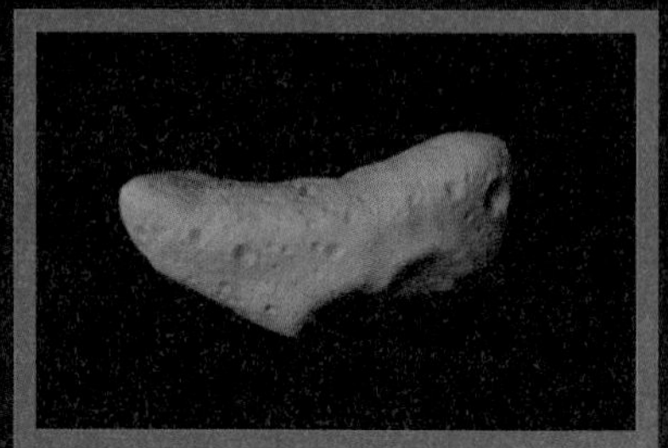

The force of gravity can change the path of an asteroid moving through the solar system.

Gravity to the rescue!

Everything in the universe—from asteroids to planets to stars—exerts gravity on every other object. This force keeps the Moon in orbit around Earth and Earth in orbit around the Sun. Gravity can also send objects on a collision course—a problem when those objects are Earth and an asteroid. Asteroids are rocky bodies found mostly in the asteroid belt between Mars and Jupiter. Jupiter's strong gravity can change the orbits of asteroids over time, occasionally sending them dangerously close to Earth.

Astronomers use powerful telescopes to track asteroids near Earth. More than a thousand asteroids are large enough to cause serious damage if they collide with Earth. If an asteroid were heading our way, how could we prevent the collision? One idea is to launch a spacecraft into the asteroid. The impact could slow it down enough to cause it to miss Earth. But if the asteroid broke apart, the pieces could rain down onto Earth!

Scientists have another idea. They propose launching a massive spacecraft into an orbit close to the asteroid. The spacecraft's gravity would exert a small tug on the asteroid. Over time, the asteroid's path would be altered enough to pass by Earth. Astronomers track objects now that are many years away from crossing paths with Earth. This gives them enough time to set a plan in motion if one of the objects appears to be on a collision course with Earth.

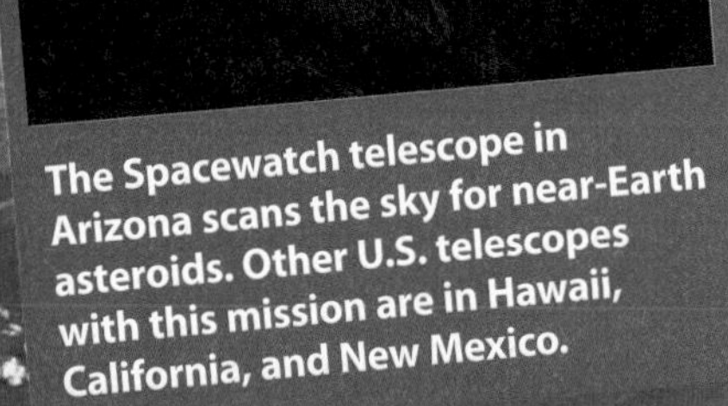

The Spacewatch telescope in Arizona scans the sky for near-Earth asteroids. Other U.S. telescopes with this mission are in Hawaii, California, and New Mexico.

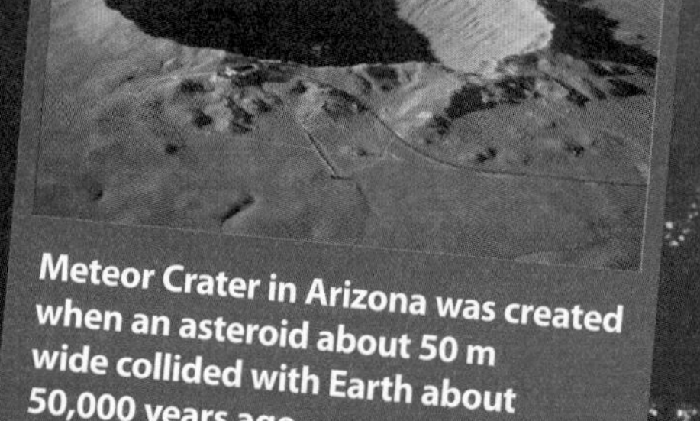

Meteor Crater in Arizona was created when an asteroid about 50 m wide collided with Earth about 50,000 years ago.

It's Your Turn

PROBLEM SOLVING With a group, come up with a plan for avoiding an asteroid's collision with Earth. Present your plan to the class. Include diagrams and details that explain exactly how your plan will work.

Lesson 2

Reading Guide

Key Concepts
ESSENTIAL QUESTIONS

- What is Newton's first law of motion?
- How is motion related to balanced and unbalanced forces?
- What effect does inertia have on the motion of an object?

Vocabulary

BrainPOP®

Newton's First Law

Inquiry How does it balance?

You probably would be uneasy standing under Balanced Rock near Buhl, Idaho. Yet this unusual rock stays in place year after year. The rock has forces acting on it. Why doesn't it fall over? The forces acting on the rock combine, and the rock does not move.

Launch Lab

10 minutes

Can you balance magnetic forces?

Magnets exert forces on each other. Depending on how you hold them, magnets either attract or repel each other. Can you balance these magnetic forces?

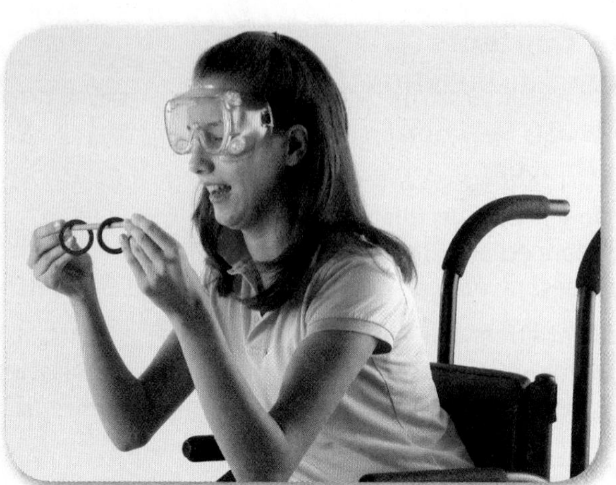

1. Read and complete a lab safety form.
2. Have your lab partner hold a **ring magnet** vertically on a **pencil,** as shown in the picture.
3. Place **another magnet** on the pencil, and use it to push the first magnet along the pencil.
4. Place a **third magnet** on the same pencil so that the outer magnets push against the middle one. Does the middle magnet still move along the pencil?

Think About This

1. Describe the forces that the other magnets exert on the first magnet in steps 3 and 4.
2. **Key Concept** Describe how the motion of the first magnet seemed to depend on whether each force on the magnet was balanced by another force.

Identifying Forces

Ospreys are birds of prey that live near bodies of water. Perhaps several minutes ago, the mother osprey in **Figure 10** was in the air in a high-speed dive. It might have plunged toward a nearby lake after seeing a fish in the water. As it neared the water, it moved its legs forward to grab the fish with its talons. It then stretched out its wings and used them to climb high into the air. Before the osprey comes to rest on its nest, it will slow its speed and land softly on the nest's edge, near the young birds waiting for food.

Figure 10 Forces change the motion of this osprey.

Forces helped the mother osprey change the speed and direction of its motion. Recall that a force is a push or a pull. Some of the forces were contact forces, such as air resistance. When soaring, the osprey spread its wings, increasing air resistance. In a dive, it held its wings close to its body, changing its shape, decreasing its surface area and air resistance. Gravity also pulled the osprey toward the ground.

To understand the motion of an object, you need to identify the forces acting on it. In this lesson you will read how forces change the motion of objects.

Combining Forces—The Net Force

Suppose you try to move a piece of heavy furniture, such as the dresser in **Figure 11.** If you push on the dresser by yourself, you have to push hard on the dresser to overcome the static friction and move it. If you ask a friend to push with you, you do not have to push as hard. When two or more forces act on an object, the forces combine. *The combination of all the forces acting on an object is the* **net force.** The way in which forces combine depends on the directions of the forces applied to an object.

Combining Forces in the Same Direction

When the forces applied to an object act in the same direction, the net force is the sum of the individual forces. In this case, the direction of the net force is the same as the direction of the individual forces.

Because forces have direction, you have to specify a **reference direction** when you combine forces. In **Figure 11,** for example, you would probably choose "to the right" as the positive reference direction. Both forces then would be positive. The net force on the dresser is the sum of the two forces pushing in the same direction. One person pushes on the dresser with a force of 200 N to the right. The other person pushes with a force of 100 N to the right. The net force on the dresser is 200 N + 100 N = 300 N to the right. The force applied to the dresser is the same as if one person pushed on the dresser with a force of 300 N to the right.

Reading Check How do you calculate the net force on an object if two forces are acting on it in the same direction?

REVIEW VOCABULARY

reference direction
a direction that you choose from a starting point to describe an object's position

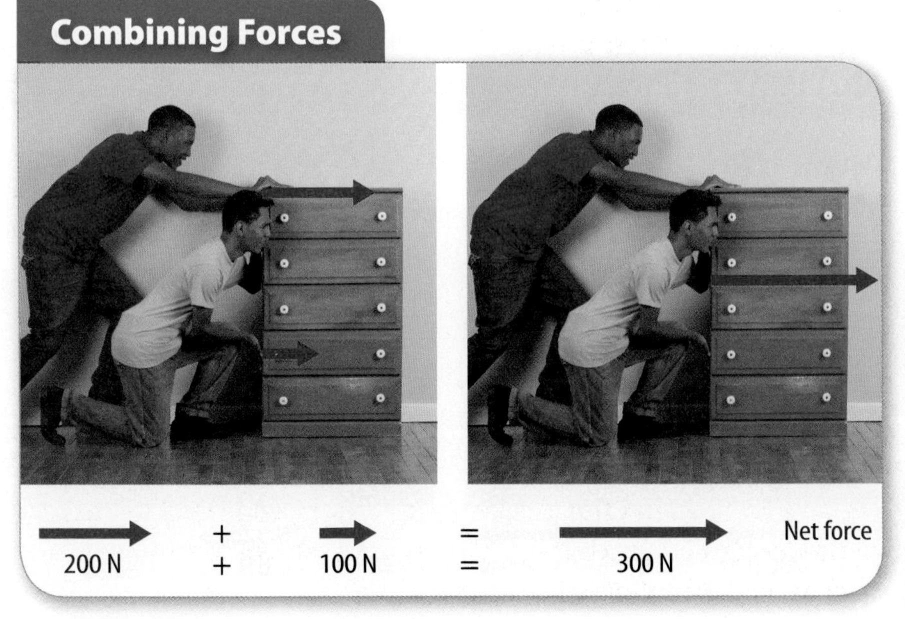

Figure 11 When forces in the same direction combine, the net force is also in the same direction. The strength of the net force is the sum of the forces.

Visual Check What would the net force be if one boy pushed with 250 N and the other boy pushed in the same direction with 180 N?

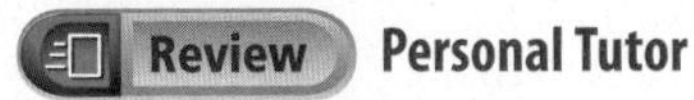

Figure 12 When two forces acting on an object in opposite directions combine, the net force is in the same direction as the larger force. The strength of the net force is the sum of the positive and negative forces. ▶

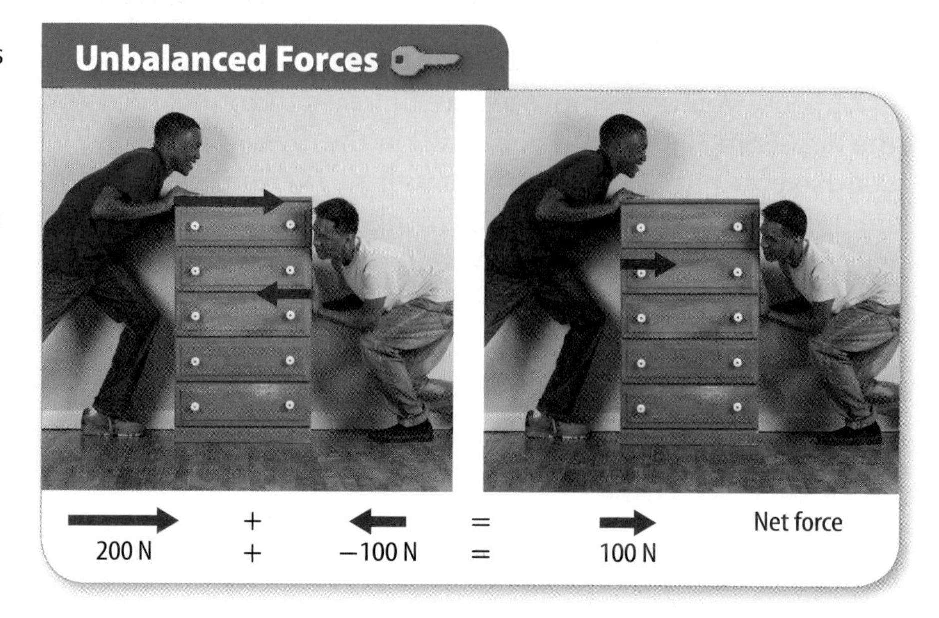

Combining Forces in Opposite Directions

When forces act in opposite directions on an object, the net force is still the sum of the forces. Suppose you choose "to the right" again as the reference direction in **Figure 12.** A force in that direction is positive, and a force in the opposite direction is negative. The net force is the sum of the positive and negative forces. The net force on the dresser is 100 N to the right.

Balanced and Unbalanced Forces

When equal forces act on an object in opposite directions, as in **Figure 13,** the net force on the object is zero. The effect is the same as if there were no forces acting on the object. *Forces acting on an object that combine and form a net force of zero are* **balanced forces.** Balanced forces do not change the motion of an object. However, the net force on the dresser in **Figure 12** is not zero. There is a net force to the right. *Forces acting on an object that combine and form a net force that is not zero are* **unbalanced forces.**

Figure 13 When two forces acting on an object in opposite directions are the same strength, the forces are balanced. ▶

Visual Check How are the force arrows for the balanced forces in the figure alike? How are they different?

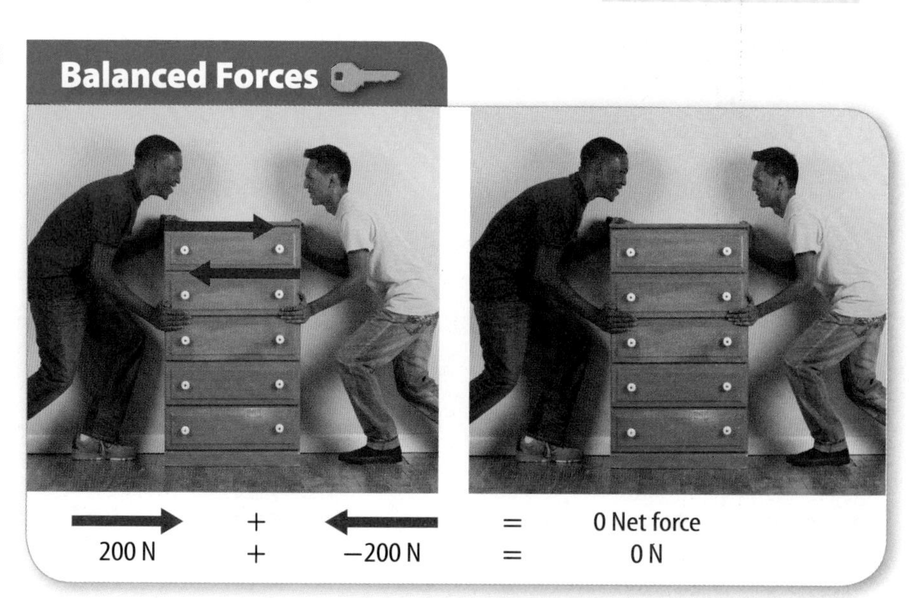

Newton's First Law of Motion

Sir Isaac Newton studied how forces affect the motion of objects. He developed three rules known as Newton's laws of motion. *According to* **Newton's first law of motion,** *if the net force on an object is zero, the motion of the object does not change.* As a result, balanced forces and unbalanced forces have different results when they act on an object.

Key Concept Check What is Newton's first law of motion?

Balanced Forces and Motion

According to Newton's first law of motion, balanced forces cause no change in an object's velocity (speed in a certain direction). This is true when an object is at rest or in motion. Look again at **Figure 13.** The dresser is at rest before the boys push on it. It remains at rest when they apply balanced forces. Similarly, because the forces in **Figure 14**—air resistance and gravity—are balanced, the parachutist moves downward at his terminal velocity. Terminal velocity is the constant velocity reached when air resistance equals the force of gravity acting on a falling object.

Reading Check What happens to the velocity of a moving car if the forces on it are balanced?

Unbalanced Forces and Motion

Newton's first law of motion only applies to balanced forces acting on an object. When unbalanced forces act on an object at rest, the object starts moving. When unbalanced forces act on an already moving object, the object's speed, direction of motion, or both change. You will read more about how unbalanced forces affect an object's motion in the next lesson.

Key Concept Check How is motion related to balanced and unbalanced forces?

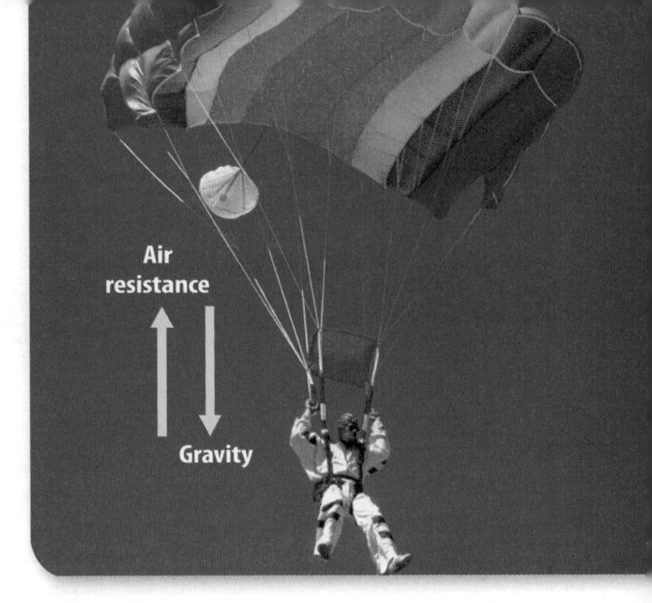

Figure 14 Balanced forces acting on an object do not change the object's speed and direction.

Inquiry MiniLab 15 minutes

How do forces affect motion?

The motion of an object depends on whether balanced or unbalanced forces act on it.

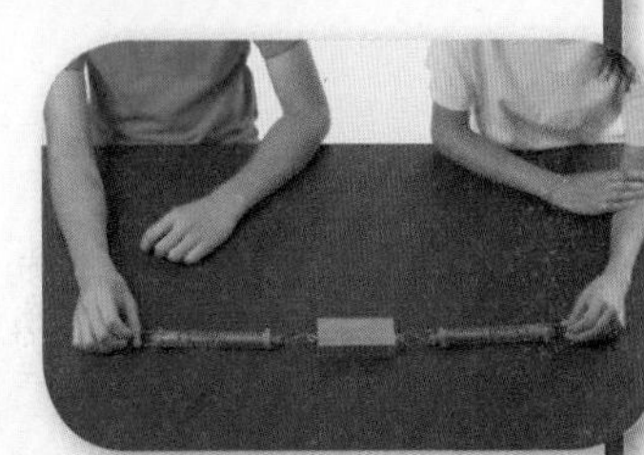

1. Read and complete a lab safety form.
2. Attach **spring scales** to opposite sides of a **wooden block** with **eyehooks.**
3. With a partner, gently pull the scales so that the block moves toward one of you. Sketch the setup in your Science Journal, including the force readings on each scale.
4. Repeat step 3, pulling on the block so that it does not move.

Analyze and Conclude

1. **Explain** Use Newton's first law of motion to explain what occurred in steps 3 and 4.
2. **Key Concept** How was the block's motion related to balanced and unbalanced forces?

Figure 15 Inertia causes the crash-test dummy to keep moving forward after the car stops.

Visual Check What effect would a shoulder belt and a lap belt have on the inertia of the crash-test dummy?

WORD ORIGIN

inertia
from Latin *iners,* means "without skill, inactive"

Inertia

According to Newton's first law, the motion of an object will not change if balanced forces act on it. *The tendency of an object to resist a change in its motion is called* **inertia** (ihn UR shuh). Inertia explains the motion of the crash-test dummy in **Figure 15.** Before the crash, the car and dummy moved with constant velocity. If no other force had acted on them, the car and dummy would have continued moving with constant velocity because of inertia. The impact with the barrier results in an unbalanced force on the car, and the car stops. The dummy continues moving forward because of its inertia.

Key Concept Check What effect does inertia have on the motion of an object?

FOLDABLES®

Make a chart with six columns and six rows. Use your chart to define and show how this lesson's vocabulary terms are related. Afterward, fold your chart in half and label the outside *Newton's First Law.*

	Net Force	Balanced Forces	Unbalanced Forces	Newton's First Law	Inertia
Net Force					
Balanced Forces					
Unbalanced Forces					
Newton's First Law					
Inertia					

Why do objects stop moving?

Think about how friction and inertia together affect an object's movement. A book sitting on a table, for example, stays in place because of inertia. When you push the book, the force you apply to the book is greater than static friction between the book and the table. The book moves in the direction of the greater force. If you stop pushing, friction stops the book.

What would happen if there were no friction between the book and the table? Inertia would keep the book moving. According to Newton's first law, the book would continue to move at the same speed in the same direction as your push.

On Earth, friction can be reduced but not totally removed. For an object to start moving, a force greater than static friction must be applied to it. To keep the object in motion, a force at least as strong as friction must be applied continuously. Objects stop moving because friction or another force acts on them.

Lesson 2 Review

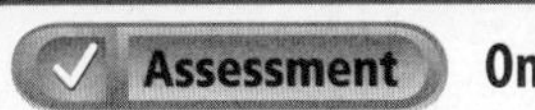

Visual Summary

Unbalanced forces cause an object to move.

According to Newton's first law of motion, if the net force on an object is zero, the object's motion does not change.

Inertia is a property that resists a change in the motion of an object.

FOLDABLES®

Use your lesson Foldable to review the lesson. Save your Foldable for the project at the end of the chapter.

What do you think NOW?

You first read the statements below at the beginning of the chapter.

3. Forces acting on an object cannot be added.

4. A moving object will stop if no forces act on it.

Did you change your mind about whether you agree or disagree with the statements? Rewrite any false statements to make them true.

Use Vocabulary

1. **Define** *net force* in your own words.
2. **Distinguish** between balanced forces and unbalanced forces.

Understand Key Concepts 

3. Which causes an object in motion to remain in motion?
   A. friction
   B. gravity
   C. inertia
   D. velocity
4. **Apply** You push a coin across a table. The coin stops. How does this motion relate to balanced and unbalanced forces?
5. **Explain** Use Newton's first law to explain why a book on a desk does not move.

Interpret Graphics

6. **Analyze** What is the missing force?

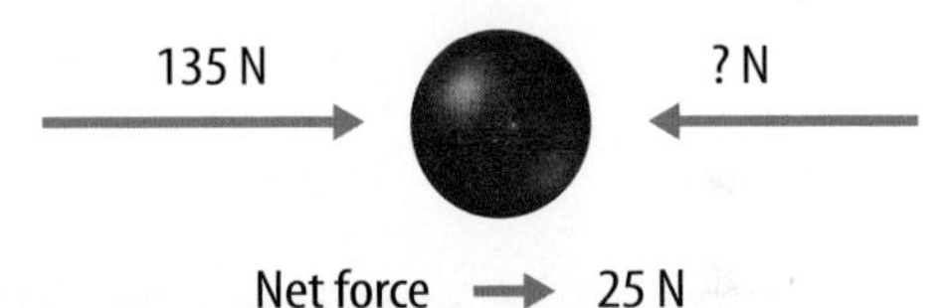

7. **Organize Information** Copy and fill in the graphic organizer below to explain Newton's first law of motion in each case.

Object at rest	
Object in motion	

Critical Thinking

8. **Extend** Three people push a piano on wheels with forces of 130 N to the right, 150 N to the left, and 165 N to the right. What are the strength and direction of the net force on the piano?
9. **Assess** A child pushes down on a box lid with a force of 25 N. At the same time, her friend pushes down on the lid with a force of 30 N. The spring on the box lid pushes upward with a force of 60 N. Can the children close the box? Why or why not?

Inquiry Skill Practice **Model** **20 minutes**

How can you model Newton's first law of motion?

Materials

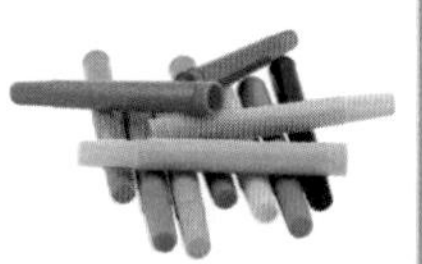

markers (red, blue, black, green)

According to Newton's first law of motion, balanced forces do not change an object's motion. Unbalanced forces change the motion of objects at rest or in motion. You can model different forces and their effects on the motion of an object.

Learn It

When you **model** a concept in science, you act it out, or imitate it. You can model the effect of balanced and unbalanced forces on motion by using movements on a line.

Try It

1. Draw a line across a sheet of lined notebook paper lengthwise. Place an X at the center. Each space to the right of the X will model a force of 1 N east, and each space to the left will model 1 N west.
2. Suppose a force of 3 N east and a force of 11 N west act on a moving object. Model these forces by starting at X and drawing a red arrow three spaces to the right. Then, start at that point and draw a blue arrow 11 spaces to the left. The net force is modeled by how far this point is from X, 8 N west.
3. Are the forces you modeled balanced or unbalanced? Will the forces change the object's motion?

Apply It

4. Suppose a force of 8 N east, a force of 12 N west, and a force of 4 N east act on a moving object. Use different colors of markers to model the forces on the object.
5. What is the net force on the object? Are the forces you modeled balanced or unbalanced? Will the forces change the object's motion?
6. **Model** other examples of balanced and unbalanced forces acting on an object. In each case, decide which forces will act on the object.
7. **Key Concept** For each of the forces you modeled, determine the net force, and decide if the forces are balanced or unbalanced. Then, decide if the forces will change the object's motion.

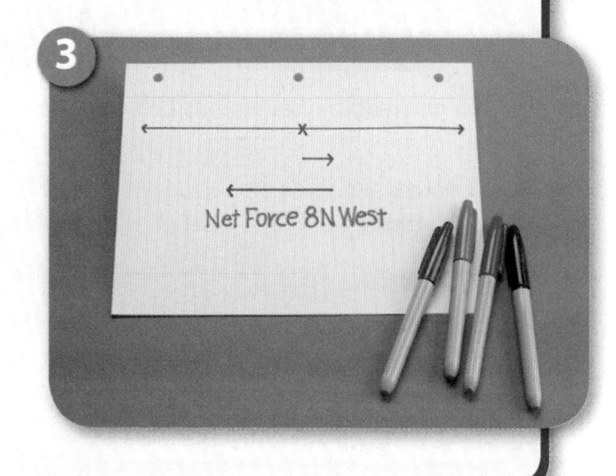

Lesson 3

Reading Guide

Key Concepts
ESSENTIAL QUESTIONS

- What is Newton's second law of motion?
- How does centripetal force affect circular motion?

Vocabulary

Newton's second law of motion p. 65
circular motion p. 66
centripetal force p. 66

g Multilingual eGlossary

Newton's Second Law

Inquiry What makes it go?

The archer pulls back the string and takes aim. When she releases the string, the arrow soars through the air. To reach the target, the arrow must quickly reach a high speed. How is it able to move so fast? The force from the string determines the arrow's speed.

Launch Lab

10 minutes

What forces affect motion along a curved path?

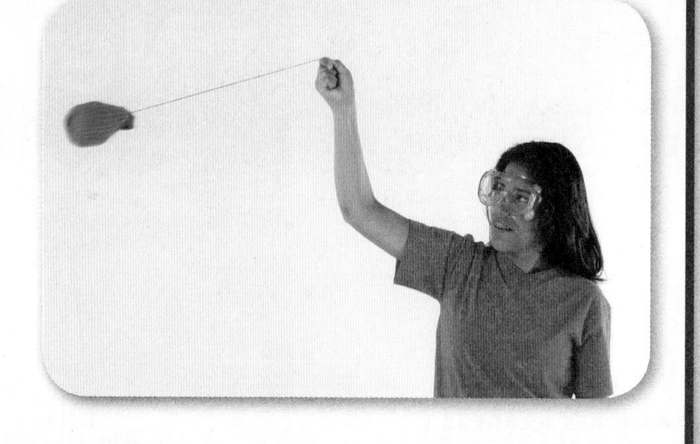

When traveling in a car or riding on a roller coaster, you can feel different forces acting on you as you move along a curved path. What are these forces? How do they affect your motion?

1. Read and complete a lab safety form.
2. Attach a piece of **string** about 1 m long to a rolled-up **sock.**

   WARNING: Find a spot away from your classmates for the next steps.

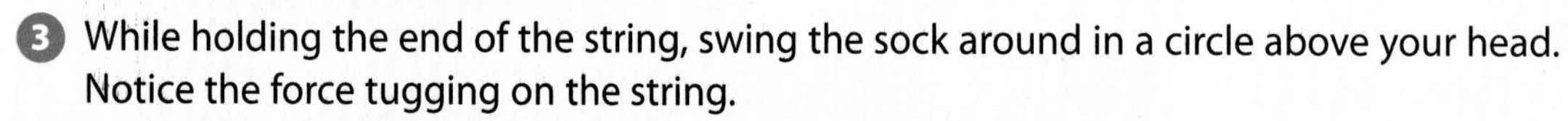

3. While holding the end of the string, swing the sock around in a circle above your head. Notice the force tugging on the string.
4. Repeat step 3 with two socks rolled together. In your Science Journal, compare the force of swinging one sock to the force of swinging two socks.

Think About This

1. Describe the forces acting along the string while you were swinging it. Classify each force as balanced or unbalanced.
2. **Key Concept** How does the force from the string seem to affect the sock's motion?

How do forces change motion?

Think about different ways that forces can change an object's motion. For example, how do forces change the motion of someone riding a bicycle? The forces of the person's feet on the pedals cause the wheels of the bicycle to turn faster and the bicycle's speed to increase. The speed of a skater slowly sliding across ice gradually decreases because of friction between the skates and the ice. Suppose you are pushing a wheelbarrow across a yard. You can change its speed by pushing with more or less force. You can change its direction by pushing it in the direction you want to move. Forces change an object's motion by changing its speed, its direction, or both its speed and its direction.

Unbalanced Forces and Velocity

Velocity is speed in a certain direction. Only unbalanced forces change an object's velocity. A bicycle's speed will not increase unless the forces of the person's feet on the pedals is greater than friction that slows the wheels. A skater's speed will not decrease if the skater pushes back against the ice with a force greater than the friction against the skates. If someone pushes the wheelbarrow with the same force but in the opposite direction that you are pushing, the wheelbarrow's direction will not change.

In the previous lesson, you read about Newton's first law of motion—balanced forces do not change an object's velocity. In this lesson you will read about how unbalanced forces affect the velocity of an object.

Unbalanced Forces on an Object at Rest

An example of how unbalanced forces affect an object at rest is shown in **Figure 16.** At first the ball is not moving. The hand holds the ball up against the downward pull of gravity. Because the forces on the ball are balanced, the ball remains at rest. When the hand moves out of the way, the ball falls downward. You know that the forces on the ball are now unbalanced because the ball's motion changed. The ball moves in the direction of the net force. When unbalanced forces act on an object at rest, the object begins moving in the direction of the net force.

Unbalanced Forces on an Object in Motion

Unbalanced forces change the velocity of a moving object. Recall that one way to change an object's velocity is to change its speed.

Speeding Up If the net force acting on a moving object is in the direction that the object is moving, the object will speed up. For example, a net force acts on the sled in **Figure 17.** Because the net force is in the direction of motion, the sled's speed increases.

Slowing Down Think about what happens if the direction of the net force on an object is opposite to the direction the object moves. The object slows down. When the boy sliding on the sled in **Figure 17** pushes his foot against the snow, friction acts in the direction opposite to his motion. Because the net force is in the direction opposite to the sled's motion, the sled's speed decreases.

Reading Check What happens to the speed of a wagon rolling to the right if a net force to the right acts on it?

Balanced Forces

Force exerted by hand

Force due to gravity

Unbalanced Forces

Force due to gravity

▲ **Figure 16** When unbalanced forces act on a ball at rest, it moves in the direction of the net force.

Speeding up

Slowing down

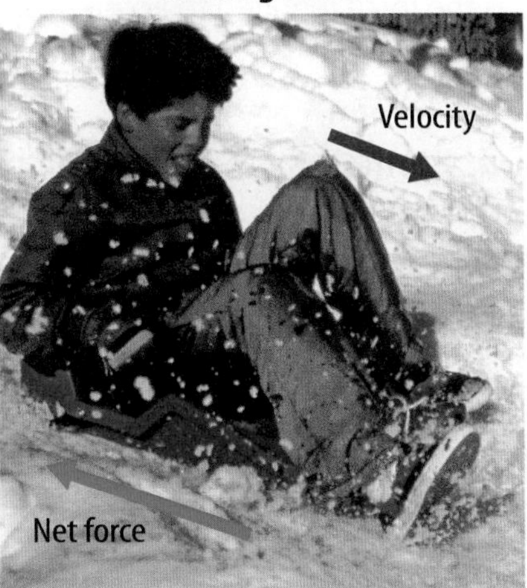

◀ **Figure 17** Unbalanced forces can cause an object to speed up or slow down.

Visual Check How would the net force and velocity arrows in the left photo change if the girl pushed harder?

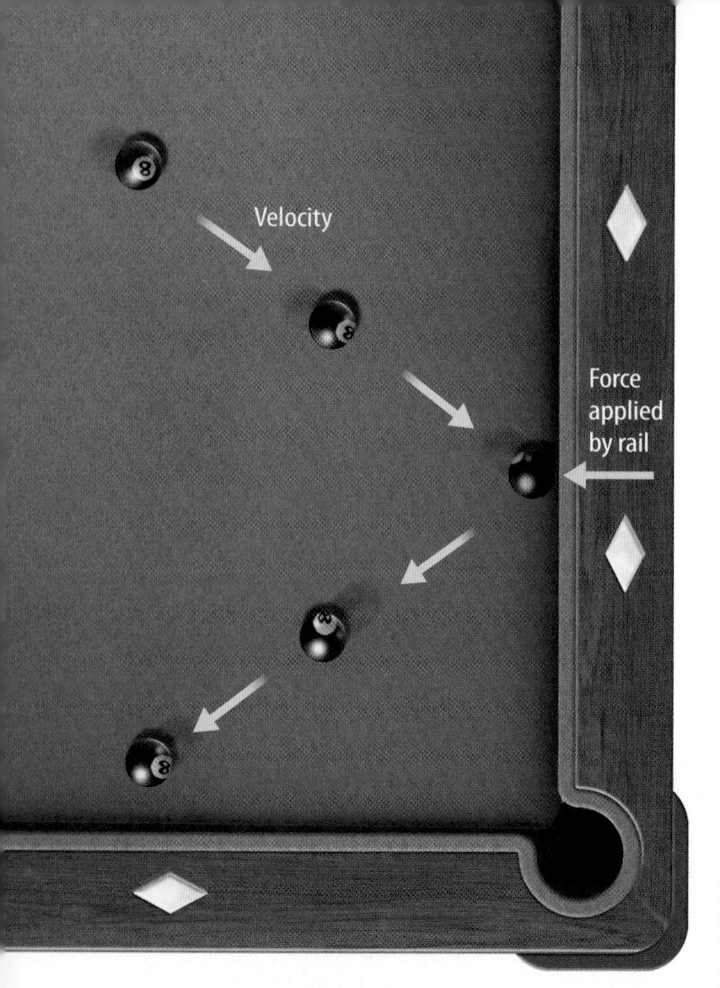

Figure 18 Unbalanced forces act on the billiard ball, causing its direction to change.

Changes in Direction of Motion

Another way that unbalanced forces can change an object's velocity is to change its direction. The ball in **Figure 18** moved at a constant velocity until it hit the rail of the billiard table. The force applied by the rail changed the ball's direction. Likewise, unbalanced forces change the direction of Earth's crust. Recall that the crust is broken into moving pieces called plates. The direction of one plate changes when another plate pushes against it with an unbalanced force.

Unbalanced Forces and Acceleration

You have read how unbalanced forces can change an object's velocity by changing its speed, its direction, or both its speed and its direction. Another name for a change in velocity over time is acceleration. When the girl in **Figure 17** pushed the sled, the sled accelerated because its speed changed. When the billiard ball in **Figure 18** hit the side of the table, the ball accelerated because its direction changed. Unbalanced forces can make an object accelerate by changing its speed, its direction, or both.

Reading Check How do unbalanced forces affect an object at rest or in motion?

Inquiry MiniLab

10 minutes

How are force and mass related?

Unbalanced forces cause an object to accelerate. If the mass of the object increases, how does the force required to accelerate the object change?

1. Read and complete a lab safety form.
2. Tie a **string** to a **small box.** Pull the box about 2 m across the floor. Notice the force required to cause the box to accelerate.
3. Put **clay** in the box to increase its mass. Pull the box so that its acceleration is about the same as before. Notice the force required.

Analyze and Conclude

1. **Compare** the strength of the force needed to accelerate the box each time.
2. **Key Concept** How did the mass affect the force needed to accelerate the box?

Newton's Second Law of Motion

Isaac Newton also described the relationship between an object's acceleration (change in velocity over time) and the net force that acts on an object. *According to* **Newton's second law of motion,** *the acceleration of an object is equal to the net force acting on the object divided by the object's mass.* The direction of acceleration is the same as the direction of the net force.

Key Concept Check What is Newton's second law of motion?

Newton's Second Law Equation

$$\text{acceleration (in m/s}^2) = \frac{\text{net force (in N)}}{\text{mass (in kg)}}$$

$$a = \frac{F}{m}$$

Notice that the equation for Newton's second law has SI units. Acceleration is expressed in meters per second squared (m/s^2), mass in kilograms (kg), and force in newtons (N). From this equation, it follows that a newton is the same as $kg \cdot m/s^2$.

Make a half-book from a sheet of notebook paper. Use it to organize your notes on Newton's second law.

Newton's Second Law

Math Skills Newton's Second Law Equation

Solve for Acceleration You throw a 0.5-kg basketball with a force of 10 N. What is the acceleration of the ball?

1. **This is what you know:** mass: $m = 0.5$ kg; force: $F = 10$ N or $10\ kg \cdot m/s^2$
2. **This is what you need to find:** acceleration: a
3. **Use this formula:** $a = \frac{F}{m}$
4. **Substitute:** the values for F and m into the formula and divide. $a = \frac{10\ N}{0.5\ kg} = 20\ \frac{kg \cdot m/s^2}{kg} = 20\ m/s^2$

Answer: The acceleration of the ball is $20\ m/s^2$.

- Math Practice
- Personal Tutor

Practice

1. A 24-N net force acts on an 8-kg rock. What is the acceleration of the rock?
2. A 30-N net force on a skater produces an acceleration of $0.6\ m/s^2$. What is the mass of the skater?
3. What net force acting on a 14-kg wagon produces an acceleration of $1.5\ m/s^2$?

Circular Motion

Newton's second law of motion describes the relationship between an object's change in velocity over time, or acceleration, and unbalanced forces acting on the object. You already read how this relationship applies to motion along a line. It also applies to circular motion. **Circular motion** *is any motion in which an object is moving along a curved path.*

Centripetal Force

The ball in **Figure 19** is in circular motion. The velocity arrows show that the ball has a tendency to move along a straight path. Inertia—not a force—causes this motion. The ball's path is curved because the string pulls the ball inward. *In circular motion, a force that acts perpendicular to the direction of motion, toward the center of the curve, is* **centripetal** (sen TRIH puh tuhl) **force.** The figure also shows that the ball accelerates in the direction of the centripetal force.

Word Origin

centripetal
from Latin *centripetus,* means "toward the center"

Key Concept Check How does centripetal force affect circular motion?

The Motion of Satellites and Planets

Another object that experiences centripetal force is a satellite. A satellite is any object in space that orbits a larger object. Like the ball in **Figure 19,** a satellite tends to move along a straight path because of inertia. But just as the string pulls the ball inward, gravity pulls a satellite inward. Gravity is the centripetal force that keeps a satellite in orbit by changing its direction. The Moon is a satellite of Earth. As shown in **Figure 19,** Earth's gravity changes the Moon's direction. Similarly, the Sun's gravity changes the direction of its satellites, including Earth.

Figure 19 Inertia of the moving object and the centripetal force acting on the object produce the circular motion of the ball and the Moon.

Visual Check How does the direction of the velocity of a satellite differ from the direction of its acceleration?

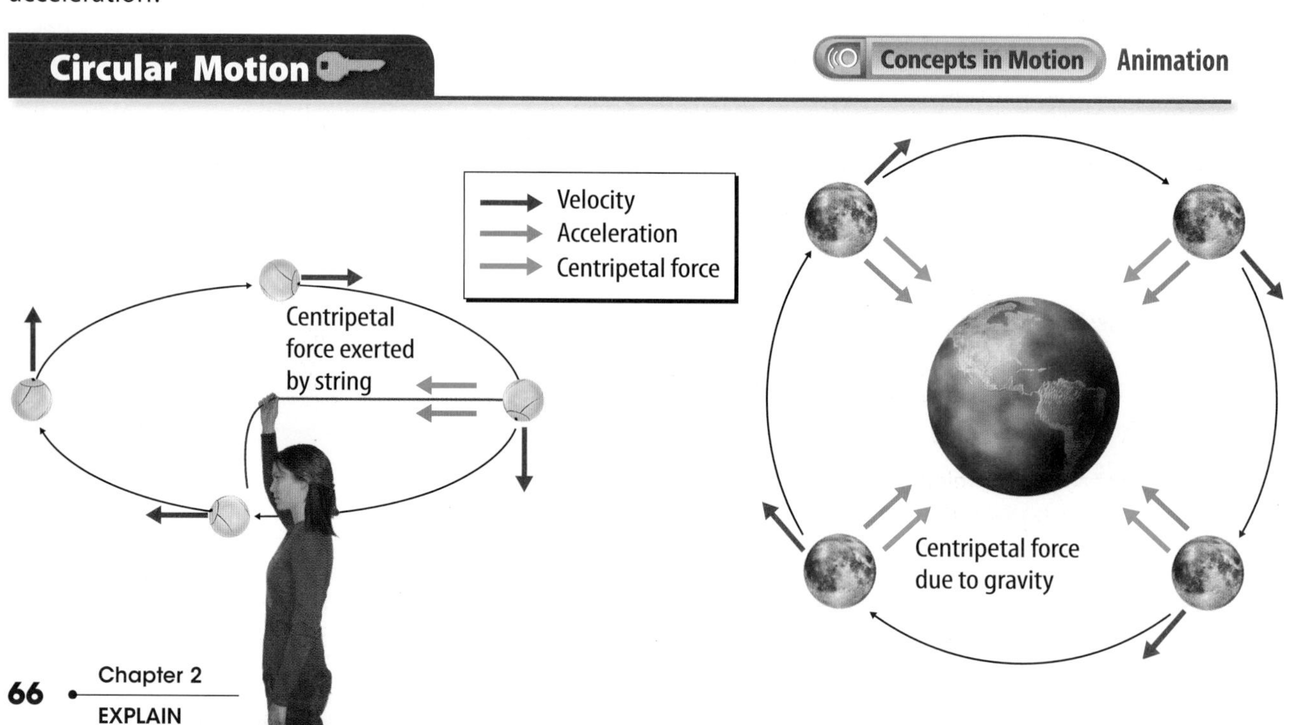

Lesson 3 Review

Visual Summary

Unbalanced forces cause an object to speed up, slow down, or change direction.

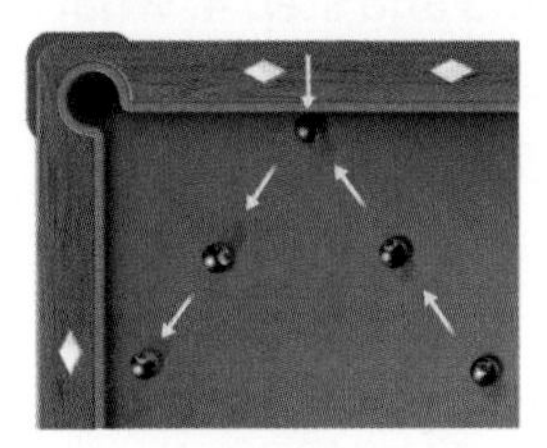

Newton's second law of motion relates an object's acceleration to its mass and the net force on the object.

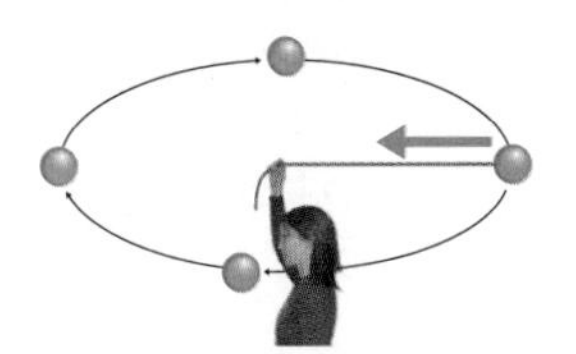

Any motion in which an object is moving along a curved path is circular motion.

FOLDABLES®

Use your lesson Foldable to review the lesson. Save your Foldable for the project at the end of the chapter.

What do you think NOW?

You first read the statements below at the beginning of the chapter.

5. When an object's speed increases, the object accelerates.

6. If an object's mass increases, its acceleration also increases if the net force acting on the object stays the same.

Did you change your mind about whether you agree or disagree with the statements? Rewrite any false statements to make them true.

Use Vocabulary

1. **Explain** Newton's second law of motion in your own words.

2. **Use the term** *circular motion* in a sentence.

Understand Key Concepts

3. A cat pushes a 0.25-kg toy with a net force of 8 N. According to Newton's second law what is the acceleration of the ball?

   A. 2 m/s^2
   B. 4 m/s^2
   C. 16 m/s^2
   D. 32 m/s^2

4. **Describe** how centripetal force affects circular motion.

Interpret Graphics

5. **Apply** Copy and fill in the graphic organizer below. Give examples of unbalanced forces that could cause an object to accelerate.

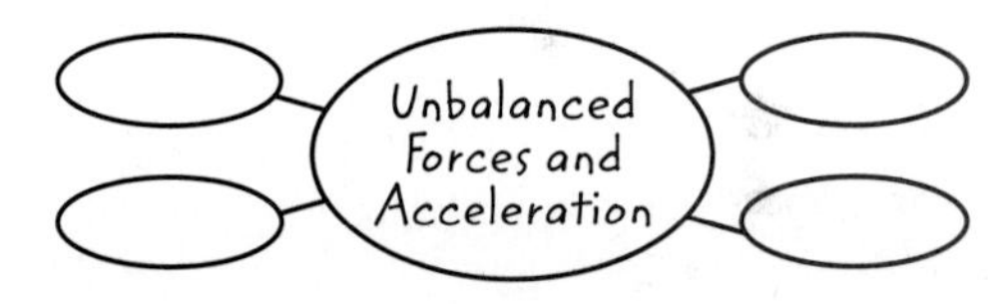

6. **Complete** Copy the graphic organizer below and complete each equation according to Newton's second law.

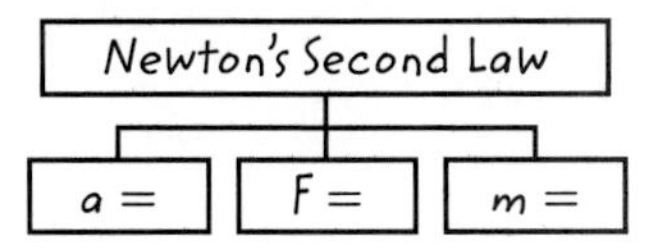

Critical Thinking

7. **Design** You need to lift up a 45-N object. Draw an illustration that explains the strength and direction of the force used to lift the object.

Math Skills

Math Practice

8. The force of Earth's gravity is about 10 N downward. What is the acceleration of a 15-kg backpack if lifted with a 15-N force?

Inquiry **Skill Practice** Use Variables

25 minutes

How does a change in mass or force affect acceleration?

Force, mass, and acceleration are all related variables. In this activity, you will use these variables to study Newton's second law of motion.

Materials

baseball

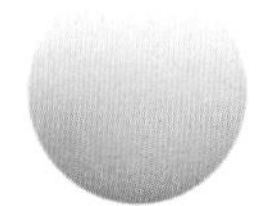

foam ball

meterstick

Safety

Learn It

Vary means "to change." A **variable** is a quantity that can be changed. For example, the variables related to Newton's second law of motion are force, mass, and acceleration. You can find the relationship between any two of these variables by changing one of them and keeping the third variable the same.

Try It

1. Read and complete a lab safety form.
2. Hold a baseball in one hand and a foam ball in your other hand. Compare the masses of the two balls.
3. Lay both balls on a flat surface. Push a meterstick against the balls at the same time with the same force. Compare the accelerations of the ball.
4. Using only the baseball and the meterstick, lightly push the ball and observe its acceleration. Again observe the acceleration as you push the baseball with a stronger push. Compare the accelerations of the ball when you used a weak force and when you used a strong force.

Apply It

5. Answer the following questions for both step 3 and step 4. What variable did you change? What variable changed as a result? What variable did you keep the same?
6. Using your results, state the relationship between acceleration and mass if the net force on an object does not change. Then, state the relationship between acceleration and force if mass does not change.
7. **Key Concept** How do your results support Newton's second law of motion?

Lesson 4

Reading Guide

Key Concepts
ESSENTIAL QUESTIONS

- What is Newton's third law of motion?
- Why don't the forces in a force pair cancel each other?
- What is the law of conservation of momentum?

Vocabulary

Newton's third law of motion p. 71
force pair p. 71
momentum p. 73
law of conservation of momentum p. 74

g Multilingual eGlossary

Newton's Third Law

Inquiry Why move up?

To reach the height she needs for her dive, this diver must move up into the air. Does she just jump up? No, she doesn't. She pushes down on the diving board and the diving board propels her into the air. How does pushing down cause the diver to move up?

Inquiry Launch Lab

10 minutes

How do opposite forces compare? 

If you think about forces you encounter every day, you might notice forces that occur in pairs. For example, if you drop a rubber ball, the falling ball pushes against the floor. The ball bounces because the floor pushes with an opposite force against the ball. How do these opposite forces compare?

1. Read and complete a lab safety form.
2. Stand so that you face your lab partner, about half a meter away. Each of you should hold a **spring scale.**
3. Hook the two scales together, and gently pull them away from each other. Notice the force reading on each scale.
4. Pull harder on the scales, and again notice the force readings on the scales.
5. Continue to pull on both scales, but let the scales slowly move toward your lab partner and then toward you at a constant speed.

Think About This

1. Identify the directions of the forces on each scale. Record this information in your Science Journal.
2. **Key Concept** Describe the relationship you noticed between the force readings on the two scales.

Opposite Forces

Have you ever been on in-line skates and pushed against a wall? When you pushed against the wall, like the boy is doing in **Figure 20,** you started moving away from it. What force caused you to move?

You might think the force of the muscles in your hands moved you away from the wall. But think about the direction of your push. You pushed against the wall in the opposite direction from your movement. It might be hard to imagine, but when you pushed against the wall, the wall pushed back in the opposite direction. The push of the wall caused you to accelerate away from the wall. When an object applies a force on another object, the second object applies a force of the same strength on the first object, but the force is in the opposite direction.

Figure 20 When the skater pushes against the wall, the wall applies a force to the skater that pushes him away from the wall.

 **Reading Check** When you are standing, you push on the floor, and the floor pushes on you. How do the directions and strengths of these forces compare?

Newton's Third Law of Motion

Newton's first two laws of motion describe the effects of balanced and unbalanced forces on one object. Newton's third law relates forces between two objects. *According to* **Newton's third law of motion,** *when one object exerts a force on a second object, the second object exerts an equal force in the opposite direction on the first object.* An example of forces described by Newton's third law is shown in **Figure 21.** When the gymnast pushes against the vault, the vault pushes back against the gymnast. Notice that the lengths of the force arrows are the same, but the directions are opposite.

Key Concept Check What is Newton's third law of motion?

Force Pairs

The forces described by Newton's third law depend on each other. *A* **force pair** *is the forces two objects apply to each other.* Recall that you can add forces to calculate the net force. If the forces of a force pair always act in opposite directions and are always the same strength, why don't they cancel each other? The answer is that each force acts on a different object. In **Figure 22,** the girl's feet act on the boat. The force of the boat acts on the girl's feet. The forces do not result in a net force of zero because they act on different objects. Adding forces can only result in a net force of zero if the forces act on the same object.

Key Concept Check Why don't the forces in a force pair cancel each other?

Action and Reaction

In a force pair, one force is called the action force. The other force is called the reaction force. The girl in **Figure 22** applies an action force against the boat. The reaction force is the force that the boat applies to the girl. For every action force, there is a reaction force that is equal in strength but opposite in direction.

Newton's Third Law

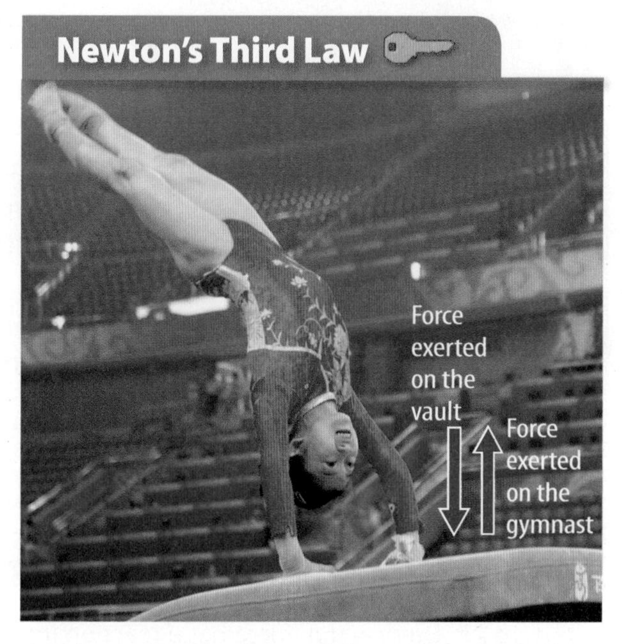

▲ **Figure 21** The force of the vault propels the gymnast upward.

Force Pairs

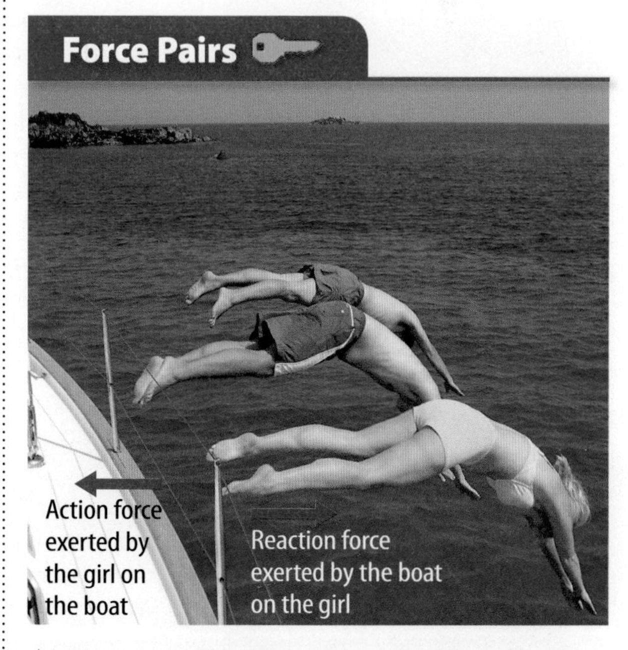

▲ **Figure 22** The force pair is the force the girl applies to the boat and the force that the boat applies to the girl.

Visual Check How can you tell that the forces don't cancel each other?

FOLDABLES®

Make a half-book from a sheet of notebook paper. Use it to summarize how Newton's third law explains the motion of a variety of common activities.

Using Newton's Third Law of Motion

When you push against an object, the force you apply is called the action force. The object then pushes back against you. The force applied by the object is called the reaction force. According to Newton's second law of motion, when the reaction force results in an unbalanced force, there is a net force, and the object accelerates. As shown in **Figure 23,** Newton's third law explains how you can swim and jump. It also explains how rockets can be launched into space.

Reading Check How does Newton's third law apply to the motion of a bouncing ball?

Action and Reaction Forces

Figure 23 Every action force has a reaction force in the opposite direction.

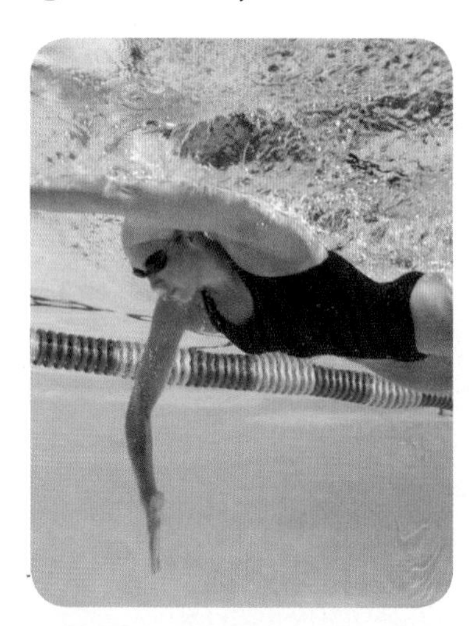

◀ **Swimming** When you swim, you push your arms against the water in the pool. The water in the pool pushes back on you in the opposite (forward) direction. If your arms push the water back with enough force, the reaction force of the water on your body is greater than the force of fluid friction. The net force is forward. You accelerate in the direction of the net force and swim forward through the water.

▶ **Jumping** When you jump, you push down on the ground, and the ground pushes up on you. The upward force of the ground combines with the downward force of gravity to form the net force acting on you. If you push down hard enough, the upward reaction force is greater than the downward force of gravity. The net force is upward. According to Newton's second law, your acceleration is in the same direction as the net force, so you accelerate upward.

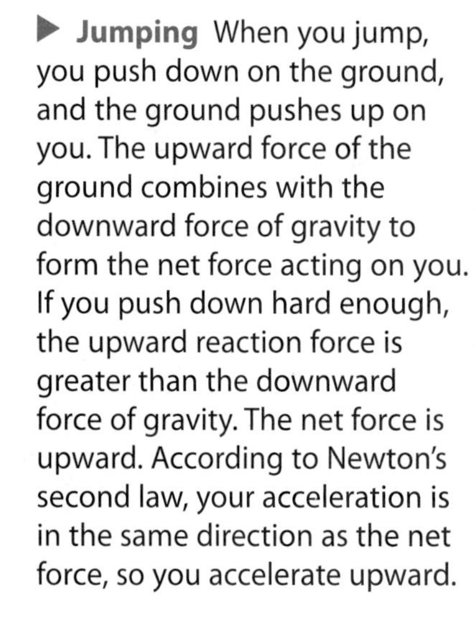

◀ **Rocket Motion** The burning fuel in a rocket engine produces a hot gas. The engine pushes the hot gas out in a downward direction. The gas pushes upward on the engine. When the upward force of the gas pushing on the engine becomes greater than the downward force of gravity on the rocket, the net force is in the upward direction. The rocket then accelerates upward.

Visual Check On what part of the swimmer's body does the water's reaction force push?

Momentum

Because action and reaction forces do not cancel each other, they can change the motion of objects. A useful way to describe changes in velocity is by describing momentum. **Momentum** *is a measure of how hard it is to stop a moving object.* It is the product of an object's mass and velocity. An object's momentum is in the same direction as its velocity.

Word Origin

momentum
from Latin *momentum,* means "movement, impulse"

Momentum Equation

momentum (in kg•m/s) = **mass** (in kg) × **velocity** (in m/s)

$$p = m \times v$$

If a large truck and a car move at the same speed, the truck is harder to stop. Because it has more mass, it has more momentum. If cars of equal mass move at different speeds, the faster car has more momentum and is harder to stop.

Newton's first two laws relate to momentum. According to Newton's first law, if the net force on an object is zero, its velocity does not change. This means its momentum does not change. Newton's second law states that the net force on an object is the product of its mass and its change in velocity. Because momentum is the product of mass and velocity, the force on an object equals its change in momentum.

Math Skills — Finding Momentum

Solve for Momentum What is the momentum of a 12-kg bicycle moving at 5.5 m/s?

1. **This is what you know:** mass: $m = 12$ kg; velocity: $v = 5.5$ m/s
2. **This is what you need to find:** momentum: p
3. **Use this formula:** $p = m \times v$
4. **Substitute:** the values for m and v into the formula and multiply. $p = \mathbf{12\ kg} \times \mathbf{5.5\ m/s} = 66$ kg•m/s

Answer: The momentum of the bicycle is 66 kg•m/s in the direction of the velocity.

Practice

1. What is the momentum of a 1.5-kg ball rolling at 3.0 m/s?
2. A 55-kg woman has a momentum of 220 kg•m/s. What is her velocity?

Review
- Math Practice
- Personal Tutor

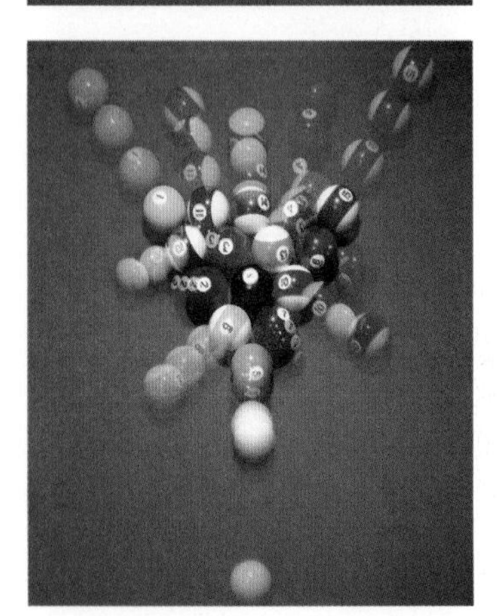

Figure 24 The total momentum of all the balls is the same before and after the collision.

Conservation of Momentum

You might have noticed that if a moving ball hits another ball that is not moving, the motion of both balls changes. The cue ball in **Figure 24** has momentum because it has mass and is moving. When it hits the other balls, the cue ball's velocity and momentum decrease. Now the other balls start moving. Because these balls have mass and velocity, they also have momentum.

The Law of Conservation of Momentum

In any collision, one object transfers momentum to another object. The billiard balls in **Figure 24** gain the momentum lost by the cue ball. The total momentum, however, does not change. *According to the* **law of conservation of momentum,** *the total momentum of a group of objects stays the same unless outside forces act on the objects.* Outside forces include friction. Friction between the balls and the billiard table decreases their velocities, and they lose momentum.

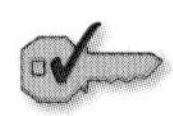

Key Concept Check What is the law of conservation of momentum?

Types of Collisions

Objects collide with each other in different ways. When colliding objects bounce off each other, it is an elastic collision. If objects collide and stick together, such as when one football player tackles another, the collision is inelastic. No matter the type of collision, the total momentum will be the same before and after the collision.

Inquiry MiniLab

15 minutes

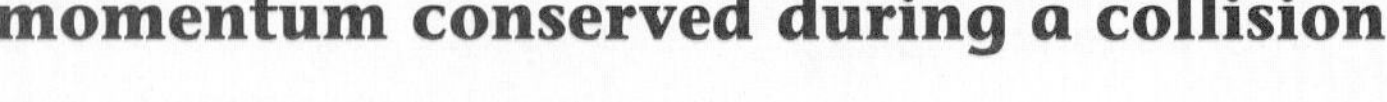

Is momentum conserved during a collision?

1. Read and complete a lab safety form.
2. Make a track by using **masking tape** to secure two **metersticks** side by side on a table, about 4 cm apart.
3. Place two **tennis balls** on the track. Roll one ball against the other. Then, roll the balls at about the same speed toward each other.
4. Place the balls so that they touch. Observe the collision as you gently roll **another ball** against them.

Analyze and Conclude

1. **Explain** how you know that momentum was transferred from one ball to another.
2. **Key Concept** What could you measure to show that momentum is conserved?

Lesson 4 Review

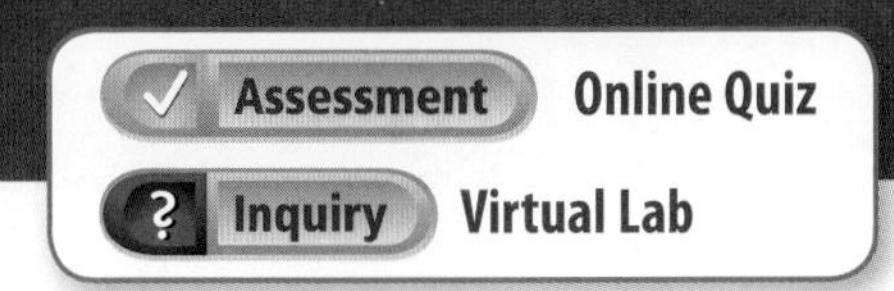

Visual Summary

Newton's third law of motion describes the force pair between two objects.

For every action force, there is a reaction force that is equal in strength but opposite in direction.

In any collision, momentum is transferred from one object to another.

FOLDABLES®

Use your lesson Foldable to review the lesson. Save your Foldable for the project at the end of the chapter.

What do you think NOW?

You first read the statements below at the beginning of the chapter.

7. If objects collide, the object with more mass applies more force.

8. Momentum is a measure of how hard it is to stop a moving object.

Did you change your mind about whether you agree or disagree with the statements? Rewrite any false statements to make them true.

Use Vocabulary

1. **Define** *momentum* in your own words.
2. The force of a bat on a ball and the force of a ball on a bat are a(n) ________.

Understand Key Concepts 

3. **State** Newton's third law of motion.
4. A ball with momentum 16 kg•m/s strikes a ball at rest. What is the total momentum of both balls after the collision?
   A. –16 kg•m/s **C.** 8 kg•m/s
   B. –8 kg•m/s **D.** 16 kg•m/s
5. **Identify** A child jumps on a trampoline. The trampoline bounces her up. Why don't the forces cancel each other?

Interpret Graphics

6. **Predict** what will happen to the velocity and momentum of each ball when the small ball hits the heavier large ball?

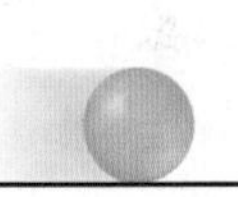

7. **Organize** Copy and fill in the table.

Event	Action Force	Reaction Force
A girl kicks a soccer ball.		
A book sits on a table.		

Critical Thinking

8. **Decide** How is it possible for a bicycle to have more momentum than a truck?

Math Skills

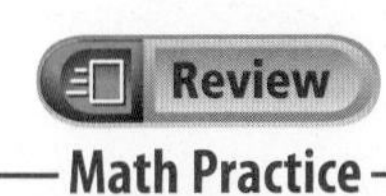

Math Practice

9. A 2.0-kg ball rolls to the right at 3.0 m/s. A 4.0-kg ball rolls to the left at 2.0 m/s. What is the momentum of the system after a head-on collision of the two balls?

Inquiry Lab **2 class periods**

Modeling Newton's Laws of Motion

Materials

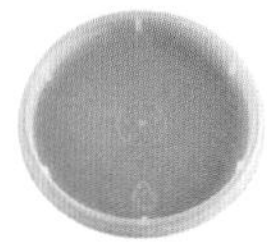

plastic lid

golf ball

modeling clay

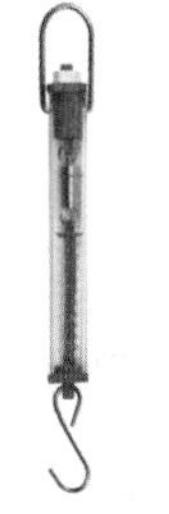

2.5-N spring scales (2)

Safety

Newton's first and second laws of motion describe the relationship between unbalanced forces and motion. These laws relate to forces acting on one object. Newton's third law describes the strength and direction of force pairs. This law relates to forces on two different objects. You can learn about all three of Newton's laws of motion by modeling them.

Question

How can you model Newton's laws of motion?

Procedure

1. Read and complete a lab safety form.
2. Attach a spring scale to a plastic lid. Add mass to the lid by placing a ball of modeling clay on it.
3. Slowly pull the lid along a table with the spring scale. Record the force reading on the scale in your Science Journal.
4. Try to use the spring scale to pull the lid with constant force and constant speed.
5. Try pulling the lid with increasing force and constant speed.
6. Pull the lid so that it accelerates quickly.
7. Increase the mass of the lid by adding more modeling clay. Repeat steps 3–6.
8. Replace the modeling clay with a golf ball. Try pulling the lid slowly. Then, try pulling it from a standstill quickly. What happens to the ball in each case? Record your results.

7

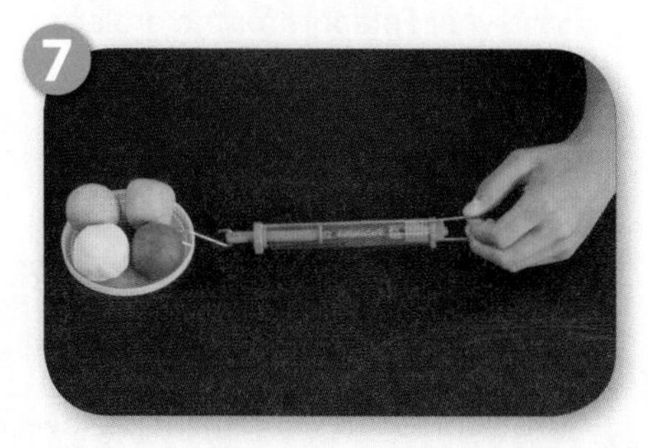

9. To model Newton's first law of motion, design an activity using the lid and the spring scales that shows that a net force of zero does not change the motion of an object.
10. To model Newton's second law of motion, design an activity that shows that if the net force acting on an object is not zero, the object accelerates.
11. To model Newton's third law of motion, plan an activity that shows action and reaction forces on an object.
12. After your teacher approves your plan, perform the activities.

Analyze and Conclude

13. **Identify** the variables in each of your models. Which variables changed and which remained constant?
14. **The Big Idea** For each law of motion that you modeled, how did the force applied to the lid relate to the motion of the lid?

Communicate Your Results

Choose one of the laws of motion, and model it for the class. Compare your model with the method of modeling used by other lab groups.

Inquiry Extension

Describe another way you could model Newton's three laws of motion using materials other than those used in this lab. For example, for Newton's first law of motion, you could pedal a bicycle at a constant speed.

Lab Tips

- ☑ Use a smooth surface so that the lid moves easily.
- ☑ You might want to make a data table in which you can record your observations and the force readings.

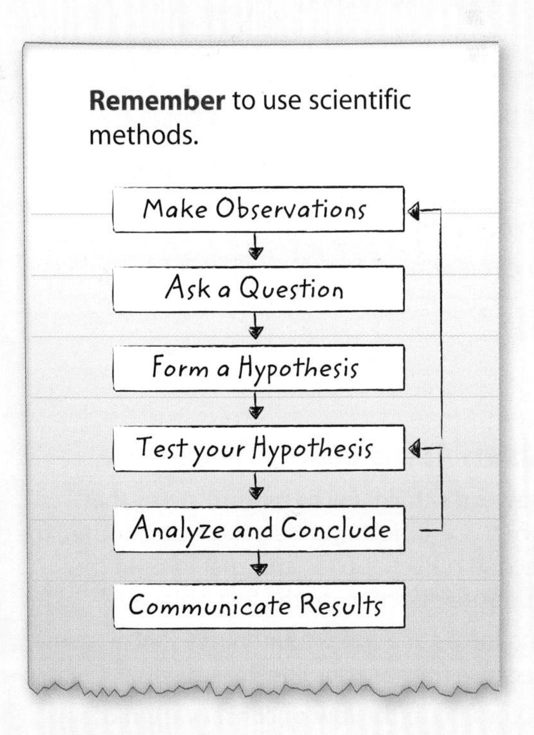

Chapter 2 Study Guide

An object's motion changes if a net force acts on the object.

Key Concepts Summary	Vocabulary
Lesson 1: Gravity and Friction • Friction is a **contact force.** Magnetism is a **noncontact force.** • The law of universal gravitation states that all objects are attracted to each other by **gravity.** • **Friction** can stop or slow down objects sliding past each other. 	**force** p. 45 **contact force** p. 45 **noncontact force** p. 46 **gravity** p. 47 **mass** p. 47 **weight** p. 48 **friction** p. 49
Lesson 2: Newton's First Law • An object's motion can only be changed by **unbalanced forces.** • According to **Newton's first law of motion,** the motion of an object is not changed by **balanced forces** acting on it. • **Inertia** is the tendency of an object to resist a change in its motion. 	**net force** p. 55 **balanced forces** p. 56 **unbalanced forces** p. 56 **Newton's first law of motion** p. 57 **inertia** p. 58
Lesson 3: Newton's Second Law • According to **Newton's second law of motion,** an object's acceleration is the net force on the object divided by its mass. • In **circular motion,** a **centripetal force** pulls an object toward the center of the curve. 	**Newton's second law of motion** p. 65 **circular motion** p. 66 **centripetal force** p. 66
Lesson 4: Newton's Third Law • **Newton's third law of motion** states that when one object applies a force on another, the second object applies an equal force in the opposite direction on the first object. • The forces of a **force pair** do not cancel because they act on different objects. • According to the **law of conservation of momentum,** momentum is conserved during a collision unless an outside force acts on the colliding objects. 	**Newton's third law of motion** p. 71 **force pair** p. 71 **momentum** p. 73 **law of conservation of momentum** p. 74

Review
- Personal Tutor
- Vocabulary eGames
- Vocabulary eFlashcards

FOLDABLES® Chapter Project

Ass emble your lesson Foldables as shown to make a Chapter Project. Use the project to review what you have learned in this chapter.

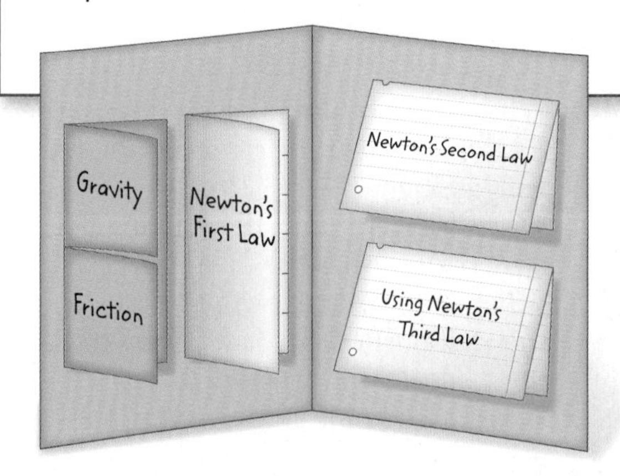

Use Vocabulary

1. The kilogram is the SI unit for ________.
2. The force of gravity on an object is its ________.
3. The sum of all the forces on an object is the ________.
4. An object that has ________ acting on it acts as if there were no forces acting on it at all.
5. A car races around a circular track. Friction on the tires is the ________ that acts toward the center of the circle and keeps the car on the circular path.
6. A heavy train requires nearly a mile to come to a complete stop because it has a lot of ________.

Link Vocabulary and Key Concepts

Concepts in Motion Interactive Concept Map

Copy this concept map, and then use vocabulary terms from the previous page to complete the concept map.

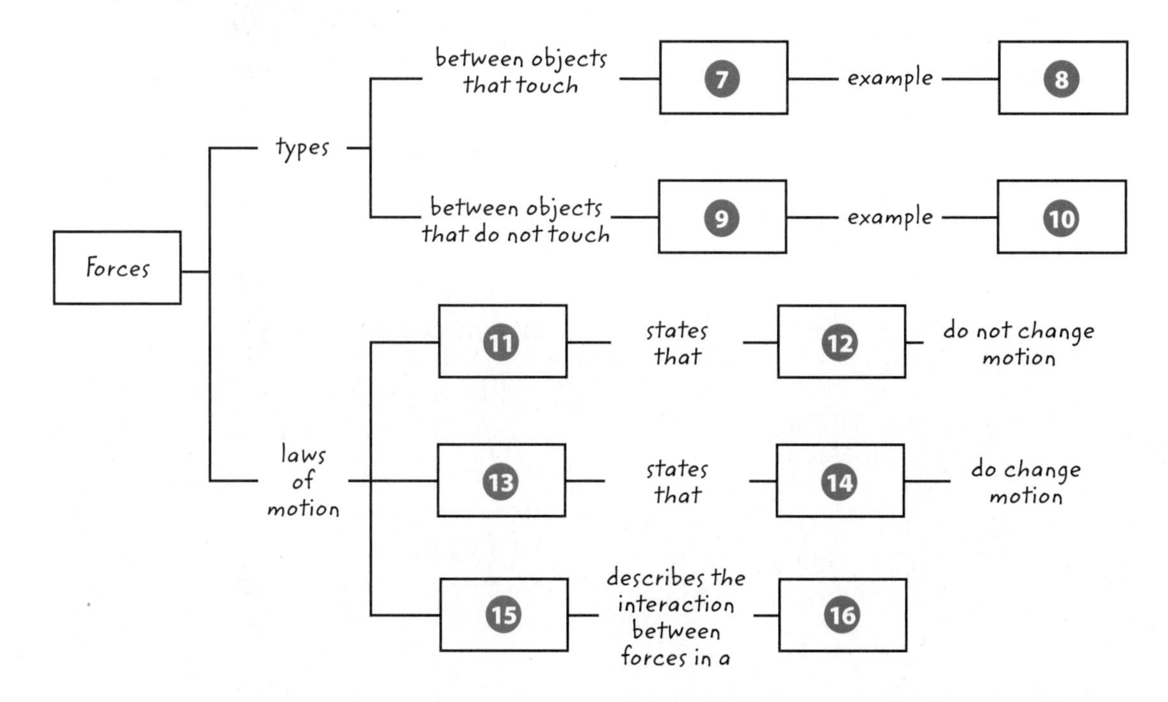

Chapter 2 Review

Understand Key Concepts 

1 The arrows in the figure represent the gravitational force between marbles that have equal mass.

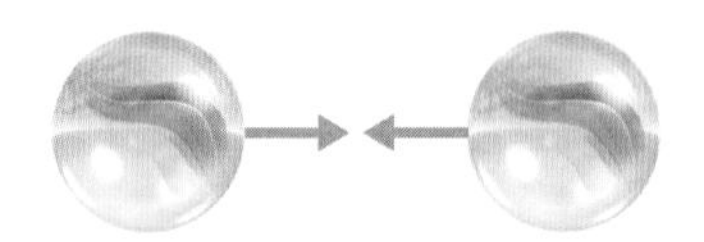

How should the force arrows look if a marble that has greater mass replaces one of these marbles?

A. Both arrows should be drawn longer.
B. Both arrows should stay the same length.
C. The arrow from the marble with less mass should be longer than the other arrow.
D. The arrow from the marble with less mass should be shorter than the other arrow.

2 A person accelerates a box across a flat table with a force less than the weight of the box. Which force is weakest?

A. the force of gravity on the box
B. the force of the table on the box
C. the applied force against the box
D. the sliding friction against the box

3 A train moves at a constant speed on a straight track. Which statement is true?

A. No horizontal forces act on the train as it moves.
B. The train moves only because of its inertia.
C. The forces of the train's engine balances friction.
D. An unbalanced force keeps the train moving.

4 The Moon orbits Earth in a nearly circular orbit. What is the centripetal force?

A. the push of the Moon on Earth
B. the outward force on the Moon
C. the Moon's inertia as it orbits Earth
D. Earth's gravitational pull on the Moon

5 A 30-kg television sits on a table. The acceleration due to gravity is 10 m/s^2. What force does the table exert on the television?

A. 0.3 N
B. 3 N
C. 300 N
D. 600 N

6 Which does NOT describe a force pair?

A. When you push on a bike's brakes, the friction between the tires and the road increases.
B. When a diver jumps off a diving board, the board pushes the diver up.
C. When an ice skater pushes off a wall, the wall pushes the skater away from the wall.
D. When a boy pulls a toy wagon, the wagon pulls back on the boy.

7 A box on a table has these forces acting on it.

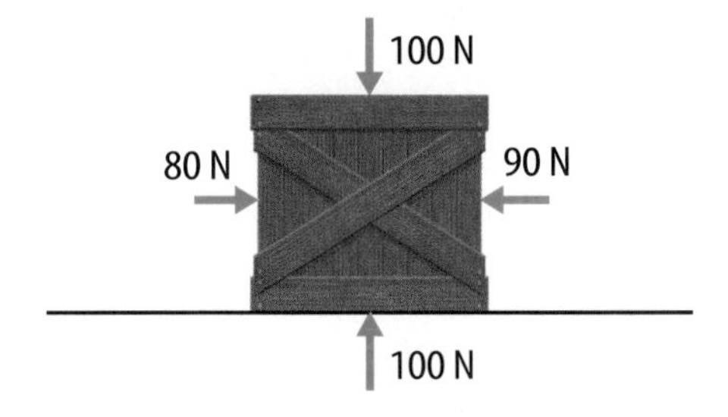

What is the static friction between the box and the table?

A. 0 N
B. 10 N
C. greater than 10 N
D. between 0 and 10 N

8 A 4-kg goose swims with a velocity of 1 m/s. What is its momentum?

A. 4 N
B. 4 $kg \cdot m/s^2$
C. 4 $kg \cdot m/s$
D. 4 m/s^2

Assessment

Online Test Practice

Critical Thinking

9 **Predict** If an astronaut moved away from Earth in the direction of the Moon, how would the gravitational force between Earth and the astronaut change? How would the gravitational force between the Moon and the astronaut change?

10 **Analyze** A box is on a table. Two people push on the box from opposite sides. Which of the labeled forces make up a force pair? Explain your answer.

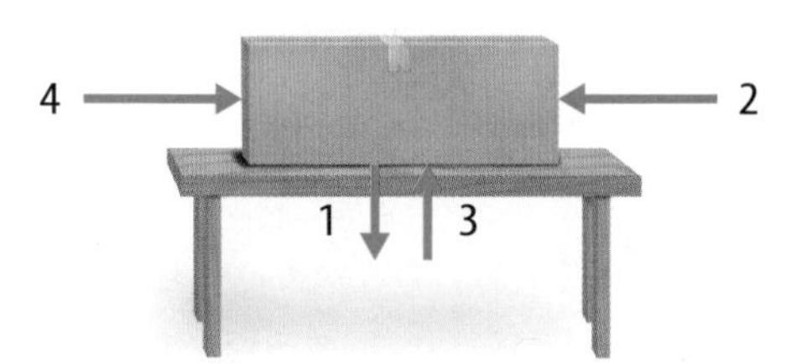

11 **Conclude** A refrigerator has a maximum static friction force of 250 N. Sam can push the refrigerator with a force of 130 N. Amir and André can each push with a force of 65 N. How could they all move the refrigerator? Will the refrigerator move with constant velocity? Why or why not?

12 **Give an example** of unbalanced forces acting on an object.

13 **Infer** Two skaters stand on ice. One weighs 250 N, and the other weighs 500 N. They push against each other and move in opposite directions. Describe the momentum of each skater after they push away from each other.

Writing in Science

14 Imagine that you are an auto designer. Your job is to design brakes for different automobiles. Write a four-sentence plan that explains what you need to consider about momentum when designing brakes for a heavy truck, a light truck, a small car, and a van.

REVIEW THE BIG IDEA

15 Explain how balanced and unbalanced forces affect objects that are not moving and those that are moving.

16 The photo below shows people on a carnival swing ride. How do forces change the motion of the riders?

Math Skills

Math Practice

Solve One-Step Equations

17 A net force of 17 N is applied to an object, giving it an acceleration of 2.5 m/s^2. What is the mass of the object?

18 A tennis ball's mass is about 0.60 kg. Its velocity is 2.5 m/s. What is the momentum of the ball?

19 A box with a mass of 0.82 kg has these forces acting on it.

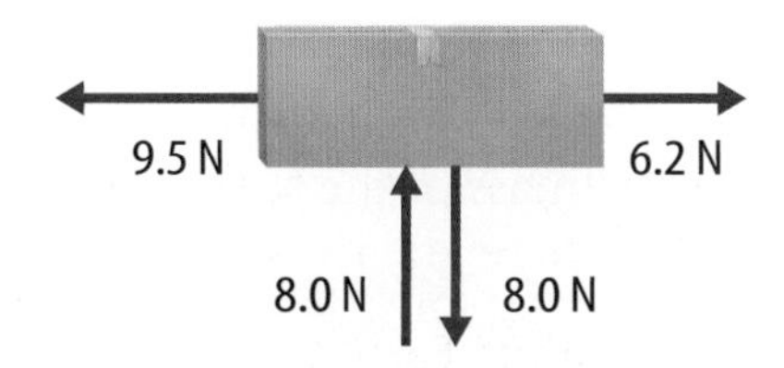

What is the strength and direction of the acceleration of the box?

Standardized Test Practice

Record your answers on the answer sheet provided by your teacher or on a sheet of paper.

Multiple Choice

1 A baseball has an approximate mass of 0.15 kg. If a bat strikes the baseball with a force of 6 N, what is the acceleration of the ball?

A 4 m/s^2

B 6 m/s^2

C 40 m/s^2

D 60 m/s^2

Use the diagram below to answer question 2.

2 The person in the diagram above is unable to move the crate from its position. Which is the opposing force?

A gravity

B normal force

C sliding friction

D static friction

3 The mass of a person on Earth is 72 kg. What is the mass of the same person on the Moon where gravity is one-sixth that of Earth?

A 12 kg

B 60 kg

C 72 kg

D 432 kg

4 A swimmer pushing off from the wall of a pool exerts a force of 1 N on the wall. What is the reaction force of the wall on the swimmer?

A 0 N

B 1 N

C 2 N

D 10 N

Use the diagram below to answer questions 5 and 6.

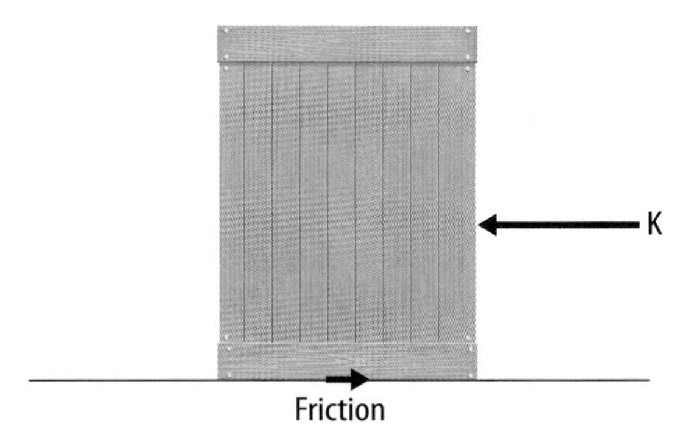

5 Which term applies to the forces in the diagram above?

A negative

B positive

C reference

D unbalanced

6 In the diagram above, what happens when force *K* is applied to the crate at rest?

A The crate remains at rest.

B The crate moves back and forth.

C The crate moves to the left.

D The crate moves to the right.

7 What is another term for change in velocity?

A acceleration

B inertia

C centripetal force

D maximum speed

Use the diagram below to answer question 8.

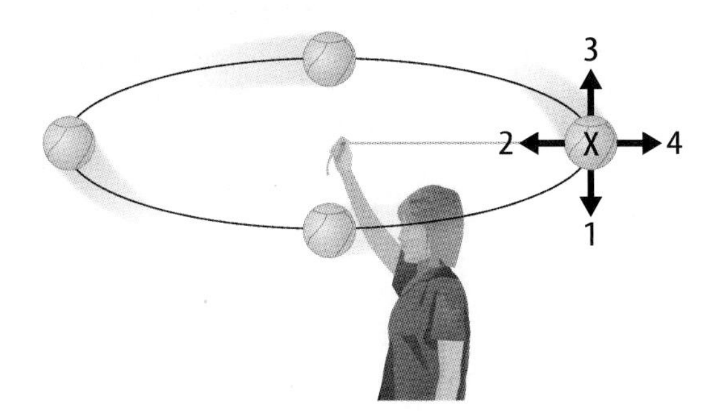

8 The person in the diagram is spinning a ball on a string. When the ball is in position *X*, what is the direction of the centripetal force?

A 1

B 2

C 3

D 4

9 Which is ALWAYS a contact force?

A electric

B friction

C gravity

D magnetic

10 When two billiard balls collide, which is ALWAYS conserved?

A acceleration

B direction

C force

D momentum

Constructed Response

Use the table below to answer question 11.

Newton's Laws of Motion	Explanation
First	
Second	
Third	

11 Explain each of Newton's laws of motion. What is one practical application of each law?

Use the diagram below to answer questions 12 and 13.

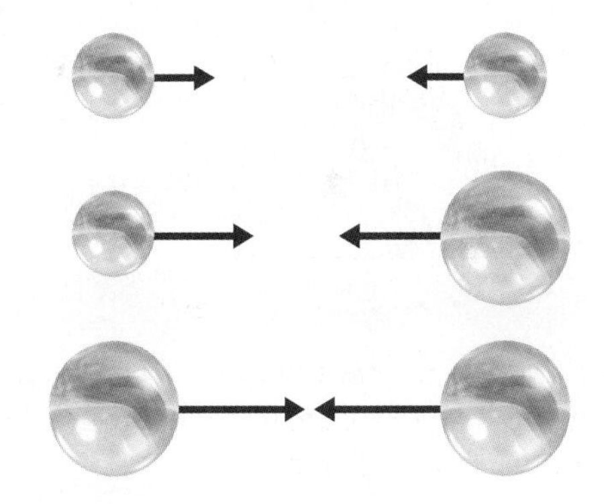

12 The arrows in the diagram above represent forces. What scientific law does the diagram illustrate? What does the law state?

13 Using the diagram, explain how marble mass affects gravitational attraction.

NEED EXTRA HELP?													
If You Missed Question...	1	2	3	4	5	6	7	8	9	10	11	12	13
Go to Lesson...	3	1	1	4	2	2	3	3	1	4	2–4	1	1

Chapter 3

Energy, Work, and Simple Machines

THE BIG IDEA How does energy cause change?

Inquiry Simple Machines?

Sailing is one activity in which simple machines are used. The pulleys on this deck help raise and lower the heavy sails on the boat. Without simple machines, transferring energy can be a much harder task.

- What are simple machines?
- How do simple machines transfer energy?
- How does energy cause change?

Get Ready to Read

What do you think?

Before you read, decide if you agree or disagree with each of these statements. As you read this chapter, see if you change your mind about any of the statements.

1. Energy is the ability to produce motion.
2. Waves transfer energy from place to place.
3. Energy cannot be created or destroyed, but it can be transformed.
4. Work describes how much energy it takes for a force to push or to pull an object.
5. All machines are 100 percent efficient.
6. Simple machines do work using one motion.

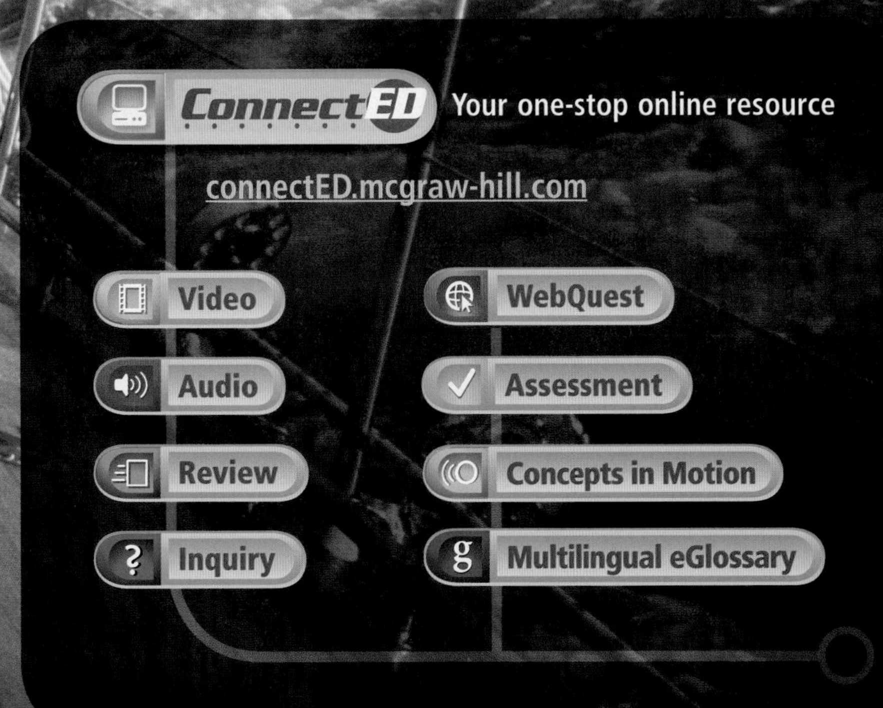

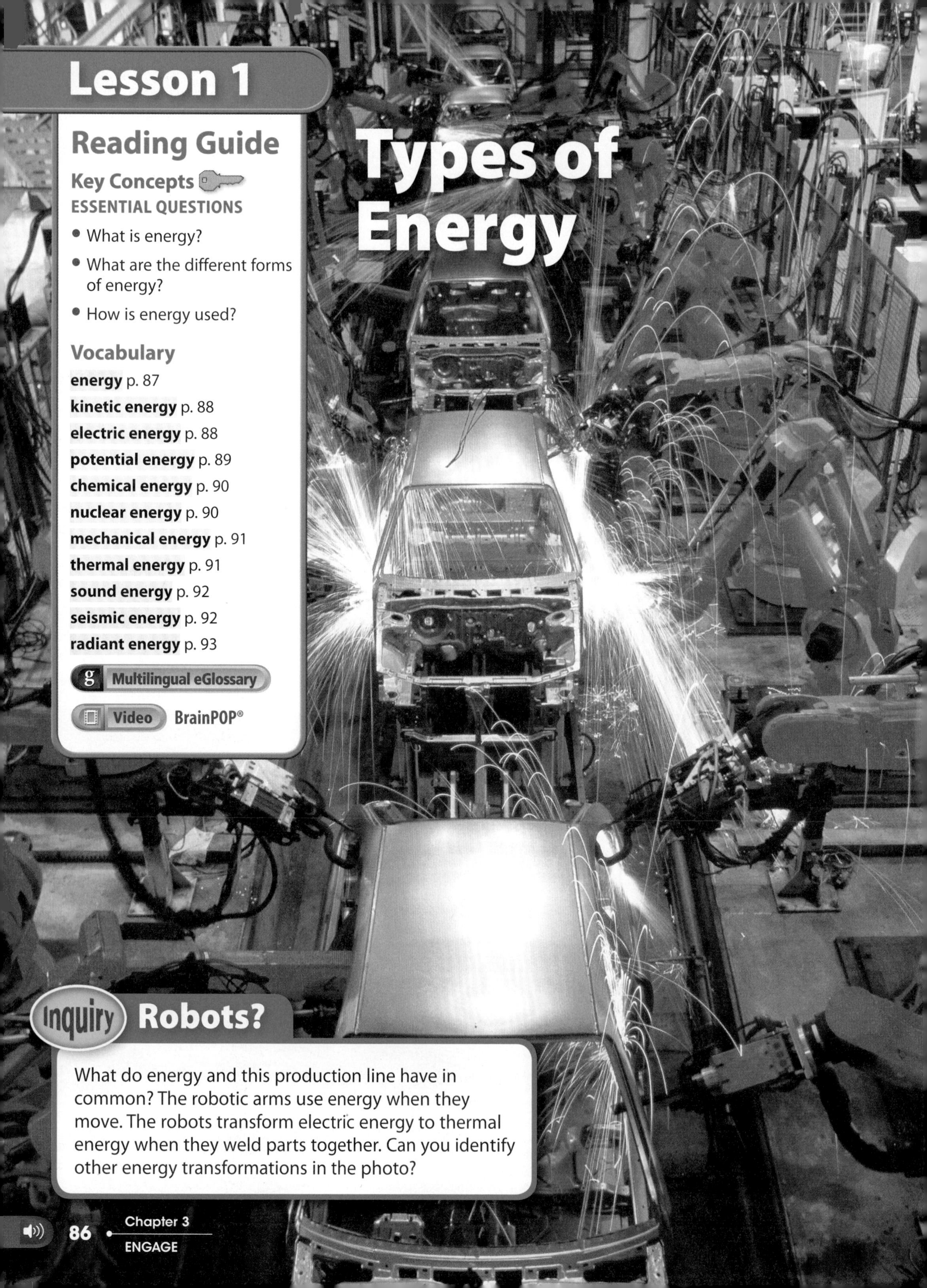

Lesson 1

Types of Energy

Reading Guide

Key Concepts
ESSENTIAL QUESTIONS

- What is energy?
- What are the different forms of energy?
- How is energy used?

Vocabulary

g Multilingual eGlossary

Video BrainPOP®

Inquiry Robots?

What do energy and this production line have in common? The robotic arms use energy when they move. The robots transform electric energy to thermal energy when they weld parts together. Can you identify other energy transformations in the photo?

Launch Lab

20 minutes

Where does energy come from?

How can you heat your hands when they are cold? You could rub them together, put them in your pockets, or hold them near a heater. What makes your hands get warmer?

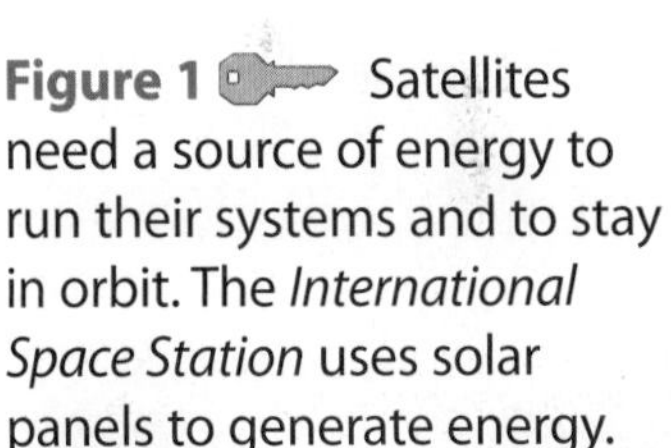

1. Read and complete a lab safety form.
2. As you complete each of the following steps, observe and record any changes in your Science Journal. Discuss the changes with your lab group. In each case, ask: What caused this change to occur? Record your ideas.
3. Rub your hands together. What do you feel?
4. Use a **match** to light a **candle.** Holding your hands near the flame, what do you see and feel?

⚠ *Use caution around an open flame.*

5. Turn on a **flashlight.** Where did the light come from?
6. Observe the overhead lights in your classroom. What is the source of the light?

Think About This

1. Where did the light and the heat come from in steps 3, 4, 5, and 6?
2. **Key Concept** How many different sources of energy can you recall? Briefly explain each one and tell how they differ from one another.

What is energy?

You probably have heard the word *energy* used on the television, the radio, or the Internet. Commercials claim that the newest models of cars are energy efficient. What is energy? Scientists define **energy** as *the ability to cause a change.*

Using this definition, what does energy have to do with the cars in production on the previous page? Most cars use some type of fuel such as gasoline or diesel as their energy source. The car's engine transforms the energy stored in the fuel to a form of energy that moves the car. Compared to other cars, an energy-efficient car uses less fuel to move the car a certain distance.

Gasoline and diesel fuel are not the only sources of energy. Food is an energy source for your body. The solar panels shown in **Figure 1** provide energy for the *International Space Station.* As you will read, wind, coal, nuclear fuel, Earth's interior, and the Sun also are sources of energy. Energy from each of these sources can be transformed into other forms of energy, such as electric energy. Every time you turn on a light, you use energy that was transformed from one form to another.

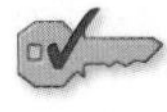

Key Concept Check What is energy?

Figure 1 Satellites need a source of energy to run their systems and to stay in orbit. The *International Space Station* uses solar panels to generate energy.

FOLDABLES®

Make a vertical 3×4 folded table. Label it as shown. Use it to organize your notes about the different types of energy in each category.

Types of Energy	Examples	Notes
Kinetic		
Potential		
Energy for Waves		

Kinetic Energy

You just turned the page of this book. As the page was moving it had **kinetic energy**—*the energy an object has because it is in motion.* Anything that is in motion has kinetic energy, including large objects that you can see as well as small particles, such as molecules, ions, atoms, and electrons.

Kinetic Energy of Objects

When the wind blows, the blades of the wind turbines in **Figure 2** turn. Because they are moving, they have kinetic energy. Kinetic energy depends on mass. If the turbine blades were smaller and had less mass, they would have less kinetic energy. Kinetic energy also depends on speed. When the wind blows harder, the blades move faster and have more kinetic energy. When the wind stops, the blades stop. When the blades are not moving, the kinetic energy of the blades is zero. One of the drawbacks of using wind-generated energy is that wind does not always blow, making the supply of energy inconsistent.

Reading Check What is one drawback of wind energy?

WORD ORIGIN

electric
from Greek *electrum*, means "amber"; because electricity was first generated by rubbing pieces of amber together

Electric Energy

When you turn on a lamp or use a cell phone, you are using a type of kinetic energy—electric energy. Recall that all objects are composed of atoms. Electrons move around the nucleus of an atom, and they move from one atom to another. When electrons move, they have kinetic energy and create an electric current. *The energy that an electric current carries is a form of kinetic energy called* **electric energy.**

Electric energy can be produced by moving objects. When the blades of wind turbines rotate, they turn a generator that changes the kinetic energy of the moving blades into electric energy. Electric energy generated from the kinetic energy of wind creates no waste products.

Figure 2 Wind turbines convert kinetic energy in the wind to electric energy.

Visual Check Why does the kinetic energy of the blades change?

Figure 3 Hydroelectric energy plants use the gravitational potential energy stored in water to produce electricity.

Potential Energy

Suppose you hold up a piece of paper. When the paper is held above the ground, it has potential energy. **Potential energy** *is stored energy that depends on the interaction of objects, particles, or atoms.*

Gravitational Potential Energy

Gravitational potential energy is a type of potential energy stored in an object due to its height above Earth's surface. The water at the top of the dam in **Figure 3** has gravitational potential energy. Gravitational potential energy depends on the mass of an object and its distance from Earth's surface. The more mass an object has and the greater its distance from Earth, the greater its gravitational potential energy.

In a hydroelectric energy plant, water above a dam flows through turbines as it falls. Generators connected to the spinning turbines convert the gravitational potential energy of the water into electric energy.

Hydroelectric power plants are a very clean source of energy. About 7 percent of all electric power in the United States is produced from hydroelectric energy. However, hydroelectric plants can interrupt the movement of animals in streams and rivers.

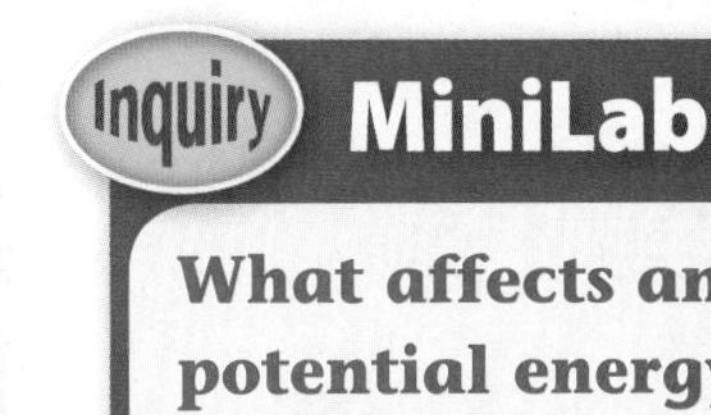

Inquiry MiniLab 20 minutes

What affects an object's potential energy?

Have you ever accidentally dropped a dish? Did it break? Why is the dish more likely to break if it falls all the way to the floor than if it falls just a short distance to a table?

1. Read and complete a lab safety form.
2. Copy the table into your Science Journal.
3. Use a **balance** to find the mass of a **ball bearing** and a **marble.** Record the masses.
4. Stand a **meterstick** on the table next to a flat pad of **clay.** Drop the ball bearing onto the clay from heights of 20 cm, 60 cm, and 100 cm. Drop the ball so that it forms three separate craters.
5. Observe the differences in the craters. Use a **dropper** to measure the number of drops of water it takes to fill each crater.
6. Flatten the clay again. Repeat steps 4 and 5 with the marble.

		Drop Height (20 cm)	Drop Height (60 cm)	Drop Height (100 cm)
Object	Mass (g)	Crater Volume (drops)	Crater Volume (drops)	Crater Volume (drops)
Ball bearing				
Marble				

Analyze and Conclude

1. **Recognize Relationships** What is the relationship between the mass of the object and the volume of the crater and between the drop height and the volume of the crater?
2. **Key Concept** What caused the differences in the sizes of the craters? Explain in terms of gravitational potential energy and kinetic energy.

Figure 4 Chemical energy and nuclear energy are two forms of potential energy.

Chemical Energy

Most electric energy in the United States comes from fossil fuels such as petroleum, natural gas, and coal. The atoms that make up these fossil fuels are joined by chemical bonds. Chemical bonds have the potential to break apart. Therefore, chemical bonds have a form of potential energy called chemical energy. **Chemical energy** *is energy that is stored in and released from the bonds between atoms.*

When fossil fuels burn, the chemical bonds between the atoms that make up the fossil fuel break apart. When this happens, chemical energy transforms to thermal energy. This energy is used to heat water and form steam. The steam is used to turn a turbine, which is connected to a generator that generates electric energy.

A drawback of fossil fuels is that they introduce harmful waste products, such as sulfur dioxide and carbon dioxide, into the environment. Sulfur dioxide in the air creates acid rain. Most scientists suspect that increased levels of carbon dioxide in the atmosphere contribute to climate change. Scientists are searching for replacement fuels that do not harm the environment.

Fossil fuels are not the only source of chemical energy. Chemical energy also is stored in the foods you eat. Your body converts the energy stored in chemical bonds in food into the kinetic energy of your moving muscles and into the electric energy that sends signals through your nerves to your brain.

 **Reading Check** What is chemical energy?

Nuclear Energy

The majority of energy on Earth comes from the Sun. A process, called nuclear fusion, in the Sun joins the nuclei of atoms and, in the process, releases large amounts of energy. On Earth, nuclear energy plants, such as the one shown in **Figure 4,** break apart the nuclei of certain atoms using a process called nuclear fission. Both nuclear fusion and nuclear fission release **nuclear energy**—*energy stored in and released from the nucleus of an atom.*

Nuclear fission produces a large amount of energy from just a small amount of fuel. However, the process produces radioactive waste that is hazardous and difficult to dispose of safely.

Kinetic and Potential Energies Combined

Recall that a moving object has kinetic energy. Objects such as wind turbine blades and particles, such as molecules, ions, atoms, and electrons, often have kinetic and potential energies.

Mechanical Energy

The sum of potential energy and kinetic energy in a system of objects is **mechanical energy.** Mechanical energy is the energy a system has because of the movement of its parts (kinetic energy) and because of the position of its parts (potential energy). An object, such as the wind turbine shown in **Figure 5,** has mechanical energy because the parts that make up the system have both potential energy and kinetic energy. A rotating blade has kinetic energy because of its motion, and it has gravitational potential energy because of its distance from Earth's surface.

Thermal Energy

The particles that make the wind turbine also have thermal energy. **Thermal energy** *is the sum of the kinetic energy and potential energy of the particles that make up an object.* Although you cannot see the individual particles move, they vibrate back and forth in place. This movement gives the particles kinetic energy. The particles also have potential energy because of the distance between particles and the charge of the particles.

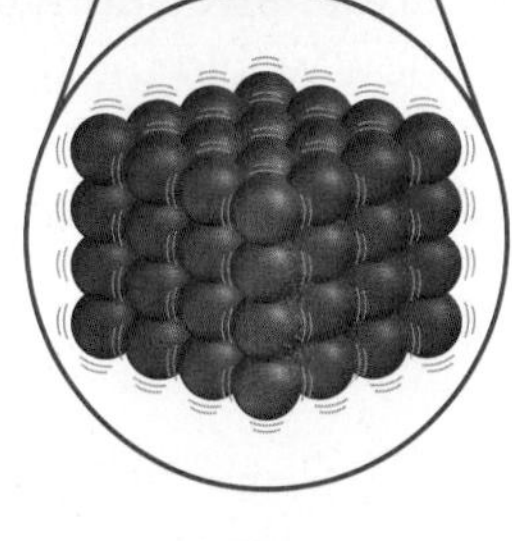

▲ **Figure 5** The entire wind turbine has mechanical energy. The particles that make up the wind turbine have thermal energy.

Geothermal Energy

The particles in Earth's interior contain great amounts of thermal energy. This energy is called geothermal energy. In geothermal energy plants, such as the one shown in **Figure 6,** thermal energy is used to heat water and turn it to steam. The steam turns turbines in electric generators, converting the geothermal energy to electric energy. Geothermal energy produces almost no pollution. However, geothermal plants must be built in places where molten rock is close to Earth's surface.

◀ **Figure 6** Geothermal energy plants convert thermal energy of the particles deep inside Earth to electric energy. The states with the most geothermal plants are Alaska, Hawaii, and California.

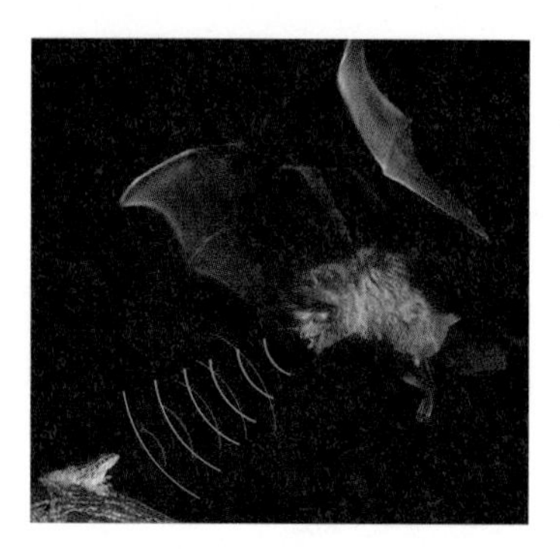

▲ **Figure 7** Bats use sound energy to detect the location of their prey.

Visual Check If the bat was farther away from the frog, how would the time it takes for the bat to receive the bounced wave change?

Energy from Waves

Have you ever seen waves crash on a beach? When a big wave crashes, you hear the sound of the impact. The movement and the sound result from the energy carried by the wave. Waves are disturbances that carry energy from one place to another. Waves move only energy, not matter.

Sound Energy

If you clap your hands together, you create a sound wave in the air. Sound waves move through matter. **Sound energy** *is energy carried by sound waves.* Some animals, such as the bat shown in **Figure 7,** emit sound waves to find their prey. The length of time it takes sound waves to travel to their prey and echo back tells the bat the location of the prey it is hunting.

Seismic Energy

You probably have seen news reports showing photographs of damage caused by earthquakes, similar to that in **Figure 8.** Earthquakes occur when Earth's tectonic plates, or large portions of Earth's crust, suddenly shift position. The kinetic energy of the plate movement is carried through the ground by seismic waves. **Seismic energy** *is the energy transferred by waves moving through the ground.* Seismic energy can destroy buildings and roads.

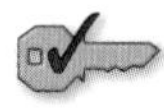

Key Concept Check What are the different forms of energy?

SCIENCE USE V. COMMON USE

vacuum

Science Use a space that contains no matter

Common Use to clean with a vacuum cleaner or sweeper

▲ **Figure 8** The seismic energy of a large earthquake caused severe damage to this building in San Francisco, California. In some locations, newly constructed homes and buildings are built to withstand many earthquakes.

Radiant Energy

When you listen to the radio, use a lamp to read, or call someone on your cell phone, do you think of waves? Electromagnetic waves are electric and magnetic waves that move perpendicular to each other, as shown in **Figure 9.** Radio waves, light waves, and microwaves are all electromagnetic waves, as shown in **Figure 10.** Some electromagnetic waves can travel through solids, liquids, gases, and **vacuums.** *The energy carried by electromagnetic waves is* **radiant energy.**

The Sun's energy is transmitted to Earth by electromagnetic waves. Photovoltaic (foh toh vohl TAY ihk) cells, also called solar cells, are made of special material that transforms the radiant energy of light into electric energy. You might have used a solar calculator. It does not need batteries because it has a photovoltaic cell. Photovoltaic cells also are used to provide energy to satellites, offices, and homes. Because so much sunlight hits the surface of Earth, the supply of solar energy is plentiful. Also, using solar energy as a source for electric energy produces almost no waste or pollution. However, only about 0.1 percent of the electric energy used in the United States comes directly from the Sun.

Key Concept Check How is radiant energy used?

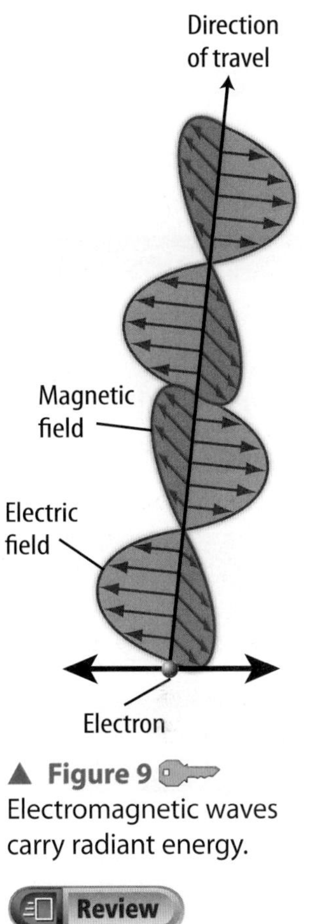

▲ **Figure 9** Electromagnetic waves carry radiant energy.

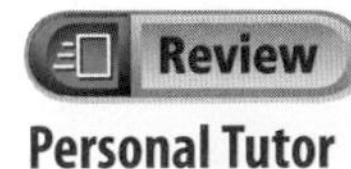

Personal Tutor

Figure 10 Radiant energy is carried by different forms of electromagnetic waves. ▼

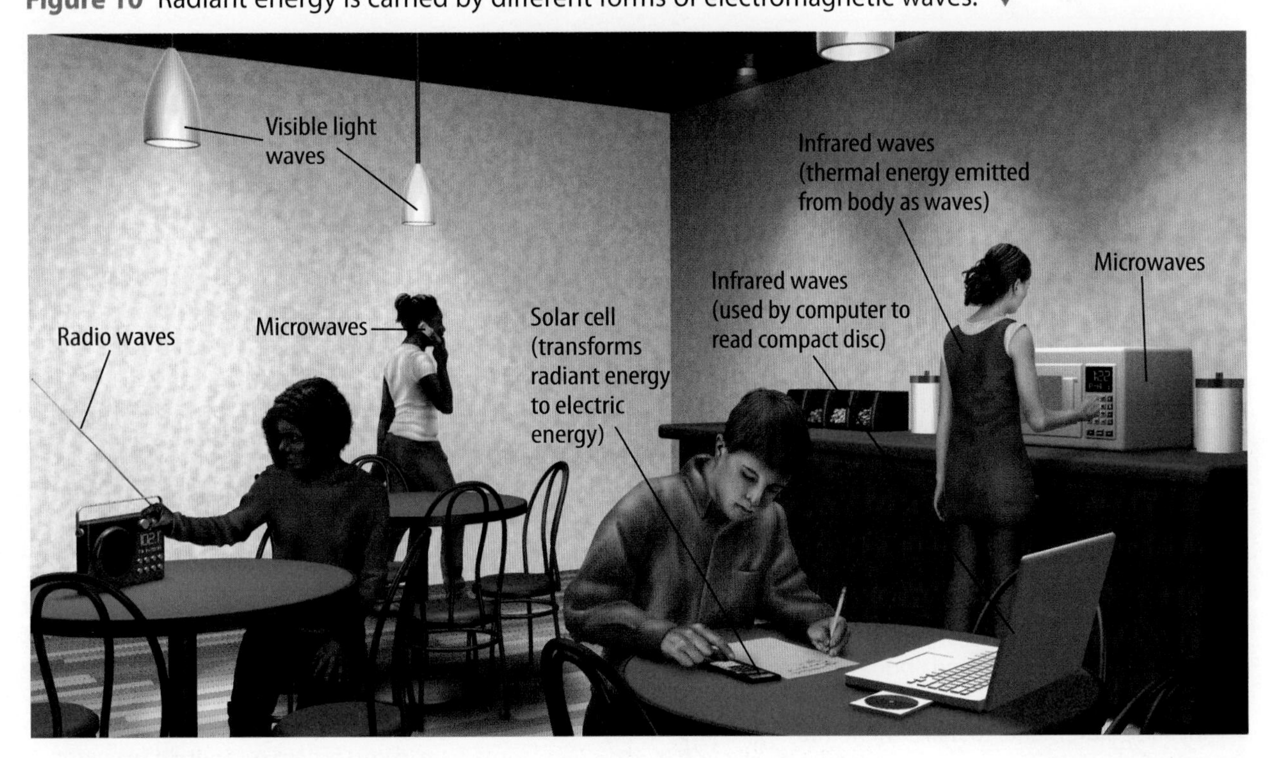

Lesson 1 Review

Visual Summary

There are different forms of energy, including solar energy.

Wind turbines have different kinds of energy including kinetic, mechanical, potential, and thermal.

Nuclear fuel pellets contain potential energy that is stored in the nuclei of atoms.

Use your lesson Foldable to review the lesson. Save your Foldable for the project at the end of the chapter.

What do you think NOW?

You first read the statements below at the beginning of the chapter.

1. Energy is the ability to produce motion.

2. Waves transfer energy from place to place.

Did you change your mind about whether you agree or disagree with the statements? Rewrite any false statements to make them true.

Use Vocabulary

1. **Define** *energy* in your own words.
2. **Distinguish** between kinetic energy and potential energy.
3. Energy carried by electromagnetic waves is _________.

Understand Key Concepts 

4. **Compare** seismic and sound energies.
5. Which of the following is NOT a form of stored energy?
   A. chemical energy
   B. electric energy
   C. gravitational potential energy
   D. nuclear energy
6. **Explain** how hydroelectric energy plants convert potential energy into kinetic energy.

Interpret Graphics

7. **Summarize** Fill in the following graphic organizer to show forms of potential energy.

Critical Thinking

8. **Apply** At graduation a student throws a cap into the air. During which part of the cap's journey does it have the most kinetic energy? When does it have the most potential energy? Explain your answer.
9. **Assess** Which forms of energy are involved when you turn on a desk lamp and the bulb becomes hot?
10. **Summarize** List the different types of energy plants mentioned in this lesson and identify which type of energy (kinetic energy, potential energy, or radiant energy) is converted into electric energy in each.

AMERICAN MUSEUM of NATURAL HISTORY

SCIENCE & SOCIETY

Using Solar Panels

Energy from Sunlight

A home's roof does more than keep the rain out! It's equipped with solar panels that supply some of the home's energy needs. Solar panels make electricity without using fossil fuels.

Large solar panels, such as those on this house, are made up of many individual photovoltaic cells. The term *photovoltaic* refers to an energy transformation from light to electricity.

Solar panels have a variety of components. Each has an important function. Most solar panels have a top layer of glass that protects the parts inside the panel. Under the glass is an anti-reflective layer that helps the panel absorb sunlight rather than reflect it. On the back, is a layer made to keep the solar panel from getting too hot.

▲ These solar panels contain materials that can transform energy from one form to another.

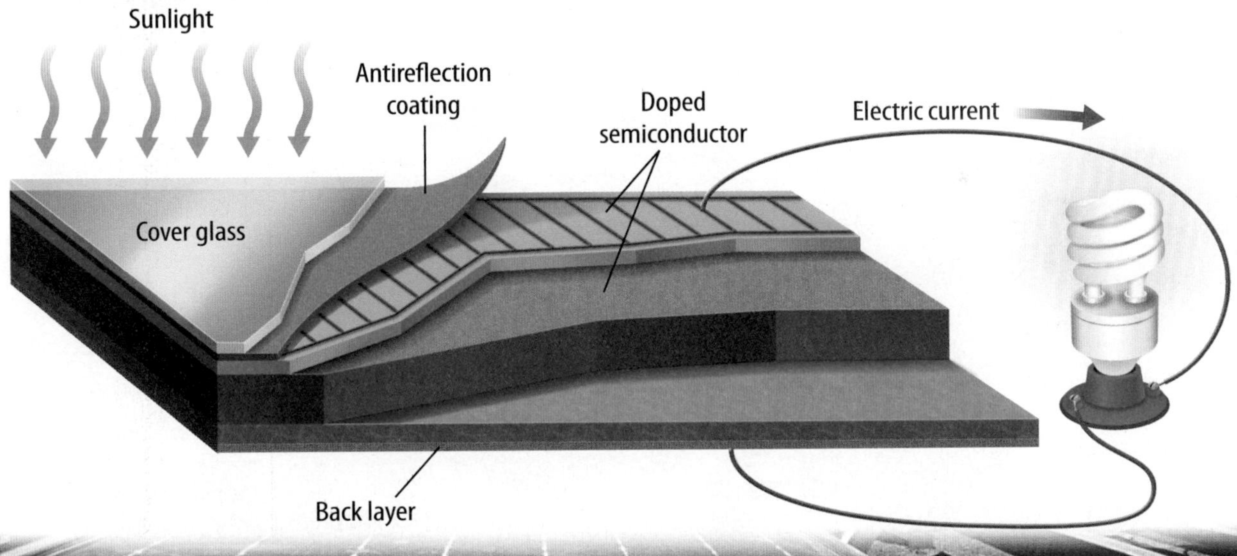

▲ Electric current flows from the solar panel to objects in the home that use electricity, such as lightbulbs, and back to the solar panel in a complete circuit.

The middle of the solar panel contains a large number of individual photovoltaic cells. That's where the energy happens! In a photovoltaic cell, sunlight strikes a doped semiconductor, or a semiconductor with atoms of other elements that increase conductivity. The energy in the sunlight knocks electrons in the doped semiconductor out of their positions and gives them energy to move. Recall that when electrons move, they create an electric current. Wires attached to the doped semiconductor allow the flowing electrons, or electric current, to travel to the electric circuits within the home and back again.

It's Your Turn

RESEARCH AND REPORT How might solar panels affect your life? How is new technology making solar panels less expensive to make and more efficient to use? Research to find out, and then share what you learn with in the rest of your class.

Lesson 2

Reading Guide

Key Concepts 
ESSENTIAL QUESTIONS

- What is the law of conservation of energy?
- In what ways can energy be transformed?
- How are energy and work related?

Vocabulary

energy transformation p. 97
law of conservation of energy p. 97
work p. 99

g Multilingual eGlossary

Video

- Science Video
- What's Science Got to do With It?

Energy Transformations and Work

Inquiry Space Aliens?

It might look like an invasion from space, but these solar-powered cars are in a race. Large solar panels across the width of the cars transform radiant energy from the Sun into electric energy that moves the cars.

Launch Lab

15 minutes

How far will it go?

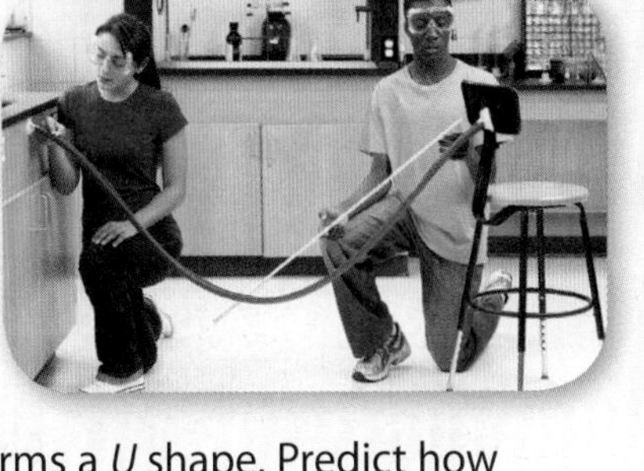

Suppose you are hired to design a roller coaster. Could you make it any shape you wanted? Could a hill in the middle of the ride be higher than the starting point?

1. Read and complete a lab safety form.
2. **Tape** one end of a **foam track** to the wall or other vertical object so that the end is 70–100 cm above the floor.
3. Tape the other end of the track to a chair so that the track forms a *U* shape. Predict how far a **marble** will travel if you release it at the top of the track on the wall side. Record your prediction in your Science Journal. Then test your prediction. Use a **meterstick** to measure the height from which you drop the marble and the height to which it rises.
4. Repeat step 3 several times using different heights above and below the starting point.

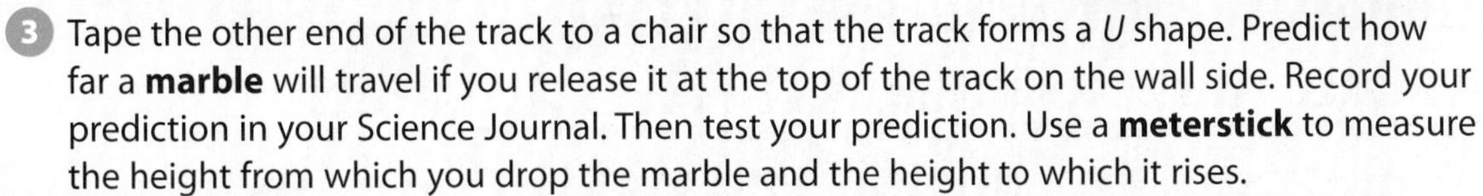

Think About This

1. How does the height to which the marble rises relate to the height at which it started?
2. **Key Concept** Do you think a hill at the end of the roller coaster ride could be higher than the starting point of the coaster car? Why or why not? Explain in terms of potential and kinetic energy.

Energy Transformations

As you read in Lesson 1, different types of electric energy plants supply the energy you use in your home and school. **Energy transformation** *is the conversion of one form of energy to another,* as shown in **Figure 11.** The electric energy in the wiring of the heat lamp is transformed into thermal energy.

Energy also is transferred when it moves from one object to another. When energy is transferred, the form of energy does not have to change. For example, the thermal energy in the heat lamp is transferred to the air and then to the zebra.

Academic Vocabulary

transform
(verb) to change form or structure

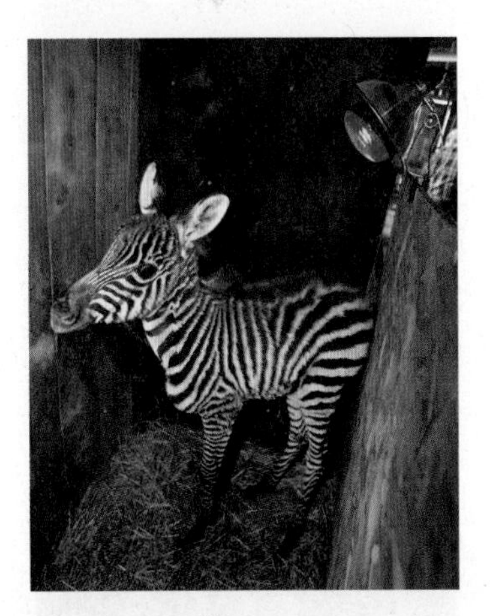

Figure 11 Electric energy is transformed into thermal energy in the heat lamp. Thermal energy from the lamp is transferred to the zebra.

Energy Conservation

Suppose you turn on a light switch. The radiant energy coming from the bulb had many other forms before it shined in your eyes. It was electric energy in the lamp's wiring, chemical energy in the fuel at the electric energy plant. The **law of conservation of energy** *says that energy can be transformed from one form to another, but it cannot be created or destroyed.* Even though energy can change forms, the total amount of energy in the universe does not change. It just changes form.

Key Concept Check What is the law of conservation of energy?

Review

Personal Tutor

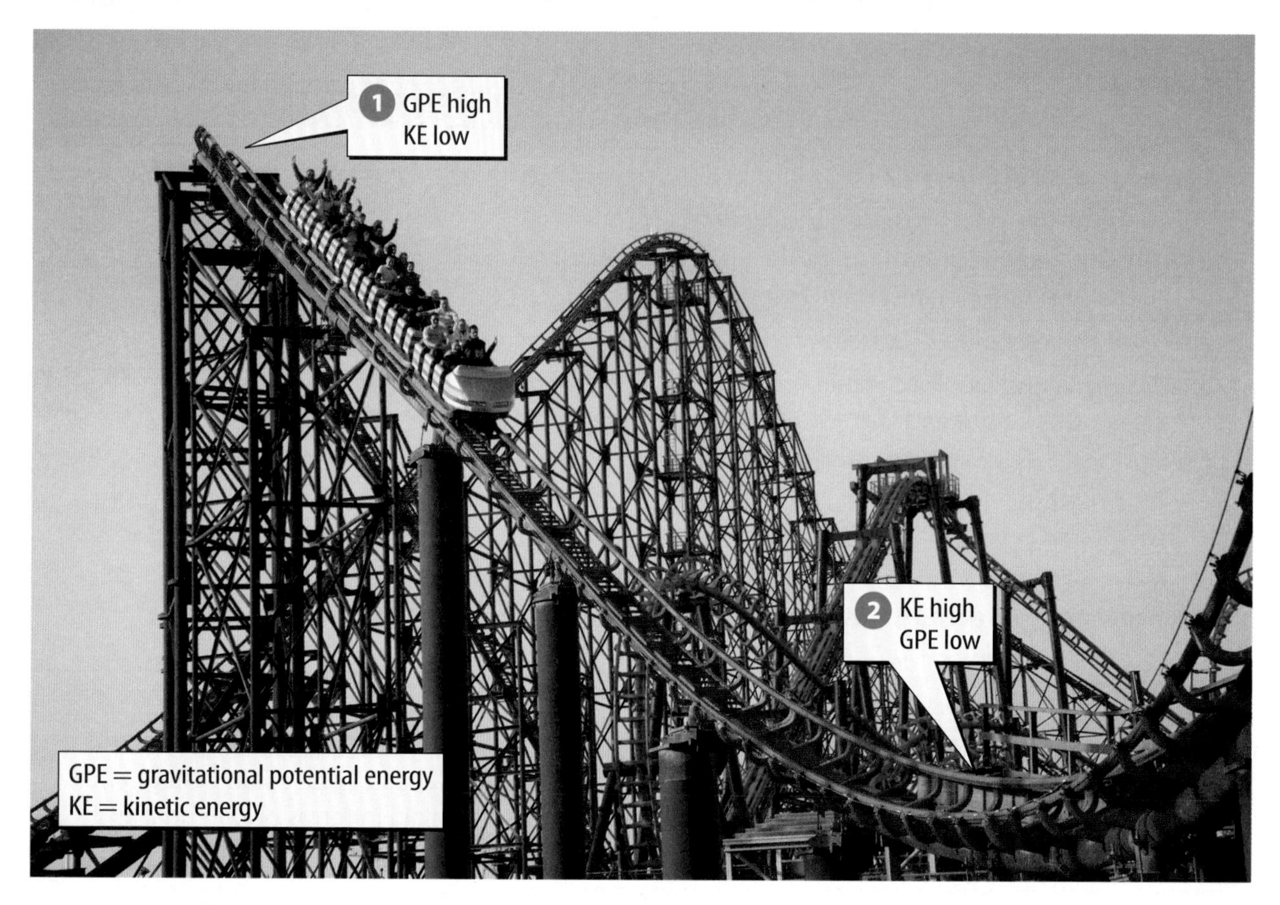

▲ **Figure 12** When you ride a roller coaster, your gravitational potential energy is transformed to kinetic energy and back to gravitational potential energy.

Roller Coasters

Have you ever thought about the energy transformations that occur on a roller coaster? Most roller coasters start off by pulling you to the top of a big hill. When you go up a hill, the distance between you and Earth increases and so does your potential energy. Next, you race down the hill. You move faster and faster. The gravitational potential energy is transformed to kinetic energy. At the bottom of the hill, your gravitational potential energy is small, but you have a lot of kinetic energy. This kinetic energy is transformed back to gravitational potential energy as you move up the next hill.

Figure 13 To carry out life processes, humans and other animals transform the chemical energy of plants into other forms of energy. ▼

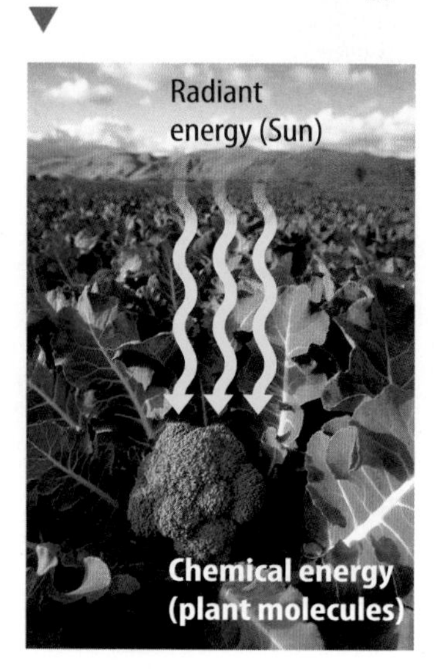

Plants and the Body

When a plant carries on photosynthesis, as shown in **Figure 13,** it transforms radiant energy from the Sun into chemical energy. The chemical energy is stored in the bonds of the plant's molecules. When you eat the broccoli, your body breaks apart the chemical bonds in the molecules that make up the broccoli. This releases chemical energy that your body transforms to energy your body needs, such as energy for movement, temperature control, and other life processes.

Electric Energy Plants

About 300 million years ago, plants carried out photosynthesis, just like plants do today. These ancient plants stored radiant energy from the Sun as chemical energy in their molecular bonds. After they died, the plants became buried under sediment. After much time and pressure from the sediments above them, these plants turned into fossil fuels. When electric energy plants burn fossil fuels, they transform the chemical energy from the molecules that were made by plants that lived millions of years ago. That chemical energy is transformed to the electric energy that you use in your home and school.

As you read in Lesson 1, other forms of energy, such as solar, wind, geothermal, and hydroelectric energy, also are transformed to electric energy by electric energy plants.

Key Concept Check Identify three energy transformations that occur to make electric energy.

Energy and Work

When you study for a test, do you do work? It might seem like it, but it would not be work as defined by science. **Work** *is the transfer of energy that occurs when a force makes an object move in the direction of the force while the force acts on the object.* Recall that forces are pushes or pulls. When you lift an object, you transfer energy from your body to the object. As the boy lifts the drums in **Figure 14**, they move and have kinetic energy. As the drums get higher off the ground, they gain gravitational potential energy. The boy has done work on the drums.

On the right in **Figure 14,** the boy is standing still with his drums lifted in place. Because he is not moving the drums, he is not doing work. To do work on an object, an object must move in the direction of the force. Work is done only while the force is moving the object.

Key Concept Check If you do work on an object, how will its energy change?

Figure 14 The boy does work on the drums when he lifts them. Once the drums are in place, no work is being done.

Visual Check What energy transformations occur as the drums are lifted?

The drummer does work on the drums as he lifts them. The drums' kinetic energy and gravitational potential energy increase.

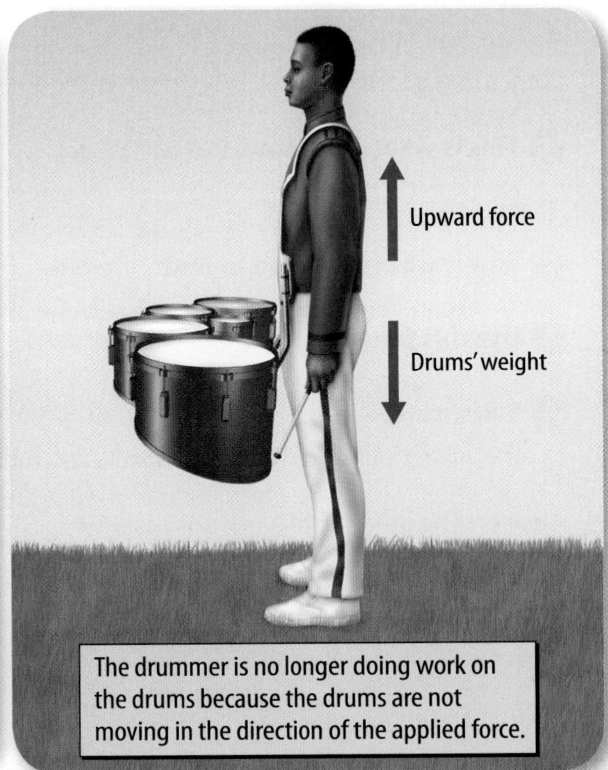

The drummer is no longer doing work on the drums because the drums are not moving in the direction of the applied force.

Word Origin

work
from Greek *ergon*, means "activity"

Doing Work

How much **work** do you do when you lift your backpack off the ground? If you lift a backpack with a force of 20 N, you do less work than if you lift a backpack with a force of 40 N. Work depends on the amount of force applied to the object.

Work also depends on the distance the object moves during the time the force is applied. If you lift a backpack 1 m you do less work than if you lift it 2 m. Suppose you toss a backpack in the air. When you release it, it continues moving upward. Even though the backpack is still moving when you let go, no work is being done. This is because you are no longer applying a force to the backpack while it is in the air.

Foldables®

Create a vertical half-book. Label it as shown. Use it to summarize, in your own words, the relationship between work and energy.

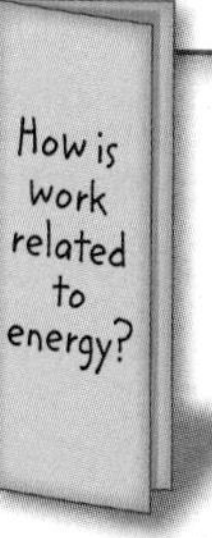

Calculating Work

The equation for work is shown below. *Force* is the force applied to the object. *Distance* is the distance the object moves in the direction of the force while the force is acting on it.

Work Equation

work (in joules) = **force** (in newtons) × **distance** (in meters)

$$W = Fd$$

The force in the equation is in newtons (N), and distance is in meters (m). The product of newtons and meters is newton-meter (N•m). A newton-meter is also called a joule (J).

Math Skills Work Equation

Solve for Work A student lifts a bag from the floor to his or her shoulder 1.2 m above the floor, using a force of 50 N. How much work does the student do on the bag?

1. **This is what you know:** force: $F = 50$ N; distance: $d = 1.2$ m
2. **This is what you need to find:** work: W
3. **Use this formula:** $W = Fd$
4. **Substitute:** the values for F and d into the formula and multiply: $W = (50 \text{ N}) \times (1.2 \text{ m}) = 60 \text{ N·m} = 60 \text{ J}$

Answer: The amount of work done is **60 J.**

Practice

A student pulls out his or her chair in order to sit down. The student pulls the chair 0.75 m with a force of 20 N. How much work does he or she do on the chair?

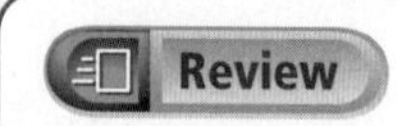

- **Math Practice**
- **Personal Tutor**

Energy and Heat

Have you ever heard the phrase *burning rubber?* The tires of race cars are made of rubber. The tires and the road are in contact, and they move past each other very fast. Recall that friction is a force between two surfaces in contact with each other. The direction of friction is in the opposite direction of the motion.

Friction between a car's tires and the road causes some of the kinetic energy of the tires to transform into thermal energy. If race cars are going really fast, thermal energy in the tires causes the rubber to give off a burnt odor.

Reading Check What is friction?

In every energy transformation and every energy transfer, some energy is transformed into thermal energy, as shown in **Figure 15.** This thermal energy is transferred to the surroundings. Thermal energy moving from a region of higher temperature to a region of lower temperature is called heat. Scientists sometimes call this heat *waste energy* because it is not easily used to do useful work.

Figure 15 Thermal energy is released to the surroundings during energy transformations and energy transfers in the engines of race cars.

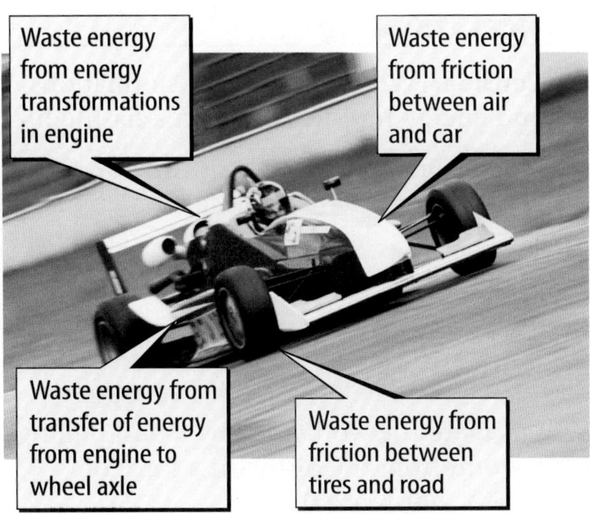

Inquiry MiniLab — 20 minutes

How do energy transformations work for you?

Every time you turn on a light, comb your hair, or ride your bike, you are transforming energy. What forms of energy are involved?

1. Copy the table into your Science Journal.
2. On the table, record an energy chain for each object, similar to the example. Include all types of energy used to make the object function, as well as the types of energy produced. Use the following abbreviations:

   R = radiant; T = thermal; C = chemical; N = nuclear; E = electric; S = sound; Mp = potential mechanical; Mk = kinetic mechanical

Object	Energy Chain	Object	Energy Chain
Flashlight	C-E-T & R	Nuclear power plant	
Sun		Auto-mobile	
Microwave oven		Television	

Analyze and Conclude

1. **Apply** Give three different examples in which the following energy transformations would take place: electric to thermal, chemical to thermal, and mechanical to electric.
2. **Generalize** Which type of energy that is no longer useful is produced in all transformations? Explain.
3. **Key Concept** Select three of the energy changes, and explain how each one does work.

Lesson 2 Review

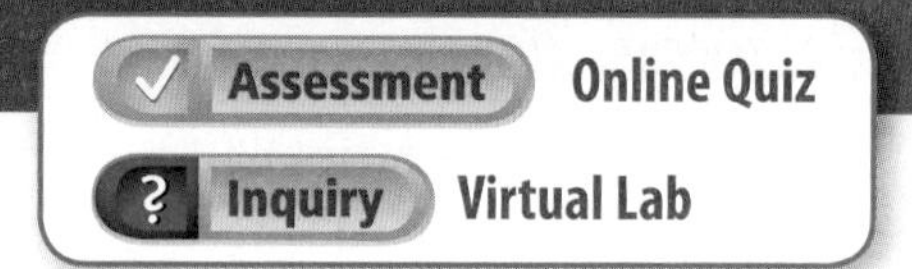

Visual Summary

Energy is always conserved.

Energy can be transformed into different kinds of energy.

Work and energy are related.

FOLDABLES®

Use your lesson Foldable to review the lesson. Save your Foldable for the project at the end of the chapter.

What do you think NOW?

You first read the statements below at the beginning of the chapter.

3. Energy cannot be created or destroyed, but it can be transformed.

4. Work describes how much energy it takes for a force to push or to pull an object.

Did you change your mind about whether you agree or disagree with the statements? Rewrite any false statements to make them true.

Use Vocabulary

1 A(n) ________ occurs when energy is converted from one form to another.

Understand Key Concepts 

2 **Distinguish** between work and energy.

3 **Define** the law of conservation of energy in your own words.

4 Which is NOT an example of work?

A. holding books in your arms
B. lifting a box from a table
C. placing a bowl on a high shelf
D. pushing a cart across the room

5 **Describe** the energy transformations that occur when a piece of wood is burned.

Interpret Graphics

6 **Explain** the gravitational potential energy transformations that occur when the object at right is in motion.

7 **Summarize** Copy and fill in the graphic organizer below to show what work is the product of.

work	

Critical Thinking

8 **Consider** Which energy transformations and energy transfers occur in a flashlight?

9 **Model** Draw a picture showing how energy is transferred to a sidewalk on a hot summer day. Label the different forms of energy in your drawing.

Math Skills

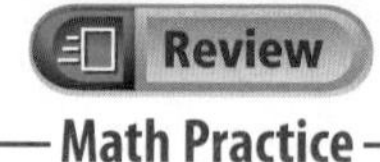

Math Practice

10 **Calculate** the work done by a bird pulling a worm from the dirt with a force of 0.05 N a distance 0.07 m.

Inquiry **Skill Practice** **Follow a Procedure** 45 minutes

How can you transfer energy to make a vehicle move?

Materials

corrugated cardboard

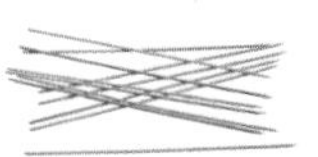

bamboo skewer

meterstick

rubber bands of different lengths and widths

masking tape

Also needed: string, scissors, compact disks, 1.5 cm faucet washers, adhesive putty

Safety

You have learned how energy can transform from one type to another. How can you and your classmates use this information to build a vehicle that uses an unusual source of energy?

Learn It

It is helpful to **follow a procedure** when you are doing something for the first time. A procedure tells you how to use the materials and what steps to take.

Try It

1. Read and complete a lab safety form.
2. Obtain instructions for building a vehicle. Gather the materials for your vehicle.
3. Build your vehicle according to the directions on your instruction sheet. When you are finished, have your teacher check your vehicle before testing it.
4. Test your vehicle several times. Try to begin with the same amount of potential energy for each trial.
5. Discuss with your teammates how you could make your vehicle go faster or farther. If the vehicle doesn't travel in a straight line, modify the design.
6. Modify your vehicle so that it uses only gravitational potential energy as its energy source. You may not use the original energy source and the vehicle must run on a straight, level course. You may not provide any external energy to get your vehicle started.

Apply It

7. Which energy transformations moved your vehicle? Be sure to describe both potential and kinetic types of energy.
8. Compare the energy sources that you used to power your vehicle. What are the advantages and the disadvantages of each energy source?
9. What variables affect the amount of gravitational potential energy you can use to move your vehicle?
10. **Key Concept** Was energy conserved as it moved your vehicle? Explain your answer. Why did your vehicle stop?

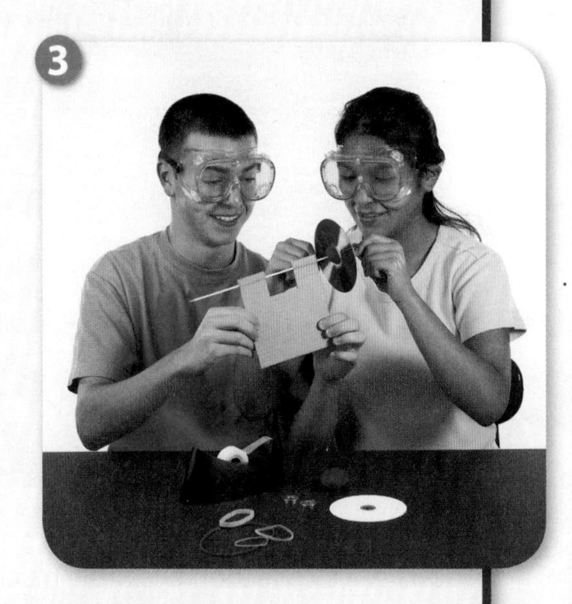

Lesson 3

Reading Guide

Key Concepts 
ESSENTIAL QUESTIONS

- What are simple machines?
- In what ways can machines make work easier?

Vocabulary

simple machine p. 105
inclined plane p. 106
screw p. 106
wedge p. 106
lever p. 106
wheel and axle p. 106
pulley p. 107
complex machine p. 107
efficiency p. 109

g Multilingual eGlossary

Video BrainPOP®

Machines

Inquiry A Machine?

When you look at a unicycle, you probably don't see a collection of simple machines. However just like the bicycle that you will read about in this lesson, a unicycle contains simple machines.

Launch Lab

15 minutes

Can you make work easier?

Have you ever tried to pull a nail from a board without a claw hammer? The claw hammer makes an impossible task quite easy. What are some other ways to make work easier?

1. Read and complete a lab safety form.
2. Try to press the tip of a piece of **wire** into a **pine block** with your fingers. Then press a **thumbtack** with the same diameter into the block. Describe in your Science Journal how the amount of force you used differed in each case.
3. Screw an **eyehook** into the block as far as it will go. Start a **second eyehook** and then run your **pencil** through the hole in the eyehook. Use the pencil to screw in the eyehook. Compare the force you used in each case.
4. Tie a length of **string** around a **book.** Hook a **spring scale** through the string and lift the book to a height of 30 cm. Record the reading on the scale. Then use the spring scale to slide the book along a **ramp** to a height of 30 cm. Record the reading on the scale as you pull the book.

Think About This

1. How did the force needed in the first attempt of each task differ from the second attempt? What caused this difference?
2. **Key Concept** How did the amount of work you did using the two methods in each step compare? What was the same? What was different? Explain.

Machines Transfer Mechanical Energy

Suppose you want to open a bottle like the one in **Figure 16.** If you use a bottle opener, you can easily pry off the top. A bottle opener is a machine. Many machines transfer mechanical energy from one object to another. The bottle opener transfers mechanical energy from your hand to the bottle cap. In this lesson, you will read about the ways in which machines transfer mechanical energy to other objects.

Simple Machines

Did you walk up a ramp this morning? Did you cut food with a knife? If so, you used a simple machine. **Simple machines** *are machines that do work using one movement.* As shown in **Figure 17** on the next page, a simple machine can be an inclined plane, a screw, a wedge, a lever, a pulley, or a wheel and axle. Simple machines do not change the amount of work required to do a task; they only change the way work is done.

Reading Check What is a simple machine?

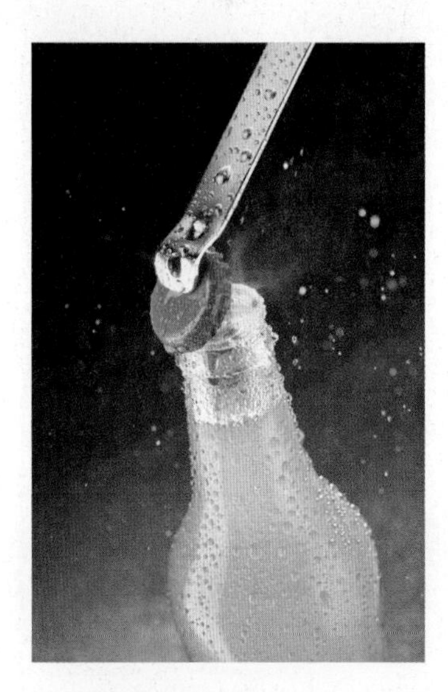

Figure 16 The bottle opener is a machine that transfers energy from your hand to the bottle cap.

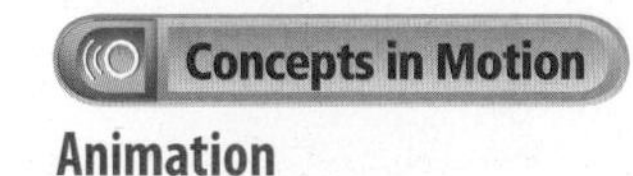

Animation

Figure 17 Simple machines do work using one movement. They can change the direction of a force or the amount of force required to perform a task.

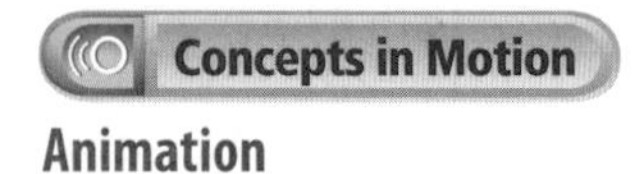

Visual Check Identify another example for each simple machine.

Animation

Inclined Plane Furniture movers often use ramps to move furniture into a truck. It is easier to slide a sofa up a ramp than to lift it straight up into the truck. An **inclined plane**, such as the ramp shown in **Figure 17**, *is a flat, sloped surface.* Ramps with gentle slopes require less force to move an object than steeper ramps, but you have to move the object a greater distance.

Screw A screw, such as screw-top bottle, is a special type of inclined plane. A **screw** *is an inclined plane wrapped around a cylinder.* A screw changes the direction of the force from one that acts in a straight line to one that rotates.

Wedge Like all knives, pizza cutters are a special type of inclined plane. A **wedge** *is an inclined plane that moves.* Notice how the wedge changes the direction of the input force.

REVIEW VOCABULARY

plane
a flat, level surface

Lever The tab in **Figure 17** on the next page, is a **lever**, *which is a simple machine that pivots around a fixed point.* The fixed point on a beverage can is where the finger tab attaches to the can. Bottle openers, scissors, seesaws, tennis racquets, and wheel barrows are other examples of levers. Levers decrease the amount of force required to complete a task, but the force must be applied over a longer distance.

Wheel and Axle A doorknob, a car's steering wheel, and a screwdriver are a type of simple machine called a **wheel and axle**—*a shaft attached to a wheel of a larger diameter so that both rotate together.* The wheel and the axle are usually both circular objects. The object with the larger diameter is the wheel, and the object with the smaller diameter is the axle. When you use a wheel and axle, such as a screwdriver, you apply a small input force over a large distance to the wheel (screwdriver handle). This causes the axle (screwdriver shaft) to rotate a smaller distance with a greater output force.

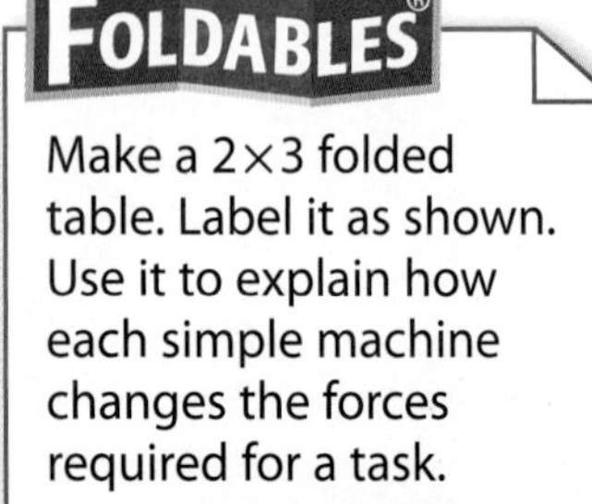

Make a 2×3 folded table. Label it as shown. Use it to explain how each simple machine changes the forces required for a task.

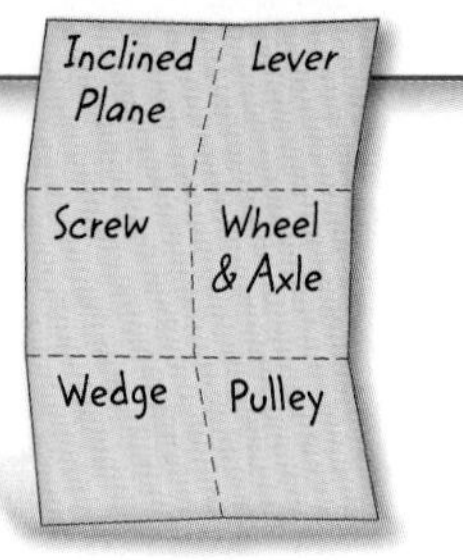

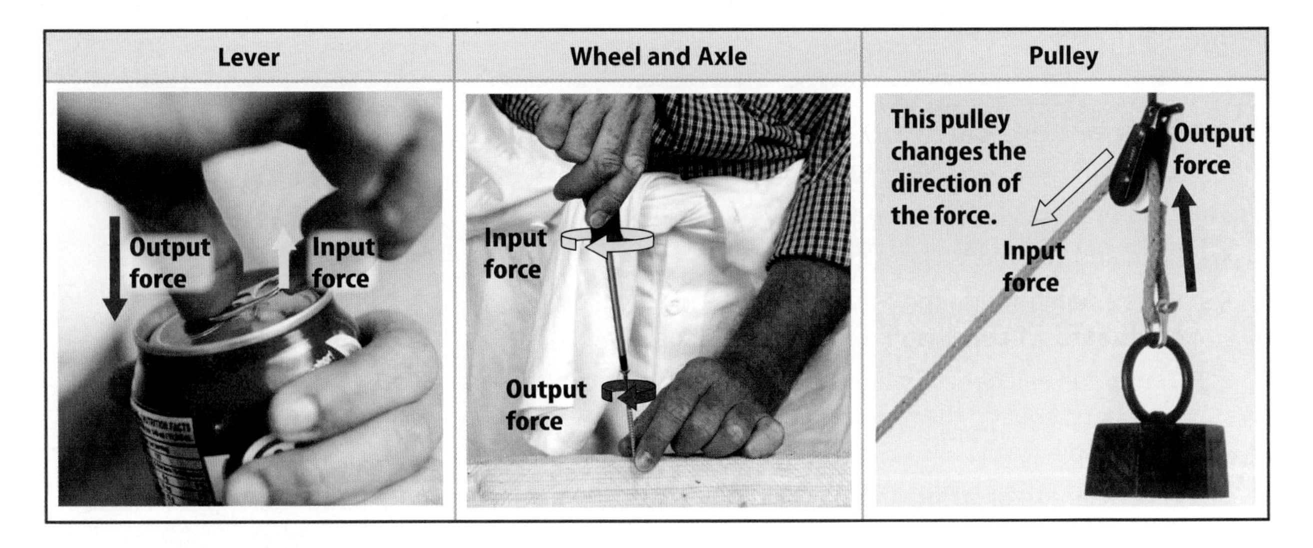

Pulley Have you ever raised a flag on a flagpole or watched someone raise a flag? The rope that you pull goes through a **pulley,** *which is a grooved wheel with a rope or cable wrapped around it.* A single pulley, such as the kind on a flagpole, changes the direction of a force. A series of pulleys decreases the force you need to lift an object because the number of ropes or cables supporting the object increases.

Key Concept Check What are examples of simple machines?

Complex Machines

Bicycles, such as the one in **Figure 18,** are made up of many different simple machines. The pedal stem is a lever. The pedal and gears together act as a wheel and axle. The chain around the gear acts as a pulley system. *Two or more simple machines working together are a* **complex machine.** Complex machines, such as bicycles, use more than one motion to accomplish tasks.

Reading Check How is a complex machine different from a simple machine?

Figure 18 A bicycle is a complex machine that is made of many simple machines.

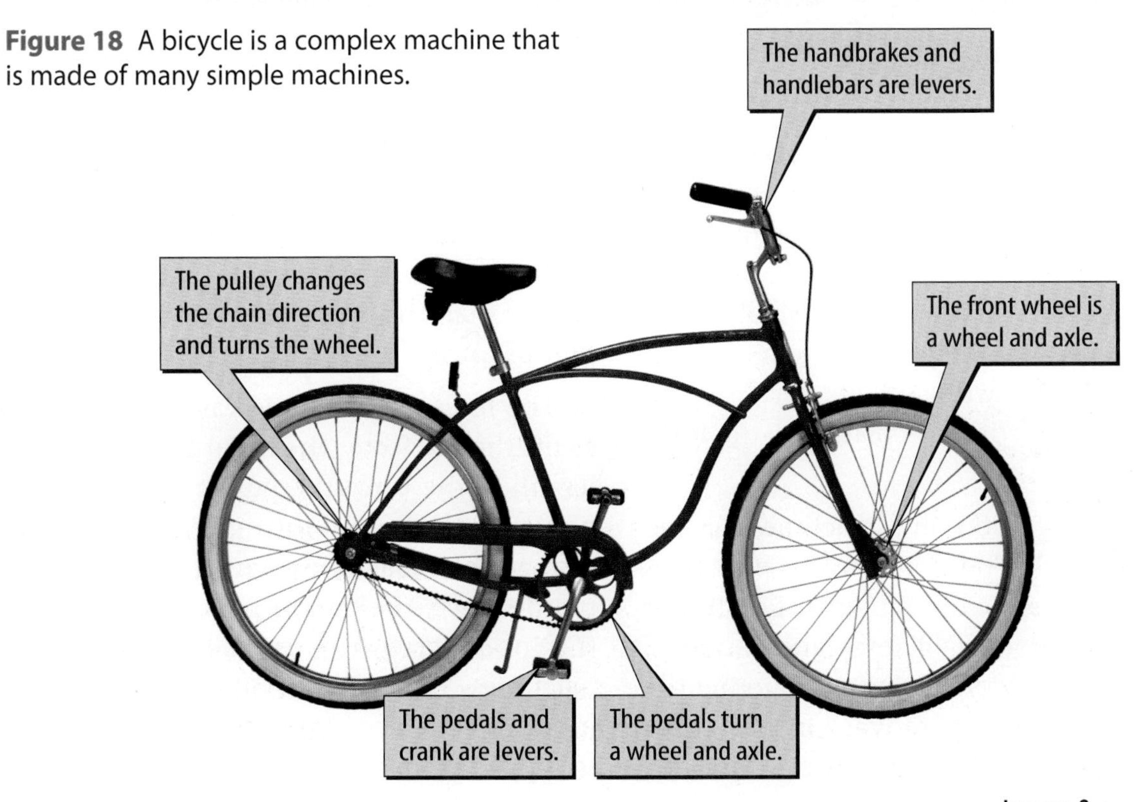

MiniLab

20 minutes

Does a wheel and axle make work easier?  

Can you shift gears on a bike? Why is it easier to push the pedals in one gear than in another? How do the sizes of the two wheels affect one another?

1. Read and complete a lab safety form.
2. Press a **nail** through the center of 5-cm and 10-cm diameter **cardboard wheels** that have been **glued** together.
3. **Clamp** the nail horizontally at the top of a **ring stand.** If necessary, enlarge the nail hole in the wheels slightly so that the wheels spin freely.
4. Wrap a length of **string** around the groove in each wheel. Attach one end of each string to the wheel groove with a **pin.** The strings should hang down on opposite sides of the two wheels. Attach four **clothespins** to the free end of the string on the smaller wheel. Add clothespins to the free end of the string on the larger wheel until the four clothespins are lifted. Record the number of clothespins on each string in your Science Journal.
5. Repeat step 4, using other combinations of clothespins on the larger and smaller wheels. Record the combinations.

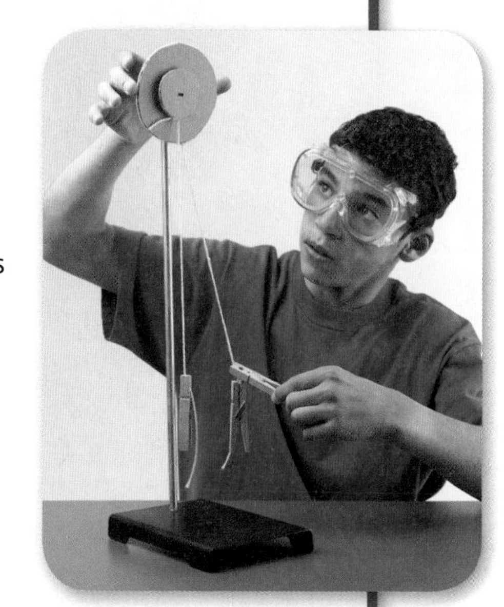

Analyze and Conclude

1. **Infer** In terms of work, what do the clothespins represent?
2. **Apply** What simple machines are used in this investigation?
3. **Key Concept** Explain why a smaller force on the large wheel was able to lift a larger force on the small wheel.

Machines and Work

Think of a window washer like the one in **Figure 19** on the next page. It takes a great amount of work to lift the washer's own weight plus the weight of buckets of water, window-washing tools, and the platform up in the air. The window washer is able to do this work because the pulley system that lifts him makes the work easier. Because two ropes are supporting the platform, the force required is half.

The work you do on a machine is called the input work. The work the machine does on an object is the output work. Recall that work is the product of force and distance. Machines make work easier by changing the distance the object moves or the force required to do work on an object.

Changing Distance and Force

To pull himself toward the top of the building, the window washer pulls down on a rope. The rope runs through a pulley system. The distance the window washer must pull the rope (the input distance) is much greater than the distance he moves (the output distance).

The force the window washer has to use to lift the platform (the input force) is much less than the force the pulley exerts on the platform (the output force). When the input distance of a machine is larger than the output distance, the output force is larger than the input force. This is true for all simple machines. Like other simple machines, the input force is decreased, but the distance it is applied is increased.

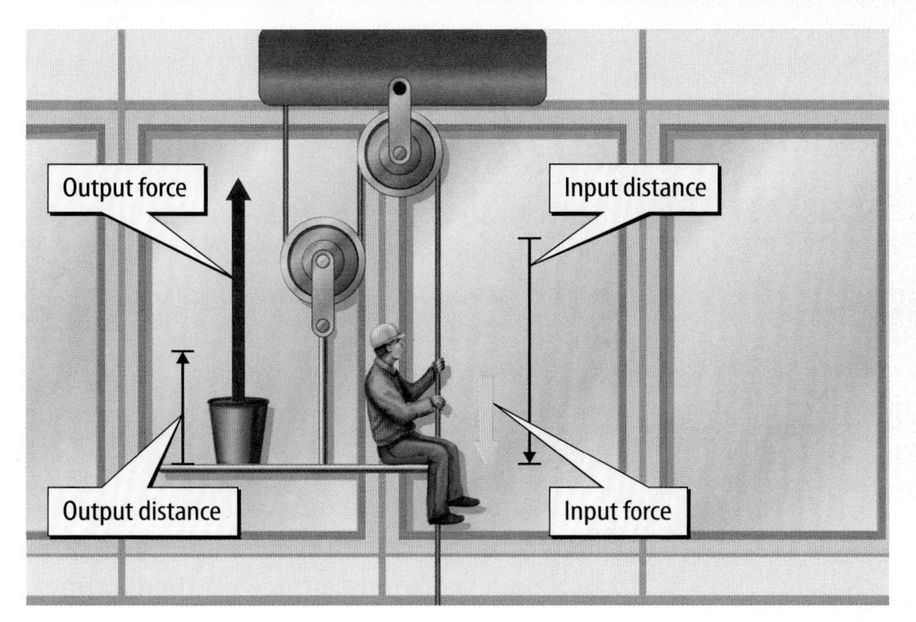

Figure 19 The window washer lifts his platform using a pulley system that increases the distance over which the force is exerted, decreases the input force needed, and changes the direction of the force.

Visual Check How does the pulley make raising the platform easier for the window washer?

Changing Direction

Machines also can change the direction of a force. A window washer pulls down on the rope. The pulley system changes the direction of the force, which pulls the platform up.

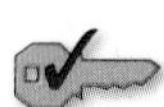 **Key Concept Check** How can machines make work easier?

Efficiency

Suppose the window washer wants to buy a new pulley system. One way to compare machines is to calculate each machine's efficiency. **Efficiency** *is the ratio of output work to input work.* In other words, it is a measure of how much work put into the machine is changed into useful output work. Input and output work are measured in joules (J). Efficiency is expressed as a percentage by multiplying the ratio by 100%.

WORD ORIGIN
efficiency
from Latin *efficere,* means "work out, accomplish"

Efficiency Equation

$$\text{efficiency (in \%)} = \frac{\text{output work (in J)}}{\text{input work (in J)}} \times 100\% = \frac{W_{out}}{W_{in}} \times 100\%$$

The window washer considers two systems that require 100 J of input work. The first one does 90 J of output work on his platform. The other pulley system does 95 J of output work. The efficiency of the first pulley system is (90 J/100 J) × 100% = 90%. The efficiency of the second one is (95 J/100 J) × 100% = 95%. The window washer decides to buy the second pulley system.

The efficiency of a machine is never 100%. Some work is always transformed into wasted thermal energy because of friction. One way to improve the efficiency of a machine is to lubricate the moving parts by applying a substance, such as oil, to them. This reduces the friction between the moving parts so that less input work is transformed to waste energy.

Figure 20 Newton's laws of motion help explain the forces applied by machines.

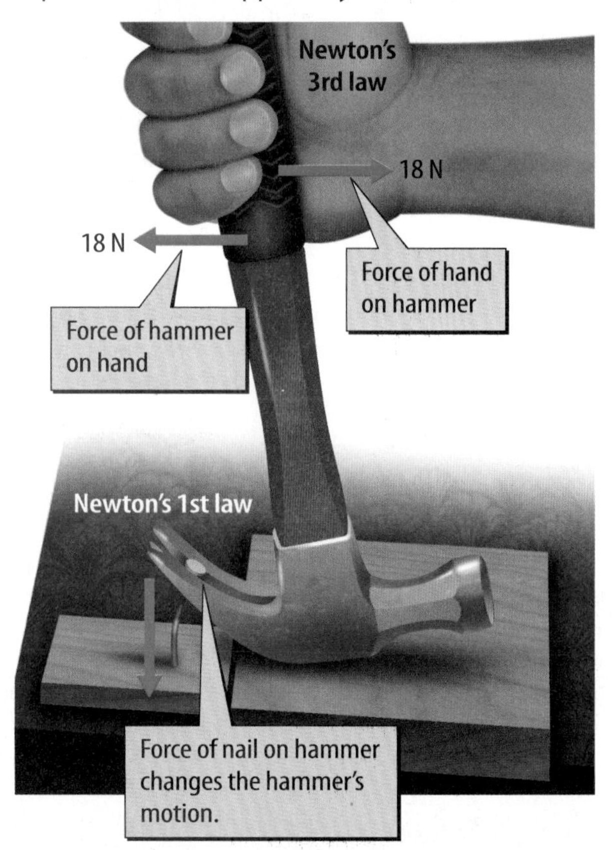

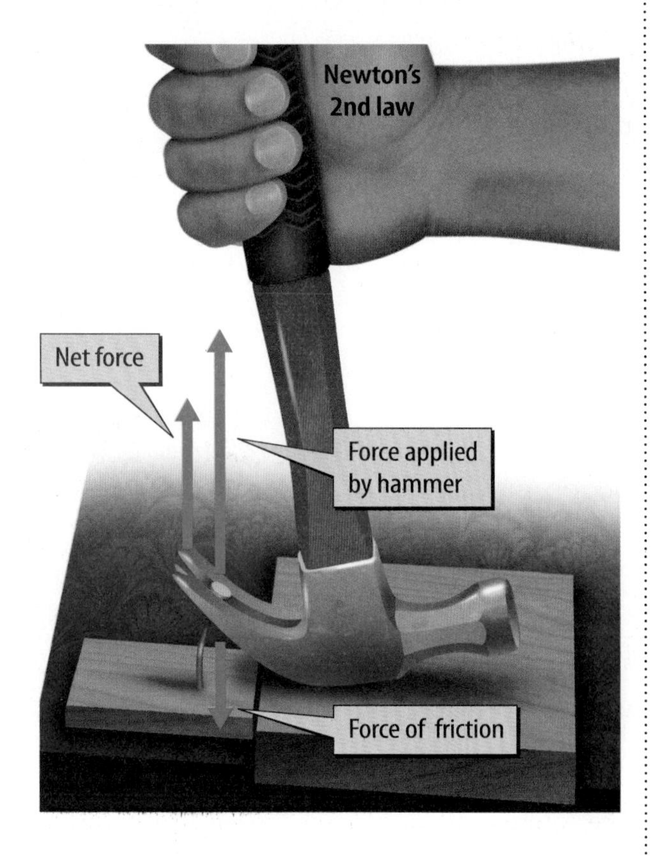

Newton's Laws and Simple Machines

Recall that Newton's laws of motion tell you how forces change the motion of objects. As you have read, machines apply forces on objects. For example, Newton's third law says that if one object applies a force on a second object, the second object applies an equal and opposite force on the first object.

As shown in the top part of **Figure 20,** when you use a hammer as a lever to pull out a nail, you apply a force on the hammer. The hammer applies an equal force in the opposite direction on your hand.

Reading Check What is Newton's third law?

According to Newton's first law, the motion of an object changes when the forces that act on the object are unbalanced. When you pull on the hammer handle, the claws of the hammer apply a force on the nail. However, unless you pull hard enough, the nail does not move.

The nail does not move because there is another force acting on the nail—the force due to friction between the nail and the wood. Unless you pull hard enough, the force of friction balances the force the hammer exerts on the nail. As a result, the motion of the nail does not change—the nail does not move.

If you pull hard enough, then the upward force the hammer applies on the nail is greater than the force of friction on the nail, as shown in the bottom part of **Figure 20.** Then the forces on the nail are unbalanced and the motion of the nail changes—the nail moves upward.

According to Newton's second law of motion, the change in motion of an object is in the same direction as the total, or net, force on the object. The nail moves upward because the net force on the nail is upward.

Lesson 3 Review

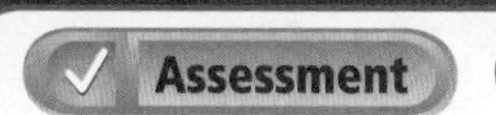

Visual Summary

A bottle opener is a simple machine.

There are six types of simple machines, and a ramp is one example.

A bicycle is an example of a complex machine that is made up of different simple machines.

Use your lesson Foldable to review the lesson. Save your Foldable for the project at the end of the chapter.

What do you think NOW?

You first read the statements below at the beginning of the chapter.

5. All machines are 100 percent efficient.

6. Simple machines do work using one motion.

Did you change your mind about whether you agree or disagree with the statements? Rewrite any false statements to make them true.

Use Vocabulary

1 **Contrast** simple and complex machines.

2 **Define** *efficiency* in your own words.

3 **Explain** the six simple machines discussed in this lesson.

Understand Key Concepts 

4 **Identify** What kind of simple machine is a thumbtack?

5 How does an inclined plane affect the work that is done on an object?

- **A.** It decreases the input distance.
- **B.** It increases the input distance.
- **C.** It changes the direction of the input force.
- **D.** It changes the direction of the output force.

Interpret Graphics

6 **Explain** which simple machine the object shown below represents.

7 **Summarize** Copy and complete the following graphic organizer showing the ways that simple machines can change the work done on an object.

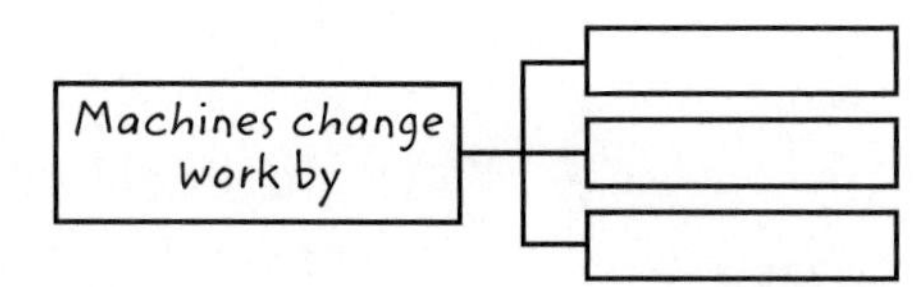

Critical Thinking

8 **Design** a machine that you could use to lift a bag of groceries from the floor to the counter using less force than if you lifted the bag with just your hands. Which simple machine would you use?

3 class periods

Build a Powered Vehicle

Materials

antacid tablets

plastic bottle

vinegar

baking soda

office supplies

Also needed: craft supplies, creative building materials, creative construction tools

Safety

In this chapter, you read about various forms of energy, how energy is transferred to do work, and how simple machines change how work is done. Now you can put it all together to design your own powered vehicle. Your teacher will give you the rules for this challenge.

Ask a Question

How can you design and construct a powered vehicle that will meet the criteria for the challenge? Consider the possible sources of energy you might use. How will you transfer energy to power the vehicle? Will you design a vehicle for speed, distance, or both? Consider the materials you have available. What materials could you bring from home?

Make Observations

1. Read and complete a lab safety form.
2. Brainstorm ideas with your teammates. Generate ideas by asking questions such as a) What are some possible energy sources? Consider possibilities such as wind, solar, chemical, electric, elastic, or magnetic. b) How will you convert potential energy to kinetic energy? c) What energy transformations will you use? d) How will you reduce the loss of energy due to friction? e) What will you use for the body of the vehicle? f) Will the vehicle have wheels? If so, how large will they be? What will they be made of? Record all of your ideas in your Science Journal.
3. Outline the steps in constructing your vehicle. Draw a diagram of the vehicle. Decide who will obtain which materials before the next lab period. Make sure each person on the team has a task in the design and building of the vehicle. You must be able to explain how your vehicle works and how it applies the ideas found in this chapter.
4. On the second lab day, follow the steps you outlined and build your vehicle. Be sure you have followed all the rules.
5. Test your vehicle and make any needed modifications. Make sure that your vehicle is sturdy enough to make a number of runs.

Form a Hypothesis

6. Formulate a hypothesis that explains why your vehicle will move due to energy transformations.

Test Your Hypothesis

7. On day 3, place your vehicle in competition with vehicles from other teams. Points will be awarded for distance, speed, and creativity in the application and transformation of energy sources. Be prepared to explain and answer questions about your design.
8. Write a report describing the design and scientific principles that went into making your vehicle.

Analyze and Conclude

9. **Identify Cause and Effect** What caused your vehicle to begin moving? What caused it to stop?
10. **Analyze** How could you increase the distance traveled by your vehicle without changing the source of energy?
11. **The Big Idea** Describe the relationship between the energy from your source and the work done on the vehicle.

Lab Tips

- ☑ Don't ignore less-obvious sources of energy in your discussions. Could you use a chemical reaction to power your vehicle? Could you produce portable electricity without a battery?
- ☑ Think outside the box. For example, instead of pushing the car, how could a power source pull the car?

Communicate Your Results

Work with your team to describe the design process and explain how you arrived at your choices for powering and building your team's vehicle. Compare and contrast the processes used by each team and discuss why some might have been more effective than others.

Inquiry Extension

If you could hold the challenge outdoors, how might you use wind energy, solar energy, or water to power your vehicle? Draw a design showing the parts of your vehicle and how it would work.

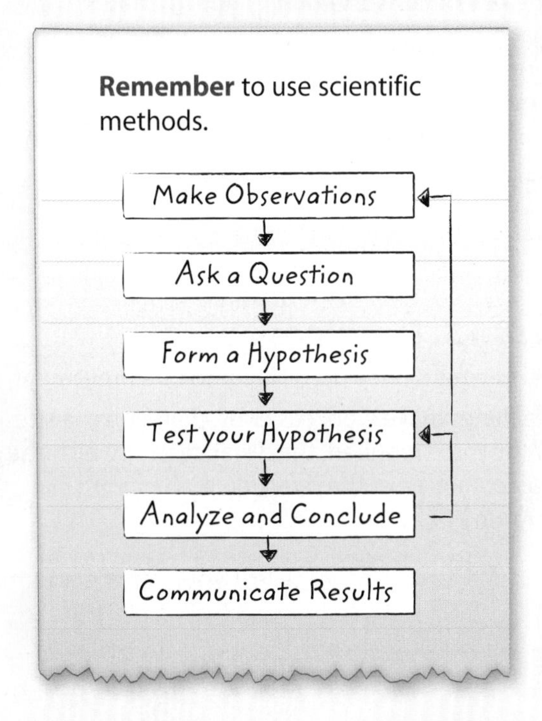

Chapter 3 Study Guide

Energy causes change by affecting the movement and position of objects. Energy can be transformed from one form to another and transferred from object to object.

Key Concepts Summary

Lesson 1: Types of Energy

- **Energy** is the ability to cause change.
- **Kinetic energy** is the energy of objects in motion, including **electric energy.** The forms of **potential energy** include gravitational potential energy, **chemical energy,** and **nuclear energy. Thermal energy** and **mechanical energy** are forms of energy involving both kinetic and potential energies. **Sound energy, seismic energy,** and **radiant energy** are all transferred by waves.
- Energy is used to move cars, heat homes, produce light, move muscles, catch prey, and cook food among many other examples.

Vocabulary

energy p. 87
kinetic energy p. 88
electric energy p. 88
potential energy p. 89
chemical energy p. 90
nuclear energy p. 90
mechanical energy p. 91
thermal energy p. 91
sound energy p. 92
seismic energy p. 92
radiant energy p. 93

Lesson 2: Energy Transformations and Work

- The **law of conservation of energy** states that energy can be transformed from one form to another, but it can never be created or destroyed.
- Energy can be transformed from one form to another in a variety of ways.
- Doing **work** on an object transfers energy to the object.

Vocabulary

energy transformation p. 97
law of conservation of energy p. 97
work p. 99

Lesson 3: Machines

- **Simple machines** do work using one movement.
- Machines make work easier by changing the size of the force required, the distance over which the object moves, or the direction of the input and output forces.

Vocabulary

simple machine p. 105
inclined plane p. 106
screw p. 106
wedge p. 106
lever p. 106
wheel and axle p. 106
pulley p. 107
complex machine p. 107
efficiency p. 109

Study Guide

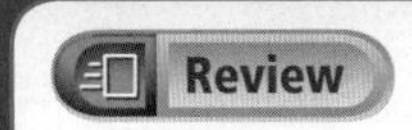

- Personal Tutor
- Vocabulary eGames
- Vocabulary eFlashcards

FOLDABLES® Chapter Project

Assemble your lesson Foldables as shown to make a Chapter Project. Use the project to review what you have learned in this chapter.

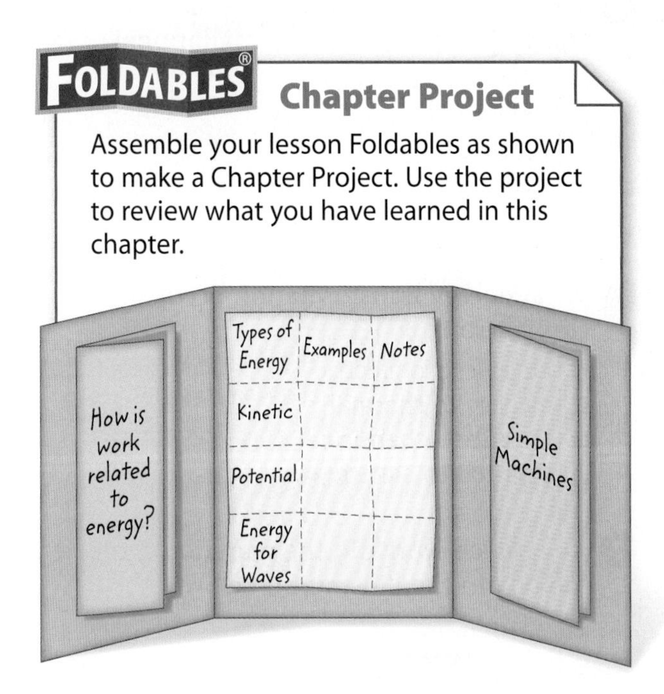

Use Vocabulary

1. Use the term *thermal energy* in a sentence.
2. The ________ of an object increases as it moves faster.
3. Define the term *energy transformation* in your own words.
4. The product of force and distance is ________.
5. Define the term *radiant energy* in your own words.
6. A(n) ________ is made of more than one simple machine.

Link Vocabulary and Key Concepts

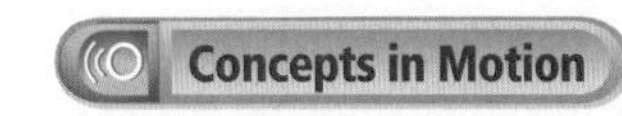

Interactive Concept Map

Copy this concept map, and then use vocabulary terms from the previous page to complete the concept map.

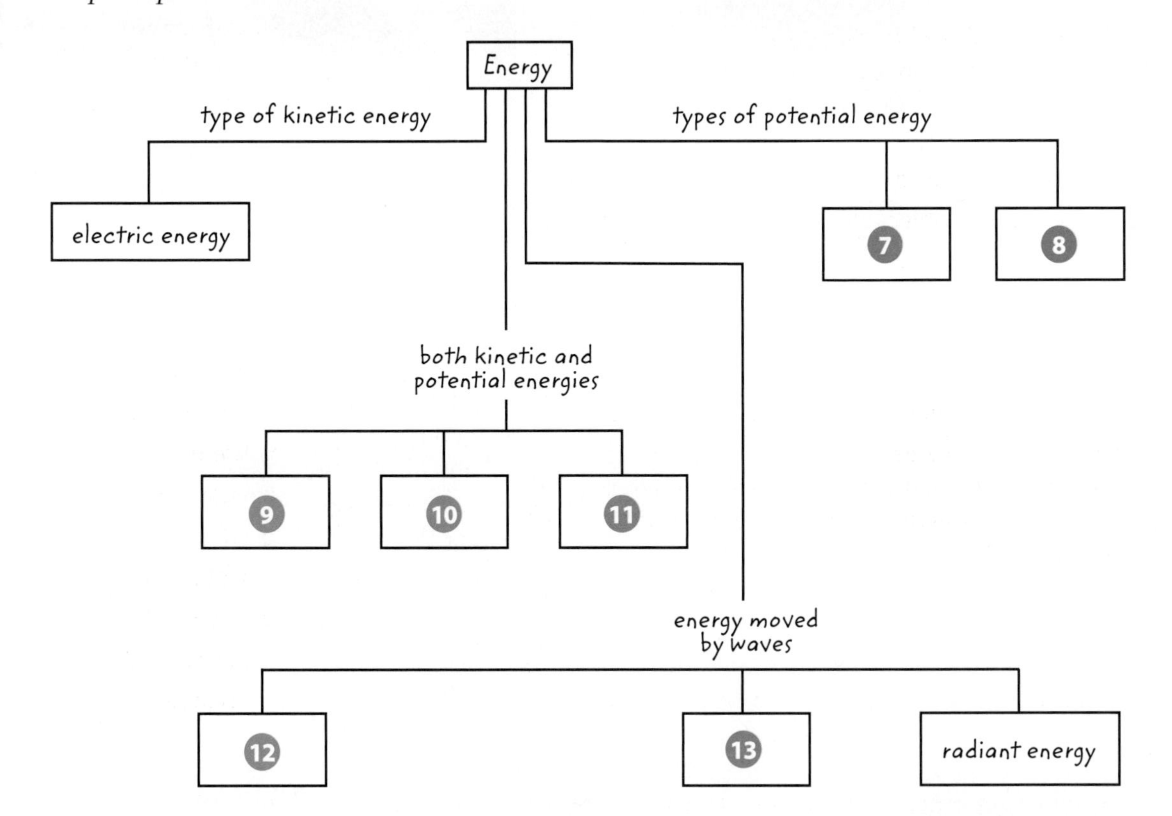

Chapter 3 Review

Understand Key Concepts

1. Which of the following is gravitational potential energy?
   - A. the energy stored in an object that is 10 m above the ground
   - B. the energy of an electron moving through a copper wire
   - C. the energy stored in the bonds of a carbohydrate molecule
   - D. the energy stored in the nucleus of a uranium atom

2. Which of the following increases the kinetic energy of an object?
   - A. decreasing the mass of the object
   - B. decreasing the volume of the object
   - C. increasing the object's height
   - D. increasing the object's speed

3. At which point in the photo below is the gravitational potential energy the greatest?
   - A. I
   - B. II
   - C. III
   - D. IV

4. The input work Shelly does on a rake is 80 J. The output work the rake does on the leaves is 70 J. What is the efficiency of the rake?
   - A. 70 percent
   - B. 80 percent
   - C. 87.5 percent
   - D. 95.4 percent

5. Which of the following types of electric energy plants transforms gravitational potential energy to electric energy?
   - A. fossil fuel
   - B. geothermal
   - C. hydroelectric
   - D. nuclear

6. What energy transformation occurs in a clothes iron?
   - A. chemical to electric
   - B. electric to thermal
   - C. kinetic to chemical
   - D. thermal to electric

7. How much work did the man do on the toolbox in the illustration below?

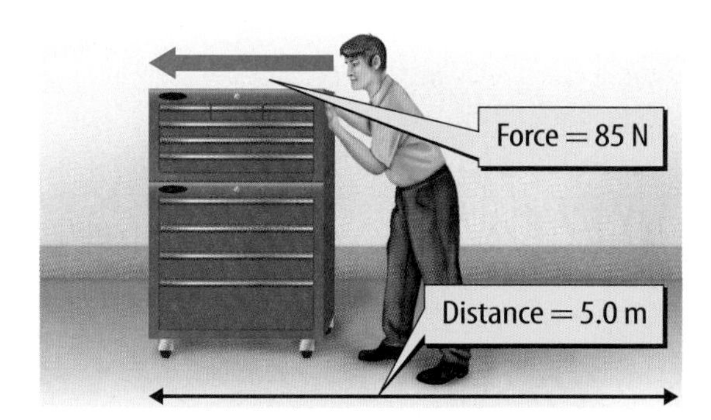

   - A. 0.06 m/N
   - B. 17 N/m
   - C. 425 J
   - D. 2,125 J

8. Which form of energy is NOT carried by waves?
   - A. chemical energy
   - B. radiant energy
   - C. seismic energy
   - D. sound energy

9. Which is NOT a simple machine?
   - A. inclined plane
   - B. lever
   - C. loop and hook
   - D. wheel and axle

Assessment
Online Test Practice

Critical Thinking

10 **Infer** How does an airplane's kinetic energy and potential energy change as it takes off and lands?

11 **Critique** You overhear someone say, "I'm going to nuke it" when referring to cooking food in a microwave. Explain why this terminology is incorrect.

12 **Consider** You are going to turn a screw using a wrench. Will the work you do on the wrench be more or less than the work done by the wrench on the screw? Explain.

13 **Compare** Describe the energy transformations that are similar in the human body and in fossil fuel electric energy plants.

14 **Explain** A coach sets up a tug-of-war between two evenly matched teams. Both teams pull against the rope as hard as they can, but the rope does not move. Is any work being done? Why or why not?

15 **Consider** You pull a nail out of a piece of wood using the back of a hammer. When you feel the nail, it is warm. Why?

16 **Explain** at least two reasons why the spatula pictured below is considered a simple machine.

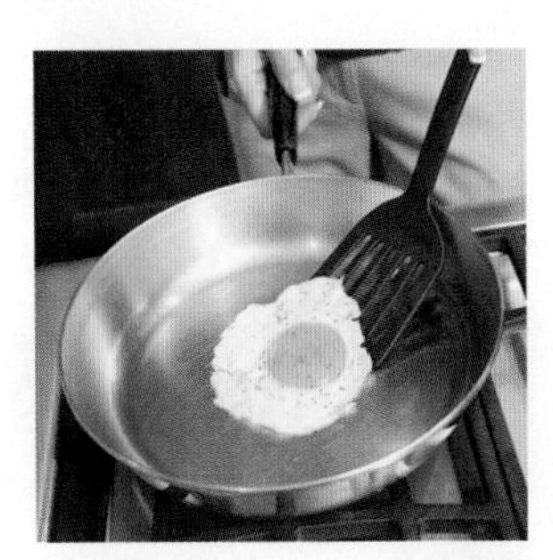

Writing in Science

17 **Write** Find a complex machine around your house or your school, and write a paragraph describing the different simple machines that it contains.

REVIEW THE BIG IDEA

18 How is energy transformed in electric energy plants, in roller coasters, and by machines?

19 The photo below shows the deck of a sailboat. How do the pulleys make rasing the sails easier?

Math Skills

Math Practice

Calculate Work

20 Humpty Dumpty weighs 400 N. He falls off a wall 3 m high. How much work was done by gravity on Humpty Dumpty?

21 A mover lifts a 12-kg box straight up 1.5 m. How much work is done on the box?

Standardized Test Practice

Record your answers on the answer sheet provided by your teacher or on a sheet of paper.

Multiple Choice

1 What does all energy have?

A size and shape

B mass and volume

C the ability to cause change

D the ability to transport matter

Use the figure below to answer question 2.

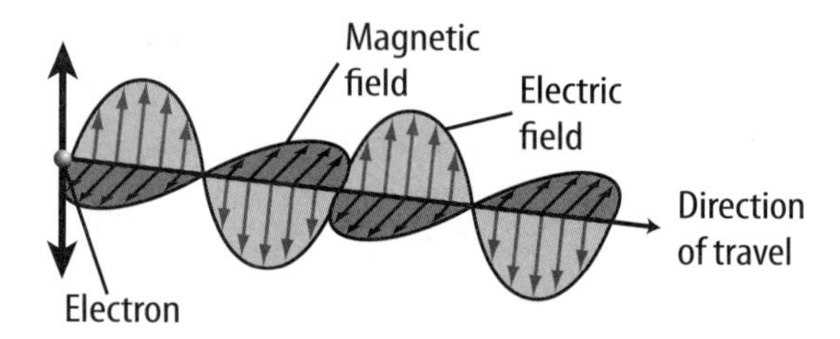

2 Which form of energy is being transmitted in the picture?

A chemical energy

B electric energy

C radiant energy

D sound energy

3 How do people use the nuclear energy produced from nuclear fission?

A to produce electric energy

B to power handheld machines

C to grow and maintain body cells

D to cook food in a microwave oven

4 Which is true of energy?

A It cannot be destroyed.

B It cannot be transmitted.

C It cannot change matter.

D It cannot be transformed.

Use the figure below to answer questions 5 and 6.

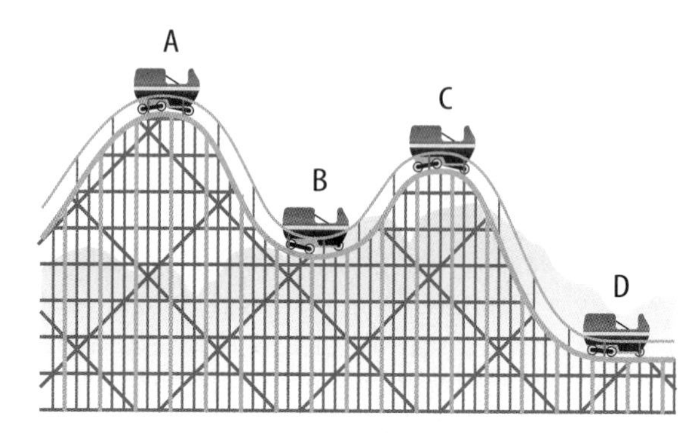

5 The figure shows four cars on a roller coaster track. At which point is gravitational potential energy the greatest?

A point A

B point B

C point C

D point D

6 What happens to the roller-coaster car's energy as it moves from point A to point B?

A New energy is created.

B The energy is destroyed.

C New energy transforms from the car's mass.

D The energy transforms from one kind to another.

7 Which equation shows how work and force are related?

A work = force + distance

B work = force − distance

C work = force × distance

D work = force ÷ distance

Online Standardized Test Practice

Use the figure below to answer question 8.

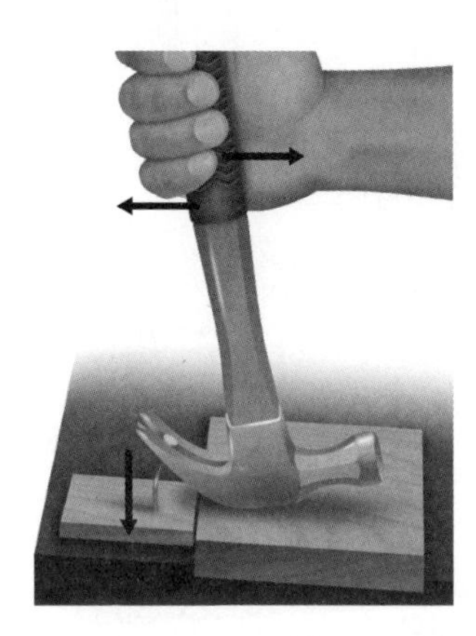

8 The figure shows a person using a hammer to remove a nail from a board. Which simple machine describes how the hammer is being used in this picture?

A inclined plane

B lever

C pulley

D wedge

9 How can simple machines make work easier?

A by increasing the amount of work done

B by decreasing the amount of work done

C by changing the distance or the force needed to do work

D by getting rid of the work needed to move an object

Constructed Response

10 A softball has more mass than a baseball. Compare the kinetic energy of a softball with that of a baseball moving at the same speed.

11 What is an energy transformation? Give an example of an energy transformation used to cook food.

Use the figure to answer questions 13 and 14.

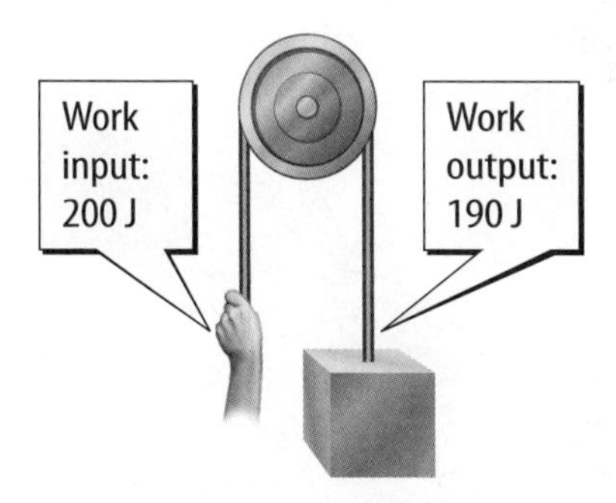

12 What simple machine is shown? What is the efficiency of this machine?

13 How could the efficiency of this machine be improved? Could it ever by 100%? Explain.

NEED EXTRA HELP?													
If You Missed Question...	1	2	3	4	5	6	7	8	9	10	11	12	13
Go to Lesson...	1	1	1	2	2	2	2	3	3	1	2	3	3

Chapter 4

Sound and Light

THE BIG IDEA

How do sound and light waves travel and interact with matter?

Inquiry Why is this reflection weird?

The mirror's curved surface forms unusual images. You can see these images because of reflected light waves.

- How are light waves reflected from a surface?
- How does the shape of a shiny surface affect the image you see?
- How do your eyes see the reflection?

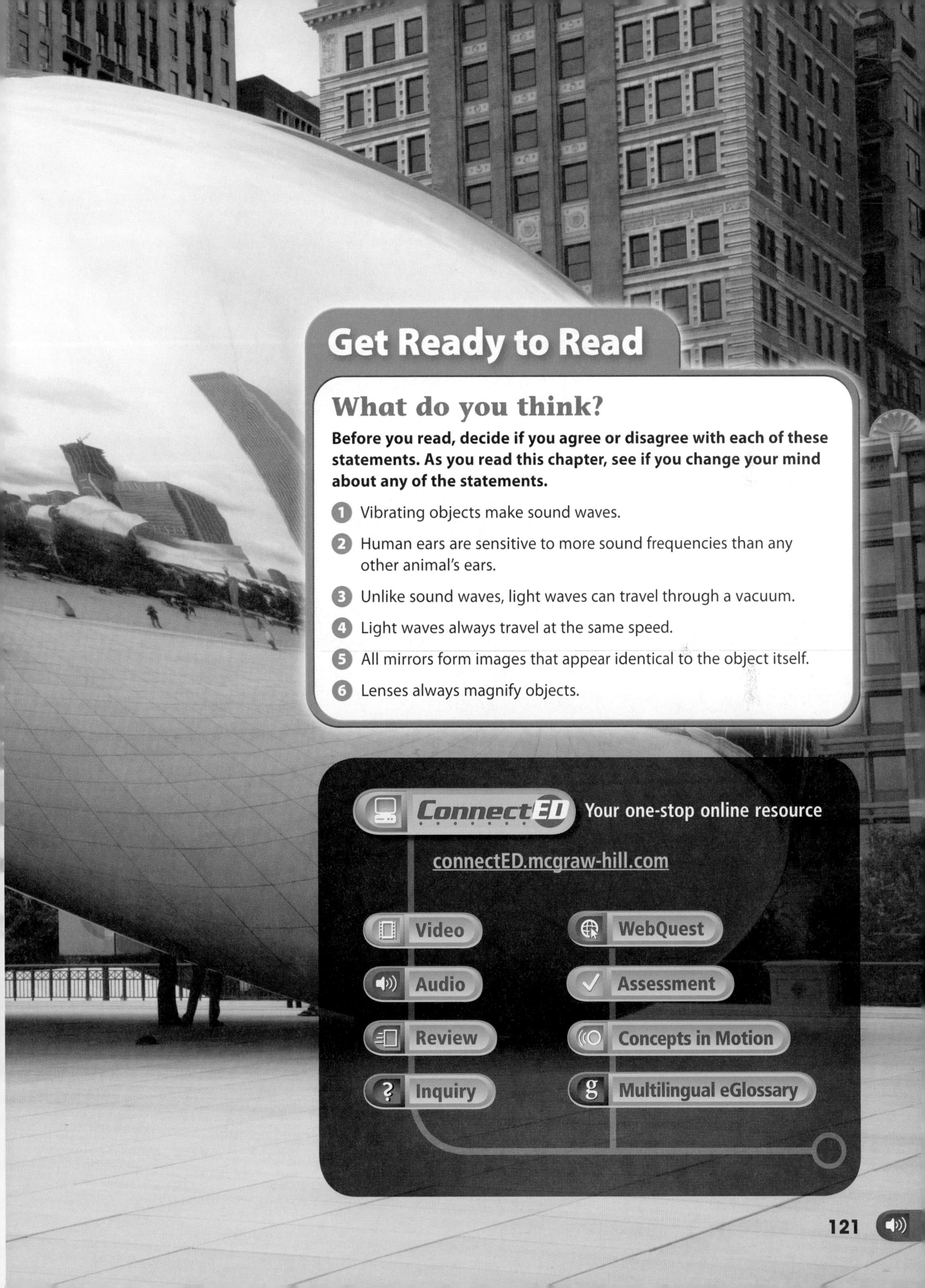

Get Ready to Read

What do you think?

Before you read, decide if you agree or disagree with each of these statements. As you read this chapter, see if you change your mind about any of the statements.

1. Vibrating objects make sound waves.
2. Human ears are sensitive to more sound frequencies than any other animal's ears.
3. Unlike sound waves, light waves can travel through a vacuum.
4. Light waves always travel at the same speed.
5. All mirrors form images that appear identical to the object itself.
6. Lenses always magnify objects.

ConnectED Your one-stop online resource

connectED.mcgraw-hill.com

- Video
- Audio
- Review
- Inquiry
- WebQuest
- Assessment
- Concepts in Motion
- Multilingual eGlossary

Lesson 1

Reading Guide

Key Concepts
ESSENTIAL QUESTIONS

- How are sound waves produced?
- Why does the speed of sound waves vary in different materials?
- How do your ears enable you to hear sounds?

Vocabulary

sound wave p. 123
pitch p. 127
echo p. 129

Multilingual eGlossary

Video Science Video

Sound

Inquiry Why are its ears so big?

The ears of this brown long-eared bat are nearly as long as its body. This bat finds its next meal by listening for the faint sounds that come from spiders and insects. How do large ears help a long-eared bat hear these sounds?

15 minutes

How is sound produced?

When an object vibrates, it produces sound. How does the sound produced depend on how the object vibrates?

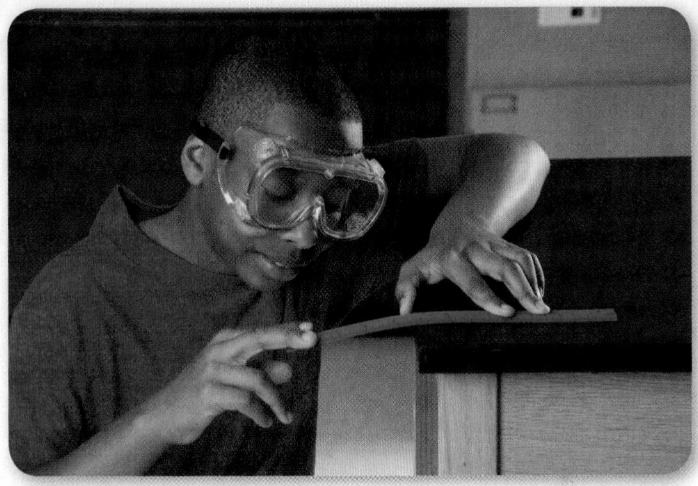

1. Read and complete a lab safety form.
2. Place a **ruler** on a table so it extends over the table edge. Hold the ruler firmly on the table with one hand.
3. With the other hand, lightly bend the protruding end of the ruler down and then release it. Observe the ruler's motion and note the sound it produces.
4. Move the ruler back 2 cm so there is less of it extending over the edge of the table. Repeat step 3.

Think About This

1. How did the vibration rate and the sound change as the length of the ruler over the side of the table decreased?
2. **Key Concept** Were the sound and the ruler's vibration rate related? Explain.

What is sound?

Have you ever walked down a busy city street and noticed all the sounds? You might hear many sounds every day, such as the music from an MP3 player, as shown in **Figure 1.** All sounds have one thing in common. The sounds travel from one place to another as sound waves. *A* **sound wave** *is a* ***longitudinal wave*** *that can travel only through matter.*

REVIEW VOCABULARY

longitudinal wave
a wave in which particles in a material move along the same direction that the wave travels

Sound waves can travel only through matter—solids, liquids, and gases. The sounds you might hear now are traveling through air—a mixture of solids and gases. You might have dived underwater and heard someone call to you. Then the sound waves traveled through a liquid. Sound waves travel through a solid when you knock on a door. As you will read, vibrating objects produce sound waves.

All sounds might have something else in common, too. Vibrating objects produce sound waves. For example, when you knock on a door, you produce sound waves by making the door vibrate. How do vibrating objects make sound waves?

Figure 1 Earbuds produce sound waves that travel into the listener's ears.

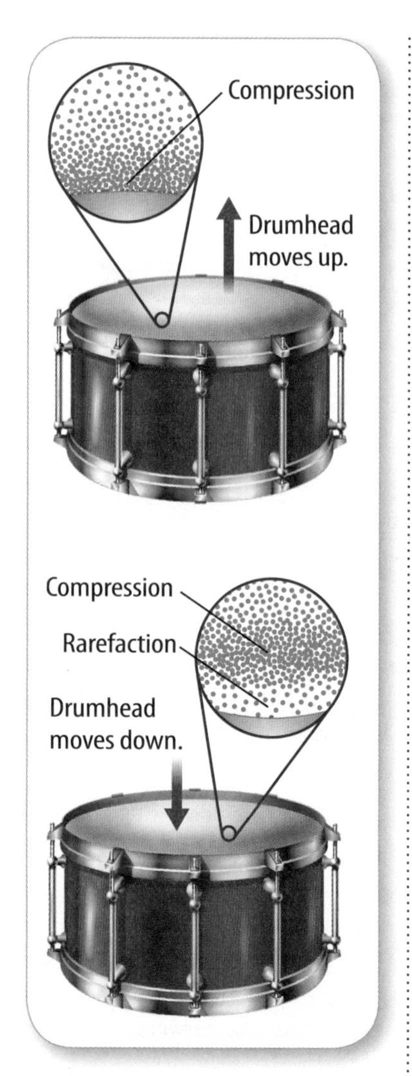

▲ **Figure 2** Vibrations of the drumhead produce sound waves.

Vibrations and Sound

Objects such as doors or drums vibrate when you hit them. For example, when you hit a drum, the drumhead moves up and down, or vibrates, as shown in **Figure 2.** These vibrations produce sound waves by moving molecules in air.

Compressions and Rarefactions

As the drumhead moves up, it pushes the molecules in the air above it closer together. The region where molecules are closer together is a compression, as shown in **Figure 2.** When the drumhead moves down, it produces a rarefaction. This is a region where molecules are farther apart. As the drumhead vibrates down and up, it produces a series of rarefactions and compressions that travels away from the drumhead. This series of rarefactions and compressions is a sound wave.

The vibrating drumhead causes molecules in the air to move closer together and then farther apart. The molecules in air move back and forth in the same direction that the sound wave travels. As a result, a sound wave is a longitudinal wave.

Key Concept Check How do vibrating objects produce sound waves?

Wavelength and Frequency

A sound wave can be described by its wavelength and frequency. Wavelength is the distance between a point on a wave and the nearest point just like it, as shown in **Figure 3.** A sound wave's frequency is the number of wavelengths that pass a given point in one second. Recall that the SI unit of frequency is hertz (Hz). The faster an object vibrates, the higher the frequency of the sound wave produced.

Figure 3 Wavelength is the distance between one compression and the next compression or the distance between a rarefaction and the next rarefaction. ▼

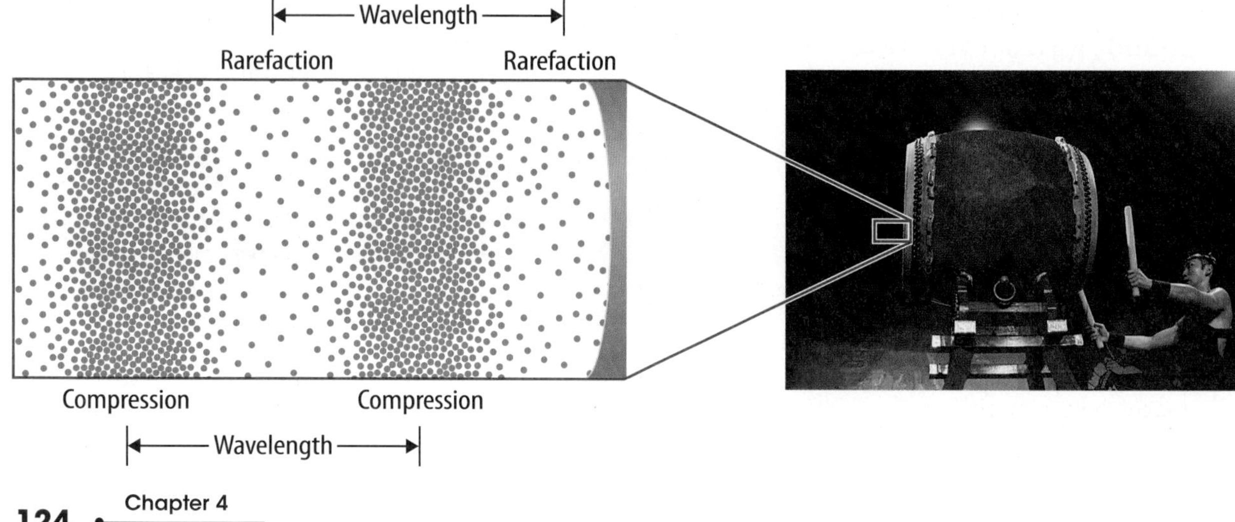

Table 1 The Speed of Sound Waves in Different Materials					
Gases (0°C)		**Liquids (25°C)**		**Solids**	
Material	**Speed (m/s)**	**Material**	**Speed (m/s)**	**Material**	**Speed (m/s)**
Carbon dioxide	259	Ethanol	1,207	Brick	3,480
Dry Air	331	Mercury	1,450	Ice	3,850
Water vapor	405	Water	1,500	Aluminum	6,420
Helium	965	Glycerine	1,904	Diamond	17,500

Table 1 Sound waves travel at different speeds in different materials. Sound waves usually travel fastest in solids and slowest in gases.

Concepts in Motion Interactive Table

Speeds of Sound Waves

Sound waves traveling through air cause the sounds you might hear every day. Like all types of waves, the speed of a sound wave depends on the material in which it travels.

Sound in Gases, Liquids, and Solids

Table 1 lists the speeds of sound waves in different materials. A sound wave's speed increases when the material's density increases. Solids and liquids are usually more dense than gases.

In addition, a sound wave's speed increases when the strengths of the forces between the particles—atoms or molecules—in the material increase. These forces are usually strongest in solids and weakest in gases. Overall, sound waves usually travel faster in solids than in liquids or gases.

Key Concept Check Why is the speed of sound waves faster in solids than in liquids or gases?

Temperature and Sound Waves

The temperature of a material also affects the speed of a sound wave. The speed of a sound wave in a material increases as the temperature of the material increases. For example, the speed of a sound wave in dry air increases from 331 m/s to 343 m/s as the air temperature increases from 0°C to 20°C. A sound wave in air travels faster on a warm, summer day than on a cold, winter day.

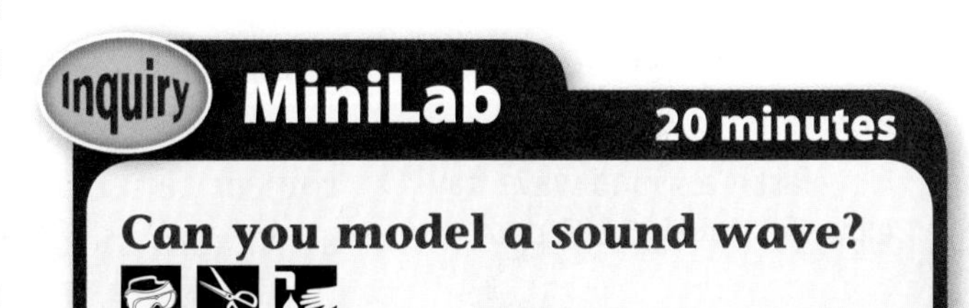

Inquiry MiniLab **20 minutes**

Can you model a sound wave?

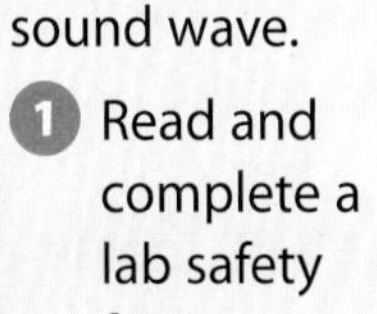

A wave on a coiled spring toy is similar to a sound wave.

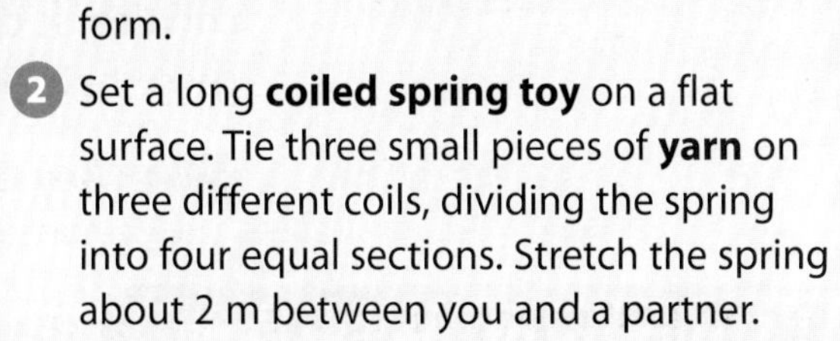

1. Read and complete a lab safety form.
2. Set a long **coiled spring toy** on a flat surface. Tie three small pieces of **yarn** on three different coils, dividing the spring into four equal sections. Stretch the spring about 2 m between you and a partner.
3. Squeeze together about one-fourth of the coils and hold the end of the spring with the other hand. While holding the end of the spring tightly, release the group of coils. Observe the wave.

Analyze and Conclude

1. **Draw** three sketches of the spring in your Science Journal, showing how the wave traveled through the spring. Label the compressions and rarefactions.
2. **Key Concept** Explain how the wave on the spring is similar to a sound wave.

Math Skills 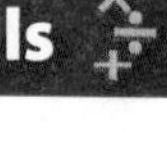

Use a Simple Equation

Speed (s) is equal to the distance (d) something travels divided by the time (t) it takes to cover that distance:

$$s = \frac{d}{t}$$

You can use this equation to calculate the speed of sound waves. For example, if a sound wave travels a distance of 662 meters in 2 seconds in air, its speed is:

$$s = \frac{d}{t} = \frac{662 \text{ m}}{2 \text{ s}} = 331 \text{ m/s}$$

Practice

How fast is a sound wave traveling if it travels 5,000 m in 5 s?

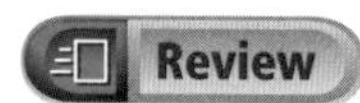

Review

- **Math Practice**
- **Personal Tutor**

The Human Ear

When you think about your ears, you probably only think about the structure on each side of your head. However, there is more to your ears than those structures. The human ear has three parts—the outer ear, the middle ear, and the inner ear, as shown in **Figure 4.**

1 The Outer Ear

The outer ear collects sound waves. The structure on each side of your head is part of the outer ear. The ear canal is also part of the outer ear, as shown in **Figure 4.** The visible part of the outer ear funnels sound waves into the ear canal. The ear canal channels sound waves into the middle ear.

2 The Middle Ear

The middle ear amplifies sound waves. As shown in **Figure 4,** the middle ear includes the eardrum and three tiny bones. The eardrum is a thin membrane that stretches across the ear canal. The three tiny bones are called the hammer, the anvil, and the stirrup. A sound wave hitting the eardrum causes it to vibrate. The vibrations travel to the three bones, which amplify the sound.

3 The Inner Ear

The inner ear converts vibrations to nerve signals that travel to the brain. The inner ear consists of a small, fluid-filled chamber called the cochlea (KOH klee uh). Tiny hairlike cells line the inside of the cochlea. As a sound wave travels into the cochlea, it causes some hair cells to vibrate. The movements of these hair cells produce nerve signals that travel to the brain.

Key Concept Check What is the function of each of the three parts of the ear?

Structure of the Ear

Concepts in Motion **Animation**

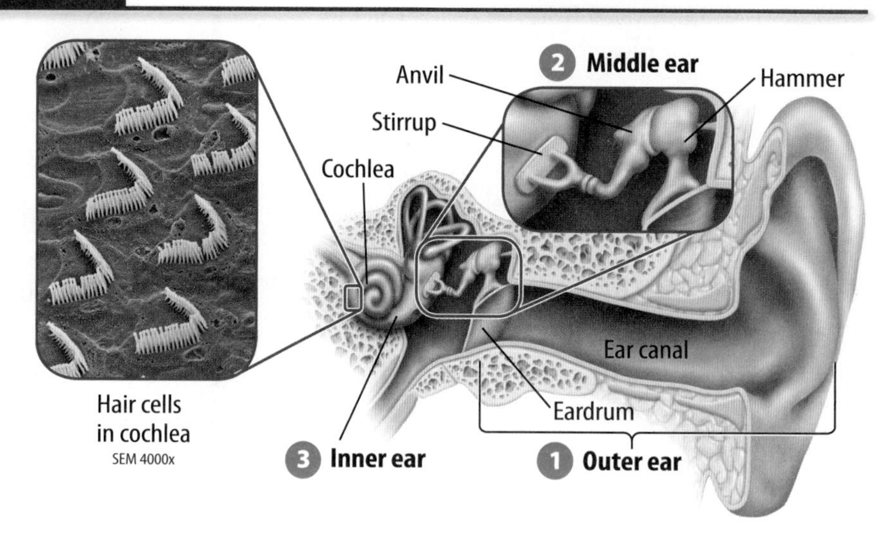

Figure 4 The human ear has three parts. The outer ear collects sound waves, the middle ear amplifies these waves, and the inner ear converts them to nerve signals.

Visual Check Which parts of the ear are considered the middle ear?

Frequencies and the Human Ear

As you can see in **Table 2,** humans hear sounds with frequencies between about 20 and 20,000 Hz. Some mammals can hear sounds with frequencies greater than 100,000 Hz.

Table 2 Frequencies Different Mammals Can Hear

Creature	Frequency Range (Hz)
Human	20–20,000
Dog	67–45,000
Cat	45–64,000
Bat	2,000–110,000
Beluga whale	1,000–123,000
Porpoise	75–150,000

▲ **Table 2** Different mammals can hear sound waves over different ranges of frequencies.

Sound and Pitch

Have you ever played a guitar? A guitar has strings with different thicknesses. If you pluck a thick string, you hear a low note. If you pluck a thin string, you hear a higher note. The sound a thick string makes has a lower pitch than the sound a thin string makes. *The* **pitch** *of a sound is the perception of how high or low a sound seems.* A sound wave with a higher frequency has a higher pitch. A sound wave with a lower frequency has a lower pitch.

Reading Check How does the pitch of a sound wave depend on the frequency of the sound wave?

You can produce sounds of different pitches by using your vocal cords. The vocal cords, shown in **Figure 5,** are two membranes in your neck above your windpipe, or trachea (TRAY kee uh). When you speak, you force air from your lungs through the space between the vocal cords. This causes the vocal cords to vibrate, creating sound waves you and other people hear as your voice.

Muscles connected to your vocal cords enable you to change the pitch of your voice. When these muscles contract, they pull on your vocal cords. This stretches the vocal cords and they become longer and thinner. The pitch of your voice is then higher, just as a thinner guitar string has a higher pitch than a thicker guitar string. When these muscles relax, the vocal cords become shorter and thicker and the pitch of your voice becomes lower.

Make a two-tab concept map book. Label it as shown. Use it to organize information about pitch and loudness.

The Ear
Pitch | Loudness

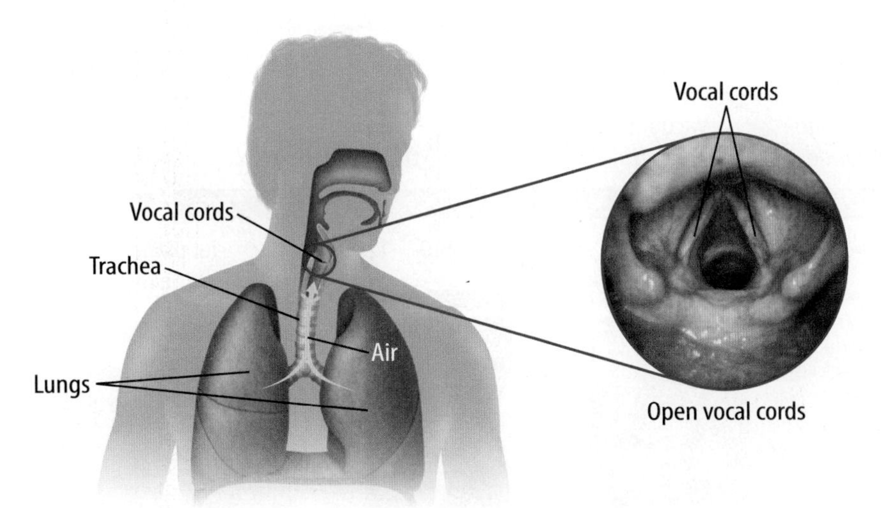

◀ **Figure 5** The vocal cords vibrate by opening and closing when air is forced through them. These vibrations produce the sounds of the human voice.

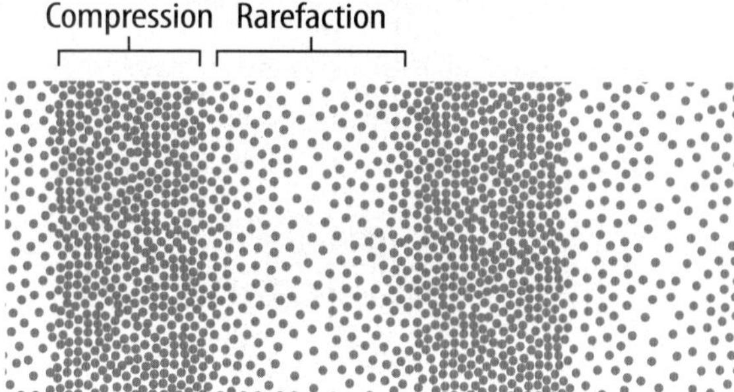

Low amplitude sound wave

Compression Rarefaction

High amplitude sound wave

▲ **Figure 6** The amplitude of a sound wave depends on how close together or far apart the particles are in the compressions and rarefactions.

Visual Check How do distances between particles differ in high- and low-amplitude sound waves?

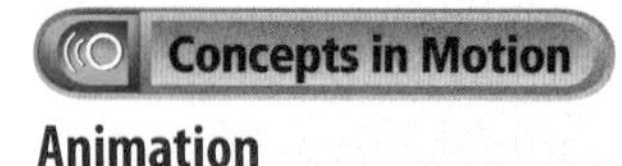

Animation

Sound and Loudness

Why is a shout louder than a whisper? Loudness is the human sensation of how much energy a sound wave carries. Sound waves produced by shouting carry more energy than sound waves produced by whispering. As a result, a shout sounds louder than a whisper.

Amplitude and Energy

The amplitude of a wave depends on the amount of energy that the wave carries. The more energy a wave has, the greater the amplitude. **Figure 6** shows the difference between a high-amplitude sound wave and a low-amplitude sound wave. High-amplitude sound waves have particles that are closer together in the compressions and farther apart in the rarefactions.

The Decibel Scale

The decibel scale, shown in **Figure 7,** is one way to compare the loudness of sounds. On this scale, the softest sound a person can hear is about 0 decibels (dB), and a jet plane taking off is at about 150 dB. Normal conversation is at about 50 dB. A sound wave that is 10 dB higher than another sound wave carries 10 times more energy. However, people hear the higher-energy sound wave as being only twice as loud.

Figure 7 The loudness of sounds can be compared on the decibel scale. ▼

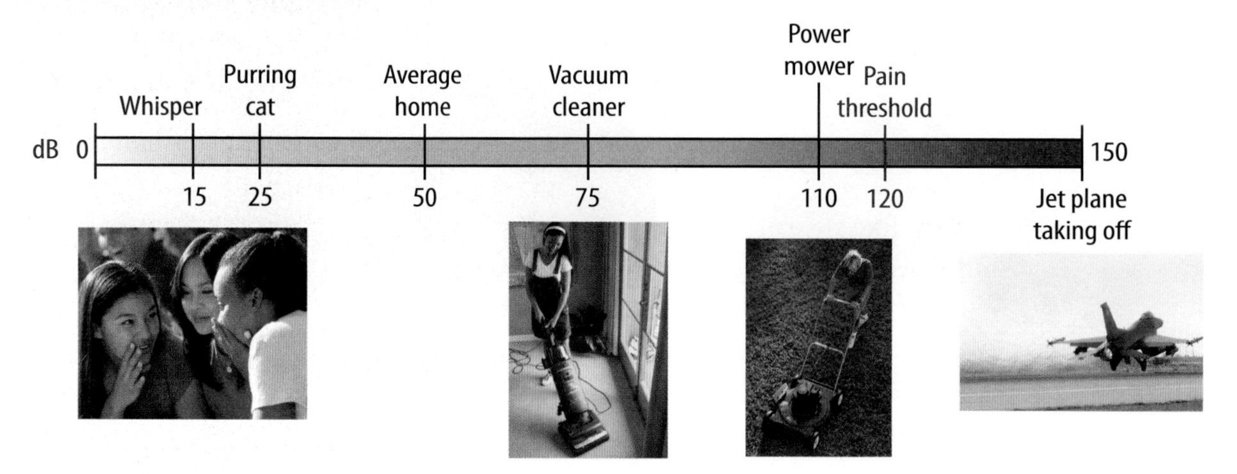

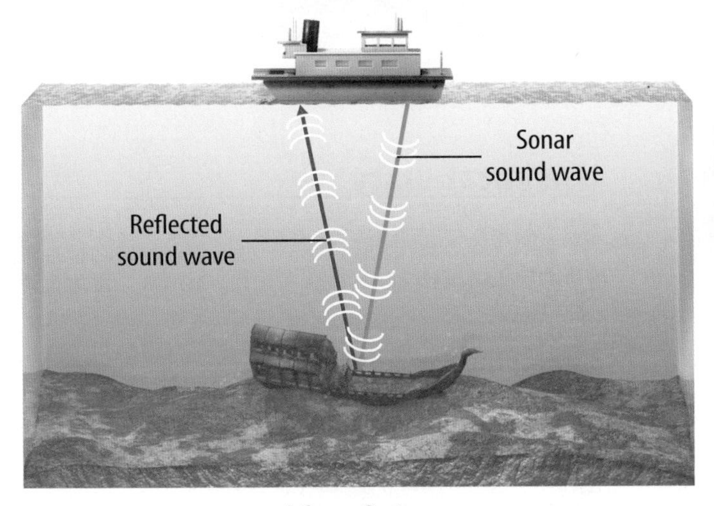

A Sonar System

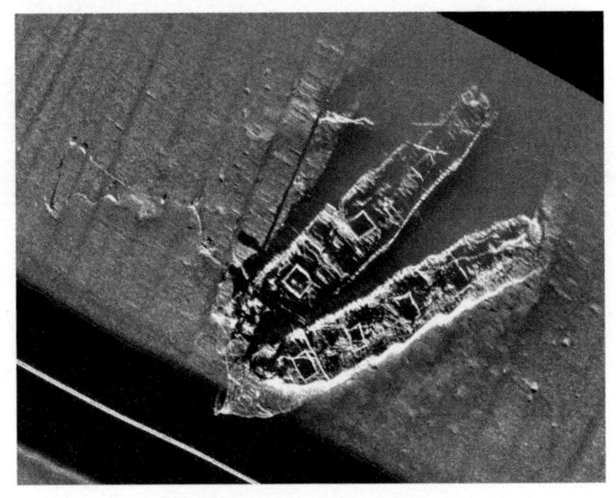

Image Made by Sonar System

▲ **Figure 8** A sonar system locates underwater objects by sending out sound waves and detecting the reflected sound waves. The photo on the right is a sonar image of two sunken ships.

Using Sound Waves

Have you ever yelled in a cave or a big, empty room? You might have heard the echo of your voice. *An* **echo** *is a reflected sound wave.* You may be able to hear echoes in a gymnasium or a cafeteria. You probably can't tell how far away a wall is by hearing an echo. However, sonar systems and some animals use reflected sound waves to determine how far away objects are.

Sonar and Echolocation

Sonar systems use reflected sound waves to locate objects under water, as shown in **Figure 8.** The sonar system emits a sound wave that reflects off an underwater object. The distance to the object can be calculated from the time difference between when the sound leaves the ship and when the sound returns to the ship. Sonar is used to map the ocean floor and to detect submarines, schools of fish, and other objects under water.

Word Origin
echo
from Greek *ekhe*,
means "sound"

 **Reading Check** How do sonar systems use sound waves?

Some animals use a method called echolocation to navigate and hunt. Echolocation is a type of sonar. Bats and dolphins, for example, emit high-pitched sounds and interpret the echoes reflected from objects. Echolocation enables bats and dolphins to locate prey and detect objects.

Ultrasound

Ultrasound scanners, like the one shown in **Figure 9,** convert high-frequency sound waves to images of internal body parts. The sound waves reflect from structures within the body. The scanner analyzes the reflected waves and produces images, called sonograms, of the body structures. These images can be used to help diagnose disease or other medical conditions.

Figure 9 Ultrasound scanners produce images that doctors can use to diagnose diseases. ▼

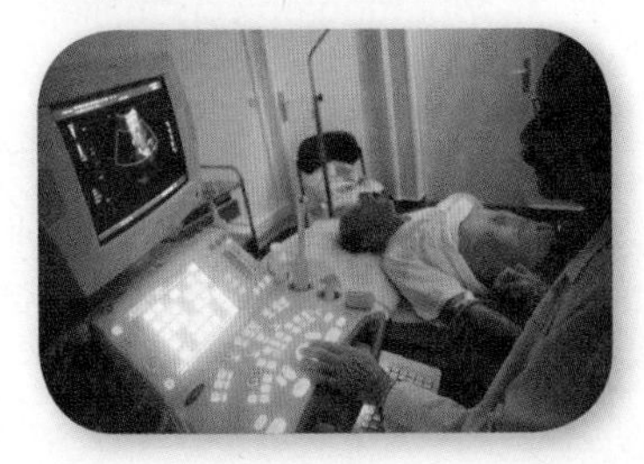

Lesson 1 Review

Online Quiz

Visual Summary

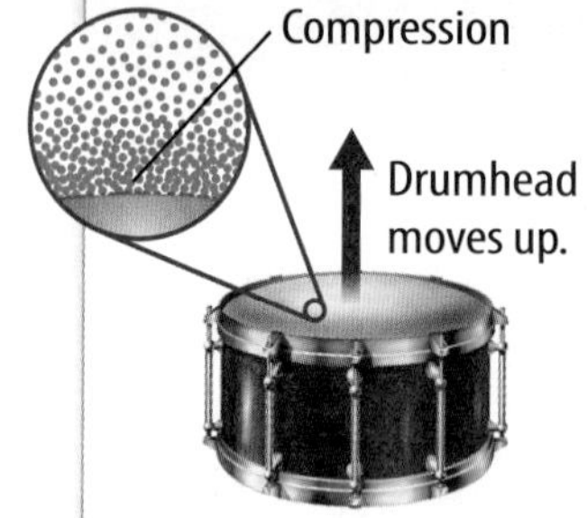

A sound wave is a longitudinal wave that can travel only through matter.

Vocal cords

The pitch is how high or low the frequency of a sound wave is. You create different pitches using your vocal cords.

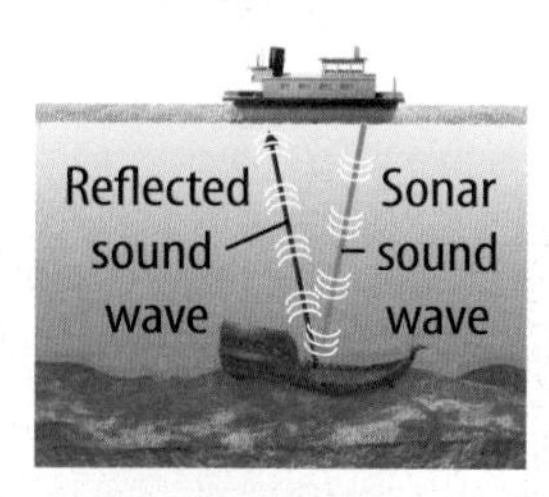

An echo is a reflected sound wave. Ships use sonar to find underwater objects.

FOLDABLES®

Use your lesson Foldable to review the lesson. Save your Foldable for the project at the end of the chapter.

What do you think NOW?

You first read the statements below at the beginning of the chapter.

1. Vibrating objects make sound waves.
2. Human ears are sensitive to more sound frequencies than any other animal's ears.

Did you change your mind about whether you agree or disagree with the statements? Rewrite any false statements to make them true.

Use Vocabulary

1. **Define** *echo* in your own words.
2. **Distinguish** between sound and a sound wave.
3. **Use the word** *pitch* in a sentence.

Understand Key Concepts

4. In which material do sound waves travel fastest?
   - **A.** aluminum
   - **B.** carbon dioxide
   - **C.** ethanol
   - **D.** water
5. **Predict** Would a barking dog produce sound waves that travel faster during the day or at night? Explain your answer.

Interpret Graphics

6. **Describe** The image below shows part of a sound wave. Describe how this image would change if the wavelength of the sound wave decreased.

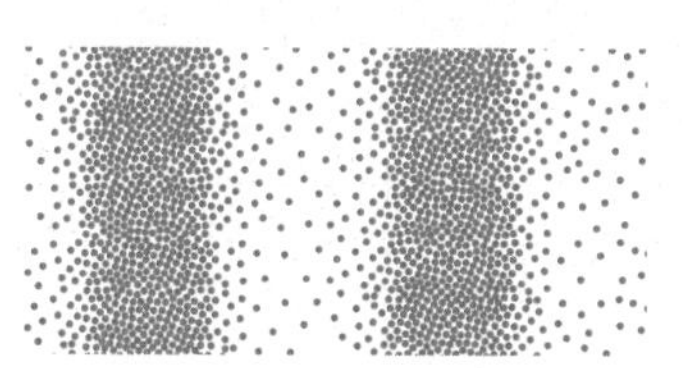

7. **Sequence** Copy and fill in a graphic organizer like the one below that shows the path a sound wave travels from the air until it is interpreted by the brain.

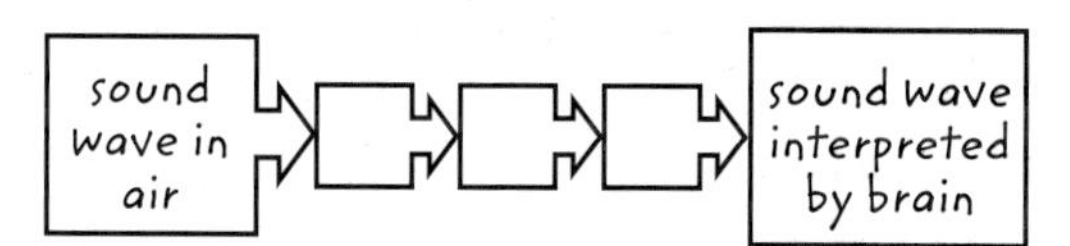

Critical Thinking

8. **Infer** A string vibrates with a frequency of 10 Hz. Why can't a person hear the sound waves produced by the vibrating string, no matter how large the amplitude of the waves?

Math Skills

Math Practice

9. If a sound wave travels 1,620 m in 3 s, what is its speed?

HOW IT WORKS

How do ultrasound machines work?

Using Ultrasound to Safely Monitor a Human Fetus

Ultrasound waves are sound waves with frequencies so high that humans cannot hear them. An ultrasound machine has a hand-held device, called a transducer, that emits and receives ultrasound waves. This enables medical professionals to see inside a human body.

1 **A technician passes the transducer over the patient's skin, transmitting ultrasound waves into the patient's body.**

2 **When the ultrasound waves strike different surfaces, some of those waves reflect back to the transducer.**

3 **The transducer detects the reflected ultrasound waves and converts them into electronic signals.**

4 **A computer receives the signals from the transducer and produces an image.**

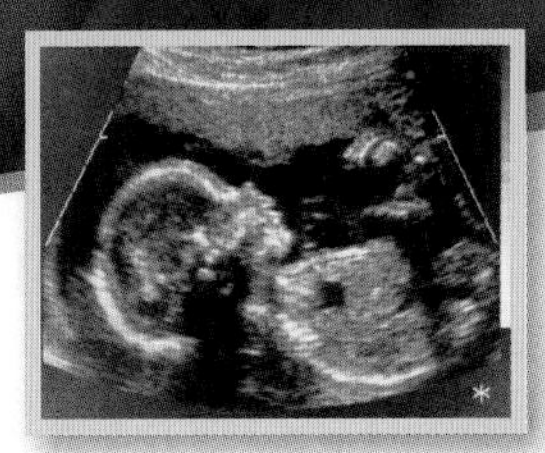

It's Your Turn

RESEARCH other ways that medical professionals use ultrasound machines.

Lesson 2

Reading Guide

Key Concepts
ESSENTIAL QUESTIONS

- How are light waves different from sound waves?
- How do waves in the electromagnetic spectrum differ?
- What happens to light waves when they interact with matter?

Vocabulary

light source p. 135
light ray p. 135
transparent p. 136
translucent p. 136
opaque p. 136

- BrainPOP®
- Science Video
- What's Science Got to do With It?

Light

Are both men real?

No, the man on the right is a hologram. A hologram is a type of image that seems to be three-dimensional. Light from a laser is reflected from the person and is used to create the life-like image. What are light waves and how do they interact with matter?

Launch Lab

15 minutes

What happens when light waves pass through water?

Do light waves always travel in a straight line? What happens to light waves when they travel through water?

1. Read and complete a lab safety form.
2. Add **distilled water** to a **500-mL beaker** until it is two-thirds full.
3. Use **scissors** to cut a thin slit in a sheet of **paper. Tape** the paper over the lens of a **flashlight.**
4. Turn on the flashlight and tilt it slightly downward so the light beam is visible on the tabletop. Place the water-filled beaker in the light beam. Record your observations in your Science Journal.

Think About This

Key Concept Compare the direction of the light beam before it entered the water to after it left the water.

What is light?

As you read these words, you are probably looking at a page in a book. You might also see the desk on which the book is resting as well as the light from a lamp. What do your eyes detect when you see something?

Your eyes sense light waves. You see books and desks when light waves reflect off these objects and enter your eyes. Some objects, like a candle flame and a lightbulb that is lit, also emit light waves. You see a candle flame or a glowing lightbulb because the light waves they emit enter your eyes.

Light—An Electromagnetic Wave

Light is a type of wave called an electromagnetic wave. Like sound waves, electromagnetic waves can travel through matter. But they can also travel through a vacuum, where no matter is present. For example, light can travel through the space between Earth and the Sun.

Light travels through a vacuum at a speed of about 300,000 km/s. However, light waves slow down when they travel through matter. The speed of light in some different materials is listed in **Table 3.** Light waves travel much faster than sound waves. For example, in air the speed of light is about 900,000 times faster than the speed of sound.

Key Concept Check How are light waves different from sound waves?

Table 3 Light waves travel fastest in empty space. When light waves travel in matter, they move fastest in gases and slowest in solids.

Table 3 Speed of Light Waves in Some Materials

Material	Wave Speed (km/s)
Vacuum	300,000
Air	299,920
Water	225,100
Glass	193,000

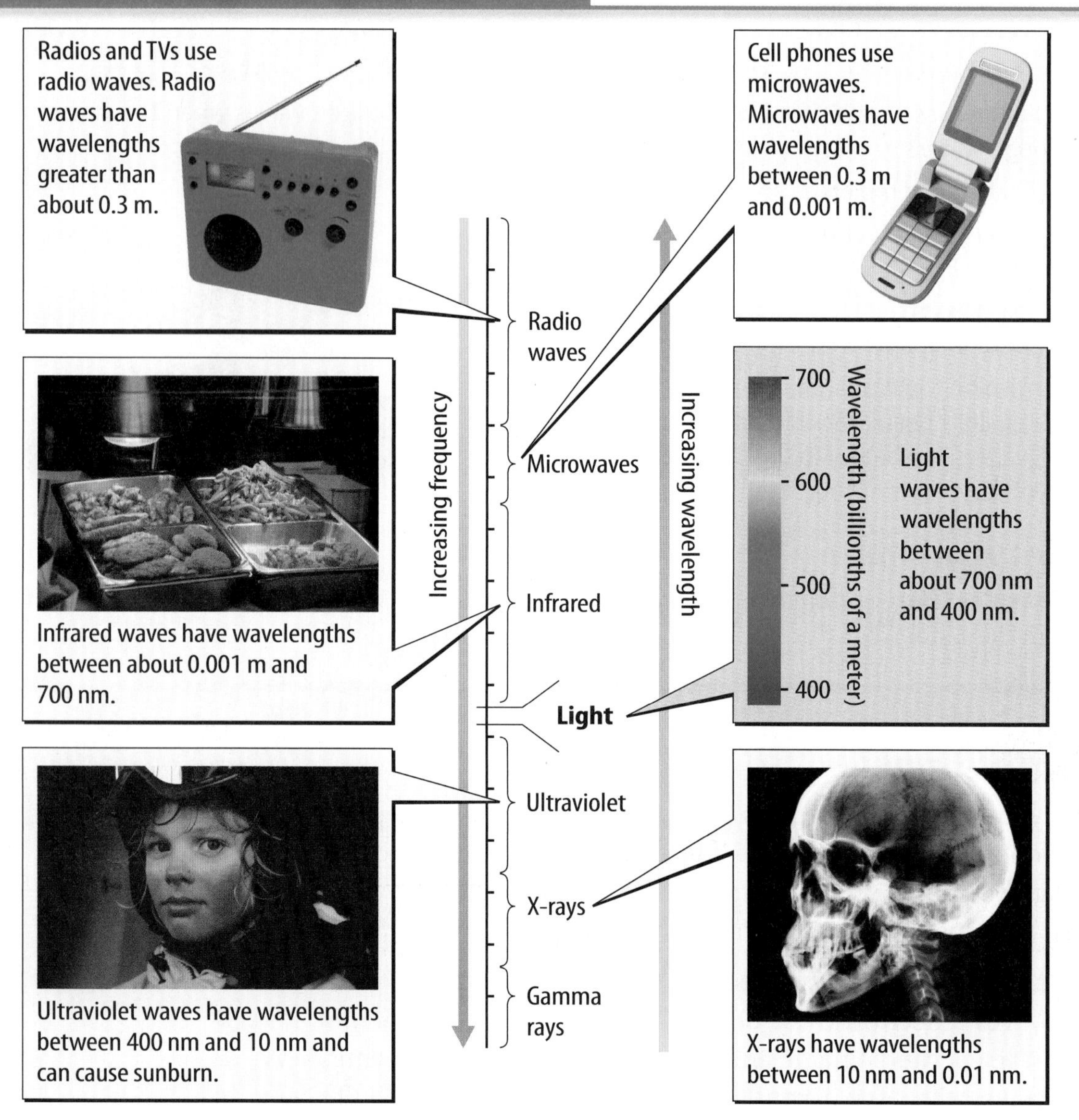

Figure 10 Electromagnetic waves are classified according to their wavelength or frequency. Visible light waves are part of the electromagnetic spectrum.

Visual Check Which type of electromagnetic waves have the longest wavelengths?

The Electromagnetic Spectrum

Besides visible light waves, there are other types of electromagnetic waves, such as X-rays and radio waves. Scientists classify electromagnetic waves into groups based on their wavelengths, as shown in **Figure 10.** The entire range of electromagnetic waves is called the electromagnetic spectrum.

Key Concept Check How are waves in the electromagnetic spectrum different?

Light waves are only a small part of the electromagnetic spectrum. The wavelengths of light waves are so short they are usually measured in nanometers (nm). One nanometer equals one-billionth of a meter. The wavelengths of light waves are from about 700 nm to about 400 nm. This is about one-hundredth the width of a human hair. You see different colors when different wavelengths of light waves enter your eyes.

Light-Emitting Objects

Think about walking into a dark room and turning on a light. The lightbulb produces light waves that travel away from the bulb in all directions, as shown in **Figure 11.** *A* **light source** *is something that emits light.* In order to emit light, the lightbulb transforms electric energy into light energy. Other examples of light sources are the Sun and burning candles. The Sun transforms nuclear energy into light energy. Burning candles transform chemical energy into light energy. Light sources convert other forms of energy into light energy.

▲ **Figure 11** Light travels in all directions away from its source.

Light Rays

As you just read, light waves spread out in all directions from a light source. You also can think of light in terms of light rays. *A* **light ray** *is a narrow beam of light that travels in a straight line.* The arrows in **Figure 11** represent some of the light rays moving away from the light source. Unless light rays come in contact with a surface or pass through a different material, they travel in straight lines.

 **Reading Check** What is a light ray?

Make a layered book from two sheets of paper. Use it to summarize information about light and how light waves interact with matter.

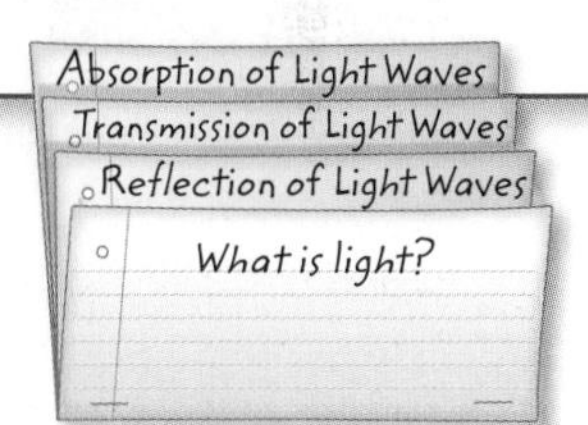

Light Reflection

Suppose you are in a dark room. Do you see anything? Now you turn on a light. What do you see now? Light sources emit light. But other objects, like books, reflect light. In order to see an object that is not a light source, light waves must reflect from an object and enter your eyes.

Seeing Objects

When you see a light source, light rays travel directly from the light source into your eye. Light rays also reflect off objects, as shown in **Figure 12.** Light rays reflect from an object in many directions. Some of the light rays that reflect from an object enter your eye, enabling you to see the object.

◀ **Figure 12** Some light waves from a light source reach the page and reflect off it. The girl sees the page when some of the reflected light enters her eyes.

The Interaction of Light and Matter

Like all waves, when light waves interact with matter they can be transmitted, absorbed, or reflected.

- Reflection occurs when light waves come in contact with the surface of a material and bounce off.
- Transmission occurs when light waves travel through a material.
- Absorption occurs when interactions with a material convert light energy into other forms such as thermal energy.

In many materials, reflection, transmission, and absorption occur at the same time. For example, the tinted glass of an office building reflects some light, transmits some light, and absorbs some light.

Key Concept Check What can happen to light waves when they interact with matter?

Figure 13 The butterfly's wing, the frosted glass, and the curtains interact with light waves differently.

Transparent

Translucent

Opaque

Depending on how they interact with light, materials can be classified as transparent, translucent, or opaque, as shown in **Figure 13.** *A material is* **transparent** *if it allows almost all light that strikes it to pass through and forms a clear image. A material is* **translucent** *if it allows most of the light that strikes it to pass but forms a blurry image. A material is* **opaque** *if light does not pass through it.* An opaque material, such as light-blocking cloth curtains, does not transmit light.

WORD ORIGIN

opaque
from Latin *opacus,* means "shady, dark"

The Reflection of Light Waves

Figure 14 shows what happens when a surface reflects light waves. All waves, including light waves, obey the law of reflection. According to the law of reflection, the angle of incidence always equals the angle of reflection. In **Figure 14,** the line perpendicular to a surface is called the normal. The angle between the normal and the incoming light ray is the angle of incidence. The angle between the reflected light ray and the normal is the angle of reflection.

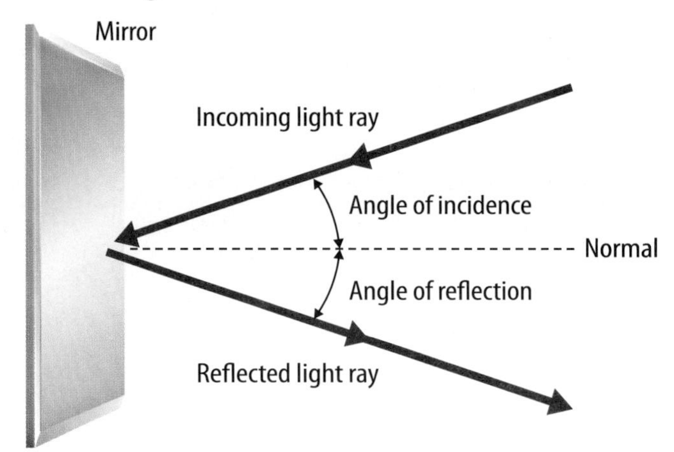

Figure 14 When a surface reflects a light ray, the angle of incidence equals the angle of reflection.

Visual Check How will the angle of reflection change if the angle of incidence increases?

Figure 15 Particles of dust floating in the air scatter light rays in a sunbeam. When light rays strike these particles, light rays reflect in many different directions. ▼

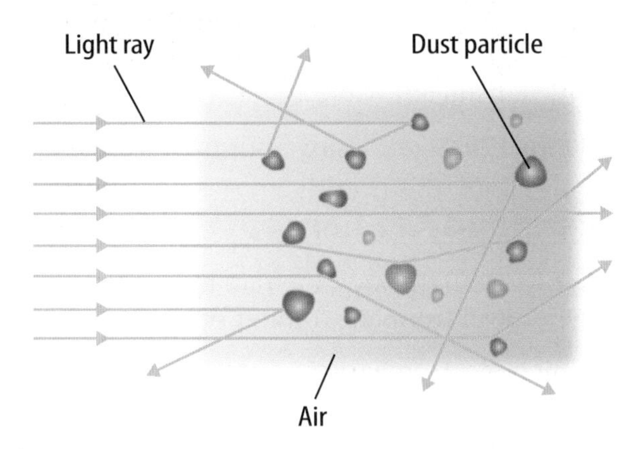

Scattering

When a beam of sunlight shines through a window, you might notice tiny particles of dust. You see the dust particles because they reflect light waves. As **Figure 15** shows, dust particles reflect light waves in many different directions because they have different shapes. This is an example of scattering. Scattering occurs when light waves traveling in one direction are made to travel in many directions. The dust particles scatter the light waves in the sunbeam.

The Refraction of Light Waves

Like all types of waves, light waves can change direction when they travel from one material to another. The light beam in **Figure 16** changes direction as it goes from the air into the glass and from the glass into the air. A wave that changes direction as it travels from one material into another is refracting.

Recall that light waves travel at different speeds in different materials. Refraction occurs when a wave changes speed. The greater the change in speed, the more the light wave refracts or changes direction.

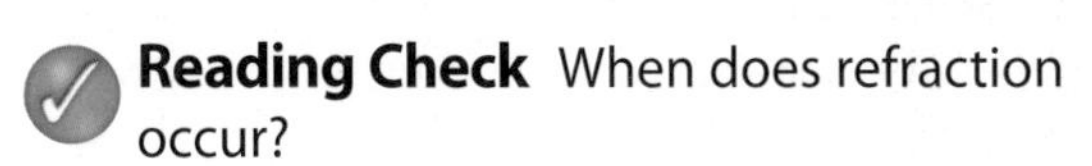
Reading Check When does refraction occur?

MiniLab **15 minutes**

Can you see a light beam in water?

Scattering by water droplets in air enables you to see a car's headlight beams on a foggy day or night. What could enable you to see a light beam traveling in water?

1. Read and complete a lab safety form.
2. Add **distilled water** to a **clear glass jar** until it is two-thirds full.
3. Shine the light from a **flashlight** through the water and record your observations in your Science Journal.
4. Add a few drops of **milk** to the water and swirl the jar to mix the milk and water. Repeat step 3.

Analyze and Conclude

1. **Compare** the appearance of the light beams in steps 3 and 4.
2. **Key Concept** Hypothesize how the milk enabled the light beam to be visible in the milk-water mixture.

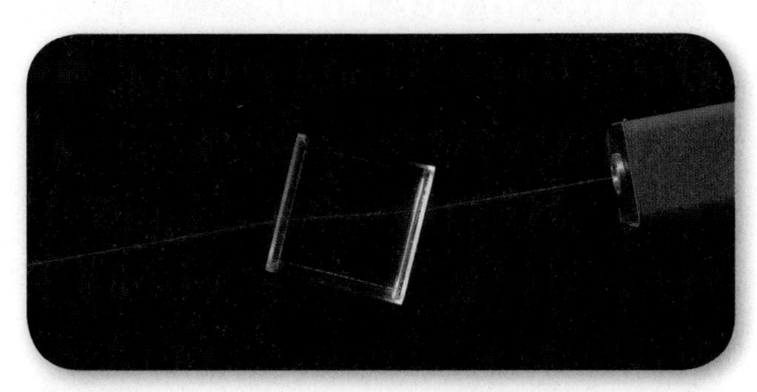

▲ **Figure 16** The red beam of light slows down as it enters the glass rectangle. It speeds up as it leaves the rectangle and enters the air.

Lesson 2 Review

Visual Summary

An object is seen when light waves emitted by the object or reflected by the object enter the eye.

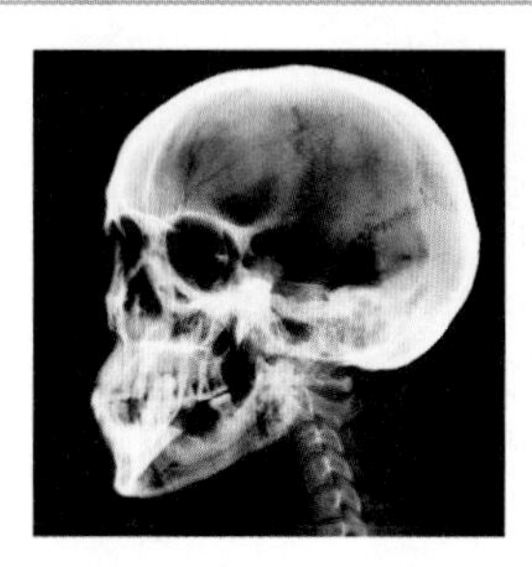

The electromagnetic spectrum includes electromagnetic waves of different wavelengths, such as X-rays.

When light waves interact with matter, they can be absorbed, reflected, or transmitted.

FOLDABLES®

Use your lesson Foldable to review the lesson. Save your Foldable for the project at the end of the chapter.

What do you think NOW?

You first read the statements below at the beginning of the chapter.

3. Unlike sound waves, light waves can travel through a vacuum.

4. Light waves always travel at the same speed.

Did you change your mind about whether you agree or disagree with the statements? Rewrite any false statements to make them true.

Use Vocabulary

1. **Explain** the difference between transparent and translucent.
2. **Define** *light source* using your own words.

Understand Key Concepts

3. **Apply** Do light waves refract more when they travel from air to water or air to glass?
4. Which electromagnetic wave has the shortest wavelength?
   A. gamma
   B. infrared
   C. radio
   D. ultraviolet

Interpret Graphics

5. **Evaluate** If a light wave has an angle of incidence of 30° as shown below, what is its angle of reflection?

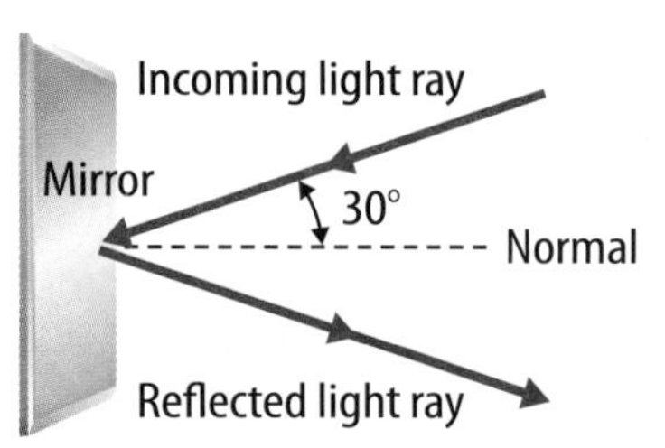

6. **Organize Information** Copy and fill in the table below.

Interaction	Description
Absorption	
	Light wave bounces off a surface.
Transmission	

Critical Thinking

7. **Describe** how the speed of light waves changes when they travel from air into a water-filled aquarium and back into air.
8. **Draw** You turn on a light in a dark room and see a chair. Sketch the path that the light waves traveled that enabled you to see the chair.

Skill Practice Model

30 minutes

How are light rays reflected from a plane mirror?

Scientists use **models** to describe how the laws of science might work in real-life applications. How does the law of reflection look in a real-life application?

Materials

flashlight

protractor

metric ruler

scissors

tape

small plane mirror, at least 10 cm on a side

modeling clay

Also needed: white, unlined paper; black construction paper

Safety

Learn It

In science, **models** are used to demonstrate ideas and concepts that often are difficult to grasp. A model that you can see and manipulate with your hands helps you understand a difficult concept and observe how the natural process works.

Try It

1. Read and complete a lab safety form.
2. With the scissors, cut a narrow slit in the construction paper and tape it over the flashlight lens.
3. Place the mirror at one end of the unlined paper. Push the mirror into lumps of clay so it stands vertically, and tilt the mirror so it leans slightly toward the table.
4. Mark the center of the mirror on the unlined paper. Then draw a line on the paper perpendicular to the mirror from the mark. Label this line *N* for normal.
5. Draw lines on the paper from the center mark at angles 30°, 45°, and 60° to line *N*.
6. Turn on the flashlight and place it so the beam is along the 45° line. This is the angle of incidence. Have a partner draw a short line tracing the reflected light beam and mark the line *45*.
7. Repeat step 6 for the 30°, 60°, and *N* lines.
8. Remove the paper from the setup. Measure and record in your Science Journal the angles that the reflected beams made with *N*. These are the angles of reflection.

Apply It

9. **Key Concept** Infer from your results the relationship between the angle of incidence and the angle of reflection.

Lesson 3

Reading Guide

Key Concepts
ESSENTIAL QUESTIONS

- What is the difference between regular and diffuse reflection?
- What types of images are formed by mirrors and lenses?
- How does the human eye enable a person to see?

Vocabulary

mirror p. 142
lens p. 143
cornea p. 144
iris p. 145
pupil p. 145
retina p. 146

Multilingual eGlossary

Video BrainPOP®

Mirrors, Lenses, and the Eye

Are there two mountains?

Have you ever seen an image on the surface of a lake? If so, you have observed light waves reflecting from a mirrorlike surface. Are all reflected images the same? How does the shape of a mirror's surface affect the image that you see?

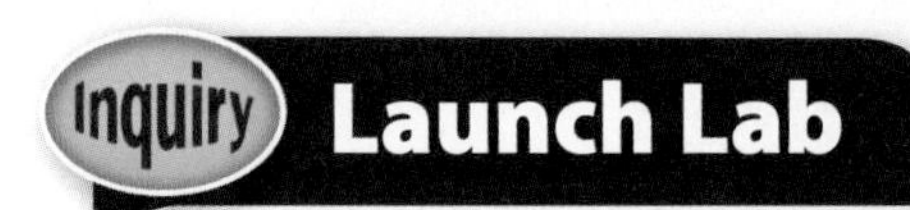

Launch Lab

15 minutes

Are there different types of reflections?

Some surfaces are like mirrors. Other surfaces do not form reflected images you can see. What is the difference between a surface that forms a sharp reflected image and one that does not?

1. Read and complete a lab safety form.
2. Place a **black bowl** on a **paper towel.** Look straight down at the bottom of the bowl. Record your observations in your Science Journal.
3. Carefully add about 3 cm of water to the bowl. Look at the bottom of the bowl. Record your observations.
4. Tap the side of the bowl gently and look again. Record your observations.

Think About This

1. Compare your observations in steps 2, 3, and 4.
2. **Key Concept** What do you think caused the differences in the images you observed?

Why are some surfaces mirrors?

When you look at a smooth pond, you can see a sharp image of yourself reflected off the water surface. When you look at a lake on a windy day, you do not see a sharp image. Why are these images different? A smooth surface reflects light rays traveling in the same direction at the same angle. This is called regular reflection, as shown in **Figure 17.** Because the light rays travel the same way relative to each other before and after reflection, the reflected light rays form a sharp image.

When a surface is not smooth, light rays still follow the law of reflection. However, light rays traveling in the same direction hit the rough surface at different angles. The reflected light rays travel in many different directions, as shown in **Figure 17.** This is called diffuse reflection. You do not see a clear image when diffuse reflection occurs.

Key Concept Check Contrast regular and diffuse reflection.

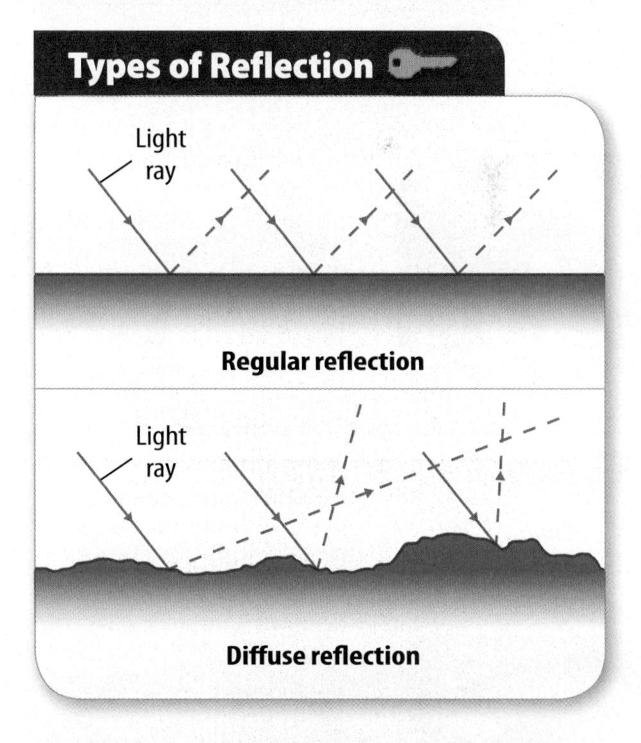

Figure 17 Light waves always obey the law of reflection, whether the surface is smooth or rough. Regular reflection occurs from a smooth surface and forms a sharp image. Diffuse reflection occurs from a rough surface and doesn't form a clear image.

Table 4 Images and Mirrors

Concave Mirror

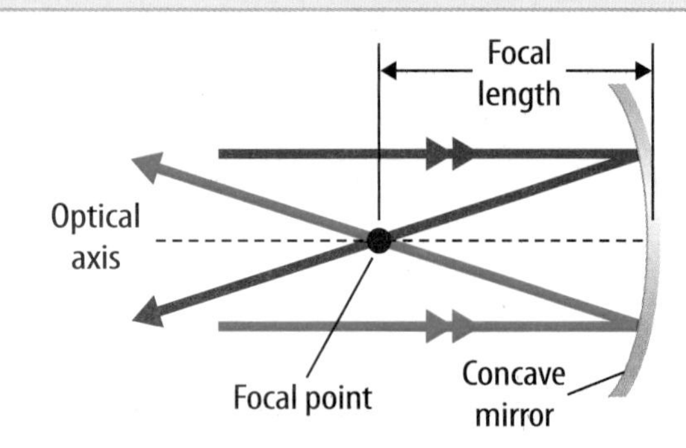

The focal length is the distance from the center of the mirror to the focal point.

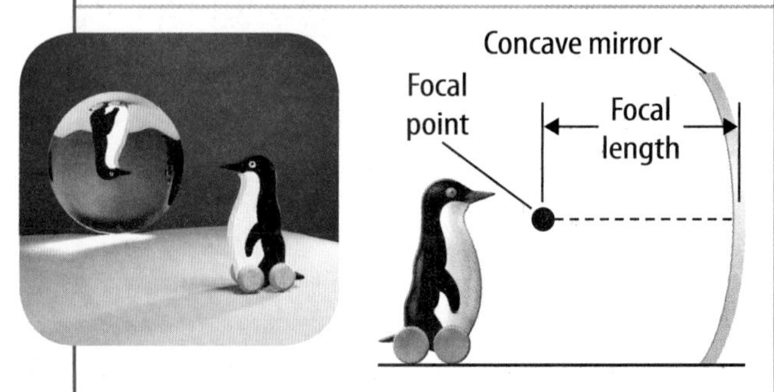

The image in a concave mirror is upside down when an object is more than one focal length from the mirror.

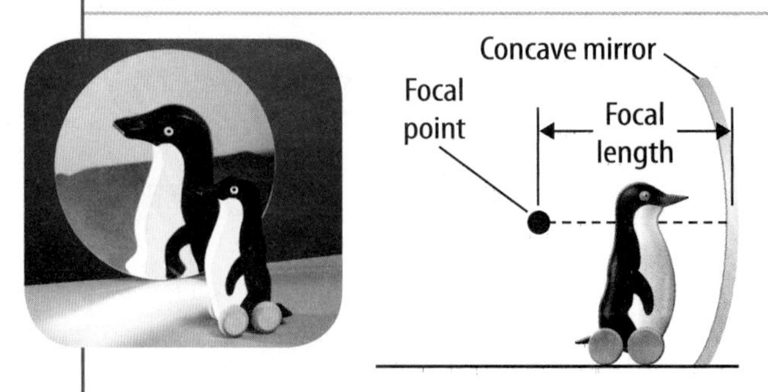

The image in a concave mirror is right-side up when an object is less than one focal length from the mirror.

Convex Mirror

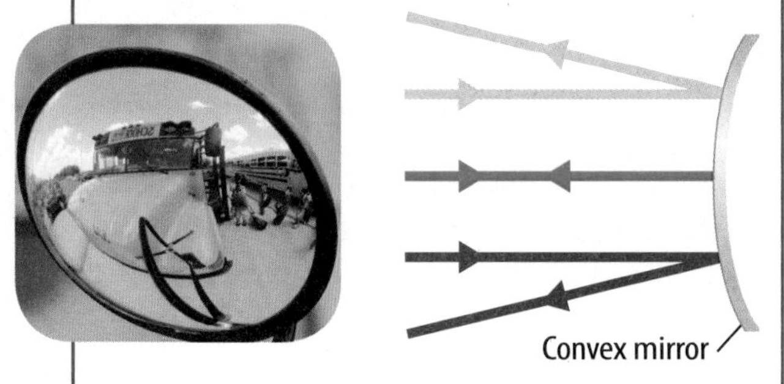

The image in a convex mirror is always right-side up and smaller than the object.

Types of Mirrors

When you look at a wall mirror, the image you see is about the same size that you are and right-side up. *A* **mirror** *is any reflecting surface that forms an image by regular reflection.* The image formed by a mirror depends on the shape of the mirror's surface.

Plane Mirrors

A plane mirror is a mirror that has a flat reflecting surface. The image formed by the mirror looks like a photograph of the object except that the image is reversed left to right. The size of the image in the mirror depends on how far the object is from the mirror. The image gets smaller as the object gets farther from the mirror.

Concave Mirrors

Reflecting surfaces that are curved inward are concave mirrors, as shown in **Table 4.** Light rays that are parallel to a line called the optical axis, which is shown in **Table 4,** are reflected through one point—the focal point. The distance from the mirror to the focal point is called the focal length.

The type of image formed depends on where the object is, as shown in **Table 4.** If an object is more than one focal length from a concave mirror, the image will be upside down. If the object is closer than one focal length, the image will be right-side up. If an object is placed exactly at the focal point, no image forms.

Convex Mirrors

A convex mirror has a reflecting surface that is curved outward, as shown in **Table 4.** The image is always right-side up and smaller than the object. Store security mirrors and passenger-side car mirrors are usually convex mirrors.

Key Concept Check How do the images formed by plane mirrors, concave mirrors, and convex mirrors depend on the distance of an object from the mirror?

Types of Lenses

Have you ever used a magnifying lens or binoculars? Or you might wear glasses that help you see more clearly. All of these items use lenses to change the way an image of an object forms. *A* **lens** *is a transparent object with at least one curved side that causes light to change direction.* The more curved the sides of a lens, the more the light changes direction as it passes through the lens.

Convex Lenses

A convex lens is curved outward on at least one side so it is thicker in the middle than at its edges. Just like a concave mirror, a convex lens has a focal point and a focal length, as shown in **Table 5.** The more curved the lens is, the shorter the focal length.

The image formed by a convex lens depends on where the object is just like it does for a concave mirror. When an object is farther than one focal length from a convex lens, the image is upside down, as shown to the right.

When an object is less than one focal length from a convex lens, the image is larger and right side up. For example, the dollar bill in **Table 5** is less than one focal length from the lens. As a result, the image of the dollar bill is larger than the actual dollar bill and is right-side up. Both a magnifying lens and a camera lens are convex lenses.

Key Concept Check How does the image formed by a convex lens depend on the distance of the object from the lens?

Concave Lenses

A concave lens is curved inward on at least one side and thicker at its edges. The image formed by a concave lens is upright and smaller than the object, as shown in **Table 5.** Concave lenses are usually used in combinations with other lenses in instruments such as telescopes and microscopes.

Table 5 Images and Lenses

Convex Lens

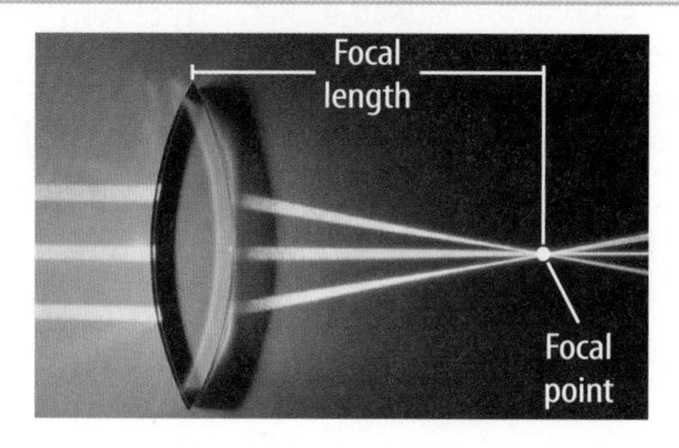

A convex lens is thicker in its middle than its edges. The focal length is the distance from the center of the lens to the focal point.

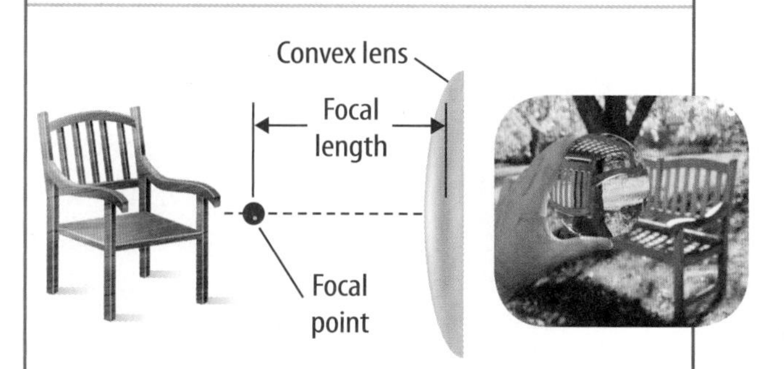

The image formed by a convex lens is upside down when an object is more than one focal length from the lens.

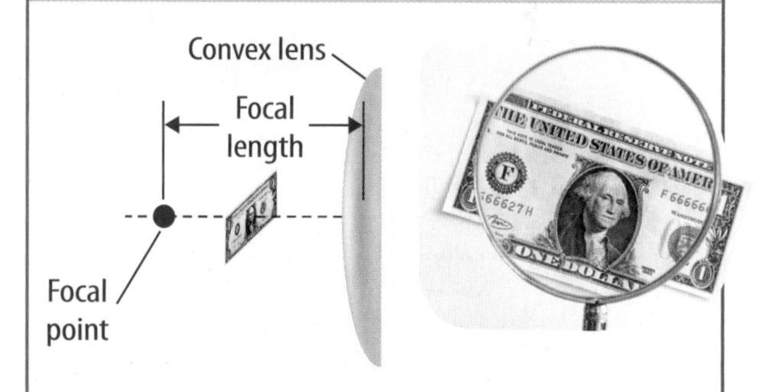

The image formed by a convex lens is right-side up when an object is less than one focal length from the lens.

Concave Lens

A concave lens is thinner in its middle than its edge. The image formed by a concave lens is always right-side up and smaller than the object.

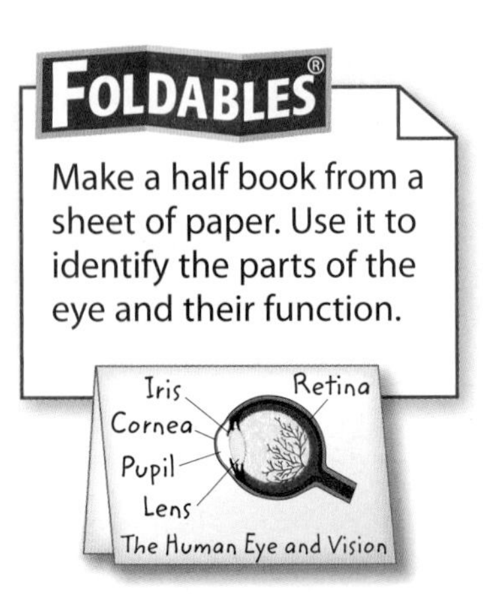

Light and the Human Eye

Microscopes, binoculars, and telescopes are instruments that contain lenses that form images of objects. Human eyes also contain lenses, as well as other parts, that can enable a person to see.

The structure of a human eye is shown in **Figure 18.** To see an object, light waves from an object travel through two convex lenses in the eye. The first of these lenses is called the cornea, and the second is simply called the lens. These lenses form an image of the object on a thin layer of tissue at the back of the eye. Special cells in this layer **convert** the image into electrical signals. Nerves carry these signals to the brain.

ACADEMIC VOCABULARY
convert
(verb) to change from one form into another

 **Reading Check** What is the function of the lenses in the eye?

Cornea

Light waves first travel through the cornea (KOR nee uh), as shown in **Figure 18.** *The* **cornea** *is a convex lens made of transparent tissue located on the outside of the eye.* Most of the change in direction of light rays occurs at the cornea. Some vision problems are corrected by changing the cornea's shape.

The Structure of the Human Eye

Figure 18 The eye is made of a number of parts that have different functions.

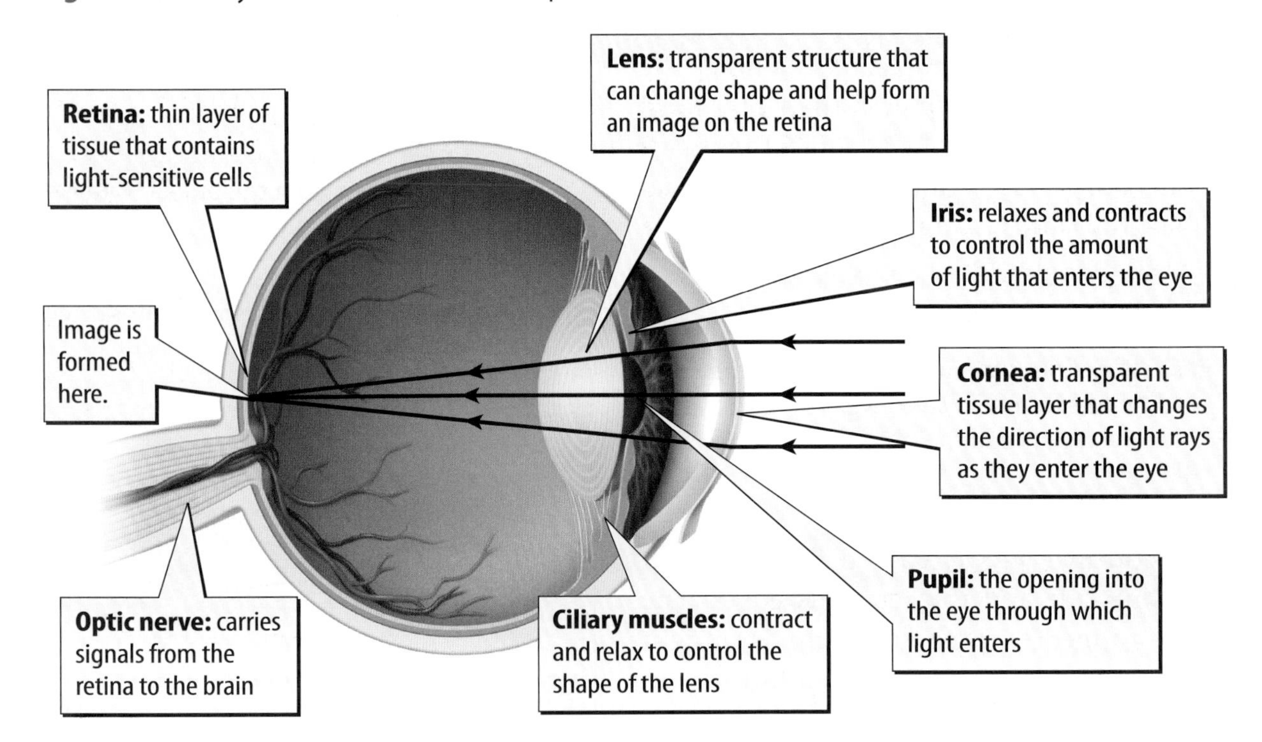

Visual Check On which part of the eye is an image formed?

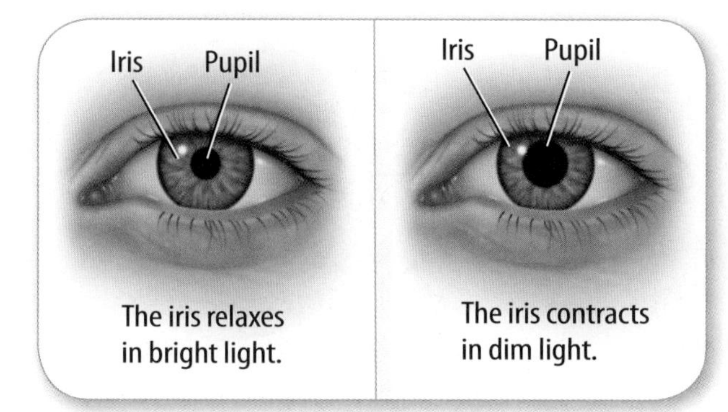

▲ **Figure 19** The iris controls the amount of light that enters the eye.

Iris and Pupil

The **iris** *is the colored part of the eye. The* **pupil** *is an opening into the interior of the eye at the center of the iris.* When the iris changes size, the amount of light that enters the eye changes. As shown in **Figure 19,** in bright light, the iris relaxes and the pupil becomes smaller. Then less light enters the eye. In dim light, the iris contracts and the pupil becomes larger. Then more light enters the eye.

Lens

Behind the iris is the lens, as shown in **Figure 18.** It is made of flexible, transparent tissue. The lens enables the eye to form a sharp image of nearby and distant objects. The muscles surrounding the lens change the lens's shape. To focus on nearby objects, these muscles relax and the lens becomes more curved, as shown in **Figure 20.** To focus on distant objects, these muscles pull on the lens and make it flatter.

Inquiry MiniLab — 20 minutes

How does the size of an image change?

The size of the image divided by the size of the object is called the magnification. How do the magnification and the image size change as an object gets farther from a convex lens?

Data Table

Distance of Object from Lens (cm)	Magnification
20.0	3.00
30.0	1.00
45.0	0.50
75.0	0.25

1. Using **graph paper,** plot the data in the data table. Plot the distance of the object from the lens on the *x*-axis and the magnification on the *y*-axis.
2. Draw a smooth curve through the points.

Analyze and Conclude

1. **Estimate** from your graph how far the object is from the lens if the magnification is 2.00.
2. **Key Concept** If the object is 60 cm from the lens, would you move the lens closer or farther from the object to make it look larger? Explain your answer.

Figure 20 The lens in the eye changes shape, enabling the formation of sharp images of objects that are either nearby or far away. ▼

Concepts in Motion Animation

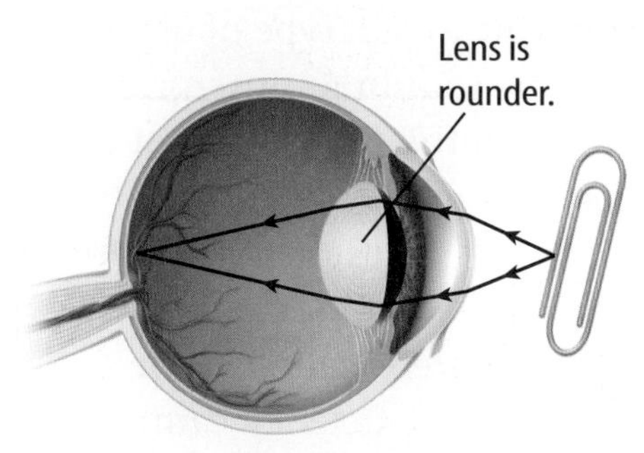

Lens becomes rounder and a sharp image forms of a nearby object.

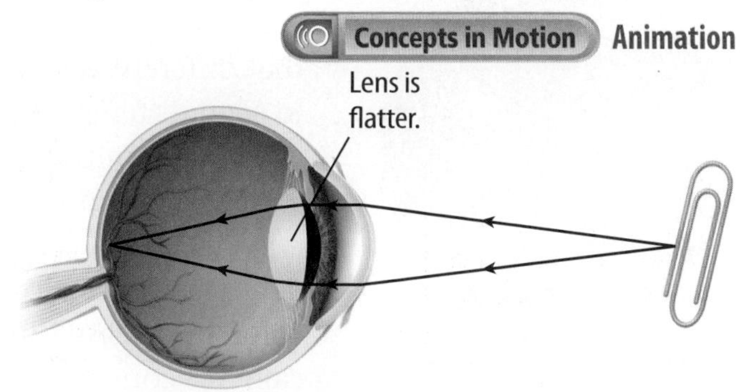

Lens becomes flatter and a sharp image forms of a distant object.

Figure 21 Rod cells in the retina respond to dim light. Cone cells in the retina enable you to see colors. ▶

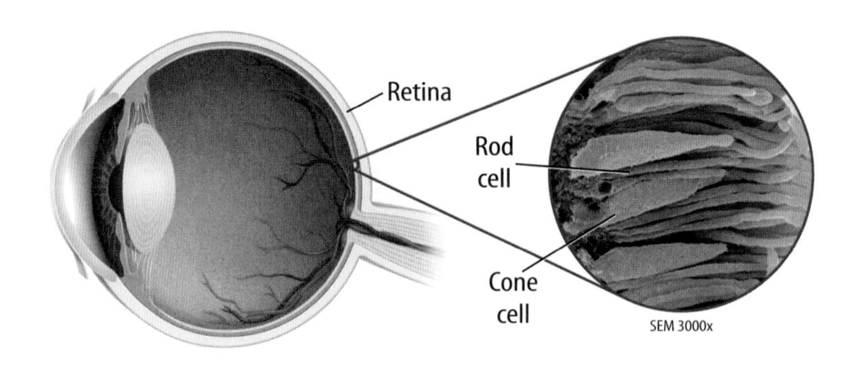

Word Origin

retina
from Latin *rete,* means "net"

Retina

The **retina** *is a layer of special light-sensitive cells in the back of the eye,* as shown in **Figure 21.** After light travels through the lens, an image forms on your retina. There, chemical reactions produce nerve signals that the optic nerve sends to your brain. There are two types of light-sensitive cells in your retina—rod cells and cone cells.

Key Concept Check Identify the parts of the eye that form a sharp image of an object and the parts that convert an image into electrical signals.

Rod Cells There are more than 100 million rod cells in a human retina. Rod cells are sensitive to low-light levels. They enable people to see objects in dim light. However, the signals rod cells send to your brain do not enable you to see colors.

Cone Cells A retina contains over 6 million cone cells. Cone cells enable a person to see colors. However, cone cells need brighter light than rod cells to function. In very dim light, only rod cells function. That is why objects seem to have no color in very dim light.

How do cone cells enable you to see colors? The responses of cone cells to light waves with different wavelengths enable you to see different colors.

The retina has three types of cone cells. Each type of cone cell responds to a different range of wavelengths. This means that different wavelengths of light cause each type of cone cell to send different signals to the brain. Your brain interprets the different combinations of signals from the three types of cone cells as different colors. However, in some people not all three types of cone cells function properly. These people cannot detect certain colors. This condition is commonly known as color blindness but is more appropriately called color deficiency. People with some kinds of color deficiency cannot see the number 74 in **Figure 22.**

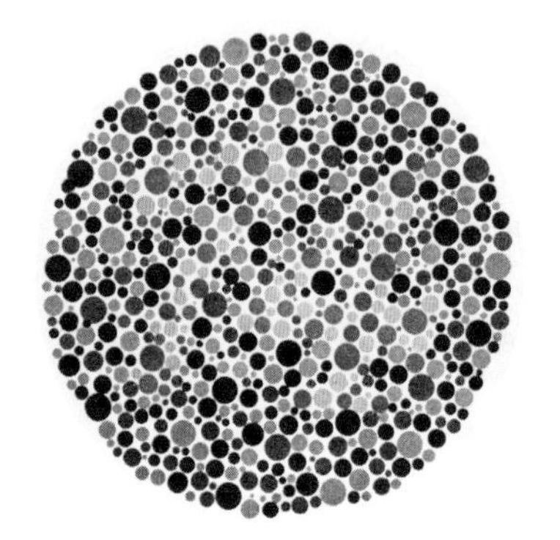

▲ **Figure 22** An image like this is used to test for color deficiency.

The Colors of Objects

The objects you see around you are different colors. A banana is mostly yellow, but a rose might be red. Why is a banana a different color from a rose? Bananas and roses do not give off, or emit, light. Instead, they reflect light. The colors of an object depend on the wavelengths of the light waves it reflects.

Reflection of Light and Color

When light waves of different wavelengths interact with an object, the object absorbs some light waves and reflects others. The wavelengths of light waves absorbed and reflected depend on the materials from which the object is made.

For example, **Figure 23** shows that the rose is red because the petals of the rose reflect light waves with certain wavelengths. When these light waves enter your eye, they cause the cone cells in your retina to send certain nerve signals to your brain. These signals cause you to see the rose as red.

A banana reflects different wavelengths of light than a rose. These different wavelengths cause cone cells in the retina to send different signals to your brain. These signals cause you to see the banana as yellow instead of red.

You might think that light waves have colors. Color, however, is a sensation produced by your brain when light waves enter your eyes. Light waves have no color as they travel from an object to your eyes.

Key Concept Check Why do you experience the sensation of color?

The Color of Objects that Emit Light

Some objects such as the Sun, lightbulbs, and neon lights emit light. The color of an object that emits light depends on the wavelengths of the light waves it emits. For example, a red neon light emits light waves with wavelengths that you see as red.

The rose reflects light waves with wavelengths that you see as red. It absorbs all other wavelengths of light.

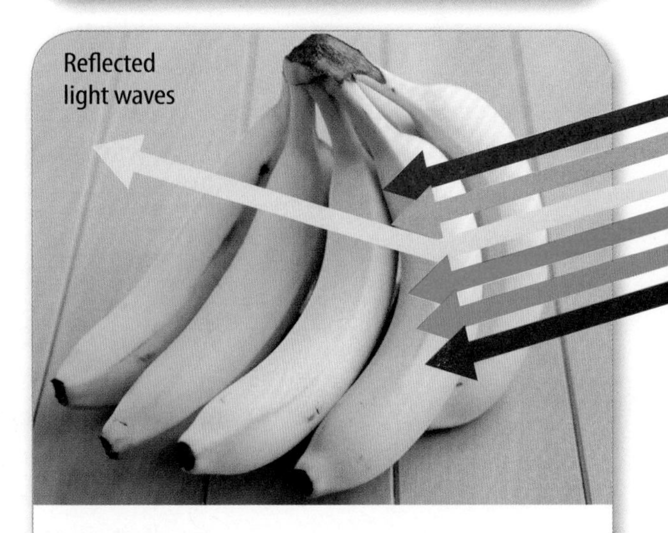

The banana reflects light waves with wavelengths that you see as yellow. It absorbs all other wavelengths of light.

Figure 23 The color of an object depends on the wavelengths of the light waves the object reflects.

Animation

Figure 24 Different wavelengths of light change direction by different amounts when they move into and out of a prism. This causes the waves to spread out. ▶

White Light—A Combination of Light Waves

You might have noticed at a concert that the colors of objects on stage depend on the colors of the spotlights. A shirt that appears blue when a white spotlight shines on it might appear black when a red spotlight shines on it. That same shirt will appear blue when a blue spotlight shines on it. How is light that is white different from light that is red or blue?

Light that you see as white is actually a combination of light waves of many different wavelengths. **Figure 24** shows what happens when white light travels through a prism. Light waves with different wavelengths spread out after passing through the prism and form a color spectrum.

Changing Colors

Figure 25 shows why the color of a blue shirt appears different when different spotlights shine on it. The shirt reflects only those wavelengths that are seen as blue. It absorbs all other wavelengths of light. When white light or blue light hits the shirt, it reflects the wavelengths you see as blue. However, when red light strikes the shirt, almost no light is reflected. This causes the shirt to appear black. An object appears black when it absorbs almost all light waves that strike it.

Figure 25 The appearance of an object changes under different colors of light. ▼

When white light strikes the shirt, only the wavelengths that you see as blue are reflected. The shirt appears blue under white light.

When blue light strikes the shirt, the blue light is reflected. This makes the shirt appear blue under blue light.

When red light strikes the shirt, the light is absorbed and no light is reflected. This makes the shirt appear black under red light.

Lesson 3 Review

Assessment — Online Quiz
Inquiry — Virtual Lab

Visual Summary

A mirror is a surface that causes a regular reflection. The shape of the reflecting surface and the position of the object determine what the image looks like.

A lens is a transparent object with at least one curved side that causes light waves to change direction. The shape of the lens and the position of the object determine how the image appears.

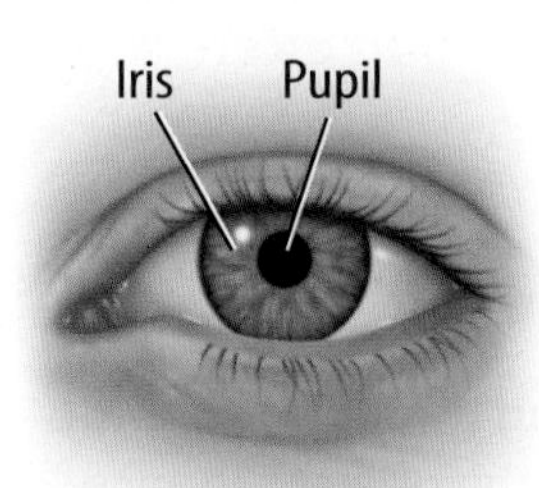

The eye has different parts with different functions. The iris is the colored part of your eye. The iris opens and closes, controlling the amount of light that enters the eye.

FOLDABLES®

Use your lesson Foldable to review the lesson. Save your Foldable for the project at the end of the chapter.

What do you think NOW?

You first read the statements below at the beginning of the chapter.

5. All mirrors form images that appear identical to the object itself.

6. Lenses always magnify objects.

Did you change your mind about whether you agree or disagree with the statements? Rewrite any false statements to make them true.

Use Vocabulary

1. The layer of tissue in the eye that contains cells sensitive to light is the ________.
2. **Define** *cornea* in your own words.
3. **Distinguish** between a lens and a mirror.

Understand Key Concepts 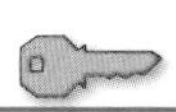

4. **Draw** a picture of two light rays that reflect off a smooth surface.
5. **Compare** the function of the cornea and the lens of the eye.
6. Which statement describes the image formed by a convex mirror?
   A. It will be caused by refraction.
   B. It will be smaller than the object.
   C. It will be upside down.
   D. It will produce a beam of light.

Interpret Graphics

7. **Describe** what is occurring in the figure below.

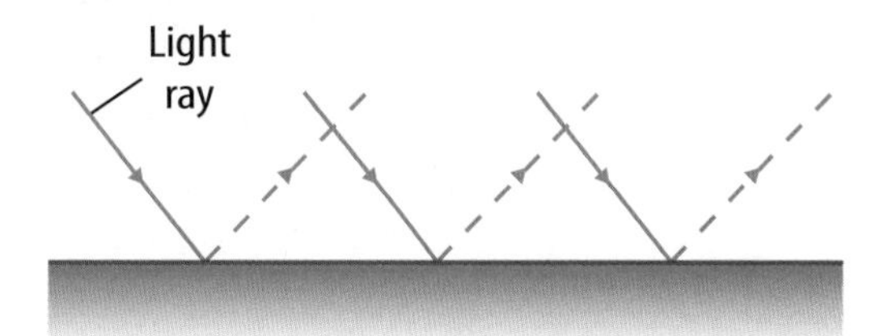

8. **Sequence** Copy and fill in the graphic organizer below to trace the path of light through the different parts of the human eye.

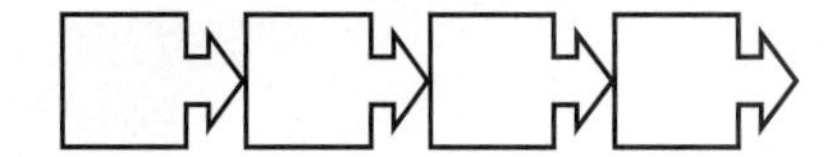

Critical Thinking

9. **Infer** A person cannot see well in dim light. What part of his or her eye is damaged?

45 minutes

The Images Formed by a Lens

Materials

convex lens

modeling clay

meterstick

flashlight

masking tape

20-cm square piece of white posterboard

Safety

The type of image formed by a convex lens is related to the distance of the object from the lens—the object distance. The distance from the lens to the image is called the image distance.

Ask a Question

How do the images formed by a convex lens depend on the distance of an object from the lens?

Make Observations

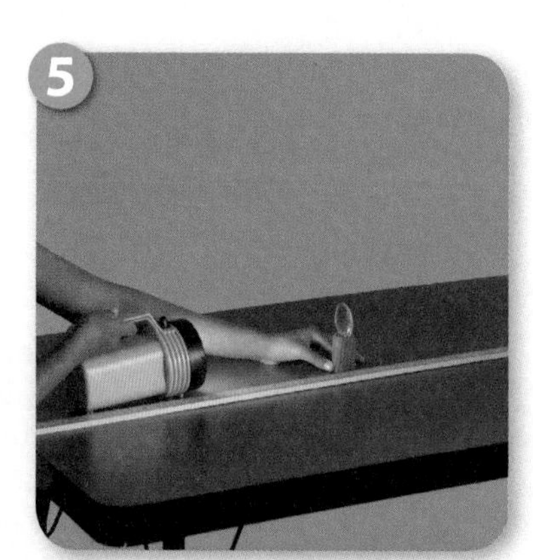

1. Read and complete a lab safety form.
2. Make a data table like the one shown below in your Science Journal to record your collected data.
3. Use the modeling clay to make the lens stand upright on the lab table.
4. Using masking tape, form the letter F on the surface of the flashlight lens. Turn on the flashlight and place it 0.25 m from the lens.
5. Position the flashlight so the flashlight beam is shining through the lens. Record the distance from the flashlight to the lens in the object distance column in your data table.
6. Hold the posterboard upright on the other side of the lens, and move it back and forth until a sharp image of the letter F is obtained.
7. Measure the distance of the card from the lens using a meterstick. Record this distance in the image distance column in your data table.
8. In the third column of your data table, record whether the image is upright or inverted, and smaller or larger.
9. Repeat steps 5 through 8 for object distances of 0.50 m, 1.5 m, and 2.0 m and record your data in your data table.

Convex Lens Data		
Object Distance (m)	Image Distance (m)	Image Type
0.25		
0.50		
1.5		
2.0		

Form a Hypothesis

10. Use your data to form a hypothesis about the type of image that would be formed if the object were far from the lens.

Test Your Hypothesis

11. Place the object far from the lens. Then repeat steps 6 through 8 to test your hypothesis.

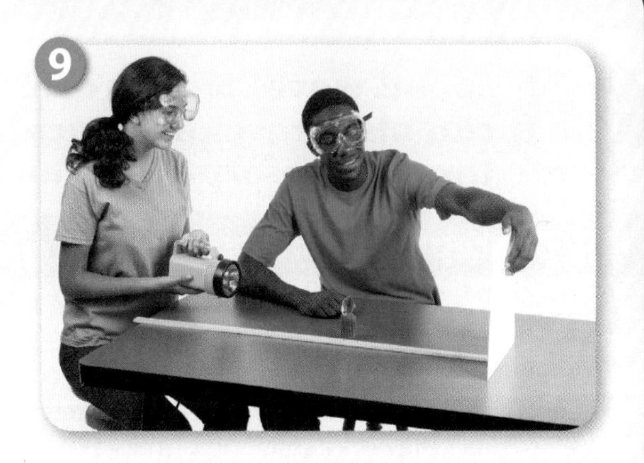

Analyze and Conclude

12. **Make and Use Graphs** Plot your data with the object distance on the *x*-axis and the image distance on the *y*-axis. Using your graph, determine the image distance of an object that is placed at 1.25 m.
13. **The Big Idea** Obtain the focal length of the lens from your teacher. Mark the focal length of the lens on your graph with a dotted line. Explain how focal length, object distance, and image distance are related.

Lab Tips

☑ Be sure the flashlight's position does not change when the cardboard is being moved.

Communicate Your Results

Write a report explaining the steps you took in this lab. Include information on your initial observations, your hypothesis, information on how you tested your hypothesis, and your final conclusions.

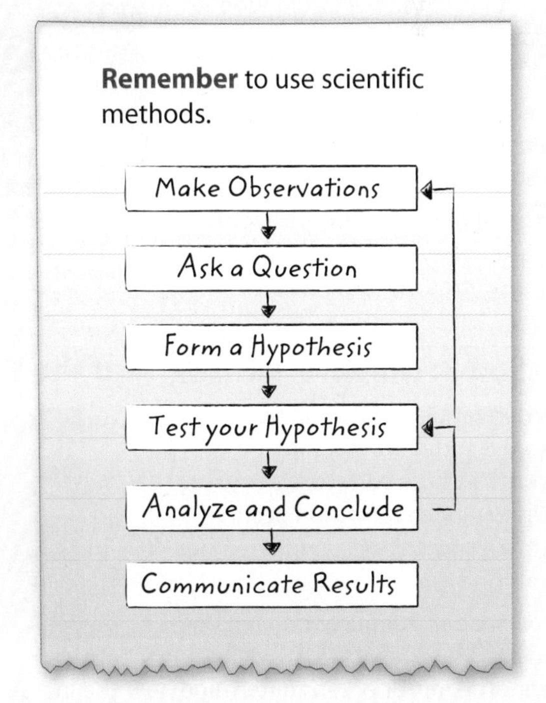

Use your graph to predict the image distance of an object that is placed at an object distance of 0.1 m. If possible, test your prediction.

Chapter 4 Study Guide

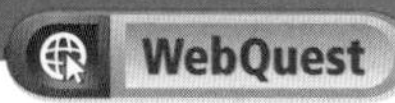

Sound waves must travel through matter, while light waves can also travel in a vacuum. Waves interact with matter through absorption, transmission, and reflection.

Key Concepts Summary

Lesson 1: Sound

- Vibrating objects produce **sound waves.**
- Sound waves travel at different speeds in different materials. Sound waves usually travel fastest in solids and slowest in gases.
- The outer ear collects sound waves. The middle ear amplifies sound waves. The inner ear converts sound waves to nerve signals.

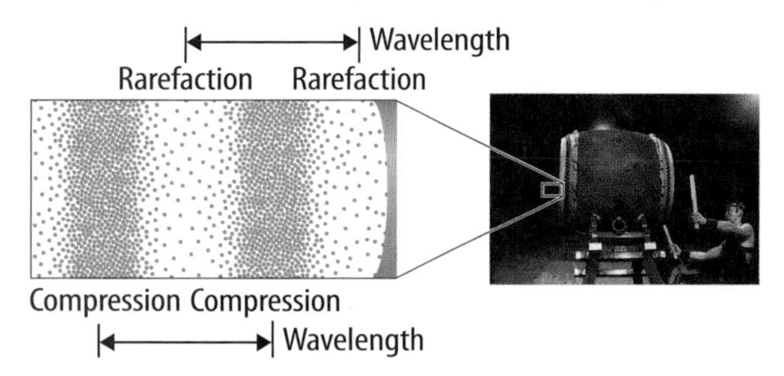

Vocabulary

sound wave p. 123
pitch p. 127
echo p. 129

Lesson 2: Light

- Light waves are electromagnetic waves that can travel in matter and through a vacuum.
- Electromagnetic waves have different wavelengths and frequencies.
- When light waves interact with matter, they are reflected, transmitted, or absorbed.

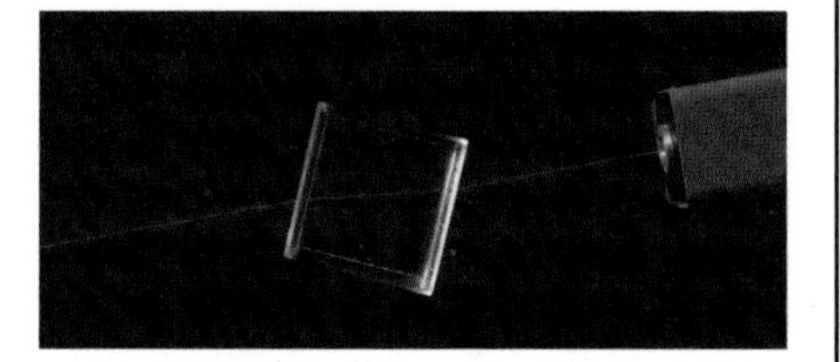

Vocabulary

light source p. 135
light ray p. 135
transparent p. 136
translucent p. 136
opaque p. 136

Lesson 3: Mirrors, Lenses, and the Eye

- When regular reflection occurs from a surface, a clear image forms and the surface is a **mirror.** When diffuse reflection occurs from a surface, a clear image does not form.
- The shape of a mirror or a **lens** and the distance of an object from the mirror or lens determine how the image appears.
- When light rays enter the eye through the **cornea** and pass through the **pupil,** an image forms on the **retina.** Rod and cone cells convert the image to nerve signals that travel to the brain.

Vocabulary

mirror p. 142
lens p. 143
cornea p. 144
iris p. 145
pupil p. 145
retina p. 146

- Personal Tutor
- Vocabulary eGames
- Vocabulary eFlashcards

FOLDABLES® Chapter Project

Assemble your lesson Foldables as shown to make a Chapter Project. Use the project to review what you have learned in this chapter.

Use Vocabulary

1. A vibrating object produces ________.
2. A sonar system detects the ________ of a sound wave.
3. A(n) ________ travels in a straight path until it is refracted, reflected, or absorbed.
4. A window is a(n) ________ object because you can see objects clearly through it.
5. The Sun is a(n) ________ ________ because it emits light waves.
6. The ________ of the eye controls how much light enters it.
7. A(n) ________ is a surface that produces a regular reflection.

Link Vocabulary and Key Concepts

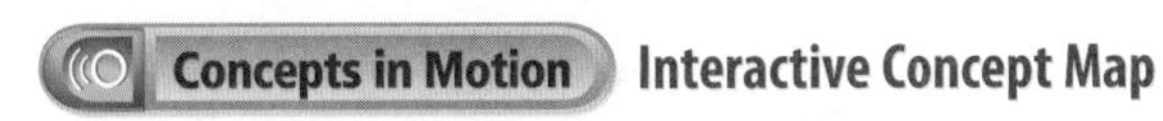

Copy this concept map, and then use vocabulary terms from the previous page to complete the concept map.

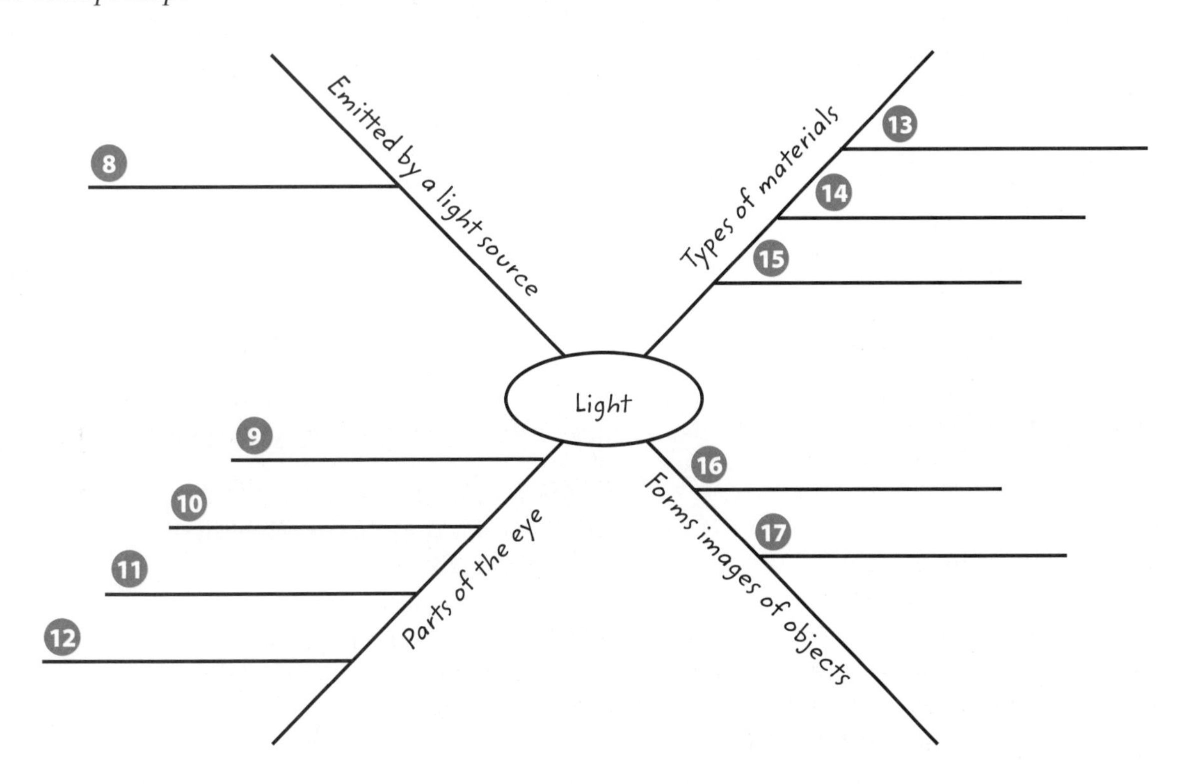

Chapter 4 Review

Understand Key Concepts 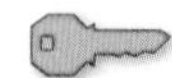

1. Which part of the ear acts as the amplifier?
   A. cochlea
   B. nerves
   C. inner ear
   D. middle ear

2. Identify the part of the ear that serves as the sound-wave collector.

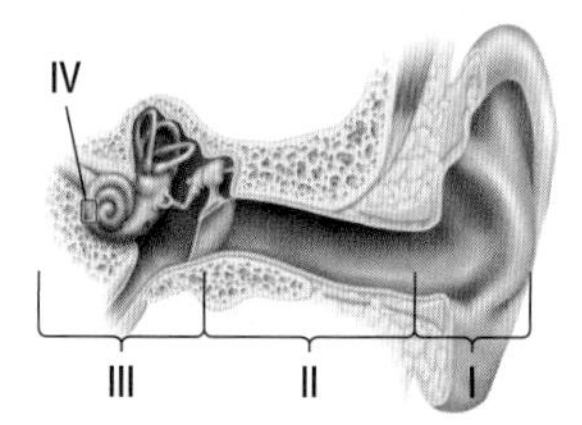

   A. I
   B. II
   C. III
   D. IV

3. Which produces sound that has the highest pitch?
   A. a tuba
   B. emergency siren
   C. lion's roar
   D. thunder

4. The speed of light is slowest in which medium?
   A. cold air
   B. outer space
   C. pond water
   D. window glass

5. A shirt appears red under white light. What color would it appear under blue light?
   A. black
   B. blue
   C. red
   D. white

6. Which enables a person to see color?
   A. cone cells
   B. cornea
   C. iris
   D. rod cells

7. The loudness of a sound wave depends on which of these?
   A. amplitude
   B. frequency
   C. pitch
   D. wavelength

8. What is the function of the cornea?
   A. changes the direction of light waves that enter the eye
   B. controls the amount of light that enters the eye
   C. controls the shape of the lens
   D. enables eye to focus on near and distant objects

9. On the electromagnetic spectrum, which waves have wavelengths longer than the wavelengths of visible light?
   A. gamma rays
   B. radio waves
   C. ultraviolet waves
   D. X-rays

10. What type of image will form in the mirror shown below?

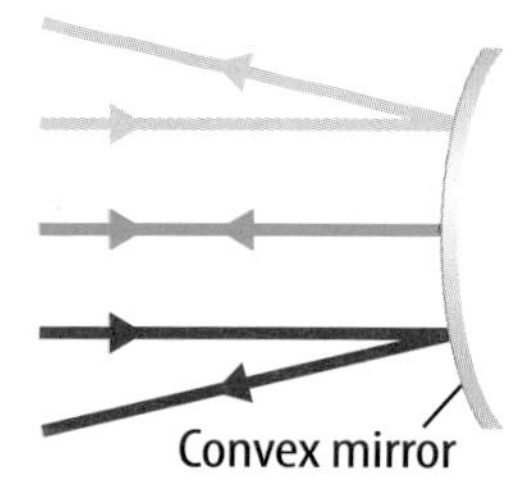

   A. Image is larger and right-side up.
   B. Image is larger and upside down.
   C. Image is smaller and right-side up.
   D. Image is smaller and upside down.

11. A reflected light ray makes an angle of 60° to the normal. At what angle did the light ray strike the surface?
   A. 30° to the normal
   B. 60° to the normal
   C. parallel to the normal
   D. perpendicular to the normal

Assessment

Online Test Practice

Critical Thinking

12. **Compare** an echo to light that hits a mirror.

13. **Summarize** Listen to the sounds around you. Choose one sound and describe its path from its source to the point where the nerve signal is sent to your brain.

14. **Judge** Frosted glass is made of glass with a scratched surface. Decide whether frosted glass is opaque, translucent, or transparent and explain your reasoning.

15. **Infer** On a hot summer day, black pavement feels much hotter than light-gray concrete. Explain why this is so.

16. **Evaluate** Stores are often equipped with mirrors like the one shown below. What type of mirror is this and why is it useful?

17. **Compare** When you enter a room filled with many hard surfaces and start talking, your voice echoes all around you. How is this similar to what happens when light rays scatter?

18. **Infer** A film camera forms an upside-down image of a tree on the film. What is the location of the tree relative to the focal length of the lens?

Writing in Science

19. **Write** a 500–700-word essay about an object such as a telescope, a microscope, or a periscope that uses mirrors or lenses to create an image. Describe how and why the mirrors and lenses are used in the object.

REVIEW THE BIG IDEA

20. What are three possible results when light waves strike matter?

21. In many movies, you can both see and hear explosions that happen in outer space. Explain how this is inaccurate.

22. The photo below shows a Chicago sculpture named "Cloud Gate." What happens to light waves when they strike the surface of the sculpture? What type of surface is this?

Math Skills

Math Practice

Use Equations

23. What is the speed of a sound wave if it travels 2,500 m in 2 s?

24. A sound wave travels through air at 331 m/s. How far would it travel at 0°C in 5 s?

25. A sound wave travels through water at 1,500 m/s. How far would it travel in 5 s?

Standardized Test Practice

Record your answers on the answer sheet provided by your teacher or on a sheet of paper.

Multiple Choice

1 Which statement about sound waves is false?

A They are longitudinal waves.

B They consist of compressions and rarefactions.

C They travel in empty space.

D They result from the vibrations of objects.

Use the diagram below to answer question 2.

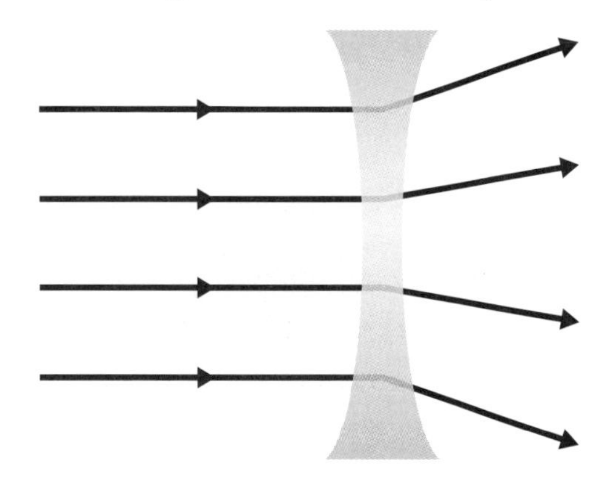

2 Which describes the image formed by the lens shown above?

A right-side up, larger than the object

B right-side up, smaller than the object

C upside down, larger than the object

D upside down, smaller than the object

3 A shirt that appears red under red light will also appear red under which color light?

A blue

B green

C yellow

D white

Use the table below to answer question 4.

Wave Type	Wavelength
Radio waves	More than 0.001 m
Microwaves	0.3 m to 0.001 m
Visible light	700 nm to 400 nm
Ultraviolet light	400 nm to 10 nm
X-rays	10 nm to 0.01 nm

4 Which electromagnetic waves can have wavelengths of 500 nm?

A microwaves

B ultraviolet light

C visible light

D X-rays

5 When does the refraction of light waves occur?

A when they bounce off a reflecting surface

B when they are absorbed

C when they move far from their source

D when they pass from one material to another and change speed

6 On which property of the particles in a material does the speed of sound depend?

A the dimensions of the particles

B the forces between the particles

C the shape of the particles

D the number of particles

7 Which property of sound waves does the decibel scale compare?

A frequency

B volume

C pitch

D speed

Assessment

Online Standardized Test Practice

Use the diagram below to answer question 8.

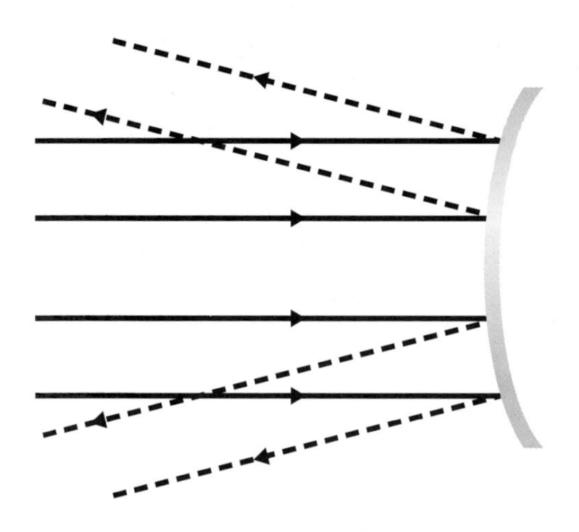

8 Which describes the image formed by the mirror shown above?

A larger than the object, right-side up

B larger than the object, upside down

C smaller than the object, right-side up

D smaller than the object, upside down

9 Which surface produces a diffuse reflection of light?

A a bathroom mirror

B the surface of a calm lake

C the hood of a newly waxed car

D a white painted wall

10 Which illustrates how the ear responds to sound waves?

A amplify → collect → transmit

B collect → amplify → convert

C convert → transmit → collect

D transmit → convert → amplify

Constructed Response

Use the table below to answer question 11.

Part of the Eye	Function
Cornea	
Iris	
Pupil	
Lens	
Retina	

11 Complete the table above. Which parts of the eye form a sharp image inside the eye? Explain how a person is able to see colors.

Use the table below to answer question 12.

	Light Waves	Sound Waves
Type of wave		
Material in which wave moves slowest		
Speed in air		
Human structure that detects these waves		

12 Complete the table above to compare and contrast light and sound waves.

13 When you turn on a lamp, you are able to see everything in a room. Explain why you see the lamp. Then, explain why you see other objects in the room.

NEED EXTRA HELP?													
If You Missed Question...	1	2	3	4	5	6	7	8	9	10	11	12	13
Go to Lesson...	1	3	3	2	2	1	1	3	3	1	3	1,2	2

Unit 2

INTERACTIONS OF MATTER

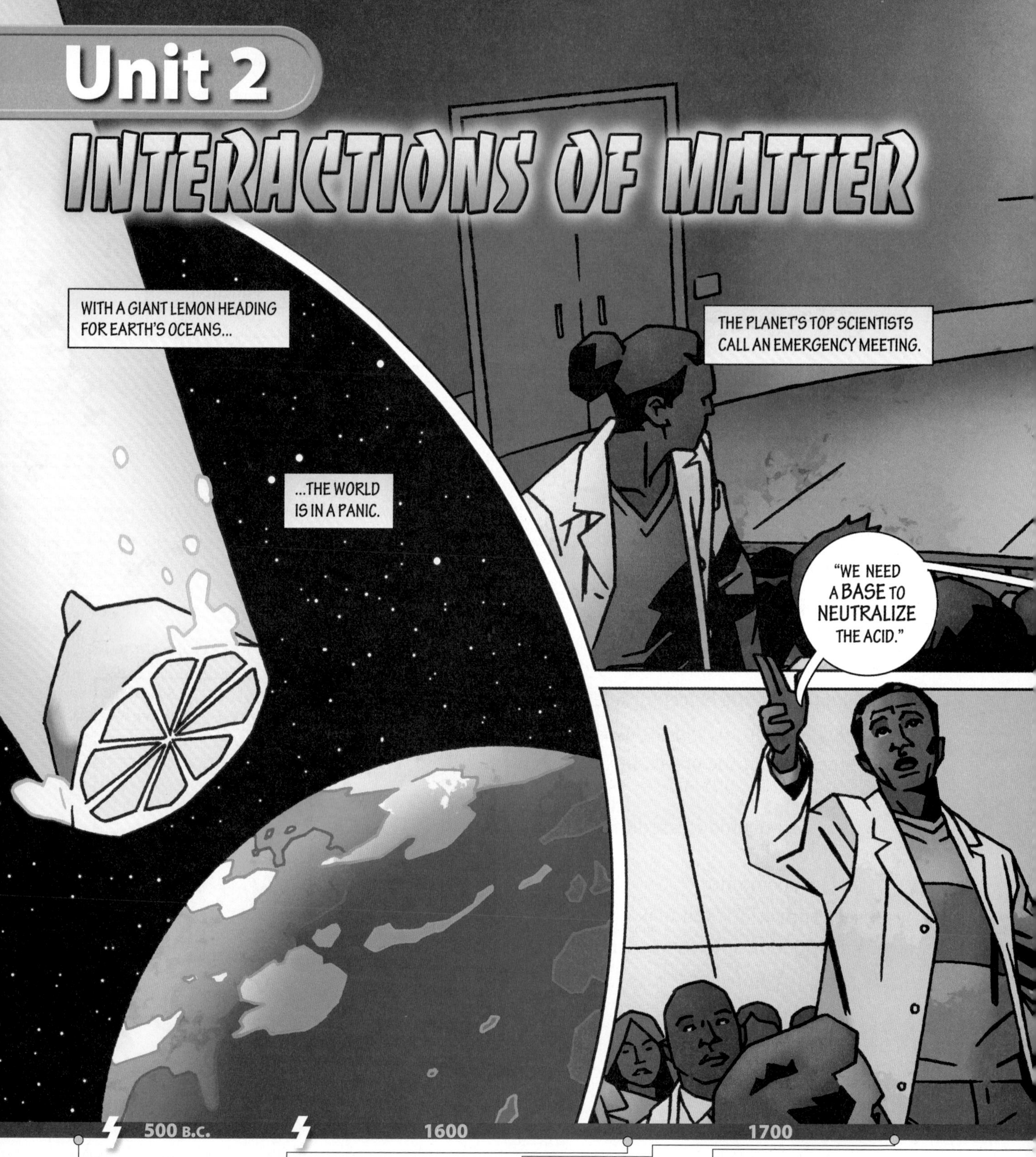

500 B.C. | 1600 | 1700

1000 B.C.
Chemistry is considered more of an art than a science. Chemical arts include the smelting of metals and the making of drugs, dyes, iron, and bronze.

1661
A clear distinction is made between chemistry and alchemy when *The Sceptical Chymist* is published by Robert Boyle. Modern chemistry begins to emerge.

1789
Antoine Lavoisier, the "father of modern chemistry," clearly outlines the law of conservation of mass.

1803
John Dalton publishes his atomic theory, which states that all matter is composed of atoms, which are small and indivisible and can join together to form chemical compounds. Dalton is considered the originator of modern atomic theory.

1800 1900

1869
The first periodic table is published by Dmitri Mendeleev. The table arranges elements into vertical columns and horizontal rows and is arranged by atomic number.

1953
James Watson and Francis Crick develop the double-helix model of DNA. This discovery leads to a spike in research of the biochemistry of life.

1983
Kary Mullis devises the polymerase chain reaction (PCR), a technique for copying a small portion of DNA in a lab environment. PCR can be used to synthesize specific pieces of DNA and makes the sequencing of DNA of organisms possible.

Inquiry
Visit ConnectED for this unit's STEM activity.

Unit 2

Nature of SCIENCE

Health and Science

Have an upset stomach? Chew on some charcoal. Have a headache? Rub a little peppermint oil on your temples. As shown in **Figure 1,** people have used chemicals to fix physical ailments for thousands of years, long before the development of the first medicines. Many cures were discovered by accident. People did not understand why the cures worked, only that they did work.

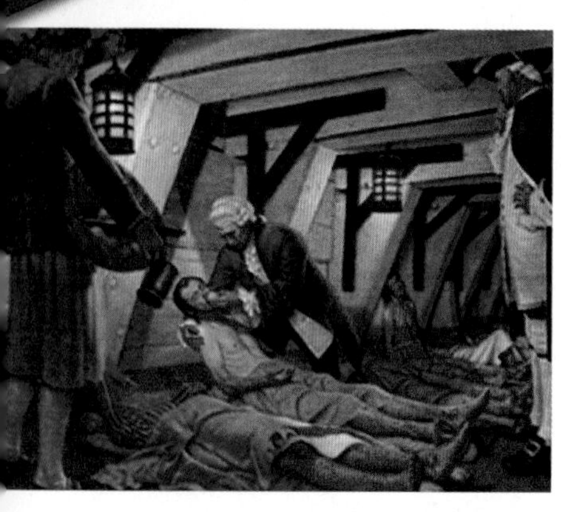

▲ **Figure 1** Thousands of years ago, people believed that evil spirits were responsible for illness. They often treated the physical symptoms with herbs or other natural materials.

Asking Questions About Health

Over time, people asked questions about which cures worked and which cures did not work. They made observations, recorded their findings, and had discussions about cures with other people. This process was the start of the scientific investigation of health. **Health** is the overall condition of an organism or group of organisms at any given time. Early studies focused on treating the physical parts of the body. The study of how chemicals interact in organisms did not come until much later. Recognizing that chemicals can affect health opened a whole new field of study known as biochemistry. The time line in **Figure 2** shows some of the medical and chemical discoveries people made that led to the development of medicines that save lives.

Figure 2 The time line shows several significant discoveries and developments in the history of medicine. ▼

4,200 years ago Clay tablets describe using sesame oil on wounds to treat infection.

3,500 years ago An ancient papyrus described how Egyptians applied moldy bread to wounds to prevent infection.

More than 3,300 years later, scientists found that a chemical in mold broke down the cell membranes of bacteria, killing them. Similar discoveries led to the development of antibiotics.

2,500 years ago Hippocrates, known as the "Father of Medicine," is the first physician known to separate medical knowledge from myth and superstition.

Year 900 The first pharmacy opened in Persia, which is now Iraq.

1740s A doctor found that the disease called scurvy was caused by a lack of Vitamin C.

Early explorers on long sea voyages often lost their teeth or developed deadly sores. Ships could not carry many fruits and vegetables, which contain Vitamin C, because they spoil quickly. Scientists suspect that many early explorers might have died because their diets did not include the proper vitamins.

Benefits and Risks of Medicines

Scientists might recognize that a person's body is missing a necessary chemical, but that does not mean they can always fix the problem. For example, people used to get necessary vitamins and minerals by eating natural, whole foods. Today, food processing destroys many nutrients. Foods last longer, but they do not provide all the nutrients the body needs.

Researchers still do not understand the role of many chemicals in the body. Taking a medicine to fix one problem sometimes causes others, called side effects. Some side effects can be worse than the original problem. For example, antibiotics kill some disease-causing bacteria. However, widespread use of antibiotics has resulted in "super bugs"—bacteria that are resistant to treatment.

Histamines are chemicals that have many functions in the body, including regulating sleep and decreasing sensitivity to allergens. However, low levels of histamines have been linked to some serious illnesses. Many medicines have long-term effects on health. Before you take a medicine, you should recognize that you are adding a chemical to your body. You should be as informed as possible about any possible side effects.

Inquiry MiniLab

15 minutes

Is everyone's chemistry the same?

Each person's body is a unique "chemical factory." Why might using the same medicine to treat illness not work exactly the same way in everyone?

1. Read and complete a lab safety form.
2. Place a strip of **pH paper** on your tongue. Immediately place the paper in a **self-sealing plastic bag.**
3. Compare the color of your paper to the **color guide.** Record the pH in your Science Journal.
4. Record your pH on a class chart for comparison.

Analyze and Conclude

1. **Organize Data** What was the range of pH values among your classmates?
2. **Predict** How might differences in pH affect how well a medicine works in different people?

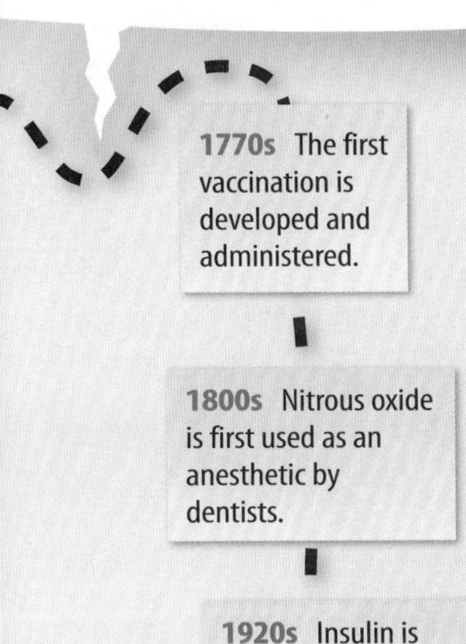

1770s The first vaccination is developed and administered.

1800s Nitrous oxide is first used as an anesthetic by dentists.

1920s Insulin is identified as the missing hormone in people with diabetes.

Scientists studying digestion in dogs noticed that ants were attracted to the urine of a dog whose pancreas had been removed. They determined the dog's urine contained sugar, which attracted ants. Eventually, scientists discovered that diabetes resulted from a lack of insulin, a chemical produced in the pancreas that regulates blood sugar. Today, some people with diabetes wear an insulin pump that monitors their blood sugar and delivers insulin to their bodies.

1920s Penicillin is discovered, but not developed for treatment of disease until the mid-1940s.

2000s First vaccine to target a cause of cancer

Chapter 5

Thermal Energy

THE BIG IDEA **How can thermal energy be used?**

Inquiry What are these colors?

This image shows the thermal energy of cars moving in traffic. The white indicates regions of high thermal energy, and the dark blue indicates regions of low thermal energy.

- What is thermal energy?
- How does thermal energy relate to temperature and heat?
- How can thermal energy be used?

Get Ready to Read

What do you think?

Before you read, decide if you agree or disagree with each of these statements. As you read this chapter, see if you change your mind about any of the statements.

1. Temperature is the same as thermal energy.
2. Heat is the movement of thermal energy from a hotter object to a cooler object.
3. It takes a large amount of energy to significantly change the temperature of an object with a low specific heat.
4. The thermal energy of an object can never be increased or decreased.
5. Car engines create energy.
6. Refrigerators cool food by moving thermal energy from inside the refrigerator to the outside.

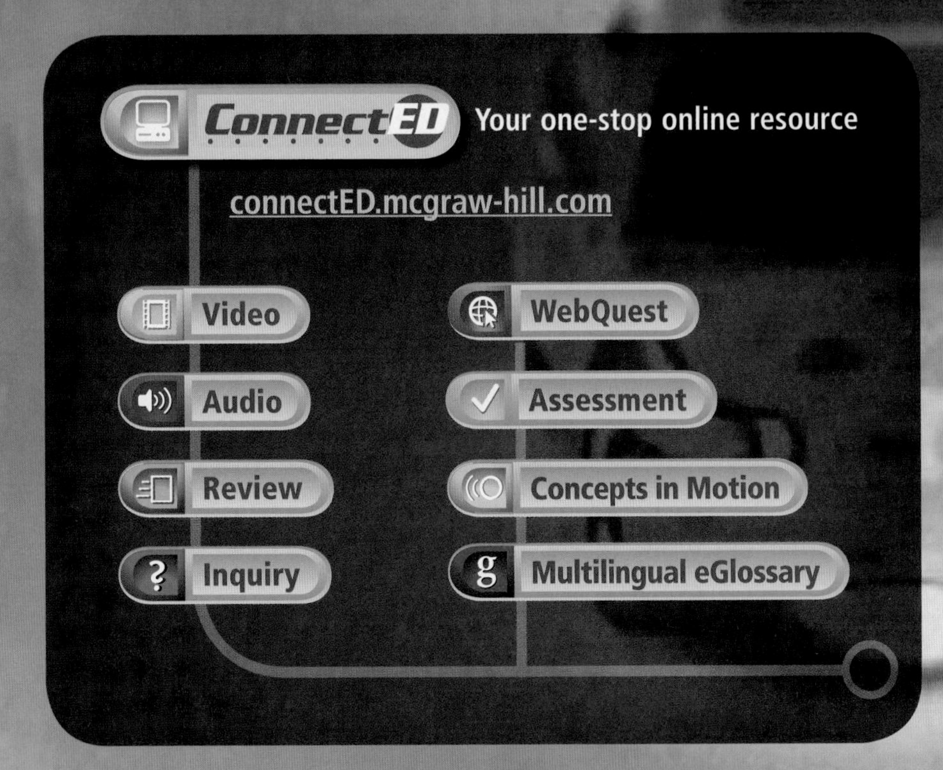

Lesson 1

Reading Guide

Key Concepts
ESSENTIAL QUESTIONS

- How are temperature and kinetic energy related?
- How do heat and thermal energy differ?

Vocabulary

thermal energy p. 166
temperature p. 167
heat p. 169

g Multilingual eGlossary

Thermal Energy, Temperature, and Heat

Inquiry How hot is it?

Forty gallons of sugar-maple sap must be heated to a very high temperature for several days to produce 1 gallon of maple syrup. What kind of energy is needed to achieve this very high temperature? Is there a difference between heat, temperature, and thermal energy?

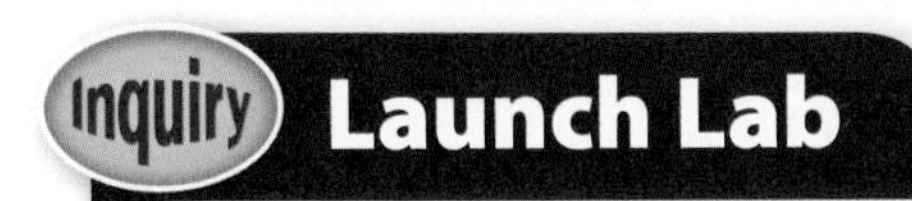

15 minutes

How can you describe temperature?

Have you ever used Fahrenheit or Celsius to describe the temperature? Why can't you just make up your own temperature scale?

1. Read and complete a lab safety form.
2. Use a **ruler** and a **permanent marker** to divide a **clear plastic straw** into 12 equal parts. Number the lines. Give your scale a name.
3. Add a room-temperature **colored alcohol-water mixture** to an **empty plastic water bottle** until it is about $\frac{1}{4}$ full.
4. Place one end of the straw into the bottle with the tip just below the surface of the liquid. Seal the straw onto the bottle top with **clay.**
5. Place the bottle in a **hot water bath**, and observe the liquid in your straw.

Think About This

1. Why is it important for scientists to use the same scale to measure temperature?
2. **Key Concept** What are some ways to make the liquid in your thermometer rise or fall?

Kinetic and Potential Energy

What do a soaring soccer ball and the particles that make up hot maple syrup have in common? They all have energy, or the ability to cause change. What type of energy does a moving soccer ball have? Recall that any moving object has kinetic energy. When the athlete in **Figure 1** kicks the ball and puts it in motion, the ball has kinetic energy.

In addition to kinetic energy, when the soccer ball is in the air, it has potential energy. Potential energy is stored energy due to the interaction between two objects. For example, think of Earth as one object and the ball as another. When the ball is in the air, it is attracted to Earth due to gravity. This attraction is called gravitational potential energy. In other words, since the ball has the potential to change, it has potential energy. And, the higher the ball is in the air, the greater the potential energy of the ball.

You also might recall that the potential energy plus the kinetic energy of an object is the mechanical energy of the object. When a soccer ball is flying through the air, you could describe the mechanical energy of the ball by describing both its kinetic and potential energy. On the next page, you will read about how the particles that make up maple syrup have energy, just like a soaring soccer ball.

Reading Check How could you describe the energy of a moving object?

REVIEW VOCABULARY

kinetic energy
the energy an object or a particle has because it is moving

potential energy
stored energy

Figure 1 This soccer ball has both kinetic energy and potential energy.

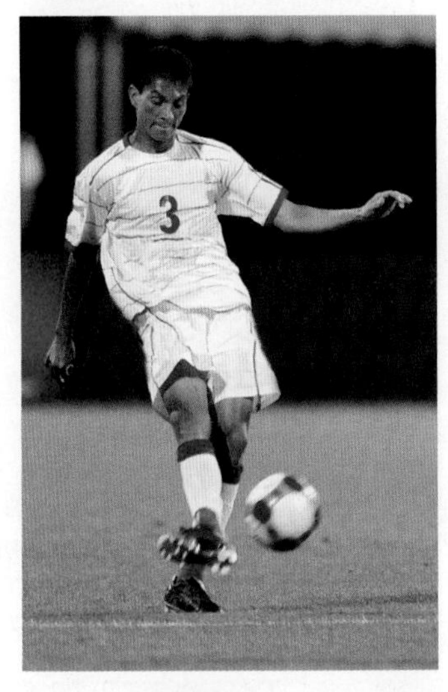

What is thermal energy?

Every solid, liquid, and gas is made up of trillions of tiny particles that are constantly moving. Moving particles make up the books you read, the air you breathe, and the maple syrup you pour on your pancakes. For example, the particles that make up a book, or any solid, vibrate in place. The particles that make up the air around you, or any gas, are spread out and move freely and quickly. Because particles are in motion, they have kinetic energy, like the soaring soccer ball in **Figure 2.** The faster particles move, the more kinetic energy they have.

The particles that make up matter also have potential energy. Like the interaction between a soccer ball and Earth, particles that make up matter interact with and are attracted to one another. The particles that make up solids usually are held very close together by attractive forces. The particles that make up a liquid are slightly farther apart than those that make up a solid. And, the particles that make up a gas are much more spread out than those that make up either a solid or a liquid. The greater the average distance between particles, the greater the potential energy of the particles.

Recall that a flying soccer ball has mechanical energy, which is the sum of its potential energy and its kinetic energy. The particles that make up the soccer ball, or any material, have a similar kind of energy called thermal energy. **Thermal energy** *is the sum of the kinetic energy and the potential energy of the particles that make up a material.* Thermal energy describes the energy of the particles that make up a solid, a liquid, or a gas.

Reading Check How are thermal energy and mechanical energy similar? How are they different?

Figure 2 The potential energy of the soccer ball depends on the distance between the ball and Earth. The potential energy of the particles of matter depends on their distance from one another.

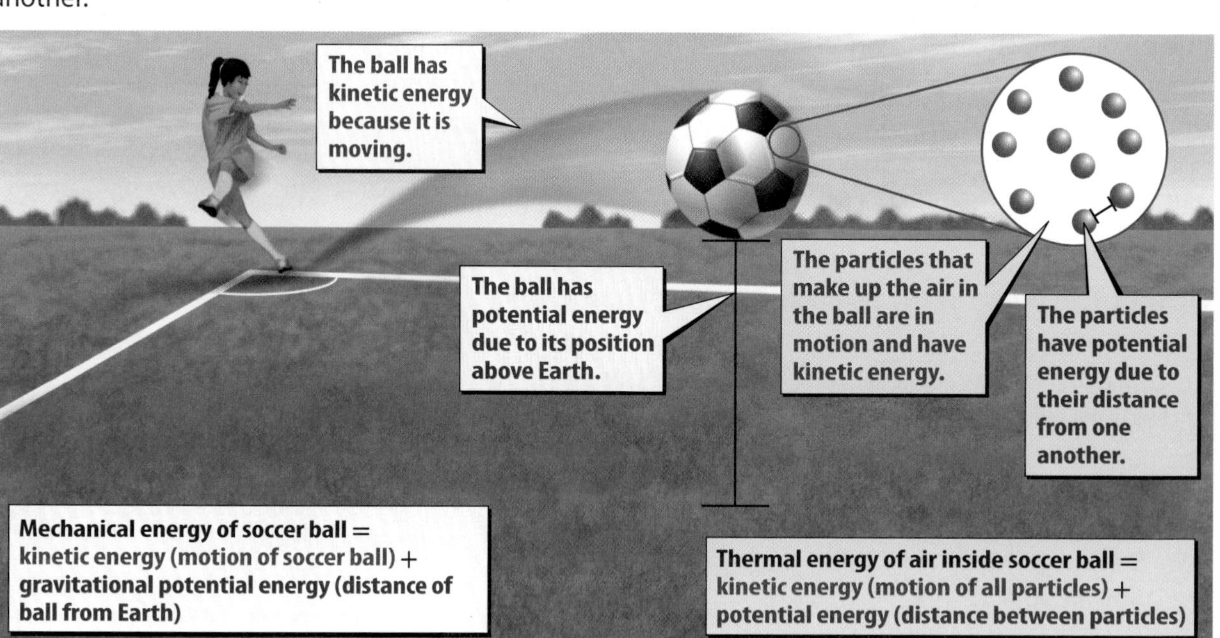

Figure 3 The air's temperature depends on how fast the particles in the air move.

Visual Check What happens to the motion of the particles in the air as temperature increases?

Concepts in Motion **Animation**

What is temperature?

When you think of temperature, you probably think of it as a measurement of how warm or cold something is. However, scientists define temperature in terms of kinetic energy.

Average Kinetic Energy and Temperature

The particles that make up the air inside and outside the house in **Figure 3** are moving. However, they are not moving at the same speed. The particles in the air in the warm house move faster and have more kinetic energy than those outside on a cold winter evening. **Temperature** *represents the average kinetic energy of the particles that make up a material.*

WORD ORIGIN
temperature
from Latin *temperatura*, means "moderating, tempering"

The greater the average kinetic energy of particles, the greater the temperature. The temperature of the air inside the house is higher than the temperature of the air outside the house. This is because the particles that make up the air inside the house have greater average kinetic energy than those outside. In other words, the particles of air inside the house are moving at a greater average speed than those outside.

Key Concept Check How are temperature and kinetic energy related?

Thermal Energy and Temperature

Temperature and thermal energy are related, but they are not the same. For example, as a frozen pond melts, both ice and water are present and they have the same temperature. Therefore, the particles that make up the ice and the water have the same average kinetic energy, or speed. However, the particles do not have the same thermal energy. This is because the average distance of the particles that make up liquid water and ice are different. The particles that make up the liquid and the solid water have different potential energies and, therefore, different thermal energies.

FOLDABLES®

Make a vertical three-column chart book. Label it as shown. Use it to organize your notes on the properties of heat, temperature, and thermal energy.

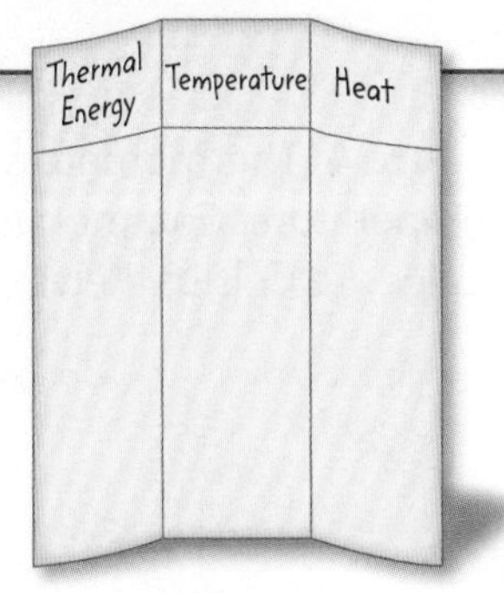

Math Skills 

Convert Between Temperature Scales

To convert Fahrenheit to Celsius, use the following equation:

$$°C = \frac{(°F - 32)}{1.8}$$

For example, to convert **176°F** to Celsius:

1. Always perform the operation in parentheses first.

$$176 - 32 = 144$$

2. Divide the answer from Step 1 by 1.8.

$$\frac{144}{1.8} = \mathbf{80°C}$$

To convert Celsius to Fahrenheit, follow the same steps using the following equation:

$$°F = (°C \times 1.8) + 32$$

Practice

1. Convert 86°F to Celsius.
2. Convert 37°C to Fahrenheit.

- **Math Practice**
- **Personal Tutor**

Measuring Temperature

How can you measure temperature? It would be impossible to measure the kinetic energy of individual particles and then calculate their average kinetic energy to determine the temperature. Instead, you can use thermometers, such as the ones in **Figure 4**, to measure temperature.

A common type of thermometer is a bulb thermometer. A bulb thermometer is a glass tube connected to a bulb that contains a liquid such as alcohol. When the temperature of the alcohol increases, the alcohol expands and rises in the glass tube. When the temperature of the alcohol decreases, the alcohol contracts back into the bulb. The height of the alcohol in the tube indicates the temperature.

There are other types of thermometers too, such as an electronic thermometer. This thermometer measures changes in the resistance of an electric circuit and converts this measurement to a temperature.

Temperature Scales

You might have seen the temperature in a weather report given in degrees Fahrenheit and degrees Celsius. On the Fahrenheit scale, water freezes at 32° and boils at 212°. On the Celsius scale, water freezes at 0° and boils at 100°. The Celsius scale is used by scientists worldwide.

Scientists also use the Kelvin scale. On the Kelvin scale, water freezes at 273 K and boils at 373 K. The lowest possible temperature for any material is 0 K. This is known as absolute zero. If a material were at 0 K, the particles in that material would not be moving and would no longer have kinetic energy. Scientists have not been able to cool any material to 0 K.

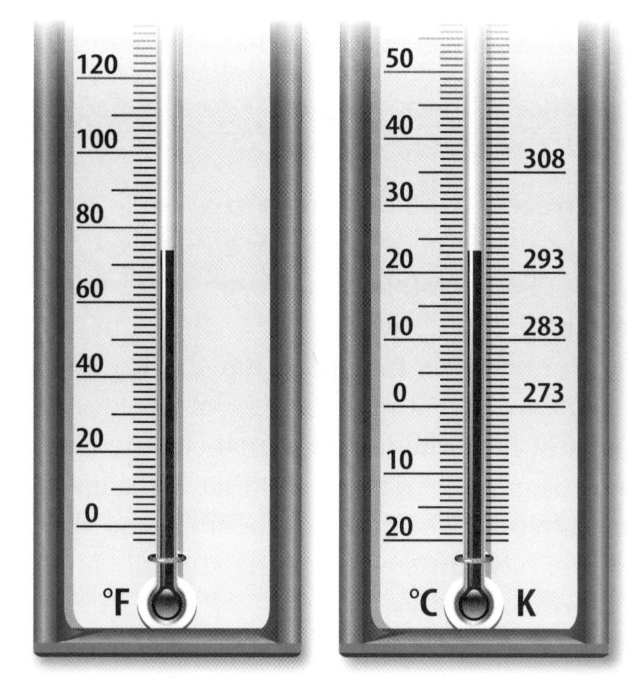

Figure 4 Thermometers measure temperature. Common temperature scales are Celsius, Kelvin, and Fahrenheit.

Figure 5 The hot cocoa heats the air and the girl's hands.

What is heat?

Have you ever held a cup of hot cocoa on a cold day like the girl in **Figure 5**? When you do, thermal energy moves from the warm cup to your hands. *The movement of thermal energy from a warmer object to a cooler object is called* **heat.** Another way to say this is that thermal energy from the cup heats your hands, or the cup is heating your hands.

Just as temperature and thermal energy are not the same thing, neither are heat and thermal energy. All objects have thermal energy. However, you heat something when thermal energy transfers from one object to another. The girl in **Figure 5** heats her hands because thermal energy transfers from the hot cocoa to her hands.

Key Concept Check How do heat and thermal energy differ?

The rate at which heating occurs depends on the difference in temperatures between the two objects. The difference in temperatures between the hot cocoa and the air is greater than the difference in temperatures between the hot cocoa and the cup. The hot cocoa heats the air more than it heats the cup. Heating continues until all objects in contact are the same temperature.

Inquiry MiniLab — 10 minutes

How do temperature scales compare?

If someone told you it was 2°C or 300 K outside, would you know whether it was warm or cold?

	Celsius (°C)	Fahrenheit (°F)	Kelvin (K)
Room temperature			
Light jacket weather			
Hot summer day			

1. Copy the table into your Science Journal.
2. Lay a **ruler** across **Figure 4** so that it lines up with the temperatures at which water freezes. Record the temperatures
3. Repeat step 2 for the three values in the table.

Analyze and Conclude

1. **Estimate** Imagine that it is snowing outside. What might the temperature be in degrees Celsius? In kelvin?
2. **Key Concept** Why doesn't the Kelvin scale include negative numbers?

Lesson 1 Review

Visual Summary

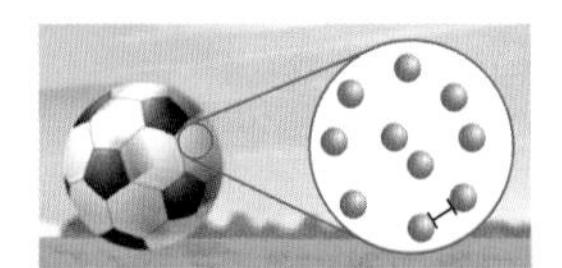

The greater the distance between two particles or two objects, the greater the potential energy.

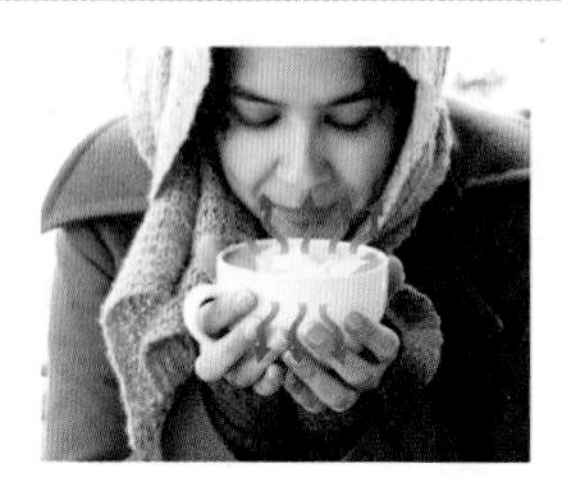

Heat is the movement of thermal energy from a warmer object to a cooler object.

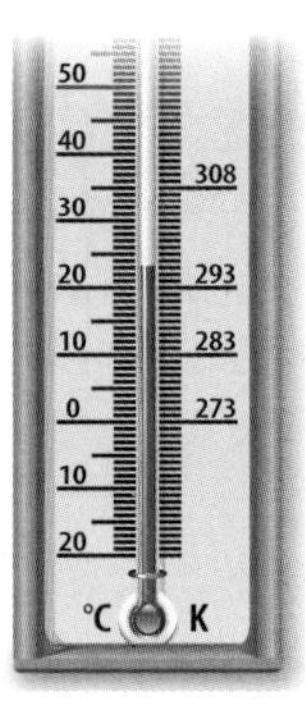

When thermal energy moves between a material and its environment, the material's temperature changes.

FOLDABLES®

Use your lesson Foldable to review the lesson. Save your Foldable for the project at the end of the chapter.

What do you think NOW?

You first read the statements below at the beginning of the chapter.

1. Temperature is the same as thermal energy.
2. Heat is the movement of thermal energy from a hotter object to a cooler object.

Did you change your mind about whether you agree or disagree with the statements? Rewrite any false statements to make them true.

Use Vocabulary

1. The sum of kinetic energy and potential energy of the particles in a material is ________.
2. **Relate** temperature to the average kinetic energy in a material.

Understand Key Concepts

3. **Differentiate** between thermal energy and heat.
4. Which increases the kinetic energy of the particles that make up a bowl of soup?
   A. dividing the soup in half
   B. putting the soup in a refrigerator
   C. heating the soup for 1 min on a stove
   D. decreasing the distance between the particles that make up the soup
5. **Infer** Suppose a friend tells you he has a temperature of 38°C. Your temperature 37°C. Do the particles that make up your body or your friend's body have a greater average kinetic energy? Explain.

Interpret Graphics

6. **Identify** Copy and fill in the following graphic organizer to show the forms of energy that make up thermal energy.

Critical Thinking

7. **Explain** How could you increase the kinetic thermal energy of a liquid?

Math Skills

Math Practice

8. Maple sap boils at 104°C. At what Fahrenheit temperature does the sap boil?

How do different materials affect thermal energy transfer?

Materials

cardboard

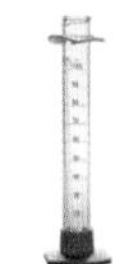

100-mL graduated cylinder

2 thermometers

Also Needed
1-L square plastic container, test containers (metal, polystyrene, ceramic, glass, plastic), large rubber band, hot water

Safety

You might have noticed that thermal energy moves more easily through some substances than others. For example, juice stays colder in a foam cup than in a can. How does the container's material affect how quickly thermal energy moves through it?

Learn It

To **form a hypothesis** is to propose an explanation for an observation. The explanation should be testable. One way to **test a hypothesis** is by gathering data that shows whether the hypothesis is correct.

Try It

1. Read and complete a lab safety form.
2. Observe the test containers. Write a hypothesis in your Science Journal that explains why you think a certain material will slow the transfer of thermal energy more than others.
3. Copy the table below.
4. Each lab group will test one container. Stand your test container in the center of a 1-L plastic container.
5. Add 125 mL of hot water to the test container. Measure and record the water's temperature.
6. Add room-temperature water to the plastic container until the level in both containers is equal. Measure and record the room-temperature water's temperature.
7. Place a cardboard square over the test container. Use two thermometers to take the temperature of the water in both containers every 2 min for 20 min. Record your data in your table.
8. Compare your data with the data gathered by the other teams. Rank the test containers from slowest to fastest thermal energy transfer in your Science Journal.

Apply It

9. **Analyze Data** Did your data support your hypothesis? Why or why not?
10. **Key Concept** What happened to the thermal energy of the water in the test container? Why did this happen?

°C	0 min	2 min	4 min	6 min	8 min	10 min	12 min	14 min	16 min	18 min	20 min
Temp in test container											
Temp in outer container											

Lesson 2

Reading Guide

Key Concepts
ESSENTIAL QUESTIONS

- What is the effect of having a small specific heat?
- What happens to a material when it is heated?
- In what ways can thermal energy be transferred?

Vocabulary

radiation p. 173
conduction p. 174
thermal conductor p. 174
thermal insulator p. 174
specific heat p. 175
thermal contraction p. 176
thermal expansion p. 176
convection p. 178
convection current p. 179

Science Video

Thermal Energy Transfers

Inquiry Keeping Warm?

Imagine camping in the mountains on a cold winter night. Your survival could depend on keeping warm. There are many things you could do to get warm and stay warm. In this picture, how is thermal energy transferred from the fire to the camper? Why does his coat keep him from losing thermal energy?

Launch Lab

15 minutes

How hot is it?

When you touch an ice cube, you sense that it is cold. When you get inside a car on a warm day, you sense that it is hot. How accurate is your sense of touch in predicting temperature?

1. Read and complete a lab safety form.
2. Place the palm of one hand flat against a piece of **metal** and the other hand against a piece of **wood.** Observe which material feels colder, and record it in your Science Journal.
3. Repeat step 2 with other materials, including **cardboard, glass, plastic,** and **foam.**
4. Rank the materials from coldest to warmest in your Science Journal.
5. Place a **liquid crystal thermometer** on each material. Record the temperature of each material in your Science Journal.

Think About This

1. Were you able to accurately rank the materials by temperature only by touching them?
2.  **Key Concept** Why might some of the materials in this experiment feel cooler than others even though they are in the same room?

How is thermal energy transferred?

Have you ever gotten into a car, such as the one in **Figure 6,** on a hot summer day? You can guess that the inside of the car is hot even before you touch the door handle. You open the door and hot air seems to pour out of the car. When you touch the metal safety-belt buckle, it is hot. How is thermal energy transferred between objects? Thermal energy is transferred in three ways—by radiation, conduction, and convection.

SCIENCE USE v. COMMON USE

vacuum

Science Use a space that contains little or no matter

Common Use a device for cleaning carpets and rugs that uses suction

Radiation

The transfer of thermal energy from one material to another by electromagnetic waves is called **radiation.** All matter, including the Sun, fire, you, and even ice, transfers thermal energy by radiation. Warm objects emit more radiation than cold objects do. For example, when you place your hands near a fire, you can more easily feel the transfer of thermal energy by radiation than when you place your hands near a block of ice.

Thermal energy from the Sun heats the inside of the car in **Figure 6** by radiation. In fact, radiation is the only way thermal energy can travel from the Sun to Earth. This is because space is a **vacuum.** However, radiation also transfers thermal energy through solids, liquids, and gases.

 **Reading Check** How does the Sun heat the inside of a car?

Figure 6 The Sun heats this car by radiation.

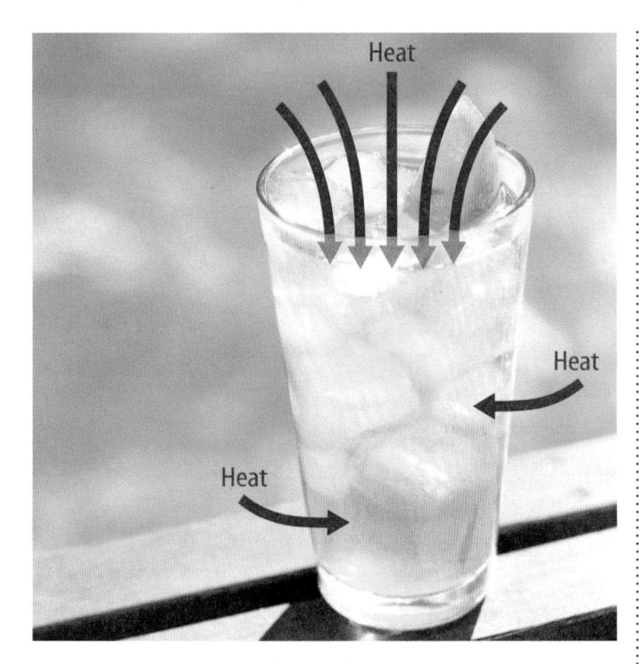

Figure 7 The hot air transfers thermal energy to, or heats, the cool lemonade by conduction. Eventually the kinetic thermal energy and temperature of the air and the lemonade will be equal.

Animation

Conduction

Suppose it's a hot day and you have a cold glass of lemonade, such as the one in **Figure 7.** The lemonade has a lower temperature than the surrounding air. Therefore, the particles that make up the lemonade have less kinetic energy than the particles that make up the air. When particles with different kinetic energies collide, the particles with higher kinetic energy transfer energy to particles with lower kinetic energy.

In **Figure 7,** the particles that make up the air collide with and transfer kinetic energy to the particles that make up the lemonade. As a result, the average kinetic energy, or temperature, of the particles that make up the lemonade increases. Since kinetic energy is being transferred, thermal energy is being transferred. *The transfer of thermal energy between materials by the collisions of particles is called* **conduction.** Conduction continues until the thermal energy of all particles in contact is equal.

Thermal Conductors and Insulators

On a hot day, why does a metal safety-belt buckle in a car feel hotter than the safety belt? Both the buckle and safety belt receive the same amount of thermal energy from the Sun. The metal that makes up the buckle is a good thermal conductor. *A* **thermal conductor** *is a material through which thermal energy flows easily.* Atoms in good thermal conductors have electrons that move easily. These electrons transfer kinetic energy when they collide with other electrons and atoms. Metals are better thermal conductors than nonmetals. The cloth that makes up a safety belt is a good thermal insulator. *A* **thermal insulator** *is a material through which thermal energy does not flow easily.* The electrons in the atoms of a good thermal insulator do not move easily. These materials do not transfer thermal energy easily because fewer collisions occur between electrons and atoms.

FOLDABLES®

Make a vertical three-column chart book. Label it as shown. Use it to describe the ways thermal energy is transferred.

Conduction | Radiation | Convection

Specific Heat

The amount of thermal energy required to increase the temperature of 1 kg of a material by 1°C is called **specific heat.** Every material has a specific heat. It does not take much energy to change the temperature of a material with a low specific heat but it can take a lot of energy to change the temperature of a material with high specific heat.

Academic Vocabulary
specific
(adjective) precise and detailed; belonging to a distinct category

Thermal conductors, such as the metal safety-belt buckle in **Figure 8,** have a lower specific heat than thermal insulators, such as the cloth safety belt. This means it takes less thermal energy to increase the buckle's temperature than it takes to increase the temperature of the cloth safety belt by the same amount.

The specific heat of water is particularly high. It takes a large amount of energy to increase the temperature of water. The high specific heat of water has many beneficial effects. For example, much of your body is water. Water's high specific heat helps prevent your body from overheating. The high specific heat of water is one of the reasons why pools, lakes, and oceans stay cool in summer. Water's high specific heat also makes it ideal for cooling machinery, such as car engines and rock-cutting saws.

Key Concept Check What does it mean if a material has a low specific heat?

Specific Heat

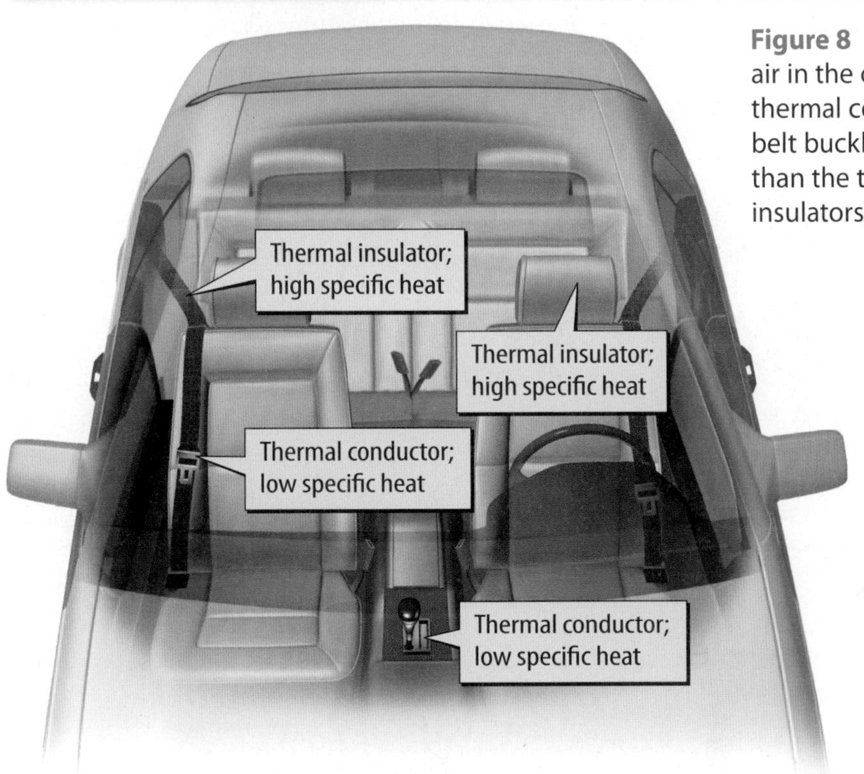

Figure 8 On a hot summer day, the air in the car is hot. The temperature of thermal conductors, such as the safety-belt buckles, increases more quickly than the temperature of thermal insulators, such as the seat material.

Thermal Expansion and Contraction

What happens if you take an inflated balloon outside on a cold day? Thermal energy transfers from the particles that make up the air inside the balloon to the particles that make up the balloon material and then to the cold outside air. As the particles that make up the air in the balloon lose thermal energy, which includes kinetic energy, they slow down and get closer together. This causes the volume of the balloon to decrease. **Thermal contraction** *is a decrease in a material's volume when its temperature decreases.*

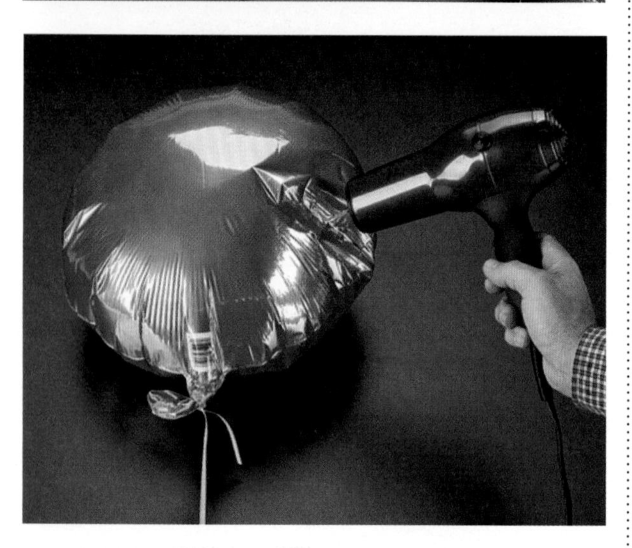

▲ **Figure 9** Air inside the balloon increases in volume when the temperature increases.

How could you reinflate the balloon? You could heat the air inside the balloon with a hair dryer, like in **Figure 9.** The particles that make up the hot air coming out of the hair dryer transfer thermal energy, which includes kinetic energy, to the particles that make up the air inside the balloon. As the average kinetic energy of the particles increases, the air temperature increases. Also, as the average kinetic energy of the particles increases, they speed up and spread out, increasing the volume of air inside the balloon. **Thermal expansion** *is an increase in a material's volume when its temperature increases.*

Thermal expansion and contraction are most noticeable in gases, less noticeable in liquids, and the least noticeable in solids.

Key Concept Check What happens to the volume of a gas when it is heated?

Sidewalk Gaps

In many places, outdoor temperatures become very hot in the summer. High temperatures can cause thermal expansion in structures, such as concrete sidewalks. If the concrete expands too much or expands unevenly, it could crack. Therefore, control joints are cut into sidewalks, as shown in **Figure 10.** If the sidewalk does crack, it should crack smoothly at the control joint.

▲ **Figure 10** Sidewalks can withstand thermal expansion and contraction because of control joints.

Hot-Air Balloons

How do hot-air balloons work? As shown in **Figure 11,** a burner heats the air in the balloon, causing thermal expansion. The particles that make up the air inside the balloon move faster and faster. As the particles collide with one another, some are forced outside the balloon through the opening at the bottom. Now, there are fewer particles in the balloon than in the same volume of air outside the balloon. The balloon is less dense, and it begins to rise through denser outside air.

To land a hot-air balloon, the balloonist allows the air inside the balloon to gradually cool. The air undergoes thermal contraction. However, the balloon itself does not contract. Instead, denser air from outside the balloon fills the space inside. As the density of the balloon increases, it slowly descends.

Figure 11 Hot-air balloonists control their balloons using thermal expansion and contraction.

Ovenproof Glass

If you put an ordinary drinking glass into a hot oven, the glass might break or shatter. However, an ovenproof glass dish would not be damaged in a hot oven. Why is this so?

Different parts of ordinary glass expand at different rates when heated. This causes it to crack or shatter. Ovenproof glass is designed to expand less than ordinary glass when heated, which means that it usually does not crack in the oven.

Inquiry MiniLab

20 minutes

How does adding thermal energy affect a wire?

How could thermal energy help you remove a metal lid from a glass jar?

1. Read and complete a lab safety form.
2. Set up **two ring stands** so that the rings are 1–2 m apart. Tie the ends of a **2-m length of wire** to the rings so that the wire is straight and tight.
3. Use **thread** to tie a **weight** to the middle of the wire.
4. Use a **ruler** to measure the distance from the bottom of the weight to the table. Record your data in your Science Journal.
5. Using **matches** light two **candles.** Move the candle flames back and forth under the wire. Repeat step 4 every minute for 5 min. Blow out the candles.
6. Repeat step 4 again every minute for 5 min as the wire cools.

Analyze and Conclude

1. **Predict** What would happen if you continued to heat the wire? Explain.
2. **Apply** How could you use this idea to help you remove a metal lid from a glass jar?
3. **Key Concept** What happens to the particles that make up the wire when the wire is heated? How do you know?

Figure 12 This cycle of cooler water sinking and forcing warmer water upward is an example of convection.

Personal Tutor

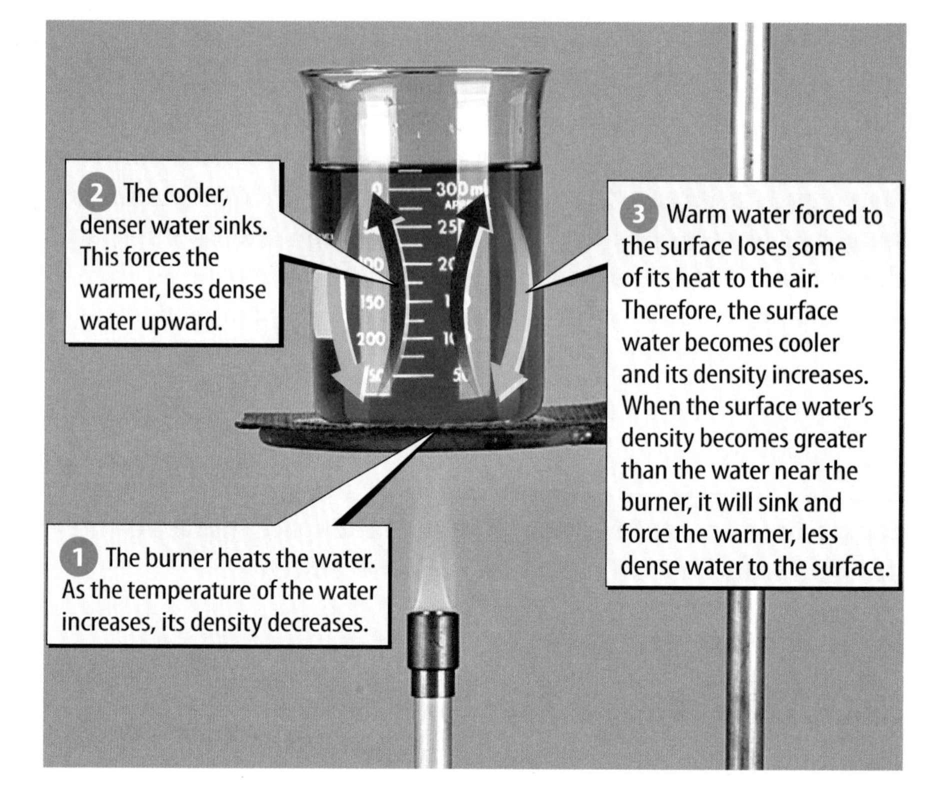

Convection

When you heat a pan of water on the stove, the burner heats the pan by conduction. This process, shown in **Figure 12,** involves the movement of thermal energy within a fluid. The particles that make up liquids and gases move around easily. As they move, they transfer thermal energy from one location to another. **Convection** *is the transfer of thermal energy by the movement of particles from one part of a material to another.* Convection only occurs in fluids, such as water, air, magma, and maple syrup.

WORD ORIGIN
convection
from Latin *convectionem,* means "the act of carrying"

 **Key Concept Check** What are the three processes that transfer thermal energy?

Density, Thermal Expansion, and Thermal Contraction

In **Figure 12,** the burner transfers thermal energy to the beaker, which transfers thermal energy to the water. Thermal expansion occurs in water nearest the bottom of the beaker. Heating increases the water's volume making it less dense.

At the same time, water molecules at the water's surface transfer thermal energy to the air. This causes cooling and thermal contraction of the water on the surface. The denser water at the surface sinks to the bottom, forcing the less dense water upward. This cycle continues until all the water in the beaker is at the same temperature.

Convection Currents in Earth's Atmosphere

The movement of fluids in a cycle because of convection is a **convection current.** Convection currents circulate the water in Earth's oceans and other bodies of water. They also circulate the air in a room, and the materials in Earth's interior. Convection currents also move matter and thermal energy from inside the Sun to its surface.

On Earth, convection currents move air between the equator and latitudes near 30°N and 30°S. This plays an important role in Earth's climates, as shown in **Figure 13.**

Convection Currents in Earth's Atmosphere

Figure 13 Convection currents in the atmosphere influence the locations of rain forests and deserts.

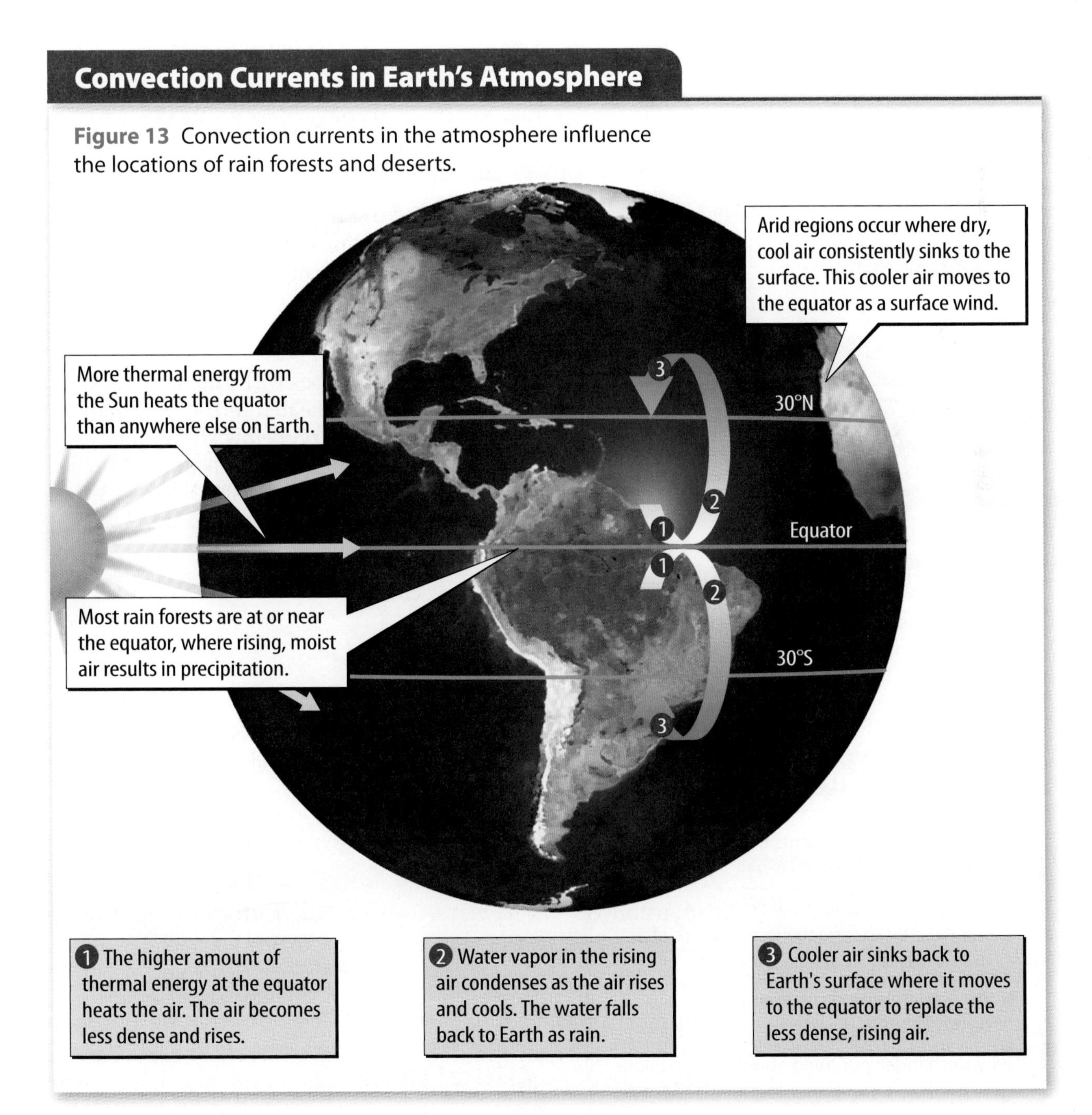

Lesson 2 Review

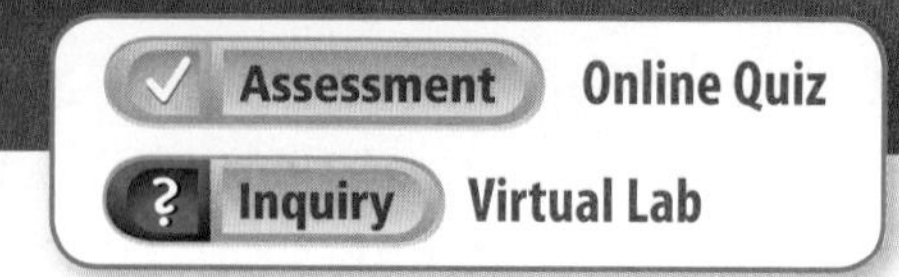

Visual Summary

When a material has a low specific heat, transferring a small amount of energy to the material increases its temperature significantly.

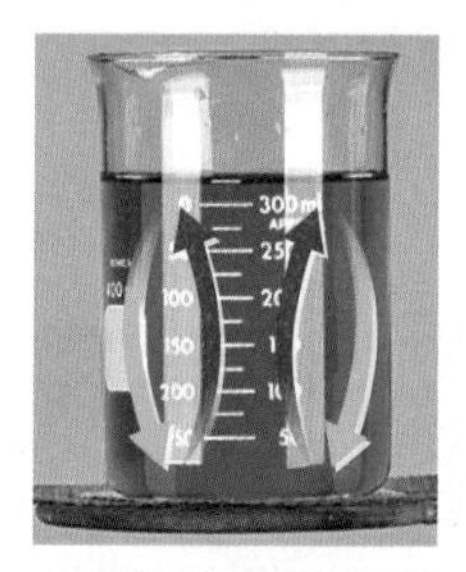

Thermal energy can be transferred through radiation, conduction, or convection.

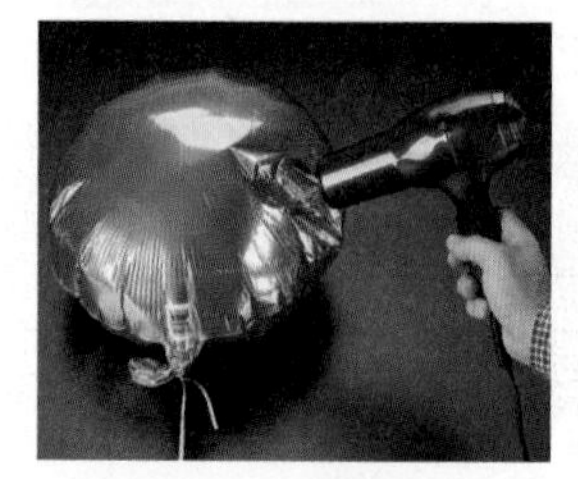

When a material is heated, the thermal energy of the material increases and the material expands.

FOLDABLES®

Use your lesson Foldable to review the lesson. Save your Foldable for the project at the end of the chapter.

What do you think NOW?

You first read the statements below at the beginning of the chapter.

3. It takes a large amount of energy to significantly change the temperature of an object with a low specific heat.

4. The thermal energy of an object can never be increased or decreased.

Did you change your mind about whether you agree or disagree with the statements? Rewrite any false statements to make them true.

Use Vocabulary

1. The transfer of thermal energy by electromagnetic waves is ________.
2. **Define** *convection* in your own words.

Understand Key Concepts

3. **Contrast** radiation with conduction.
4. Why do hot-air balloons rise?
   - **A.** thermal conduction
   - **B.** thermal convection
   - **C.** thermal expansion
   - **D.** thermal radiation
5. **Infer** why the sauce on a hot pizza burns your mouth but the crust of the pizza does not burn your mouth.

Interpret Graphics

6. **Analyze** Two cubes with the same mass and volume are heated in the same pan of water. The graph below shows the change in temperature with time. Which cube has the higher specific heat?

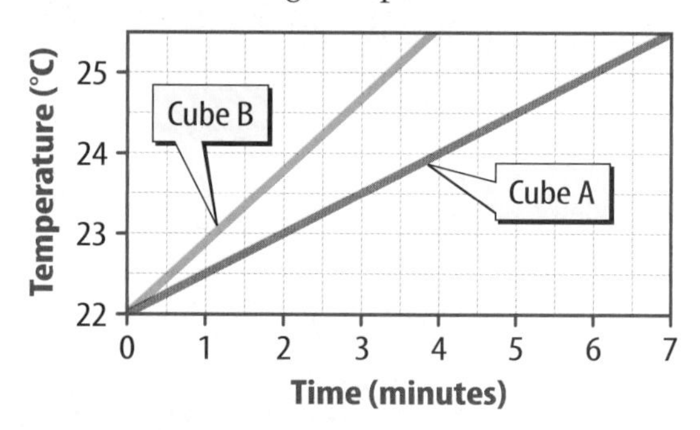

7. **Organize** Copy and fill in the graphic organizer to show how thermal energy is transferred.

Critical Thinking

8. **Explain** Why do you use a pot holder when taking hot food out of the oven?

SCIENCE & SOCIETY

Insulating the Home

It's what's between the walls that matters.

The first requirement of a shelter is to protect you from the weather. If the weather where you live is mild year-round, almost any kind of shelter will do. However, basic shelters, such as huts and tents, are not comfortable during cold winters or hot summers. Over many centuries, societies have experimented with thermal insulators that keep the inside of a shelter warm during winter and cool during summer.

One of the first thermal insulators used in shelters was air. Because air is a poor conductor of thermal energy, using cavity walls became a common form of insulating homes in the United States.

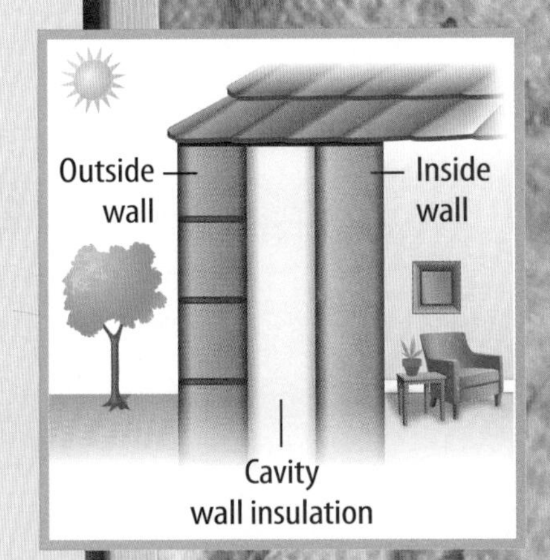

An air gap was not the perfect solution, however. Convection currents in the cavity carried some thermal energy across the gap. At first, no one seemed to mind. But, in the 1970s, the cost of heating and cooling homes suddenly increased. People began looking for a better way to reduce the transfer of thermal energy between the outside and the inside.

To meet the growing demand for better insulation, scientists began researching to find better insulation materials. If you could stop the convection currents, they reasoned, you could stop the transfer of thermal energy. One way was to use materials such as polymer foam or fiberglass. Each of these trapped air between the walls and held it there.

But how do you install insulation if your house is already built? You poke holes in the walls and blow it in! This process has little effect on your home's structure, and it decreases the cost of heating and cooling the house.

It's Your Turn

MAKE A POSTER A material's insulating ability is rated with an R-value. Find out what an R-value is. Then make a poster showing the ratings of common materials.

Lesson 3

Reading Guide

Key Concepts
ESSENTIAL QUESTIONS

- How does a thermostat work?
- How does a refrigerator keep food cold?
- What are the energy transformations in a car engine?

Vocabulary

heating appliance p. 183
thermostat p. 184
refrigerator p. 184
heat engine p. 186

Multilingual eGlossary

Using Thermal Energy

Inquiry Concentrating Energy?

This power plant uses mirrors to focus light toward a tower. The tower then transforms some of the light into thermal energy. In what ways do we use thermal energy?

15 minutes

How can you transform energy?

If you rub your hands together very quickly, do they become warm? Where does the thermal energy come from?

1. Read and complete a lab safety form.
2. Copy the table into your Science Journal.
3. Place a **thermometer strip** on the surface of a **block of wood.** Record the temperature after the thermometer stops changing color.
4. Remove the thermometer and rub the wood vigorously with **sandpaper** for 30 seconds. Quickly replace the thermometer, and record the temperature.
5. Repeat steps 3 and 4 on another part of the wood. This time, sand the wood for 60 seconds.

	Starting temp (°C)	Ending temp (°C)
30 s		
60 s		

Think About This

1. Did the temperature of the wood change? Why or why not?
2. When did the wood have the highest temperature? Explain this result.
3. **Key Concept** What energy transformations take place in this activity?

Thermal Energy Transformations

You can convert other forms of energy into thermal energy. Repeatedly stretching a rubber band makes it hot. Burning wood heats the air. A toaster gets hot when you turn it on.

You also can convert thermal energy into other forms of energy. Burning coal can generate electricity. Thermostats transform thermal energy into mechanical energy that switch heaters on and off. When you convert energy from one form to another, you can use the energy to perform useful tasks.

Remember that energy cannot be created or destroyed. Even though many devices transform energy from one form to another or transfer energy from one place to another, the total amount of energy does not change.

FOLDABLES®

Make a vertical four-tab book. Label it as shown. Use it to explain the energy transformation that occurs in each device.

Heating Appliances

A device that converts electric energy into thermal energy is a **heating appliance.** Curling irons, coffeemakers, and clothes irons are some examples of heating appliances.

Other devices, such as computers and cell phones, also become warm when you use them. This is because some electric energy always is converted to thermal energy in an electronic device. However, the thermal energy that most electronic devices generate is not used for any purpose.

Figure 14 The coil in a thermostat contains two different metals that expand at two different rates.

Thermostats

WORD ORIGIN

thermostat
from Greek *therme,* meaning "heat"; and *statos,* meaning "a standing"

You might have heard the furnace in your house or in your classroom turn on in the winter. After the room warms, the furnace turns off. *A* **thermostat** *is a device that regulates the temperature of a system.* Kitchen refrigerators, toasters, and ovens are all equipped with thermostats.

Most thermostats used in home heating systems contain a bimetallic coil. A bimetallic coil is made of two types of metal joined together and bent into a coil, as shown in **Figure 14.** The metal on the inside of the coil expands and contracts more than the metal on the outside of the coil. After the room warms, the thermal energy in the air causes the bimetallic coil to uncurl slightly. This moves a switch that turns off the furnace. As the room cools, the metal on the inside of the coil contracts more than the metal on the outside, curling the coil tighter. This moves the switch in the other direction, turning on the furnace.

 **Key Concept Check** How does the bimetallic coil in a thermostat respond to heating and cooling?

Refrigerators

A device that uses electric energy to transfer thermal energy from a cooler location to a warmer location is called a **refrigerator.** Recall that thermal energy naturally flows from a warmer area to a cooler area. The opposite might seem impossible. But, that is exactly how your refrigerator works. So, how does a refrigerator move thermal energy from its cold inside to the warm air outside? Pipes that surround the refrigerator are filled with a fluid, called a coolant, that flows through the pipes. Thermal energy from inside the refrigerator transfers to the coolant, keeping the inside of the refrigerator cold.

Vaporizing the Coolant

A coolant is a substance that evaporates at a low temperature. In a refrigerator, a coolant is pumped through pipes on the inside and the outside of the refrigerator. The coolant, which begins as a liquid, passes through an expansion valve and cools. As the cold gas flows through pipes inside the refrigerator, it absorbs thermal energy from the refrigerator compartment and vaporizes. The coolant gas becomes warmer, and the inside of the refrigerator becomes cooler.

Condensing the Coolant

The coolant flows to an electric compressor at the bottom of the refrigerator. Here, the coolant is compressed, or forced into a smaller space, which increases its thermal energy. Then, the gas is pumped through condenser coils. In the coils, the thermal energy of the gas is greater than that of the surrounding air. This causes thermal energy to flow from the coolant gas to the air behind the refrigerator. As thermal energy is removed from the gas, it condenses, or becomes liquid. Then, the liquid coolant is pumped up through the expansion valve. The cycle repeats.

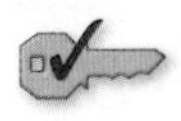
Key Concept Check How does a refrigerator keep food cold?

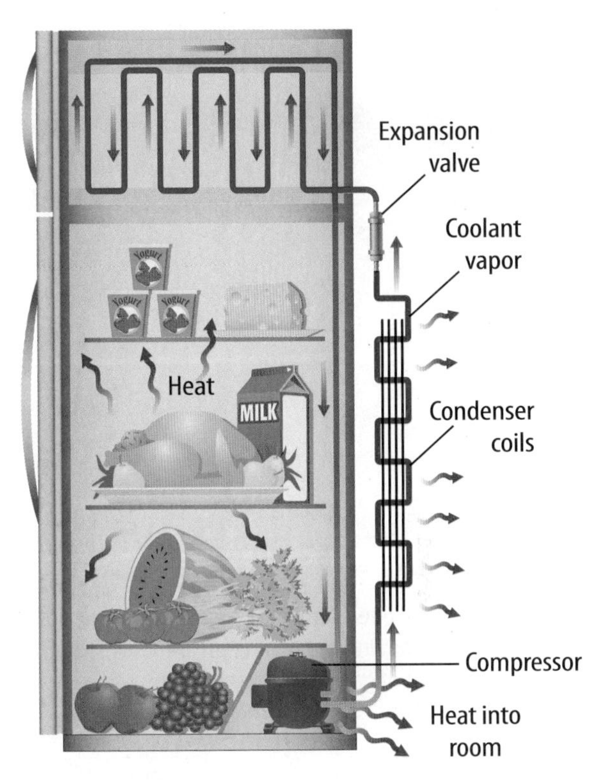

Figure 15 Coolant in a refrigerator moves thermal energy from inside to outside the refrigerator.

Inquiry MiniLab — 10 minutes

Can thermal energy be used to do work?

You know you can raise the thermal energy of a substance by doing work on it. Is the opposite true? Can thermal energy cause something to move?

1. Read and complete a lab safety form.
2. Add 10 mL of water to a **100-mL beaker.**
3. Place a **small square of aluminum foil** over the top of the beaker.
4. Place the beaker on a **hot plate,** and turn it on. Observe the results and record them in your Science Journal.

Analyze and Conclude

1. **Infer** Is thermal energy used to do work in this lab? Explain your answer.
2. **Key Concept** Is thermal energy transformed into another form of energy in this experiment? If so, what is the other form of energy?

Internal Combustion Engine

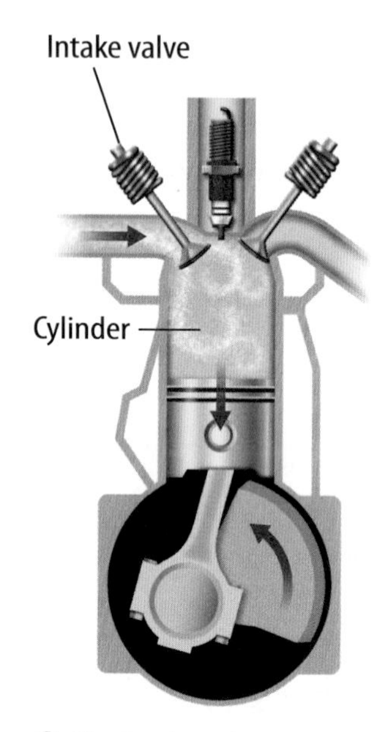

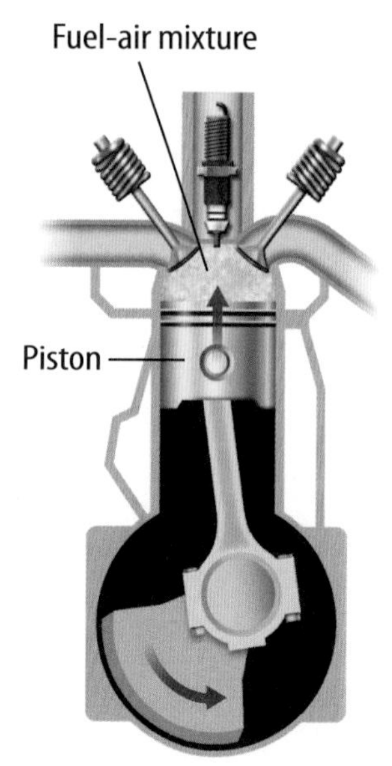

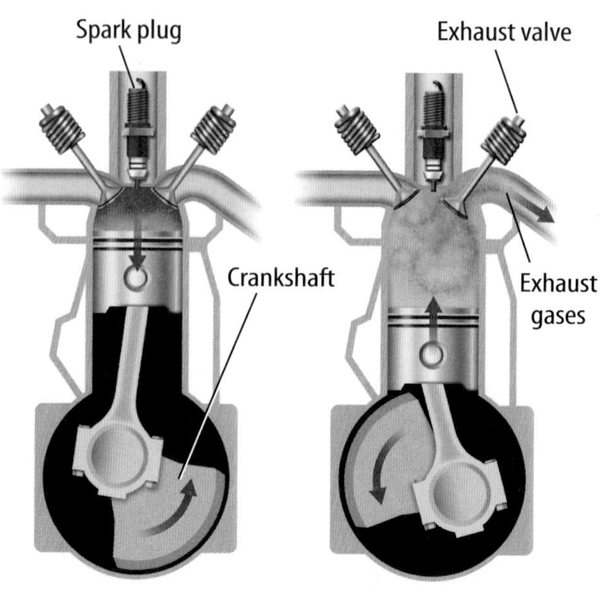

1. The intake valve opens as the piston moves downward, drawing a mixture of gasoline and air into the cylinder.
2. The intake valve closes as the piston moves upward, compressing the fuel-air mixture.
3. A spark plug ignites the fuel-air mixture. As the mixture burns, hot gases expand, pushing the piston down.
4. As the piston moves up, the exhaust valve opens, and the hot gases are pushed out of the cylinder.

Figure 16 Internal combustion engines transform the chemical energy from fuel to thermal energy, which then produces mechanical energy.

Heat Engines

A typical automobile engine is a heat engine. *A* **heat engine** *is a machine that converts thermal energy into mechanical energy.* When a heat engine converts thermal energy into mechanical energy, the mechanical energy moves the vehicle. Most cars, buses, boats, trucks, and lawn mowers use a type of heat engine called an internal combustion engine. **Figure 16** shows how one type of internal combustion engine converts thermal energy into mechanical energy.

Perhaps you have heard someone refer to a car as having a six-cylinder engine. A cylinder is a tube with a piston that moves up and down. At one end of the cylinder a spark ignites a fuel-air mixture. The ignited fuel-air mixture expands and pushes the piston down. This action occurs because the fuel's chemical energy converts to thermal energy. Some of the thermal energy immediately converts to mechanical energy.

A heat engine is not efficient. Most automobile engines only convert about 20 percent of the chemical energy in gasoline into mechanical energy. The remaining energy from the gasoline is lost to the environment.

Key Concept Check What is one form of energy that is output from a heat engine?

Lesson 3 Review

Visual Summary

A bimetallic coil inside a thermostat controls a switch that turns a heating or cooling device on or off.

A refrigerator keeps food cold by moving thermal energy from the inside of the refrigerator out to the refrigerator's surroundings.

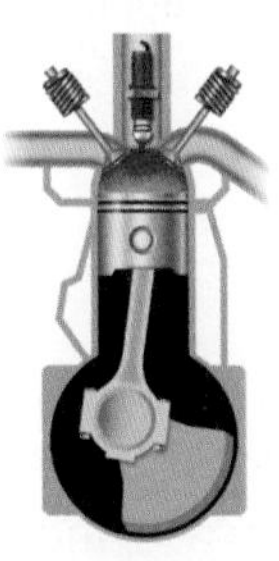

In a car engine, chemical energy in fuel is transformed into thermal energy. Some of this thermal energy is then transformed into mechanical energy.

FOLDABLES®

Use your lesson Foldable to review the lesson. Save your Foldable for the project at the end of the chapter.

What do you think NOW?

You first read the statements below at the beginning of the chapter.

5. Car engines create energy.

6. Refrigerators cool food by moving thermal energy from inside the refrigerator to the outside.

Did you change your mind about whether you agree or disagree with the statements? Rewrite any false statements to make them true.

Use Vocabulary

1. A ________ is a device that converts electric energy into thermal energy.
2. **Explain** how an internal combustion engine works.

Understand Key Concepts

3. **Describe** the path of thermal energy in a refrigerator.
4. Which sequence describes the energy transformation in an automobile engine?
   - **A.** chemical→thermal→mechanical
   - **B.** thermal→kinetic→potential
   - **C.** thermal→mechanical→potential
   - **D.** thermal→chemical→mechanical
5. **Explain** how a thermostat uses electric energy, mechanical energy, and thermal energy.

Interpret Graphics

6. **Predict** Suppose you pointed a hair dryer at the device pictured below and turned on the hair dryer. What would happen?

7. **Sequence** Copy the graphic organizer below. Use it to show the steps involved in one cycle of an internal combustion engine.

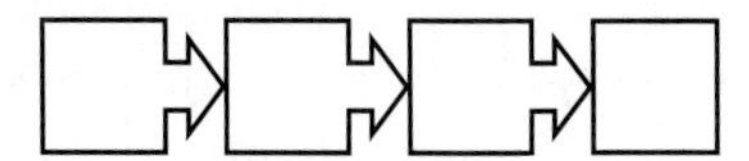

Critical Thinking

8. **Explain** how two of the devices you read about in this chapter could be used in one appliance.

3 class periods

Design an Insulated Container

Materials

aluminum foil

self-sealing plastic bag

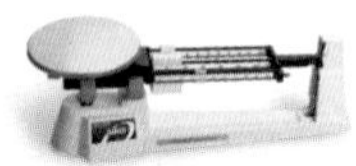

triple-beam balance

creative building materials

office supplies

Also Needed
frozen fruit pop, foam packing peanuts, rubber bands

Safety

Many refrigerated or frozen food products must be kept cold as they are transported long distances. Meat or fresh fruits might travel from South America to grocery stores in the United States. Imagine that you have been hired to design a container that will keep a frozen fruit pop from melting for as long as possible.

Ask a Question

How can you construct a container that will prevent a frozen fruit pop inside a plastic bag from melting? Think about thermal energy transfer by conduction, convection, and radiation. You will begin with a shoe box, but you can modify it in any way. Consider the materials you have available. Ask yourself what material you can bring from home that might slow the melting of a frozen fruit pop.

Make Observations

1. Read and complete a lab safety form.
2. In your Science Journal, write your ideas about
   - how you can you reduce the amount of thermal energy moving by conduction, convection, and radiation;
   - what materials you will use inside and outside your box;
   - what materials you will need to bring from home.
3. Outline the steps in preparing for your box. Have your teacher check your procedures. Decide who will obtain which materials before the next lab period. Design a logo for your container.
4. As a class, decide how many hours you will wait before checking the condition of your frozen fruit pop.

Form a Hypothesis

5. Formulate a hypothesis explaining why the materials you use inside your bag will be effective in insulating the frozen fruit pop. Remember, your hypothesis should be a testable explanation based on observations.

Test Your Hypothesis

6. On the second lab day, follow the steps you have outlined and prepare your container. Check it over one more time to be sure you have accounted for all ways that thermal energy could enter or leave the box.
7. Obtain a frozen fruit pop. Place it inside a self-sealing plastic bag. Seal the bag. Quickly measure and record its mass. Attach your logo and return the pop to the freezer.
8. On the third lab day, remove your frozen fruit pop from the freezer. Do not open the plastic bag. Place your frozen fruit pop in your container and seal it. Place your container in a location assigned by your teacher.
9. After the set amount of time, remove the fruit pop from the container. Open the plastic bag, and pour off any melted juice. Reseal the bag. Measure and record the mass.

Lab Tips

☑ Keep in mind that you are trying to keep thermal energy out of the package.

☑ The length of the test time you decide on should be long enough to allow some of the fruit pop to melt.

Analyze and Conclude

10. **Calculate** What percentage of your fruit pop remained frozen? How long do you think it would take for the fruit pop to completely melt in your container? Justify your answer.
11. **Analyze** What are some possible ways thermal energy entered your bag? How could you improve the package on another try?
12. **The Big Idea** How would you modify your design to keep something hot inside the bag? Explain your answer.

Communicate Your Results

Make a class graph showing the percentages of the different frozen fruit pops remaining. Discuss why some packages were more or less effective.

Explore designs for portable coolers. What are the most effective portable packages that keep things hot or cold without external cooling or heating?

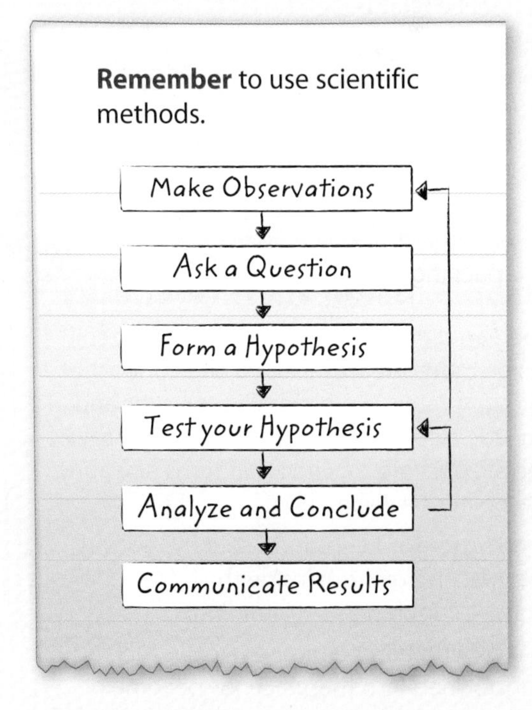

Chapter 5 Study Guide

Thermal energy can be transferred by conduction, radiation, and convection. Thermal energy also can be transformed into other forms of energy and used in devices such as thermostats, refrigerators, and automobile engines.

Key Concepts Summary

Lesson 1: Thermal Energy, Temperature, and Heat

- The **temperature** of a material is the average kinetic energy of the particles that make up the material.
- **Heat** is the movement of **thermal energy** from a material or area with a higher temperature to a material or area with a lower temperature.
- When a material is heated, the material's temperature changes.

Vocabulary

thermal energy p. 166
temperature p. 167
heat p. 169

Lesson 2: Thermal Energy Transfers

- When a material has a low **specific heat,** transferring a small amount of energy to the material increases its temperature significantly.
- When a material is heated, the thermal energy of the material increases and the material expands.
- Thermal energy can be transferred by **conduction, radiation,** or **convection.**

Vocabulary

radiation p. 173
conduction p. 174
thermal conductor p. 174
thermal insulator p. 174
specific heat p. 175
thermal contraction p. 176
thermal expansion p. 176
convection p. 178
convection current p. 179

Lesson 3: Using Thermal Energy

- The two different metals in a bimetallic coil inside a **thermostat** expand and contract at different rates. The bimetallic coil curls and uncurls, depending on the thermal energy of the air, pushing a switch that turns a heating or cooling device on or off.
- A **refrigerator** keeps food cold by moving thermal energy from inside the refrigerator out to the refrigerator's surroundings.
- In a car engine, chemical energy in fuel is transformed into thermal energy. Some of this thermal energy is then transformed into mechanical energy.

Vocabulary

heating appliance p. 183
thermostat p. 184
refrigerator p. 184
heat engine p. 186

- Personal Tutor
- Vocabulary eGames
- Vocabulary eFlashcards

FOLDABLES® Chapter Project

Assemble your lesson Foldables as shown to make a Chapter Project. Use the project to review what you have learned in this chapter.

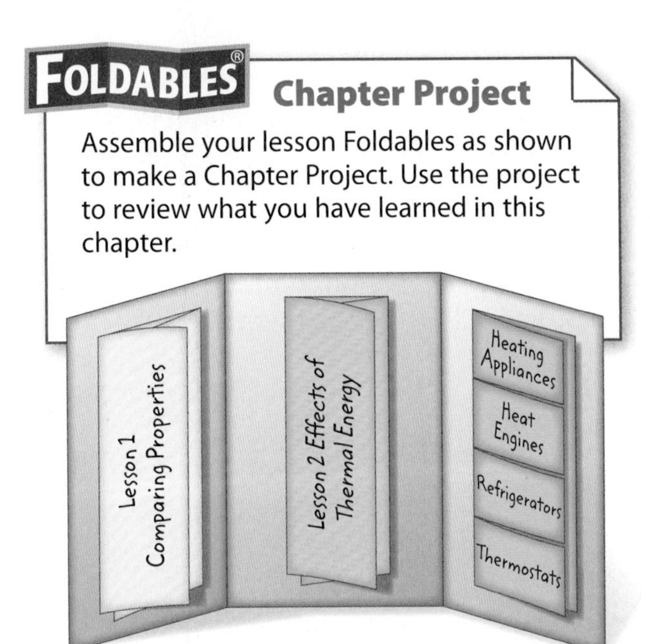

Use Vocabulary

1. When you increase the ________ of a cup of hot cocoa, you increase the average kinetic energy of the particles that make up the hot cocoa.
2. The increase in volume of a material when heated is ________ ________.
3. A(n) ________ is used to control the temperature in a room.
4. Thermal energy is transferred by ________ between two objects that are touching.
5. A fluid moving in a circular pattern because of changes in density is a ________.
6. Define *heating appliance* in your own words.

Link Vocabulary and Key Concepts

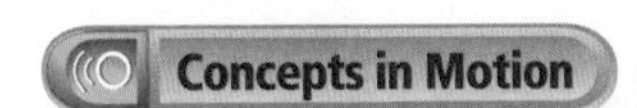

Interactive Concept Map

Copy this concept map, and then use vocabulary terms from the previous page to complete the concept map.

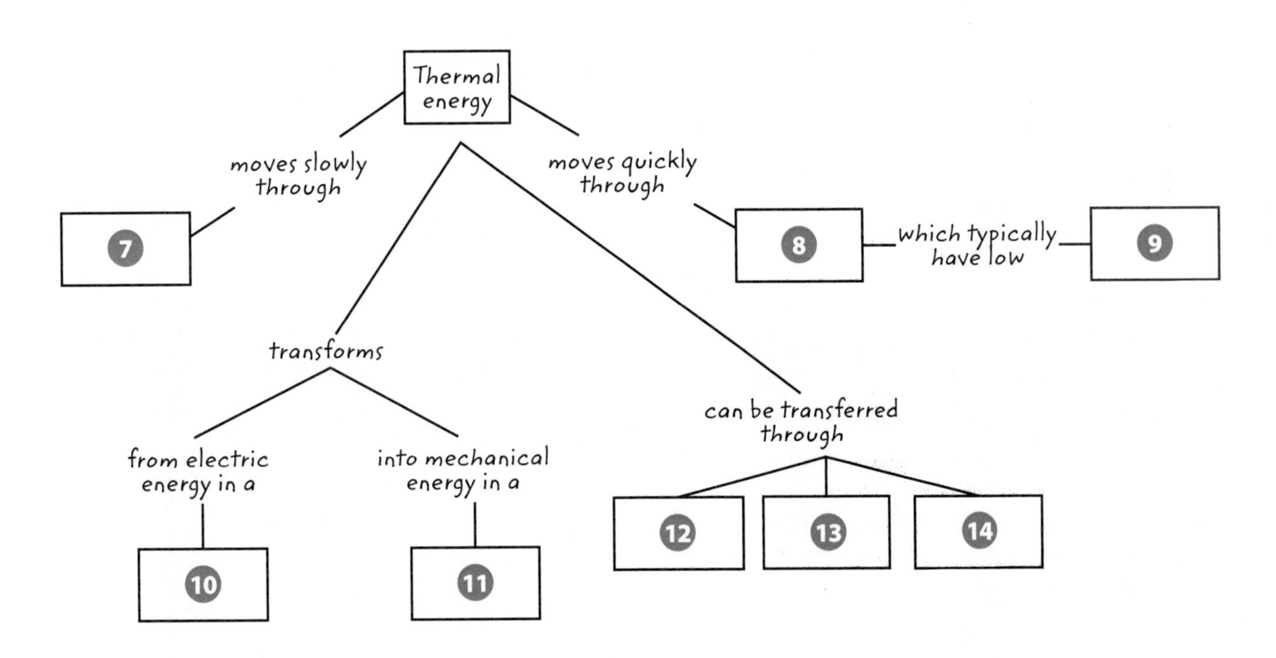

Chapter 5 Review

Understand Key Concepts 

1 Which would decrease a material's thermal energy?

A. heating the material
B. increasing the kinetic energy of the particles that make up the material
C. increasing the temperature of the material
D. moving the material to a location where the temperature is lower

2 You put a metal spoon in a bowl of hot soup. Why does the spoon feel hotter than the outside of the bowl?

A. The bowl is a better conductor than the spoon.
B. The bowl has a lower specific heat than the spoon.
C. The spoon is a good thermal insulator.
D. The spoon transfers thermal energy better than the bowl does.

3 In the picture to the right, thermal energy moves from the

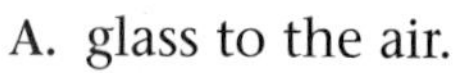

A. glass to the air.
B. lemonade to the air.
C. ice to the lemonade.
D. air to the lemonade.

4 Which has the lowest specific heat?

A. an object that is made out of metal
B. an object that does not transfer thermal energy easily
C. an object with electrons that do not move easily
D. an object that requires a lot of energy to change its temperature

5 Which does NOT occur in an internal combustion engine?

A. Most of the thermal energy is wasted.
B. Thermal energy forces the piston downward.
C. Thermal energy is converted into chemical energy.
D. Thermal energy is converted into mechanical energy.

6 Which statement about radiation is correct?

A. In solids, radiation transfers electromagnetic energy, but not thermal energy.
B. Cooler objects radiate the same amount of thermal energy as warmer objects.
C. Radiation occurs in fluids such as gas and water, but not solids such as metals.
D. Radiation transfers thermal energy from the Sun to Earth.

7 The device below detects an increase in room temperature as

A. an increase in thermal energy causes a bimetallic coil to curl.
B. an increase in thermal energy causes a bimetallic coil to uncurl.
C. a switch causes a bimetallic coil to curl.
D. a switch causes a bimetallic coil to uncurl

8 Which is the lowest temperature?

A. 0°C
B. 0°F
C. 32°F
D. 273 K

9 Which energy conversion typically occurs in a heating appliance?

A. chemical energy to thermal energy
B. electric energy to thermal energy
C. thermal energy to chemical energy
D. thermal energy to mechanical energy

Assessment
Online Test Practice

Critical Thinking

10 **Compare** A swimming pool with a temperature of 30°C has more thermal energy than a cup of soup with a temperature of 60°C. Explain why this is so.

11 **Contrast** A spoon made of aluminum and a spoon made of steel have the same mass. The aluminum spoon has a higher specific heat than the steel spoon. Which spoon becomes hotter more quickly when placed in a pan of boiling water?

12 **Describe** How do convection currents influence Earth's climate?

13 **Diagram** A room has a heater on one side and an open window letting in cool air on the opposite side. Diagram the convection current in the room. Label the warm air and the cool air.

14 **Evaluate** When engineers build bridges, they separate sections of the roadway with expansion joints such as the one below that allow movement between the sections. Why are expansion joints necessary?

15 **Explain** Why is conduction slower in a gas than in a liquid or a solid?

Writing in Science

16 **Research** various types of heat engines that have been developed throughout history. Write 3–5 paragraphs explaining the energy transformations in one of these engines.

REVIEW THE BIG IDEA

17 **Describe** each of the three ways thermal energy can be transferred. Give an example of each.

18 What do the different colors in this photograph indicate?

Math Skills

Math Practice

Convert Between Temperature Scales

19 If water in a bath is at 104°F, then what is the temperature of the water in degrees Celsius?

20 Convert −40°C to degrees Fahrenheit.

Standardized Test Practice

Record your answers on the answer sheet provided by your teacher or on a sheet of paper.

Multiple Choice

1 Which statement describes the thermal energy of an object?

A kinetic energy of particles + potential energy of particles

B kinetic energy of particles ÷ number of particles

C potential energy of particles ÷ number of particles

D kinetic energy of particles ÷ (kinetic energy of particles + potential energy of particles)

2 Which term describes a transfer of thermal energy?

A heat

B specific heat

C temperature

D thermal energy

Use the figures below to answer question 3.

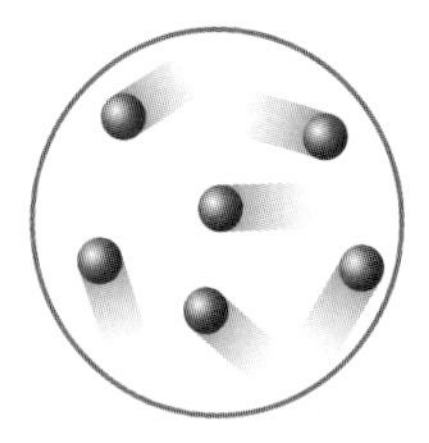

Sample X

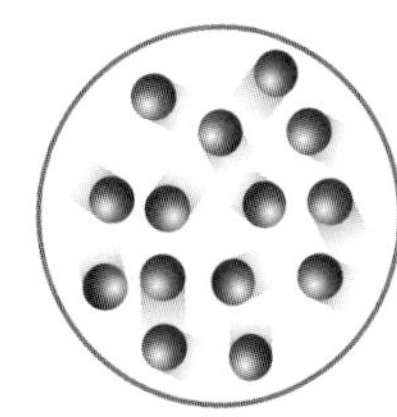

Sample Y

3 The figures show two different samples of air. In what way do they differ?

A Sample X is at a higher temperature than sample Y.

B Sample X has a higher specific heat than sample Y.

C Particles of sample Y have a higher average kinetic energy than those of sample X.

D Particles of sample Y have a higher average thermal energy than those of sample X.

Use the table below to answer question 4.

Material	Specific Heat (in J/g•K)
Air	1.0
Copper	0.4
Water	4.2
Wax	2.5

4 The table shows the specific heat of four materials. Which statement can be concluded from the information in the table?

A Copper is a thermal insulator.

B Wax is a thermal conductor.

C Air takes the most thermal energy to change its temperature.

D Water takes the most thermal energy to change its temperature.

5 Which term describes what happens to a cold balloon when placed in a hot car?

A thermal conduction

B thermal contraction

C thermal expansion

D thermal insulation

6 A girl stirs soup with a metal spoon. Which process causes her hand to get warmer?

A conduction

B convection

C insulation

D radiation

7 In a thermostat's coil, what causes the two metals in the strip to curl and uncurl?

A They contract at the same rate when cooled.

B They expand at different rates when heated.

C They have the same specific heat.

D They melt at different temperatures.

Use the figure to below to answer questions 8–10.

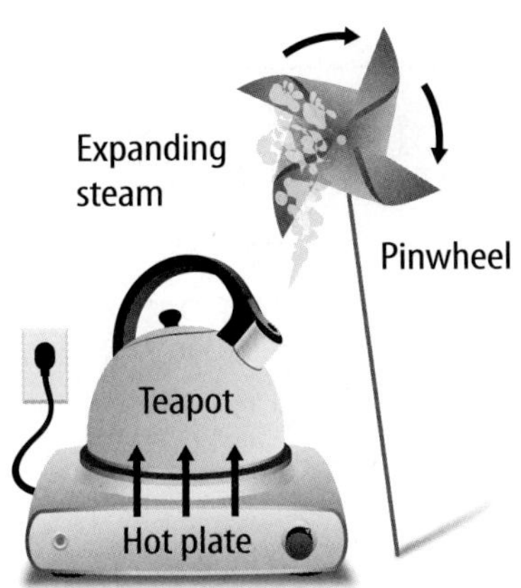

8 Which term describes the transfer of thermal energy between the hot plate and the teapot?

A conduction

B convection

C insulation

D radiation

9 Which energy transformations are taking place in this system?

A electrical → thermal → chemical

B electrical → thermal → mechanical

C thermal → electrical → chemical

D thermal → electrical → mechanical

10 What kind of machine is represented by the hot plate, the teapot, the steam, and the pinwheel working together?

A bimetallic coil

B heat engine

C refrigerator

D thermostat

Constructed Response

Use the figure to answer questions 11 and 12.

11 The foam cooler and the metal pan both contain ice. Describe the energy transfers that cause the ice to melt in each container.

12 After 1 hour, the ice in the metal pan had melted more than the ice in the foam cooler. What is it about the containers that could explain the difference in the melting rates?

13 What causes the air around a refrigerator to become warmer as the refrigerator is cooling the air inside it?

14 How does a car's internal combustion engine convert thermal energy to mechanical energy?

NEED EXTRA HELP?														
If You Missed Question...	1	2	3	4	5	6	7	8	9	10	11	12	13	14
Go to Lesson...	1	1	1	2	2	2	3	2	3	3	2	2	3	3

Chapter 6

States of Matter

THE BIG IDEA

What physical changes and energy changes occur as matter goes from one state to another?

Inquiry Liquid Glass?

When you look at this blob of molten glass, can you envision it as a beautiful vase? The solid glass was heated in a furnace until it formed a molten liquid. Air is blown through a pipe to make the glass hollow and give it form.

- Can you identify a solid, a liquid, and a gas in the photo?
- What physical changes and energy changes do you think occurred when the glass changed state?

Get Ready to Read

What do you think?

Before you read, decide if you agree or disagree with each of these statements. As you read this chapter, see if you change your mind about any of the statements.

1. Particles moving at the same speed make up all matter.
2. The particles in a solid do not move.
3. Particles of matter have both potential energy and kinetic energy.
4. When a solid melts, thermal energy is removed from the solid.
5. Changes in temperature and pressure affect gas behavior.
6. If the pressure on a gas increases, the volume of the gas also increases.

Lesson 1

Reading Guide

Key Concepts

ESSENTIAL QUESTIONS

- How do particles move in solids, liquids, and gases?
- How are the forces between particles different in solids, liquids, and gases?

Vocabulary

solid p. 201
liquid p. 202
viscosity p. 202
surface tension p. 203
gas p. 204
vapor p. 204

g Multilingual eGlossary

Solids, Liquids, and Gases

Inquiry Giant Bubbles?

Giant bubbles can be made from a solution of water, soap, and a syrupy liquid called glycerine. These liquids change the properties of water. Soap changes water's surface tension. Glycerine changes the evaporation rate. How do surface tension and evaporation work?

10 minutes

How can you see particles in matter?

It's sometimes difficult to picture how tiny objects, such as the particles that make up matter, move. However, you can use other objects to model the movement of these particles.

1. Read and complete a lab safety form.
2. Place about 50 **copper pellets** into a **plastic petri dish.** Place the cover on the dish, and secure it with **tape.**
3. Hold the dish by the edges. Gently vibrate the dish from side to side no more than 1–2 mm. Observe the pellets. Record your observations in your Science Journal.
4. Repeat step 3, vibrating the dish less than 1 cm from side to side.
5. Repeat step 3, vibrating the dish 3–4 cm from side to side.

Think About This

1. If the pellets represent particles in matter, what do you think the shaking represents?
2. In which part of the experiment do you think the pellets were like a liquid? Explain.
3. **Key Concept** If the pellets represent molecules of water, what do you think are the main differences among molecules of ice, water, and vapor?

Describing Matter

Take a closer look at the photo on the previous page. Do you see **matter?** The three most common forms, or states, of matter on Earth are solids, liquids, and gases. The giant bubble contains air, which is a mixture of gases. The ocean water and the soap mixture used to make the bubble are liquids. The sand, sign, and walkway are a few of the solids in the photo.

REVIEW VOCABULARY
matter
anything that takes up space and has mass

There is a fourth state of matter, plasma, that is not shown in this photo. Plasma is high-energy matter consisting of positively and negatively charged particles. Plasma is the most common state of matter in space. It also is in lightning flashes, fluorescent lighting, and stars, such as the Sun.

There are many ways to describe matter. You can describe the state, the color, the texture, and the odor of matter using your senses. You also can describe matter using measurements, such as mass, volume, and density. Mass is the amount of matter in an object. The units for mass are often grams (g) or kilograms (kg). Volume is the amount of space that a sample of matter occupies. The units for liquid volume are usually liters (L) or milliliters (mL). The units for solid volume are usually cubic centimeters (cm^3) or cubic meters (m^3). Density is the mass per unit volume of a substance. The units are usually g/cm^3 or g/mL. Density of a given substance remains constant, regardless of the size of the sample.

Particles in Motion

Have you ever wondered what makes something a solid, a liquid, or a gas? Two main factors that determine the state of matter are particle motion and particle forces.

Particles, such as atoms, ions, or molecules, moving in different ways make up all matter. The particles that make up some matter are close together and vibrate back and forth. In other types of matter, the particles are farther apart, move freely, and can spread out. Regardless of how close particles are to each other, they all move in random motion—movement in all directions and at different speeds. However, particles will move in straight lines until they collide with something. Collisions usually change the speed and direction of the particles' movements.

FOLDABLES®

Use a sheet of notebook paper to make a three-tab Foldable as shown. Record information about each state of matter under the tabs.

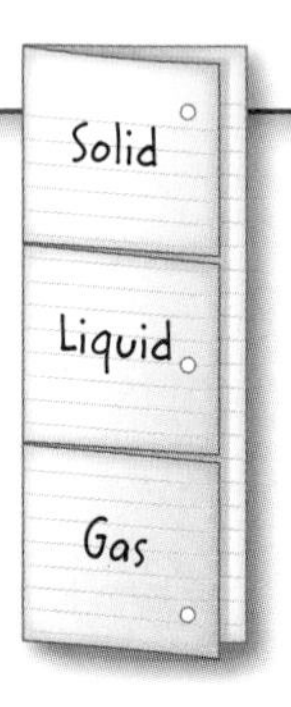

Forces Between Particles

Recall that atoms that make up matter contain positively charged protons and negatively charged electrons. There is a force of attractions between these oppositely charged particles, as shown in **Figure 1.**

You just read that the particles that make up matter move at all speeds and in all directions. If the motion of particles slows, the particles move closer together. This is because the attraction between them pulls them toward each other. Strong attractive forces hold particles close together. As the motion of particles increases, particles move farther apart. The attractive forces between particles get weaker. The spaces between them increase and the particles can slip past one another. As the motion of particles continues to increase, they move even farther apart. Eventually, the distance between particles is so great that there are little or no attractive forces between the particles. The particles move randomly and spread out. As you continue to read, you will learn how particle motion and particle forces determine whether matter is a solid, a liquid, or a gas.

Figure 1 The forces between particles of matter and the movement of particles determine the physical state of matter.

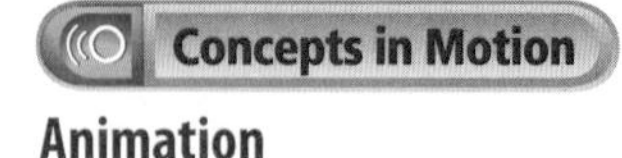

Animation

Particles move slowly and can only vibrate in place. Therefore, the attractive forces between particles are strong.

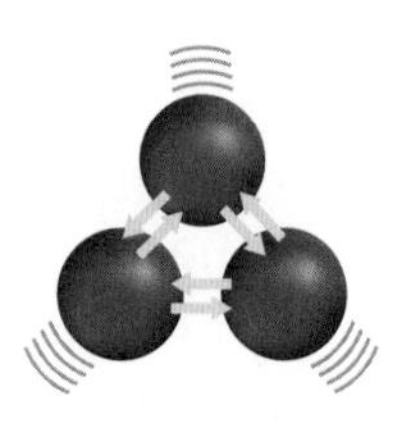

Particles move faster and slip past each other. The distance between particles increases. Therefore, the attractive forces between particles are weaker.

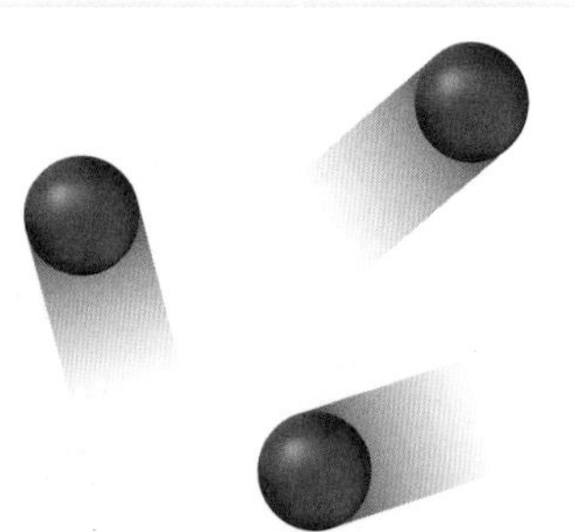

Particles move fast. The distance between the particles is great, and therefore, the attractive forces between particles are very weak.

Solids

If you had to describe a solid, what would you say? You might say, a **solid** *is matter that has a definite shape and a definite volume.* For example, if the skateboard in **Figure 2** moves from one location to another, the shape and volume of it do not change.

Particles in a Solid

Why doesn't a solid change its shape and volume? Notice in **Figure 2** how the particles in a solid are close together. The particles are very close to their neighboring particles. That's because the attractive forces between the particles are strong and hold them close together. The strong attractive forces and slow motion of the particles keep them tightly held in their positions. The particles simply vibrate back and forth in place. This arrangement gives solids a definite shape and volume.

Key Concept Check Describe the movement of particles in a solid and the forces between them.

Types of Solids

All solids are not the same. For example, a diamond and a piece of charcoal don't look alike. However, they are both solids made of only carbon atoms. A diamond and a lump of charcoal both contain particles that strongly attract each other and vibrate in place. What makes them different is the arrangement of their particles. Notice in **Figure 3** that the arrangement of particles in a diamond is different from that in charcoal. A diamond is a crystalline solid. It has particles arranged in a specific, repeating order. Charcoal is an amorphous solid. It has particles arranged randomly. Different particle arrangements give these materials different properties. For example, a diamond is a hard material, and charcoal is a brittle material.

Reading Check What is the difference between crystalline and amorphous solids?

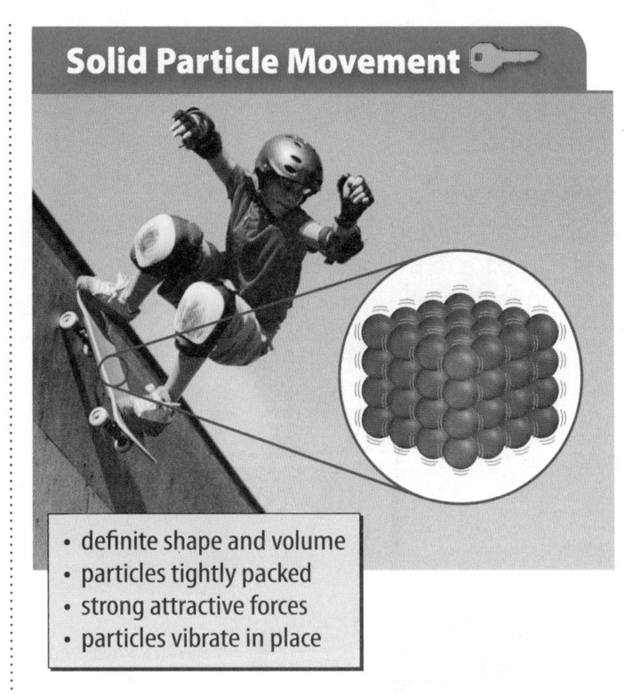

▲ **Figure 2** The particles in a solid have strong attractive forces and vibrate in place.

Figure 3 Carbon is a solid that can have different particle arrangements. ▼

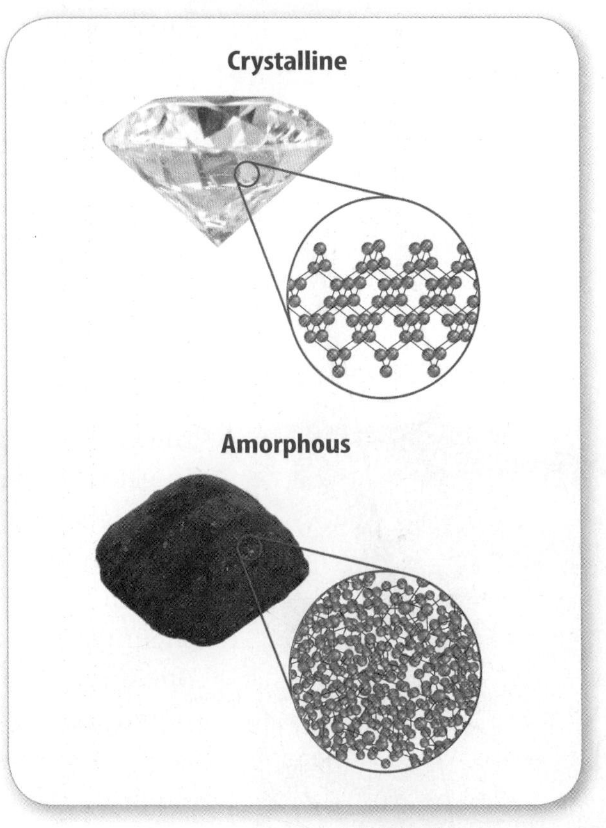

Figure 4 The motion of particles in a liquid causes the particles to move slightly farther apart. ▶

Visual Check How does the spacing among these particles compare to the particle spacing in Figure 2?

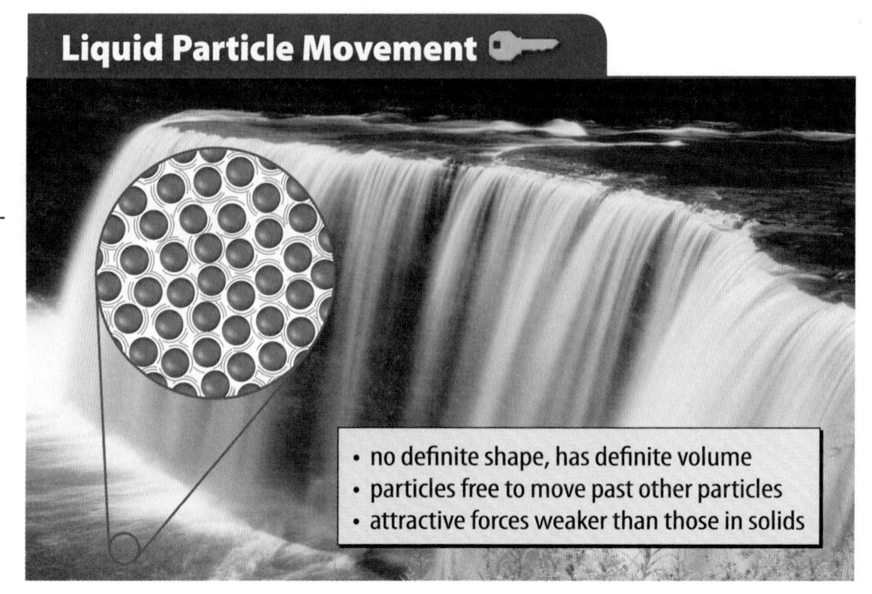

Liquids

You have probably seen a waterfall, such as the one in Figure 4. Water is a liquid. *A* **liquid** *is matter with a definite volume but no definite shape.* Liquids flow and can take the shape of their containers. The container for this water is the riverbed.

WORD ORIGIN
viscosity
from Latin *viscum*, means "sticky"

Particles in a Liquid

How can liquids change their shape? The particle motion in the liquid state of a material is faster than the particle motion in the solid state. This increased particle motion causes the particles to move slightly farther apart. As the particles move farther apart, the attractive forces between the particles decrease. The weaker attractive forces allow particles to slip past one another. The weather forces also enable liquids to flow and take the shape of their containers.

Viscosity

If you have ever poured or dipped honey, as shown in Figure 5, you have experienced a liquid with a high viscosity. **Viscosity** (vihs KAW sih tee) *is a measurement of a liquid's resistance to flow.* Honey has high viscosity, while water has low viscosity. Viscosity is due to particle mass, particle shape, and the strength of the attraction between the particles of a liquid. In general, the stronger the attractive forces between particles, the higher the viscosity. For many liquids, viscosity decreases as the liquid becomes warmer. As a liquid becomes warmer, particles begin to move faster and the attractive forces between them get weaker. This allows particles to more easily slip past one another. The mass and shape of particles that make up a liquid also affect viscosity. Large particles or particles with complex shapes tend to move more slowly and have difficulty slipping past one another.

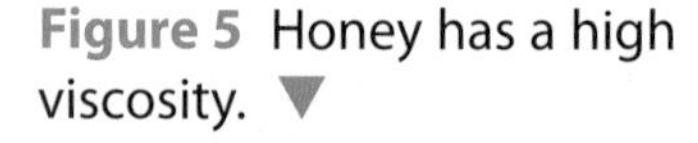

Figure 5 Honey has a high viscosity. ▼

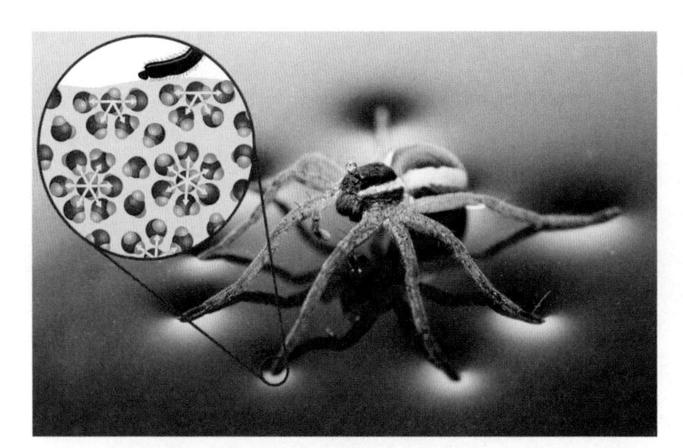

Figure 6 The surface tension of water enables this spider to walk on the surface of a lake.

Surface Tension

How can the nursery web spider in **Figure 6** walk on water? Believe it or not, it is because of the interactions between molecules.

The blowout in **Figure 6** shows the attractive forces between water molecules. Water molecules below the surface are surrounded on all sides by other water molecules. Therefore, they have attractive forces, or pulls, in all directions. The attraction between similar molecules, such as water molecules, is called cohesion.

Water molecules at the surface of a liquid do not have liquid water molecules above them. As a result, they experience a greater downward pull, and the surface particles become tightly stretched like the head of a drum. Molecules at the surface of a liquid have **surface tension,** *the uneven forces acting on the particles on the surface of a liquid.* Surface tension allows a spider to walk on water. In general, the stronger the attractive forces between particles, the greater the surface tension of the liquid.

Recall the giant bubbles at the beginning of the chapter. The thin water-soap film surrounding the bubbles forms because of surface tension between the particles.

Key Concept Check Describe the movement of particles in a liquid and the forces between them.

Inquiry MiniLab 20 minutes

How can you make bubble films?

Have you ever observed surface tension? Which liquids have greater surface tension?

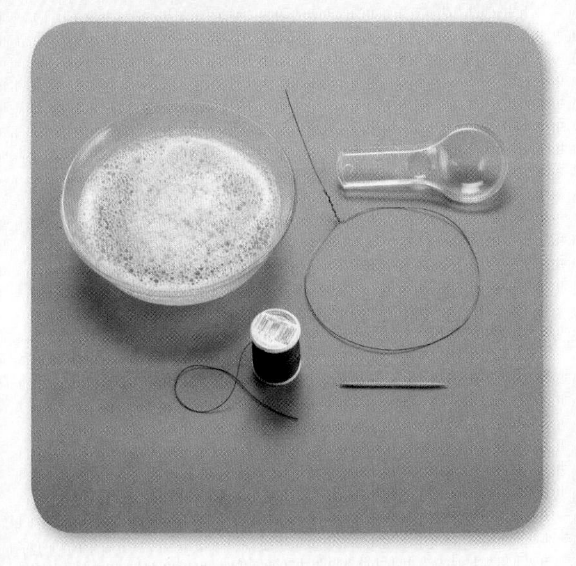

1. Read and complete a lab safety form.
2. Place about 100 mL of cool water in a **small bowl.** Lower a **wire bubble frame** into the bowl, and gently lift it. Use a **magnifying lens** to observe the edges of the frame. Write your observations in your Science Journal.
3. Add a full **dropper** of **liquid dishwashing soap** to the water. Stir with a **toothpick** until mixed. Lower the frame into the mixture and lift it out. Record your observations.
4. Use a toothpick to break the bubble film on one side of the thread. Observe.

Analyze and Conclude

1. **Recognize Cause and Effect** Explain what caused the thread to form an arc when half the bubble film broke.
2. **Key Concept** Explain why pure water doesn't form bubbles. What happens to the forces between water molecules when you add soap?

Gas Particle Movement

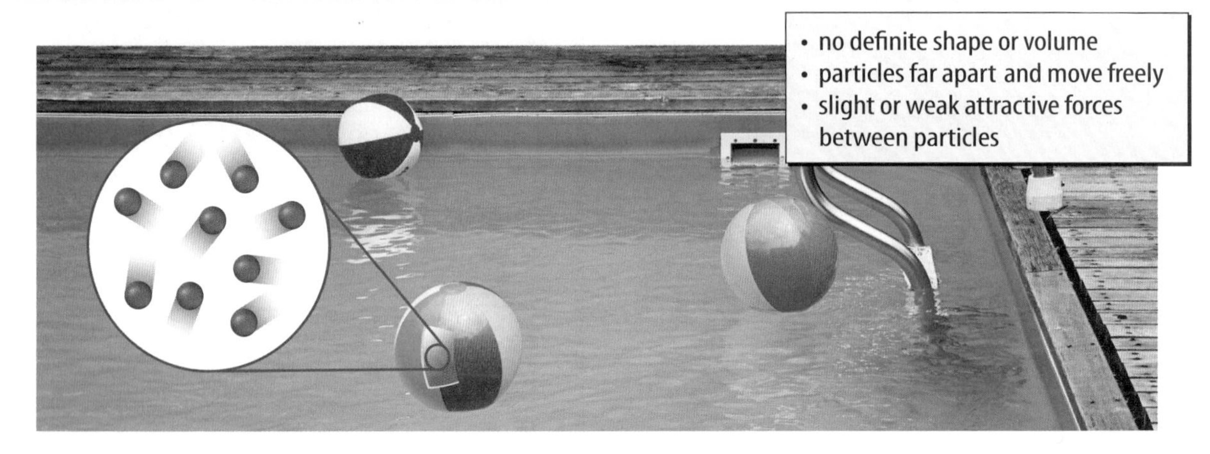

Figure 7 The particles in a gas are far apart, and there are little or no attractive forces between particles.

Visual Check What are gas particles likely to hit as they move?

Gases

Look at the photograph in **Figure 7.** Where is the gas? *A* **gas** *is matter that has no definite volume and no definite shape.* It is not easy to identify the gas because you cannot see it. However, gas particles are inside and outside the inflatable balls. Air is a mixture of gases, including nitrogen, oxygen, argon, and carbon dioxide.

Reading Check What is a gas, and what is another object that contains a gas?

Particles in a Gas

Why don't gases have definite volumes or definite shapes like solids and liquids? Compare the particles in **Figures 2, 4,** and **7.** Notice how the distance between particles differs. As the particles move faster, such as when matter goes from the solid state to the liquid state, the particles move farther apart. When the particles in matter move even faster, such as when matter goes from the liquid state to the gas state, the particles move even farther apart. When the distances between particles change, the attractive forces between the particles also change.

Forces Between Particles

As a type of matter goes from the solid state to the liquid state, the distance between the particles increases and the attractive forces between the particles decrease. When the same matter goes from the liquid state to the gas state, the particles are even farther apart and the attractive forces between the particles are weak or absent. As a result, the particles spread out to fill their container. Because gas particles lack attractive forces between particles, they have no definite shape or definite volume.

Vapor

Have you ever heard the term *vapor? The gas state of a substance that is normally a solid or a liquid at room temperature is called* **vapor.** For example, water is normally a liquid at room temperature. When it is in a gas state, such as in air, it is called water vapor. Other substances that can form a vapor are rubbing alcohol, iodine, mercury, and gasoline.

Key Concept Check How do particles move and interact in a gas?

Lesson 1 Review

Assessment Online Quiz

Visual Summary

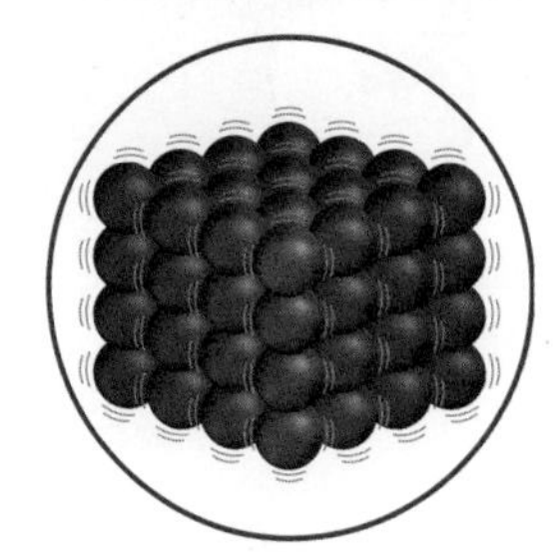

The particles that make up a solid can only vibrate in place. The particles are close together, and there are strong forces among them.

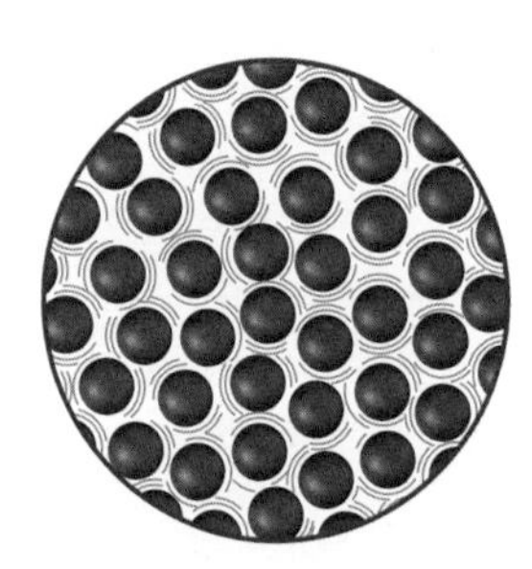

The particles that make up a liquid are far enough apart that particles can flow past other particles. The forces among these particles are weaker than those in a solid.

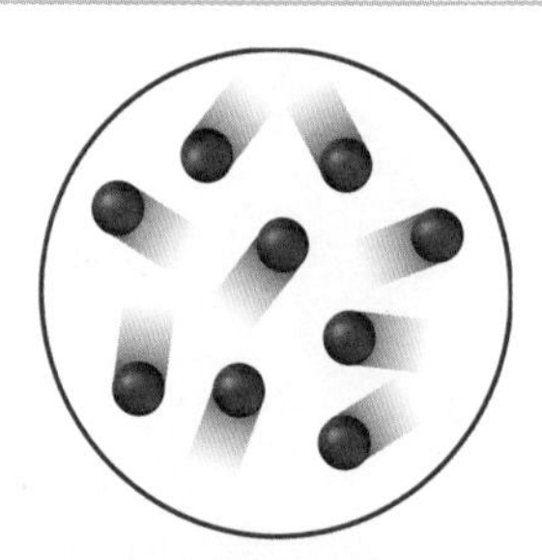

The particles that make up a gas are far apart. There is little or no attraction between the particles.

FOLDABLES®

Use your lesson Foldable to review the lesson. Save your Foldable for the project at the end of the chapter.

What do you think NOW?

You first read the statements below at the beginning of the chapter.

1. Particles moving at the same speed make up all matter.
2. The particles in a solid do not move.

Did you change your mind about whether you agree or disagree with the statements? Rewrite any false statements to make them true.

Use Vocabulary

1. A measurement of how strongly particles attract one another at the surface of a liquid is ________.
2. **Define** *solid, liquid,* and *gas* in your own words.
3. A measurement of a liquid's resistance to flow is known as ________.

Understand Key Concepts 

4. Which state of matter rarely is found on Earth?
   - **A.** gas
   - **B.** liquid
   - **C.** plasma
   - **D.** solid
5. **Compare** particle movement in solids, liquids, and gases.
6. **Compare** the forces between particles in a liquid and in a gas.

Interpret Graphics

7. **Explain** why the particles at the surface in the image below have surface tension while the particles below the surface do not.

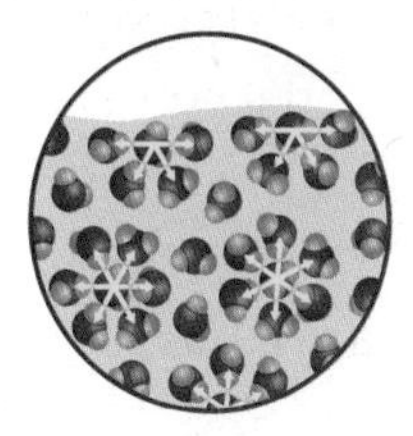

8. **Summarize** Copy and fill in the graphic organizer to compare two types of solids.

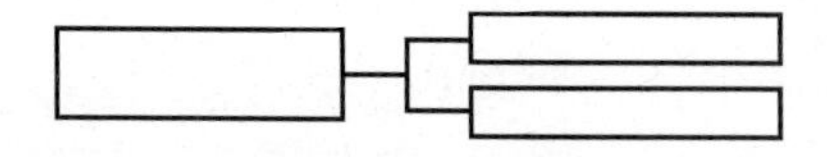

Critical Thinking

9. **Hypothesize** how you could change the viscosity of a cold liquid, and explain why your idea would work.
10. **Summarize** the relationship between the motion of particles and attractive forces between particles.

HOW IT WORKS

Freeze-Drying Foods

Have you noticed that the berries you find in some breakfast cereals are lightweight and dry—much different from the berries you get from the market or the garden?

Fresh fruit would spoil quickly if it were packaged in breakfast cereal, so fruits in cereals are often freeze-dried. When liquid is returned to the freeze-dried fruit, its physical properties more closely resemble fresh fruit. Freeze-drying, or lyophilization (lie ah fuh luh ZAY shun), is the process in which a solvent (usually water) is removed from a solid. During this process, a frozen solvent changes to a gas without going through the liquid state. Freeze-dried foods are lightweight and long-lasting. Astronauts have been using freeze-dried food during space travel since the 1960s.

How Freeze-Drying Works

1 Machines called freeze-dryers are used to freeze-dry foods and other products. Fresh or cooked food is flash-frozen, changing moisture in the food to a solid.

2 The frozen food is placed in a large vacuum chamber, where moisture is removed. Heat is applied to accelerate moisture removal. Condenser plates remove vaporized solvent from the chamber and convert the frozen food to a freeze-dried solid.

3 Freeze-dried food is sealed in oxygen- and moisture-proof packages to ensure stability and freshness. When the food is rehydrated, it returns to its near-normal state of weight, color, and texture.

It's Your Turn

PREDICT/DISCOVER What kinds of products besides food are freeze-dried? Use library or internet resources to learn about other products that undergo the freeze-drying process. Discuss the benefits or drawbacks of freeze-drying.

Lesson 2

Changes in State

Reading Guide

Key Concepts
ESSENTIAL QUESTIONS

- How is temperature related to particle motion?
- How are temperature and thermal energy different?
- What happens to thermal energy when matter changes from one state to another?

Vocabulary

kinetic energy p. 208
temperature p. 208
thermal energy p. 209
vaporization p. 211
evaporation p. 212
condensation p. 212
sublimation p. 212
deposition p. 212

Multilingual eGlossary

Video

- BrainPOP®
- What's Science Got to do With It?

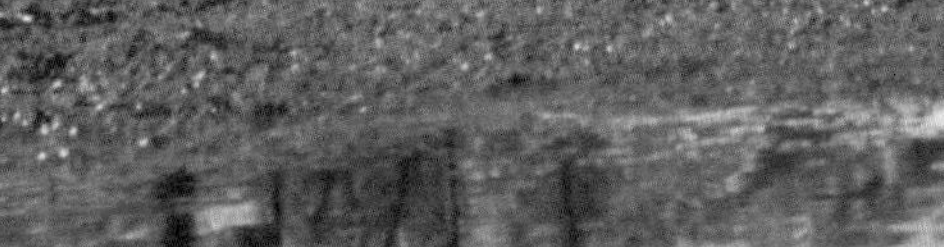

Inquiry Spring Thaw?

When you look at a snowman, you probably don't think about states of matter. However, water is one of the few substances that you frequently observe in three states of matter at Earth's temperatures. What energy changes are involved when matter changes state?

Inquiry Launch Lab **10 minutes**

Do liquid particles move?

If you look at a glass of milk sitting on a table, it appears to have no motion. But appearances can be deceiving!

1. Read and complete a lab safety form.
2. Use a **dropper,** and place one drop of **2 percent milk** on a **glass slide.** Add a **cover slip.**
3. Place the slide on a **microscope** stage, and focus on low power. Focus on a single globule of fat in the milk. Observe the motion of the globule for several minutes. Record your observations in your Science Journal.

Think About This

1. Describe the motion of the fat globule.
2. What do you think caused the motion of the globule?
3. **Key Concept** What do you think would happen to the motion of the fat globule if you warmed the milk? Explain.

Kinetic and Potential Energy

When snow begins to melt after a snowstorm, all three states of water are present. The snow is a solid, the melted snow is a liquid, and the air above the snow and ice contains water vapor, a gas. What causes particles to change state?

Kinetic Energy

Recall that the particles that make up matter are in constant motion. These particles have **kinetic energy,** *the energy an object has due to its motion.* The faster particles move, the more kinetic energy they have. Within a given substance, such as water, particles in the solid state have the least amount of kinetic energy. This is because they only vibrate in place. Particles in the liquid state move faster than particles in the solid state. Therefore, they have more kinetic energy. Particles in the gaseous state move very quickly and have the most kinetic energy of particles of a given substance.

Temperature *is a measure of the average kinetic energy of all the particles in an object.* Within a given substance, a temperature increase means that the particles, on average, are moving at greater speeds, or have a greater average kinetic energy. For example, water molecules at 25°C are generally moving faster and have more kinetic energy than water molecules at 10°C.

Key Concept Check How is temperature related to particle motion?

Potential Energy

In addition to kinetic energy, particles have potential energy. Potential energy is stored energy due to the interactions between particles or objects. For example, when you pick up a ball and then let it go, the gravitational force between the ball and Earth causes the ball to fall toward Earth. Before you let the ball go, it has potential energy.

Potential energy typically increases when objects get farther apart and decreases when they get closer together. The basketball in the top part of **Figure 8** is farther off the ground than it is in the bottom part of the figure. The farther an object is from Earth's surface, the greater the gravitational potential energy. As the ball gets closer to the ground, the potential energy decreases.

You can think of the potential energy of particles in a similar way. The chemical potential energy is due to the position of the particles relative to other particles. The chemical potential energy of particles increases and decreases as the distances between particles increase or decrease. The particles in the top part of **Figure 8** are farther apart than the particles in the bottom part. The particles that are farther apart have greater chemical potential energy.

Thermal Energy

Thermal energy *is the total potential and kinetic energies of an object.* You can change an object's state of matter by adding or removing thermal energy. When you add thermal energy to an object, the particles either move faster (increased kinetic energy) or get farther apart (increased potential energy) or both. The opposite is true when you remove thermal energy from an object. If enough thermal energy is added or removed, a change of state can occur.

Key Concept Check How do thermal energy and temperature differ?

Figure 8 The potential energy of the ball depends on the distance between the ball and Earth. The potential energy of particles in matter depends on the distances between the particles.

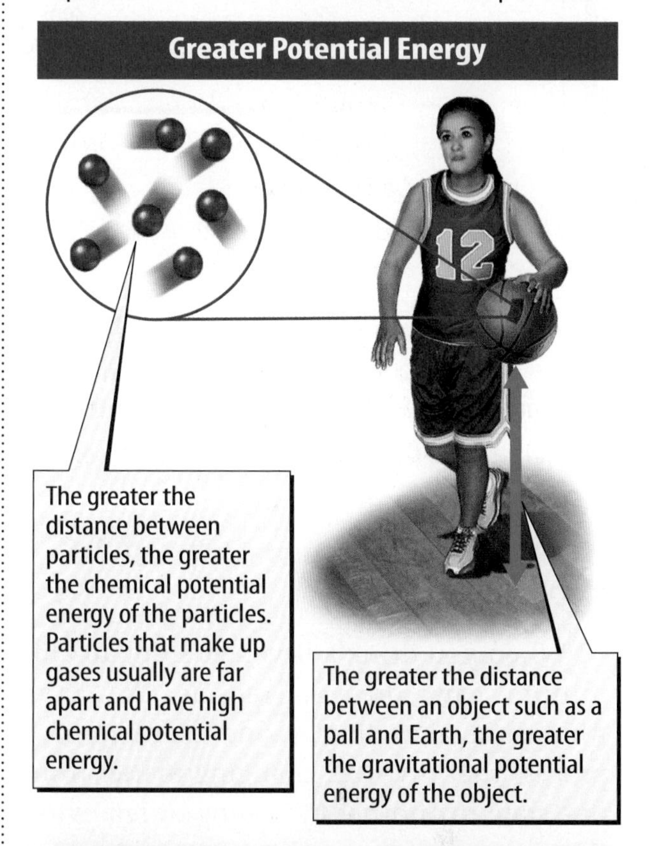

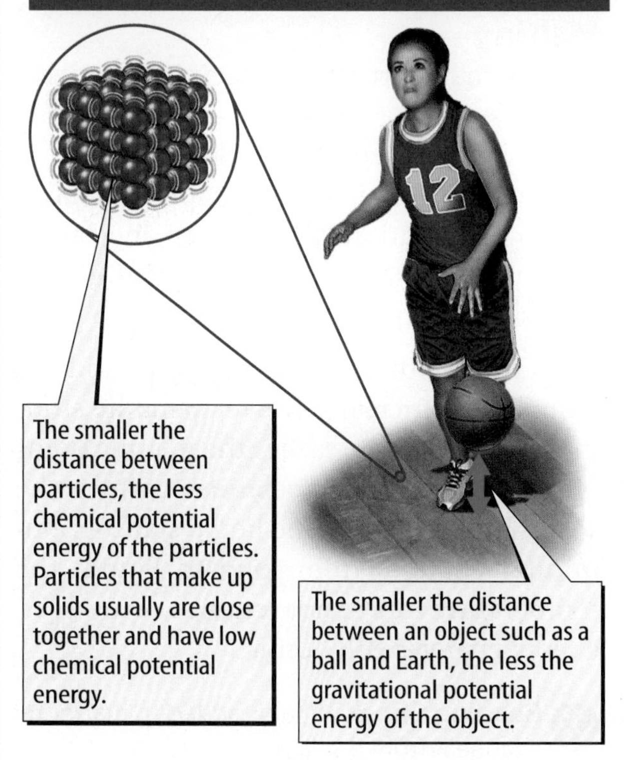

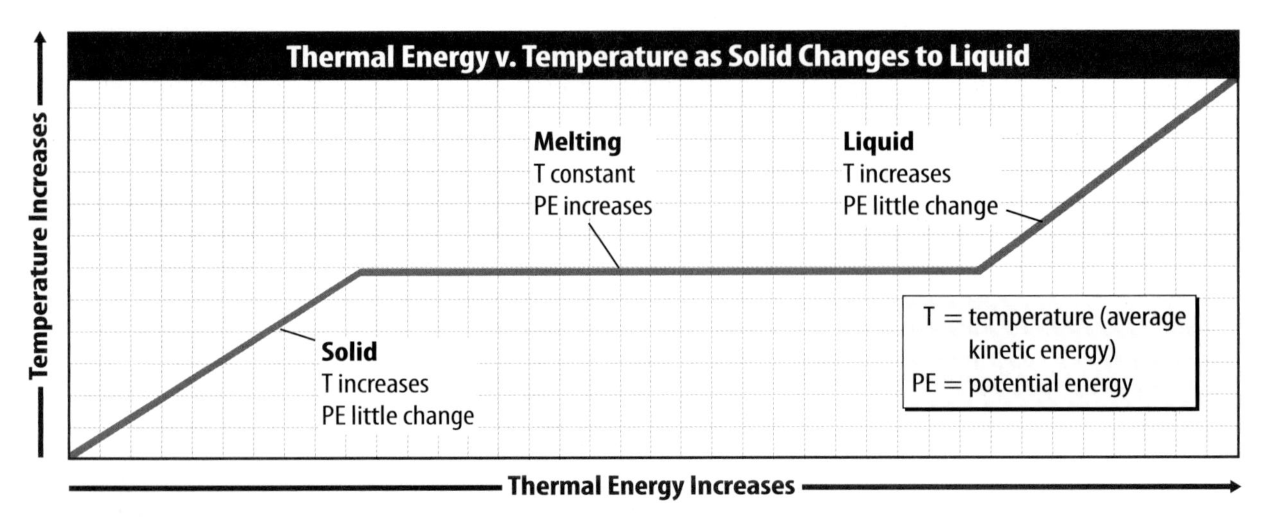

Figure 9 Adding thermal energy to matter causes the particles that make up the matter to increase in kinetic energy, potential energy, or both.

Visual Check During melting, which factor remains constant?

Solid to Liquid or Liquid to Solid

When you drink a beverage from an aluminum can, do you recycle the can? Aluminum recycling is one example of a process that involves changing matter from one state to another by adding or removing thermal energy.

Melting

The first part of the recycling process involves melting aluminum cans. To change matter from a solid to a liquid, thermal energy must be added. The graph in **Figure 9** shows the relationship between increasing temperature and increasing thermal energy (potential energy + kinetic energy).

At first, both the thermal energy and the temperature increase. The temperature stops increasing when it reaches the melting point of the matter, the temperature at which the solid state changes to the liquid state. As aluminum changes from solid to liquid, the temperature does not change. However, energy changes still occur.

Reading Check What is added to matter to change it from a solid to a liquid?

Energy Changes

What happens when a solid reaches its melting point? Notice the line on the graph is horizontal. This means that the temperature, or average kinetic energy, stops increasing. However, the amount of thermal energy continues to increase. How is this possible?

Once a solid reaches its melting point, the average speed of particles does not change, but the distance between the particles does change. The particles move farther apart and potential energy increases. Once a solid completely melts, the addition of thermal energy will cause the kinetic energy of the particles to increase again, as shown by a temperature increase.

Freezing

After the aluminum melts, it is poured into molds to cool. As the aluminum cools, thermal energy leaves it. Freezing is a process that is the reverse of melting. The temperature at which matter changes from the liquid state to the solid state is its freezing point. To observe the temperature and thermal energy changes that occur to hot aluminum blocks, move from right to left on the graph in **Figure 9.**

▲ **Figure 10** Boiling and evaporation are two kinds of vaporization.

Visual Check Why doesn't the evaporation flask have bubbles below the surface?

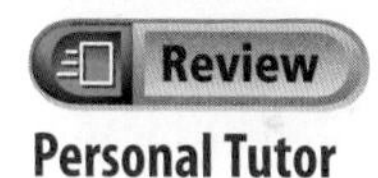

Personal Tutor

Liquid to Gas or Gas to Liquid

When you heat water, do you ever notice how bubbles begin to form at the bottom and rise to the surface? The bubbles contain water vapor, a gas. *The change in state of a liquid into a gas is* **vaporization.** **Figure 10** shows two types of vaporization—evaporation and boiling.

Boiling

Vaporization that occurs within a liquid is called boiling. The temperature at which boiling occurs in a liquid is called its boiling point. In **Figure 11,** notice the energy changes that occur during this process. The kinetic energy of particles increases until the liquid reaches its boiling point.

At the boiling point, the potential energy of particles begins increasing. The particles move farther apart until the attractive forces no longer hold them together. At this point, the liquid changes to a gas. When boiling ends, if thermal energy continues to be added, the kinetic energy of the gas particles begins to increase again. Therefore, the temperature begins to increase again as shown on the graph.

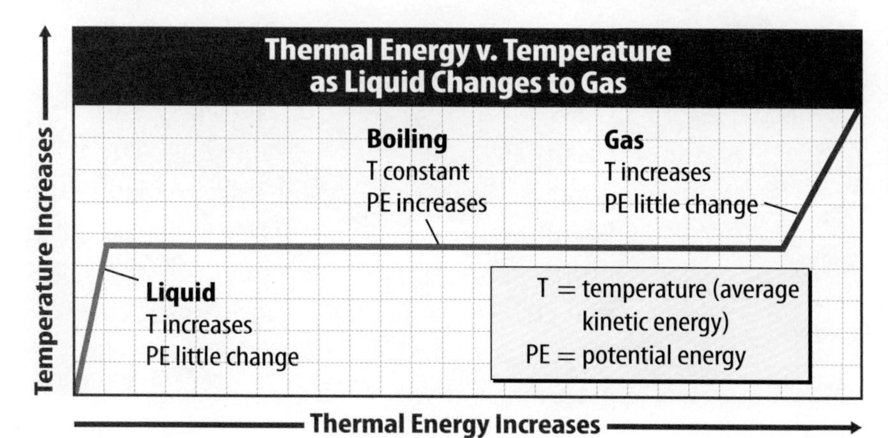

◀ **Figure 11** When thermal energy is added to a liquid, kinetic energy and potential energy changes occur.

WORD ORIGIN

evaporation
from Latin *evaporare,* means "disperse in steam or vapor"

Evaporation

Unlike boiling, **evaporation** *is vaporization that occurs only at the surface of a liquid.* Liquid in an open container will vaporize, or change to a gas, over time due to evaporation.

Condensation

Boiling and evaporation are processes that change a liquid to a gas. A reverse process also occurs. When a gas loses enough thermal energy, the gas changes to a liquid, or condenses. *The change of state from a gas to a liquid is called* **condensation.** Overnight, water vapor often condenses on blades of grass, forming dew.

Solid to Gas or Gas to Solid

Is it possible for a solid to become a gas without turning to a liquid first? Yes, in fact, dry ice does. Dry ice, as shown in **Figure 12,** is solid carbon dioxide. It turns immediately into a gas when thermal energy is added to it. The process is called sublimation. **Sublimation** *is the change of state from a solid to a gas without going through the liquid state.* As dry ice sublimes, it cools and condenses the water vapor in the surrounding air, creating a thick fog.

SCIENCE USE V. COMMON USE

deposition
Science Use the change of state of a gas to a solid without going through the liquid state

Common Use giving a legal testimony under oath

The opposite of sublimation is deposition. **Deposition** *is the change of state of a gas to a solid without going through the liquid state.* For deposition to occur, thermal energy has to be removed from the gas. You might see deposition in autumn when you wake up and there is frost on the grass. As water vapor loses thermal energy, it changes into a solid known as frost.

Reading Check Why are sublimation and deposition unusual changes of state?

Figure 12 Dry ice sublimes—goes directly from the solid state to the gas state—when thermal energy is added. Frost is an example of the opposite process—deposition.

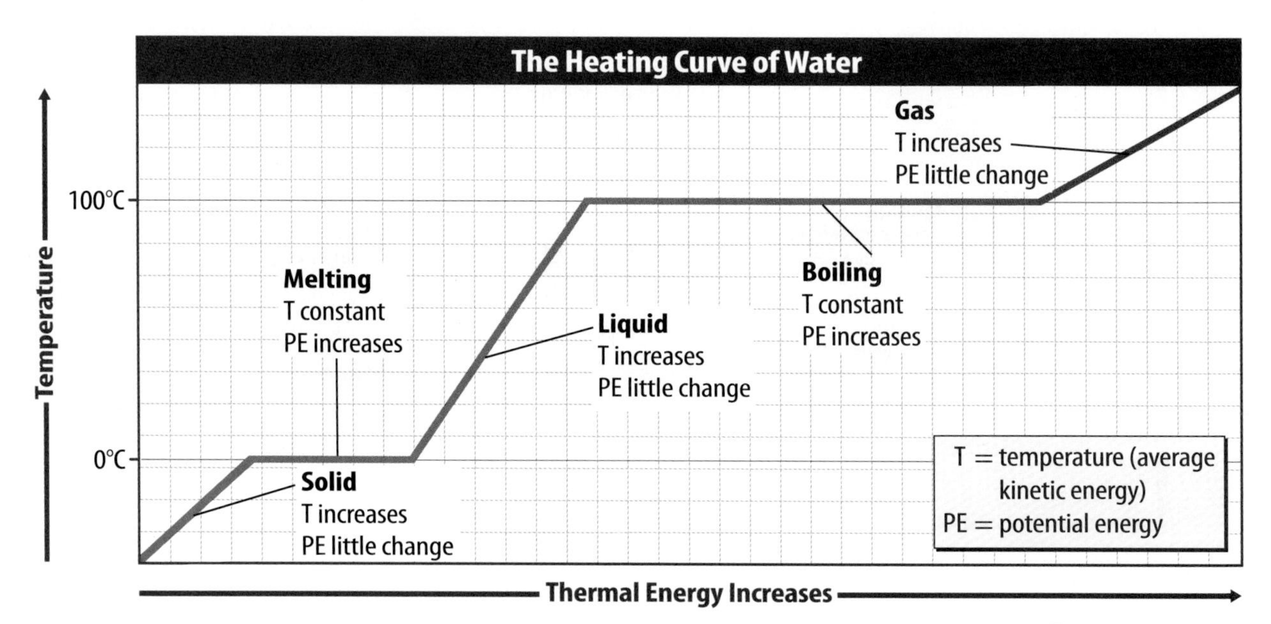

Figure 13 Water undergoes energy changes and state changes as thermal energy is added and removed.

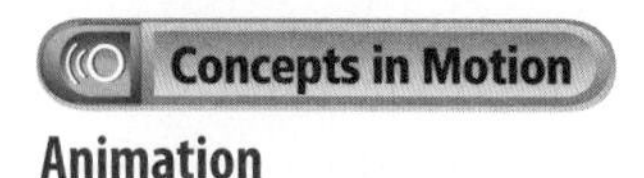

Animation

States of Water

Water is the only substance that exists naturally as a solid, a liquid, and a gas within Earth's temperature range. To better understand the energy changes during a change in state, it is helpful to study the heating curve of water, as shown in **Figure 13.**

Adding Thermal Energy

Suppose you place a beaker of ice on a hot plate. The hot plate transfers thermal energy to the beaker and then to the ice. The temperature of the ice increases. Recall that this means the average kinetic energy of the water molecules increases.

At 0°C, the melting point of water, the water molecules vibrate so rapidly that they begin to move out of their places. At this point, added thermal energy only increases the distance between particles and decreases attractive forces—melting occurs. Once melting is complete, the average kinetic energy of the particles (temperature) begins to increase again as more thermal energy is added.

When water reaches 100°C, the boiling point, liquid water begins to change to water vapor. Again, kinetic energy is constant as vaporization occurs. When the change of state is complete, the kinetic energy of molecules increases once more, and so does the temperature.

Key Concept Check Describe the changes in thermal energy as water goes from a solid to a liquid.

Removing Thermal Energy

The removal of thermal energy is the reverse of the process shown in **Figure 13.** Cooling water vapor changes the gas to a liquid. Cooling the water further changes it to ice.

FOLDABLES®

Fold a sheet of notebook paper to make a four-tab Foldable as shown. Label the tabs, define the terms, and record what you learn about each term under the tabs.

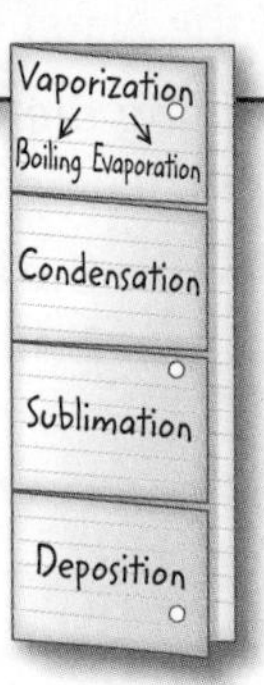

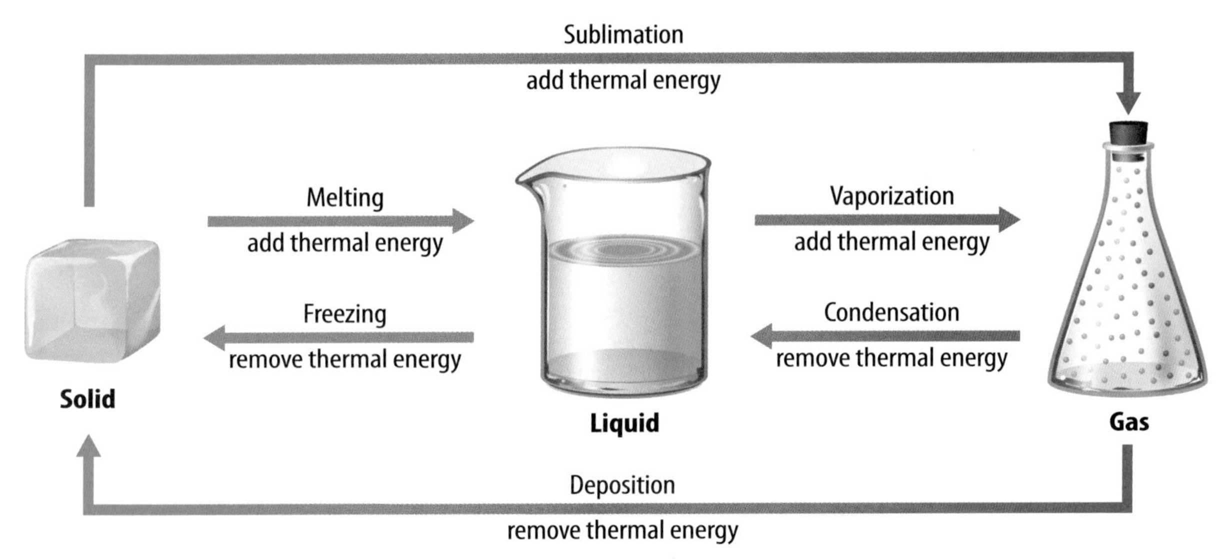

Figure 14 For a change of state to occur, thermal energy must move into or out of matter.

Conservation of Mass and Energy

The diagram in **Figure 14** illustrates the energy changes that occur as thermal energy is added or removed from matter. Notice that opposite processes, melting and freezing and vaporization and condensation, are shown. When matter changes state, matter and energy are always conserved.

When water vaporizes, it appears to disappear. If the invisible gas is captured and its mass added to the remaining mass of the liquid, you would see that matter is conserved. This is also true for energy. Surrounding matter, such as air, often absorbs thermal energy. If you measured all the thermal energy, you would find that energy is conserved.

Inquiry MiniLab

20 minutes

How can you make a water thermometer?

What causes liquid in a thermometer to rise and fall?

1. Read and complete a lab safety form.
2. Place one drop of **food coloring** in a **flask.** Fill the flask to the top with room temperature tap water. Over a **sink or pan,** insert a **one-holed stopper fitted with a glass tube** into the flask. Press down gently. The liquid should rise partway into the tube. Mark the level of the water with a **grease pencil.**
3. Holding the tube by its neck, lower the flask into a pan of hot water. Observe the water level for 3 min. Record your observations in your Science Journal.
4. Remove the flask from the hot water, and lower it into a pan of **ice water.** Observe the water level for 3 min, and record your observations.

Analyze and Conclude

Key Concept Explain what happens to the column of water and the water particles as they are heated and cooled.

Lesson 2 Review

Visual Summary

All matter has thermal energy. Thermal energy is the sum of potential and kinetic energy.

When thermal energy is added to a liquid, vaporization can occur.

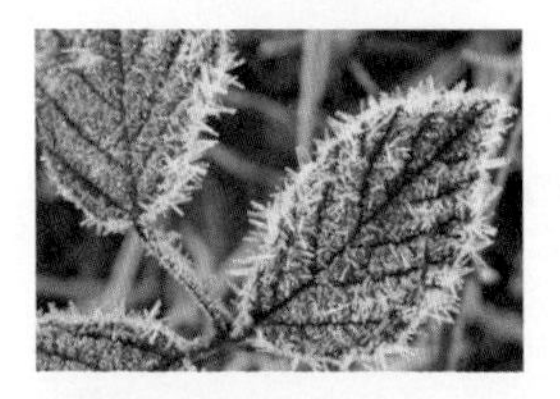

When enough thermal energy is removed from matter, a change of state can occur.

Use your lesson Foldable to review the lesson. Save your Foldable for the project at the end of the chapter.

What do you think NOW?

You first read the statements below at the beginning of the chapter.

3. Particles of matter have both potential energy and kinetic energy.

4. When a solid melts, thermal energy is removed from the solid.

Did you change your mind about whether you agree or disagree with the statements? Rewrite any false statements to make them true.

Use Vocabulary

1. The measure of average kinetic energy of the particles in a material is __________.
2. **Define** *kinetic energy* and *thermal energy* in your own words.
3. The change of a liquid into a gas is known as __________.

Understand Key Concepts

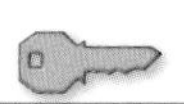

4. The process that is opposite of condensation is known as
   A. deposition.
   B. freezing.
   C. melting.
   D. vaporization.
5. **Explain** how temperature and particle motion are related.
6. **Describe** the relationship between temperature and thermal energy.
7. **Generalize** the changes in thermal energy when matter increases in temperature and then changes state.

Interpret Graphics

8. **Describe** what is occurring below.

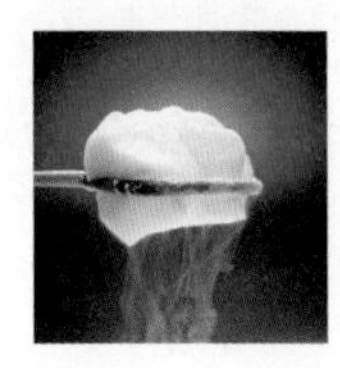

9. **Summarize** Copy and fill in the graphic organizer below to identify the two types of vaporization that can occur in matter.

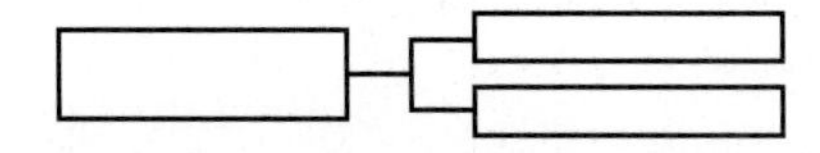

Critical Thinking

10. **Summarize** the energy and state changes that occur when freezing rain falls and solidifies on a wire fence.
11. **Compare** the amount of thermal energy needed to melt a solid and the amount of thermal energy needed to freeze the same liquid.

How does dissolving substances in water change its freezing point?

You know that when thermal energy is removed from a liquid, the particles move more slowly. At the freezing point, the particles move so slowly that the attractive forces pull them together to form a solid. What happens if the water contains particles of another substance, such as salt? You will form a hypothesis and test the hypothesis to find out.

Materials

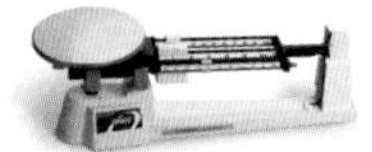

triple-beam balance

beaker

foam cup

50-mL graduated cylinder

distilled water

Also needed: ice-salt slush, test tubes, thermometers

Safety

Learn It

To **form a hypothesis** is to propose a possible explanation for an observation that is testable by a scientific investigation. You **test the hypothesis** by conducting a scientific investigation to see whether the hypothesis is supported.

Try It

1. Read and complete a lab safety form.
2. Form a hypothesis that answers the question in the title of the lab. Write your hypothesis in your Science Journal.
3. Copy the data table in your Science Journal.
4. Use a triple-beam balance to measure 5 g of table salt (NaCl). Dissolve the 5 g of table salt in 50 mL of distilled water.
5. Place 40 mL of distilled water in one large test tube. Place 40 mL of the salt-water mixture in a second large test tube.
6. Measure and record the temperature of the liquids in each test tube.
7. Place both test tubes into a large foam cup filled with crushed ice-salt slush. Gently rotate the thermometers in the test tubes. Record the temperature in each test tube every minute until the temperature remains the same for several minutes.

Apply It

8. How does the data tell you when the freezing point of the liquid has been reached?
9. Was your hypothesis supported? Why or why not?
10. **Key Concept** Explain your observations in terms of how temperature affects particle motion and how a liquid changes to a solid.

Water	Time (min)	0	1	2	3	4	5	6	7	8
	Temperature (°C)									
Salt water	Time (min)	0	1	2	3	4	5	6	7	8
	Temperature (°C)									

Lesson 3

The Behavior of Gases

Reading Guide

Key Concepts
ESSENTIAL QUESTIONS

- How does the kinetic molecular theory describe the behavior of a gas?
- How are temperature, pressure, and volume related in Boyle's law?
- How is Boyle's law different from Charles's law?

Vocabulary

kinetic molecular theory p. 218

pressure p. 219

Boyle's law p. 220

Charles's law p. 221

g Multilingual eGlossary

Video
What's Science Got to do With It?

Inquiry Survival Gear?

Why do some pilots wear oxygen masks? Planes fly at high altitudes where the atmosphere has a lower pressure and gas molecules are less concentrated. If the pressure is not adjusted inside the airplane, a pilot must wear an oxygen mask to inhale enough oxygen to keep the body functioning.

Launch Lab

15 minutes

Are volume and pressure of a gas related?

Pressure affects gases differently than it affects solids and liquids. How do pressure changes affect the volume of a gas?

1. Read and complete a lab safety form.
2. Stretch and blow up a **small balloon** several times.
3. Finally, blow up the balloon to a diameter of about 5 cm. Twist the neck, and stretch the mouth of the balloon over the opening of a **plastic bottle. Tape** the neck of the balloon to the bottle.
4. Squeeze and release the bottle several times while observing the balloon. Record your observations in your Science Journal.

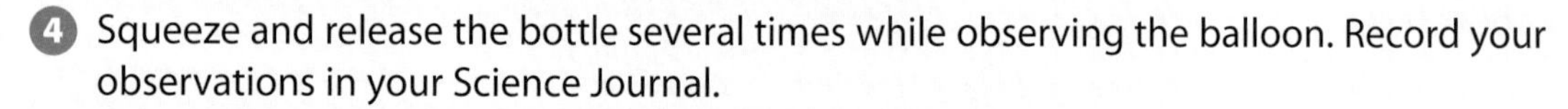

Think About This

1. Why doesn't the balloon deflate when you attach it to the bottle?
2. What caused the balloon to inflate when you squeezed the bottle?
3. **Key Concept** Using this lab as a reference, do you think pressure and volume of a gas are related? Explain.

Understanding Gas Behavior

Pilots do not worry as much about solids and liquids at high altitudes as they do gases. That is because gases behave differently than solids and liquids. Changes in temperature, pressure, and volume affect the behavior of gases more than they affect solids and liquids.

The explanation of particle behavior in solids, liquids, and gases is based on the kinetic molecular theory. The **kinetic molecular theory** *is an explanation of how particles in matter behave.* Some basic ideas in this theory are

ACADEMIC VOCABULARY
theory
(noun) an explanation of things or events that is based on knowledge gained from many observations and investigations

- small particles make up all matter;
- these particles are in constant, random motion;
- the particles collide with other particles, other objects, and the walls of their container;
- when particles collide, no energy is lost.

You have read about most of these, but the last two statements are very important in explaining how gases behave.

Key Concept Check How does the kinetic molecular theory describe the behavior of a gas?

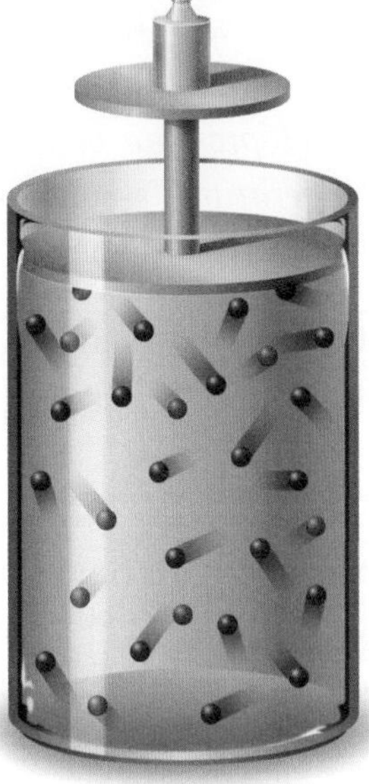

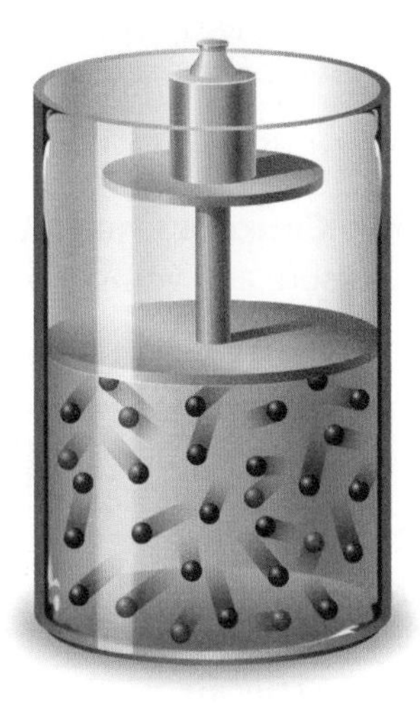

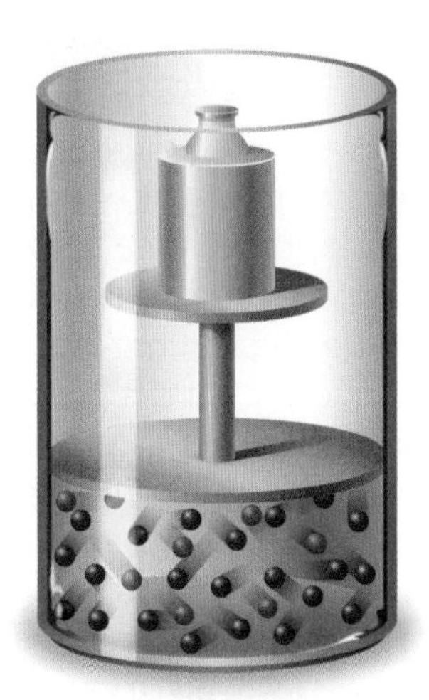

Figure 15 As pressure increases, the volume of the gas decreases.

What is pressure?

Particles in gases move constantly. As a result of this movement, gas particles constantly collide with other particles and their container. When particles collide with their container, pressure results. **Pressure** *is the amount of force applied per unit of area.* For example, gas in a cylinder, as shown in **Figure 15,** might contain trillions of gas particles. These particles exert forces on the cylinder each time they strike it. When a weight is added to the plunger, the plunger moves down, compressing the gas in the cylinder. With less space to move around, the particles that make up the gas collide with each other more frequently, causing an increase in pressure. The more the particles are compressed, the more often they collide, increasing the pressure.

WORD ORIGIN

pressure
from Latin *pressura,* means "to press"

Pressure and Volume

Figure 15 also shows the relationship between pressure and volume of gas at a constant temperature. What happens to pressure if the volume of a container changes? Notice that when the volume is greater, the particles have more room to move. This additional space results in fewer collisions within the cylinder, and pressure is less. The gas particles in the middle cylinder have even less volume and more pressure. In the cylinder on the right, the pressure is greater because the volume is less. The particles collide with the container more frequently. Because of the greater number of collisions within the container, pressure is greater.

FOLDABLES®

Fold a sheet of notebook paper to make a three-tab Foldable and label as shown. Use your Foldable to compare two important gas laws.

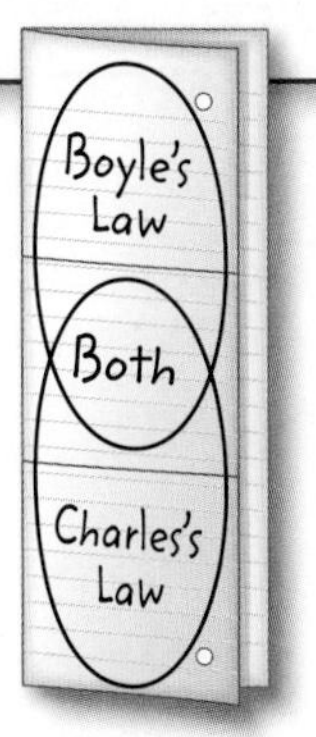

Math Skills

Solve Equations

Boyle's law can be stated by the equation

$$V_2 = \frac{P_1 V_1}{P_2}$$

P_1 and V_1 represent the pressure and volume before a change. P_2 and V_2 are the pressure and volume after a change. Pressure is often measured in kilopascals (kPa). For example, what is the final volume of a gas with an initial volume of **50.0 mL** if the pressure increases from **600.0 kPa** to **900.0 kPa**?

1. Replace the terms in the equation with the actual values.

$$V_2 = \frac{(600.0\ \text{kPa})(50.0\ \text{mL})}{(900.0\ \text{kPa})}$$

2. Cancel units, multiply, and then divide.

$$V_2 = \frac{(600.0\ \cancel{\text{kPa}})(50.0\ \text{mL})}{(900.0\ \cancel{\text{kPa}})}$$

$$V_2 = 33.3\ \text{mL}$$

Practice

What is the final volume of a gas with an initial volume of 100.0 mL if the pressure decreases from 500.0 kPa to 250.0 kPa?

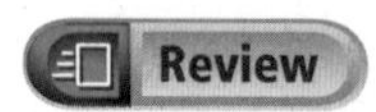

- **Math Practice**
- **Personal Tutor**

Boyle's Law

You read that the pressure and volume of a gas are related. Robert Boyle (1627–1691), a British scientist, was the first to describe this property of gases. **Boyle's law** *states that pressure of a gas increases if the volume decreases and pressure of a gas decreases if the volume increases, when temperature is constant.* This law can be expressed mathematically as shown to the left.

Key Concept Check What is the relationship between pressure and volume of a gas if temperature is constant?

Boyle's Law in Action

You have probably felt Boyle's law in action if you have ever traveled in an airplane. While on the ground, the air pressure inside your middle ear and the pressure of the air surrounding you are equal. As the airplane takes off and begins to increase in altitude, the air pressure of the surrounding air decreases. However, the air pressure inside your middle ear does not decrease. The trapped air in your middle ear increases in volume, which can cause pain. These pressure changes also occur when the plane is landing. You can equalize this pressure difference by yawning or chewing gum.

Graphing Boyle's Law

This relationship is shown in the graph in **Figure 16.** Pressure is on the *x*-axis, and volume is on the *y*-axis. Notice that the line decreases in value from left to right. This shows that as the pressure of a gas increases, the volume of the gas decreases.

Figure 16 The graph shows that as pressure increases, volume decreases. This is true only if the temperature of the gas is constant.

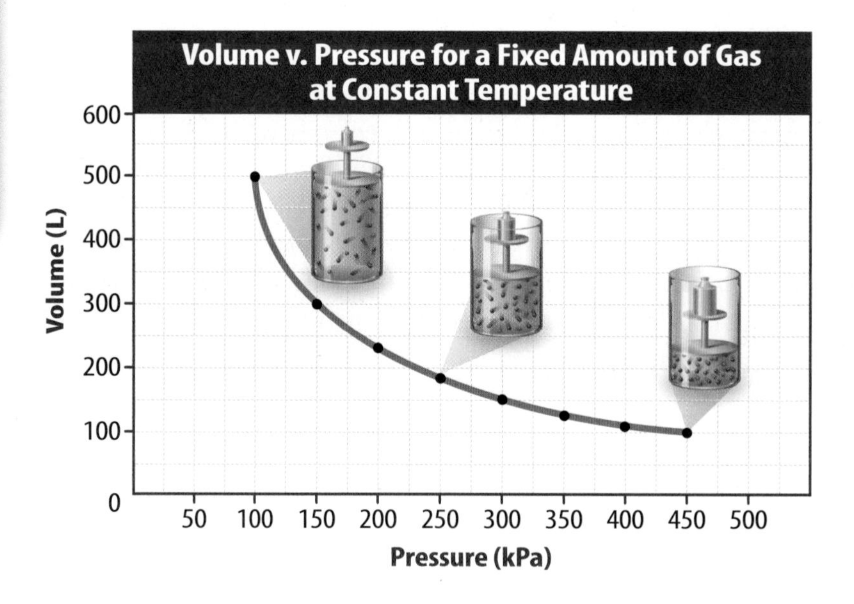

Figure 17 As the temperature of a gas increases, the kinetic energy of the particles increases. The particles move farther apart, and volume increases.

Temperature and Volume

Pressure and volume changes are not the only factors that affect gas behavior. Changing the temperature of a gas also affects its behavior, as shown in **Figure 17.** The gas in the cylinder on the left has a low temperature. The average kinetic energy of the particles is low, and they move closer together. The volume of the gas is less. When thermal energy is added to the cylinder, the gas particles move faster and spread farther apart. This increases the pressure from gas particles, which push up the plunger. This increases the volume of the container.

Charles's Law

Jacque Charles (1746–1823) was a French scientist who described the relationship between temperature and volume of a gas. **Charles's law** *states that the volume of a gas increases with increasing temperature, if the pressure is constant.* Charles's practical experience with gases was most likely the result of his interest in balloons. Charles and his colleague were the first to pilot and fly a hydrogen-filled balloon in 1783.

Key Concept Check How is Boyle's law different from Charles's law?

Inquiry MiniLab — 20 minutes

How does temperature affect the volume?

You can observe Charles's law in action using a few lab supplies.

1. Read and complete a lab safety form.
2. Stretch and blow up a **small balloon** several times.
3. Finally, blow up the balloon to a diameter of about 5 cm. Twist the neck and stretch the mouth of the balloon over the opening of an **ovenproof flask.**
4. Place the flask on a cold **hot plate.** Turn on the hot plate to low, and gradually heat the flask. Record your observations in your Science Journal.
5. ⚠ Use **tongs** to remove the flask from the hot plate. Allow the flask to cool for 5 min. Record your observations.
6. Place the flask in a **bowl of ice water.** Record your observations.

Analyze and Conclude

Key Concept What is the effect of temperature changes on the volume of a gas?

Charles's Law in Action

You have probably seen Charles's law in action if you have ever taken a balloon outside on a cold winter day. Why does a balloon appear slightly deflated when you take it from a warm place to a cold place? When the balloon is in cold air, the temperature of the gas inside the balloon decreases. Recall that a decrease in temperature is a decrease in the average kinetic energy of particles. As a result, the gas particles slow down and begin to get closer together. Fewer particles hit the inside of the balloon. The balloon appears partially deflated. If the balloon is returned to a warm place, the kinetic energy of the particles increases. More particles hit the inside of the balloon and push it out. The volume increases.

 **Reading Check** What happens when you warm a balloon?

Graphing Charles's Law

The relationship described in Charles's law is shown in the graph of several gases in **Figure 18.** Temperature is on the *x*-axis and volume is on the *y*-axis. Notice that the lines are straight and represent increasing values. Each line in the graph is extrapolated to −273°C. *Extrapolated* means the graph is extended beyond the observed data points. This temperature also is referred to as 0 K (kelvin), or absolute zero. This temperature is theoretically the lowest possible temperature of matter. At absolute zero, all particles are at the lowest possible energy state and do not move. The particles contain a minimal amount of thermal energy (potential energy + kinetic energy).

 **Key Concept Check** Which factors must be constant in Boyle's law and in Charles's law?

Figure 18 The volume of a gas increases when the temperature increases at constant pressure.

Visual Check What do the dashed lines mean?

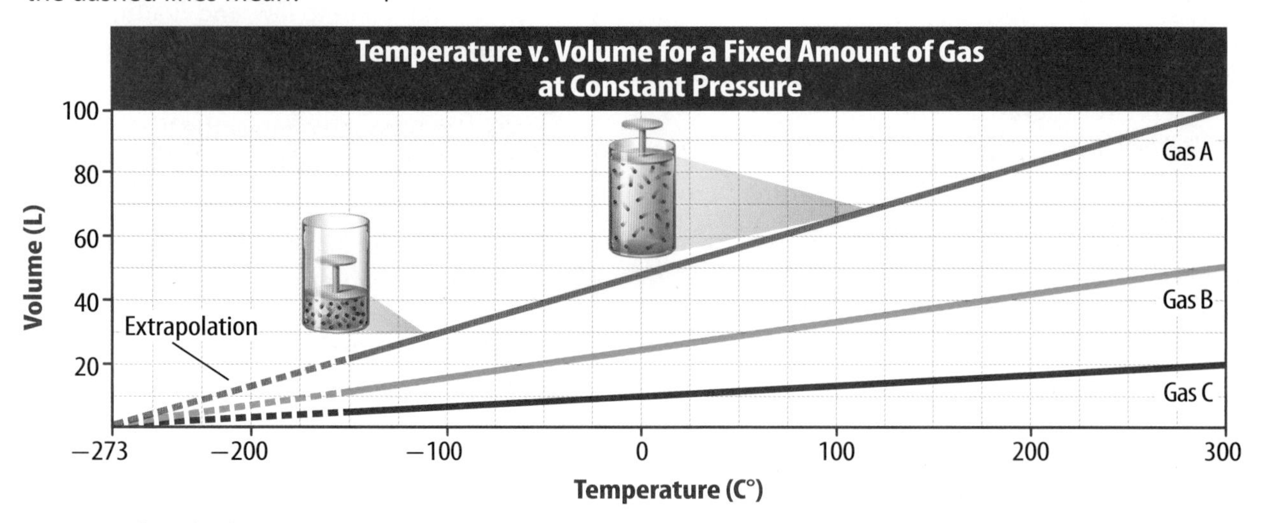

Lesson 3 Review

✓ Assessment **Online Quiz**

? Inquiry **Virtual Lab**

Visual Summary

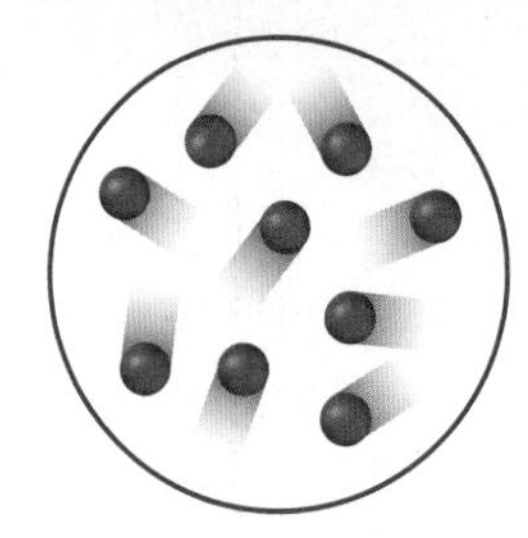

The explanation of particle behavior in solids, liquids, and gases is based on the kinetic molecular theory.

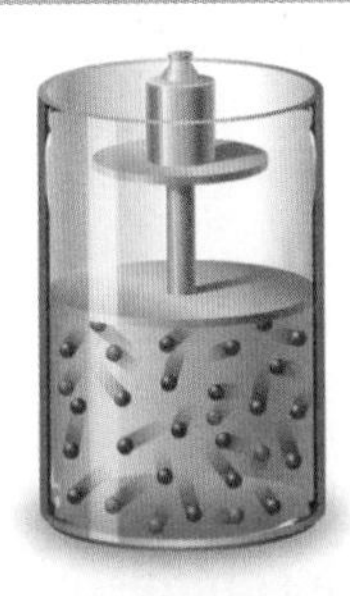

As volume of a gas decreases, the pressure increases when at constant temperature.

At constant pressure, as the temperature of a gas increases, the volume also increases.

Use your lesson Foldable to review the lesson. Save your Foldable for the project at the end of the chapter.

What do you think NOW?

You first read the statements below at the beginning of the chapter.

5. Changes in temperature and pressure affect gas behavior.

6. If the pressure on a gas increases, the volume of the gas also increases.

Did you change your mind about whether you agree or disagree with the statements? Rewrite any false statements to make them true.

Use Vocabulary

1 **List** the basic ideas of the kinetic molecular theory.

2 ________ is force applied per unit area.

Understand Key Concepts

3 Which is held constant when a gas obeys Boyle's law?

A. motion

B. pressure

C. temperature

D. volume

4 **Describe** how the kinetic molecular theory explains the behavior of a gas.

5 **Contrast** Charles's law with Boyle's law.

6 **Explain** how temperature, pressure, and volume are related in Boyle's law.

Interpret Graphics

7 **Explain** what happens to the particles to the right when more weights are added.

8 **Identify** Copy and fill in the graphic organizer below to list three factors that affect gas behavior.

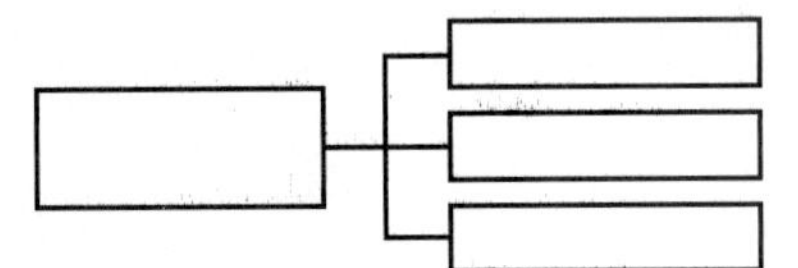

Critical Thinking

9 **Describe** what would happen to the pressure of a gas if the volume of the gas doubles while at a constant temperature.

Math Skills

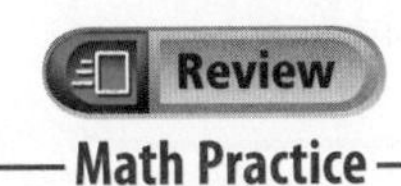

Math Practice

10 **Calculate** The pressure on 400 mL of a gas is raised from 20.5 kPa to 80.5 kPa. What is the final volume of the gas?

Design an Experiment to Collect Data

Materials

triple-beam balance

50-mL graduated cylinders

beakers

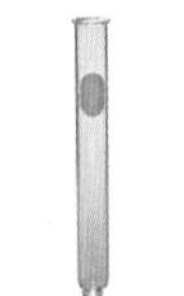

test tubes

thermometers

distilled water

Also needed: ice, salt

Safety

In this chapter, you have learned about the relationship between the motion of particles in matter and change of state. How might you use your knowledge of particles in real life? Suppose that you work for a state highway department in a cold climate. Your job is to test three products. You must determine which is the most effective in melting existing ice, the best at keeping melted ice from refreezing, and the best product to buy.

Question

How can you compare the products? What might make one product better than another? Consider how you can describe and compare the effect of each product on both existing ice and the freezing point of water. Think about controls, variables, and the equipment you have available.

Procedure

1. Read and complete a lab safety form.
2. In your Science Journal, write a set of procedures you will use to answer your questions. Include the materials and steps you will use to test the effect of each product on existing ice and on the freezing point of water. How will you record your data? Draw any data tables, such as the example below, that you might need. Have your teacher approve your procedures.

Distilled Water	Time (min)	0	1	2	3	4	5	6	7	8
	Temperature (°C)									
Product A	Time (min)	0	1	2	3	4	5	6	7	8
	Temperature (°C)									
Product B	Time (min)	0	1	2	3	4	5	6	7	8
	Temperature (°C)									
Product C	Time (min)	0	1	2	3	4	5	6	7	8
	Temperature (°C)									

3. Begin by observing and recording your observations on how each product affects ice. Does it make ice melt or melt faster?
4. Test the effect of each product on the freezing point of water. Think about how you will ensure that each product is tested in the same way.
5. Add any additional tests you think you might need to make your recommendation.

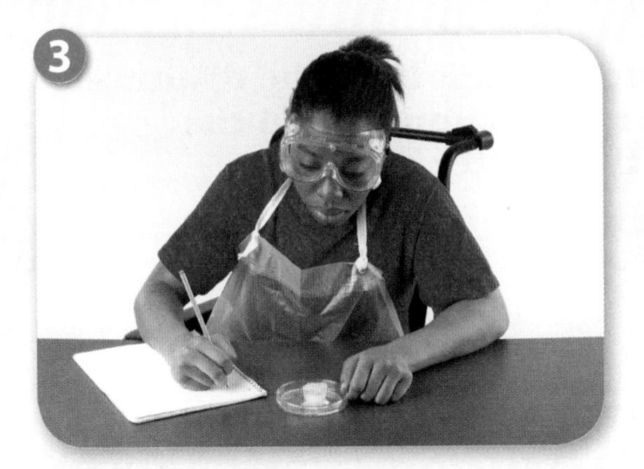

Analyze and Conclude

6. **Analyze the data** you have collected. Which product was most effective in melting existing ice? How do you know?
7. **Determine** which product was most effective in lowering the freezing point of water.
8. **Draw or make a model** to show the effect of dissolved solids on water molecules.
9. **Recognize Cause and Effect** In terms of particles, what causes dissolved solids to lower the freezing point of water?
10. **Draw Conclusions** In terms of particles, why are some substances more effective than others in lowering the freezing point of water?
11. THE BIG IDEA **The Big Idea** Why is the kinetic molecular theory important in understanding how and why matter changes state?

Communicate Your Results

You are to present your recommendations to the road commissioners. Create a graphic presentation that clearly displays your results and justifies your recommendations about which product to buy.

In some states, road crews spray liquid deicer on the roads. If your teacher approves, you may enjoy testing liquids, such as alcohol, corn syrup, or salad oil.

Lab Tips

- To ensure fair testing, add the same mass of each product to the ice cubes at the same time.
- Be sure to add the same mass of each solid to the same volume of water. About 1 g of solid in 10 mL of water is a good ratio.
- Keep adding crushed ice/salt slush to the cup so that the liquid in the test tubes remains below the surface.

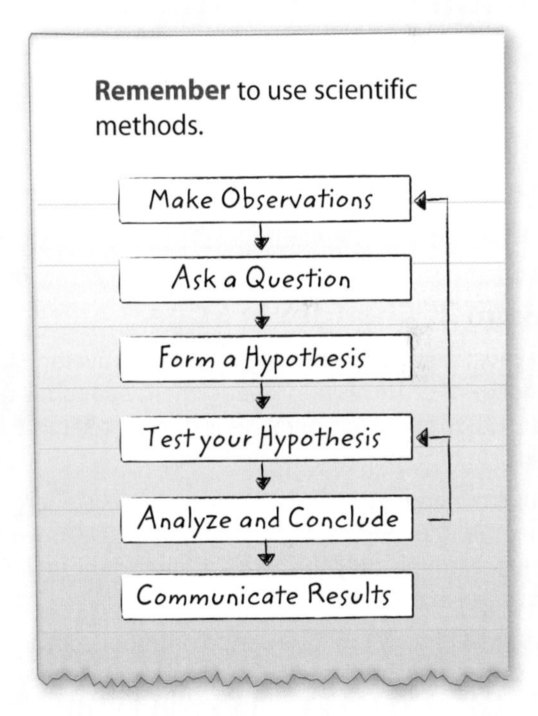

Chapter 6 Study Guide

As matter changes from one state to another, the distances and the forces between the particles change, and the amount of thermal energy in the matter changes.

Key Concepts Summary

Lesson 1: Solids, Liquids, and Gases

- Particles vibrate in **solids.** They move faster in **liquids** and even faster in **gases.**
- The force of attraction among particles decreases as matter goes from a solid, to a liquid, and finally to a gas.

Solid

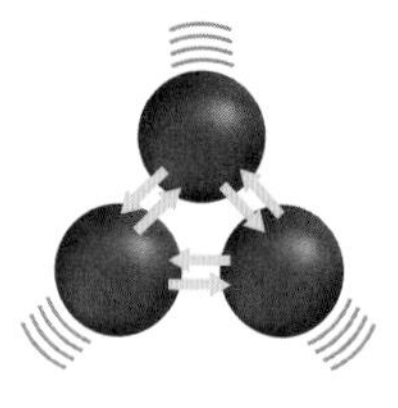

Liquid

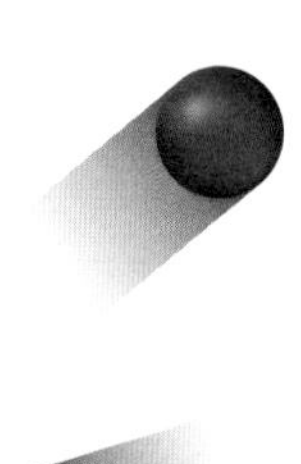
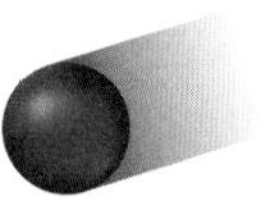

Gas

Vocabulary

solid p. 201
liquid p. 202
viscosity p. 202
surface tension p. 203
gas p. 204
vapor p. 204

Lesson 2: Changes in State

- Because **temperature** is defined as the average **kinetic energy** of particles and kinetic energy depends on particle motion, temperature is directly related to particle motion.
- **Thermal energy** includes both the kinetic energy and the potential energy of particles in matter. However, temperature is only the average kinetic energy of particles in matter.
- Thermal energy must be added or removed from matter for a change of state to occur.

Vocabulary

kinetic energy p. 208
temperature p. 208
thermal energy p. 209
vaporization p. 211
evaporation p. 212
condensation p. 212
sublimation p. 212
deposition p. 212

Lesson 3: The Behavior of Gases

- The **kinetic molecular theory** states basic assumptions that are used to describe particles and their interactions in gases and other states of matter.
- **Pressure** of a gas increases if the volume decreases, and pressure of a gas decreases if the volume increases, when temperature is constant.
- **Boyle's law** describes the behavior of a gas when pressure and volume change at constant temperature. **Charles's law** describes the behavior of a gas when temperature and volume change, and pressure is constant.

Vocabulary

kinetic molecular theory p. 218
pressure p. 219
Boyle's law p. 220
Charles's law p. 221

- Personal Tutor
- Vocabulary eGames
- Vocabulary eFlashcards

FOLDABLES® Chapter Project

Assemble your lesson Foldables as shown to make a Chapter Project. Use the project to review what you have learned in this chapter.

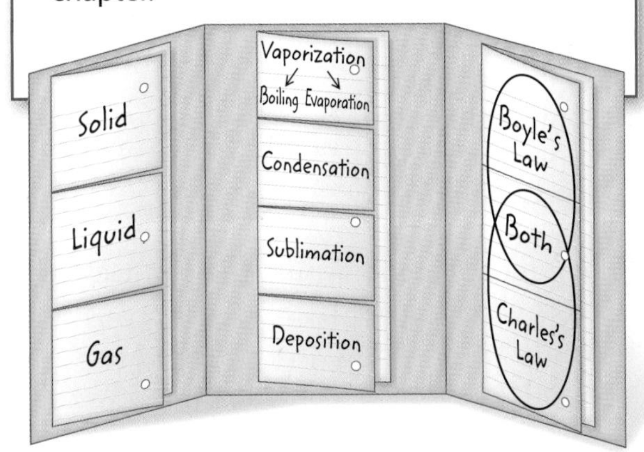

Use Vocabulary

Replace the underlined word with the correct term.

1. Matter with a definite shape and a definite volume is known as a gas.
2. Surface tension is a measure of a liquid's resistance to flow.
3. The gas state of a substance that is normally a solid or a liquid at room temperature is a pressure.
4. Boiling is vaporization that occurs at the surface of a liquid.
5. Boyle's law is an explanation of how particles in matter behave.
6. When graphing a gas obeying Boyle's law, the line will be a straight line with a positive slope.

Link Vocabulary and Key Concepts

Concepts in Motion Interactive Concept Map

Copy this concept map, and then use vocabulary terms from the previous page to complete the concept map.

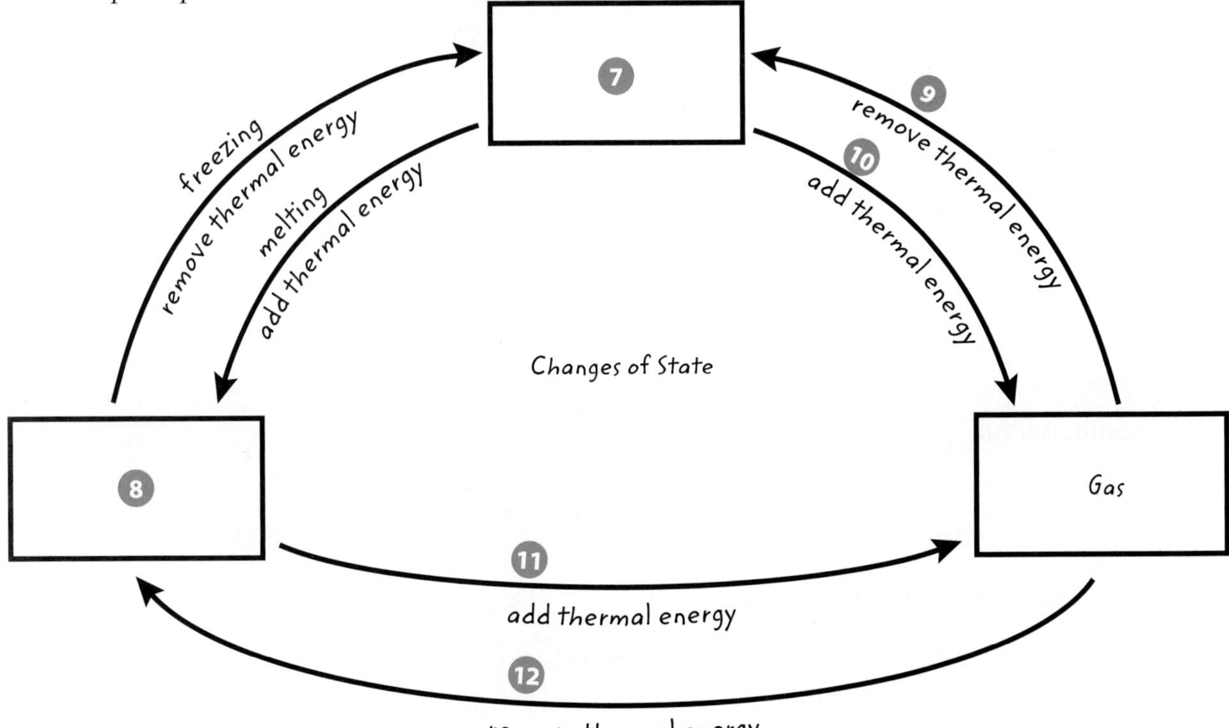

Chapter 6 Review

Understand Key Concepts 

1. What would happen if you tried to squeeze a gas into a smaller container?
   A. The attractive forces between the particles would increase.
   B. The force of the particles would prevent you from doing it.
   C. The particles would have fewer collisions with the container.
   D. The repulsive forces of the particles would pull on the container.

2. Which type of motion in the figure below best represents the movement of gas particles?

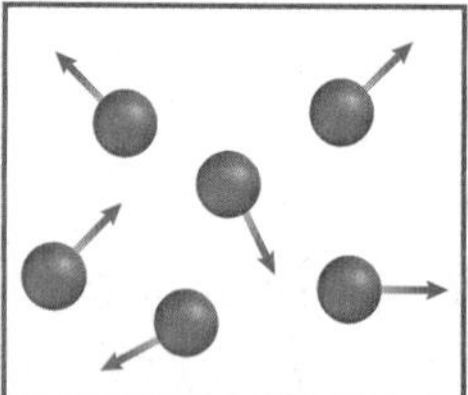

Motion 1

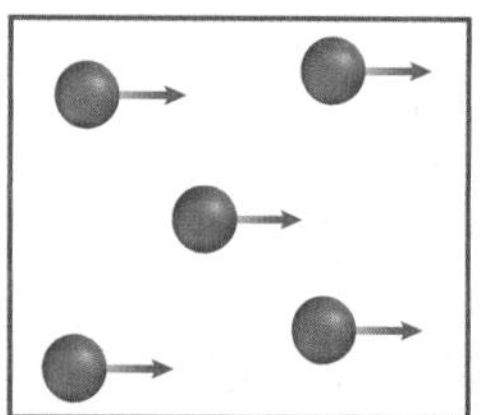

Motion 2

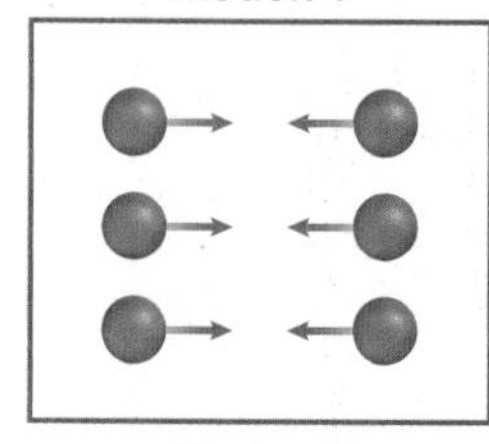

Motion 3

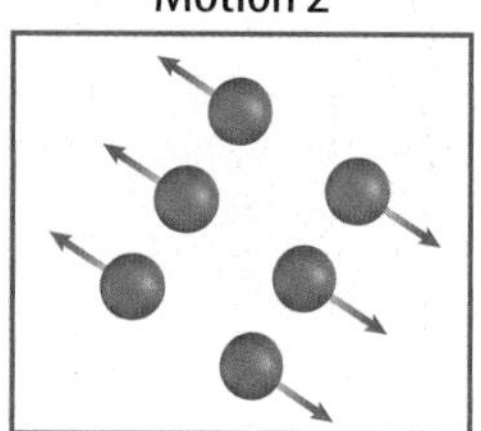

Motion 4

   A. motion 1
   B. motion 2
   C. motion 3
   D. motion 4

3. A pile of snow slowly disappears into the air, even though the temperature remains below freezing. Which process explains this?
   A. condensation
   B. deposition
   C. evaporation
   D. sublimation

4. Which unit is a density unit?
   A. cm^3
   B. cm^3/g
   C. g
   D. g/cm^3

5. Which is a form of vaporization?
   A. condensation
   B. evaporation
   C. freezing
   D. melting

6. When a needle is placed on the surface of water, it floats. Which idea best explains why this happens?
   A. Boyle's law
   B. molecular theory
   C. surface tension
   D. viscosity theory

7. In which material would the particles be most closely spaced?
   A. air
   B. brick
   C. syrup
   D. water

Use the graph below to answer questions 8 and 9.

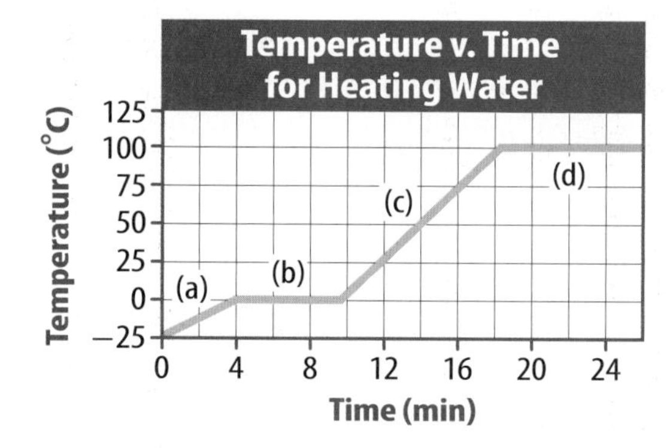

8. Which area of the graph above shows melting of a solid?
   A. a
   B. b
   C. c
   D. d

9. Which area or areas of the graph above shows a change in the potential energy of the particles?
   A. a
   B. a and c
   C. b and d
   D. c

Critical Thinking

10. **Explain** how the distances between particles in a solid, a liquid, and a gas help determine the densities of each.

11. **Describe** what would happen to the volume of a balloon if it were submerged in hot water.

12. **Assess** The particles of an unknown liquid have very weak attractions for other particles in the liquid. Would you expect the liquid to have a high or low viscosity? Explain your answer.

13. **Rank** these liquids from highest to lowest viscosity: honey, rubbing alcohol, and ketchup.

14. **Evaluate** Each beaker below contains the same amount of water. The thermometers show the temperature in each beaker. Explain the kinetic energy differences in each beaker.

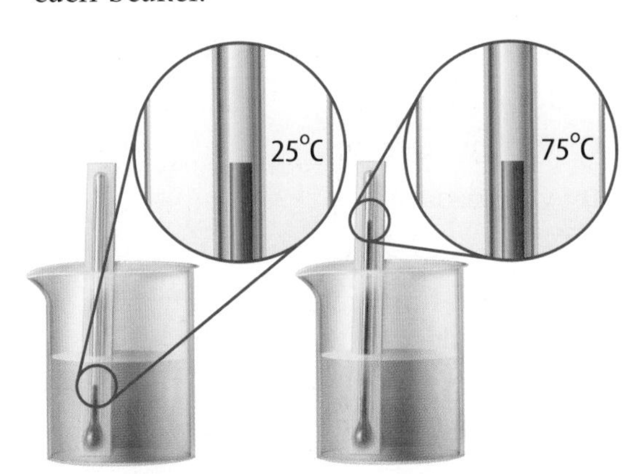

15. **Summarize** A glass with a few milliliters of water is placed on a counter. No one touches the glass. Explain what happens to the water after a few days.

Writing in Science

16. **Write** a paragraph that describes how you could determine the melting point of a substance from its heating or cooling curve.

REVIEW THE BIG IDEA

17. During springtime in Alaska, frozen rivers thaw and boats can navigate the rivers again. What physical changes and energy changes occur to the ice molecules when ice changes to water? Explain the process in which water in the river changes to water vapor.

18. In the photo below, explain how the average kinetic energy of the particles changes as the molten glass cools. What instrument could you use to verify the change in the average kinetic energy of the particles?

Math Skills

Math Practice

Solve Equations

19. The pressure on 1 L of a gas at a pressure of 600 kPa is lowered to 200 kPa. What is the final volume of the gas?

20. A gas has a volume of 30 mL at a pressure of 5000 kPa. What is the volume of the gas if the pressure is lowered to 1,250 kPa?

Standardized Test Practice

Record your answers on the answer sheet provided by your teacher or on a separate sheet of paper.

Multiple Choice

1 Which property applies to matter that consists of particles vibrating in place?

A has a definite shape

B takes the shape of the container

C flows easily at room temperature

D particles far apart

Use the figure below to answer questions 2 and 3.

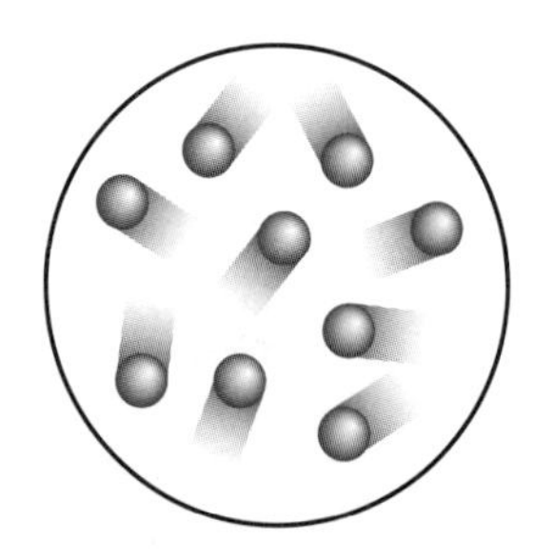

2 Which state of matter is represented above?

A amorphous solid

B crystalline solid

C gas

D liquid

3 Which best describes the attractive forces between particles shown in the figure?

A The attractive forces keep the particles vibrating in place.

B The particles hardly are affected by the attractive forces.

C The attractive forces keep the particles close together but still allow movement.

D The particles are locked in their positions because of the attractive forces between them.

4 What happens to matter as its temperature increases?

A The average kinetic energy of its particles decreases.

B The average thermal energy of its particles decreases.

C The particles gain kinetic energy.

D The particles lose potential energy.

Use the figure to answer question 5.

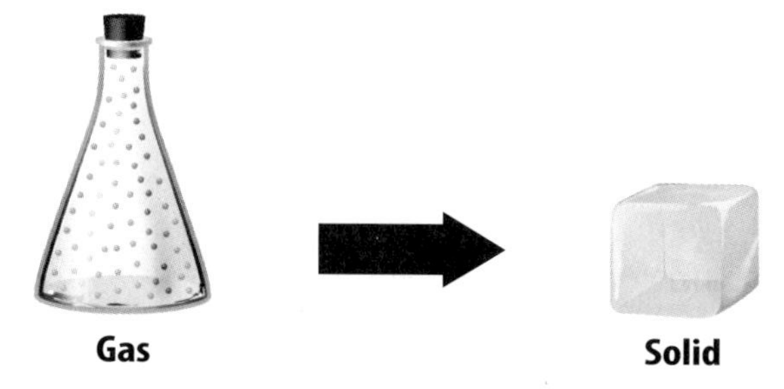

5 Which process is represented in the figure?

A deposition

B freezing

C sublimation

D vaporization

6 Which is a fundamental assumption of the kinetic molecular theory?

A All atoms are composed of subatomic particles.

B The particles of matter move in predictable paths.

C No energy is lost when particles collide with one another.

D Particles of matter never come into contact with one another.

Assessment

Online Standardized Test Practice

7 Which is true of the thermal energy of particles?

A Thermal energy includes the potential and the kinetic energy of the particles.

B Thermal energy is the same as the average kinetic energy of the particles.

C Thermal energy is the same as the potential energy of particles.

D Thermal energy is the same as the temperature of the particles.

Use the graph below to answer question 8.

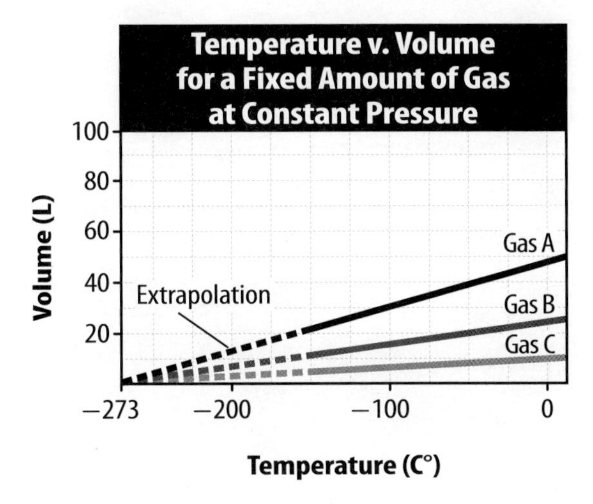

8 Which relationship is shown in the graph?

A Boyle's law

B Charles's law

C kinetic molecular theory

D definition of thermal energy

Constructed Response

9 Some people say that something that does not move very quickly is "as slow as molasses in winter." What property of molasses is described by the saying? Based on the saying, how do you think this property changes with temperature?

Use the graph to answer questions 10 and 11.

A scientist measured the temperature of a sample of frozen mercury as thermal energy is added to the sample. The graph below shows the results.

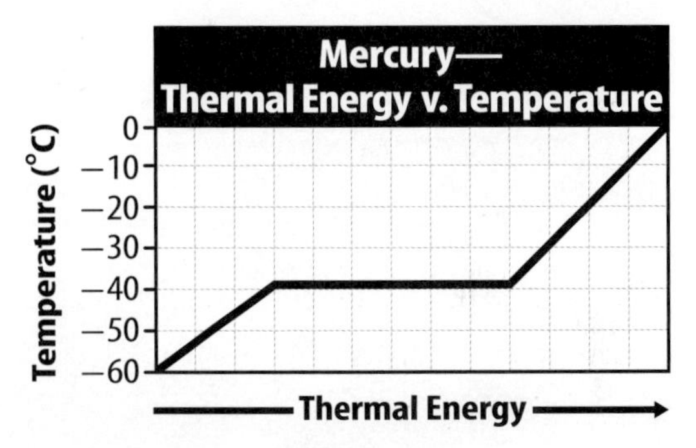

10 At what temperature does mercury melt? How do you know?

11 Describe the motion and arrangement of mercury atoms while the temperature is constant.

12 Atmospheric pressure is greater at the base of a mountain than at its peak. A hiker drinks from a water bottle at the top of a mountain. The bottle is capped tightly. At the base of the mountain, the water bottle has collapsed slightly. What happened to the gas inside the bottle? Assume constant temperature. Explain.

NEED EXTRA HELP?												
If You Missed Question...	1	2	3	4	5	6	7	8	9	10	11	12
Go to Lesson...	1	1	1	2	2	3	2	3	1	1	2	3

Chapter 7

Understanding the Atom

THE BIG IDEA **What are atoms, and what are they made of?**

Inquiry All This to Study Tiny Particles?

This huge machine is called the Large Hadron Collider (LHC). It's like a circular racetrack for particles and is about 27 km long. The LHC accelerates particle beams to high speeds and then smashes them into each other. The longer the tunnel, the faster the beams move and the harder they smash together. Scientists study the tiny particles produced in the crash.

- How might scientists have studied matter before colliders were invented?
- What do you think are the smallest parts of matter?
- What are atoms, and what are they made of?

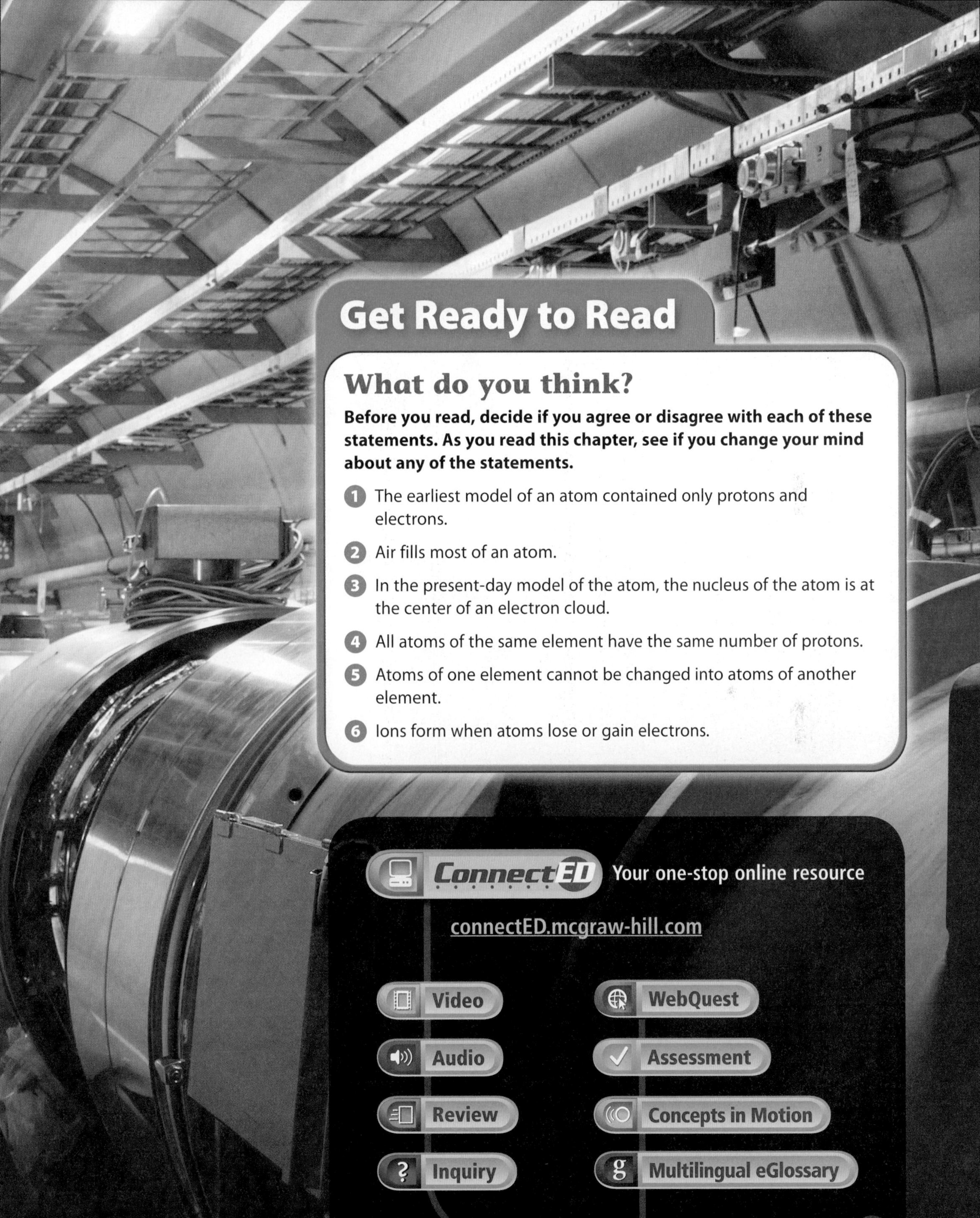

Get Ready to Read

What do you think?

Before you read, decide if you agree or disagree with each of these statements. As you read this chapter, see if you change your mind about any of the statements.

1. The earliest model of an atom contained only protons and electrons.
2. Air fills most of an atom.
3. In the present-day model of the atom, the nucleus of the atom is at the center of an electron cloud.
4. All atoms of the same element have the same number of protons.
5. Atoms of one element cannot be changed into atoms of another element.
6. Ions form when atoms lose or gain electrons.

ConnectED Your one-stop online resource

connectED.mcgraw-hill.com

- Video
- Audio
- Review
- Inquiry
- WebQuest
- Assessment
- Concepts in Motion
- Multilingual eGlossary

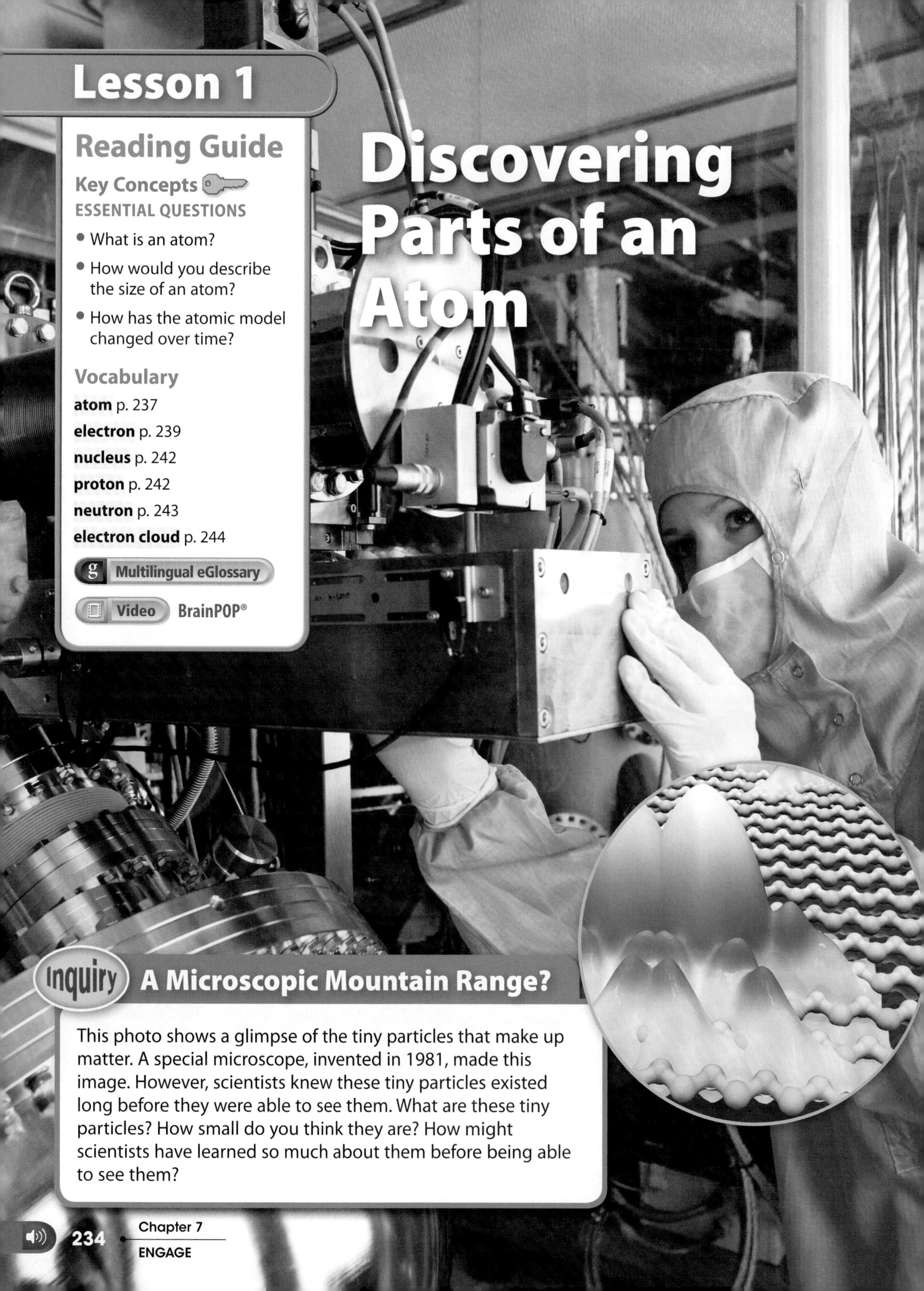

Lesson 1

Reading Guide

Key Concepts
ESSENTIAL QUESTIONS

- What is an atom?
- How would you describe the size of an atom?
- How has the atomic model changed over time?

Vocabulary

atom p. 237
electron p. 239
nucleus p. 242
proton p. 242
neutron p. 243
electron cloud p. 244

g Multilingual eGlossary

Video BrainPOP®

Discovering Parts of an Atom

Inquiry A Microscopic Mountain Range?

This photo shows a glimpse of the tiny particles that make up matter. A special microscope, invented in 1981, made this image. However, scientists knew these tiny particles existed long before they were able to see them. What are these tiny particles? How small do you think they are? How might scientists have learned so much about them before being able to see them?

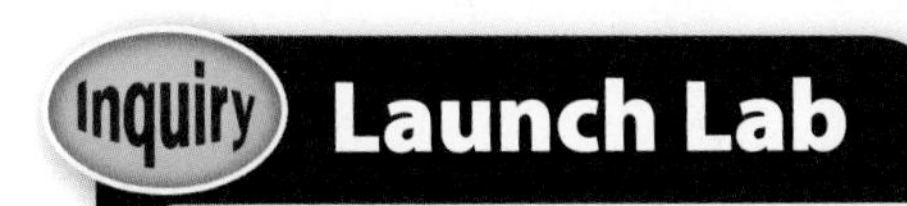

Inquiry Launch Lab

10 minutes

What's in there?

When you look at a sandy beach from far away, it looks like a solid surface. You can't see the individual grains of sand. What would you see if you zoomed in on one grain of sand?

1. Read and complete a lab safety form.
2. Have your partner hold a **test tube** of **a substance,** filled to a height of 2–3 cm.
3. Observe the test tube from a distance of at least 2 m. Write a description of what you see in your Science Journal.
4. Pour about 1 cm of the substance onto a piece of **waxed paper.** Record your observations.
5. Use a **toothpick** to separate out one particle of the substance. Suppose you could zoom in. What do you think you would see? Record your ideas in your Science Journal.

Think About This

1. Do you think one particle of the substance is made of smaller particles? Why or why not?
2. **Key Concept** Do you think you could use a microscope to see what the particles are made of? Why or why not?

Early Ideas About Matter

Look at your hands. What are they made of? You might answer that your hands are made of things such as skin, bone, muscle, and blood. You might recall that each of these is made of even smaller structures called cells. Are cells made of even smaller parts? Imagine dividing something into smaller and smaller parts. What would you end up with?

Greek philosophers discussed and debated questions such as these more than 2,000 years ago. At the time, many thought that all matter is made of only four elements—fire, water, air, and earth, as shown in **Figure 1.** However, they weren't able to test their ideas because scientific tools and methods, such as experimentation, did not exist yet. The ideas proposed by the most influential philosophers usually were accepted over the ideas of less influential philosophers. One philosopher, Democritus (460–370 B.C.), challenged the popular idea of matter.

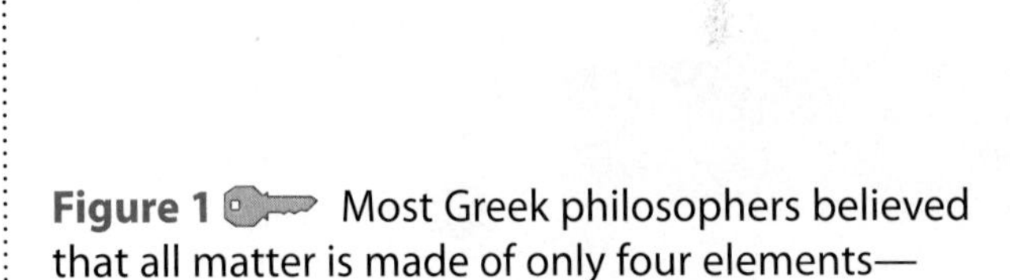

Figure 1 Most Greek philosophers believed that all matter is made of only four elements—fire, water, air, and earth.

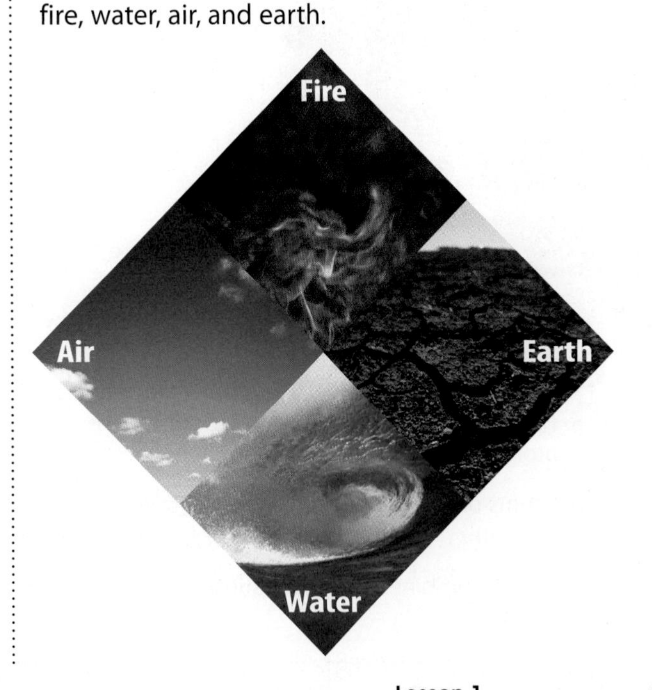

Democritus

Democritus believed that matter is made of small, solid objects that cannot be divided, created, or destroyed. He called these objects *atomos,* from which the English word *atom* is derived. Democritus proposed that different types of matter are made from different types of atoms. For example, he said that smooth matter is made of smooth atoms. He also proposed that nothing is between these atoms except empty space. **Table 1** summarizes Democritus's ideas.

Although Democritus had no way to test his ideas, many of his ideas are similar to the way scientists describe the atom today. Because Democritus's ideas did not conform to the popular opinion and because they could not be tested scientifically, they were open for debate. One philosopher who challenged Democritus's ideas was Aristotle.

Reading Check According to Democritus, what might atoms of gold look like?

Aristotle

Aristotle (384–322 B.C.) did not believe that empty space exists. Instead, he favored the more popular idea—that all matter is made of fire, water, air, and earth. Because Aristotle was so influential, his ideas were accepted. Democritus's ideas about atoms were not studied again for more than 2,000 years.

Dalton's Atomic Model

In the late 1700s, English schoolteacher and scientist John Dalton (1766–1844) revisited the idea of atoms. Since Democritus's time, advancements had been made in technology and scientific methods. Dalton made careful observations and measurements of chemical reactions. He combined data from his own scientific research with data from the research of other scientists to propose the atomic theory. **Table 1** lists ways that Dalton's atomic theory supported some of the ideas of Democritus.

Table 1 Similarities Between Democritus's and Dalton's Ideas

Democritus

1. Atoms are small solid objects that cannot be divided, created, or destroyed.
2. Atoms are constantly moving in empty space.
3. Different types of matter are made of different types of atoms.
4. The properties of the atoms determine the properties of matter.

John Dalton

1. All matter is made of atoms that cannot be divided, created, or destroyed.
2. During a chemical reaction, atoms of one element cannot be converted into atoms of another element.
3. Atoms of one element are identical to each other but different from atoms of another element.
4. Atoms combine in specific ratios.

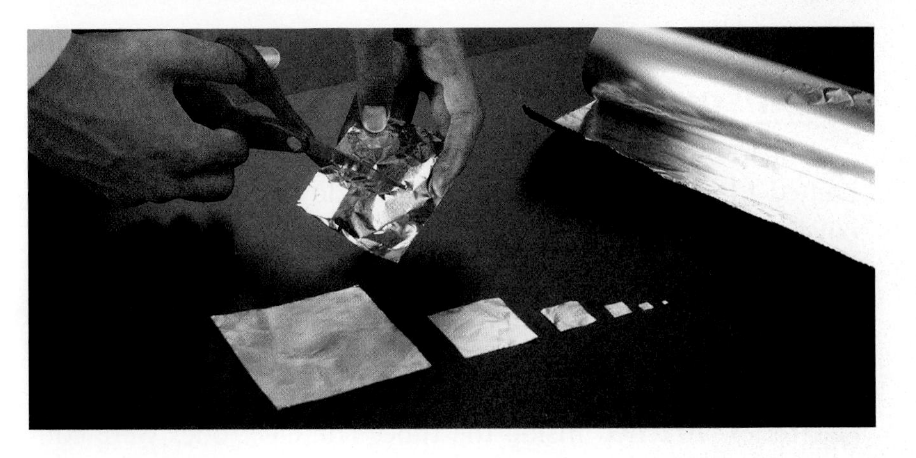

◀ **Figure 2** If you could keep dividing a piece of aluminum, you eventually would have the smallest possible piece of aluminum—an aluminum atom.

The Atom

Today, scientists agree that matter is made of atoms with empty space between and within them. What is an atom? Imagine dividing the piece of aluminum shown in **Figure 2** into smaller and smaller pieces. At first you would be able to cut the pieces with scissors. But eventually you would have a piece that is too small to see—much smaller than the smallest piece you could cut with scissors. This small piece is an aluminum atom. An aluminum atom cannot be divided into smaller aluminum pieces. *An* **atom** *is the smallest piece of an element that still represents that element.*

 **Key Concept Check** What is a copper atom?

The Size of Atoms

Just how small is an atom? Atoms of different elements are different sizes, but all are very, very small. You cannot see atoms with just your eyes or even with most microscopes. Atoms are so small that about 7.5 trillion carbon atoms could fit into the period at the end of this sentence.

 **Key Concept Check** How would you describe the size of an atom?

Seeing Atoms

Scientific experiments verified that matter is made of atoms long before scientists were able to see atoms. However, the 1981 invention of a high-powered microscope, called a scanning tunneling microscope (STM), enabled scientists to see individual atoms for the first time. **Figure 3** shows an STM image. An STM uses a tiny, metal tip to trace the surface of a piece of matter. The result is an image of atoms on the surface.

Even today, scientists still cannot see inside an atom. However, scientists have learned that atoms are not the smallest particles of matter. In fact, atoms are made of much smaller particles. What are these particles, and how did scientists discover them if they could not see them?

Figure 3 A scanning tunneling microscope created this image. The yellow sphere is a manganese atom on the surface of gallium arsenide. ▼

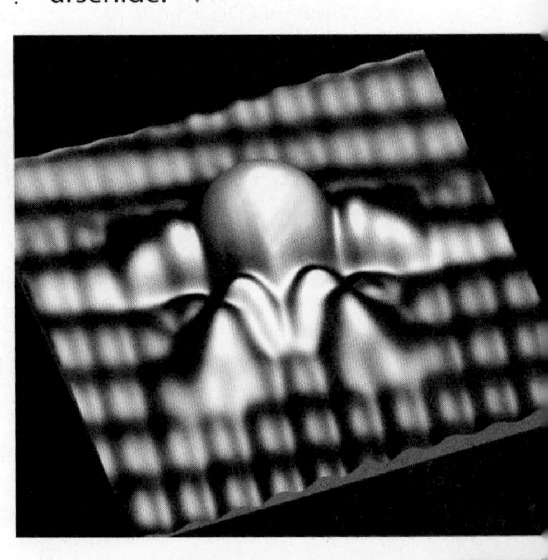

Thomson—Discovering Electrons

Not long after Dalton's findings, another English scientist, named J.J. Thomson (1856–1940), made some important discoveries. Thomson and other scientists of that time worked with cathode ray tubes. If you ever have seen a neon sign, an older computer monitor, or the color display on an ATM screen, you have seen a cathode ray tube. Thomson's cathode ray tube, shown in **Figure 4,** was a glass tube with pieces of metal, called electrodes, attached inside the tube. The electrodes were connected to wires, and the wires were connected to a battery. Thomson discovered that if most of the air was removed from the tube and electricity was passed through the wires, greenish-colored rays traveled from one electrode to the other end of the tube. What were these rays made of?

Negative Particles

Scientists called these rays cathode rays. Thomson wanted to know if these rays had an electric charge. To find out, he placed two plates on opposite sides of the tube. One plate was positively charged, and the other plate was negatively charged, as shown in **Figure 4.** Thomson discovered that these rays bent toward the positively charged plate and away from the negatively charged plate. Recall that opposite charges attract each other, and like charges repel each other. Thomson concluded that cathode rays are negatively charged.

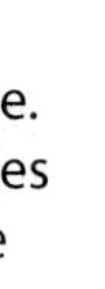

Reading Check If the rays were positively charged, what would Thomson have observed as they passed between the plates?

Figure 4 As the cathode rays passed between the plates, they were bent toward the positive plate. Because opposite charges attract, the rays must be negatively charged.

Thomson's Cathode Ray Tube Experiment

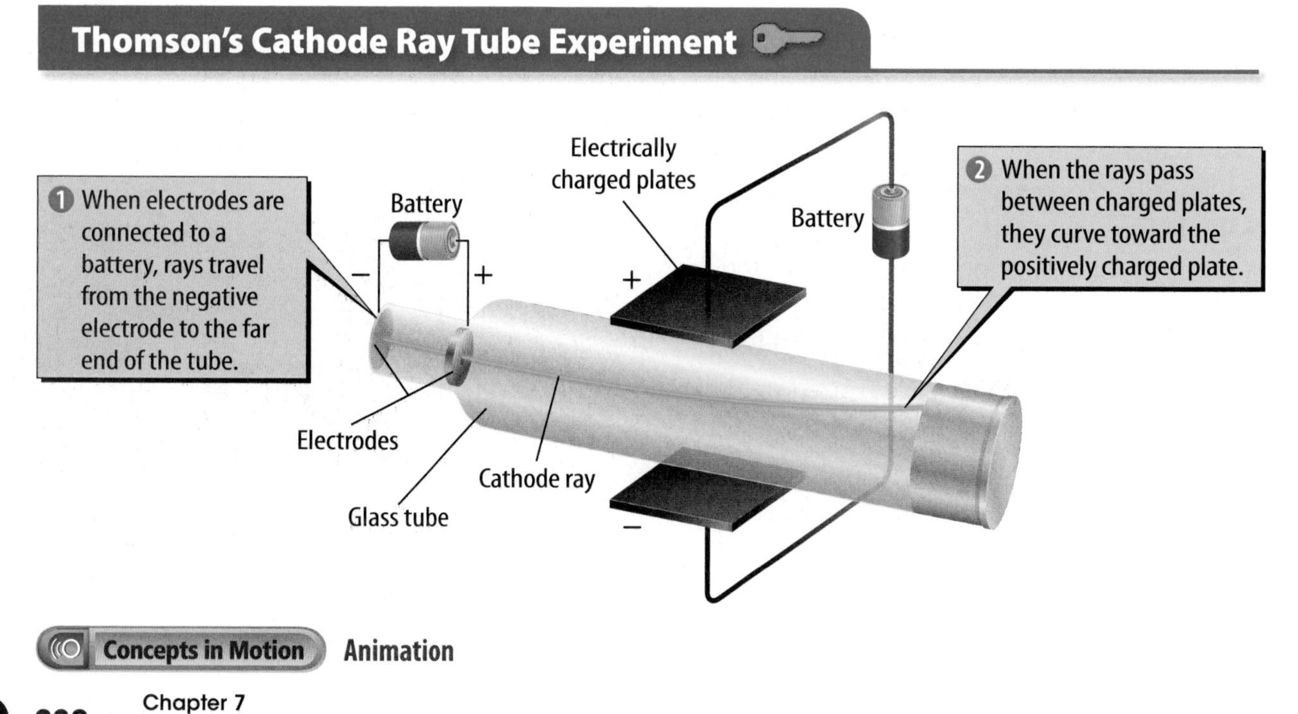

Concepts in Motion Animation

Parts of Atoms

Through more experiments, Thomson learned that these rays were made of particles that had mass. The mass of one of these particles was much smaller than the mass of the smallest atoms. This was surprising information to Thomson. Until then, scientists understood that the smallest particle of matter is an atom. But these rays were made of particles that were even smaller than atoms.

Where did these small, negatively charged particles come from? Thomson proposed that these particles came from the metal atoms in the electrode. Thomson discovered that identical rays were produced regardless of the kind of metal used to make the electrode. Putting these clues together, Thomson concluded that cathode rays were made of small, negatively charged particles. He called these particles electrons. *An* **electron** *is a particle with one negative charge (1−).* Because atoms are neutral, or not electrically charged, Thomson proposed that atoms also must contain a positive charge that balances the negatively charged electrons.

Thomson's Atomic Model

Thomson used this information to propose a new model of the atom. Instead of a solid, neutral sphere that was the same throughout, Thomson's model of the atom contained both positive and negative charges. He proposed that an atom was a sphere with a positive charge evenly spread throughout. Negatively charged electrons were mixed through the positive charge, similar to the way chocolate chips are mixed in cookie dough. **Figure 5** shows this model.

Reading Check How did Thomson's atomic model differ from Dalton's atomic model?

FOLDABLES®

Use two sheets of paper to make a layered book. Label it as shown. Use it to organize your notes and diagrams on the parts of an atom.

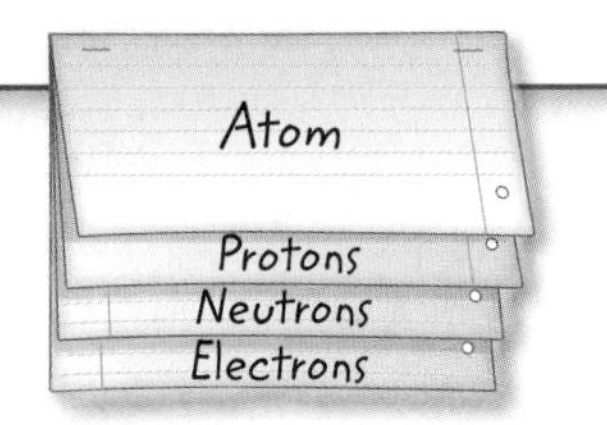

WORD ORIGIN

electron
from Greek *electron,* means "amber," the physical force so called because it first was generated by rubbing amber. Amber is a fossilized substance produced by trees.

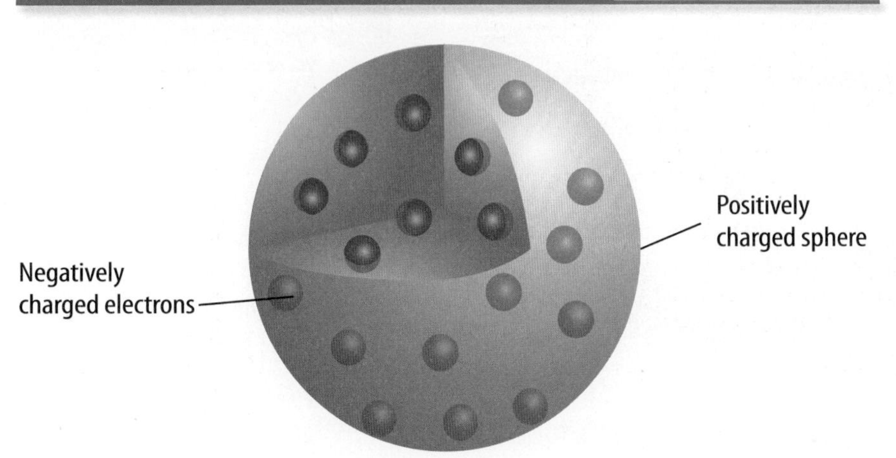

Figure 5 Thomson's model of the atom contained a positively charged sphere with negatively charged electrons within it.

Rutherford—Discovering the Nucleus

The discovery of electrons stunned scientists. Ernest Rutherford (1871–1937) was a student of Thomson's who eventually had students of his own. Rutherford's students set up experiments to test Thomson's atomic model and to learn more about what atoms contain. They discovered another surprise.

Rutherford's Predicted Result

Imagine throwing a baseball into a pile of table tennis balls. The baseball likely would knock the table tennis balls out of the way and continue moving in a relatively straight line. This is similar to what Rutherford's students expected to see when they shot alpha particles into atoms. Alpha particles are dense and positively charged. Because they are so dense, only another dense particle could deflect the path of an alpha particle. According to Thomson's model, the positive charge of the atom was too spread out and not dense enough to change the path of an alpha particle. Electrons wouldn't affect the path of an alpha particle because electrons didn't have enough mass. The result that Rutherford's students expected is shown in **Figure 6.**

 **Reading Check** Explain why Rutherford's students did not think an atom could change the path of an alpha particle.

Figure 6 The Thomson model of the atom did not contain a charge that was dense enough to change the path of an alpha particle. Rutherford expected the positive alpha particles to travel straight through the foil without changing direction.

Rutherford's Predicted Result

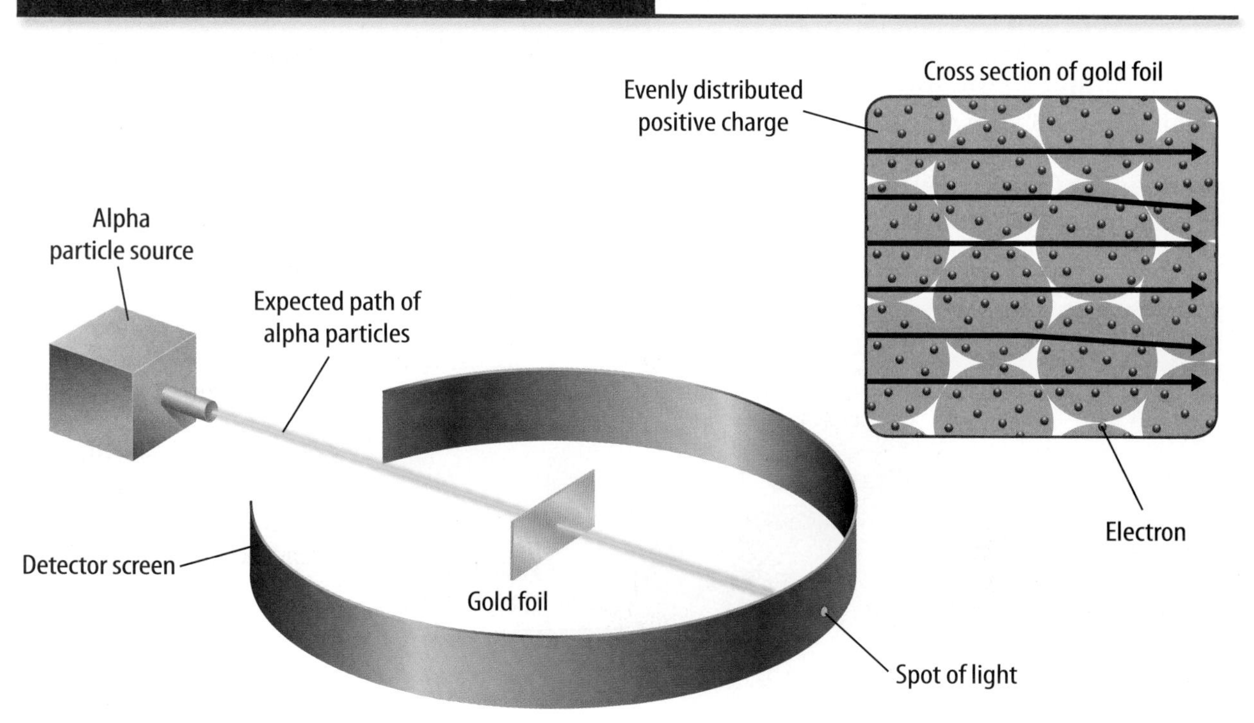

The Gold Foil Experiment

Rutherford's students went to work. They placed a source of alpha particles near a very thin piece of gold foil. Recall that all matter is made of atoms. Therefore, the gold foil was made of gold atoms. A screen surrounded the gold foil. When an alpha particle struck the screen, it created a spot of light. Rutherford's students could determine the path of the alpha particles by observing the spots of light on the screen.

The Surprising Result

Figure 7 shows what the students observed. Most of the particles did indeed travel through the foil in a straight path. However, a few particles struck the foil and bounced off to the side. And one particle in 10,000 bounced straight back! Rutherford later described this surprising result, saying it was almost as incredible as if you had fired a 38-cm shell at a piece of tissue paper and it came back and hit you. The alpha particles must have struck something dense and positively charged inside the nucleus. Thomson's model had to be refined.

Key Concept Check Given the results of the gold foil experiment, how do you think an actual atom differs from Thomson's model?

Figure 7 Some alpha particles traveled in a straight path, as expected. But some changed direction, and some bounced straight back.

Visual Check What do the dots on the screen indicate?

The Surprising Result

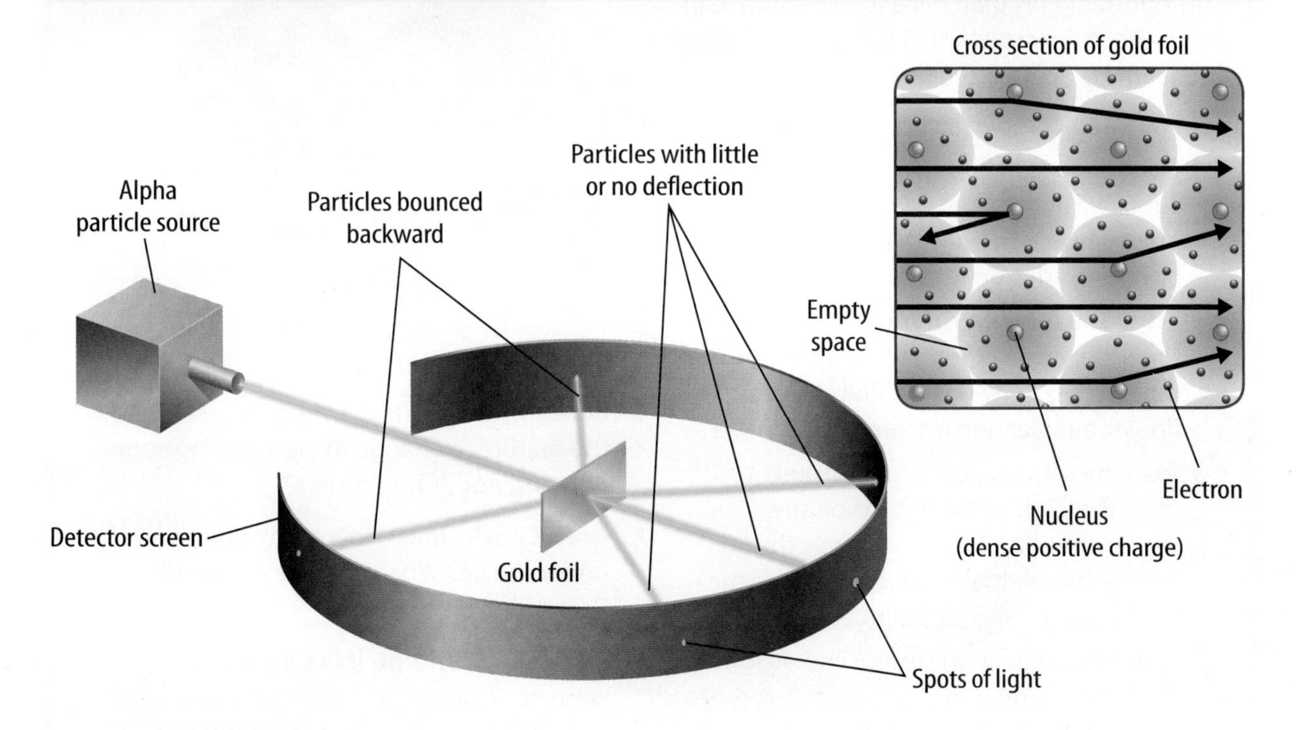

Concepts in Motion Animation

Rutherford's Atomic Model

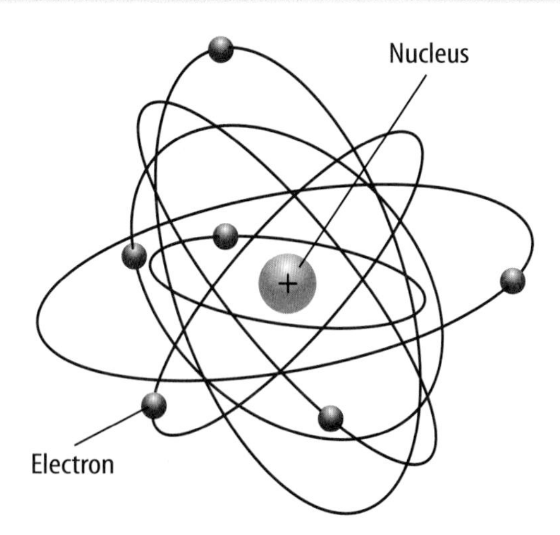

Figure 8 Rutherford's model contains a small, dense, positive nucleus. Tiny, negatively charged electrons travel in empty space around the nucleus.

Rutherford's Atomic Model

Because most alpha particles traveled through the foil in a straight path, Rutherford concluded that atoms are made mostly of empty space. The alpha particles that bounced backward must have hit a dense, positive mass. Rutherford concluded that *most of an atom's mass and positive charge is concentrated in a small area in the center of the atom called the* **nucleus.** **Figure 8** shows Rutherford's atomic model. Additional research showed that the positive charge in the nucleus was made of positively charged particles called protons. *A* **proton** *is an atomic particle that has one positive charge (1+).* Negatively charged electrons move in the empty space surrounding the nucleus.

 **Reading Check** How did Rutherford explain the observation that some of the alpha particles bounced directly backward?

Inquiry MiniLab

20–30 minutes

How can you gather information about what you can't see?

Rutherford did his gold foil experiment to learn more about the structure of the atom. What can you learn by doing a similar investigation?

1. Read and complete a lab safety form.
2. Place a piece of white **newsprint** on a flat surface. Your teacher will place an upside-down **shoe box lid** with holes cut on opposite sides on the newsprint.
3. Place one end of a **ruler** on a **book,** with the other end pointing toward one of the holes in the shoe box lid. Roll a **marble** down the ruler and into one of the holes.
4. Team members should use **markers** to draw the path of the marble on the newsprint as it enters and leaves the lid. Predict the path of the marble under the lid. Draw it on the lid. Number the path *1.*
5. Take turns repeating steps 3 and 4 eight to ten times, numbering each path *2, 3, 4,* etc. Move the ruler and aim it in a slightly different direction each time.

Analyze and Conclude

1. **Recognize Cause and Effect** What caused the marble to change its path during some rolls and not during others?
2. **Draw Conclusions** How many objects are under the lid? Where are they located? Draw your answer.
3. **Key Concept** If the shoe box lid were an accurate model of the atom, what hypothesis would you make about the atom's structure?

Discovering Neutrons

The modern model of the atom was beginning to take shape. Rutherford's colleague, James Chadwick (1891–1974), also researched atoms and discovered that, in addition to protons, the nucleus also contained neutrons. *A* **neutron** *is a neutral particle that exists in the nucleus of an atom.*

Bohr's Atomic Model

Rutherford's model explained much of his students' experimental evidence. However, there were several observations that the model could not explain. For example, scientists noticed that if certain elements were heated in a flame, they gave off specific colors of light. Each color of light had a specific amount of energy. Where did this light come from? Niels Bohr (1885–1962), another student of Rutherford's, proposed an answer. Bohr studied hydrogen atoms because they contain only one electron. He experimented with adding electric energy to hydrogen and studying the energy that was released. His experiments led to a revised atomic model.

Electrons in the Bohr Model

Bohr's model is shown in **Figure 9.** Bohr proposed that electrons move in circular orbits, or energy levels, around the nucleus. Electrons in an energy level have a specific amount of energy. Electrons closer to the nucleus have less energy than electrons farther away from the nucleus. When energy is added to an atom, electrons gain energy and move from a lower energy level to a higher energy level. When the electrons return to the lower energy level, they release a specific amount of energy as light. This is the light that is seen when elements are heated.

Limitations of the Bohr Model

Bohr reasoned that if his model were accurate for atoms with one electron, it would be accurate for atoms with more than one electron. However, this was not the case. More research showed that, although electrons have specific amounts of energy, energy levels are not arranged in circular orbits. How do electrons move in an atom?

Key Concept Check How did Bohr's atomic model differ from Rutherford's?

Figure 9 In Bohr's atomic model, electrons move in circular orbits around the atom. When an electron moves from a higher energy level to a lower energy level, energy is released—sometimes as light. Further research showed that electrons are not arranged in orbits.

Concepts in Motion Animation

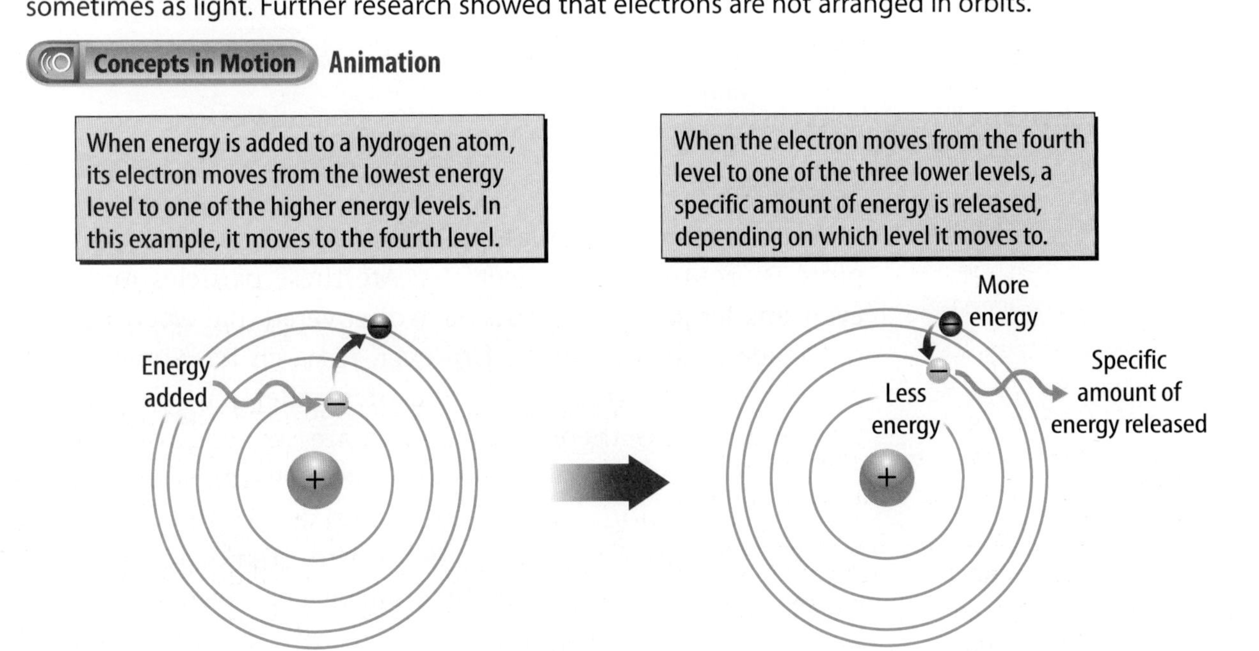

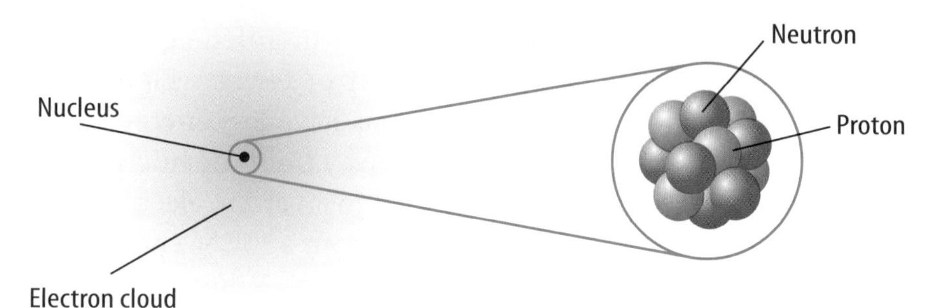

Figure 10 In this atom, electrons are more likely to be found closer to the nucleus than farther away.

Visual Check Why do you think this model of the atom doesn't show the electrons?

The Modern Atomic Model

In the modern atomic model, electrons form an electron cloud. *An* **electron cloud** *is an area around an atomic nucleus where an electron is most likely to be located.* Imagine taking a time-lapse photograph of bees around a hive. You might see a blurry cloud. The cloud might be denser near the hive than farther away because the bees spend more time near the hive.

In a similar way, electrons constantly move around the nucleus. It is impossible to know both the speed and exact location of an electron at a given moment in time. Instead, scientists only can predict the likelihood that an electron is in a particular location. The electron cloud shown in **Figure 10** is mostly empty space but represents the likelihood of finding an electron in a given area. The darker areas represent areas where electrons are more likely to be.

Key Concept Check How has the model of the atom changed over time?

Quarks

You have read that atoms are made of smaller parts—protons, neutrons, and electrons. Are these particles made of even smaller parts? Scientists have discovered that electrons are not made of smaller parts. However, research has shown that protons and neutrons are made of smaller particles called quarks. Scientists theorize that there are six types of quarks. They have named these quarks up, down, charm, strange, top, and bottom. Protons are made of two up quarks and one down quark. Neutrons are made of two down quarks and one up quark. Just as the model of the atom has changed over time, the current model might also change with the invention of new technology that aids the discovery of new information.

Lesson 1 Review

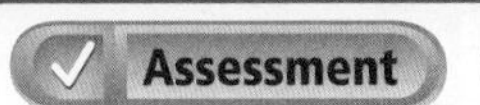

Online Quiz

Visual Summary

If you were to divide an element into smaller and smaller pieces, the smallest piece would be an atom.

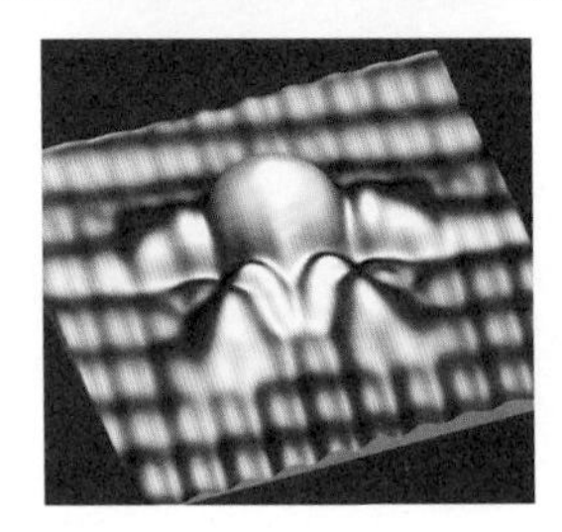

Atoms are so small that they can be seen only by using very powerful microscopes.

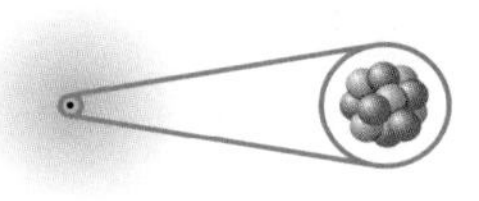

Scientists now know that atoms contain a dense, positive nucleus surrounded by an electron cloud.

FOLDABLES®

Use your lesson Foldable to review the lesson. Save your Foldable for the project at the end of the chapter.

What do you think NOW?

You first read the statements below at the beginning of the chapter.

1. The earliest model of an atom contained only protons and electrons.

2. Air fills most of an atom.

3. In the present-day model of the atom, the nucleus of the atom is at the center of an electron cloud.

Did you change your mind about whether you agree or disagree with the statements? Rewrite any false statements to make them true.

Use Vocabulary

1 The smallest piece of the element gold is a gold ________.

2 **Write** a sentence that describes the nucleus of an atom.

3 **Define** *electron cloud* in your own words.

Understand Key Concepts 

4 What is an atom mostly made of?

A. air
B. empty space
C. neutrons
D. protons

5 Why have scientists only recently been able to see atoms?

A. Atoms are too small to see with ordinary microscopes.
B. Early experiments disproved the idea of atoms.
C. Scientists didn't know atoms existed.
D. Scientists were not looking for atoms.

6 **Draw** Thomson's model of the atom, and label the parts of the drawing.

7 **Explain** how Rutherford's students knew that Thomson's model of the atom needed to change.

Interpret Graphics

8 **Contrast** Copy the graphic organizer below and use it to contrast the locations of electrons in Thomson's, Rutherford's, Bohr's, and the modern models of the atom.

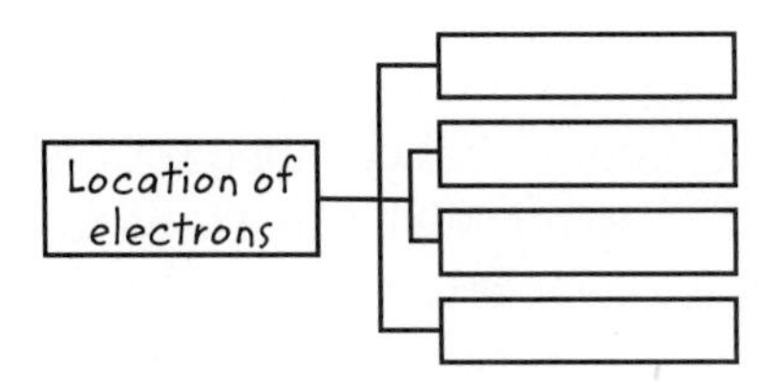

Critical Thinking

9 **Explain** what might have happened in Rutherford's experiment if he had used a thin sheet of copper instead of a thin sheet of gold.

Subatomic Particles

Welcome To The Particle Zoo

QUARKS

BOSONS

LEPTONS

Much has changed since Democritus and Aristotle studied atoms.

When Democritus and Aristotle developed ideas about matter, they probably never imagined the kinds of research being performed today! From the discovery of electrons, protons, and neutrons to the exploration of quarks and other particles, the atomic model continues to change.

You've learned about quarks, which make up protons and neutrons. But quarks are not the only kind of particles! In fact, some scientists call the collection of particles that have been discovered the particle zoo, because different types of particles have unique characteristics, just like the different kinds of animals in a zoo.

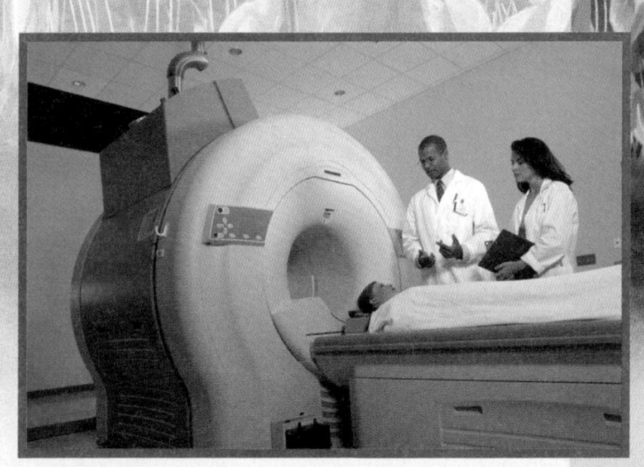

▲ **MRIs are just one way in which particle physics technology is applied.**

In addition to quarks, scientists have discovered a group of particles called leptons, which includes the electron. Gluons and photons are examples of bosons—particles that carry forces. Some particles, such as the Higgs Boson, have been predicted to exist but have yet to be observed in experiments.

Identifying and understanding the particles that make up matter is important work. However, it might be difficult to understand why time and money are spent to learn more about tiny subatomic particles. How can this research possibly affect everyday life? Research on subatomic particles has changed society in many ways. For example, magnetic resonance imaging (MRI), a tool used to diagnose medical problems, uses technology that was developed to study subatomic particles. Cancer treatments using protons, neutrons, and X-rays are all based on particle physics technology. And, in the 1990s, the need for particle physicists to share information with one another led to the development of the World Wide Web!

It's Your Turn

RESEARCH AND REPORT Learn more about research on subatomic particles. Find out about one recent discovery. Make a poster to share what you learn with your classmates.

Lesson 2

Protons, Neutrons, and Electrons—How Atoms Differ

Reading Guide

Key Concepts
ESSENTIAL QUESTIONS

- What happens during nuclear decay?
- How does a neutral atom change when its number of protons, electrons, or neutrons changes?

Vocabulary

atomic number p. 249
isotope p. 250
mass number p. 250
average atomic mass p. 251
radioactive p. 252
nuclear decay p. 253
ion p. 254

Multilingual eGlossary

BrainPOP®

Inquiry Is this glass glowing?

Under natural light, this glass vase is yellow. But when exposed to ultraviolet light, it glows green! That's because it is made of uranium glass, which contains small amounts of uranium, a radioactive element. Under ultraviolet light, the glass emits radiation.

Inquiry Launch Lab 15 minutes

How many different things can you make?

Many buildings are made of just a few basic building materials, such as wood, nails, and glass. You can combine those materials in many different ways to make buildings of various shapes and sizes. How many things can you make from three materials?

1. Read and complete a lab safety form.
2. Use **colored building blocks** to make as many different objects as you can with the following properties:
   - Each object must have a different number of red blocks.
   - Each object must have an equal number of red and blue blocks.
   - Each object must have at least as many yellow blocks as red blocks but can have no more than two extra yellow blocks.
3. As you complete each object, record in your Science Journal the number of each color of block used to make it. For example, R = 1; B = 1; Y = 2.
4. When time is called, compare your objects with others in the class.

Think About This

1. How many different objects did you make? How many different objects did the class make?
2. How many objects do you think you could make out of the three types of blocks?
3. **Key Concept** In what ways does changing the number of building blocks change the properties of the objects?

The Parts of the Atom

If you could see inside any atom, you probably would see the same thing—empty space surrounding a very tiny nucleus. A look inside the nucleus would reveal positively charged protons and neutral neutrons. Negatively charged electrons would be whizzing by in the empty space around the nucleus.

Table 2 compares the properties of protons, neutrons, and electrons. Protons and neutrons have about the same mass. The mass of electrons is much smaller than the mass of protons or neutrons. That means most of the mass of an atom is found in the nucleus. In this lesson, you will learn that, while all atoms contain protons, neutrons, and electrons, the numbers of these particles are different for different types of atoms.

Table 2 Properties of Protons, Neutrons, and Electrons

	Electron	Proton	Neutron
Symbol	e−	p	n
Charge	1−	1+	0
Location	electron cloud around the nucleus	nucleus	nucleus
Relative mass	1/1,840	1	1

Different Elements—Different Numbers of Protons

Look at the periodic table on the inside back cover of this book. Notice that there are more than 115 different elements. Recall that an element is a substance made from atoms that all have the same number of protons. For example, the element carbon is made from atoms that all have six protons. Likewise, all atoms that have six protons are carbon atoms. *The number of protons in an atom of an element is the element's* **atomic number.** The atomic number is the whole number listed with each element on the periodic table.

What makes an atom of one element different from an atom of another element? Atoms of different elements contain different numbers of protons. For example, oxygen atoms contain eight protons; nitrogen atoms contain seven protons. Different elements have different atomic numbers. **Figure 11** shows some common elements and their atomic numbers.

Neutral atoms of different elements also have different numbers of electrons. In a neutral atom, the number of electrons equals the number of protons. Therefore, the number of positive charges equals the number of negative charges.

Reading Check What two numbers can be used to identify an element?

FOLDABLES®

Create a three-tab book and label it as shown. Use it to organize the three ways that atoms can differ.

Different Numbers of:
Protons | Neutrons | Electrons

Figure 11 Atoms of different elements contain different numbers of protons.

Visual Check Explain the difference between an oxygen atom and a carbon atom.

Different Elements Review Personal Tutor

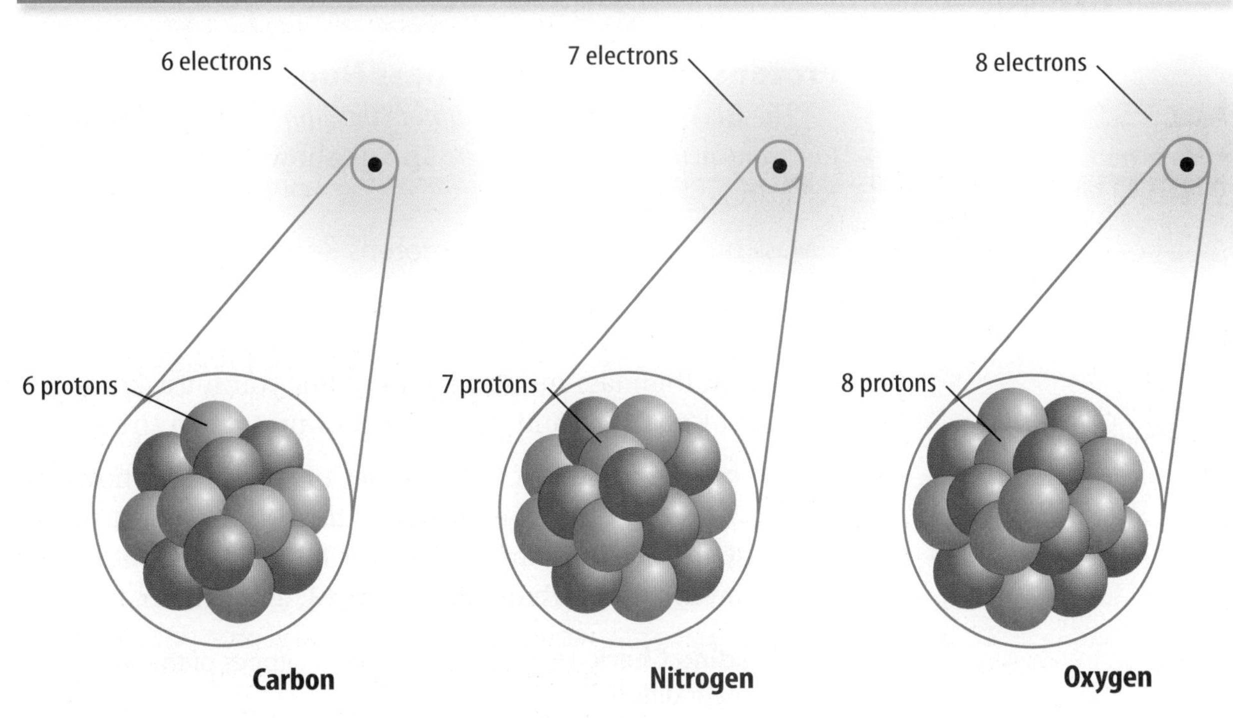

Math Skills 

Use Percentages

You can calculate the average atomic mass of an element if you know the percentage of each isotope in the element. Lithium (Li) contains 7.5% Li-6 and 92.5% Li-7. What is the average atomic mass of Li?

1. Divide each percentage by 100 to change to decimal form.
   $\frac{7.5\%}{100} = 0.075$
   $\frac{92.5\%}{100} = 0.925$
2. Multiply the mass of each isotope by its decimal percentage.
   $6 \times 0.075 = 0.45$
   $7 \times 0.925 = 6.475$
3. Add the values together to get the average atomic mass.
   $0.45 + 6.475 = 6.93$

Practice

Nitrogen (N) contains 99.63% N-14 and 0.37% N-15. What is the average atomic mass of nitrogen?

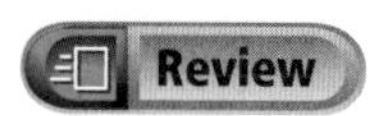

- **Math Practice**
- **Personal Tutor**

Word Origin

isotope
from Greek *isos*, means "equal"; and *topos*, means "place"

Table 3 Naturally Occurring Isotopes of Carbon

Isotope	Carbon-12 Nucleus	Carbon-13 Nucleus	Carbon-14 Nucleus
Abundance	98.89%	<1.11%	<0.01%
Protons	6	6	6
Neutrons	+6	+7	+8
Mass Number	12	13	14

Neutrons and Isotopes

You have read that atoms of the same element have the same numbers of protons. However, atoms of the same element can have different numbers of neutrons. For example, carbon atoms all have six protons, but some carbon atoms have six neutrons, some have seven neutrons, and some have eight neutrons. These three different types of carbon atoms, shown in **Table 3,** are called isotopes. **Isotopes** *are atoms of the same element that have different numbers of neutrons.* Most elements have several isotopes.

Protons, Neutrons, and Mass Number

The **mass number** *of an atom is the sum of the number of protons and neutrons in an atom.* This is shown in the following equation.

$$\text{Mass number} = \text{number of protons} + \text{number of neutrons}$$

Any one of these three quantities can be determined if you know the value of the other two quantities. For example, to determine the mass number of an atom, you must know the number of neutrons and the number of protons in the atom.

The mass numbers of the isotopes of carbon are shown in **Table 3.** An isotope often is written with the element name followed by the mass number. Using this method, the isotopes of carbon are written carbon-12, carbon-13, and carbon-14.

Reading Check How do two different isotopes of the same element differ?

Average Atomic Mass

You might have noticed that the periodic table does not list mass numbers or the numbers of neutrons. This is because a given element can have several isotopes. However, you might notice that there is a decimal number listed with most elements, as shown in **Figure 12.** This decimal number is the average atomic mass of the element. The **average atomic mass** *of an element is the average mass of the element's isotopes, weighted according to the abundance of each isotope.*

Table 3 shows the three isotopes of carbon. The average atomic mass of carbon is 12.01. Why isn't the average atomic mass 13? After all, the average of the mass numbers 12, 13, and 14 is 13. The average atomic mass is weighted based on each isotope's abundance—how much of each isotope is present on Earth. Almost 99 percent of Earth's carbon is carbon-12. That is why the average atomic mass is close to 12.

 **Reading Check** What does the term *weighted average* mean?

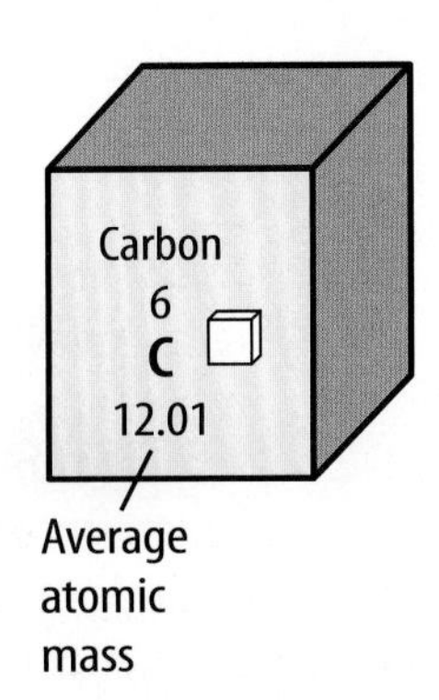

Figure 12 The element carbon has several isotopes. The decimal number 12.01 is the average atomic mass of these isotopes.

Inquiry MiniLab

20 minutes

How many penny isotopes do you have?

All pennies look similar, and all have a value of one cent. But do they have the same mass? Let's find out.

1. Read and complete a lab safety form.
2. Copy the data table into your Science Journal.
3. Use a **balance** to find the mass of **10 pennies minted before 1982.** Record the mass in the data table.
4. Divide the mass by 10 to find the average mass of one penny. Record the answer.
5. Repeat steps 3 and 4 with **10 pennies minted after 1982.**
6. Have a team member combine pre- and post-1982 pennies for a total of 10 pennies. Find the mass of the ten pennies and the average mass of one penny. Record your observations.

Penny Sample	Mass of 10 pennies (g)	Average mass of 1 penny (g)
Pre-1982		
Post-1982		
Unknown mix		

Analyze and Conclude

1. **Compare and Contrast** How did the average mass of pre- and post-1982 pennies compare?
2. **Draw Conclusions** How many pennies of each type were in the 10 pennies assembled by your partner? How do you know?
3. **Key Concept** How does this activity relate to the way in which scientists calculate the average atomic mass of an element?

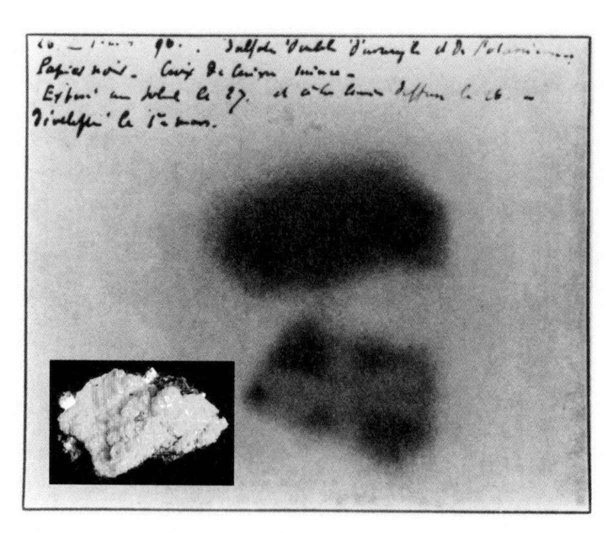

▲ **Figure 13** The black and white photo shows Henri Becquerel's photographic plate. The dark area on the plate was exposed to radiation given off by uranium in the mineral even though the mineral was not exposed to sunlight.

ACADEMIC VOCABULARY

spontaneous
(adjective) occurring without external force or cause

Figure 14 Marie Curie studied radioactivity and discovered two new radioactive elements—polonium and radium. ▼

Radioactivity

More than 1,000 years ago, people tried to change lead into gold by performing chemical reactions. However, none of their reactions were successful. Why not? Today, scientists know that a chemical reaction does not change the number of protons in an atom's nucleus. If the number of protons does not change, the element does not change. But in the late 1800s, scientists discovered that some elements change into other elements **spontaneously.** How does this happen?

An Accidental Discovery

In 1896, a scientist named Henri Becquerel (1852–1908) studied minerals containing the element uranium. When these minerals were exposed to sunlight, they gave off a type of energy that could pass through paper. If Becquerel covered a photographic plate with black paper, this energy would pass through the paper and expose the film. One day, Becquerel left the mineral next to a wrapped, unexposed plate in a drawer. Later, he opened the drawer, unwrapped the plate, and saw that the plate contained an image of the mineral, as shown in **Figure 13.** The mineral spontaneously emitted energy, even in the dark! Sunlight wasn't required. What was this energy?

Radioactivity

Becquerel shared his discovery with fellow scientists Pierre and Marie Curie. Marie Curie (1867–1934), shown in **Figure 14,** called *elements that spontaneously emit radiation* **radioactive.** Becquerel and the Curies discovered that the radiation released by uranium was made of energy and particles. This radiation came from the nuclei of the uranium atoms. When this happens, the number of protons in one atom of uranium changes. When uranium releases radiation, it changes to a different element!

Types of Decay

Radioactive elements contain unstable nuclei. **Nuclear decay** *is a process that occurs when an unstable atomic nucleus changes into another more stable nucleus by emitting radiation.* Nuclear decay can produce three different types of radiation—alpha particles, beta particles, and gamma rays. **Figure 15** compares the three types of nuclear decay.

Alpha Decay An alpha particle is made of two protons and two neutrons. When an atom releases an alpha particle, its atomic number decreases by two. Uranium-238 decays to thorium-234 through the process of alpha decay.

Beta Decay When beta decay occurs, a neutron in an atom changes into a proton and a high-energy electron called a beta particle. The new proton becomes part of the nucleus, and the beta particle is released. In beta decay, the atomic number of an atom increases by one because it has gained a proton.

Gamma Decay Gamma rays do not contain particles, but they do contain a lot of energy. In fact, gamma rays can pass through thin sheets of lead! Because gamma rays do not contain particles, the release of gamma rays does not change one element into another element.

Key Concept Check What happens during radioactive decay?

Uses of Radioactive Isotopes

The energy released by radioactive decay can be both harmful and beneficial to humans. Too much radiation can damage or destroy living cells, making them unable to function properly. Some organisms contain cells, such as cancer cells, that are harmful to the organism. Radiation therapy can be beneficial to humans by destroying these harmful cells.

Figure 15 Alpha and beta decay change one element into another element.

Visual Check Explain the change in atomic number for each type of decay.

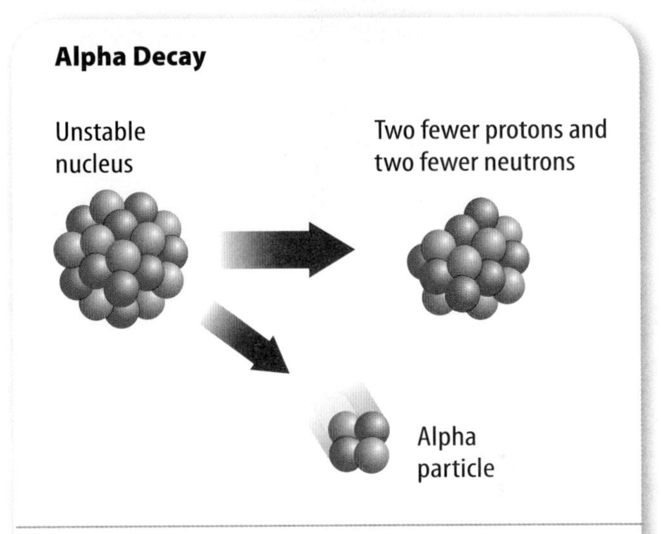

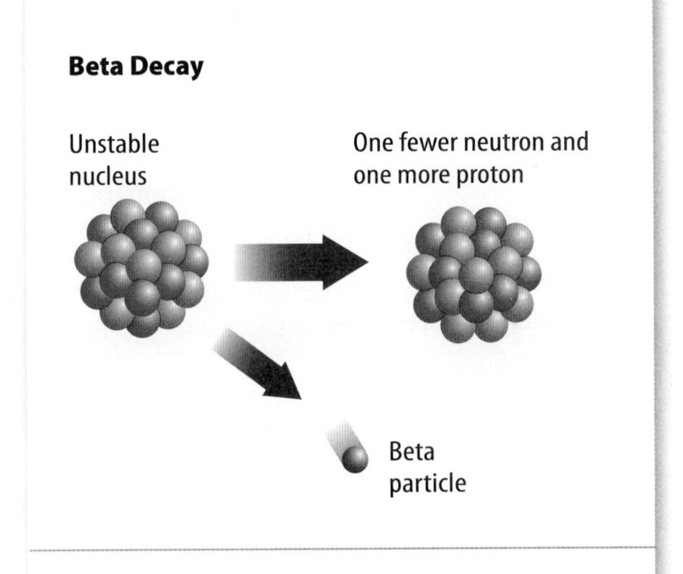

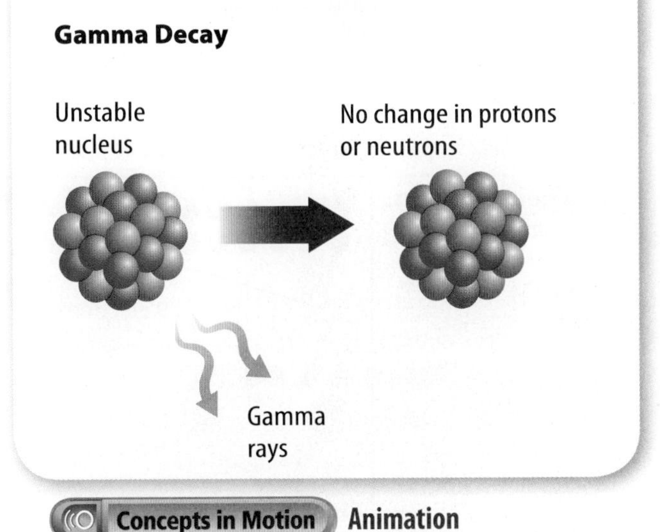

Concepts in Motion Animation

Ions—Gaining or Losing Electrons

What happens to a neutral atom if it gains or loses electrons? Recall that a neutral atom has no overall charge. This is because it contains equal numbers of positively charged protons and negatively charged electrons. When electrons are added to or removed from an atom, that atom becomes an ion. *An* **ion** *is an atom that is no longer neutral because it has gained or lost electrons.* An ion can be positively or negatively charged depending on whether it has lost or gained electrons.

Positive Ions

When a neutral atom loses one or more electrons, it has more protons than electrons. As a result, it has a positive charge. An atom with a positive charge is called a positive ion. A positive ion is represented by the element's symbol followed by a superscript plus sign ($^+$). For example, **Figure 16** shows how sodium (Na) becomes a positive sodium ion (Na^+).

Negative Ions

When a neutral atom gains one or more electrons, it now has more electrons than protons. As a result, the atom has a negative charge. An atom with a negative charge is called a negative ion. A negative ion is represented by the element's symbol followed by a superscript negative sign ($^-$). **Figure 16** shows how fluorine (F) becomes a fluoride ion (F^-).

Key Concept Check How does a neutral atom change when its number of protons, electrons, or neutrons changes?

Figure 16 An ion is formed when a neutral atom gains or loses an electron.

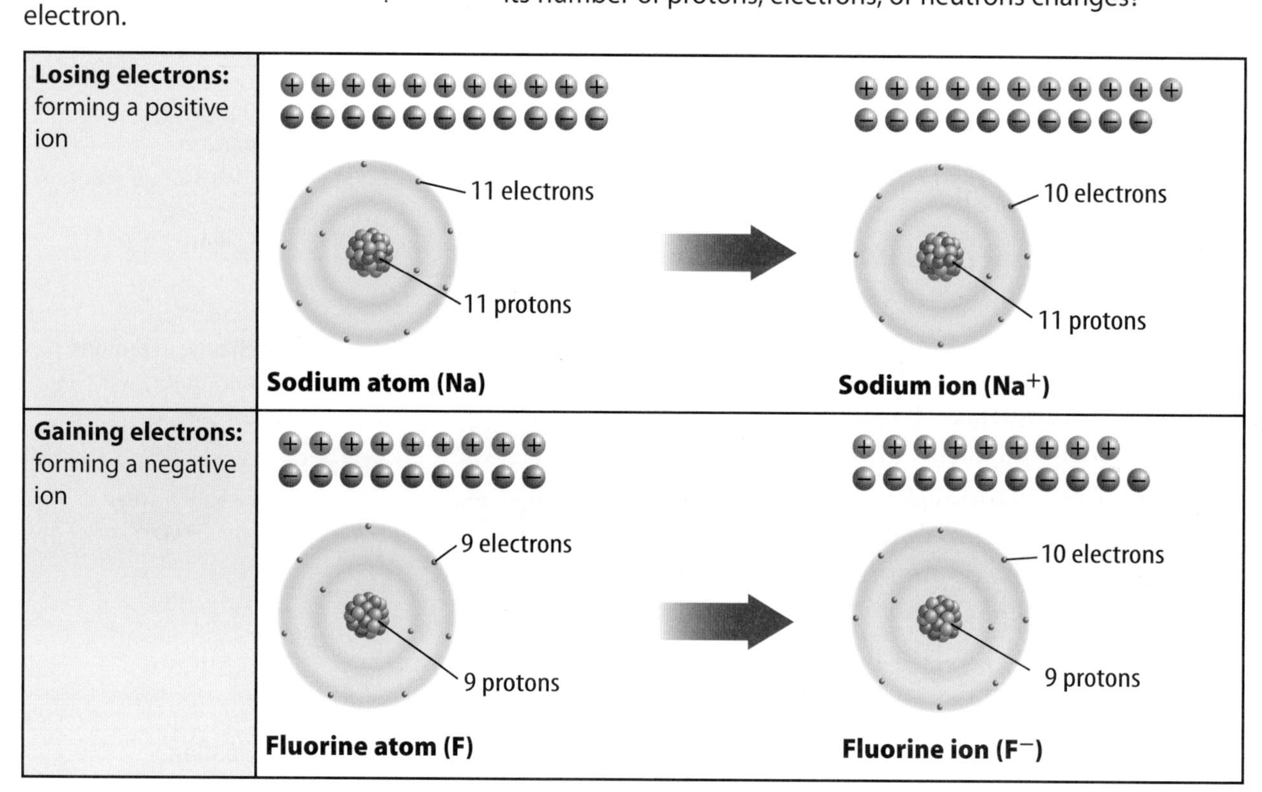

Lesson 2 Review

Assessment **Online Quiz**

Inquiry **Virtual Lab**

Visual Summary

Carbon Nitrogen

Different elements contain different numbers of protons.

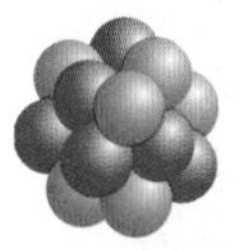
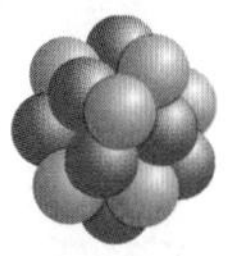

Isotopes

Two isotopes of a given element contain different numbers of neutrons.

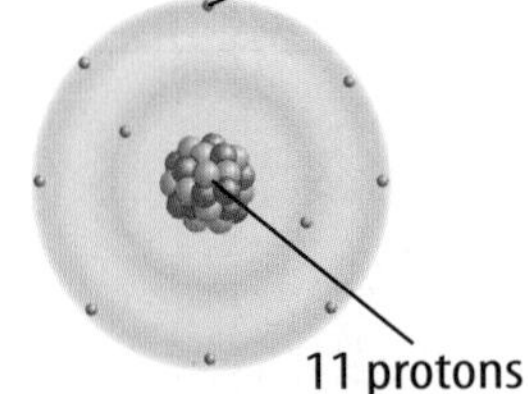

Sodium ion (Na^+)

When a neutral atom gains or loses an electron, it becomes an ion.

FOLDABLES®

Use your lesson Foldable to review the lesson. Save your Foldable for the project at the end of the chapter.

What do you think NOW?

You first read the statements below at the beginning of the chapter.

4. All atoms of the same element have the same number of protons.

5. Atoms of one element cannot be changed into atoms of another element.

6. Ions form when atoms lose or gain electrons.

Did you change your mind about whether you agree or disagree with the statements? Rewrite any false statements to make them true.

Use Vocabulary

1 The number of protons in an atom of an element is its ________.

2 Nuclear decay occurs when an unstable atomic nucleus changes into another nucleus by emitting ________.

3 **Describe** how two isotopes of nitrogen differ from two nitrogen ions.

Understand Key Concepts 

4 An element's average atomic mass is calculated using the masses of its

A. electrons. **C.** neutrons.
B. isotopes. **D.** protons.

5 **Compare and contrast** oxygen-16 and oxygen-17.

6 **Show** what happens to the electrons of a neutral calcium atom (Ca) when it is changed into a calcium ion (Ca^{2+}).

Interpret Graphics

7 **Contrast** Copy and fill in this graphic organizer to contrast how different elements, isotopes, and ions are produced.

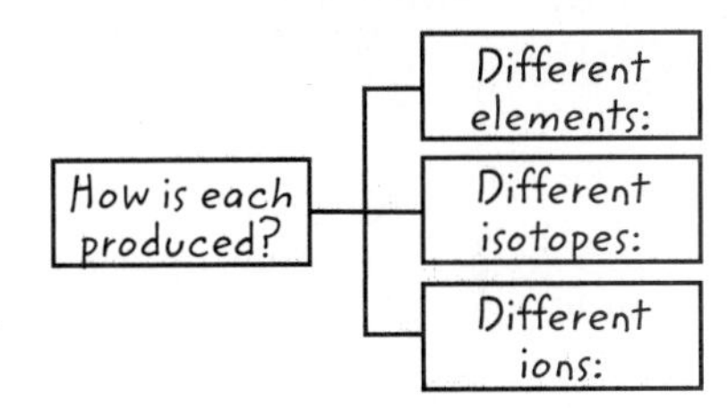

Critical Thinking

8 **Consider** Find two neighboring elements on the periodic table whose positions would be reversed if they were arranged by atomic mass instead of atomic number.

9 **Infer** Can an isotope also be an ion?

Math Skills

Math Practice

10 A sample of copper (Cu) contains 69.17% Cu-63. The remaining copper atoms are Cu-65. What is the average atomic mass of copper?

2 class periods

Communicate Your Knowledge About the Atom

Materials

computer

creative building materials

drawing and modeling materials

office supplies

Also needed: recording devices, software, or other equipment for multimedia presentations

In this chapter, you have learned many things about atoms. Suppose that you are asked to take part in an atom fair. Each exhibit in the fair will help visitors understand something new about atoms in an exciting and interesting way. What will your exhibit be like? Will you hold a mock interview with Democritus or Rutherford? Will you model a famous experiment and explain its conclusion? Can visitors assemble or make models of their own atoms? Is there a multimedia presentation? Will your exhibit be aimed at children or adults? The choice is yours!

Question

Which concepts about the atom did you find most interesting? How can you present the information in exciting, creative, and perhaps unexpected ways? Think about whether you will present the information yourself or have visitors interact with the exhibit.

Procedure

1. In your Science Journal, write your ideas about the following questions:

- What specific concepts about the atom do you want your exhibit to teach?
- How will you present the information to your visitors?
- How will you make the information exciting and interesting to keep your visitors' attention?

2. Outline the steps in preparing for your exhibit.

- What materials and equipment will you need?
- How much time will it take to prepare each part of your exhibit?
- Will you involve anyone else? For example, if you are going to interview a scientist about an early model of the atom, who will play the scientist? What questions will you ask?

3. Have your teacher approve your plan.
4. Follow the steps you outlined, and prepare your exhibit.
5. Ask family members and/or several friends to view your exhibit. Invite them to tell you what they've learned from your exhibit. Compare this with what you had expected to teach in your exhibit.
6. Ask your friends for feedback about what could be more effective in teaching the concepts you intend to teach.
7. Modify your exhibit to make it more effective.
8. If you can, present your exhibit to visitors of various ages, including teachers and students from other classes. Have visitors fill out a comment form.

Analyze and Conclude

9. **Infer** What did visitors to your exhibit find the most and least interesting? How do you know?
10. **Predict** What would you do differently if you had a chance to plan your exhibit again? Why?
11. **The Big Idea** In what ways did your exhibit help visitors understand the current model of the atom?

Communicate Your Results

After the fair is over, discuss the visitors' comments and how you might improve organization or individual experiences if you were to do another fair.

Inquiry Extension

Design and describe an interactive game or activity that would teach the same concept you used for your exhibit. Invite other students to comment on your design.

Lab Tips

- For very young visitors, you might draw a picture book about the three parts of the atom and make them into cartoon characters.
- Plan your exhibit so visitors spend about 5 minutes there. Don't try to present too much or too little information.
- Remember that a picture is worth a thousand words!

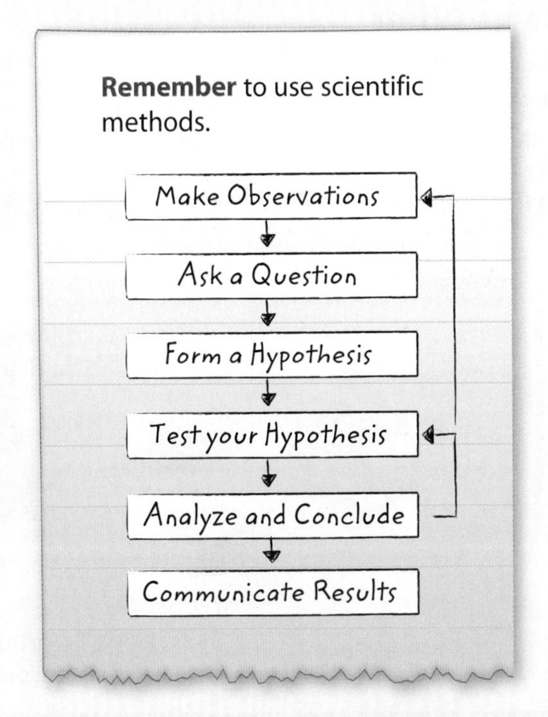

Chapter 7 Study Guide

An atom is the smallest unit of an element and is made mostly of empty space. It contains a tiny nucleus surrounded by an electron cloud.

Key Concepts Summary

Lesson 1: Discovering Parts of the Atom

- If you were to divide an element into smaller and smaller pieces, the smallest piece would be an **atom.**
- Atoms are so small that they can be seen only by powerful scanning microscopes.
- The first model of the atom was a solid sphere. Now, scientists know that an atom contains a dense positive **nucleus** surrounded by an **electron cloud.**

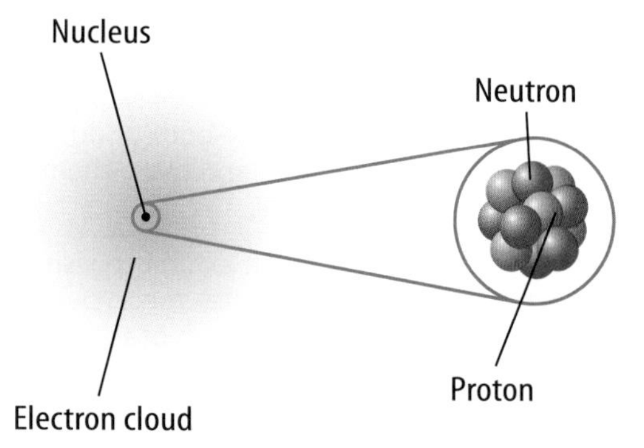

Vocabulary

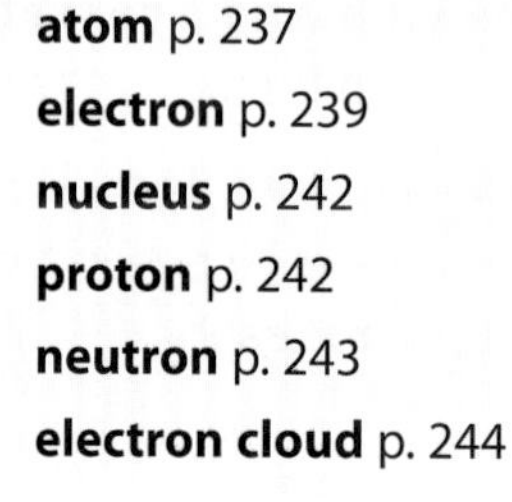

atom p. 237
electron p. 239
nucleus p. 242
proton p. 242
neutron p. 243
electron cloud p. 244

Lesson 2: Protons, Neutrons, and Electrons—How Atoms Differ

- **Nuclear decay** occurs when an unstable atomic nucleus changes into another more stable nucleus by emitting radiation.
- Different elements contain different numbers of protons. Two **isotopes** of the same element contain different numbers of neutrons. When a neutral atom gains or loses an electron, it becomes an **ion.**

Nuclear Decay

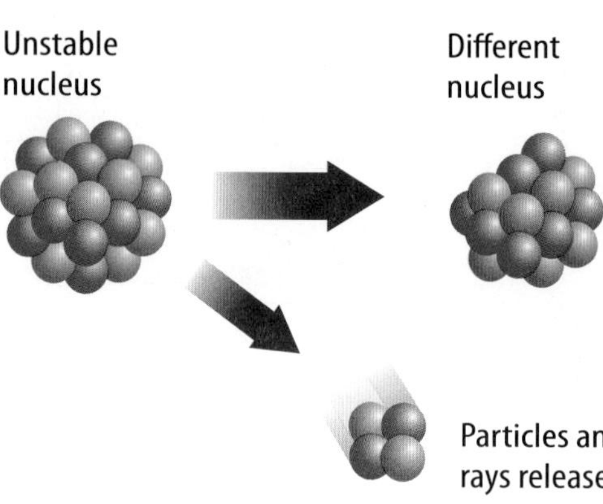

Vocabulary

atomic number p. 249
isotope p. 250
mass number p. 250
average atomic mass p. 251
radioactive p. 252
nuclear decay p. 253
ion p. 254

- Personal Tutor
- Vocabulary eGames
- Vocabulary eFlashcards

FOLDABLES® Chapter Project

Assemble your lesson Foldables as shown to make a Chapter Project. Use the project to review what you have learned in this chapter.

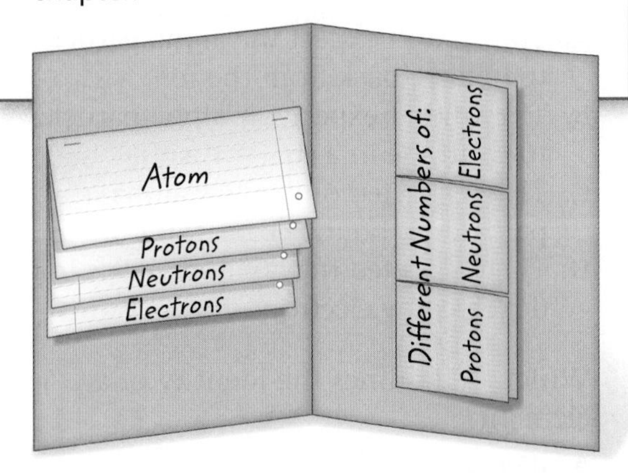

Use Vocabulary

1. A(n) ________ is a very small particle that is the basic unit of matter.
2. Electrons in an atom move throughout the ________ surrounding the nucleus.
3. ________ is the weighted average mass of all of an element's isotopes.
4. All atoms of a given element have the same number of ________.
5. When ________ occurs, one element is changed into another element.
6. Isotopes have the same ________, but different mass numbers.

Link Vocabulary and Key Concepts

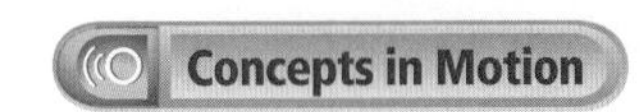

Interactive Concept Map

Copy this concept map, and then use vocabulary terms from the previous page to complete the concept map.

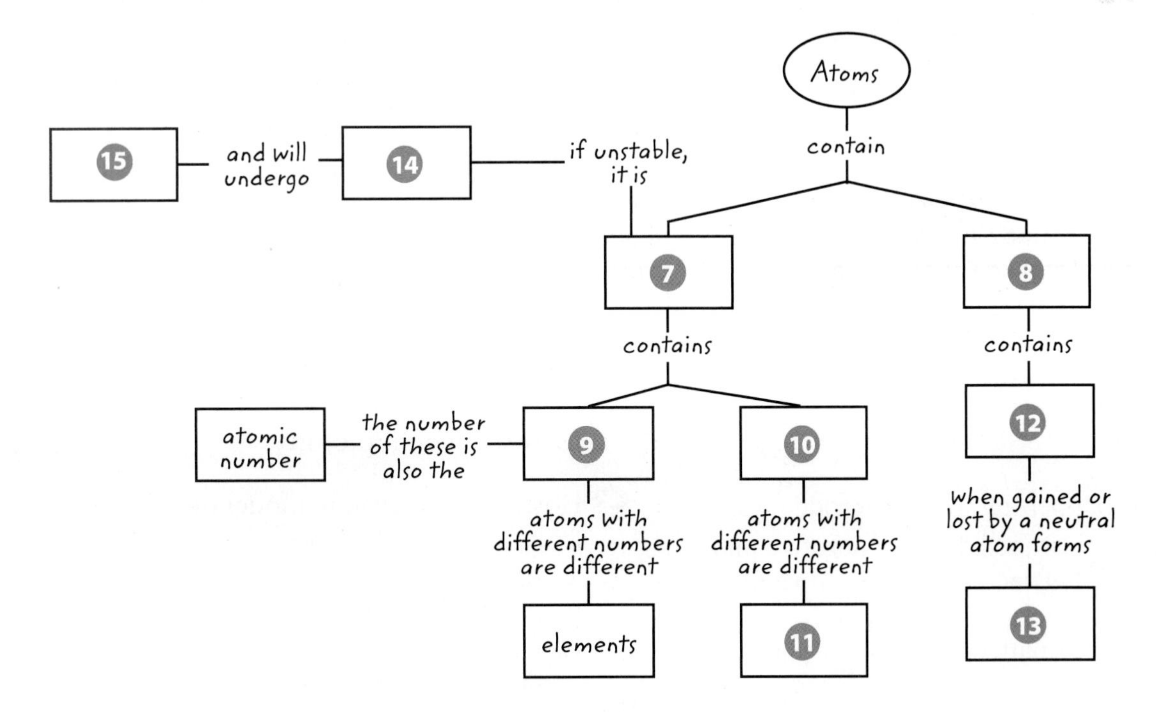

Chapter 7 Review

Understand Key Concepts 

1. Which part of an atom makes up most of its volume?
   A. its electron cloud
   B. its neutrons
   C. its nucleus
   D. its protons

2. What did Democritus believe an atom was?
   A. a solid, indivisible object
   B. a tiny particle with a nucleus
   C. a nucleus surrounded by an electron cloud
   D. a tiny nucleus with electrons surrounding it

3. If an ion contains 10 electrons, 12 protons, and 13 neutrons, what is the ion's charge?
   A. 2−
   B. 1−
   C. 2+
   D. 3+

4. J.J. Thomson's experimental setup is shown below.

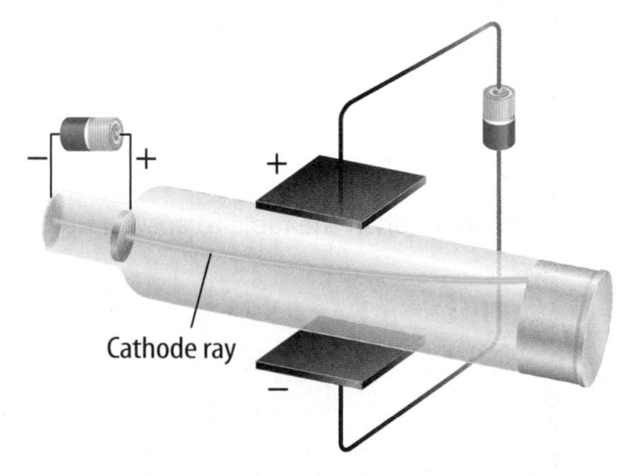

What is happening to the cathode rays?
   A. They are attracted to the negative plate.
   B. They are attracted to the positive plate.
   C. They are stopped by the plates.
   D. They are unaffected by either plate.

5. How many neutrons does iron-59 have?
   A. 30
   B. 33
   C. 56
   D. 59

6. Why were Rutherford's students surprised by the results of the gold foil experiment?
   A. They didn't expect the alpha particles to bounce back from the foil.
   B. They didn't expect the alpha particles to continue in a straight path.
   C. They expected only a few alpha particles to bounce back from the foil.
   D. They expected the alpha particles to be deflected by electrons.

7. Which determines the identity of an element?
   A. its mass number
   B. the charge of the atom
   C. the number of its neutrons
   D. the number of its protons

8. The figure below shows which of the following?

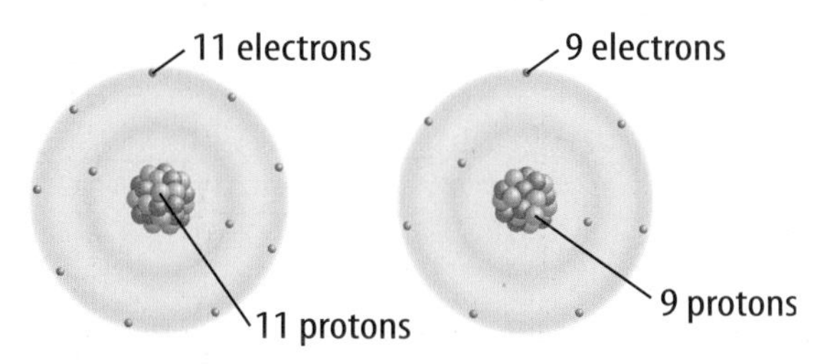

   A. two different elements
   B. two different ions
   C. two different isotopes
   D. two different protons

9. How is Bohr's atomic model different from Rutherford's model?
   A. Bohr's model has a nucleus.
   B. Bohr's model has electrons.
   C. Electrons in Bohr's model are located farther from the nucleus.
   D. Electrons in Bohr's model are located in circular energy levels.

Assessment
Online Test Practice

Critical Thinking

10. **Consider** what would have happened in the gold foil experiment if Dalton's theory had been correct.

11. **Contrast** How does Bohr's model of the atom differ from the present-day atomic model?

12. **Describe** the electron cloud using your own analogy.

13. **Summarize** how radioactive decay can produce new elements.

14. **Hypothesize** What might happen if a negatively charged ion comes into contact with a positively charged ion?

15. **Infer** Why isn't mass number listed with each element on the periodic table?

16. **Explain** How is the average atomic mass calculated?

17. **Infer** Oxygen has three stable isotopes.

Isotope	Average Atomic Mass
Oxygen-16	0.99757
Oxygen-17	0.00038
Oxygen-18	0.00205

What can you determine about the average atomic mass of oxygen without calculating it?

Writing in Science

18. **Write** a newspaper article that describes how the changes in the atomic model provide an example of the scientific process in action.

REVIEW THE BIG IDEA

19. **Describe** the current model of the atom. Explain the size of atoms. Also explain the charge, the location, and the size of protons, neutrons, and electrons.

20. **Summarize** The Large Hadron Collider, shown below, is continuing the study of matter and energy. Use a set of four drawings to summarize how the model of the atom changed from Thomson, to Rutherford, to Bohr, to the modern model.

Math Skills

Math Practice

Use Percentages

Use the information in the table to answer questions 21 and 22.

Magnesium (Mg) Isotope	Percent Found in Nature
Mg-24	78.9%
Mg-25	10.0%
Mg-26	

21. What is the percentage of Mg-26 found in nature?

22. What is the average atomic mass of magnesium?

Standardized Test Practice

Record your answers on the answer sheet provided by your teacher or on a sheet of paper.

Multiple Choice

1 Which best describes an atom?

A a particle with a single negative charge

B a particle with a single positive charge

C the smallest particle that still represents a compound

D the smallest particle that still represents an element

Use the figure below to answer questions 2 and 3.

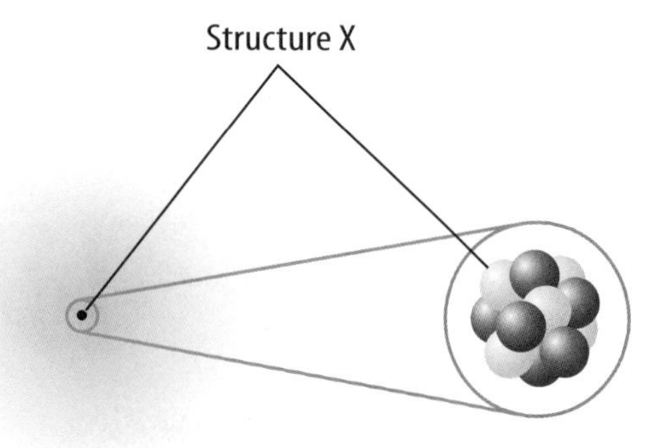

2 What is Structure X?

A an electron

B a neutron

C a nucleus

D a proton

3 Which best describes Structure X?

A most of the atom's mass, neutral charge

B most of the atom's mass, positive charge

C very small part of the atom's mass, negative charge

D very small part of the atom's mass, positive charge

4 Which is true about the size of an atom?

A It can only be seen using a scanning tunneling microscope.

B It is about the size of the period at the end of this sentence.

C It is large enough to be seen using a magnifying lens.

D It is too small to see with any type of microscope.

Use the figure below to answer question 5.

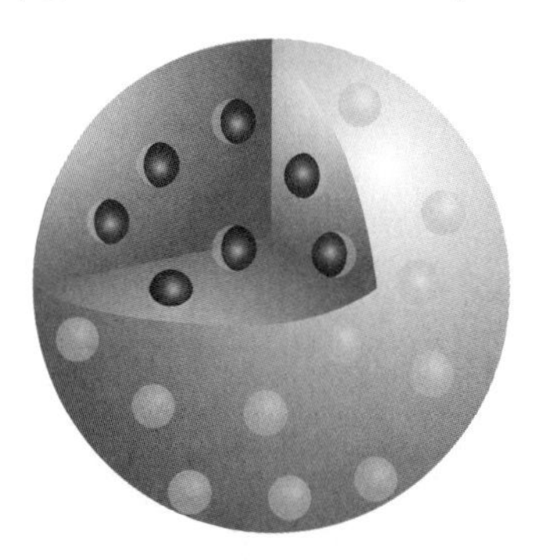

5 Whose model for the atom is shown?

A Bohr's

B Dalton's

C Rutherford's

D Thomson's

6 What structure did Rutherford discover?

A the atom

B the electron

C the neutron

D the nucleus

Use the table below to answer questions 7–9.

Particle	Number of Protons	Number of Neutrons	Number of Electrons
1	4	5	2
2	5	5	5
3	5	6	5
4	6	6	6

7 What is atomic number of particle 3?

A 3

B 5

C 6

D 11

8 Which particles are isotopes of the same element?

A 1 and 2

B 2 and 3

C 2 and 4

D 3 and 4

9 Which particle is an ion?

A 1

B 2

C 3

D 4

10 Which reaction starts with a neutron and results in the formation of a proton and a high-energy electron?

A alpha decay

B beta decay

C the formation of positive ion

D the formation of negative ion

Constructed Response

Use the figure below to answer questions 11 and 12.

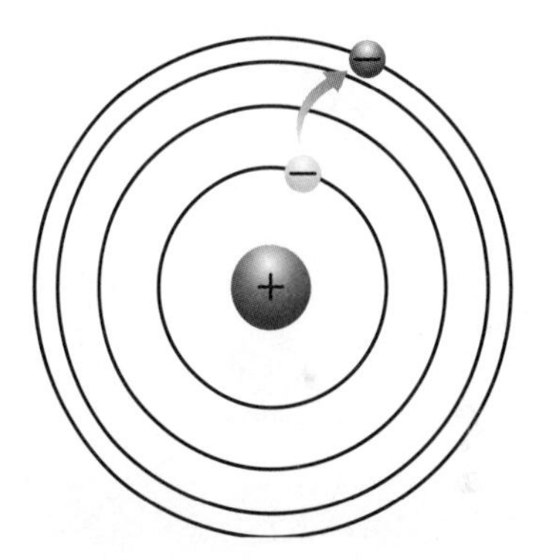

11 Identify the atomic model shown in the figure, and describe its characteristics.

12 How does this atomic model differ from the modern atomic model?

13 Compare two different neutral isotopes of the same element. Then compare two different ions of the same element. What do all of these particles have in common?

14 How does nuclear decay differ from the formation of ions? What parts of the atom are affected in each type of change?

NEED EXTRA HELP?														
If You Missed Question...	1	2	3	4	5	6	7	8	9	10	11	12	13	14
Go to Lesson...	1	1	1	1	1	1	2	2	2	2	1	1	2	2

Chapter 8

Elements and Chemical Bonds

THE BIG IDEA How do elements join together to form chemical compounds?

Inquiry How do they combine?

How many different words could you type using just the letters on a keyboard? The English alphabet has only 26 letters, but a dictionary lists hundreds of thousands of words using these letters! Similarly only about 115 different elements make all kinds of matter.

- How do so few elements form so many different kinds of matter?
- Why do you think different types of matter have different properties?
- How are atoms held together to produce different types of matter?

Get Ready to Read

What do you think?

Before you read, decide if you agree or disagree with each of these statements. As you read this chapter, see if you change your mind about any of the statements.

1. Elements rarely exist in pure form. Instead, combinations of elements make up most of the matter around you.
2. Chemical bonds that form between atoms involve electrons.
3. The atoms in a water molecule are more chemically stable than they would be as individual atoms.
4. Many substances dissolve easily in water because opposite ends of a water molecule have opposite charges.
5. Losing electrons can make some atoms more chemically stable.
6. Metals are good electrical conductors because they tend to hold onto their valence electrons very tightly.

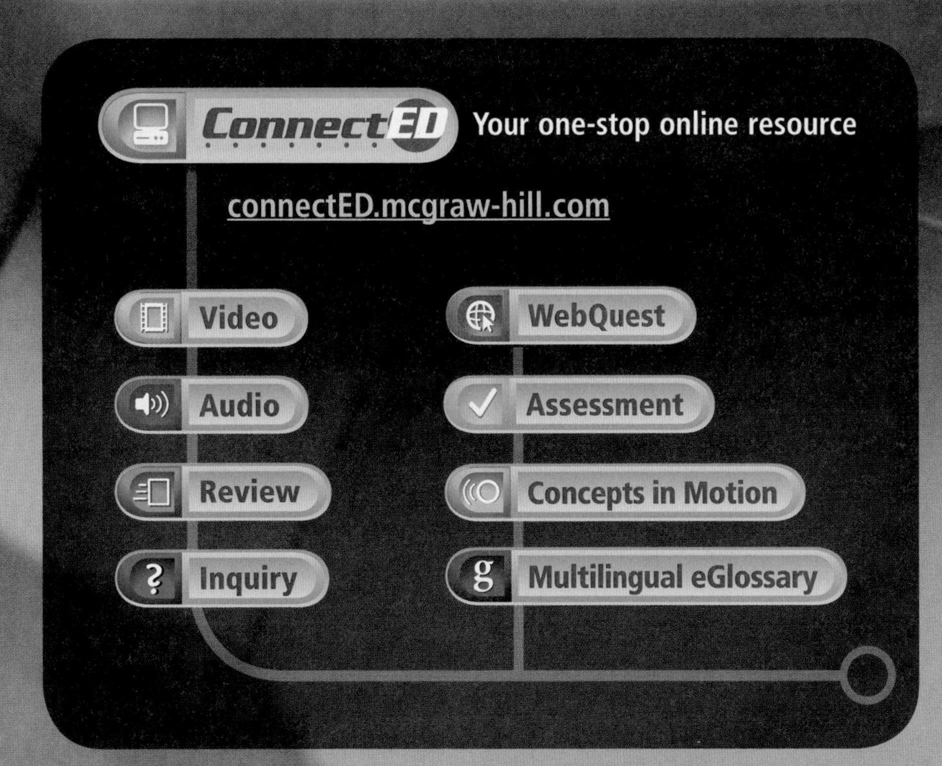

Lesson 1

Electrons and Energy Levels

Reading Guide

Key Concepts
ESSENTIAL QUESTIONS

- How is an electron's energy related to its distance from the nucleus?
- Why do atoms gain, lose, or share electrons?

Vocabulary

chemical bond p. 268
valence electron p. 270
electron dot diagram p. 271

Multilingual eGlossary

Video BrainPOP®

Inquiry Are pairs more stable?

Rowing can be hard work, especially if you are part of a racing team. The job is made easier because the rowers each pull on the water with a pair of oars. How do pairs make the boat more stable?

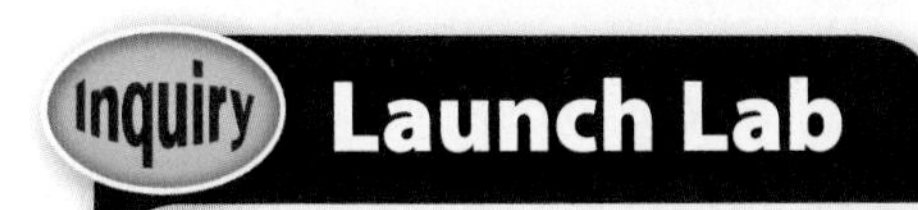

Launch Lab

20 minutes

How is the periodic table organized?

How do you begin to put together a puzzle of a thousand pieces? You first sort similar pieces into groups. All edge pieces might go into one pile. All blue pieces might go into another pile. Similarly, scientists placed the elements into groups based on their properties. They created the periodic table, which organizes information about all the elements.

1. Obtain six **index cards** from your teacher. Using one card for each element name, write the names *beryllium, sodium, iron, zinc, aluminum,* and *oxygen* at the top of a card.
2. Open your textbook to the periodic table printed on the inside back cover. Locate the element key for each element written on your cards.
3. For each element, find the following information and write it on the index card: symbol, atomic number, atomic mass, state of matter, and element type.

Think About This

1. What do the elements in the blue blocks have in common? In the green blocks? In the yellow blocks?
2. **Key Concept** Each element in a column on the periodic table has similar chemical properties and forms bonds in similar ways. Based on this, for each element you listed on a card, name another element on the periodic table that has similar chemical properties.

The Periodic Table

Imagine trying to find a book in a library if all the books were unorganized. Books are organized in a library to help you easily find the information you need. The periodic table is like a library of information about all chemical elements.

A copy of the periodic table is on the inside back cover of this book. The table has more than 100 blocks—one for each known element. Each block on the periodic table includes basic properties of each element such as the element's state of matter at room temperature and its atomic number. The atomic number is the number of protons in each atom of the element. Each block also lists an element's atomic mass, or the average mass of all the different isotopes of that element.

Periods and Groups

You can learn about some properties of an element from its position on the periodic table. Elements are organized in periods (rows) and groups (columns). The periodic table lists elements in order of atomic number. The atomic number increases from left to right as you move across a period. Elements in each group have similar chemical properties and react with other elements in similar ways. In this lesson, you will read more about how an element's position on the periodic table can be used to predict its properties.

Reading Check How is the periodic table organized?

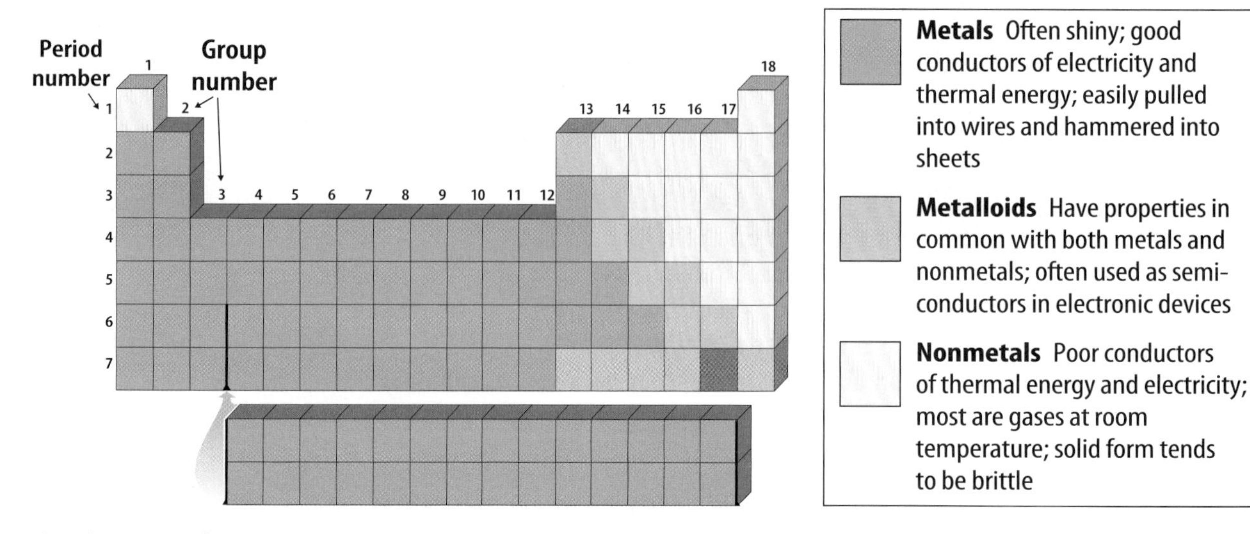

▲ **Figure 1** Elements on the periodic table are classified as metals, nonmetals, or metalloids.

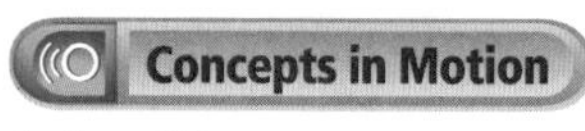

Animation

Metals, Nonmetals, and Metalloids

The three main regions of elements on the periodic table are shown in **Figure 1.** Except for hydrogen, elements on the left side of the table are metals. Nonmetals are on the right side of the table. Metalloids form the narrow stair-step region between metals and nonmetals.

Reading Check Where are metals, nonmetals, and metalloids on the periodic table?

REVIEW VOCABULARY

compound
matter that is made up of two or more different kinds of atoms joined together by chemical bonds

Atoms Bond

In nature, pure elements are rare. Instead, atoms of different elements chemically combine and form **compounds.** Compounds make up most of the matter around you, including living and nonliving things. There are only about 115 elements, but these elements combine and form millions of compounds. Chemical bonds hold them together. *A* **chemical bond** *is a force that holds two or more atoms together.*

Electron Number and Arrangement

Recall that atoms contain protons, neutrons, and electrons, as shown in **Figure 2.** Each proton has a positive charge; each neutron has no charge; and each electron has a negative charge. The atomic number of an element is the number of protons in each atom of that element. In a neutral (uncharged) atom, the number of protons equals the number of electrons.

The exact position of electrons in an atom cannot be determined. This is because electrons are in constant motion around the nucleus. However, each electron is usually in a certain area of space around the nucleus. Some are in areas close to the nucleus, and some are in areas farther away.

Figure 2 Protons and neutrons are in an atom's nucleus. Electrons move around the nucleus. ▼

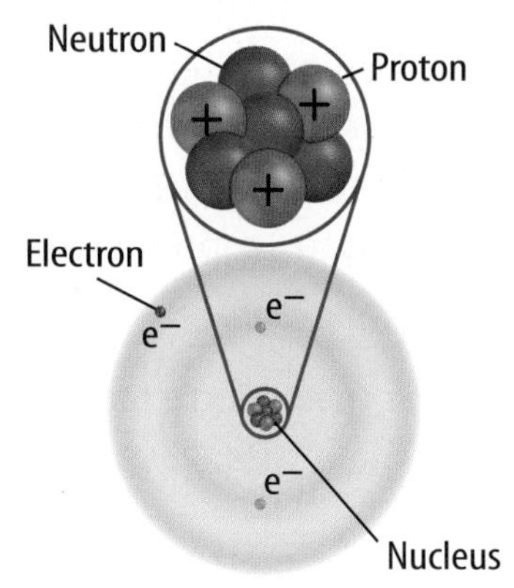

Electrons and Energy Different electrons in an atom have different amounts of energy. An electron moves around the nucleus at a distance that corresponds to its amount of energy. Areas of space in which electrons move around the nucleus are called energy levels. Electrons closest to the nucleus have the least amount of energy. They are in the lowest energy level. Electrons farthest from the nucleus have the greatest amount of energy. They are in the highest energy level. The energy levels of an atom are shown in **Figure 3.** Notice that only two electrons can be in the lowest energy level. The second energy level can hold up to eight.

Key Concept Check How is an electron's energy related to its position in an atom?

Electrons and Bonding Imagine two magnets. The closer they are to each other, the stronger the attraction of their opposite ends. Negatively charged electrons have a similar attraction to the positively charged nucleus of an atom. The electrons in energy levels closest to the nucleus of the same atom have a strong attraction to that nucleus. However, electrons farther from that nucleus are weakly attracted to it. These outermost electrons can easily be attracted to the nucleus of other atoms. This attraction between the positive nucleus of one atom and the negative electrons of another is what causes a chemical bond.

FOLDABLES®

Make two quarter-sheet note cards from a sheet of paper. Use them to organize your notes on valence electrons and electron dot diagrams.

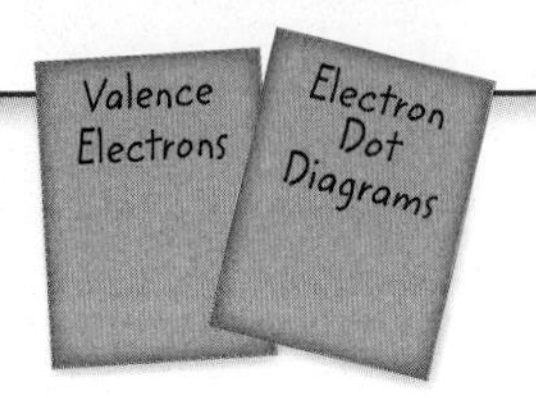

Figure 3 Electrons are in certain energy levels within an atom.

Electron Energy Levels

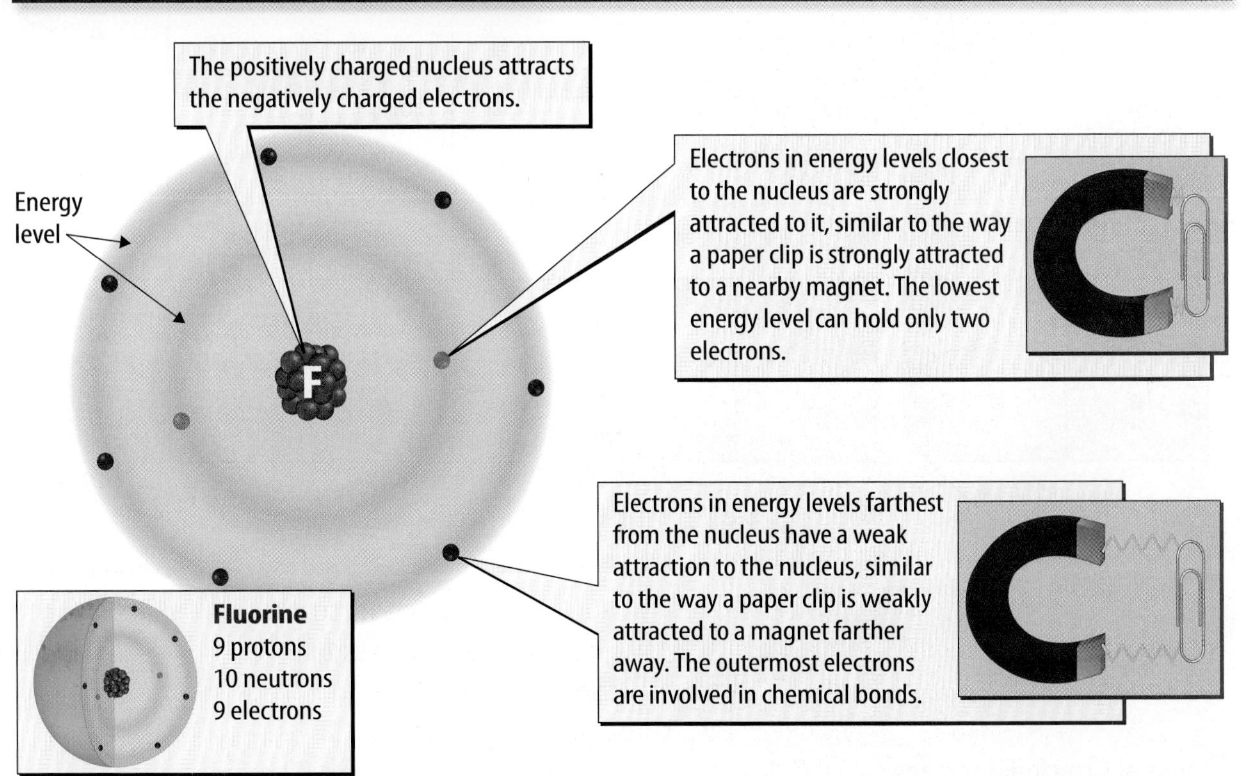

Valence Electrons

You have read that electrons farthest from their nucleus are easily attracted to the nuclei of nearby atoms. These outermost electrons are the only electrons involved in chemical bonding. Even atoms that have only a few electrons, such as hydrogen or lithium, can form chemical bonds. This is because these electrons are still the outermost electrons and are exposed to the nuclei of other atoms. A **valence electron** *is an outermost electron of an atom that participates in chemical bonding.* Valence electrons have the most energy of all electrons in an atom.

Word Origin
valence
from Latin *valentia,* means "strength, capacity"

The number of valence electrons in each atom of an element can help determine the type and the number of bonds it can form. How do you know how many valence electrons an atom has? The periodic table can tell you. Except for helium, elements in certain groups have the same number of valence electrons. **Figure 4** illustrates how to use the periodic table to determine the number of valence electrons in the atoms of groups 1, 2, and 13–18. Determining the number of valence electrons for elements in groups 3–12 is more complicated. You will learn about these groups in later chemistry courses.

Figure 4 You can use the group numbers at the top of the columns to determine the number of valence electrons in atoms of groups 1, 2, and 13–18.

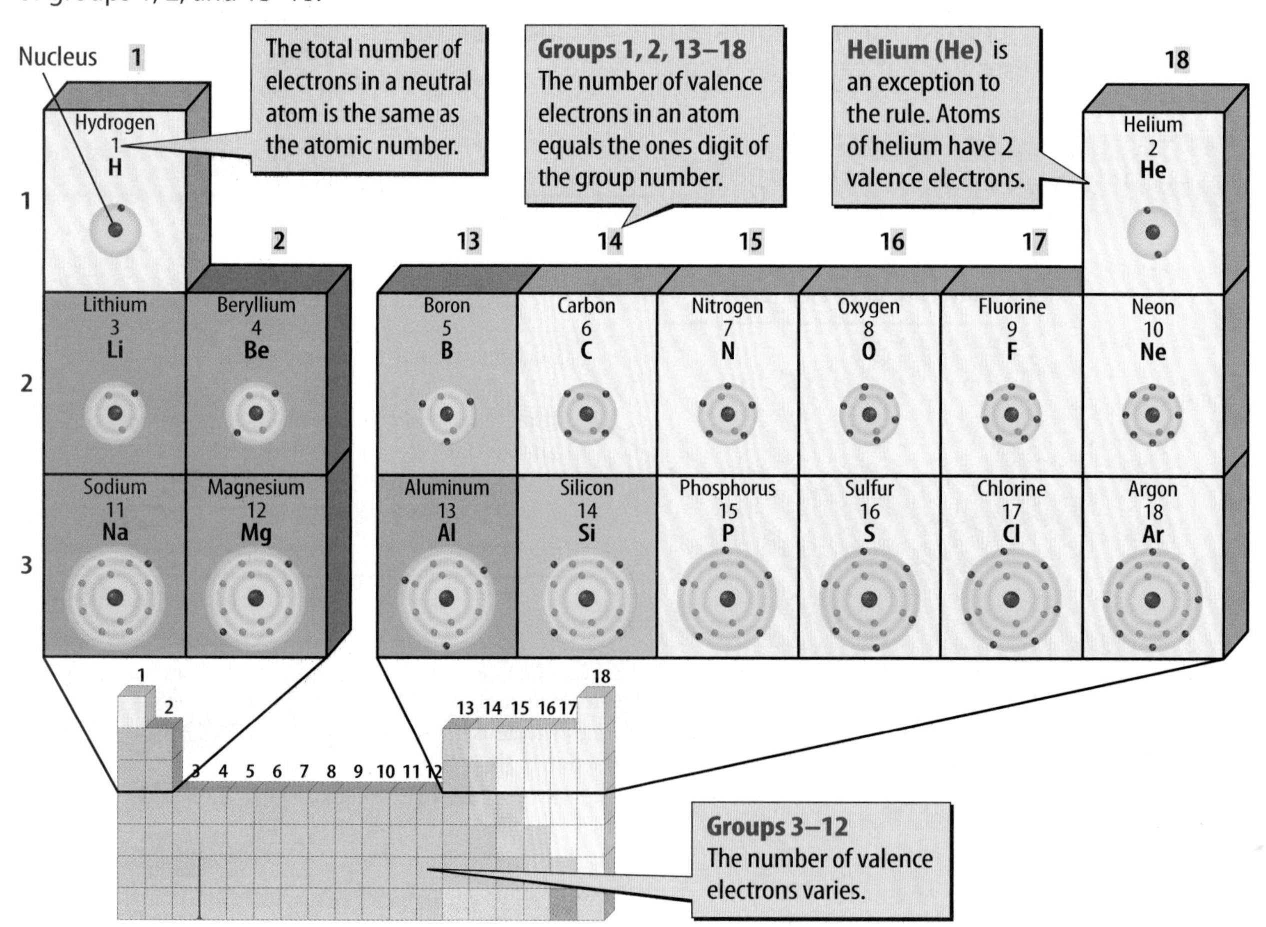

Visual Check How many valence electrons does an atom of phosphorous (P) have?

Writing and Using Electron Dot Diagrams

Figure 5 Electron dot diagrams show the number of valence electrons in an atom.

Steps for writing a dot diagram	Beryllium	Carbon	Nitrogen	Argon
1 Identify the element's group number on the periodic table.	2	14	15	18
2 Identify the number of valence electrons. • This equals the ones digit of the group number.	2	4	5	8
3 Draw the electron dot diagram. • Place one dot at a time on each side of the symbol (top, right, bottom, left). Repeat until all dots are used.	Be·	·Ċ·	·N̈·	:Är:
4 Determine if the atom is chemically stable. • An atom is chemically stable if all dots on the electron dot diagram are paired.	Chemically Unstable	Chemically Unstable	Chemically Unstable	Chemically Stable
5 Determine how many bonds this atom can form. • Count the dots that are unpaired.	2	4	3	0

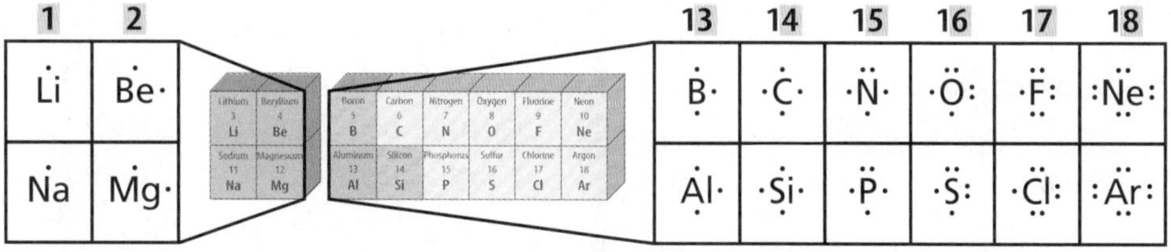

1	2	13	14	15	16	17	18
Li	Be	B	C	N	O	F	Ne
Na	Mg	Al	Si	P	S	Cl	Ar

Electron Dot Diagrams

In 1916 an American Chemist named Gilbert Lewis developed a method to show an element's valence electrons. He developed the **electron dot diagram**, *a model that represents valence electrons in an atom as dots around the element's chemical symbol.*

Electron dot diagrams can help you predict how an atom will bond with other atoms. Dots, representing valence electrons, are placed one-by-one on each side of an element's chemical symbol until all the dots are used. Some dots will be paired up, others will not. The number of unpaired dots is often the number of bonds an atom can form. The steps for writing dot diagrams are shown in **Figure 5.**

Reading Check Why are electron dot diagrams useful?

Recall that each element in a group has the same number of valence electrons. As a result, every element in a group has the same number of dots in its electron dot diagram.

Notice in **Figure 5** that an argon atom, Ar, has eight valence electrons, or four pairs of dots, in the diagram. There are no unpaired dots. Atoms with eight valence electrons do not easily react with other atoms. They are chemically stable. Atoms that have between one and seven valence electrons are reactive, or chemically unstable. These atoms easily bond with other atoms and form chemically stable compounds.

Atoms of hydrogen and helium have only one energy level. These atoms are chemically stable with two valence electrons.

Inquiry MiniLab 20 minutes

How does an electron's energy relate to its position in an atom?

Electrons in energy levels closest to the nucleus are strongly attracted to it. You can use paper clips and a magnet to model a similar attraction.

1. Read and complete a lab safety form.
2. Pick up a **paper clip** with a **magnet.** Use the first paper clip to pick up another one.
3. Continue picking up paper clips in this way until you have a chain of paper clips and no more will attach.
4. Gently pull off the paper clips one by one.

Analyze and Conclude

1. **Observe** Which paper clip was the easiest to remove? Which was the most difficult?
2. **Use Models** In what way do the magnet and the paper clips act as a model for an atom?
3. **Key Concept** How does an electron's position in an atom affect its ability to take part in chemical bonding?

Noble Gases

The elements in Group 18 are called noble gases. With the exception of helium, noble gases have eight valence electrons and are chemically stable. Chemically stable atoms do not easily react, or form bonds, with other atoms. The electron structures of two noble gases—neon and helium—are shown in **Figure 6.** Notice that all dots are paired in the dot diagrams of these atoms.

Stable and Unstable Atoms

Atoms with unpaired dots in their electron dot diagrams are reactive, or chemically unstable. For example, nitrogen, shown in **Figure 6,** has three unpaired dots in its electron dot diagram, and it is reactive. Nitrogen, like many other atoms, becomes more stable by forming chemical bonds with other atoms.

When an atom forms a bond, it gains, loses, or shares valence electrons with other atoms. By forming bonds, atoms become more chemically stable. Recall that atoms are most stable with eight valence electrons. Therefore, atoms with less than eight valence electrons form chemical bonds and become stable. In Lessons 2 and 3, you will read which atoms gain, lose, or share electrons when forming stable compounds.

Key Concept Check Why do atoms gain, lose, or share electrons?

Figure 6 Atoms gain, lose, or share valence electrons and become chemically stable.

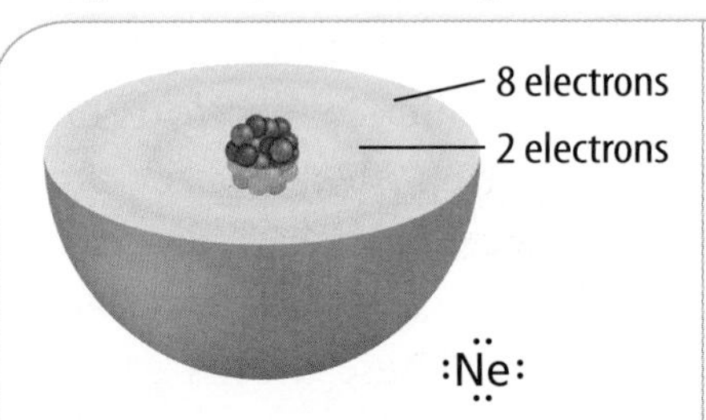

Neon has 10 electrons: 2 inner electrons and 8 valence electrons. A neon atom is chemically stable because it has 8 valence electrons. All dots in the dot diagram are paired.

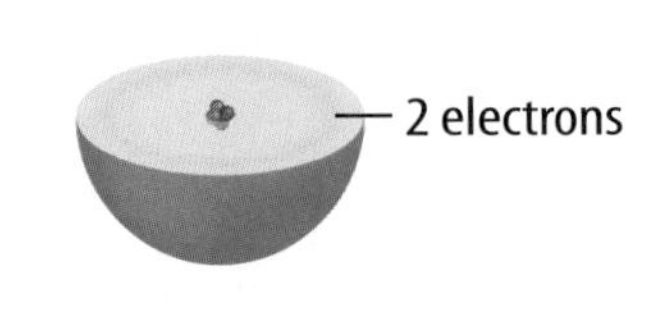

Helium has 2 electrons. Because an atom's lowest energy level can hold only 2 electrons, the 2 dots in the dot diagram are paired. Helium is chemically stable.

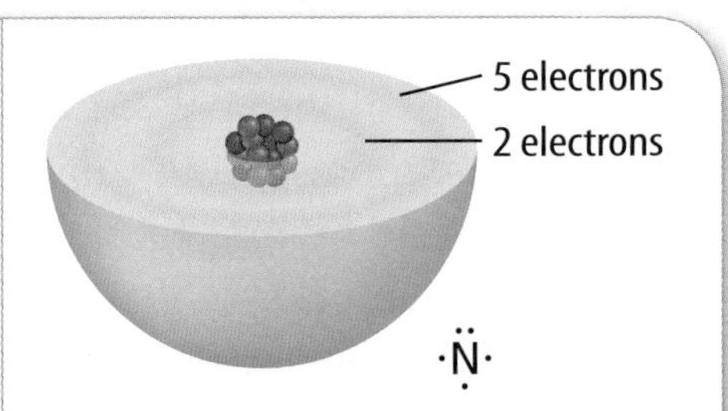

Nitrogen has 7 electrons: 2 inner electrons and 5 valence electrons. Its dot diagram has 1 pair of dots and 3 unpaired dots. Nitrogen atoms become more stable by forming chemical bonds.

Lesson 1 Review

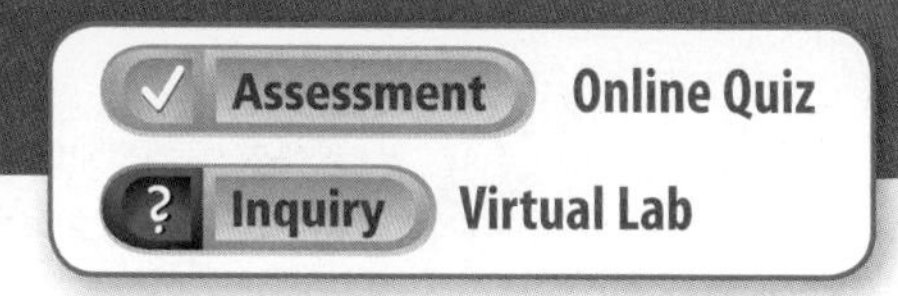

Visual Summary

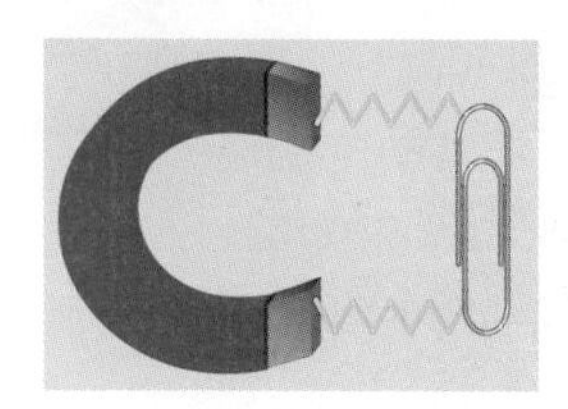

Electrons are less strongly attracted to a nucleus the farther they are from it, similar to the way a magnet attracts a paper clip.

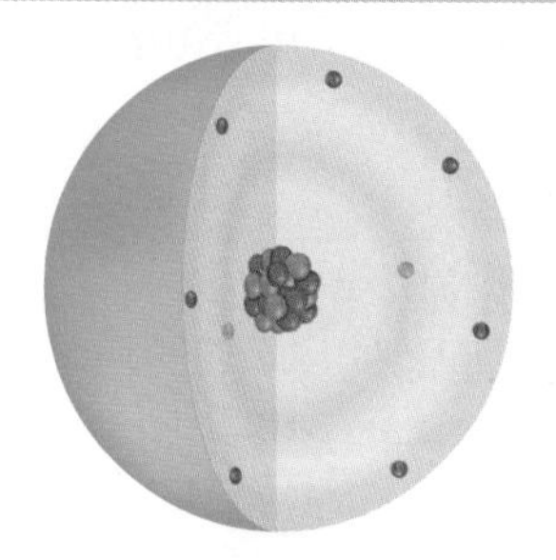

Electrons in atoms are in energy levels around the nucleus. Valence electrons are the outermost electrons.

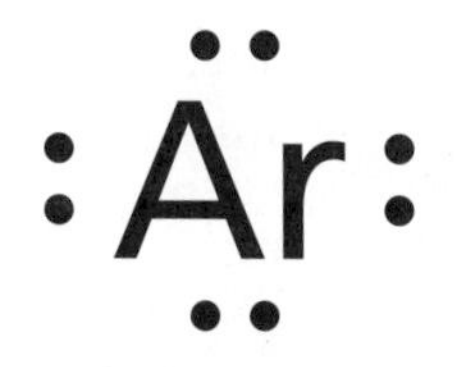

All noble gases, except He, have four pairs of dots in their electron dot diagrams. Noble gases are chemically stable.

FOLDABLES®

Use your lesson Foldable to review the lesson. Save your Foldable for the project at the end of the chapter.

What do you think NOW?

You first read the statements below at the beginning of the chapter.

1. Elements rarely exist in pure form. Instead, combinations of elements make up most of the matter around you.
2. Chemical bonds that form between atoms involve electrons.

Did you change your mind about whether you agree or disagree with the statements? Rewrite any false statements to make them true.

Use Vocabulary

1. **Use the term** *chemical bond* in a complete sentence.
2. **Define** *electron dot diagram* in your own words.
3. The electrons of an atom that participate in chemical bonding are called ________.

Understand Key Concepts 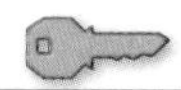

4. **Identify** the number of valence electrons in each atom: calcium, carbon, and sulfur.
5. Which part of the atom is shared, gained, or lost when forming a chemical bond?
   - **A.** electron
   - **B.** neutron
   - **C.** nucleus
   - **D.** proton
6. **Draw** electron dot diagrams for oxygen, potassium, iodine, nitrogen, and beryllium.

Interpret Graphics

7. **Determine** the number of valence electrons in each diagram shown below.

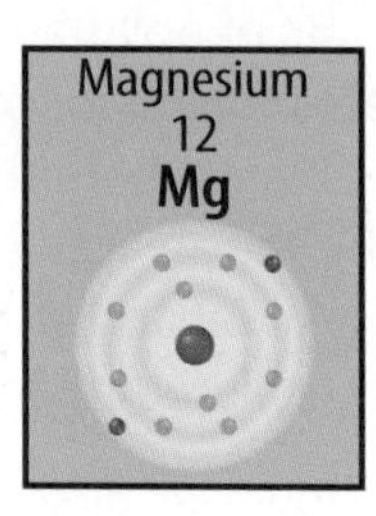

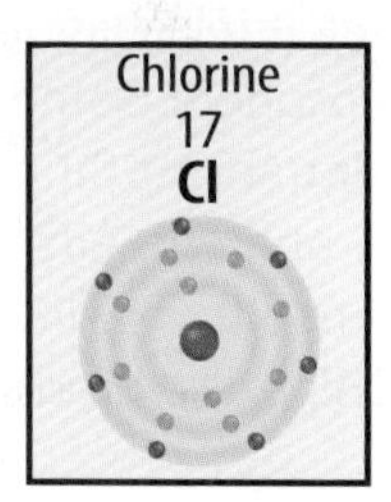

8. **Organize Information** Copy and fill in the graphic organizer below to describe one or more details for each concept: electron energy, valence electrons, stable atoms.

Concept	Description

Critical Thinking

9. **Compare** krypton and bromine in terms of chemical stability.
10. **Decide** An atom of nitrogen has five valence electrons. How could a nitrogen atom become more chemically stable?

GREEN SCIENCE

New Green Airships

The Difference of One Valence Electron

Faster than ocean liners and safer than airplanes, airships used to be the best way to travel. The largest, the *Hindenburg*, was nearly the size of the *Titanic*. To this day, no larger aircraft has ever flown. So, what happened to the giant airship? The answer lies in a valence electron.

The builders of the *Hindenburg* filled it with a lighter-than-air gas, hydrogen, so that it would float. Their plan was to use helium, a noble gas. However, helium was scarce. They knew hydrogen was explosive, but it was easier to get. For nine years, hydrogen airships floated safely back and forth across the Atlantic. But in 1937, disaster struck. Just before it landed, the *Hindenburg* exploded in flames. The age of the airship was over.

H

Since the *Hindenburg,* airplanes have become the main type of air transportation. A big airplane uses hundreds of gallons of fuel to take off and fly. As a result, it releases large amounts of pollutants into the atmosphere. Some people are looking for other types of air transportation that will be less harmful to the environment. Airships may be the answer. An airship floats and needs very little fuel to take off and stay airborne. Airships also produce far less pollution than other aircraft.

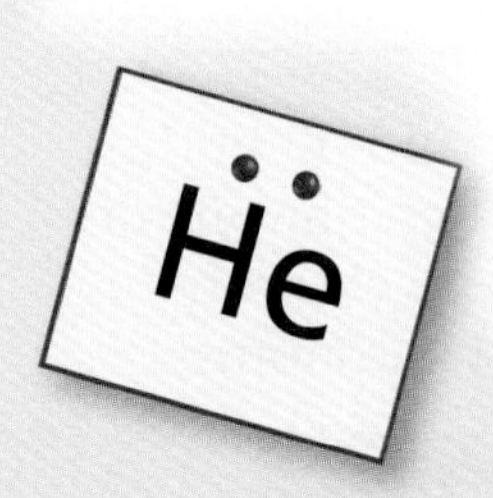

Today, however, airships use helium not hydrogen. With two valence electrons instead of one, as hydrogen has, helium is unreactive. Thanks to helium's chemical stability, someday you might be a passenger on a new, luxurious, but not explosive, version of the *Hindenburg*.

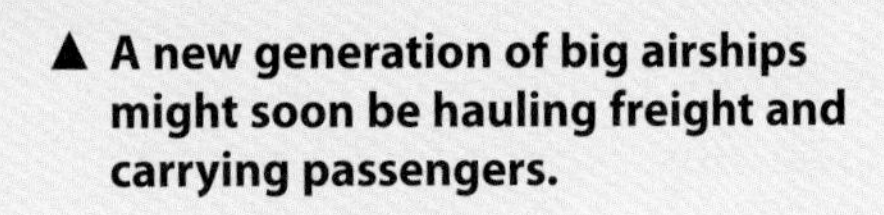

▲ **A new generation of big airships might soon be hauling freight and carrying passengers.**

It's Your Turn

RESEARCH Precious documents deteriorate with age as their surfaces react with air. Parchment turns brown and crumbles. Find out how our founding documents have been saved from this fate by noble gases.

Lesson 2

Reading Guide

Key Concepts
ESSENTIAL QUESTIONS

- How do elements differ from the compounds they form?
- What are some common properties of a covalent compound?
- Why is water a polar compound?

Vocabulary

covalent bond p. 277
molecule p. 278
polar molecule p. 279
chemical formula p. 280

g Multilingual eGlossary

Compounds, Chemical Formulas, and Covalent Bonds

Inquiry How do they combine?

A jigsaw puzzle has pieces that connect in a certain way. The pieces fit together by sharing tabs with other pieces. All of the pieces combine and form a complete puzzle. Like pieces of a puzzle, atoms can join together and form a compound by sharing electrons.

Inquiry Launch Lab

20 minutes

How is a compound different from its elements?

The sugar you use to sweeten foods at home is probably sucrose. Sucrose contains the elements carbon, hydrogen, and oxygen. How does table sugar differ from the elements that it contains?

1. Read and complete a lab safety form.
2. Air is a mixture of several gases, including oxygen and hydrogen. Charcoal is a form of carbon. Write some properties of oxygen, hydrogen, and carbon in your Science Journal.
3. Obtain from your teacher a piece of **charcoal** and a **beaker** with **table sugar** in it.
4. Observe the charcoal. In your Science Journal, describe the way it looks and feels.
5. Observe the table sugar in the beaker. What does it look and feel like? Record your observations.

Think About This

1. Compare and contrast the properties of charcoal, hydrogen, and oxygen.
2. **Key Concept** How do you think the physical properties of carbon, hydrogen, and oxygen change when they combined to form sugar?

From Elements to Compounds

Have you ever baked cupcakes? First, combine flour, baking soda, and a pinch of salt. Then, add sugar, eggs, vanilla, milk, and butter. Each ingredient has unique physical and chemical properties. When you mix the ingredients together and bake them, a new product results—cupcakes. The cupcakes have properties that are different from the ingredients.

In some ways, compounds are like cupcakes. Recall that a compound is a substance made up of two or more different elements. Just as cupcakes are different from their ingredients, compounds are different from their elements. An element is made of one type of atom, but compounds are chemical combinations of different types of atoms. Compounds and the elements that make them up often have different properties.

Chemical **bonds** join atoms together. Recall that a chemical bond is a force that holds atoms together in a compound. In this lesson, you will learn that one way that atoms can form bonds is by sharing valence electrons. You will also learn how to write and read a chemical formula.

Science Use v. Common Use

bond

Science Use a force that holds atoms together in a compound

Common Use a close personal relationship between two people

 **Key Concept Check** How is a compound different from the elements that compose it?

Figure 7 A covalent bond forms when two nonmetal atoms share electrons.

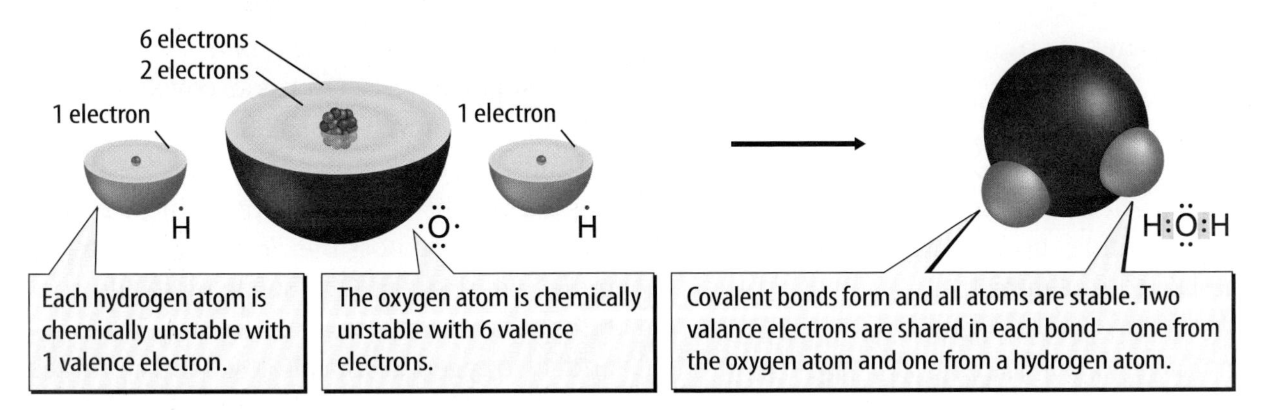

Covalent Bonds—Electron Sharing

As you read in Lesson 1, one way that atoms can become more chemically stable is by sharing valence electrons. When unstable, nonmetal atoms bond together, they bond by sharing valence electrons. *A* **covalent bond** *is a chemical bond formed when two atoms share one or more pairs of valence electrons.* The atoms then form a stable covalent compound.

A Noble Gas Electron Arrangement

Look at the reaction between hydrogen and oxygen in **Figure 7.** Before the reaction, each hydrogen atom has one valence electron. The oxygen atom has six valence electrons. Recall that most atoms are chemically stable with eight valence electrons—the same electron arrangement as a noble gas. An atom with less than eight valence electrons becomes stable by forming chemical bonds until it has eight valence electrons. Therefore, an oxygen atom forms two bonds to become stable. A hydrogen atom is stable with two valence electrons. It forms one bond to become stable.

Shared Electrons

If the oxygen atom and each hydrogen atom share their unpaired valence electrons, they can form two covalent bonds and become a stable covalent compound. Each covalent bond contains two valence electrons—one from the hydrogen atom and one from the oxygen atom. Since these electrons are shared, they count as valence electrons for both atoms in the bond. Each hydrogen atom now has two valence electrons. The oxygen atom now has eight valence electrons, since it bonds to two hydrogen atoms. All three atoms have the electron arrangement of a noble gas and the compound is stable.

FOLDABLES®

Make three quarter-sheet note cards from a sheet of paper to organize information about single, double, and triple covalent bonds.

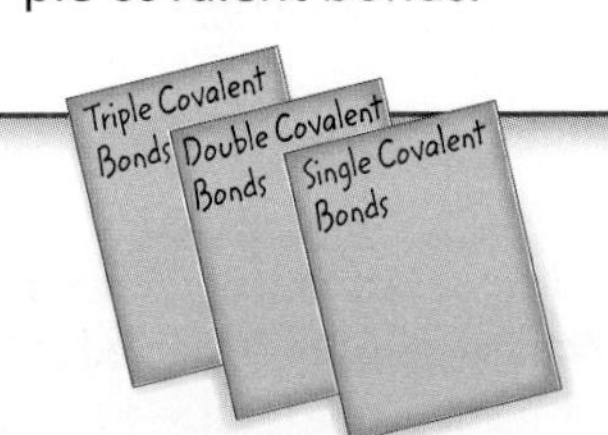

Double and Triple Covalent Bonds

As shown in **Figure 8,** a single covalent bond exists when two atoms share one pair of valence electrons. A double covalent bond exists when two atoms share two pairs of valence electrons. Double bonds are stronger than single bonds. A triple covalent bond exists when two atoms share three pairs of valence electrons. Triple bonds are stronger than double bonds. Multiple bonds are explained in **Figure 8.**

Covalent Compounds

When two or more atoms share valence electrons, they form a stable covalent compound. The covalent compounds carbon dioxide, water, and sugar are very different, but they also share similar properties. Covalent compounds usually have low melting points and low boiling points. They are usually gases or liquids at room temperature, but they can also be solids. Covalent compounds are poor conductors of thermal energy and electricity.

Molecules

The chemically stable unit of a covalent compound is a molecule. *A* **molecule** *is a group of atoms held together by covalent bonding that acts as an independent unit.* Table sugar ($C_{12}H_{22}O_{11}$) is a covalent compound. One grain of sugar is made up of trillions of sugar molecules. Imagine breaking a grain of sugar into the tiniest microscopic particle possible. You would have a molecule of sugar. One sugar molecule contains 12 carbon atoms, 22 hydrogen atoms, and 11 oxygen atoms all covalently bonded together. The only way to further break down the molecule would be to chemically separate the carbon, hydrogen, and oxygen atoms. These atoms alone have very different properties from the compound sugar.

Key Concept Check What are some common properties of covalent compounds?

Multiple Bonds

Figure 8 The more valence electrons that two atoms share, the stronger the covalent bond is between the atoms.

One Single Covalent Bond

When two hydrogen atoms bond, they form a single covalent bond.

H· + ·H ⟶ H:H

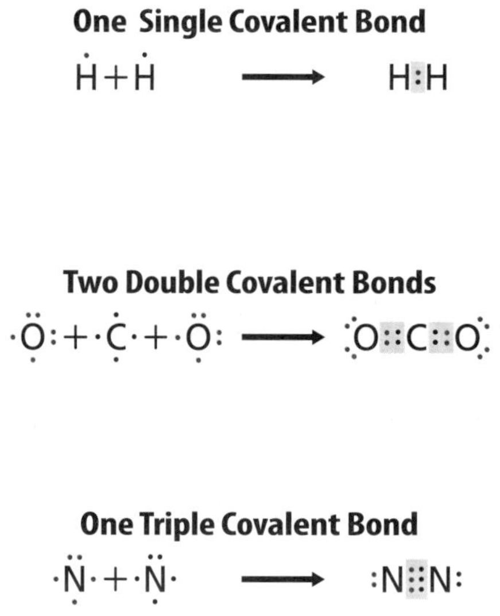

In a single covalent bond, 1 pair of electrons is shared between two atoms. Each H atom shares 1 valence electron with the other.

Two Double Covalent Bonds

When one carbon atom bonds with two oxygen atoms, two double covalent bonds form.

·Ö: + ·Ċ· + ·Ö: ⟶ :O::C::O:

In a double covalent bond, 2 pairs of electrons are shared between two atoms. One O atom and the C atom each share 2 valence electrons with the other.

One Triple Covalent Bond

When two nitrogen atoms bond, they form a triple covalent bond.

·N· + ·N· ⟶ :N⋮⋮N:

In a triple covalent bond, 3 pairs of electrons are shared between two atoms. Each N atom shares 3 valence electrons with the other.

Visual Check Is the bond stronger between atoms in hydrogen gas (H_2) or nitrogen gas (N_2)? Why?

Water and Other Polar Molecules

In a covalent bond, one atom can attract the shared electrons more strongly than the other atom can. Think about the valence electrons shared between oxygen and hydrogen atoms in a water molecule. The oxygen atom attracts the shared electrons more strongly than each hydrogen atom does. As a result, the shared electrons are pulled closer to the oxygen atom, as shown in **Figure 9.** Since electrons have a negative charge, the oxygen atom has a partial negative charge. The hydrogen atoms have a partial positive charge. *A molecule that has a partial positive end and a partial negative end because of unequal sharing of electrons is a* **polar molecule.**

WORD ORIGIN

polar
from Latin *polus,* means "pole"

The charges on a polar molecule affect its properties. Sugar, for example, dissolves easily in water because both sugar and water are polar. The negative end of a water molecule pulls on the positive end of a sugar molecule. Also, the positive end of a water molecule pulls on the negative end of a sugar molecule. This causes the sugar molecules to separate from one another and mix with the water molecules.

 **Key Concept Check** Why is water a polar compound?

Nonpolar Molecules

A hydrogen molecule, H_2, is a nonpolar molecule. Because the two hydrogen atoms are identical, their attraction for the shared electrons is equal. The carbon dioxide molecule, CO_2, in **Figure 9** is also nonpolar. A nonpolar compound will not easily dissolve in a polar compound, but it will dissolve in other nonpolar compounds. Oil is an example of a nonpolar compound. It will not dissolve in water. Have you ever heard someone say "like dissolves like"? This means that polar compounds can dissolve in other polar compounds. Similarly, nonpolar compounds can dissolve in other nonpolar compounds.

Figure 9 Atoms of a polar molecule share their valence electrons unequally. Atoms of a nonpolar molecule share their valence electrons equally.

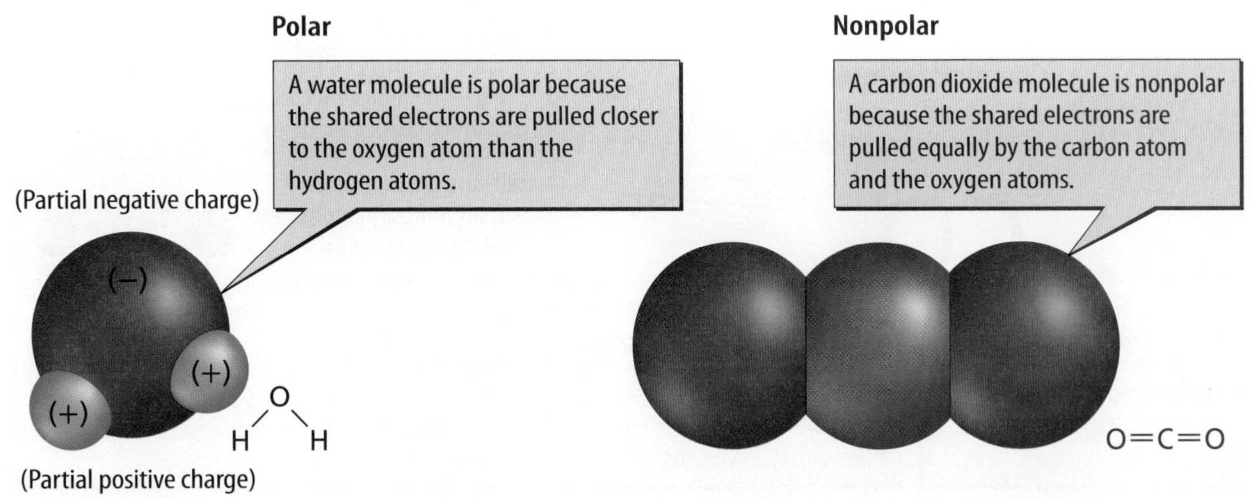

Inquiry MiniLab 20 minutes

How do compounds form?

Use building blocks to model ways in which elements combine to form compounds.

1. Examine various types of **interlocking plastic blocks.** Notice that the blocks have different numbers of holes and pegs. Attaching one peg to one hole represents a shared pair of electrons. 
2. Draw the electron dot diagrams for carbon, nitrogen, oxygen, and hydrogen in your Science Journal. Based on the diagrams, decide which block should represent an atom of each element.
3. Use the blocks to make models of H_2, CO_2, NH_3, H_2O, and CH_4. All pegs on the largest block must fit into a hole, and no blocks can stick out over the edge of a block, either above or below it.

Analyze and Conclude

1. **Explain** how you decided which type of block should be assigned to each type of atom.
2. **Key Concept** Name at least one way that your models show the difference between a compound and the elements that combine and form the compound.

Chemical Formulas and Molecular Models

How do you know which elements make up a compound? *A* **chemical formula** *is a group of chemical symbols and numbers that represent the elements and the number of atoms of each element that make up a compound.* Just as a recipe lists ingredients, a chemical formula lists the elements in a compound. For example, the chemical formula for carbon dioxide shown in **Figure 10** is CO_2. The formula uses chemical symbols that show which elements are in the compound. Notice that CO_2 is made up of carbon (C) and oxygen (O). A subscript, or small number after a chemical symbol, shows the number of atoms of each element in the compound. Carbon dioxide (CO_2) contains two atoms of oxygen bonded to one atom of carbon.

A chemical formula describes the types of atoms in a compound or a molecule, but it does not explain the shape or appearance of the molecule. There are many ways to model a molecule. Each one can show the molecule in a different way. Common types of models for CO_2 are shown in **Figure 10.**

Reading Check What information is given in a chemical formula?

Figure 10 Chemical formulas and molecular models provide information about molecules.

Chemical Formula

A carbon dioxide molecule is made up of carbon (C) and oxygen (O) atoms.

$$CO_2$$

A symbol without a subscript indicates one atom. Each molecule of carbon dioxide has one carbon atom.

The subscript 2 indicates two atoms of oxygen. Each molecule of carbon dioxide has two oxygen atoms.

Dot Diagram
- Shows atoms and valence electrons

:Ö::C::Ö:

Structural Formula
- Shows atoms and lines; each line represents one shared pair of electrons

O=C=O

Ball-and-Stick Model
- Balls represent atoms and sticks represent bonds; used to show bond angles

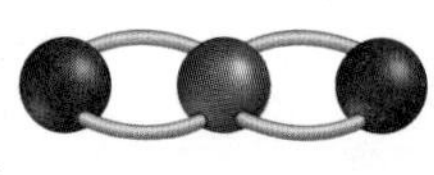

Space-Filling Model
- Spheres represent atoms; used to show three-dimensional arrangement of atoms

Lesson 2 Review

Assessment Online Quiz

Visual Summary

CO_2

A chemical formula is one way to show the elements that make up a compound.

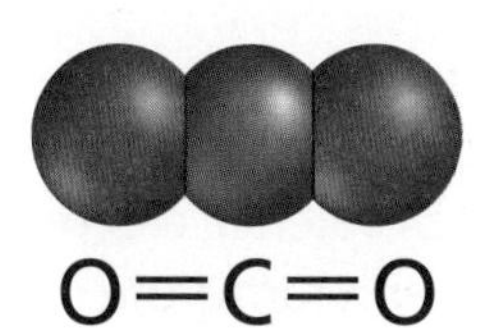

O=C=O

A covalent bond forms when atoms share valence electrons. The smallest particle of a covalent compound is a molecule.

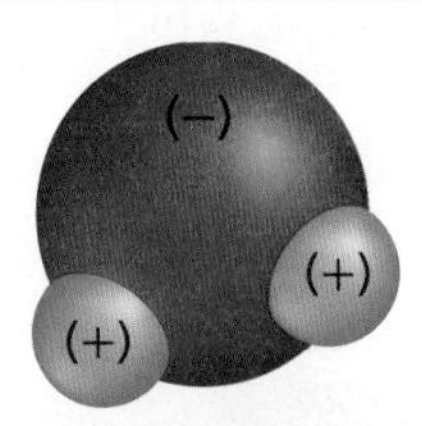

Water is a polar molecule because the oxygen and hydrogen atoms unequally share electrons.

FOLDABLES®

Use your lesson Foldable to review the lesson. Save your Foldable for the project at the end of the chapter.

What do you think NOW?

You first read the statements below at the beginning of the chapter.

3. The atoms in a water molecule are more chemically stable than they would be as individual atoms.

4. Many substances dissolve easily in water because opposite ends of a water molecule have opposite charges.

Did you change your mind about whether you agree or disagree with the statements? Rewrite any false statements to make them true.

Use Vocabulary

1. **Define** *covalent bond* in your own words.
2. The group of symbols and numbers that shows the types and numbers of atoms that make up a compound is a ______.
3. **Use the term** *molecule* in a complete sentence.

Understand Key Concepts 

4. **Contrast** Name at least one way water (H_2O) is different from the elements that make up water.
5. **Explain** why water is a polar molecule.
6. A sulfur dioxide molecule has one sulfur atom and two oxygen atoms. Which is its correct chemical formula?

   A. SO_2 **C.** S_2O_2
   B. $(SO)_2$ **D.** S_2O

Interpret Graphics

7. **Examine** the electron dot diagram for chlorine below.

In chlorine gas, two chlorine atoms join to form a Cl_2 molecule. How many pairs of valence electrons do the atoms share?

8. **Compare and Contrast** Copy and fill in the graphic organizer below to identify at least one way polar and nonpolar molecules are similar and one way they are different.

Polar and Nonpolar Molecules	
Similarities	
Differences	

Critical Thinking

9. **Develop** an analogy to explain the unequal sharing of valence electrons in a water molecule.

Inquiry **Skill Practice** **Model** 25 minutes

How can you model compounds?

Materials

colored pencils

Chemists use models to explain how electrons are arranged in an atom. Electron dot diagrams are models used to show how many valence electrons an atom has. Electron dot diagrams are useful because they can help predict the number and type of bond an atom will form.

Learn It

In science, **models** are used to help you visualize objects that are too small, too large, or too complex to understand. A model is a representation of an object, idea, or event.

Try It

1. Use the periodic table to write the electron dot diagrams for hydrogen, oxygen, carbon, and silicon.
2. Using your electron dot diagrams from step 1, write electron dot diagrams for the following compounds: H_2O, CO, CO_2, SiO_2, C_2H_2, and CH_4. Use colored pencils to differentiate the electrons for each atom. Remember that all the above atoms, except hydrogen and helium, are chemically stable when they have eight valence electrons. Hydrogen and helium are chemically stable with two valence electrons.

Apply It

3. Based on your model, describe silicon's electron dot diagram and arrangement of valence electrons before and after it forms the compound SiO_2.
4. **Key Concept** Which of the covalent compounds you modeled contain double bonds? Which contain triple bonds?

2

1	2
Hydrogen 1 H	
Lithium 3 Li	Beryllium 4 Be
Sodium 11 Na	Magnesium 12 Mg
Potassium 19 K	Calcium 20 Ca
Rubidium 37 Rb	Strontium 38 Sr
Cesium 55 Cs	Barium 56 Ba
Francium 87 Fr	Radium 88 Ra

13	14	15	16	17	18
					Helium 2 He
Boron 5 B	Carbon 6 C	Nitrogen 7 N	Oxygen 8 O	Fluorine 9 F	Neon 10 Ne
Aluminum 13 Al	Silicon 14 Si	Phosphorus 15 P	Sulfur 16 S	Chlorine 17 Cl	Argon 18 Ar
Gallium 31 Ga	Germanium 32 Ge	Arsenic 33 As	Selenium 34 Se	Bromine 35 Br	Krypton 36 Kr
Indium 49 In	Tin 50 Sn	Antimony 51 Sb	Tellurium 52 Te	Iodine 53 I	Xenon 54 Xe
Thallium 81 Tl	Lead 82 Pb	Bismuth 83 Bi	Polonium 84 Po	Astatine 85 At	Radon 86 Rn

Lesson 3

Reading Guide

Key Concepts 
ESSENTIAL QUESTIONS

- What is an ionic compound?
- How do metallic bonds differ from covalent and ionic bonds?

Vocabulary

ion p. 284
ionic bond p. 286
metallic bond p. 287

Multilingual eGlossary

Ionic and Metallic Bonds

Inquiry What is this?

This scene might look like snow along a shoreline, but it is actually thick deposits of salt on a lake. Over time, tiny amounts of salt dissolved in river water that flowed into this lake and built up as water evaporated. Salt is a compound that forms when elements form bonds by gaining or losing valence electrons, not sharing them.

Launch Lab

15 minutes

How can atoms form compounds by gaining and losing electrons?

Metals often lose electrons when forming stable compounds. Nonmetals often gain electrons.

1. Read and complete a lab safety form.
2. Make two model atoms of sodium, and one model atom each of calcium, chlorine, and sulfur. To do this, write each element's chemical symbol with a **marker** on a **paper plate.** Surround the symbol with small balls of **clay** to represent valence electrons. Use one color of clay for the metals (groups 1 and 2 elements) and another color of clay for nonmetals (groups 16 and 17 elements).
3. To model sodium sulfide (Na_2S), place the two sodium atoms next to the sulfur atom. To form a stable compound, move each sodium atom's valence electron to the sulfur atom.
4. Form as many other compound models as you can by removing valence electrons from the groups 1 and 2 plates and placing them on the groups 16 and 17 plates.

Think About This

1. What other compounds were you able to form?
2. **Key Concept** How do you think your models are different from covalent compounds?

FOLDABLES®

Make two quarter-sheet note cards as shown. Use the cards to summarize information about ionic and metallic compounds.

WORD ORIGIN

ion
from Greek *ienai,* means "to go"

Understanding Ions

As you read in Lesson 2, the atoms of two or more nonmetals form compounds by sharing valence electrons. However, when a metal and a nonmetal bond, they do not share electrons. Instead, one or more valence electrons transfers from the metal atom to the nonmetal atom. After electrons transfer, the atoms bond and form a chemically stable compound. Transferring valence electrons results in atoms with the same number of valence electrons as a noble gas.

When an atom loses or gains a valence electron, it becomes an ion. *An* **ion** *is an atom that is no longer electrically neutral because it has lost or gained valence electrons.* Because electrons have a negative charge, losing or gaining an electron changes the overall charge of an atom. An atom that loses valence electrons becomes an ion with a positive charge. This is because the number of electrons is now less than the number of protons in the atom. An atom that gains valence electrons becomes an ion with a negative charge. This is because the number of protons is now less than the number of electrons.

Reading Check Why do atoms that a gain electrons become an ion with a negative charge?

Losing Valence Electrons

Look at the periodic table on the inside back cover of this book. What information about sodium (Na) can you infer from the periodic table? Sodium is a metal. Its atomic number is 11. This means each sodium atom has 11 protons and 11 electrons. Sodium is in group 1 on the periodic table. Therefore, sodium atoms have one valence electron, and they are chemically unstable.

Metal atoms, such as sodium, become more stable when they lose valence electrons and form a chemical bond with a nonmetal. If a sodium atom loses its one valence electron, it would have a total of ten electrons. Which element on the periodic table has atoms with ten electrons? Neon (Ne) atoms have a total of ten electrons. Eight of these are valence electrons. When a sodium atom loses one valence electron, the electrons in the next lower energy level are now the new valence electrons. The sodium atom then has eight valence electrons, the same as the noble gas neon and is chemically stable.

Gaining Valence Electrons

In Lesson 2, you read that nonmetal atoms can share valence electrons with other nonmetal atoms. Nonmetal atoms can also gain valence electrons from metal atoms. Either way, they achieve the electron arrangement of a noble gas. Find the nonmetal chlorine (Cl) on the periodic table. Its atomic number is 17. Atoms of chlorine have seven valence electrons. If a chlorine atom gains one valence electron, it will have eight valence electrons. It will also have the same electron arrangement as the noble gas argon (Ar).

When a sodium atom loses a valence electron, it becomes a positively charged ion. This is shown by a plus (+) sign. When a chlorine atom gains a valence electron, it becomes a negatively charged ion. This is shown by a negative (–) sign. **Figure 11** illustrates the process of a sodium atom losing an electron and a chlorine atom gaining an electron.

Reading Check Are atoms of a group 16 element more likely to gain or lose valence electrons?

Losing and Gaining Electrons

Figure 11 Sodium atoms have a tendency to lose a valence electron. Chlorine atoms have a tendency to gain a valence electron.

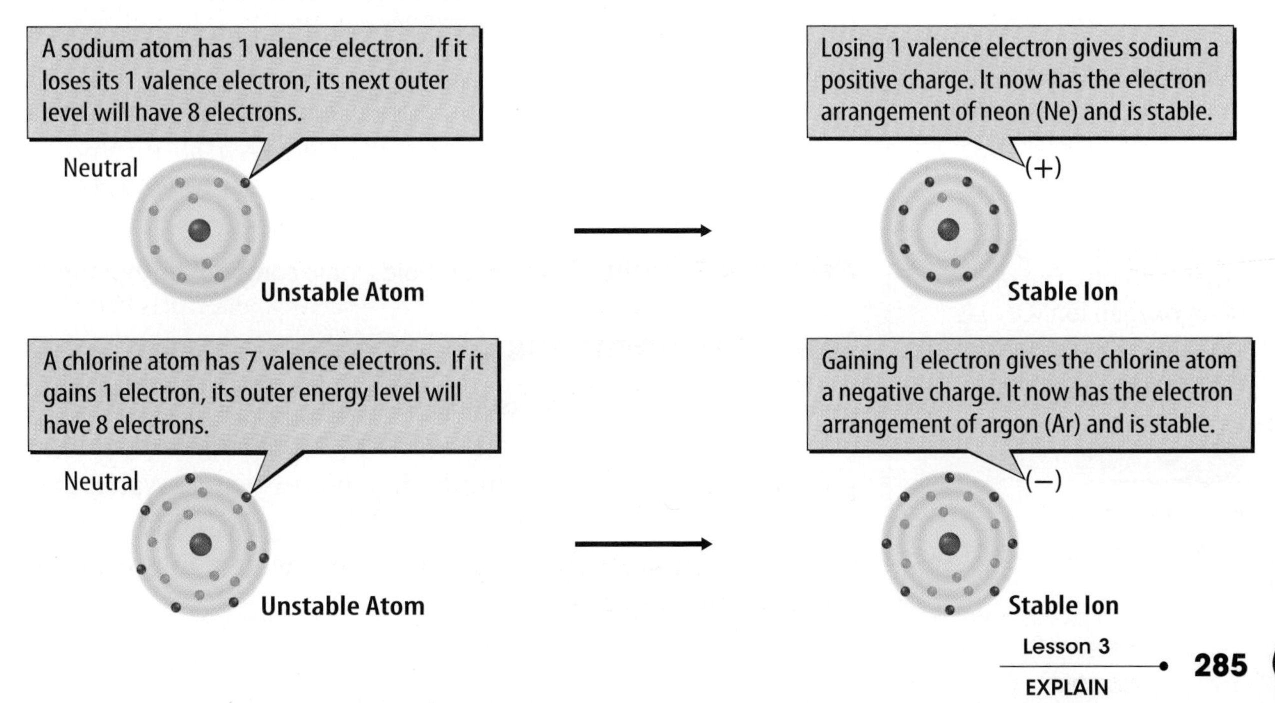

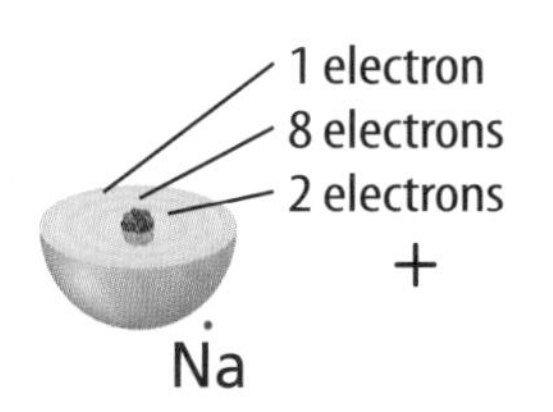

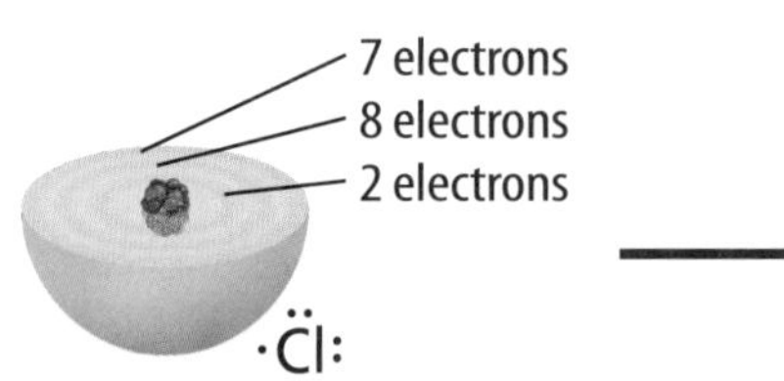

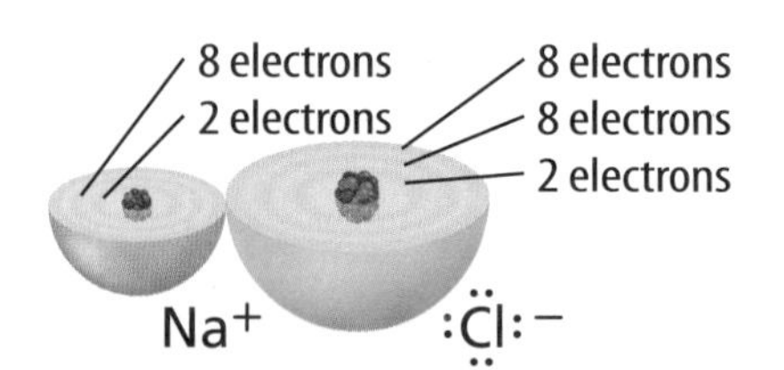

Sodium and chlorine atoms are stable when they have eight valence electrons. A sodium atoms loses one valence electron and becomes stable. A chlorine atom gains one valence electron and becomes stable.

The positively charged sodium ion and the negatively charged chlorine ion attract each other. Together they form a strong ionic bond.

Figure 12 An ionic bond forms between Na and Cl when an electron transfers from Na to Cl.

Concepts in Motion **Animation**

Math Skills

Use Percentage

An atom's radius is measured in picometers (pm), 1 trillion times smaller than a meter. When an atom becomes an ion, its radius increases or decreases. For example, a Na atom has a radius of **186 pm.** A Na^+ ion has a radius of **102 pm.** By what percentage does the radius change?

Subtract the atom's radius from the ion's radius.

$$102 \text{ pm} - 186 \text{ pm} = -84 \text{ pm}$$

Divide the difference by the atom's radius.

$$-84 \text{ pm} \div 186 \text{ pm} = -0.45$$

Multiply the answer by 100 and add a % sign.

$$-0.45 \times 100 = -45\%$$

A negative value is a decrease in size. A positive value is an increase.

Practice

The radius of an oxygen (O) atom is 73 pm. The radius of an oxygen ion (O^{2-}) is 140 pm. By what percentage does the radius change?

- **Math Practice**
- **Personal Tutor**

Determining an Ion's Charge

Atoms are electrically neutral because they have the same number of protons and electrons. Once an atom gains or loses electrons, it becomes a charged ion. For example, the atomic number for nitrogen (N) is 7. Each N atom has 7 protons and 7 electrons and is electrically neutral. However, an N atom often gains 3 electrons when forming an ion. The N ion then has 10 electrons. To determine the charge, subtract the number of electrons in the ion from the number of protons.

$$7 \text{ protons} - 10 \text{ electrons} = -3 \text{ charge}$$

A nitrogen ion has a −3 charge. This is written as N^{3-}.

Ionic Bonds—Electron Transferring

Recall that metal atoms typically lose valence electrons and nonmetal atoms typically gain valence electrons. When forming a chemical bond, the nonmetal atoms gain the electrons lost by the metal atoms. Take a look at **Figure 12.** In NaCl, or table salt, a sodium atom loses a valence electron. The electron is transferred to a chlorine atom. The sodium atom becomes a positively charged ion. The chlorine atom becomes a negatively charged ion. These ions attract each other and form a stable ionic compound. *The attraction between positively and negatively charged ions in an ionic compound is an* **ionic bond.**

 **Key Concept Check** What holds ionic compounds together?

Ionic Compounds

Ionic compounds are usually solid and brittle at room temperature. They also have relatively high melting and boiling points. Many ionic compounds dissolve in water. Water that contains dissolved ionic compounds is a good conductor of electricity. This is because an electrical charge can pass from ion to ion in the solution.

Comparing Ionic and Covalent Compounds

Recall that in a covalent bond, two or more nonmetal atoms share electrons and form a unit, or molecule. Covalent compounds, such as water, are made up of many molecules. However, when nonmetal ions bond to metal ions in an ionic compound, there are no molecules. Instead, there is a large collection of oppositely charged ions. All of the ions attract each other and are held together by ionic bonds.

Metallic Bonds—Electron Pooling

Recall that metal atoms typically lose valence electrons when forming compounds. What happens when metal atoms bond to other metal atoms? Metal atoms form compounds with one another by combining, or pooling, their valence electrons. A **metallic bond** *is a bond formed when many metal atoms share their pooled valence electrons.*

The pooling of valence electrons in aluminum is shown in **Figure 13.** The aluminum atoms lose their valence electrons and become positive ions, indicated by the plus (+) signs. The negative (−) signs indicate the valence electrons, which move from ion to ion. Valence electrons in metals are not bonded to one atom. Instead, a "sea of electrons" surrounds the positive ions.

Key Concept Check How do metal atoms bond with one another?

Figure 13 Valence electrons move among all the aluminum (Al) ions.

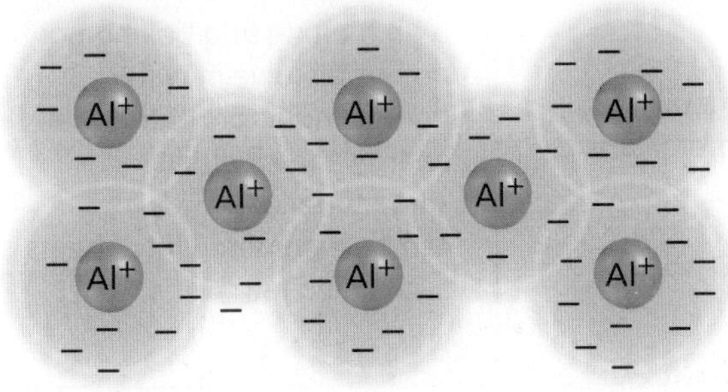

Inquiry MiniLab — 20 minutes

How many ionic compounds can you make?

You have read that in ionic bonding, metal atoms transfer electrons to nonmetal atoms.

1. Copy the table below into your Science Journal.

Group	Elements	Type	Dot Diagram
1	Li, Na, K	Metal	$\dot{X}$
2	Be, Mg, Ca	Metal	
14	C	Nonmetal	
15	N, P	Nonmetal	
16	O, S	Nonmetal	
17	F, Cl	Nonmetal	

2. Fill in the last column with the correct dot diagram for each group. Color the dots of the metal atoms with a **red marker** and the dots of the nonmetal atoms with a **blue marker.**
3. Using the information in your table, create five different ionic bonds. Write (a) the equation for the electron transfer and (b) the formula for each compound. For example:

a. $\dot{Na} + \dot{Na} + \cdot\ddot{O}: \longrightarrow Na^{+} + Na^{+} + :\ddot{O}:^{2-}$

b. Na_2O

Analyze and Conclude

1. **Explain** What happens to the metal and nonmetal ions after the electrons have been transferred?
2. **Key Concept** Describe the ionic bonds that hold the ions together in your compounds.

Properties of Metallic Compounds

Metals are good conductors of thermal energy and electricity. Because the valence electrons can move from ion to ion, they can easily **conduct** an electric charge. When a metal is hammered into a sheet or drawn into a wire, it does not break. The metal ions can slide past one another in the electron sea and move to new positions. Metals are shiny because the valence electrons at the surface of a metal interact with light. **Table 1** compares the covalent, ionic, and metallic bonds that you studied in this chapter.

Reading Check How does valence electron pooling explain why metals can be hammered into a sheet?

ACADEMIC VOCABULARY

conduct
(verb) to serve as a medium through which something can flow

Table 1 Bonds can form when atoms share valence electrons, transfer valence electrons, or pool valence electrons.

Concepts in Motion

Interactive Table

Table 1 Covalent, Ionic, and Metallic Bonds

Type of Bond	What is bonding?	Properties of Compounds
Covalent 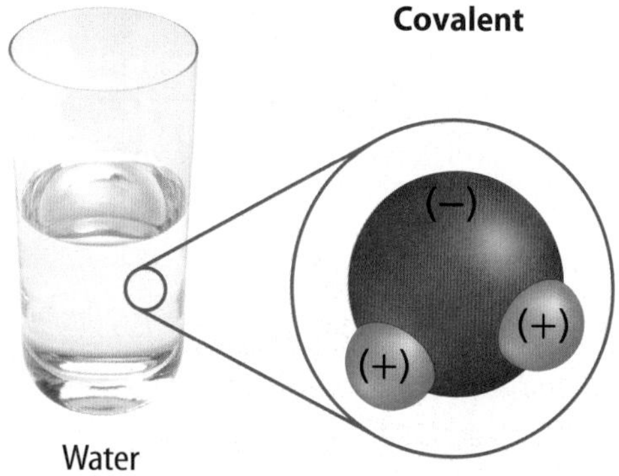 Water	nonmetal atoms; nonmetal atoms	• gas, liquid, or solid • low melting and boiling points • often not able to dissolve in water • poor conductors of thermal energy and electricity • dull appearance
Ionic 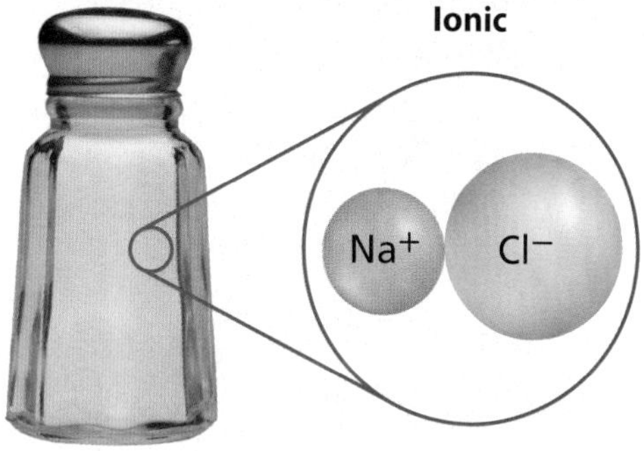 Salt	nonmetal ions; metal ions	• solid crystals • high melting and boiling points • dissolves in water • solids are poor conductors of thermal energy and electricity • ionic compounds in water solutions conduct electricity
Metallic 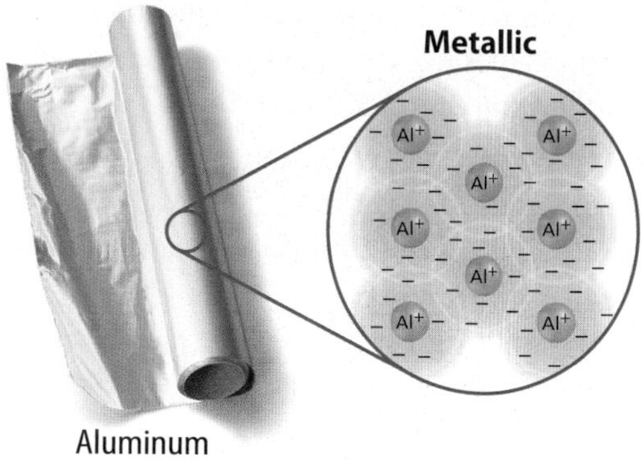 Aluminum	metal ions; metal ions	• usually solid at room temperature • high melting and boiling points • do not dissolve in water • good conductors of thermal energy and electricity • shiny surface • can be hammered into sheets and pulled into wires

Lesson 3 Review

Assessment Online Quiz

Visual Summary

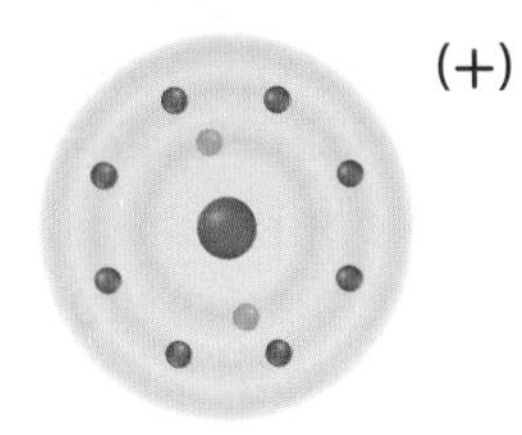

Metal atoms lose electrons and nonmetal atoms gain electrons and form stable compounds. An atom that has gained or lost an electron is an ion.

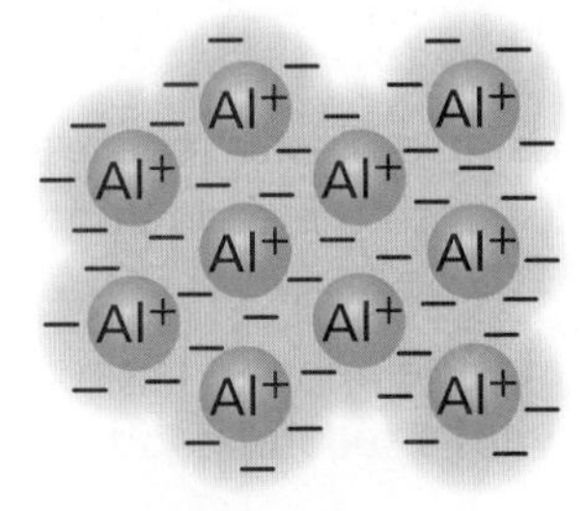

An ionic bond forms between positively and negatively charged ions.

A metallic bond forms when many metal atoms share their pooled valence electrons.

FOLDABLES

Use your lesson Foldable to review the lesson. Save your Foldable for the project at the end of the chapter.

What do you think 

You first read the statements below at the beginning of the chapter.

5. Losing electrons can make some atoms more chemically stable.

6. Metals are good electrical conductors because they tend to hold onto their valence electrons very tightly.

Did you change your mind about whether you agree or disagree with the statements? Rewrite any false statements to make them true.

Use Vocabulary

1. **Define** *ionic bond* in your own words.
2. An atom that changes so that it has an electrical charge is a(n) ________.
3. **Use the term** *metallic bond* in a sentence.

Understand Key Concepts 

4. **Recall** What holds ionic compounds together?
5. Which element would most likely bond with lithium and form an ionic compound?
   A. beryllium **C.** fluorine
   B. calcium **D.** sodium
6. **Contrast** Why are metals good conductors of electricity while covalent compounds are poor conductors?

Interpret Graphics

7. **Organize** Copy and fill in the graphic organizer below. In each oval, list a common property of an ionic compound.

Critical Thinking

8. **Design** a poster to illustrate how ionic compounds form.
9. **Evaluate** What type of bonding does a material most likely have if it has a high melting point, is solid at room temperature, and easily dissolves in water?

Math Skills

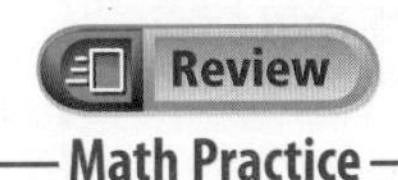

Math Practice

10. The radius of the aluminum (Al) atom is 143 pm. The radius of the aluminum ion (Al^{3+}) is 54 pm. By what percentage did the radius change as the ion formed?

Inquiry Lab

45 minutes

Ions in Solution

Materials

250-mL beaker

plastic spoon

dull pennies (20)

stopwatch

white vinegar

table salt

iron nails (2)

sandpaper

Safety

You know that ions can combine and form stable ionic compounds. Ions can also separate in a compound and dissolve in solution. For example, pennies become dull over time because the copper ions on the surface of the pennies react with oxygen in the air and form copper(II) oxide. When you place dull pennies in a vinegar-salt solution, the copper ions separate from the oxygen ions. These ions dissolve in the solution.

Question

How do elements join together to make chemical compounds?

Procedure

1. Read and complete a lab safety form.
2. Pour 50 mL of white vinegar into a 250-mL beaker. Using a plastic spoon, add a spoonful of table salt to the vinegar. Stir the mixture with the spoon until the salt dissolves.
3. Add 20 dull pennies to the vinegar-salt solution. Leave the pennies in the solution for 10 minutes. Use a stopwatch or a clock with a second hand to measure the time.
4. After 10 minutes, use the plastic spoon to remove the pennies from the solution. Rinse the pennies in tap water. Place them on paper towels to dry. Record the change to the pennies in your Science Journal.

3

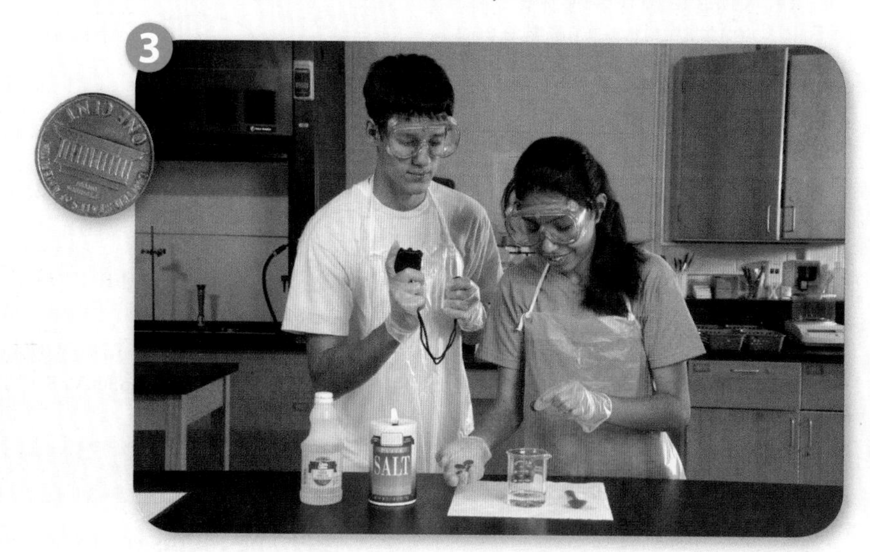

Form a Hypothesis

5. If you place an iron nail in the vinegar-salt solution, predict what changes will occur to the nail.

Test Your Hypothesis

6. Use sandpaper to clean two nails. Place one nail in the vinegar-salt solution, and place the other nail on a clean paper towel. You will compare the dry nail to the one in the solution and observe changes as they occur.
7. Every 5 minutes observe the nail in the solution and record your observations in your Science Journal. Remember to use the dry nail to help detect changes in the wet nail. Use a stopwatch or a clock with a second hand to measure the time. Keep the nail in the solution for 25 minutes
8. After 25 minutes, use a plastic spoon to remove the nail from the solution. Dispose of all materials as directed by your teacher.

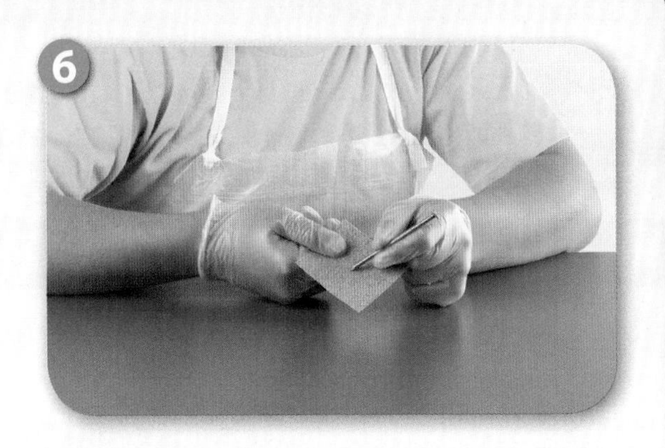

Lab Tips

☑ Be sure the pennies are separated when they are in the vinegar-salt solution. You may need to stir them with the plastic spoon.

☑ Use the plastic spoon to bring the nail out of the solution when checking for changes.

Analyze and Conclude

9. **Compare and Contrast** What changes occurred when you placed the dull pennies in the vinegar-salt solution?
10. **Recognize Cause and Effect** What changes occurred to the nail in the leftover solution? Infer why these changes occurred.
11. **The Big Idea** Give two examples of how elements chemically combine and form compounds in this lab.

Communicate Your Results

Create a chart suitable for display summarizing this lab and your results.

The Statue of Liberty is made of copper. Research why the statue is green.

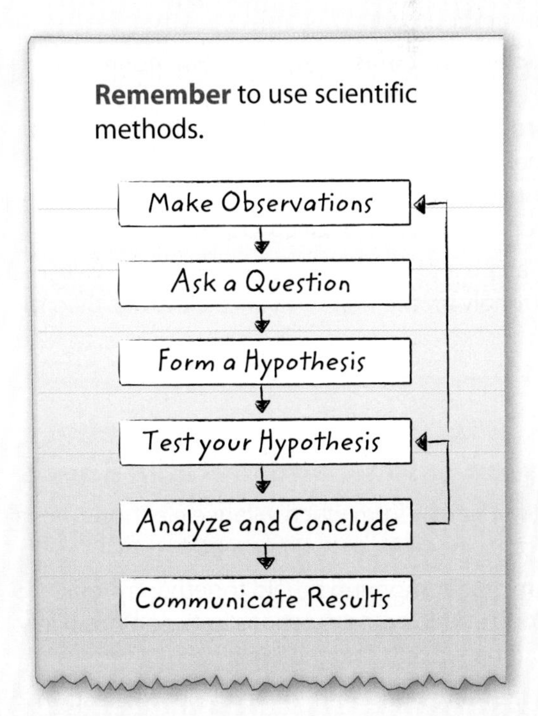

Chapter 8 Study Guide

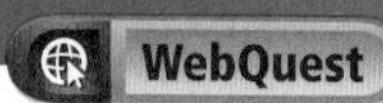

Elements can join together by sharing, transferring, or pooling electrons to make chemical compounds.

Key Concepts Summary

Lesson 1: Electrons and Energy Levels

- Electrons with more energy are farther from the atom's nucleus and are in a higher energy level.
- Atoms with fewer than eight **valence electrons** gain, lose, or share valence electrons and form stable compounds. Atoms in stable compounds have the same electron arrangement as a noble gas.

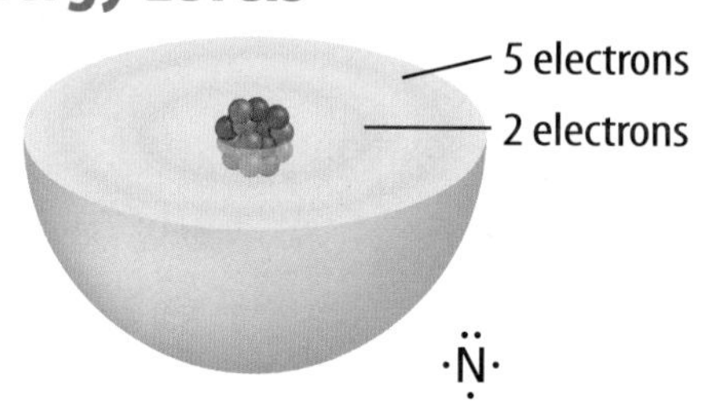

Vocabulary

chemical bond p. 268
valence electron p. 270
electron dot diagram p. 271

Lesson 2: Compounds, Chemical Formulas, and Covalent Bonds

- A compound and the elements it is made from have different chemical and physical properties.
- A **covalent bond** forms when two nonmetal atoms share valence electrons. Common properties of covalent compounds include low melting points and low boiling points. They are usually gas or liquid at room temperature and poor conductors of electricity.
- Water is a polar compound because the oxygen atom pulls more strongly on the shared valence electrons than the hydrogen atoms do.

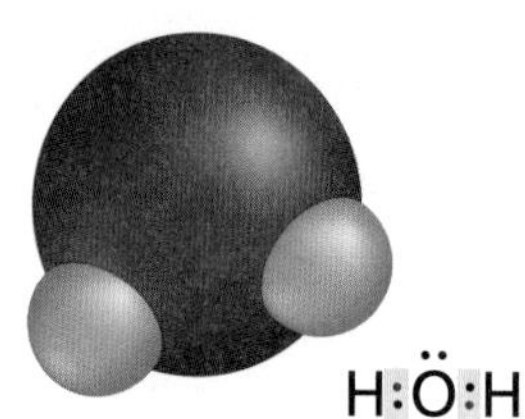

Vocabulary

covalent bond p. 277
molecule p. 278
polar molecule p. 279
chemical formula p. 280

Lesson 3: Ionic and Metallic Bonds

- **Ionic bonds** form when valence electrons move from a metal atom to a nonmetal atom.
- An ionic compound is held together by ionic bonds, which are attractions between positively and negatively charged **ions.**
- A **metallic bond** forms when valence electrons are pooled among many metal atoms.

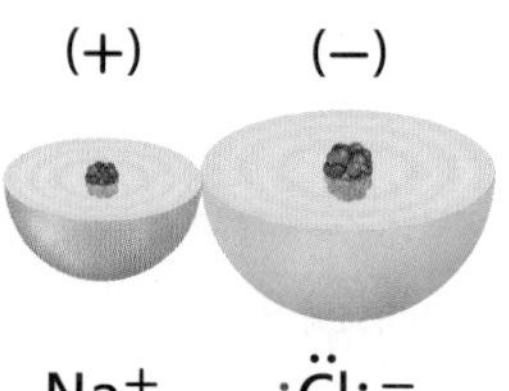

Vocabulary

ion p. 284
ionic bond p. 286
metallic bond p. 287

- Personal Tutor
- Vocabulary eGames
- Vocabulary eFlashcards

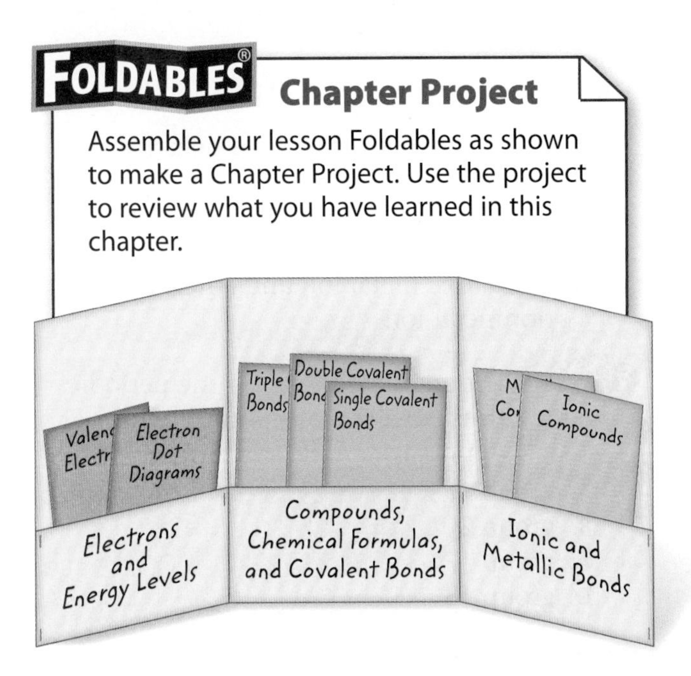

Chapter Project

Assemble your lesson Foldables as shown to make a Chapter Project. Use the project to review what you have learned in this chapter.

Use Vocabulary

1. The force that holds atoms together is called a(n) ________.
2. You can predict the number of bonds an atom can form by drawing its ________.
3. The nitrogen and hydrogen atoms that make up ammonia (NH_3) are held together by a(n) ________ because the atoms share valence electrons unequally.
4. Two hydrogen atoms and one oxygen atom together are a ________ of water.
5. A positively charged sodium ion and a negatively charged chlorine ion are joined by a(n) ________ to form the compound sodium chloride.

Link Vocabulary and Key Concepts

Copy this concept map, and then use vocabulary terms from the previous page and other terms from the chapter to complete the concept map.

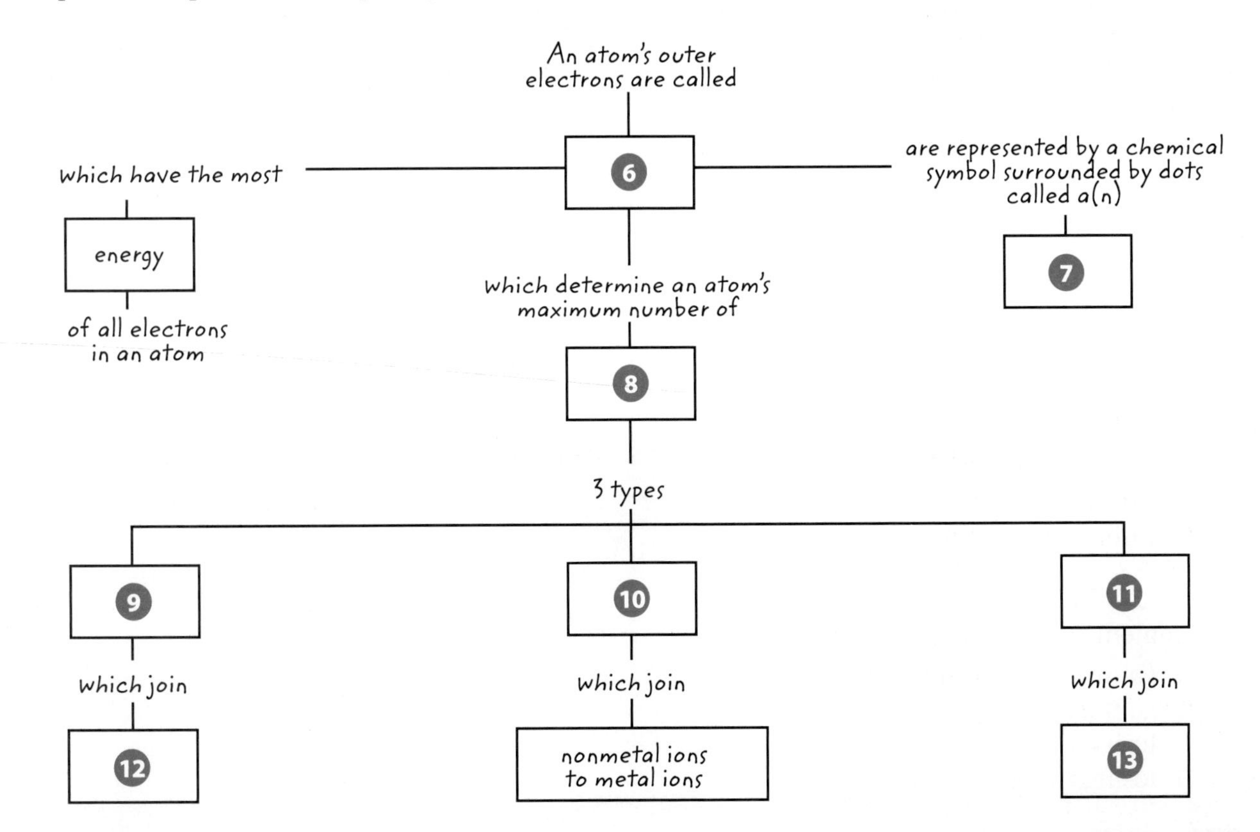

Chapter 8 Review

Understand Key Concepts

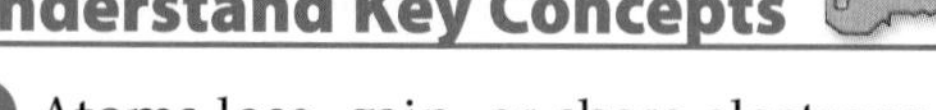

1. Atoms lose, gain, or share electrons and become as chemically stable as
   A. an electron.
   B. an ion.
   C. a metal.
   D. a noble gas.

2. Which is the correct electron dot diagram for boron, one of the group 13 elements?
   A. $\overset{\cdot}{\underset{\cdot}{B}}\cdot$
   B. $\cdot\overset{\cdot\cdot}{\underset{\cdot\cdot}{B}}:$
   C. $:\overset{\cdot\cdot}{\underset{\cdot\cdot}{B}}:$
   D. $\cdot\overset{\cdot\cdot}{\underset{\cdot}{B}}\cdot$

3. If an electron transfers from one atom to another atom, what type of bond will most likely form?
   A. covalent
   B. ionic
   C. metallic
   D. polar

4. What change would make an atom represented by this diagram have the same electron arrangement as a noble gas?

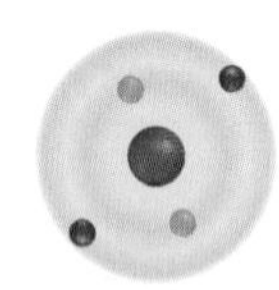

   A. gaining two electrons
   B. gaining four electrons
   C. losing two electrons
   D. losing four electrons

5. What would make bromine, a group 17 element, more similar to a noble gas?
   A. gaining one electron
   B. gaining two electrons
   C. losing one electron
   D. losing two electrons

6. Which would most likely be joined by an ionic bond?
   A. a positive metal ion and a positive nonmetal ion
   B. a positive metal ion and a negative nonmetal ion
   C. a negative metal ion and a positive nonmetal ion
   D. a negative metal ion and a negative nonmetal ion

7. Which group of elements on the periodic table forms covalent compounds with other nonmetals?
   A. group 1
   B. group 2
   C. group 17
   D. group 18

8. Which best describes an atom represented by this diagram?

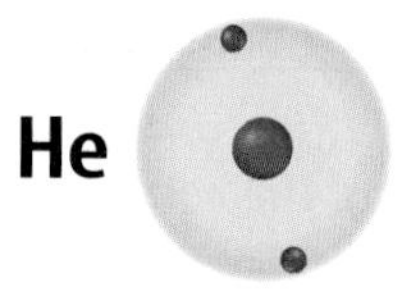

   A. It is likely to bond by gaining six electrons.
   B. It is likely to bond by losing two electrons.
   C. It is not likely to bond because it is already stable.
   D. It is not likely to bond because it has too few electrons.

9. How many dots would a dot diagram for selenium, one of the group 16 elements, have?
   A. 6
   B. 8
   C. 10
   D. 16

Assessment
Online Test Practice

Critical Thinking

10 **Classify** Use the periodic table to classify the elements potassium (K), bromine (Br), and argon (Ar) according to how likely their atoms are to do the following.

a. lose electrons to form positive ions
b. gain electrons to form negative ions
c. neither gain nor lose electrons

11 **Describe** the change that is shown in this illustration. How does this change affect the stability of the atom?

$$\cdot\ddot{\underset{\cdot}{N}}\cdot \longrightarrow :\ddot{\underset{\cdot\cdot}{N}}:^{3-}$$

12 **Analyze** One of your classmates draws an electron dot diagram for a helium atom with two dots. He tells you that these dots mean each helium atom has two unpaired electrons and can gain, lose, or share electrons to have four pairs of valence electrons and become stable. What is wrong with your classmate's argument?

13 **Explain** why the hydrogen atoms in a hydrogen gas molecule (H_2) form nonpolar covalent bonds but the oxygen and hydrogen atoms in water molecules (H_2O) form polar covalent bonds.

14 **Contrast** Why is it possible for an oxygen atom to form a double covalent bond, but it is not possible for a chlorine atom to form a double covalent bond?

Writing in Science

15 **Compose** a poem at least ten lines long that explains ionic bonding, covalent bonding, and metallic bonding.

REVIEW THE BIG IDEA

16 Which types of atoms pool their valence electrons to form a "sea of electrons"?

17 Describe a way in which elements joining together to form chemical compounds is similar to the way the letters on a computer keyboard join together to form words.

Math Skills

Math Practice

Element	Atomic Radius	Ionic Radius
Potassium (K)	227 pm	133 pm
Iodine (I)	133 pm	216 pm

18 What is the percent change when an iodine atom (I) becomes an ion (I^-)?

19 What is the percent change when a potassium atom (K) becomes an ion (K^+)?

Standardized Test Practice

Record your answers on the answer sheet provided by your teacher or on a sheet of paper.

Multiple Choice

1 Which information does the chemical formula CO_2 NOT give you?

A number of valence electrons in each atom

B ratio of atoms in the compound

C total number of atoms in one molecule of the compound

D type of elements in the compound

Use the diagram below to answer question 2.

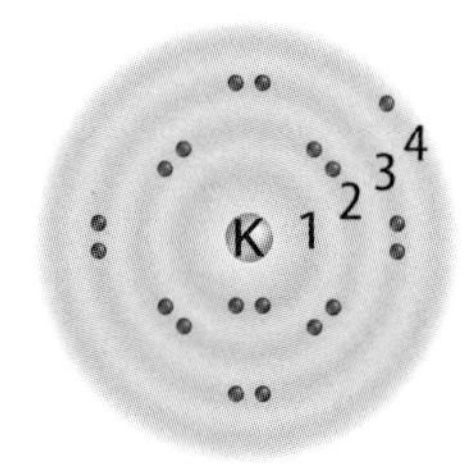

2 The diagram above shows a potassium atom. Which is the second-highest energy level?

A 1

B 2

C 3

D 4

3 What is shared in a metallic bond?

A negatively charged ions

B neutrons

C pooled valence electrons

D protons

4 Which is a characteristic of most nonpolar compounds?

A conduct electricity poorly

B dissolve easily in water

C solid crystals

D shiny surfaces

Use the diagram below to answer question 5.

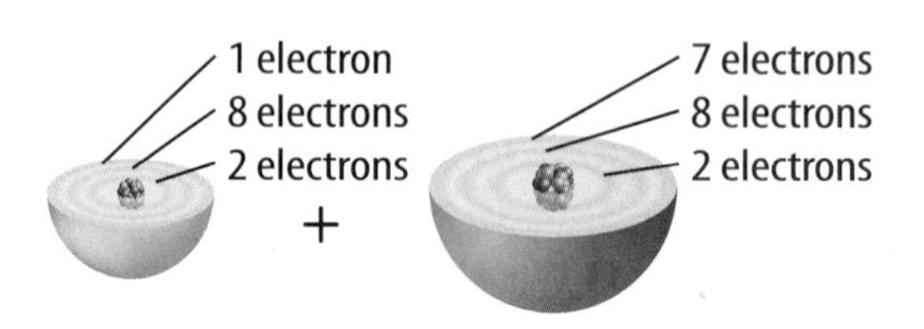

5 The atoms in the diagram above are forming a bond. Which represents that bond?

A

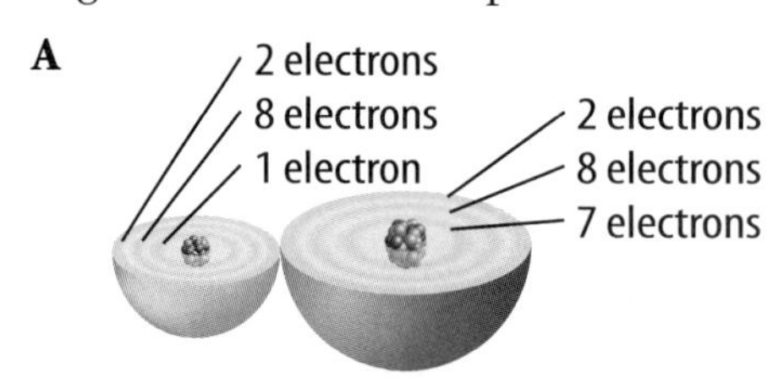

B

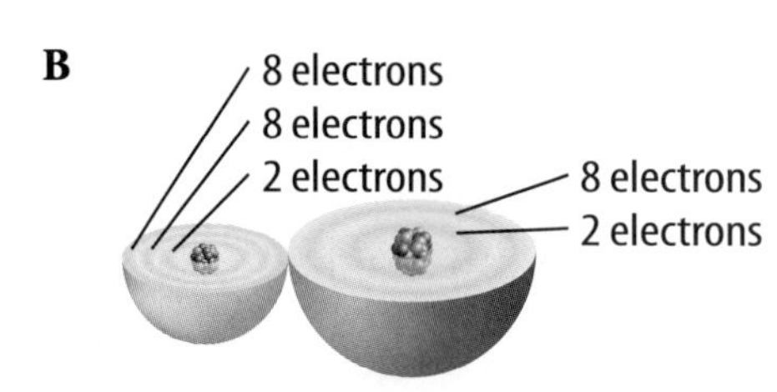

C

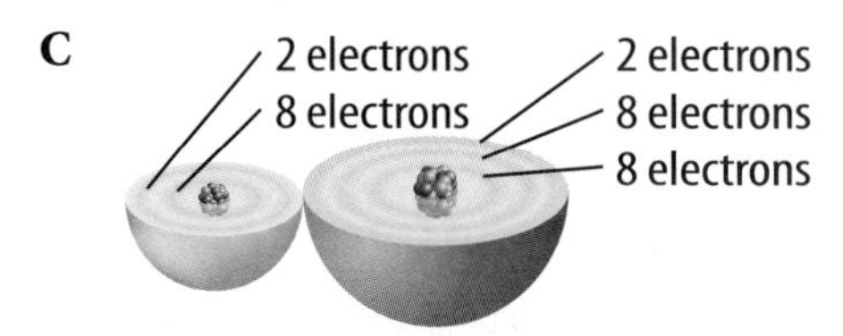

D

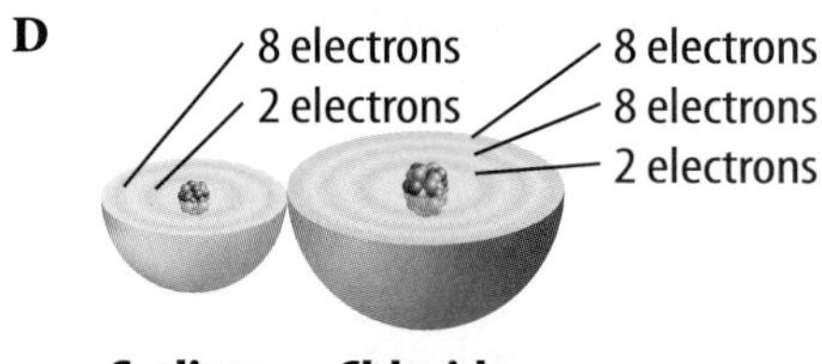

6 Covalent bonds typically form between the atoms of elements that share

A nuclei.

B oppositely charged ions.

C protons.

D valence electrons.

Use the diagram below to answer question 7.

Water Molecule

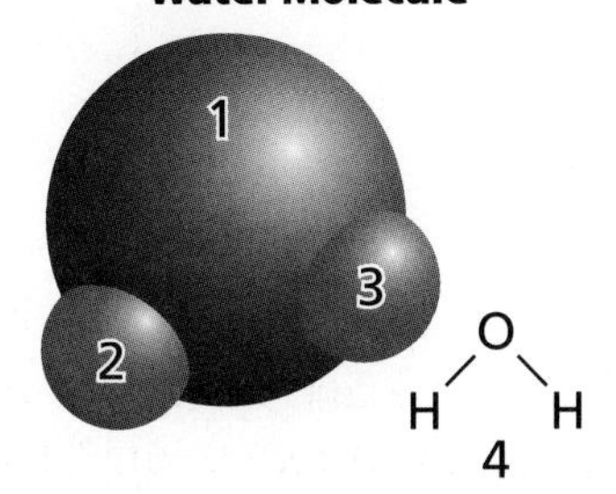

7 In the diagram above, which shows an atom with a partial negative charge?

A 1

B 2

C 3

D 4

8 Which compound is formed by the attraction between negatively and positively charged ions?

A bipolar

B covalent

C ionic

D nonpolar

9 The atoms of noble gases do NOT bond easily with other atoms because their valence electrons are

A absent.

B moving.

C neutral.

D stable.

Constructed Response

Use the table below to answer question 10.

Property	Rust	Iron	Oxygen
Color			Clear
Solid, liquid, or gas			
Strength		Strong	Does NOT apply
Usefulness			

10 Rust is a compound of iron and oxygen. Compare the properties of rust, iron, and oxygen by filling in the missing cells in the table above. What can you conclude about the properties of compounds and their elements?

Use the diagram below to answer questions 11 and 12.

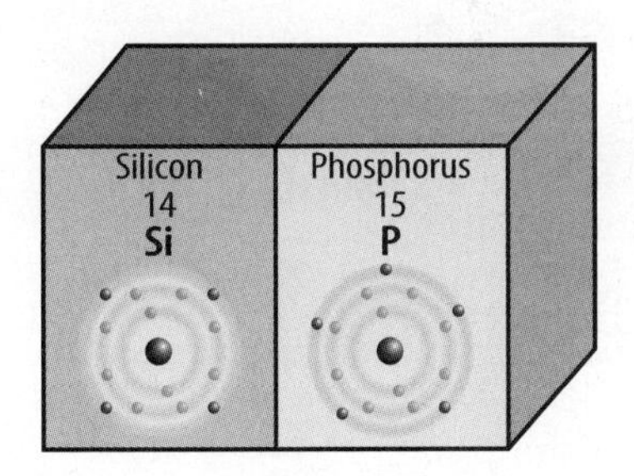

11 In the diagram, how are valence electrons illustrated? How many valence electrons does each element have?

12 Describe a stable electron configuration. For each element above, how many electrons are needed to make a stable electron configuration?

NEED EXTRA HELP?												
If You Missed Question...	1	2	3	4	5	6	7	8	9	10	11	12
Go to Lesson...	2	1	3	3	3	2	2	3	1	2	1	1

Chapter 9

Chemical Reactions and Equations

THE BIG IDEA What happens to atoms and energy during a chemical reaction?

Inquiry How does it work?

An air bag deploys in less than the blink of an eye. How does the bag open so fast? At the moment of impact, a sensor triggers a chemical reaction between two chemicals. This reaction quickly produces a large amount of nitrogen gas. This gas inflates the bag with a pop.

- A chemical reaction can produce a gas. How is this different from a gas produced when a liquid boils?
- Where do you think the nitrogen gas that is in an air bag comes from? Do you think any of the chemicals in the air bag contain the element nitrogen?
- What do you think happens to atoms and energy during a chemical reaction?

Get Ready to Read

What do you think?

Before you read, decide if you agree or disagree with each of these statements. As you read this chapter, see if you change your mind about any of the statements.

1. If a substance bubbles, you know a chemical reaction is occurring.
2. During a chemical reaction, some atoms are destroyed and new atoms are made.
3. Reactions always start with two or more substances that react with each other.
4. Water can be broken down into simpler substances.
5. Reactions that release energy require energy to get started.
6. Energy can be created in a chemical reaction.

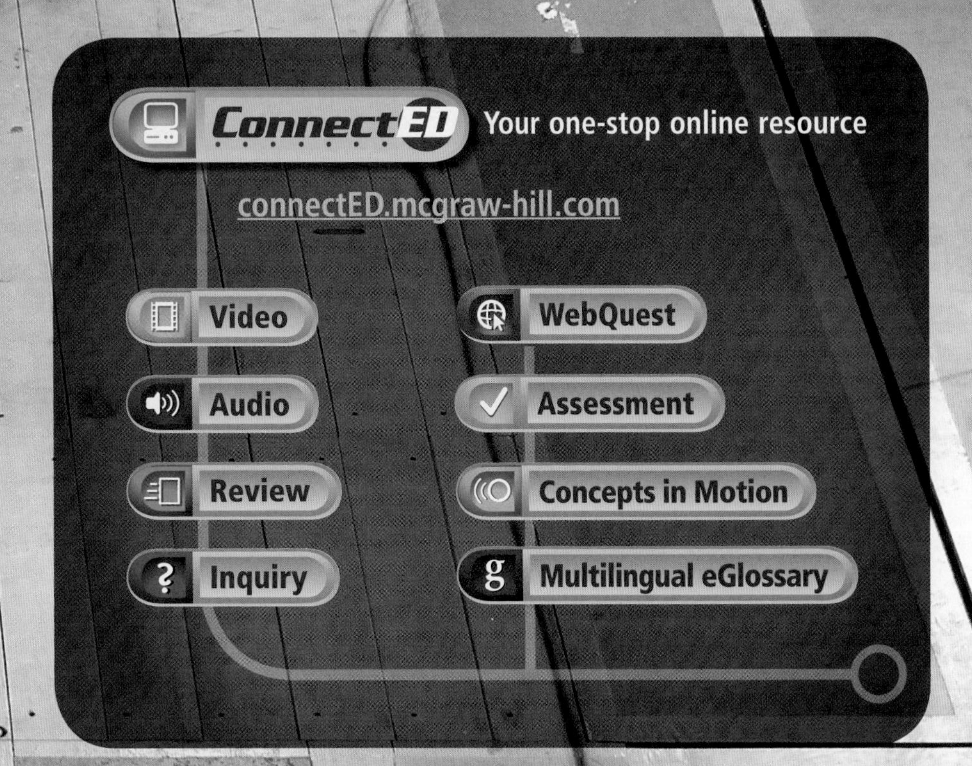

Lesson 1

Reading Guide

Key Concepts
ESSENTIAL QUESTIONS

- What are some signs that a chemical reaction might have occurred?
- What happens to atoms during a chemical reaction?
- What happens to the total mass in a chemical reaction?

Vocabulary

chemical reaction p. 301
chemical equation p. 304
reactant p. 305
product p. 305
law of conservation of mass p. 306
coefficient p. 308

Multilingual eGlossary

Understanding Chemical Reactions

Inquiry Does it run on batteries?

Flashes of light from fireflies dot summer evening skies in many parts of the United States. But, firefly light doesn't come from batteries. Fireflies make light using a process called bioluminescence (bi oh lew muh NE cents). In this process, chemicals in the firefly's body combine in a two-step process and make new chemicals and light.

Launch Lab

15–20 minutes

Where did it come from?

Does a boiled egg have more mass than a raw egg? What happens when liquids change to a solid?

1. Read and complete a lab safety form.
2. Use a **graduated cylinder** to add 25 mL of **solution A** to a **self-sealing plastic bag.** Place a **stoppered test tube** containing **solution B** into the bag. Be careful not to dislodge the stopper.
3. Seal the bag completely, and wipe off any moisture on the outside with a **paper towel.** Place the bag on the **balance.** Record the total mass in your Science Journal.
4. Without opening the bag, remove the stopper from the test tube and allow the liquids to mix. Observe and record what happens.
5. Place the sealed bag and its contents back on the balance. Read and record the mass.

Think About This

1. What did you observe when the liquids mixed? How would you account for this observation?
2. Did the mass of the bag's contents change? If so, could the change have been due to the precision of the balance, or did the matter in the bag change its mass? Explain.
3. **Key Concept** Do you think matter was gained or lost in the bag? How can you tell?

Changes in Matter

When you put liquid water in a freezer, it changes to solid water, or ice. When you pour brownie batter into a pan and bake it, the liquid batter changes to a solid, too. In both cases, a liquid changes to a solid. Are these changes the same?

Physical Changes

Recall that matter can undergo two types of changes—chemical or physical. A physical change does not produce new substances. The substances that exist before and after the change are the same, although they might have different physical properties. This is what happens when liquid water freezes. Its physical properties change from a liquid to a solid, but the water, H_2O, does not change into a different substance. Water molecules are always made up of two hydrogen atoms bonded to one oxygen atom regardless of whether they are solid, liquid, or gas.

Chemical Changes

Recall that during a chemical change, one or more substances change into new substances. The starting substances and the substances produced have different physical and chemical properties. For example, when brownie batter bakes, a chemical change occurs. Many of the substances in the baked brownies are different from the substances in the batter. As a result, baked brownies have physical and chemical properties that are different from those of brownie batter.

A chemical change also is called a chemical reaction. These terms mean the same thing. *A* **chemical reaction** *is a process in which atoms of one or more substances rearrange to form one or more new substances.* In this lesson, you will read what happens to atoms during a reaction and how these changes can be described using equations.

Reading Check What types of properties change during a chemical reaction?

Signs of a Chemical Reaction

How can you tell if a chemical reaction has taken place? You have read that the substances before and after a reaction have different properties. You might think that you could look for changes in properties as a sign that a reaction occurred. In fact, changes in the physical properties of color, state of matter, and odor are all signs that a chemical reaction might have occurred. Another sign of a chemical reaction is a change in energy. If substances get warmer or cooler or if they give off light or sound, it is likely that a reaction has occurred. Some signs that a chemical reaction might have occurred are shown in **Figure 1.**

However, these signs are not proof of a chemical change. For example, bubbles appear when water boils. But, bubbles also appear when baking soda and vinegar react and form carbon dioxide gas. How can you be sure that a chemical reaction has taken place? The only way to know is to study the chemical properties of the substances before and after the change. If they have different chemical properties, then the substances have undergone a chemical reaction.

Key Concept Check What are some signs that a chemical reaction might have occurred?

Figure 1 You can detect a chemical reaction by looking for changes in properties and changes in energy of the substances that reacted.

Change in Properties	
Change in color Bright copper changes to green when the copper reacts with certain gases in the air. 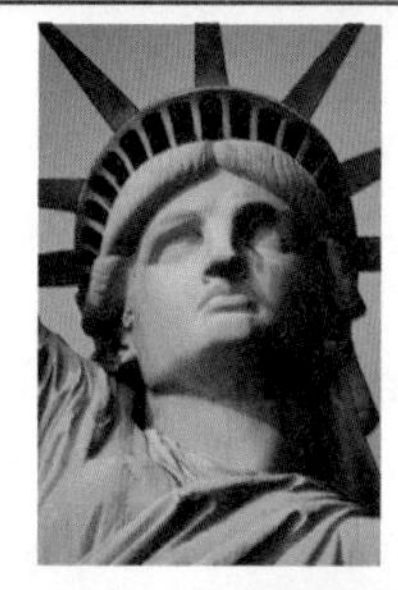	**Formation of bubbles** Bubbles of carbon dioxide form when baking soda is added to vinegar. 
Change in odor When food burns or rots, a change in odor is a sign of chemical change. 	**Formation of a precipitate** A precipitate is a solid formed when two liquids react. 
Change in Energy	
Warming or cooling Thermal energy is either given off or absorbed during a chemical change. 	**Release of light** A firefly gives off light as the result of a chemical change. 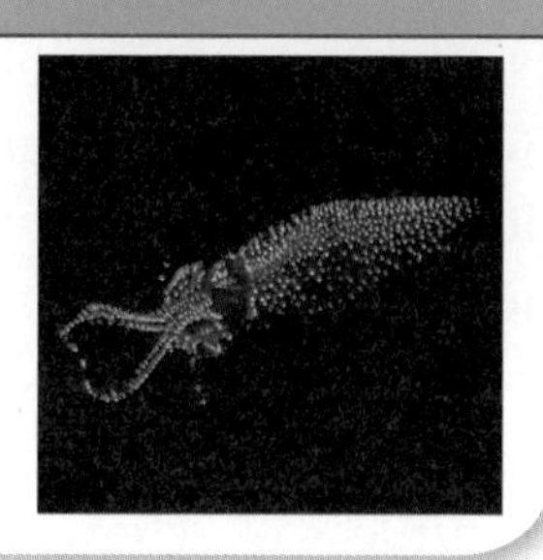

What happens in a chemical reaction?

During a chemical reaction, one or more substances react and form one or more new substances. How are these new substances formed?

Atoms Rearrange and Form New Substances

To understand what happens in a reaction, first review substances. Recall that there are two types of substances—elements and compounds. Substances have a fixed arrangement of atoms. For example, in a single drop of water, there are trillions of oxygen and hydrogen atoms. However, all of these atoms are arranged in the same way—two atoms of hydrogen are bonded to one atom of oxygen. If this arrangement changes, the substance is no longer water. Instead, a different substance forms with different physical and chemical properties. This is what happens during a chemical reaction. Atoms of elements or compounds rearrange and form different elements or compounds.

Bonds Break and Bonds Form

How does the rearrangement of atoms happen? Atoms rearrange when **chemical bonds** between atoms break. Recall that constantly moving particles make up all substances, including solids. As particles move, they collide with one another. If the particles collide with enough energy, the bonds between atoms can break. The atoms separate, rearrange, and new bonds can form. The reaction that forms hydrogen and oxygen from water is shown in **Figure 2.** Adding electric energy to water molecules can cause this reaction. The added energy causes bonds between the hydrogen atoms and the oxygen atoms to break. After the bonds between the atoms in water molecules break, new bonds can form between pairs of hydrogen atoms and between pairs of oxygen atoms.

REVIEW VOCABULARY
chemical bond
an attraction between atoms when electrons are shared, transferred, or pooled

 **Key Concept Check** What happens to atoms during a chemical reaction?

Figure 2 Notice that no new atoms are created in a chemical reaction. The existing atoms rearrange and form new substances.

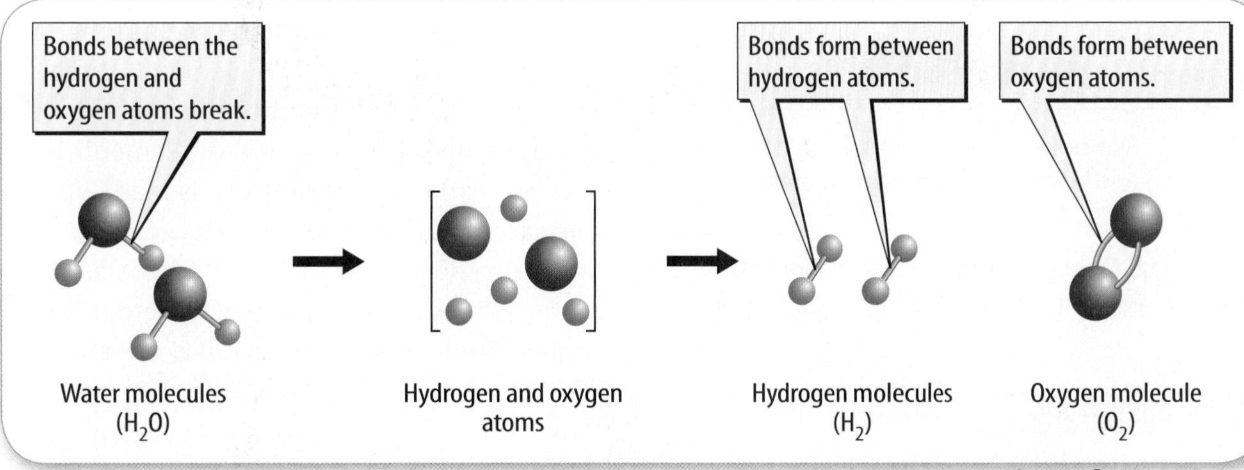

Table 1 Symbols and Formulas of Some Elements and Compounds

Substance	Formula	# of atoms
Carbon	C	C: 1
Copper	Cu	Cu: 1
Cobalt	Co	Co: 1
Oxygen	O_2	O: 2
Hydrogen	H_2	H: 2
Chlorine	Cl_2	Cl: 2
Carbon dioxide	CO_2	C: 1 O: 2
Carbon monoxide	CO	C: 1 O: 1
Water	H_2O	H: 2 O: 1
Hydrogen peroxide	H_2O_2	H: 2 O: 2
Glucose	$C_6H_{12}O_6$	C: 6 H: 12 O: 6
Sodium chloride	NaCl	Na: 1 Cl: 1
Magnesium hydroxide	$Mg(OH)_2$	Mg: 1 O: 2 H: 2

Table 1 Symbols and subscripts describe the type and number of atoms in an element or a compound.

Visual Check Describe the number of atoms in each element in the following: C, Co, CO, and CO_2.

Concepts in Motion **Interactive Table**

Chemical Equations

Suppose your teacher asks you to produce a specific reaction in your science laboratory. How might your teacher describe the reaction to you? He or she might say something such as "react baking soda and vinegar to form sodium acetate, water, and carbon dioxide." It is more likely that your teacher will describe the reaction in the form of a chemical equation. *A* **chemical equation** *is a description of a reaction using element symbols and chemical formulas.* Element symbols represent elements. Chemical formulas represent compounds.

Element Symbols

Recall that symbols of elements are shown in the periodic table. For example, the symbol for carbon is C. The symbol for copper is Cu. Each element can exist as just one atom. However, some elements exist in nature as diatomic molecules—two atoms of the same element bonded together. A formula for one of these diatomic elements includes the element's symbol and the subscript *2*. A subscript describes the number of atoms of an element in a compound. Oxygen (O_2) and hydrogen (H_2) are examples of diatomic molecules. Some element symbols are shown above the blue line in **Table 1.**

Chemical Formulas

When atoms of two or more different elements bond, they form a compound. Recall that a chemical formula uses elements' symbols and subscripts to describe the number of atoms in a compound. If an element's symbol does not have a subscript, the compound contains only one atom of that element. For example, carbon dioxide (CO_2) is made up of one carbon atom and two oxygen atoms. Remember that two different formulas, no matter how similar, represent different substances. Some chemical formulas are shown below the blue line in **Table 1.**

Writing Chemical Equations

A chemical equation includes both the substances that react and the substances that are formed in a chemical reaction. *The starting substances in a chemical reaction are* **reactants.** *The substances produced by the chemical reaction are* **products.** **Figure 3** shows how a chemical equation is written. Chemical formulas are used to describe the reactants and the products. The reactants are written to the left of an arrow, and the products are written to the right of the arrow. Two or more reactants or products are separated by a plus sign. The general structure for an equation is:

reactant + reactant → product + product

When writing chemical equations, it is important to use correct chemical formulas for the reactants and the products. For example, suppose a certain chemical reaction produces carbon dioxide and water. The product carbon dioxide would be written as CO_2 and not as CO. CO is the formula for carbon monoxide, which is not the same compound as CO_2. Water would be written as H_2O and not as H_2O_2, the formula for hydrogen peroxide.

Inquiry MiniLab — 10 minutes

How does an equation represent a reaction?

Sulfur dioxide (SO_2) and oxygen (O_2) react and form sulfur trioxide (SO_3). How does an equation represent the reaction?

1. Read and complete a lab safety form.
2. Use **yellow modeling clay** to model two atoms of sulfur. Use **red modeling clay** to model six atoms of oxygen.
3. Make two molecules of SO_2 with a sulfur atom in the middle of each molecule. Make one molecule of O_2. Sketch the models in your Science Journal.
4. Rearrange atoms to form two molecules of SO_3. Place a sulfur atom in the middle of each molecule. Sketch the models in your Science Journal.

Analyze and Conclude

1. **Identify** the reactants and the products in this chemical reaction.
2. **Write** a chemical equation for this reaction.
3. **Explain** What do the letters represent in the equation? The numbers?
4. **Key Concept** In terms of chemical bonds, what did you model by pulling molecules apart and building new ones?

Figure 3 An equation is read much like a sentence. This equation is read as "carbon plus oxygen produces carbon dioxide."

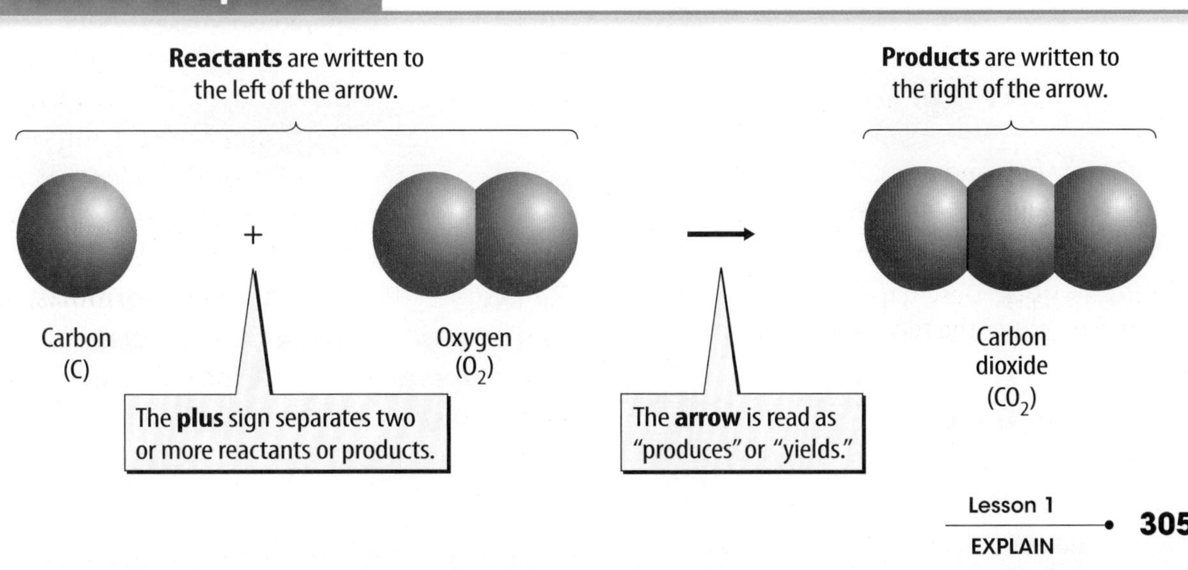

Word Origin

product
from Latin *producere,* means "bring forth"

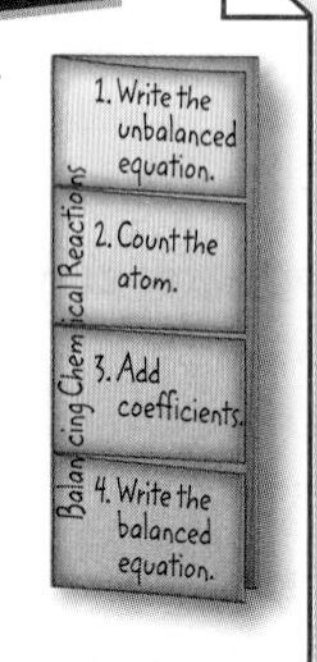

Foldables®

Make a vertical four-tab book. Label it as shown. Use it to study the steps of balancing equations.

Conservation of Mass

A French chemist named Antoine Lavoisier (AN twan • luh VWAH see ay) (1743–1794) discovered something interesting about chemical reactions. In a series of experiments, Lavoisier measured the masses of substances before and after a chemical reaction inside a closed container. He found that the total mass of the reactants always equaled the total mass of the **products.** Lavoisier's results led to the law of conservation of mass. *The* **law of conservation of mass** *states that the total mass of the reactants before a chemical reaction is the same as the total mass of the products after the chemical reaction.*

Atoms are conserved.

The discovery of atoms provided an explanation for Lavoisier's observations. Mass is conserved in a reaction because atoms are conserved. Recall that during a chemical reaction, bonds break and new bonds form. However, atoms are not destroyed, and no new atoms form. All atoms at the start of a chemical reaction are present at the end of the reaction. **Figure 4** shows that mass is conserved in the reaction between baking soda and vinegar.

Key Concept Check What happens to the total mass of the reactants in a chemical reaction?

Figure 4 As this reaction takes place, the mass on the balance remains the same, showing that mass is conserved.

Conservation of Mass

The baking soda is contained in a balloon. The balloon is attached to a flask that contains vinegar.

When the balloon is tipped up, the baking soda pours into the vinegar. The reaction forms a gas that is collected in the balloon.

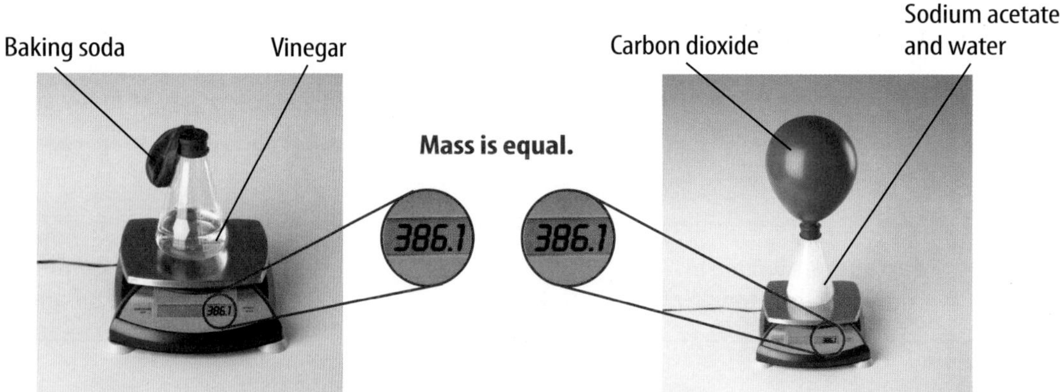

baking soda + **vinegar** → **sodium acetate** + **water** + **carbon dioxide**

$NaHCO_3$ + $HC_2H_3O_2$ → $NaC_2H_3O_2$ + H_2O + CO_2

$NaHCO_3$	$HC_2H_3O_2$		$NaC_2H_3O_2$	H_2O	CO_2
1 Na: ●	4 H: ●●●●	**Atoms are equal.**	1 Na: ●	2 H: ●●	1 C: ●
1 H: ●	2 C: ●●		2 C: ●●	1 O: ●	2 O: ●●
1 C: ●	2 O: ●●		3 H: ●●●		
3 O: ●●●			2 O: ●●		

Is an equation balanced?

How does a chemical equation show that atoms are conserved? An equation is written so that the number of atoms of each element is the same, or balanced, on each side of the arrow. The equation showing the reaction between carbon and oxygen that produces carbon dioxide is shown below. Remember that oxygen is written as O_2 because it is a diatomic molecule. The formula for carbon dioxide is CO_2.

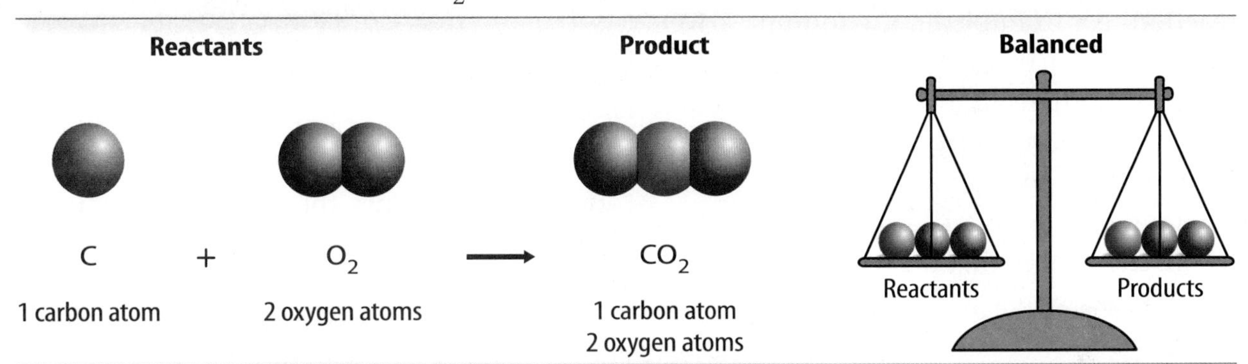

Is there the same number of carbon atoms on each side of the arrow? Yes, there is one carbon atom on the left and one on the right. Carbon is balanced. Is oxygen balanced? There are two oxygen atoms on each side of the arrow. Oxygen also is balanced. The atoms of all elements are balanced. Therefore, the equation is balanced.

You might think a balanced equation happens automatically when you write the symbols and formulas for reactants and products. However, this usually is not the case. For example, the reaction between hydrogen (H_2) and oxygen (O_2) that forms water (H_2O) is shown below.

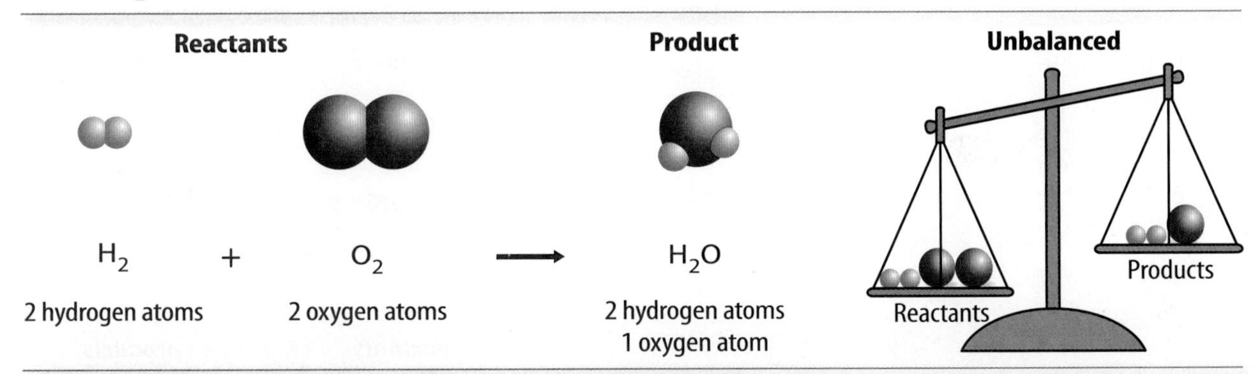

Count the number of hydrogen atoms on each side of the arrow. There are two hydrogen atoms in the product and two in the reactants. They are balanced. Now count the number of oxygen atoms on each side of the arrow. Did you notice that there are two oxygen atoms in the reactants and only one in the product? Because they are not equal, this equation is not balanced. To accurately represent this reaction, the equation needs to be balanced.

Balancing Chemical Equations

When you balance a chemical equation, you count the atoms in the reactants and the products and then add coefficients to balance the number of atoms. A **coefficient** *is a number placed in front of an element symbol or chemical formula in an equation.* It is the number of units of that substance in the reaction. For example, in the formula $2H_2O$, the *2* in front of H_2O is a coefficient, This means that there are two molecules of water in the reaction. Only coefficients can be changed when balancing an equation. Changing subscripts changes the identities of the substances that are in the reaction.

If one molecule of water contains two hydrogen atoms and one oxygen atom, how many H and O atoms are in two molecules of water ($2H_2O$)? Multiply each by 2.

2 × 2 H atoms = 4 H atoms
2 × 1 O atom = 2 O atoms

When no coefficient is present, only one unit of that substance takes part in the reaction. **Table 2** shows the steps of balancing a chemical equation.

Table 2 Balancing a Chemical Equation

Step	Example
1 Write the unbalanced equation. Make sure that all chemical formulas are correct.	$H_2 + O_2 \rightarrow H_2O$ — *reactants* / *products*
2 Count atoms of each element in the reactants and in the products. **a.** Note which, if any, elements have a balanced number of atoms on each side of the equation. Which atoms are not balanced? **b.** If all of the atoms are balanced, the equation is balanced.	$H_2 + O_2 \rightarrow H_2O$ — *reactants* H = 2, O = 2; *products* H = 2, O = 1
3 Add coefficients to balance the atoms. **a.** Pick an element in the equation that is not balanced, such as oxygen. Write a coefficient in front of a reactant or a product that will balance the atoms of that element. **b.** Recount the atoms of each element in the reactants and the products. Note which atoms are not balanced. Some atoms that were balanced before might no longer be balanced. **c.** Repeat step 3 until the atoms of each element are balanced.	$H_2 + O_2 \rightarrow 2H_2O$ — *reactants* H = 2, O = 2; *products* H = 4, O = 2 $2H_2 + O_2 \rightarrow 2H_2O$ — *reactants* H = 4, O = 2; *products* H = 4, O = 2
4 Write the balanced chemical equation including the coefficients.	$2H_2 + O_2 = 2H_2O$

Visual Check In row 2 above, which element is not balanced? In the top of row 3, which element is not balanced?

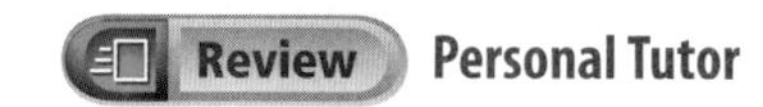

Lesson 1 Review

Assessment Online Quiz

Visual Summary

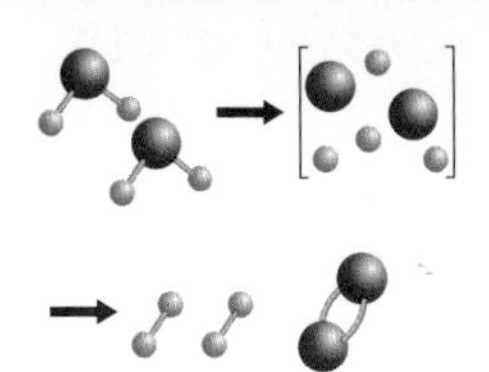

A chemical reaction is a process in which bonds break and atoms rearrange, forming new bonds.

$2H_2 + O_2 \rightarrow 2H_2O$

A chemical equation uses symbols to show reactants and products of a chemical reaction.

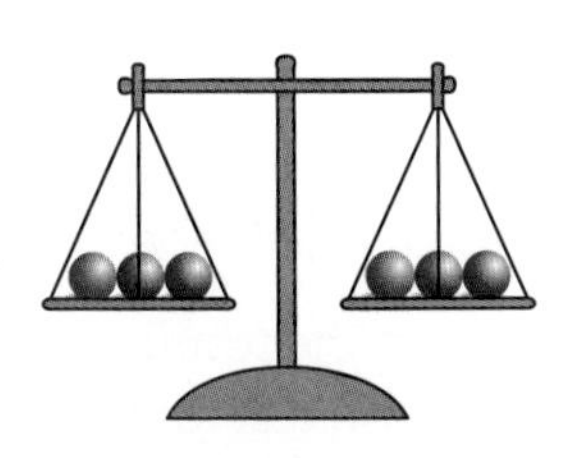

The mass and the number of each type of atom do not change during a chemical reaction. This is the law of conservation of mass.

FOLDABLES®

Use your lesson Foldable to review the lesson. Save your Foldable for the project at the end of the chapter.

What do you think NOW?

You first read the statements below at the beginning of the chapter.

1. If a substance bubbles, you know a chemical reaction is occurring.

2. During a chemical reaction, some atoms are destroyed and new atoms are made.

Did you change your mind about whether you agree or disagree with the statements? Rewrite any false statements to make them true.

Use Vocabulary

1 **Define** *reactants* and *products.*

Understand Key Concepts

2 Which is a sign of a chemical reaction?

- **A.** chemical properties change
- **B.** physical properties change
- **C.** a gas forms
- **D.** a solid forms

3 **Explain** why subscripts cannot change when balancing a chemical equation.

4 **Infer** Is the reaction below possible? Explain why or why not.

$$H_2O + NaOH \rightarrow NaCl + H_2$$

Interpret Graphics

5 **Describe** the reaction below by listing the bonds that break and the bonds that form.

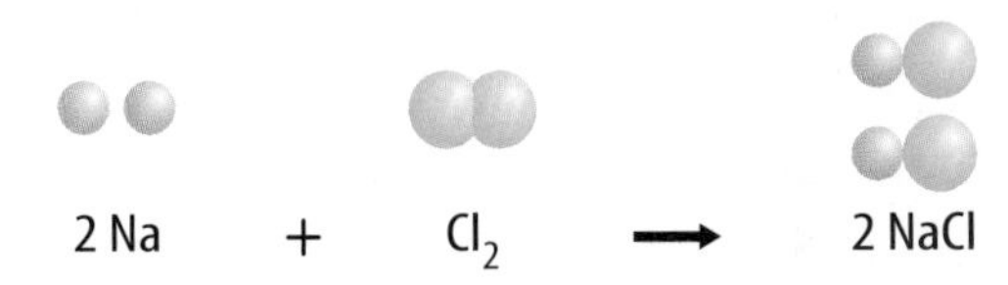

6 **Interpret** Copy and complete the table to determine if this equation is balanced:

$$CH_4 + 2O_2 \rightarrow CO_2 + 2H_2O$$

Is this reaction balanced? Explain.

Type of Atom	Number of Atoms in the Balanced Chemical Equation	
	Reactants	Products

Critical Thinking

7 Balance this chemical equation. Hint: Balance Al last and then use a multiple of 2 and 3.

$$Al + HCl \rightarrow AlCl_3 + H_2$$

Follow a Procedure

40 minutes

What can you learn from an experiment?

Materials

test tubes and rack

ammonium hydroxide (NH_4OH)

aluminum foil

sodium bicarbonate ($NaHCO_3$)

Also needed: copper foil, tongs, salt water, copper sulfate solution ($CuSO_4$), 25-mL graduated cylinder, Bunsen burner, plastic spoon, toothpick, ring stand and clamp, splints, matches, paper towel

Safety

Observing reactions allows you to compare different types of changes that can occur. You can then design new experiments to learn more about reactions.

Learn It

If you have never tested for a chemical reaction before, it is helpful to **follow a procedure.** A procedure tells you which materials to use and what steps to take.

Try It

1. Read and complete a safety form.
2. Copy the table into your Science Journal. During each procedure, record observations in the table.

3a. Dip a strip of aluminum foil into salt water in a test tube for about 1 min to remove the coating.

3b. Place 5 mL of copper sulfate solution in a test tube. Lift the aluminum foil from the salt water. Drop it into the test tube of copper sulfate so that the bottom part is in the liquid. Look for evidence of a chemical change. Set the test tube in a rack, and do the other procedures.

4. Use tongs to hold a small piece of copper foil in a flame for 3 min. Set the foil on a heat-proof surface, and allow it to cool. Use a toothpick to examine the product.
5. Place a spoonful of sodium bicarbonate in a dry test tube. Clamp the tube to a ring stand at a 45° angle. Point the mouth of the tube away from people. Move a burner flame back and forth under the tube. Observe the reaction. Test for carbon dioxide with a lighted wood splint.
6. Add 1 drop of ammonium hydroxide to a test tube containing 5 mL of copper sulfate solution.
7. Pour the liquid from the test tube in step 3b into a clean test tube. Dump the aluminum onto a paper towel. Record your observations of both the liquid and the solid.

Apply It

8. Using the table, write a balanced equation for each reaction.
9. Why did the color of the copper sulfate disappear in step 3b?
10. **Key Concept** How can you tell the difference between types of reactions by the number and type of reactants and products?

Step	Reactants	Products	Observations and Evidence of Chemical Reaction
3 + 7	$Al + CuSO_4$	$Cu + Al_2(SO_4)_3$	
4	$Cu + O_2$	CuO	
5	$NaHCO_3$	$CO_2 + Na_2CO_3 + H_2O$	
6	$NH_4OH + CuSO_4$	$(NH_4)_2SO_4 + Cu(OH)_2$	

Lesson 2

Reading Guide

Key Concepts
ESSENTIAL QUESTIONS

- How can you recognize the type of chemical reaction by the number or type of reactants and products?
- What are the different types of chemical reactions?

Vocabulary

synthesis p. 313
decomposition p. 313
single replacement p. 314
double replacement p. 314
combustion p. 314

Video
What's Science Got to do With It?

Types of Chemical Reactions

Inquiry Where did it come from?

When lead nitrate, a clear liquid, combines with potassium iodide, another clear liquid, a yellow solid appears instantly. Where did it come from? Here's a hint—the name of the solid is lead iodide. Did you guess that parts of each reactant combined and formed it? You'll learn about this and other types of reactions in this lesson.

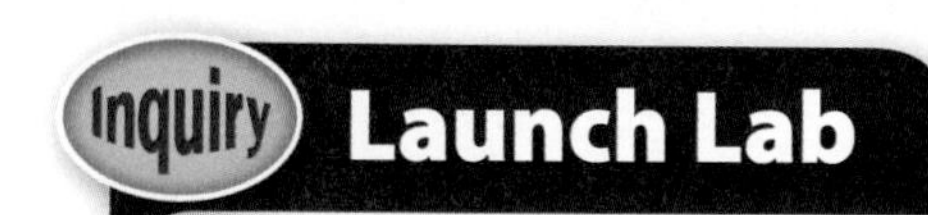

15 minutes

What combines with what?

The reactants and the products in a chemical reaction can be elements, compounds, or both. In how many ways can these substances combine?

1. Read and complete a lab safety form.
2. Divide a **sheet of paper** into four equal sections labeled *A, B, Y,* and *Z.* Place **red paper clips** in section A, **yellow clips** in section B, **blue clips** in section Y, and **green clips** in section Z.
3. Use another sheet of paper to copy the table shown to the right. Turn the paper so that a long edge is at the top. Print *REACTANTS → PRODUCTS* across the top then complete the table.
4. Using the paper clips, model the equations listed in the table. Hook the clips together to make diatomic elements or compounds. Place each clip model onto your paper over the matching written equation.
5. As you read this lesson, match the types of equations to your paper clip equations.

	REACTANTS → PRODUCTS
1	$AY \rightarrow A + Y$
2	$B + Z \rightarrow BZ$
3	$2A_2 + Y_2 \rightarrow 2A_2Y$
4	$A + BY \rightarrow B + AY$
5	$Z + BY \rightarrow Y + BZ$
6	$AY + BZ \rightarrow AZ + BY$

Think About This

1. Which equation represents hydrogen combining with oxygen and forming water? How do you know?
2. **Key Concept** How could you use the number and type of reactants to identify a type of chemical reaction?

Patterns in Reactions

If you have ever used hydrogen peroxide, you might have noticed that it is stored in a dark bottle. This is because light causes hydrogen peroxide to change into other substances. Maybe you have seen a video of an explosion demolishing an old building, like in **Figure 5.** How is the reaction with hydrogen peroxide and light similar to a building demolition? In both, one reactant breaks down into two or more products.

The breakdown of one reactant into two or more products is one of four major types of chemical reactions. Each type of chemical reaction follows a unique pattern in the way atoms in reactants rearrange to form products. In this lesson, you will read how chemical reactions are classified by recognizing patterns in the way the atoms recombine.

Figure 5 When dynamite explodes, it chemically changes into several products and releases energy.

Types of Chemical Reactions

There are many different types of reactions. It would be impossible to memorize them all. However, most chemical reactions fit into four major categories. Understanding these categories of reactions can help you predict how compounds will react and what products will form.

Synthesis

A **synthesis** (SIHN thuh sus) *is a type of chemical reaction in which two or more substances combine and form one compound.* In the synthesis reaction shown in **Figure 6,** magnesium (Mg) reacts with oxygen (O_2) in the air and forms magnesium oxide (MgO). You can recognize a synthesis reaction because two or more reactants form only one product.

Decomposition

In a **decomposition** *reaction, one compound breaks down and forms two or more substances.* You can recognize a decomposition reaction because one reactant forms two or more products. For example, hydrogen peroxide (H_2O_2), shown in **Figure 6,** decomposes and forms water (H_2O) and oxygen gas (O_2). Notice that decomposition is the reverse of synthesis.

Key Concept Check How can you tell the difference between synthesis and decomposition reactions?

FOLDABLES®

Make a horizontal four-door book. Label it as shown. Use it to organize your notes about the different types of chemical reactions.

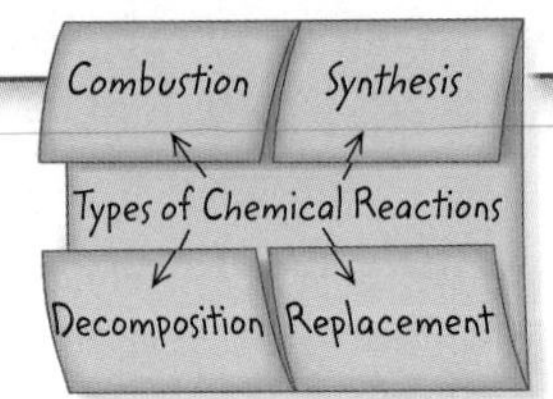

WORD ORIGIN

synthesis
from Greek *syn-*, means "together"; and *tithenai*, means "put"

Figure 6 Synthesis and decomposition reactions are opposites of each other.

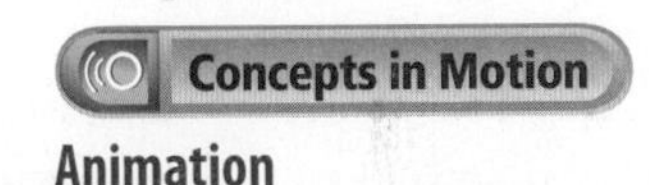

Animation

Synthesis and Decomposition Reactions

Synthesis Reactions

Examples:
$2Na + Cl_2 \rightarrow 2NaCl$
$2H_2 + O_2 \rightarrow 2H_2O$
$H_2O + SO_3 \rightarrow H_2SO_4$

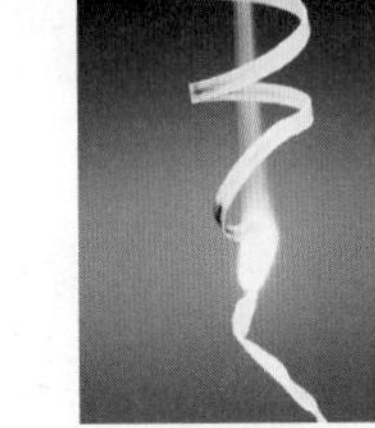

$2Mg + O_2 \rightarrow 2MgO$
magnesium + oxygen → magnesium oxide

Decomposition Reactions

Examples:
$CaCO_3 \rightarrow CaO + CO_2$
$2H_2O \rightarrow 2H_2 + O_2$
$2KClO_3 \rightarrow 2KCl + 3O_2$

$2H_2O_2 \rightarrow 2H_2O + O_2$
hydrogen peroxide → water + oxygen

Replacement Reactions

Single Replacement

Examples:
$Fe + CuSO_4 \rightarrow FeSO_4 + Cu$
$Zn + 2HCl + ZnCl_2 + H_2$

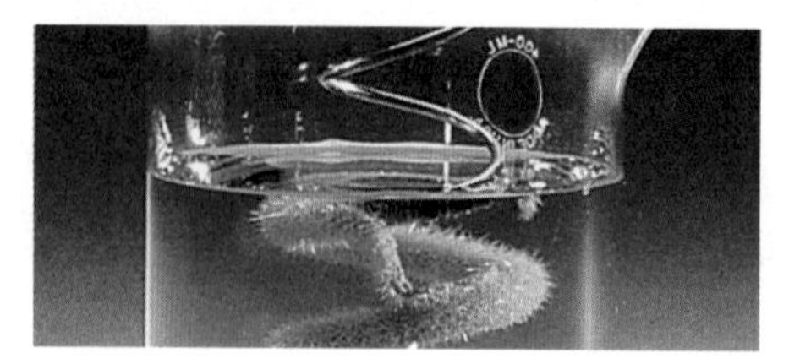

$2AgNO_3 + Cu \rightarrow Cu(NO_3)_2 + 2Ag$
silver nitrate, copper, copper nitrate, silver

Double Replacement

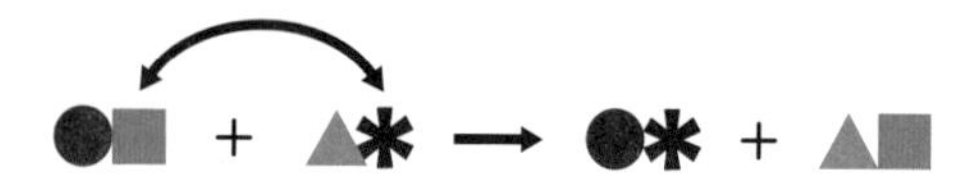

Examples:
$NaCl + AgNO_3 \rightarrow NaNO_3 + AgCl$
$HCl + FeS \rightarrow FeCl_2 + H_2S$

$Pb(NO_3)_2 + 2KI \rightarrow 2KNO_3 + PbI_2$
lead nitrate, potassium iodide, potassium nitrate, lead iodide

▲ **Figure 7** In each of these reactions, an atom or group of atoms replaces another atom or group of atoms.

Replacement

In a replacement reaction, an atom or group of atoms replaces part of a compound. There are two types of replacement reactions. *In a* **single-replacement** *reaction, one element replaces another element in a compound.* In this type of reaction, an element and a compound react and form a different element and a different compound. *In a* **double-replacement** *reaction, the negative ions in two compounds switch places, forming two new compounds.* In this type of reaction, two compounds react and form two new compounds. **Figure 7** describes these replacement reactions.

Combustion Reactions 

substance + O_2 → substance(s)

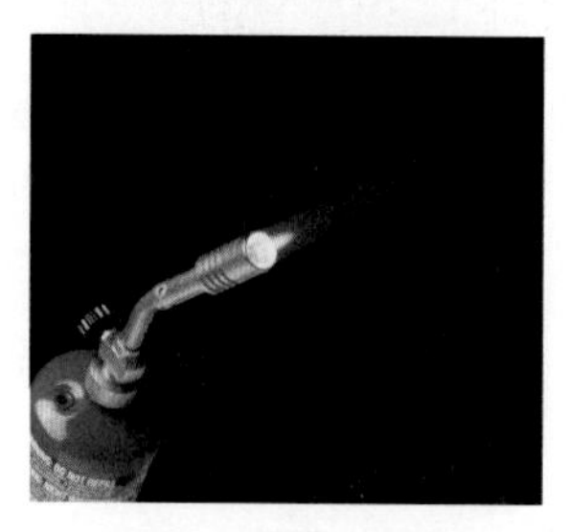

$C_3H_8 + 5O_2 \rightarrow 3CO_2 + 4H_2O$
propane, oxygen, carbon dioxide, water

Example:
$2C_4H_{10} + 13O_2 \rightarrow 8CO_2 + 10H_2O$

▲ **Figure 8** Combustion reactions always contain oxygen (O_2) as a reactant and often produce carbon dioxide (CO_2) and water (H_2O).

Combustion

Combustion *is a chemical reaction in which a substance combines with oxygen and releases energy.* This energy usually is released as thermal energy and light energy. For example, burning is a common combustion reaction. The burning of fossil fuels, such as propane (C_3H_8) shown in **Figure 8**, produces the energy we use to cook food, power vehicles, and light cities.

 **Key Concept Check** What are the different types of chemical reactions?

Lesson 2 Review

Assessment Online Quiz

Inquiry Virtual Lab

Visual Summary

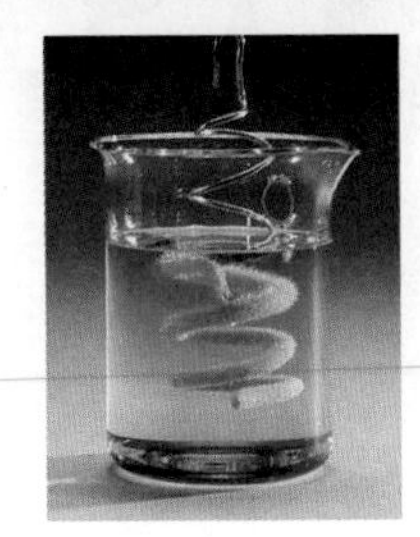

Chemical reactions are classified according to patterns seen in their reactants and products.

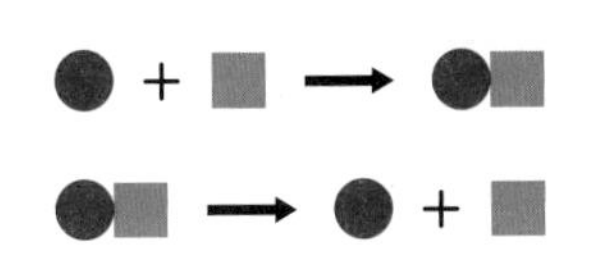

In a synthesis reaction, there are two or more reactants and one product. A decomposition reaction is the opposite of a synthesis reaction.

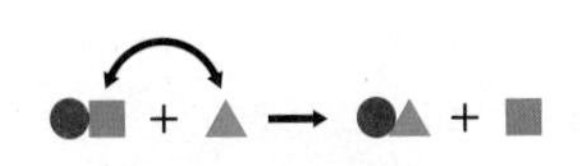

In replacement reactions, an element, or elements, in a compound is replaced with another element or elements.

FOLDABLES®

Use your lesson Foldable to review the lesson. Save your Foldable for the project at the end of the chapter.

What do you think NOW?

You first read the statements below at the beginning of the chapter.

3. Reactions always start with two or more substances that react with each other.

4. Water can be broken down into simpler substances.

Did you change your mind about whether you agree or disagree with the statements? Rewrite any false statements to make them true.

Use Vocabulary

1. **Contrast** synthesis and decomposition reactions using a diagram.

2. A reaction in which parts of two substances switch places and make two new substances is a(n) ______.

Understand Key Concepts 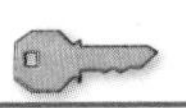

3. **Classify** the reaction shown below.

$$2Na + Cl_2 \rightarrow 2NaCl$$

A. combustion
B. decomposition
C. single replacement
D. synthesis

4. Write a balanced equation that produces H_2 and O_2 from H_2O. Classify this reaction.

5. **Classify** In which two groups of reactions can this reaction be classified?

$$2SO_2 + O_2 \rightarrow 2SO_3$$

Interpret Graphics

6. **Complete** this table to identify four types of chemical reactions and the patterns shown by the reactants and the products.

Type of Reaction	Pattern of Reactants and Products
Synthesis	at least two reactants; one product

Critical Thinking

7. **Design** a poster to illustrate single- and double-replacement reactions.

8. **Infer** The combustion of methane (CH_4) produces energy. Where do you think this energy comes from?

HOW IT WORKS

How does a light stick work?

What makes it glow?

Glowing neon necklaces, bracelets, or sticks—chances are you've worn or used them. Light sticks—also known as glow sticks—come in brilliant colors and provide light without electricity or batteries. Because they are lightweight, portable, and waterproof, they provide an ideal light source for campers, scuba divers, and other activities in which electricity is not readily available. Light sticks also are useful in emergency situations in which an electric current from battery-powered lights could ignite a fire.

Light sticks give off light because of a chemical reaction that happens inside the tube. During the reaction, energy is released as light. This is known as chemiluminescence (ke mee lew muh NE sunts).

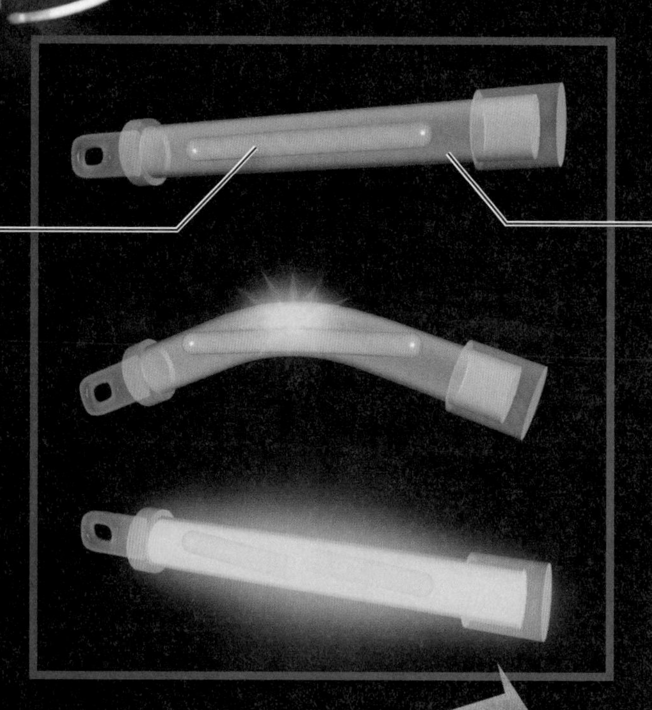

A light stick consists of a plastic tube with a glass tube inside it. Hydrogen peroxide fills the glass tube..

A solution of phenyl oxalate ester and fluorescent dye surround the glass tube.

When you bend the outer plastic tube, the inner glass tube breaks, causing the hydrogen peroxide, ester, and dye to mix together.

When the solutions mix together, they react. Energy produced by the reaction causes the electrons in the dye to produce light.

It's Your Turn

RESEARCH AND REPORT Research bioluminescent organisms, such as fireflies and sea animals. How is the reaction that occurs in these organisms similar to or different from that in a glow stick? Work in small groups, and present your findings to the class.

Lesson 3

Reading Guide

Key Concepts
ESSENTIAL QUESTIONS

- Why do chemical reactions always involve a change in energy?
- What is the difference between an endothermic reaction and an exothermic reaction?
- What factors can affect the rate of a chemical reaction?

Vocabulary

endothermic p. 319
exothermic p. 319
activation energy p. 320
catalyst p. 322
enzyme p. 322
inhibitor p. 322

g Multilingual eGlossary

Energy Changes and Chemical Reactions

Inquiry Energy from Bonds?

A deafening roar, a blinding light, and the power to lift 2 million kg—what is the source of all this energy? Chemical bonds in the fuel store all the energy needed to launch a space shuttle. Chemical reactions release the energy in these bonds.

20 minutes

Where's the heat?

Does a chemical change always produce a temperature increase?

1. Read and complete a lab safety form.
2. Copy the table into your Science Journal.
3. Use a **graduated cylinder** to measure 25 mL of **citric acid solution** into a **foam cup.** Record the temperature with a **thermometer**.
4. Use a **plastic spoon** to add a rounded spoonful of **solid sodium bicarbonate** to the cup. Stir.
5. Use a **clock** or **stopwatch** to record the temperature every 15 s until it stops changing. Record your observations during the reaction.
6. Add 25 mL of **sodium bicarbonate solution** to a **second foam cup.** Record the temperature. Add a spoonful of **calcium chloride**. Repeat step 5.

Time	Temperature (°C)	
	Citric Acid Solution	Sodium Bicarbonate Solution
Starting temp.		
15 s		
30 s		
45 s		
1 min		
1 min, 15 s		
1 min, 30 s		
1 min, 45 s		
2 min		
2 min, 15 sec		

Think About This

1. What evidence do you have that the changes in the two cups were chemical reactions?
2. What happened to the temperature in the two cups? How would you explain the changes?
3. **Key Concept** Based on your observations and past experience, would a change in temperature be enough to convince you that a chemical change had taken place? Why or why not? What else could cause a temperature change?

Energy Changes

What is about 1,500 times heavier than a typical car and 300 times faster than a roller coaster? Do you need a hint? The energy it needs to move this fast comes from a chemical reaction that produces water. If you guessed a space shuttle, you are right!

It takes a large amount of energy to launch a space shuttle. The shuttle's main engines burn almost 2 million L of liquid hydrogen and liquid oxygen. This chemical reaction produces water vapor and a large amount of energy. The energy produced heats the water vapor to high temperatures, causing it to expand rapidly. When the water expands, it pushes the shuttle into orbit. Where does all this energy come from?

Chemical Energy in Bonds

Recall that when a chemical reaction occurs, chemical bonds in the reactants break and new chemical bonds form. Chemical bonds contain a form of energy called chemical energy. Breaking a bond absorbs energy from the surroundings. The formation of a chemical bond releases energy to the surroundings. Some chemical reactions release more energy than they absorb. Some chemical reactions absorb more energy than they release. You can feel this energy change as a change in the temperature of the surroundings. Keep in mind that in all chemical reactions, energy is conserved.

Key Concept Check Why do chemical reactions involve a change in energy?

Endothermic Reactions—Energy Absorbed

Have you ever heard someone say that the sidewalk was hot enough to fry an egg? To fry, the egg must absorb energy. *Chemical reactions that absorb thermal energy are* **endothermic** *reactions.* For an endothermic reaction to continue, energy must be constantly added.

reactants + thermal energy → products

In an endothermic reaction, more energy is required to break the bonds of the reactants than is released when the products form. Therefore, the overall reaction absorbs energy. The reaction on the left in **Figure 9** is an endothermic reaction.

Exothermic Reactions—Energy Released

Most chemical reactions release energy as opposed to absorbing it. *An* **exothermic** *reaction is a chemical reaction that releases thermal energy.*

reactants → products + thermal energy

In an exothermic reaction, more energy is released when the products form than is required to break the bonds in the reactants. Therefore, the overall reaction releases energy. The reaction shown on the right in **Figure 9** is exothermic.

Key Concept Check What is the difference between an endothermic reaction and an exothermic reaction?

FOLDABLES®

Make a vertical three-tab Venn book. Label it as shown. Use it to compare and contrast energy in chemical reactions.

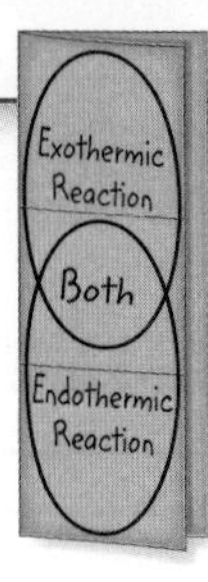

WORD ORIGIN

exothermic
from Greek *exo-*, means "outside"; and *therm*, means "heat"

Figure 9 Whether a reaction is endothermic or exothermic depends on the amount of energy contained in the bonds of the reactants and the products.

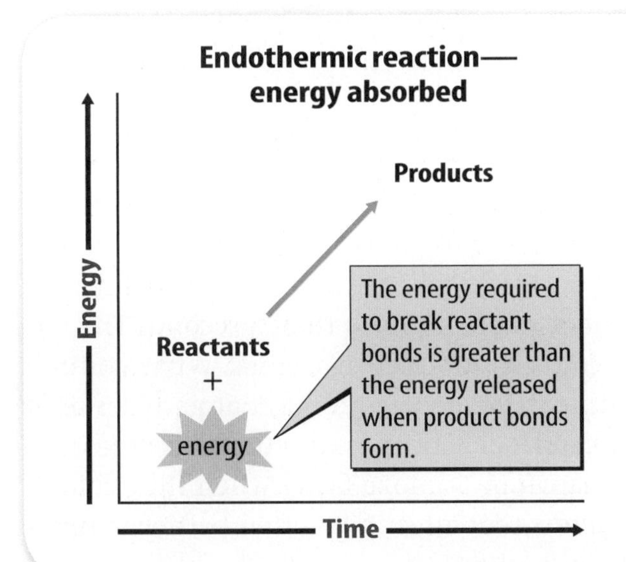

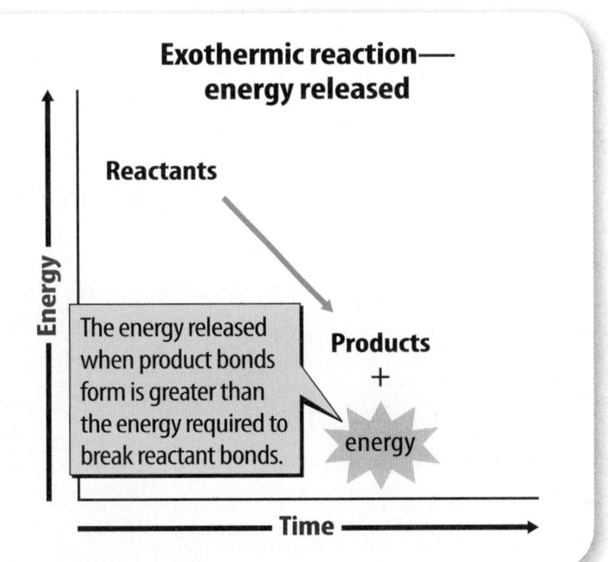

Visual Check Why does one arrow point upward and the other arrow point downward in these diagrams?

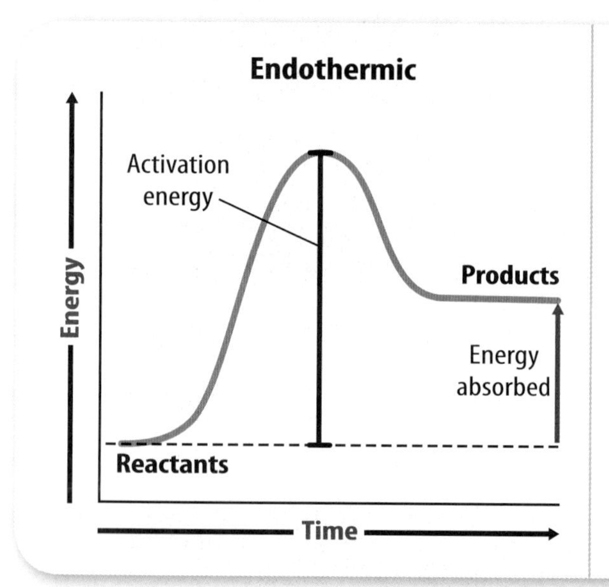

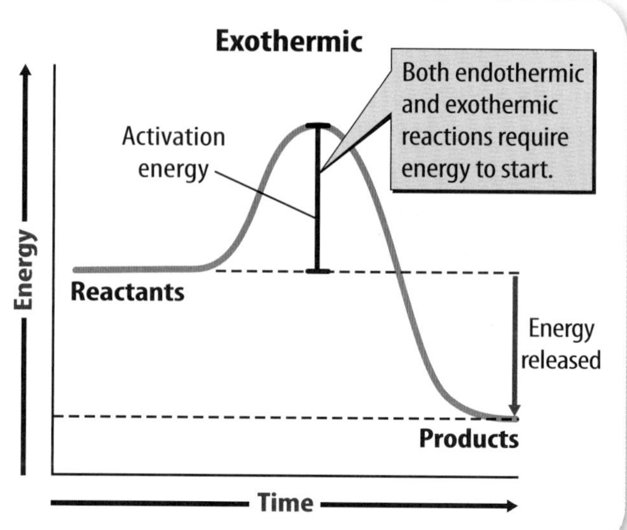

Figure 10 Both endothermic and exothermic reactions require activation energy to start the reaction.

Visual Check How can a reaction absorb energy to start but still be exothermic?

Activation Energy

You might have noticed that some chemical reactions do not start by themselves. For example, a newspaper does not burn when it comes into contact with oxygen in air. However, if a flame touches the paper, it starts to burn.

All reactions require energy to start the breaking of bonds. This energy is called activation energy. **Activation energy** *is the minimum amount of energy needed to start a chemical reaction.* Different reactions have different activation energies. Some reactions, such as the rusting of iron, have low activation energy. The energy in the surroundings is enough to start these reactions. If a reaction has high activation energy, more energy is needed to start the reaction. For example, wood requires the thermal energy of a flame to start burning. Once the reaction starts, it releases enough energy to keep the reaction going. **Figure 10** shows the role activation energy plays in endothermic and exothermic reactions.

Reaction Rates

Some chemical reactions, such as the rusting of a bicycle wheel, happen slowly. Other chemical reactions, such as the explosion of fireworks, happen in less than a second. The rate of a reaction is the speed at which it occurs. What controls how fast a chemical reaction occurs? Recall that particles must collide before they can react. Chemical reactions occur faster if particles collide more often or move faster when they collide. There are several factors that affect how often particles collide and how fast particles move.

Reading Check How do particle collisions relate to reaction rate?

Surface Area

Surface area is the amount of exposed, outer area of a solid. Increased surface area increases reaction rate because more particles on the surface of a solid come into contact with the particles of another substance. For example, if you place a piece of chalk in vinegar, the chalk reacts slowly with the acid. This is because the acid contacts only the particles on the surface of the chalk. But, if you grind the chalk into powder, more chalk particles contact the acid, and the reaction occurs faster.

Temperature

Imagine a crowded hallway. If everyone in the hallway were running, they would probably collide with each other more often and with more energy than if everyone were walking. This is also true when particles move faster. At higher temperatures, the average speed of particles is greater. This speeds reactions in two ways. First, particles collide more often. Second, collisions with more energy are more likely to break chemical bonds.

Concentration and Pressure

Think of a crowded hallway again. Because the concentration of people is higher in the crowded hallway than in an empty hallway, people probably collide more often. Similarly, increasing the concentration of one or more reactants increases collisions between particles. More collisions result in a faster reaction rate. In gases, an increase in pressure pushes gas particles closer together. When particles are closer together, more collisions occur. Factors that affect reaction rate are shown in **Figure 11.**

Math Skills

Use Geometry

The surface area (SA) of one side of a 1-cm cube is 1 cm $\times$ 1 cm, or 1 cm^2. The cube has 6 equal sides. Its total SA is $6 \times 1\ cm^2$, or 6 cm^2. What is the total SA of the two solids made when the cube is cut in half?

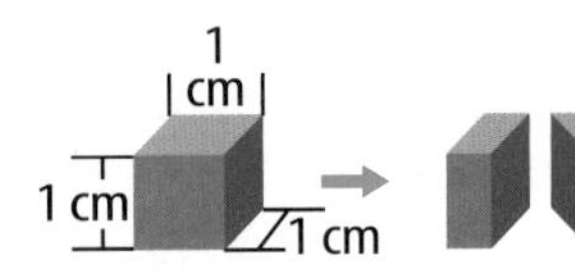

1. The new surfaces made each have an area of $1\ cm \times 1\ cm = 1\ cm^2$.
2. Multiply the area by the number of new surfaces. $2 \times 1 = 2\ cm^2$
3. Add the SA of the original cube to the new SA. $6\ cm^2 + 2\ cm^2$ The total SA is 8 cm^2.

Practice

Calculate the amount of SA gained when a 2-cm cube is cut in half.

Figure 11 Several factors can affect reaction rate.

Slower Reaction Rate

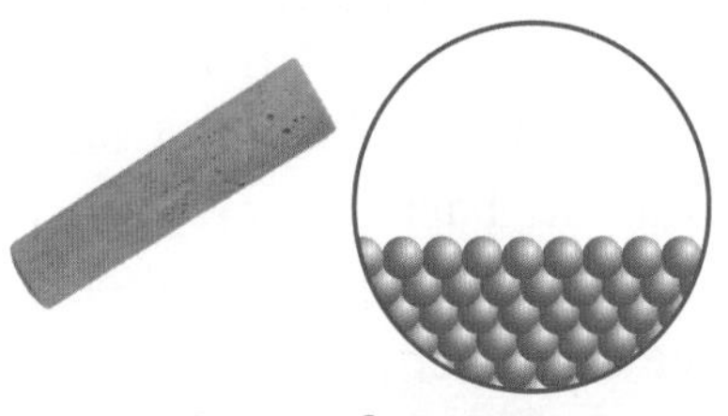

Less surface area

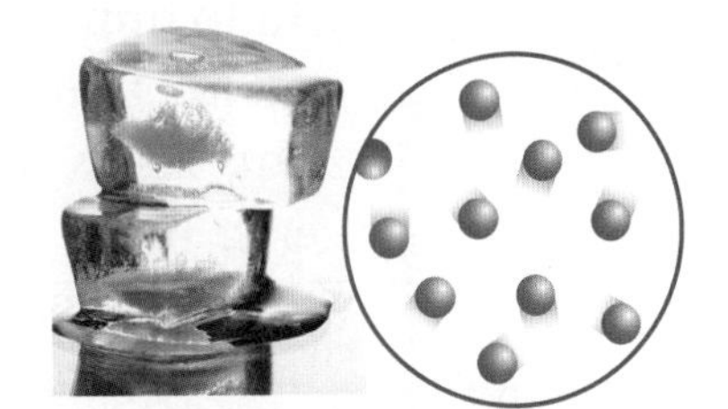

Lower temperature

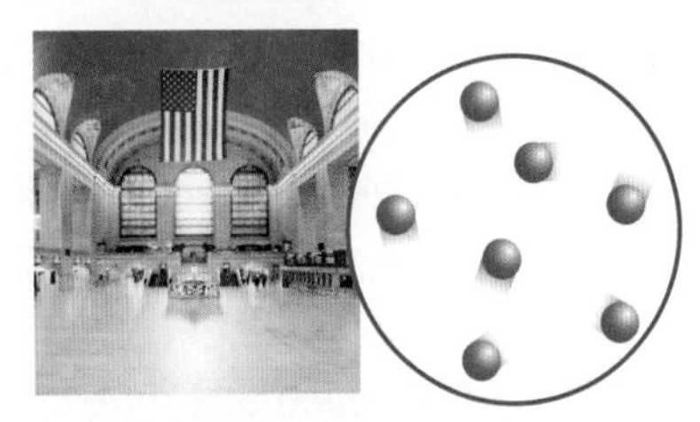

Lower concentration

Faster Reaction Rate

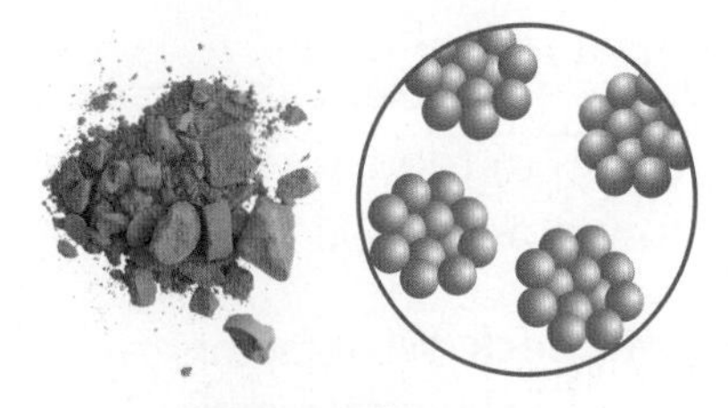

More surface area

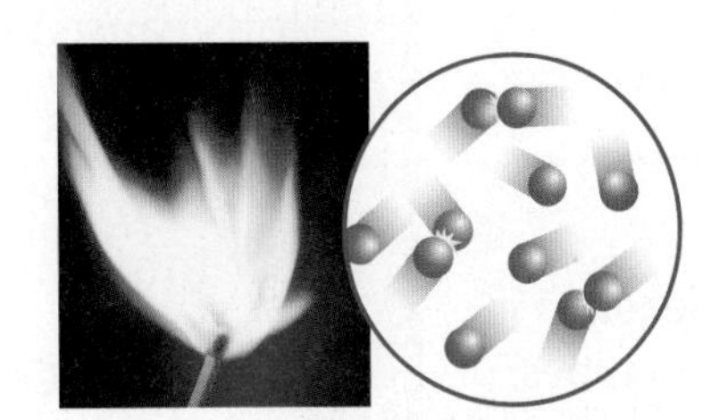

Higher temperature

Higher concentration

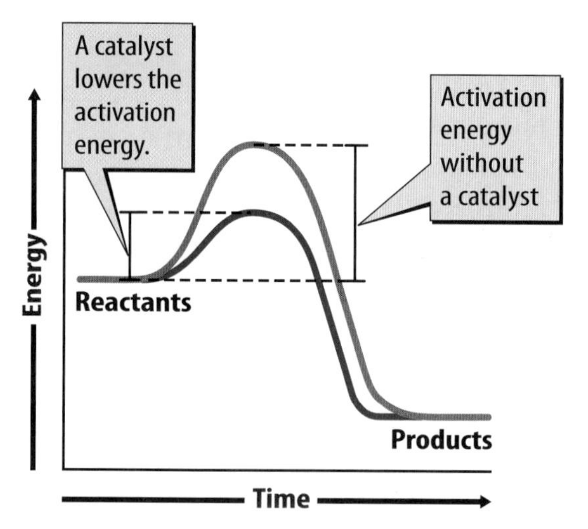

Figure 12 The blue line shows how a catalyst can increase the reaction rate.

Inquiry MiniLab

20 minutes

Can you speed up a reaction?

Can you speed up the decomposition of hydrogen peroxide (H_2O_2)? The reaction is $H_2O_2 \rightarrow H_2O$ and O_2.

1. Read and complete a lab safety form.
2. Use **tape** to label three **test tubes** *1, 2,* and *3.* Place the tubes in a **test-tube rack.**
3. Add 10 mL of **hydrogen peroxide** to each test tube.
4. Observe tube 1 for changes. Add a small piece of **raw potato** to tube 2. Record observations in your Science Journal.
5. Add a pinch of **dry yeast** to tube 3. Shake the tube gently. Record observations.
6. Use **matches** to light a **wood splint,** then blow it out, leaving a glowing tip. One at a time, hold each test tube at a 45° angle and insert the glowing splint into the tube just above the liquid. Record your observations.

Analyze and Conclude

1. **Draw Conclusions** What was the chemical reaction when the potato and yeast were added?
2. **Key Concept** Why is the reaction in tube 3 faster than in the other two tubes?

Catalysts

A **catalyst** *is a substance that increases reaction rate by lowering the activation energy of a reaction.* One way catalysts speed reactions is by helping reactant particles contact each other more often. Look at **Figure 12.** Notice that the activation energy of the reaction is lower with a catalyst than it is without a catalyst. A catalyst isn't changed in a reaction, and it doesn't change the reactants or products. Also, a catalyst doesn't increase the amount of reactant used or the amount of product that is made. It only makes a given reaction happen faster. Therefore, catalysts are not considered reactants in a reaction.

You might be surprised to know that your body is filled with catalysts called enzymes. *An* **enzyme** *is a catalyst that speeds up chemical reactions in living cells.* For example, the enzyme protease (PROH tee ays) breaks the protein molecules in the food you eat into smaller molecules that can be absorbed by your intestine. Without enzymes, these reactions would occur too slowly for life to exist.

Inhibitors

Recall than an enzyme is a molecule that speeds reactions in organisms. However, some organisms, such as bacteria, are harmful to humans. Some medicines contain molecules that attach to enzymes in bacteria. This keeps the enzymes from working properly. If the enzymes in bacteria can't work, the bacteria die and can no longer infect a human. The active ingredients in these medicines are called inhibitors. *An* **inhibitor** *is a substance that slows, or even stops, a chemical reaction.* Inhibitors can slow or stop the reactions caused by enzymes.

Inhibitors are also important in the food industry. Preservatives in food are substances that inhibit, or slow down, food spoilage.

Key Concept Check What factors can affect the rate of a chemical reaction?

Lesson 3 Review

Assessment Online Quiz

Visual Summary

Endothermic

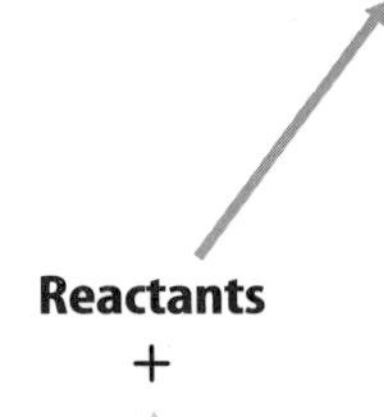

Chemical reactions that release energy are exothermic, and those that absorb energy are endothermic.

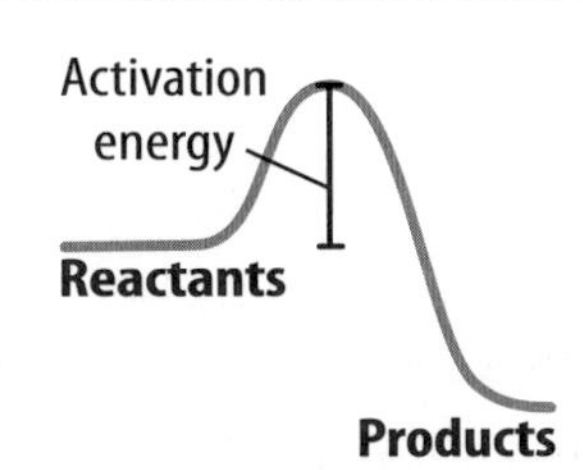

Activation energy must be added to a chemical reaction for it to proceed.

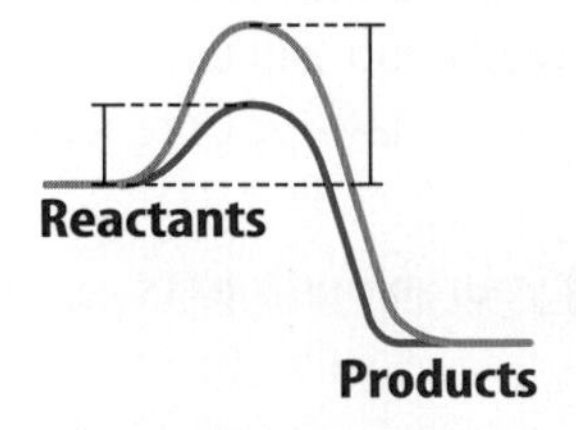

Catalysts, including enzymes, speed up chemical reactions. Inhibitors slow them down.

FOLDABLES®

Use your lesson Foldable to review the lesson. Save your Foldable for the project at the end of the chapter.

What do you think NOW?

You first read the statements below at the beginning of the chapter.

5. Reactions that release energy require energy to get started.

6. Energy can be created in a chemical reaction.

Did you change your mind about whether you agree or disagree with the statements? Rewrite any false statements to make them true.

Use Vocabulary

1 The smallest amount of energy required by reacting particles for a chemical reaction to begin is the ________.

Understand Key Concepts 

2 How does a catalyst increase reaction rate?

A. by increasing the activation energy
B. by increasing the amount of reactant
C. by increasing the contact between particles
D. by increasing the space between particles

3 **Contrast** endothermic and exothermic reactions in terms of energy.

4 **Explain** When propane burns, heat and light are produced. Where does this energy come from?

Interpret Graphics

5 **List** Copy and complete the graphic organizer to describe four ways to increase the rate of a reaction.

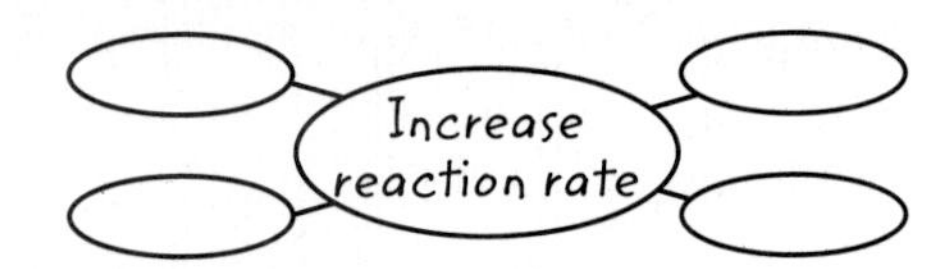

Critical Thinking

6 **Infer** Explain why keeping a battery in a refrigerator can extend its life.

7 **Infer** Explain why a catalyst does not increase the amount of product that can form.

Math Skills

Math Practice

8 An object measures 4 cm × 4 cm × 4 cm.

a. What is the surface area of the object?
b. What is the total surface area if you cut the object into two equal pieces?

2 class periods

Design an Experiment to Test Advertising Claims

Materials

graduated cylinder

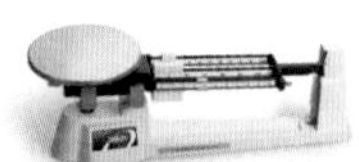

balance

droppers

baking soda

plastic spoon

Also needed: various brands of liquid and solid antacids (both regular and maximum strength), beakers, universal indicator in dropper bottle, 0.1M HCl solution, stirrers

Safety

Antacids contain compounds that react with excess acid in your stomach and prevent a condition called heartburn. Suppose you work for a laboratory that tests advertising claims about antacids. What kinds of procedures would you follow? How would you decide which antacid is the most effective?

Ask a Question

Ask a question about the claims that you would like to investigate. For example: what does *most effective* mean? What would make an antacid the strongest?

Make Observations

1. Read and complete a lab safety form.
2. Study the selection of antacids available for testing. You will use a 0.1M HCl solution to simulate stomach acid. Use the questions below to discuss with your lab partners which advertising claim you might test and how you might test it.
3. In your Science Journal, write a procedure for each variable that you will test to answer your question. Include the materials and steps you will use to test each variable. Place the steps of each procedure in order. Have your teacher approve your procedures.
4. Make a chart or table to record observations during your experiments.

Questions
Which advertising claim will I test? What question am I trying to answer?
What will be the independent and the dependent variables for each test? Recall that the independent variable is the variable that is changed. A dependent variable changes when you change the independent variable.
What variables will be held constant in each test?
How many different procedures will I use, and what equipment will I need?
How much of each antacid will I use? How many antacids will I test?
How will I use the indicator?
How many times will I do each test?
How will I record the data and observations?
What will I analyze to form a conclusion?

Form a Hypothesis

5. Write a hypothesis for each variable. Your hypothesis should identify the independent variable and state why you think changing the variable will alter the effectiveness of an antacid tablet.

Test Your Hypothesis

6. On day 2, use the available materials to perform your experiments. Accurately record all observations and data for each test.
7. Add any additional tests you think you need to answer your questions.
8. Examine the data you have collected. If the data are not conclusive, what other tests can you do to provide more information?
9. Write all your observations and measurements in your Science Journal. Use tables to record any quantitative data.

Analyze and Conclude

10. **Infer** What do you think advertisers mean when they say their product is most effective?
11. **Draw Conclusions** If you needed an antacid, which one would you use, based on the limited information provided from your experiments? Explain your reasoning.
12. **Analyze** Would breaking an antacid tablet into small pieces before using it make it more effective? Why or why not?
13. **The Big Idea** How does understanding chemical reactions enable you to analyze products and their claims?

Communicate Your Results

Combine your data with other teams. Compare the results and conclusions. Discuss the validity of advertising claims for each brand of antacid.

Research over-the-counter antacids that were once available by prescription only. Do they work in the same way as the antacids you tested? Explain.

Lab TIPS

- ☑ Think about how you might measure the amount of acid the tablet neutralizes. Would you add the tablet to the acid or the acid to the tablet? What does the indicator show you?
- ☑ Try your tests on a small scale before using the full amounts to see how much acid you might need.
- ☑ Always get your teacher's approval before trying any new test.

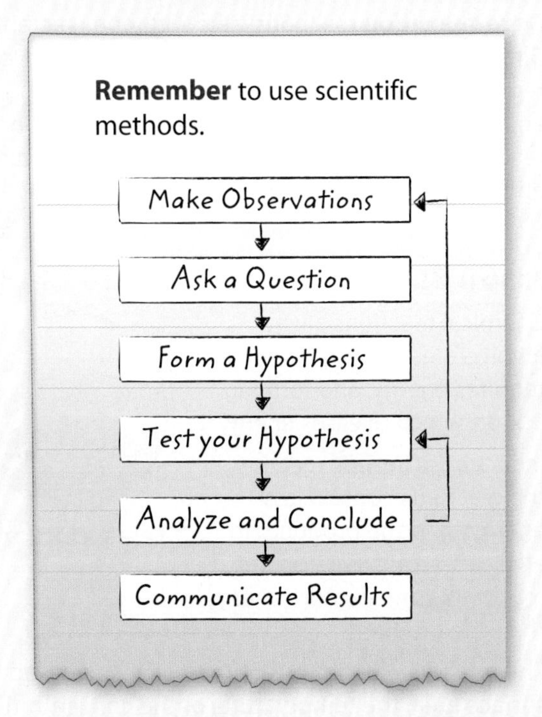

Chapter 9 Study Guide

Atoms are neither created nor destroyed in chemical reactions. Energy can be released when chemical bonds form or absorbed when chemical bonds are broken.

Key Concepts Summary

Lesson 1: Understanding Chemical Reactions

- There are several signs that a **chemical reaction** might have occurred, including a change in temperature, a release of light, a release of gas, a change in color or odor, and the formation of a solid from two liquids.
- In a chemical reaction, atoms of **reactants** rearrange and form **products.**
- The total mass of all the reactants is equal to the total mass of all the products in a reaction.

Reactants			Products		
1 Na: ●		**Atoms are equal.**	1 Na: ●		
1 H: ●	4 H: ●●●●		2 C: ●●		1 C: ●
1 C: ●	2 C: ●●		3 H: ●●●	2 H: ●●	
3 O: ●●●	2 O: ●●		2 O: ●●	1 O: ●	2 O: ●●

Vocabulary

chemical reaction p. 301
chemical equation p. 304
reactant p. 305
product p. 305
law of conservation of mass p. 306
coefficient p. 308

Lesson 2: Types of Chemical Reactions

- Most chemical reactions fit into one of a few main categories—synthesis, decomposition, combustion, and single- or double-replacement.
- **Synthesis** reactions create one product. **Decomposition** reactions start with one reactant. **Single-** and **double-replacement** reactions involve replacing one element or group of atoms with another element or group of atoms. **Combustion** reactions involve a reaction between one reactant and oxygen, and they release thermal energy.

Vocabulary

synthesis p. 313
decomposition p. 313
single replacement p. 314
double replacement p. 314
combustion p. 314

Lesson 3: Energy Changes and Chemical Reactions

- Chemical reactions always involve breaking bonds, which requires energy, and forming bonds, which releases energy.
- In an **endothermic** reaction, the reactants contain less energy than the products. In an **exothermic** reaction, the reactants contain more energy than the products.
- The rate of a chemical reaction can be increased by increasing the surface area, the temperature, or the concentration of the reactants or by adding a **catalyst.**

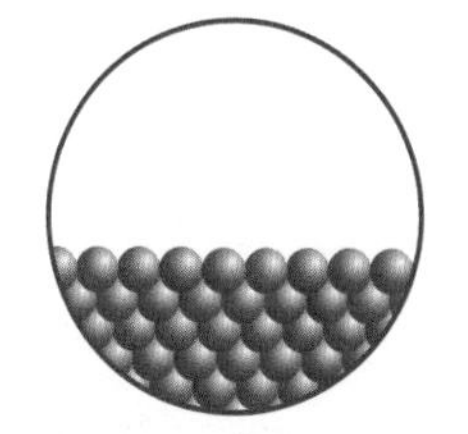

Less surface area

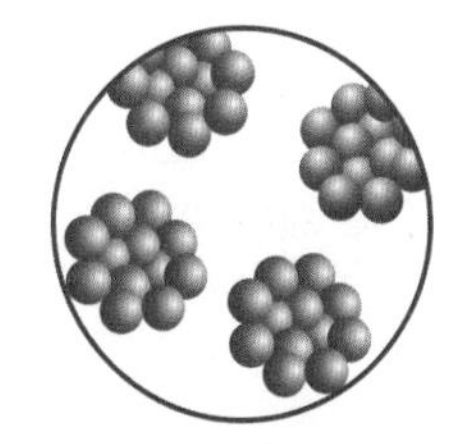

More surface area

Vocabulary

endothermic p. 319
exothermic p. 319
activation energy p. 320
catalyst p. 322
enzyme p. 322
inhibitor p. 322

Review
- Personal Tutor
- Vocabulary eGames
- Vocabulary eFlashcards

FOLDABLES® Chapter Project

Assemble your lesson Foldables as shown to make a Chapter Project. Use the project to review what you have learned in this chapter.

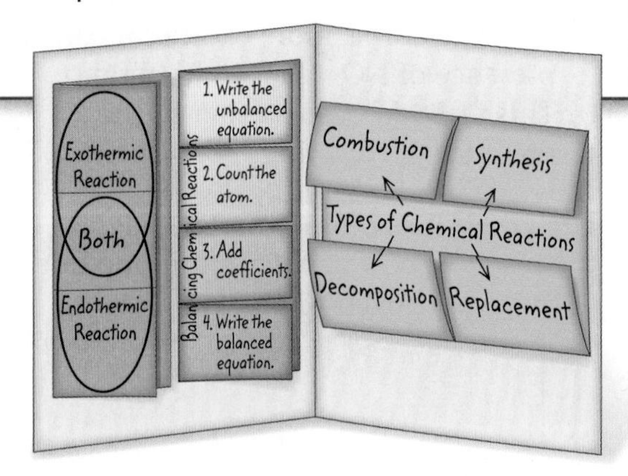

Use Vocabulary

1. When water forms from hydrogen and oxygen, water is the ________.
2. A(n) ________ uses symbols instead of words to describe a chemical reaction.
3. In a(n) ________ reaction, one element replaces another element in a compound.
4. When Na_2CO_3 is heated, it breaks down into CO_2 and Na_2O in a(n) ________ reaction.
5. The chemical reactions that keep your body warm are ________ reactions.
6. Even exothermic reactions require ________ to start.

Link Vocabulary and Key Concepts

Concepts in Motion Interactive Concept Map

Copy this concept map, and then use vocabulary terms from the previous page and other terms from the chapter to complete the concept map.

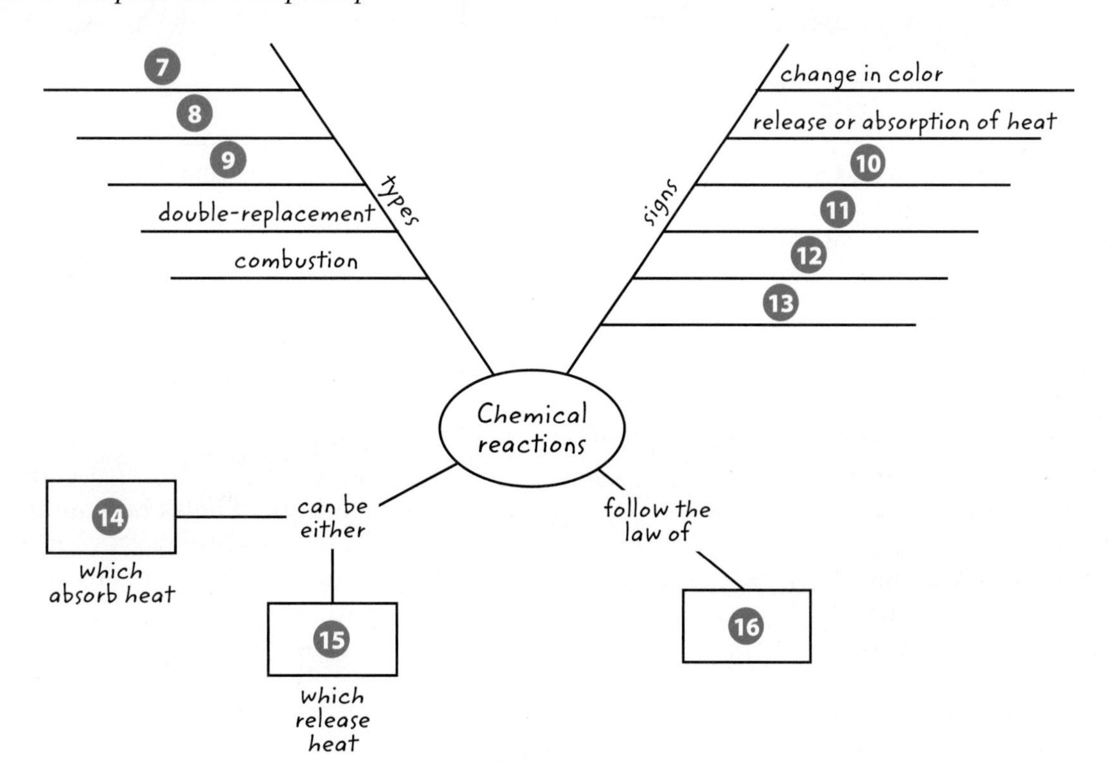

Chapter 9 Review

Understand Key Concepts 

1 How many carbon atoms react in this equation?

$$2C_4H_{10} + 13O_2 \rightarrow 8CO_2 + 10H_2O$$

A. 2
B. 4
C. 6
D. 8

2 The chemical equation below is unbalanced.

$$Zn + HCl \rightarrow ZnCl_2 + H_2$$

Which is the correct balanced chemical equation?

A. $Zn + H_2Cl_2 \rightarrow ZnCl_2 + H_2$
B. $Zn + HCl \rightarrow ZnCl + H$
C. $2Zn + 2HCl \rightarrow ZnCl_2 + H_2$
D. $Zn + 2HCl \rightarrow ZnCl_2 + H_2$

3 When iron combines with oxygen gas and forms rust, the total mass of the products

A. depends on the reaction conditions.
B. is less than the mass of the reactants.
C. is the same as the mass of the reactants.
D. is greater than the mass of the reactants.

4 Potassium nitrate forms potassium oxide, nitrogen, and oxygen in certain fireworks.

$$4KNO_3 \rightarrow 2K_2O + 2N_2 + 5O_2$$

This reaction is classified as a

A. combustion reaction.
B. decomposition reaction.
C. single-replacement reaction.
D. synthesis reaction.

5 Which type of reaction is the reverse of a decomposition reaction?

A. combustion
B. synthesis
C. double-replacement
D. single-replacement

6 The compound NO_2 can act as a catalyst in the reaction that converts ozone (O_3) to oxygen (O_2) in the upper atmosphere. Which statement is true?

A. More oxygen is created when NO_2 is present.
B. NO_2 is a reactant in the chemical reaction that converts O_3 to O_2.
C. This reaction is more exothermic in the presence of NO_2 than in its absence.
D. This reaction occurs faster in the presence of NO_2 than in its absence.

7 The graph below is an energy diagram for the reaction between carbon monoxide (CO) and nitrogen dioxide (NO_2).

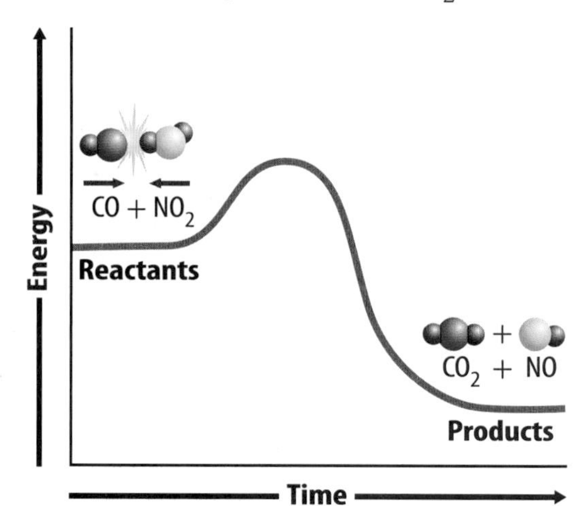

Which is true about this reaction?

A. More energy is required to break reactant bonds than is released when product bonds form.
B. Less energy is required to break reactant bonds than is released when product bonds form.
C. The bonds of the reactants do not require energy to break because the reaction releases energy.
D. The bonds of the reactants require energy to break, and therefore the reaction absorbs energy.

Assessment

Online Test Practice

Critical Thinking

8 **Predict** The diagram below shows two reactions—one with a catalyst (blue) and one without a catalyst (orange).

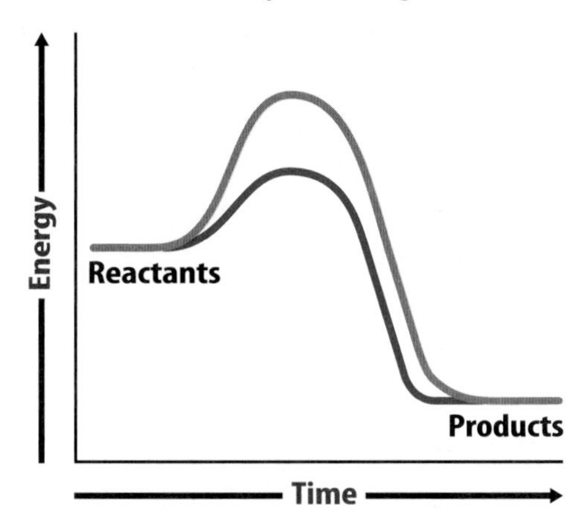

How would the blue line change if an inhibitor were used instead of a catalyst?

9 **Analyze** A student observed a chemical reaction and collected the following data:

Observations before the reaction	A white powder was added to a clear liquid.
Observations during the reaction	The reactants bubbled rapidly in the open beaker.
Mass of reactants	4.2 g
Mass of products	4.0 g

The student concludes that mass was not conserved in the reaction. Explain why this is not a valid conclusion. What might explain the difference in mass?

10 **Explain Observations** How did the discovery of atoms explain the observation that the mass of the products always equals the mass of the reactants in a reaction?

Writing in Science

11 **Write instructions** that explain the steps in balancing a chemical equation. Use the following equation as an example.

$$MnO_2 + HCl \rightarrow MnCl_2 + H_2O + Cl_2$$

REVIEW THE BIG IDEA

12 Explain how atoms and energy are conserved in a chemical reaction.

13 When a car air bag inflates, sodium azide (NaN_3) decomposes and produces nitrogen gas (N_2) and another product. What element does the other product contain? How do you know?

Math Skills

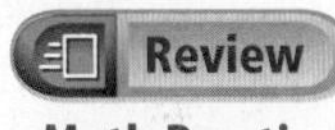

Math Practice

Use Geometry

14 What is the surface area of the cube shown below? What would the total surface area be if you cut the cube into 27 equal cubes?

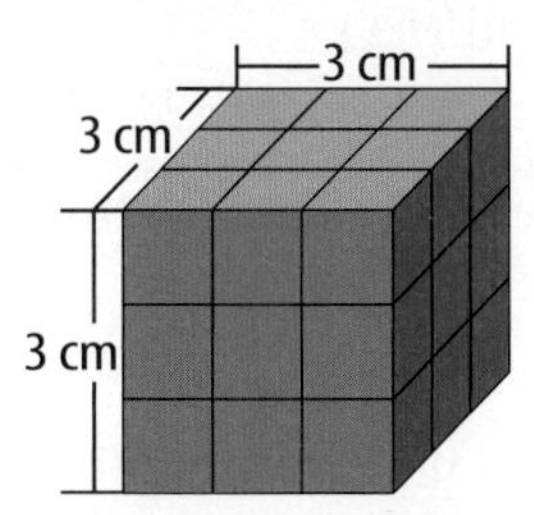

15 Suppose you have ten cubes that measure 2 cm on each side.

a. What is the total surface area of the cubes?

b. What would the surface area be if you glued the cubes together to make one object that is two cubes wide, one cube high, and five cubes long? Hint: draw a picture of the final cube and label the length of each side.

Standardized Test Practice

Record your answers on the answer sheet provided by your teacher or on a sheet of paper.

Multiple Choice

1 How can you verify that a chemical reaction has occurred?

A Check the temperature of the starting and ending substances.

B Compare the chemical properties of the starting substances and ending substances.

C Look for a change in state.

D Look for bubbling of the starting substances.

Use the figure below to answer questions 2 and 3.

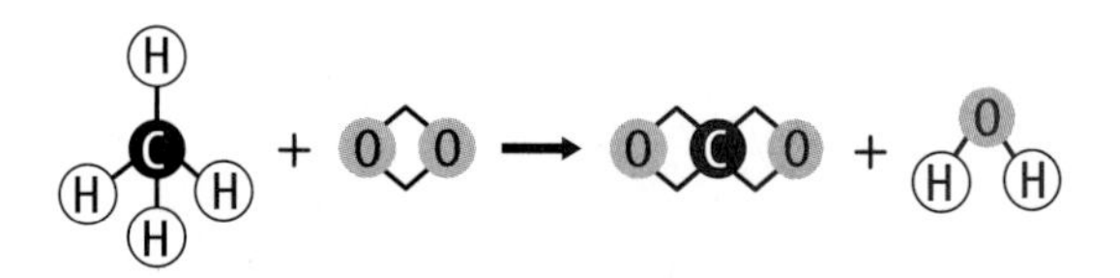

2 The figure above shows models of molecules in a chemical reactions. Which substances are reactants in this reaction?

A CH_4 and CO_2

B CH_4 and O_2

C CO_2 and H_2O

D O_2 and H_2O

3 Which equation shows that atoms are conserved in the reaction?

A $CH_4 + O_2 \longrightarrow CO_2 + H_2O$

B $CH_4 + O_2 \longrightarrow CO_2 + 2H_2O$

C $CH_4 + 2O_2 \longrightarrow CO_2 + 2H_2O$

D $2CH_4 + O_2 \longrightarrow 2CO_2 + H_2O$

4 Which occurs before new bonds can form during a chemical reaction?

A The atoms in the original substances are destroyed.

B The bonds between atoms in the original substances are broken.

C The atoms in the original substances are no longer moving.

D The bonds between atoms in the original substances get stronger.

Use the figure below to answer question 5.

5 The figure above uses shapes to represent a chemical reaction. What kind of chemical reaction does the figure represent?

A decomposition

B double-replacement

C single-replacement

D synthesis

6 Which type of chemical reaction has only one reactant?

A decomposition

B double-replacement

C single-replacement

D synthesis

7 Which element is always a reactant in a combustion reaction?

A carbon

B hydrogen

C nitrogen

D oxygen

Use the figure below to answer question 8.

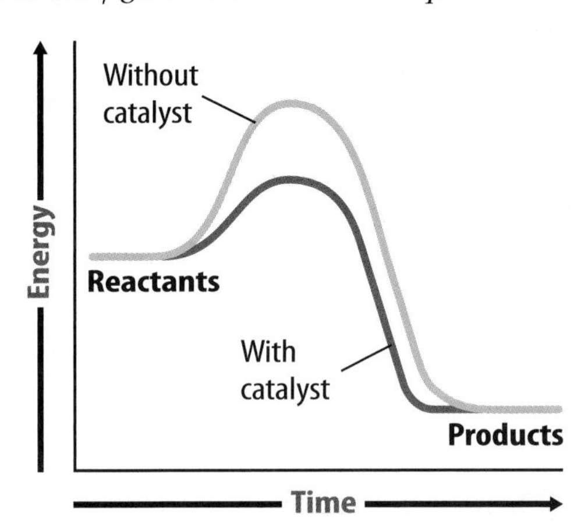

8 The figure above shows changes in energy during a reaction. The lighter line shows the reaction without a catalyst. The darker line shows the reaction with a catalyst. Which is true about these two reactions?

A The reaction with the catalyst is more exothermic than the reaction without the catalyst.

B The reaction with the catalyst requires less activation energy than the reaction without the catalyst.

C The reaction with the catalyst requires more reactants than the reaction without the catalyst.

D The reaction with the catalyst takes more time than the reaction without the catalyst.

Constructed Response

9 Explain the role of energy in chemical reactions.

10 How does a balanced chemical equation illustrate the law of conservation of mass?

11 Many of the reactions that occur when something decays are decomposition reactions. What clues show that this type of reaction is taking place? What happens during a decomposition reaction?

Use the figure below to answer questions 12 and 13.

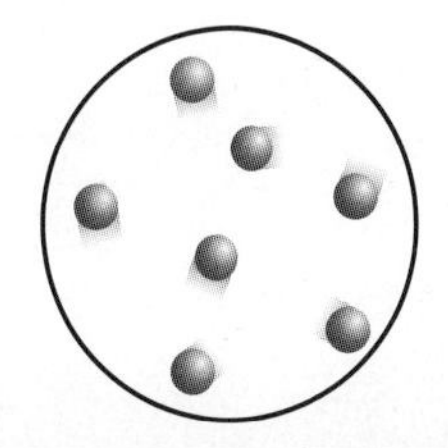

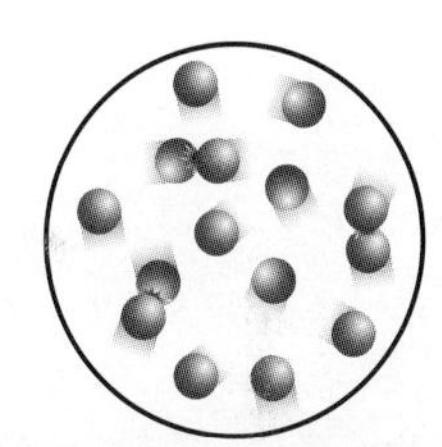

12 Compare the two gas samples represented in the figure in terms of pressure and concentration.

13 Describe the conditions that would increase the rate of a reaction.

NEED EXTRA HELP?													
If You Missed Question...	1	2	3	4	5	6	7	8	9	10	11	12	13
Go to Lesson...	1	1	1	1	2	2	2	3	3	1	2	3	3

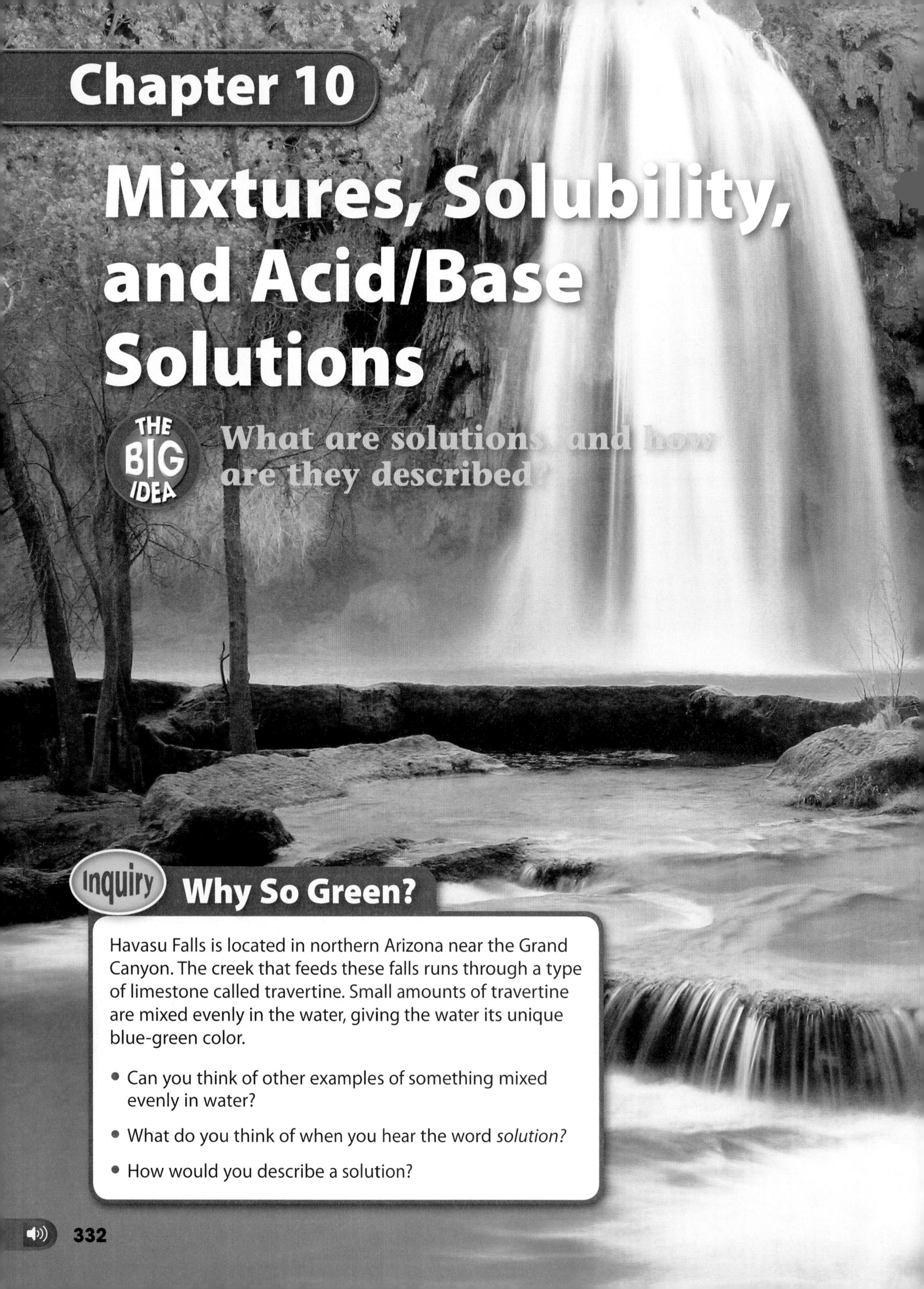

Chapter 10

Mixtures, Solubility, and Acid/Base Solutions

THE BIG IDEA What are solutions, and how are they described?

Inquiry Why So Green?

Havasu Falls is located in northern Arizona near the Grand Canyon. The creek that feeds these falls runs through a type of limestone called travertine. Small amounts of travertine are mixed evenly in the water, giving the water its unique blue-green color.

- Can you think of other examples of something mixed evenly in water?
- What do you think of when you hear the word *solution?*
- How would you describe a solution?

Get Ready to Read

What do you think?

Before you read, decide if you agree or disagree with each of these statements. As you read this chapter, see if you change your mind about any of the statements.

1. You can identify a mixture by looking at it without magnification.
2. A solution is another name for a homogeneous mixture.
3. Solutions can be solids, liquids, or gases.
4. A teaspoon of soup is less concentrated than a cup of the same soup.
5. Acids are found in many foods.
6. You can determine the exact pH of a solution by using pH paper.

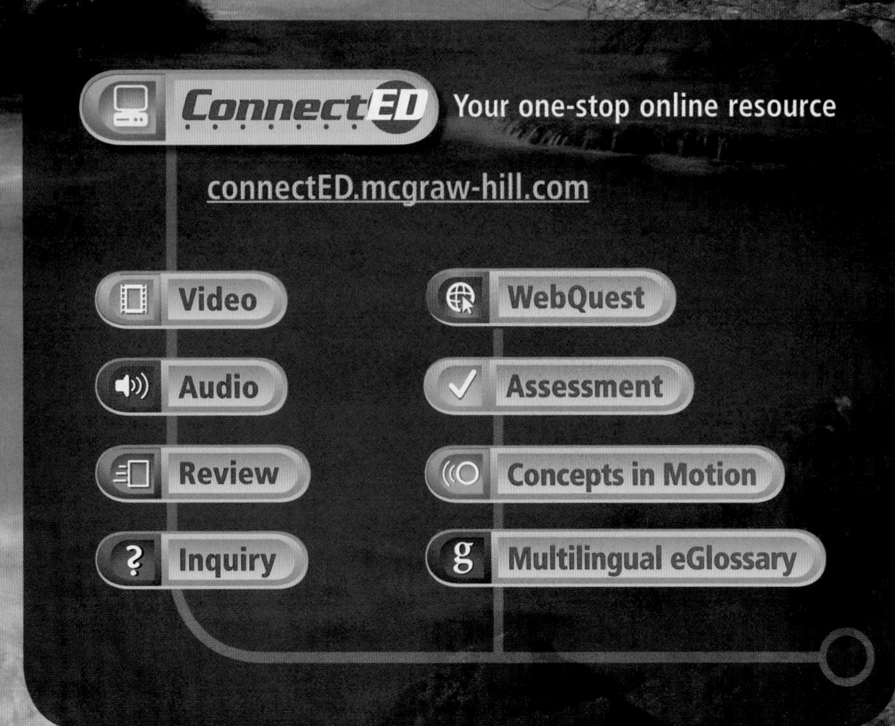

Lesson 1

Substances and Mixtures

Reading Guide

Key Concepts
ESSENTIAL QUESTIONS

- How do substances and mixtures differ?
- How do solutions compare and contrast with heterogeneous mixtures?
- In what three ways do compounds differ from mixtures?

Vocabulary

substance p. 336
mixture p. 336
heterogeneous mixture p. 337
homogeneous mixture p. 337
solution p. 337

g Multilingual eGlossary

Inquiry What's in the water?

The water in which these fish live is so clear that you might think it is pure water. But if it were pure water (H_2O), the fish could not survive. This is because fish need oxygen. What looks like pure water is actually a mixture of water and other substances, including oxygen. Fish obtain the oxygen from the water.

Launch Lab

15 minutes

What makes black ink black?

Many of the products we use every day are mixtures. How can you tell if something is a mixture?

1. Read and complete a lab safety form.
2. Lay a **coffee filter** on your table.
3. Find the center of the coffee filter, and mark it lightly with a **pencil.**
4. Use a **permanent marker** to draw a circle with a diameter of 5 cm around the center of the coffee filter. Do not fill this circle in.
5. Pour **rubbing alcohol** to a depth of 1 cm into a **beaker.**

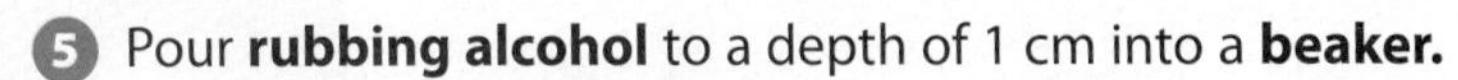

6. Using the eraser end of your pencil, push the center of the coffee filter down into the beaker until the center, but not the ink, touches the liquid. Keep the ink above the surface of the liquid.
7. Observe the liquid in the bottom of the beaker and the circle on the coffee filter. Record your observations in your Science Journal.
8. Dispose of rubbing alcohol and used coffee filters as instructed by your teacher.

Think About This

1. What happened to the black ink circle on the coffee filter?
2. What was the purpose of the rubbing alcohol?
3. What do you think you would see if you used green ink instead?
4. **Key Concept** How do you think this shows that black ink is a mixture?

Matter: Substances and Mixtures

Think about the journey you take to get to school. How many different types of matter do you see? You might see metal, plastic, rocks, concrete, bricks, plants, fabric, water, skin, and hair. So many different types of matter exist around you that it's hard to imagine them all. You might notice that you can group types of matter into categories. For example, keys, coins, and paper clips are all made of metal. Grouping matter into categories helps you understand how some things are similar to each other, but different from other things.

You might be surprised to know that nearly all types of matter can be sorted into just two major categories—substances and mixtures. What are substances and mixtures, and how are they different from each other?

FOLDABLES®

Make horizontal two- and four-tab books. Assemble, staple, and label as shown. Use it to organize your notes on matter.

Two Types of Matter
Substances | Mixtures
Elements | Compounds | Heterogeneous | Homogeneous

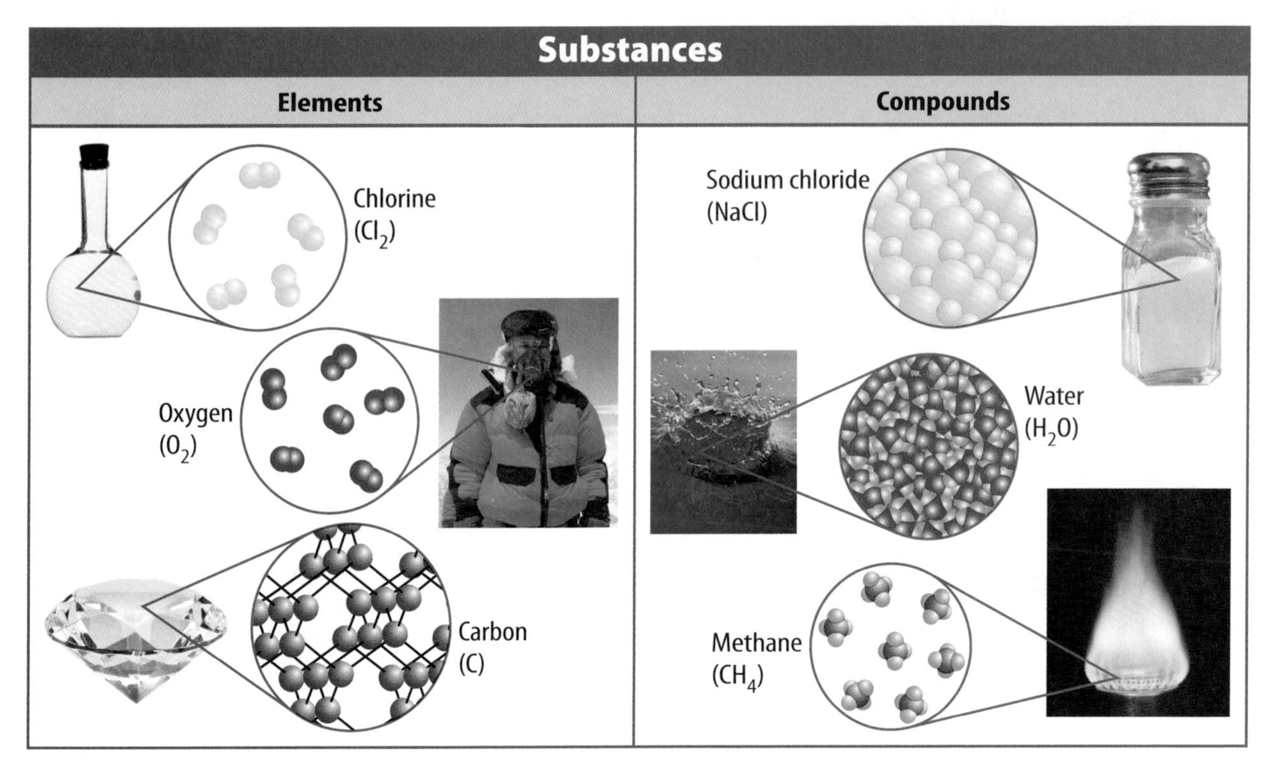

Figure 1 Elements are substances made of only one type of atom. Compounds are made of atoms of two or more elements bonded together.

What is a substance?

A **substance** *is matter that is always made up of the same combination of atoms.* Some substances are shown in **Figure 1.** There are two types of substances—elements and compounds. Recall that an element is matter made of only one type of atom, such as oxygen. A **compound** is matter made of atoms of two or more elements chemically bonded together, such as water (H_2O). Because the compositions of elements and compounds do not change, all elements and compounds are substances.

Science Use v. Common Use

substance

Science Use matter that is always made of the same combination of atoms

Common Use any physical material from which something is made

What is a mixture?

A **mixture** *is two or more substances that are physically blended but are not chemically bonded together.* The amounts of each substance in a mixture can vary. Granite, a type of rock, is a mixture. If you look at a piece of granite, you can see bits of white, black, and other colors. Another piece of granite will have different amounts of each color. The composition of rocks varies.

Air is a mixture, too. Air contains about 78 percent nitrogen, 21 percent oxygen, and 1 percent other substances. However, this composition varies. Air in a scuba tank can have more than 21 percent oxygen and less of the other substances.

Review Vocabulary

compound

matter made of two or more types of atoms bonded together

It's not always easy to identify a mixture. A piece of granite looks like a mixture, but air does not. Granite and air are examples of the two different types of mixtures—heterogeneous (he tuh roh JEE nee us) and homogeneous (hoh muh JEE nee us).

Key Concept Check How do substances and mixtures differ?

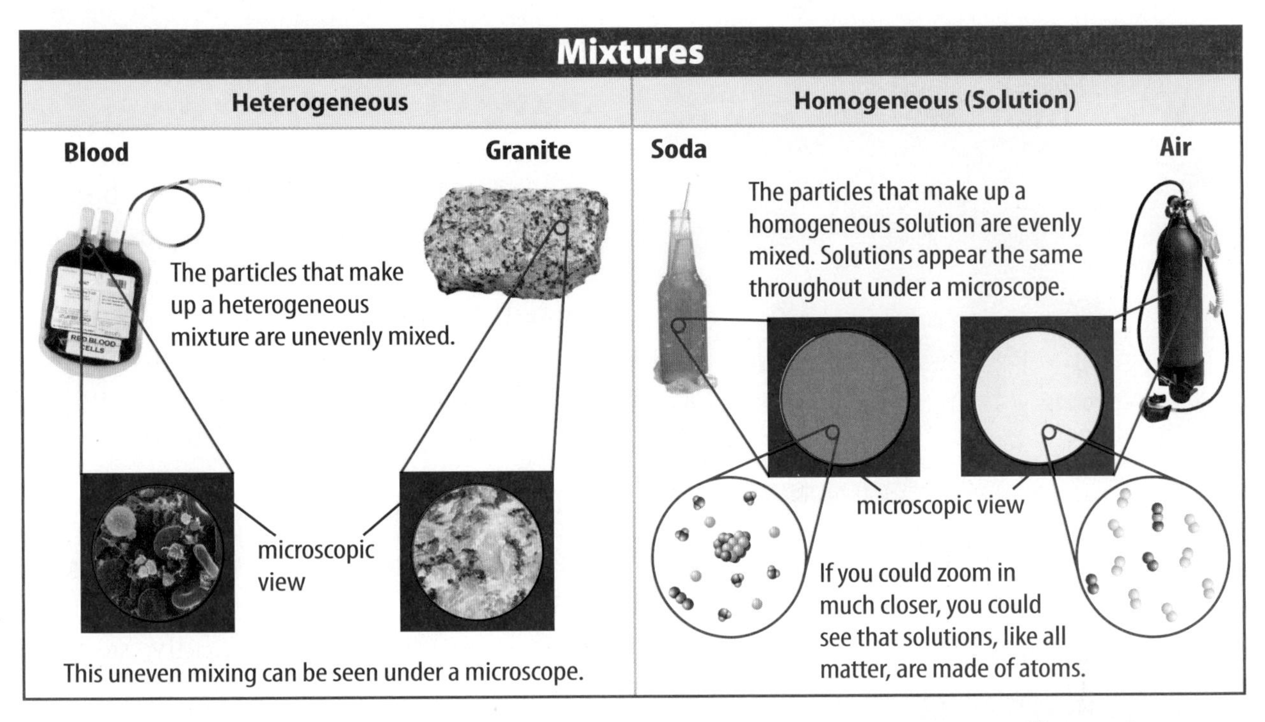

Figure 2 Solutions look evenly mixed even under a microscope. But if you could zoom in even closer to see individual atoms, you would see that solutions are made of two or more substances mixed evenly but not bonded.

Visual Check Describe what you might see if you were to look at a homogeneous mixture under an ordinary microscope.

Heterogeneous Mixtures *A* **heterogeneous mixture** *is a mixture in which substances are not evenly mixed.* For example, the substances that make up granite, a heterogeneous mixture, are unevenly mixed. When you look at a piece of granite, you can easily see the different parts. Often, you can see the different substances and parts of a heterogeneous mixture with unaided eyes, but sometimes you can only see them with a microscope. For example, blood looks evenly mixed—its color and texture are the same throughout. But, when you view blood with a microscope, as shown in **Figure 2,** you can see areas with more of one component and less of another.

Solutions—Homogeneous Mixtures Many mixtures look evenly mixed even when you view them with a powerful microscope. These mixtures are homogeneous. *A* **homogeneous mixture** *is a mixture in which two or more substances are evenly mixed on the atomic level but not bonded together.* The individual atoms or compounds of each substance are mixed. The mixture looks the same throughout under a microscope because individual atoms and compounds are too small to see.

Air is a homogeneous mixture. If you view air under a microscope, you can't see the individual substances that make it up. This is shown in **Figure 2.** *Another name for a homogeneous mixture is* **solution.** As you read about solutions in this chapter, remember that the term *solution* means "homogeneous mixture."

Key Concept Check How can you determine whether a mixture is homogeneous or heterogeneous?

Word Origin

heterogeneous
from Greek *heteros,* means "different"; and *genos,* means "kind"

homogeneous
from Greek *homos,* means "same"; and *genos,* means "kind"

Figure 3 Some properties of both the sugar and the water are observed in the mixture.

Inquiry MiniLab — 20 minutes

Which one is the mixture?

Can you tell the difference between pure water and a mixture of water and salt?

1. Read and complete a lab safety form.
2. Label two **beakers** *A* and *B* as shown.
3. Pour 50 mL of **liquid A** into beaker A.
4. Pour 50 mL of **liquid B** into beaker B.
5. Place both beakers on a **hot plate.**
6. Allow the water to boil until the beakers are nearly dry. Record your observations in your Science Journal.
   ⚠ *Use a hot mitt to handle hot glassware.*

Analyze and Conclude

1. **Compare and Contrast** How did the appearance of beakers A and B change?
2. **Key Concept** Which liquid, A or B, was the mixture? How do you know?
3. **Describe** how the water could be collected and used as a freshwater source.

How do compounds and mixtures differ?

You have read that a compound contains two or more elements chemically bonded together. In contrast, the substances that make up a mixture are not chemically bonded. So, mixing is a physical change. The substances that exist before mixing still exist in the mixture. This leads to two important differences between compounds and mixtures.

Substances keep their properties.

Because substances that make up a mixture are not changed chemically, some of their properties are observed in the mixture. Sugar water, shown in **Figure 3,** is a mixture of two compounds—sugar and water. After the sugar is mixed in, you can't see the sugar in the water, but you can still taste it. Some properties of the water, such as its liquid state, are also observed in the mixture.

In contrast, the properties of a compound can be different from the properties of the elements that make it up. Sodium and chlorine bond to form table salt. Sodium is a soft, opaque, silvery metal. Chlorine is a greenish, poisonous gas. None of these properties are observed in table salt.

Mixtures can be separated.

Because the substances that make up a mixture are not bonded together, they can be separated from each other using physical methods. The physical properties of one substance are different from those of another. These differences can be used to separate the substances. In contrast, the only way compounds can be separated is by a chemical change that breaks the bonds between the elements. **Figure 4** summarizes the characteristics of substances and mixtures.

Key Concept Check In what three ways do compounds differ from mixtures?

Figure 4 This organizational chart shows how different types of matter are classified. All matter can be classified as either a substance or a mixture.

Visual Check Can a mixture be made only of elements? Explain.

Matter

- anything that has mass and takes up space
- Most matter on Earth is made up of atoms.

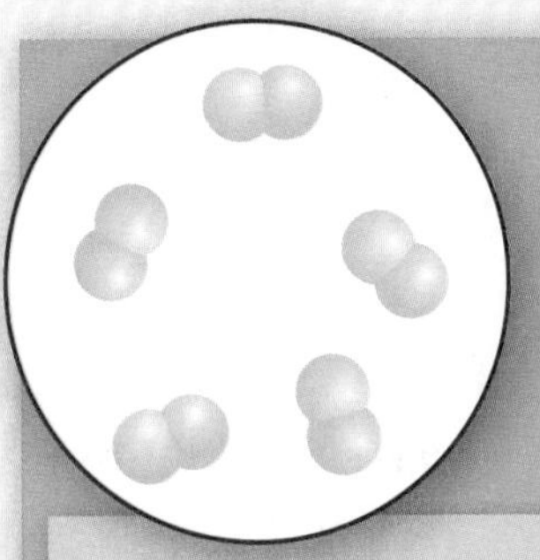

Substances

- matter with a composition that is always the same
- two types of substances: elements and compounds

Elements

- consist of just one type of atom
- organized on the periodic table
- Elements can exist as single atoms or as diatomic molecules—two atoms bonded together.

Compounds

- two or more types of atoms bonded together
- can't be separated by physical methods
- Properties of a compound are different from the properties of the elements that make it up.
- two types: ionic and covalent

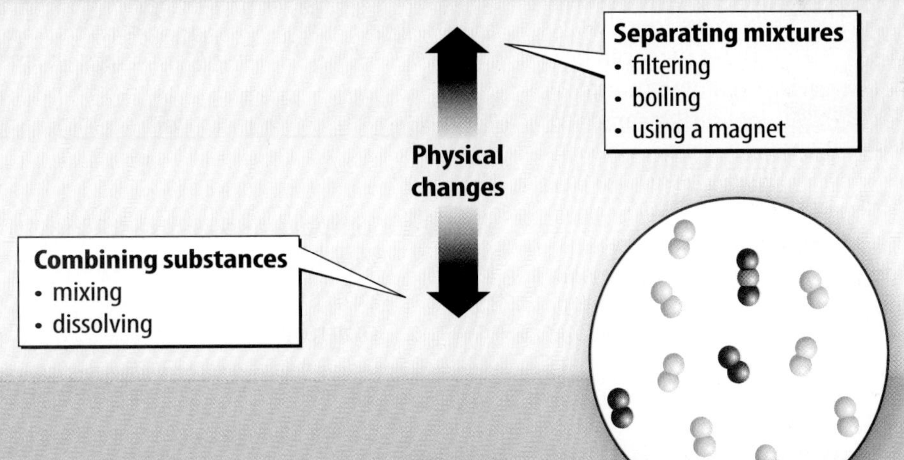

Mixtures

- matter that can vary in composition
- made of two or more substances mixed but not bonded together
- can be separated into substances by physical methods
- two types of mixtures: heterogeneous and homogeneous

Heterogeneous mixtures

- two or more substances unevenly mixed
- Uneven mixing is visible with unaided eyes or a microscope.

Homogeneous mixtures (solutions)

- two or more substances evenly mixed
- Homogeneous mixtures appear uniform under a microscope.

Lesson 1 Review

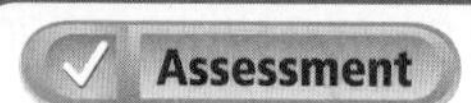

Visual Summary

Substance **Mixtures**

Substances have a composition that does not change. The composition of mixtures can vary.

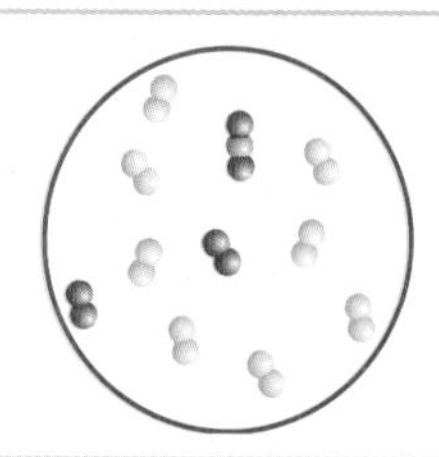

Solutions (homogeneous mixtures) are mixed at the atomic level.

Mixtures contain parts that are not bonded together. These parts can be separated using physical means.

FOLDABLES®

Use your lesson Foldable to review the lesson. Save your Foldable for the project at the end of the chapter.

What do you think NOW?

You first read the statements below at the beginning of the chapter.

1. You can identify a mixture by looking at it without magnification.
2. A solution is another name for a homogeneous mixture.

Did you change your mind about whether you agree or disagree with the statements? Rewrite any false statements to make them true.

Use Vocabulary

1. **Identify** What is another name for a homogeneous mixture?
2. **Contrast** homogeneous and heterogeneous mixtures.

Understand Key Concepts 

3. **Explain** why a compound is classified as a substance.
4. **Describe** two tests that you can run to determine if something is a substance or a mixture.

Interpret Graphics

5. **Classify** each group of particles below as an element, a compound, or a mixture.

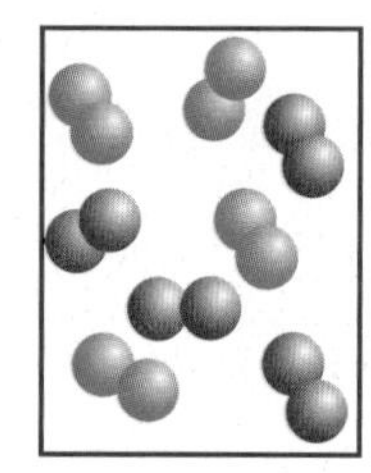

A

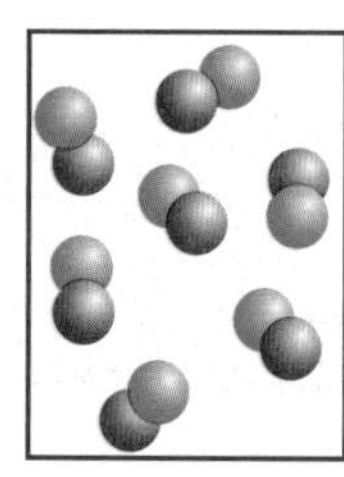

B

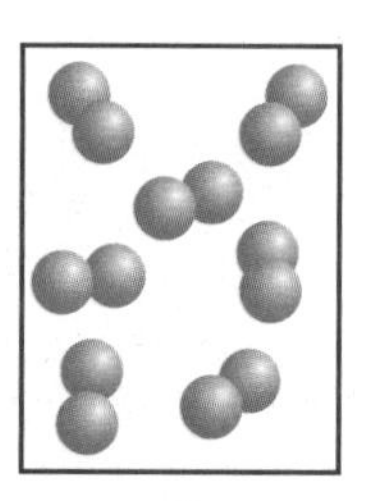

C

6. **Compare and contrast** Copy the graphic organizer below and use it to compare and contrast heterogeneous mixtures and solutions.

Heterogeneous mixtures | Solutions

Critical Thinking

7. **Explain** the following statement: All compounds are substances, but not all substances are compounds.
8. **Suppose** you have found an unknown substance in a laboratory. It has the formula H_2O_2 written on the bottle. Is it water? How do you know?

Sports Drinks and Your Body

HOW NATURE WORKS

Electrolytes and the Conduction of Nerve Impulses

When you exercise, you sweat. When you're thirsty, do you drink water or a sports drink? Chances are you have seen ads that claim sports drinks are better than water because they contain electrolytes. What are electrolytes, and how are they used in your body?

What are electrolytes?

Many elements, including potassium, sodium, chlorine, and calcium form ions. An electrolyte is a charged particle, scientifically known as an ion. Electrolytes can conduct electric charges. Pure water cannot conduct electric charges, but water containing electrolytes can.

Electrolytes in Your Body

Why does your body need electrolytes? Solutions of water and electrolytes surround all of the cells in your body. Electrolytes enable these solutions to carry nerve impulses from one cell to another. Your body's voluntary movements, such as walking, and involuntary movements, such as your heart beating, are caused by nerve impulses. Without electrolytes, these nerve impulses cannot move normally.

Nutrient	Function	Sex/Age	Adequate Intake	Food Sources
Sodium	maintains fluid volume outside of cells and aids in normal cell function	males/females 14–18	1.5 g/day	Sodium is added to many processed foods.
Potassium	maintains fluid volume inside/outside of cells and aids in normal cell function; supports blood and kidney health	males/females 14–18	4.7 g/day	fresh fruits and vegetables, dried peas, dairy products, meats, and nuts
Water	allows transport of nutrients to cells and removal of waste products	males females	3.3 L/day 2.3 L/day	all beverages, including water, as well as moisture in foods

Replenishing Fluids

Have you ever noticed that sweat is salty? Sweat is a solution of water and electrolytes, including sodium. Sports drinks can replace water and electrolytes. Some foods, such as bananas, oranges, and lima beans, also contain electrolytes. However, many sports drinks contain ingredients your body doesn't need, such as caffeine, sugar, and artificial colors. Unless you are sweating for an extended period of time, you don't need to replace electrolytes, but replacing water is always essential.

It's Your Turn

RESEARCH Study three different sports drinks. Create a bar graph that compares ingredients such as water, sugar, electrolytes, and caffeine in each brand. Draw a conclusion about which type of drink is best for your body.

Lesson 2

Reading Guide

Key Concepts
ESSENTIAL QUESTIONS

- Why do some substances dissolve in water and others do not?
- How do concentration and solubility differ?
- How can the solubility of a solute be changed?

Vocabulary

solvent p. 343
solute p. 343
polar molecule p. 344
concentration p. 346
solubility p. 348
saturated solution p. 348
unsaturated solution p. 348

Multilingual eGlossary

What's Science Got to do With It?

Properties of Solutions

Inquiry Stairs?

These stair-like formations, called terraces, are located in Mammoth Hot Springs, a part of Yellowstone National Park in Wyoming. Hot spring water is a mixture of water and dissolved carbon dioxide and limestone. When this mixture reaches Earth's surface, the pressure on the mixture is reduced. This causes the limestone to leave the mixture, forming the terraces.

Launch Lab

15 minutes

How are they different?

If you have ever looked at a bottle of Italian salad dressing, you know that some substances do not easily form solutions. The oil and vinegar do not mix, and the spices sink to the bottom. However, the salt in salad dressing does mix evenly with the other substances and forms a solution. How can we describe the difference quantitatively?

1. Read and complete a lab safety form.
2. Label one **beaker** *A* and **another beaker** *B.*
3. Measure 100 mL of water and pour it into beaker A.
4. Measure 100 mL of water and pour it into beaker B.
5. Add 10 g of **baking soda** to beaker A, and stir with a **plastic spoon** for 2 min or until all the baking soda dissolves, whichever happens first.
6. Add 25 g of **sugar** to beaker B and stir with a plastic spoon for 2 min or until all of the dissolves, whichever happens first.
7. Observe the mixtures in each beaker. Record your observations in your Science Journal.

Think About This

1. What substance dissolved better in water? How do you know?
2. Predict what would happen if you were to use 200 mL of water instead of 100 mL.
3. Do you think more baking soda might dissolve if you stirred the solution longer?
4. **Key Concept** Why do you think one substance dissolved more easily in water than the other substance? What factors do you think contribute to this difference?

Parts of Solutions

You've read that a solution is a homogeneous mixture. Recall that in a solution, substances are evenly mixed on the atomic level. How does this mixing occur? Dissolving is the process of mixing one substance into another to form a solution. Scientists use two terms to refer to the substances that make up a solution. Generally, the **solvent** *is the substance that exists in the greatest quantity in a solution. All other substances in a solution are* **solutes.** Recall that air is a solution of 78 percent nitrogen, 21 percent oxygen, and 1 percent other substances. Which substance is the solvent? In air, nitrogen exists in the greatest quantity. Therefore, it is the solvent. The oxygen and other substances are solutes. In this lesson, you will read the terms *solute* and *solvent* often. Refer back to this page if you forget what these terms mean.

Reading Check How do a solute and a solvent differ?

Make a four-tab shutter-fold. Label it as shown. Collect information about which solvents dissolve which solutes.

Polar solvents dissolve:
Nonpolar solvents dissolve:
Like Dissolves Like
Polar solvents do not dissolve:
Nonpolar solvents do not dissolve:

State of Solution	Solvent Is:	Solute Can Be:
Solid	solid	**gas or solid (called alloys)** This saxophone is a solid solution of solid copper and solid zinc.
Liquid	liquid	**solid, liquid, and/or gas** Soda is a liquid solution of liquid water, gaseous carbon dioxide, and solid sugar and other flavorings.
Gas	gas	**gas** This lighted sign contains a gaseous mixture of gaseous argon and gaseous mercury.

Table 1 Types of Solutions

Concepts in Motion Interactive Table

The electrons spend more time near the oxygen atom. This makes the end with the oxygen atom slightly negative (–).

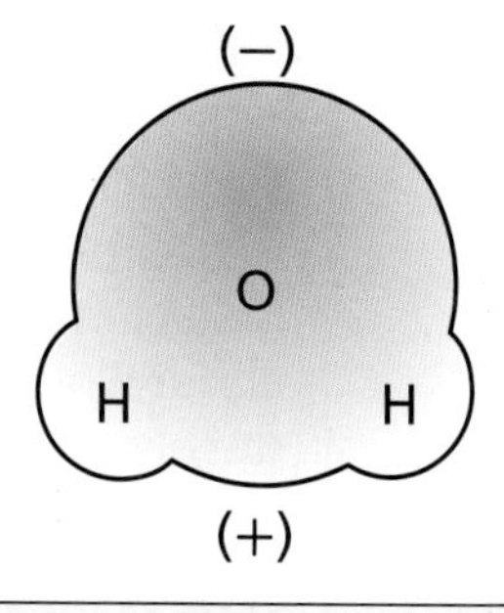

The end with the hydrogen atoms is slightly positive (+).

▲ **Figure 5** Water is a polar molecule. The end with the oxygen atom is slightly negative, and the end with the hydrogen atoms is slightly positive.

Personal Tutor

Types of Solutions

When you think of a solution, you might think of a liquid. However, solutions can exist in all three states of matter—solid, liquid, or gas. The state of the solvent, because it exists in the greatest quantity, determines the state of the solution. **Table 1** contrasts solid, liquid, and gaseous solutions.

Water as a Solvent

Did you know that over 75 percent of your brain and almost 90 percent of your lungs are made of water? Water is one of the few substances on Earth that exists naturally in all three states—solid, liquid, and gas. However, much of this water is not pure water. In nature, water almost always exists as a solution; it contains dissolved solutes. Why does nearly all water on Earth contain dissolved solutes? The answer has to do with the structure of the water molecule.

The Polarity of Water

A water molecule, such as the one shown in **Figure 5,** is a covalent compound. Recall that atoms are held together with covalent bonds when sharing electrons. In a water molecule, one oxygen atom shares electrons with two hydrogen atoms. However, these electrons are not shared equally. The electrons in the oxygen-hydrogen bonds more often are closer to the oxygen atom than they are to the hydrogen atoms. This unequal sharing of electrons gives the end with the oxygen atom a slightly negative charge. And, it gives the end with the hydrogen atoms a slightly positive charge. Because of the unequal sharing of electrons, a water molecule is said to be polar. *A* **polar molecule** *is a molecule with a slightly negative end and a slightly positive end.* Nonpolar molecules have an even distribution of charge. Solutes and solvents can be polar or nonpolar.

Like Dissolves Like

Water is often called the universal solvent because it dissolves many different substances. But water can't dissolve everything. Why does water dissolve some substances but not others? Water is a polar solvent. Polar solvents dissolve polar solutes easily. Nonpolar solvents dissolve nonpolar solutes easily. This is summarized by the phrase "like dissolves like." Because water is a polar solvent, it dissolves most polar and ionic solutes.

Key Concept Check Why do some substances dissolve in water and others do not?

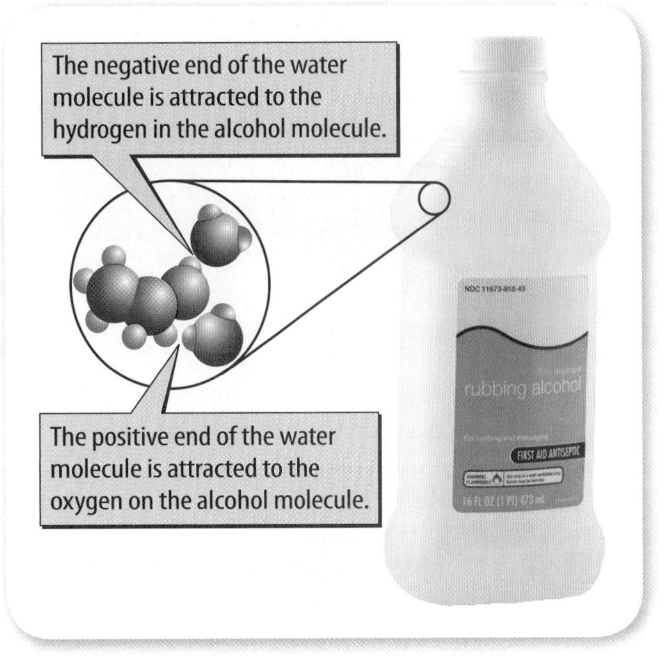

▲ **Figure 6** When a polar solute, such as rubbing alcohol, dissolves in a polar solvent, such as water, the poles of the solvent are attracted to the oppositely charged poles of the solute.

Polar Solvents and Polar Molecules

Because water molecules are polar, water dissolves groups of other polar molecules. **Figure 6** shows what rubbing alcohol, a substance used as a disinfectant, looks like when it is in solution with water. Molecules of rubbing alcohol also are polar. Therefore, when rubbing alcohol and water mix, the positive ends of the water molecules are attracted to the negative ends of the alcohol molecules. Similarly, the negative ends of the water molecules are attracted to the positive ends of the alcohol. In this way, alcohol molecules dissolve in the solvent.

Polar Solvents and Ionic Compounds

Many ionic compounds are also soluble in water. Recall that ionic compounds are composed of alternating positive and negative ions. Sodium chloride (NaCl) is an ionic compound composed of sodium ions (Na^+) and chloride ions (Cl^-). When sodium chloride dissolves, these ions are pulled apart by the water molecules. This is shown in **Figure 7.** The negative ends of the water molecules attract the positive sodium ions. The positive ends of the water molecules attract the negative chloride ions.

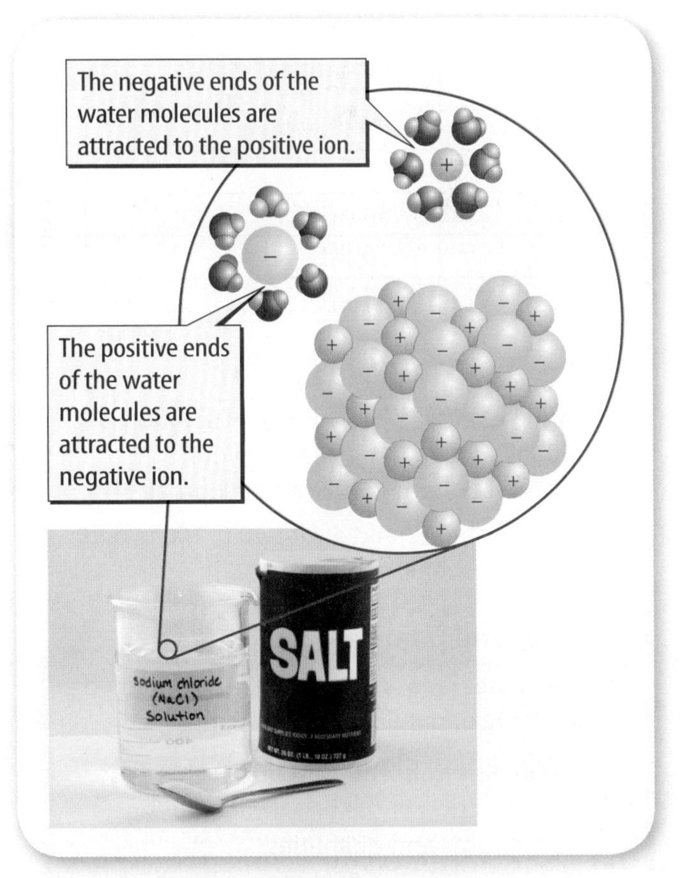

▲ **Figure 7** When ionic solutes dissolve, the positive poles of the solvent are attracted to the negative ions. The negative poles of the solvent are attracted to the positive ions.

Figure 8 The volumes of both drinks are the same, but the glass on the left contains more solute than the solution on the right.

Inquiry MiniLab 15 minutes

How much is dissolved?

How can you make a glass and a pitcher of lemonade with the same sweetness?

1. Read and complete a lab safety form.
2. Make a table similar to the one below.

Mass of sugar (g)	Volume of solution (mL)	Observation
25 g	50 mL	
50 g	50 mL	
100 g	50 mL	
125 g	50 mL	

3. Add 25 g of **sugar** to a **beaker.**
4. Add water until the volume of solution is 50 mL. Stir for 2 min. Record your observations in your Science Journal.
5. Repeat steps 3 and 4 with 50 g, 100 g, and 125 g of sugar.

Analyze and Conclude

1. **Calculate** What mass of sugar is contained in 25 mL of the first three solutions?
2. **Key Concept** Describe the concentration of the first three solutions using words and quantities.
3. **Infer** How might you make 100 mL of the first solution so that it has the same concentration as 50 mL of this solution?

Concentration—How much is dissolved?

Have you ever tasted a spoonful of soup and wished it had more salt in it? In a way, your taste buds were measuring the amount, or concentration, of salt in the soup. **Concentration** *is the amount of a particular solute in a given amount of solution.* In the soup, salt is a solute. Saltier soup has a higher concentration of salt. Soup with less salt has a lower concentration of salt. Look at the glasses of fruit drink in **Figure 8.** Which drink has a higher concentration of solute? The darker blue drink has a higher concentration of solute.

Concentrated and Dilute Solutions

One way to describe the saltier soup is to say that it is more concentrated. The less salty soup is more dilute. The terms *concentrated* and *dilute* are one way to describe how much solute is dissolved in a solution. However, these terms don't state the exact amount of solute dissolved. What one person thinks is concentrated might be what another person thinks is dilute. Soup that tastes too salty to you might be perfect for someone else. How can concentration be described more precisely?

Reading Check Why is the term *dilute* not a precise way to describe concentration?

Describing Concentration Using Quantity

A more precise way to describe concentration is to state the quantity of solute in a given quantity of solution. When a solution is made of a solid dissolved in a liquid, such as salt in water, concentration is the mass of solute in a given volume of solution. Mass usually is stated in grams, and volume usually is stated in liters. For example, concentration can be stated as grams of solute per 1 L of solution. However, concentration can be stated using any units of mass or volume.

Calculating Concentration—Mass per Volume

One way that concentration can be calculated is by the following equation:

$$\text{Concentration } (C) = \frac{\text{mass of solute } (m)}{\text{volume of solution } (V)}$$

To calculate concentration, you must know both the mass of solute and the volume of solution that contains this mass. Then divide the mass of solute by the volume of solution.

Reading Check If more solvent is added to a solution, what happens to the concentration of the solution?

Concentration—Percent by Volume

Not all solutions are made of a solid dissolved in a liquid. If a solution contains only liquids or gases, its concentration is stated as the volume of solute in a given volume of solution. In this case, the units of volume must be the same—usually mL or L. Because the units match, the concentration can be stated as a percentage. Percent by volume is calculated by dividing the volume of solute by the total volume of solution and then multiplying the quotient by 100. For example, if a container of orange drink contains 3 mL of acetic acid in a 1,000-mL container, the concentration is 0.3 percent.

Math Skills — Calculate Concentration

Solve for Concentration

Suppose you want to calculate the concentration of salt in a **0.4 L** can of soup. The back of the can says it contains **1.6 g** of salt. What is its concentration in g/L? In other words, how much salt would be contained in 1 L of soup?

1 This is what you know: mass: **1.6 g**; volume: **0.4 L**

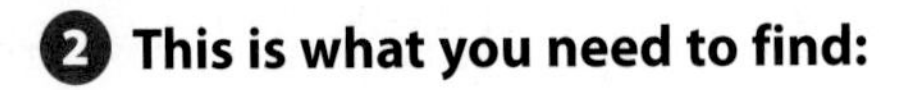

2 This is what you need to find: concentration: C

3 Use this formula: $C = \frac{m}{V}$

4 Substitute: the values for m and V into the formula and divide. $C = \frac{1.6\text{ g}}{0.4\text{ L}} = 4\text{ g/L}$

Answer: The concentration is 4 g/L. As you might expect, 0.4 L of soup contains less salt (1.6 g) than 1 L of soup (4 g). However, the concentration of both amounts of soup is the same—4 g/L.

Practice

1. What is the concentration of 5 g of sugar in 0.2 L of solution?
2. How many grams of salt are in 5 L of a solution with a concentration of 3 g/L?
3. Suppose you add water to 6 g of sugar to make a solution with a concentration of 3 g/L. What is the total volume of the solution?

- **Math Practice**
- **Personal Tutor**

Solubility—How much can dissolve?

Have you ever put too much sugar into a glass of iced tea? What happens? Not all of the sugar dissolves. You stir and stir, but there is still sugar at the bottom of the glass. That is because there is a limit to how much solute (sugar) can be dissolved in a solvent (water). **Solubility** (sahl yuh BIH luh tee) *is the maximum amount of solute that can dissolve in a given amount of solvent at a given temperature and pressure.* If a substance has a high solubility, more of it can dissolve in a given solvent.

Word Origin
solubility
from Latin solvere, means "to loosen"

Key Concept Check How do concentration and solubility differ?

Saturated and Unsaturated Solutions

If you add water to a dry sponge, the sponge absorbs the water. If you keep adding water, the sponge becomes saturated. It can't hold any more water. This is **analogous** (uh NA luh gus), or similar, to what happens when you stir too much sugar into iced tea. Some sugar dissolves, but the excess sugar does not dissolve. The solution is saturated. *A* **saturated solution** *is a solution that contains the maximum amount of solute the solution can hold at a given temperature and pressure. An* **unsaturated solution** *is a solution that can still dissolve more solute at a given temperature and pressure.*

Academic Vocabulary
analogous
(adjective) showing a likeness in some ways between things that are otherwise different

Factors that Affect How Much Can Dissolve

Can you change the amount of a particular solute that can dissolve in a solvent? Yes. Recall the definition of solubility—the maximum amount of solute that can dissolve in a given amount of solvent at a given temperature and pressure. Changing either temperature or pressure changes how much solute can dissolve in a solvent.

Figure 9 Generally, solids are more soluble at warmer temperatures.

Visual Check How many grams of KNO_3 will dissolve in 100 g of water at 10°C?

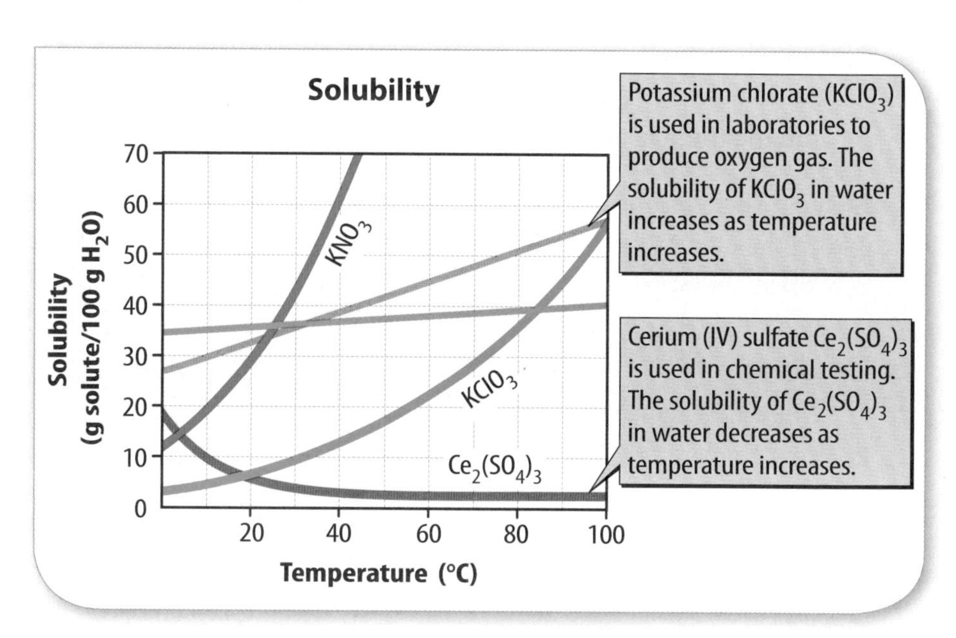

Effect of Temperature Have you noticed that more sugar dissolves in hot tea than in iced tea? The solubility of sugar in water increases with temperature. This is true for many solid solutes, as shown in **Figure 9.** Notice that some solutes become less soluble when temperature is increased.

How does temperature affect the solubility of a gas in a liquid? Recall that soda, or soft drinks, contains carbon dioxide, a gaseous solute, dissolved in liquid water. The bubbles you see in soda are made of undissolved carbon dioxide. Have you ever noticed that more carbon dioxide bubbles out when you open a warm can of soda than when you open a cold can? This is because the solubility of a gas in a liquid decreases when the temperature of the solution increases.

Effect of Pressure What keeps carbon dioxide dissolved in an unopened can of soda? In a can, the carbon dioxide in the space above the liquid soda is under pressure. This causes the gas to move to an area of lower pressure—the solvent. The gas moves into the solvent, and a solution is formed. When the can is opened, as shown in **Figure 10,** this pressure is released, and the carbon dioxide gas leaves the solution. Pressure does not affect the solubility of a solid solute in a liquid.

Key Concept Check How can the solubility of a solute be changed?

How Fast a Solute Dissolves

Temperature and pressure can affect how much solute dissolves. If solute and solvent particles come into contact more often, the solute dissolves faster. **Figure 11** shows three ways to increase how often solute particles contact solvent particles. Each of these methods will make a solute dissolve faster. However, it is important to note that stirring the solution or crushing the solute will not make more solute dissolve.

▲ **Figure 10** The solubility of a gas in a liquid is directly proportional to the pressure of the gas above the solution. When the soft drink can is opened, the reduced pressure inside the can forces carbon dioxide gas to come out of the solution.

Stirring the solution

Crushing the solute

Increasing the temperature

Figure 11 Several factors can affect how quickly a solute will dissolve in a solution. However, dissolving more quickly won't necessarily make more solute dissolve.

Lesson 2 Review

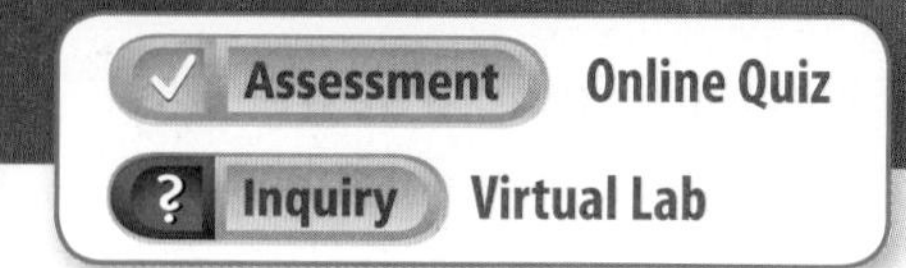

Visual Summary

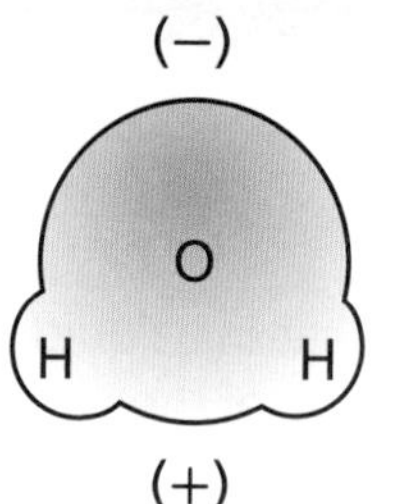

Polar molecule

Substances dissolve in other substances that have similar polarity. In other words, like dissolves like.

Concentration is the amount of substance that is dissolved. Solubility is the maximum amount that can dissolve.

Both temperature and pressure affect the solubility of solutes in solutions.

FOLDABLES®

Use your lesson Foldable to review the lesson. Save your Foldable for the project at the end of the chapter.

What do you think NOW?

You first read the statements below at the beginning of the chapter.

3. Solutions can be solids, liquids, or gases.

4. A teaspoon of soup is less concentrated than a cup of the same soup.

Did you change your mind about whether you agree or disagree with the statements? Rewrite any false statements to make them true.

Use Vocabulary

1. **Define** *polar molecule* in your own words.

Understand Key Concepts 

2. **Explain** how you could use the solubility of a substance to make a saturated solution.
3. **Predict** whether an ionic compound will dissolve in a nonpolar solvent.

Interpret Graphics

4. **Read a Graph** Use the graph to determine what you would observe in a solution of 30 g of $KClO_3$ in 100 g of water at 10°C.

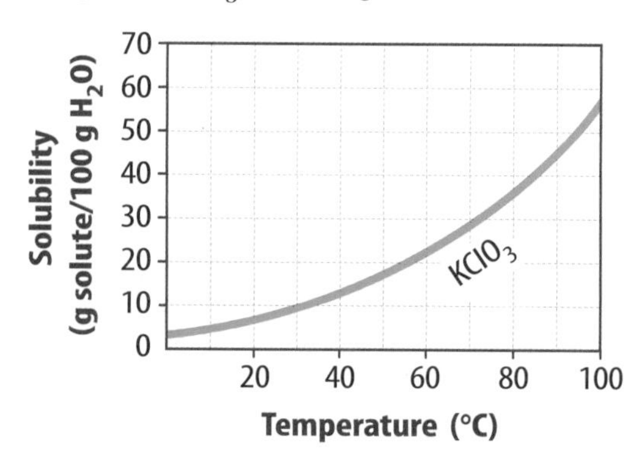

5. **Organize** Copy the graphic organizer, and use it to organize three factors that increase the speed a solute dissolves in a liquid.

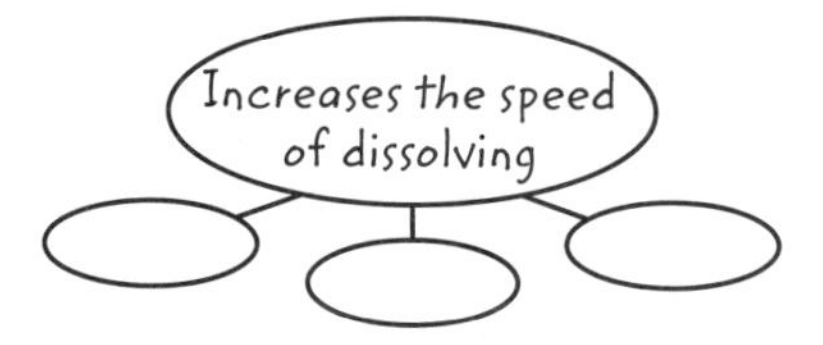

Critical Thinking

6. **Explain** A student wants to increase the maximum amount of sugar that can dissolve in water. She crushes the sugar and then stirs it into the water. Does this work?

Math Skills

Math Practice

7. Use ratios to explain how a tablespoon of soup and a cup of the same soup have the same concentration.

Inquiry **Skill Practice** **Manipulate Variables** **45 minutes**

How does a solute affect the conductivity of a solution?

When some substances dissolve in water, they form ions, or charged particles. Other substances do not. Solutions that contain ions conduct electricity. In this lab, you will determine how one variable affects another.

Materials

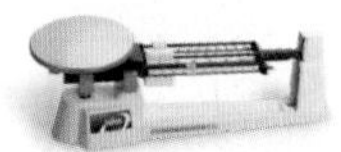

triple-beam balance

250-mL beaker

stirring rod

salt

Also needed: 6-V battery, wires, sugar, miniature lightbulb with base distilled water

Safety

Learn It

In an experiment, you can manipulate variables or factors. The factor you change is called the independent variable. The factor or factors that change as a result of a change to the independent variable are called the dependent variables. All other factors should be kept constant.

Try It

1. Read and complete a lab safety form.
2. Dissolve 20 g of salt in 100 mL of distilled water to make a saltwater solution with a concentration of 200 g/L. Label the beaker *salt water, 200 g/L.*
3. Prepare a sugar solution with a concentration of 200 g/L. Label the beaker *sugar water, 200 g/L.*
4. Create a circuit as shown in the photograph below using the salt-water solution, wires, a 6-V battery, and a lightbulb with base.

5. Once the circuit is complete, record your observations in your Science Journal.
6. Remove the wires from the solution, and rinse them with plain water.
7. Repeat steps 4–6 using the sugar-water solution.
8. Repeat steps 4–6 using only distilled water.

Apply It

9. Based on your observations, what happened when the circuit was made using each of the solutions?
10. What would happen if the concentration of the solution that conducted the electricity were changed to a more dilute solution? What if it was changed to a more concentrated solution?
11. **Key Concept** Test your hypothesis by creating two more saltwater solutions, one with a concentration of 100 g/L and one with a concentration of 300 g/L. Use them, one at a time, to complete the circuit. Record your observations in your Science Journal.

Lesson 3

Acid and Base Solutions

Reading Guide

Key Concepts
ESSENTIAL QUESTIONS

- What happens when acids and bases dissolve in water?
- How does the concentration of hydronium ions affect pH?
- What methods can be used to measure pH?

Vocabulary

acid p. 354
hydronium ion p. 354
base p. 354
pH p. 356
indicator p. 358

Multilingual eGlossary

Video BrainPOP®

Inquiry What's eating her?

When this statue was first carved, it didn't have any of these odd-shaped marks on its surface. This damage was caused by acid rain—precipitation that contains water and dissolved substances called acids. When acid rain falls on this statue, the acid reacts with the stone, dissolving it and then carrying it away.

15 minutes

What color is it?

Did you know that all rain is naturally acidic? As raindrops fall through the air, they pick up molecules of carbon dioxide. An acid called carbonic acid is formed when the water molecules react with the carbon dioxide molecules. An indicator is a substance that can be used to tell if a solution is acidic, basic, or neutral.

1. Read and complete a lab safety form.
2. Half fill a **beaker** with the **colored solution.**
3. Place one end of a **straw** into the solution.
   ⚠ **Caution:** *Do not suck liquid through the straw.*
4. Blow through the straw, making bubbles in the solution. Continue blowing, and count how many times you have to blow bubbles until you observe a change.
5. Record your observations in your Science Journal.

Think About This

1. Describe what change you saw take place.
2. What do you think made this change occur?
3. How do you think the results would have been different if you had held your breath for several seconds before blowing through the straw?
4. **Key Concept** Recall the change you observed in this Launch Lab. What did the color of the indicator solution tell you as your breath dissolved in the solution?

What are acids and bases?

Would someone ever drink an acid? At first thought, you might answer no. After all, when people think of acids, they often think of acids such as those found in batteries or in acid rain. However, acids are found in other items, including milk, vinegar, fruits, and green leafy vegetables. Some examples of acids that you might eat are shown in **Figure 12.** Along with the word *acid,* you might have heard the word *base.* Like acids, you can also find bases in your home. Detergent, antacids, and baking soda are examples of items that contain bases. But acids and bases are found in more than just household goods. As you will learn in this lesson, they are necessities for our daily life.

Figure 12 You might be surprised to learn that acids are common in the foods you eat. All of the foods pictured here contain acids.

Figure 13 Acids, such as hydrochloric acid, produce hydronium ions when they dissolve in water. ▶

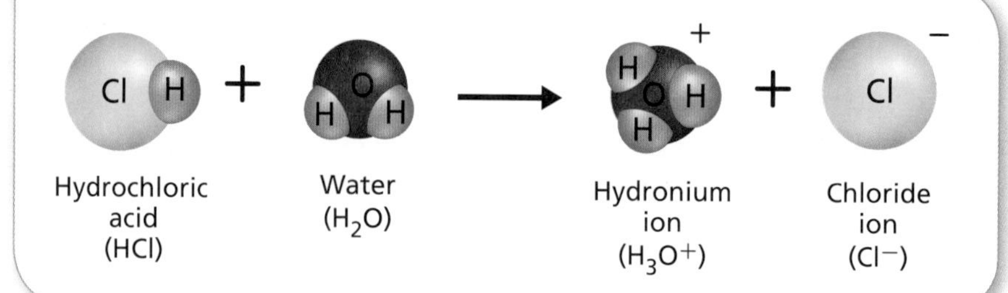

Acids

Have you ever tasted the sourness of a lemon or a grapefruit? This sour taste is due to the acid in the fruit. *An* **acid** *is a substance that produces a hydronium ion (H_3O^+) when dissolved in water.* Nearly all acid molecules contain one or more hydrogen atoms (H). When an acid mixes with water, this hydrogen atom separates from the acid. It quickly combines with a water molecule, resulting in a hydronium ion. This process is shown in **Figure 13.** A **hydronium ion,** H_3O^+, *is a positively charged ion formed when an acid dissolves in water.*

WORD ORIGIN

acid
from Latin *acidus,* means "sour"

Bases

A **base** *is a substance that produces hydroxide ions (OH^-) when dissolved in water.* When a hydroxide compound such as sodium hydroxide (NaOH) mixes with water, hydroxide ions separate from the base and form hydroxide ions (OH^-) in water. Some bases, such as ammonia (NH_3), do not contain hydroxide ions. These bases produce hydroxide ions by taking hydrogen atoms away from water, leaving hydroxide ions (OH^-). This process is shown in **Figure 14.** Some properties and uses of acids and bases are shown in **Table 2.**

Figure 14 Bases, such as sodium hydroxide and ammonia, produce hydroxide ions when they dissolve in water. ▼

Visual Check How is dissolving an acid, shown above, similar to dissolving ammonia, shown below?

Key Concept Check What happens when acids and bases dissolve in water?

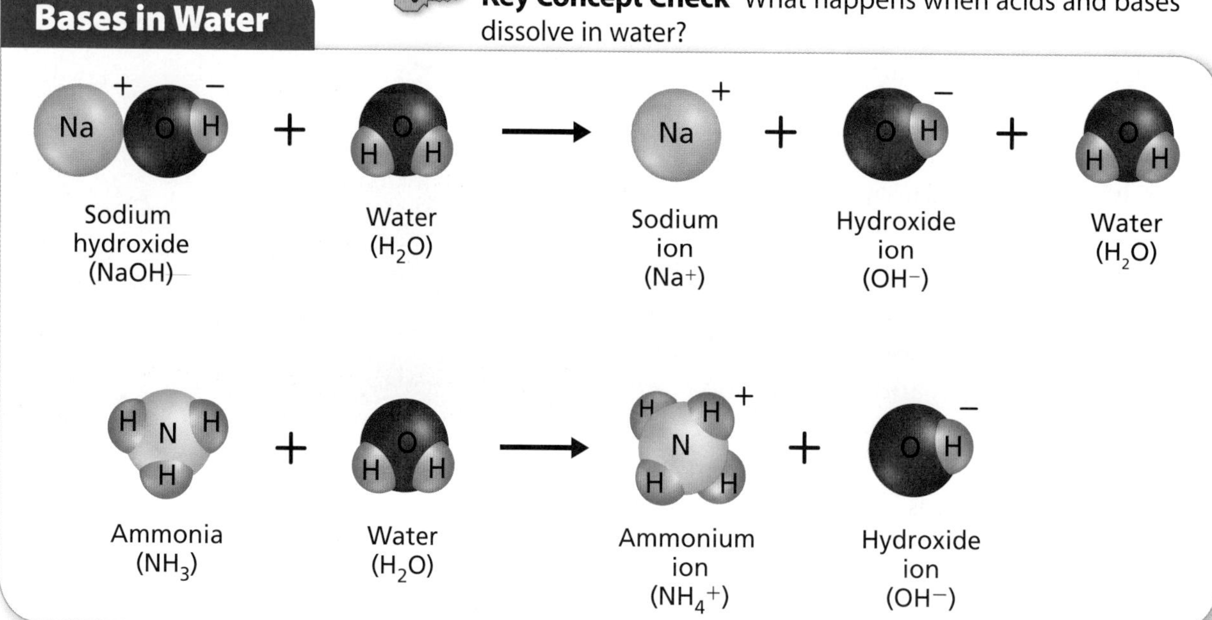

Table 2 Properties and Uses of Acids and Bases

	Acids	Bases
Ions produced	H, O, H, H, + Acids produce H_3O^+ in water.	O, H, − Bases produce OH^- ions in water.
Examples	• hydrochloric acid, HCl • acetic acid, CH_3COOH • citric acid, $H_3C_6H_5O_7$ • lactic acid, $C_3H_6O_3$	• sodium hydroxide, $NaOH$ • ammonia, NH_3 • sodium carbonate, Na_2CO_3 • calcium hydroxide, $Ca(OH)_2$
Some properties	• Acids provide the sour taste in food (never taste acids in the laboratory). • Most can damage skin and eyes. • Acids react with some metals to produce hydrogen gas. • H_3O^+ ions can conduct electricity in water. • Acids react with bases to form neutral solutions.	• Bases provide the bitter taste in food (never taste bases in the laboratory). • Most can damage skin and eyes. • Bases are slippery when mixed with water. • OH^- ions can conduct electricity in water. • Bases react with acids to form neutral solutions.
Some uses	• Acids are responsible for for natural and artificial flavoring in foods, such as fruits. • Lactic acid is found in milk. • Acid in your stomach breaks down food. • Blueberries, strawberries, and many vegetable crops grow better in acidic soil. • Acids are used to make products such as fertilizers, detergents, and plastics.	• Bases are found in natural and artificial flavorings in food, such as cocoa beans. • Antacids neutralize stomach acid, alleviating heartburn. • Bases are found in cleaners such as shampoo, dish detergent, and window cleaner. • Many flowers grow better in basic soil. • Bases are used to make products such as rayon and paper.

What is pH?

Have you ever seen someone test the water in a swimming pool? It is likely that the person was testing the pH of the water. Swimming pool water should have a pH around 7.4. If the pH of the water is higher or lower than 7.4, the water might become cloudy, burn swimmers' eyes, or contain too many bacteria. What does a pH of 7.4 mean?

Hydronium Ions

The **pH** *is an inverse measure of the concentration of hydronium ions (H_3O^+) in a solution.* What does *inverse* mean? It means that as one thing increases, another thing decreases. In this case, as the concentration of hydronium ions increases, pH decreases. A solution with a lower pH is more acidic. As the concentration of hydronium ions decreases, the pH increases. A solution with a higher pH is more basic. This relationship is shown in **Figure 15.**

Balance of Hydronium and Hydroxide Ions

All acid and base solutions contain both hydronium and hydroxide ions. In a neutral solution, such as water, the concentrations of hydronium and hydroxide ions are equal. What distinguishes an acid from a base is which of the two ions is present in the greater concentration. Acids have a greater concentration of hydronium ions (H_3O^+) than hydroxide ions (OH^-). Bases have a greater concentration of hydroxide ions than hydronium ions. Brackets around a chemical formula mean *concentration.*

Acids	$[H_3O^+] > [OH^-]$
Neutral	$[H_3O^+] = [OH^-]$
Bases	$[H_3O^+] < [OH^-]$

Key Concept Check How does the concentration of hydronium ions affect pH?

pH Scale

Figure 15 Notice that as hydronium concentration increases, the pH decreases.

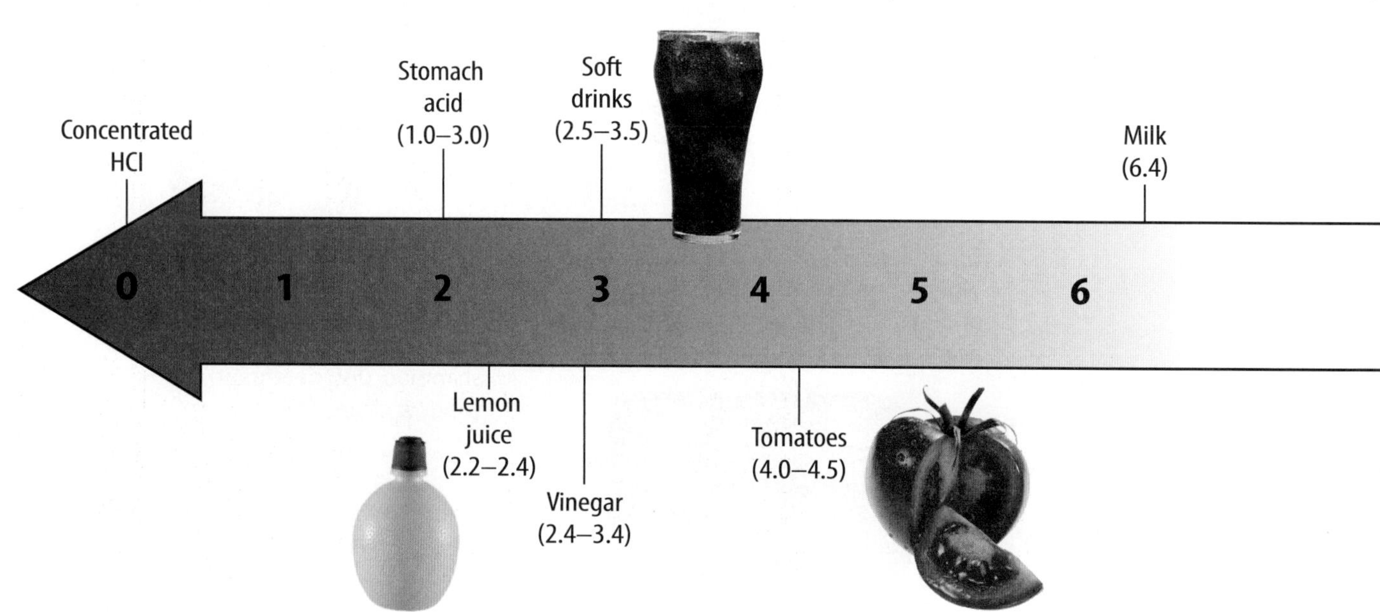

The pH Scale

The pH scale is used to indicate how acidic or basic a solution is. Notice in **Figure 15** that the pH scale contains values that range from below 0 to above 14. Acids have a pH below 7. Bases have a pH above 7. Solutions that are neutral have a pH of 7—they are neither acidic nor basic.

You might be wondering what the numbers on the pH scale mean. How is the concentration of hydronium ions different in a solution with a pH of 1 from the concentration in a solution with a pH of 2? A change in one pH unit represents a tenfold change in the acidity or basicity of a solution. For example, if one solution has a pH of 1 and a second solution has a pH of 2, then the first solution is not twice as acidic as the second solution; it is ten times more acidic.

The difference in acidity or basicity between two solutions is represented by 10^n, where n is the difference between the two pH values. For example, how much more acidic is a solution with a pH of 1 than a solution with pH of 3? First, calculate the difference, n, between the two pH values: $n = 3 - 1 = 2$. Then use the formula, 10^n, to calculate the difference in acidity: $10^2 = 100$. A solution with a pH of 1 is 100 times more acidic than a solution with a pH of 3.

Reading Check How much more acidic is a solution with a pH of 1 than a solution with a pH of 4?

FOLDABLES®

Make a small horizontal shutterfold. Label and draw a pH scale as shown. Shade the scale with colored pencils to differentiate between acids and bases. Use the foldable to compare acid and base solutions.

Review Personal Tutor

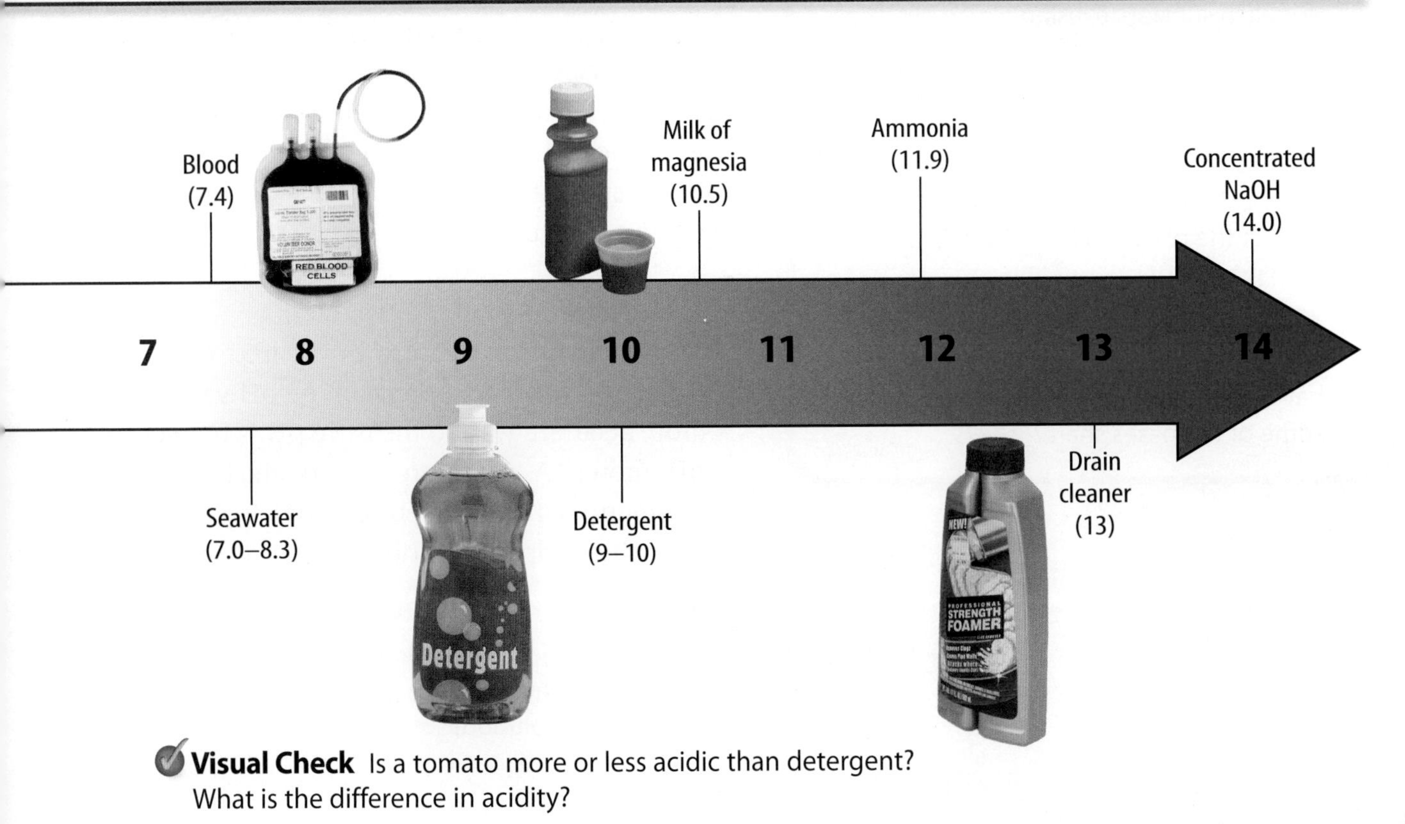

Visual Check Is a tomato more or less acidic than detergent? What is the difference in acidity?

MiniLab 30 minutes

Is it an acid or a base?

Acids and bases are found in many of the products people use in their homes every day. Properties of acids and bases make them desirable for use as cleaners, preservatives, or even flavorings.

1. Read and complete a lab safety form.
2. Get samples of **household products** from your teacher.
3. Dip a **pH strip** into the first product. Blot off excess liquid with a **paper towel.**
4. Compare the color of the pH strip to the **pH strip color chart,** and read the pH value.
5. Make a table in your Science Journal that shows the products and their pH values. Identify each product as acidic, neutral, or basic.
6. Repeat steps 3–5 for each substance, using a new pH strip for each substance.

Analyze and Conclude

1. **Predict** Using your table, think of three other products that could be tested. Predict the pH of each product. Explain your predictions.
2. **Key Concept** Based on your table, what characteristics do some of the acids have in common? What characteristics do some of the bases share?

How is pH measured?

How is the pH of a solution, such as swimming pool water, measured? Water test kits contain chemicals that change color when an acid or a base is added to them. These chemicals are called indicators.

pH Indicators

Indicators can be used to measure the approximate pH of a solution. *An* **indicator** *is a compound that changes color at different pH values when it reacts with acidic or basic solutions.* The pH of a solution is measured by adding a drop or two of the indicator to the solution. When the solution changes color, this color is matched to a set of standard colors that correspond to certain pH values. There are many different indicators—each indicator changes color over a specific range of pH values. For example, bromthymol blue is an indicator that changes from yellow to green to blue between pH 6 and pH 7.6.

pH Testing Strips

pH also can be measured using pH testing strips. The strips contain an indicator that changes to a variety of colors over a range of pH values. To use pH strips, dip the strip into the solution. Then match the resulting color to the list of standard colors that represent specific pH values.

pH Meters

Although pH strips are quick and easy, they provide only an approximate pH value. A more accurate way to measure pH is to use a pH meter. A pH meter is an electronic instrument with an electrode that is sensitive to the hydronium ion concentration in solution.

Key Concept Check What are two methods that can be used to measure the pH of a solution?

Lesson 3 Review

Assessment Online Quiz

Visual Summary

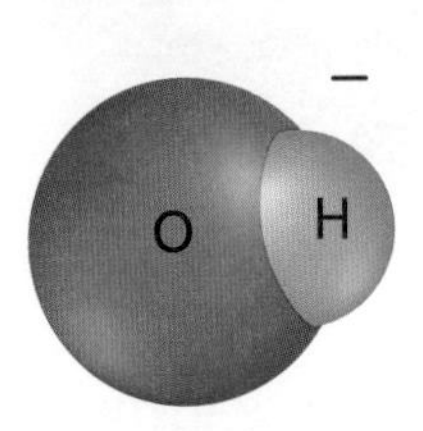

Acids contain hydrogen ions that are released and form hydronium ions in water. Bases are substances that form hydroxide ions when dissolved in water.

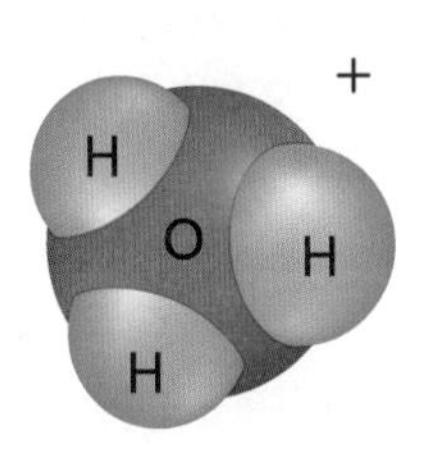

Hydronium ion concentration changes inversely with pH. This means that as hydronium ion concentration increases, the pH decreases.

pH can be measured using indicators or digital pH meters.

FOLDABLES®

Use your lesson Foldable to review the lesson. Save your Foldable for the project at the end of the chapter.

What do you think NOW?

You first read the statements below at the beginning of the chapter.

5. Acids are found in many foods.

6. You can determine the exact pH of a solution using pH paper.

Did you change your mind about whether you agree or disagree with the statements? Rewrite any false statements to make them true.

Use Vocabulary

1 A measure of the concentration of hydronium ions (H_3O^+) in a solution is ___________.

2 A(n) ___________ is used to determine the approximate pH of a solution.

Understand Key Concepts

3 **Describe** What happens to a hydrogen atom in an acid when the acid is dissolved in water?

4 **Explain** How does pH vary with hydronium ion and hydroxide ion concentrations in water?

5 **Show** Does an acidic solution contain hydroxide ions? Explain your answer with a diagram.

Interpret Graphics

6 **Predict** what is produced when hydrofluoric acid (HF) is dissolved in water in the equation below.

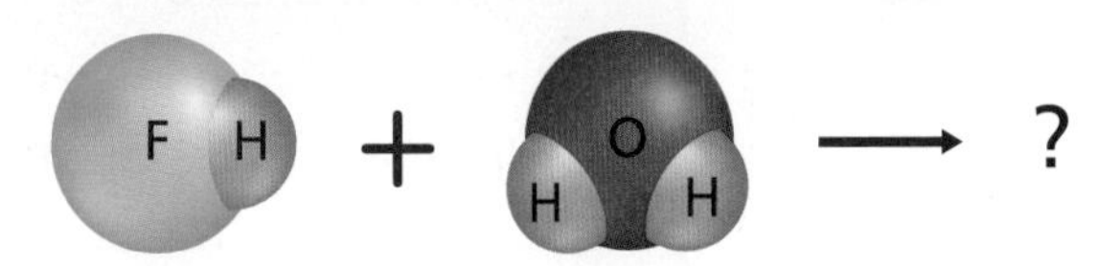

7 **Contrast** Copy the graphic organizer below, and use it to describe and contrast three ways to measure pH. In the organizer, describe which methods are most and least accurate.

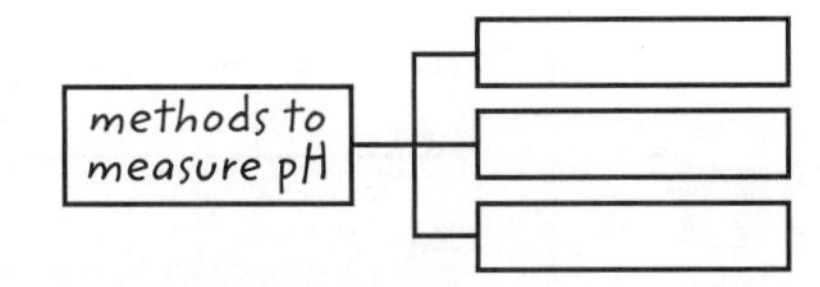

Critical Thinking

8 **Describe** the concentration of hydronium ions and hydroxide ions when a base is added slowly to a white vinegar solution. The pH of white vinegar is 3.1.

Lab

1 class period

Can the pH of a solution be changed?

Materials

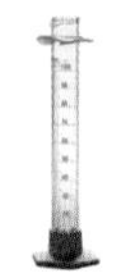

100-mL graduated cylinder

triple-beam balance

beakers

dropper

white vinegar

Also needed: baking soda, stirring rod, universal indicator, water

Safety

Many foods that people eat come from plants that grow best in soils that have a pH within a particular range. Potatoes grow best in soils that have a pH between 4.5 and 6.0. Blueberry plants grow best in soils with a pH of 4.0 to 4.5. Strawberry plants grow best in soils with a pH of 5.3 to 6.2. How might someone grow all of these plants in the same garden? The pH of soil often has to be changed to make it suitable to grow these plants. Using the knowledge you have gained from this chapter and the techniques that you have practiced, you will discover whether the pH of a solution can be changed.

Question

Can the pH of a solution be changed? If so, how quickly can it happen, and how does one tell?

Procedure

1. Read and complete a lab safety form.
2. Using the materials provided, make a baking soda solution with a concentration of 5 g/L.
3. Add 15 drops of universal indicator to the solution.
4. In your Science Journal, make a table like the one shown here.

Number of Drops of Vinegar Added	Solution pH According to pH Paper	Color of Solution	Solution pH According to Indicator
0			
10			
20			
30			

5. Determine and record the pH and color of the solution.

6. Add 10 drops of vinegar to the solution and stir.
7. Determine and record the pH and color of the solution.
8. Repeat steps 7 and 8 until you have added 200 drops of vinegar.
9. Dispose of your solutions according to your teacher's instructions.

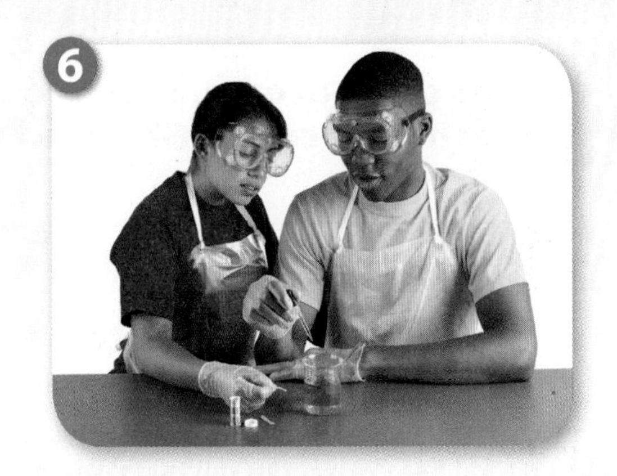

Analyze and Conclude

10. **Organize Information** Create a line graph that shows the data you have collected. Remember to put your independent variable on the *x*-axis.
11. **Analyze Data** Look at your table and the graph you made and discuss what, if any, relationships you see in your data.
12. **Predict** What do you think would happen if you performed this experiment again using an acid stronger than vinegar?
13. **The Big Idea** How can the pH of a solution or substance be changed?

Communicate Your Results

Compare your results with those of other groups. Look for differences, and discuss possible causes for discrepancies among the data collected.

Inquiry Extension

Perform some additional research in one of the following areas:

- Why is soil pH important?
- Why are acids and bases added to products, such as pH-balanced shampoo, to adjust their pH?
- Why is the pH of blood important to a person's health? How can blood pH become too high or too low? How can it be corrected?

Lab Tips

- ☑ The concentration of the starting solution is important. Zero your balance, and make accurate measurements.
- ☑ Stir the solution thoroughly after adding the vinegar each time.
- ☑ Always get your teacher's approval before trying any new test.

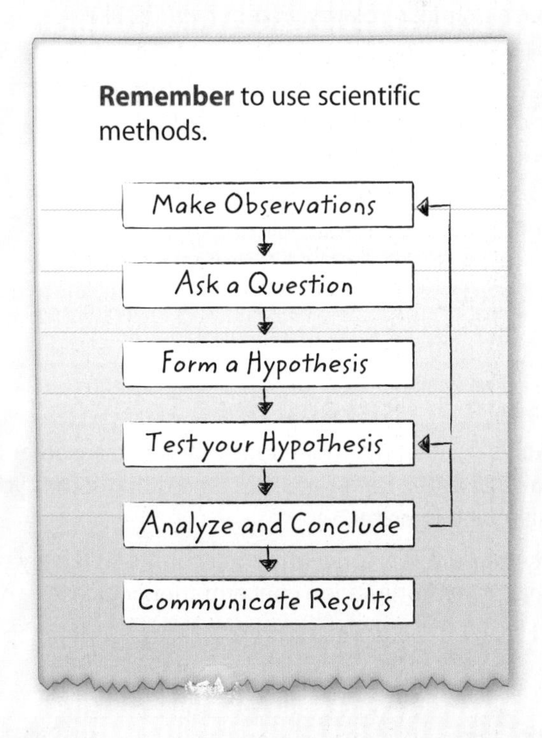

Chapter 10 Study Guide

Solutions are homogeneous mixtures. They can be described by the concentration and type of solute they contain.

Key Concepts Summary

Lesson 1: Substances and Mixtures

- **Substances** have a fixed composition. The composition of **mixtures** can vary.
- **Solutions** and **heterogeneous mixtures** are both types of mixtures. Solutions are mixed at the atomic level.
- Mixtures contain parts that are not bonded together. These parts can be separated using physical means, and their properties can be seen in the solution.

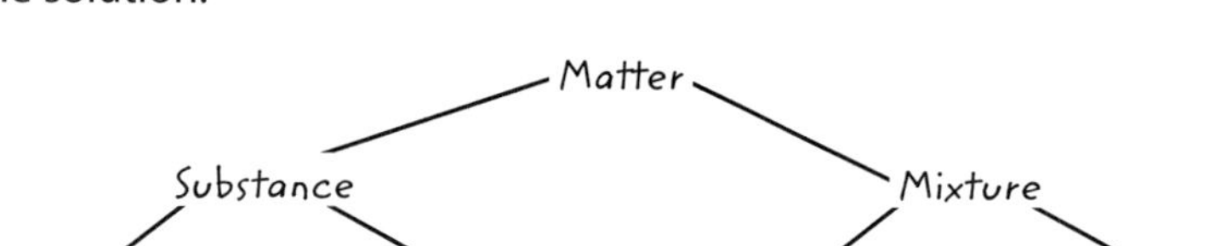

Vocabulary

substance p. 336
mixture p. 336
heterogeneous mixture p. 337
homogeneous mixture p. 337
solution p. 337

Lesson 2: Properties of Solutions

- Substances dissolve other substances that have a similar polarity. In other words, like dissolves like.
- **Concentration** is the amount of a **solute** that is dissolved. **Solubility** is the maximum amount of a solute that can dissolve.
- Both temperature and pressure affect the solubility of solutes in solutions.

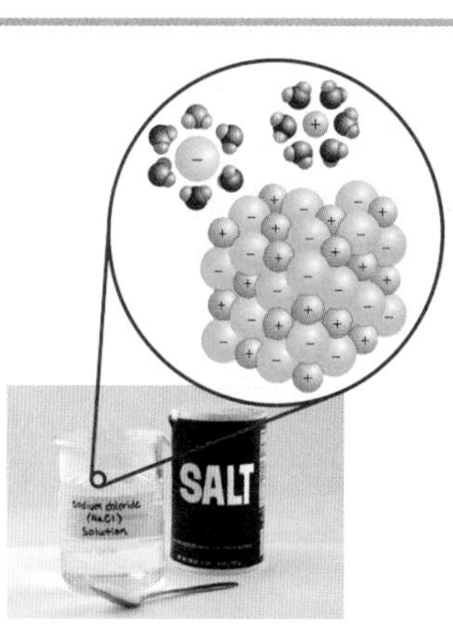

Vocabulary

solvent p. 343
solute p. 343
polar molecule p. 344
concentration p. 346
solubility p. 348
saturated solution p. 348
unsaturated solution p. 348

Lesson 3: Acid and Base Solutions

- **Acids** contain hydrogen ions that are released and form **hydronium ions** in water. **Bases** are substances that form hydroxide ions when dissolved in water.
- Hydronium ion concentration changes inversely with **pH.** This means that as hydronium ion concentration increases, the pH decreases.
- pH can be measured using **indicators** or digital pH meters.

acidic basic
0 1 2 3 4 5 6 7 8 9 10 11 12 13 14

Vocabulary

acid p. 354
hydronium ion p. 354
base p. 354
pH p. 356
indicator p. 358

Study Guide

- Personal Tutor
- Vocabulary eGames
- Vocabulary eFlashcards

FOLDABLES® Chapter Project

Assemble your lesson Foldables as shown to make a Chapter Project. Use the project to review what you have learned in this chapter.

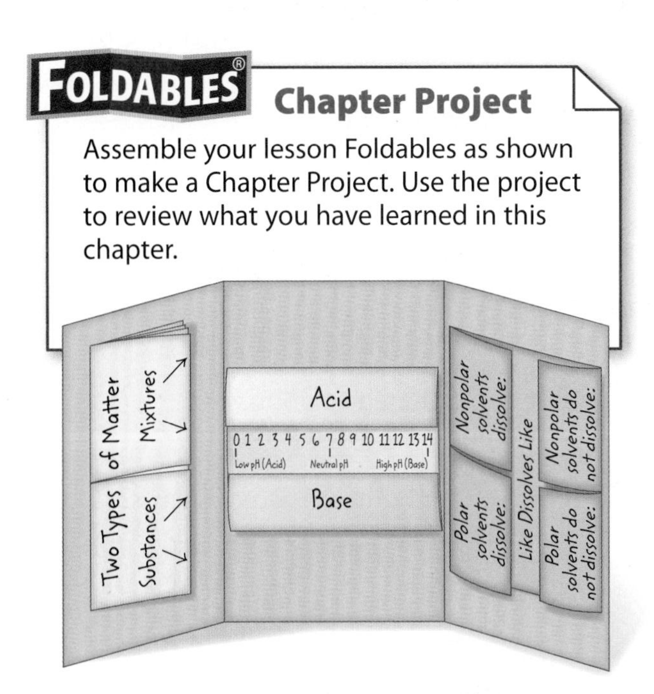

Use Vocabulary

1. The parts of a __________ can be seen with unaided eyes or with a microscope.
2. It is impossible to tell the difference between a solution and a __________ just by looking at them.
3. Water dissolves other __________ easily.
4. Two equal volumes of a solution that contain different amounts of the same solute have a different __________.
5. As __________ concentration decreases, pH increases.
6. A(n) __________ can be added to milk to neutralize it.

Link Vocabulary and Key Concepts

Concepts in Motion **Interactive Concept Map**

Use vocabulary terms from the previous page to complete the concept map.

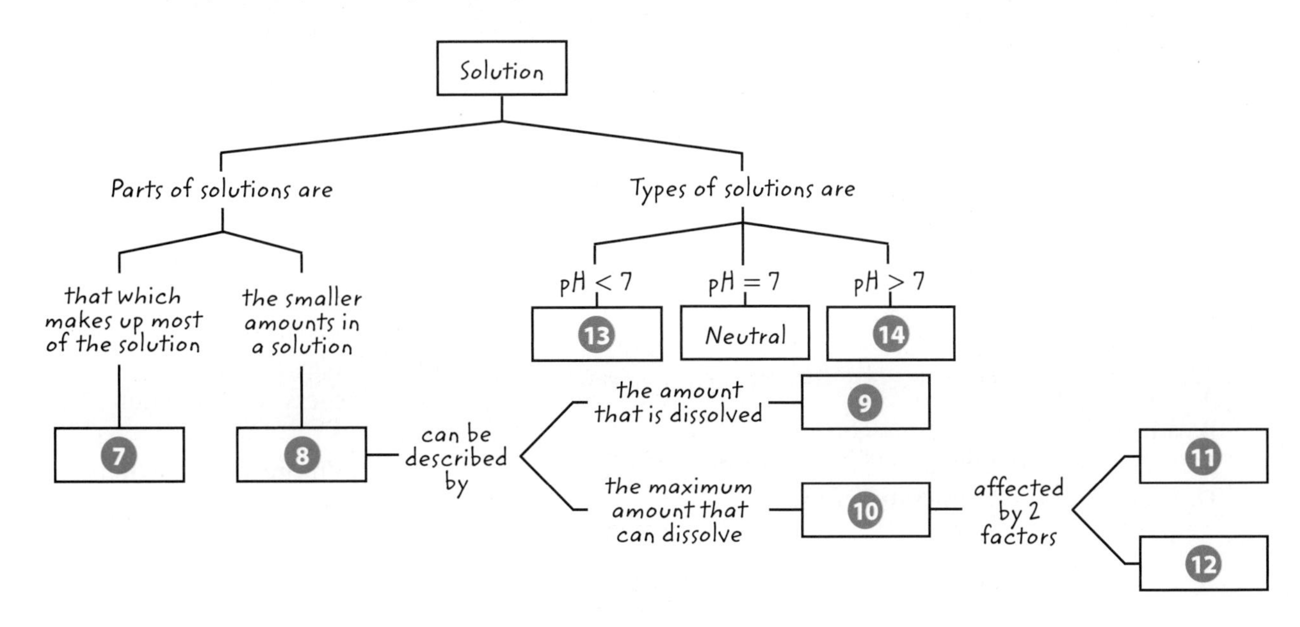

Chapter 10 Review

Understand Key Concepts 

1. Which is a solution?
   A. copper
   B. vinegar
   C. pure water
   D. a raisin cookie

2. The graph below shows the solubility of sodium chloride (NaCl) in water.

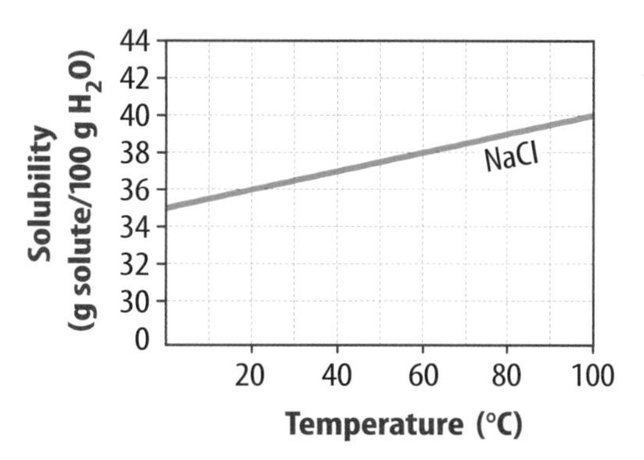

   What mass of sodium chloride must be added to 100 g of water at 80°C to form a saturated salt solution?
   A. 36 g
   B. 39 g
   C. 40 g
   D. 100 g

3. What would you add to a solution with a pH of 1.5 to obtain a solution with a pH of 7?
   A. milk (pH 6.4)
   B. vinegar (pH 3.0)
   C. lye (pH 13.0)
   D. coffee (pH 5.0)

4. Which can change the solubility of a solid in a liquid?
   A. crushing the solute
   B. stirring the solute
   C. increasing the pressure of the solution
   D. increasing the temperature of the solution

5. Which ions are present in the greatest amount in a solution with a pH of 8.5?
   A. hydrogen ions
   B. hydronium ions
   C. hydroxide ions
   D. oxygen ions

6. Which best describes a solution that contains the maximum dissolved solute?
   A. It is a concentrated solution.
   B. It is a dilute solution.
   C. It is a saturated solution.
   D. It is an unsaturated solution.

7. Which is a mixture of two elements?

A. 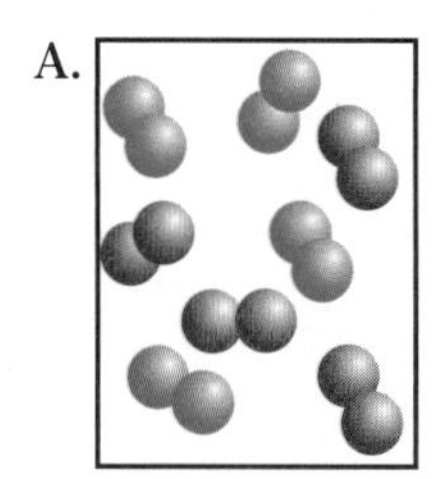

C. 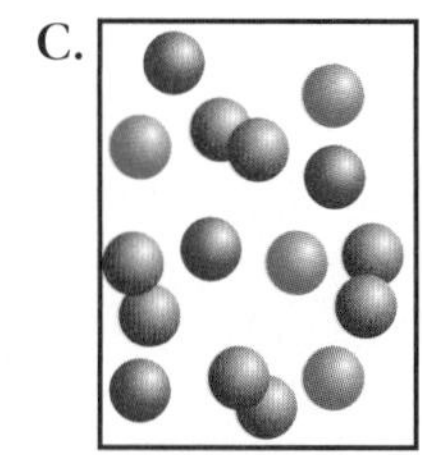

B. 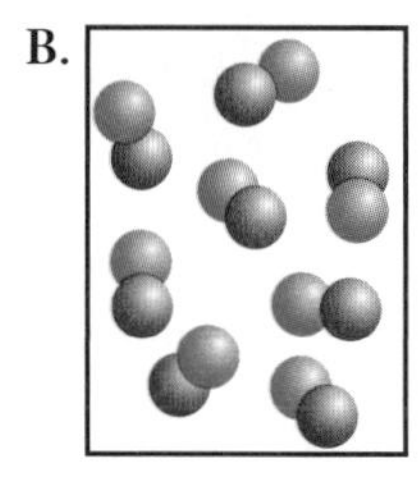

D. 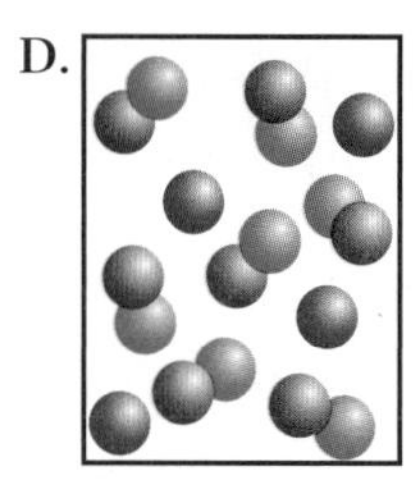

8. Which explains why a soft drink bubbles when the cap is released?
   A. The gas becomes less soluble when temperature decreases.
   B. The gas becomes more soluble when temperature decreases.
   C. The gas becomes less soluble when pressure decreases.
   D. The gas becomes more soluble when pressure decreases.

Assessment

Online Test Practice

Critical Thinking

9. **Infer** How can you tell which component in a solution is the solvent?

10. **Predict** The graph below shows the solubility of potassium chloride (KCl) in water.

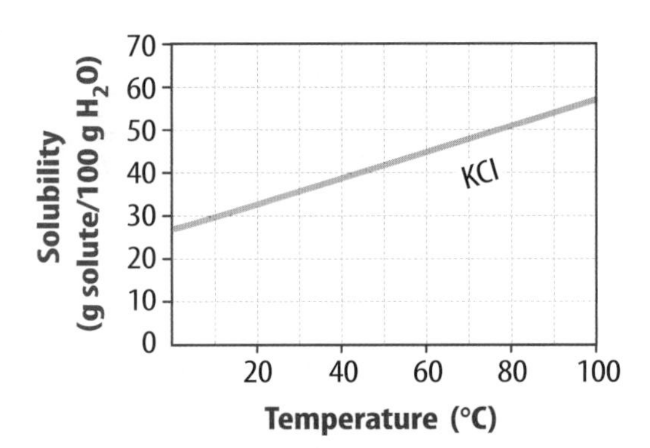

Imagine you have made a solution that contains 50 g of potassium chloride (KCl) in 100 g of solution. Predict what you would observe as you gradually increased the temperature from 0°C to 100°C.

11. **Organize** The pH of three solutions is shown below.

    Milk (pH 6.7)
    Coffee (pH 5)
    Ammonia (pH 11.6)

    Place these solutions in order of
    a. most acidic to least acidic
    b. most basic to least basic
    c. highest OH^- concentration to lowest OH^- concentration

12. **Explain** The pH of a solution is inversely related to the concentration of hydronium ions in solution. Explain what this means.

13. **Design** a method to determine the solubility of an unknown substance at 50°C.

Writing in Science

14. **Compose** A haiku is a poem containing three lines of five, seven, and five syllables, respectively. Write a haiku describing what happens when an acid is dissolved in water.

REVIEW THE BIG IDEA

15. What are solutions? List at least three ways a solution can be described.

16. How do solutions differ from other types of matter?

Math Skills

Review

Math Practice

Calculate Concentration

17. Calculate the concentration of sugar in g/L in a solution that contains 40 g of sugar in 100 mL of solution. There are 1,000 mL in 1 L.

18. There are many ways to make a solution of a given concentration. What are two ways you could make a sugar solution with a concentration of 100 g/L?

19. A salt solution has a concentration of 200 g/L. How many grams of salt are contained in 500 mL of this solution? How many grams of salt would be contained in 2 L of this solution?

Standardized Test Practice

Record your answers on the answer sheet provided by your teacher or on a sheet of paper.

Multiple Choice

Use the figures below to answer question 1.

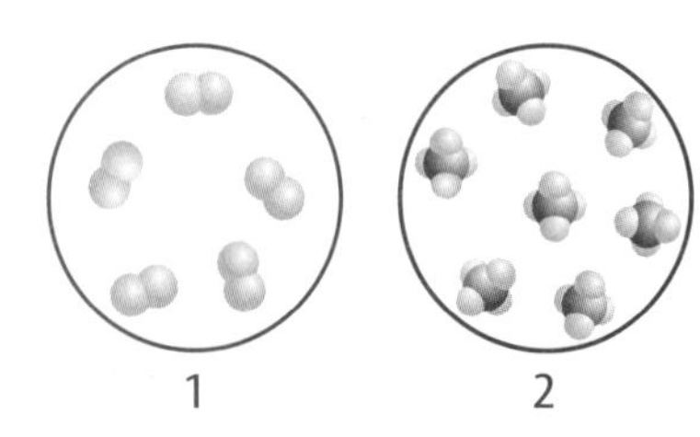

1 Which statement describes the two figures?

A Both 1 and 2 are mixtures.

B Both 1 and 2 are substances.

C 1 is a mixture and 2 is a substance.

D 1 is a substance and 2 is a mixture.

2 Which statement is an accurate comparison of solutions and homogeneous mixtures.

A They are the same.

B They are opposites.

C Solutions are more evenly mixed than homogeneous mixtures.

D Homogeneous mixtures are more evenly mixed than solutions.

3 A worker uses a magnet to remove bits of iron from a powdered sample. Which describes the sample before the worker used the magnet to remove the iron?

A The sample is a compound because the iron was removed using a physical method.

B The sample is a compound because the iron was removed using a chemical change.

C The sample is a mixture because the iron was removed using a chemical change.

D The sample is a mixture because the iron was removed using a physical method.

4 A beaker contains a mixture of sand and small pebbles. What kind of mixture is this?

A compound

B heterogeneous

C homogeneous

D solution

5 Which type of substance would best dissolve in a solvent that was made of nonpolar molecules?

A a water-based solvent

B an ionic compound

C a solute made of polar molecules

D a solute made of nonpolar molecules

Use the figure to answer question 6.

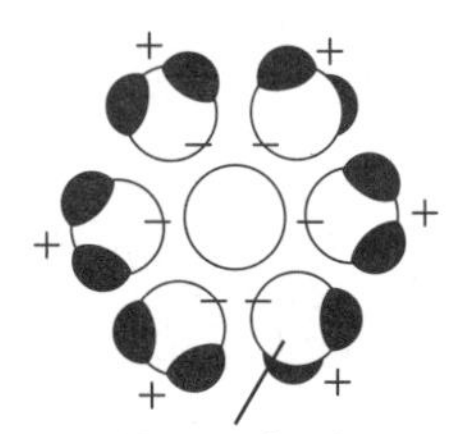

6 The figure shows how water molecules surround an ion in a solution. What can you conclude about the ions?

A It is negative because the negative ends of the water molecule are attracted to it.

B It is negative because the positive ends of the water molecule are attracted to it.

C It is positive because the negative ends of the water molecule are attracted to it.

D It is positive because the positive ends of the water molecule are attracted to it.

Assessment

Online Standardized Test Practice

7 A girl makes two glasses of lemonade using a powder mix. She pours one cup of water into each glass. She adds one spoonful of powder to the first glass and two spoonfuls of powder to the second glass. How do the solutions in the two glasses compare?

A The first glass has a greater concentration of powder mix.

B The first glass has a greater solubility.

C The second glass has a greater concentration of powder mix.

D The second glass has a greater solubility.

Use the table below to answer question 8.

Sample solution	Change in blue litmus	Change in red litmus
1	turns red	no change
2	no change	turns blue
3	turns red	no change
4	no change	no change

8 A scientist collects the data above using litmus paper. Blue litmus paper is a type of pH indicator that turns red when placed in an acidic solution. Red litmus paper is an indicator that turns blue when placed in an basic solution. Neutral solutions cause no change in either color of litmus paper. Which sample solution must be a base?

A solution 1

B solution 2

C solution 3

D solution 4

Constructed Response

9 Explain how the concentration of hydronium ions and the concentration of hydroxide ions change when a base is dissolved in water.

10 A researcher mixes a solution that is 40 percent helium gas and 60 percent nitrogen gas. Which gas is the solute and which is the solvent? What would the mixture look like through a microscope? What would the mixture look like at the atomic level?

11 A student is dissolving rock salt in water. Describe three ways to increase the rate of dissolving.

Use the figure to answer questions 12 and 13.

12 The figure shows what happens when hydrogen iodide (HI) dissolves in water. Is hydrogen iodide an acid, a base, or a neutral substance? Explain.

13 What can you conclude about the pH of the aqueous solution of hydrogen iodide?

NEED EXTRA HELP?													
If You Missed Question...	1	2	3	4	5	6	7	8	9	10	11	12	13
Go to Lesson...	1	1	1	1	2	2	2	2	3	3	1	2	3

Unit 3

UNDERSTANDING THE UNIVERSE

"IF YOU LOOK TO YOUR RIGHT, YOU'LL SEE OUR CLOSEST NEIGHBOR, THE ANDROMEDA GALAXY...

"ONLY 2.5 MILLION LIGHT-YEARS AWAY."

"NEXT, WE HAVE A DYING STAR THAT HAS EXPANDED INTO A RED GIANT."

2000 B.C. 1600 1700 1800

1600 B.C. Babylonian texts show records of people observing Venus without the aid of technology. Its appearance is recorded for 21 years.

265 B.C. Greek astronomer Timocharis makes the first recorded observation of Mercury.

1610 Galileo Galilei observes the four largest moons of Jupiter through his telescope.

1613 Galileo records observations of the planet Neptune but mistakes it for a star.

1655 Astronomer Christiaan Huygens observes Saturn and discovers its rings, which were previously thought to be large moons on each side.

1781 William Herschel discovers the planet Uranus.

1930
Clyde Tombaugh discovers Pluto, making him the first American to discover a planet.

1971
Mariner 9 visits Mars and becomes the first human-made object to orbit a planet other than Earth.

2006
After research and consideration, the International Astronomical Union votes to remove Pluto from the list of planets in the solar system.

Inquiry
Visit ConnectED for this unit's **STEM** activity.

Patterns

You might sometimes see the Moon as a large, glowing disk in the night sky. At other times, the Moon appears as different shapes. These shapes are the Moon's phases, or the changing portions of the Moon that are seen from Earth. The Moon's phases occur as a repeating pattern about every 28 days. A **pattern** is a consistent plan or model used as a guide for understanding and predicting things. You can predict the next phase of the Moon or you can determine the previous phase of the Moon if you know the pattern.

Patterns in Earth Science

Patterns help scientists understand observations. This allows them to predict future events or understand past events. For instance, geologists are Earth scientists who measure the chemical composition, age, and location of rocks. They look for patterns in these measurements. The patterns allow geologists to propose what processes formed the rocks millions or billions of years ago. Geologists also use patterns to draw conclusions about how Earth has changed over time and to estimate how it will change in the future.

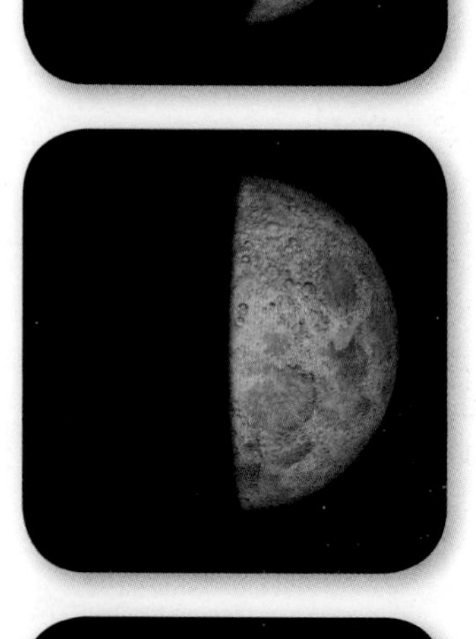

Meteorologists are scientists who study weather and climate. They study patterns of fronts, winds, cloud formation, precipitation, and ocean temperatures to make weather forecasts. For example, meteorologists track patterns in hurricanes, such as wind speed, movements, and rotation velocity. These patterns help meteorologists understand the conditions under which a hurricane can form. Predicting the strength and the path of a storm can help save lives, buildings, and property. When meteorologists see weather patterns similar to those of past hurricanes, they can predict the severity of the storm and when and where it will hit. Then meteorologists can send advance warnings for people to safely prepare.

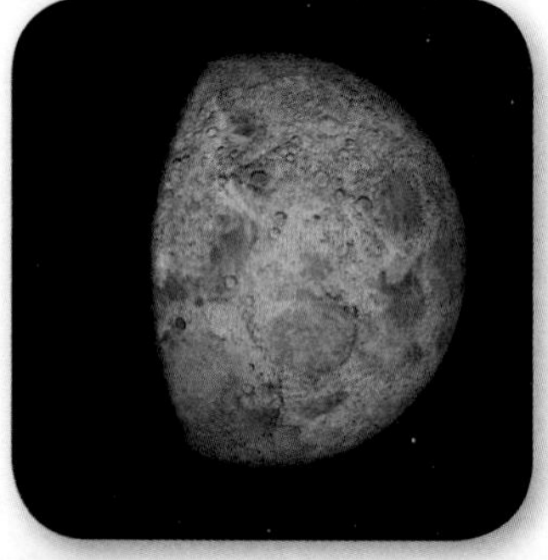

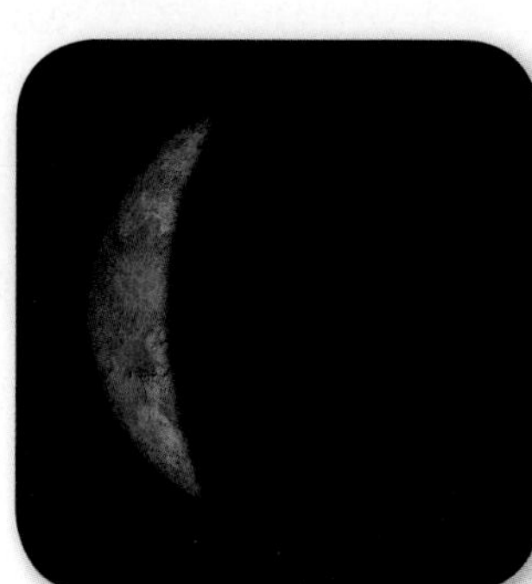

Types of Patterns

Physical Patterns

A pattern you observe using your eyes or other instruments is a physical pattern. Earthquake uplift and erosion reveal physical patterns in layers of rock, as shown in the photo. Patterns in exposed rock layers tell geologists many things, including the order in which the rocks formed, the different minerals and fossils the rocks contain, and the age and movement of landforms.

Patterns in Graphs

Scientists plot data on graphs and then analyze the graphs for patterns. Patterns on graphs can appear as straight lines, curved lines, or waves. The graph to the right shows a pattern in sea level as it increased between 1994 and 2008. Scientists analyze graphic patterns to predict events in the future. For example, a scientist might predict that in 2014, the sea level will be 30 mm higher than it was in 2008.

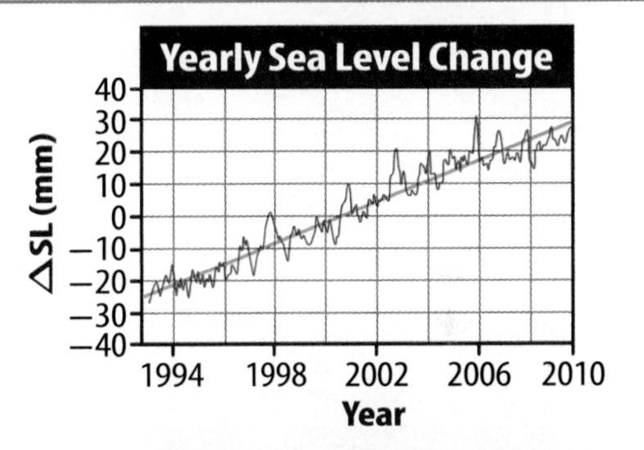

Cyclic Patterns

An event that repeats many times in a predictable order, such as the phases of the Moon, has a cyclic pattern. As shown in the graph, water temperatures in both the North and South Atlantic Ocean rise and fall equally each year. The annual changes in the water temperature follow a cyclic pattern. How do the temperature patterns in the two oceans differ?

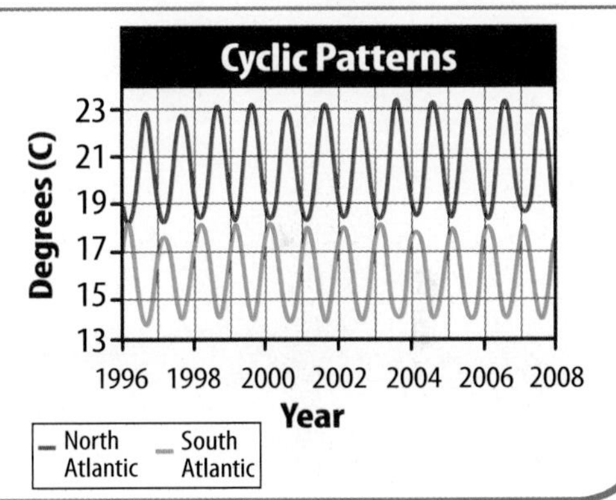

Inquiry MiniLab

15 minutes

What patterns are in your year?

What are some of the cyclic patterns in your life?

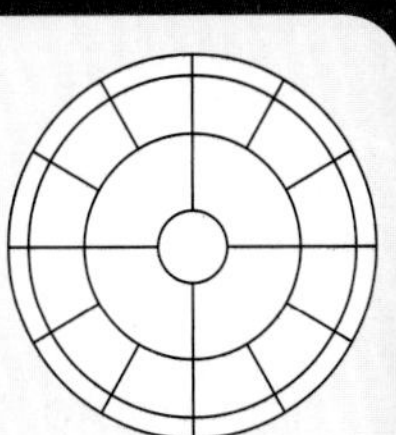

1. On a sheet of **notebook paper,** draw four concentric circles with diameters of 20 cm, 18 cm, 10 cm, and 4 cm. Write your name in the innermost circle.
2. Divide the two outermost circles into 12 equal sections. Write each month of the year (one month per section) in each section of the outermost circle. In the next circle, write personal events or activities that take place in each corresponding month.
3. Divide the 10-cm circle into four sections. Write the weather conditions and the plant patterns that correspond to the months in the outermost circle.

Analyze and Conclude

Observe If you start at one month and move inward through the rings, what patterns do you observe? How do these observations fit into the yearly cycle?

Chapter 11

The Solar System

THE BIG IDEA

What kinds of objects are in the solar system?

One, Two, or Three Planets?

This photo, taken by the Cassini spacecraft, shows part of Saturn's rings and two of its moons. Saturn is a planet that orbits the Sun. The moons, tiny Epimetheus and much larger Titan, orbit Saturn. Besides planets and moons, many other objects are in the solar system.

- How would you describe a planet such as Saturn?
- How do astronomers classify the objects they discover?
- What types of objects do you think make up the solar system?

Get Ready to Read

What do you think?

Before you read, decide if you agree or disagree with each of these statements. As you read this chapter, see if you change your mind about any of the statements.

1. Astronomers measure distances between space objects using astronomical units.
2. Gravitational force keeps planets in orbit around the Sun.
3. Earth is the only inner planet that has a moon.
4. Venus is the hottest planet in the solar system.
5. The outer planets also are called the gas giants.
6. The atmospheres of Saturn and Jupiter are mainly water vapor.
7. Asteroids and comets are mainly rock and ice.
8. A meteoroid is a meteor that strikes Earth.

Your one-stop online resource

connectED.mcgraw-hill.com

Lesson 1

Reading Guide

Key Concepts
ESSENTIAL QUESTIONS

- How are the inner planets different from the outer planets?
- What is an astronomical unit and why is it used?
- What is the shape of a planet's orbit?

Vocabulary

asteroid p. 377
comet p. 377
astronomical unit p. 378
period of revolution p. 378
period of rotation p. 378

Video

- BrainPOP®
- Science Video

The Structure of the Solar System

Inquiry Are these stars?

Did you know that shooting stars are not actually stars? The bright streaks are small, rocky particles burning up as they enter Earth's atmosphere. These particles are part of the solar system and are often associated with comets.

Launch Lab

10 minutes

How do you know which distance unit to use?

You can use different units to measure distance. For example, millimeters might be used to measure the length of a bolt, and kilometers might be used to measure the distance between cities. In this lab, you will investigate why some units are easier to use than others for certain measurements.

1. Read and complete a lab safety form.
2. Use a **centimeter ruler** to measure the length of a **pencil** and the thickness of this **book.** Record the distances in your Science Journal.
3. Use the centimeter ruler to measure the width of your classroom. Then measure the width of the room using a **meterstick.** Record the distances in your Science Journal.

Think About This

1. Why are meters easier to use than centimeters for measuring the classroom?
2.  **Key Concept** Why do you think astronomers might need a unit larger than a kilometer to measure distances in the solar system?

What is the solar system?

Have you ever made a wish on a star? If so, you might have wished on a planet instead of a star. Sometimes, as shown in **Figure 1,** the first starlike object you see at night is not a star at all. It's Venus, the planet closest to Earth.

It's hard to tell the difference between planets and stars in the night sky because they all appear as tiny lights. Thousands of years ago, observers noticed that a few of these tiny lights moved, but others did not. The ancient Greeks called these objects planets, which means "wanderers." Astronomers now know that the planets do not wander about the sky; the planets move around the Sun. The Sun and the group of objects that move around it make up the solar system.

When you look at the night sky, a few of the tiny lights that you can see are part of our solar system. Almost all of the other specks of light are stars. They are much farther away than any objects in our solar system. Astronomers have discovered that some of those stars also have planets moving around them.

Reading Check What object do the planets in the solar system move around?

Figure 1 When looking at the night sky, you will likely see stars and planets. In the photo below, the planet Venus is the bright object seen above the Moon.

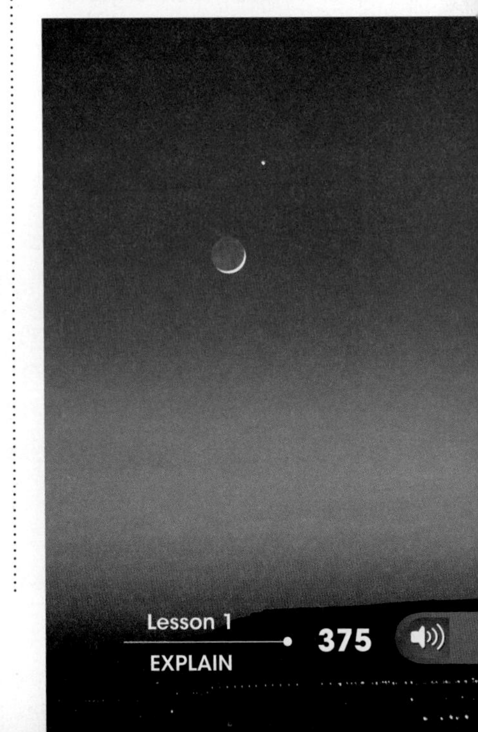

Objects in the Solar System

Ancient observers looking at the night sky saw many stars but only five planets—Mercury, Venus, Mars, Jupiter, and Saturn. The invention of the telescope in the 1600s led to the discovery of additional planets and many other space objects.

The Sun

The largest object in the solar system is the Sun, a **star.** Its diameter is about 1.4 million km—ten times the diameter of the largest planet, Jupiter. The Sun is made mostly of hydrogen gas. Its mass makes up about 99 percent of the entire solar system's mass.

SCIENCE USE V. COMMON USE

star

Science Use an object in space made of gases in which nuclear fusion reactions occur that emit energy

Common Use a shape that usually has five or six points around a common center

Inside the Sun, a process called nuclear fusion produces an enormous amount of energy. The Sun emits some of this energy as light. The light from the Sun shines on all of the planets every day. The Sun also applies gravitational forces to objects in the solar system. Gravitational forces cause the planets and other objects to move around, or **orbit,** the Sun.

REVIEW VOCABULARY

orbit

(noun) the path an object follows as it moves around another object

(verb) to move around another object

Objects That Orbit the Sun

Different types of objects orbit the Sun. These objects include planets, dwarf planets, asteroids, and comets. Unlike the Sun, these objects don't emit light but only reflect the Sun's light.

Planets Astronomers classify some objects that orbit the Sun as planets, as shown in **Figure 2.** An object is a planet only if it orbits the Sun and has a nearly spherical shape. Also, the mass of a planet must be much larger than the total mass of all other objects whose orbits are close by. The solar system has eight objects classified as planets.

Reading Check What is a planet?

Figure 2 The orbits of the inner and outer planets are shown to scale. The Sun and the planets are not to scale. The outer planets are much larger than the inner planets.

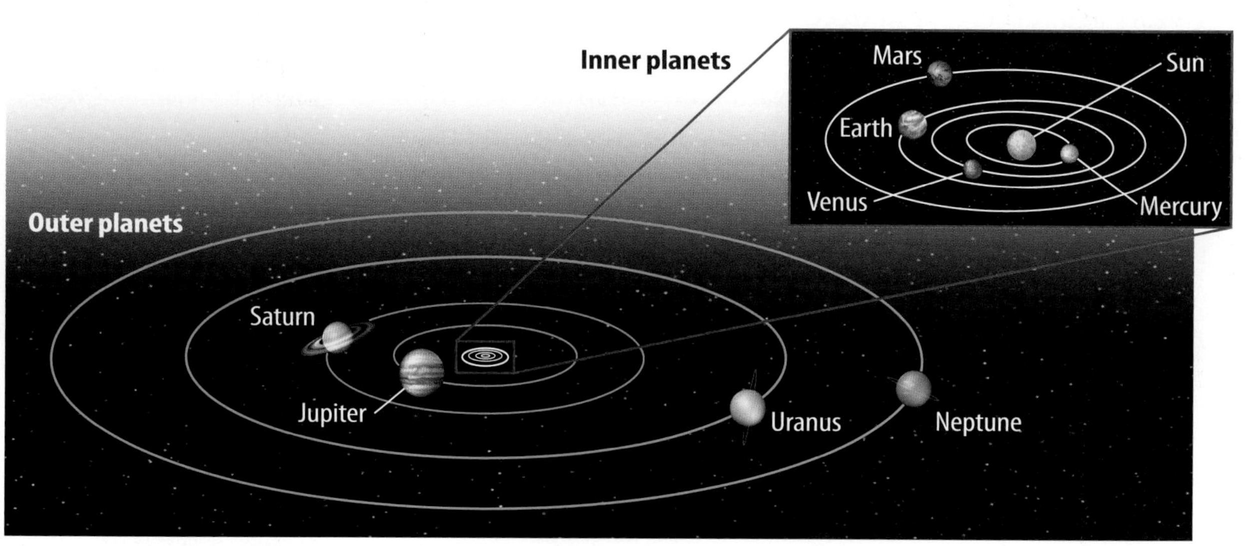

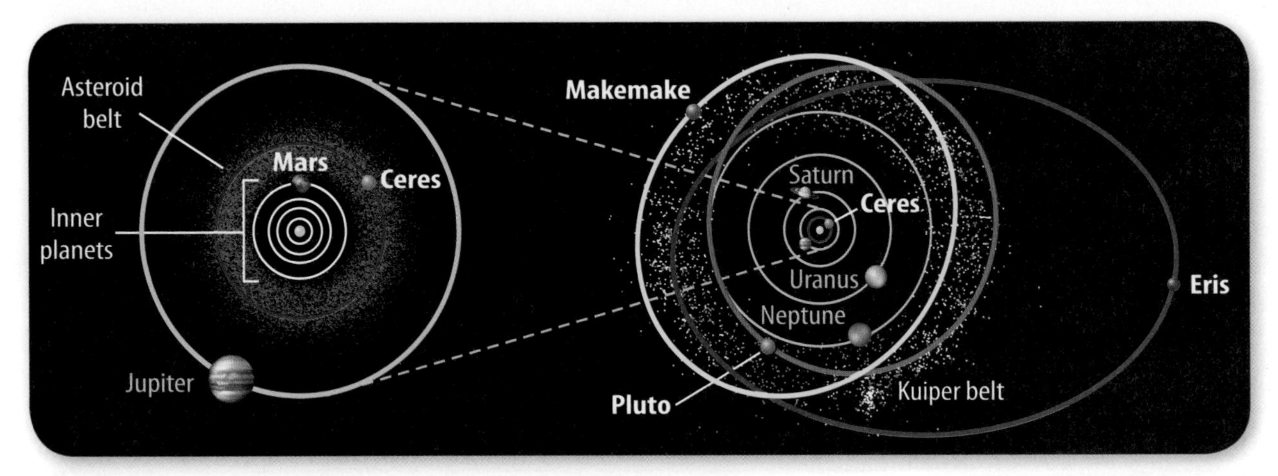

▲ **Figure 3** Ceres, a dwarf planet, orbits the Sun as planets do. The orbit of Ceres is in the asteroid belt between Mars and Jupiter.

Visual Check Which dwarf planet is farthest from the Sun?

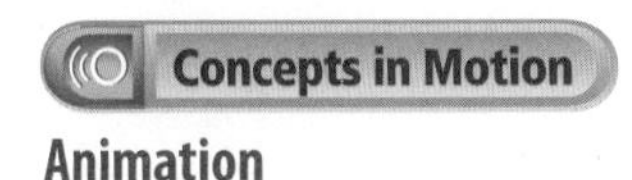

Animation

Inner Planets and Outer Planets As shown in **Figure 2,** the four planets closest to the Sun are the inner planets. The inner planets are Mercury, Venus, Earth, and Mars. These planets are made mainly of solid rocky materials. The four planets farthest from the Sun are the outer planets. The outer planets are Jupiter, Saturn, Uranus (YOOR uh nus), and Neptune. These planets are made mainly of ice and gases such as hydrogen and helium. The outer planets are much larger than Earth and are sometimes called gas giants.

Key Concept Check Describe how the inner planets differ from the outer planets.

Dwarf Planets Scientists classify some objects in the solar system as dwarf planets. A dwarf planet is a spherical object that orbits the Sun. It is not a moon of another planet and is in a region of the solar system where there are many objects orbiting near it. But, unlike a planet, a dwarf planet does not have more mass than objects in nearby orbits. **Figure 3** shows the locations of the dwarf planets Ceres (SIHR eez), Eris (IHR is), Pluto, and Makemake (MAH kay MAH kay). Dwarf planets are made of rock and ice and are much smaller than Earth.

WORD ORIGIN

asteroid
from Greek *asteroeides,* means "resembling a star"

Asteroids *Millions of small, rocky objects called* **asteroids** *orbit the Sun in the asteroid belt between the orbits of Mars and Jupiter.* The asteroid belt is shown in **Figure 3.** Asteroids range in size from less than a meter to several hundred kilometers in length. Unlike planets and dwarf planets, asteroids, such as the one shown in **Figure 4,** usually are not spherical.

Figure 4 The asteroid Gaspra orbits the Sun in the asteroid belt. Its odd shape is about 19 km long and 11 km wide. ▼

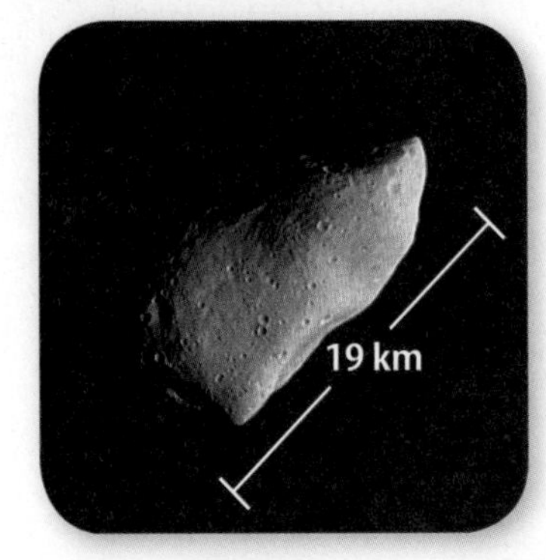

Comets You might have seen a picture of a comet with a long, glowing tail. *A* **comet** *is made of gas, dust, and ice and moves around the Sun in an oval-shaped orbit.* Comets come from the outer parts of the solar system. There might be 1 trillion comets orbiting the Sun. You will read more about comets, asteroids, and dwarf planets in Lesson 4.

The Astronomical Unit

On Earth, distances are often measured in meters (m) or kilometers (km). Objects in the solar system, however, are so far apart that astronomers use a larger distance unit. *An* **astronomical unit** *(AU) is the average distance from Earth to the Sun—about 150 million km.* **Table 1** lists each planet's average distance from the Sun in km and AU.

 **Key Concept Check** Define what an astronomical unit is and explain why it is used.

Table 1 Because the distances of the planets from the Sun are so large, it is easier to express these distances using astronomical units rather than kilometers.

Interactive Table

Table 1 Average Distance of the Planets from the Sun

Planet	Average Distance (km)	Average Distance (AU)
Mercury	57,910,000	0.39
Venus	108,210,000	0.72
Earth	149,600,000	1.00
Mars	227,920,000	1.52
Jupiter	778,570,000	5.20
Saturn	1,433,530,000	9.58
Uranus	2,872,460,000	19.20
Neptune	4,495,060,000	30.05

The Motion of the Planets

Have you ever swung a ball on the end of a string in a circle over your head? In some ways, the motion of a planet around the Sun is like the motion of that ball. As shown in **Figure 5** on the next page, the Sun's gravitational force pulls each planet toward the Sun. This force is similar to the pull of the string that keeps the ball moving in a circle. The Sun's gravitational force pulls on each planet and keeps it moving along a curved path around the Sun.

 **Reading Check** What causes planets to orbit the Sun?

Revolution and Rotation

Objects in the solar system move in two ways. They orbit, or revolve, around the Sun. *The time it takes an object to travel once around the Sun is its* **period of revolution.** Earth's period of revolution is one year. The objects also spin, or rotate, as they orbit the Sun. *The time it takes an object to complete one rotation is its* **period of rotation.** Earth has a period of rotation of one day.

FOLDABLES®

Make a tri-fold book from a sheet of paper and label it as shown. Use it to summarize information about the types of objects that make up the solar system.

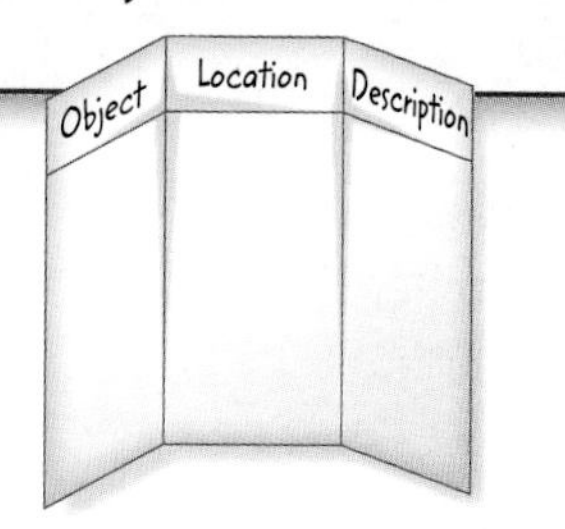

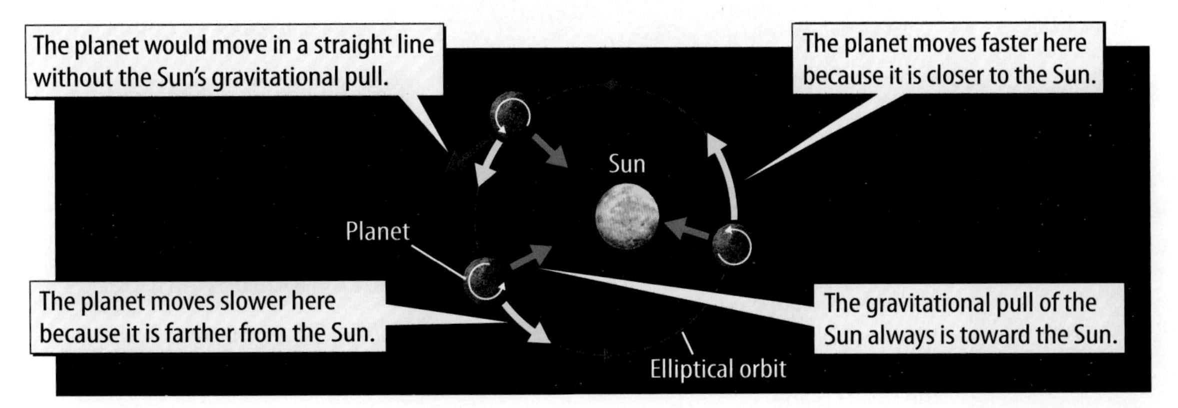

Figure 5 Planets and other objects in the solar system revolve around the Sun because of its gravitational pull on them.

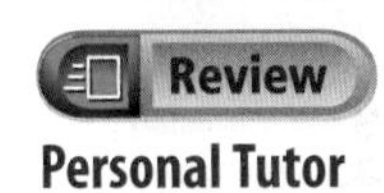

Personal Tutor

Planetary Orbits and Speeds

Unlike a ball swinging on the end of a string, planets do not move in circles. Instead, a planet's orbit is an ellipse—a stretched-out circle. Inside an ellipse are two special points, each called a focus. These focus points, or foci, determine the shape of the ellipse. The foci are equal distances from the center of the ellipse. As shown in **Figure 5,** the Sun is at one of the foci; the other foci is empty space. As a result, the distance between the planet and the Sun changes as the planet moves.

A planet's speed also changes as it orbits the Sun. The closer the planet is to the Sun, the faster it moves. This also means that planets farther from the Sun have longer periods of revolution. For example, Jupiter is more than five times farther from the Sun than Earth. Not surprisingly, Jupiter takes 12 times longer than Earth to revolve around the Sun.

Key Concept Check Describe the shape of a planet's orbit.

Inquiry MiniLab

20 minutes

How can you model an elliptical orbit?

In this lab you will explore how the locations of foci affect the shape of an ellipse.

1. Read and complete a lab safety form.
2. Place a sheet of **paper** on a **corkboard.** Insert two **push pins** 8 cm apart in the center of the paper.
3. Use **scissors** to cut a 24-cm piece of **string.** Tie the ends of the string together.
4. Place the loop of string around the pins. Use a pencil to draw an ellipse as shown.
5. Measure the maximum width and length of the ellipse. Record the data in your Science Journal.
6. Move one of the push pins so that the pins are 5 cm apart. Repeat steps 4 and 5.

Analysis

1. **Compare and contrast** the two ellipses.
2. **Key Concept** How are the shapes of the ellipses you drew similar to the orbits of the inner and outer planets?

Lesson 1 Review

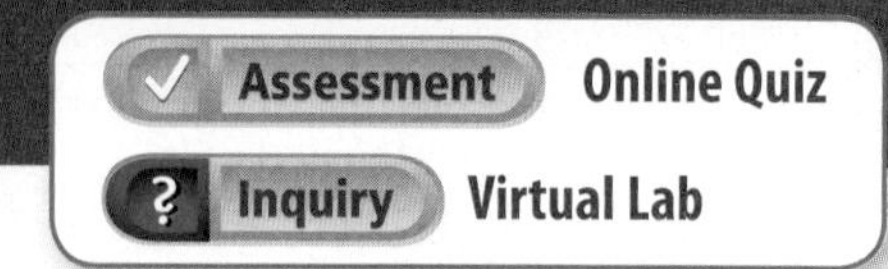

Visual Summary

The solar system contains the Sun, the inner planets, the outer planets, the dwarf planets, asteroids, and comets.

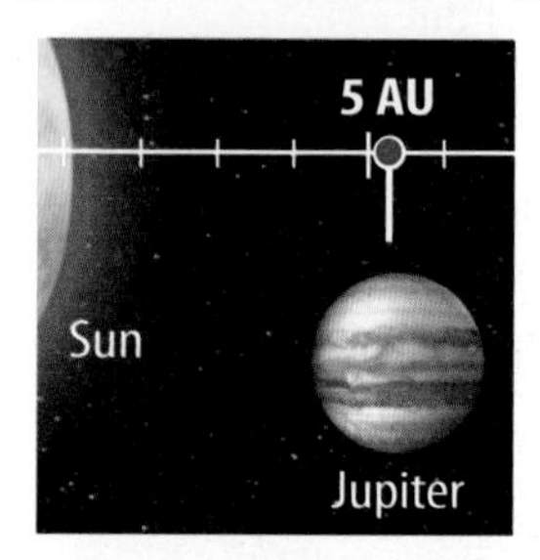

An astronomical unit (AU) is a unit of distance equal to about 150 million km.

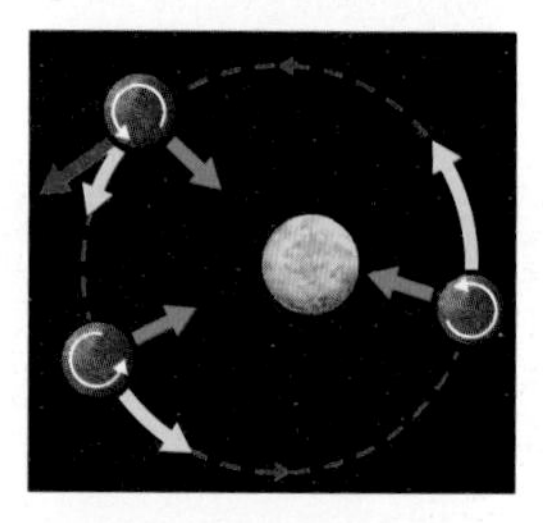

The speeds of the planets change as they move around the Sun in elliptical orbits.

FOLDABLES®

Use your lesson Foldable to review the lesson. Save your Foldable for the project at the end of the chapter.

What do you think NOW?

You first read the statements below at the beginning of the chapter.

1. Astronomers measure distances between space objects using astronomical units.
2. Gravitational force keeps planets in orbit around the Sun.

Did you change your mind about whether you agree or disagree with the statements? Rewrite any false statements to make them true.

Use Vocabulary

1. **Compare and contrast** a period of revolution and a period of rotation.
2. **Define** *dwarf planet* in your own words.
3. **Distinguish** between an asteroid and a comet.

Understand Key Concepts 

4. **Summarize** how and why planets orbit the Sun and how and why a planet's speed changes in orbit.
5. **Infer** why an astronomical unit is not used to measure distances on Earth.
6. Which distinguishes a dwarf planet from a planet?
   - **A.** mass
   - **B.** the object it revolves around
   - **C.** shape
   - **D.** type of orbit

Interpret Graphics

7. **Explain** what each arrow in the diagram represents.

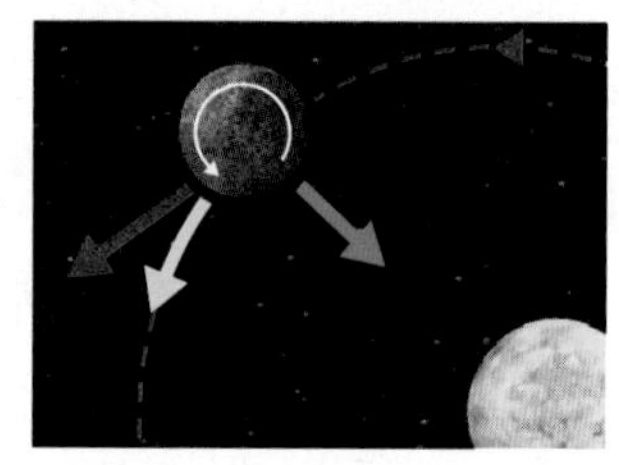

8. **Take Notes** Copy the table below. List information about each object or group of objects in the solar system mentioned in the lesson. Add additional lines as needed.

Object	Description
Sun	
Planets	

Critical Thinking

9. **Evaluate** How would the speed of a planet be different if its orbit were a circle instead of an ellipse?

Meteors are pieces of a comet or an asteroid that heat up as they fall through Earth's atmosphere. Meteors that strike Earth are called meteorites. ▶

History from Space

CAREERS in SCIENCE

Meteorites give a peek back in time.

About 4.6 billion years ago, Earth and the other planets did not exist. In fact, there was no solar system. Instead, a large disk of gas and dust, known as the solar nebula, swirled around a forming Sun, as shown in the top picture to the right. How did the planets and other objects in the solar system form?

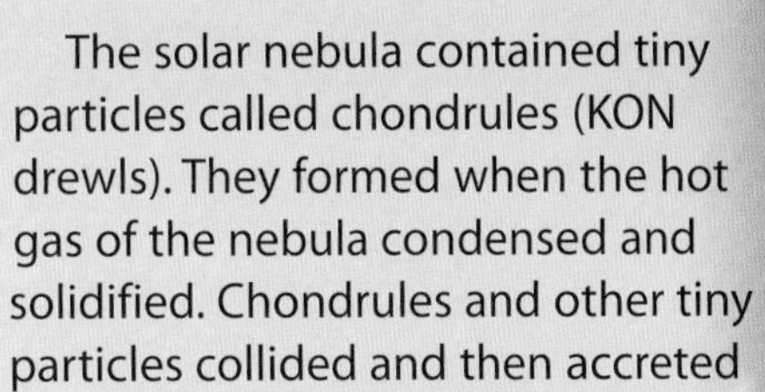

▲ **Denton Ebel holds a meteorite that broke off the Vesta asteroid.**

Denton Ebel is looking for the answer. He is a geologist at the American Museum of Natural History in New York City. Ebel explores the hypothesis that over millions of years, tiny particles in the solar nebula clumped together and formed the asteroids, comets, and planets that make up our solar system.

The solar nebula contained tiny particles called chondrules (KON drewls). They formed when the hot gas of the nebula condensed and solidified. Chondrules and other tiny particles collided and then accreted (uh KREET ed) or clumped together. This process eventually formed asteroids, comets, and planets. Some of the asteroids and comets have not changed much in over 4 billion years. Chondrite meteorites are pieces of asteroids that fell to Earth. The chondrules within the meteorites are the oldest solid material in our solar system.

For Ebel, chondrite meteorites contain information about the formation of the solar system. Did the materials in the meteorite form throughout the solar system and then accrete? Or did asteroids and comets form and accrete near the Sun, drift outward to where they are today, and then grow larger by accreting ice and dust? Ebel's research is helping to solve the mystery of how our solar system formed.

Accretion Hypothesis

According to the accretion hypothesis, the solar system formed in stages.

First there was a solar nebula. The Sun formed when gravity caused the nebula to collapse.

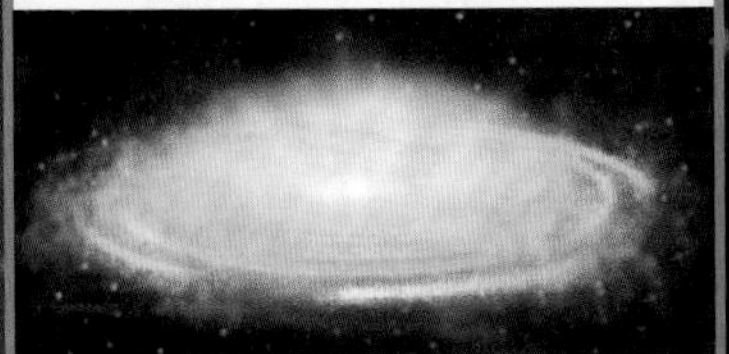

The rocky inner planets formed from accreted particles.

The gaseous outer planets formed as gas, ice, and dust condensed and accreted.

It's Your Turn

TIME LINE Work in groups. Learn more about the history of Earth from its formation until life began to appear. Create a time line showing major events. Present your time line to the class.

Lesson 2

Reading Guide

Key Concepts 

ESSENTIAL QUESTIONS

- How are the inner planets similar?
- Why is Venus hotter than Mercury?
- What kind of atmospheres do the inner planets have?

Vocabulary

terrestrial planet p. 383

greenhouse effect p. 385

 Multilingual eGlossary

Video

What's Science Got to do With It?

The Inner Planets

Where is this?

This spectacular landscape is the surface of Mars, one of the inner planets. Other inner planets have similar rocky surfaces. It might surprise you to learn that there are planets in the solar system that have no solid surface on which to stand.

Launch Lab

20 minutes

What affects the temperature on the inner planets?

Mercury and Venus are closer to the Sun than Earth. What determines the temperature on these planets? Let's find out.

1. Read and complete a lab safety form.
2. Insert a **thermometer** into a **clear 2-L plastic bottle.** Wrap **modeling clay** around the lid to hold the thermometer in the center of the bottle. Form an airtight seal with the clay.
3. Rest the bottle against the side of a **shoe box** in direct sunlight. Lay a second **thermometer** on top of the box next to the bottle so that the bulbs are at about the same height. The thermometer bulb should not touch the box. Secure the thermometer in place using **tape.**

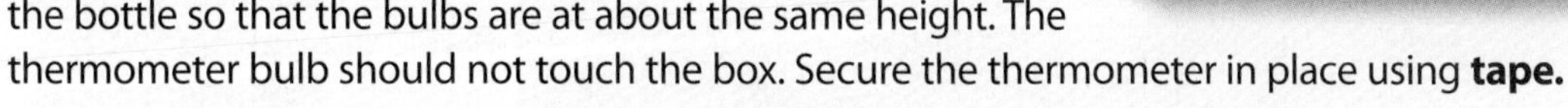

4. Read the thermometers and record the temperatures in your Science Journal.
5. Wait 15 minutes and then read and record the temperature on each thermometer.

Think About This

1. How did the temperature of the two thermometers compare?
2. **Key Concept** What do you think caused the difference in temperature?

Planets Made of Rock

Imagine that you are walking outside. How would you describe the ground? You might say it is dusty or grassy. If you live near a lake or an ocean, you might say the ground is sandy or wet. But beneath the ground or lake or ocean is a layer of solid rock.

The inner planets—Mercury, Venus, Earth, and Mars—are also called terrestrial planets. **Terrestrial planets** *are the planets closest to the Sun, are made of rock and metal, and have solid outer layers*. Like Earth, the other inner planets also are made of rock and metallic materials and have a solid outer layer. However, as shown in **Figure 6,** the inner planets have different sizes, atmospheres, and surfaces.

WORD ORIGIN

terrestrial
from Latin *terrestris,* means "earthly"

INNER PLANETS

Visual Check Which is the smallest inner planet?

Figure 6 The inner planets are roughly similar in size. Earth is about two and half times larger than Mercury. All inner planets have a solid outer layer.

Figure 7 The *Messenger* space probe flew by Mercury in 2008 and photographed the planet's cratered surface.

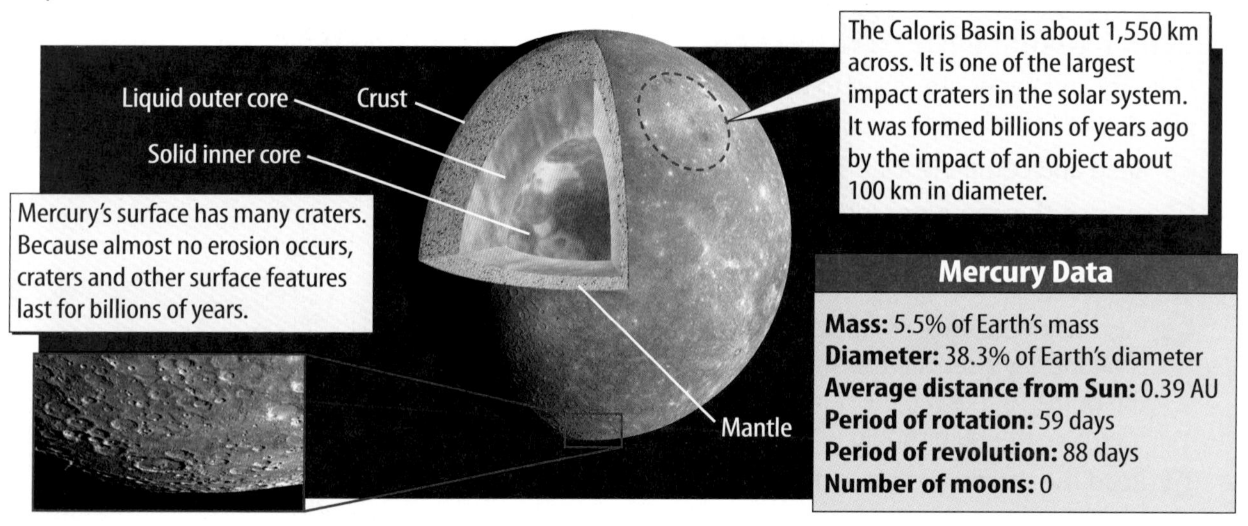

Mercury

The smallest planet and the planet closest to the Sun is Mercury, shown in **Figure 7.** Mercury has no atmosphere. A planet has an atmosphere when its gravity is strong enough to hold gases close to its surface. The strength of a planet's gravity depends on the planet's mass. Because Mercury's mass is so small, its gravity is not strong enough to hold onto an atmosphere. Without an atmosphere there is no wind that moves energy from place to place across the planet's surface. This results in temperatures as high as 450°C on the side of Mercury facing the Sun and as cold as −170°C on the side facing away from the Sun.

FOLDABLES®

Make a four-door book. Label each door with the name of an inner planet. Use the book to organize your notes on the inner planets.

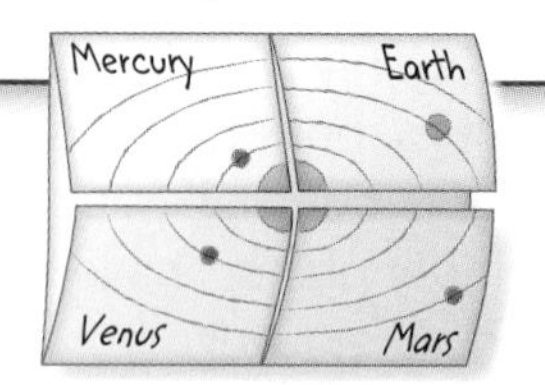

Mercury's Surface

Impact craters, depressions formed by collisions with objects from space, cover the surface of Mercury. There are smooth plains of solidified lava from long-ago eruptions. There are also high cliffs that might have formed when the planet cooled quickly, causing the surface to wrinkle and crack. Without an atmosphere, almost no erosion occurs on Mercury's surface. As a result, features that formed billions of years ago have changed very little.

Mercury's Structure

The structures of the inner planets are similar. Like all inner planets, Mercury has a core made of iron and nickel. Surrounding the core is a layer called the mantle. The mantle is mainly made of silicon and oxygen. The crust is a thin, rocky layer above the mantle. Mercury's large core might have been formed by a collision with a large object during Mercury's formation.

 **Key Concept Check** How are the inner planets similar?

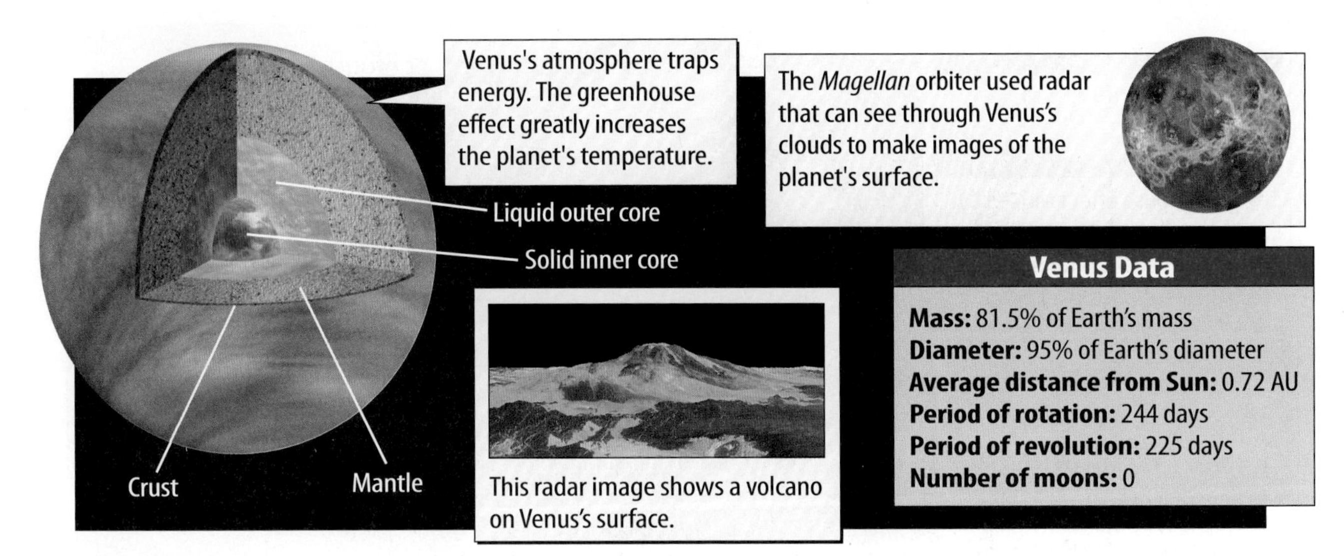

Figure 8 Because a thick layer of clouds covers Venus, its surface has not been seen. Between 1990 and 1994, the *Magellan* space probe mapped the surface using radar.

Venus

The second planet from the Sun is Venus, as shown in **Figure 8.** Venus is about the same size as Earth. It rotates so slowly that its period of rotation is longer than its period of revolution. This means that a day on Venus is longer than a year. Unlike most planets, Venus rotates from east to west. Several space probes have flown by or landed on Venus.

Venus's Atmosphere

The atmosphere of Venus is about 97 percent carbon dioxide. It is so dense that the atmospheric pressure on Venus is about 90 times greater than on Earth. Even though Venus has almost no water in its atmosphere or on its surface, a thick layer of clouds covers the planet. Unlike the clouds of water vapor on Earth, the clouds on Venus are made of acid.

The Greenhouse Effect on Venus

With an average temperature of about 460°C, Venus is the hottest planet in the solar system. The high temperatures are caused by the greenhouse effect. *The* **greenhouse effect** *occurs when a planet's atmosphere traps solar energy and causes the surface temperature to increase.* Carbon dioxide in Venus's atmosphere traps some of the solar energy that is absorbed and then emitted by the planet. This heats up the planet. Without the greenhouse effect, Venus would be almost 450°C cooler.

 **Key Concept Check** Why is Venus hotter than Mercury?

Venus's Structure and Surface

Venus's internal structure, as shown in **Figure 8,** is similar to Earth's. Radar images show that more than 80 percent of Venus's surface is covered by solidified lava. Much of this lava might have been produced by volcanic eruptions that occurred about half a billion years ago.

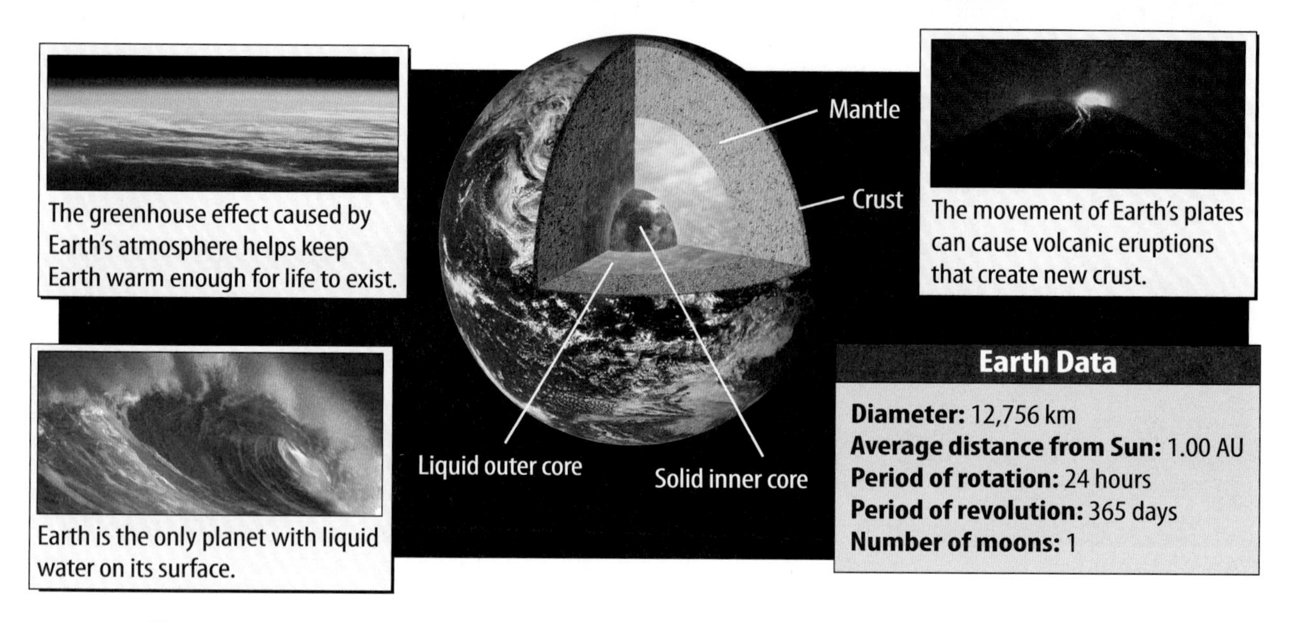

Figure 9 Earth has more water in its atmosphere and on its surface than the other inner planets. Earth's surface is younger than the surfaces of the other inner planets because new crust is constantly forming.

Inquiry MiniLab — 20 minutes

How can you model the inner planets?

In this lab, you will use modeling clay to make scale models of the inner planets.

Planet	Actual Diameter (km)	Model Diameter (cm)
Mercury	4,886	
Venus	12,118	
Earth	12,756	8.0
Mars	6,786	

1. Use the data above for Earth to calculate in your Science Journal each model's diameter for the other three planets.
2. Use **modeling clay** to make a ball that represents the diameter of each planet. Check the diameter with a **centimeter ruler**.

Analyze Your Results

1. **Explain** how you converted actual diameters (km) to model diameters (cm).
2. **Key Concept** How do the inner planets compare? Which planets have approximately the same diameter?

Earth

Earth, shown in **Figure 9**, is the third planet from the Sun. Unlike Mercury and Venus, Earth has a moon.

Earth's Atmosphere

A mixture of gases and a small amount of water vapor make up most of Earth's atmosphere. They produce a greenhouse effect that increases Earth's average surface temperature. This effect and Earth's distance from the Sun warm Earth enough for large bodies of liquid water to exist. Earth's atmosphere also absorbs much of the Sun's radiation and protects the surface below. Earth's protective atmosphere, the presence of liquid water, and the planet's moderate temperature range support a variety of life.

Earth's Structure

As shown in **Figure 9**, Earth has a solid inner core surrounded by a liquid outer core. The mantle surrounds the liquid outer core. Above the mantle is Earth's crust. It is broken into large pieces, called plates, that constantly slide past, away from, or into each other. The crust is made mostly of oxygen and silicon and is constantly created and destroyed.

Reading Check Why is there life on Earth?

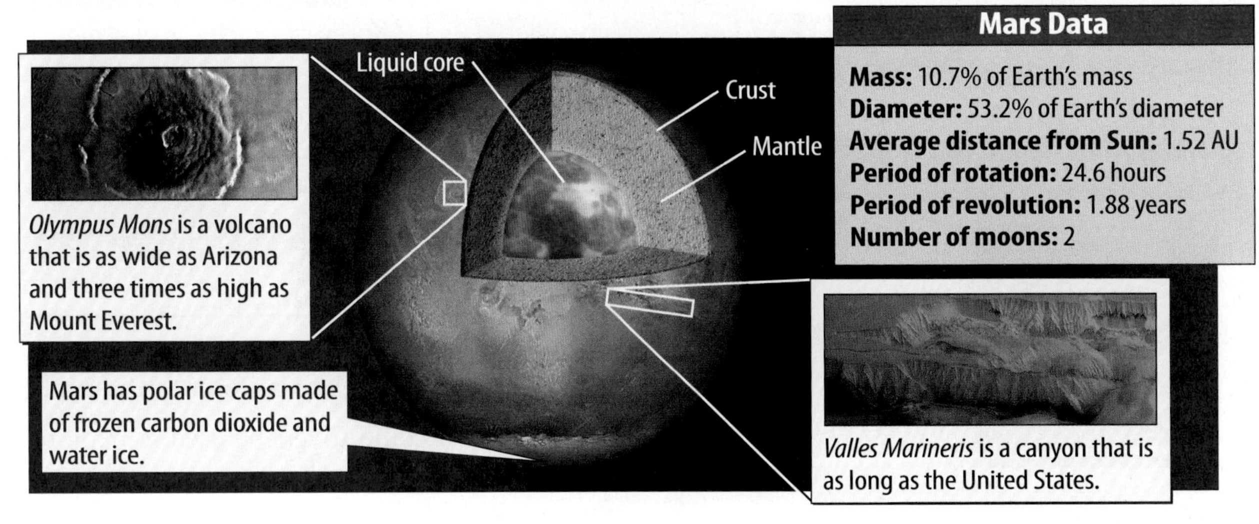

▲ **Figure 10** Mars is a small, rocky planet with deep canyons and tall mountains.

Mars

The fourth planet from the Sun is Mars, shown in **Figure 10.** Mars is about half the size of Earth. It has two very small and irregularly shaped moons. These moons might be asteroids that were captured by Mars's gravity.

Many space probes have visited Mars. Most of them have searched for signs of water that might indicate the presence of living organisms. Images of Mars show features that might have been made by water, such as the gullies in **Figure 11.** So far no evidence of liquid water or life has been found.

Mars's Atmosphere

The atmosphere of Mars is about 95 percent carbon dioxide. It is thin and much less dense than Earth's atmosphere. Temperatures range from about −125°C at the poles to about 20°C at the equator during a martian summer. Winds on Mars sometimes produce great dust storms that last for months.

Mars's Surface

The reddish color of Mars is because its soil contains iron oxide, a compound in rust. Some of Mars's major surface features are shown in **Figure 10.** The enormous canyon Valles Marineris is about 4,000 km long. The Martian volcano Olympus Mons is the largest known mountain in the solar system. Mars also has polar ice caps made of frozen carbon dioxide and ice.

The southern hemisphere of Mars is covered with craters. The northern hemisphere is smoother and appears to be covered by lava flows. Some scientists have proposed that the lava flows were caused by the impact of an object about 2,000 km in diameter.

Key Concept Check Describe the atmosphere of each inner planet.

Figure 11 Gullies such as these might have been formed by the flow of liquid water. ▼

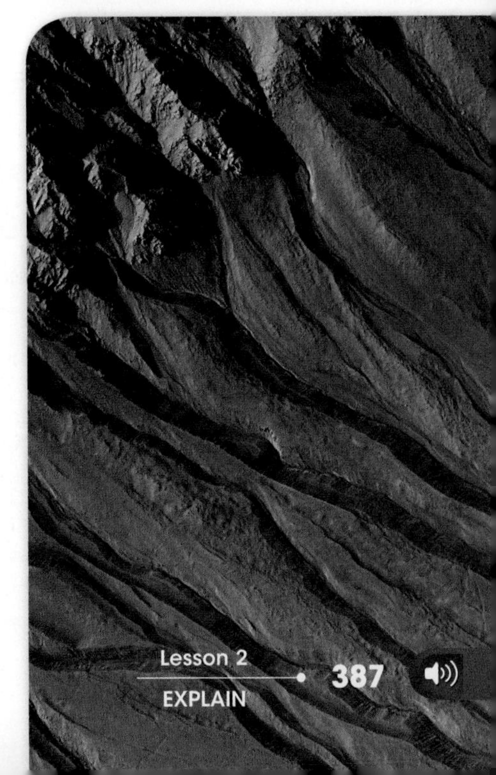

Lesson 2 Review

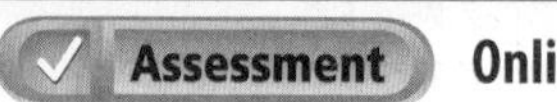

Visual Summary

The terrestrial planets include Mercury, Venus, Earth, and Mars.

The inner planets all are made of rocks and minerals, but they have different characteristics. Earth is the only planet with liquid water.

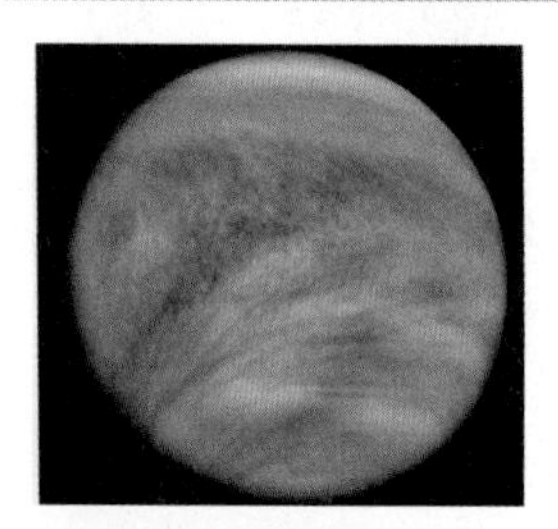

The greenhouse effect greatly increases the surface temperature of Venus.

FOLDABLES®

Use your lesson Foldable to review the lesson. Save your Foldable for the project at the end of the chapter.

What do you think NOW?

You first read the statements below at the beginning of the chapter.

3. Earth is the only inner planet that has a moon.

4. Venus is the hottest planet in the solar system.

Did you change your mind about whether you agree or disagree with the statements? Rewrite any false statements to make them true.

Use Vocabulary

1 **Define** *greenhouse effect* in your own words.

Understand Key Concepts 

2 **Explain** why Venus is hotter than Mercury, even though Mercury is closer to the Sun.

3 **Infer** Why could rovers be used to explore Mars but not Venus?

4 Which of the inner planets has the greatest mass?

A. Mercury
B. Venus
C. Earth
D. Mars

5 **Relate** Describe the relationship between an inner planet's distance from the Sun and its period of revolution.

Interpret Graphics

6 **Infer** Which planet shown below is most likely able to support life now or was able to in the past? Explain your reasoning.

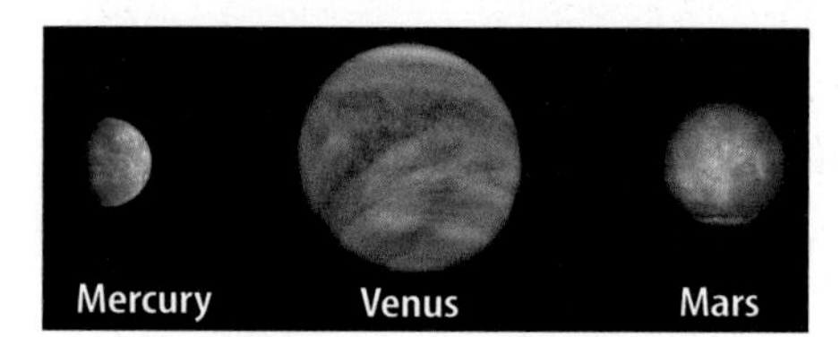

7 **Compare and Contrast** Copy and fill in the table below to compare and contrast properties of Venus and Earth.

Planet	Similarities	Differences
Venus		
Earth		

Critical Thinking

8 **Imagine** How might the temperatures on Mercury be different if it had the same mass as Earth? Explain.

9 **Judge** Do you think the inner planets should be explored or should the money be spent on other things? Justify your opinion.

Inquiry **Skill Practice** **Graphing Data** **25 minutes**

What can we learn about planets by graphing their characteristics?

Scientists collect and analyze data, and draw conclusions based on data. They are particularly interested in finding trends and relationships in data. One commonly used method of finding relationships is by graphing data. Graphing allows different types of data be to seen in relation to one another.

Learn It

Scientists know that some properties of the planets are related. **Graphing data** makes the relationships easy to identify. The graphs can show mathematical relationships such as direct and inverse relationships. Often, however, the graphs show that there is no relationship in the data.

Try It

1. You will plot two graphs that explore the relationships in data. The first graph compares a planet's distance from the Sun and its orbital period. The second graph compares a planet's distance from the Sun and its radius. Make a prediction about how these two sets of data are related, if at all. The data is shown in the table below.

Planet	Average Distance From the Sun (AU)	Orbital Period (yr)	Planet Radius (km)
Mercury	0.39	0.24	2440
Venus	0.72	0.62	6051
Earth	1.00	1.0	6378
Mars	1.52	1.9	3397
Jupiter	5.20	11.9	71,492
Saturn	9.58	29.4	60,268
Uranus	19.2	84.0	25,559
Neptune	30.1	164	24,764

2. Use the data in the table to plot a line graph showing orbital period versus average distance from the Sun. On the *x*-axis, plot the planet's distance from the Sun. On the *y*-axis, plot the planet's orbital period. Make sure the range of each axis is suitable for the data to be plotted, and clearly label each planet's data point.

3. Use the data in the table to plot a line graph showing planet radius versus average distance from the Sun. On the *y*-axis, plot the planet's radius. Make sure the range of each axis is suitable for the data to be plotted, and clearly label each planet's data point.

Apply It

4. Examine the *Orbital Period v. Distance from the Sun* graph. Does the graph show a relationship? If so, describe the relationship between a planet's distance from the Sun and its orbital period in your Science Journal.

5. Examine the *Planet Radius v. Distance from the Sun* graph. Does the graph show a relationship? If so, describe the relationship between a planet's distance from the Sun and its radius.

6. **Key Concept** Identify one or two characteristics the inner planets share that you learned from your graphs.

Lesson 3

Reading Guide

Key Concepts 
ESSENTIAL QUESTIONS

- How are the outer planets similar?
- What are the outer planets made of?

Vocabulary

Galilean moons p. 393

g Multilingual eGlossary

The Outer Planets

What's below?

Clouds often prevent airplane pilots from seeing the ground below. Similarly, clouds block the view of Jupiter's surface. What do you think is below Jupiter's colorful cloud layer? The answer might surprise you—Jupiter is not at all like Earth.

Launch Lab

15 minutes

How do we see distant objects in the solar system?

Some of the outer planets were discovered hundreds of years ago. Why weren't all planets discovered?

1. Read and complete a lab safety form.
2. Use a **meterstick, masking tape,** and the **data table** to mark and label the position of each object on the tape on the floor along a straight line.
3. Shine a **flashlight** from "the Sun" horizontally along the tape.
4. Have a partner hold a page of this **book** in the flashlight beam at each planet location. Record your observations in your Science Journal.

Object	Distance from Sun (cm)
Sun	0
Jupiter	39
Saturn	71
Uranus	143
Neptune	295

Think About This

1. What happens to the image of the page as you move away from the flashlight?
2. **Key Concept** Why do you think it is more difficult to observe the outer planets than the inner planets?

The Gas Giants

Have you ever seen water drops on the outside of a glass of ice? They form because water vapor in the air changes to a liquid on the cold glass. Gases also change to liquids at high pressures. These properties of gases affect the outer planets.

The outer planets, shown in **Figure 12,** are called the gas giants because they are primarily made of hydrogen and helium. These elements are usually gases on Earth.

The outer planets have strong gravitational forces due to their large masses. The strong gravity creates tremendous atmospheric pressure that changes gases to liquids. Thus, the outer planets mainly have liquid interiors. In general, the outer planets have a thick gas and liquid layer covering a small, solid core.

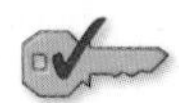

Key Concept Check How are the outer planets similar?

Figure 12 The outer planets are primarily made of gases and liquids.

Visual Check Which outer planet is the largest?

Jupiter

Figure 13 describes Jupiter, the largest planet in the solar system. Jupiter's diameter is more than 11 times larger than the diameter of Earth. Its mass is more than twice the mass of all the other planets combined. One way to understand just how big Jupiter is is to realize that more than 1,000 Earths would fit within this gaseous planet's volume.

Jupiter takes almost 12 Earth years to complete one orbit. Yet, it rotates faster than any other planet. Its period of rotation is less than 10 hours. Jupiter and all the outer planets have a ring system.

FOLDABLES®

Make a four-door book. Label each door with the name of an outer planet. Use the book to organize your notes on the outer planets.

Jupiter's Atmosphere

The atmosphere on Jupiter is about 90 percent hydrogen and 10 percent helium and is about 1,000 km deep. Within the atmosphere are layers of dense, colorful clouds. Because Jupiter rotates so quickly, these clouds stretch into colorful, swirling bands. The Great Red Spot on the planet's surface is a storm of swirling gases.

Jupiter's Structure

Overall, Jupiter is about 80 percent hydrogen and 20 percent helium with small amounts of other materials. The planet is a ball of gas swirling around a thick liquid layer that conceals a solid core. About 1,000 km below the outer edge of the cloud layer, the pressure is so great that the hydrogen gas changes to liquid. This thick layer of liquid hydrogen surrounds Jupiter's core. Scientists do not know for sure what makes up the core. They suspect that the core is made of rock and iron. The core might be as large as Earth and could be 10 times more massive.

Key Concept Check Describe what makes up each of Jupiter's three distinct layers.

Figure 13  Jupiter is mainly hydrogen and helium. Throughout most of the planet, the pressure is high enough to change the hydrogen gas into a liquid.

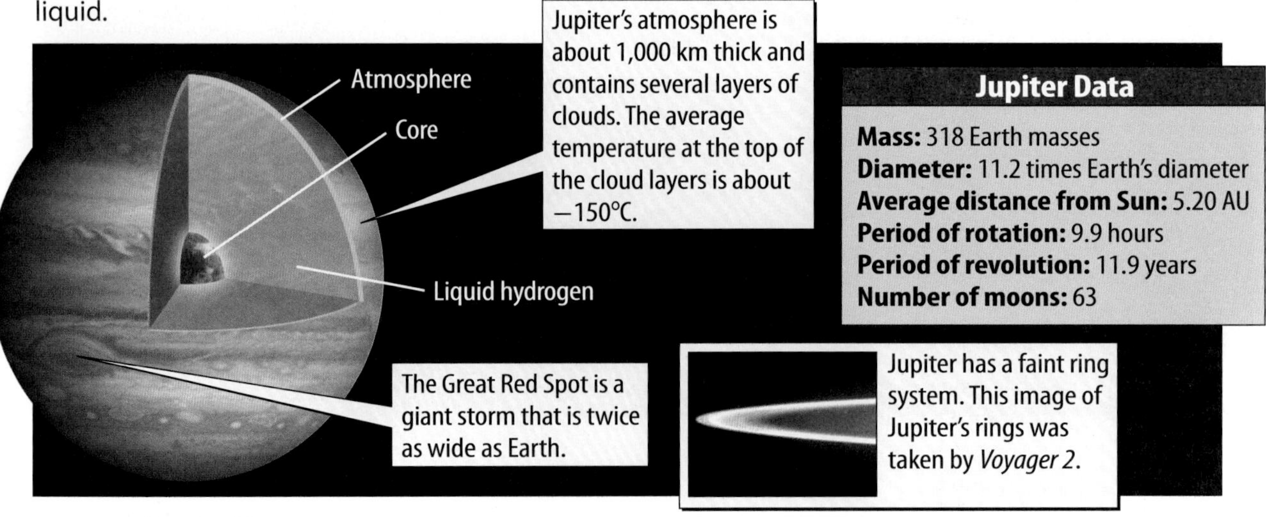

The Moons of Jupiter

Jupiter has at least 63 moons, more than any other planet. Jupiter's four largest moons were first discovered by Galileo Galilei in 1610. *The four largest moons of Jupiter—Io, Europa, Ganymede, and Callisto—are known as the* **Galilean moons.** The Galilean moons all are made of rock and ice. The moons Ganymede, Callisto, and Io are larger than Earth's Moon. Collisions between Jupiter's moons and meteorites likely resulted in the particles that make up the planet's faint rings.

Saturn

Saturn is the sixth planet from the Sun. Like Jupiter, Saturn rotates rapidly and has horizontal bands of clouds. Saturn is about 90 percent hydrogen and 10 percent helium. It is the least dense planet. Its density is less than that of water.

Saturn's Structure

Saturn is made mostly of hydrogen and helium with small amounts of other materials. As shown in **Figure 14,** Saturn's structure is similar to Jupiter's structure—an outer gas layer, a thick layer of liquid hydrogen, and a solid core.

The ring system around Saturn is the largest and most complex in the solar system. Saturn has seven bands of rings, each containing thousands of narrower ringlets. The main ring system is over 70,000 km wide, but it is likely less than 30 m thick. The ice particles in the rings are possibly from a moon that was shattered in a collision with another icy object.

Key Concept Check Describe what makes up Saturn and its ring system.

Math Skills 

Ratios

A ratio is a quotient—it is one quantity divided by another. Ratios can be used to compare distances. For example, Jupiter is 5.20 AU from the Sun, and Neptune is 30.05 from the Sun. Divide the larger distance by the smaller distance:

$$\frac{30.05\ \cancel{AU}}{5.20\ \cancel{AU}} = 5.78$$

Neptune is 5.78 times farther from the Sun than Jupiter.

Practice

How many times farther from the Sun is Uranus (distance = 19.20 AU) than Saturn (distance = 9.58 AU)?

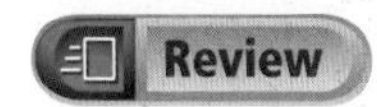

- **Math Practice**
- **Personal Tutor**

Figure 14 Like Jupiter, Saturn is mainly hydrogen and helium. Saturn's rings are one of the most noticeable features of the solar system.

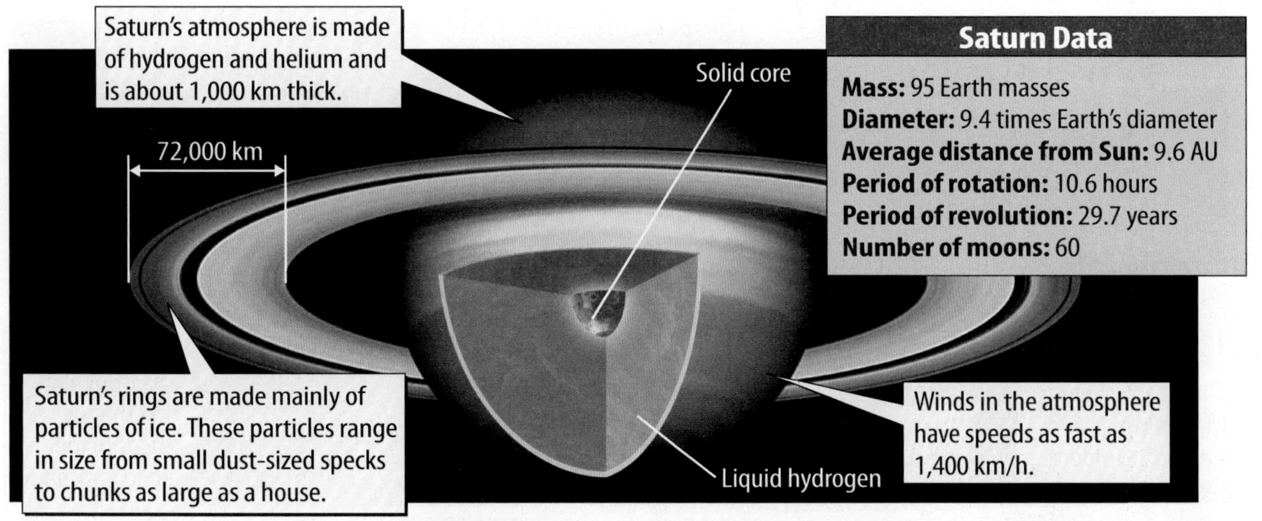

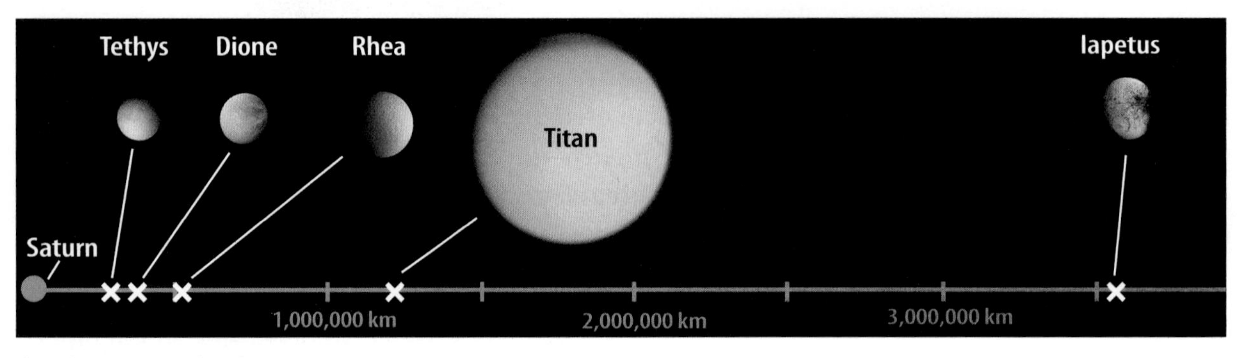

▲ **Figure 15** The five largest moons of Saturn are shown above drawn to scale. Titan is Saturn's largest moon.

Saturn's Moons

Saturn has at least 60 moons. The five largest moons, Titan, Rhea, Dione, Iapetus, and Tethys, are shown in **Figure 15.** Most of Saturn's moons are chunks of ice less than 10 km in diameter. However, Titan is larger than the planet Mercury. Titan is the only moon in the solar system with a dense atmosphere. In 2005, the *Cassini* orbiter released the *Huygens* (HOY guns) **probe** that landed on Titan's surface.

WORD ORIGIN

probe
from Medieval Latin *proba*, means "examination"

Uranus

Uranus, shown in **Figure 16,** is the seventh planet from the Sun. It has a system of narrow, dark rings and a diameter about four times that of Earth. *Voyager 2* is the only space probe to explore Uranus. The probe flew by the planet in 1986.

Uranus has a deep atmosphere composed mostly of hydrogen and helium. The atmosphere also contains a small amount of methane. Beneath the atmosphere is a thick, slushy layer of water, ammonia, and other materials. Uranus might also have a solid, rocky core.

Key Concept Check Identify the substances that make up the atmosphere and the thick slushy layer on Uranus.

Figure 16  Uranus is mainly gas and liquid, with a small solid core. Methane gas in the atmosphere gives Uranus a bluish color. ▼

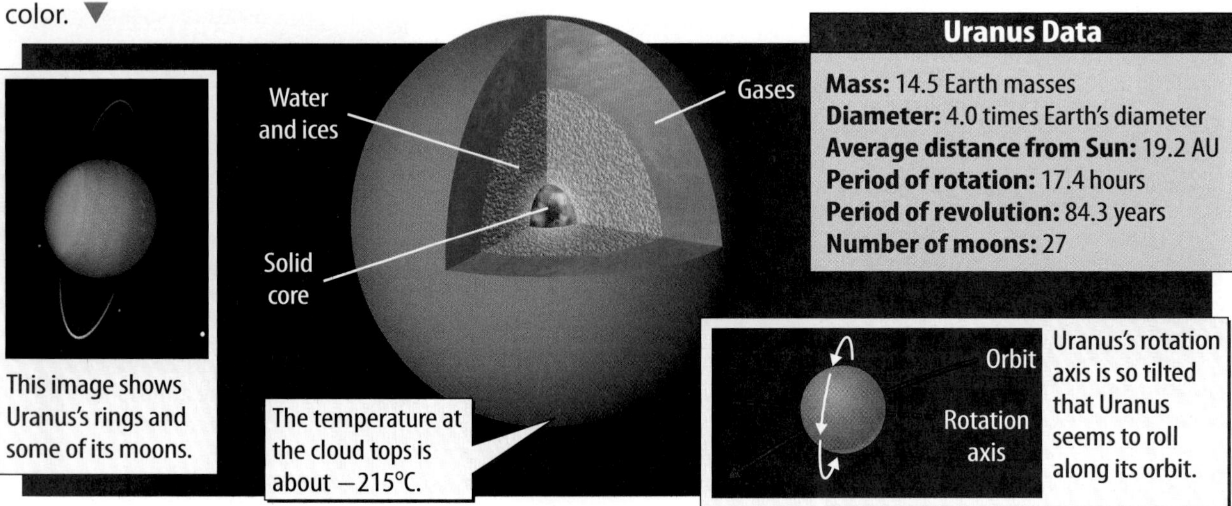

Uranus Data
Mass: 14.5 Earth masses
Diameter: 4.0 times Earth's diameter
Average distance from Sun: 19.2 AU
Period of rotation: 17.4 hours
Period of revolution: 84.3 years
Number of moons: 27

Uranus's Axis and Moons

Figure 16 shows that Uranus has a tilted axis of rotation. In fact, it is so tilted that the planet moves around the Sun like a rolling ball. This sideways tilt might have been caused by a collision with an Earth-sized object.

Uranus has at least 27 moons. The two largest moons, Titania and Oberon, are considerably smaller than Earth's moon. Titania has an icy cracked surface that once might have been covered by an ocean.

Neptune

Neptune, shown in **Figure 17,** was discovered in 1846. Like Uranus, Neptune's atmosphere is mostly hydrogen and helium, with a trace of methane. Its interior also is similar to the interior of Uranus. Neptune's interior is partially frozen water and ammonia with a rock and iron core.

Neptune has at least 13 moons and a faint, dark ring system. Its largest moon, Triton, is made of rock with an icy outer layer. It has a surface of frozen nitrogen and geysers that erupt nitrogen gas.

Key Concept Check How does the atmosphere and interior of Neptune compare with that of Uranus?

MiniLab — 15 minutes

How do Saturn's moons affect its rings?

In this lab, sugar models Saturn's rings. How might Saturn's moons affect its rings?

1. Read and complete a lab safety form.
2. Hold two **sharpened pencils** with their points even and then **tape** them together.
3. Insert a third pencil into the hole in a **record.** Hold the pencil so the record is in a horizontal position.
4. Have your partner sprinkle **sugar** evenly over the surface of the record. Hold the taped pencils vertically over the record so that the tips rest in the record's grooves.
5. Slowly turn the record. In your Science Journal, record what happens to the sugar.

Analyze and Conclude

1. **Compare and Contrast** What feature of Saturn's rings do the pencils model?
2. **Infer** What do you think causes the spaces between the rings of Saturn?
3. **Key Concept** What would have to be true for a moon to interact in this way with Saturn's rings?

Figure 17 The atmosphere of Neptune is similar to that of Uranus—mainly hydrogen and helium with a trace of methane. The dark circular areas on Neptune are swirling storms. Winds on Neptune sometimes exceed 1,000 km/h.

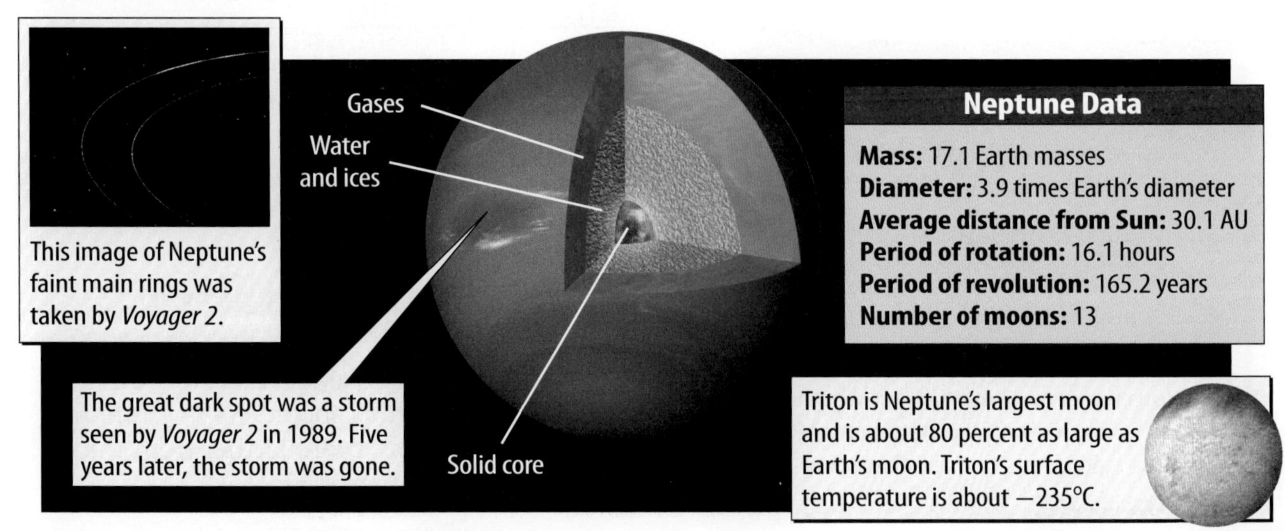

This image of Neptune's faint main rings was taken by *Voyager 2*.

The great dark spot was a storm seen by *Voyager 2* in 1989. Five years later, the storm was gone.

Neptune Data

Mass: 17.1 Earth masses
Diameter: 3.9 times Earth's diameter
Average distance from Sun: 30.1 AU
Period of rotation: 16.1 hours
Period of revolution: 165.2 years
Number of moons: 13

Triton is Neptune's largest moon and is about 80 percent as large as Earth's moon. Triton's surface temperature is about −235°C.

Lesson 3 Review

Online Quiz

Visual Summary

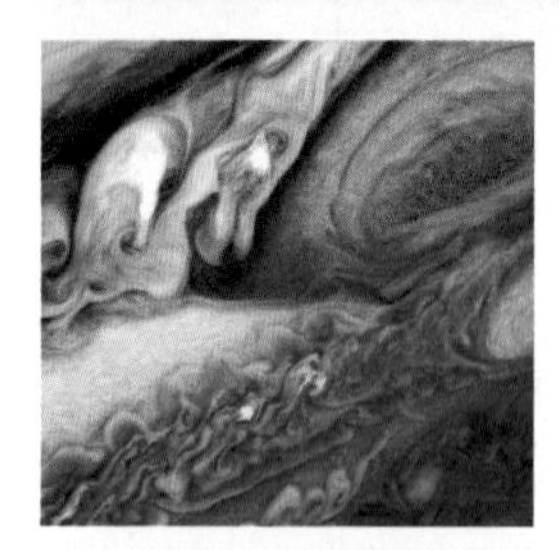

All of the outer planets are primarily made of materials that are gases on Earth. Colorful clouds of gas cover Saturn and Jupiter.

Jupiter is the largest outer planet. Its four largest moons are known as the Galilean moons.

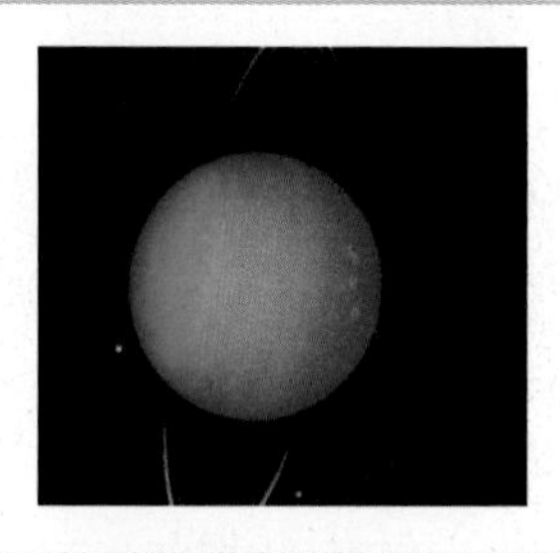

Uranus has an unusual tilt, possibly due to a collision with a large object.

FOLDABLES®

Use your lesson Foldable to review the lesson. Save your Foldable for the project at the end of the chapter.

What do you think NOW?

You first read the statements below at the beginning of the chapter.

5. The outer planets also are called the gas giants.

6. The atmospheres of Saturn and Jupiter are mainly water vapor.

Did you change your mind about whether you agree or disagree with the statements? Rewrite any false statements to make them true.

Use Vocabulary

1 **Identify** What are the four Galilean moons of Jupiter?

Understand Key Concepts 

2 **Contrast** How are the rings of Saturn different from the rings of Jupiter?

3 Which planet's rings probably formed from a collision between an icy moon and another icy object?

A. Jupiter
B. Neptune
C. Saturn
D. Uranus

4 **List** the outer planets by increasing mass.

Interpret Graphics

5 **Infer** from the diagram below how Uranus's tilted axis affects its seasons.

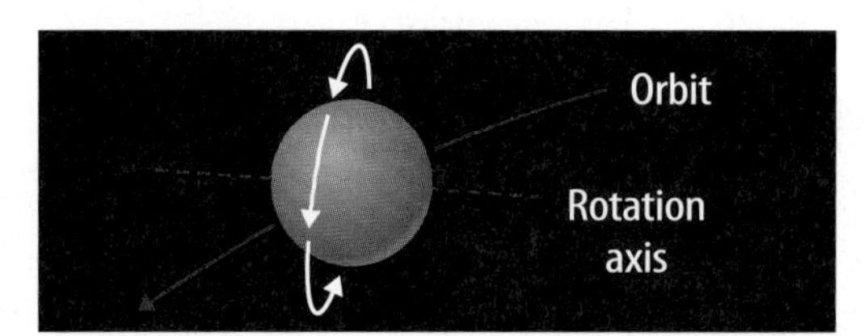

6 **Organize Information** Copy the organizer below and use it to list the outer planets.

Critical Thinking

7 **Predict** what would happen to Jupiter's atmosphere if its gravitational force suddenly decreased. Explain.

8 **Evaluate** Is life more likely on a dry and rocky moon or on an icy moon? Explain.

Math Skills

Math Practice

9 **Calculate** Mars is about 1.52 AU from the Sun, and Saturn is about 9.58 AU from the Sun. How many times farther from the Sun is Saturn than Mars?

Pluto

CAREERS in SCIENCE

What in the world is it?

Since Pluto's discovery in 1930, students have learned that the solar system has nine planets. But in 2006, the number of planets was changed to eight. What happened?

Neil deGrasse Tyson is an astrophysicist at the American Museum of Natural History in New York City. He and his fellow Museum scientists were among the first to question Pluto's classification as a planet. One reason was that Pluto is smaller than six moons in our solar system, including Earth's moon. Another reason was that Pluto's orbit is more oval-shaped, or elliptical, than the orbits of other planets. Also, Pluto has the most tilted orbit of all planets—17 degrees out of the plane of the solar system. Finally, unlike other planets, Pluto is mostly ice.

Tyson also questioned the definition of a planet—an object that orbits the Sun. Then shouldn't comets be planets? In addition, he noted that when Ceres, an object orbiting the Sun between Jupiter and Mars, was discovered in 1801, it was classified as a planet. But, as astronomers discovered more objects like Ceres, it was reclassified as an asteroid. Then, during the 1990s, many space objects similar to Pluto were discovered. They orbit the Sun beyond Neptune's orbit in a region called the Kuiper belt.

These new discoveries led Tyson and others to conclude that Pluto should be reclassified. In 2006, the International Astronomical Union agreed. Pluto was reclassified as a dwarf planet—an object that is spherical in shape and orbits the Sun in a zone with other objects. Pluto lost its rank as smallest planet, but became "king of the Kuiper belt."

Pluto TIME LINE

1930
Astronomer Clyde Tombaugh discovers a ninth planet, Pluto.

1992
The first object is discovered in the Kuiper belt.

July 2005
Eris—a Pluto-sized object—is discovered in the Kuiper belt.

January 2006
NASA launches *New Horizons* spacecraft, expected to reach Pluto in 2015.

August 2006
Pluto is reclassified as a dwarf planet.

Neil deGrasse Tyson is director of the Hayden Planetarium at the American Museum of Natural History. ▶

This illustration shows what Pluto might look like if you were standing on one of its moons.

It's Your Turn

RESEARCH With a group, identify the different types of objects in our solar system. Consider size, composition, location, and whether the objects have moons. Propose at least two different ways to group the objects.

Lesson 4

Reading Guide

Key Concepts 

ESSENTIAL QUESTIONS

- What is a dwarf planet?
- What are the characteristics of comets and asteroids?
- How does an impact crater form?

Vocabulary

meteoroid p. 402

meteor p. 402

meteorite p. 402

impact crater p. 402

g Multilingual eGlossary

Dwarf Planets and Other Objects

Will it return?

You would probably remember a sight like this. This image of comet C/2006 P1 was taken in 2007. The comet is no longer visible from Earth. Believe it or not, many comets appear then reappear hundreds to millions of years later.

Launch Lab

15 minutes

How might asteroids and moons form?

In this activity, you will explore one way moons and asteroids might have formed.

1. Read and complete a lab safety form.
2. Form a small ball from **modeling clay** and roll it in **sand.**
3. Press a thin layer of modeling clay around a **marble.**
4. Tie equal lengths of **string** to each ball. Hold the strings so the balls are above a **sheet of paper.**
5. Have someone pull back the marble so that its string is parallel to the tabletop and then release it. Record the results in your Science Journal.

Think About This

1. If the collision you modeled occurred in space, what would happen to the sand?
2. **Key Concept** Infer one way scientists propose moons and asteroids formed.

Dwarf Planets

Ceres was discovered in 1801 and was called a planet until similar objects were discovered near it. Then it was called an asteroid. For decades after Pluto's discovery in 1930, it was called a planet. Then, similar objects were discovered, and Pluto lost its planet classification. What type of object is Pluto?

Pluto once was classified as a planet, but it is now classified as a dwarf planet. In 2006, the International Astronomical Union (IAU) adopted "dwarf planet" as a new category. The IAU defines a dwarf planet as an object that orbits the Sun, has enough mass and gravity to form a sphere, and has objects similar in mass orbiting near it or crossing its orbital path. Astronomers classify Pluto, Ceres, Eris, MakeMake (MAH kay MAH kay), and Haumea (how MAY uh) as dwarf planets. **Figure 18** shows four dwarf planets.

Key Concept Check Describe the characteristics of a dwarf planet.

Dwarf Planets

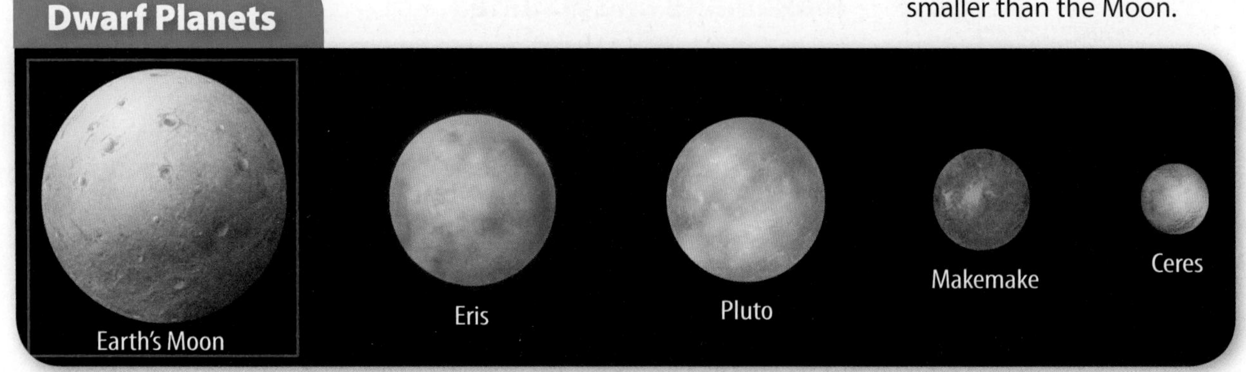

Figure 18 Four dwarf planets are shown to scale. All dwarf planets are smaller than the Moon.

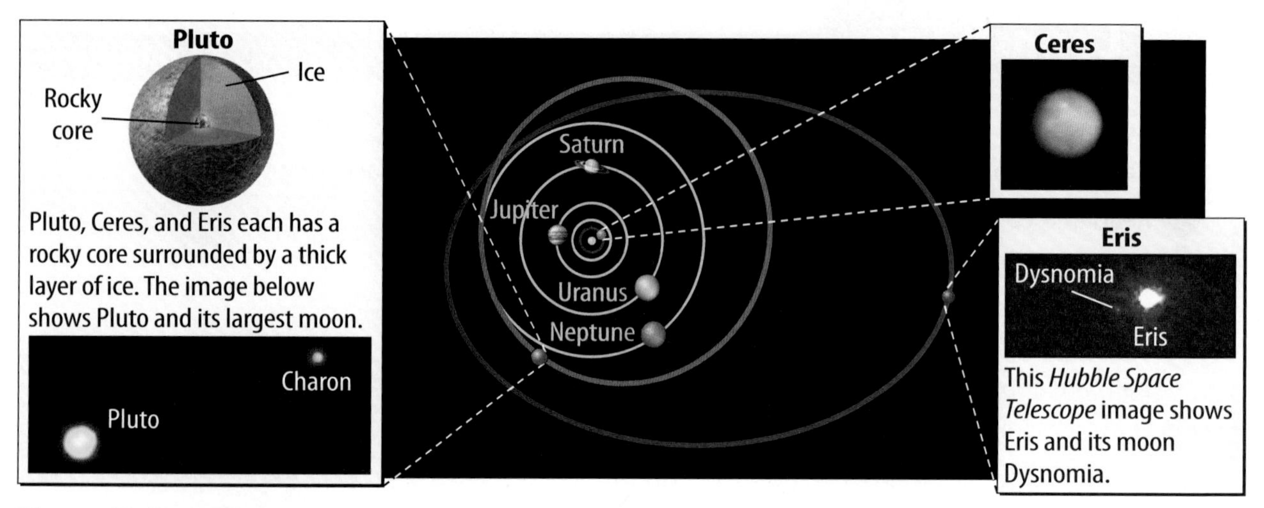

Figure 19 Because most dwarf planets are so far from Earth, astronomers do not have detailed images of them.

Visual Check Which dwarf planet orbits closest to Earth?

Ceres

Ceres, shown in **Figure 19,** orbits the Sun in the asteroid belt. With a diameter of about 950 km, Ceres is about one-fourth the size of the Moon. It is the smallest dwarf planet. Ceres might have a rocky core surrounded by a layer of water ice and a thin, dusty crust.

Pluto

Pluto is about two-thirds the size of the Moon. Pluto is so far from the Sun that its period of revolution is about 248 years. Like Ceres, Pluto has a rocky core surrounded by ice. With an average surface temperature of about −230°C, Pluto is so cold that it is covered with frozen nitrogen.

Pluto has three known moons. The largest moon, Charon, has a diameter that is about half the diameter of Pluto. Pluto also has two smaller moons, Hydra and Nix.

Eris

The largest dwarf planet, Eris, was discovered in 2003. Its orbit lasts about 557 years. Currently, Eris is three times farther from the Sun than Pluto is. The structure of Eris is probably similar to Pluto. Dysnomia (dis NOH mee uh) is the only known moon of Eris.

Makemake and Haumea

In 2008, the IAU designated two new objects as dwarf planets: Makemake and Haumea. Though smaller than Pluto, Makemake is one of the largest objects in a region of the solar system called the Kuiper (KI puhr) belt. The Kuiper belt extends from about the orbit of Neptune to about 50 AU from the Sun. Haumea is also in the Kuiper belt and is smaller than Pluto.

Reading Check Which dwarf planet is the largest? Which dwarf planet is the smallest?

FOLDABLES®

Make a layered book from two sheets of paper. Label it as shown. Use it to organize your notes on other objects in the solar system.

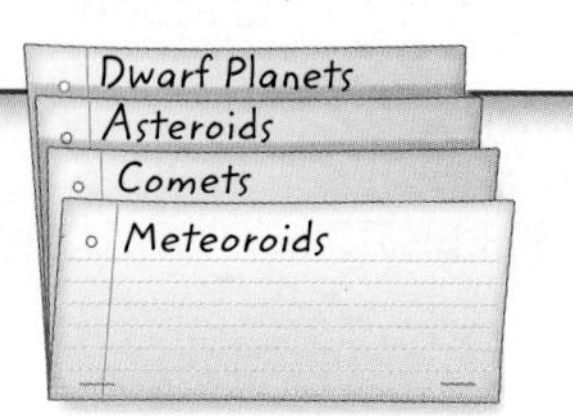

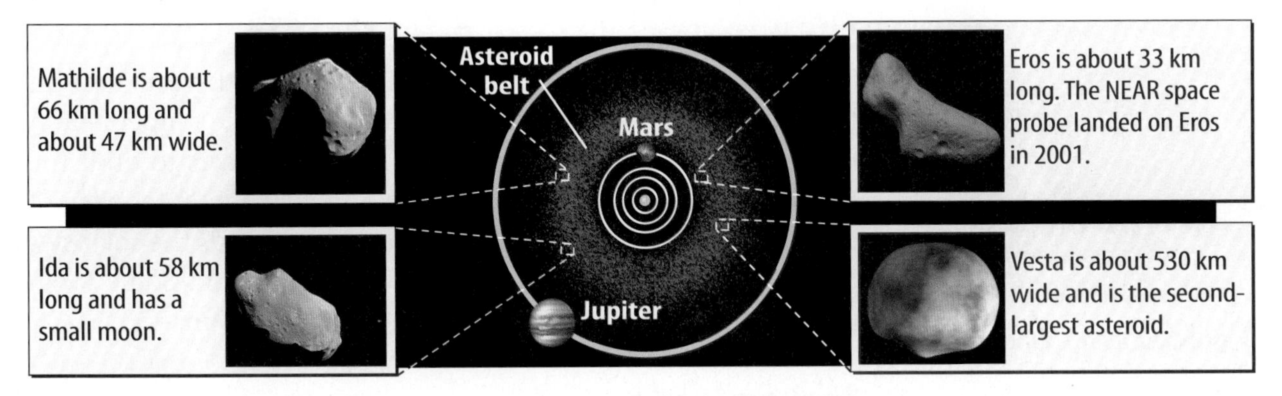

Figure 20 The asteroids that orbit the Sun in the asteroid belt are many sizes and shapes.

Asteroids

Recall from Lesson 1 that asteroids are pieces of rock and ice. Most asteroids orbit the Sun in the asteroid belt. The asteroid belt is between the orbits of Mars and Jupiter, as shown in **Figure 20.** Hundreds of thousands of asteroids have been discovered. The largest asteroid, Pallas, is over 500 km in diameter.

Asteroids are chunks of rock and ice that never clumped together like the rocks and ice that formed the inner planets. Some astronomers suggest that the strength of Jupiter's gravitational field might have caused the chunks to collide so violently, and they broke apart instead of sticking together. This means that asteroids are objects left over from the formation of the solar system.

 **Key Concept Check** Where do the orbits of most asteroids occur?

Comets

Recall that comets are mixtures of rock, ice, and dust. The particles in a comet are loosely held together by the gravitational attractions among the particles. As shown in **Figure 21,** comets orbit the Sun in long elliptical orbits.

The Structure of Comets

The solid, inner part of a comet is its nucleus, as shown in **Figure 21.** As a comet moves closer to the Sun, it absorbs thermal energy and can develop a bright tail. Heating changes the ice in the comet into a gas. Energy from the Sun pushes some of the gas and dust away from the nucleus and makes it glow. This produces the comet's bright tail and glowing nucleus, called a coma.

 **Key Concept Check** Describe the characteristics of a comet.

Figure 21 When energy from the Sun strikes the gas and dust in the comet's nucleus, it can create a two-part tail. The gas tail always points away from the Sun.

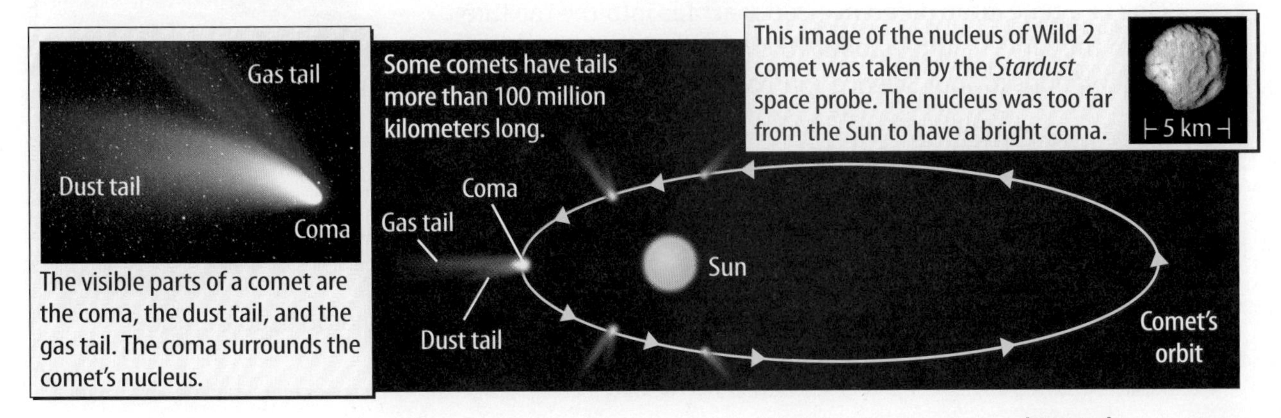

Figure 22 When a large meteorite strikes, it can form a giant impact crater like this 1.2-km wide crater in Arizona.

Short-Period and Long-Period Comets

A short-period comet takes less than 200 Earth years to orbit the Sun. Most short-period comets come from the Kuiper belt. A long-period comet takes more than 200 Earth years to orbit the Sun. Long-period comets come from a area at the outer edge of the solar system, called the Oort cloud. It surrounds the solar system and extends about 100,000 AU from the Sun. Some long-period comets take millions of years to orbit the Sun.

Meteoroids

Every day, many millions of particles called meteoroids enter Earth's atmosphere. *A* **meteoroid** *is a small, rocky particle that moves through space.* Most meteoroids are only about as big as a grain of sand. As a meteoroid passes through Earth's atmosphere, friction makes the meteoroid and the air around it hot enough to glow. *A* **meteor** *is a streak of light in Earth's atmosphere made by a glowing meteoroid.* Most meteoroids burn up in the atmosphere. However, some meteoroids are large enough that they reach Earth's surface before they burn up completely. When this happens, it is called a meteorite. *A* **meteorite** *is a meteoroid that strikes a planet or a moon.*

Word Origin

meteor
from Greek *meteoros,* means "high up"

When a large meteoroite strikes a moon or planet, it often forms a bowl-shaped depression such as the one shown in **Figure 22.** *An* **impact crater** *is a round depression formed on the surface of a planet, moon, or other space object by the impact of a meteorite.* There are more than 170 impact craters on Earth.

 **Key Concept Check** What causes an impact crater to form?

Inquiry MiniLab

20 minutes

How do impact craters form?

In this lab, you will model the formation of an impact crater.

1. Pour a layer of **flour** about 3 cm deep in a **cake pan.**
2. Pour a layer of **cornmeal** about 1 cm deep on top of the flour.
3. One at a time, drop different-sized **marbles** into the mixture from the same height—about 15 cm. Record your observations in your Science Journal.

Analyze and Conclude

1. **Describe** the mixture's surface after you dropped the marbles.
2. **Recognize Cause and Effect** Based on your results, explain why impact craters on moons and planets differ.
3. **Key Concept** Explain how the marbles used in the activity could be used to model meteoroids, meteors, and meteorites.

Lesson 4 Review

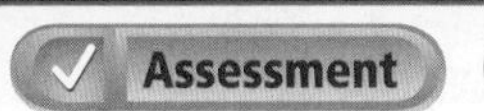

Visual Summary

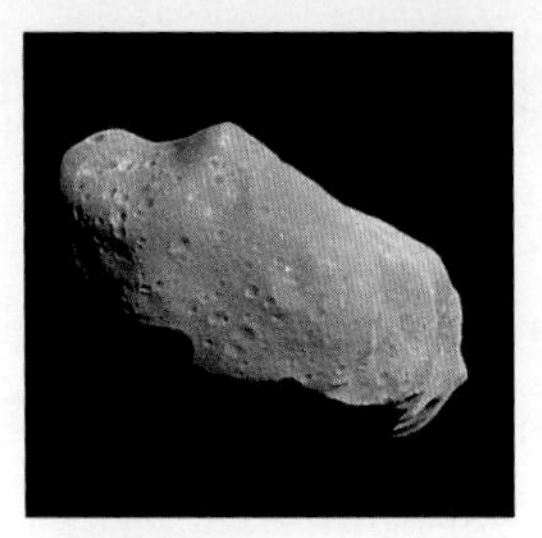

An asteroid, such as Ida, is a chunk of rock and ice that orbits the Sun.

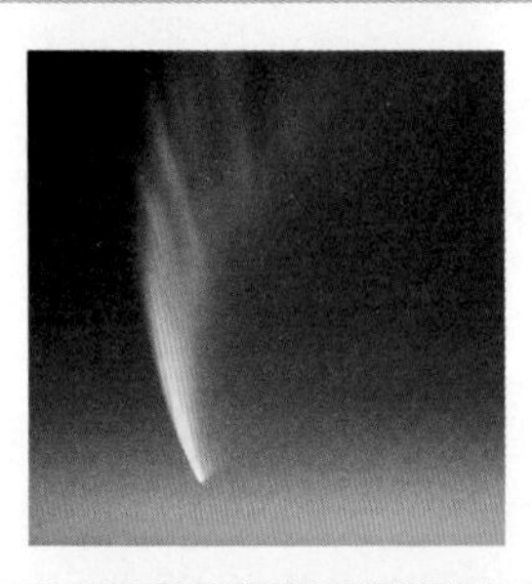

Comets, which are mixture of rock , ice, and dust, orbit the Sun. A comet's tail is caused by its interaction with the Sun.

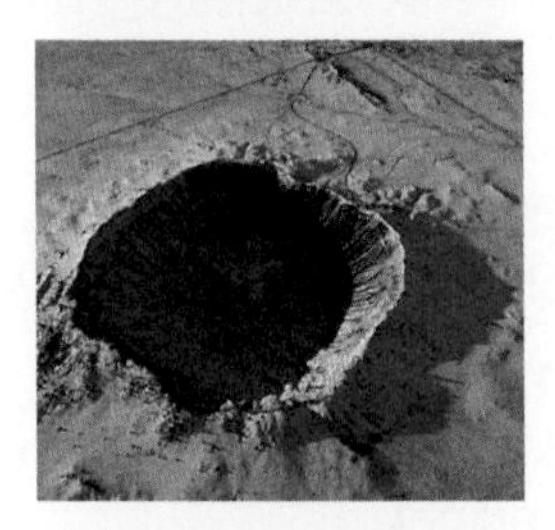

When a large meteorite strikes a planet or moon, it often makes an impact crater.

FOLDABLES®

Use your lesson Foldable to review the lesson. Save your Foldable for the project at the end of the chapter.

What do you think NOW?

You first read the statements below at the beginning of the chapter.

7. Asteroids and comets are mainly rock and ice.

8. A meteoroid is a meteor that strikes Earth.

Did you change your mind about whether you agree or disagree with the statements? Rewrite any false statements to make them true.

Use Vocabulary

1. **Define** *impact crater* in your own words.
2. **Distinguish** between a meteorite and a meteoroid.
3. **Use the term** *meteor* in a complete sentence.

Understand Key Concepts 

4. Which produces an impact crater?
   A. comet
   B. meteor
   C. meteorite
   D. planet
5. **Reason** Are you more likely to see a meteor or a meteoroid? Explain.
6. **Differentiate** between objects located in the asteroid belt and objects located in the Kuiper belt.

Interpret Graphics

7. **Explain** why some comets have a two-part tail during portions of their orbit.

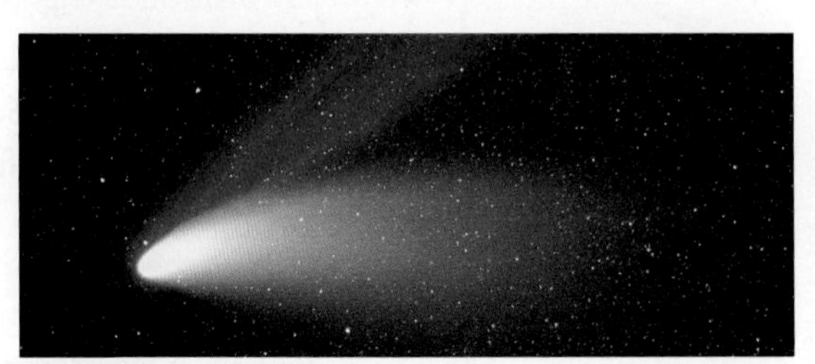

8. **Organize Information** Copy the table below and list the major characteristics of a dwarf planet.

Object	Defining Characteristic
Dwarf Planet	

Critical Thinking

9. **Compose** a paragraph describing what early sky observers might have thought when they saw a comet.
10. **Evaluate** Do you agree with the decision to reclassify Pluto as a dwarf planet? Defend your opinion.

40 minutes

Scaling down the Solar System

Materials

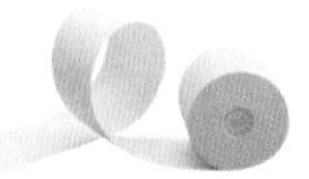

2.25 in–wide register tape (several rolls)

meterstick

masking tape

colored markers

Safety

A scale model is a physical representation of something that is much smaller or much larger. Reduced-size scale models are made of very large things, such as the solar system. The scale used must reduce the actual size to a size reasonable for the model.

Question

What scale can you use to represent the distances between solar system objects?

Procedure

1. First, decide how big your solar system will be. Use the data given in the table to figure out how far apart the Sun and Neptune would be if a scale of 1 meter = 1 AU is used. Would a solar system based on that scale fit in the space you have available?
2. With your group determine the scale that results in a model that fits the available space. Larger models are usually more accurate, so choose a scale that produces the largest model that fits in the available space.
3. Once you have decided on a scale, copy the table in your Science Journal. Replace the word *(Scale)* in the third column of the table with the unit you have chosen. Then fill in the scaled distance for each planet.

Planet	Distance from the Sun (AU)	Distance from the Sun (Scale)
Mercury	0.39	
Venus	0.72	
Earth	1.00	
Mars	1.52	
Jupiter	5.20	
Saturn	9.54	
Uranus	19.18	
Neptune	30.06	

4. On register tape, mark the positions of objects in the solar system based on your chosen scale. Use a length of register tape that is slightly longer than the scaled distance between the Sun and Neptune.
5. Tape the ends of the register tape to a table or the floor. Mark a dot at one end of the paper to represent the Sun. Measure along the tape from the center of the dot to the location of Mercury. Mark a dot at this position and label it *Mercury.* Repeat this process for the remaining planets.

Analyze and Conclude

6. **Critique** There are many objects in the solar system. These objects have different sizes, structures, and orbits. Examine your scale model of the solar system. How accurate is the model? How could the model be changed to be more accurate?
7. **The Big Idea** Pluto is a dwarf planet located beyond Neptune. Based on the pattern of distance data for the planets shown in the table, approximately how far from the Sun would you expect to find Pluto? Explain you reasoning.
8. **Calculate** What length of register tape is needed if a scale of 30 cm = 1 AU is used for the solar system model?

Communicate Your Results

Compare your model with other groups in your class by taping them all side-by-side. Discuss any major differences in your models. Discuss the difficulties in making the scale models much smaller.

Inquiry Extension

How can you build a scale model of the solar system that accurately shows both planetary diameters and distances? Describe how you would go about figuring this out.

Lab Tips

- ☑ A scale is the ratio between the actual size of something and a representation of it.
- ☑ The distances between the planets and the Sun are average distances because planetary orbits are not perfect circles.

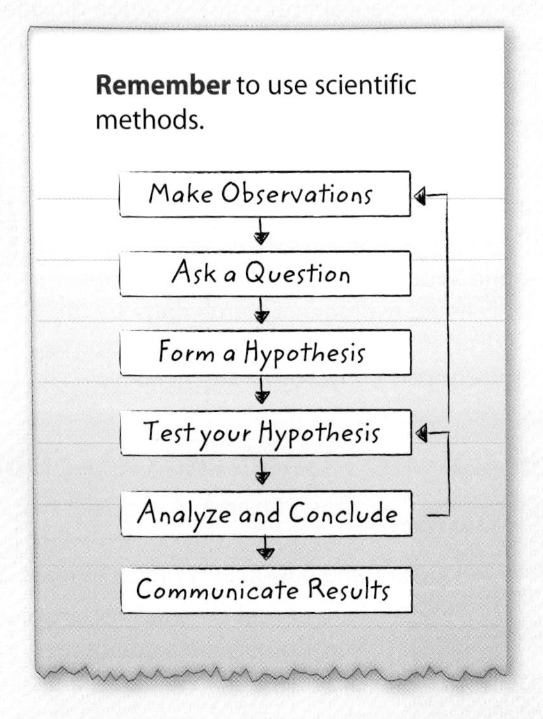

Chapter 11 Study Guide

The solar system contains planets, dwarf planets, comets, asteroids, and other small solar system bodies.

Key Concepts Summary

Lesson 1: The Structure of the Solar System

- The inner planets are made mainly of solid materials. The outer planets, which are larger than the inner planets, have thick gas and liquid layers covering a small solid core.
- Astronomers measure vast distances in space in **astronomical units;** an astronomical unit is about 150 million km.
- The speed of each planet changes as it moves along its elliptical orbit around the Sun.

Vocabulary

asteroid p. 377
comet p. 377
astronomical unit p. 378
period of revolution p. 378
period of rotation p. 378

Lesson 2: The Inner Planets

- The inner planets—Mercury, Venus, Earth, and Mars—are made of rock and metallic materials.
- The **greenhouse effect** makes Venus the hottest planet.
- Mercury has no atmosphere. The atmospheres of Venus and Mars are almost entirely carbon dioxide. Earth's atmosphere is a mixture of gases and a small amount of water vapor.

Vocabulary

terrestrial planet p. 383
greenhouse effect p. 385

Lesson 3: The Outer Planets

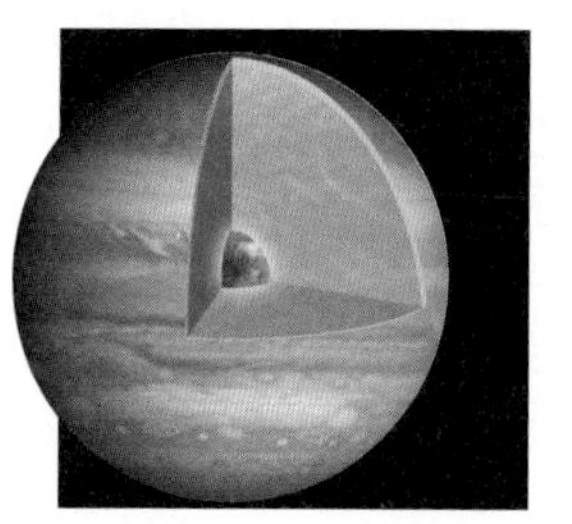

- The outer planets—Jupiter, Saturn, Uranus, and Neptune—are primarily made of hydrogen and helium.
- Jupiter and Saturn have thick cloud layers, but are mainly liquid hydrogen. Saturn's rings are largely particles of ice. Uranus and Neptune have thick atmospheres of hydrogen and helium.

Vocabulary

Galilean moons p. 393

Lesson 4: Dwarf Planets and Other Objects

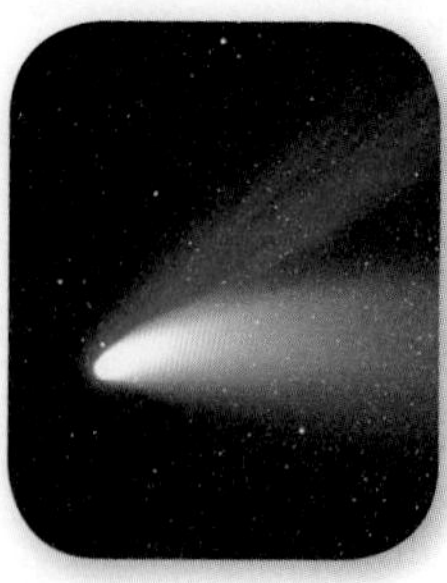

- A dwarf planet is an object that orbits a star, has enough mass to pull itself into a spherical shape, and has objects similar in mass orbiting near it.
- An asteroid is a small rocky object that orbits the Sun. Comets are made of rock, ice, and dust and orbit the Sun in highly elliptical paths.
- The impact of a **meteorite** forms an **impact crater.**

Vocabulary

meteoroid p. 402
meteor p. 402
meteorite p. 402
impact crater p. 402

- Personal Tutor
- Vocabulary eGames
- Vocabulary eFlashcards

FOLDABLES® Chapter Project

Assemble your lesson Foldables as shown to make a Chapter Project. Use the project to review what you have learned in this chapter.

Use Vocabulary

Match each phrase with the correct vocabulary term from the Study Guide.

1. the time it takes an object to complete one rotation on its axis
2. the average distance from Earth to the Sun
3. the time it takes an object to travel once around the Sun
4. an increase in temperature caused by energy trapped by a planet's atmosphere
5. an inner planet
6. the four largest moons of Jupiter
7. a streak of light in Earth's atmosphere made by a glowing meteoroid

Interactive Concept Map

Link Vocabulary and Key Concepts

Copy this concept map, and then use vocabulary terms to complete the concept map.

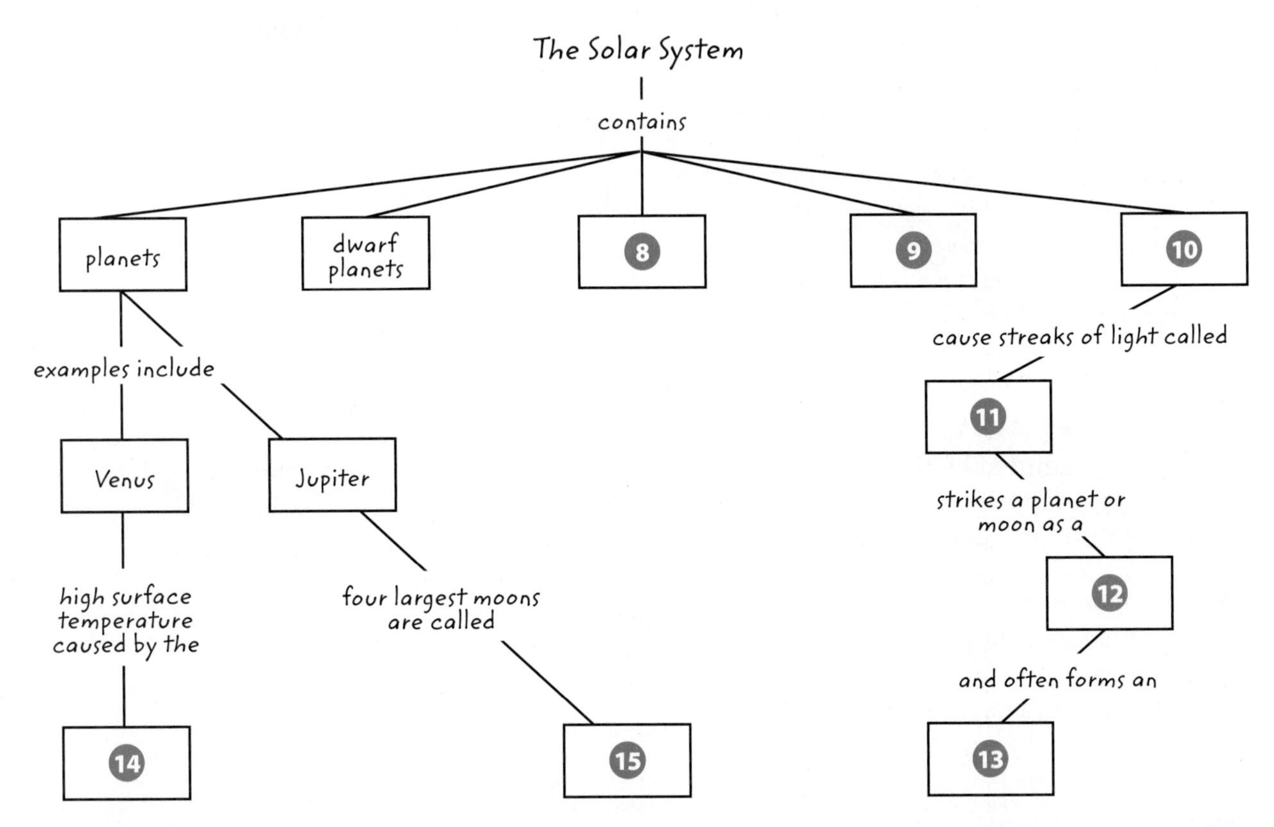

Chapter 11 Review

Understand Key Concepts 

1. Which solar system object is the largest?
   A. Jupiter
   B. Neptune
   C. the Sun
   D. Saturn

2. Which best describes the asteroid belt?
   A. another name for the Oort cloud
   B. the region where comets originate
   C. large chunks of gas, dust, and ice
   D. millions of small rocky objects

3. Which describes a planet's speed as it orbits the Sun?
   A. It constantly decreases.
   B. It constantly increases.
   C. It does not change.
   D. It increases then decreases.

4. The diagram below shows a planet's orbit around the Sun. What does the blue arrow represent?
   A. the gravitational pull of the Sun
   B. the planet's orbital path
   C. the planet's path if Sun did not exist
   D. the planet's speed

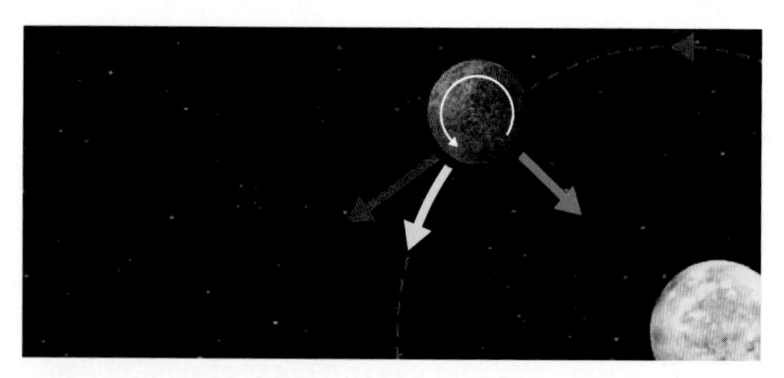

5. Which describes the greenhouse effect?
   A. effect of gravity on temperature
   B. energy emitted by the Sun
   C. energy trapped by atmosphere
   D. reflection of light from a planet

6. How are the terrestrial planets similar?
   A. similar densities
   B. similar diameters
   C. similar periods of rotation
   D. similar rocky surfaces

7. Which inner planet is the hottest?
   A. Earth
   B. Mars
   C. Mercury
   D. Venus

8. The photograph below shows how Earth appears from space. How does Earth differ from other inner planets?

   A. Its atmosphere contains large amounts of methane.
   B. Its period of revolution is much greater.
   C. Its surface is covered by large amounts of liquid water.
   D. Its surface temperature is higher.

9. Which two gases make up most of the outer planets?
   A. ammonia and helium
   B. ammonia and hydrogen
   C. hydrogen and helium
   D. methane and hydrogen

10. Which is true of the dwarf planets?
    A. more massive than nearby objects
    B. never have moons
    C. orbit near the Sun
    D. spherically shaped

11. Which is a bright streak of light in Earth's atmosphere?
    A. a comet
    B. a meteor
    C. a meteorite
    D. a meteoroid

12. Which best describes an asteroid?
    A. icy
    B. rocky
    C. round
    D. wet

Critical Thinking

13. **Relate** changes in speed during a planet's orbit to the shape of the orbit and the gravitational pull of the Sun.

14. **Compare** In what ways are planets and dwarf planets similar?

15. **Apply** Like Venus, Earth's atmosphere contains carbon dioxide. What might happen on Earth if the amount of carbon dioxide in the atmosphere increases? Explain.

16. **Defend** A classmate states that life will someday be found on Mars. Defend the statement and offer a reason why life might exist on Mars.

17. **Infer** whether a planet with active volcanoes would have more or fewer craters than a planet without active volcanoes. Explain.

18. **Support** Use the diagram of the asteroid belt to support the explanation of how the belt formed.

19. **Evaluate** The *Huygens* probe transmitted data about Titan for only 90 min. In your opinion, was this worth the effort of sending the probe?

20. **Explain** why Jupiter's moon Ganymede is not considered a dwarf planet, even though it is bigger than Mercury.

Writing in Science

21. **Compose** a pamphlet that describes how the International Astronomical Union classifies planets, dwarf planets, and small solar system objects.

REVIEW THE BIG IDEA

22. What kinds of objects are in the solar system? Summarize the types of space objects that make up the solar system and give at least one example of each.

23. The photo below shows part of Saturn's rings and two of its moons. Describe what Saturn and its rings are made of and explain why the other two objects are moons.

Math Skills

Math Practice

Use Ratios

Inner Planet Data			
Planet	Diameter (% of Earth's diameter)	Mass (% of Earth's mass)	Average Distance from Sun (AU)
Mercury	38.3	5.5	0.39
Venus	95	81.5	0.72
Earth	100	100	1.00
Mars	53.2	10.7	1.52

24. Use the table above to calculate how many times farther from the Sun Mars is compared to Mercury.

25. Calculate how much greater Venus's mass is compared to Mercury's mass.

Standardized Test Practice

Record your answers on the answer sheet provided by your teacher or on a sheet of paper.

Multiple Choice

1 Which is a terrestrial planet?

A Ceres
B Neptune
C Pluto
D Venus

2 An astronomical unit (AU) is the average distance

A between Earth and the Moon.
B from Earth to the Sun.
C to the nearest star in the galaxy.
D to the edge of the solar system.

3 Which is NOT a characteristic of ALL planets?

A exceed the total mass of nearby objects
B have a nearly spherical shape
C have one or more moons
D make an elliptical orbit around the Sun

Use the diagram below to answer question 4.

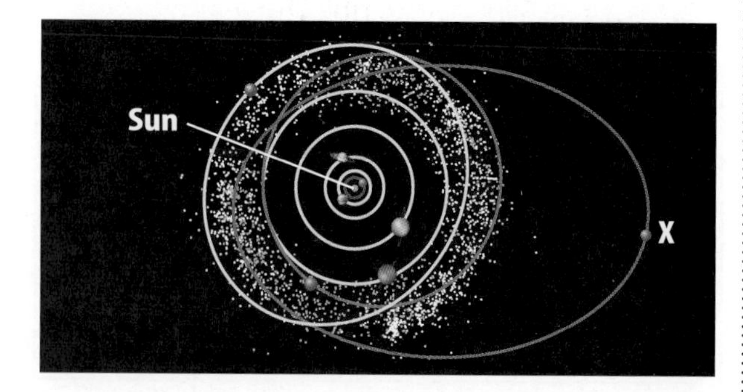

4 Which object in the solar system is marked by an *X* in the diagram?

A asteroid
B meteoroid
C dwarf planet
D outer planet

Use the diagram of Saturn below to answer questions 5 and 6.

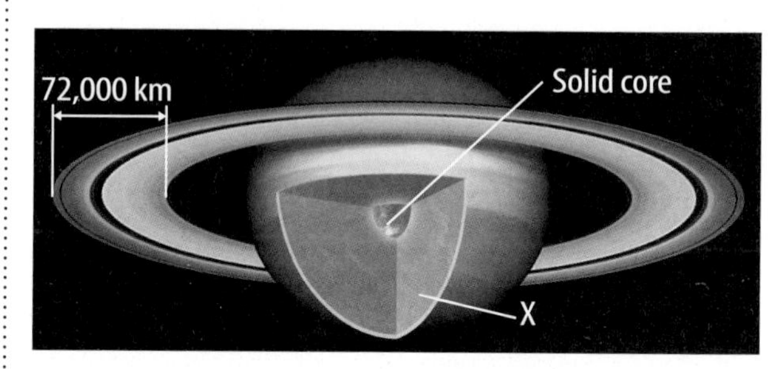

5 The thick inner layer marked *X* in the diagram above is made of which material?

A carbon dioxide
B gaseous helium
C liquid hydrogen
D molten rock

6 In the diagram, Saturn's rings are shown to be 72,000 km in width. Approximately how thick are Saturn's rings?

A 30 m
B 1,000 km
C 14,000 km
D 1 AU

7 Which are NOT found on Mercury's surface?

A high cliffs
B impact craters
C lava flows
D sand dunes

8 What is the primary cause of the extremely high temperatures on the surface of Venus?

A heat rising from the mantle
B lack of an atmosphere
C proximity to the Sun
D the greenhouse effect

Use the diagram below to answer question 9.

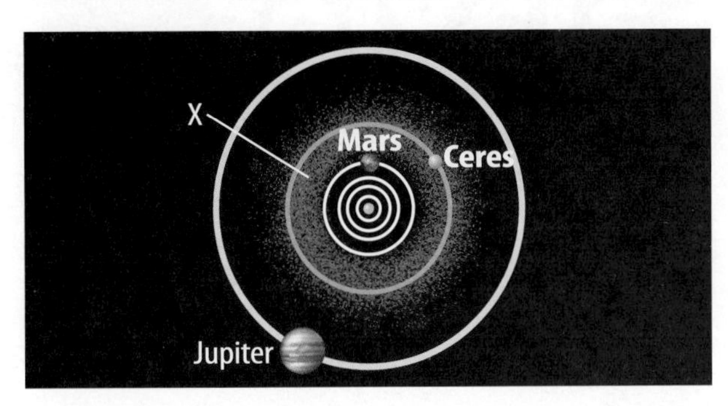

9 In the diagram above, which region of the solar system is marked by an *X*?

- **A** the asteroid belt
- **B** the dwarf planets
- **C** the Kuiper belt
- **D** the Oort cloud

10 What is a meteorite?

- **A** a surface depression formed by collision with a rock from space
- **B** a fragment of rock that strikes a planet or a moon
- **C** a mixture of ice, dust, and gas with a glowing tail
- **D** a small rocky particle that moves through space

11 What gives Mars its reddish color?

- **A** ice caps of frozen carbon dioxide
- **B** lava from Olympus Mons
- **C** liquid water in gullies
- **D** soil rich in iron oxide

Constructed Response

Use the table below to answer questions 12 and 13.

	Inner Planets	Outer Planets
Also called		
Relative size		
Main materials		
General structure		
Number of moons		

12 Copy the table and complete the first five rows to compare the features of the inner planets and outer planets.

13 In the blank row of the table, add another feature of the inner planets and outer planets. Then, describe the feature you have chosen.

14 What features of Earth make it suitable for supporting life as we know it?

15 How are planets, dwarf planets, and asteroids both similar and different?

NEED EXTRA HELP?															
If You Missed Question...	1	2	3	4	5	6	7	8	9	10	11	12	13	14	15
Go to Lesson...	2	1	1	1	3	3	2	2	1,4	4	2	2,3	2,3	2	1,4

Chapter 12

Stars and Galaxies

What makes up the universe, and how does gravity affect the universe?

Inquiry What can't you see?

This photograph shows a small part of the universe. You can see many stars and galaxies in this image. But the universe also contains many things you cannot see.

- How do scientists study the universe?
- What makes up the universe?
- How does gravity affect the universe?

Get Ready to Read

What do you think?

Before you read, decide if you agree or disagree with each of these statements. As you read this chapter, see if you change your mind about any of the statements.

1. The night sky is divided into constellations.
2. A light-year is a measurement of time.
3. Stars shine because there are nuclear reactions in their cores.
4. Sunspots appear dark because they are cooler than nearby areas.
5. The more matter a star contains, the longer it is able to shine.
6. Gravity plays an important role in the formation of stars.
7. Most of the mass in the universe is in stars.
8. The Big Bang theory is an explanation of the beginning of the universe.

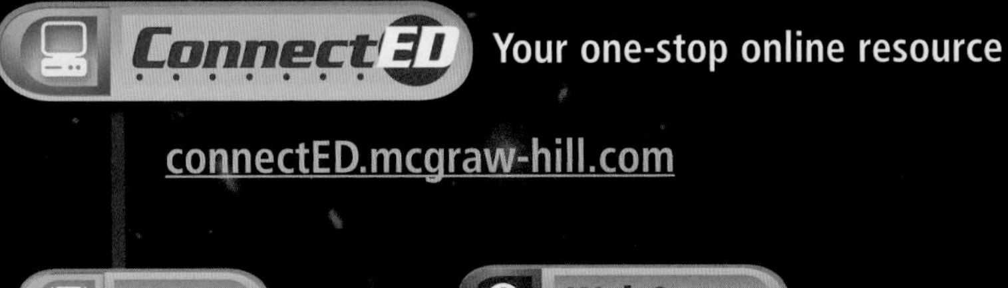

Lesson 1

The View from Earth

Reading Guide

Key Concepts

ESSENTIAL QUESTIONS

- How do astronomers divide the night sky?
- What can astronomers learn about stars from their light?
- How do scientists measure the distance and the brightness of objects in the sky?

Vocabulary

spectroscope p. 417
astronomical unit p. 418
light-year p. 418
apparent magnitude p. 419
luminosity p. 419

g Multilingual eGlossary

Inquiry Where is this?

Unless you have visited a remote part of the country, you have probably never seen the sky look like this. It is similar to what the night sky looked like to your ancestors—before towns and cities brightened the sky.

Launch Lab

20 minutes

How can you "see" invisible energy?

You see because of the Sun's light. You feel the heat of the Sun's energy. The Sun produces other kinds of energy that you can't directly see or feel.

1. Read and complete a lab safety form.
2. Put 5–6 **beads** into a **clear container.** Observe the color of the beads.
3. In a darkened room, shine light from a **flashlight** onto the beads for several seconds. Record your observations in your Science Journal. Repeat this step, exposing the beads to light from an **incandescent lightbulb** and a **fluorescent light.** Record your observations.
4. Stand outside in a shady spot for several seconds. Then expose the beads to direct sunlight. Record your observations.

Think About This

1. How did the light from the different light sources affect the color of the beads?
2. What do you think made the beads change color?
3. **Key Concept** How do you think invisible forms of light help scientists understand stars and other objects in the sky?

Looking at the Night Sky

Have you ever looked up at the sky on a clear, dark night and seen countless stars? If you have, you are lucky. Few people see a sky like that shown on the previous page. Lights from towns and cities make the night sky too bright for faint stars to be seen.

If you look at a clear night sky for a long time, the stars seem to move. But what you are really seeing is Earth's movement. Earth spins, or rotates, once every 24 hours. Day turns to night and then back to day as Earth rotates. Because Earth rotates from west to east, objects in the sky rise in the east and set in the west.

Earth spins on its axis, an imaginary line from the North Pole to the South Pole. The star Polaris is almost directly above the North Pole. As Earth spins, stars near Polaris appear to travel in a circle around Polaris, as shown in **Figure 1.** These stars never set when viewed from the northern hemisphere. They are always present in the night sky.

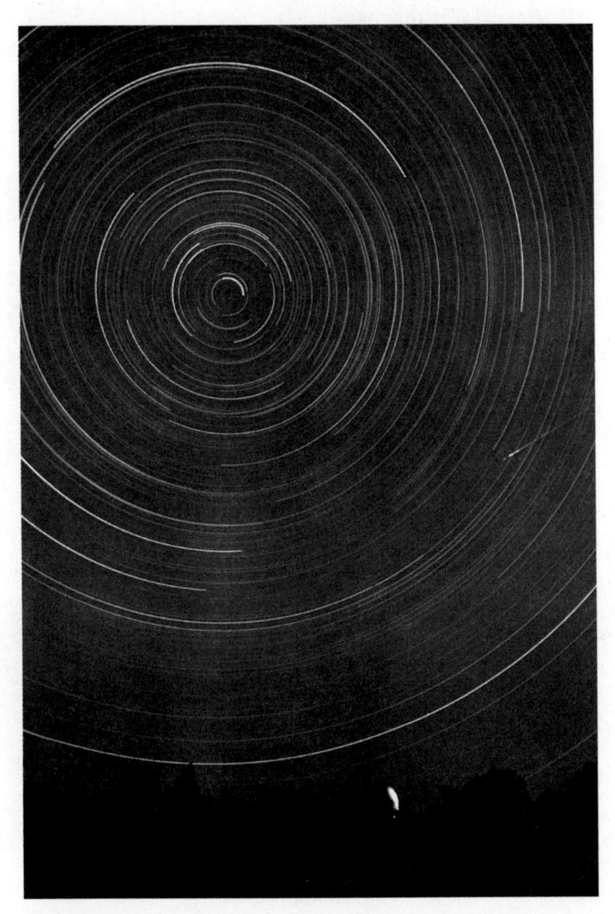

Figure 1 The stars around Polaris appear as streaks of light in this time-lapse photograph.

FOLDABLES®

Make a horizontal two-tab book. Label it as shown. Use it to organize your notes on astronomy.

How Scientists Divide the Night Sky | What Observations of the Stars Tell Scientists

Naked-Eye Astronomy

You don't need expensive equipment to view the sky. *Naked-eye astronomy* means gazing at the sky with just your eyes, without binoculars or a telescope. Long before the telescope was invented, people observed stars to tell time and find directions. They learned about planets, seasons, and astronomical events merely by watching the sky. As you practice naked-eye astronomy, remember never to look directly at the Sun—it could damage your eyes.

Constellations

As people in ancient cultures gazed at the night sky, they saw patterns. The patterns resembled people, animals, or objects, such as the hunter and the dragon shown in **Figure 2.** The Greek astronomer Ptolemy (TAH luh mee) identified dozens of star patterns nearly 2,000 years ago. Today, these patterns and others like them are known as ancient constellations.

Present-day astronomers use many ancient constellations to divide the sky into 88 regions. Some of these regions, which are also called constellations, are shown in the sky map in **Figure 2.** Dividing the sky helps scientists communicate to others what area of sky they are studying.

 **Key Concept Check** How do astronomers divide the night sky?

Figure 2 Most modern constellations contain an ancient constellation.

Visual Check Why does east appear on the left and west appear on the right on the sky map?

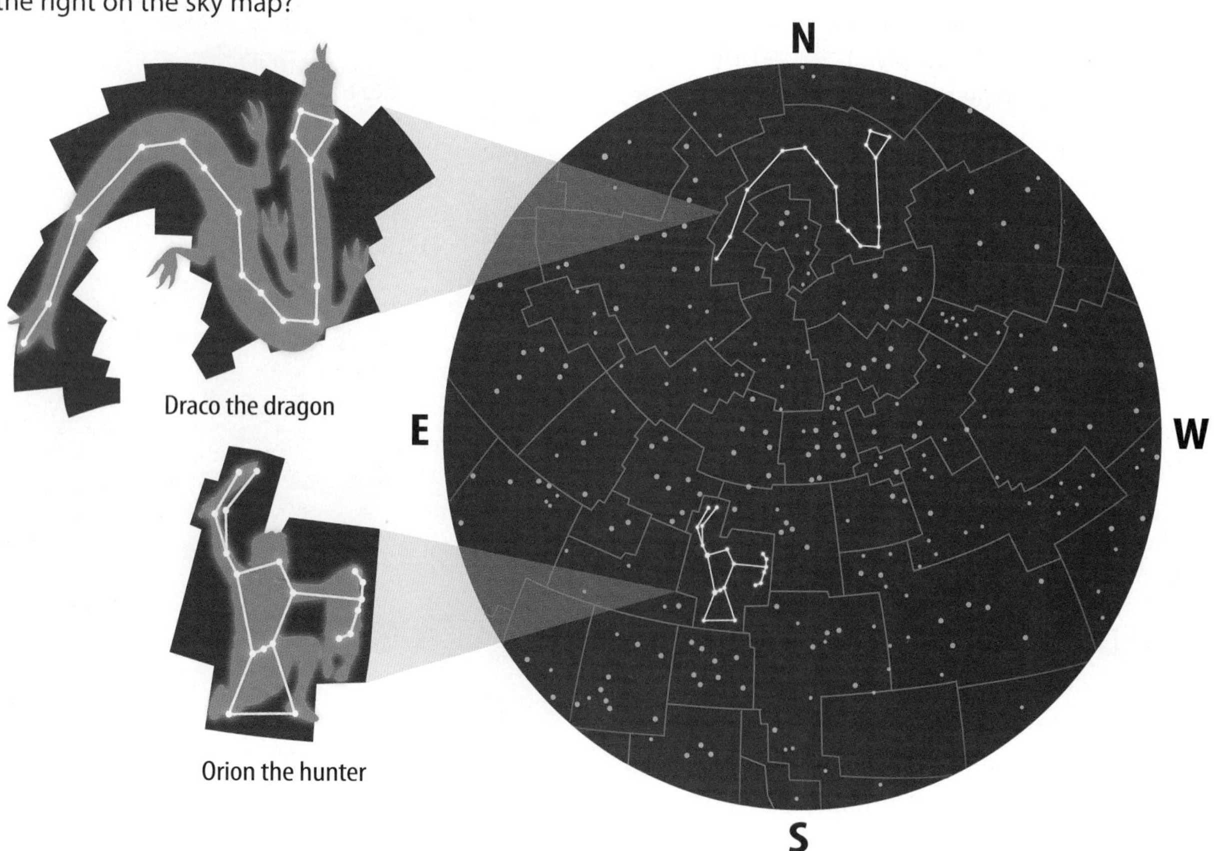

Electromagnetic Spectrum

Review Personal Tutor

Figure 3 Different parts of the electromagnetic spectrum have different wavelengths and different energies. You can see only a small part of the energy in these wavelengths.

Visual Check Which wavelength has the highest energy?

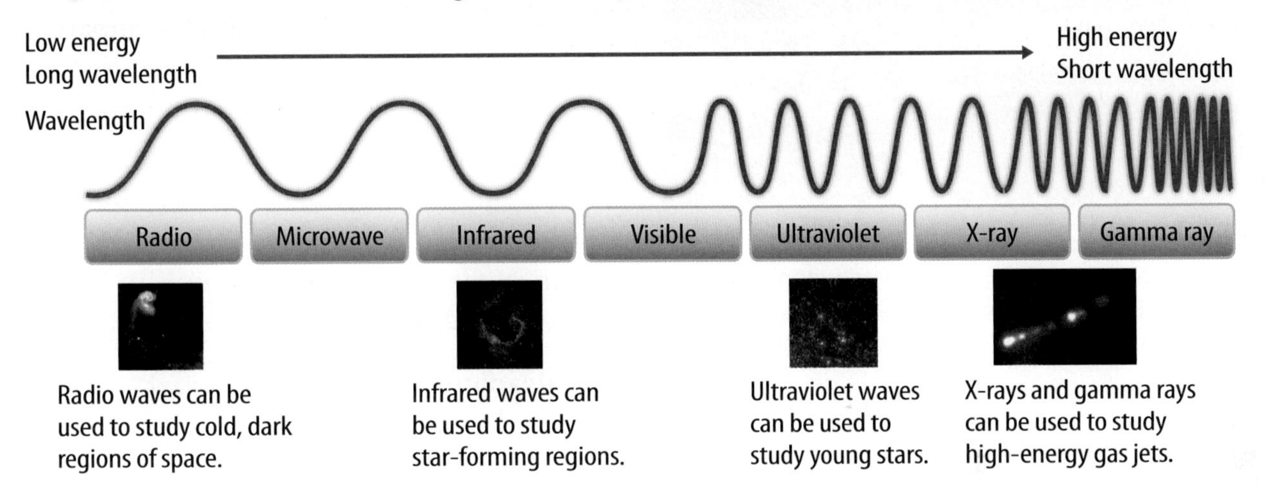

Telescopes

Telescopes can collect much more light than the human eye can detect. Visible light is just one part of the electromagnetic spectrum. As shown in **Figure 3,** the electromagnetic spectrum is a continuous range of wavelengths. Longer wavelengths have low energy. Shorter wavelengths have high energy. Different objects in space emit different ranges of wavelengths. The range of wavelengths that a star emits is the star's spectrum (plural, spectra).

Spectroscopes

Scientists study the spectra of stars using an instrument called a spectroscope. *A* **spectroscope** *spreads light into different wavelengths.* Using spectroscopes, astronomers can study stars' characteristics, including temperatures, compositions, and energies. For example, newly formed stars emit mostly radio and infrared waves, which have low energy. Exploding stars emit mostly high-energy ultraviolet waves and X-rays.

Key Concept Check What can astronomers learn from a star's spectrum?

Inquiry MiniLab — 20 minutes

How does light differ?

Light from the Sun is different from light from a lightbulb. How do the light sources differ?

1. Read and complete a lab safety form.
2. Follow instructions included with your **spectroscope.** Use it to observe various **light sources** around the classroom. Then use it to look at a bright part of the sky. ⚠ Do not look directly at the Sun.
3. Use **colored pencils** to draw what you see for each light source in your Science Journal.

Analyze and Conclude

1. **Compare and Contrast** What colors did you see for each light source? How did the colors differ?
2. **Key Concept** How might a spectroscope be used to learn about stars?

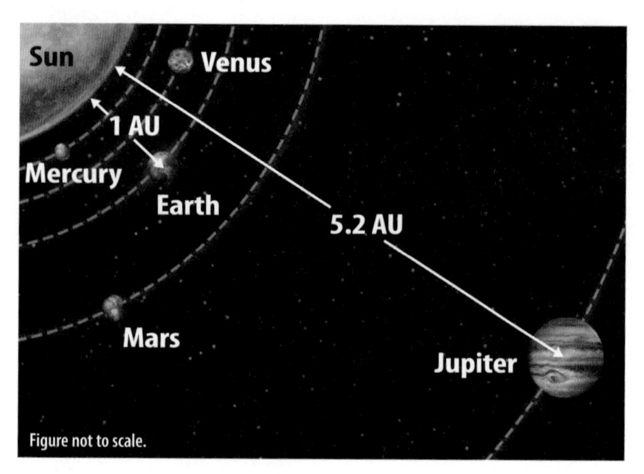

Figure 4 Measurements in the solar system are based on the average distance between Earth and the Sun—1 astronomical unit (AU). The most distant planet, Neptune, is 30 AU from the Sun.

WORD ORIGIN

parallax
from Greek *parallaxis,* means "alteration"

Math Skills

Use Proportions

Proportions can be used to calculate distances to astronomical objects. Light can travel nearly 10 trillion km in 1 year (y). How many years would it take light to reach Earth from a star that is 100 trillion km away?

1. Set up a proportion.

$$\frac{10 \text{ trillion km}}{1 \text{ y}} = \frac{100 \text{ trillion km}}{x \text{ y}}$$

2. Cross multiply.

$$10 \text{ trillion km} \times (x) \text{ y} = 100 \text{ trillion km} \times 1 \text{ y}$$

3. Solve for x by dividing both sides by 10 trillion km.

$$x = \frac{100 \text{ trillion km}}{10 \text{ trillion km}} = 10 \text{ y}$$

Practice

How many years would it take light to reach Earth from a star 60 trillion km away?

- **Math Practice**
- **Personal Tutor**

Measuring Distances

Hold up your thumb at arm's length. Close one eye, and look at your thumb. Now open that eye, and close the other eye. Did your thumb seem to jump? This is an example of parallax. Parallax is the apparent change in an object's position caused by looking at it from two different points.

Astronomers use angles created by parallax to measure how far objects are from Earth. Instead of the eyes being the two points of view, they use two points in Earth's orbit around the Sun.

 **Reading Check** What is parallax?

Distances Within the Solar System

Because the universe is too large to be measured easily in meters or kilometers, astronomers use other units of measurement. For distances within the solar system, they use astronomical units (AU). *An* **astronomical unit** *is the average distance between Earth and the Sun, about 150 million km.* Astronomical units are convenient to use in the solar system because distances easily can be compared to the distance between Earth and the Sun, as shown in **Figure 4.**

Distances Beyond the Solar System

Astronomers measure distances to objects beyond the solar system using a larger distance unit—the light-year. Despite its name, a light-year measures distance, not time. *A* **light-year** *is the distance light travels in 1 year.* Light travels at a rate of about 300,000 km/s. That means 1 light-year is about 10 trillion km! Proxima Centauri, the nearest star to the Sun, is 4.2 light-years away.

Looking Back in Time

Because it takes time for light to travel, you see a star not as it is today, but as it was when light left it. At 4.2 light-years away, Proxima Centauri appears as it was 4.2 years ago. The farther away an object, the longer it takes for its light to reach Earth.

Measuring Brightness

When you look at stars, you can see that some are dim and some are bright. Astronomers measure the brightness of stars in two ways: by how bright they appear from Earth and by how bright they actually are.

Apparent Magnitude

Scientists measure how bright stars appear from Earth using a scale developed by the ancient Greek astronomer Hipparchus (hi PAR kus). Hipparchus assigned a number to every star he saw in the night sky, based on the star's brightness. Astronomers today call these numbers magnitudes. *The* **apparent magnitude** *of an object is a measure of how bright it appears from Earth.*

As shown in **Figure 5,** some objects have negative apparent magnitudes. That is because Hipparchus assigned a value of 1 to all of the brightest stars. He also did not assign values to the Sun, the Moon, or Venus. Astronomers later assigned negative numbers to the Sun, the Moon, Venus, and a few bright stars.

Academic Vocabulary

apparent
(adjective) appearing to the eye or mind

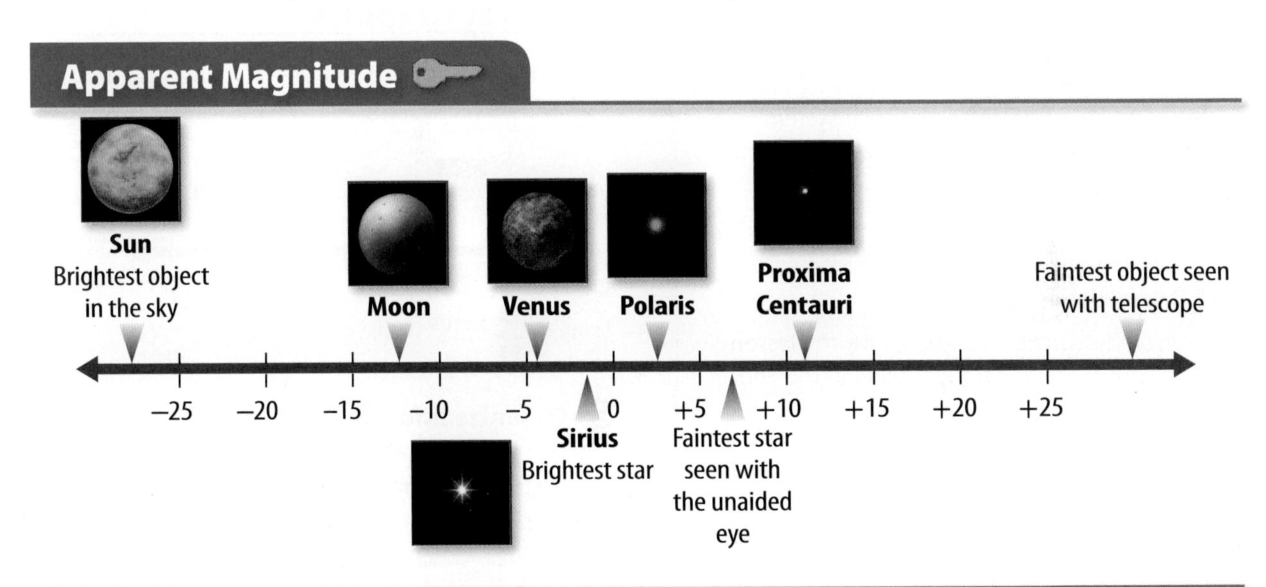

Figure 5 The fainter a star or other object in the sky appears, the greater its apparent magnitude.

Visual Check What is the apparent magnitude of Sirius?

Absolute Magnitude

Stars can appear bright or dim depending on their distances from Earth. But stars also have actual, or absolute, magnitudes. **Luminosity** (lew muh NAH sih tee) *is the true brightness of an object.* The luminosity of a star, measured on an absolute magnitude scale, depends on the star's temperature and size, not its distance from Earth. A star's luminosity, apparent magnitude, and distance are related. If scientists know two of these factors, they can determine the third using mathematical formulas.

Key Concept Check How do scientists measure the brightness of stars?

Lesson 1 Review

Visual Summary

Ancient people recognized patterns in the night sky. These patterns are known as the ancient constellations.

Different wavelengths of the electromagnetic spectrum carry different energies.

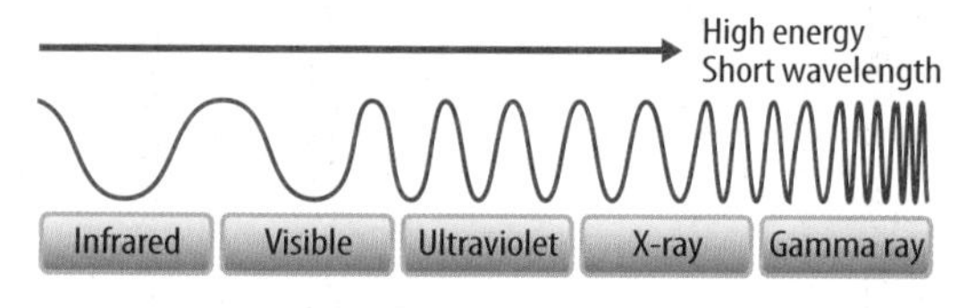

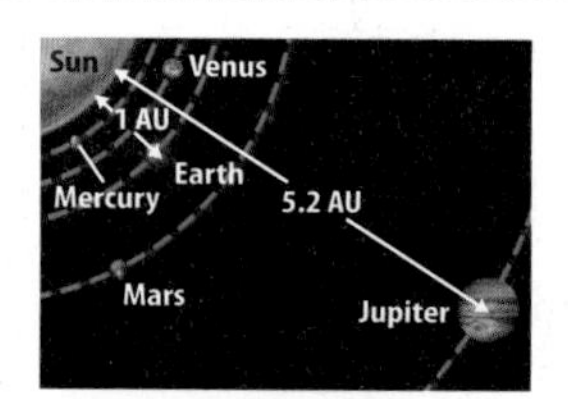

Astronomers measure distances within the solar system using astronomical units.

FOLDABLES®

Use your lesson Foldable to review the lesson. Save your Foldable for the project at the end of the chapter.

What do you think NOW?

You first read the statements below at the beginning of the chapter.

1. The night sky is divided into constellations.

2. A light-year is a measurement of time.

Did you change your mind about whether you agree or disagree with the statements? Rewrite any false statements to make them true.

Use Vocabulary

1. A device that spreads light into different wavelengths is a(n) ________.
2. **Define** *astronomical unit* and *light-year* in your own words.
3. **Distinguish** between apparent magnitude and luminosity.

Understand Key Concepts

4. Which does a light-year measure?
   A. brightness
   B. distance
   C. time
   D. wavelength
5. **Describe** how scientists divide the sky.

Interpret Graphics

6. **Analyze** Which star in the diagram below appears the brightest from Earth?

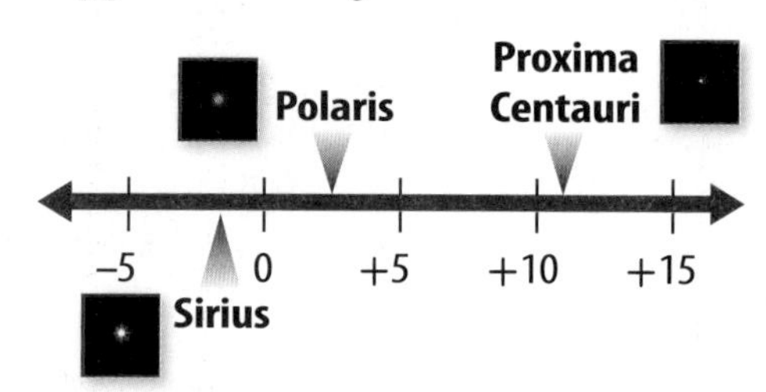

7. **Organize Information** Copy and fill in the graphic organizer below to list three things astronomers can learn from a star's light.

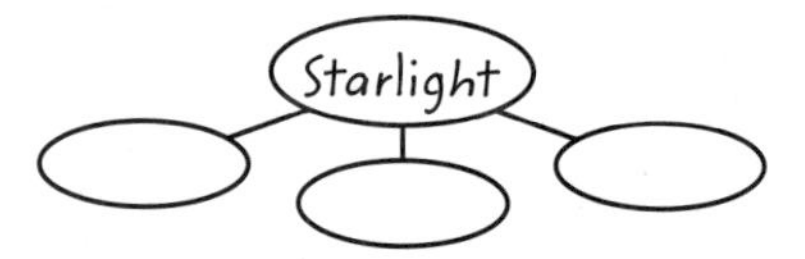

Critical Thinking

8. **Evaluate** why astronomers use modern constellation regions instead of ancient constellation patterns to divide the sky.

Math Skills

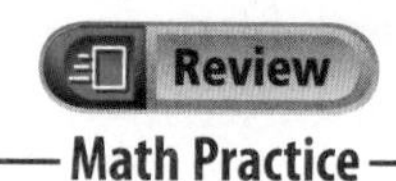

Math Practice

9. The Andromeda galaxy is about 25,000,000,000,000,000,000 km from Earth. How long does it take light to reach Earth from the Andromeda galaxy?

Interpret Scientific Illustrations **30 minutes**

How can you use scientific illustrations to locate constellations?

You might have heard that stars in the Big Dipper point to Polaris. The Big Dipper is a small star pattern in the larger constellation of Ursa Major. *Ursa Major* means "big bear" in Latin. It is the third-largest of the 88 modern constellations in the sky. Study the image of Ursa Major. Can you find the seven stars that form the Big Dipper? You can use a star finder to locate stars on any clear night of the year. The star finder also helps you see how constellations move across the sky.

Materials

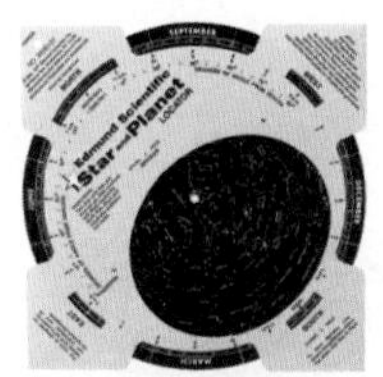

star chart

adhesive stars

graph paper

Learn It

Scientific illustrations can help you understand difficult or complicated subjects. **Interpret scientific illustrations** on the star finder to learn about the night sky.

Try It

1. Read and complete a lab safety form.
2. Read the user information provided with the star finder.
3. Rotate the wheel until the star finder is set to the day and time when you will be viewing the night sky. Observe how the ancient constellations marked on the star finder move.

4. Make a list of the bright stars, constellations, and planets you might be able to see in the sky.
5. Use the star finder outdoors on a clear night. As you hold the star finder overhead, be sure the arrows are pointing in the correct directions.

Apply It

6. What ancient constellations, planets, and stars were you able to see?
7. Did you locate Polaris? Why will you be able to see Polaris 6 months from now?
8. Which constellations won't you be able to see 6 months from now?
9. Why do stars appear to move?
10. How might ancient constellations have helped people in the past?
11. **Key Concept** How does dividing the sky into constellations help scientists study the sky?

Lesson 2

Reading Guide

Key Concepts
ESSENTIAL QUESTIONS

- How do stars shine?
- How are stars layered?
- How does the Sun change over short periods of time?
- How do scientists classify stars?

Vocabulary

nuclear fusion p. 423
star p. 423
radiative zone p. 424
convection zone p. 424
photosphere p. 424
chromosphere p. 424
corona p. 424
Hertzsprung-Russell diagram p. 427

g Multilingual eGlossary

The Sun and Other Stars

Inquiry Volcanoes on the Sun?

No, it's the Sun's atmosphere! The Sun's atmosphere can extend millions of kilometers into space. Sometimes the atmosphere becomes so active it disrupts communication systems and power grids on Earth.

Launch Lab

15 minutes

What are those spots on the Sun?

If you could see the Sun up close, what would it look like? Does it look the same all the time?

1. Examine a **collage of Sun images.** Notice the dates on which the pictures were taken.
2. Discuss with a partner what the dark spots might be and why they change position.
3. Select one spot. Estimate how long it took the spot to move completely across the surface of the Sun. Record your estimate in your Science Journal.

Think About This

1. What do you think the spots are?
2. Why do you think the spots move across the surface of the Sun?
3. **Key Concept** How do you think the Sun changes over days, months, and years?

How Stars Shine

The hotter something is, the more quickly its atoms move. As atoms move, they collide. If a gas is hot enough and its atoms move quickly enough, the nuclei of some of the atoms combine. **Nuclear fusion** *is a process that occurs when the nuclei of several atoms combine into one larger nucleus.*

Nuclear fusion releases a great amount of energy. This energy powers stars. *A* **star** *is a large ball of gas held together by gravity with a core so hot that nuclear fusion occurs.* A star's core can reach millions or hundreds of millions of degrees Celsius. When energy leaves a star's core, it travels throughout the star and radiates into space. As a result, the star shines.

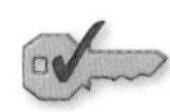

Key Concept Check How do stars shine?

FOLDABLES®

Make a vertical four-tab book. Label it as shown. Use it to organize your notes about the changing features of the Sun.

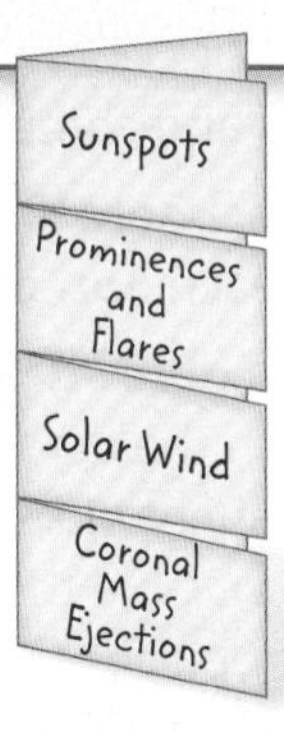

Composition and Structure of Stars

The Sun is the closest star to Earth. Because it is so close, scientists can easily observe it. They can send probes to the Sun, and they can study its spectrum using spectroscopes on Earth-based telescopes. Spectra of the Sun and other stars provide information about **stellar** composition. The Sun and most stars are made almost entirely of hydrogen and helium gas. A star's composition changes slowly over time as hydrogen in its core fuses into more complex nuclei.

SCIENCE USE V. COMMON USE

stellar

Science Use anything related to stars

Common Use outstanding, exemplary

Layers of the Sun

Convection zone
Photosphere
Radiative zone
Core
Corona
Chromosphere

Figure 6 The Sun is divided into six layers.

Visual Check Where is the photosphere located in relation to the Sun's other layers?

Inquiry MiniLab

Can you model the Sun's structure?

Making a two-dimensional model of the Sun can help you visualize its parts.

1. Read and complete a lab safety form.
2. Use **scissors** to cut out each **Sun part.**
3. Use a **glue stick** to attach the corona to a sheet of **black paper.** Glue the other pieces to the corona in this order: chromosphere, convection zone, radiative zone, core.

4. Glue only the top edge of the photosphere over the convection zone.

Analyze and Conclude

1. Draw the path a particle of light would follow from the core to the photosphere.
2. **Key Concept** How does this activity model a star's ability to shine?

Interior of Stars

When first formed, all stars fuse hydrogen into helium in their cores. Helium is denser than hydrogen, so it sinks to the inner part of the core after it forms.

The core is one of three interior layers of a typical star, as shown in the drawing of the Sun in **Figure 6.** *The* **radiative zone** *is a shell of cooler hydrogen above a star's core.* Hydrogen in this layer is dense. Light energy bounces from atom to atom as it gradually makes its way upward, out of the radiative zone.

Above the radiative zone is the **convection zone,** *where hot gas moves up toward the surface and cooler gas moves deeper into the interior.* Light energy moves quickly upward in the convection zone.

Key Concept Check What are the interior layers of a star?

Atmosphere of Stars

Beyond the convection zone are the three outer layers of a star. These layers make up a star's atmosphere. The **photosphere** *is the apparent surface of a star.* In the Sun, it is the dense, bright part you can see, where light energy radiates into space. From Earth, the Sun's photosphere looks smooth. But like the rest of the Sun, it is made of gas.

Above the photosphere are the two outer layers of a star's atmosphere. *The* **chromosphere** *is the orange-red layer above the photosphere,* as shown in **Figure 6.** *The* **corona** *is the wide, outermost layer of a star's atmosphere.* The temperature of the corona is higher than the photosphere or the chromosphere. It has an irregular shape and can extend outward for several million kilometers.

The Sun's Changing Features

The interior features of the Sun are stable over millions of years. But the Sun's atmosphere can change over years, months, or even minutes. Some of these features are illustrated in **Table 1** on the following page.

Table 1 The Sun is dynamic. It changes over years, months, hours, and minutes.

Key Concept Check Which parts of the Sun change over short periods of time?

Table 1 Changing Features of the Sun

Sunspots Regions of strong magnetic activity are called sunspots. Cooler than the rest of the photosphere, sunspots appear as dark splotches on the Sun. They seem to move across the Sun as the Sun rotates. The number of sunspots changes over time. They follow a cycle, peaking in number every 11 years. An average sunspot is about the size of Earth.	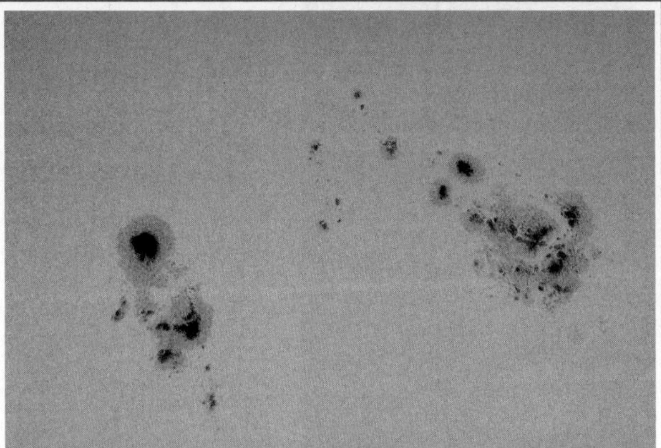
Prominences and Flares The loop shown here is a prominence. Prominences are clouds of gas that make loops and jets extending into the corona. They sometimes last for weeks. Flares are sudden increases in brightness often found near sunspots or prominences. They are violent eruptions that last from minutes to hours. Both prominences and flares begin at or just above the photosphere.	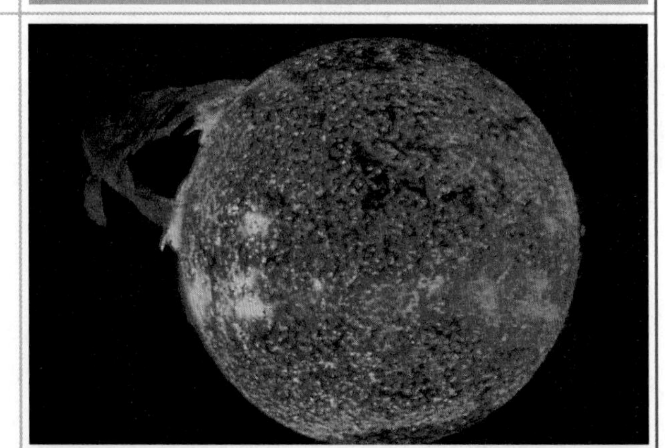
Coronal Mass Ejections (CMEs) Huge bubbles of gas ejected from the corona are coronal mass ejections (CMEs). They are much larger than flares and occur over the course of several hours. Material from a CME can reach Earth, occasionally causing a radio blackout or a malfunction in an orbiting satellite.	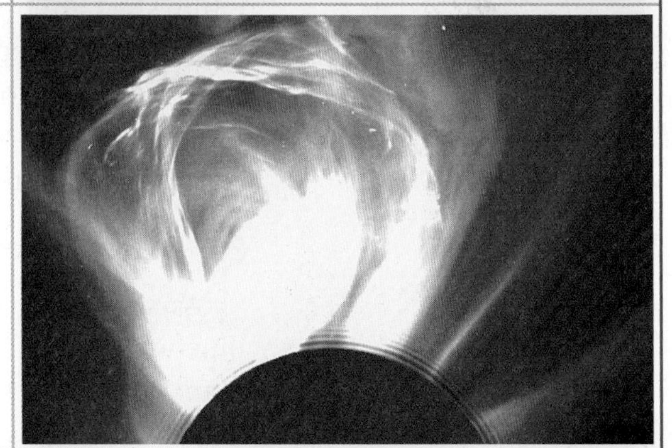
The Solar Wind Charged particles that stream continually away from the Sun create the solar wind. The solar wind passes Earth and extends to the edge of the solar system. Auroras are curtains of light created when particles from the solar wind or a CME interact with Earth's magnetic field. Auroras occur in both the northern and southern hemispheres. The northern lights are shown here.	

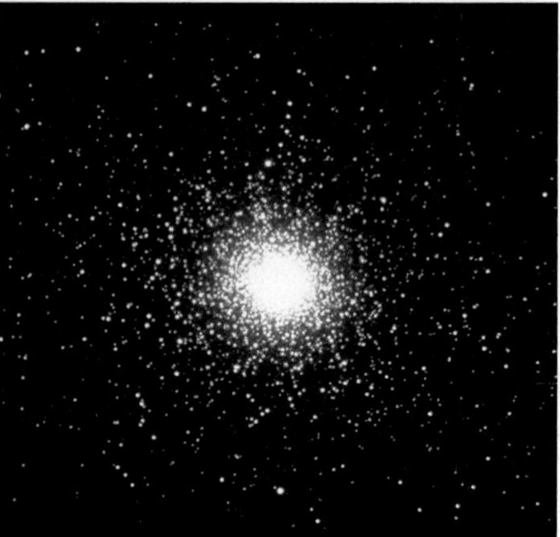

Figure 7 Open clusters (top) contain fewer than 1,000 stars. Globular clusters (bottom) can contain hundreds of thousands of stars.

Groups of Stars

The Sun has no stellar companion. The star closest to the Sun is 4.2 light-years away. Many stars are single stars, such as the Sun. Most stars exist in multiple star systems bound by gravity.

The most common star system is a binary system, where two stars orbit each other. By studying the orbits of binary stars, astronomers can determine the stars' masses. Many stars exist in large groupings called clusters. Two types of star clusters—open clusters and **globular** clusters—are shown in **Figure 7.** Stars in a cluster all formed at about the same time and are the same distance from Earth. If astronomers determine the distance to or the age of one star in a cluster, they know the distance to or the age of every star in that cluster.

Word Origin
globular
from Latin *globus*, means "round mass, sphere"

Classifying Stars

How do you classify a star? Which properties are important? Scientists classify stars according to their spectra. Recall that a star's spectrum is the light it emits spread out by wavelength. Stars have different spectra and different colors depending on their surface temperatures.

Temperature, Color, and Mass

Have you ever seen coals in a fire? Red coals are the coolest, and blue-white coals are the hottest. Stars are similar. Blue-white stars are hotter than red stars. Orange, yellow, and white stars are intermediate in temperature. Though there are exceptions, color in most stars is related to mass, as shown in **Figure 8.** Blue-white stars tend to have the most mass, followed by white stars, yellow stars, orange stars, and red stars.

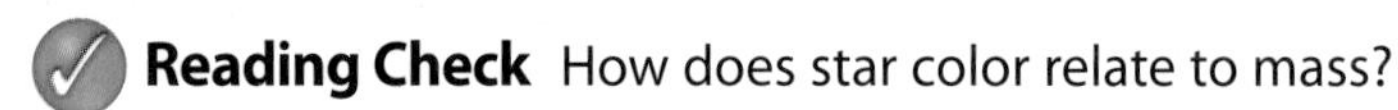

Reading Check How does star color relate to mass?

As shown in **Figure 8,** the Sun is tiny compared to large, blue-white stars. However, scientists suspect that most stars—as many as 90 percent—are smaller than the Sun. These stars are called red dwarfs. The smallest star in **Figure 8** is a red dwarf.

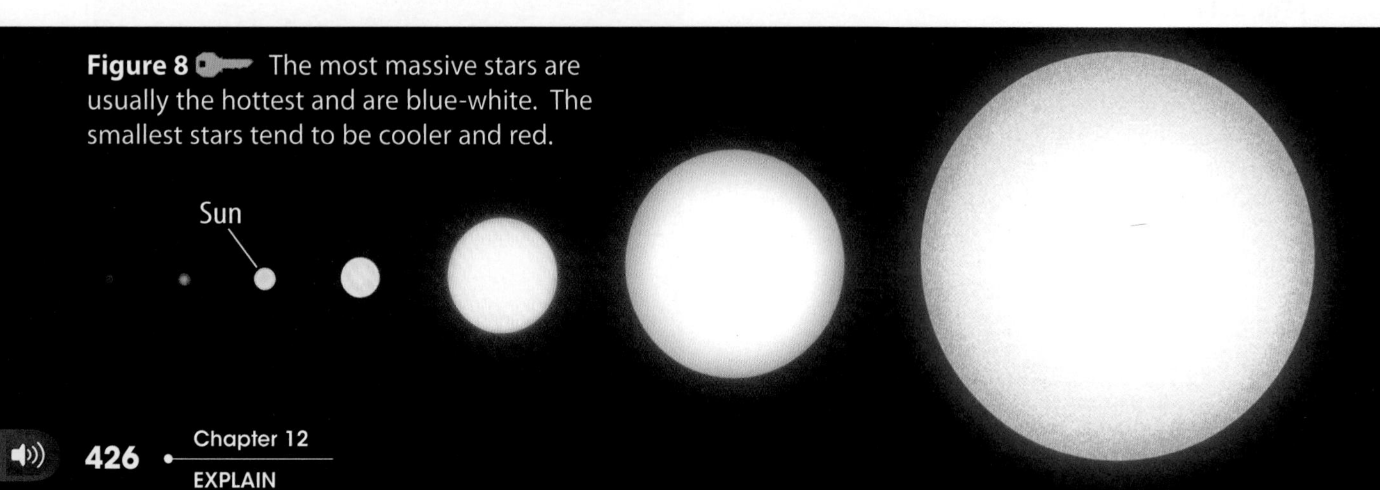

Figure 8 The most massive stars are usually the hottest and are blue-white. The smallest stars tend to be cooler and red.

Hertzsprung-Russell Diagram 

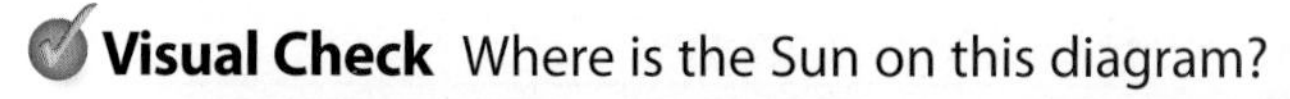

Figure 9 The H-R diagram plots luminosity against temperature. Most stars exist along the main sequence, the band that stretches from the upper left to the lower right.

Visual Check Where is the Sun on this diagram?

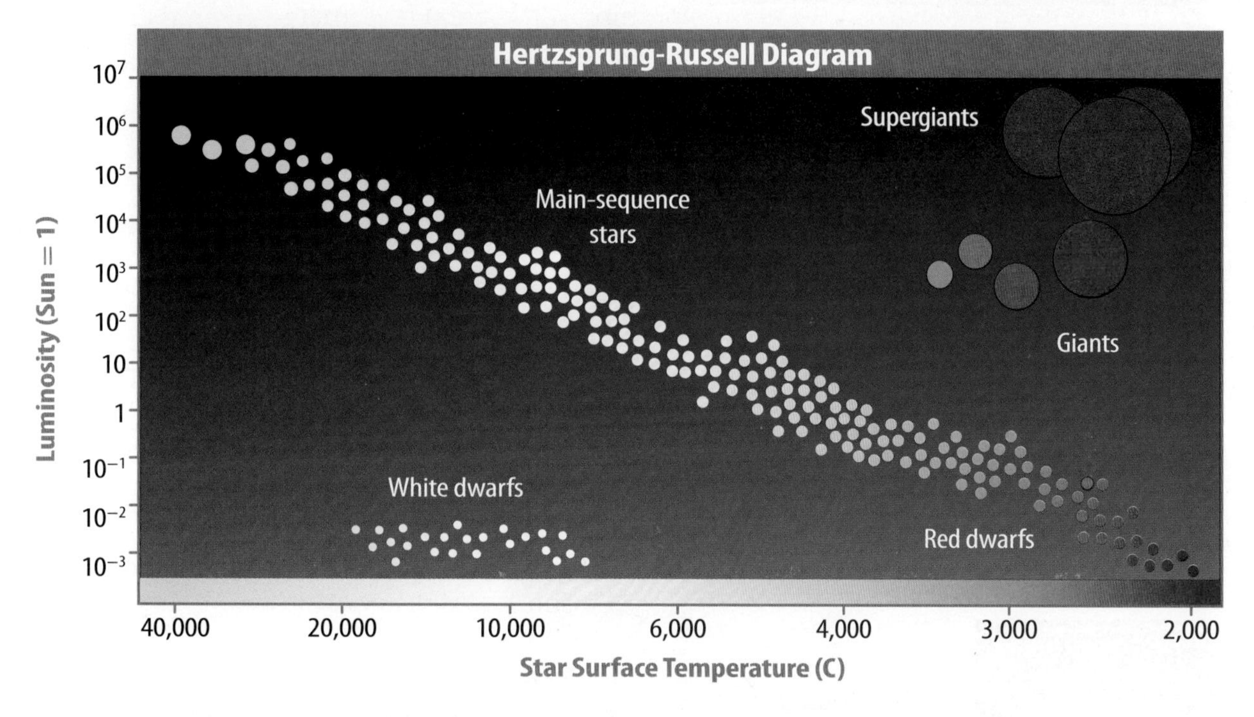

Hertzsprung-Russell Diagram

When scientists plot the temperatures of stars against their luminosities, the result is a graph like that shown in **Figure 9.** The **Hertzsprung-Russell diagram** (or H-R diagram) *is a graph that plots luminosity v. temperature of stars.* The *y*-axis of the H-R diagram displays increasing luminosity. The *x*-axis displays decreasing temperature.

The H-R diagram is named after two astronomers who developed it in the early 1900s. It is an important tool for categorizing stars. It also is an important tool for determining distances of some stars. If a star has the same temperature as a star on the H-R diagram, astronomers often can determine its luminosity. As you read earlier, if scientists know a star's luminosity, they can calculate its distance from Earth.

Key Concept Check What is the Hertzsprung-Russell diagram?

The Main Sequence

Most stars spend the majority of their lives on the main sequence. On the H-R diagram, main sequence stars form a curved line from the upper left corner to the lower right corner of the graph. The mass of a star determines both its temperature and its luminosity; the higher the mass the hotter and brighter the star. Because high-mass stars have more gravity pulling inward than low-mass stars, their cores have higher temperatures and produce and use more energy through fusion. High-mass stars have a shorter life span than low-mass stars. High-mass stars burn through their hydrogen much faster and move off the main sequence. A downside to a large-mass star is that the life span of the star is much shorter than average- or low-mass stars.

As shown in **Figure 9,** some groups of stars on the H-R diagram lie outside of the main sequence. These stars are no longer fusing hydrogen into helium in their cores. Some of these stars are cooler, but brighter and larger, such as supergiants. Other stars are dimmer and smaller, but much hotter, such as white dwarfs.

Lesson 2 Review

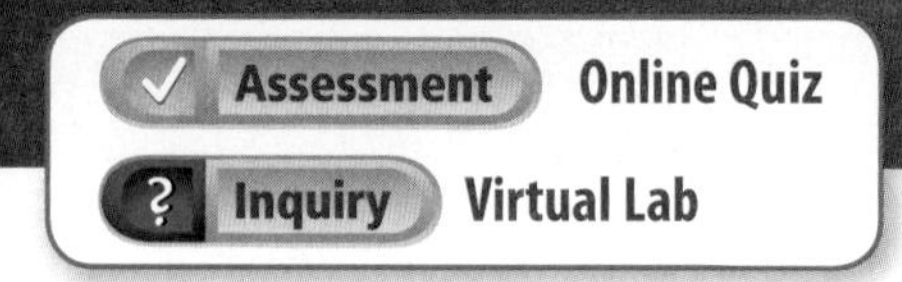

Visual Summary

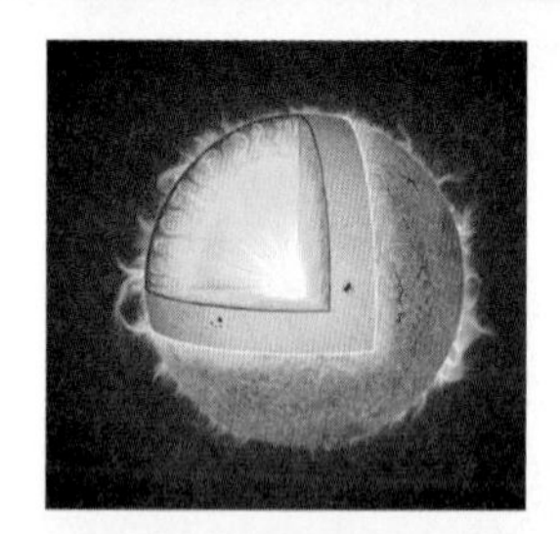

Hot gas moves up and cool gas moves down in the Sun's convection zone.

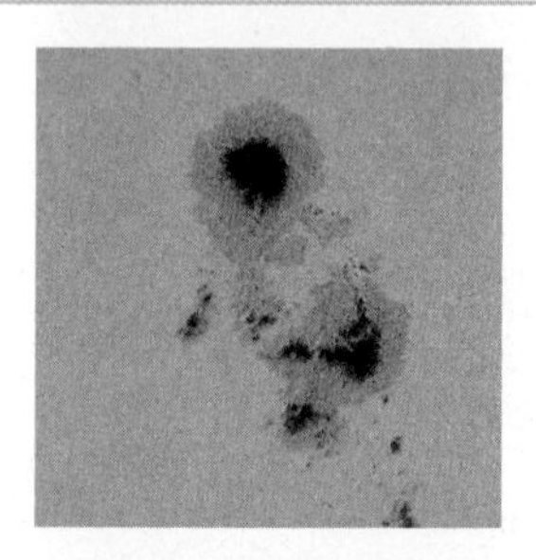

Sunspots are relatively dark areas on the Sun that have strong magnetic activity.

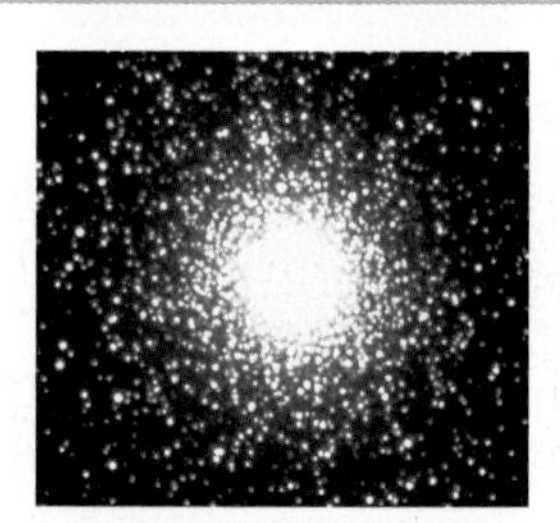

Globular clusters contain hundreds of thousands of stars.

FOLDABLES®

Use your lesson Foldable to review the lesson. Save your Foldable for the project at the end of the chapter.

What do you think NOW?

You first read the statements below at the beginning of the chapter.

3. Stars shine because there are nuclear reactions in their cores.

4. Sunspots are dark because they are cooler than nearby areas.

Did you change your mind about whether you agree or disagree with the statements? Rewrite any false statements to make them true.

Use Vocabulary

1. The ________ is a graph that plots luminosity v. temperature.
2. **Use the term** *photosphere* in a sentence.
3. **Define** *star* in your own words.

Understand Key Concepts 

4. Which part of a star extends millions of kilometers into space?
   A. chromosphere
   B. corona
   C. photosphere
   D. radiative zone
5. **Explain** how stars produce and release energy.
6. **Construct** an H-R diagram, and show the positions of the main sequence and the Sun.

Interpret Graphics

7. **Identify** Which star on the diagram below is hottest? Which is coolest? Which star represents the Sun?

8. **Organize Information** Copy and fill in the graphic organizer below to list the Sun's radiative zone, corona, convection zone, chromosphere, and photosphere in order outward from the core.

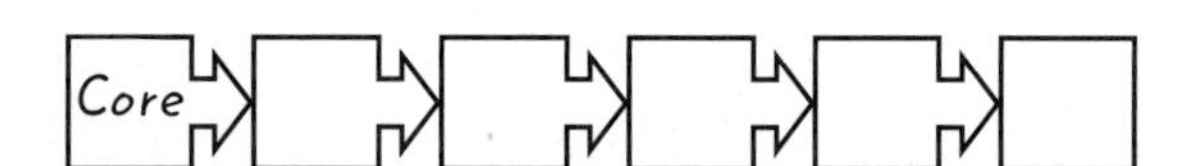

Critical Thinking

9. **Assess** why scientists monitor the Sun's changing features.
10. **Evaluate** In what way is the Sun an average star? In what way is it not an average star?

HOW IT WORKS

Viewing the Sun in 3-D

NASA's Solar Terrestrial Relations Observatory

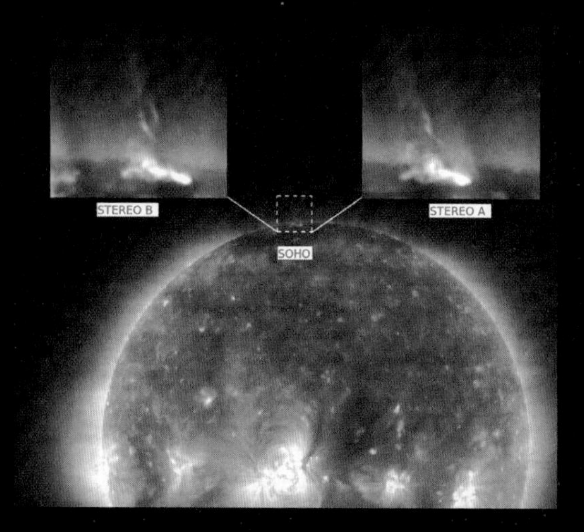

You might have used a telescope to look at objects far in the distance or to look at stars and planets. Although telescopes allow you to see a distant object in closer detail, you cannot see a three-dimensional view of objects in space. To get a three-dimensional view of the Sun, astronomers use two space telescopes. NASA's *Solar Terrestrial Relations Observatory* (STEREO) telescopes orbit the Sun in front of and behind Earth and give astronomers a 3-D view of the Sun. Why is this important?

STEREO B is in orbit around the Sun behind Earth.

In January 2009, the telescopes were 90 degrees apart.

In February 2011, the craft will be 180 degrees apart.

Earth

STEREO A is in orbit around the Sun ahead of Earth.

If a coronal mass ejection (CME) erupts from the Sun, it can blast more than a billion tons of material into space. The powerful energy in a CME can damage satellites and power grids if Earth happens to be in its way. Before STEREO, scientists had only a straight-on view of CMEs approaching Earth. With STEREO, they have two different views. Each STEREO telescope carries several cameras that can detect many wavelengths. Scientists combine the pictures from each type of camera to make one 3-D image. In this way, they can track a CME from its emergence on the Sun all the way to its impact with Earth.

It's Your Turn

RESEARCH AND REPORT How can power and satellite companies prepare for an approaching CME? Find out and write a short report on what you find. Share your findings with the class.

Lesson 3

Evolution of Stars

Reading Guide

Key Concepts
ESSENTIAL QUESTIONS

- How do stars form?
- How does a star's mass affect its evolution?
- How is star matter recycled in space?

Vocabulary

nebula p. 431
white dwarf p. 433
supernova p. 433
neutron star p. 434
black hole p. 434

g Multilingual eGlossary

Inquiry Exploding Star?

No, this is a cloud of gas and dust where stars form. How do you think stars form? Do you think stars ever stop shining?

Inquiry Launch Lab

20 minutes

Do stars have life cycles?

You might have learned about the life cycles of plants or animals. Do stars, such as the Sun, have life cycles? Before you find out, review the life cycle of a sunflower.

1. Read and complete a lab safety form.
2. Obtain an **envelope containing slips of paper** that explain the life cycle of a sunflower.
3. Use **colored pencils** to draw a sunflower in the middle of a piece of **paper,** or use a **glue stick** to glue a sunflower picture on the paper.
4. Using your knowledge of plant life cycles, arrange the slips of paper around the sunflower in the order in which the events listed on them occur. Draw arrows to show how the steps form a cycle.

Think About This

1. Does the life cycle of a sunflower have a beginning and an end? Explain your answer.
2. Do you think that every stage in the life cycle takes the same amount of time? Why or why not?
3. **Key Concept** How do you think the life cycle of a star compares to the life cycle of a sunflower? Do you think all stars have the same life cycle?

Life Cycle of a Star

Like living things, stars have life cycles. They are "born," and after millions or billions of years, they "die." Stars die in different ways, depending on their masses. But all stars—from white dwarfs to supergiants—form in the same way.

WORD ORIGIN
nebula
from Latin *nebula,* means "mist" or "little cloud"

Nebulae and Protostars

Stars form deep inside clouds of gas and dust. *A cloud of gas and dust is a* **nebula** (plural, nebulae). Star-forming nebulae are cold, dense, and dark. Gravity causes the densest parts to collapse, forming regions called protostars. Protostars continue to contract, pulling in surrounding gas, until their cores are hot and dense enough for nuclear fusion to begin. As they contract, protostars produce enormous amounts of thermal energy.

FOLDABLES®
Make a vertical five-tab book. Label it as shown. Use it to organize your notes on the life cycle of a star.

Protostar
Main Sequence
Red Giant
Red Supergiant
Supernova

Birth of a Star

Over many thousands of years, the energy produced by protostars heats the gas and dust surrounding them. Eventually, the surrounding gas and dust blows away, and the protostars become visible as stars. Some of this material might later become planets or other objects that orbit the star. During the star-formation process, nebulae glow brightly, as shown in the photograph on the previous page.

 **Key Concept Check** How do stars form?

Main-Sequence Stars

Recall the main sequence of the Hertzsprung-Russell diagram. Stars spend most of their lives on the main sequence. A star becomes a main-sequence star as soon as it begins to fuse hydrogen into helium in the core. It remains on the main sequence for as long as it continues to fuse hydrogen into helium. Average-mass stars such as the Sun remain on the main sequence for billions of years. High-mass stars remain on the main sequence for only a few million years. Even though massive stars have more hydrogen than lower-mass stars, they process it at a much faster rate.

When a star's hydrogen supply is nearly gone, the star moves off the main sequence. It begins the next stage of its life cycle, as shown in **Figure 10.** Not all stars go through all phases in **Figure 10.** Lower-mass stars do not have enough mass to become supergiants.

Figure 10 Massive stars become red giants, then larger red giants, then red supergiants.

Visual Check Which element forms in only the most massive stars?

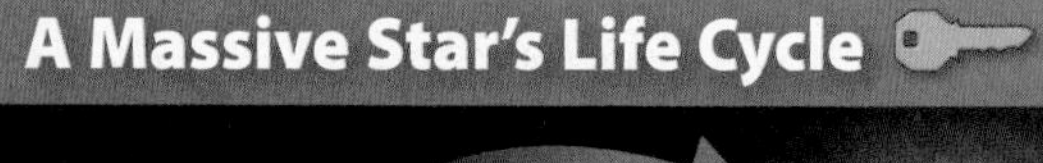

Massive star

Helium fuses and forms carbon.

Hydrogen fuses and forms helium.

Red giant

When the star's hydrogen supply is gone, gravity causes the core to contract and heat up. Thermal energy in the star's center causes the star's outer layers to expand and cool. The star becomes a red giant. Eventually, the interior becomes hot enough to resume nuclear fusion. The outer layers contract, and the star begins to fuse helium nuclei and form carbon.

Carbon fuses and forms other elements.

Helium fuses and forms carbon.

Hydrogen fuses and forms helium.

Larger red giant

When helium in the core runs low, the core again collapses under gravity and the outer layers expand. The star becomes a red giant for a second time, but this time it is even larger. It again contracts when it begins to fuse carbon and form other elements.

Hydrogen

Helium

Carbon

Neon

Oxygen

Silicon

Iron

Red supergiant

The process repeats again and again. The star becomes a red supergiant as different elements are formed during fusion. Iron nuclei form in the most massive stars.

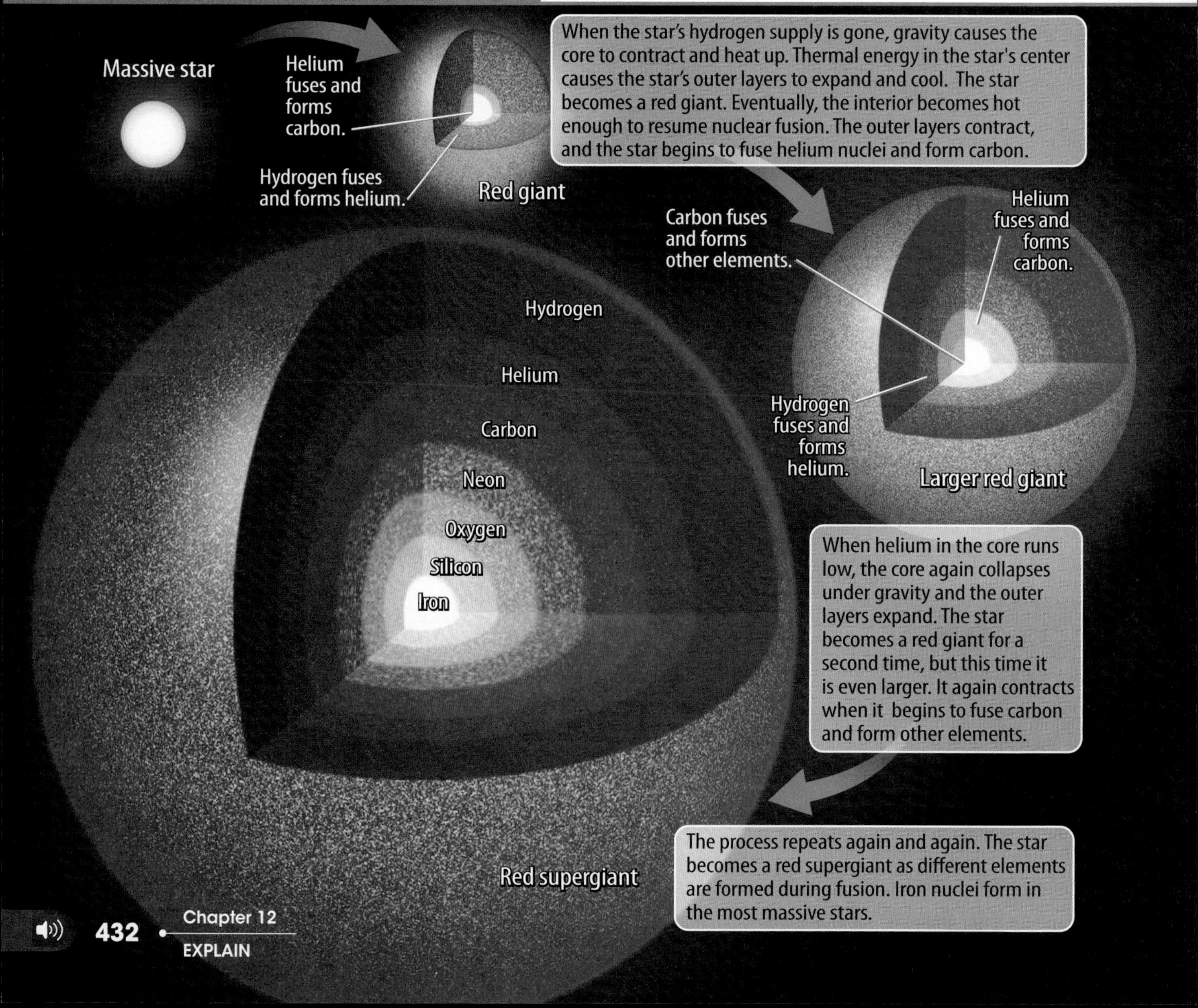

End of a Star

All stars form in the same way. But stars die in different ways, depending on their masses. Massive stars collapse and explode. Lower-mass stars die more slowly.

White Dwarfs

Average-mass stars, such as the Sun, do not have enough mass to fuse elements beyond helium. They do not get hot enough. After helium in their cores is gone, the stars cast off their gases, exposing their cores. The core becomes a **white dwarf**, *a hot, dense, slowly cooling sphere of carbon.*

What will happen to Earth and the solar system when the Sun runs out of fuel? When the Sun runs out of hydrogen, in about 5 billion years, it will become a red giant. Once helium fusion begins, the Sun will contract. When the helium is gone, the Sun will expand again, probably absorbing Mercury, Venus, and Earth and pushing Mars outward, as shown in **Figure 11.** Eventually, the Sun will become a white dwarf. Imagine the mass of the Sun squeezed a million times until it is the size of Earth. That's the size of a white dwarf. Scientists expect that all stars with masses less than 8–10 times that of the Sun will eventually become white dwarfs.

Reading Check What will happen to Earth when the Sun runs out of fuel?

Supernovae

Stars with more than 10 times the mass of the Sun do not become white dwarfs. Instead, they explode. *A* **supernova** (plural, supernovae) *is an enormous explosion that destroys a star.* In the most massive stars, a supernova occurs when iron forms in the star's core. Iron is stable and does not fuse. After a star forms iron, it loses its internal energy source, and the core collapses quickly under the force of gravity. So much energy is released that the star explodes. When it explodes, a star can become one billion times brighter and form elements even heavier than iron.

Figure 11 In about 5 billion years, the Sun will become a red giant and then a white dwarf.

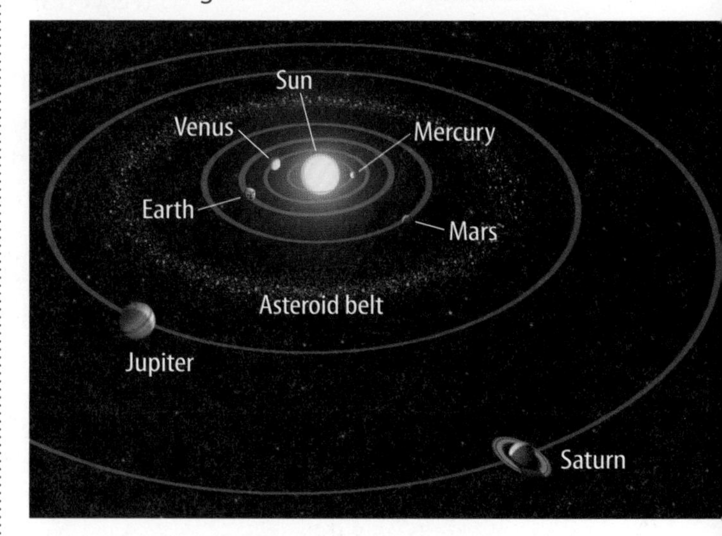

The Sun will remain on the main sequence for 5 billion more years.

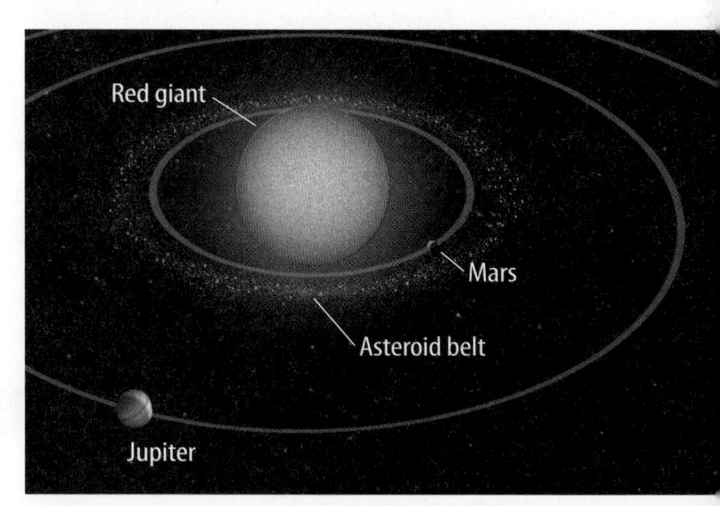

When the Sun becomes a red giant for the second time, the outer layers will probably absorb Earth and push Mars and Jupiter outward.

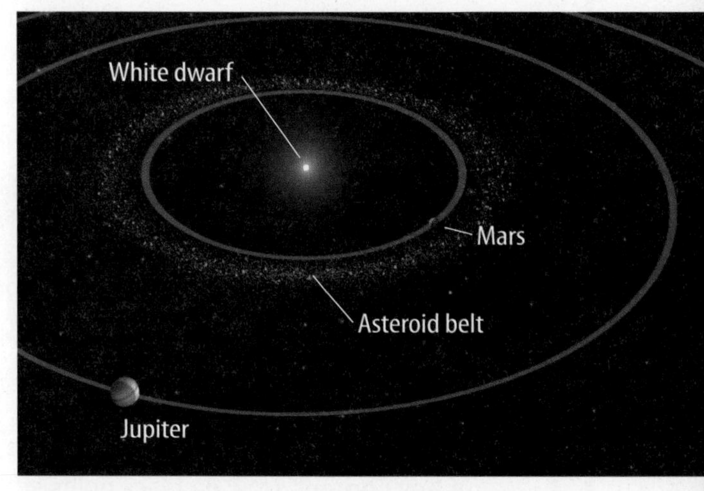

When the Sun becomes a white dwarf, the solar system will be a cold, dark place.

Neutron Stars

Have you ever eaten cotton candy? A bag of cotton candy is made from just a few spoonfuls of spun sugar. Cotton candy is mostly air. Similarly, atoms are mostly empty space. During a supernova, the outer layers of the star are blown away and the core collapses under the heavy force of gravity. The space in atoms disappears as protons and electrons combine to form neutrons. *A* **neutron star** *is a dense core of neutrons that remains after a supernova.* Neutron stars are only about 20 km wide, with cores so dense that a teaspoonful would weigh more than 1 billion tons.

REVIEW VOCABULARY
neutron
a neutral particle in the nucleus of an atom

Black Holes

For the most massive stars, atomic forces holding neutrons together are not strong enough to overcome so much mass in such a small volume. Gravity is too strong, and the matter crushes into a black hole. *A* **black hole** *is an object whose gravity is so great that no light can escape.*

A black hole does not suck matter in like a vacuum cleaner. But its gravity is very strong because all of its mass is concentrated in a single point. Because astronomers cannot see a black hole, they only can infer its existence. For example, if they detect a star circling around something, but they cannot see what that something is, they infer it is a black hole.

Key Concept Check How does a star's mass determine if it will become a white dwarf, a neutron star, or a black hole?

Inquiry MiniLab

15 minutes

How do astronomers detect black holes?

The only way astronomers can detect black holes is by studying the movement of objects nearby. How do black holes affect nearby objects?

1. Read and complete a lab safety form.
2. With a partner, make two stacks of **books** of equal height about 25 cm apart. Place a piece of **thin cardboard** on top of the books.
3. Spread some **staples** over the cardboard. Hold a **magnet** under the cardboard. Observe what happens to the staples.
4. While one student holds the magnet in place beneath the cardboard, the other student gently rolls a **small magnetic marble** across the cardboard. Repeat several times, rolling the marble in different pathways. Record your observations in your Science Journal.

Analyze and Conclude

1. **Infer** What did the pull of the magnet represent?
2. **Cause and Effect** How did the magnet affect the staples and the movement of the marble?
3. **Key Concept** How do black holes affect nearby objects?

Recycling Matter

At the end of a star's life cycle, much of its gas escapes into space. This gas is recycled. It becomes the building blocks of future generations of stars and planets.

Planetary Nebulae

You read that average-mass stars, such as the Sun, become white dwarfs. When a star becomes a white dwarf, it casts off hydrogen and helium gases in its outer layers, as shown in **Figure 12.** The expanding, cast-off matter of a white dwarf is a planetary nebula. Most of the star's carbon remains locked in the white dwarf. But the gases in the planetary nebula can be used to form new stars.

Planetary nebulae have nothing to do with planets. They are so named because early astronomers thought they were regions where planets were forming.

Supernova Remnants

During a supernova, a massive star comes apart. This sends a shock wave into space. The expanding cloud of dust and gas is called a supernova remnant. A supernova remnant is shown in **Figure 13.** Like a snowplow pushing snow in its path, a supernova remnant pushes on the gas and dust it encounters.

In a supernova, a star releases the elements that formed inside it during nuclear fusion. Almost all of the elements in the universe other than hydrogen and helium were created by nuclear reactions inside the cores of massive stars and released in supernovae. This includes the oxygen in air, the silicon in rocks, and the carbon in you.

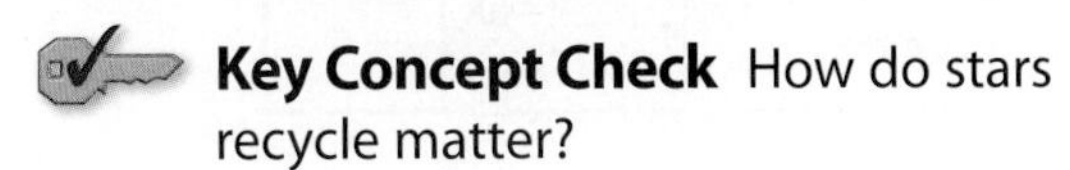

Key Concept Check How do stars recycle matter?

Gravity causes recycled gases and other matter to clump together in nebulae and form new stars and planets. As you will read in the next lesson, gravity also causes stars to clump together into even larger structures called galaxies.

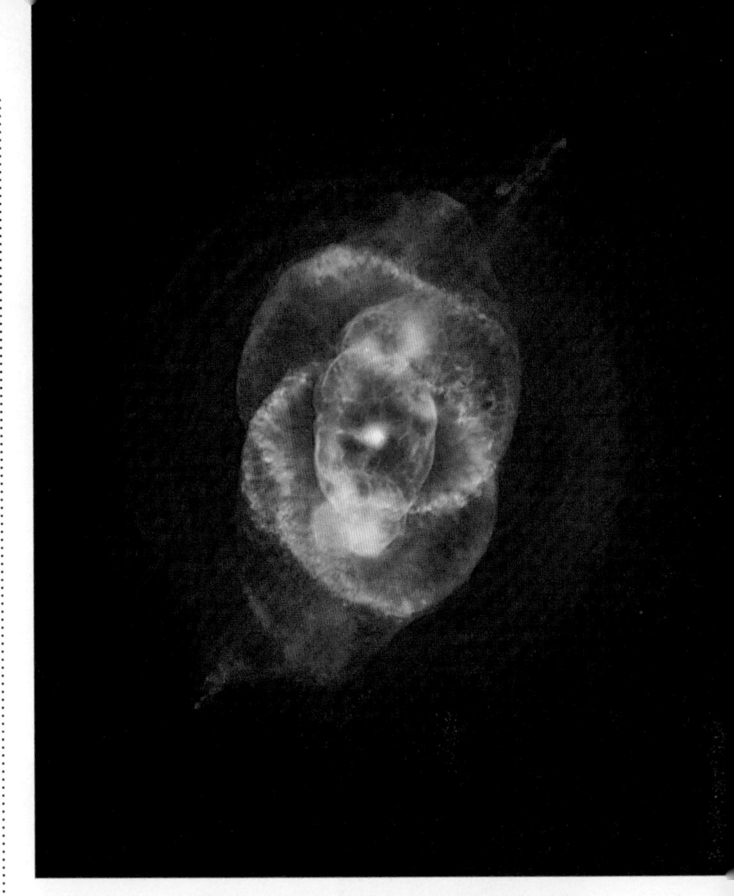

▲ **Figure 12** White dwarfs cast off helium and hydrogen as planetary nebulae. The gases can be used by new generations of stars.

▲ **Figure 13** Many of the elements in you and in matter all around you were formed inside massive stars and released in supernovae.

Lesson 3 Review

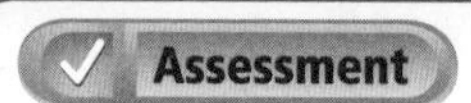

Visual Summary

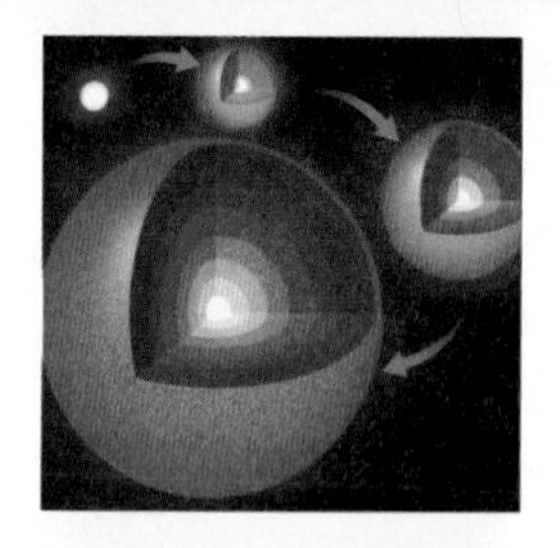

Iron is formed in the cores of the most massive stars.

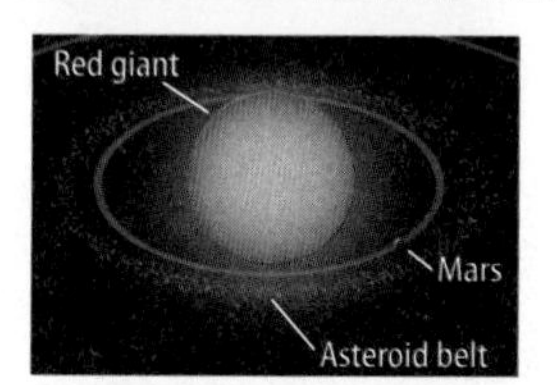

The Sun will become a red giant in about 5 billion years.

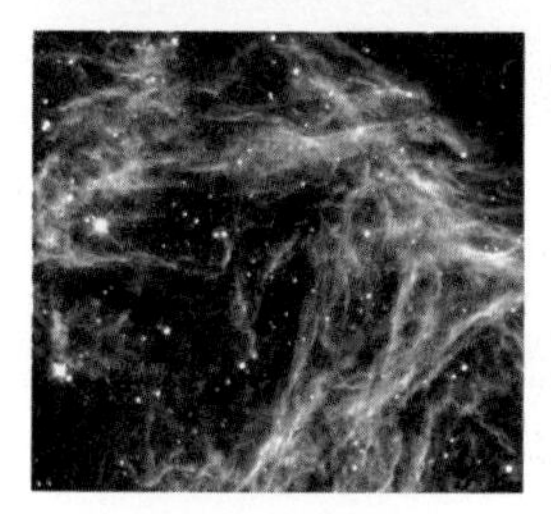

Matter is recycled in supernovae.

FOLDABLES®

Use your lesson Foldable to review the lesson. Save your Foldable for the project at the end of the chapter.

What do you think NOW?

You first read the statements below at the beginning of the chapter.

5. The more matter a star contains, the longer it is able to shine.

6. Gravity plays an important role in the formation of stars.

Did you change your mind about whether you agree or disagree with the statements? Rewrite any false statements to make them true.

Use Vocabulary

1. Planetary nebulae are the expanding outer layers of a(n) ________.
2. **Define** *supernova* in your own words.
3. **Use the terms** *neutron star* and *black hole* in a sentence.

Understand Key Concepts 

4. Which type of star will the Sun eventually become?
   - **A.** neutron star
   - **B.** red dwarf
   - **C.** red supergiant
   - **D.** white dwarf
5. **Explain** how supernovae recycle matter.
6. **Rank** black holes, neutron stars, and white dwarfs from smallest to largest. Then rank them from most massive to least massive.

Interpret Graphics

7. **Describe** details of the process occurring in the photo below.

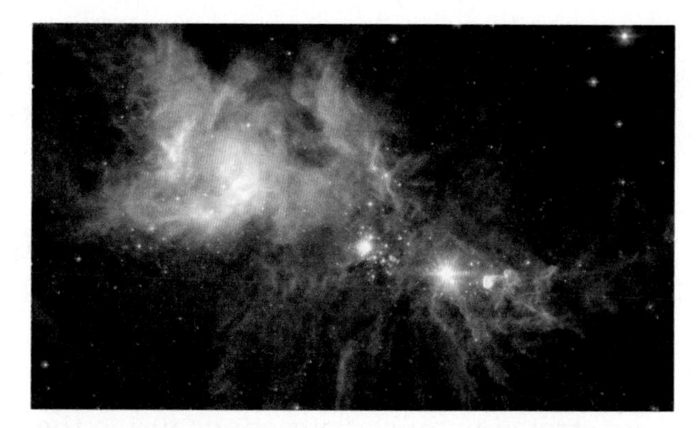

8. **Organize Information** Copy and fill in the graphic organizer below to list what happens to a star following a supernova.

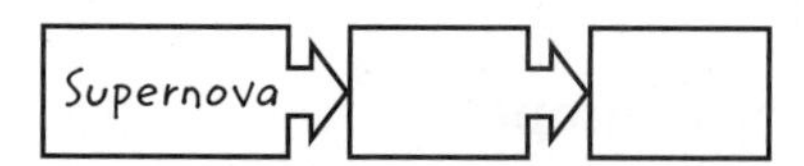

Critical Thinking

9. **Predict** whether the Sun will eventually become a black hole. Why or why not?
10. **Evaluate** why mass is so important in determining the evolution of a star.

Make and Use Graphs

45 minutes

How can graphing data help you understand stars?

How can you make sense of everything in the universe? Graphs help you organize information. The Hertzsprung-Russell diagram is a graph that plots the color, or temperature, of stars against their luminosities. What can you learn about stars by plotting them on a graph similar to the H-R diagram?

Materials

graph paper

Learn It

Displaying information on graphs makes it easier to see how objects are related. Lines on graphs show you patterns and enable you to make predictions. Graphs display a lot of information in an easily understandable form. In this activity, you will **make and use graphs,** plotting the temperature, the color, and the mass of stars.

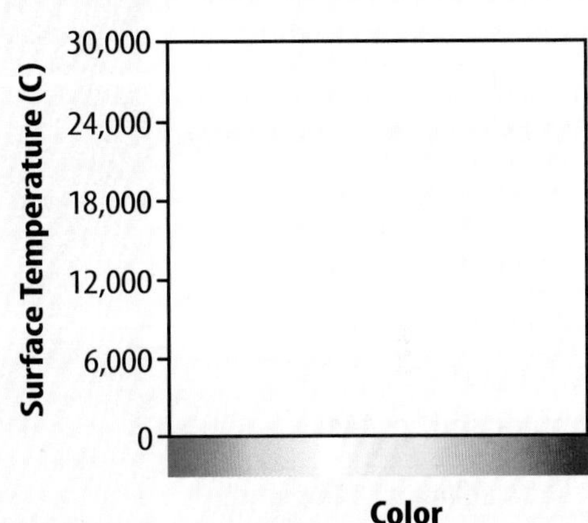

Try It

1. Using graph paper or your Science Journal, draw a graph like the one shown at right.
2. Use the color and temperature data in the table below to plot the position of each star on your graph. Mark the points by attaching adhesive stars to the graph.
3. If stars have similar data, plot them in a cluster. Label each star with its name.
4. Draw a curve that joins the data points as smoothly as possible.
5. Make another graph and plot temperature v. mass of the stars in the table.

Star	Color	Temperature (K)	Mass in solar masses
Sun	Yellow	5,700	1
Alnilam	Blue-white	27,000	40
Altair	White	8,000	1.9
Alpha Centauri A	Yellow	6,000	1.08
Alpha Centauri B	Orange	4,370	0.68
Barnard's Star	Red	3,100	0.1
Epsilon Eridani	Orange	4,830	0.78
Hadar	Blue-white	25,500	10.5
Proxima Centauri	Red	3,000	0.12
Regulus	White	11,000	8
Sirius A	White	9,500	2.6
Spica	Blue-white	22,000	10.5
Vega	White	9,900	3

Apply It

6. All of the stars on your graph are main-sequence stars. What is the relationship between the color and the temperature of a main-sequence star?
7. What is the relationship between the mass and the temperature of a main-sequence star? How are color and mass related?
8. **Key Concept** Which star would be the most likely to eventually form a black hole? Why?

Lesson 4

Galaxies and the Universe

Reading Guide

Key Concepts
ESSENTIAL QUESTIONS

- What are the major types of galaxies?
- What is the Milky Way, and how is it related to the solar system?
- What is the Big Bang theory?

Vocabulary

galaxy p. 439
dark matter p. 439
Big Bang theory p. 444
Doppler shift p. 444

g Multilingual eGlossary

Inquiry Disk in Space?

Yes, this is the disk of a galaxy—a huge collection of stars. You see this galaxy on its edge. If you were to look down on it from above, it would look like a two-armed spiral. Do you think all galaxies are shaped like spirals? What about the galaxy you live in?

20 minutes

Does the universe move?

Scientists think the universe is expanding. What does that mean? Are stars and galaxies moving away from each other? Is the universe moving?

1. Read and complete a lab safety form.
2. Copy the table at right into your Science Journal.
3. Use a **marker** to make three dots 5–7 cm apart on one side of a **large round balloon.** Label the dots *A, B,* and *C.* The dots represent galaxies.
4. Blow up the balloon to a diameter of about 8 cm. Hold the balloon closed as your partner uses a **measuring tape** to measure the distance between each galaxy on the balloon's surface. Record the distances on the table.
5. Repeat step 4 two more times, blowing up the balloon a little more each time.

Balloon size	A–B (cm)	B–C (cm)	A–C (cm)
Small			
Medium			
Large			

Think About This

1. What happened to the distances between galaxies as the balloon expanded?
2. If you were standing in one of the galaxies, what would you observe about the other galaxies?
3. **Key Concept** If the balloon were a model of the universe, what do you think might have caused galaxies to move in this way?

Galaxies

Most people live in towns or cities where houses are close together. Not many houses are found in the wilderness. Similarly, most stars exist in galaxies. **Galaxies** *are huge collections of stars.* The universe contains hundreds of billions of galaxies, and each galaxy can contain hundreds of billions of stars.

 **Reading Check** What are galaxies?

Dark Matter

Gravity holds stars together. Gravity also holds galaxies together. When astronomers examine how galaxies, such as those in **Figure 14,** rotate and gravitationally interact, they find that most of the matter in galaxies is invisible. *Matter that emits no light at any wavelength is* **dark matter.** Scientists hypothesize that more than 90 percent of the universe's mass is dark matter. Scientists do not fully understand dark matter. They do not know what composes it.

Figure 14 By examining interacting galaxies such as these, astronomers hypothesize that most mass in the universe is invisible dark matter.

Types of Galaxies

There are three major types of galaxies: spiral, elliptical, and irregular. **Table 2** gives a brief description of each type.

Key Concept Check What are the major types of galaxies?

Table 2 Types of Galaxies

	**Spiral Galaxies** The stars, gas, and dust in a spiral galaxy exist in spiral arms that begin at a central disk. Some spiral arms are long and symmetrical; others are short and stubby. Spiral galaxies are thicker near the center, a region called the central bulge. A spherical halo of globular clusters and older, redder stars surrounds the disk. NGC 5679, shown here, contains a pair of spiral galaxies.
Elliptical Galaxies Unlike spiral galaxies, elliptical galaxies do not have internal structure. Some are spheres, like basketballs, while others resemble footballs. Elliptical galaxies have higher percentages of old, red stars than spiral galaxies do. They contain little or no gas and dust. Scientists suspect that many elliptical galaxies form by the gravitational merging of two or more spiral galaxies. The elliptical galaxy pictured here is NGC 5982, part of the Draco Group.	
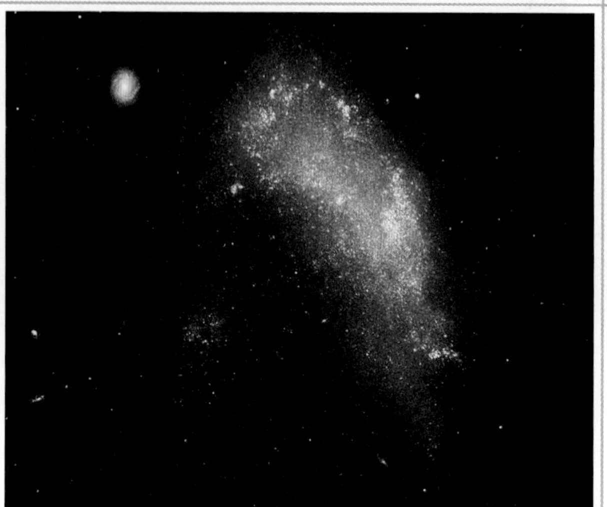	**Irregular Galaxies** Irregular galaxies are oddly shaped. Many form from the gravitational pull of neighboring galaxies. Irregular galaxies contain many young stars and have areas of intense star formation. Shown here is the irregular galaxy NGC 1427A.

MiniLab

20 minutes

Can you identify a galaxy?

The *Hubble Space Telescope,* shown below, is an orbiting telescope that gives astronomers clear pictures of the night sky. What kinds of galaxies can you see in pictures taken by the *Hubble Telescope*?

1. Study each image on the ***Hubble Space Telescope* image sheet.** For each image, identify at least two galaxies. Are they spiral, elliptical, or irregular? Write your observations in your Science Journal, labeled with the letter of the image.

Analyze and Conclude

1. **Draw Conclusions** Why are some galaxies easier to identify than others?
2. **Infer** What interactions do you see among some of the galaxies?
3. **Key Concept** Do you think the shapes of galaxies can change over time? Why or why not?

Groups of Galaxies

Galaxies are not distributed evenly in the universe. Gravity holds them together in groups called clusters. Some clusters of galaxies are enormous. The Virgo Cluster is 60 million light-years from Earth. It contains about 2,000 galaxies. Most clusters exist in even larger structures called superclusters. Between superclusters are voids, which are regions of nearly empty space. Scientists hypothesize that the large-scale structure of the universe resembles a sponge.

Reading Check What holds clusters of galaxies together?

FOLDABLES®

Make a horizontal single-tab matchbook. Label it as shown. Use it to describe the contents of the Milky Way.

Milky Way

Galaxy

The Milky Way

The solar system is in the Milky Way, a spiral galaxy that contains gas, dust, and almost 200 billion stars. The Milky Way is a member of the Local Group, a cluster of about 30 galaxies. Scientists expect the Milky Way will begin to merge with the Andromeda Galaxy, the largest galaxy in the Local Group, in about 3 billion years. Because stars are far apart in galaxies, it is not likely that many stars will actually collide during this event.

Where is Earth in the Milky Way? **Figure 15** on the next two pages shows an artist's drawing of the Milky Way and Earth's place in it.

WORD ORIGIN

galaxy
from Greek *galactos,* means "milk"

The Milky Way

Figure 15 The Milky Way is shown here in two separate views, from the top (left page) and on edge (right page). Because Earth is located inside the disk of the Milky Way, people cannot see beyond the central bulge to the other side.

Key Concept Check Where is Earth in the Milky Way?

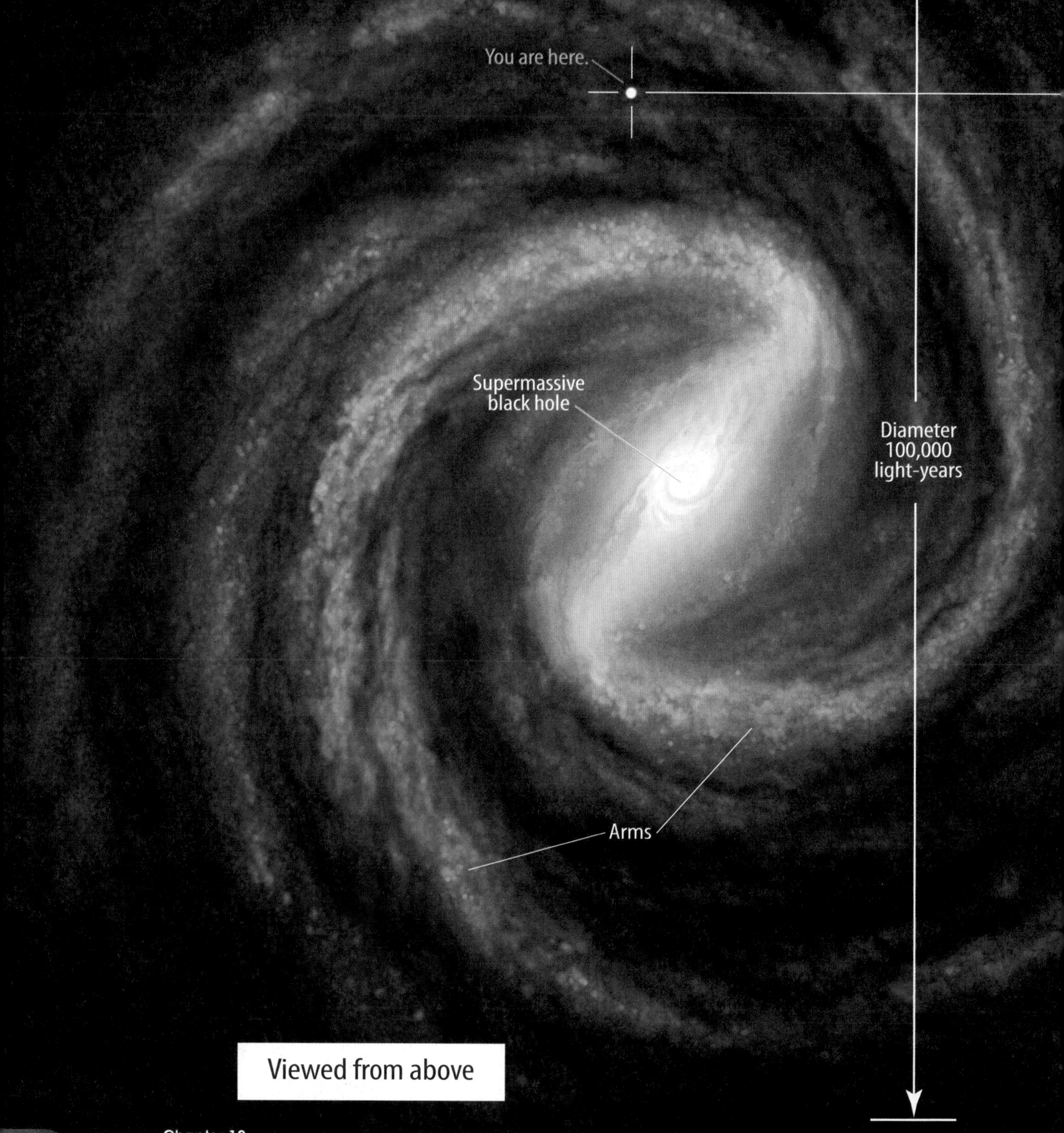

You are here.

Supermassive black hole

Diameter 100,000 light-years

Arms

Viewed from above

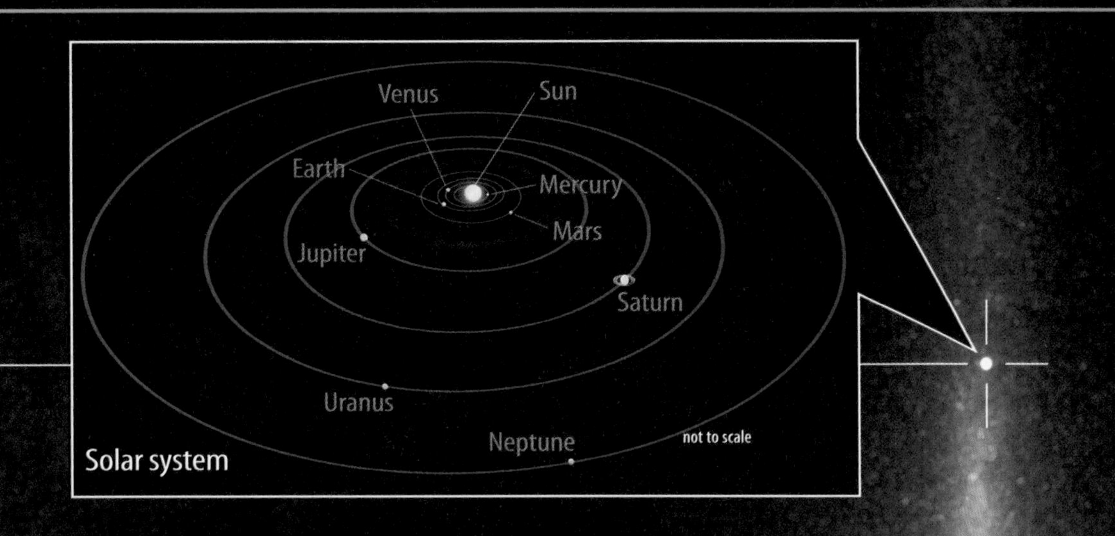

Disk

Globular clusters

Bulge

Open clusters

Halo

Arms

Viewed on edge

The Big Bang Theory

When astronomers look into space, they look back in time. Is there a beginning to time? According to the **Big Bang theory,** *the universe began from one point billions of years ago and has been expanding ever since.*

Key Concept Check What is the Big Bang theory?

Origin and Expansion of the Universe

Most scientists agree that the universe is 13 – 14 billion years old. When the universe began, it was dense and hot—so hot that even atoms didn't exist. After a few hundred thousand years, the universe cooled enough for atoms to form. Eventually, stars formed, and gravity pulled them into galaxies.

As the universe expands, space stretches and galaxies move away from one another. The same thing happens in a loaf of unbaked raisin bread. As the dough rises, the raisins move apart. Scientists observe how space stretches by measuring the speed at which galaxies move away from Earth. As the galaxies move away, their wavelengths lengthen and stretch out. How does light stretch?

Doppler Shift

You have probably heard the siren of a speeding police car. As **Figure 16** illustrates, when the car moves toward you, the sound waves compress. As the car moves away, the sound waves spread out. Similarly, when visible light travels toward you, its wavelength compresses. When light travels away from you, its wavelength stretches out. It shifts to the red end of the visual light portion of the electromagnetic spectrum. *The shift to a different wavelength is called the* **Doppler shift.** Because the universe is expanding, light from galaxies is red-shifted. The more distant a galaxy is, the faster it moves away from Earth, and the more it is red-shifted.

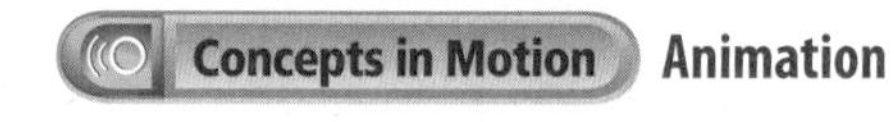

Dark Energy

Will the universe expand forever? Or will gravity cause the universe to contract? Scientists have observed that galaxies are moving away from Earth faster over time. To explain this, they suggest a force called dark energy is pushing the galaxies apart.

Dark energy, like dark matter, is an active area of research. There is still much to learn about the universe and all it contains.

Doppler Shift

Concepts in Motion **Animation**

Figure 16 The sound waves from an approaching police car are compressed. As the car speeds away, the sound waves are stretched out. Similarly, when an object is moving away, its light is stretched out. The light's wavelength shifts toward a longer wavelength.

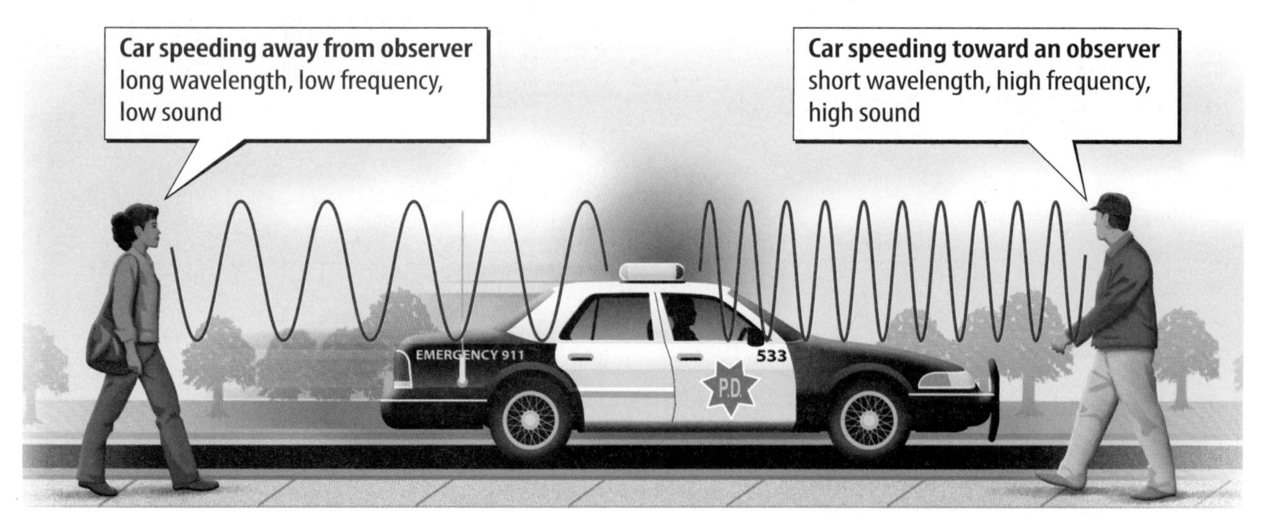

Lesson 4 Review

Assessment Online Quiz

Visual Summary

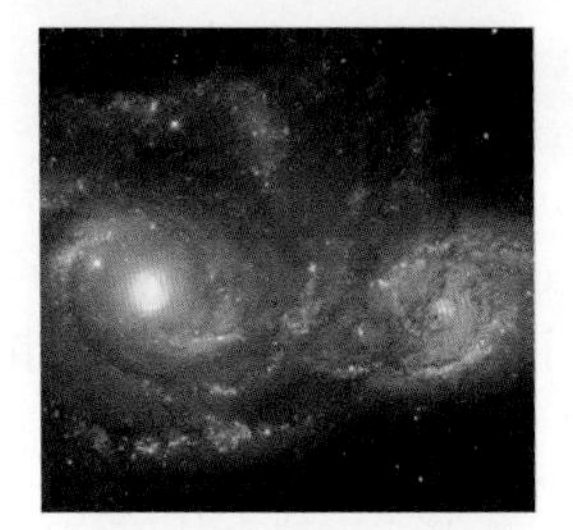

By studying interacting galaxies, scientists have determined that most mass in the universe is dark matter.

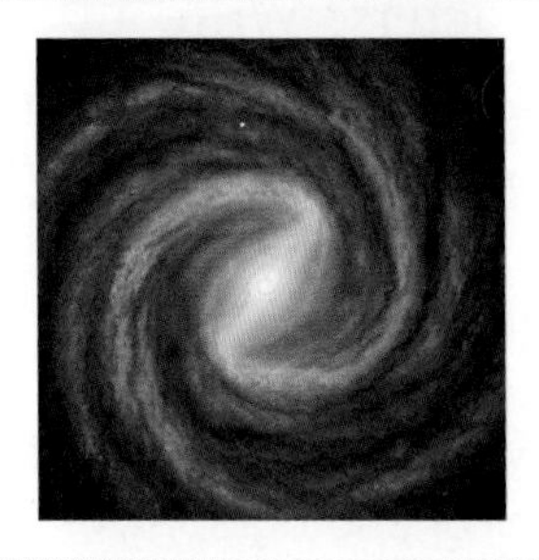

The Sun is one of billions of stars in the Milky Way.

When an object moves away, its light stretches out, just as a siren's sound waves stretch out as the siren moves away.

FOLDABLES®

Use your lesson Foldable to review the lesson. Save your Foldable for the project at the end of the chapter.

What do you think NOW?

You first read the statements below at the beginning of the chapter.

7. Most of the mass in the universe is in stars.

8. The Big Bang theory is an explanation of the beginning of the universe.

Did you change your mind about whether you agree or disagree with the statements? Rewrite any false statements to make them true.

Use Vocabulary

1. Stars exist in huge collections called ________.
2. **Use the term** *dark matter* in a sentence.
3. **Define** the *Big Bang theory*.

Understand Key Concepts 

4. Which is NOT a major galaxy type?
   - **A.** dark
   - **B.** elliptical
   - **C.** irregular
   - **D.** spiral
5. **Identify** Sketch the Milky Way, and identify the location of the solar system.
6. **Explain** how scientists know the universe is expanding.

Interpret Graphics

7. **Identify** Sketch the Milky Way, shown below. Identify the bulge, the halo, and the disk.

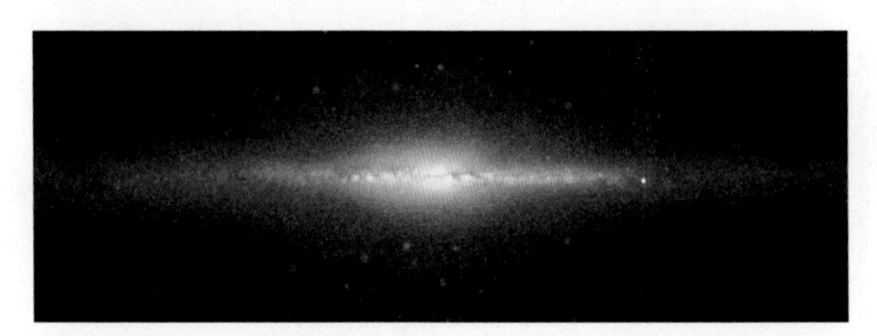

8. **Organize Information** Copy and fill in the graphic organizer below. List the three major types of galaxies and some characteristics of each.

Galaxy Type	Characteristics

Critical Thinking

9. **Assess** the role of gravity in the structure of the universe.
10. **Predict** what the solar system and the universe might be like in 10 billion years.

Describe a Trip Through Space

Materials

paper

colored pencils

astronomy magazines

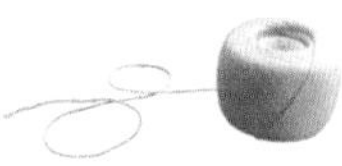
string

glue

scissors

Safety

Imagine you could travel through space at speeds even faster than light. Based on what you have learned in this chapter, where would you choose to go? What would you like to see? What would it be like to move through the Milky Way and out into distant galaxies? Would you travel with anyone or meet any characters? Write a book describing your trip through space.

Question

Where will your trip take you, and how will you describe it? How can you write a fictional, but scientifically accurate, story about your trip? Will you make a picture book? If so, will you sketch your own pictures, use diagrams or photographs, or both? Will your book be mostly words, or will it be like a graphic novel? How can you draw your readers into the story?

Procedure

1. In your Science Journal, write ideas about where your trip will take you, how you will travel, what will happen along the way, and who or what you might meet.
2. Draw a graphic organizer, such as the one below, in your Science Journal. Use it to help you organize your ideas.
3. Write an outline of your story. Use it to guide you as you write the story.
4. List things you will need to research, pictures you will need to find or draw, and any other materials you will need. How will you bind your book? Will you make more than one copy?

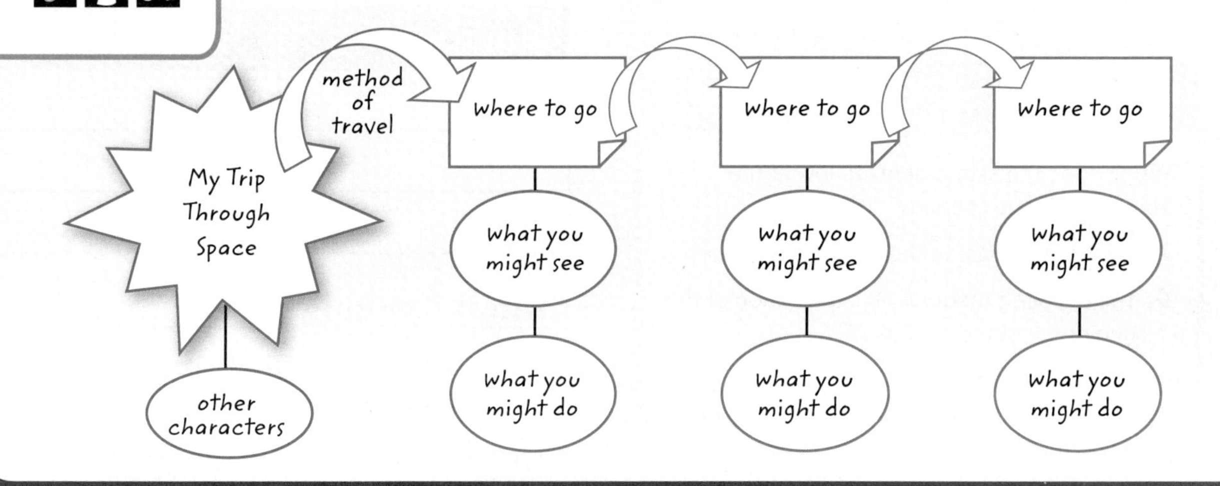

5. Write your book. Add pictures or illustrations. Bind the pages together into book form.
6. Have a friend read your book and tell you if you succeeded in telling your story in an engaging way. What suggestions does your friend have for improvement?
7. Revise and improve the book based on your friend's suggestions.

Analyze and Conclude

8. **Research Information** What new information did you learn as you did research for your book?
9. **Calculate** how many light-years you traveled from Earth.
10. **Draw Conclusions** How would your story be limited if you could only travel at the speed of light?
11. **The Big Idea** How does your story help people understand the size of the universe, what it contains, and how gravity affects it?

Communicate Your Results

You may wish to make a copy of your book and give it to the school library or add it to a library of books in your classroom.

Combine your book with books written by other students in your class to make an almanac of the universe. Add pages that give statistics and other interesting facts about the universe.

Lab Tips

- Think about your audience as you plan your book. Are you writing it for young children or for students your own age? What kinds of books do you and your friends enjoy reading?
- What metaphors or other kinds of figurative language can you add to your writing that will draw readers into your story?

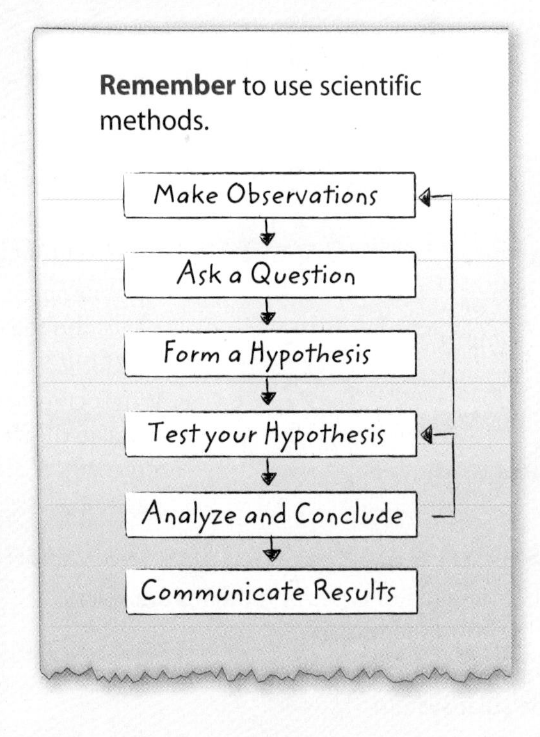

Chapter 12 Study Guide

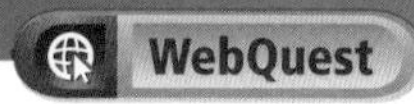

The universe is made up of stars, gas, and dust, as well as invisible dark matter. Material in the universe is not randomly arranged, but is pulled by gravity into galaxies.

Key Concepts Summary	Vocabulary
Lesson 1: The View from Earth • The sky is divided into 88 constellations. • Astronomers learn about the energy, distance, temperature, and composition of stars by studying their light. • Astronomers measure distances in space in **astronomical units** and in **light-years.** They measure star brightness as **apparent magnitude** and as **luminosity.** 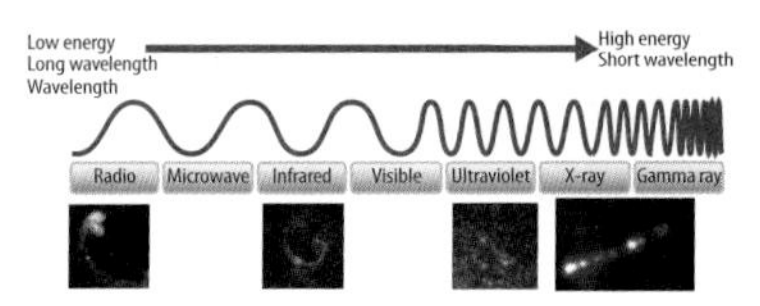 	**spectroscope** p. 417 **astronomical unit** p. 418 **light-year** p. 418 **apparent magnitude** p. 419 **luminosity** p. 419
Lesson 2: The Sun and Other Stars • **Stars** shine because of **nuclear fusion** in their cores. • Stars have a layered structure—they conduct energy through their **radiative zones** and their **convection zones** and release the energy at their **photospheres.** • Sunspots, prominences, flares, and coronal mass ejections are temporary phenomena on the Sun. • Astronomers classify stars by their temperatures and luminosities. 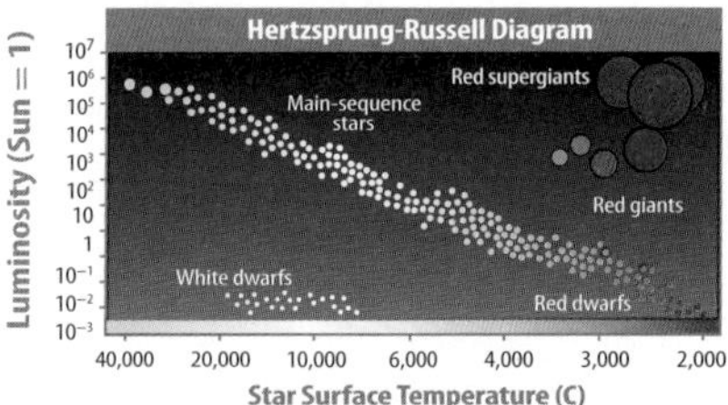 	**nuclear fusion** p. 423 **star** p. 423 **radiative zone** p. 424 **convection zone** p. 424 **photosphere** p. 424 **chromosphere** p. 424 **corona** p. 424 **Hertzsprung-Russell diagram** p. 427
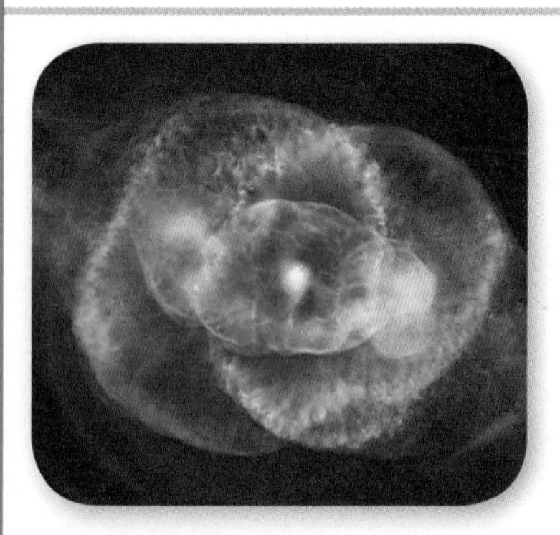 **Lesson 3: Evolution of Stars** • Stars are born in clouds of gas and dust called **nebulae.** • What happens to a star when it leaves the main sequence depends on its mass. • Matter is recycled in the planetary nebulae of **white dwarfs** and the remnants of **supernovae.**	**nebula** p. 431 **white dwarf** p. 433 **supernova** p. 433 **neutron star** p. 434 **black hole** p. 434
Lesson 4 Galaxies and the Universe • The three major types of **galaxies** are spiral, elliptical, and irregular. • The Milky Way is the spiral galaxy that contains the solar system. • The **Big Bang theory** explains the origin of the universe. 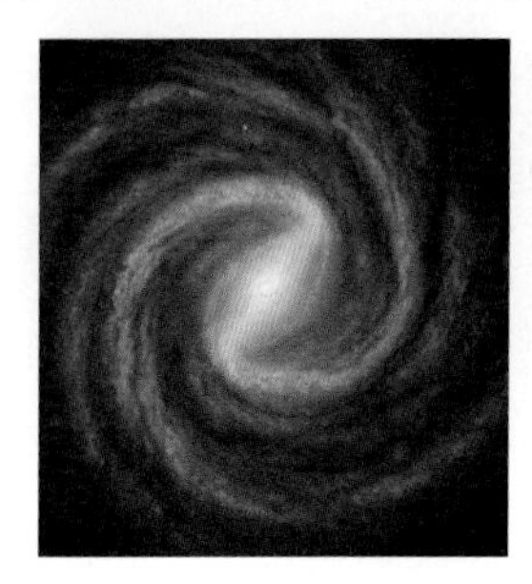	**galaxy** p. 439 **dark matter** p. 439 **Big Bang theory** p. 444 **Doppler shift** p. 444

- Personal Tutor
- Vocabulary eGames
- Vocabulary eFlashcards

FOLDABLES® Chapter Project

Assemble your lesson Foldables as shown to make a Chapter Project. Use the project to review what you have learned in this chapter.

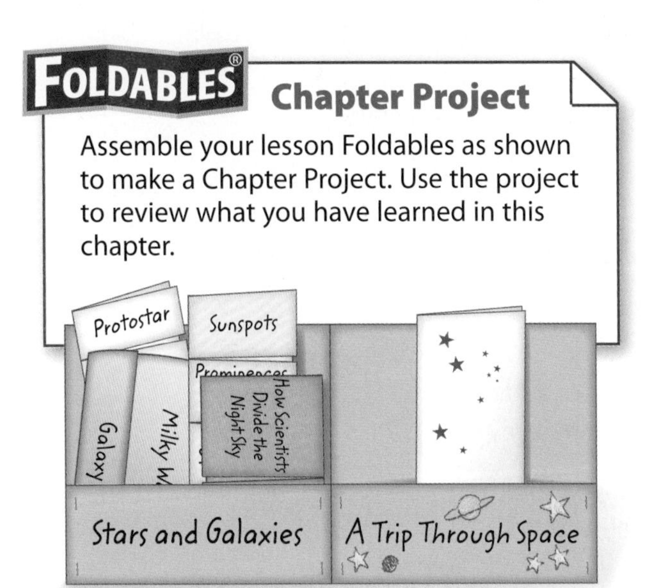

Use Vocabulary

1. **Explain** how nebulae are related to stars.
2. **Define** *Doppler shift*.
3. **Compare** neutron stars and black holes.
4. **Explain** the role of white dwarfs in recycling matter.
5. **Distinguish** between an astronomical unit and a light-year.
6. How does a convection zone transfer energy?
7. **Use the term** *dark matter* in a sentence.
8. On what diagram would you find a plot of stellar luminosity v. temperature?
9. **Compare** photosphere and corona.

Interactive Concept Map

Link Vocabulary and Key Concepts

Copy this concept map, and then use vocabulary terms from the previous page to complete the concept map.

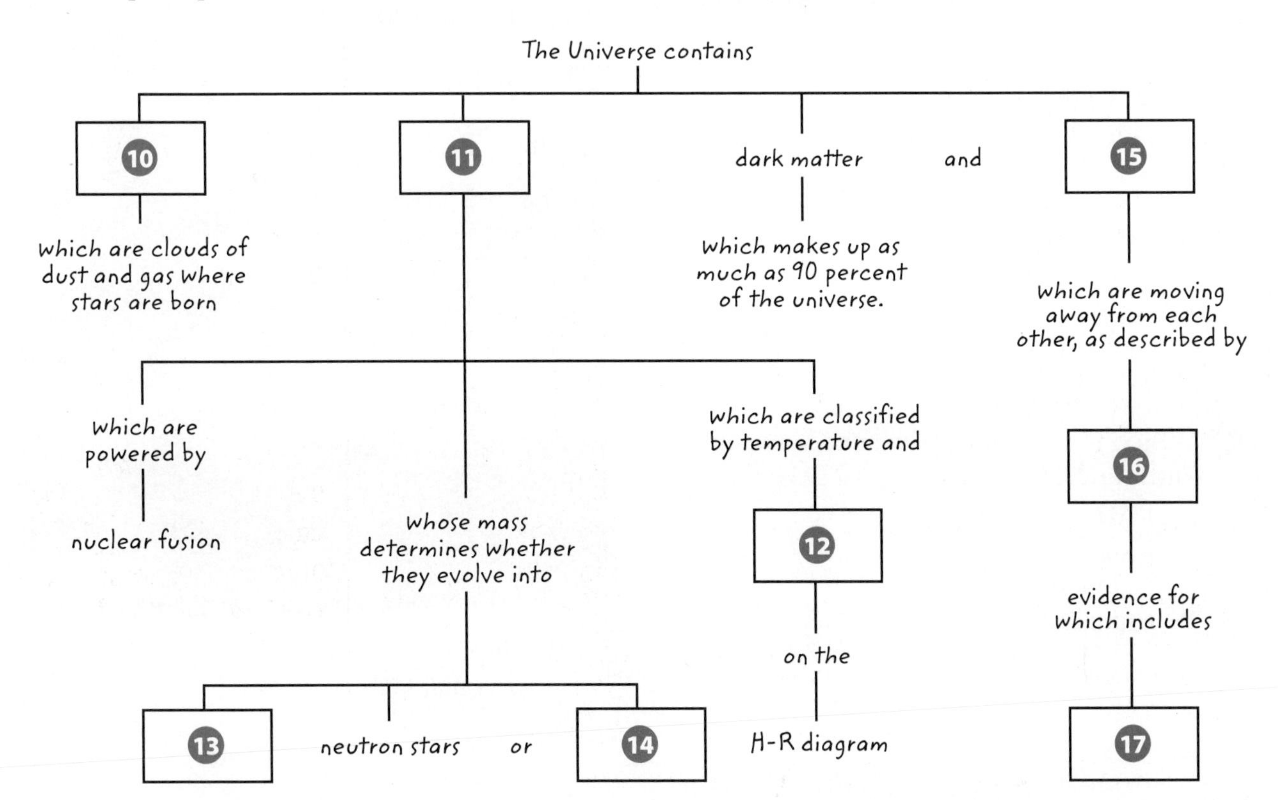

Chapter 12 Review

Understand Key Concepts 

1. Scientists divide the sky into
   A. astronomical units.
   B. clusters.
   C. constellations.
   D. light-years.

2. Which part of the Sun is marked with an *X* on the diagram below?

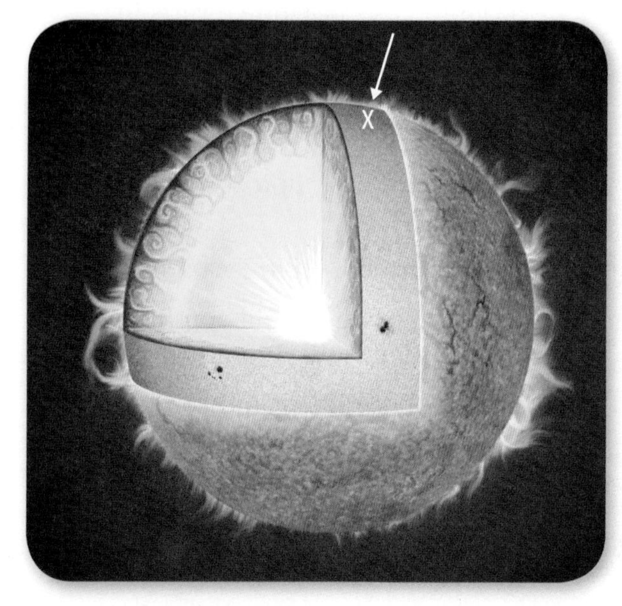

   A. convection zone
   B. corona
   C. photosphere
   D. radiative zone

3. Which might change, depending on the distance to a star?
   A. absolute magnitude
   B. apparent magnitude
   C. composition
   D. luminosity

4. Which is the average distance between Earth and the Sun?
   A. 1 AU
   B. 1 km
   C. 1 light-year
   D. 1 magnitude

5. Which is most important in determining the fate of a star?
   A. the star's color
   B. the star's distance
   C. the star's mass
   D. the star's temperature

6. What star along the main sequence will likely end in a supernova?
   A. blue-white
   B. orange
   C. red
   D. yellow

7. Which term does NOT belong with the others?
   A. black hole
   B. neutron star
   C. red dwarf
   D. supernova

8. What does the Big Bang theory state?
   A. The universe is ageless.
   B. The universe is collapsing.
   C. The universe is expanding.
   D. The universe is infinite.

9. Which type of galaxy is illustrated below?

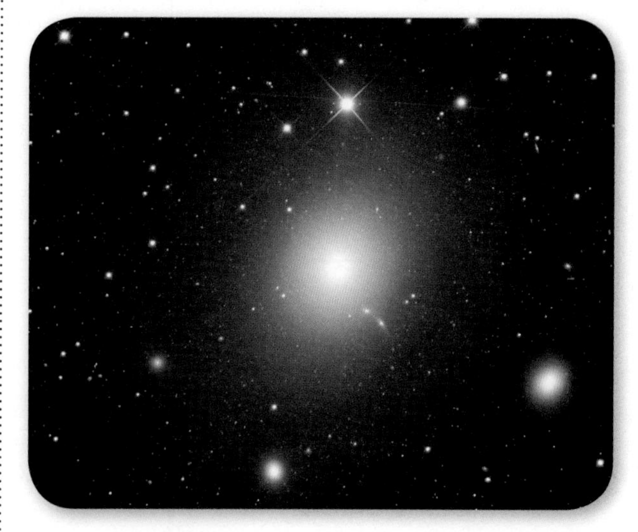

   A. elliptical
   B. irregular
   C. peculiar
   D. spiral

Assessment

Online Test Practice

Critical Thinking

10 **Explain** how energy is released in a star.

11 **Assess** how the invention of the telescope changed people's views of the universe.

12 **Imagine** you are asked to classify 10,000 stars. Which properties would you measure?

13 **Deduce** why supernovae are needed for life on Earth.

14 **Predict** how the Sun would be different if it were twice as massive.

15 **Imagine** that you are writing to a friend who lives in the Virgo Cluster of galaxies. What would you write as your return address? Be specific.

16 **Interpret** The figure below shows part of the solar system. Explain what is happening.

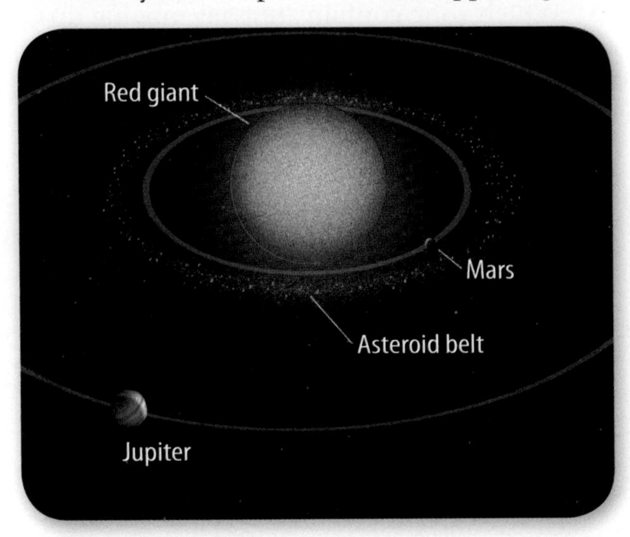

Writing in Science

17 **Write** You are a scientist being interviewed by a magazine on the topic of black holes. Write three questions an interviewer might ask, as well as your answers.

REVIEW THE BIG IDEA

18 What makes up the universe, and how does gravity affect the universe?

19 The photo below shows an image of the early universe obtained with the *Hubble Space Telescope*. Identify the objects you see. Make a list of other objects in the universe that you cannot see on this image.

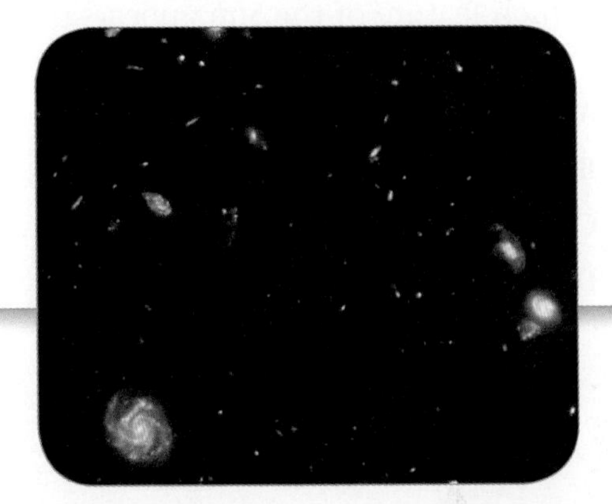

Math Skills

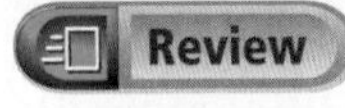

Math Practice

Use Proportions

20 The Milky Way galaxy is about 100,000 light-years across. What is this distance in kilometers?

21 Astronomers sometimes use a distance unit called a parsec. One parsec is 3.3 light-years. What is the distance, in parsecs, of a nebula that is 82.5 light-years away?

22 The distance to the Orion nebula is about 390 parsecs. What is this distance in light-years?

Standardized Test Practice

Record your answers on the answer sheet provided by your teacher or on a sheet of paper.

Multiple Choice

1 Which characteristics can by studied by analyzing a star's spectrum?

A absolute and apparent magnitudes
B formation and evolution
C movement and luminosity
D temperature and composition

2 Which feature of the Sun appears in cycles of about 11 years?

A coronal mass ejections
B solar flares
C solar wind
D sunspots

Use the graph below to answer question 3.

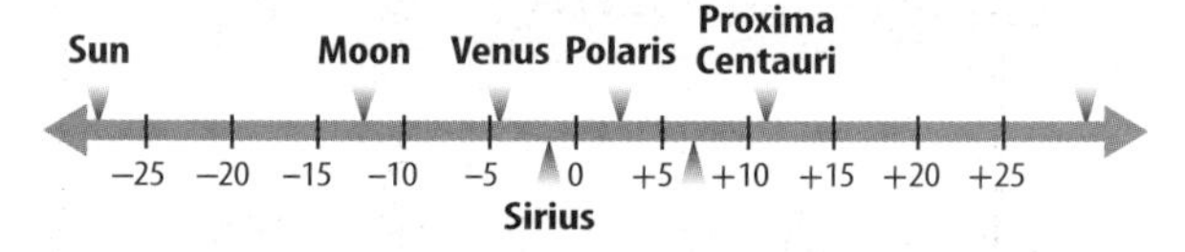

3 Which star on the graph has the greatest apparent magnitude?

A Polaris
B Proxima Centauri
C Sirius
D the Sun

4 Where in the Milky Way is the solar system located?

A at the edge of the disk
B inside a globular cluster
C near the supermassive black hole
D within the central bulge

Use the figure below to answer question 5.

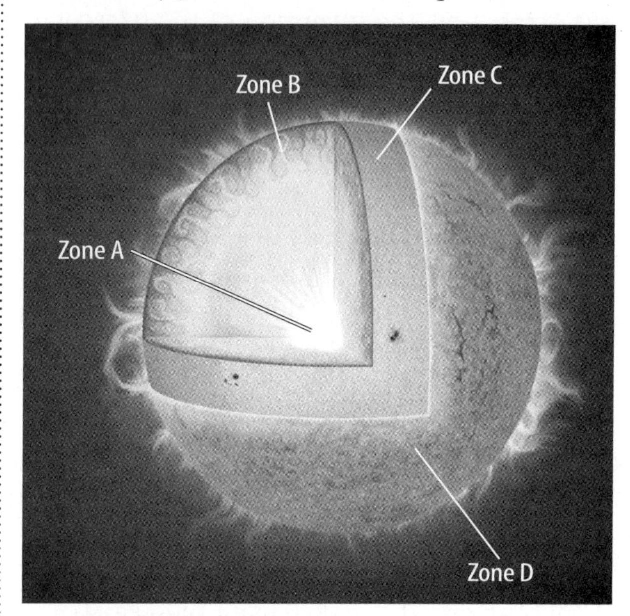

5 Which zone contains hot gas moving up toward the surface and cooler gas moving down toward the center of the Sun?

A zone A
B zone B
C zone C
D zone D

6 Which contains most of the mass of the universe?

A black holes
B dark matter
C gas and dust
D stars

7 Which stellar objects eventually form from the most massive stars?

A black holes
B diffuse nebulae
C planetary nebulae
D white dwarfs

Online Standardized Test Practice

Use the figure below to answer question 8.

8 Which is a characteristic for this type of galaxy?

A It contains no dust.

B It contains little gas.

C It contains many young stars.

D It contains mostly old stars.

9 Where do stars form?

A in black holes

B in constellations

C in nebulae

D in supernovae

10 What term describes the process that causes a star to shine?

A binary fission

B coronal mass ejection

C nuclear fusion

D stellar composition

11 What ancient star grouping do modern astronomers use to divide the sky into regions?

A astronomical unit

B constellation

C galaxy

D star cluster

Constructed Response

Use the diagram below to answer question 12.

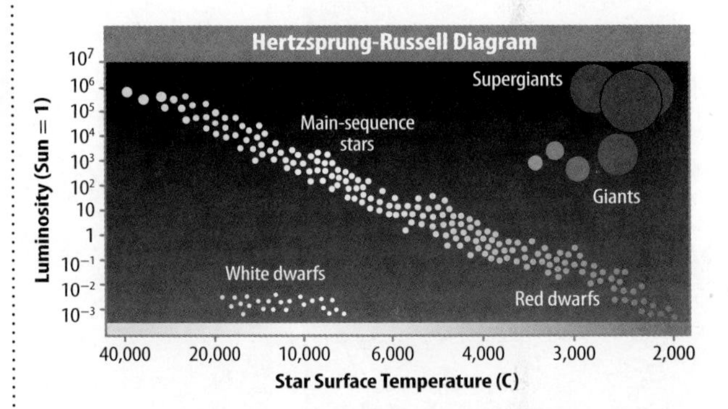

12 Use the information in the diagram above to describe red giants and white dwarfs based on their sizes, temperatures, and luminosities.

13 Describe the life cycle of a main-sequence star. What event causes the star to leave the main sequence?

14 How does the red shift of galaxies support the Big Bang theory?

15 Explain how planetary nebula recycle matter.

NEED EXTRA HELP?															
If You Missed Question...	1	2	3	4	5	6	7	8	9	10	11	12	13	14	15
Go to Lesson...	1	2	1	4	2	4	3	4	3	2	1	2	3	4	3

Unit 4

Earth and Geologic Changes

5 Billion B.C. — 1700 — 1800

4.57 billion years ago
The Sun forms.

4.54 billion years ago
Earth forms.

1778
French naturalist Comte du Buffon creates a small globe resembling Earth and measures its cooling rate to estimate Earth's age. He concludes that Earth is approximately 75,000 years old.

1830
Geologist Charles Lyell begins publishing *The Principles of Geology*; his work popularizes the concept that the features of Earth are perpetually changing, eroding, and reforming.

1862
Physicist William Thomson publishes calculations that Earth is approximately 20 million years old. He claims that Earth had formed as a completely molten object, and he calculates the amount of time it would take for the surface to cool to its present temperature.

1900

2000

1899–1900
John Joly releases his findings from calculating the age of Earth using the rate of oceanic salt accumulation. He determines that the oceans are about 80–100 million years old.

1905
Ernest Rutherford and Bertrand Boltwood use radiometric dating to determine the age of rock samples. This technique would later be used to determine the age of Earth.

1956
Today's accepted age of Earth is determined by C.C. Patterson using uranium-lead isotope dating on several meteorites.

Inquiry
Visit ConnectED for this unit's  activity.

Unit 4

Nature of SCIENCE

Science and History

About 500,000 years ago, early humans used stone to make tools, weapons, and small decorative items. Then, about 8,000 years ago, someone might have spied a shiny object among the rocks. It was gold—thought to be the first metal discovered by humans. Gold was very different from stone. It did not break when it was struck. It could easily be shaped into useful and beautiful objects. Over time, other metals were discovered. Each metal helped advance human civilization. Metals from Earth's crust have helped humans progress from the Stone Age to the Moon, to Mars, and beyond.

Gold

Since the time of its discovery, gold has been a symbol of wealth and power. It is used mainly in jewelry, coins, and other valuable objects. King Tut's coffin was made of pure gold. Tut's body was surrounded by the largest collection of gold objects ever discovered—chariots, statues, jewelry, and a golden throne. Because gold is so valuable, much of it is recycled. If you own a piece of gold jewelry, it might contain gold that was mined thousands of years ago!

Lead

Ancient Egyptians used the mineral lead sulfide, also called galena, as eye paint. About 5,500 years ago, metalworkers found that galena melts at a low temperature, forming puddles of the lead. Lead bends easily, and the Romans shaped it into pipes for carrying water. Over the years, the Romans realized that lead was entering the water and was toxic to humans. Despite possible danger, lead water pipes were common in modern homes for decades. Finally, however, in 2004 the use of lead pipes in home construction was banned.

Copper

The first metal commonly traded was copper. About 5,000 years ago, Native Americans mined more than half a million tons of copper from the area that is now Michigan. Copper is stronger than gold. Back then, it was shaped into saws, axes, and other tools. Stronger saws made it easier to cut down trees. The wood from trees then could be used to build boats, which allowed trade routes to expand. Many cultures today still use methods to shape copper that are similar to those used by ancient peoples.

Tin and Bronze

Around 4,500 years ago, the Sumerians noticed differences in the copper they used. Some flowed more easily when it melted and was stronger after it hardened. They discovered that this harder copper contained another metal—tin. Metalworkers began combining tin and copper to produce a metal called bronze. Bronze eventually replaced copper as the most important metal to society. Bronze was strong and cheap enough to make everyday tools. It could easily be shaped into arrowheads, armor, axes, and sword blades. People admired the appearance of bronze. It continues to be used in sculptures. Bronze, along with gold and silver, is used in Olympic medals as a symbol of excellence.

Iron and Steel

Although iron-containing rock was known centuries ago, people couldn't build fires hot enough to melt the rock and separate out the iron. As fire-building methods improved, iron use became more common. It replaced bronze for all uses except art. Iron farm tools revolutionized agriculture. Iron weapons became the choice for war. Like metals used by earlier civilizations, iron increased trade and wealth, and improved people's lives.

In the 17th century, metalworkers developed a way to mix iron with carbon. This process formed steel. Steel quickly became valued for its strength, resistance to rusting, and ease of use in welding. Besides being used in the construction of skyscrapers, bridges, and highways, steel is used to make tools, ships, vehicles, machines, and appliances.

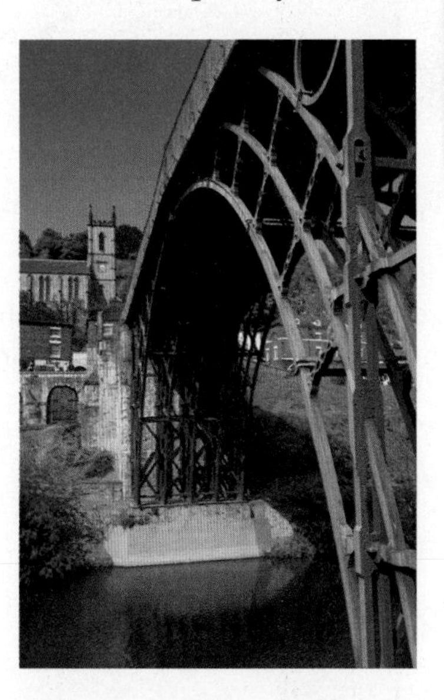

Try to imagine your world without metals. Throughout history, metals changed society as people learned to use them.

Inquiry MiniLab
20 minutes

How do a metal's properties affect its uses?

Why are different common objects made of a variety of different metals?

1. Read and complete a lab safety form.
2. Examine a **lead fishing weight,** a piece of **copper tubing,** and an **iron bolt.**
3. Create a table comparing characteristics of the objects in your Science Journal.
4. Use a **hammer** to tap on each item. Record your observations in your table.

Analyze and Conclude

1. **Infer** Why was lead, not copper or iron, used to make the fishing weight?
2. **Compare** What similarities do all three objects share?
3. **Infer** Why do you think ancient peoples used lead for pipes and iron for weapons?

Chapter 13

Minerals and Rocks

How are minerals and rocks formed, identified, classified, and used?

Inquiry What sort of staircase is this?

This is Giant's Causeway in Ireland. Columns of rocks, such as these, are present in several places on Earth. Some look like staircases, such as the rocks pictured here. Others look like a pile of telephone poles that have been knocked over.

- Did they form naturally?
- Are they always the same shape?
- What caused these rocks to form in this way?

Get Ready to Read

What do you think?

Before you read, decide if you agree or disagree with each of these statements. As you read this chapter, see if you change your mind about any of the statements.

1. Minerals generally are identified by observing their color.
2. Minerals are made of crystals.
3. Once a rock forms, it lasts forever.
4. All rocks form when melted rock cools and changes into a solid.
5. All rock types are related through the rock cycle.
6. Rocks move at a slow and constant rate through the rock cycle.

ConnectED Your one-stop online resource

connectED.mcgraw-hill.com

- Video
- Audio
- Review
- Inquiry
- WebQuest
- Assessment
- Concepts in Motion
- Multilingual eGlossary

Lesson 1

Minerals

Reading Guide

Key Concepts
ESSENTIAL QUESTIONS

- How do minerals form?
- What properties can be used to identify minerals?
- What are some uses of minerals in everyday life?

Vocabulary

mineral p. 461
crystal structure p. 462
crystallization p. 463
streak p. 465
luster p. 465
cleavage p. 465
fracture p. 465
ore p. 467

Multilingual eGlossary

Video
- BrainPOP®
- Science Video

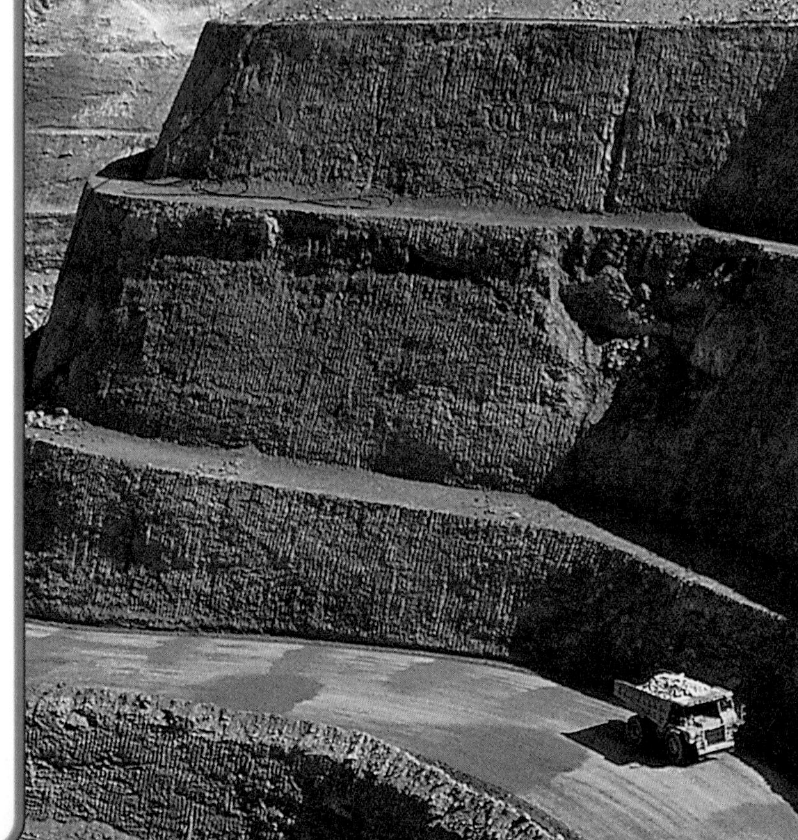

Inquiry Where on Earth are minerals?

Minerals are all around you every day. Sometimes you can even pick up a mineral right from the ground! But most minerals need to mined from beneath Earth's surface, which requires large equipment and areas of land. What is a mineral? Where can you find minerals? What can you use them for?

Launch Lab

20 minutes

Is everything crystal clear?

Do you have several shirts that are nearly the same color or style? If so, you probably know the subtle differences that make each unique. The same is true for the thousands of minerals on Earth. To most people, many of these minerals look exactly the same. In this lab, you'll examine four transparent minerals and demonstrate that not everything is crystal clear.

1. Read and complete a lab safety form.
2. In your Science Journal, draw a data table in which to record your observations. Record your observations after each step.
3. Use a **magnifying lens** to examine each **mineral.**
4. Place each mineral over this sentence, and observe the words.
5. Place the minerals in a **small bowl** of warm water for 2–3 minutes. Take the minerals out of the water and dry them with **paper towels.** Examine each mineral.
6. Carefully place one drop of **dilute hydrochloric acid** on each mineral. Record your observations. Use the paper towels to wipe the minerals dry.

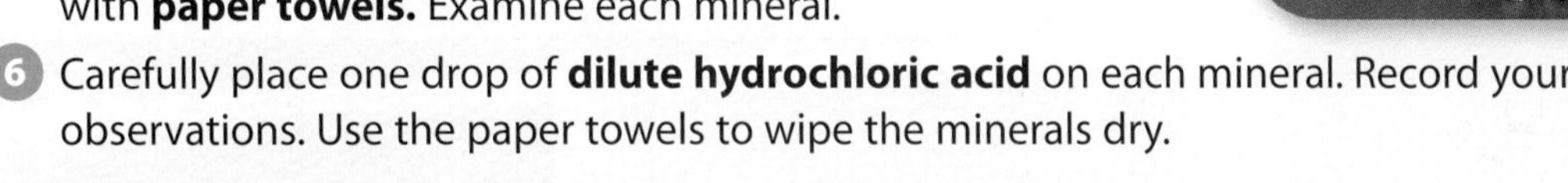

Think About This

1. How are the minerals the same? How are they different?
2. **Key Concept** How do you think each mineral might be used in everyday life?

What is a mineral?

Do you ever drink mineral water? Maybe you take vitamins and minerals to stay healthy. The word *mineral* has many common meanings, but for geologists, scientists who study Earth and the materials of which it is made, this word has a very specific definition.

A **mineral** *is a solid that is naturally occurring, inorganic, and has a crystal structure, and definite chemical composition.* In order for a substance to be classified as a mineral, it must have all five of the characteristics listed in the definition above. Samples of pyrite and coal are shown in **Figure 1.** Coal is made of ancient compressed plant material. Pyrite crystals are made of the elements iron and sulfur. One of these is a mineral, but the other is not. By considering each of the five characteristics of minerals, you can determine which sample is the mineral. The information on the next page will help you do this.

Figure 1 Both coal, on the left, and pyrite, on the right, are shiny, hard substances that form deep inside Earth. But only one is a mineral.

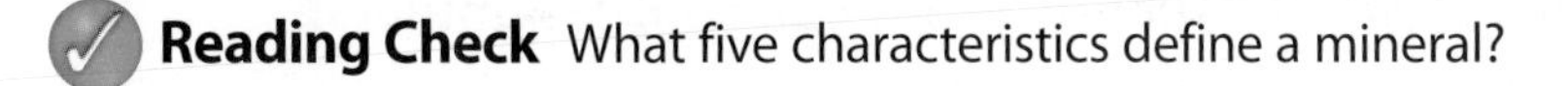

Reading Check What five characteristics define a mineral?

FOLDABLES®

Make a horizontal five-tab book, and label it as shown. Use it to record your notes on mineral characteristics.

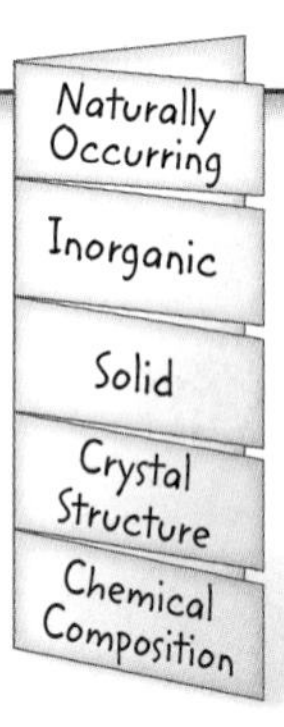

Characteristics of Minerals

To be classified as a mineral, a substance must form naturally. Materials made by people are not considered minerals. Diamonds that form deep beneath Earth's surface are minerals, but diamonds that are made in a laboratory are not. As shown in **Figure 2,** these two types of diamonds can look very similar.

Materials that contain carbon and were once alive are organic. Minerals cannot be organic. This means that a mineral cannot have once been alive, and it cannot contain anything that was once alive, such as plant parts.

A mineral must be solid. Liquids and gases are not considered to be minerals. So, while solid ice is a mineral, water is not.

Figure 2 Natural and artificial diamonds look very much alike, but only the natural diamond on the right is a mineral.

A mineral must have a crystal structure. *The atoms in a crystal are arranged in an orderly, repeating pattern called a* **crystal structure.** This organized structure produces smooth faces and sharp edges on a crystal. The faces and edges of the pyrite crystals shown in **Figure 1** are produced by this internal atomic structure.

A mineral is made of specific amounts of elements. A chemical formula shows how much of each element is present in the mineral. For example, pyrite is made of the elements iron (Fe) and sulfur (S). There always must be one iron atom for every two sulfur atoms. Therefore, the chemical formula for pyrite is FeS_2.

Look again at **Figure 1.** Because the plants that turned into the coal were once alive, coal is not a mineral. The pyrite has all five characteristics of a mineral, so it is a mineral.

Use Ratios

A ratio compares numbers. For example, in the chemical formula for water, H_2O, the number *2* is called a subscript. The subscript tells you how many atoms of that element are in the formula. A symbol with no subscript means that element has one atom. So, the ratio of hydrogen (H) atoms to oxygen (O) atoms in H_2O is 2:1. This is read *two to one.*

Practice

Quartz has the formula SiO_2. What is the ratio of silicon (Si) atoms to oxygen (O) atoms in quartz?

- **Math Practice**
- **Personal Tutor**

Mineral Formation

How do atoms form minerals? Atoms within a liquid join together to form a solid. **Crystallization** *is the process by which atoms form a solid with an orderly, repeating pattern.* Crystallization can happen in two main ways.

Crystallization from Magma When melted rock material—called magma—cools, some of the atoms join together and form solid crystals. As the liquid continues to cool, atoms are added to the surface of the crystals. The longer it takes the magma to cool, the more atoms are added to the crystal. Large crystals grow when the magma cools slowly. If the magma cools quickly, there is only enough time for small crystals to grow.

Crystallization from Water Many substances, such as sugar and salt, dissolve in water, especially if the water is warm. When water cools or evaporates, the particles of the dissolved substances come together again in the solution and crystallize. The gold shown in **Figure 3** formed this way. The orderly arrangement of atoms in this mineral is visible using a very powerful microscope.

Key Concept Check How do minerals form?

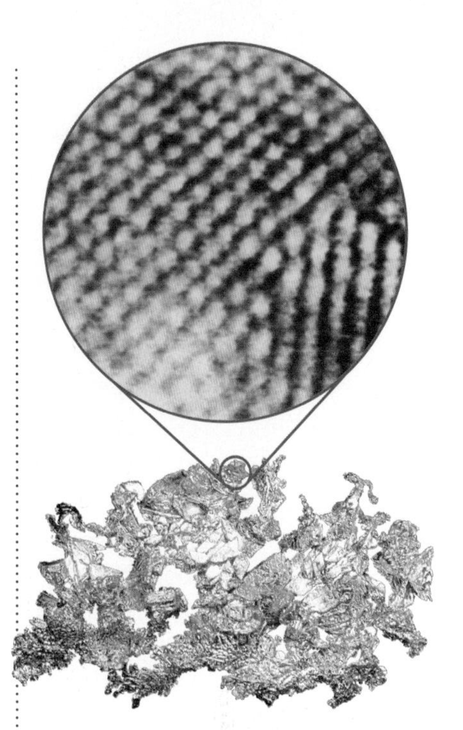

Figure 3 The orderly atomic structure of gold crystals produces these neat rows of atoms.

Inquiry MiniLab

20 minutes

How does your garden grow?

Just like plants in a garden, minerals grow, or form, in different ways. Some minerals form when magma or lava cools and hardens. Other minerals form when atoms and molecules dissolved in water join and crystallize. Can you grow a "garden" of mineral crystals?

1. Read and complete a lab safety form.
2. Pour 500 mL of hot water into a **beaker.** Using a **spoon,** slowly add the **solid** to the water. Stir with a **stirring rod** until no more of the solid dissolves.
3. Use **hot mitts** to pick up the beaker and carefully pour about 100 mL of the solution into each **shallow glass dish.**
4. Put the dishes where they will not be disturbed. Use a **magnifying lens** to observe the crystals that form in your dish over the next couple of days. Record your observations in your Science Journal.

Analyze and Conclude

1. **Observe** What do your crystals look like? Be specific in your description.
2. **Measure** How big are the largest crystals? The smallest ones?
3. **Recognize Cause and Effect** What factors affected the size of the crystals that formed?
4. **Key Concept** How did the minerals form in this activity?

SCIENCE USE V. COMMON USE

property

Science Use a quality or characteristic of an individual or thing

Common Use something owned or possessed

Mineral Identification

Every mineral has a unique set of physical **properties,** or characteristics. These properties are used to identify minerals. Generally, it is necessary to test several properties in order to distinguish between similar minerals.

Density

If you pick up two mineral samples that are about the same size, one might feel heavier than the other. The heavier mineral has a higher density. It has more mass in the same volume. The densities of many minerals are quite similar, but a very high or a very low density can be useful for identifying a mineral.

Hardness

The hardness of a mineral is measured by observing how easily it is scratched or how easily it scratches something else. The Mohs hardness scale, shown in **Table 1,** ranks hardness from 1 to 10. On this scale, diamond is the hardest mineral, with a hardness value of 10. The softest mineral is talc, with a hardness of 1.

Table 1 The minerals on the Mohs hardness scale are classified according to their relative hardness. Some common materials that can be used to test mineral samples also are included.

Visual Check Which minerals are harder than glass?

Table 1 Mohs Hardness Scale

Hardness	Mineral or Ordinary Object	Hardness	Mineral or Ordinary Object
10	diamond	5	apatite
9	corundum	4.5	wire nail
8	topaz	4	fluorite
7	quartz	3.5	copper wire or copper coin (penny)
6.5	streak plate	3	calcite
6	feldspar	2.5	fingernail
5.5	glass, steel knife blade	2	gypsum
		1	talc

Color and Streak

Some minerals have a unique color that can be used for identification. The mineral malachite always has a distinctive green color. But the colors of most minerals vary. Quartz is a common mineral that has many different colors.

Even though the colors of most minerals vary from specimen to specimen, the color of a mineral's powder does not. *The color of a mineral's powder is called its* **streak.** Streak is observed by scratching the mineral across a tile of unglazed porcelain. Sometimes, the color of a mineral and the color of its streak are different, as illustrated by the hematite shown in **Figure 4.**

Luster

Minerals reflect light in different ways. **Luster** *describes the way that a mineral's surface reflects light.* Many terms are used to describe mineral luster. Some of these terms are *metallic, glassy, earthy,* or *pearly.* **Figure 4** contains an example of the mineral hematite, which can have either a metallic luster or a dull luster. The muscovite mica shown in **Figure 4** has a pearly luster, and the quartz has a glassy luster.

Cleavage and Fracture

Two properties are used to describe the ways that minerals break. *If a mineral breaks along smooth, flat surfaces, it displays* **cleavage.** A mineral can break along a single cleavage direction, or it can have more than one direction. As shown in **Figure 4,** muscovite mica has one cleavage direction and peels off in sheets. Halite has three distinct cleavage directions and breaks into cubes.

A mineral that breaks along rough or irregular surfaces displays **fracture.** The quartz in **Figure 4** shows fracture.

Reading Check How does cleavage differ from fracture?

Figure 4 Minerals can be described in many different ways. The minerals pictured here have a variety of colors, streaks, and luster, and they break in different ways.

Visual Check Which mineral displays cleavage? Which displays fracture?

Hematite can have a red, brown, or black color, but its streak is always a dark, rusty red. Hematite's luster can be dull or metallic.

Muscovite displays cleavage in one direction. Parts of the crystals can peel off in flakes or sheets.

Quartz forms in a variety of colors—clear, purple, orange, or pink, such as the quartz pictured above.

Academic Vocabulary

exhibit
(verb) to display, to present for the public to see

Crystal Shape

Minerals **exhibit** many different crystal shapes. A mineral's atomic structure determines its crystal shape. Crystal shapes can vary greatly. The crystals of hematite, shown in **Figure 4,** are relatively shapeless, so they are described as massive. Muscovite mica has diamond-shaped or six-sided crystals, but muscovite commonly occurs in flat, sheetlike layers, as shown in **Figure 4.** Amethyst, a type of quartz, has crystals shaped like pyramids, as shown in **Figure 5.**

Sometimes crystals grow so close to each other that the crystal shape is not visible. If there is room for large crystals to grow, the crystal shape can be useful for identifying the mineral. A comparison of small, crowded amethyst crystals and large, well-formed quartz crystals is shown in **Figure 5.**

▲ **Figure 5** The smaller amethyst crystals on the top formed in a small, crowded space. More space allowed the quartz crystals on the bottom to grow large and distinct.

Key Concept Check What are the common properties used to identify minerals?

Unusual Properties

Some minerals have unusual properties that make them easy to identify. For example, halite tastes salty. Magnetite is magnetic and attracts steel objects. Calcite fizzes when acid is dropped on it. A variety of calcite, called Iceland spar, also has a property called double refraction. As shown in **Figure 6,** images viewed through the calcite crystal appear doubled.

Quartz crystals can produce an electric current when compressed. This property makes quartz crystals useful in radios, microphones, and watches. Several minerals display the property of fluorescence. As shown in **Figure 6,** calcite and quartz glow under ultraviolet light.

Iceland spar, a form of calcite, displays double refraction.

Calcite (orange) and quartz (green) glow under ultraviolet light.

Figure 6 Some minerals have unique properties.

Figure 7 Minerals have a variety of uses in our daily lives. Toothpaste, cosmetics, and table salt are just a few everyday items that contain minerals.

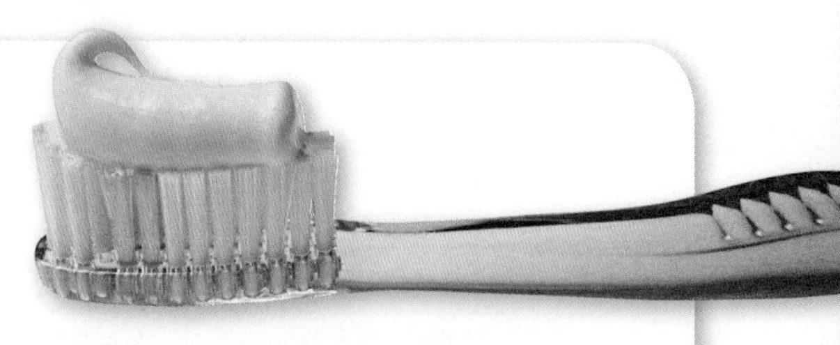

▲ Toothpaste contains calcite or silica.

◀ Common table salt contains the mineral halite.

Some cosmetics contain mica. ▶

Minerals in Everyday Life

You might not be aware of how important minerals are in your life. From the moment you wake in the morning until you fall asleep at night, you use materials made from minerals. A few common examples are shown in **Figure 7.** Some minerals are valuable because we use them every day. We appreciate others simply for their beauty.

Did you know that beverage cans and car batteries are made from minerals? These items are made of metals. Most metals combine with other elements in the formation of a mineral. For example, aluminum can be extracted from the rock bauxite. The minerals must be processed to remove the metals from them. *Deposits of metallic or nonmetallic minerals that can be produced at a profit are called* **ores.**

Some minerals, such as gemstones, are valuable because of their appearance. Specific physical properties make the gemstones valuable. They usually are harder than quartz. Their luster is generally brilliant, and gemstones often have intense colors. The natural crystals are cut and polished, such as the emeralds shown in **Figure 8.**

Key Concept Check How are minerals used in everyday life?

Figure 8 When rough emeralds, such as the one on the left, are cut and polished, they look like the gem to the right. ▼

Lesson 1 Review

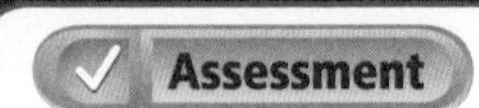

Visual Summary

Hardness varies from mineral to mineral and can be used to help identify certain minerals.

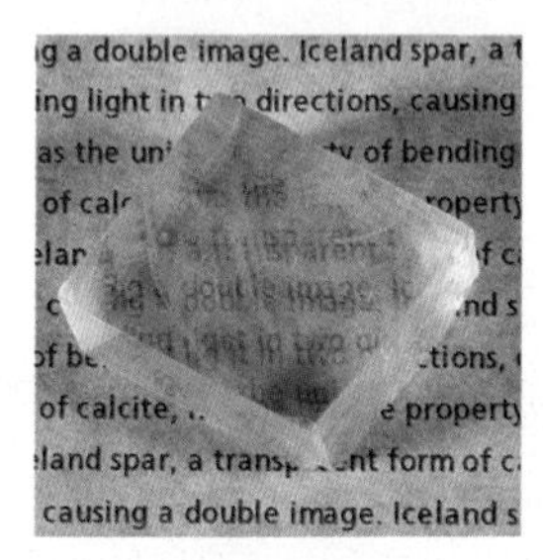

All minerals have specific crystal shapes and properties that can be used to help identify them.

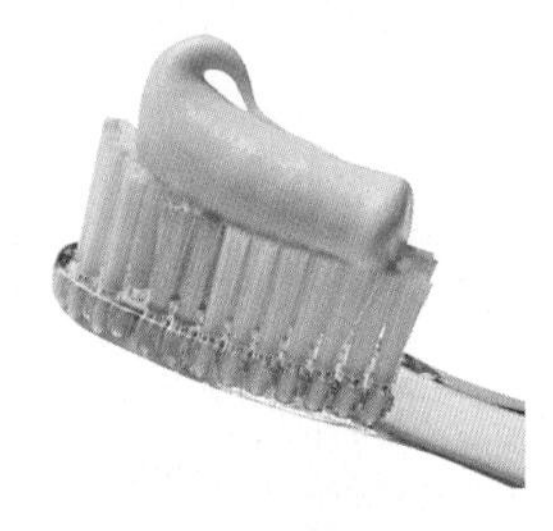

Minerals are present in everyday items such as toothpaste, makeup, and household items.

FOLDABLES®

Use your lesson Foldable to review the lesson. Save your Foldable for the project at the end of the chapter.

What do you think NOW?

You first read the statements below at the beginning of the chapter.

1. Minerals generally are identified by observing their color.
2. Minerals are made of crystals.

Did you change your mind about whether you agree or disagree with the statements? Rewrite any false statements to make them true.

Use Vocabulary

1. **Distinguish** between the terms *crystal structure* and *crystallization.*
2. The color of a mineral's powder is called its ________.
3. **Define** *luster* in your own words.

Understand Key Concepts

4. **Explain** Bones contain elements found in minerals and have a definite structure. Why is bone NOT a mineral?
5. Which mineral property is least reliable in mineral identification?
   A. color
   B. streak
   C. hardness
   D. luster
6. **Give an example** of an object you use daily that is made from a mineral.

Interpret Graphics

7. **Draw** a graphic organizer like the one below, and list the five characteristics used to define a mineral.

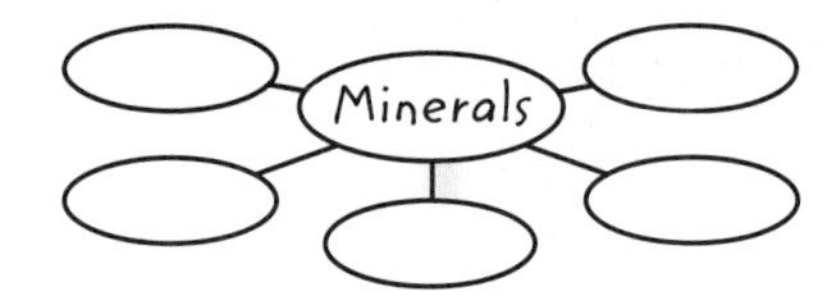

Critical Thinking

8. **Suggest** a plan to identify unknown mineral samples.
9. **Justify** why natural diamonds are worth much more than manufactured diamonds.

Math Skills

Math Practice

10. Calcite has the formula $CaCO_3$. What is the ratio of calcium (Ca) to carbon (C) to oxygen (O) atoms in calcite?
11. What is the formula for hematite if it's ratio of iron (Fe) to oxygen (O) atoms is 3:4?

Cave of Crystals

HOW NATURE WORKS

An Amazing Sight!

For any rock collector, a geode is a thrilling find. Splitting these hollow rocks in half reveals perfectly formed crystals inside. In 2000, two brothers found themselves inside something like a giant geode. They discovered the Cave of Crystals 300 m under the Chihuahuan Desert in Mexico. The limestone cave is filled with some of the largest crystals ever found—gigantic, shimmering beams the size of trees. Inside the cave, the temperature is about 43°C, and it is very humid. When researchers are in the cave, they wear special suits so they can work in the steaming hot conditions. Scientists are now working to identify and understand the natural conditions that allowed the crystals to become so large.

AMERICAN MUSEUM OF NATURAL HISTORY

What are the crystals made of? The crystals are made of the mineral selenite, a transparent and colorless form of gypsum. Gypsum has a hardness of 2 on the Mohs hardness scale. So even though some of the crystals have edges that are as sharp as blades, the crystals are soft enough to be scratched by your fingernail.

How did the crystals form? Hundreds of thousands of years ago, a magma chamber below the caves heated groundwater in Earth's crust. The hot water became mineral-rich by dissolving minerals from the surrounding rock. It seeped through cracks in the limestone caves. As the water cooled over time, the dissolved minerals crystallized.

◀ These selenite crystals are the same shape as the giants in the Cave of Crystals. They are much smaller, though, because they did not have as much time to grow.

How did the crystals get so big? The more time crystals have to grow, the bigger they become. The Cave of Crystals was left undisturbed for thousands of years, so the crystals grew very large. The crystals stopped growing only because local miners drained the groundwater, emptying the cave of the mineral-rich water.

It's Your Turn

JOURNAL ENTRY Suppose you are in charge of tourism for the Cave of Crystals. Write a tourist brochure describing what visitors can expect to see and experience in the cave. Be sure to include a list of dos and don'ts.

Lesson 2

Reading Guide

Key Concepts
ESSENTIAL QUESTIONS

- What characteristics can be used to classify rocks?
- How do the different types of rocks form?
- What are some uses of rocks in everyday life?

Vocabulary

rock p. 471
grain p. 471
magma p. 472
lava p. 472
texture p. 472
sediment p. 473
lithification p. 473
foliation p. 475

Multilingual eGlossary

Video BrainPOP®

Rocks

Inquiry Are these rocks?

These rocks are at the bottom of the Grand Canyon and are very old. They have certain characteristics that allow geologists to identify exactly what type of rock they are. All rocks do not look the same. Geologists use the different characteristics of rocks to categorize the different rock types on Earth.

Inquiry Launch Lab

15 minutes

What does a rock's texture tell you about the rock?

Do you have a sweater or a shirt that is so soft you want to wear it all the time? Softness is a texture. The word texture is also used to describe rocks. But, it does not refer to the way a rock's surface feels. A rock's texture refers to the sizes of the grains, or particles, in the rock and how they are arranged. What can a rock's texture tell you about how the rock formed?

1. Read and complete a lab safety form.
2. Make a data table in your Science Journal in which to record your observations. Your table should include columns for three different rocks, drawings of their textures, and detailed descriptions of the textures.
3. Use a **magnifying lens** to closely examine sets of **rocks.** Note the sizes of the grains and how the grains in each rock are arranged. Record your observations.

Think About This

1. Which rock might have formed from magma? Which might have formed in water? Which might have formed under pressure?
2. Contrast the three different rock textures.
3. **Key Concept** How do you think texture might be used to classify rocks?

What is a rock?

Sometimes you can tell how an object was made simply by looking at the finished product. If someone serves you eggs for breakfast, you can tell whether they were fried or scrambled. In much the same way, a geologist can tell how a rock was formed just by looking at it. The two rocks in **Figure 9** mostly contain quartz, feldspar and biotite mica. But the rocks look different because they formed in different ways.

Figure 9 These rocks contain the same minerals, but, because they formed differently, they have different appearances.

Visual Check How are these two rock samples different?

A **rock** *is a naturally occurring solid mixture composed of minerals, smaller rock fragments, organic matter, or glass. The individual particles in a rock are called* **grains.** Both rocks shown in **Figure 9** are made of mineral grains. The arrangement of the grains give clues to understanding how the rocks formed.

Reading Check What types of grains make up rocks?

Classifying Rocks

Most of Earth's surface is made of rocks. The different kinds of rocks are classified based on the way they form. There are three major types of rocks: igneous, metamorphic, and sedimentary.

Igneous Rocks

Igneous rocks are the most abundant rocks on Earth. Most of them form deep below Earth's surface, but some form on Earth's surface. Igneous rocks might form in different places, but they all form in a similar way.

Formation of Igneous Rocks *Molten rock is called* **magma** *when it is inside Earth. Molten rock that erupts onto Earth's surface is called* **lava.** As magma or lava cools, mineral crystals begin to form. These minerals form the grains of a new igneous rock.

Texture and Composition Geologists classify igneous rocks according to texture and mineral composition. For rocks, **texture** *refers to grain size and how the grains are arranged.*

Lava at Earth's surface cools quickly, so crystals do not have much time to increase in size. The crystals are small, like the crystals in the basalt shown in **Figure 10.** Geologists describe the texture of igneous rocks with small crystals as fine-grained. Deep below Earth's surface, magma cools slowly, and crystals have more time to grow. The crystals are larger, like the crystals in the granite shown in **Figure 10.** Geologists describe the texture of igneous rocks with large crystals as coarse-grained.

Igneous rocks such as granite and basalt differ in texture. They also differ in composition, or the minerals they contain. Granite contains mostly light-colored minerals such as quartz and potassium feldspar. Basalt is made of dark minerals such as pyroxene [pi RAHK seen] and olivine [AHL ih veen]. Note the differences in color between the two rocks in **Figure 10.**

Key Concept Check What characteristics are used to classify igneous rocks?

Environments of Igneous Rock Formation 

Figure 10 Igneous rocks form as magma or lava cools. Magma that cools deep beneath Earth's surface cools slowly and forms large crystals, such as the ones in granite. Lava that cools at Earth's surface cools quickly and forms small crystals, such as the crystals in basalt.

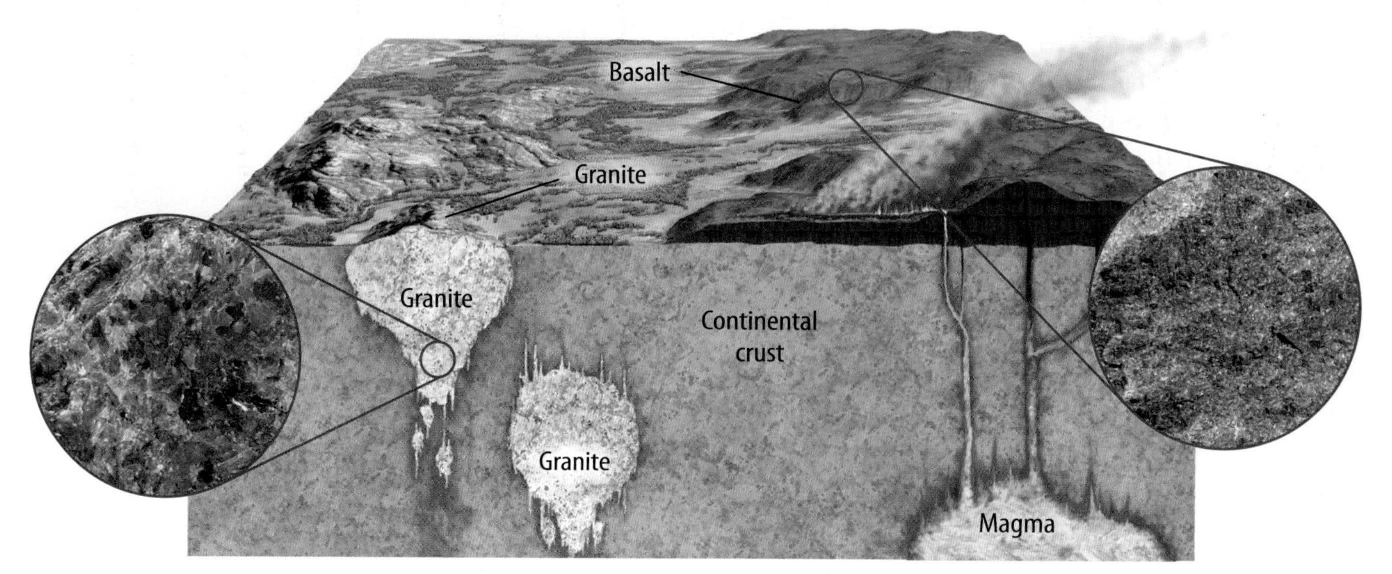

Environments of Sedimentary Rock Formation

Figure 11 Sediment travels downhill and settles in layers within basins. Eventually, the sediment becomes compacted and cemented together, forming sedimentary rocks.

Sedimentary Rocks

Natural processes break down rocks exposed at Earth's surface. *Rock and mineral fragments that are loose or suspended in water are called* **sediment.** Just as magma is a source material for igneous rocks, sediment is the source material for sedimentary rocks.

Formation of Sedimentary Rocks **Lithification** *is the process through which sediment turns into rock.* Study **Figure 11.** How can sediment become solid rock? Usually, sediment is formed through weathering by water, ice, or wind. These agents also remove, or erode, the sediment. Sediment eventually is deposited in low areas called basins. Layers of sediment build up, and the weight of the upper layers compacts the lower layers.

Dissolved minerals, usually quartz or calcite in water, cement the grains together and form sedimentary rocks, such as the sandstone and shale in **Figure 11.** Dissolved solids also can crystallize directly from a water solution and form sedimentary rocks such as rock salt.

Texture and Composition Similar to igneous rocks, sedimentary rocks can be described as fine-grained or coarse-grained. The shape of the grains also can be described as rounded or angular. Grains usually are angular when first broken but often become rounded during transport. Rounded grains can help distinguish sedimentary rocks from some igneous rocks.

The composition of a sedimentary rock depends on the minerals in the sediment from which it formed. Sandstone is a sedimentary rock that usually is made of quartz grains. Shale is made from much smaller grains of quartz and clay minerals.

FOLDABLES®

Use a sheet of paper to make a folded table. Label it as shown. Use it to compare types of rocks.

Rocks	Formation	Texture	Composition
Igneous			
Sedimentary			
Metamorphic			

Figure 12 Each metamorphic rock has a parent rock.

Visual Check Which rock does not show the direction of increased pressure?

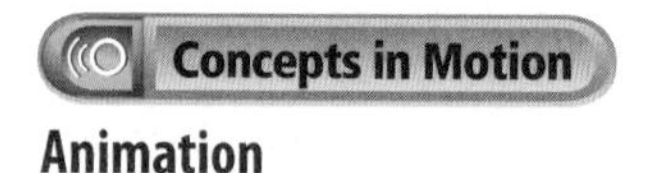

Animation

Metamorphic Rocks

Sometimes rocks can change into new and different rocks without erosion or melting. Extreme high temperatures and pressure cause these changes. The original rocks are called parent rocks, and the new rocks are called metamorphic rocks.

Formation of Metamorphic Rocks Metamorphic rocks form when parent rocks are squeezed, heated, or exposed to hot fluids. The rocks do not melt. They remain solid, but the texture and, sometimes, the mineral composition of the parent rock change. This process is metamorphism.

Texture and Composition The textures of most metamorphic rocks result from increases in temperature and pressure. This process is illustrated in **Figure 12.** The mineral composition of metamorphic rocks might be the result of minerals that are present in the parent rock, or they might grow in the new metamorphic rock. You read about gemstones in Lesson 1. Many gemstones are minerals that formed as a result of metamorphism.

Key Concept Check How do metamorphic rocks form?

Inquiry MiniLab

20 minutes

How do heat and pressure change rocks?

Have you ever discovered a forgotten peanut butter sandwich at the bottom of your backpack? If so, you know well that heat and pressure can change things! When rocks are exposed to heat and pressure, they can become metamorphic rocks. In this lab, you'll model two types of metamorphism.

1. Read and complete a lab safety form.
2. Drop a handful of **uncooked spaghetti noodles** onto your desk. Observe and record how they are arranged relative to each other.
3. In one quick motion, use the edges of your hands to bring the noodles back together. Observe and record how the noodles now are arranged relative to each other.
4. Put an **egg white** in a **shallow glass dish.**
5. Use a **hot plate** to heat a **beaker** half-filled with water.
   ⚠ Do not touch the hot plate or boil the water.
6. Use **tongs** to place the beaker on the egg white. Observe what happens and record your observations in your Science Journal.

Analyze and Conclude

1. **Recognize Cause and Effect** How did the spaghetti noodles and the egg white change? How are these changes like metamorphism?
2. **Key Concept** How does this activity model how metamorphic rocks form?

Foliated Metamorphic Rocks Recall that crystals form in a variety of shapes. Minerals with flat shapes, such as mica, produce a foliated texture. **Foliation** [foh lee AY shun] *results when uneven pressures cause flat minerals to line up, giving the rock a layered appearance.* Eventually distinct bands of light and dark minerals form, as shown in the sample of gneiss [NISE] in **Figure 12.** Foliation is the most obvious characteristic of metamorphic rocks.

Nonfoliated Metamorphic Rocks Marble, another type of metamorphic rock is not foliated. The grains in the marble pictured in **Figure 12** are not flattened like the grains in gneiss. The calcite crystals that make up marble became blocklike and square when exposed to high temperatures and pressure. Marble has a nonfoliated texture.

Rocks in Everyday Life

Rocks are abundant natural resources that are used in many ways based on their physical characteristics. Some igneous rocks are hard and durable, such as the granite used to construct the fountain shown in **Figure 13.** The igneous rock pumice is soft but contains small pieces of hard glass, which makes it useful for polishing and cleaning.

Natural layering makes sedimentary rock a high-quality building stone. Both sandstone and limestone are used in buildings. The building pictured in **Figure 14** is made of sandstone. Limestone also is used to make cement, which is then used in construction applications, including building highways.

Foliated metamorphic rocks, such as slate, split into flat pieces. Slate makes durable, fireproof roofing shingles, such as the ones shown in **Figure 15.** Other metamorphic rocks are used in art. Marble is soft enough to carve and often is used for making detailed sculptures.

Key Concept Check What are some everyday uses for rocks?

Figure 13 Granite was used to build this fountain. ▶

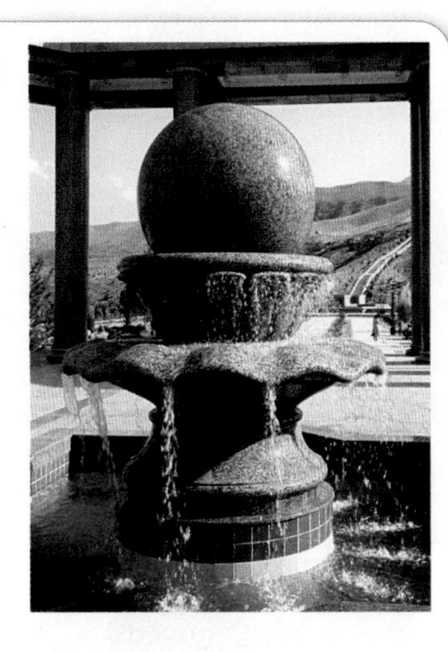

Figure 14 This building in Jordan was carved and constructed out of sandstone. ▼

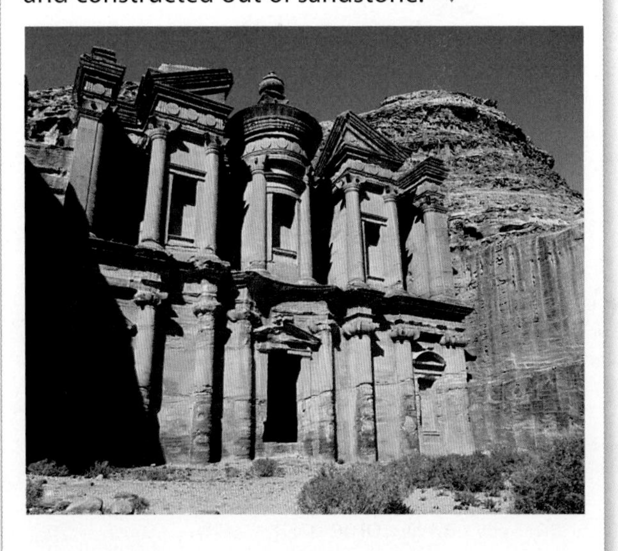

Figure 15 Slate sometimes is used on roofs like the ones on this house. ▼

Lesson 2 Review

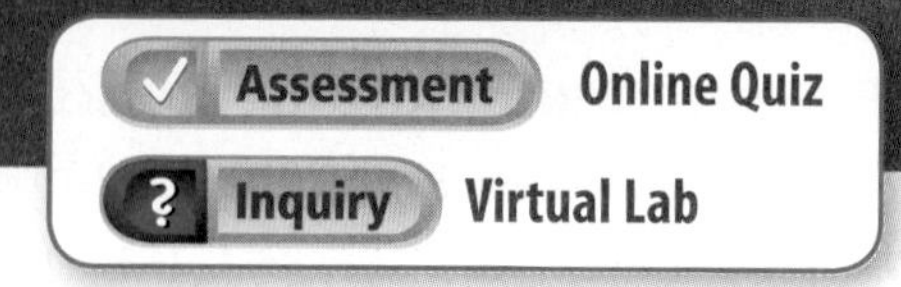

Visual Summary

Interlocking crystals of different sizes are common in igneous rocks.

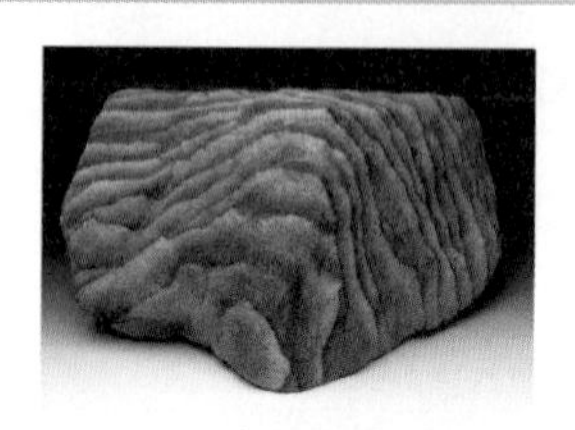

The individual grains that form sedimentary rocks can be mineral grains or fragments of other rocks.

Increases in temperature and pressure cause minerals to change in size and shape.

FOLDABLES®

Use your lesson Foldable to review the lesson. Save your Foldable for the project at the end of the chapter.

What do you think NOW?

You first read the statements below at the beginning of the chapter.

3. Once a rock forms, it lasts forever.

4. All rocks form when melted rock cools and changes into a solid.

Did you change your mind about whether you agree or disagree with the statements? Rewrite any false statements to make them true.

Use Vocabulary

1. **Distinguish** between lava and magma.
2. Loose grains of rock material are called ________.
3. **Use the term** *lithification* in a sentence.

Understand Key Concepts 

4. **Explain** how sedimentary rocks form.
5. Which list shows the correct sequence?
   A. shale, foliation, basalt
   B. gneiss, lithification, shale
   C. granite, metamorphism, gneiss
   D. sandstone, lithification, sediment
6. **Compare** the formation of sandstone to the formation of gneiss.

Interpret Graphics

7. **Analyze** the figure below. Identify the parent rock, the metamorphic rock, and what the red arrow represents.

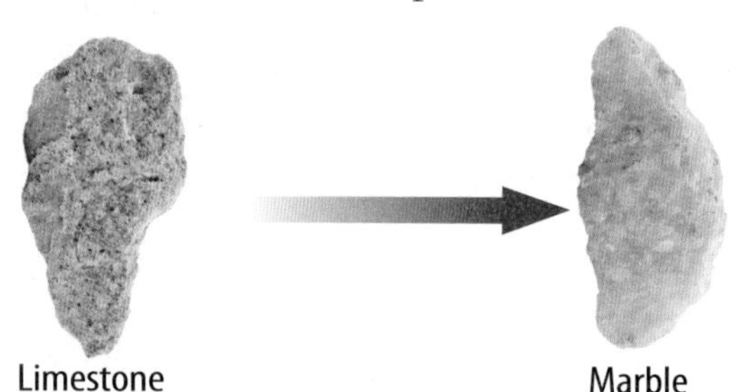

8. **Draw** a graphic organizer like the one below. List the two major textures of metamorphic rocks and why they form.

Texture	Formation

Critical Thinking

9. **Invent** a new way to classify the major rock types.
10. **Classify** the following rock and justify your answer: light brown in color; small, sand-sized pieces of quartz cemented together in layers.

Inquiry **Skill Practice** **Compare and Contrast** **30 minutes**

Materials

rock samples

magnifying lens

glass plate

Safety

How are rocks similar and different?

Do you have any collections? If so, you might have the objects in the collections sorted according to how they are the same or how they are different. In this lab, you will use a key to **compare and contrast** various rock samples in order to identify them.

Learn It

Comparing objects means finding similarities. Contrasting objects means finding differences. How can you **compare and contrast** rock samples to correctly identify different rocks?

Try It

1. Read and complete a lab safety form.
2. In your Science Journal, draw a table like the one shown below.
3. Use the rock identification key as a guide while you examine your rocks.
4. Study your rock samples. Organize them into groups based on common characteristics. In your Science Journal, record what characteristics you used to group the rocks.
5. Use the key and a magnifying lens, as needed, to identify each rock.
6. Record each rock type and name in your data table.

Apply It

7. **Compare and Contrast** How are the rocks similar? How are they different?
8. **Communicate** Were there any rocks that you could not identify? Why do you think this happened?
9. **Key Concept** What characteristics can be used to classify rocks?

Data for Ten Different Rock Samples

Sample	Type of Rock	Name of Rock
1.		
2.		
3.		
4.		
5.		
6.		
7.		
8.		
9.		
10.		

Lesson 3

The Rock Cycle

Reading Guide

Key Concepts
ESSENTIAL QUESTIONS

- How do surface processes contribute to the rock cycle?
- How is the rock cycle related to plate tectonics?

Vocabulary

rock cycle p. 479
extrusive rock p. 480
intrusive rock p. 480
uplift p. 480
deposition p. 481

g Multilingual eGlossary

Inquiry Where did all the rock layers go?

Monument Valley, Utah, is home to these towering rock formations. Over millions of years, processes on Earth's surface have worn away the surrounding rock layers, leaving the more resistant rocks behind. What do you think this landscape will look like in another few million years?

Inquiry Launch Lab

20 minutes

Do you "rock"?

Have you ever walked across a gravel road in your bare feet? If so, you know that rocks are hard. However, even though they are hard, rocks can change. How can you make a model of a rock that allows you to observe some of the changes that can turn it from one type into another?

1. Read and complete a lab safety form.
2. Break **small candles** in half over a piece of **waxed paper.**
3. Drop the pieces of candle into very warm water.
4. After 10–20 seconds, use **forceps** to remove all the candle pieces and stack them back on the waxed paper.
5. Wrap the candles in the waxed paper and squeeze tightly to press the warm pieces.

Think About This

1. What type of rock did you model? Explain.
2. What changed the model rocks in step 2 to those in step 5? What type of rock formed?
3. **Key Concept** How might different processes contribute to the rock cycle?

What is the rock cycle?

Do you have a recycling program at school? Or do you recycle at home? When materials such as paper or metal are recycled, they are used over again but not always for the same things. The metal from the beverage can you recycled yesterday might end up in a baseball bat.

Recycling also occurs naturally on Earth. The rock material that formed Earth 4.6 billion years ago is still here, but much of it has changed many times throughout Earth's history. *The series of processes that continually change one rock type into another is called the* **rock cycle.**

As materials move through the rock cycle, they can take the form of igneous rocks, sedimentary rocks, or metamorphic rocks. At times, the material might not be rock at all. It might be sediment, magma, or lava, such as that pictured in **Figure 16.**

Reading Check How is the rock cycle similar to recycling?

Figure 16 Earth materials move through the rock cycle, changing both form and their location on Earth.

Figure 17 As rocks and rock material move slowly through the rock cycle, they are continually transformed from one rock type to another.

Visual Check Describe a path through the rock cycle that would result in the formation of a metamorphic rock.

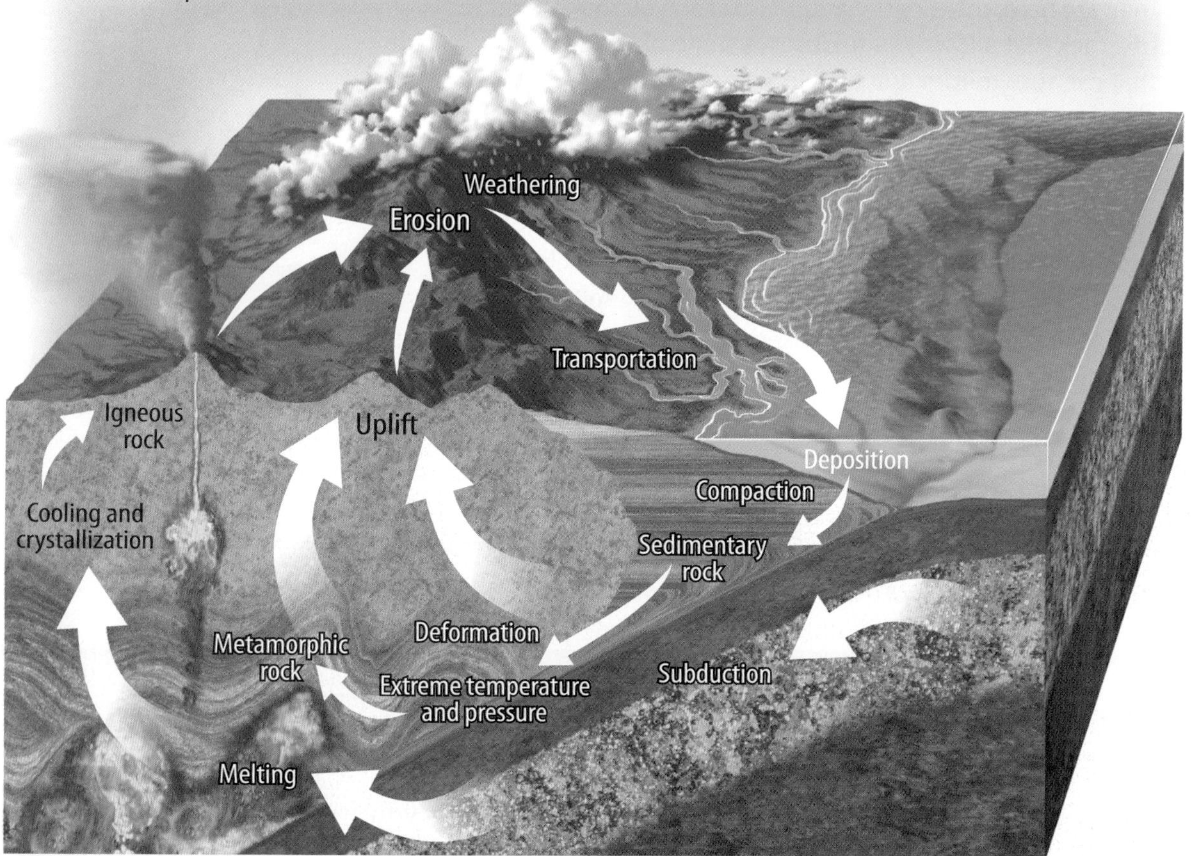

Processes of the Rock Cycle

Mineral and rock formation are important processes in the rock cycle. The rock cycle is continuous, with no beginning or end. As shown in **Figure 17,** some processes take place on Earth's surface, and others take place deep beneath Earth's surface.

Cooling and Crystallization

Melted rock material is present both on and below Earth's surface. *When lava cools and crystallizes on Earth's surface, the igneous rock that forms is called* **extrusive rock.** *When magma cools and crystallizes inside Earth, the igneous rock that forms is called* **intrusive rock.**

Science Use v. Common Use

intrusive

Science Use igneous rock that forms as a result of injecting magma into an existing rock body

Common Use the condition of being not welcome or invited

Uplift

If intrusive rocks form deep within Earth, how are they ever exposed at the surface? **Uplift** *is the process that moves large amounts of rock up to Earth's surface and to higher elevations.* Uplift is driven by Earth's tectonic activity and often is associated with mountain building.

Weathering and Erosion

Uplift brings rocks to Earth's surface where they are exposed to the environment. Glaciers, wind, and rain, along with the activities of some organisms, start to break down exposed rocks. The same glaciers, wind, and rain also carry sediment to low-lying areas, called basins, by the process of erosion.

Deposition

Eventually, glaciers, wind, and water slow down enough that they can no longer transport the sediment. *The process of laying down sediment in a new location is called* **deposition.** Deposition forms layers of sediment. As time passes, more and more layers are deposited.

Key Concept Check How are surface processes involved in the rock cycle?

Compaction and Cementation

The weight of overlying layers of sediment pushes the grains of the bottom layers closer together. This process is called compaction. Sedimentary rocks have tiny spaces, called pores, between the grains. Pores sometimes contain water and dissolved minerals. When these minerals crystallize, they cement the grains together. **Figure 17** shows the path of sediment from weathering and erosion to compaction and cementation.

FOLDABLES®

Make a horizontal two-column chart book. Label it as shown. Use it to organize your notes on rock formation.

Internal Processes | External Processes

Inquiry MiniLab **15 minutes**

How can you turn one sedimentary rock into another?

The formation of a rock involves many changes. How can you model some of the changes that turn one sedimentary rock into another?

1. Read and complete a lab safety form.
2. Stack some **craft sticks** on the table. Prop up one end of a **baking dish** by laying it on the craft sticks.
3. Rub two **rocks** together over the dish. Observe how the rocks change.
4. Observe what happens to the sediment. Record your observations in your Science Journal.
5. Add some **white glue** to the sediment that collects at the bottom of the dish. Allow the glue to dry. Observe.
6. Predict what will happen to the sediment as the glue dries. Record your prediction.

Analyze and Conclude

1. **Model** What rock cycle process did you model in step 3? In step 4? In step 5?
2. **Predict** How did your prediction compare to the actual outcome?
3. **Key Concept** How are surface processes involved in the rock cycle?

Word Origin

metamorphism
from Greek *metamorphoun,* means "to transform"

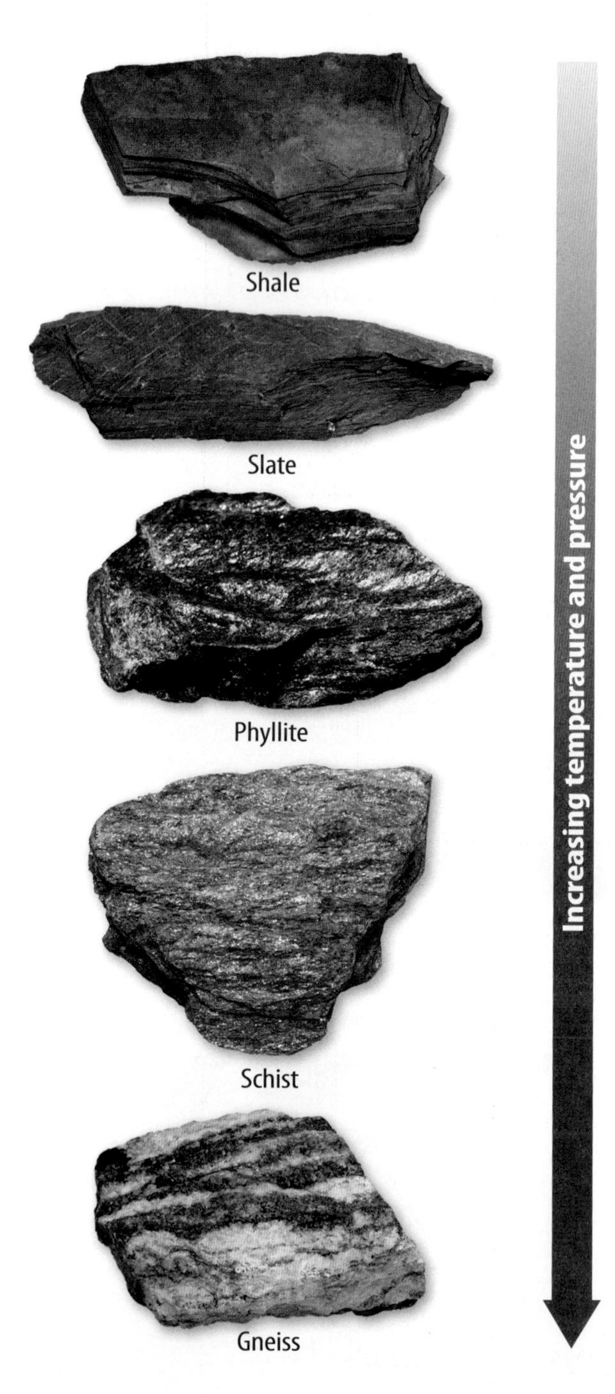

Figure 18 Rocks change form under high temperatures and pressure.

Visual Check How do the characteristics of the rocks change with increased pressure and temperature?

 **Personal Tutor**

Temperature and Pressure

Recall that rocks subjected to high temperature and pressure undergo **metamorphism.** This usually occurs far below Earth's surface. The progression of metamorphism can be observed in some rocks, as shown in **Figure 18.** As temperature and pressure increase, the sedimentary rock shale, shown at the top of **Figure 18,** changes to the metamorphic rock slate.

The rocks shown in **Figure 18,** slate, phyllite, schist, and gneiss, form from shale with increasing temperature and pressure. If the temperature is high enough, the rock melts and becomes magma. Igneous rocks form as the magma cools, and the material continues through the rock cycle.

Rocks and Plate Tectonics

The theory of plate tectonics states that Earth's surface is broken into rigid plates. The plates move as a result of Earth's internal thermal energy and convection in the mantle. The theory explains the movement of continents. It also explains earthquakes, volcanoes, and the formation of new crust. These events occur at plate boundaries, where tectonic plates interact.

Igneous rock forms where volcanoes occur and where plates move apart. Where plates collide, rocks are subjected to intense pressure and can undergo metamorphism. Colliding plates also can cause uplift or can push rock deep below Earth's surface, where it melts and forms magma. At Earth's surface, uplifted rocks are exposed and weathered. Weathered rock forms sediment, which eventually can form sedimentary rock.

Processes within Earth that move tectonic plates also drive part of the rock cycle. The rock cycle also includes surface processes. As long as these processes exist, the rock cycle will continue.

 **Key Concept Check** How is the rock cycle related to plate tectonics?

Lesson 3 Review

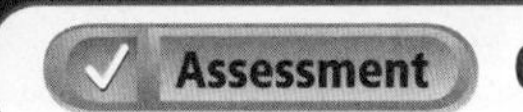

Visual Summary

Weathering and erosion are important processes in the rock cycle.

Uplift contributes to rock cycle processes on Earth's surface.

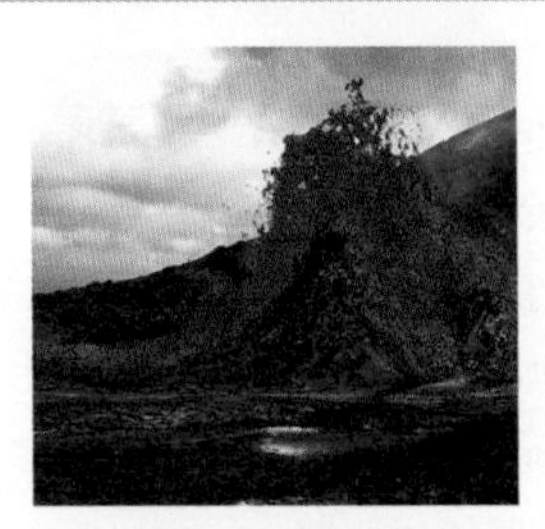

Plate tectonic activity contributes to rock cycle processes beneath Earth's surface.

FOLDABLES®

Use your lesson Foldable to review the lesson. Save your Foldable for the project at the end of the chapter.

What do you think NOW?

You first read the statements below at the beginning of the chapter.

5. All rocks are related through the rock cycle.

6. Rocks move at a slow and constant rate through the rock cycle.

Did you change your mind about whether you agree or disagree with the statements? Rewrite any false statements to make them true.

Use Vocabulary

1. **Distinguish** between intrusive igneous rocks and extrusive igneous rocks.
2. **Define** *deposition* in your own words.
3. **Use the term** *rock cycle* in a sentence.

Understand Key Concepts 

4. Which term refers to breaking rocks apart?
   A. cementation
   B. crystallization
   C. deposition
   D. weathering
5. **Give an example** of a rock cycle process that occurs on Earth's surface and one that occurs below Earth's surface.
6. **Classify** each of the following terms as Earth materials or rock cycle processes: magma, crystallization, sedimentary rocks, sediment, uplift, cementation.

Interpret Graphics

7. **Determine** what process must occur between the magma chamber and the intrusive rock in the figure below.

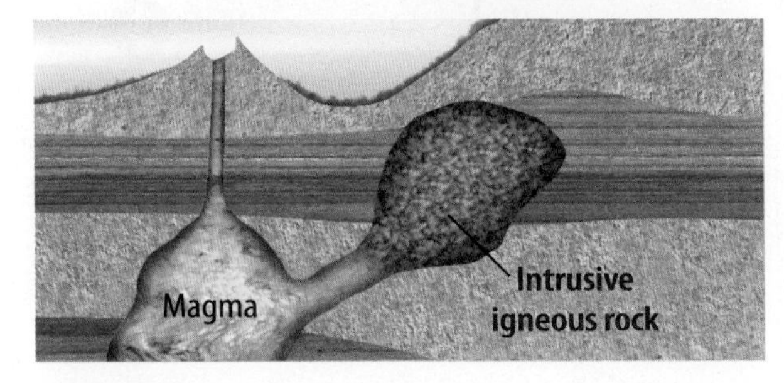

8. **Draw** a graphic organizer like the one below and sequence the following terms: erosion, compaction, sedimentary rock, deposition, uplift, weathering, cementation.

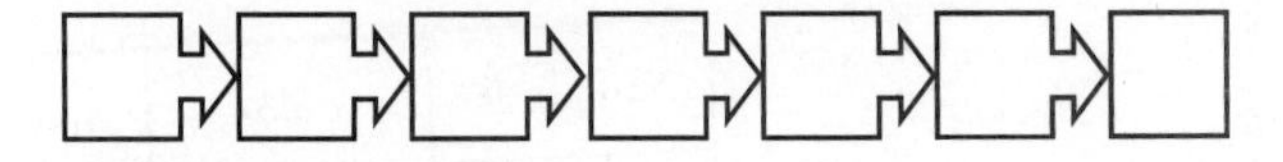

Critical Thinking

9. **Infer** where you would look along a tectonic plate to find the youngest rock.

 Lab

2 class periods

Design a Forensic Investigation

Materials

bag of "loot"

mineral-testing tools

magnifying lens

paper towels

Safety

The local jewelry store has been robbed! The thief had planned to leave common minerals at the scene to replace the valuable ones. However, she was caught in the act, and both the valuable and the common minerals ended up mixed together on the store's floor. The police have called you in as a forensic geologist to help them sort through the rubble and identify the valuable minerals.

Question

Think about what you learned about minerals in this chapter. Use that information and the paragraph above to formulate a question to help you solve this case.

Procedure

1. Read and complete a lab safety form.
2. Copy the data table shown below into your Science Journal. Complete the table as you test each mineral. Add columns and rows as needed.
3. Remove the loot from its bag. Use a magnifying lens to examine each mineral. Record any interesting or unique observations about each mineral in your data table.

Sample #	color	hardness	reaction to HCL	streak	luster	other
1.						
2.						

4. Use the mineral-testing tools to determine the hardness, the streak, and the reaction to dilute hydrochloric acid of each piece of loot. Record your observations in your data table.
5. Use your data table and the Minerals table in the Reference Handbook to identify each of the minerals involved in this imaginary heist.

Analyze and Conclude

6. **Compare and Contrast** How are minerals alike and how do they differ?
7. **Evaluate** Were any of the samples valuable minerals such as gold, silver, rubies, diamonds, or emeralds? How do you know?
8. **Draw Conclusions** What properties of minerals make them useful or desirable for jewelry?
9. **The Big Idea** How are minerals and rocks formed, identified, classified, and used?

Communicate Your Results

Choose your favorite piece of loot from the bag. On a small piece of poster board, write the name of this mineral and its properties. Then, with two or more other students, "combine" your minerals to form a rock. Write the name of the rock you made on another piece of poster board. Compose a two-minute skit to demonstrate and explain how the rock formed and how it, and the minerals in it, might be used in everyday life.

Inquiry Extension

Use your observations from this lab to visually examine jewelry worn by your classmates or family members. Try to identify the minerals in each piece if jewelry. Be careful not to damage the jewelry!

Lab Tips

☑ **CAUTION:** Use only one drop of hydrochloric acid to test each mineral. Use paper towels to completely remove any acid from the minerals.

☑ When testing hardness, use the mineral to try to scratch the testing materials. Do not scratch the mineral itself unless instructed to do so.

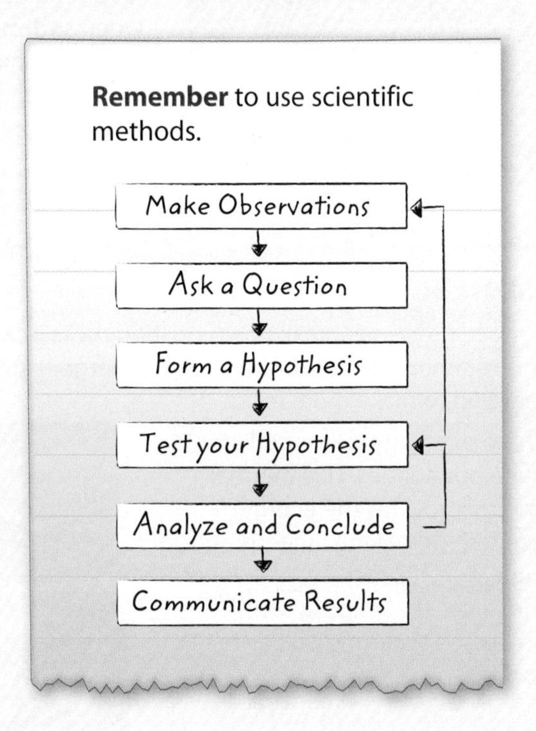

Chapter 13 Study Guide

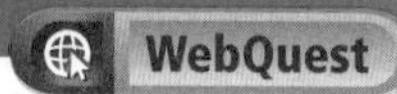

Minerals and rocks form through natural processes, have practical uses in everyday life, and are valued for their beauty. Minerals can be identified based on their physical properties. Rocks are classified based on their physical characteristics and how they formed.

Key Concepts Summary	Vocabulary
Lesson 1: Minerals • **Minerals** form when solids crystallize from molten material or from solutions. • Properties such as color, **streak,** hardness, and **cleavage** are used to identify minerals. Unique properties such as magnetism, reaction to acid, and fluorescence can also be used to identify certain minerals. • Minerals are used to make everyday products such as toothpaste and makeup. Metals are used in cars and buildings. Gemstones are valued for their beauty. 	**mineral** p. 461 **crystal structure** p. 462 **crystallization** p. 463 **streak** p. 465 **luster** p. 465 **cleavage** p. 465 **fracture** p. 465 **ore** p. 467
Lesson 2: Rocks • Rocks are classified based on their **texture** and composition.  • Igneous rocks form when **magma** or **lava** solidifies. Sedimentary rocks form when **sediments** are **lithified.** Metamorphic rocks form when parent rocks are changed by thermal energy, pressure, or hot fluids. • Rocks are used in construction, abrasives, and art.	**rock** p. 471 **grain** p. 471 **magma** p. 472 **lava** p. 472 **texture** p. 472 **sediment** p. 473 **lithification** p. 473 **foliation** p. 475
Lesson 3: The Rock Cycle • Surface processes break down existing rocks into sediment. They transport this sediment to locations where it undergoes **deposition** and can be recycled to make more rocks. • Thermal energy is released at plate boundaries. This thermal energy provides the energy needed for making igneous and metamorphic rocks. It also drives the forces that expose rocks to processes occurring on Earth's surface. 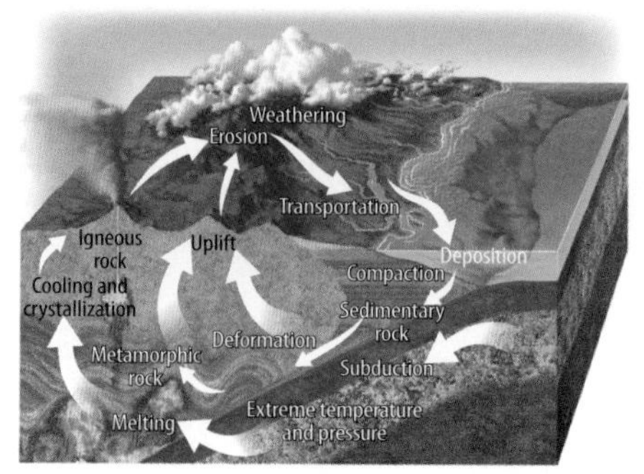 	**rock cycle** p. 479 **extrusive rock** p. 480 **intrusive rock** p. 480 **uplift** p. 480 **deposition** p. 481

- Personal Tutor
- Vocabulary eGames
- Vocabulary eFlashcards

FOLDABLES® Chapter Project

Assemble your lesson Foldables as shown to make a Chapter Project. Use the project to review what you have learned in this chapter.

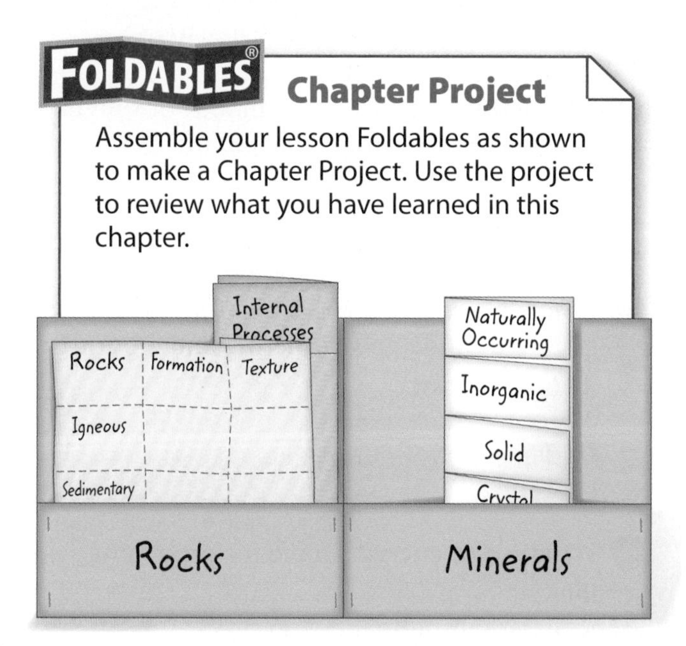

Use Vocabulary

1. A mineral deposit that can be mined for a profit is a(n) ________.
2. How does color differ from streak?
3. Loose rock and mineral fragments are called ________.
4. Define the word *rock* in your own words.
5. The process that brings rocks formed deep within Earth to the surface is called ________.
6. Relate the words *deposition* and *lithification.*

Link Vocabulary and Key Concepts

Interactive Concept Map

Copy this concept map, and then use vocabulary terms from the previous page to complete the concept map.

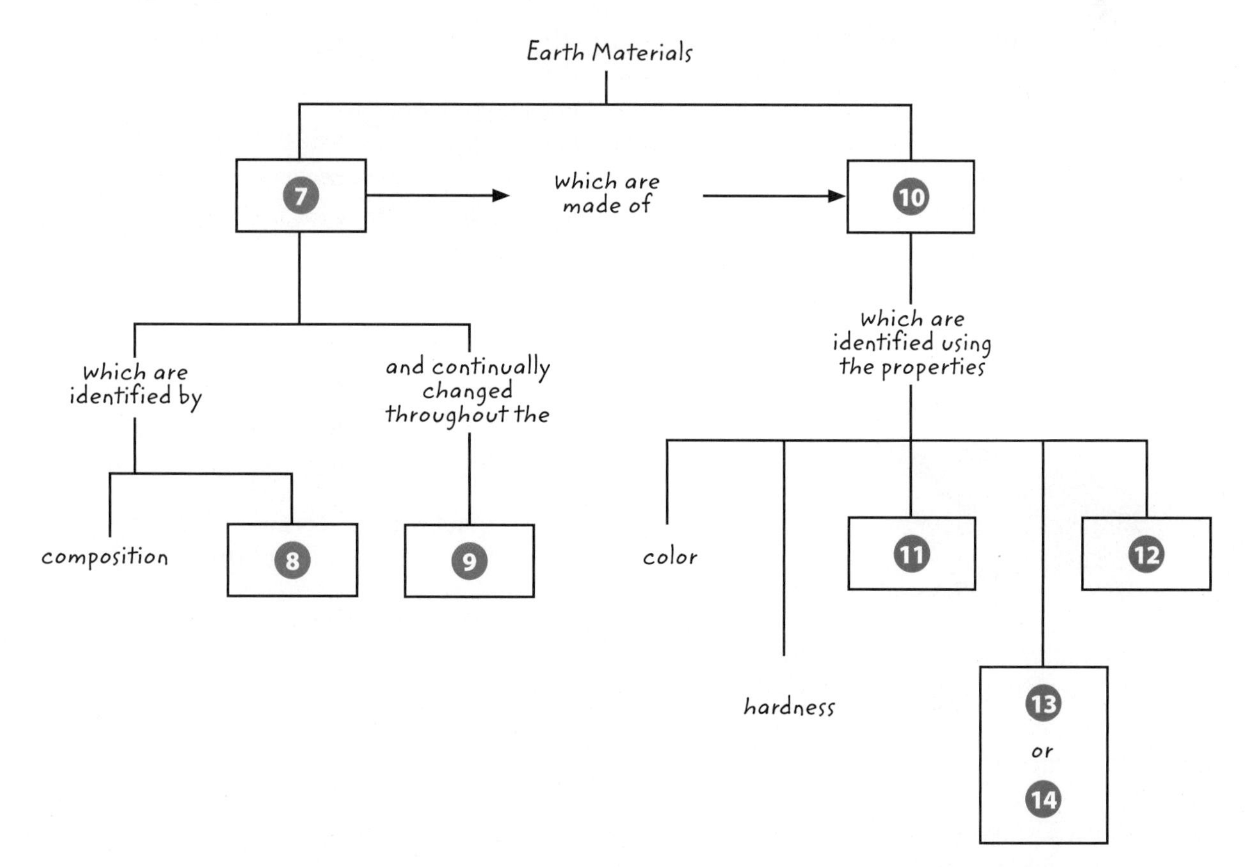

Chapter 13 Review

Understand Key Concepts 

1. Based on what you know about the Mohs hardness scale and mineral hardness, which mineral would make a good sandpaper?
   A. fluorite
   B. gypsum
   C. quartz
   D. talc

2. Which describes a way that minerals form?
   A. changing from gas to liquid
   B. changing from liquid to gas
   C. changing from liquid to solid
   D. changing from solid to gas

3. Which are mineral resources?
   A. gemstones and wood
   B. metals and cotton
   C. metals and gemstones
   D. wood and cotton

4. What can you learn from a mineral's chemical formula?
   A. composition
   B. crystal structure
   C. texture
   D. hardness

5. Which type of rock forms from magma and contains large interlocking crystals?
   A. extrusive igneous
   B. intrusive igneous
   C. foliated metamorphic
   D. nonfoliated metamorphic

6. What characteristic can be used to identify the mineral pictured above?
   A. color
   B. crystal structure
   C. density
   D. hardness

7. Texture of sedimentary rock refers to
   A. whether the rock cooled slowly or quickly.
   B. whether the rock feels smooth or rough.
   C. whether the rock is foliated or nonfoliated.
   D. whether the rock has coarse or fine grains.

8. Which process is necessary in order for granite to undergo weathering and erosion?
   A. cementation
   B. compaction
   C. deposition
   D. uplift

9. What is the energy source for producing magma?
   A. external thermal energy
   B. internal thermal energy
   C. pressure
   D. the Sun

10. What processes are necessary in order to turn the material in the picture below into rock?

   A. compaction and cementation
   B. cooling and crystallization
   C. uplift and deposition
   D. weathering and erosion

Assessment
Online Test Practice

Critical Thinking

11. **Compare** the textures of the three main types of rocks.

12. **Infer** why there is no longer a rock cycle on the Moon.

13. **Evaluate** the relative worth of minerals that are valued for their beauty and those that are used for practical purposes.

14. **Analyze** which rock is more useful for building a roof—slate or granite? Explain your reasoning.

15. **Predict** what would happen to a sample of gneiss if it were heated enough to melt the mineral grains.

16. **Construct** a flow chart showing the formation of quartzite starting with a mountain and ending with quartzite. (Quartzite is metamorphosed sandstone.)

17. **Critique** the rock cycle diagram below. Include one feature of the diagram that you find useful and one feature that could be improved.

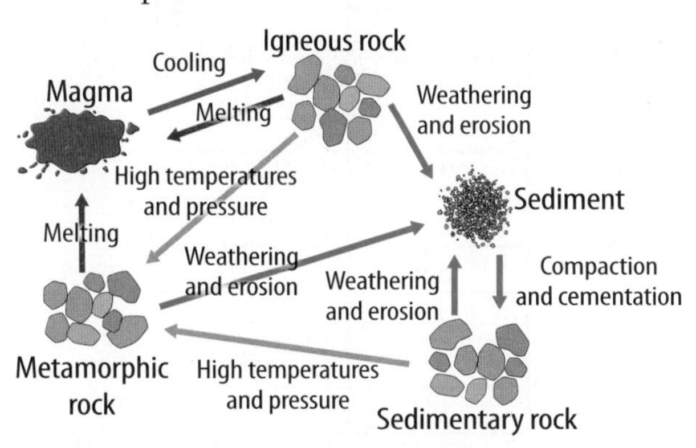

18. **Summarize** How do geologists identify minerals?

Writing in Science

19. **Write** an acrostic poem based on the term *rock cycle.* Acrostic poems are written without rhyming. The letters of the given term form the first letter of each line of the poem, which, when read downward, spells out the term.

REVIEW THE BIG IDEA

20. How is rock classification related to the rock cycle?

21. What are some minerals and rocks that you use every day?

Math Skills

Math Practice

Use Ratios

22. The ratio of iron (Fe) to chromium (Cr) to oxygen (O) in the mineral chromite is 1:2:4. What is the formula for chromite?

23. One form of the mineral feldspar has the formula $KAlSi_3O_8$. What is the ratio of potassium (K) to aluminum (Al) to silicon (Si) to oxygen (O) atoms in feldspar?

24. The ratio of aluminum (Al) to oxygen (O) atoms in the mineral corundum is 2:3. Write the formula for corundum.

Standardized Test Practice

Record your answers on the answer sheet provided by your teacher or on a sheet of paper.

Multiple Choice

1 Which is NOT a characteristic of minerals?

A They are only solids.

B They form from decaying materials.

C They have a crystal structure.

D They have a definite composition.

2 Which property is NOT used to identify minerals?

A color

B fracture

C streak

D weight

Use the figure below to answer question 3.

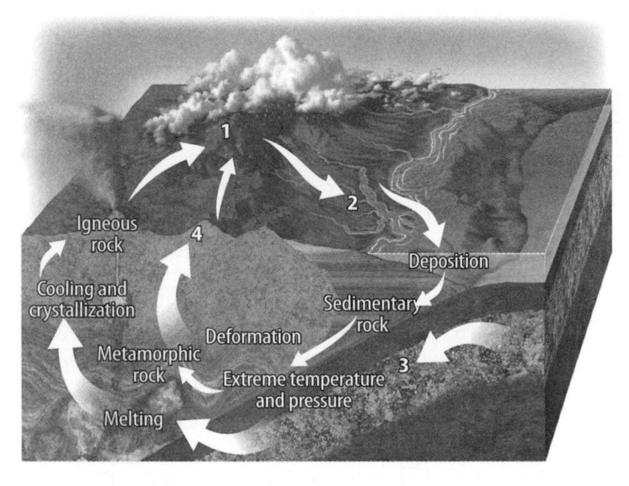

3 Which arrow on the diagram represents uplift?

A 1

B 2

C 3

D 4

4 Which process produces features that distinguish gneiss from marble?

A crystallization

B foliation

C lithification

D sedimentation

5 Which characteristic is unique to metamorphic rock?

A It can form from erosion.

B It can form from lithification.

C It can form from parent rocks.

D It can have a layered appearance.

Use the table below to answer question 6.

10	diamond
9	corundum
8	topaz
7	quartz
6	feldspar
5.5	glass
5	apatite
4.5	nail
4	fluorite
3.5	penny
3	calcite
2.5	fingernail
2	gypsum
1	talc

6 Use the Mohs hardness scale to determine which statement is correct.

A A fingernail will scratch talc.

B A nail will scratch apatite.

C Corundum will scratch diamond.

D Quartz will not scratch glass.

Assessment
Online Standardized Test Practice

7 Which is NOT present in sediment?

A magma

B minerals

C rock fragments

D organic matter

8 Which is the last step in the formation of sedimentary rock?

A cementation

B compaction

C deposition

D erosion

Use the figure below to answer question 9.

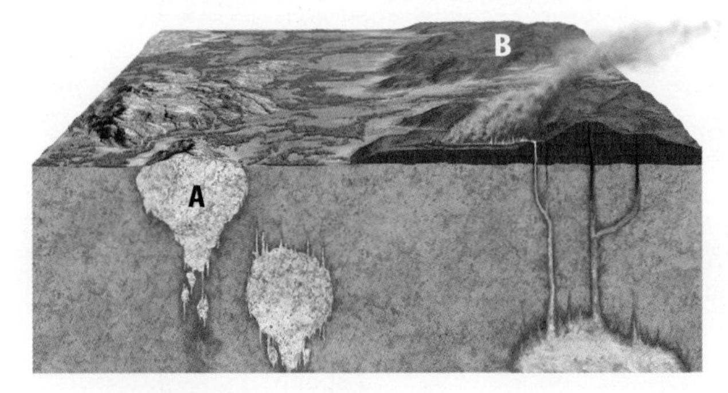

9 Which statement about the igneous rocks in the diagram is correct?

A Rock A is extrusive.

B Rock B is intrusive.

C Rock A cooled slowly and formed small crystals.

D Rock B cooled quickly and formed small crystals.

Constructed Response

Use the figures below to answer questions 10 and 11.

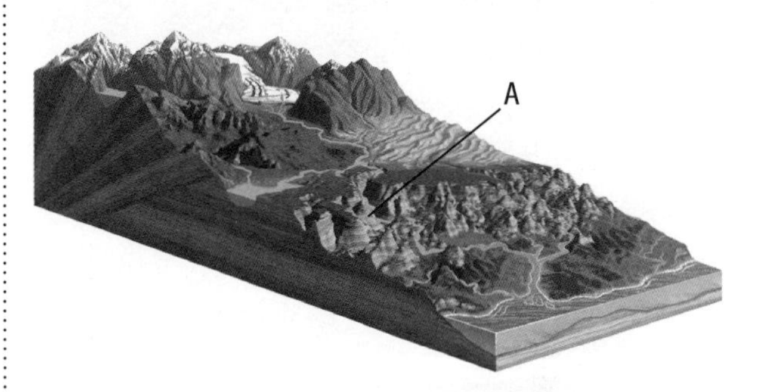

10 Identify the kind of rock that is labeled *A* in the figure. Then describe how it forms.

11 Describe a use for the type of rocks formed by the process in the figure. Give at least two examples of rocks used in this way.

12 Identify three surface processes that are part of the rock cycle. Explain the roles of each process in the cycle.

13 Which properties of a mineral can you observe using a penny, a pocketknife, and a magnifying lens? Explain your answers.

NEED EXTRA HELP?													
If You Missed Question...	1	2	3	4	5	6	7	8	9	10	11	12	13
Go to Lesson...	1	1	3	2	3	1	2	2	2	2	2	2	1

Chapter 14

Plate Tectonics

What is the theory of plate tectonics?

Is this a volcano?

Iceland is home to many active volcanoes like this one. This eruption is called a fissure eruption. This occurs when lava erupts from a long crack, or fissure, in Earth's crust.

- Why is the crust breaking apart here?
- What factors determine where a volcano will form?
- How are volcanoes associated with plate tectonics?

Get Ready to Read

What do you think?

Before you read, decide if you agree or disagree with each of these statements. As you read this chapter, see if you change your mind about any of the statements.

1. India has always been north of the equator.
2. All the continents once formed one supercontinent.
3. The seafloor is flat.
4. Volcanic activity occurs only on the seafloor.
5. Continents drift across a molten mantle.
6. Mountain ranges can form when continents collide.

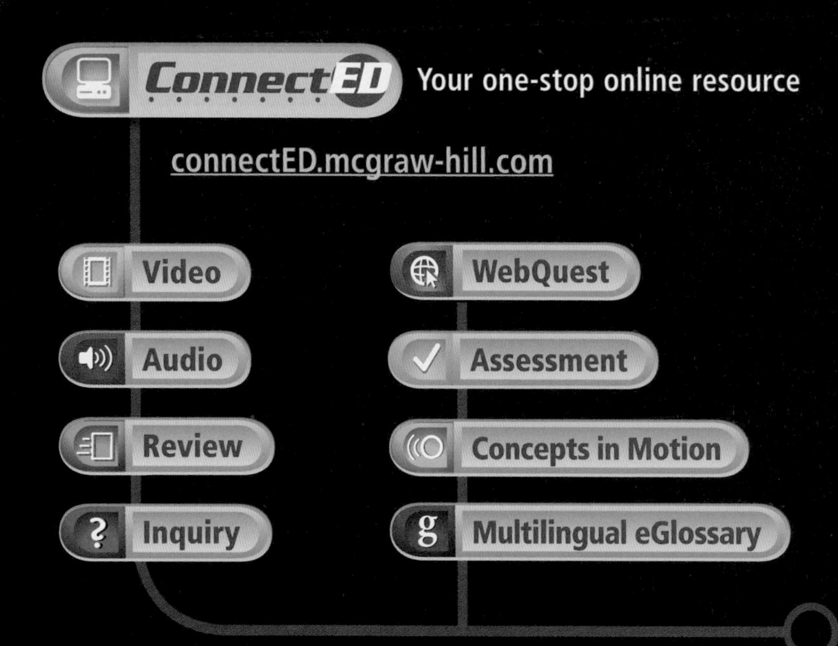

Lesson 1

Reading Guide

Key Concepts
ESSENTIAL QUESTIONS

- What evidence supports continental drift?
- Why did scientists question the continental drift hypothesis?

Vocabulary

Pangaea p. 495

continental drift p. 495

 Multilingual eGlossary

 Video BrainPOP®

The Continental Drift Hypothesis

Inquiry How did this happen?

In Iceland, elongated cracks called rift zones are easy to find. Why do rift zones occur here? Iceland is above an area of the seafloor where Earth's crust is breaking apart. Earth's crust is constantly on the move. Scientists realized this long ago, but they could not prove how or why this happened.

Inquiry Launch Lab

20 minutes

Can you put together a peel puzzle?

Early map makers observed that the coastlines of Africa and South America appeared as if they could fit together like pieces of a puzzle. Scientists eventually discovered that these continents were once part of a large landmass. Can you use an orange peel to illustrate how continents may have fit together?

1. Read and complete a lab safety form.
2. Carefully peel an **orange,** keeping the orange-peel pieces as large as possible.
3. Set the orange aside.
4. Refit the orange-peel pieces back together in the shape of a sphere.
5. After successfully reconstructing the orange peel, disassemble your pieces.
6. Trade the entire orange peel with a classmate and try to reconstruct his or her orange peel.

Think About This

1. Which orange peel was easier for you to reconstruct? Why?
2. Look at a world map. Do the coastlines of any other continents appear to fit together?
3.  **Key Concept** What additional evidence would you need to prove that all the continents might have once fit together?

Pangaea

Did you know that Earth's surface is on the move? Can you feel it? Each year, North America moves a few centimeters farther away from Europe and closer to Asia. That is several centimeters, or about the thickness of this book. Even though you don't necessarily feel this motion, Earth's surface moves slowly every day.

Nearly 100 years ago Alfred Wegener (VAY guh nuhr), a German scientist, began an important investigation that continues today. Wegener wanted to know whether Earth's continents were fixed in their positions. He proposed that *all the continents were once part of a supercontinent called* **Pangaea** (pan JEE uh). Over time Pangaea began breaking apart, and the continents slowly moved to their present positions. Wegener proposed the hypothesis of **continental drift**, *which suggested that continents are in constant motion on the surface of Earth.*

Alfred Wegener observed the similarities of continental coastlines now separated by oceans. Look at the outlines of Africa and South America in **Figure 1.** Notice how they could fit together like pieces of a puzzle. Hundreds of years ago mapmakers noticed this jigsaw-puzzle pattern as they made the first maps of the continents.

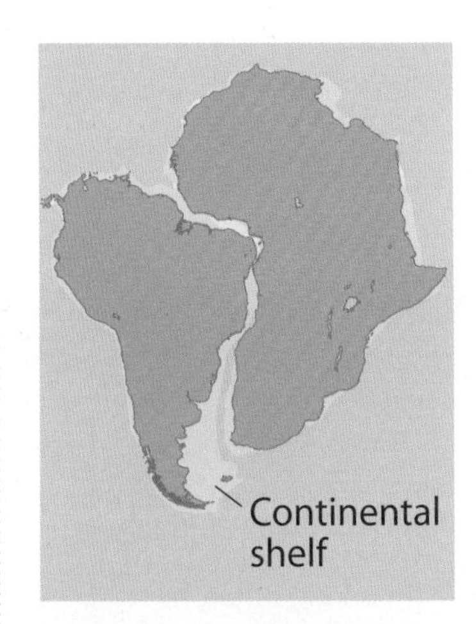

Figure 1 The eastern coast of South America mirrors the shape of the west coast of Africa.

Evidence That Continents Move

If you had discovered continental drift, how would you have tested your hypothesis? The most obvious evidence for continental drift is that the continents appear to fit together like pieces of a puzzle. But scientists were skeptical, and Wegener needed additional evidence to help support his hypothesis.

Climate Clues

When Wegener pieced Pangaea together, he proposed that South America, Africa, India, and Australia were located closer to Antarctica 280 million years ago. He suggested that the climate of the Southern Hemisphere was much cooler at the time. Glaciers covered large areas that are now parts of these continents. These glaciers would have been similar to the ice sheet that covers much of Antarctica today.

Wegener used climate clues to support his continental drift hypothesis. He studied the sediments deposited by glaciers in South America and Africa, as well as in India and Australia. Beneath these sediments, Wegener discovered glacial grooves, or deep scratches in rocks made as the glaciers moved across land. **Figure 2** shows where these glacial features are found on neighboring continents today. These continents were once part of the supercontinent Pangaea, when the climate in the Southern Hemisphere was cooler.

Climate Clues

Concepts in Motion Animation

Figure 2 If the southern hemisphere continents could be reassembled into Pangaea, the presence of an ice sheet would explain the glacial features on these continents today.

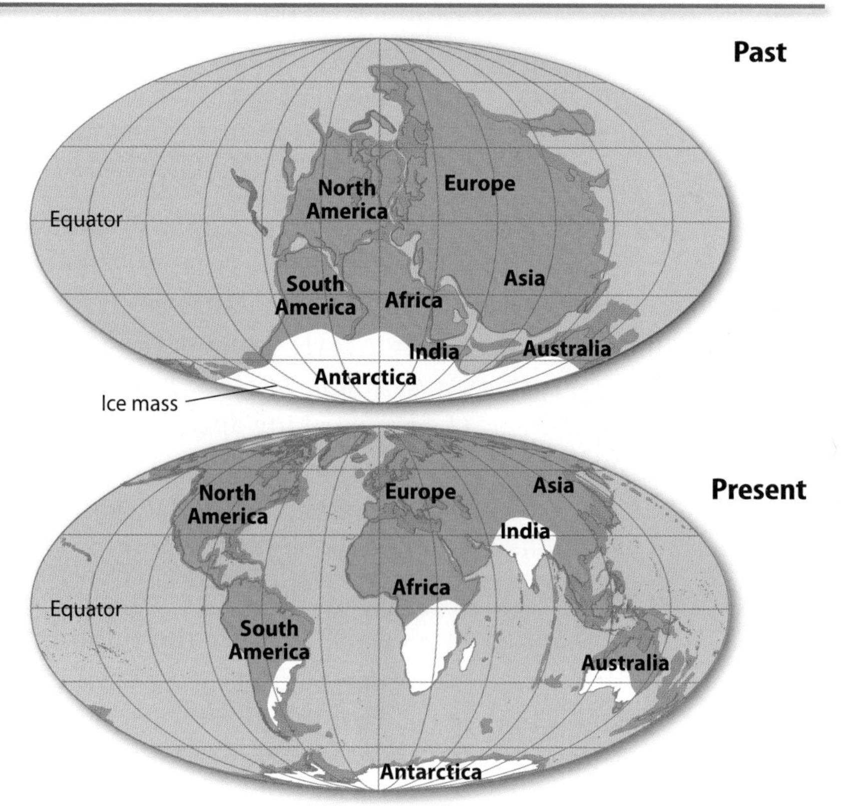

Fossil Clues

Animals and plants that live on different continents can be unique to that continent alone. Lions live in Africa but not in South America. Kangaroos live in Australia but not on any other continent. Because oceans separate continents, these animals cannot travel from one continent to another by natural means. However, **fossils** of similar organisms have been found on several continents separated by oceans. How did this happen? Wegener argued that these continents must have been connected some time in the past.

Review Vocabulary

fossil
the naturally preserved remains, imprints, or traces of organisms that lived long ago

Fossils of a plant called *Glossopteris* (glahs AHP tur us) have been discovered in rocks from South America, Africa, India, Australia, and Antarctica. These continents are far apart today. The plant's seeds could not have traveled across the vast oceans that separate them. **Figure 3** shows that when these continents were part of Pangaea 225 million years ago, *Glossopteris* lived in one region. Evidence suggests these plants grew in a swampy environment. Therefore, the climate of this region, including Antarctica, was different than it is today. Antarctica had a warm and wet climate. The climate had changed drastically from what it was 55 million years earlier when glaciers existed.

Reading Check How did climate in Antarctica change between 280 and 225 million years ago?

Fossil Clues

Figure 3 Fossils of *Glossopteris* have been found on many continents that are now separated by oceans. The orange area in the image on the right represents where *Glossopteris* fossils have been found.

Visual Check Which of the continents would not support *Glossopteris* growth today?

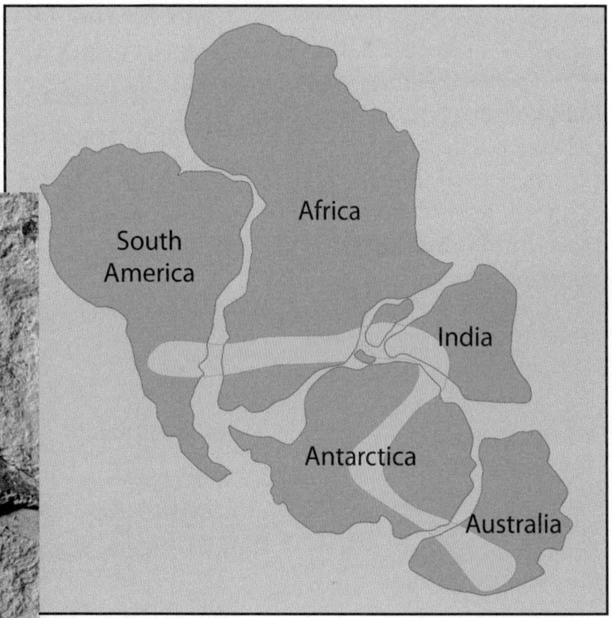

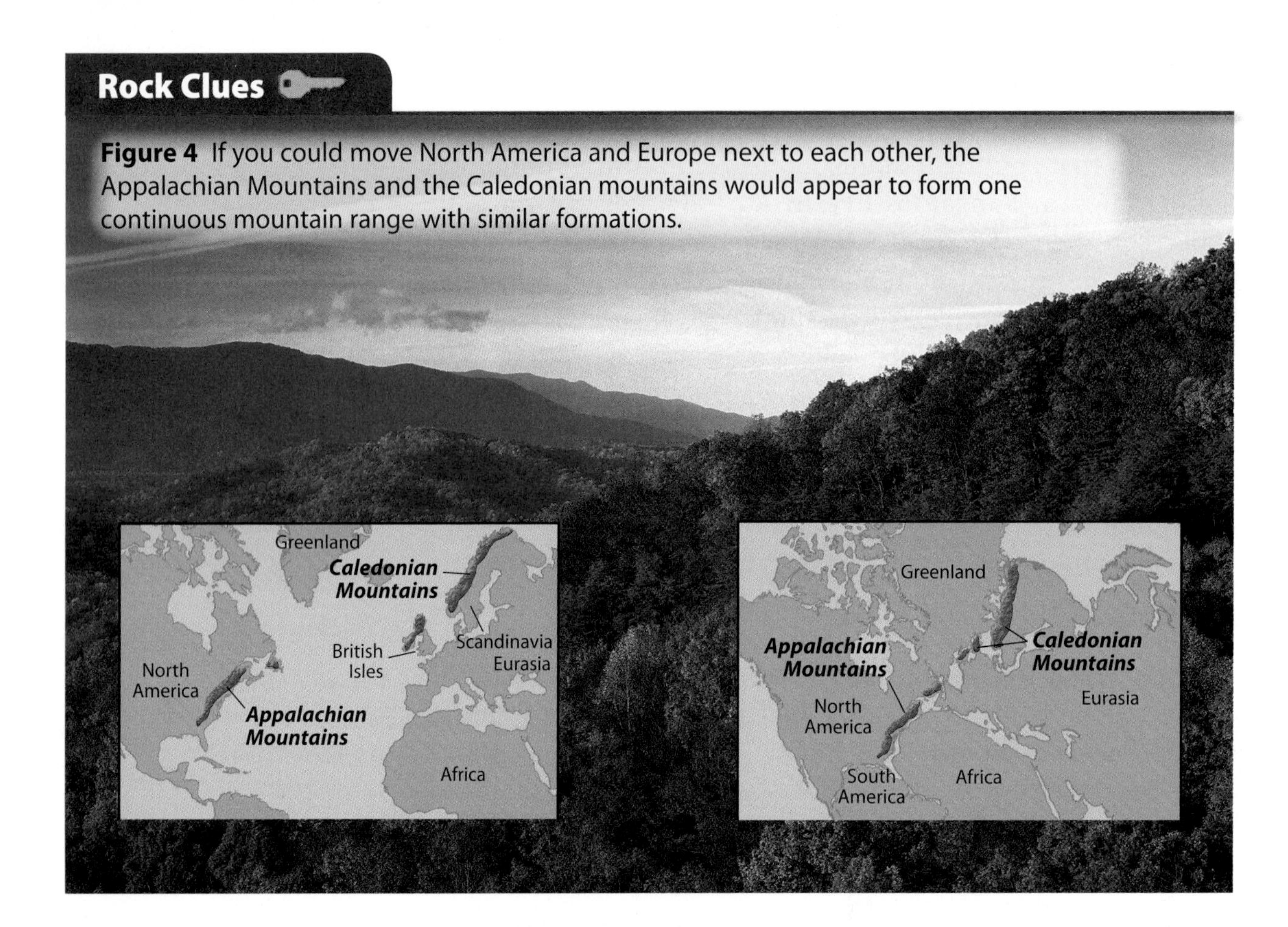

Figure 4 If you could move North America and Europe next to each other, the Appalachian Mountains and the Caledonian mountains would appear to form one continuous mountain range with similar formations.

Rock Clues

Wegener realized he needed more evidence to support the continental drift hypothesis. He observed that mountain ranges like the ones shown in **Figure 4** and rock formations on different continents had common origins. Today, geologists can determine when these rocks formed. For example, geologists suggest that large-scale volcanic eruptions occurred on the western coast of Africa and the eastern coast of South America at about the same time hundreds of millions of years ago. The volcanic rocks from the eruptions are identical in both chemistry and age. Refer back to **Figure 1.** If you could superimpose similar rock types onto the maps, these rocks would be in the area where Africa and South America fit together.

The Caledonian mountain range in northern Europe and the Appalachian Mountains in eastern North America are similar in age and structure. They are also composed of the same rock types. If you placed North America and Europe next to each other, these mountains would meet and form one long, continuous mountain belt. **Figure 4** illustrates where this mountain range would be.

Key Concept Check How were similar rock types used to support the continental drift hypothesis?

FOLDABLES®

Make a horizontal half-book and write the title as shown. Use it to organize your notes on the continental drift hypothesis.

Evidence for the Continental Drift Hypothesis

What was missing?

Wegener continued to support the continental drift hypothesis until his death in 1930. Wegener's ideas were not widely accepted until nearly four decades later. Why were scientists skeptical of Wegener's hypothesis? Although Wegener had evidence to suggest that continents were on the move, he could not explain how they moved.

One reason scientists questioned continental drift was because it is a slow process. It was not possible for Wegener to measure how fast the continents moved. The main objection to the continental drift hypothesis, however, was that Wegener could not explain what forces caused the continents to move. The mantle beneath the continents and the seafloor is made of solid rock. How could continents push their way through solid rock? Wegener needed more scientific evidence to prove his hypothesis. However, this evidence was hidden on the seafloor between the drifting continents. The evidence necessary to prove continental drift was not discovered until long after Wegener's death.

Key Concept Check Why did scientists argue against Wegener's continental drift hypothesis?

Science Use v. Common Use

mantle

Science Use the middle layer of Earth, situated between the crust above and the core below

Common Use a loose, sleeveless garment worn over other clothes

Inquiry MiniLab

20 minutes

How do you use clues to put puzzle pieces together?

When you put a puzzle together, you use clues to figure out which pieces fit next to each other. How did Wegener use a similar technique to piece together Pangaea?

1. Read and complete a lab safety form.
2. Using **scissors,** cut a piece of **newspaper** or a page from a **magazine** into an irregular shape with a diameter of about 25 cm.
3. Cut the piece of paper into at least 12 but not more than 20 pieces.
4. Exchange your puzzle with a partner and try to fit the new puzzle pieces together.
5. Reclaim your puzzle and remove any three pieces. Exchange your incomplete puzzle with a different partner. Try to put the incomplete puzzles back together.

Analyze and Conclude

1. **Summarize** Make a list of the clues you used to put together your partner's puzzle.
2. **Describe** How was putting together a complete puzzle different from putting together an incomplete puzzle?
3. **Key Concept** What clues did Wegener use to hypothesize the existence of Pangaea? What clues were missing from Wegener's puzzle?

Lesson 1 Review

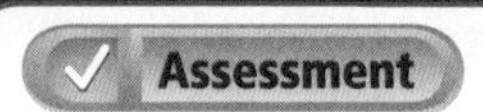

Online Quiz

Visual Summary

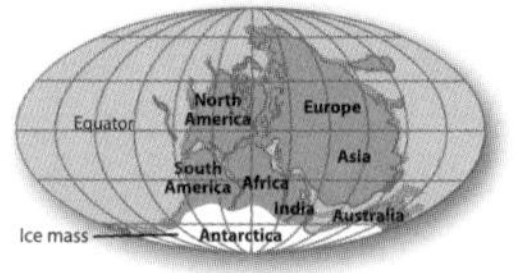

Past

All continents were once part of a supercontinent called Pangaea.

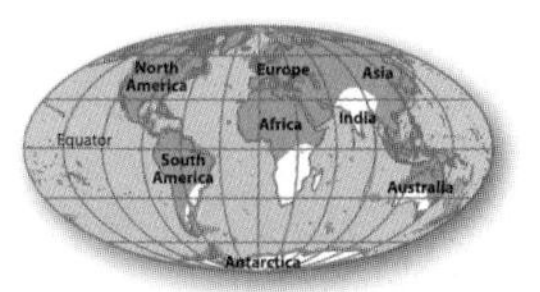

Present

Evidence found on present-day continents suggests that the continents have moved across Earth's surface.

FOLDABLES®

Use your lesson Foldable to review the lesson. Save your Foldable for the project at the end of the chapter.

What do you think NOW?

You first read the statements below at the beginning of the chapter.

1. India has always been north of the equator.
2. All the continents once formed one supercontinent.

Did you change your mind about whether you agree or disagree with the statements? Rewrite any false statements to make them true.

Use Vocabulary

1. **Define** *Pangaea.*
2. **Explain** the continental drift hypothesis and the evidence used to support it.

Understand Key Concepts 

3. **Identify** the scientist who first proposed that the continents move away from or toward each other.
4. Which can be used as an indicator of past climate?
   - **A.** fossils
   - **B.** lava flows
   - **C.** mountain ranges
   - **D.** tides

Interpret Graphics

5. **Interpret** Look at the map of the continents below. What direction has South America moved relative to Africa?

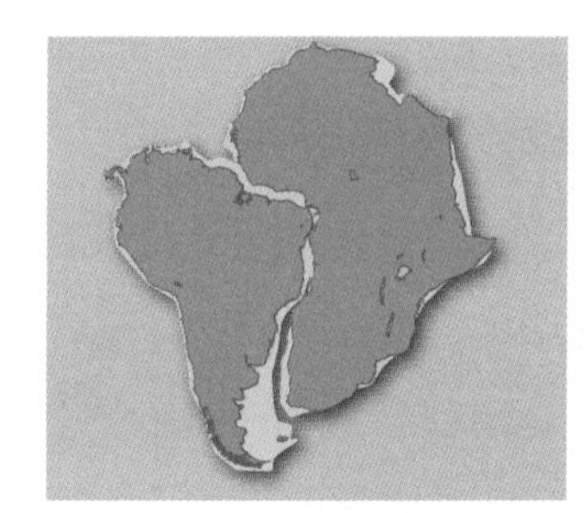

6. **Summarize** Copy and fill in the graphic organizer below to show the evidence Alfred Wegener used to support his continental drift hypothesis.

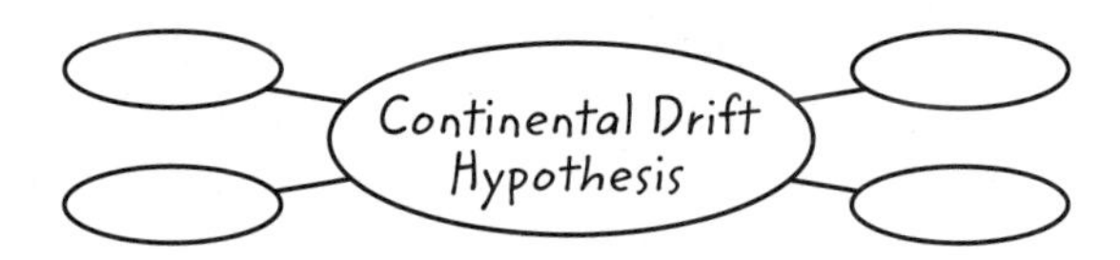

Critical Thinking

7. **Recognize** The shape and age of the Appalachian Mountains are similar to the Caledonian mountains in northern Europe. What else could be similar?
8. **Explain** If continents continue to drift, is it possible that a new supercontinent will form? Which continents might be next to each other 200 million years from now?

▼ This small mammal is a close living relative of an animal that once roamed Antarctica.

CAREERS in SCIENCE

▲ Ross MacPhee is a paleontologist working for the American Museum of Natural History in New York City. Here, he is searching for fossils in Antarctica.

Gondwana

A Fossil Clue from the Giant Landmass that Once Dominated the Southern Hemisphere

If you could travel back in time 120 million years, you would probably discover that Earth looked very different than it does today. Scientists believe that instead of seven continents, there were two giant landmasses, or supercontinents, on Earth at that time. Scientists named the landmass in the northern hemisphere *Laurasia*. The landmass in the southern hemisphere is known as *Gondwana*. It included the present-day continents of Antarctica, South America, Australia, and Africa.

How do scientists know that Gondwana existed? Ross MacPhee is a paleontologist—a scientist who studies fossils. MacPhee recently traveled to Antarctica where he discovered the fossilized tooth of a small land mammal. After carefully examining the tooth, he realized that it resembled fossils from ancient land mammals found in Africa and North America. MacPhee believes that these mammals are the ancient relatives of a mammal living today on the African island-nation of Madagascar.

How did the fossil remains and their present-day relatives become separated by kilometers of ocean? MacPhee hypothesizes that the mammal migrated across land bridges that once connected parts of Gondwana. Over millions of years, the movement of Earth's tectonic plates broke up this supercontinent. New ocean basins formed between the continents, resulting in the arrangement of landmasses that we see today.

LAURASIA
North America
Europe and Asia
GONDWANA
South America
Africa
Arabia
India
Australia
Antarctica

▲ *Gondwana* and *Laurasia* formed as the supercontinent Pangaea broke apart.

It's Your Turn

RESEARCH Millions of years ago, the island of Madagascar separated from the continent of Gondwana. In this environment, the animals of Madagascar changed and adapted. Research and report on one animal. Describe some of its unique adaptations.

Lesson 2

Reading Guide

Key Concepts 
ESSENTIAL QUESTIONS

- What is seafloor spreading?
- What evidence is used to support seafloor spreading?

Vocabulary
mid-ocean ridge p. 503
seafloor spreading p. 504
normal polarity p. 506
magnetic reversal p. 506
reversed polarity p. 506

Multilingual eGlossary

Development of a Theory

Inquiry What do the colors represent?

The colors in this satellite image show topography. The warm colors, red, pink, and yellow, represent landforms above sea level. The greens and blues indicate changes in topography below sea level. Deep in the Atlantic Ocean there is a mountain range, shown here as a linear feature in green. Is there a connection between this landform and the continental drift hypothesis?

15 minutes

Can you guess the age of the glue?

The age of the seafloor can be determined by measuring magnetic patterns in rocks from the bottom of the ocean. How can similar patterns in drying glue be used to show age relationships between rocks exposed on the seafloor?

1. Read and complete a lab safety form.
2. Carefully spread a thin layer of **rubber cement** on a sheet of **paper.**
3. Observe for 3 minutes. Record the pattern of how the glue dries in your Science Journal.
4. Repeat step 2. After 1 minute, exchange papers with a classmate.
5. Ask the classmate to observe and tell you which part of the glue dried first.

Think About This

1. What evidence helped you to determine the oldest and youngest glue layers?
2. How is this similar to a geologist trying to estimate the age of rocks on the seafloor?
3. **Key Concept** How could magnetic patterns in rock help predict a rock's age?

Mapping the Ocean Floor

During the late 1940s after World War II, scientists began exploring the seafloor in greater detail. They were able to determine the depth of the ocean using a device called an echo sounder, as shown in **Figure 5.** Once ocean depths were determined, scientists used these data to create a topographic map of the seafloor. These new topographic maps of the seafloor revealed that vast mountain ranges stretched for many miles deep below the ocean's surface. *The mountain ranges in the middle of the oceans are called* **mid-ocean ridges.** Mid-ocean ridges, shown in **Figure 5,** are much longer than any mountain range on land.

Figure 5 An echo sounder produces sound waves that travel from a ship to the seafloor and back. The deeper the ocean, the longer the time this takes. Depth can be used to determine seafloor topography.

Seafloor Topography

Mid-ocean Ridge
Sediment
Magma

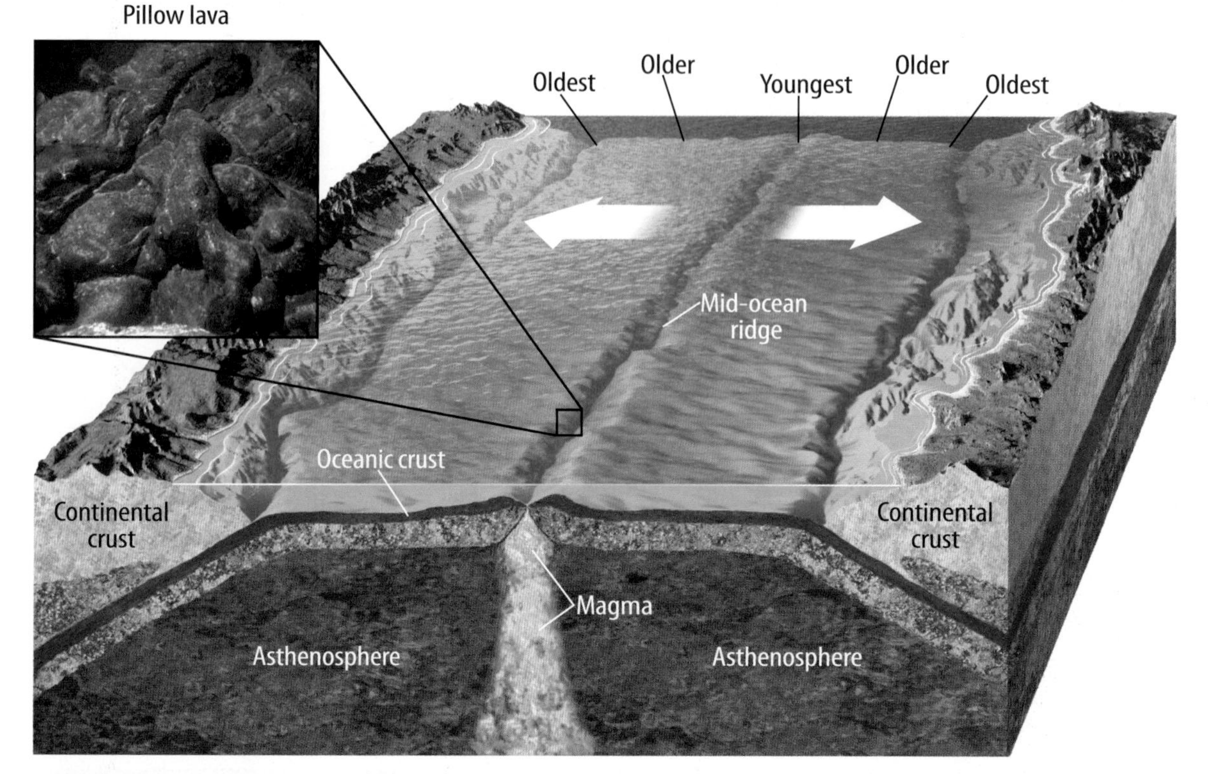

Figure 6 When lava erupts along a mid-ocean ridge, it cools and crystallizes, forming a type of rock called basalt. Basalt is the dominant rock on the seafloor. The youngest basalt is closest to the ridge. The oldest basalt is farther away from the ridge.

Visual Check Looking at the image above, can you propose a pattern that exists in rocks on either side of the mid-ocean ridge?

Seafloor Spreading

By the 1960s scientists discovered a new process that helped explain continental drift. This process, shown in **Figure 6,** is called seafloor spreading. **Seafloor spreading** *is the process by which new oceanic crust forms along a mid-ocean ridge and older oceanic crust moves away from the ridge.*

When the seafloor spreads, the mantle below melts and forms magma. Because magma is less dense than solid mantle material, it rises through cracks in the crust along the mid-ocean ridge. When magma erupts on Earth's surface, it is called lava. As this lava cools and crystallizes on the seafloor, it forms a type of rock called basalt. Because the lava erupts into water, it cools rapidly and forms rounded structures called pillow lavas. Notice the shape of the pillow lava shown in **Figure 6.**

As the seafloor continues to spread apart, the older oceanic crust moves away from the mid-ocean ridge. The closer the crust is to a mid-ocean ridge, the younger the oceanic crust is. Scientists argued that if the seafloor spreads, the continents must also be moving. A mechanism to explain continental drift was finally discovered long after Wegener proposed his hypothesis.

Key Concept Check What is seafloor spreading?

Topography of the Seafloor

The rugged mountains that make up the mid-ocean ridge system can form in two different ways. For example, large amounts of lava can erupt from the center of the ridge, cool and build up around the ridge. Or, as the lava cools and forms new crust, it cracks. The rocks move up or down along these cracks in the seafloor, forming jagged mountain ranges.

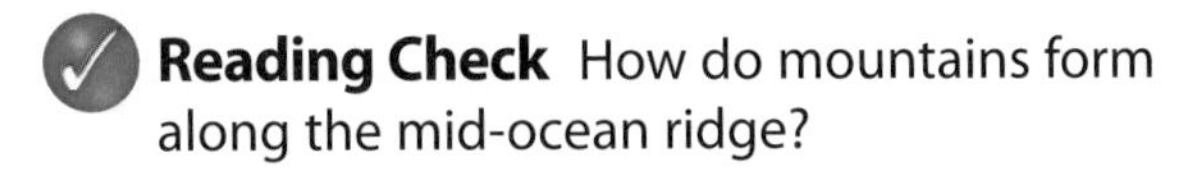

Reading Check How do mountains form along the mid-ocean ridge?

Over time, sediment accumulates on top of the oceanic crust. Close to the mid-ocean ridge there is almost no sediment. Far from the mid-ocean ridge, the layer of sediment becomes thick enough to make the seafloor smooth. This part of the seafloor, shown in **Figure 7**, is called the abyssal (uh BIH sul) plain.

Moving Continents Around

The theory of seafloor spreading provides a way to explain how continents move. Continents do not move through the solid mantle or the seafloor. Instead, continents move as the seafloor spreads along a mid-ocean ridge.

Inquiry MiniLab — 20 minutes

How old is the Atlantic Ocean?

If you measure the width of the Atlantic Ocean and you know the rate of seafloor spreading, you can calculate the age of the Atlantic.

1. Use a **ruler** to measure the horizontal distance between a point on the eastern coast of South America and a point on the western coast of Africa on a **world map.** Repeat three times and calculate the average distance in your Science Journal.
2. Use the map's legend to convert the average distance from centimeters to kilometers.
3. If Africa and South America have been moving away from each other at a rate of 2.5 cm per year, calculate the age of the Atlantic Ocean.

Analyze and Conclude

1. **Measure** Did your measurements vary?
2. **Key Concept** How does the age you calculated compare to the breakup of Pangaea 200 million years ago?

Abyssal Plain

Figure 7 The abyssal plain is flat due to an accumulation of sediments far from the ridge.

Visual Check Compare and contrast the topography of a mid-ocean ridge to an abyssal plain.

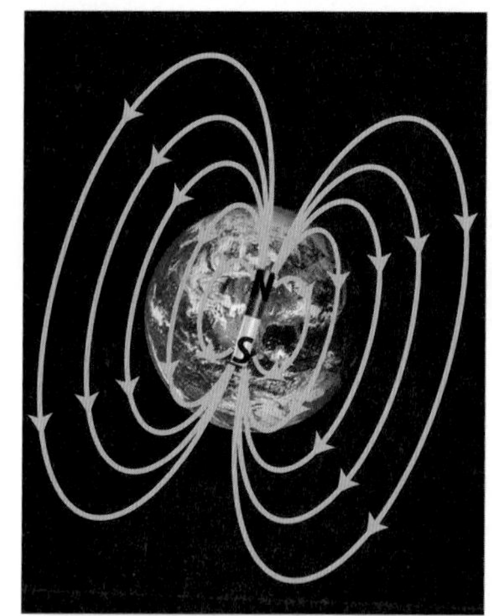

Reversed magnetic field

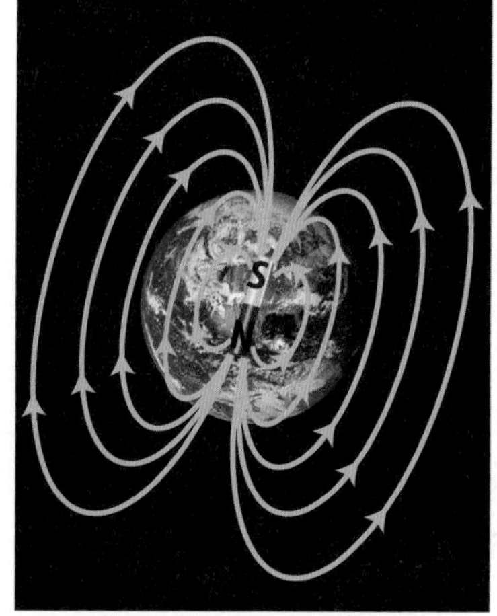

Normal magnetic field

▲ **Figure 8** Earth's magnetic field is like a large bar magnet. It has reversed direction hundreds of times throughout history.

Development of a Theory

The first evidence used to support seafloor spreading was discovered in rocks on the seafloor. Scientists studied the magnetic signature of minerals in these rocks. To understand this, you need to understand the direction and orientation of Earth's magnetic field and how rocks record magnetic information.

Magnetic Reversals

Recall that the iron-rich, liquid outer core is like a giant magnet that creates Earth's magnetic field. The direction of the magnetic field is not constant. Today's magnetic field, shown in **Figure 8,** is described as having **normal polarity**—*a state in which magnetized objects, such as compass needles, will orient themselves to point north.* Sometimes a **magnetic reversal** *occurs and the magnetic field reverses direction.* The opposite of normal polarity is **reversed polarity**—*a state in which magnetized objects would reverse direction and orient themselves to point south,* as shown in **Figure 8.** Magnetic reversals occur every few hundred thousand to every few million years.

Reading Check Is Earth's magnetic field currently normal or reversed polarity?

Rocks Reveal Magnetic Signature

Basalt on the seafloor contains iron-rich minerals that are magnetic. Each mineral acts like a small magnet. **Figure 9** shows how magnetic minerals align themselves with Earth's magnetic field. When lava erupts from a vent along a mid-ocean ridge, it cools and crystallizes. This permanently records the direction and orientation of Earth's magnetic field at the time of the eruption. Scientists have discovered parallel patterns in the magnetic signature of rocks on either side of a mid-ocean ridge.

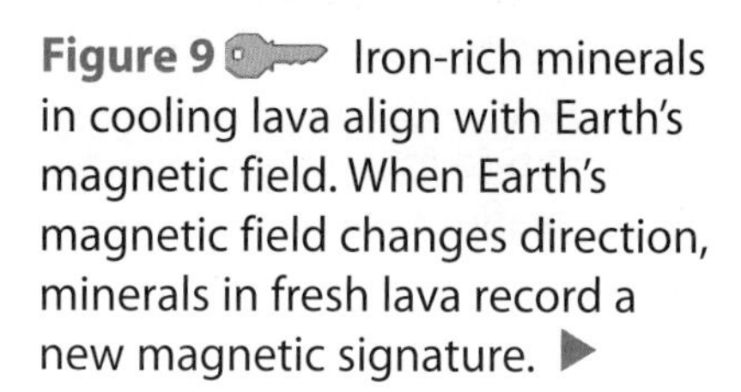

Figure 9 Iron-rich minerals in cooling lava align with Earth's magnetic field. When Earth's magnetic field changes direction, minerals in fresh lava record a new magnetic signature. ▶

Visual Check Describe the pattern in the magnetic stripes shown in the image to the right.

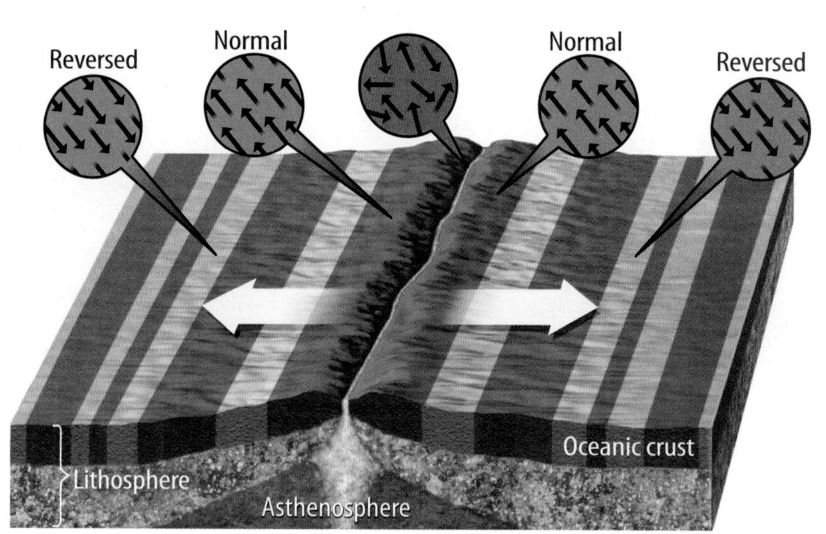

Seafloor Spreading Theory

Figure 10 A mirror image in the magnetic stripes on either side of the mid-ocean ridge shows that the crust formed at the ridge is carried away in opposite directions.

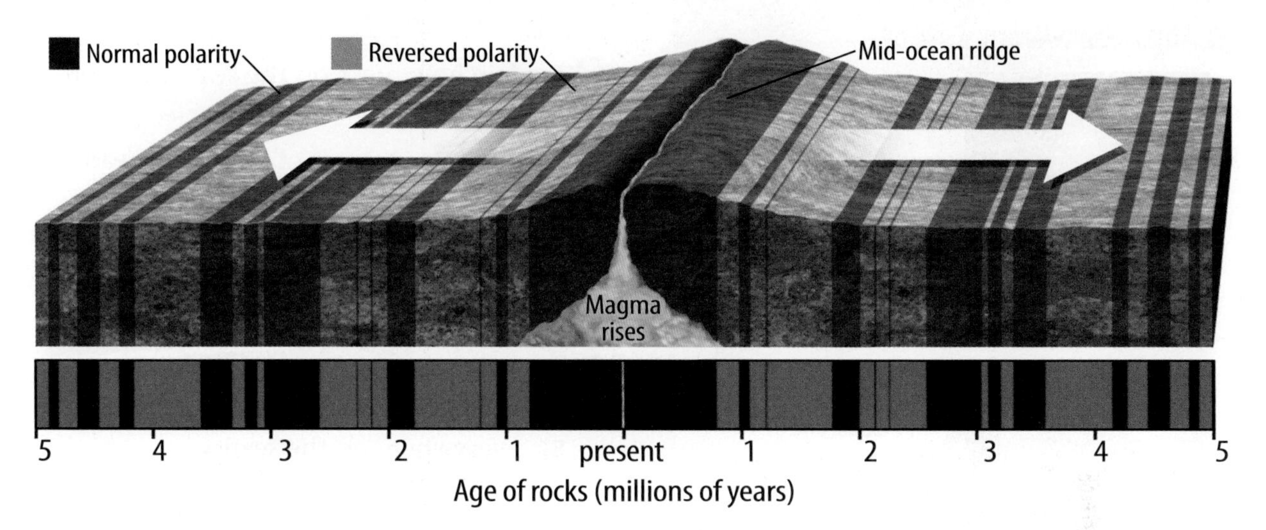

Evidence to Support the Theory

How did scientists prove the theory of seafloor spreading? Scientists studied magnetic minerals in rocks from the seafloor. They used a magnetometer (mag nuh TAH muh tur) to measure and record the magnetic signature of these rocks. These measurements revealed a surprising pattern. Scientists have discovered parallel magnetic stripes on either side of the mid-ocean ridge. Each pair of stripes has a similar composition, age, and magnetic character. Each magnetic stripe in **Figure 10** represents crust that formed and magnetized at a mid-ocean ridge during a period of either **normal** or reversed polarity. The pairs of magnetic stripes confirm that the ocean crust formed at mid-ocean ridges is carried away from the center of the ridges in opposite directions.

ACADEMIC VOCABULARY

normal
(adjective) conforming to a type, standard, or regular pattern

 **Reading Check** How do magnetic minerals help support the theory of seafloor spreading?

Other measurements made on the seafloor confirm seafloor spreading. By drilling a hole into the seafloor and measuring the temperature beneath the surface, scientists can measure the amount of thermal energy leaving Earth. The measurements show that more thermal energy leaves Earth near mid-ocean ridges than is released from beneath the abyssal plains.

Additionally, sediment collected from the seafloor can be dated. Results show that the sediment closest to the mid-ocean ridge is younger than the sediment farther away from the ridge. Sediment thickness also increases with distance away from the mid-ocean ridge.

FOLDABLES®

Make a layered book using two sheets of notebook paper. Use the two pages to record your notes and the inside to illustrate seafloor spreading.

Seafloor Spreading

Lesson 2 Review

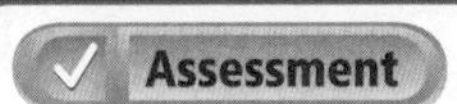

Visual Summary

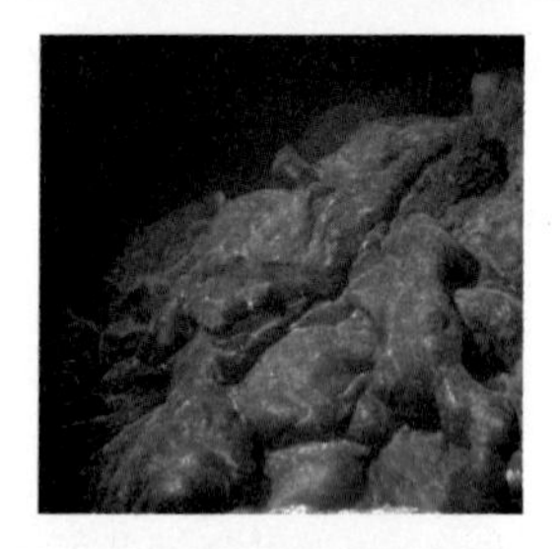

Lava erupts along mid-ocean ridges.

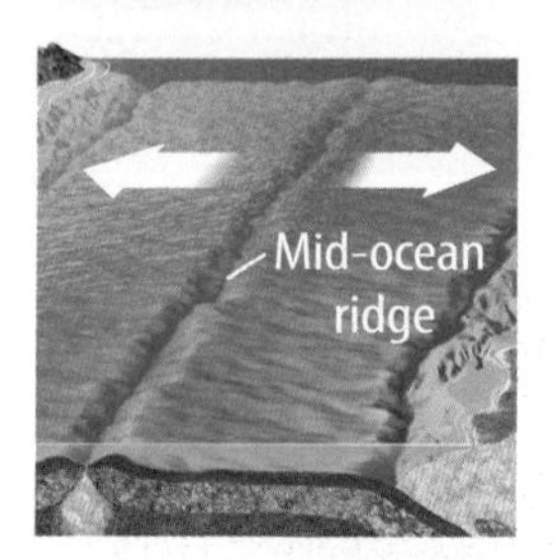

Mid-ocean ridges are large mountain ranges that extend throughout Earth's oceans.

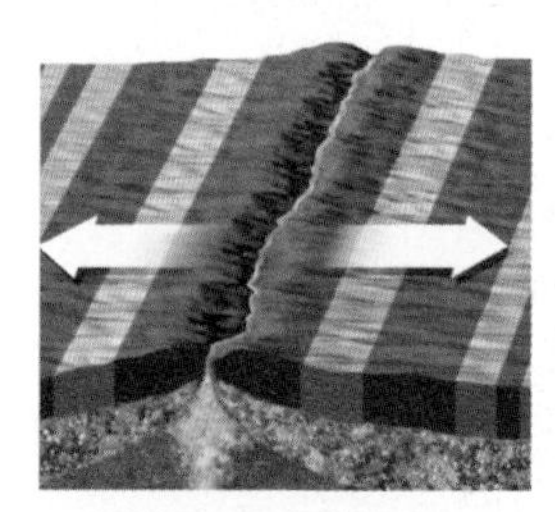

A magnetic reversal occurs when Earth's magnetic field changes direction.

FOLDABLES®

Use your lesson Foldable to review the lesson. Save your Foldable for the project at the end of the chapter.

What do you think 

You first read the statements below at the beginning of the chapter.

3. The seafloor is flat.

4. Volcanic activity occurs only on the seafloor.

Did you change your mind about whether you agree or disagree with the statements? Rewrite any false statements to make them true.

Use Vocabulary

1. **Explain** how rocks on the seafloor record magnetic reversals over time.
2. **Diagram** the process of seafloor spreading.
3. **Use the term** *seafloor spreading* to explain how a mid-ocean ridge forms.

Understand Key Concepts 

4. Oceanic crust forms
   - **A.** at mid-ocean ridges.
   - **B.** everywhere on the seafloor.
   - **C.** on the abyssal plains.
   - **D.** by magnetic reversals.
5. **Explain** why magnetic stripes on the seafloor are parallel to the mid-ocean ridge.
6. **Describe** how scientists can measure the depth to the seafloor.

Interpret Graphics

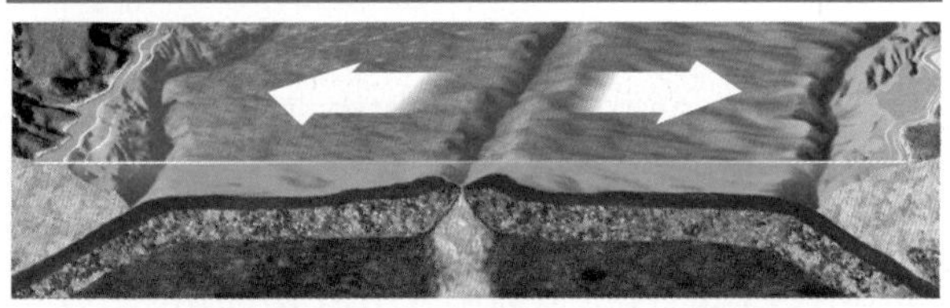

7. **Determine** Refer to the image above. Where is the youngest crust? Where is the oldest crust?
8. **Describe** how seafloor spreading helps to explain the continental drift hypothesis.
9. **Sequence Information** Copy and fill in the graphic organizer below to explain the steps in the formation of a mid-ocean ridge.

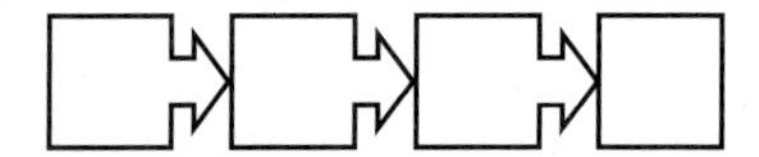

Critical Thinking

10. **Infer** why magnetic stripes in the Pacific Ocean are wider than in the Atlantic Ocean.
11. **Explain** why the thickness of seafloor sediments increases with increasing distance from the ocean ridge.

Model

30 minutes

How do rocks on the seafloor vary with age away from a mid-ocean ridge?

Scientists discovered that new ocean crust forms at a mid-ocean ridge and spreads away from the ridge slowly over time. This process is called seafloor spreading. The age of the seafloor is one component that supports this theory.

Materials

vanilla yogurt
berry yogurt

foam board
(10 cm × 4 cm)

waxed paper

plastic spoon

Safety

Do not eat anything used in this lab.

Learn It

Scientists use **models** to represent real-world science. By creating a three-dimensional model of volcanic activity along the Mid-Atlantic Ridge, scientists can model the seafloor spreading process. They can then compare this process to the actual age of the seafloor. In this skill lab, you will investigate how the age of rocks on the seafloor changes with distance away from the ridge.

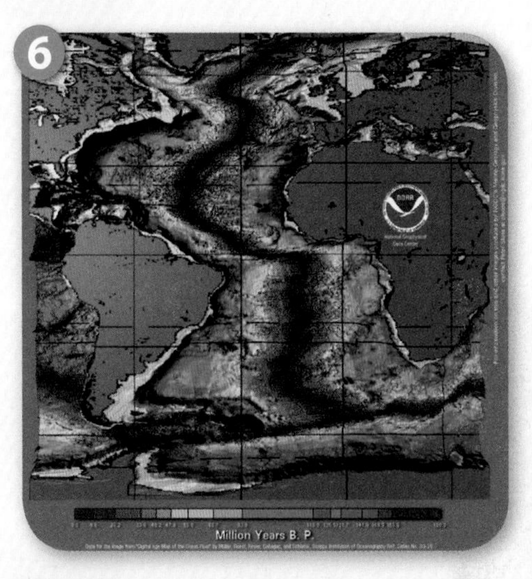

Try It

1. Read and complete a lab safety form.
2. Lay the sheet of waxed paper flat on the lab table. Place two spoonfuls of vanilla yogurt in a straight line near the center of the waxed paper, leaving it lumpy and full.
3. Lay the two pieces of foam board over the yogurt, leaving a small opening in the middle. Push the foam boards together and down, so the yogurt oozes up and over each of the foam boards.
4. Pull the foam boards apart and add a new row of two spoonfuls of berry yogurt down the middle. Lift the boards and place them partly over the new row. Push them together gently. Observe the outer edges of the new yogurt while you are moving the foam boards together.
5. Repeat step 4 with one more spoonful of vanilla yogurt. Then repeat again with one more spoonful of berry yogurt.

Apply It

6. Compare the map and the model. Where is the Mid-Atlantic Ridge on the map? Where is it represented in your model?
7. Which of your yogurt strips matches today on this map? And millions of years ago?
8. How do scientists determine the ages of different parts of the ocean floor?
9. **Conclude** What happened to the yogurt when you added more?
10. **Key Concept** What happens to the material already on the ocean floor when magma erupts along a mid-ocean ridge?

Lesson 3

Reading Guide

Key Concepts
ESSENTIAL QUESTIONS

- What is the theory of plate tectonics?
- What are the three types of plate boundaries?
- Why do tectonic plates move?

Vocabulary

g Multilingual eGlossary

The Theory of Plate Tectonics

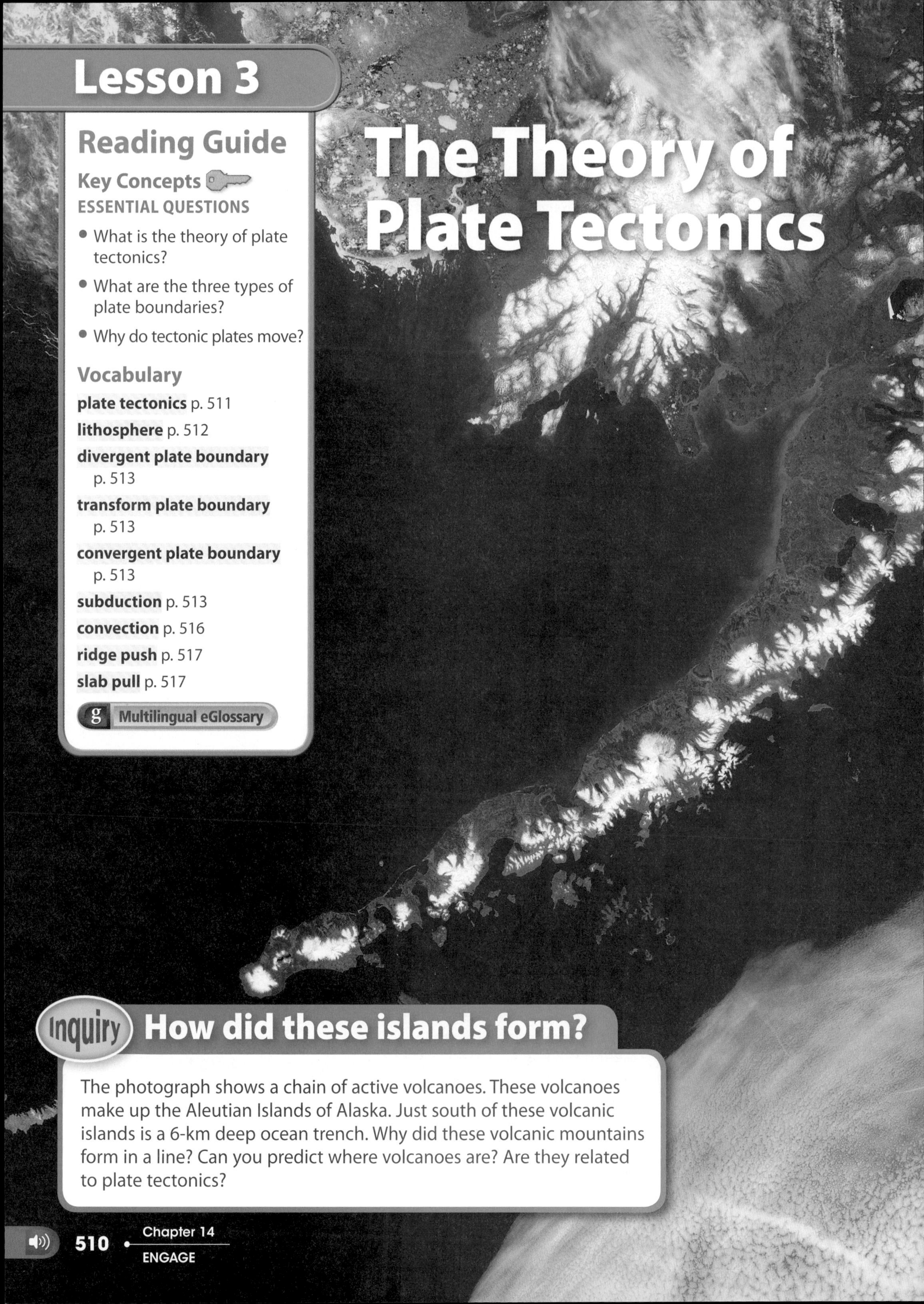

Inquiry How did these islands form?

The photograph shows a chain of active volcanoes. These volcanoes make up the Aleutian Islands of Alaska. Just south of these volcanic islands is a 6-km deep ocean trench. Why did these volcanic mountains form in a line? Can you predict where volcanoes are? Are they related to plate tectonics?

Launch Lab

15 minutes

Can you determine density by observing buoyancy?

Density is the measure of an object's mass relative to its volume. Buoyancy is the upward force a liquid places on objects that are immersed in it. If you immerse objects with equal densities into liquids that have different densities, the buoyant forces will be different. An object will sink or float depending on the density of the liquid compared to the object. Earth's layers differ in density. These layers float or sink depending on density and buoyant force.

1. Read and complete a lab safety form.
2. Obtain four **test tubes.** Place them in a **test-tube rack.** Add **water** to one test tube until it is ¾ full.
3. Repeat with the other test tubes using **vegetable oil** and **glucose syrup.** One test tube should remain empty.
4. Drop **beads** of equal density into each test tube. Observe what the object does when immersed in each liquid. Record your observations in your Science Journal.

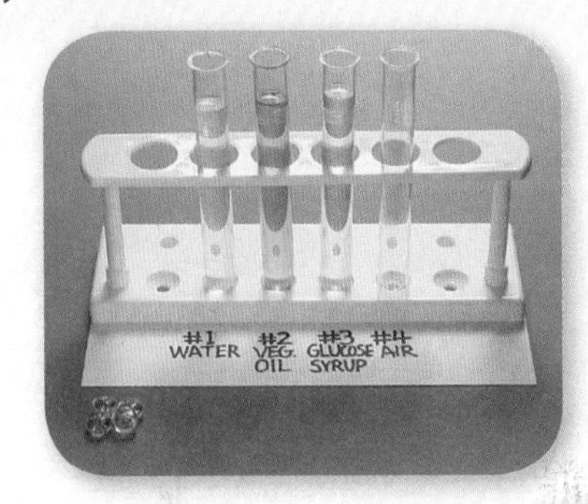

Think About This

1. How did you determine which liquid has the highest density?
2. Key Concept What happens when layers of rock with different densities collide?

The Plate Tectonics Theory

When you blow into a balloon, the balloon expands and its surface area also increases. Similarly, if oceanic crust continues to form at mid-ocean ridges and is never destroyed, Earth's surface area should increase. However, this is not the case. The older crust must be destroyed somewhere—but where?

By the late 1960s a more complete theory, called plate tectonics, was proposed. The theory of **plate tectonics** states that *Earth's surface is made of rigid slabs of rock, or plates, that move with respect to each other.* This new theory suggested that Earth's surface is divided into large plates of rigid rock. Each plate moves over Earth's hot and semi-plastic mantle.

 **Key Concept Check** What is plate tectonics?

Geologists use the word *tectonic* to describe the forces that shape Earth's surface and the rock structures that form as a result. Plate tectonics provides an explanation for the occurrence of earthquakes and volcanic eruptions. When plates separate on the seafloor, earthquakes result and a mid-ocean ridge forms. When plates come together, one plate can dive under the other, causing earthquakes and creating a chain of volcanoes. When plates slide past each other, earthquakes can result.

Earth's Tectonic Plates

North American Plate
Juan de Fuca Plate
Caribbean Plate
Cocos Plate
Pacific Plate
Nazca Plate
South American Plate
Scotia Plate
Eurasian Plate
Arabian Plate
African Plate
Antarctic Plate
Indo-Australian Plate
Philippine Plate
Pacific Plate
North American Plate

Divergent boundary
Convergent boundary
Plate boundary

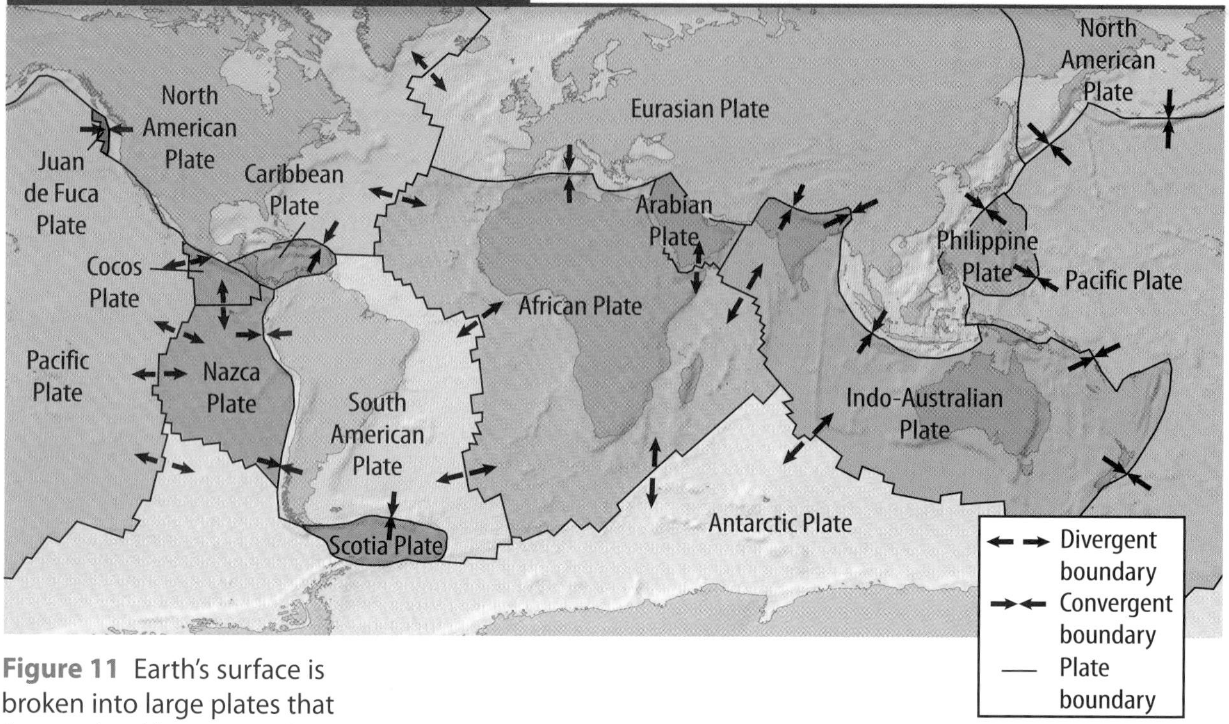

Figure 11 Earth's surface is broken into large plates that fit together like pieces of a giant jigsaw puzzle. The arrows show the general direction of movement of each plate.

Tectonic Plates

You read on the previous page that the theory of plate tectonics states that Earth's surface is divided into rigid plates that move relative to one another. These plates are "floating" on top of a hot and semi-plastic mantle. The map in **Figure 11** illustrates Earth's major plates and the boundaries that define them. The Pacific Plate is the largest plate. The Juan de Fuca Plate is one of the smallest plates. It is between the North American and Pacific Plates. Notice the boundaries that run through the oceans. Many of these boundaries mark the positions of the mid-ocean ridges.

Earth's outermost layers are cold and rigid compared to the layers within Earth's interior. *The cold and rigid outermost rock layer is called the* **lithosphere.** It is made up of the crust and the solid, uppermost mantle. The lithosphere is thin below mid-ocean ridges and thick below continents. Earth's tectonic plates are large pieces of lithosphere. These lithospheric plates fit together like the pieces of a giant jigsaw puzzle.

The layer of Earth below the lithosphere is called the asthenosphere (as THEN uh sfihr). This layer is so hot that it behaves like a **plastic** material. This enables Earth's plates to move because the hotter, plastic mantle material beneath them can flow. The interactions between lithosphere and asthenosphere help to explain plate tectonics.

SCIENCE USE V. COMMON USE

plastic

Science Use capable of being molded or changing shape without breaking

Common Use any of numerous organic, synthetic, or processed materials made into objects

Reading Check What are Earth's outermost layers called?

Plate Boundaries

Place two books side by side and imagine each book represents a tectonic plate. A plate boundary exists where the books meet. How many different ways can you move the books with respect to each other? You can pull the books apart, you can push the books together, and you can slide the books past one another. Earth's tectonic plates move in much the same way.

Divergent Plate Boundaries

Mid-ocean ridges are located along divergent plate boundaries. A **divergent plate boundary** *forms where two plates separate.* When the seafloor spreads at a mid-ocean ridge, lava erupts, cools, and forms new oceanic crust. Divergent plate boundaries can also exist in the middle of a continent. They pull continents apart and form rift valleys. The East African Rift is an example of a continental rift.

FOLDABLES®

Make a layered book using two sheets of notebook paper. Use it to organize information about the different types of plate boundaries and the features that form there.

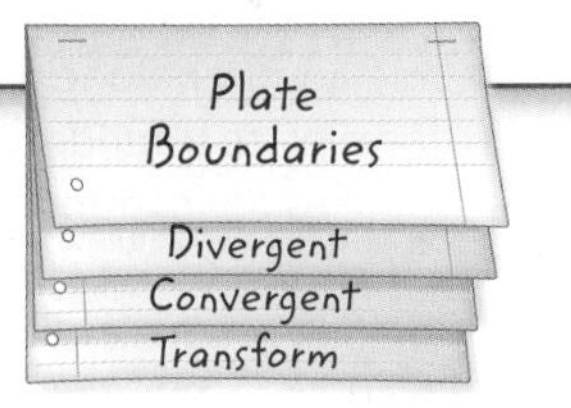

Transform Plate Boundaries

The famous San Andreas Fault in California is an example of a transform plate boundary. A **transform plate boundary** *forms where two plates slide past each other.* As they move past each other, the plates can get stuck and stop moving. Stress builds up where the plates are "stuck." Eventually, the stress is too great and the rocks break, suddenly moving apart. This results in a rapid release of energy as earthquakes.

Convergent Plate Boundaries

Convergent plate boundaries *form where two plates collide. The denser plate sinks below the more buoyant plate in a process called* **subduction.** The area where a denser plate descends into Earth along a convergent plate boundary is called a **subduction** zone.

WORD ORIGIN

subduction
from Latin *subductus,* means "to lead under, removal"

When an oceanic plate and a continental plate collide, the denser oceanic plate subducts under the edge of the continent. This creates a deep ocean trench. A line of volcanoes forms above the subducting plate on the edge of the continent. This process can also occur when two oceanic plates collide. The older and denser oceanic plate will subduct beneath the younger oceanic plate. This creates a deep ocean trench and a line of volcanoes called an island arc.

When two continental plates collide, neither plate is subducted, and mountains such as the Himalayas in southern Asia form from uplifted rock. **Table 1** on the next page summarizes the interactions of Earth's tectonic plates.

Key Concept Check What are the three types of plate boundaries?

Table 1 The direction of motion of Earth's plates creates a variety of features at the boundaries between the plates.

Concepts in Motion **Animation**

Table 1 Interactions of Earth's Tectonic Plates

Plate Boundary	Relative Motion	Example
Divergent plate boundary When two plates separate and create new oceanic crust, a divergent plate boundary forms. This process occurs where the seafloor spreads along a mid-ocean ridge, as shown to the right. This process can also occur in the middle of continents and is referred to as a continental rifting.	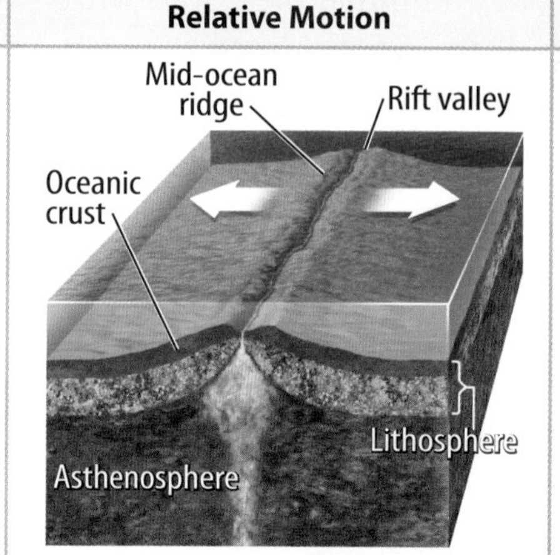	
Transform plate boundary Two plates slide horizontally past one another along a transform plate boundary. Earthquakes are common along this type of plate boundary. The San Andreas Fault, shown to the right, is part of the transform plate boundary that extends along the coast of California.	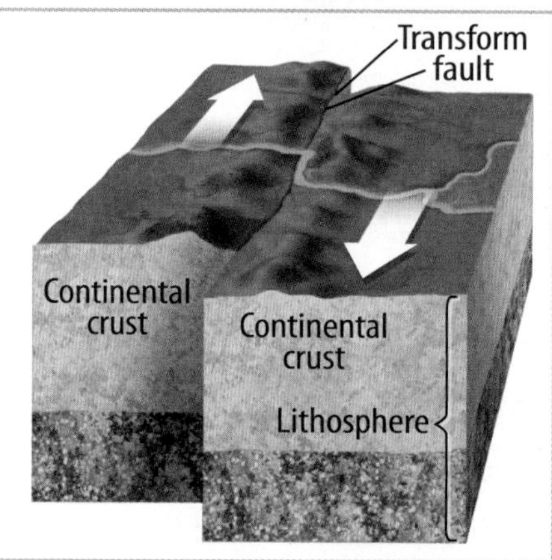	
Convergent plate boundary (ocean-to-continent) When an oceanic and a continental plate collide, they form a convergent plate boundary. The denser plate will subduct. A volcanic mountain, such as Mount Rainier in the Cascade Mountains, forms along the edge of the continent. This process can also occur where two oceanic plates collide, and the denser plate is subducted.	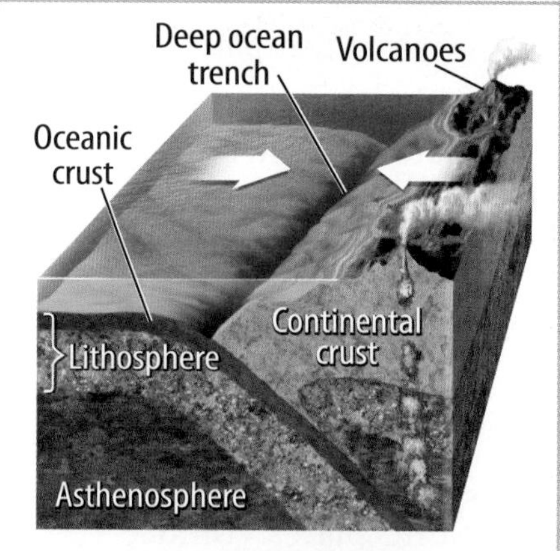	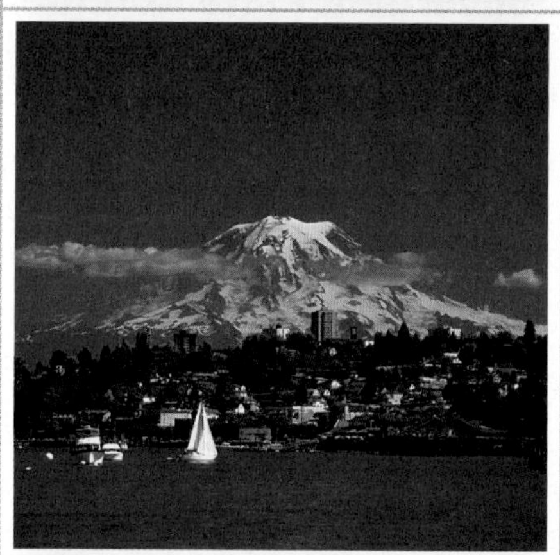
Convergent plate boundary (continent-to-continent) Convergent plate boundaries can also occur where two continental plates collide. Because both plates are equally dense, neither plate will subduct. Both plates uplift and deform. This creates huge mountains like the Himalayas, shown to the right.	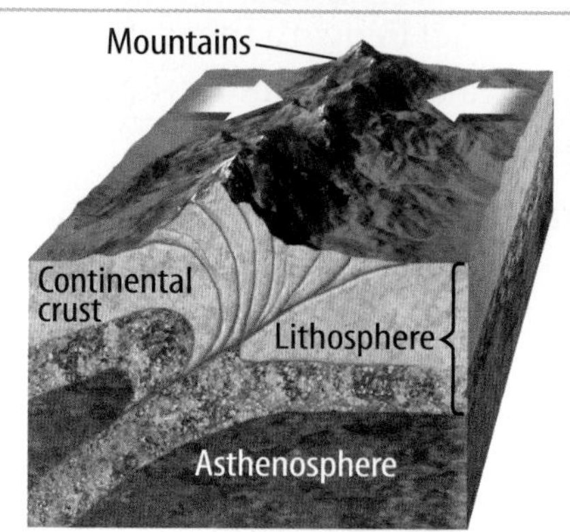	

Evidence for Plate Tectonics

When Wegener proposed the continental drift hypothesis, the technology used to measure how fast the continents move today wasn't yet available. Recall that continents move apart or come together at speeds of a few centimeters per year. This is about the length of a small paperclip.

Today, scientists can measure how fast continents move. A network of satellites orbiting Earth monitors plate motion. By keeping track of the distance between these satellites and Earth, it is possible to locate and determine how fast a tectonic plate moves. This network of satellites is called the Global Positioning System (GPS).

The theory of plate tectonics also provides an explanation for why earthquakes and volcanoes occur in certain places. Because plates are rigid, tectonic activity occurs where plates meet. When plates separate, collide, or slide past each other along a plate boundary, stress builds. A rapid release of energy can result in earthquakes. Volcanoes form where plates separate along a mid-ocean ridge or a continental rift or collide along a subduction zone. Mountains can form where two continents collide. **Figure 12** illustrates the relationship between plate boundaries and the occurrence of earthquakes and volcanoes. Refer back to the lesson opener photo. Find these islands on the map. Are they located near a plate boundary?

Key Concept Check How are earthquakes and volcanoes related to the theory of plate tectonics?

Figure 12 Notice that most earthquakes and volcanoes occur near plate boundaries.

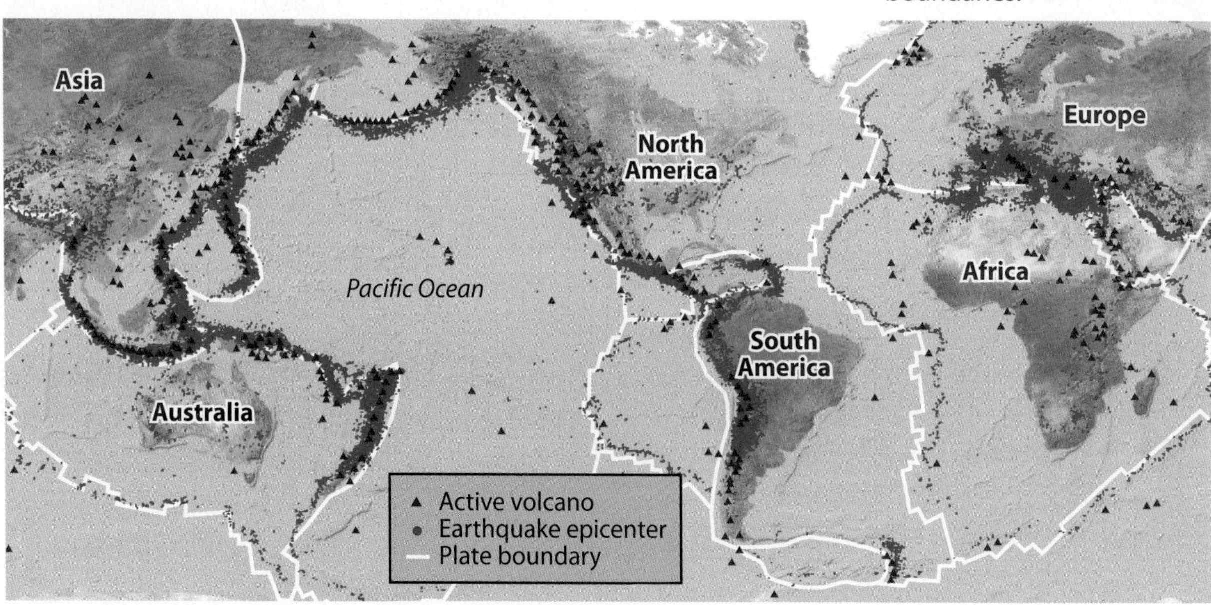

Visual Check Do earthquakes and volcanoes occur anywhere away from plate boundaries?

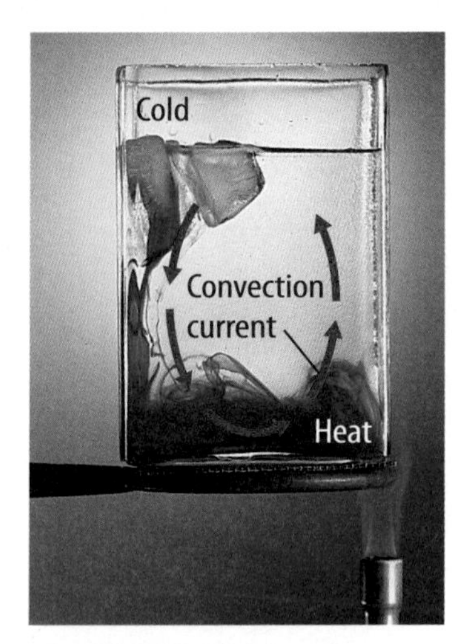

Figure 13 When water is heated, it expands. Less dense heated water rises because the colder water sinks, forming convection currents.

Personal Tutor

Plate Motion

The main objection to Wegener's continental drift hypothesis was that he could not explain why or how continents move. Scientists now understand that continents move because the asthenosphere moves underneath the lithosphere.

Convection Currents

You are probably already familiar with the process of **convection,** *the circulation of material caused by differences in temperature and density.* For example, the upstairs floors of homes and buildings are often warmer. This is because hot air rises while dense, cold air sinks. Look at **Figure 13** to see convection in action.

Reading Check What causes convection?

Plate tectonic activity is related to convection in the mantle, as shown in **Figure 14.** Radioactive elements, such as uranium, thorium, and potassium, heat Earth's interior. When materials such as solid rock are heated, they expand and become less dense. Hot mantle material rises upward and comes in contact with Earth's crust. Thermal energy is transferred from hot mantle material to the colder surface above. As the mantle cools, it becomes denser and then sinks, forming a convection current. These currents in the asthenosphere act like a conveyor belt moving the lithosphere above.

Key Concept Check Why do tectonic plates move?

Inquiry MiniLab — 20 minutes

How do changes in density cause motion?

Convection currents drive plate motion. Material near the base of the mantle is heated, which decreases its density. This material then rises to the base of the crust, where it cools, increasing in density and sinking.

1. Read and complete a lab safety form.
2. Copy the table to the right into your Science Journal and add a row for each minute. Record your observations.
3. Pour 100 mL of **carbonated water** or **clear soda** into a **beaker** or a **clear glass.**
4. Drop five **raisins** into the water. Observe the path that the raisins follow for 5 minutes.

Time Interval	Observations
First minute	
Second minute	
Third minute	

Analyze and Conclude

1. **Observe** Describe each raisin's motion.
2. **Key Concept** How does the behavior of the raisin model compare to the motion in Earth's mantle?

Convection Currents

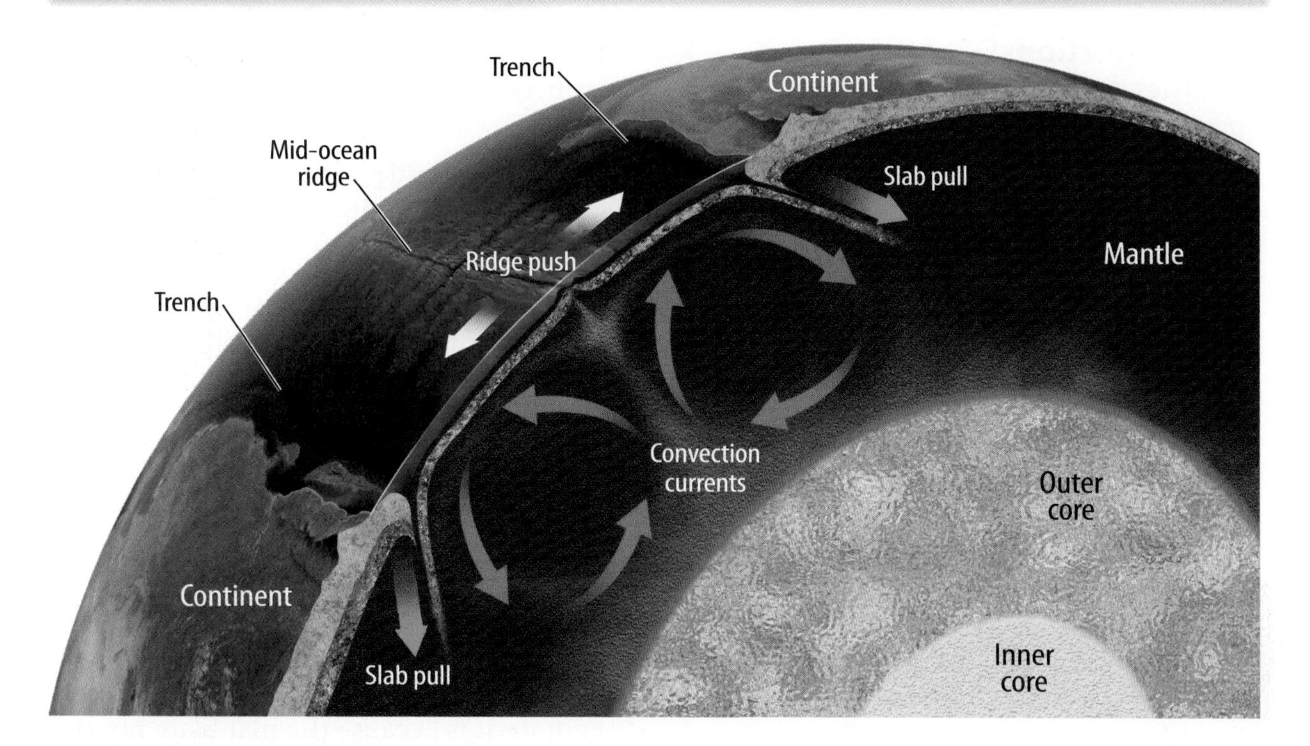

Figure 14 Convection occurs in the mantle underneath Earth's tectonic plates. Three forces act on plates to make them move: basal drag from convection currents, ridge push at mid-ocean ridges, and slab pull from subducting plates.

Visual Check What is happening to a plate that is undergoing slab pull?

Forces Causing Plate Motion

How can something as massive as the Pacific Plate move? **Figure 14** shows the three forces that interact to cause plate motion. Scientists still debate over which of these forces has the greatest effect on plate motion.

Basal Drag Convection currents in the mantle produce a force that causes motion called basal drag. Notice in **Figure 14** how convection currents in the asthenosphere circulate and drag the lithosphere similar to the way a conveyor belt moves items along at a supermarket checkout.

Ridge Push Recall that mid-ocean ridges have greater elevation than the surrounding seafloor. Because mid-ocean ridges are higher, gravity pulls the surrounding rocks down and away from the ridge. *Rising mantle material at mid-ocean ridges creates the potential for plates to move away from the ridge with a force called* **ridge push.** Ridge push moves lithosphere in opposite directions away from the mid-ocean ridge.

Slab Pull As you read earlier in this lesson, when tectonic plates collide, the denser plate will sink into the mantle along a subduction zone. This plate is called a slab. Because the slab is old and cold, it is denser than the surrounding mantle and will sink. *As a slab sinks, it pulls on the rest of the plate with a force called* **slab pull.** Scientists are still uncertain about which force has the greatest influence on plate motion.

Math Skills

Use Proportions

The plates along the Mid-Atlantic Ridge spread at an average rate of 2.5 cm/y. How long will it take the plates to spread 1 m? Use proportions to find the answer.

1. **Convert the distance to the same unit.**

$$1 \text{ m} = 100 \text{ cm}$$

2. **Set up a proportion:**

$$\frac{2.5 \text{ cm}}{1 \text{ y}} = \frac{100 \text{ cm}}{x \text{ y}}$$

3. **Cross multiply and solve for *x* as follows:**

$$2.5 \text{ cm} \times x \text{y} = 100 \text{ cm} \times 1 \text{ y}$$

4. **Divide both sides by 2.5 cm.**

$$x = \frac{100 \text{ cm y}}{2.5 \text{ cm}}$$

$$x = 40 \text{ y}$$

Practice

The Eurasian plate travels the slowest, at about 0.7 cm/y. How long would it take the plate to travel 3 m?

(1 m = 100 cm)

- **Math Practice**
- **Personal Tutor**

A Theory in Progress

Plate tectonics has become the unifying theory of geology. It explains the connection between continental drift and the formation and destruction of crust along plate boundaries. It also helps to explain the occurrence of earthquakes, volcanoes, and mountains.

The investigation that Wegener began nearly a century ago is still being revised. Several unanswered questions remain.

- Why is Earth the only planet in the solar system that has plate tectonic activity? Different hypotheses have been proposed to explain this. Extrasolar planets outside our solar system are also being studied.
- Why do some earthquakes and volcanoes occur far away from plate boundaries? Perhaps it is because the plates are not perfectly rigid. Different thicknesses and weaknesses exist within the plates. Also, the mantle is much more active than scientists originally understood.
- What forces dominate plate motion? Currently accepted models suggest that convection currents occur in the mantle. However, there is no way to measure or observe them.
- What will scientists investigate next? **Figure 15** shows an image produced by a new technique called anisotropy that creates a 3-D image of seismic wave velocities in a subduction zone. This developing technology might help scientists better understand the processes that occur within the mantle and along plate boundaries.

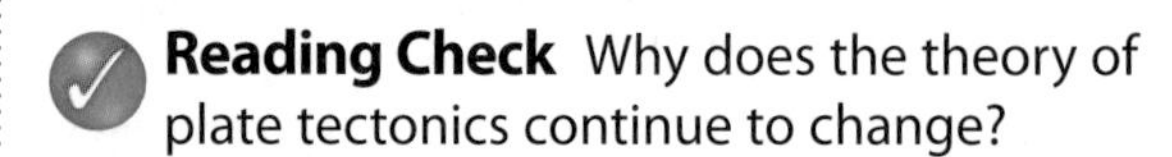
Reading Check Why does the theory of plate tectonics continue to change?

Figure 15 Seismic waves were used to produce this tomography scan. These colors show a subducting plate. The blue colors represent rigid materials with faster seismic wave velocities.

Lesson 3 Review

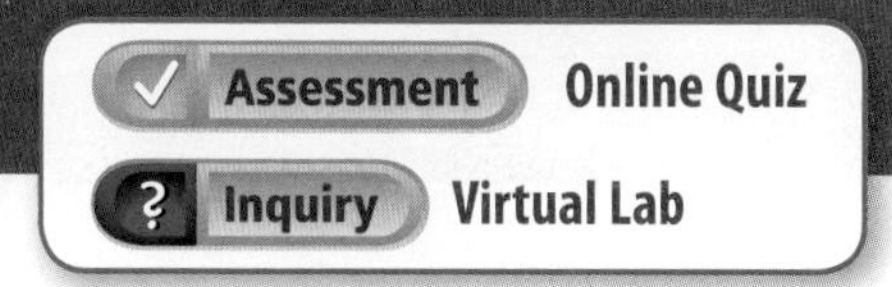

Visual Summary

Tectonic plates are made of cold and rigid slabs of rock.

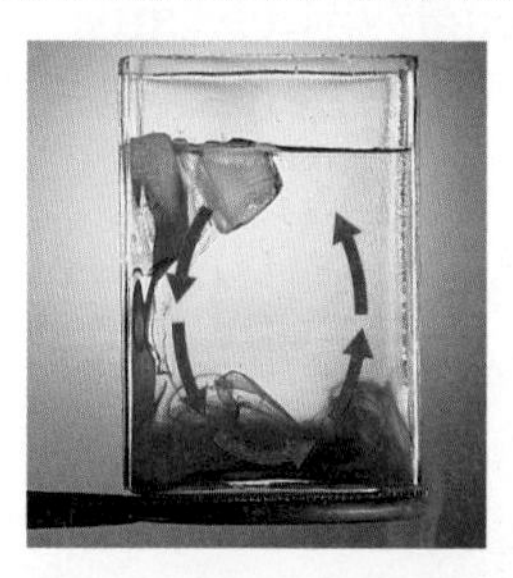

Mantle convection—the circulation of mantle material due to density differences—drives plate motion.

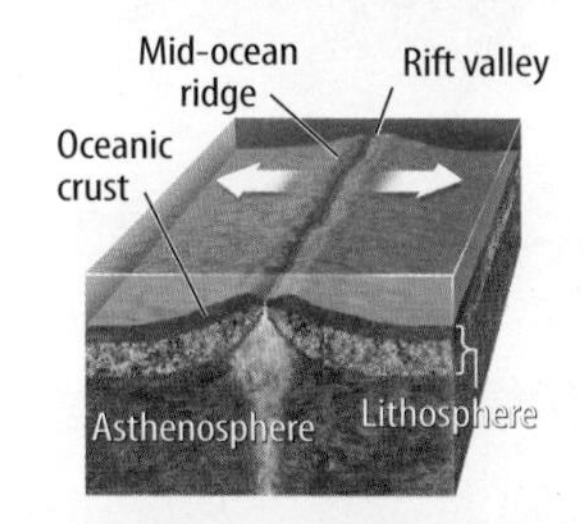

The three types of plate boundaries are divergent, convergent, and transform boundaries.

FOLDABLES®

Use your lesson Foldable to review the lesson. Save your Foldable for the project at the end of the chapter.

What do you think NOW?

You first read the statements below at the beginning of the chapter.

5. Continents drift across a molten mantle.

6. Mountain ranges can form when continents collide.

Did you change your mind about whether you agree or disagree with the statements? Rewrite any false statements to make them true.

Use Vocabulary

1. The theory that proposes that Earth's surface is broken into moving, rigid plates is called ________.

Understand Key Concepts

2. **Compare and contrast** the geological activity that occurs along the three types of plate boundaries.

3. **Explain** why mantle convection occurs.

4. Tectonic plates move because of
   - **A.** convection currents.
   - **B.** Earth's increasing size.
   - **C.** magnetic reversals.
   - **D.** volcanic activity.

Interpret Graphics

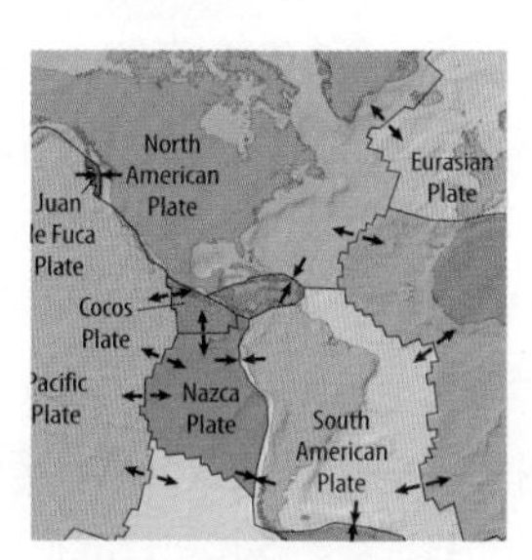

5. **Identify** Name the type of boundary between the Eurasian Plate and the North American Plate and between the Nazca Plate and South American Plate.

6. **Determine Cause and Effect** Copy and fill in the graphic organizer below to list the cause and effects of convection currents.

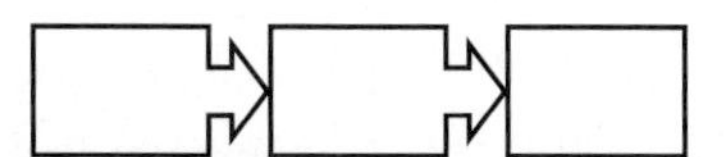

Critical Thinking

7. **Explain** why earthquakes occur at greater depths along convergent plate boundaries.

Math Skills

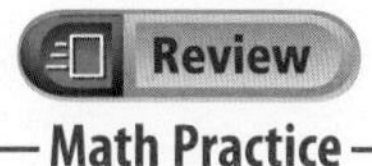

Math Practice

8. Two plates in the South Pacific separate at an average rate of 15 cm/y. How far will they have separated after 5,000 years?

Inquiry Lab

45 minutes

Movement of Plate Boundaries

Materials

graham crackers

waxed paper (four 10×10-cm squares)

dropper

frosting

plastic spoon

Safety

Earth's surface is broken into 12 major tectonic plates. Wherever these plates touch, one of four events occurs. The plates may collide and crumple or fold to make mountains. One plate may subduct under another, forming volcanoes. They may move apart and form a mid-ocean ridge, or they may slide past each other causing an earthquake. This investigation models plate movements.

Question

What happens where two plates come together?

Procedure

Part I

1. Read and complete a lab safety form.
2. Obtain the materials from your teacher.
3. Break a graham cracker along the perforation line into two pieces.
4. Lay the pieces side by side on a piece of waxed paper.
5. Slide crackers in opposite directions so that the edges of the crackers rub together.

Part II

6. Place two new graham crackers side by side but not touching.
7. In the space between the crackers, add several drops of water.
8. Slide the crackers toward each other and observe what happens.

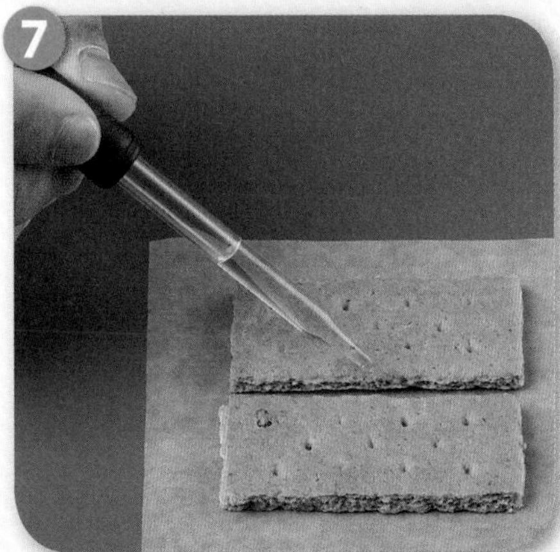

Part III

9. Place a spoonful of frosting on the waxed-paper square.
10. Place two graham crackers on top of the frosting so that they touch.
11. Push the crackers down and spread them apart in one motion.

Analyze and Conclude

12 Analyze the movement of the crackers in each of your models.

Part I

13 What type of plate boundary do the graham crackers in this model represent?

14 What do the crumbs in the model represent?

15 Did you feel or hear anything when the crackers moved past each other? Explain.

16 How does this model simulate an earthquake?

Part II

17 What does the water in this model represent?

18 What type of plate boundary do the graham crackers in this model represent?

19 Why didn't one graham cracker slide beneath the other in this model?

Part III

20 What type of plate boundary do the graham crackers in this model represent?

21 What does the frosting represent?

22 What shape does the frosting create when the crackers move?

23 What is the formation formed from the crackers and frosting?

Communicate Your Results

Create a flip book of one of the boundaries to show a classmate who was absent. Show how each boundary plate moves and the results of those movements.

Inquiry Extension

Place a graham cracker and a piece of cardboard side by side. Slide the two pieces toward each other. What type of plate boundary does this model represent? How is this model different from the three that you observed in the lab?

Lab Tips

- ☑ Use fresh graham crackers.
- ☑ Slightly heat frosting to make it more fluid for experiments.

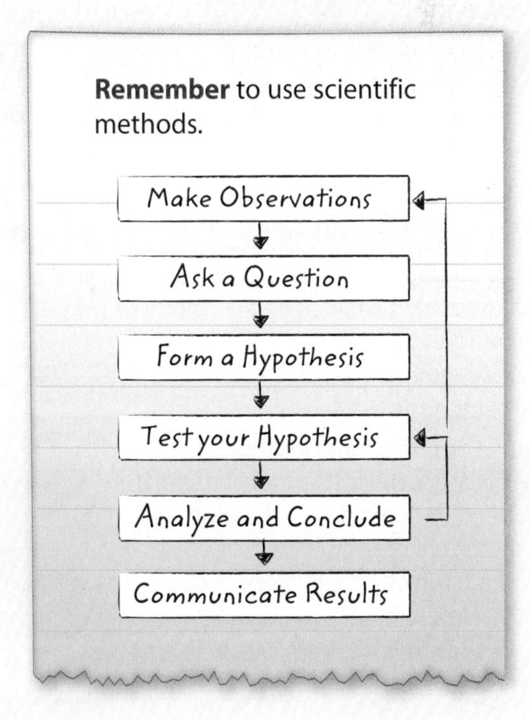

Chapter 14 Study Guide

The theory of plate tectonics states that Earth's lithosphere is broken up into rigid plates that move over Earth's surface.

Key Concepts Summary

Lesson 1: The Continental Drift Hypothesis

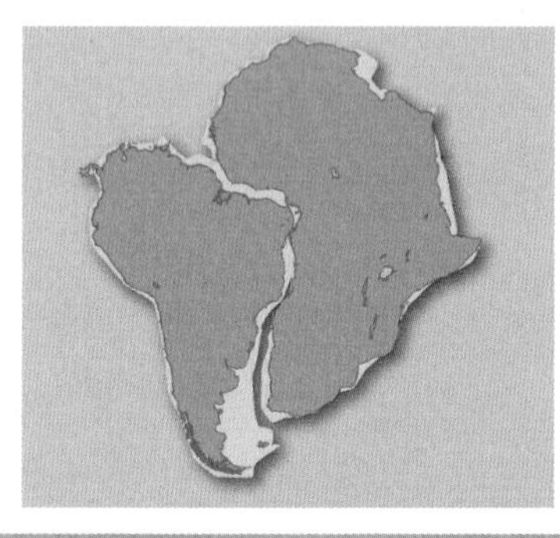

- The puzzle piece fit of continents, fossil evidence, climate, rocks, and mountain ranges supports the hypothesis of **continental drift.**
- Scientists were skeptical of continental drift because Wegener could not explain the mechanism for movement.

Vocabulary

Pangaea p. 495
continental drift p. 495

Lesson 2: Development of a Theory

- **Seafloor spreading** provides a mechanism for continental drift.
- Seafloor spreading occurs at **mid-ocean ridges.**
- Evidence of **magnetic reversal** in rock, thermal energy trends, and the discovery of seafloor spreading all contributed to the development of the theory of plate tectonics.

Vocabulary

mid-ocean ridge p. 503
seafloor spreading p. 504
normal polarity p. 506
magnetic reversal p. 506
reversed polarity p. 506

Lesson 3: The Theory of Plate Tectonics

- Types of plate boundaries, the location of earthquakes, volcanoes, and mountain ranges, and satellite measurement of plate motion support the theory of **plate tectonics.**
- Mantle **convection, ridge push,** and **slab pull** are the forces that cause plate motion. Radioactivity in the mantle and thermal energy from the core produce the energy for convection.

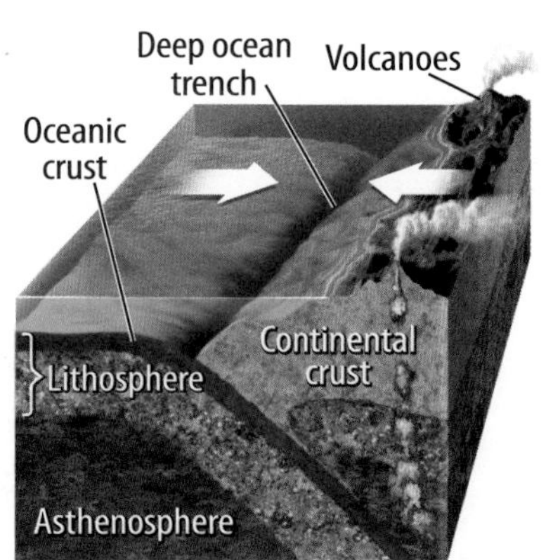

Vocabulary

plate tectonics p. 511
lithosphere p. 512
divergent plate boundary p. 513
transform plate boundary p. 513
convergent plate boundary p. 513
subduction p. 513
convection p. 516
ridge push p. 517
slab pull p. 517

- Personal Tutor
- Vocabulary eGames
- Vocabulary eFlashcards

FOLDABLES® Chapter Project

Assemble your lesson Foldables as shown to make a Chapter Project. Use the project to review what you have learned in this chapter.

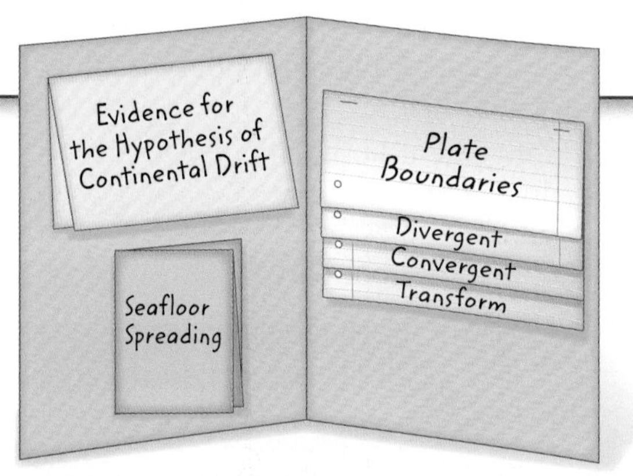

Use Vocabulary

1. The process in which hot mantle rises and cold mantle sinks is called ________.
2. What is the plate tectonics theory?
3. What was Pangaea?
4. Identify the three types of plate boundaries and the relative motion associated with each type.
5. Magnetic reversals occur when ________.
6. Explain seafloor spreading in your own words.

Link Vocabulary and Key Concepts

Interactive Concept Map

Copy this concept map, and then use vocabulary terms from the previous page to complete the concept map.

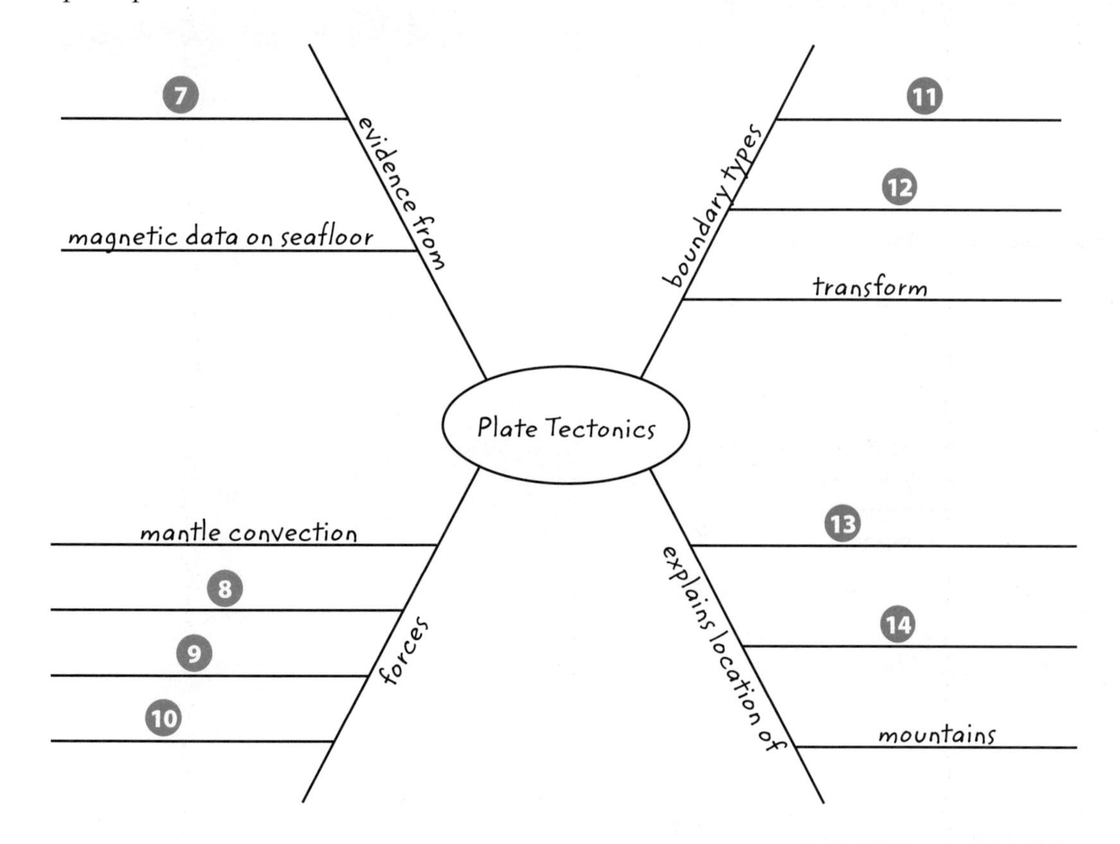

Chapter 14 Review

Understand Key Concepts 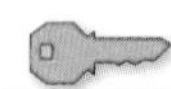

1. Alfred Wegener proposed the ________ hypothesis.
   A. continental drift
   B. plate tectonics
   C. ridge push
   D. seafloor spreading

2. Ocean crust is
   A. made from submerged continents.
   B. magnetically produced crust.
   C. produced at the mid-ocean ridge.
   D. produced at all plate boundaries.

3. What technologies did scientists NOT use to develop the theory of seafloor spreading?
   A. echo-sounding measurements
   B. GPS (global positioning system)
   C. magnetometer measurements
   D. seafloor thickness measurements

4. The picture below shows Pangaea's position on Earth approximately 280 million years ago. Where did geologists discover glacial features associated with a cooler climate?
   A. Antarctica
   B. Asia
   C. North America
   D. South America

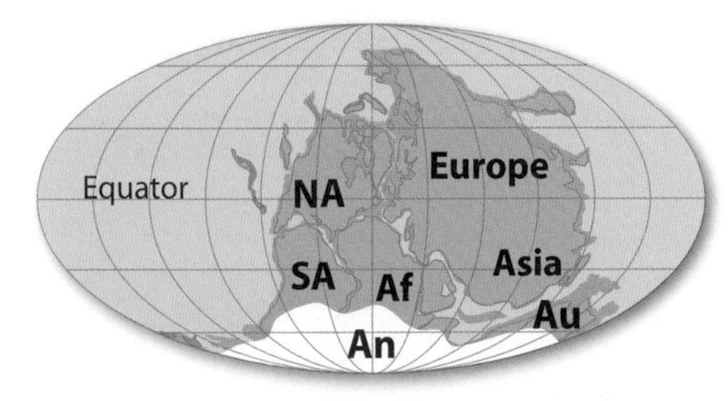

Pangaea

5. Mid-ocean ridges are associated with
   A. convergent plate boundaries.
   B. divergent plate boundaries.
   C. hot spots.
   D. transform plate boundaries.

6. Two plates of equal density form mountain ranges along
   A. continent-to-continent convergent boundaries.
   B. ocean-to-continent convergent boundaries.
   C. divergent boundaries.
   D. transform boundaries.

7. Which type of plate boundary is shown in the figure below?
   A. convergent boundary
   B. divergent boundary
   C. subduction zone
   D. transform boundary

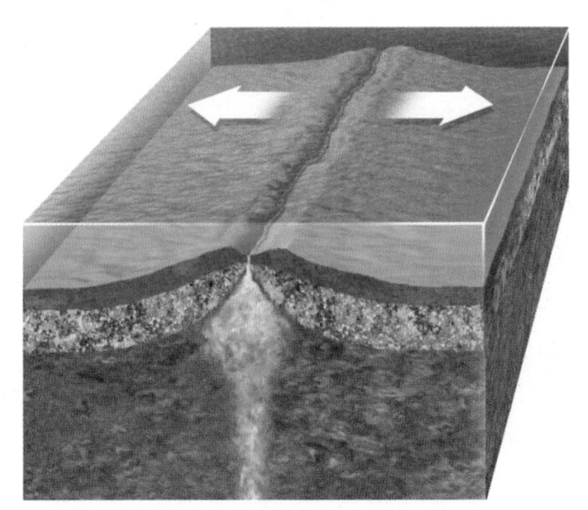

8. What happens to Earth's magnetic field over time?
   A. It changes polarity.
   B. It continually strengthens.
   C. It stays the same.
   D. It weakens and eventually disappears.

9. Which of Earth's outermost layers includes the crust and the upper mantle?
   A. asthenosphere
   B. lithosphere
   C. mantle
   D. outer core

Critical Thinking

10 **Evaluate** The oldest seafloor in the Atlantic Ocean is located closest to the edge of continents, as shown in the image below. Explain how this age can be used to figure out when North America first began to separate from Europe.

11 **Examine** the evidence used to develop the theory of plate tectonics. How has new technology strengthened the theory?

12 **Explain** Sediments deposited by glaciers in Africa are surprising because Africa is now warm. How does the hypothesis of continental drift explain these deposits?

13 **Draw** a diagram to show subduction of an oceanic plate beneath a continental plate along a convergent plate boundary. Explain why volcanoes form along this type of plate boundary.

14 **Infer** Warm peanut butter is easier to spread than cold peanut butter. How does knowing this help you understand why the mantle is able to deform in a plastic manner?

Writing in Science

15 **Predict** If continents continue to move in the same direction over the next 200 million years, how might the appearance of landmasses change? Write a paragraph to explain the possible positions of landmasses in the future. Based on your understanding of the plate tectonic theory, is it possible that new supercontinents will form in the future?

REVIEW THE BIG IDEA

16 What is the theory of plate tectonics? Distinguish between continental drift, seafloor spreading, and plate tectonics. What evidence was used to support the theory of plate tectonics?

17 Use the image below to interpret how the theory of plate tectonics helps to explain the formation of huge mountains like the Himalayas.

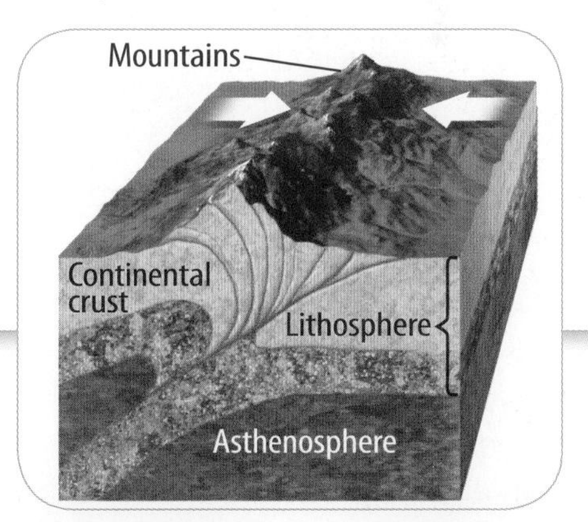

Math Skills

Math Practice

Use Proportions

18 Mountains on a convergent plate boundary may grow at a rate of 3 mm/y. How long would it take a mountain to grow to a height of 3,000 m? (1 m = 1,000 mm)

19 The North American Plate and the Pacific Plate have been sliding horizontally past each other along the San Andreas fault zone for about 10 million years. The plates move at an average rate of about 5 cm/y.

a. How far have the plates traveled, assuming a constant rate, during this time?

b. How far has the plate traveled in kilometers? (1 km = 100,000 cm)

Standardized Test Practice

Record your answers on the answer sheet provided by your teacher or on a sheet of paper.

Multiple Choice

Use the diagram below to answer questions 1 and 2.

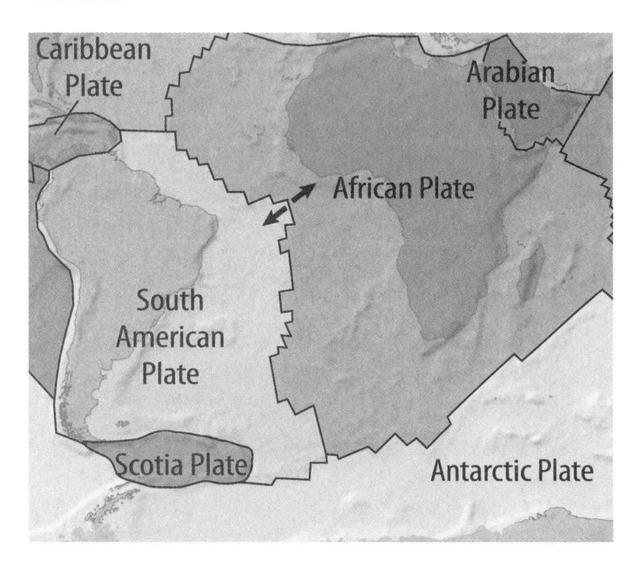

1 In the diagram above, what does the irregular line between tectonic plates represent?

A abyssal plain
B island chain
C mid-ocean ridge
D polar axis

2 What do the arrows indicate?

A magnetic polarity
B ocean flow
C plate movement
D volcanic eruption

3 What evidence helped to support the theory of seafloor spreading?

A magnetic equality
B magnetic interference
C magnetic north
D magnetic polarity

4 Which plate tectonic process creates a deep ocean trench?

A conduction
B deduction
C induction
D subduction

5 What causes plate motion?

A convection in Earth's mantle
B currents in Earth's oceans
C reversal of Earth's polarity
D rotation on Earth's axis

6 New oceanic crust forms and old oceanic crust moves away from a mid-ocean ridge during

A continental drift.
B magnetic reversal.
C normal polarity.
D seafloor spreading.

Use the diagram below to answer question 7.

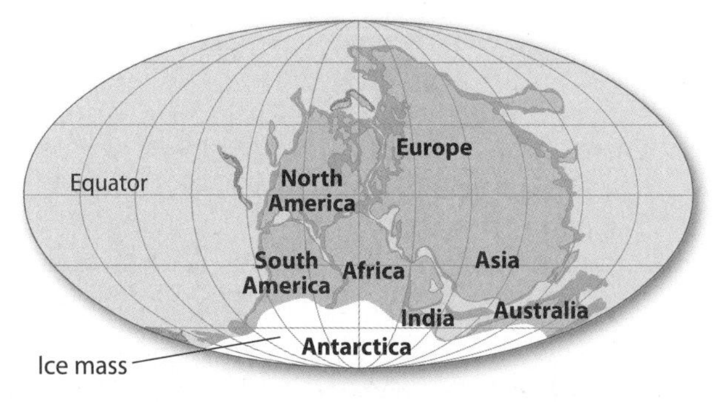

7 What is the name of Alfred Wegener's ancient supercontinent pictured in the diagram above?

A Caledonia
B continental drift
C *Glossopteris*
D Pangaea

Assessment
Online Standardized Test Practice

Use the diagram below to answer question 8.

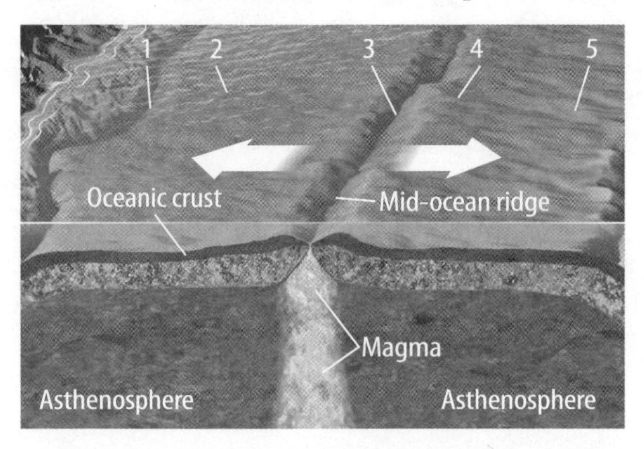

8 The numbers in the diagram represent seafloor rock. Which represent the oldest rock?

A 1 and 5

B 2 and 4

C 3 and 4

D 4 and 5

9 Which part of the seafloor contains the thickest sediment layer?

A abyssal plain

B deposition band

C mid-ocean ridge

D tectonic zone

10 What type of rock forms when lava cools and crystallizes on the seafloor?

A a fossil

B a glacier

C basalt

D magma

Constructed Response

Use the table below to answer questions 11 and 12.

Plate Boundary	Location

11 In the table above, identify the three types of plate boundaries. Then describe a real-world location for each type.

12 Create a diagram to show plate motion along one type of plate boundary. Label the diagram and draw arrows to indicate the direction of plate motion.

13 Identify and explain all the evidence that Wegener used to help support his continental drift hypothesis.

14 Why was continental drift so controversial during Alfred Wegener's time? What explanation was necessary to support his hypothesis?

15 How did scientists prove the theory of seafloor spreading?

16 If new oceanic crust constantly forms along mid-ocean ridges, why isn't Earth's total surface area increasing?

NEED EXTRA HELP?																
If You Missed Question...	1	2	3	4	5	6	7	8	9	10	11	12	13	14	15	16
Go to Lesson...	3	3	2	3	3	2	1	2	2	2	3	3	1	1	2	3

Chapter 15

Earthquakes and Volcanoes

What causes earthquakes and volcanic eruptions?

Why do volcanoes erupt?

Mount Pinatubo, a volcano in the Philippines, ejected superheated particles of ash and dust in June 1991. This truck is trying to outrun a pyroclastic flow produced during this eruption. *Pyroclastic* means "fire fragments." Why do you suppose this eruption was so dangerous?

- Why did Mount Pinatubo erupt explosively?
- Can scientists predict earthquakes and volcanic eruptions?
- What causes earthquakes and volcanic activity?

Get Ready to Read

What do you think?

Before you read, decide if you agree or disagree with each of these statements. As you read this chapter, see if you change your mind about any of the statements.

1. Earth's crust is broken into rigid slabs of rock that move, causing earthquakes and volcanic eruptions.
2. Earthquakes create energy waves that travel through Earth.
3. All earthquakes occur on plate boundaries.
4. Volcanoes can erupt anywhere on Earth.
5. Volcanic eruptions are rare.
6. Volcanic eruptions only affect people and places close to the volcano.

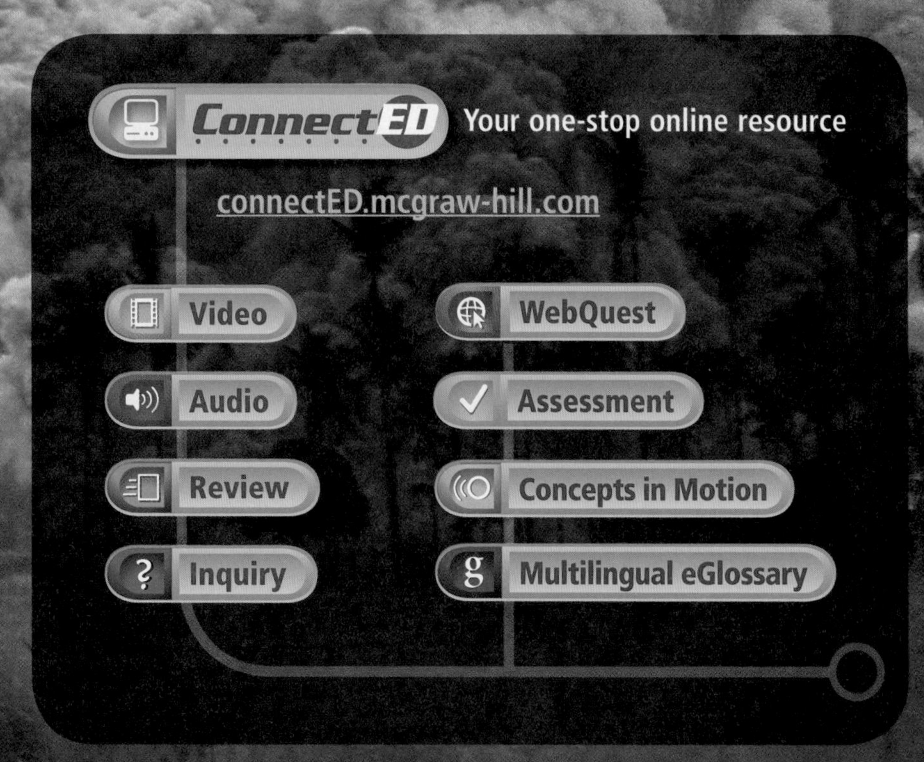

Lesson 1

Reading Guide

Key Concepts
ESSENTIAL QUESTIONS

- What is an earthquake?
- Where do earthquakes occur?
- How do scientists monitor earthquake activity?

Vocabulary

earthquake p. 531
fault p. 533
seismic wave p. 534
focus p. 534
epicenter p. 534
primary wave p. 535
secondary wave p. 535
surface wave p. 535
seismologist p. 536
seismometer p. 537
seismogram p. 537

 Multilingual eGlossary

 Video Science Video

Earthquakes

Inquiry Why did this building collapse?

This building collapsed during the Loma Prieta earthquake that shook the San Francisco Bay area of California in 1989. The magnitude 7.1 earthquake produced severe shaking and damage. Freeways and buildings collapsed and a number of injuries and fatalities occurred. Why are earthquakes common in California?

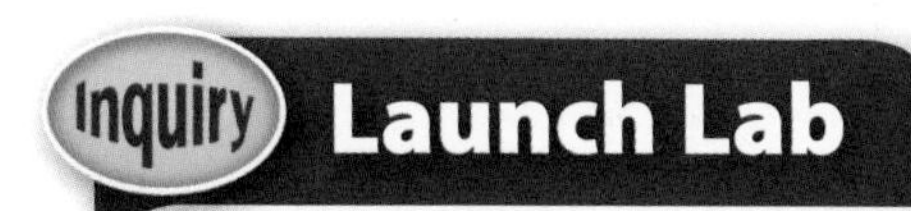

15 minutes

What causes earthquakes?

Earthquakes occur every day. On average, approximately 35 earthquakes happen on Earth every day. These earthquakes vary in severity. What causes the intense shaking of an earthquake? In this activity, you will simulate the energy released during an earthquake and observe the shaking that results.

1. Read and complete a lab safety form.
2. Tie two **large, thick rubber bands** together.
3. Loop one rubber band lengthwise around a **textbook.**
4. Use **tape** to secure a sheet of **medium-grained sandpaper** to the tabletop.
5. Tape a second sheet of sandpaper to the cover of the textbook.
6. Place the book on the table so that the sheets of sandpaper touch.
7. Slowly pull on the end of the rubber band until the book moves.
8. Observe and record what happens in your Science Journal.

Think About This

1. How does this experiment model the buildup of stress along a fault?
2. **Key Concept** Why does the rapid movement of rocks along a fault result in an earthquake?

What are earthquakes?

Have you ever tried to bend a stick until it breaks? When the stick snaps, it vibrates, releasing energy. Earthquakes happen in a similar way. **Earthquakes** *are the vibrations in the ground that result from movement along breaks in Earth's lithosphere.* These breaks are called faults.

Key Concept Check What is an earthquake?

Why do rocks move along a fault? The forces that move tectonic plates also push and pull on rocks along the fault. If these forces become large enough, the blocks of rock on either side of the fault can move horizontally or vertically past each other. The greater the force applied to a fault, the greater the chance of a large and destructive earthquake. **Figure 1** shows earthquake damage from the Northridge earthquake in 1994.

Figure 1 In 1994, the Northridge earthquake along the San Andreas Fault in California caused $20 billion in damage.

Earthquake Distribution

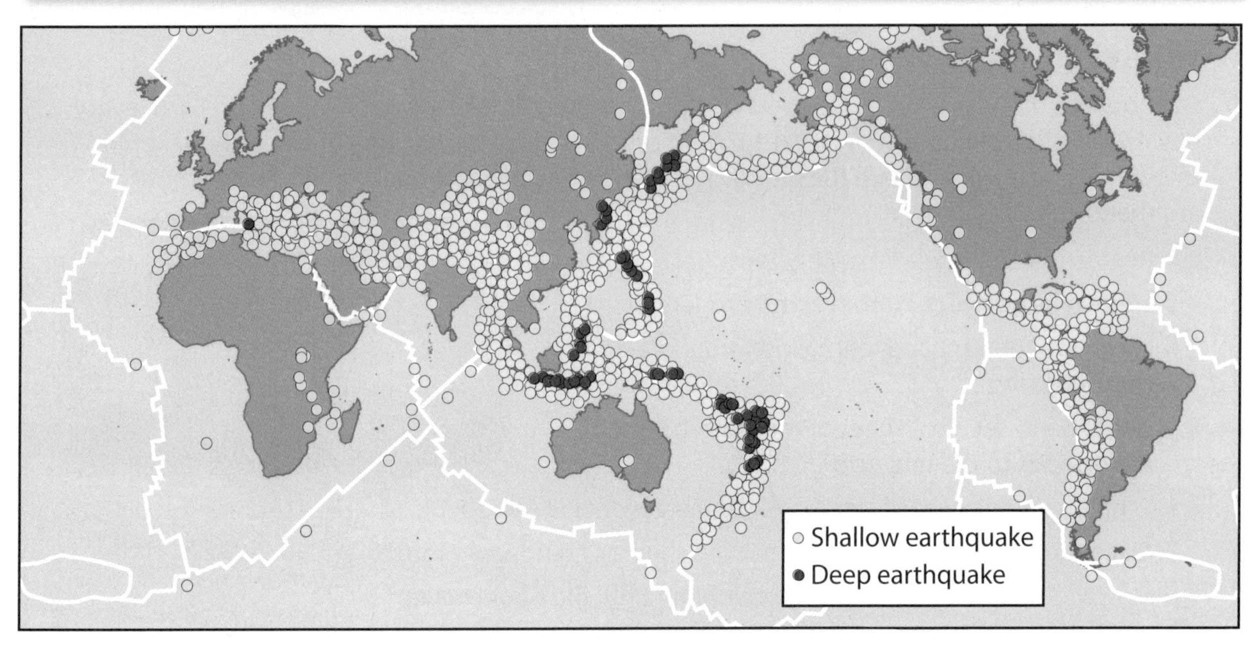

Figure 2 Notice that most earthquakes occur along plate boundaries.

Where do earthquakes occur?

The locations of major earthquakes that occurred between 2000 and 2008 are shown in **Figure 2.** Notice that only a few earthquakes occurred in the middle of a continent. Records show that most earthquakes occur in the oceans and along the edges of continents. Are there any exceptions?

Earthquakes and Plate Boundaries

Compare the location of earthquakes in **Figure 2** with tectonic **plate boundaries.** What is the relationship between earthquakes and plate boundaries? Earthquakes result from the buildup and release of stress along active plate boundaries.

REVIEW VOCABULARY

plate boundary
area where Earth's lithospheric plates move and interact with each other, producing earthquakes, volcanoes, and mountain ranges

Some earthquakes occur more than 100 km below Earth's surface, as shown in **Figure 2.** Which plate boundaries are associated with deep earthquakes? The deepest earthquakes occur where plates collide along a convergent plate boundary. Here, the denser oceanic plate subducts into the **mantle.** Earthquakes that occur along convergent plate boundaries typically release tremendous amounts of energy. They can also be disastrous.

Shallow earthquakes are common where plates separate along a divergent plate boundary, like the mid-ocean ridge system. Shallow earthquakes can also occur along transform plate boundaries like the San Andreas Fault in California. Earthquakes of varying depths occur where continents collide. Continental collisions result in the formation of large and deformed mountain ranges such as the Himalayas in Asia.

 **Key Concept Check** Where do most earthquakes occur?

Rock Deformation

At the beginning of this lesson, you read that earthquake energy is similar to bending and breaking a stick. Rocks below Earth's surface behave the same way. When a force is applied to a body of rock, depending on the properties of the rock and the force applied, the rock might bend or break.

When a force such as pressure is applied to rock along plate boundaries, the rock can change shape. This is called rock deformation. Eventually the rocks can be deformed so much that they break and move. **Figure 3** illustrates how rock deformation can result in ground displacement. Notice that rock deformation has resulted in ground displacement where the creek has been pulled in two different directions.

▲ **Figure 3** Forces at work along the San Andreas Fault in California caused displacement of this creek in two directions along a strike-slip fault.

Faults

When stress builds in places like a plate boundary, rocks can form faults. *A* **fault** *is a break in Earth's lithosphere where one block of rock moves toward, away from, or past another.* When rocks move in any direction along a fault, an earthquake occurs. The direction that rocks move on either side of the fault depends on the forces applied to the fault. **Table 1** lists three types of faults that result from motion along plate boundaries. These faults are called strike-slip, normal, and reverse faults.

 **Reading Check** What is a fault?

Table 1 The three types of faults are defined based on relative motion along the fault. ▼

Table 1 Types of Faults

Strike-slip	• Two blocks of rock slide horizontally past each other in opposite directions. • Location: transform plate boundaries	
Normal	• Forces pull two blocks of rock apart. The block of rock above the fault moves down relative to the block of rock below the fault. • Location: divergent plate boundaries	
Reverse	• Forces push two blocks of rock together. The block of rock above the fault moves up relative to the block of rock below the fault. • Location: convergent plate boundaries	

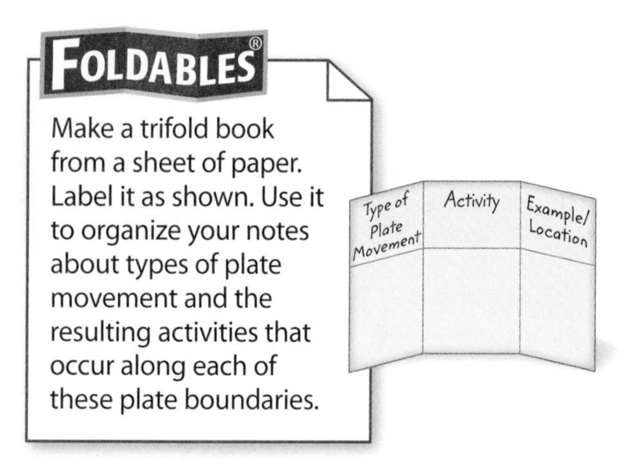

Types of Faults Strike-slip faults can form along transform plate boundaries. There, forces cause rocks to slide horizontally past each other in opposite directions. In contrast, normal faults can form when forces pull rocks apart along a divergent plate boundary. At a normal fault, one block of rock moves down relative to the other. Forces push rocks toward each other at a convergent plate boundary and a reverse fault can form. There, one block of rock moves up relative to another block of rock.

Reading Check What are the three types of faults?

Earthquake Focus and Epicenter

When rocks move along a fault, they release *energy that travels as vibrations on and in Earth called* **seismic waves.** *These waves originate where rocks first move along the fault, at a location inside Earth called the* **focus.** Earthquakes can occur anywhere between Earth's surface and depths of greater than 600 km. When you watch a news report, the reporter often will identify the earthquake's epicenter. *The* **epicenter** *is the location on Earth's surface directly above the earthquake's focus.* **Figure 4** shows the relationship between an earthquake's focus and its epicenter.

SCIENCE USE V. COMMON USE

focus

Science Use the place of origin of an earthquake

Common Use to concentrate

Figure 4 An earthquake epicenter is above a focus, where the motion along the fault first occurs.

Visual Check What is the relationship between an earthquake focus and an epicenter?

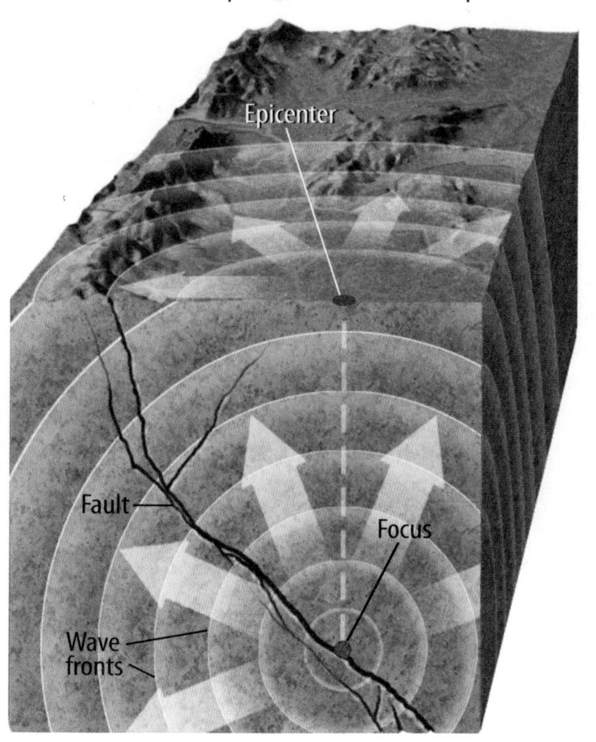

Seismic Waves

During an earthquake, a rapid release of energy along a fault produces seismic waves. Seismic waves travel outward in all directions through rock. It is similar to what happens when you drop a stone into water. When the stone strikes the water's surface, ripples move outward in circles. Seismic waves transfer energy through the ground and produce the motion that you feel during an earthquake. The energy released is strongest near the epicenter. As seismic waves move away from the epicenter, they decrease in energy and intensity. The farther you are from an earthquake's epicenter, the less the ground moves.

Types of Seismic Waves

When an earthquake occurs, particles in the ground can move back and forth, up and down, or in an elliptical motion parallel to the direction the seismic wave travels. Scientists use wave motion, wave speed, and the type of material that the waves travel through to classify seismic waves. The three types of seismic waves are **primary** waves, secondary waves, and surface waves.

WORD ORIGIN

primary
from Latin *primus,* means "first"

As shown in **Table 2, primary waves,** *also called P-waves, cause particles in the ground to move in a push-pull motion similar to a coiled spring.* P-waves are the fastest-moving seismic waves. They are the first waves that you feel following an earthquake. **Secondary waves,** *also called S-waves,* are slower than P-waves. *They cause particles to move up and down at right angles relative to the direction the wave travels.* This movement can be demonstrated by shaking a coiled spring side to side and up and down at the same time. **Surface waves** *cause particles in the ground to move up and down in a rolling motion,* similar to ocean waves. Surface waves travel only on Earth's surface closest to the epicenter. P-waves and S-waves can travel through Earth's interior. However, scientists have discovered that S-waves cannot travel through liquid.

Reading Check Describe the three types of seismic waves.

Table 2 The three types of seismic waves are classified by wave motion, wave speed, and the types of materials they can travel through.

Animation

Table 2 Properties of Seismic Waves

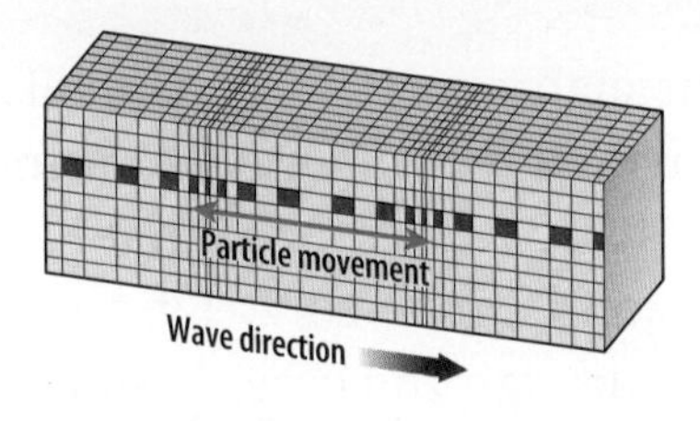	**Primary wave** • Cause rock particles to vibrate in the same direction that waves travel • Fastest seismic waves • First to be detected and recorded • Travel through solids and liquids
Secondary wave • Cause rock particles to vibrate perpendicular to the direction that waves travel • Slower than P-waves, faster than surface waves • Detected and recorded after P-waves • Only travel through solids	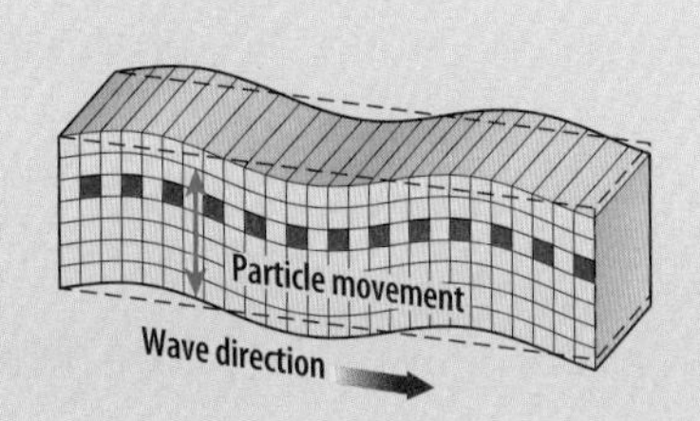
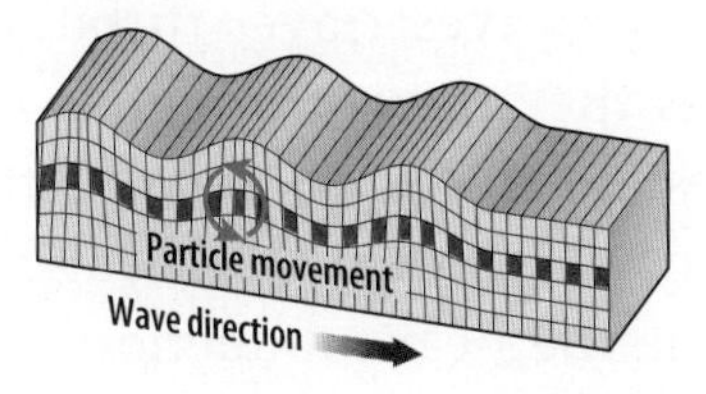	**Surface wave** • Cause rock particles to move in a rolling or elliptical motion in the same direction that waves travel • Slowest seismic wave • Generally cause the most damage at Earth's surface

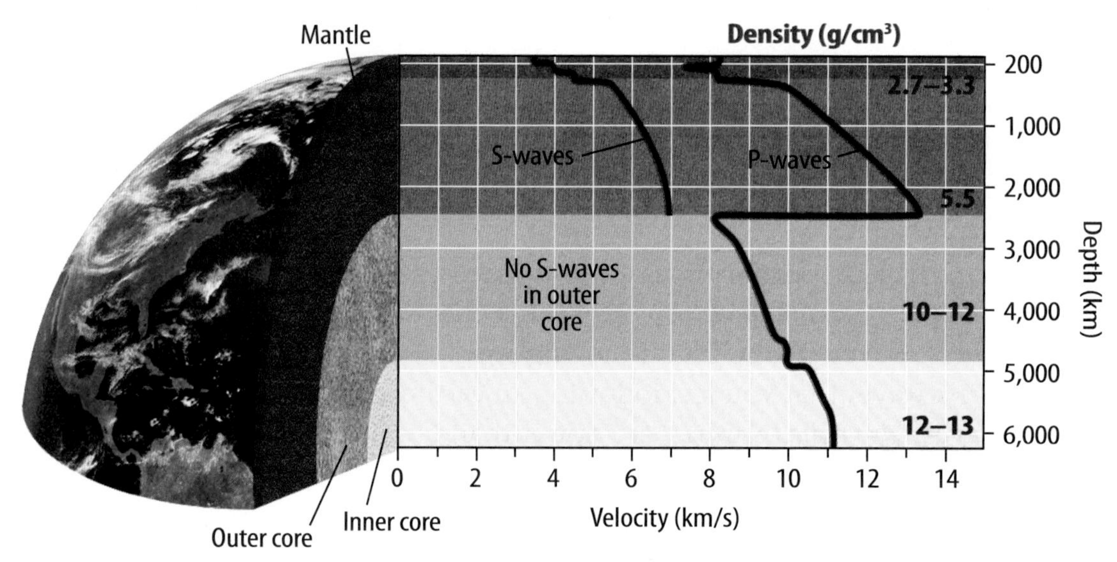

Figure 5 Seismic waves change speed and direction as they travel through Earth's interior. S-waves do not travel through Earth's outer core because it is liquid.

Visual Check What happens to P-waves and S-waves at a depth of 2500 km?

Mapping Earth's Interior

Scientists that study earthquakes are called **seismologists** (size MAH luh just). They use the properties of seismic waves to map Earth's interior. P-waves and S-waves change speed and direction depending on the material they travel through. **Figure 5** shows the speed of P-waves and S-waves at different depths within Earth's interior. By comparing these measurements to the densities of different Earth materials, scientists have determined the composition of Earth's layers.

Inner and Outer Core Through extensive earthquake studies, seismologists have discovered that S-waves cannot travel through the outer core. This discovery proved that Earth's outer core is liquid unlike the solid inner core. By analyzing the speed of P-waves traveling through the core, seismologists also have discovered that the inner and outer cores are composed of mostly iron and nickel.

Reading Check How did scientists discover that Earth's outer core is liquid?

The Mantle Seismologists also have used seismic waves to model convection currents in the mantle. The speeds of seismic waves depend on the temperature, pressure, and chemistry of the rocks that the seismic waves travel through. Seismic waves tend to slow down as they travel through hot material. For example, seismic waves are slower in areas of the mantle beneath mid-ocean ridges or near hotspots. Seismic waves are faster in cool areas of the mantle near subduction zones.

Locating an Earthquake's Epicenter

An instrument called a **seismometer** (size MAH muh ter) *measures and records ground motion and can be used to determine the distance seismic waves travel.* Ground motion is recorded as a **seismogram**, *a graphical illustration of seismic waves,* shown in **Figure 6.**

Seismologists use a method called triangulation to locate an earthquake's epicenter. This method uses the speeds and travel times of seismic waves to determine the distance to the earthquake epicenter from at least three different seismometers.

1 Find the arrival time difference.

First, determine the number of seconds between the arrival of the first P-wave and the first S-wave on the seismogram. This time difference is called lag time. Using the time scale on the bottom of the seismogram, subtract the arrival time of the first P-wave from the arrival time of the first S-wave.

2 Find the distance to the epicenter.

Next, use a graph showing the P-wave and S-wave lag time plotted against distance. Look at the *y*-axis and locate the place on the solid blue line that intersects with the lag time that you calculated from the seismogram. Then, read the corresponding distance from the epicenter on the *x*-axis.

3 Plot the distance on a map.

Next, use a ruler and a map scale to measure the distance between the seismometer and the earthquake epicenter. Draw a circle with a radius equal to this distance by placing the compass point on the seismometer location. Set the pencil at the distance measured on the scale. Draw a complete circle around the seismometer location. The epicenter is somewhere on the circle. When circles are plotted for data from at least three seismic stations, the epicenter's location can be found. This location is the point where the three circles intersect.

Triangulation

1 Find the arrival time difference.

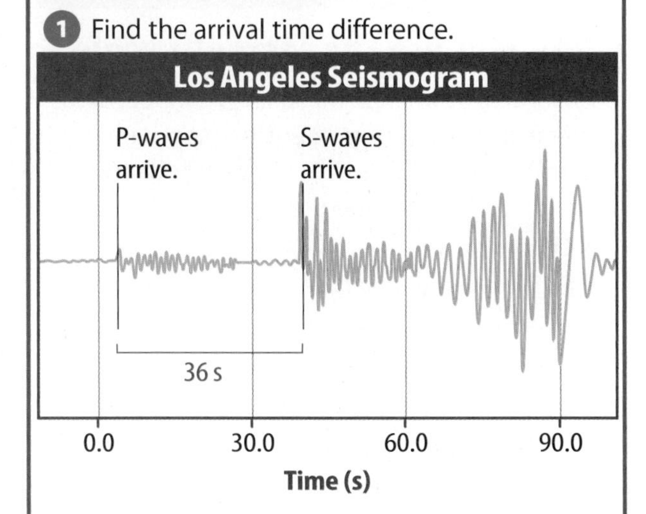

2 Find the distance to the epicenter.

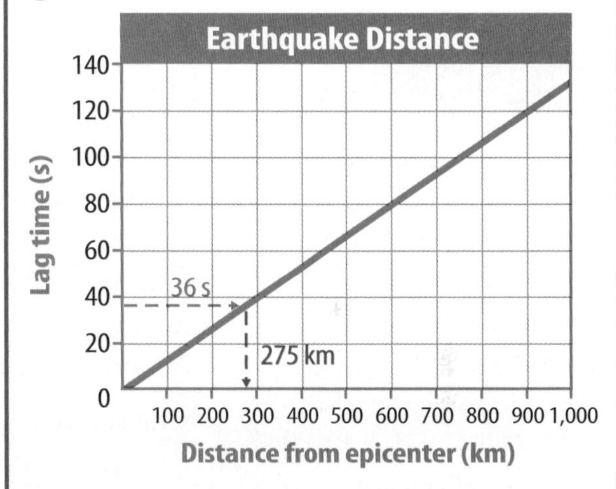

3 Plot the distance on the map.

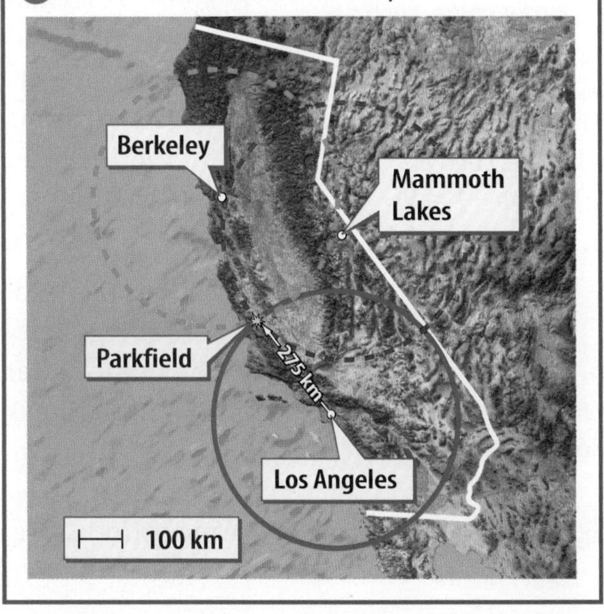

Figure 6 Seismograms provide the information necessary to locate an earthquake epicenter.

Review Personal Tutor

Inquiry MiniLab

15 minutes

Can you use the Mercalli scale to locate an epicenter?

Isoseismic (I soh SIZE mihk) lines connect areas that experience equal intensity during an earthquake. In this activity, you will observe trends in intensity and use the Mercalli scale to locate an earthquake epicenter.

1. Obtain a **map** of Mercalli ratings for the San Francisco Bay area.
2. Draw a line that connects all the points of equal intensity, making a closed loop. This is your first isoseismic line.
3. Continue drawing isoseismic lines for each Mercalli rating on the map. Just like contour lines, these lines should never cross.

Analyze and Conclude

1. **Interpret Data** Identify two cities that experienced similar effects during the earthquake.
2. **Infer** What were some of the experiences people in San Francisco might have had during the earthquake?
3.  **Key Concept** Can you identify the earthquake's epicenter on your map? Why did you choose this location?

Math Skills

Use Roman Numerals

Use the following rules to evaluate Roman numerals.

1. Values: X = 10; V = 5; I = 1
2. Add similar values that are next to one another, such as III (1 + 1 + 1 = 3)
3. Add a smaller value that comes after a larger value, such as XV (10 + 5 = 15)
4. Subtract a smaller value that precedes a larger value, such as IX (10 − 1 = 9)
5. Use the fewest possible numerals to express the value (X rather than VV)

Practice

What is the value of the Roman numeral XVI? XIV?

- Math Practice
- Personal Tutor

Determining Earthquake Magnitude

Scientists can use three different scales to measure and describe earthquakes. The Richter magnitude scale uses the amount of ground motion at a given distance from an earthquake to determine magnitude. The Richter magnitude scale is used when reporting earthquake activity to the general public.

The Richter scale begins at zero, but there is no upper limit to the scale. Each increase of 1 unit on the scale represents ten times the amount of ground motion recorded on a seismogram. For example, a magnitude 8 earthquake produces 10 times greater shaking than a magnitude 7 earthquake and 100 times greater shaking than a magnitude 6 earthquake does. The largest earthquake ever recorded was a magnitude 9.5 in Chile in 1960. The earthquake and the tsunamis that followed left nearly 2,000 people dead and 2 million people homeless.

Seismologists use the moment magnitude scale to measure the total amount of energy released by the earthquake. The energy released depends on the size of the fault that breaks, the motion that occurs along the fault, and the strength of the rocks that break during an earthquake. The units on this scale are exponential. For each increase of one unit on the scale, the earthquake releases 31.5 times more energy. That means that a magnitude 8 earthquake releases more than 992 times the amount of energy than that of a magnitude 6 earthquake.

 **Reading Check** Compare the Richter scale to the moment magnitude scale.

Describing Earthquake Intensity

Another way to measure and describe an earthquake is to evaluate the damage that results from shaking. Shaking is directly related to earthquake intensity. The Modified Mercalli scale measures earthquake intensity based on descriptions of the earthquake's effects on people and structures. The Modified Mercalli scale, shown in **Table 3,** ranges from I, when shaking is not noticeable, to XII, when everything is destroyed.

Local geology also contributes to earthquake damage. In an area covered by loose sediment, ground motion is exaggerated. The intensity of the earthquake will be greater there than in places built on solid bedrock even if they are the same distance from the epicenter. Recall the lesson opener. The 1989 Loma Prieta earthquake produced severe shaking in an area called the Marina District in the San Francisco Bay area. This area had been built on loose sediment susceptible to shaking.

Table 3 The Modified Mercalli scale is used to evaluate earthquake intensity based on the damage that results.

Table 3 Modified Mercalli Scale

I	Not felt except under unusual conditions.
II	Felt by few people; suspended objects might swing.
III	Most noticeable indoors; vibrations feel like the effects of a truck passing by.
IV	Felt by many people indoors but by few people outdoors; dishes and windows rattle; standing cars rock noticeably.
V	Felt by nearly everyone; some dishes and windows break and some walls crack.
VI	Felt by all; furniture moves; some plaster falls from walls and some chimneys are damaged.
VII	Everybody runs outdoors; some chimneys break; damage is light in well-built structures but considerable in weak structures.
VIII	Chimneys, smokestacks, and walls fall; heavy furniture is overturned; partial collapse of ordinary buildings occurs.
IX	Great general damage occurs; buildings shift off foundations; ground cracks; underground pipes break.
X	Most ordinary structures are destroyed; rails are bent; landslides are common.
XI	Few structures remain standing; bridges are destroyed; railroad rails are greatly bent; broad fissures form in the ground.
XII	Total destruction; objects are thrown upward into the air.

Earthquake Hazard Map

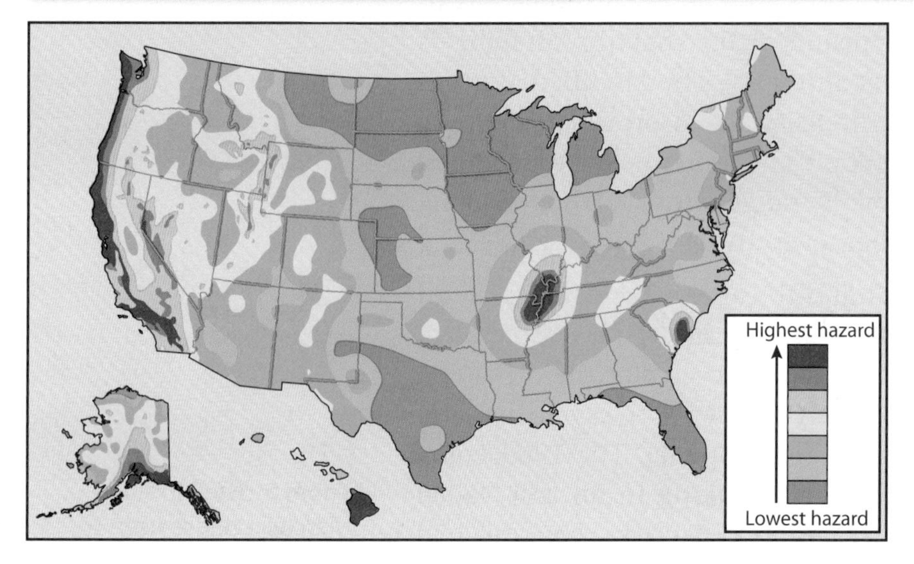

Figure 7 Areas that experienced earthquakes in the past will likely experience earthquakes again. Notice that even some parts of the central and eastern United States have high earthquake risk because of past activity.

Earthquake Risk

REVIEW VOCABULARY

convergent
tending to move toward one point or approaching each other

Recall that most earthquakes occur near tectonic plate boundaries. The transform plate boundary in California and the **convergent** plate boundaries in Oregon, Washington, and Alaska have the highest earthquake risks in the United States. However, not all earthquakes occur near plate boundaries. Some of the largest earthquakes in the United States have occurred far from plate boundaries.

From 1811–1812, three earthquakes with magnitudes between 7.8 and 8.1 occurred on the New Madrid Fault in Missouri. In contrast, the 1989 Loma Prieta earthquake had a magnitude of 7.1. **Figure 7** illustrates earthquake risk in the United States. Fortunately, high energy, destructive earthquakes are not very common. On average, only about 10 earthquakes with a magnitude greater than 7.0 occur worldwide each year. Earthquakes with magnitudes greater than 9.0, such as the Indian Ocean earthquake that caused the Asian tsunami in 2004, are rare.

Because earthquakes threaten people's lives and property, seismologists study the probability that an earthquake will occur in a given area. Probability is one of several factors that contribute to earthquake risk assessment. Seismologists also study past earthquake activity, the geology around a fault, the population density, and the building design in an area to evaluate risk. Engineers use these risk assessments to design earthquake-safe structures that are able to withstand the shaking during an earthquake. City and state governments use risk assessments to help plan and prepare for future earthquakes.

 **Key Concept Check** How do seismologists evaluate risk?

Lesson 1 Review

Visual Summary

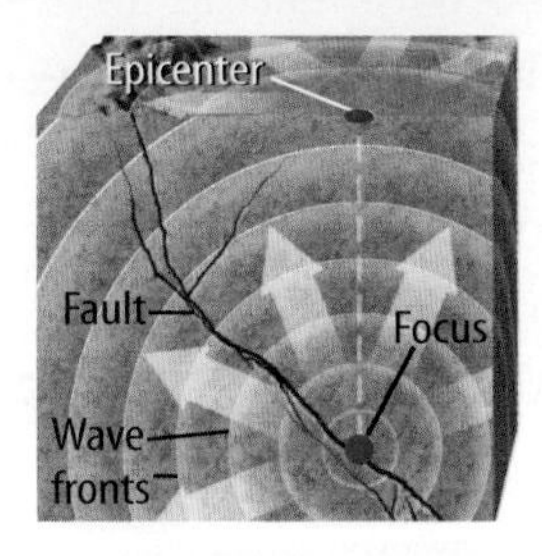

The focus is the area on a fault where an earthquake begins.

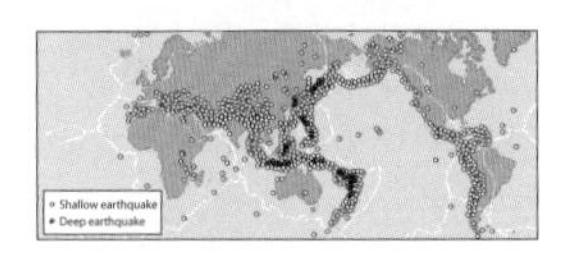

Earthquakes can occur along plate boundaries.

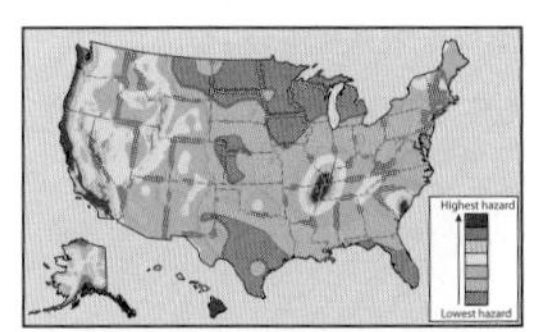

Seismologists assess earthquake risk by studying past earthquake activity and local geology.

FOLDABLES®

Use your lesson Foldable to review the lesson. Save your Foldable for the project at the end of the chapter.

What do you think NOW?

You first read the statements below at the beginning of the chapter.

1. Earth's crust is broken into rigid slabs of rock that move, causing earthquakes and volcanic eruptions.
2. Earthquakes create energy waves that travel through Earth.
3. All earthquakes occur on plate boundaries.

Did you change your mind about whether you agree or disagree with the statements? Rewrite any false statements to make them true.

Use Vocabulary

1. **Compare and contrast** the three types of faults.
2. **Distinguish** between an earthquake focus and an earthquake epicenter.
3. **Use the terms** *seismogram* and *seismometer* in a sentence.

Understand Key Concepts 

4. **Identify** areas in the United States that have the highest earthquake risk.
5. Approximately how much more energy is released in a magnitude 7 earthquake compared to a magnitude 5 earthquake?
   A. 30
   B. 60
   C. 90
   D. 1000

Interpret Graphics

6. **Compare and contrast** Create a table with the column headings for wave type, wave motion, and wave properties. Use the table to compare and contrast the three types of seismic waves.
7. **Describe** Use the image below to describe Earth's interior.

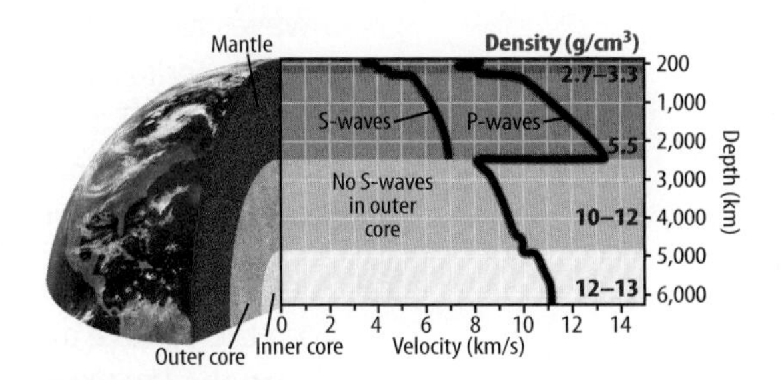

Critical Thinking

8. **Determine** what measurements you would make to evaluate earthquake risk in your hometown.

Math Skills

Math Practice

9. What is the value of Roman numeral XXVI?

Inquiry **Skill Practice** **Analyze** **45 minutes**

Can you locate an earthquake's epicenter?

Materials

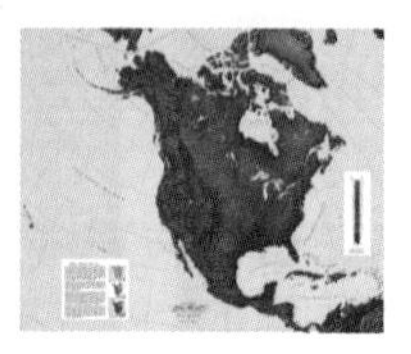

map of North America

drawing compass

Safety

Imagine the room where you are sitting suddenly begins to shake. This movement lasts for about 10 seconds. Based on the shaking that you have felt, it might seem like an earthquake happened nearby. But, to locate the epicenter, you need to analyze P-wave and S-wave data recorded for the same earthquake in at least three different locations.

Learn It

When scientists conduct experiments, they make measurements and collect and **analyze** data. For example, seismologists measure the difference in arrival times between P-waves and S-waves following an earthquake. They collect seismic wave data from at least three different locations. Using the difference in arrival times, or lag time, seismologists can determine the distance to an earthquake epicenter.

Try It

1. Read and complete a lab safety form.
2. Obtain a map of the United States from your teacher.
3. Study the three seismograms. Determine the arrival times, to the nearest second, of the P- and S-waves for each seismometer station: Berkeley, CA; Parkfield, CA; and Kanab, UT. Record the location and the arrival times for P- and S-waves in your Science Journal.
4. Subtract the P-wave arrival time from the S-wave arrival time and record the lag time in your Science Journal.
5. Use the lag time and the Earthquake Distance graph to determine the distance to the epicenter for each seismometer station.

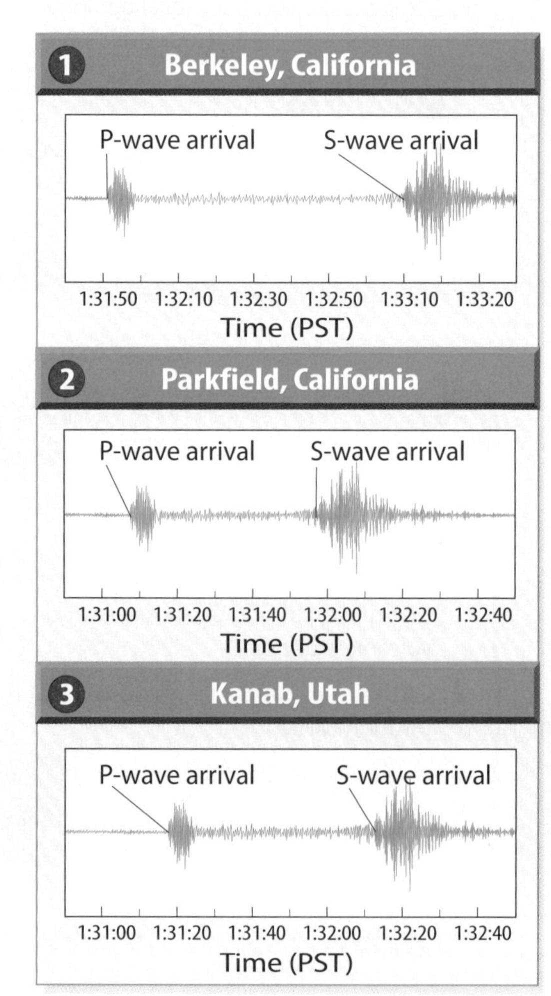

6. Use the map scale to set the spacing between the pencil and the point on the compass equal to the distance to the first seismometer. Draw a circle with a radius equal to the distance around the seismic station on the map.

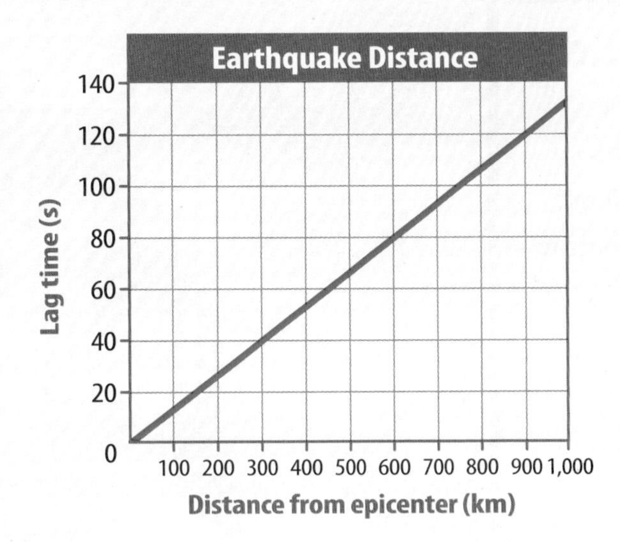

7. Repeat for the two other seismometer locations. The point where the three circles intersect marks the earthquake epicenter.

Apply It

8. Consider the difference between the arrival times of P-waves for all three seismometer locations. Why does this difference occur?

9. **Examine** the calculated lag times for all three seismograms. Why do you think the arrival-time differences are greater for the stations that are furthest from the epicenter?

10. Where did the earthquake occur?

11. **Key Concept** Why does it take three seismograms to locate an earthquake epicenter? What is this process called?

Lesson 2

Volcanoes

Reading Guide

Key Concepts
ESSENTIAL QUESTIONS

- How do volcanoes form?
- What factors contribute to the eruption style of a volcano?
- How are volcanoes classified?

Vocabulary

volcano p. 545
magma p. 545
lava p. 546
hot spot p. 546
shield volcano p. 548
composite volcano p. 548
cinder cone p. 548
volcanic ash p. 549
viscosity p. 549

g Multilingual eGlossary

Video

- Science Video
- What's Science Got to do With It?

Inquiry What makes an eruption explosive?

Notice the red, hot "fire fountain" erupting from Kilauea volcano in Hawaii. Kilauea is the most active volcano in the world. Now recall the ash eruption pictured in the chapter opener. What makes volcanoes erupt so differently? The answer can be found in magma chemistry.

Launch Lab

15 minutes

What determines the shape of a volcano?

Not all volcanoes look the same. The location of a volcano and the magma chemistry play an important part in determining the shape of a volcano.

1. Read and complete a lab safety form.
2. Obtain a **tray, a beaker of sand, a beaker with a mixture of flour and water, waxed paper,** and a **plastic spoon.**
3. Lay the waxed paper inside the tray.
4. Hold the beaker of sand about 30 cm above the tray. Slowly pour the sand onto the waxed paper and observe how it piles up.
5. Fold the paper in half and use it to carefully pour the sand back into the beaker.
6. Stir the flour and water mixture. It should be about the consistency of oatmeal. Add water if necessary.
7. Repeat steps 4 and 5 with the flour and water mixture. Record your observations for each trial in your Science Journal.

Think About This

1. What do the sand and the flour and water mixture represent?
2. **Key Concept** How do you think volcanoes get their shape?

What is a volcano?

Perhaps you have heard of some famous volcanoes such as Mount St. Helens, Kilauea, or Mount Pinatubo. All of these volcanoes have erupted within the last 30 years. A **volcano** *is a vent in Earth's crust through which melted—or molten—rock flows. Molten rock below Earth's surface is called* **magma.** Volcanoes are in many places worldwide. Some places have more volcanoes than others. In this lesson, you will learn about how volcanoes form, where they form, and about their structure and eruption style.

Reading Check What is magma?

How do volcanoes form?

Volcanic eruptions constantly shape Earth's surface. They can form large mountains, create new crust, and leave a path of destruction behind. Scientists have learned that the movement of Earth's tectonic plates causes the formation of volcanoes and the eruptions that result.

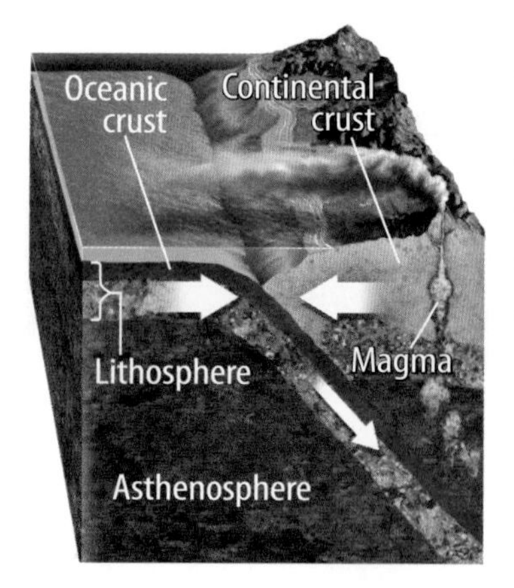

◀ Figure 8 During subduction, magma forms when one plate sinks beneath another plate.

Convergent Boundaries

Volcanoes can form along convergent plate boundaries. Recall that when two plates collide, the denser plate sinks, or subducts, into the mantle, as shown in **Figure 8.** The thermal energy below the surface and fluids driven off the subducting plate melt the mantle and form magma. Magma is less dense than the surrounding mantle and rises through cracks in the crust. This forms a volcano. *Molten rock that erupts onto Earth's surface is called* **lava.**

Figure 9 When plates spread apart, it forces magma to the surface and creates new crust. The pillow lava shown in the photograph formed at the mid-ocean ridge. ▼

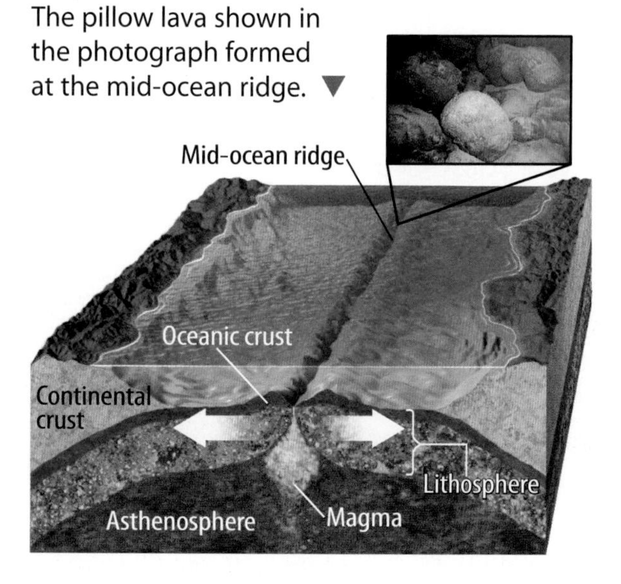

Divergent Boundaries

Lava erupts along divergent plate boundaries too. Recall that two plates spread apart along a divergent plate boundary. As the plates separate, magma rises through the vent or opening in Earth's crust that forms between them. This process commonly occurs at a mid-ocean ridge and forms new oceanic crust, as shown in **Figure 9.** More than 60 percent of all volcanic activity on Earth occurs along mid-ocean ridges.

Hot spots

Not all volcanoes form on or near plate boundaries. Volcanoes in the Hawaiian Island-Emperor Seamount chain are far from plate boundaries. *Volcanoes that are not associated with plate boundaries are called* **hot spots.** Geologists hypothesize that hot spots originate above a rising convection current from deep within Earth's mantle. They use the word *plume* to describe these rising currents of hot mantle material.

◀ Figure 10 The farther each of the Hawaiian Islands is from the hot spot, the older the island is.

Figure 10 illustrates how a new volcano forms as a tectonic plate moves over a plume. When the plate moves away from the plume, the volcano becomes dormant, or inactive. Over time, a chain of volcanoes forms as the plate moves. The oldest volcano will be farthest away from the hot spot. The youngest volcano will be directly above the hot spot.

 **Key Concept Check** How do volcanoes form?

Volcano Distribution

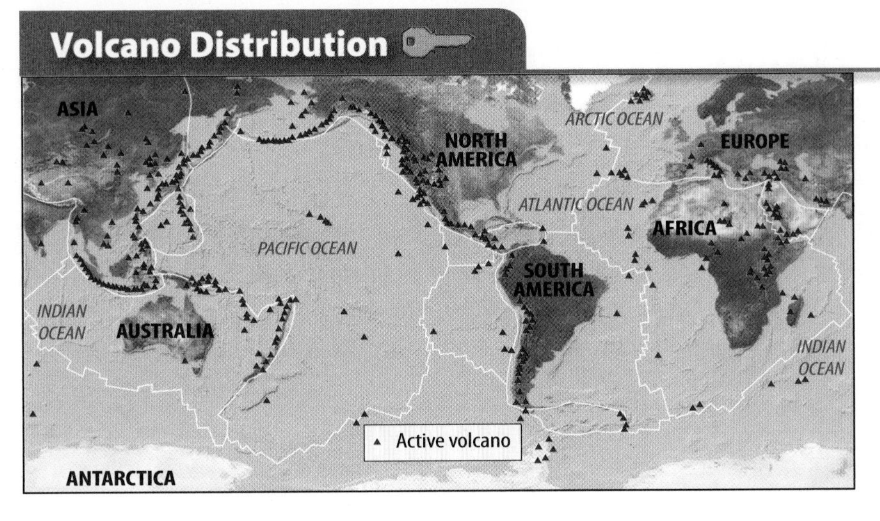

◀ **Figure 11** Most of the world's active volcanoes are located along convergent and divergent plate boundaries and hot spots.

Where do volcanoes form?

The world's active volcanoes are shown in **Figure 11.** The volcanoes all erupted within the last 100,000 years. Notice that most volcanoes are close to plate boundaries.

Ring of Fire

The Ring of Fire represents an area of earthquake and volcanic activity that surrounds the Pacific Ocean. When you compare the locations of active volcanoes and plate boundaries in **Figure 11,** you can see that volcanoes are mostly along convergent plate boundaries where plates collide. They also are located along divergent plate boundaries where plates separate. Volcanoes also can occur over hot spots, like Hawaii, the Galapagos Islands, and Yellowstone National Park in Wyoming.

Reading Check Where is the Ring of Fire?

Volcanoes in the United States

There are 60 potentially active volcanoes in the United States. Most of these volcanoes are part of the Ring of Fire. Alaska, Hawaii, Washington, Oregon, and northern California all have active volcanoes, such as Mount Redoubt in Alaska. A few of these volcanoes have produced violent eruptions, like the explosive eruption of Mount St. Helens in 1980.

The United States Geological Survey (USGS) has established three volcano observatories to monitor the potential for future volcanic eruptions in the United States. Because large populations of people live near volcanoes such as Mount Rainier in Washington, shown in **Figure 12,** the USGS has developed a hazard assessment program. Scientists monitor earthquake activity, changes in the shape of the volcano, gas emissions, and the past eruptive history of a volcano to evaluate the possibility of future eruptions.

Figure 12 Mount Rainier is an active volcano in the Cascade Mountains of the Pacific Northwest. Many people live in close proximity to the volcano. ▼

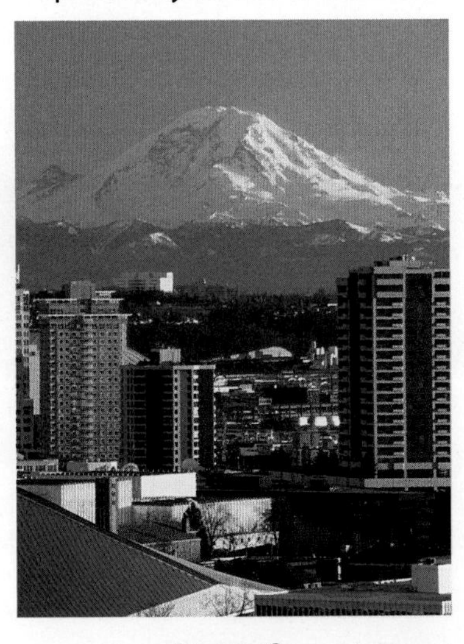

FOLDABLES®

Fold a sheet of paper to make a pyramid book. Use it to illustrate the three main types of volcanoes. Organize your notes inside the pyramid.

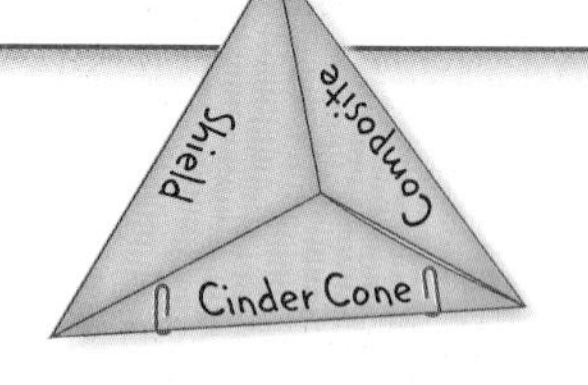

Types of Volcanoes

Volcanoes are classified based on their shapes and sizes, as shown in **Table 4.** Magma composition and eruptive style of the volcano contribute to the shape. **Shield volcanoes** *are common along divergent plate boundaries and oceanic hot spots. Shield volcanoes are large with gentle slopes of basaltic lavas.* **Composite volcanoes** *are large, steep-sided volcanoes that result from explosive eruptions of andesitic and rhyolitic lava and ash along convergent plate boundaries.* **Cinder cones** *are small, steep-sided volcanoes that erupt gas-rich, basaltic lavas.* Some volcanoes are classified as supervolcanoes—volcanoes that have very large and explosive eruptions. Approximately 630,000 years ago, the Yellowstone Caldera in Wyoming ejected more than 1,000 km^3 of rhyolitic ash and rock in one eruption. This eruption produced nearly 2,500 times the volume of material erupted from Mount St. Helens in 1980.

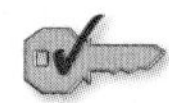

Key Concept Check What determines the shape of a volcano?

Table 4 Geologists classify volcanoes based on their size, shape, and eruptive style.

Table 4 Volcanic Features Concepts in Motion Interactive Table

Shield volcano	Composite volcano
 Large, shield-shaped volcano with gentle slopes made from basaltic lavas.	 Large, steep-sided volcano made from a mixture of andesitic and rhyolitic lava and ash.
Cinder cone volcano	**Caldera**
 Small, steep-sided volcano; made from moderately explosive eruptions of basaltic lavas.	 Large volcanic depression formed when a volcano's summit collapses or is blown away by explosive activity.

Volcanic Eruptions

When magma surfaces, it might erupt as a lava flow, such as the lava shown in **Figure 13** erupting from Kilauea volcano in Hawaii. Other times, magma might erupt explosively, sending **volcanic ash**—*tiny particles of pulverized volcanic rock and glass*—high into the atmosphere. **Figure 13** also shows Mount St. Helens in Washington, erupting violently in 1980. Why do some volcanoes erupt violently while others erupt quietly?

Eruption Style

Magma chemistry determines a volcano's eruptive style. The explosive behavior of a volcano is affected by the amount of dissolved gases, specifically the amount of water vapor, a magma contains. It is also affected by the silica, SiO_2, content of magma.

Magma Chemistry Magmas that form in different volcanic environments have unique chemical compositions. Silica is the main chemical compound in all magmas. Differences in the amount of silica affect magma thickness and its **viscosity**—*a liquid's resistance to flow.*

Magma that has a low silica content also has a low viscosity and flows easily like warm maple syrup. When the magma erupts, it flows as fluid lava that cools, crystallizes, and forms the volcanic rock basalt. This type of lava commonly erupts along mid-ocean ridges and at oceanic hot spots, such as Hawaii.

Magma that has a high silica content has a high viscosity and flows like sticky toothpaste. This type of magma forms when rocks rich in silica melt or when magma from the mantle mixes with continental crust. The volcanic rocks andesite and rhyolite form when intermediate and high silica magmas erupt from subduction zone volcanoes and continental hot spots.

Key Concept Check What factors affect eruption style?

Quiet Eruption

Violent Eruption

Figure 13 Lavas that are low in silica and the amount of dissolved gases erupt quietly. Explosive eruptions result from lava and ash that are high in silica and dissolved gases.

Dissolved Gases The presence of **dissolved** gases in magma contributes to how explosive a volcano can be. This is similar to what happens when you shake a can of soda and then open it. The bubbles come from the carbon dioxide that is dissolved in the soda. The pressure inside the can decreases rapidly when you open it. Trapped bubbles increase in size rapidly and escape as the soda erupts from the can.

ACADEMIC VOCABULARY
dissolve
(verb) to cause to disperse or disappear

All magmas contain dissolved gases. These gases include water vapor and small amounts of carbon dioxide and sulfur dioxide. As magma moves toward the surface, the pressure from the weight of the rock above decreases. As pressure decreases, the ability of gases to stay dissolved in the magma also decreases. Eventually, gases can no longer remain dissolved in the magma and bubbles begin to form. As the magma continues to rise to the surface, the bubbles increase in size and the gas begins to escape. Because gases cannot easily escape from high-viscosity lavas, this combination often results in explosive eruptions. When gases escape above ground, the lava, ash, or volcanic glass that cools and crystallizes has holes. These holes, shown in **Figure 14**, are a common feature in the volcanic rock pumice.

Figure 14 The holes in this pumice were caused by gas bubbles that escaped during a volcanic eruption.

Inquiry MiniLab

20 minutes

Can you model the movement of magma?

Magma erupts because it is less dense than Earth's crust. Similarly, oil is less dense than water and can be used to model magma.

1. Read and complete a lab safety form.
2. Half-fill a **clear plastic cup** with **pebbles.**
3. Fill the cup with **water** to a level just above the top of the pebbles.
4. Fill a **syringe** with 5 mL of **olive oil.**
5. Insert the syringe between the pebbles and the side of the cup until it touches the bottom.
6. Inject the oil slowly, 1 mL at a time.
7. Observe and record your results in your Science Journal.
8. Repeat the procedures using **motor oil.**

Analyze and Conclude

1. **Observe** What happens to the oil when you inject it into the water?
2. **Compare** How did the movement of the two oils differ?
3. **Key Concept** Which oil behaves like magma that will become basalt? Which behaves like magma that will become rhyolite? Explain.

Effects of Volcanic Eruptions

On average, about 60 different volcanoes erupt each year. The effects of lava flows, ash fall, pyroclastic flows, and mudflows can affect all life on Earth. Volcanoes enrich rock and soil with valuable nutrients and help to regulate climate. Unfortunately, they also can be destructive and sometimes even deadly.

Lava Flows Because lava flows are relatively slow moving, they are rarely deadly. But lava flows can be damaging. Mount Etna in Sicily, Italy, is Europe's most active volcano. **Figure 15** shows a fountain of fluid, hot lava erupting from one of the volcano's many vents. In May 2008, the volcano began spewing lava and ash in an eruption lasting over six months. Although lavas tend to be slow moving, they threaten communities nearby. People who live on Mount Etna's slopes are used to evacuations due to frequent eruptions.

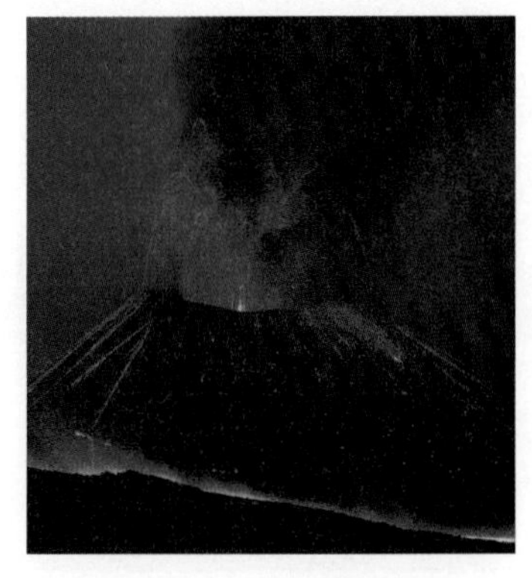

▲ **Figure 15** Mount Etna is one of the world's most active volcanoes. People that live near the volcano are accustomed to frequent eruptions of both lava and ash.

Ash Fall During an explosive eruption, volcanoes can erupt large volumes of volcanic ash. Ash columns can reach heights of more than 40 km. Recall that ash is a mixture of particles of pulverized rock and glass. Ash can disrupt air traffic and cause engines to stop mid-flight as shards of rock and ash fuse onto hot engine blades. Ash can also affect air quality and can cause serious breathing problems. Large quantities of ash erupted into the atmosphere can also affect climate by blocking out sunlight and cooling Earth's atmosphere.

Mudflows The thermal energy a volcano produces during an eruption can melt snow and ice on the summit. This meltwater can then mix with mud and ash on the mountain to form mudflows. Mudflows are also called lahars. Mount Redoubt in Alaska erupted on March 23, 2009. Snow and meltwater mixed to form the mudflows shown in **Figure 16**.

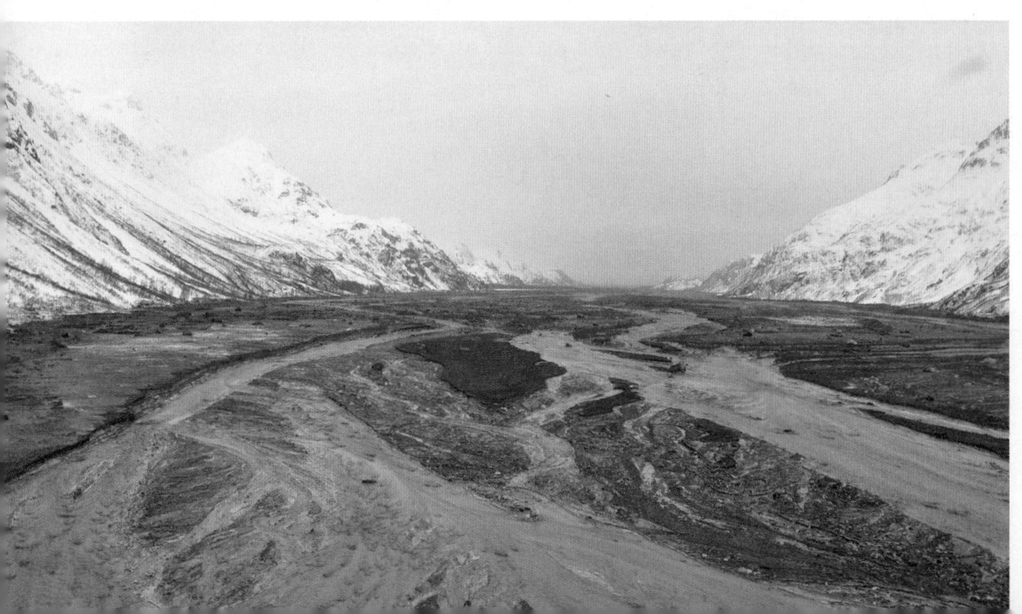

◀ **Figure 16** Many of the steep-sided composite volcanoes are covered with seasonal snow. When a volcano becomes active, the snow can melt and mix with mud and ash to form a mudflow like the one shown here in the Cook Inlet, Alaska.

▲ **Figure 17** A pyroclastic flow travels down the side of Mount Mayon in the Philippines. Pyroclastic flows are made of hot (*pyro*) volcanic particles (*clast*).

Pyroclastic Flow Explosive volcanoes can produce fast-moving avalanches of hot gas, ash, and rock called pyroclastic (pi roh KLAS tihk) flows. Pyroclastic flows travel at speeds of more than 100 km/hr and with temperatures greater than 1000°C. In 1980, Mount St. Helens produced a pyroclastic flow that killed 58 people and destroyed 1 billion km^3 of forest. Mount Mayon in the Phillippines erupts frequently producing pyroclastic flows like the one shown in **Figure 17.**

Predicting Volcanic Eruptions

Unlike earthquakes, volcanic eruptions can be predicted. Moving magma can cause ground deformation, a change in shape of the volcano, and a series of earthquakes called an earthquake swarm. Volcanic gas emissions can increase. Ground and surface water near the volcano can become more acidic. Geologists study these events, in addition to satellite and aerial photographs, to assess volcanic hazards.

Volcanic Eruptions and Climate Change

Volcanic eruptions affect climate when volcanic ash in the atmosphere blocks sunlight. High-altitude wind can move ash around the world. In addition, sulfur dioxide gases released from a volcano form sulfuric acid droplets in the upper atmosphere. These droplets reflect sunlight into space, resulting in lower temperatures as less sunlight reaches Earth's surface. **Figure 18** shows the result of sulfur dioxide gas in the atmosphere from the 1991 eruption of Mt. Pinatubo.

 **Key Concept Check** How do volcanoes affect climate?

Figure 18 In 1991, Mount Pinatubo erupted more than 20 million tons of gas and volcanic ash into the atmosphere. The greatest concentration of sulfur dioxide gas from the eruption is shown below in blue. The eruption caused temperatures to decrease by almost one degree Celsius in one year. ▼

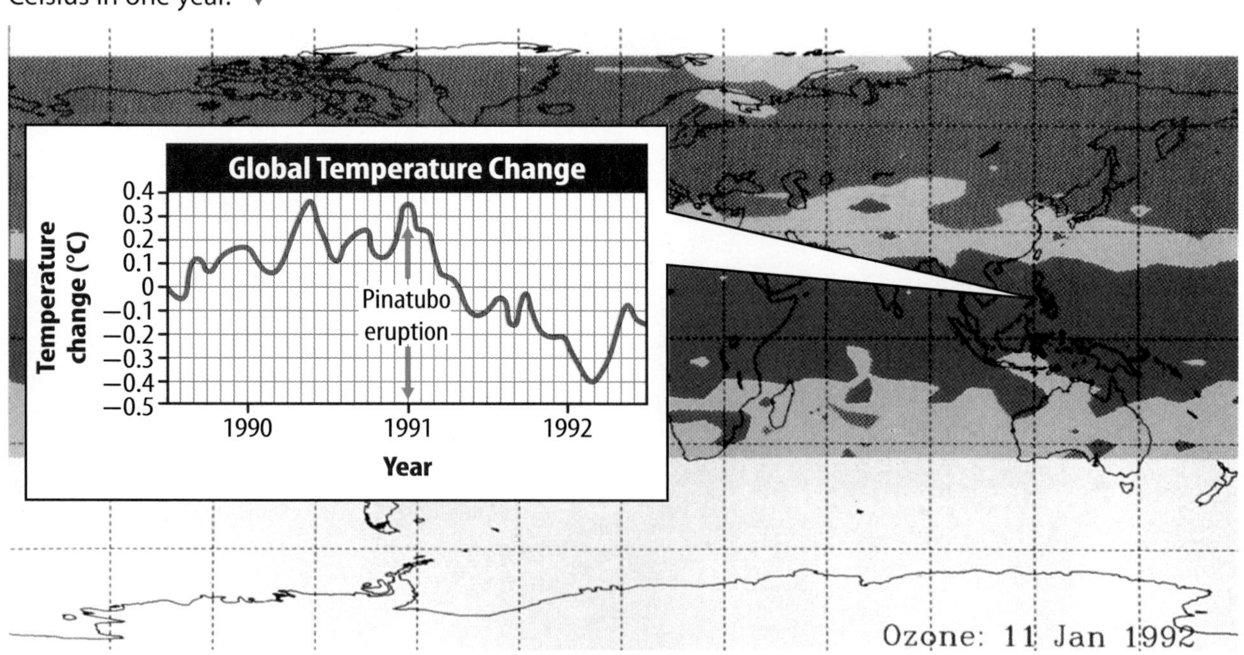

Lesson 2 Review

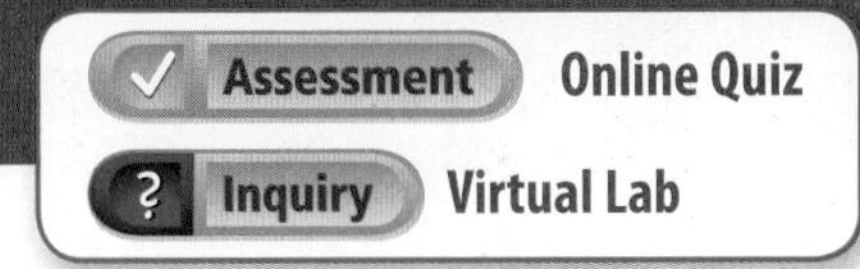

Visual Summary

Volcanoes form when magma rises through cracks in the crust and erupts from vents on Earth's surface.

Magma with low amounts of silica and low viscosity erupts to form shield volcanoes.

Magma with high amounts of silica and high viscosity erupts explosively to form composite cones.

FOLDABLES®

Use your lesson Foldable to review the lesson. Save your Foldable for the project at the end of the chapter.

What do you think NOW?

You first read the statements below at the beginning of the chapter.

4. Volcanoes can erupt anywhere on Earth.

5. Volcanic eruptions are rare.

6. Volcanic eruptions only affect people and places close to the volcano.

Did you change your mind about whether you agree or disagree with the statements? Rewrite any false statements to make them true.

Use Vocabulary

1. **Compare and contrast** lava and magma.
2. **Explain** the term *viscosity.*
3. Pulverized rock and ash that erupts from explosive volcanoes is called ________.

Understand Key Concepts 

4. **Identify** places where volcanoes form.
5. **Compare** the three main types of volcanoes.
6. What type of lava erupts from shield volcanoes?

   A. andesitic
   B. basaltic
   C. granitic
   D. rhyolitic

Interpret Graphics

7. **Analyze** the image below and explain what factors contribute to explosive eruptions.

8. **Create** a graphic organizer to illustrate the four types of eruptive products that can result from a volcanic eruption.

Critical Thinking

9. **Compare** the shapes of composite volcanoes and shield volcanoes. Why are their shapes and eruptive styles so different?
10. **Explain** how explosive volcanic eruptions can cause climate change. What might happen if Yellowstone Caldera erupted today?

Inquiry Lab 45 minutes

The Dangers of Mount Rainier

Materials

colored pencils

metric ruler

drawing compass

topographic map of Mount Rainier

Safety

If you have ever visited the area near Seattle or Tacoma, Washington, it is difficult to miss majestic snow-capped Mount Rainier on the horizon. Mount Rainier, at nearly 4.4 km, is the highest active volcano in the Cascade Mountains of western Washington. More than 3.6 million people live within 100 km to the north and west of Mount Rainier.

Mount Rainicr last erupted in 1895, but historical records show that it erupts with a frequency between 100 to 500 years. Mount Rainier's explosive past is evident from the pyroclastic flows, mudflows, and ash deposits that surround the volcano. Geologists predict that Mount Rainier will erupt in the future, but when? In this lab, you will assess the volcanic dangers of Mount Rainier.

Ask a Question

Imagine that you decide to open a mountain bike shop in either Sunrise, Longmire, or Ashford, Washington. Before you make your final decision about location, you must examine volcanic dangers of Mount Rainier. Which town is the safest choice?

Make Observations

1. Obtain a topographic map of Mount Rainier from your teacher.
2. Use the topographic map and the table below to indicate where mudflows might occur. Locate this area on your map and color it yellow.
3. Use the information in the table provided, the map scale, and a compass to identify the area that might be affected by
   - lava and pyroclastic flows; color this area orange on your map;
   - ash fall; color this area blue on your map.

   Be sure to include a legend on your map.

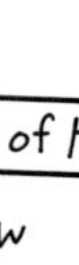

Volcanic Hazards

Type of Hazard	Range	Notes
Mudflow	Up to 64 km	Contained within river valleys
Lava and pyroclastic flow	Within 16 km of the summit	Will likely remain within boundary of Mount Rainier National Park
Falling ash	96 km downwind	Wind generally blows to the east of Mount Rainier

Form a Hypothesis

4. Use your observations to form a hypothesis about which town—Sunrise, Longmire, or Ashford—would be the safest choice for opening a bike shop. Base your hypothesis on your assessment of the volcanic hazards associated with Mount Rainier.

Test your Hypothesis

5. Compare your map to a classmate's map. If your assessments differ, explain how you developed your hypothesis.

Analyze and Conclude

6. **Calculate** If a mudflow from Mount Rainier traveled down the Nisqually River Valley, how much time would the towns of Longmire and Ashford have to prepare? *(Hint: Mudflows can move at speeds of 80 km/hr.)*
7. **Predict** Based on the extent of the volcanic hazards you mapped, would it be possible for a mudflow to reach Tacoma, Washington? Support your answer.
8. **The Big Idea** Mount Rainier is in the Cascade subduction zone. Why are hazards such as earthquakes and volcanic eruptions common along a subduction zone?

Communicate Your Results

As the owner of a bike shop, you want your clients to have a great visit to Mount Rainier. However, you want them to understand the risks associated with recreation on a volcano. Create a pamphlet that describes Mount Rainier's volcanic hazards. Include a map of the hazards. You might want to include names and contact information for local emergency response agencies.

Imagine you are riding your mountain bike on trails high above the Nisqually River Valley when a mudflow floods the valley below. Write a story describing your experience. Describe what you saw, what you heard, and what you felt. Explain how the mudflow might change the way people think about the volcanic dangers of Mount Rainier.

Lab Tips

- ☑ Use the distance scale on the map to determine the extent of volcanic hazards.
- ☑ Mudflows that originate on Mount Rainier follow topography and flow down river valleys.

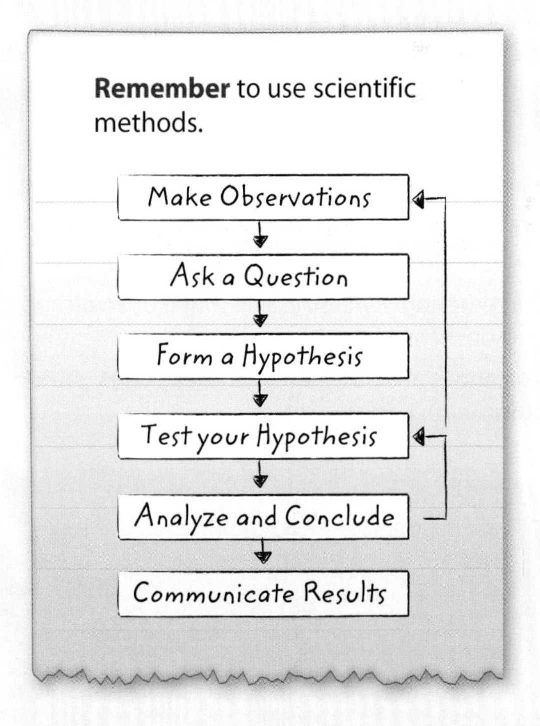

Chapter 15 Study Guide

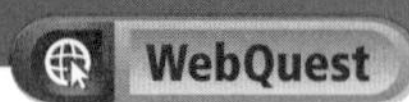

Most earthquakes occur along plate boundaries where plates slide past each other, collide, or separate. Volcanoes form at subduction zones, mid-ocean ridges, and hot spots.

Key Concepts Summary

Lesson 1: Earthquakes

- Earthquakes commonly occur on or near tectonic plate boundaries.
- Earthquakes are used to study the composition and structure of Earth's interior and to identify the location of active faults.
- Earthquakes are monitored using **seismometers** and described using the Richter magnitude scale, the moment magnitude scale, and the Modified Mercalli scale.

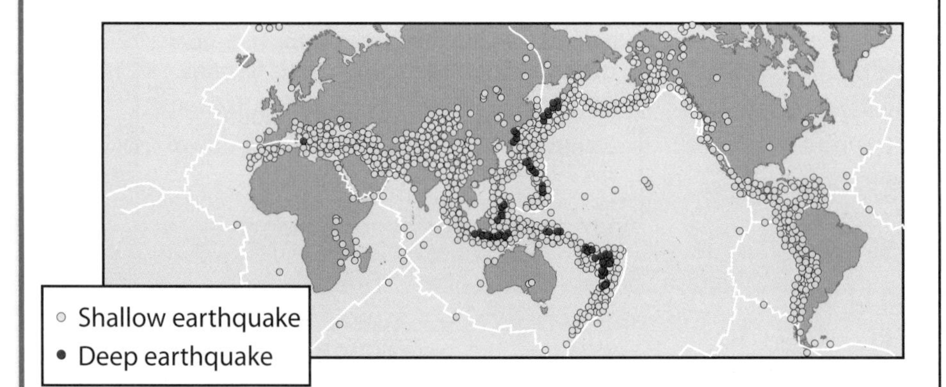

○ Shallow earthquake
● Deep earthquake

Lesson 2: Volcanoes

- Molten **magma** is forced upward through cracks in the crust, erupting from volcanoes.
- The eruption style, size, and shape of a volcano depends on the composition of the magma, including the amount of dissolved gas.
- Volcanoes are classified as **cinder cones, shield volcanoes,** and **composite cones.**

Vocabulary

Lesson 1

earthquake p. 531
fault p. 533
seismic wave p. 534
focus p. 534
epicenter p. 534
primary wave p. 535
secondary wave p. 535
surface wave p. 535
seismologist p. 536
seismometer p. 537
seismogram p. 537

Lesson 2

volcano p. 545
magma p. 545
lava p. 546
hot spot p. 546
shield volcano p. 548
composite volcano p. 548
cinder cone p. 548
volcanic ash p. 549
viscosity p. 549

Study Guide

Review
- Personal Tutor
- Vocabulary eGames
- Vocabulary eFlashcards

FOLDABLES® Chapter Project

Assemble your lesson Foldables as shown to make a Chapter Project. Use the project to review what you have learned in this chapter.

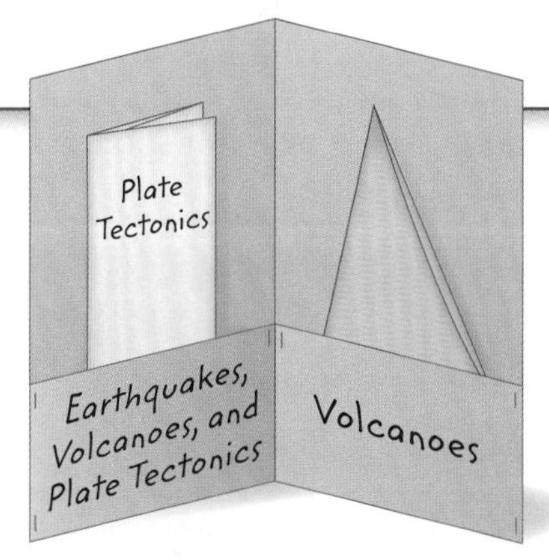

Use Vocabulary

1. A volcano with gently sloping sides is a(n) ________.
2. Write a sentence using the terms *seismic waves, P-waves,* and *S-waves.*
3. Magma that erupts quietly is ________. Magma most likely to erupt explosively is ________.
4. Volcanic activity that does not occur near a plate boundary happens at a(n) ________.
5. Molten rock inside Earth is called ________.
6. ________ are used to record ground motion during an earthquake.
7. The ________ marks the exact location where an earthquake occurs. The ________ is the place on Earth's surface directly above it.
8. A type of seismic wave that has movement similar to an ocean wave is a(n) ________.
9. A mixture of pulverized ash, rock, and gas ejected during explosive eruptions is called a(n) ________.

Link Vocabulary and Key Concepts

Concepts in Motion Interactive Concept Map

Copy this concept map, and then use vocabulary terms from the previous page to complete the concept map.

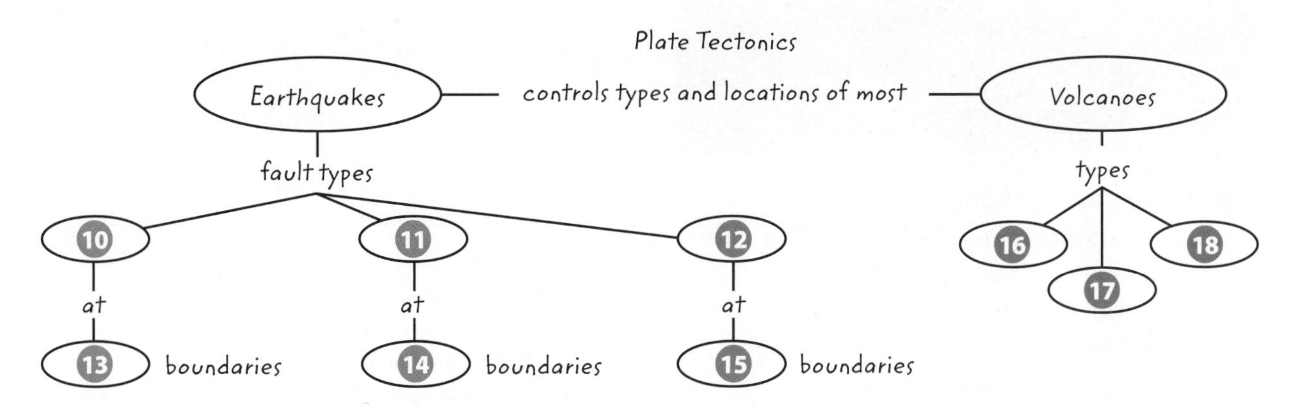

Chapter 15 Review

Understand Key Concepts 

1. Most of the volcanic activity on Earth occurs
   A. along mid-ocean ridges.
   B. along transform plate boundaries.
   C. at hot spots.
   D. within the crust.

2. At a divergent plate boundary such as a mid-ocean ridge, you should expect to find
   A. low viscosity lava and normal faults.
   B. low viscosity lava and reverse faults.
   C. high viscosity lava and normal faults.
   D. high viscosity lava and reverse faults.

3. High energy earthquakes occur
   A. away from plate boundaries.
   B. away from divergent plate boundaries.
   C. on convergent plate boundaries.
   D. on transform plate boundaries.

4. Large and explosive volcanic eruptions, such as the one shown below, can change climate because
   A. ash and gas that erupt high into the atmosphere can reflect sunlight.
   B. the magma that erupts is hot.
   C. volcanic ash keeps Earth from losing its heat.
   D. volcanic mountains block solar radiation.

5. What is an earthquake?
   A. a fault at a convergent plate boundary
   B. a wave of water in the crust
   C. energy released as rocks break and move along a fault
   D. the elastic strain stored in rocks

6. Approximately how much more ground motion is recorded on a seismogram from a magnitude 6 earthquake compared to a magnitude 4 earthquake?
   A. 10 times more
   B. 50 times more
   C. 100 times more
   D. 1,000 times more

7. The figure below shows the Hawaiian Islands, formed by a hot spot. Which island is the oldest?
   A. Hawaii
   B. Kauai
   C. Maui
   D. Oahu

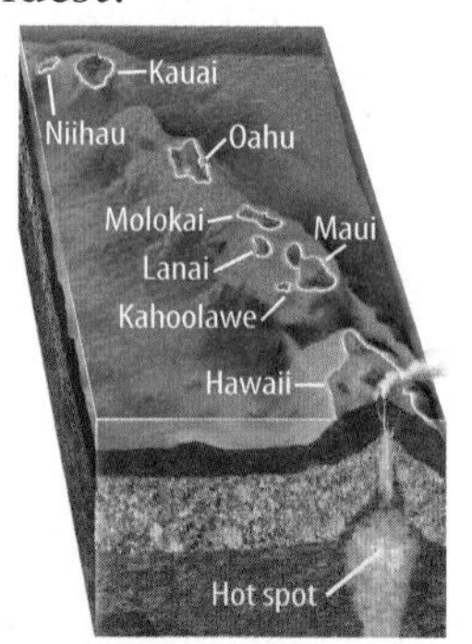

8. A lag-time graph illustrates the relationship between the time it takes a seismic wave to travel from the earthquake epicenter to a seismometer and the
   A. distance between the earthquake and the seismometer.
   B. earthquake intensity.
   C. earthquake magnitude.
   D. size of the fault.

9. Which can show the amount of energy released by an earthquake?
   A. a lag-time graph
   B. the Modified Mercalli scale
   C. the moment magnitude scale
   D. the Richter magnitude scale

10. The location of an earthquake can be determined from seismic data recorded by at least
   A. one seismometer.
   B. two seismometers.
   C. three seismometers.
   D. five seismometers.

Assessment
Online Test Practice

Critical Thinking

11 **Explain** why Alaska has such a high risk associated with earthquakes.

12 **Analyze** the various types of volcanoes shown in **Table 4.** Which type of volcano is most likely to form at a hot spot in the ocean? Explain your answer.

13 **Evaluate** the following statement: "Yellowstone is a caldera that has erupted more than 1,000 km^3 of magma three times over the past 2.2 million years." Suggest how you might test the hypothesis that there is hot molten material beneath Yellowstone today.

14 **Hypothesize** Use the map below to identify evidence to suggest that Africa is splitting into two continents.

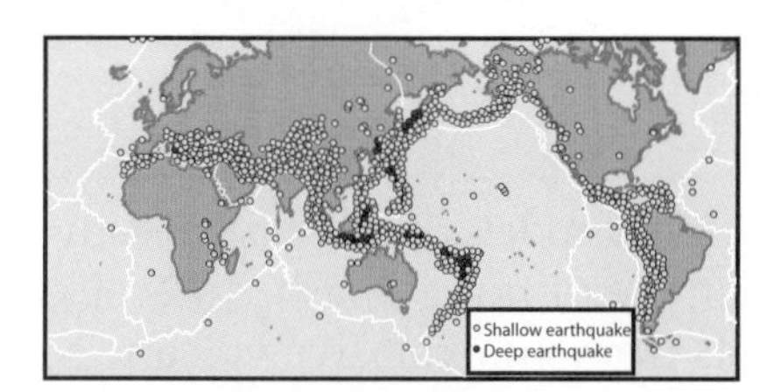

15 **Describe** how seismologists discovered that most of the mantle is solid.

16 **Identify** several reasons why a magnitude 6 earthquake in New Orleans might be more damaging than a magnitude 7 earthquake in San Francisco.

17 **Explain** why pyroclastic flows are responsible for more deaths than lava flows.

18 **Describe** Look at a map of the Hawaiian Island–Emperor Seamount chain formed by an active hot spot. Describe the relationship between these two chains. What do you think changed to form two chains instead of one?

Writing in Science

19 **Hypothesize** how scientists might be able to determine the composition of the Moon's interior given what you know about Earth's interior.

REVIEW THE BIG IDEA

20 How does the theory of plate tectonics explain the location of most earthquakes and volcanoes?

21 The photo below shows a pyroclastic flow from Mount Pinatubo in the Philippines. Why was this eruption so explosive?

Math Skills

Review

Math Practice

22 **Identify** What is the value of Roman numeral XXXIX?

23 **Evaluate** How would you write number 38 in Roman numerals?

24 **Evaluate** In Roman numerals, L = 50. What is the value of the Roman numeral XL?

25 **Determine** How would you write the number 83 in Roman numerals?

Standardized Test Practice

Record your answers on the answer sheet provided by your teacher or on a sheet of paper.

Multiple Choice

1 Along which type of plate boundary do the deepest earthquakes occur?

A convergent

B divergent

C passive

D transform

2 The Richter scale registers the magnitude of an earthquake by determining the

A amount of energy released by the earthquake.

B amount of ground motion measured at a given distance from the earthquake.

C descriptions of damage caused by the earthquake.

D type of seismic waves produced by the earthquake.

3 Which state has no active volcanoes?

A California

B Hawaii

C New York

D Washington

Use the diagram below to answer question 4.

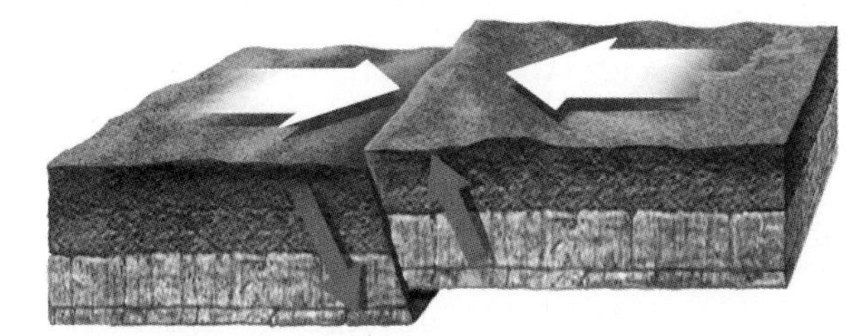

4 Which type of fault is shown in the diagram above?

A normal

B reverse

C shallow

D strike-slip

Use the diagram below to answer question 5.

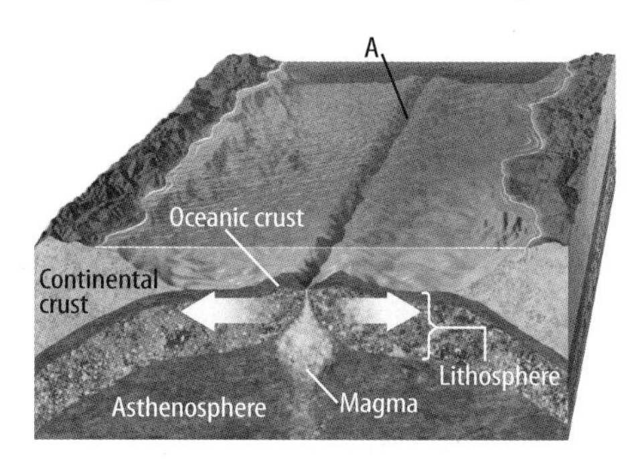

5 Which feature is labeled with the letter *A* in the diagram above?

A a caldera

B a chain of hot spot volcanoes

C a mid-ocean ridge

D a subducting tectonic plate

6 Which term describes a fast-moving avalanche of hot gas, ash, and rock that erupts from an explosive volcano?

A ash fall

B cinder cone

C lahar

D pyroclastic flow

7 Earthquakes occur along the San Andreas Fault. Which is an example of this type of plate boundary?

A convergent

B divergent

C passive

D transform

8 Hot spot volcanoes ALWAYS

A appear at plate boundaries.

B erupt in chains.

C form above mantle plumes.

D remain active.

Assessment
Online Standardized Test Practice

Use the map below to answer questions 9 and 10.

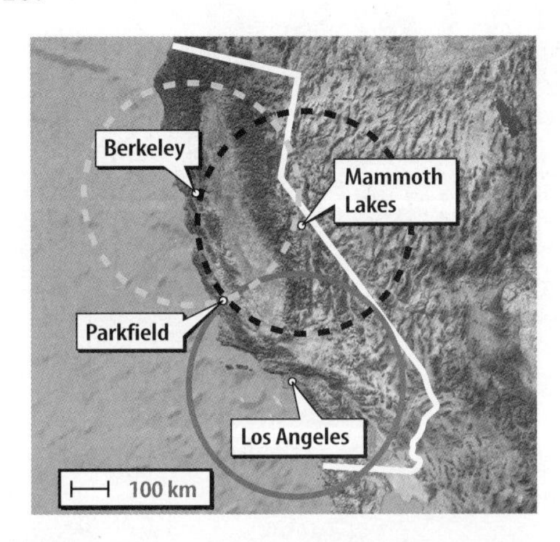

9 What do the circles represent in the map of seismic activity illustrated above?

A the distance between waves

B the distance to an earthquake epicenter

C the seismic wave speeds

D the wave travel times

10 According to the map, where is the earthquake epicenter?

A Berkeley

B Los Angeles

C Mammoth Lakes

D Parkfield

11 Where do seismic waves originate?

A above ground

B epicenter

C focus

D seismogram

Constructed Response

Use the diagram below to answer questions 12 and 13.

12 The diagram above shows one way volcanoes form. Explain the process shown in the diagram and why volcanoes form as a result of this process.

13 What type of volcano results from the process shown in the diagram? Describe it. What is the eruptive style of this type of volcano? Why?

Use the table below to answer question 14.

Wave Type	Characteristics

14 Re-create the table above and identify the three types of seismic waves. Then, describe wave characteristics such as movement, speed, and difference in arrival time for each type.

NEED EXTRA HELP?														
If You Missed Question...	1	2	3	4	5	6	7	8	9	10	11	12	13	14
Go to Lesson...	1	1	2	1	2	2	1	2	1	1	1	2	2	1

Chapter 16

Clues to Earth's Past

THE BIG IDEA

What evidence do scientists use to determine the ages of rocks?

Inquiry Always a Canyon?

The Colorado River started cutting through the rock layers of the Grand Canyon only about 6 million years ago. Hundreds of millions of years earlier, these rock layers were deposited at the bottom of an ancient sea. Even before that, a huge mountain range existed here.

- What evidence do scientists use to learn about past environments?
- What evidence do scientists use to determine the ages of rocks?

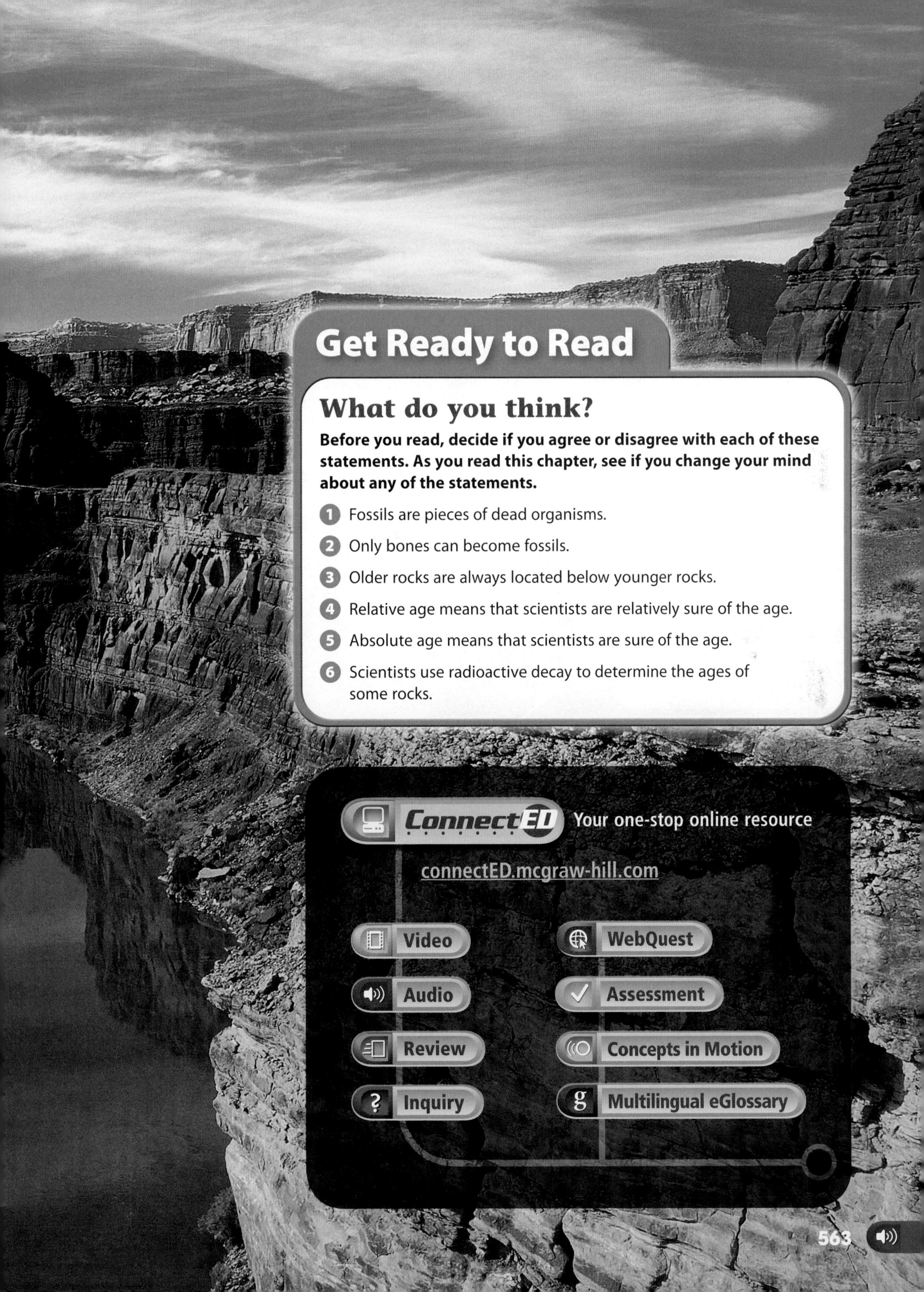

Get Ready to Read

What do you think?

Before you read, decide if you agree or disagree with each of these statements. As you read this chapter, see if you change your mind about any of the statements.

1. Fossils are pieces of dead organisms.
2. Only bones can become fossils.
3. Older rocks are always located below younger rocks.
4. Relative age means that scientists are relatively sure of the age.
5. Absolute age means that scientists are sure of the age.
6. Scientists use radioactive decay to determine the ages of some rocks.

ConnectED Your one-stop online resource

connectED.mcgraw-hill.com

- Video
- Audio
- Review
- Inquiry
- WebQuest
- Assessment
- Concepts in Motion
- Multilingual eGlossary

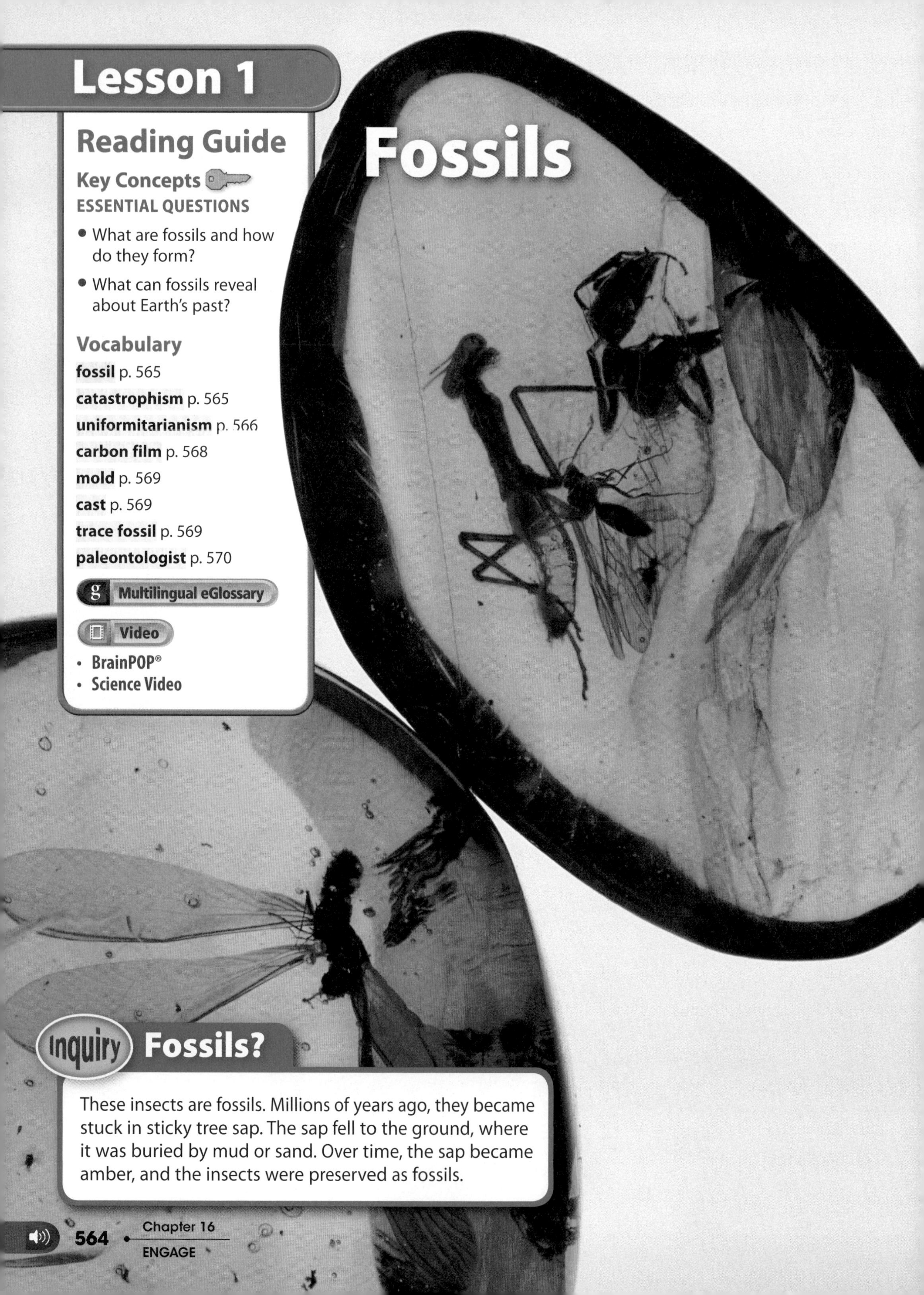

Lesson 1

Fossils

Reading Guide

Key Concepts
ESSENTIAL QUESTIONS

- What are fossils and how do they form?
- What can fossils reveal about Earth's past?

Vocabulary

fossil p. 565
catastrophism p. 565
uniformitarianism p. 566
carbon film p. 568
mold p. 569
cast p. 569
trace fossil p. 569
paleontologist p. 570

g Multilingual eGlossary

Video
- BrainPOP®
- Science Video

Inquiry Fossils?

These insects are fossils. Millions of years ago, they became stuck in sticky tree sap. The sap fell to the ground, where it was buried by mud or sand. Over time, the sap became amber, and the insects were preserved as fossils.

20 minutes

What can trace fossils show?

Did you know that a fossil can be a footprint or the imprint of an ancient nest? These are examples of trace fossils. Although trace fossils do not contain any part of an organism, they do hold clues about how organisms lived, moved, or behaved.

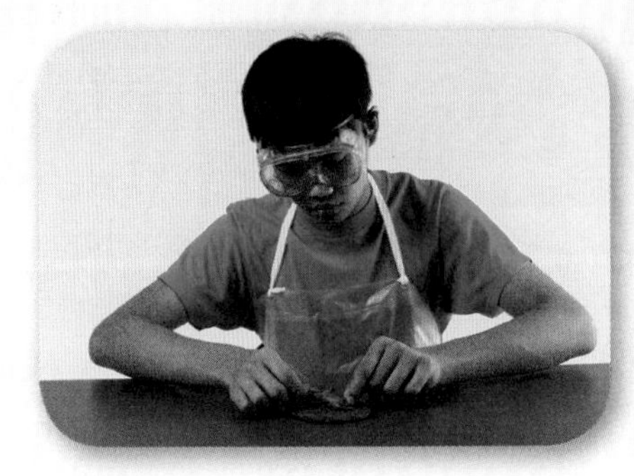

1. Read and complete a lab safety form.
2. Flatten some **clay** into a pancake shape.
3. Think about a behavior or movement you would like your fossil to model. Use available tools, such as a **plastic knife,** a **chenille stem,** or a **toothpick,** to make a fossil showing that behavior or movement.
4. Exchange your fossil with another student. Try to figure out what behavior or movement he or she modeled.

Think About This

1. Were you able to determine what behavior or movement your classmate's fossil modeled? Was he or she able to determine yours? Why or why not?
2. **Key Concept** What do you think scientists can learn by studying trace fossils?

Evidence of the Distant Past

Have you ever looked through an old family photo album? Each photo shows a little of your family's history. You might guess the age of the photographs based on the clothes people are wearing, the vehicles they are driving, or even the paper the photographs are printed on.

Just as old photos can provide clues to your family's past, rocks can provide clues to Earth's past. Some of the most obvious clues found in rocks are the remains or traces of ancient living things. **Fossils** *are the preserved remains or evidence of ancient living things.*

WORD ORIGIN
fossil
from Latin *fossilis,* means "dug up"

Catastrophism

Many fossils represent plants and animals that no longer live on Earth. Ideas about how these fossils formed have changed over time. Some early scientists thought that great, sudden, catastrophic disasters killed the organisms that became fossils. These scientists explained Earth's history as a series of disastrous events occurring over short periods of time. **Catastrophism** (kuh TAS truh fih zum) *is the idea that conditions and organisms on Earth change in quick, violent events.* The events described in catastrophism include volcanic eruptions and widespread flooding. Scientists eventually disagreed with catastrophism because Earth's history is full of violent events.

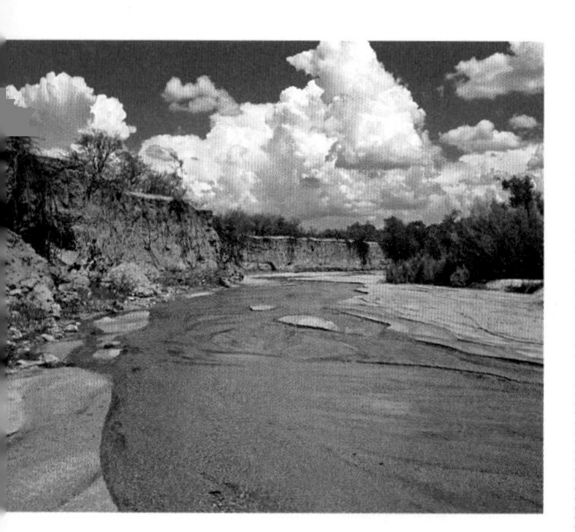

Figure 1 Hutton realized that erosion happens on small or large scales.

Uniformitarianism

Most people who supported catastrophism thought that Earth was only a few thousand years old. In the 1700s, James Hutton rejected this idea. Hutton was a naturalist and a farmer in Scotland. He observed how the landscape on his farm gradually changed over the years. Hutton thought that the processes responsible for changing the landscape on his farm could also shape Earth's surface. For example, he thought that erosion caused by streams, such as that shown in **Figure 1,** could also wear down mountains. Because he realized that this would take a long time, Hutton proposed that Earth is much older than a few thousand years.

Hutton's ideas eventually were included in a principle called uniformitarianism (yew nuh for muh TER ee uh nih zum). *The principle of* **uniformitarianism** *states that geologic processes that occur today are similar to those that have occurred in the past.* According to this view, Earth's surface is constantly being reshaped in a steady, **uniform** manner.

Academic Vocabulary

uniform

(adjective) having always the same form, manner, or degree; not varying or variable

Reading Check What is uniformitarianism?

Today, uniformitarianism is the basis for understanding Earth's past. But scientists also know that catastrophic events do sometimes occur. Huge volcanic eruptions and giant meteorite impacts can change Earth's surface very quickly. These catastrophic events can be explained by natural processes.

Inquiry MiniLab

15 minutes

How is a fossil a clue?

Fossils provide clues about once-living organisms. Sometimes those clues are hard to interpret.

1. Read and complete a lab safety form.
2. Select an **object** from a bag provided by your teacher. Do not let anyone see your object.
3. Make a fossil impression of your object by pressing only part of it into a piece of **clay.**
4. Place your clay fossil and object in separate locations indicated by your teacher.
5. Make a chart in your Science Journal that matches your classmates' objects and fossils.

Analyze and Conclude

1. Did you correctly match the objects with their fossils?
2. Why might scientists need more than one fossil of an organism to understand what it looked like?
3. **Key Concept** What do you think you could learn from fossils?

Fossil Formation

Figure 2 A fossil can form if an organism with hard parts, such as a fish, is buried quickly after it dies.

1. A dead fish falls to a river bottom during a flood. Its body is rapidly buried by mud, sand, or other sediment.

2. Over time, the body decomposes but the hard bones become a fossil.

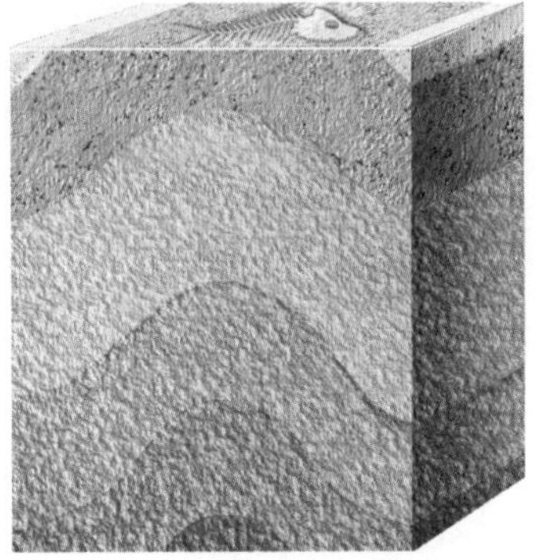

3. The sediments, hardened into rock, are uplifted and eroded, which exposes the fossil fish on the surface.

Visual Check How did the fish fossil reach the surface?

Formation of Fossils

Recall that fossils are the remains or traces of ancient living organisms. Not all dead organisms become fossils. Fossils form only under certain conditions.

Conditions for Fossil Formation

Most plants and animals are eaten or decay when they die, leaving no evidence that they ever lived. Think about the chances of an apple becoming a fossil. If it is on the ground for many months, it will decay into a soft, rotting lump. Eventually, insects and bacteria consume it.

However, some conditions increase the chances of fossil formation. An organism is more likely to become a fossil if it has hard parts, such as shells, teeth, or bones, like the fish in **Figure 2.** Unlike a soft apple, hard parts do not decay easily. Also, an organism is more likely to form a fossil if it is buried quickly after it dies. If layers of sand or mud bury an organism quickly, decay is slowed or stopped.

Key Concept Check What conditions increase the chances of fossil formation?

Fossils Come in All Sizes

You might have seen pictures of dinosaur fossils. Many dinosaurs were large animals, and large bones were left behind when they died. Not all fossils are large enough for you to see. Sometimes it is necessary to use a microscope to see fossils. Tiny fossils are called microfossils. The microfossils in **Figure 3** are each about the size of a speck of dust.

Figure 3 Details of microfossils can be seen only under a microscope.

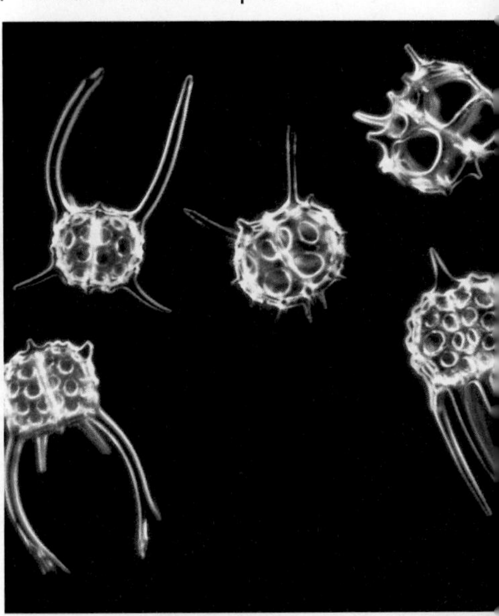

Types of Preservation

Fossils are preserved in different ways. As shown in **Figure 4,** there are many ways fossils can form.

Preserved Remains

Sometimes the actual remains of organisms are preserved as fossils. For this to happen, an organism must be completely enclosed in some material over a long period of time. This would prevent it from being exposed to air or bacteria. Generally, preserved remains are 10,000 or fewer years in age. However, insects preserved in amber—shown in the photo at the beginning of this lesson—can be millions of years old.

Carbon Films

Sometimes when an organism is buried, exposure to heat and pressure forces gases and liquids out of the organism's tissues. This leaves only the carbon behind. *A* **carbon film** *is the fossilized carbon outline of an organism or part of an organism.*

Mineral Replacement

Replicas, or copies, of organisms can form from minerals in groundwater. They fill in the pore spaces or replace the tissues of dead organisms. **Petrified** wood is an example.

SCIENCE USE V. COMMON USE

petrified

Science Use turned into stone by the replacement of tissue with minerals

Common Use made rigid with fear

Figure 4 Fossils can form in many different ways.

Types of Preservation

Preserved Remains Organisms trapped in amber, tar pits, or ice can be preserved over thousands of years. This baby mammoth was preserved in ice for more than 10,000 years before it was discovered. ▶

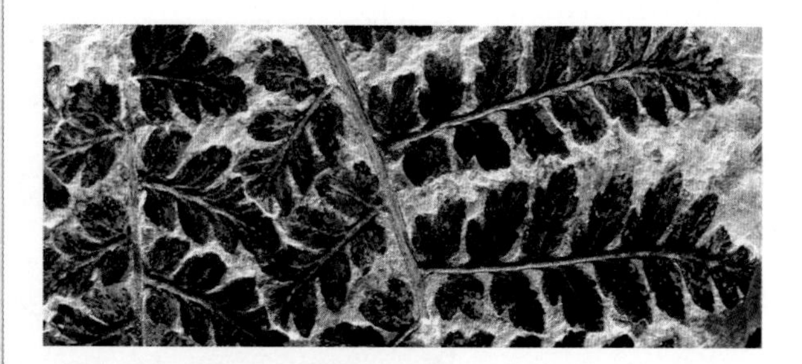

◀ **Carbon Film** Only a carbon film remains of this ancient fern. Carbon films are usually shiny black or brown. Fish, insects, and plant leaves are often preserved as carbon films.

Mineral Replacement Rock-forming minerals dissolved in groundwater can fill in pore spaces or replace the tissues of dead organisms. This petrified wood formed when silica (SiO_2) filled in the spaces between the cell walls in a dead tree. The wood petrified when the SiO_2 crystallized. ▶

Molds

Sometimes all that remains of an organism is its fossilized imprint or impression. *A* **mold** *is the impression in a rock left by an ancient organism.* A mold can form when sediment hardens around a buried organism. As the organism decays over time, an impression of its shape remains in the sediment. The sediment eventually turns to rock.

Casts

Sometimes, after a mold forms, it is filled with more sediment. *A* **cast** *is a fossil copy of an organism made when a mold of the organism is filled with sediment or mineral deposits.* The process is similar to making a gelatin dessert using a molded pan.

Trace Fossils

Some animals leave fossilized traces of their movement or activity. *A* **trace fossil** *is the preserved evidence of the activity of an organism.* Trace fossils include tracks, footprints, and nests. These fossils help scientists learn about characteristics and behaviors of animals. The dinosaur tracks in **Figure 4** reveal clues about the dinosaur's size, its speed, and whether it was traveling alone or in a group.

Reading Check What are some examples of trace fossils?

Make a tri-fold book from a sheet of paper. Label it as shown. Use it to organize your notes about the different types of fossils.

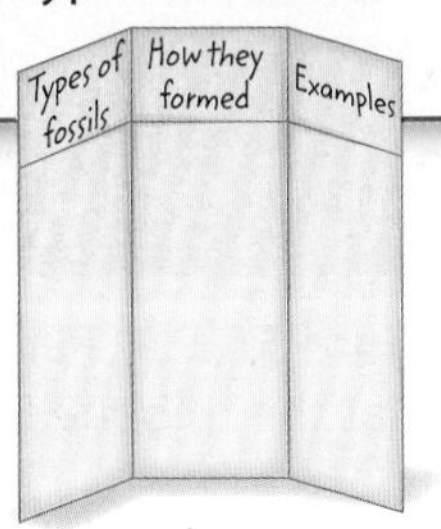

Mold This mold of an ancient mollusk formed after it was buried by sediment and then decayed. The sediment hardened, leaving an impression of its shape in the rock. ▼

▲ **Cast** This cast was formed when the mold was later filled with sediment that then hardened. Molds and casts show only the exterior, or outside, features of organisms.

Trace Fossil These trace fossils formed when dinosaur tracks in soft sediments were later filled in by other sediments, which then hardened. Trace fossils reveal information about the behavior of organisms. ▶

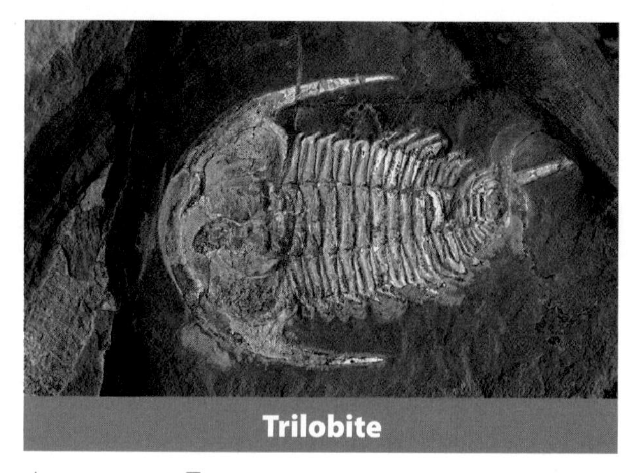

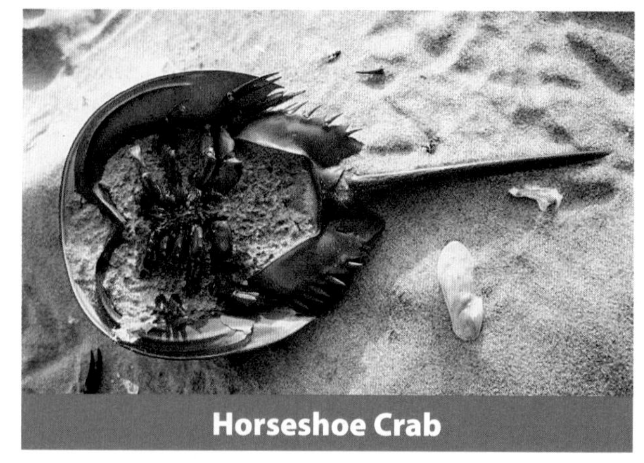

▲ **Figure 5** Partly because a trilobite fossil looks like a present-day horseshoe crab, scientists infer that the trilobite lived in an environment similar to the environment where a horseshoe crab lives.

Ancient Environments

Scientists who study fossils are called **paleontologists** (pay lee ahn TAH luh jihstz). Paleontologists use the principle of uniformitarianism to learn about ancient organisms and the environments where ancient organisms lived. For example, they can compare fossils of ancient organisms with organisms living today. The trilobite fossil and the horseshoe crab in **Figure 5** look alike. Horseshoe crabs today live in shallow water on the ocean floor. Partly because trilobite fossils look like horseshoe crabs, paleontologists infer that trilobites also lived in shallow ocean water.

Shallow Seas

Today, Earth's continents are mostly above sea level. But sea level has risen, flooding Earth's continents, many times in the past. For example, a shallow ocean covered much of North America 450 million years ago, as illustrated in the map in **Figure 6.** Fossils of organisms that lived in that shallow ocean, like those shown in **Figure 6,** help scientists reconstruct what the seafloor looked like at that time.

Key Concept Check What can fossils tell us about ancient environments?

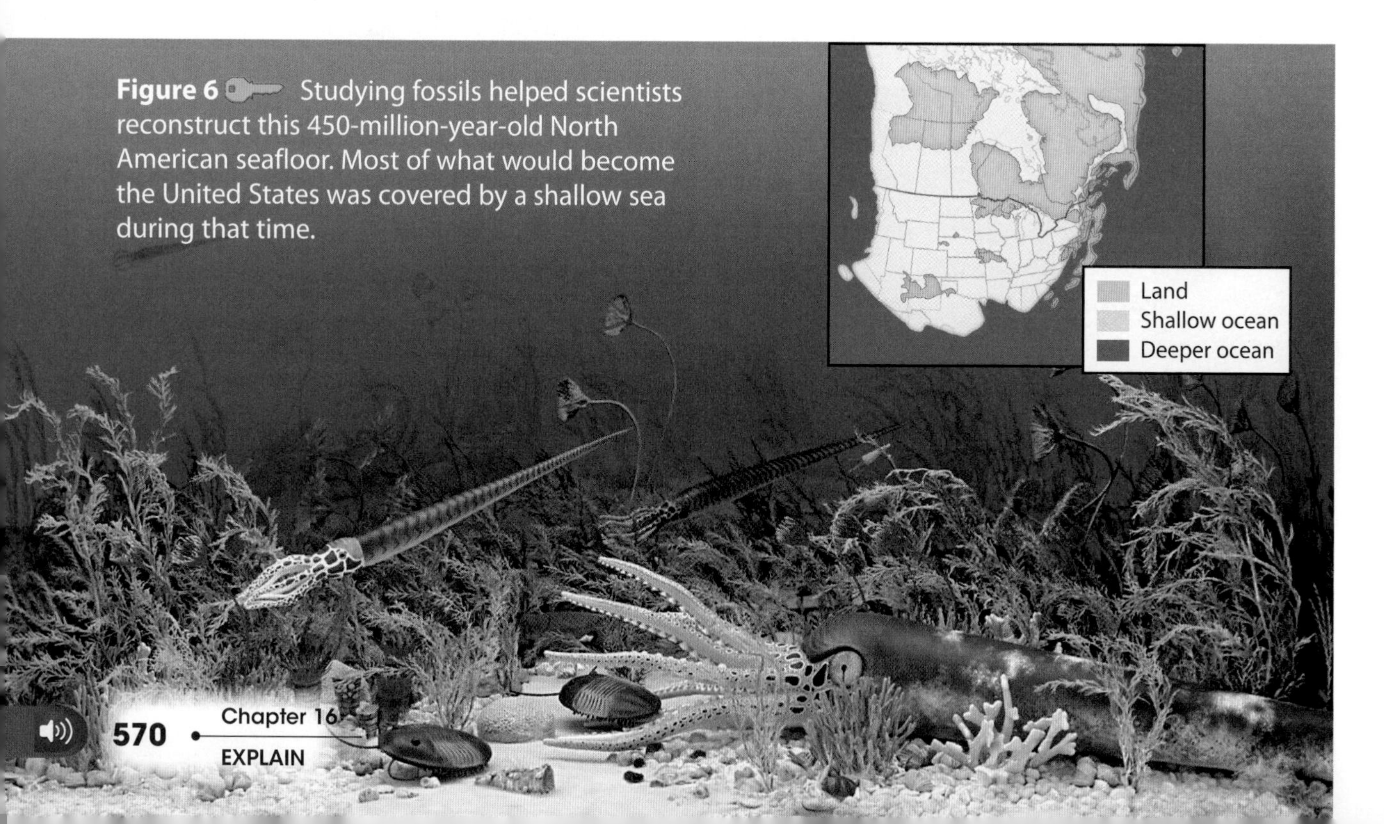

Figure 6 Studying fossils helped scientists reconstruct this 450-million-year-old North American seafloor. Most of what would become the United States was covered by a shallow sea during that time.

Figure 7 About 100 million years ago, tropical forests and swamps covered much of North America. Dinosaurs also lived on Earth at that time.

Past Climates

You might have heard people talking about global climate change, or maybe you've read about climate change. Evidence indicates that Earth's present-day climate is warming. Fossils show that Earth's climate has warmed and cooled many times in the past.

Plant fossils are especially good indicators of climate change. For example, fossils of ferns and other tropical plants dating to the time of the dinosaurs reveal that Earth was very warm 100 million years ago. Tropical forests and swamps covered much of the land, as illustrated in **Figure 7.**

Key Concept Check What was Earth's climate like when dinosaurs lived?

Millions of years later, the swamps and forests were gone, but coarse grasses grew in their place. Huge sheets of ice called glaciers spread over parts of North America, Europe, and Asia. Fossils suggest that some species that lived during this time, such as the woolly mammoth shown in **Figure 8,** were able to survive in the colder climate.

Fossils of organisms such as ferns and mammoths help scientists learn about ancient organisms and past environments. In the following lessons, you will read how scientists use fossils and other clues, such as the order of rock layers and radioactivity, to learn about the ages of Earth's rocks.

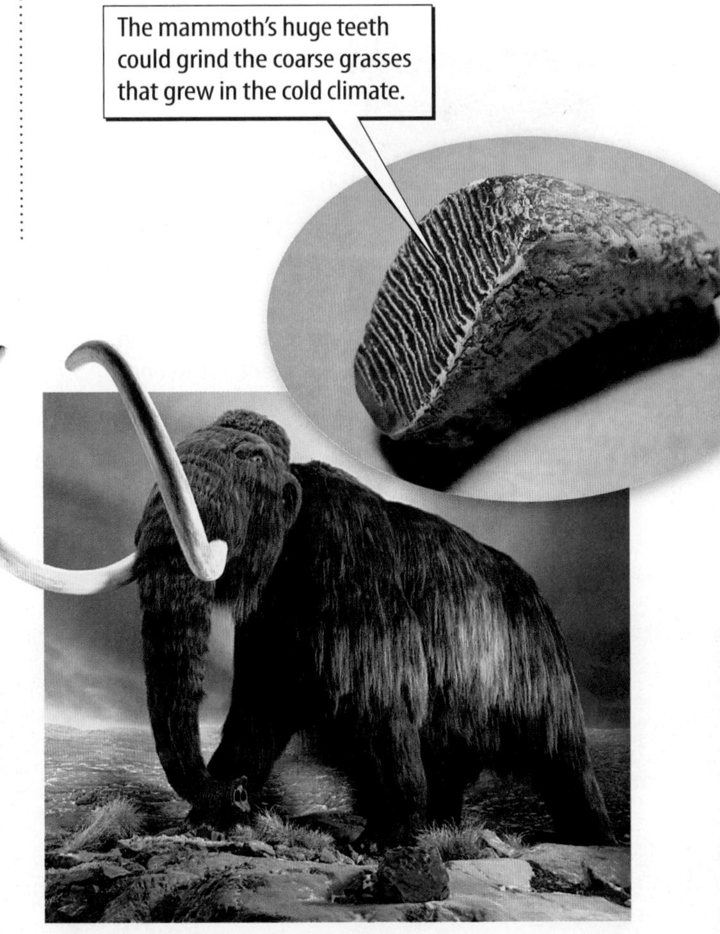

Figure 8 The woolly mammoth was well adapted to a cold climate.

Lesson 1 Review

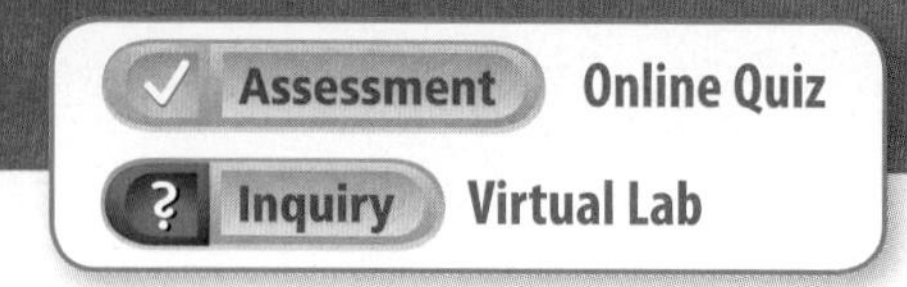

Visual Summary

The principle of uniformitarianism is the basis for understanding Earth's past.

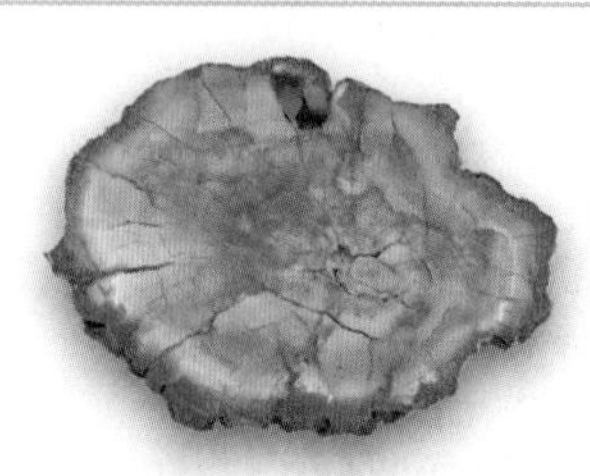

Fossils can form in many different ways.

Fossils help scientists learn about Earth's ancient organisms and past environments.

FOLDABLES®

Use your lesson Foldable to review the lesson. Save your Foldable for the project at the end of the chapter.

What do you think NOW?

You first read the statements below at the beginning of the chapter.

1. Fossils are pieces of dead organisms.

2. Only bones can become fossils.

Did you change your mind about whether you agree or disagree with the statements? Rewrite any false statements to make them true.

Use Vocabulary

1. **Distinguish** between catastrophism and uniformitarianism.
2. Plant leaves are often preserved as ______.
3. **Use the terms** *cast* and *mold* in a complete sentence.

Understand Key Concepts

4. Which conditions aid in the formation of fossils?
   - **A.** hard parts and slow burial
   - **B.** hard parts and rapid burial
   - **C.** soft parts and rapid burial
   - **D.** soft parts and slow burial
5. What human body system could be fossilized? Explain.
6. **Determine** what type of an environment a fossil palm tree would indicate.

Interpret Graphics

7. **Compare** the two sets of dinosaur footprints below. Which dinosaur was running? How can you tell?

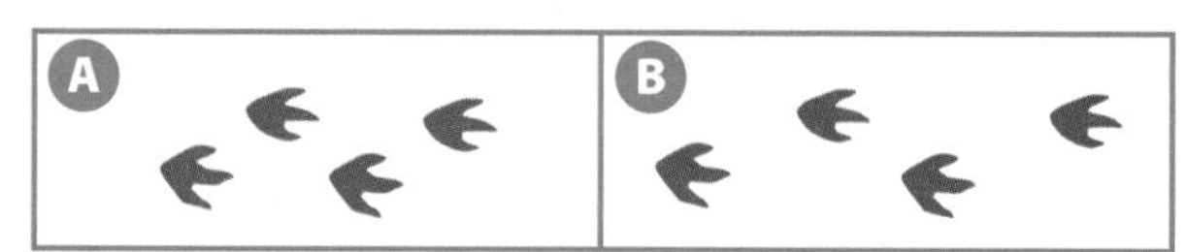

8. **Organize Information** Copy and fill in the graphic organizer below to list types of fossil preservation.

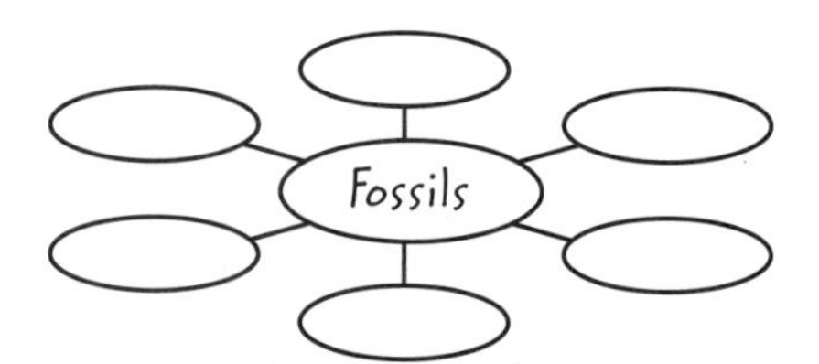

Critical Thinking

9. **Invent** a process for the formation of ocean basins consistent with catastrophism.
10. **Evaluate** how the following statement relates to what you have read in this lesson: "The present is the key to the past."

AMERICAN MUSEUM OF NATURAL HISTORY

▼ This Byronosaurus skull was discovered at Ukhaa Tolgod. It gave scientists important clues about how birds and dinosaurs are related.

Perfect Fossils— A Rare Find

Spectacular fossils formed when ancient organisms were buried quickly.

The key to a well-preserved fossil is what happens to an organism right after it dies. When organisms are buried swiftly, their remains are protected from scavengers and natural events. Over time, their bones and teeth form amazingly well-preserved fossils.

As a boy growing up in Los Angeles, Michael Novacek loved visiting the La Brea tar pits. The fossils from these tar pits are remarkably intact because animals became stuck in gooey tar puddles and quickly became submerged. These fossils inspired Novacek to become a paleontologist.

Years later, on an expedition for the American Museum of Natural History, Novacek discovered another extraordinary site. In the Gobi Desert, in Mongolia, Novacek and his team uncovered a rich collection of fossils at a site called Ukhaa Tolgod.

The fossils were astonishingly complete. One fossil was of a 2-inch mammal skeleton, still with its microscopic ear bones. Like the animals at the La Brea tar pits, those that died here were buried quickly. But after examining the evidence, scientists have determined that these animals were likely killed and covered by a devastating avalanche or landslide. Spectacular fossils formed when ancient organisms were buried quickly.

◀ Novacek and his team have found amazing fossils in Mongolia's Gobi Desert, including this velociraptor.

Today, the dunes at Ukhaa Tolgod are sandy and barren. But long ago, many plants and animals lived in the region. ▶

Discovering What Happened at Ukhaa Tolgod

Scientists have looked for clues in the rocks where the fossils were found. Most of the rocks are made of sandstone. One hypothesis was that the animals were buried alive by drifting dunes during a sandstorm. But then scientists noticed the rocks near the fossils held large pebbles that were too big to be carried by wind.

To find an explanation, they turned to Nebraska's Sand Hills—a region with giant, stable dunes similar to those that existed at Ukhaa Tolgod. In the Sand Hills, heavy rains can set off avalanches of wet sand. An enormous slab of heavy, wet sand can bury everything in its path. The current hypothesis about Ukhaa Tolgod is that heavy rains triggered an avalanche of sand that slid down the dunes and buried the animals below.

It's Your Turn

DRAW With a partner, draw a comic strip showing how animals in Ukhaa Tolgod might have died and been buried. Use the comic strip to explain how the almost-perfect fossils were preserved.

Lesson 2

Relative-Age Dating

Reading Guide

Key Concepts 

ESSENTIAL QUESTIONS

- What does relative age mean?
- How can the positions of rock layers be used to determine the relative ages of rocks?

Vocabulary

relative age p. 575
superposition p. 576
inclusion p. 577
unconformity p. 578
correlation p. 578
index fossil p. 579

g Multilingual eGlossary

Inquiry How did this happen?

Millions of years ago, hot magma from deep in Earth was forced into these red, horizontal rock layers in the Grand Canyon. As the magma cooled, it formed this dark gash. How do you think features such as this help scientists determine the relative ages of rock layers?

Launch Lab

15 minutes

Which rock layer is oldest?

Scientists study rock layers to learn about the geologic history of an area. How do scientists determine the order in which layers of rock were deposited?

1. Read and complete a lab safety form.
2. Break a **disposable polystyrene meat tray** in half. Place the two pieces on a flat surface so that the broken edges touch one another.
3. Break **another meat tray** in half. Place the two pieces directly on top of the first broken meat tray.
4. Place a **third, unbroken meat tray** on top of the two broken meat trays.

Think About This

1. If you observed rock layers that looked like your model, what would you think might have caused the break only in the two bottom layers?
2. **Key Concept** How do you think your model resembles a rock formation? Which layer in your model is youngest? Which is oldest?

Relative Ages of Rocks

You just remembered where you left the money you have been looking for. It is in the pocket of the pants you wore to the movies last Saturday. Look at your pile of dirty clothes. How can you tell where your money is? There really is some order in that pile of dirty clothes. Every time you add clothes to the pile, you place them on top, like the clothes you wore last night. And the clothes from last Saturday are on the bottom. That's where your money is.

Just as there is order in a pile of clothes, there is order in a rock formation. In the rock formation shown in **Figure 9,** the oldest rocks are in the bottom layer and the youngest rocks are in the top layer.

Maybe you have brothers and sisters. If you do, you might describe your age by saying, "I'm older than my sister and younger than my brother." In this way, you compare your age to others in your family. Geologists—the scientists who study Earth and rocks—have developed a set of principles to compare the ages of rock layers. They use these principles to organize the layers according to their relative ages. **Relative age** *is the age of rocks and geologic features compared with other rocks and features nearby.*

Key Concept Check How might you define your relative age?

Figure 9 Just as there is order in a pile of clothes, there is order in this rock formation.

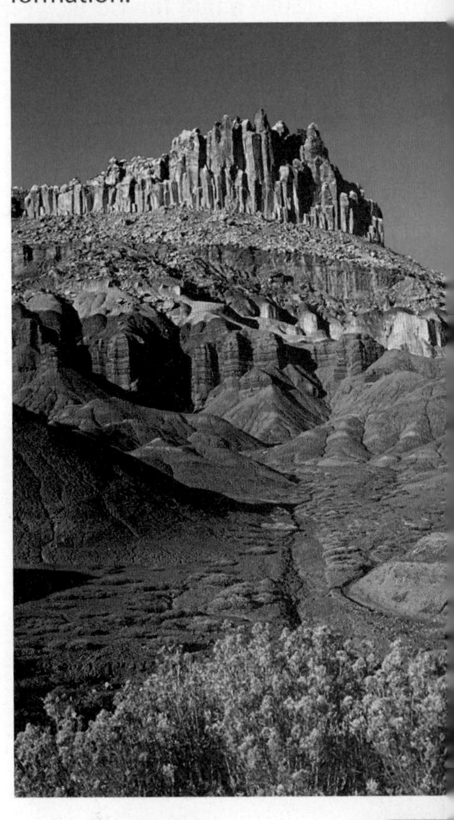

Principles of Relative-Age Dating

Figure 10 Geologic principles help scientists determine the relative order of rock layers.

Visual Check Which rock layer is the oldest?

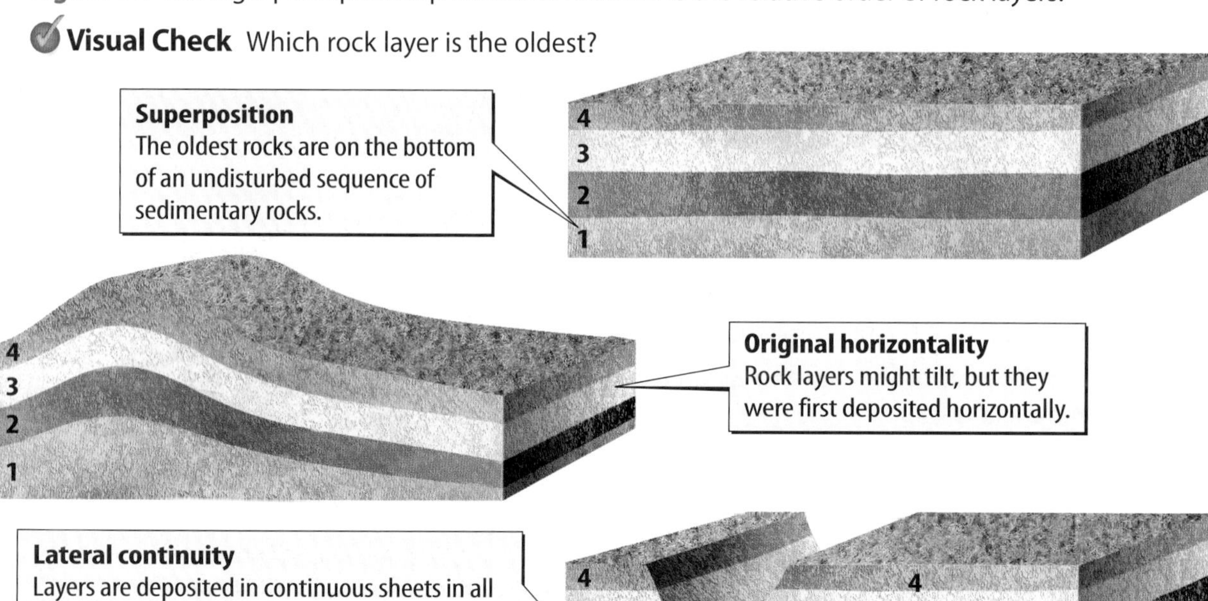

Superposition

Your pile of dirty clothes demonstrates the first principle of relative-age dating—superposition. **Superposition** *is the principle that in undisturbed rock layers, the oldest rocks are on the bottom.* Unless some force disturbs the layers after they were deposited, each layer of rocks is younger than the layer below it, as shown in **Figure 10.**

FOLDABLES®

Make a five-tab book and label it as shown. Use it to organize information about the principles of relative-age dating.

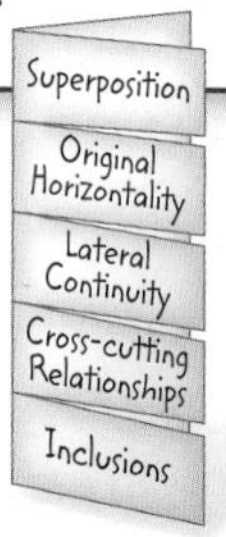

Original Horizontality

An example of the second principle of relative-age dating—original horizontality—is also shown in **Figure 10.** According to the principle of original horizontality, most rock-forming materials are deposited in horizontal layers. Sometimes rock layers are deformed or disturbed after they form. For example, the layers might be tilted or folded. Even though they might be tilted, all the layers were originally deposited horizontally.

Reading Check How might rock layers be disturbed?

Lateral Continuity

Another principle of relative-age dating is that sediments are deposited in large, continuous sheets in all **lateral** directions. The sheets, or layers, continue until they thin out or meet a barrier. This principle, called the principle of lateral continuity, is illustrated in the bottom image of **Figure 10.** A river might erode the layers, but their placements do not change.

WORD ORIGIN

lateral
from Latin *lateralis*, means "belonging to the side"

Figure 11 Dikes and faults help scientists determine the order in which rock layers were deposited.

Inclusions

Occasionally when rocks form, they contain pieces of other rocks. This can happen when part of an existing rock breaks off and falls into soft sediment or flowing magma. When the sediment or magma becomes rock, the broken piece becomes a part of it. *A piece of an older rock that becomes part of a new rock is called an* **inclusion.** According to the principle of inclusions, if one rock contains pieces of another rock, the rock containing the pieces is younger than the pieces. The vertical intrusion in **Figure 11,** called a dike, is younger than the pieces of rock inside it.

Cross-Cutting Relationships

Sometimes, forces within Earth cause rock formations to break, or fracture. When rocks move along a fracture line, the fracture is called a fault. Faults and dikes cut across existing rock. According to the principle of cross-cutting relationships, if one geologic feature cuts across another feature, the feature that it cuts across is older, as shown in **Figure 11.** This principle is illustrated in the photo at the beginning of this lesson. The black rock layer formed as magma cut across pre-existing red rock layers and crystallized.

 **Key Concept Check** What geologic principles are used in relative-age dating?

Inquiry MiniLab — 20 minutes

Can you model rock layers?

Can a classmate determine the order of your three-dimensional model of rock layers?

1. Read and complete a lab safety form.
2. Cut out a **cube template** as instructed by your teacher.
3. On the sides and top, use **colored pencils** to draw a rock formation that contains 4–5 layers. Include faults, dikes, inclusions, and other disturbances.
4. **Glue** your cube to make a three-dimensional model.
5. Exchange models with another student and determine the order of the layers.

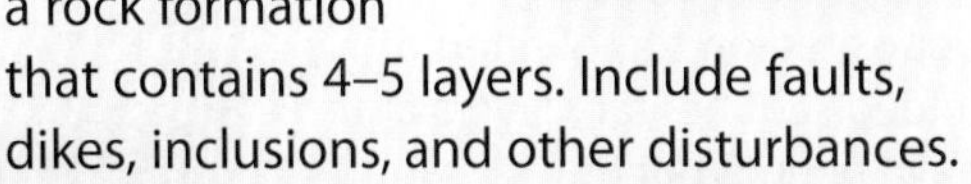

Analyze and Conclude

Key Concept Summarize how positions of rock layers can be used to determine the relative ages of rocks.

Unconformities

After rocks form, they are sometimes uplifted and exposed at Earth's surface. When rocks are exposed, wind and rain start to weather and erode them. These eroded areas represent a gap in the rock record.

Often, new rock layers are deposited on top of old, eroded rock layers. When this happens, an unconformity (un kun FOR muh tee) occurs. *An* **unconformity** *is a surface where rock has eroded away, producing a break, or gap, in the rock record.*

An unconformity is not a hollow gap in the rock. It is a surface on a layer of eroded rocks where younger rocks have been deposited. However, an unconformity does represent a gap in time. It could represent a few hundred years, a million years, or even billions of years. Three major types of unconformities are shown in **Table 1.**

Key Concept Check How does an unconformity represent a gap in time?

Correlation

You have read that rock layers contain clues about Earth. Geologists use these clues to build a record of Earth's geologic history. Many times the rock record is incomplete, such as happens in an unconformity. Geologists fill in gaps in the rock record by matching rock layers or fossils from separate locations. *Matching rocks and fossils from separate locations is called* **correlation** (kor uh LAY shun).

Matching Rock Layers

Another word for correlation is *connection.* Sometimes it is possible to connect rock layers simply by walking along rock formations and looking for similarities. At other times, soil might cover the rocks, or rocks might be eroded away. In these cases, geologists correlate rocks by matching exposed rock layers in different locations. Through correlation, geologists have established a historical record for part of the southwestern United States, as shown in **Figure 12.**

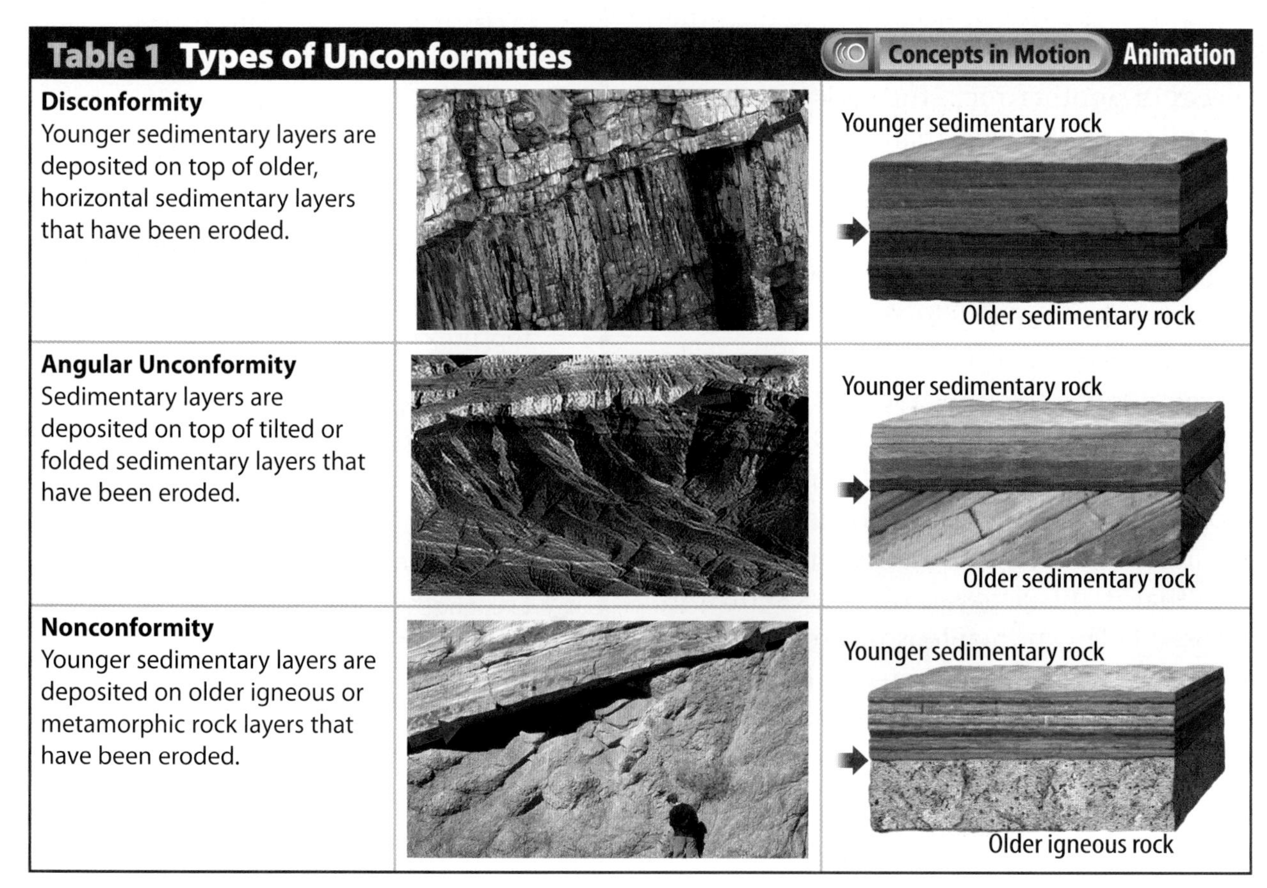

Table 1 Types of Unconformities — Concepts in Motion Animation

Type	Photo	Diagram
Disconformity Younger sedimentary layers are deposited on top of older, horizontal sedimentary layers that have been eroded.		Younger sedimentary rock; Older sedimentary rock
Angular Unconformity Sedimentary layers are deposited on top of tilted or folded sedimentary layers that have been eroded.		Younger sedimentary rock; Older sedimentary rock
Nonconformity Younger sedimentary layers are deposited on older igneous or metamorphic rock layers that have been eroded.		Younger sedimentary rock; Older igneous rock

Correlation

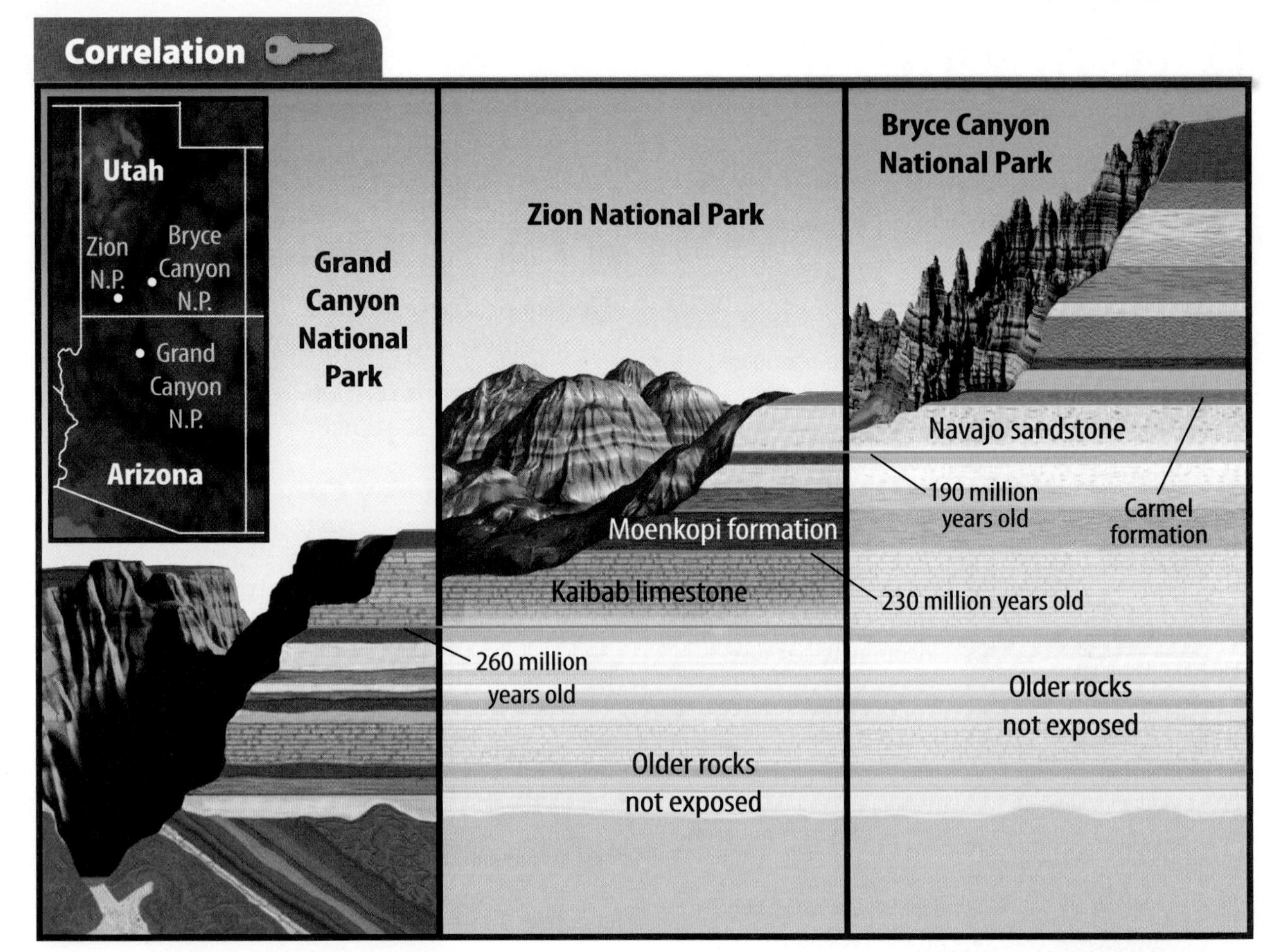

Figure 12 Exposed rock layers from three national parks have been correlated to make a historical record.

Visual Check Which geologic principles must be assumed in order to correlate these layers?

Personal Tutor

Index Fossils

The rock formations in **Figure 12** are correlated based on similarities in rock type, structure, and fossil evidence. They exist within a few hundred kilometers of one another. If scientists want to learn the relative ages of rock formations that are very far apart or on different continents, they often use fossils. If two or more rock formations contain fossils of about the same age, scientists can infer that the formations are also about the same age.

Not all fossils are useful in determining the relative ages of rock layers. Fossils of species that lived on Earth for hundreds of millions of years are not helpful. They represent time spans that are too long. The most useful fossils represent species, like certain trilobites, that existed for only a short time in many different areas on Earth. These fossils are called index fossils. **Index fossils** *represent species that existed on Earth for a short length of time, were abundant, and inhabited many locations.* When an index fossil is found in rock layers at different locations, geologists can infer that the layers are of similar age.

Key Concept Check How are index fossils useful in relative-age dating?

Lesson 2 Review

Visual Summary

Geologic principles help geologists learn the relative ages of rock layers.

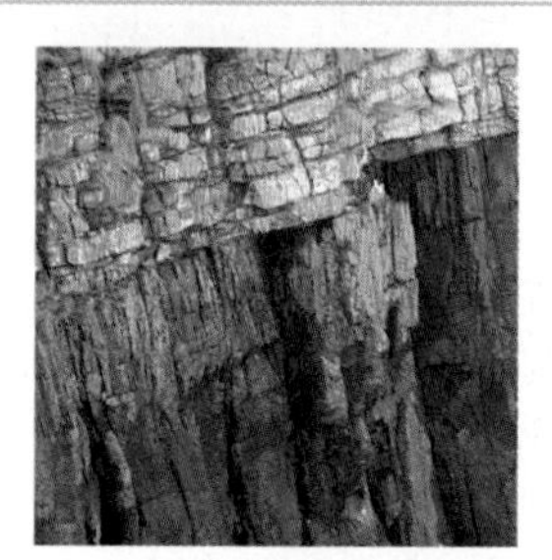

The rock record is incomplete because some of it has eroded away.

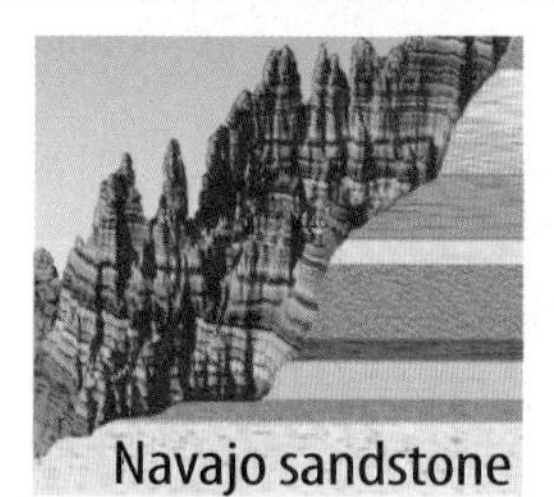

Geologists fill in gaps in the rock record by correlating rock layers.

FOLDABLES®

Use your lesson Foldable to review the lesson. Save your Foldable for the project at the end of the chapter.

What do you think NOW?

You first read the statements below at the beginning of the chapter.

3. Older rocks are always located below younger rocks.

4. Relative age means that scientists are relatively sure of the age.

Did you change your mind about whether you agree or disagree with the statements? Rewrite any false statements to make them true.

Use Vocabulary

1. A gap in the rock record is a(n) ________.
2. The principle that the oldest rocks are generally on the bottom is ________.
3. **Use the terms** *correlation* and *index fossil* in a complete sentence.

Understand Key Concepts

4. Which might be useful in correlation?
   A. amber
   B. inclusion
   C. trilobite
   D. unconformity
5. **Draw** and label a sequence of rock layers showing how an unconformity might form.
6. **Relate** uniformitarianism to principles of relative-age dating.

Interpret Graphics

Use the diagram below to answer question 7.

7. **Decide** Which is older—the rock layers or the dike? Explain which geologic principle you used to arrive at your answer.
8. **Summarize** Copy and fill in the graphic organizer below to identify five geologic principles useful in relative-age dating.

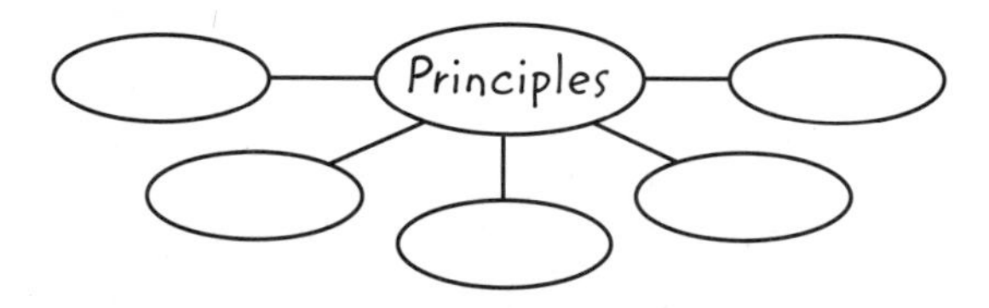

Critical Thinking

9. **Evaluate** why fossils might be more useful than rock types in correlating rock layers on two different continents.
10. **Debate** whether you think humans might be useful as index fossils in the future.

Can you correlate rock formations?

Materials

pencil

colored pencils

large, soft eraser

ruler

Most rocks have been buried in Earth for thousands, millions, or even billions of years. Occasionally, rock layers become exposed on Earth's surface. To correlate rock layers exposed at different locations, it is sometimes necessary to **interpret scientific illustrations** of the layers.

Learn It

Drawings and photos can make complex scientific data easier to understand. Use the drawings below to represent rock formations. As you correlate the layers, use the key to **interpret each illustration.**

Try It

1. As well as you can, copy the drawings of the four rock columns shown below into your Science Journal. *Do not write in this book.*
2. Color your drawings so that each rock layer is one color in each of the four rock columns. Use the key to determine what type of rocks each layer contains.
3. Carefully study your drawings. Try to determine which rock columns correlate the best.

Apply It

4. Which rock columns correlated the best? Which principle of relative-age dating did you use when correlating the rock layers?
5. Which rock layer is the oldest in rock column C? The youngest? Which geologic principle did you use to determine this?
6. Identify the type of unconformity that exists in rock column B.
7. **Key Concept** How can you use types of rocks to correlate rock layers? What other type of evidence could you use to determine the relative ages of rock layers?

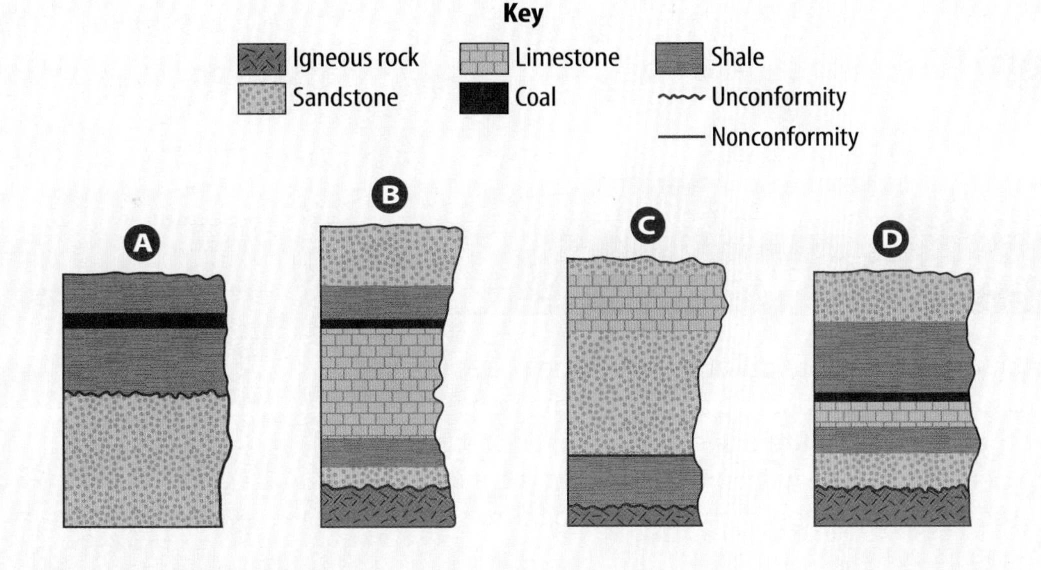

Lesson 3

Reading Guide

Key Concepts
ESSENTIAL QUESTIONS

- What does absolute age mean?
- How can radioactive decay be used to date rocks?

Vocabulary

absolute age p. 583
isotope p. 584
radioactive decay p. 584
half-life p. 585

Multilingual eGlossary

Absolute-Age Dating

Inquiry How old are they?

These mammoth bones are dry and fragile. They have not yet turned to rock. Scientists analyze samples of the bones to discover their ages. Absolute-age dating requires precise measurements in very clean laboratories. What techniques can be used to learn the age of an ancient organism simply by analyzing its bones?

Launch Lab

10 minutes

How can you describe your age?

If you described your relative age compared to your classmates', how would you do it? How do you think your actual, or absolute, age differs from your relative age?

1. One student will write down his or her birth date on an **index card.** The student will hold the card while everyone else files by and looks at it.
2. Form two groups depending on whether your birth date falls before or after the date on the card.
3. Remaining in your group, write down your own birth date on an index card. Quietly form a line in order of your birth dates.

Think About This

1. When you were in two groups, what did you know about everyone's age? When you lined up, what did you know about everyone's age? Which is your relative age? Your absolute age?
2. Can you think of a situation where it would be important to know your absolute age?
3. **Key Concept** Why do you think scientists would want to know the absolute age of a rock?

Absolute Ages of Rocks

Recall from Lesson 2 that you have a relative age. You might be older than your sister and younger than your brother—or you might be the youngest in your family. You also can describe your age by saying your age in years, such as "I am 13 years old." This is not a relative age. It is your age in numbers—your numerical age.

Similarly, scientists can describe the ages of some kinds of rocks numerically. Scientists use the term **absolute age** *to mean the numerical age, in years, of a rock or object.* By measuring the absolute ages of rocks, geologists have developed accurate historical records for many geologic formations.

Key Concept Check How is absolute age different from relative age?

Scientists have been able to determine the absolute ages of rocks and other objects only since the beginning of the twentieth century. That is when radioactivity was discovered. Radioactivity is the release of energy from unstable atoms. The image in **Figure 13** was made using X-rays. How can radioactivity be used to date rocks? In order to answer this question, you need to know about the internal structure of the atoms that make up elements.

Figure 13 The release of radioactive energy can be used to make an X-ray.

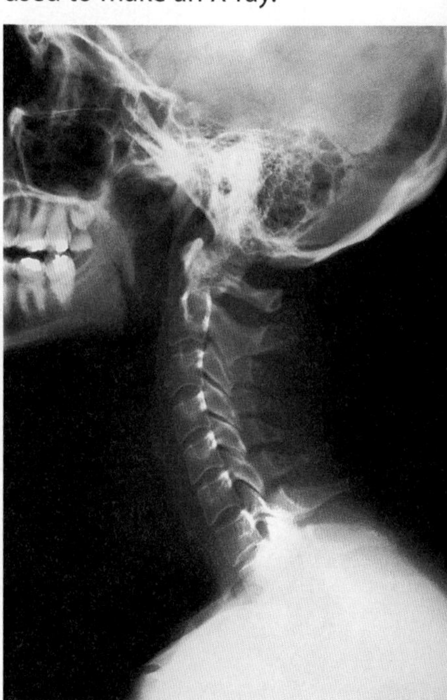

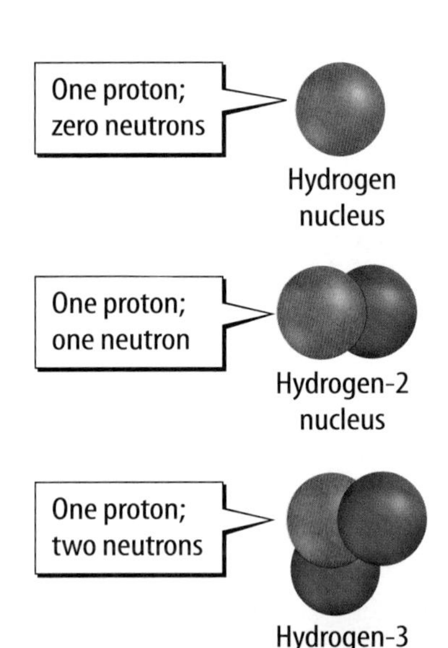

Figure 14 All forms of hydrogen contain only one proton regardless of the number of neutrons.

Atoms

You are probably familiar with the periodic table of the elements, which is shown inside the back cover of this book. Each element is made up of atoms. An atom is the smallest part of an element that has all the properties of the element. Each atom contains smaller particles called protons, neutrons, and electrons. Protons and neutrons are in an atom's nucleus. Electrons surround the nucleus.

Isotopes

All atoms of a given element have the same number of protons. For example, all hydrogen atoms have one proton. But an element's atoms can have different numbers of neutrons. The three atoms shown in **Figure 14** are all hydrogen atoms. Each has the same number of protons—one. However, one of the hydrogen atoms has no neutrons, one has one neutron, and the other has two neutrons. The three different forms of hydrogen atoms are called hydrogen **isotopes** (I suh tohps). **Isotopes** *are atoms of the same element that have different numbers of neutrons.*

Reading Check How do an element's isotopes differ?

Radioactive Decay

Most isotopes are stable. Stable isotopes do not change under normal conditions. But some isotopes are unstable. These isotopes are known as radioactive isotopes. Radioactive isotopes decay, or change, over time. As they decay, they release energy and form new, stable atoms. **Radioactive decay** *is the process by which an unstable element naturally changes into another element that is stable.* The unstable isotope that decays is called the parent isotope. The new element that forms is called the daughter isotope. **Figure 15** illustrates an example of radioactive decay. The atoms of an unstable isotope of hydrogen (parent) decay into atoms of a stable isotope of helium (daughter).

WORD ORIGIN

isotope
from Greek *isos,* means "equal"; and *topos,* means "place"

Radioactive Decay

Figure 15 An unstable parent hydrogen isotope produces the stable daughter helium isotope.

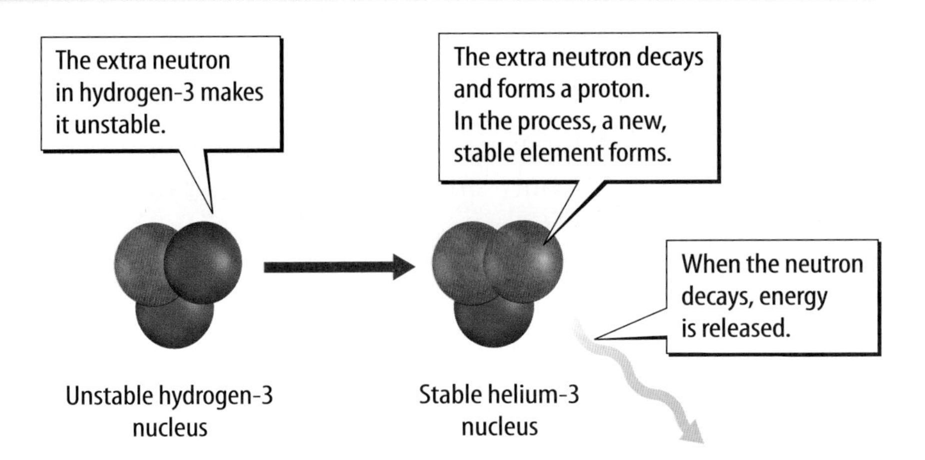

Half-Life of Radioactive Decay

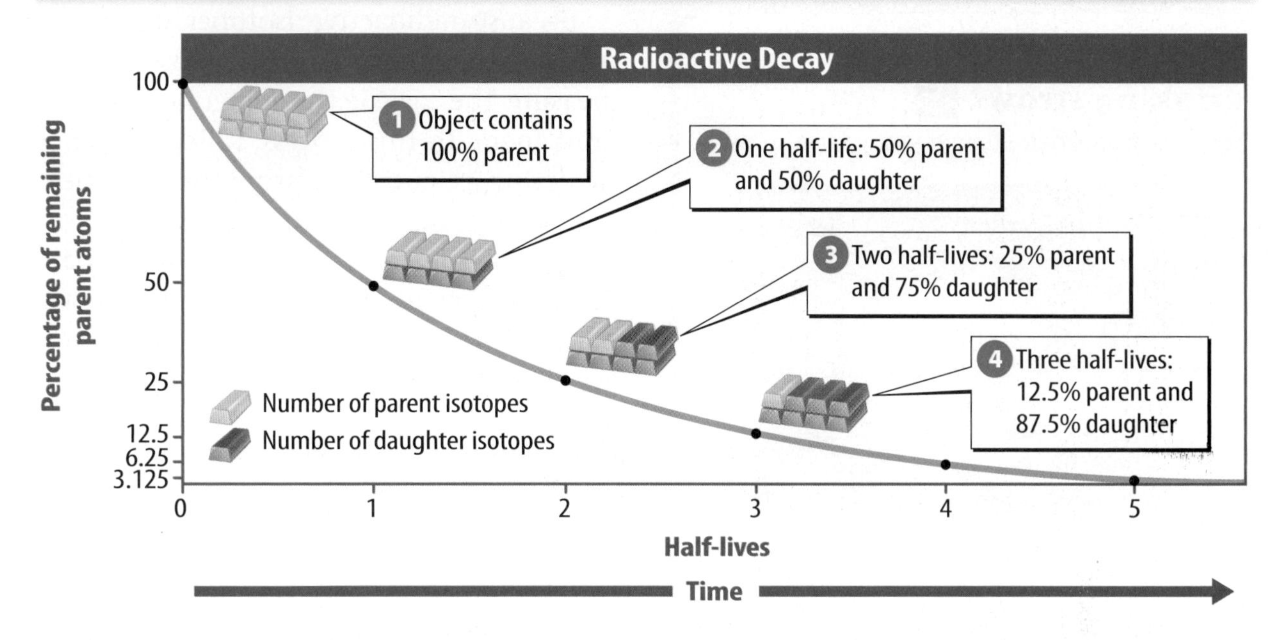

Figure 16 The half-life is the time it takes for one-half of the parent isotopes to change into daughter isotopes.

Visual Check What percentages of parent isotopes and daughter isotopes will there be after four half-lives?

Half-Life

The rate of decay from parent isotopes into daughter isotopes is different for different radioactive elements. But the rate of decay is constant for a given isotope. This rate is measured in time units called half-lives. *An isotope's* **half-life** *is the time required for half of the parent isotopes to decay into daughter isotopes.* Half-lives of radioactive isotopes range from a few microseconds to billions of years.

Reading Check What is half-life?

The graph in **Figure 16** shows how half-life is measured. As time passes, more and more unstable parent isotopes decay and form stable daughter isotopes. That means the ratio between the numbers of parent and daughter isotopes is always changing. When half the parent isotopes have decayed into daughter isotopes, the isotope has reached one half-life. At this point, 50 percent of the isotopes are parents and 50 percent of the isotopes are daughters. After two half-lives, one-half of the remaining parent isotopes have decayed so that only one-quarter as much parent remains as at the start. At this point, 25 percent of the isotopes are parent and 75 percent of the isotopes are daughter. After three half-lives, half again of the remaining parent isotopes have decayed into daughter isotopes. This process continues until nearly all parent isotopes have decayed into daughter isotopes.

Inquiry MiniLab — 10 minutes

What is the half-life of a drinking straw? 

You can model half-life with a drinking straw.

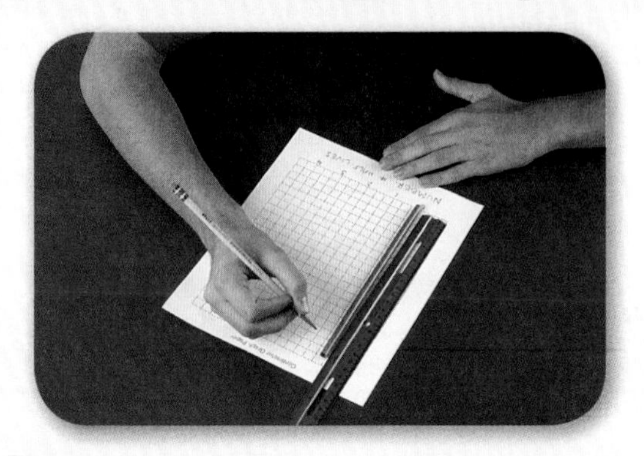

1. Read and complete a lab safety form.
2. On a piece of **graph paper,** draw an *x*-axis and a *y*-axis. Label the *x*-axis *Number of half-lives,* from 0 to 4 in equal intervals. Leave the *y*-axis blank.
3. Use a **metric ruler** to measure a **drinking straw.** Mark its height on the *y*-axis, as shown in the photo. Use **scissors** to cut the straw in half and discard half of it. Mark the height of the remaining half as the first half-life.
4. Repeat four times, each time cutting the straw in half and each time adding a measurement to your graph's *y*-axis.

Analyze and Conclude

1. Compare your graph to the graph in **Figure 16.** How is it similar? How is it different?
2. **Key Concept** Explain how your disappearing straw represents the decay of a radioactive element.

Radiometric Ages

Because radioactive isotopes decay at a constant rate, they can be used like clocks to measure the age of the material that contains them. In this process, called radiometric dating, scientists measure the amount of parent isotope and daughter isotope in a sample of the material they want to date. From this ratio, they can determine the material's age. Scientists make these very precise measurements in laboratories.

Reading Check What is measured in radiometric dating?

Radiocarbon Dating

One important radioactive isotope used for dating is an isotope of carbon called radiocarbon. Radiocarbon is also known as carbon-14, or C-14, because there are 14 particles in its nucleus—six protons and eight neutrons. Radiocarbon forms in Earth's upper atmosphere. There, it mixes in with a stable isotope of carbon called carbon-12, or C-12. The ratio of C-14 to C-12 in the atmosphere is constant.

All living things use carbon as they build and repair tissues. As long as an organism is alive, the ratio of C-14 to C-12 in its tissues is identical to the ratio in the atmosphere. However, when an organism dies, it stops taking in C-14. The C-14 already present in the organism starts to decay to nitrogen-14 (N-14). As the dead organism's C-14 decays, the ratio of C-14 to C-12 changes. Scientists measure the ratio of C-14 to C-12 in the remains of the dead organism to determine how much time has passed since the organism died.

The half-life of carbon-14 is 5,730 years. That means radiocarbon dating is useful for measuring the age of the remains of organisms that died up to about 60,000 years ago. In older remains, there is not enough C-14 left to measure accurately. Too much of it has decayed to N-14.

Radiometric Dating

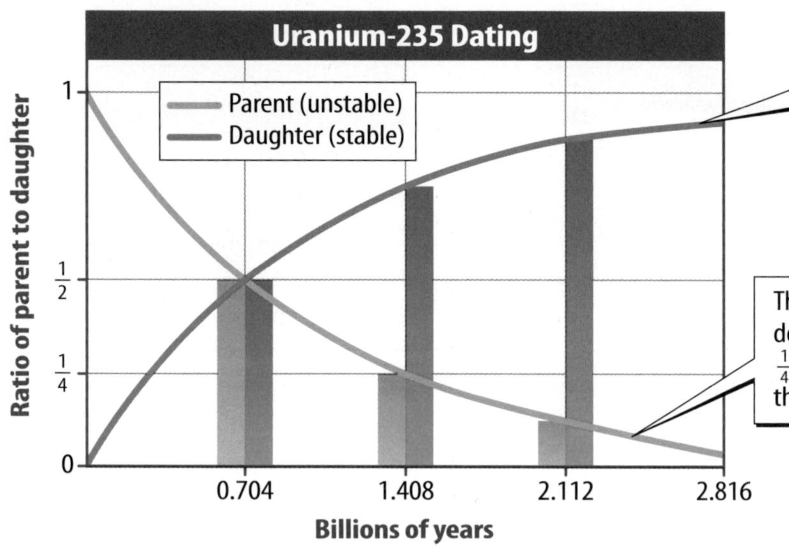

Figure 17 Scientists determine the absolute age of an igneous rock by measuring the ratio of uranium-235 isotopes (parent) to lead-207 isotopes (daughter) in the rock's minerals.

Visual Check How old is a mineral that contains 25 percent U-235?

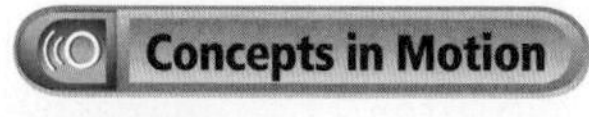

Animation

Dating Rocks

Radiocarbon dating is useful only for dating organic material—material from once-living organisms. This material includes bones, wood, parchment, and charcoal. Most rocks do not contain organic material. Even most fossils are no longer organic. In most fossils, living tissue has been replaced by rock-forming **minerals.** For dating rocks, geologists use different kinds of radioactive isotopes.

Dating Igneous Rock One of the most common isotopes used in radiometric dating is uranium-235, or U-235. U-235 is often trapped in the minerals of igneous rocks that crystallize from hot, molten magma. As soon as U-235 is trapped in a mineral, it begins to decay to lead-207, or Pb-207, as shown in **Figure 17.** Scientists measure the ratio of U-235 to Pb-207 in a mineral to determine how much time has passed since the mineral formed. This provides the age of the rock that contains the mineral.

Dating Sedimentary Rock In order to be dated by radiometric means, a rock must have U-235 or other radioactive isotopes trapped inside it. The grains in many sedimentary rocks come from a variety of weathered rocks from different locations. The radioactive isotopes within these grains generally record the ages of the grains—not the time when the sediment was deposited. For this reason, sedimentary rock is not as easily dated as igneous rock in radiometric dating.

Key Concept Check Why are radioactive isotopes not useful for dating sedimentary rocks?

REVIEW VOCABULARY

mineral
a naturally occurring, inorganic solid with a definite chemical composition and an orderly arrangement of atoms

Table 2 Radioactive Isotopes Used for Dating Rocks		
Parent Isotope	**Half-Life**	**Daughter Product**
Uranium-235	704 million years	lead-207
Potassium-40	1.25 billion years	argon-40
Uranium-238	4.5 billion years	lead-206
Thorium-232	14.0 billion years	lead-208
Rubidium-87	48.8 billion years	strontium-87

Table 2 Radioactive isotopes useful for dating rocks have long half-lives.

Math Skills

Use Significant Digits

The answer to a problem involving measurement cannot be more precise than the measurement with the fewest number of significant digits. For example, if you begin with 36 grams (2 significant digits) of U-235, how much U-235 will remain after 2 half-lives?

1. After the first half-life, $\frac{36 \text{ g}}{2} = 18 \text{ g}$ of U-235 remain.
2. After the second half-life, $\frac{18 \text{ g}}{2} = 9.0 \text{ g}$ of U-235 remain. Add the zero to retain two significant digits.

Practice

The half-life of rubidium-87 (Rb-87) is 48.8 billion years. What is the length of three half-lives of Rb-87?

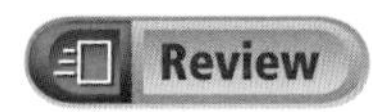

- **Math Practice**
- **Personal Tutor**

Different Types of Isotopes The half-life of uranium-235 is 704 million years. This makes it useful for dating rocks that are very old. **Table 2** lists five of the most useful radioactive isotopes for dating old rocks. All of them have long half-lives. Radioactive isotopes with short half-lives cannot be used for dating old rocks. They do not contain enough parent isotope to measure. Geologists often use a combination of radioactive isotopes to measure the age of a rock. This helps make the measurements more accurate.

Key Concept Check Why is a radioactive isotope with a long half-life useful in dating very old rocks?

The Age of Earth

The oldest known rock formation dated by geologists using radiometric means is in Canada. It is estimated to be between 4.03 billion and 4.28 billion years old. However, individual crystals of the mineral zircon in igneous rocks in Australia have been dated at 4.4 billion years.

With rocks and minerals more than 4 billion years old, scientists know that Earth must be at least that old. Radiometric dating of rocks from the Moon and meteorites indicate that Earth is 4.54 billion years old. Scientists accept this age because evidence suggests that Earth, the Moon, and meteorites all formed at about the same time.

Radiometric dating, the relative order of rock layers, and fossils all help scientists understand Earth's long history. Understanding Earth's history can help scientists understand changes occurring on Earth today—as well as changes that are likely to occur in the future.

Lesson 3 Review

Assessment Online Quiz

Visual Summary

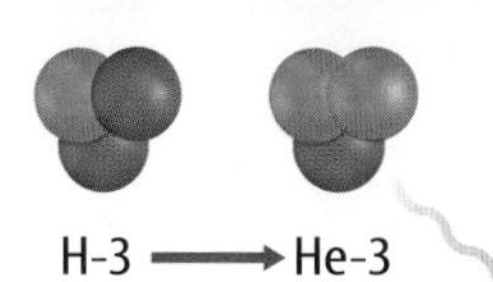

When the unstable atoms of radioactive isotopes decay, they form new, stable isotopes.

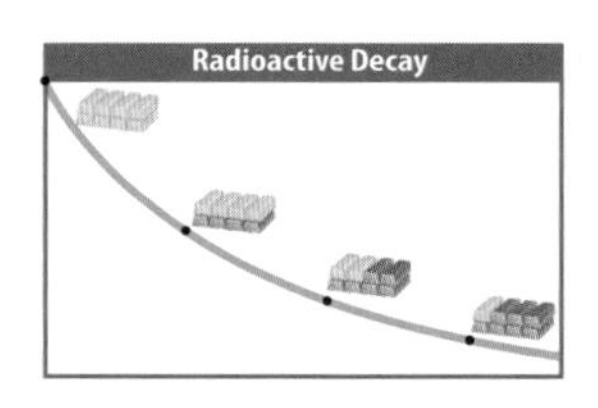

Because radioactive isotopes decay at constant rates, they can be used to determine absolute ages.

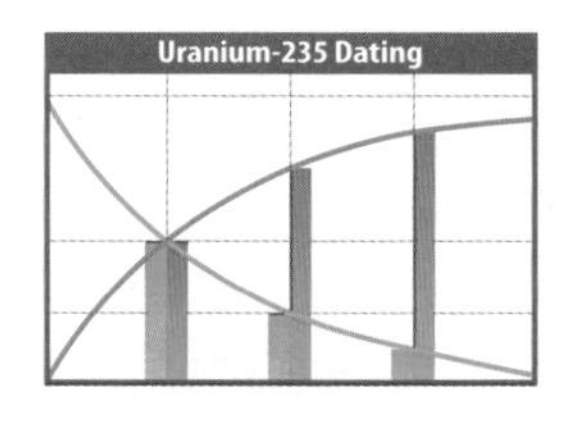

Isotopes with long half-lives are the most useful for dating old rocks.

FOLDABLES®

Use your lesson Foldable to review the lesson. Save your Foldable for the project at the end of the chapter.

What do you think NOW?

You first read the statements below at the beginning of the chapter.

5. Absolute age means that scientists are sure of the age.

6. Scientists use radioactive decay to determine the ages of some rocks.

Did you change your mind about whether you agree or disagree with the statements? Rewrite any false statements to make them true.

Use Vocabulary

1. **Compare** absolute age and relative age.
2. The rate of radioactive decay is expressed as an isotope's ________.
3. **Use the terms** *atom* and *isotope* in a complete sentence.

Understand Key Concepts 

4. Which could you date with carbon-14?
   A. a fossilized shark's tooth
   B. an arrowhead carved out of rock
   C. a petrified tree
   D. charcoal from an ancient campfire
5. **Explain** why radioactive isotopes are more useful for dating igneous rocks than they are for dating sedimentary rocks.
6. **Differentiate** between parent isotopes and daughter isotopes.

Interpret Graphics

7. **Identify** Copy and fill in the graphic organizer below to identify the three parts of an atom.

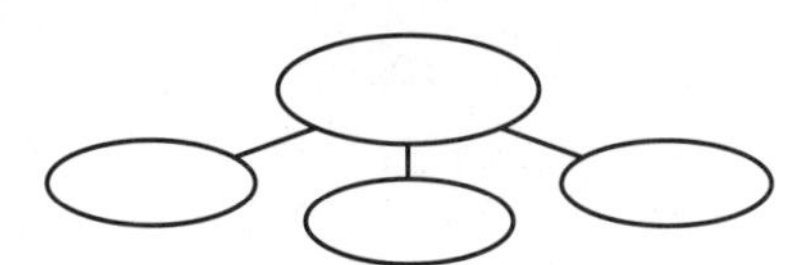

Critical Thinking

8. **Evaluate** the importance of radioactive isotopes in determining the age of Earth.

Math Skills

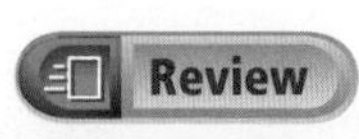

Math Practice

9. The half life of potassium-40 (K-40) is 1.25 billion years. If you begin with 130 g of K-40, how much remains after 2.5 billion years? Use the correct number of significant digits in your answer.

40 minutes

Correlate Rocks Using Index Fossils

Imagine you are a geologist and you have been asked to correlate the rock columns below in order to determine the relative ages of the layers. Recall that geologists can correlate rock layers in different ways. In this lab, use index fossils to correlate and date the layers.

Question

How can index fossils be used to determine the relative ages of Earth's rocks?

Procedure

1. Carefully examine the three rock columns on this page. Each rock layer can be identified with a letter and a number. For example, the second layer down in column A is layer A-2.
2. In your Science Journal, correlate the layers using only the fossils—not the types of rock. Before you begin, look at the fossil key on the next page. It shows the time intervals during which each organism or group of organisms lived on Earth. Refer to the key as you correlate.

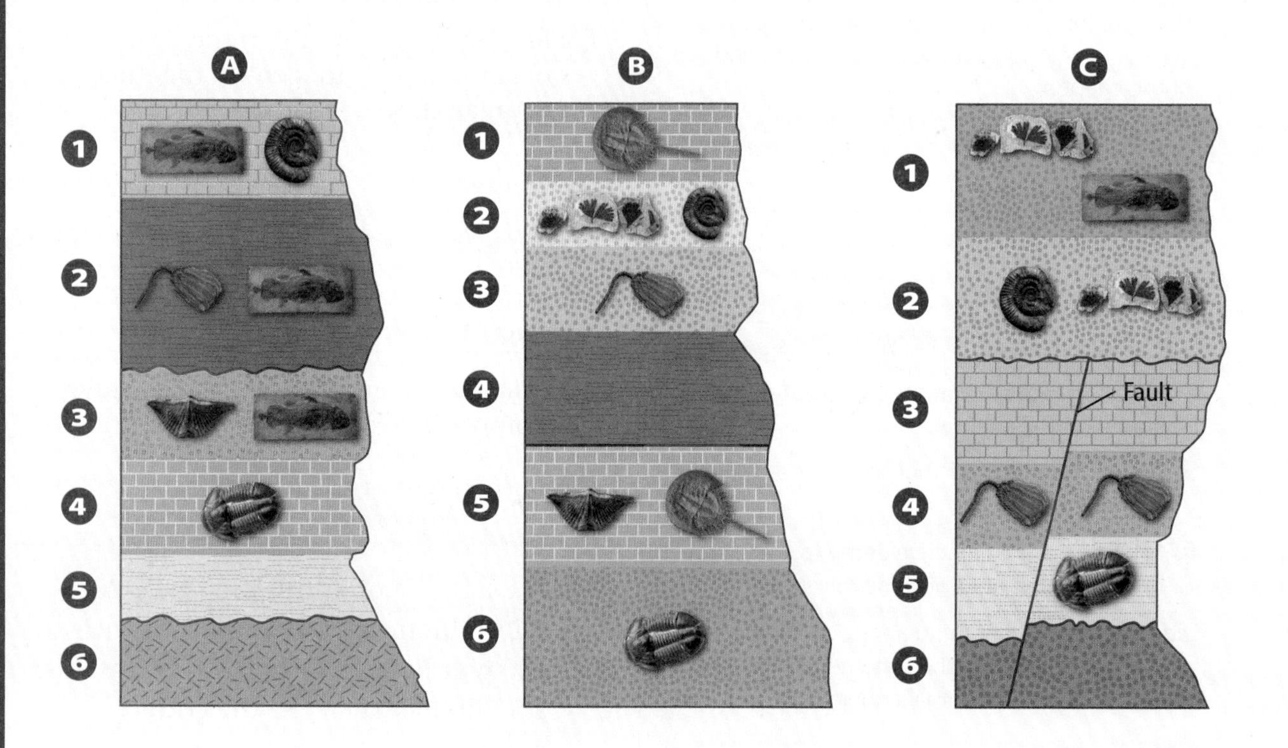

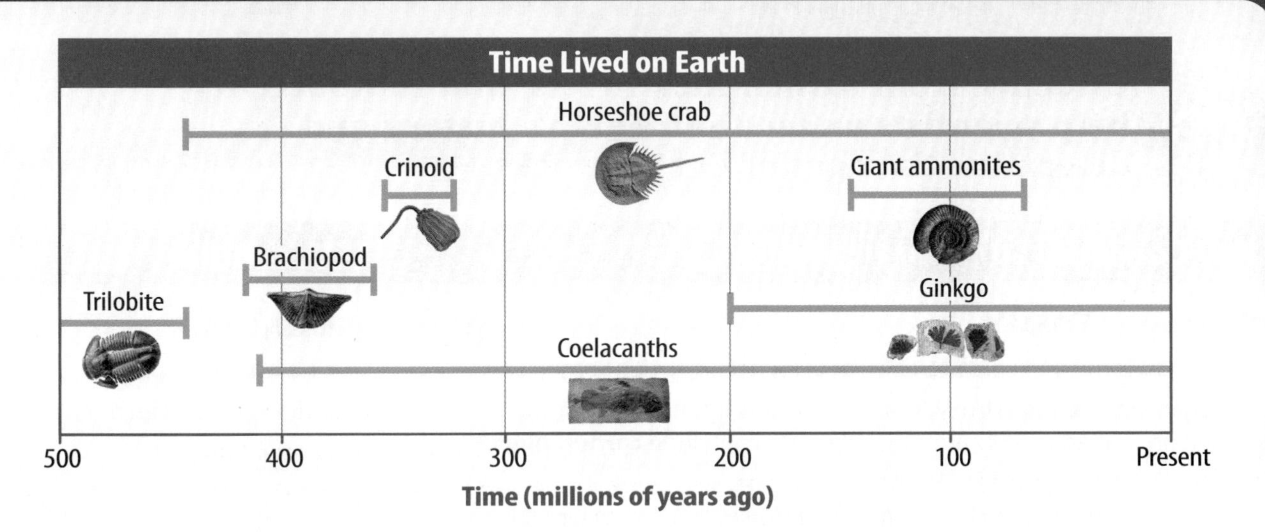

Analyze and Conclude

3. **Differentiate** Which fossils in the key appear to be index fossils? Explain your choices.
4. **Match** Correlate layer A-2 to one layer in each of the other two columns. Approximately how old are these layers? How do you know?
5. **Infer** What is the approximate age of layer B-4? *Hint: It lies between two index fossils.*
6. **Infer** How old is the fault in column C?
7. **Compare and Contrast** How does correlating rocks using fossils differ from correlating rocks using types of rock?
8. **The Big Idea** How can fossils be used to determine the relative ages of rocks?

Communicate Your Results

Choose a partner. One of you is a reporter and one is a geologist. Conduct an interview about what kinds of fossils are best used to date rocks.

Choose one of the three rock formations you correlated. Based on your results, provide a range of dates for each of the layers within it.

Lab Tips

☑ You might want to copy the rock layers in your Science Journal and correlate them by drawing lines connecting the layers.

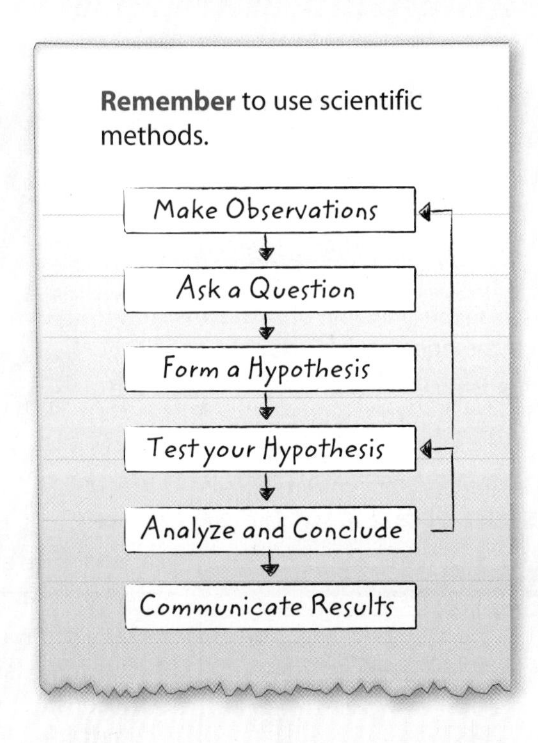

Chapter 16 Study Guide

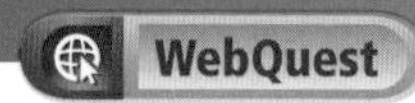

Evidence from fossils, rock layers, and radioactivity help scientists understand Earth's history and determine the ages of Earth's rocks.

Key Concepts Summary

Lesson 1: Fossils

- A **fossil** is the preserved remains or evidence of ancient organisms.
- Organisms are more likely to become fossils if they have hard parts and are buried quickly after they die. Fossils include **carbon films**, **molds**, **casts**, and **trace fossils**.
- **Paleontologists** use clues from fossils to learn about ancient life and the environments ancient organisms lived in.

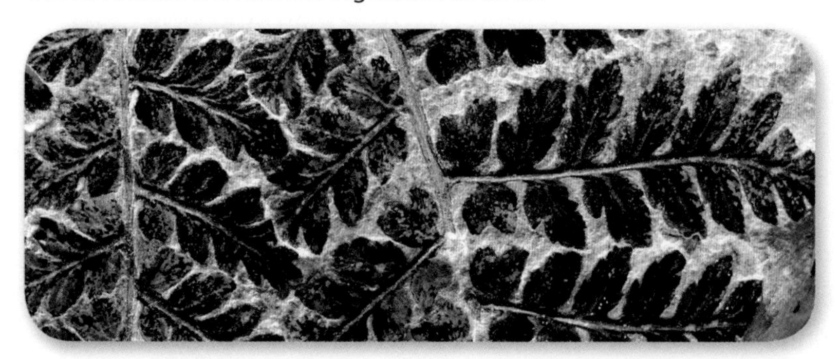

Vocabulary

fossil p. 565
catastrophism p. 565
uniformitarianism p. 566
carbon film p. 568
mold p. 569
cast p. 569
trace fossil p. 569
paleontologist p. 570

Lesson 2: Relative-Age Dating

- **Relative age** is the age of rocks and geologic features compared with rocks and features nearby.
- The relative age of rock layers can be determined using geologic principles, such as the principle of **superposition** and the principle of **inclusion. Unconformities** represent time gaps in the rock record.

Vocabulary

relative age p. 575
superposition p. 576
inclusion p. 577
unconformity p. 578
correlation p. 578
index fossil p. 579

Lesson 3: Absolute-Age Dating

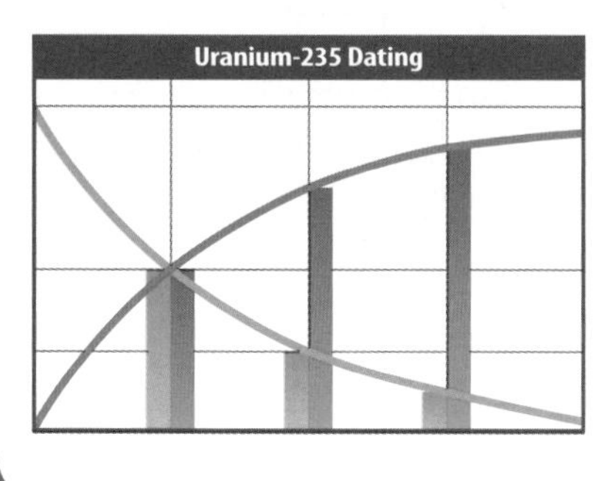

- **Absolute age** is the age in years of a rock or object.
- The **radioactive decay** of unstable **isotopes** occurs at a constant rate, measured as **half-life.** To date a rock or object, scientists measure the ratios of its parent and daughter isotopes.

Vocabulary

absolute age p. 583
isotope p. 584
radioactive decay p. 584
half-life p. 585

Study Guide

- Personal Tutor
- Vocabulary eGames
- Vocabulary eFlashcards

FOLDABLES® Chapter Project

Assemble your lesson Foldables® as shown to make a Chapter Project. Use the project to review what you have learned in this chapter.

Fossils

Superposition

Original Horizontality

Lateral Continuity

Cross-cutting Relationships

Inclusions

Dating Organic Material

Dating Rocks

Use Vocabulary

1. An ancient dinosaur track is a(n) ________ .
2. ________ use the principle of ________ to reconstruct ancient environments.
3. The principle of ________ states that the oldest layers are generally at the bottom.
4. In ________, geologists use ________ to match rock layers on separate continents.
5. A(n) ________ is an eroded surface.
6. The process of ________ can be used like a clock to determine a rock's ________ .
7. A Uranium-235 ________ decays with a constant ________ of 704 million years.

Link Vocabulary and Key Concepts

Copy this concept map, and then use vocabulary terms from the previous page to complete the concept map.

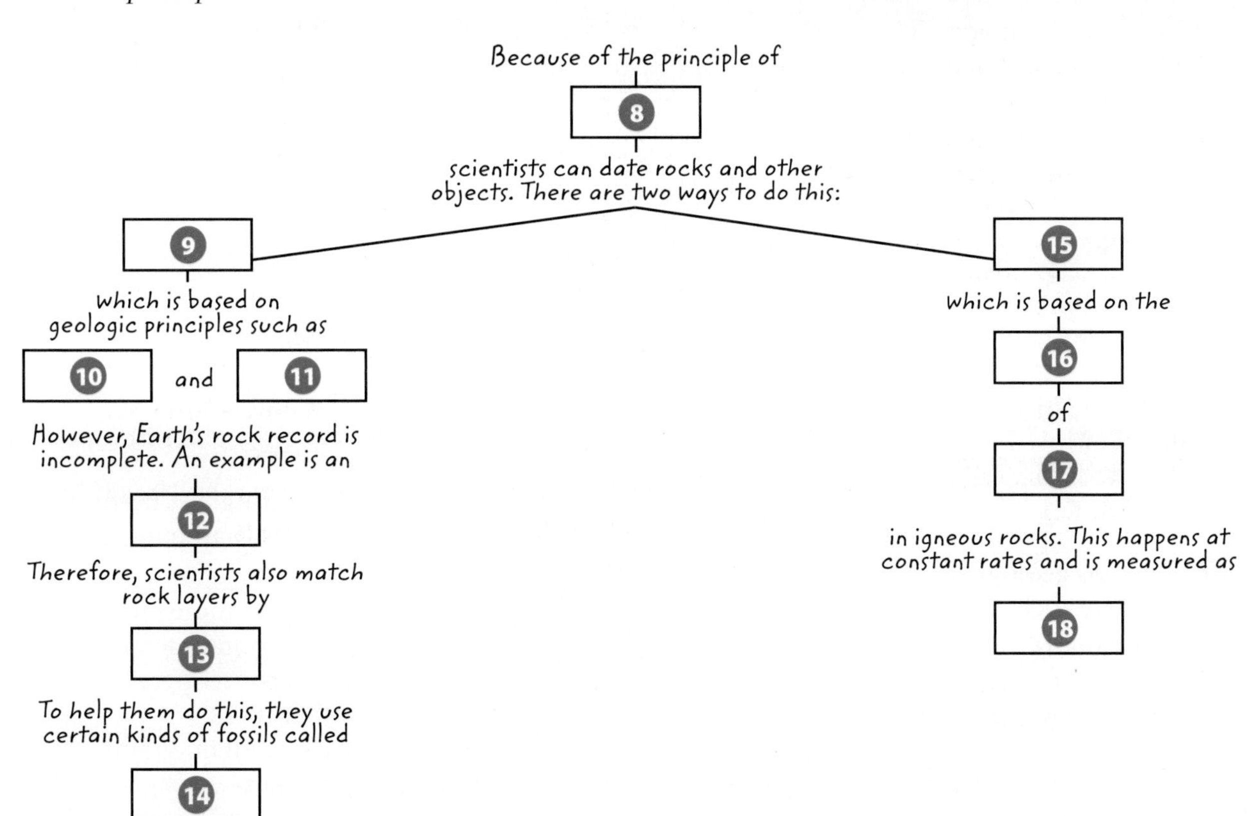

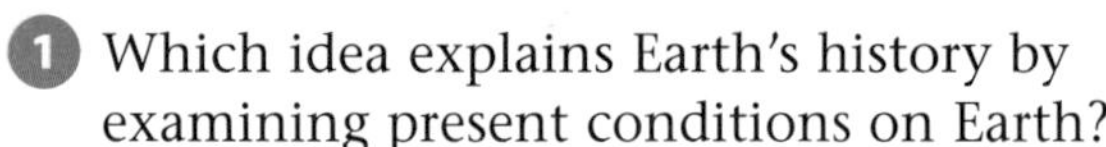

Understand Key Concepts

1. Which idea explains Earth's history by examining present conditions on Earth?
   A. absolute-age dating
   B. catastrophism
   C. relative-age dating
   D. uniformitarianism

2. Which part of a dinosaur is least likely to be fossilized?
   A. bone
   B. brain
   C. horn
   D. tooth

3. Which makes a species a good index fossil?
   A. lived a long time and was abundant
   B. lived a long time and was scarce
   C. lived a short time and was scarce
   D. lived a short time and was abundant

4. In the drawing below, what is the order of rock layers from oldest to youngest?

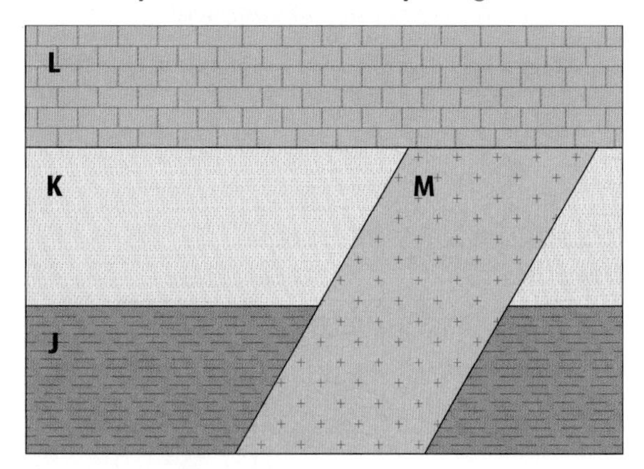

   A. J, K, L, M
   B. J, K, M, L
   C. L, K, J, M
   D. M, J, K, L

5. What do geologists look for in order to correlate rocks in different locations?
   A. different rock types and similar fossils
   B. many rock types and many fossils
   C. similar rock types and lack of fossils
   D. similar rock types and similar fossils

6. What is the half-life on the graph below?

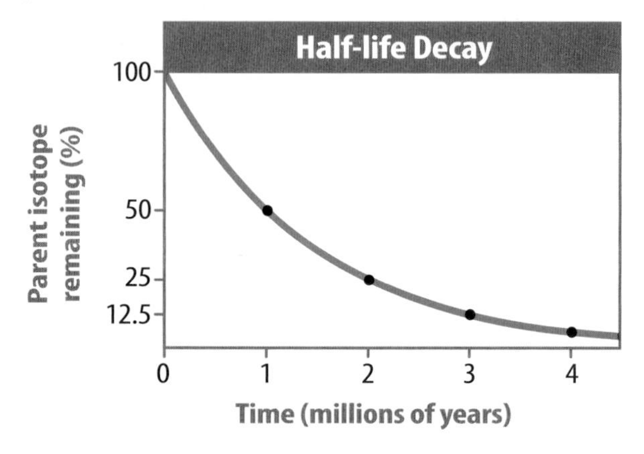

   A. 1 million years
   B. 2 million years
   C. 3 million years
   D. 4 million years

7. What are isotopes?
   A. atoms of the same element with different numbers of electrons but the same number of protons
   B. atoms of the same element with different numbers of electrons but the same number of neutrons
   C. atoms of the same element with different numbers of neutrons but the same number of protons
   D. atoms of the same element with equal numbers of neutrons and protons.

8. What do scientists measure when determining the absolute age of a rock?
   A. amount of radioactivity
   B. number of uranium atoms
   C. ratio of neutrons and electrons
   D. ratio of parent and daughter isotopes

9. Why is radiometric dating less useful to date sedimentary rocks than igneous rocks?
   A. Sedimentary rocks are more eroded.
   B. Sedimentary rocks contain fossils.
   C. Sedimentary rocks contain grains formed from other rocks.
   D. Sedimentary rocks contain grains less than 60,000 years old.

Assessment
Online Test Practice

Critical Thinking

10. **Give** an example of superposition from your own life.

11. **Suggest** a way that an ancient human might have been preserved as a fossil.

12. **Explain** why scientists use a combination of uniformitarianism and catastrophism ideas to understand Earth.

13. **Reason** You are studying a rock formation that includes layers of folded sedimentary rocks cut by faults and dikes. Describe the geologic principles you would use to determine the relative order of the layers.

14. **Construct** a graph showing the radioactive decay of an unstable isotope with a half-life of 250 years. Label three half-lives.

15. **Assess** The ash layers in the drawing below have been dated as shown. What conclusions can you draw about the ages of each of the layers A, B, and C?

C
Ash deposited 540 mya
B
Ash deposited 730 mya
A

Writing in Science

16. **Write** a paragraph of at least five sentences explaining why absolute-age dating has been more useful than relative-age dating in determining the age of Earth. Include a main idea, supporting details, and concluding sentence.

REVIEW

17. What evidence do scientists use to determine the ages of rocks?

18. The photo below shows many rock layers of the Grand Canyon. Explain how the development of the principle of uniformitarianism might have changed earlier ideas about the age of the Grand Canyon and how it formed.

Math Skills

Math Practice

Use Significant Figures

19. If you begin with 68 g of an isotope, how many grams of the original isotope will remain after four half-lives?

20. The half-life of radon-222 (Rn-222) is 3.823 days.
    a. How long would it take for three half-lives?
    b. What percentage of the original sample would remain after three half-lives?

21. The half-life of Rn-222 is 3.823 days. What was the original mass of a sample of this isotope if 0.0500 g remains after 7.646 days?

Standardized Test Practice

Record your answers on the answer sheet provided by your teacher or on a sheet of paper.

Multiple Choice

1 Which is a copy of a dead organism formed when its impression fills with mineral deposits or sediments?

A carbon film

B cast

C mold

D trace fossil

Use the diagram below to answer question 2.

2 In the diagram above, which rock layer typically is youngest?

A 1

B 2

C 3

D 4

3 Which characteristic of rocks does radioactive decay measure?

A absolute age

B lateral continuity

C relative age

D unconformity

4 Which increases the likelihood that a dead organism will be fossilized?

A fast decay of bones

B presence of few hard body parts

C quick burial after death

D vast amounts of skin

Use the diagram below to answer question 5.

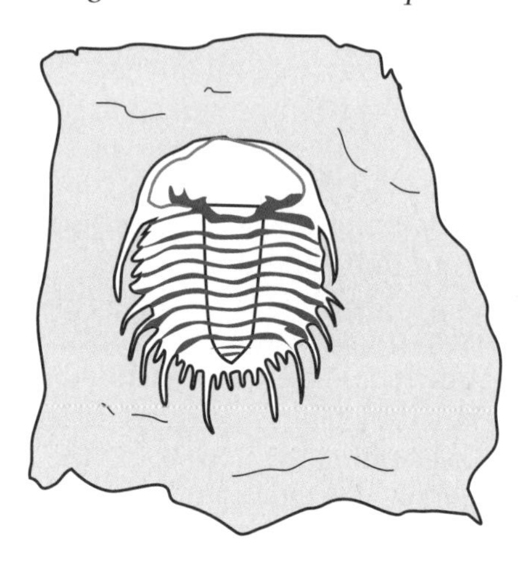

5 Which fossilized ancient organism is pictured in the diagram above?

A clam

B mammoth

C mastodon

D trilobite

6 Which explains most of Earth's geological features as a result of short periods of earthquakes, volcanoes, and meteorite impacts?

A catastrophism

B evolution

C supernaturalism

D uniformitarianism

7 Which fossil type helps geologists infer that rock layers in different geographic locations are similar in age?

A carbon film

B index fossil

C preserved remains

D trace fossil

Assessment
Online Standardized Test Practice

8 Which pie chart shows the ratio of parent to daughter atoms after four half-lives?

A

Parent
Daughter

B

Parent
Daughter

C

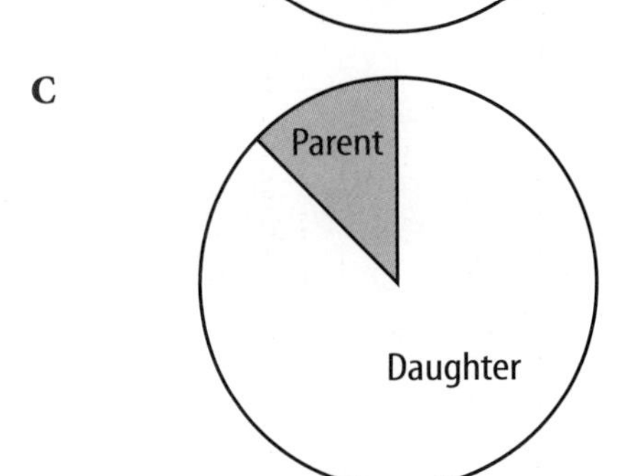

D

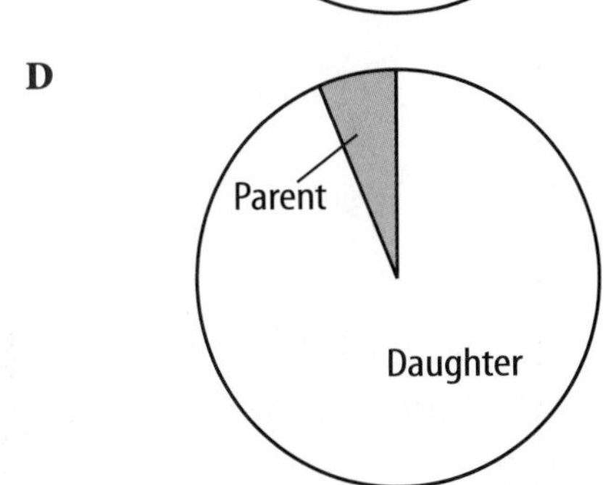

Constructed Response

Use the diagram below to answer questions 9 and 10.

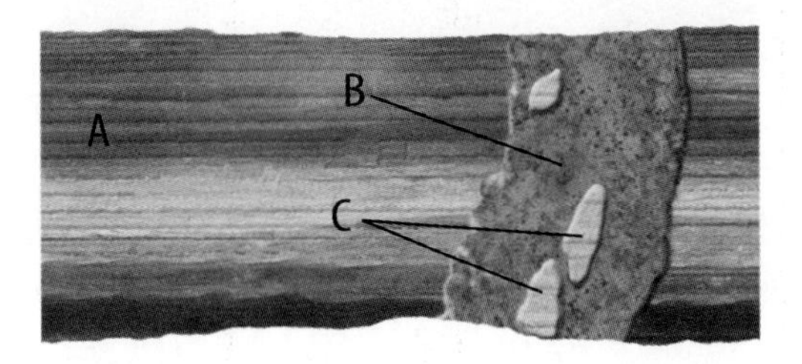

9 Are the sedimentary rock layers (A) older or younger than the dike (B)? How do you know?

10 Is the dike (B) older or younger than the inclusions (C)? How do you know?

Use the diagram below to answer question 11.

11 Identify the type of unconformity that exists in the diagram above. Hypothesize how this could have happened.

12 What is C-14? What role does it play in radiocarbon dating? Why does time limit the effectiveness of radiocarbon dating as a tool for measuring age?

NEED EXTRA HELP?												
If You Missed Question...	1	2	3	4	5	6	7	8	9	10	11	12
Go to Lesson...	1	2	3	1	1	1	2	3	2	2	2	3

Chapter 17

Geologic Time

What have scientists learned about Earth's past by studying rocks and fossils?

Inquiry What happened to the dinosaurs?

This Triceratops lived millions of years ago. Hundreds of other kinds of dinosaurs lived at the same time. Some were as big as houses; others were as small as chickens. Scientists learn about dinosaurs by studying their fossils. Like many organisms that have lived on Earth, dinosaurs disappeared suddenly. Why did the dinosaurs disappear?

- How has Earth changed over geologic time?
- How do geologic events affect life on Earth?
- What have scientists learned about Earth's past by studying rocks and fossils?

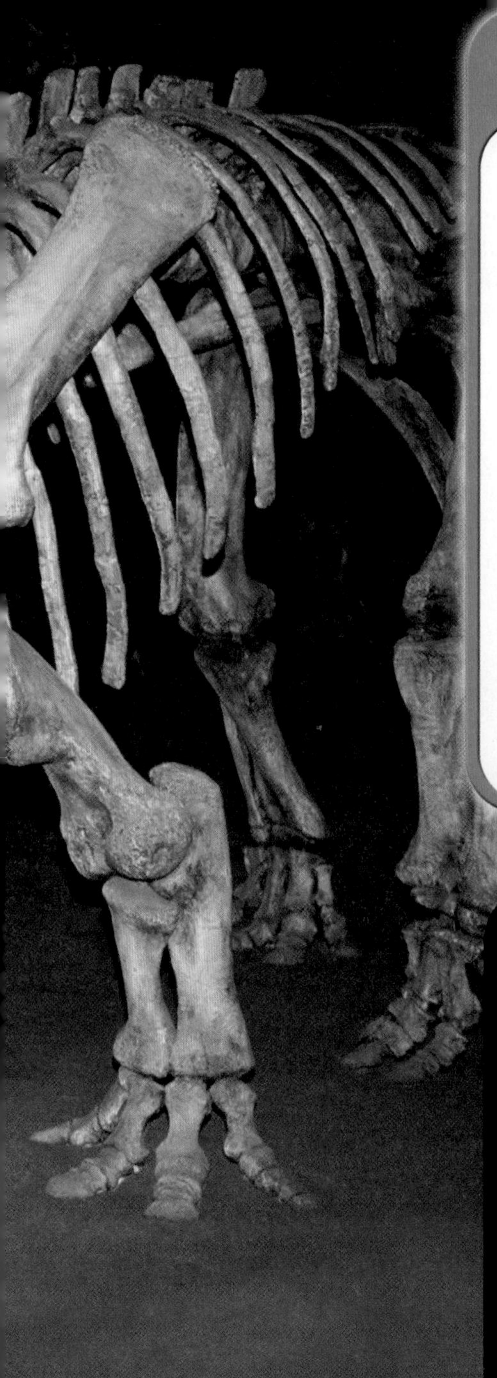

Get Ready to Read

What do you think?

Before you read, decide if you agree or disagree with each of these statements. As you read this chapter, see if you change your mind about any of the statements.

1. All geologic eras are the same length of time.
2. Meteorite impacts cause all extinction events.
3. North America was once on the equator.
4. All of Earth's continents were part of a huge supercontinent 250 million years ago.
5. All large Mesozoic vertebrates were dinosaurs.
6. Dinosaurs disappeared in a large mass extinction event.
7. Mammals evolved after dinosaurs became extinct.
8. Ice covered nearly one-third of Earth's land surface 10,000 years ago.

Your one-stop online resource

connectED.mcgraw-hill.com

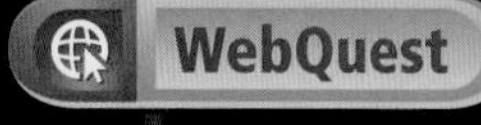

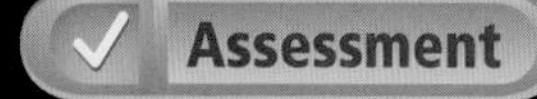

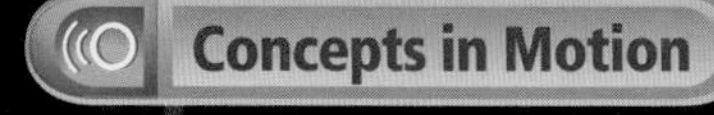

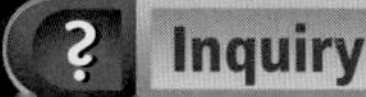

Multilingual eGlossary

Lesson 1

Reading Guide

Key Concepts
ESSENTIAL QUESTIONS

- How was the geologic time scale developed?
- What are some causes of mass extinctions?
- How is evolution affected by environmental change?

Vocabulary

eon p. 601
era p. 601
period p. 601
epoch p. 601
mass extinction p. 603
land bridge p. 604
geographic isolation p. 604

 Multilingual eGlossary

 Video

- BrainPOP®
- Science Video

Geologic History and the Evolution of Life

Inquiry What happened here?

A meteorite 50 m in diameter crashed into Earth 50,000 years ago. The force of the impact created this crater in Arizona and threw massive amounts of dust and debris into the atmosphere. Scientists hypothesize that a meteorite 200 times this size—the size of a small city—struck Earth 65 million years ago. How might it have affected life on Earth?

Launch Lab

20 minutes

Can you make a time line of your life?

How would you organize a time line of your life? You might include regular events, such as birthdays. But you might also include special events, such as a weekend camping trip or a summer vacation.

1. Read and complete a lab safety form.
2. Use **scissors** to cut two pieces of **graph paper** in half. **Tape** them together to make one long piece of paper. Write down the years of your life in horizontal sequence, marked off at regular intervals.
3. Choose up to 12 important events or periods of time in your life. Mark those events on your time line.

Think About This

1. Do the events on your time line appear at regular intervals?
2. **Key Concept** How do you think the geologic time scale is like a time line of your life?

Developing a Geologic Time Line

Think about what you did over the last year. Maybe you went on vacation during the summer or visited relatives in the fall. To organize events in your life, you use different units of time, such as weeks, months, and years. Geologists organize Earth's past in a similar way. They developed a time line of Earth's past called the geologic time scale. As shown in **Figure 1,** time units on the geologic time scale are thousands and millions of years long—much longer than the units you use to organize events in your life.

Units in the Geologic Time Scale

Eons *are the longest units of geologic time.* Earth's current eon, the Phanerozoic (fan er oh ZOH ihk) eon, began 542 million years ago (mya). *Eons are subdivided into smaller units of time called* **eras.** *Eras are subdivided into* **periods.** *Periods are subdivided into* **epochs** (EH pocks). Epochs are not shown on the time line in **Figure 1.** Notice that the time units are not equal. For example, the Paleozoic era is longer than the Mesozoic and Cenozoic eras combined.

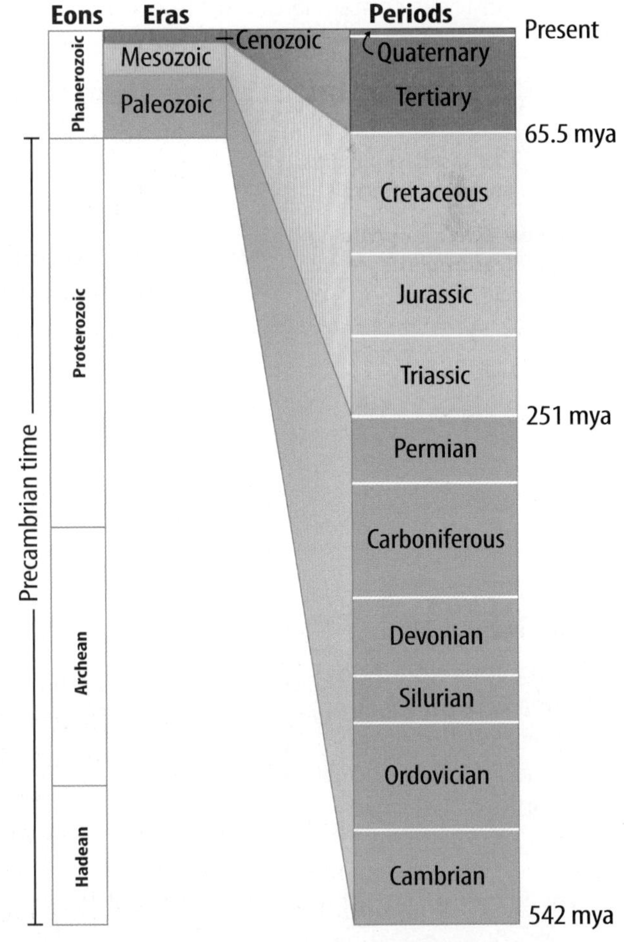

Figure 1 In the geologic time scale, the 4.6 billion years of Earth's history are divided into time units of unequal length.

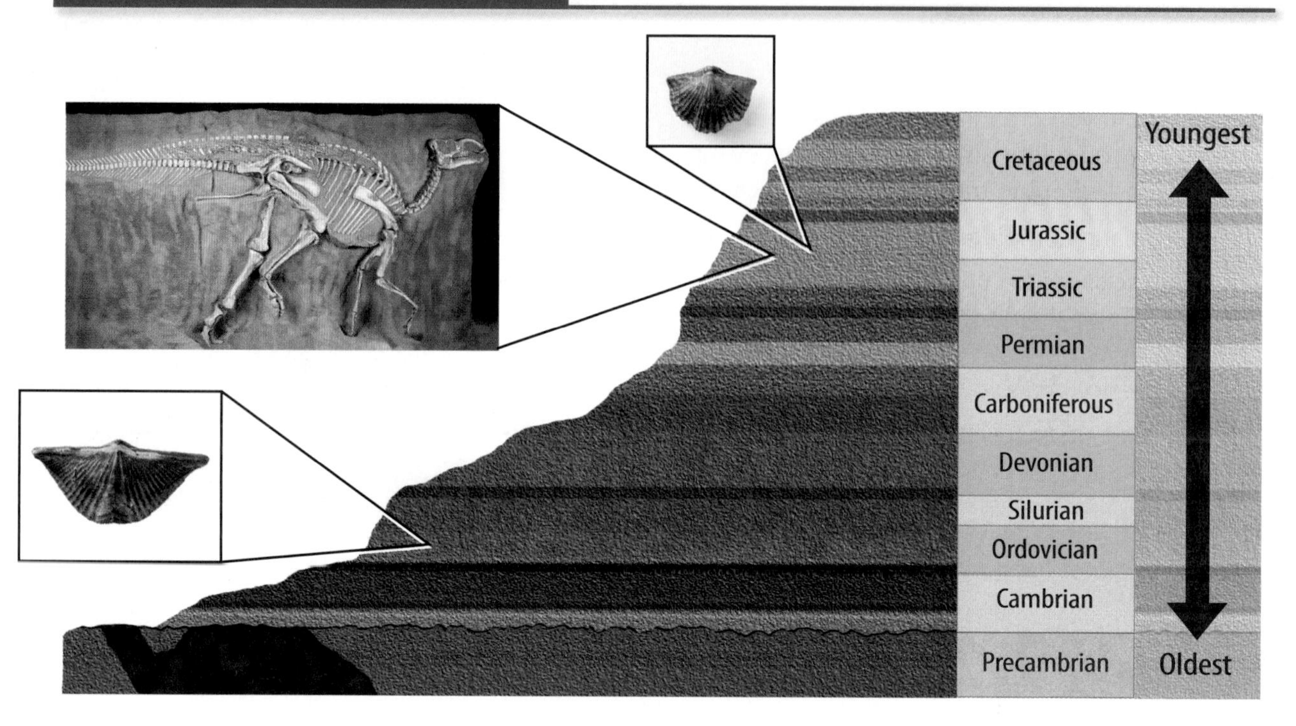

Figure 2 Both older and younger rocks contain fossils of small, relatively simple life-forms. Only younger rocks contain larger, more complex fossils.

Science Use v. Common Use
scale
Science Use a series of marks or points at known intervals
Common Use an instrument used for measuring the weight of an object

The Time Scale and Fossils

Hundreds of years ago, when geologists began developing the geologic time **scale,** they chose the time boundaries based on what they observed in Earth's rock layers. Different layers contained different fossils. For example, older rocks contained only fossils of small, relatively simple life-forms. Younger rocks contained these fossils as well as fossils of other, more complex organisms, such as dinosaurs, as illustrated in **Figure 2.**

Major Divisions in the Geologic Time Scale

While studying the fossils in rock layers, geologists often saw abrupt changes in the types of fossils within the layers. Sometimes, fossils in one rock layer did not appear in the rock layers right above it. It seemed as though the organisms that lived during that period of time had disappeared suddenly. Geologists used these sudden changes in the fossil record to mark divisions in geologic time. Because the changes did not occur at regular intervals, the boundaries between the units of time in the geologic time scale are irregular. This means the time units are of unequal length.

The time scale is a work in progress. Scientists debate the placement of the boundaries as they make new discoveries.

Foldables®

Make a four-door book from a vertical sheet of paper. Use it to organize information about the units of geologic time.

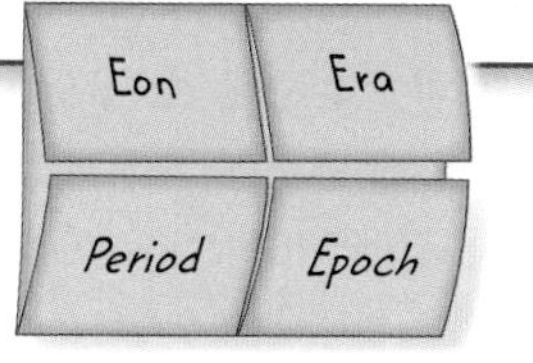

Key Concept Check Why are fossils important in the development of the geologic time scale?

Responses to Change

Sudden changes in the fossil record represent times when large populations of organisms died or became extinct. *A* **mass extinction** *is the extinction of many species on Earth within a short period of time.* As shown in **Figure 3,** there have been several mass extinction events in Earth's history.

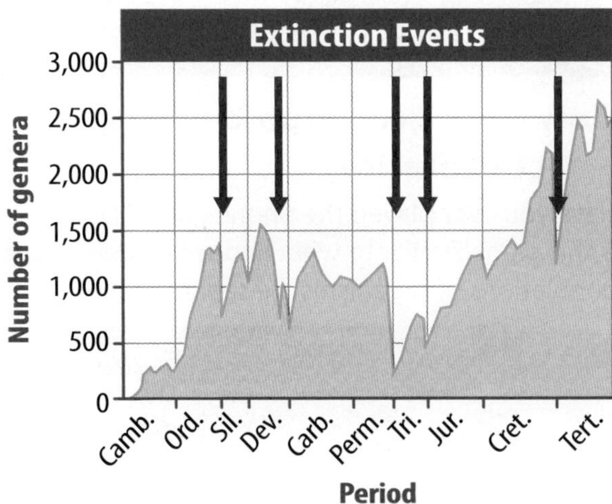

Figure 3 There have been five major mass extinctions in Earth's history. In each one, the number of genera—groups of species—decreased sharply.

Visual Check When was Earth's greatest mass-extinction event?

Changes in Climate

What could cause a mass extinction? All species of organisms depend on the environment for their survival. If the environment changes quickly and species do not adapt to the change, they die.

Many things can cause a climate change. For example, gas and dust from volcanoes can block sunlight and reduce temperatures. As you read on the first page of this lesson, the results of a meteorite crashing into Earth would block sunlight and change climate.

Scientists hypothesize that a meteorite impact might have caused the mass extinction that occurred when dinosaurs became **extinct.** Evidence for this impact is in a clay layer containing the element iridium in rocks around the world. Iridium is rare in Earth rocks but common in meteorites. No dinosaur fossils have been found in rocks above the iridium layer. A sample of rock containing this layer is shown in **Figure 4.**

Word Origin
extinct
from Latin *extinctus,* means "dying out"

Key Concept Check Describe a possible event that could cause a mass extinction.

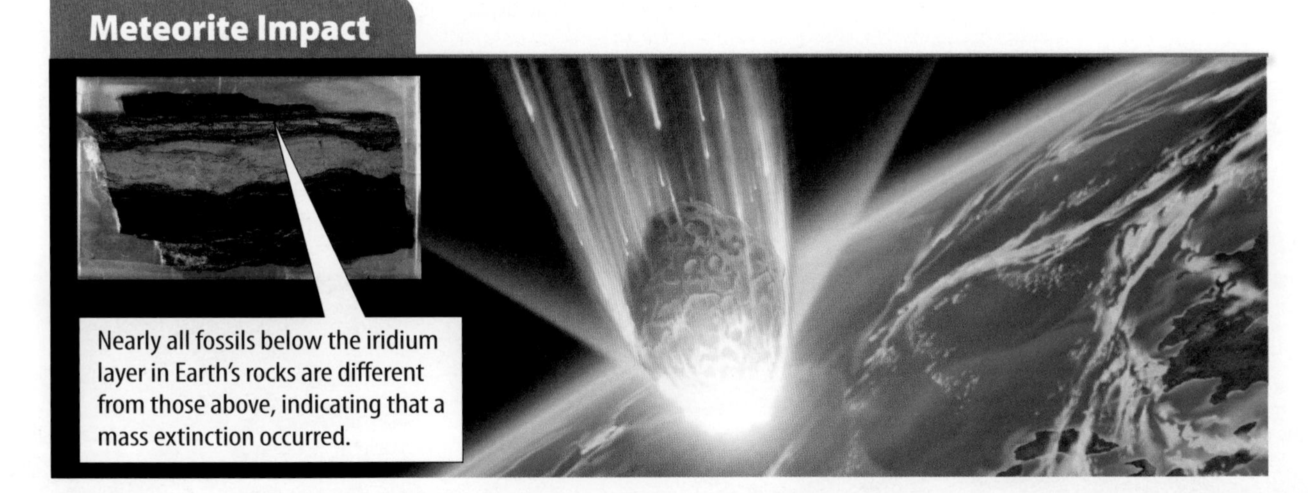

Figure 4 An iridium-enriched clay layer in Earth's rocks is evidence that a large meteorite crashed into Earth 65 million years ago. A meteorite impact can contribute to a mass extinction event.

Inquiry MiniLab

10 minutes

How does geographic isolation affect evolution?

Have you ever played the phone game? How is this game similar to what happens when populations of organisms are separated?

1. Form two groups.
2. One person in each group should whisper a sentence—provided by your teacher—into the ear of his or her neighbor. Each person in turn will whisper the sentence to his or her neighbor until it returns to the first person.

Analyze and Conclude

1. **Observe** Did the sentence change? Did it change in the same way in each group?
2. **Key Concept** How is this activity similar to organisms that are geographically isolated?

Geography and Evolution

When environments change, some species of organisms are unable to adapt. They become extinct. However, other species do adapt to environmental changes. Evolution is the change in species over time as they adapt to their environments. Sudden, catastrophic changes in the environment can affect evolution. So can the slow movement of Earth's tectonic plates.

Land Bridges When continents collide or when sea level drops, landmasses can join together. *A* **land bridge** *connects two continents that were previously separated.* Over time, organisms move across land bridges and evolve as they adapt to new environments.

Geographic Isolation The movement of tectonic plates or other slow geologic events can cause geographic areas to move apart. When this happens, populations of organisms can become isolated. **Geographic isolation** *is the separation of a population of organisms from the rest of its species due to some physical barrier, such as a mountain range or an ocean.* Separated populations of species evolve in different ways as they adapt to different environments. Even slight differences in environments can affect evolution, as shown in **Figure 5.**

Key Concept Check How can geographic isolation affect evolution?

Geographic Isolation

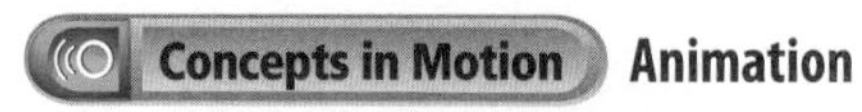

Figure 5 A population of squirrels was gradually separated as the Grand Canyon developed. Each group adapted to a slightly different environment and evolved in a different way

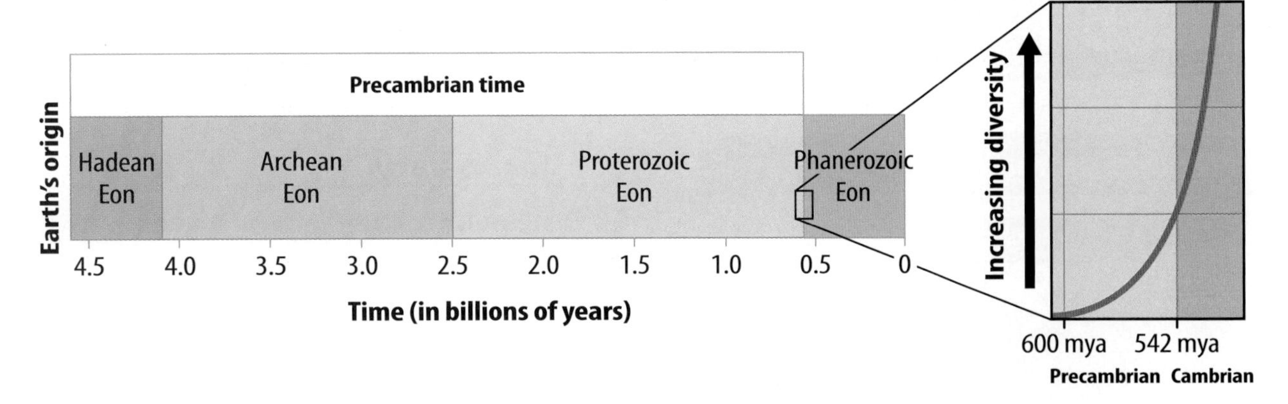

Figure 6 Precambrian time is nearly 90 percent of Earth's history. An explosion of life-forms appeared at the beginning of the Phanerozoic eon, during the Cambrian period.

Personal Tutor

Precambrian Time

Life has been evolving on Earth for billions of years. The oldest fossil evidence of life on Earth is in rocks that are about 3.5 billion years old. These ancient life-forms were simple, unicellular organisms, much like present-day bacteria. The oldest fossils of multicellular organisms are about 600 million years old. These fossils are rare, and early geologists did not know about them. They hypothesized that multicellular life first appeared in the Cambrian (KAM bree un) period, at the beginning of the Phanerozoic eon 542 mya. Time before the Cambrian was called Precambrian time. Scientists have determined that Precambrian time is nearly 90 percent of Earth's history, as shown in **Figure 6.**

Precambrian Life

The rare fossils of multicellular life-forms in Precambrian rocks are from soft-bodied organisms different from organisms on Earth today. Many of these species became extinct at the end of the Precambrian.

Cambrian Explosion

Precambrian life led to a sudden appearance of new types of multicellular life-forms in the Cambrian period. This sudden appearance of new, complex life-forms, indicated on the right in **Figure 6,** is often referred to as the Cambrian explosion. Some Cambrian life-forms, such as trilobites, were the first to have hard body parts. The trilobite fossils shown in **Figure 7,** are preserved in limestone. Because of their hard body parts, trilobites were more easily preserved. More evidence of trilobites is in the fossil record. Scientists hypothesize that some of them are distant ancestors of organisms alive today.

Reading Check What is the Cambrian explosion?

Figure 7 The hard body parts of these trilobites were preserved as fossils.

Lesson 1 Review

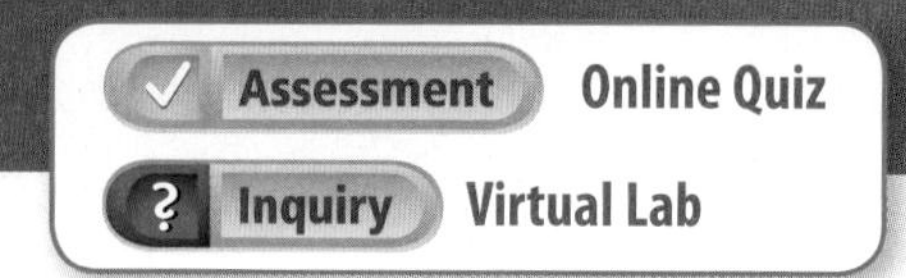

Visual Summary

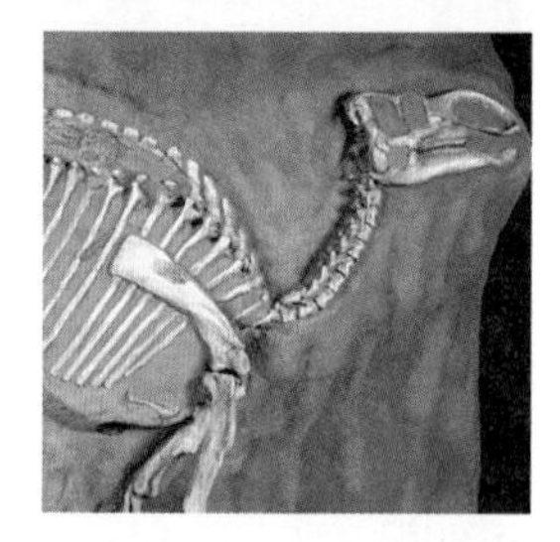

Earth's history is organized into eons, eras, periods, and epochs.

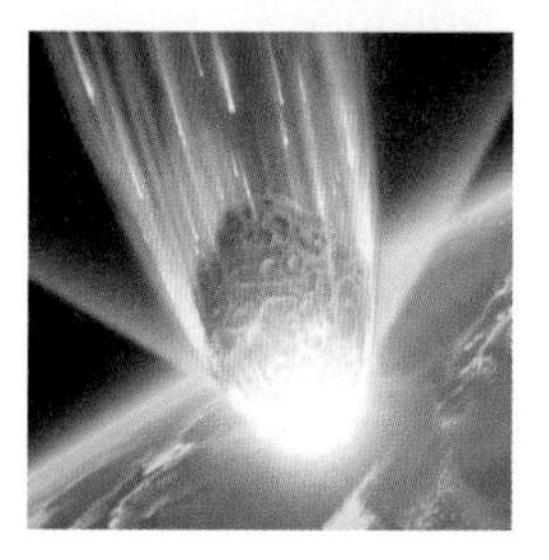

Climate change caused by the results of a meteorite impact could contribute to a mass extinction event.

Slow changes in geography affect evolution.

Use your lesson Foldable to review the lesson. Save your Foldable for the project at the end of the chapter.

What do you think NOW?

You first read the statements below at the beginning of the chapter.

1. All geologic eras are the same length of time.
2. Meteorite impacts cause all extinction events.

Did you change your mind about whether you agree or disagree with the statements? Rewrite any false statements to make them true.

Use Vocabulary

1. **Distinguish** between an eon and an era.
2. A(n) ________ might form when continents move close together.
3. A(n) ________ might occur if an environment changes suddenly.

Understand Key Concepts 

4. Which could contribute to a mass-extinction event?
   - **A.** an earthquake
   - **B.** a hot summer
   - **C.** a hurricane
   - **D.** a volcanic eruption
5. **Explain** how geographic isolation can affect evolution.
6. **Distinguish** between a calendar and the geologic time scale.

Interpret Graphics

7. **Explain** what the graph below represents. What happened at this time in Earth's past?

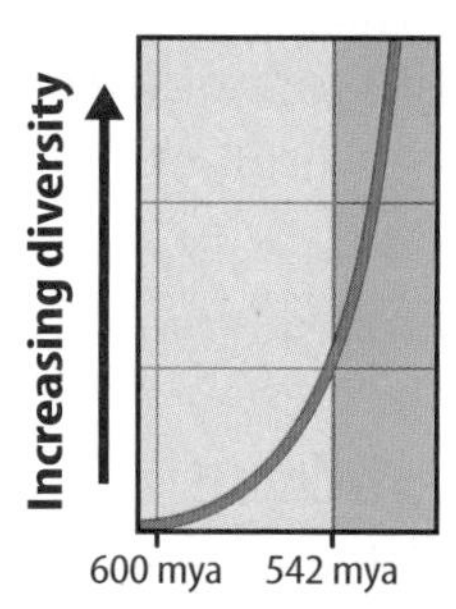

8. **Organize Information** Copy and fill in the graphic organizer below to show units of the geologic time scale from longest to shortest.

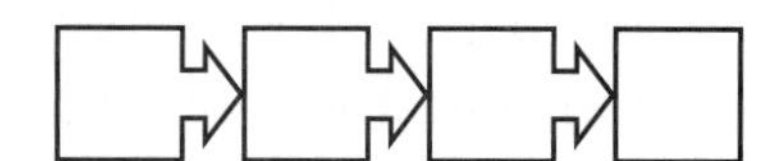

Critical Thinking

9. **Suggest** how humans might contribute to a mass extinction event.
10. **Propose** why Precambrian rocks contain few fossils.

How has life changed over time?

Fossil evidence indicates that there have been wide fluctuations in the types, or diversity, of organisms that have lived on Earth over geologic time.

Learn It

Line graphs compare two variables and show how one variable changes in response to another variable. Line graphs are particularly useful in presenting data that change over time. The first line graph below shows how the diversity of genera has changed over time. The second graph shows how extinction rates, presented as percentages of genera, have changed over time. **Interpret data** in these graphs to learn how they relate to each another.

Try It

1. Carefully study each graph. Note that time, the independent variable, is plotted on the *x*-axis of each graph. The dependent variable of each graph—the diversity, or number of genera, in one graph and the extinction rate in the other graph—are plotted on the *y*-axes.
2. Use the graphs to answer questions 3–7.

Apply It

3. According to the graph on the left, at what time in Earth's past was diversity the lowest? At what time was diversity the highest?
4. Approximately what percentage of genera became extinct 250 million years ago?
5. Approximately when did each of Earth's major mass extinctions take place?
6. What is the relationship between diversity and extinction rate?
7. **Key Concept** How have mass extinctions helped scientists develop the geologic time scale?

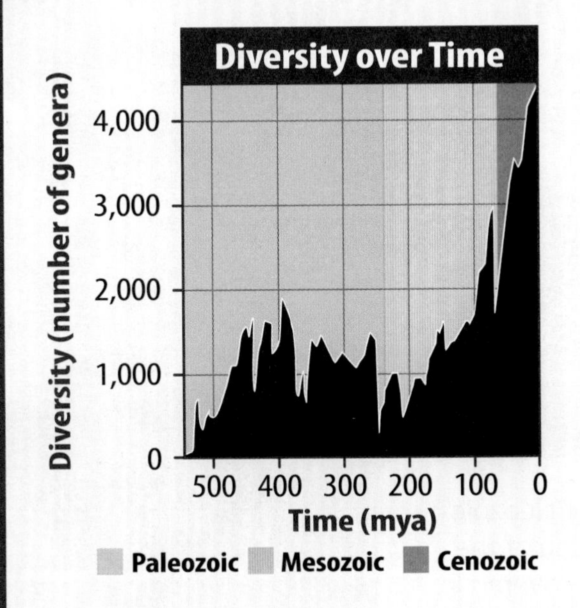

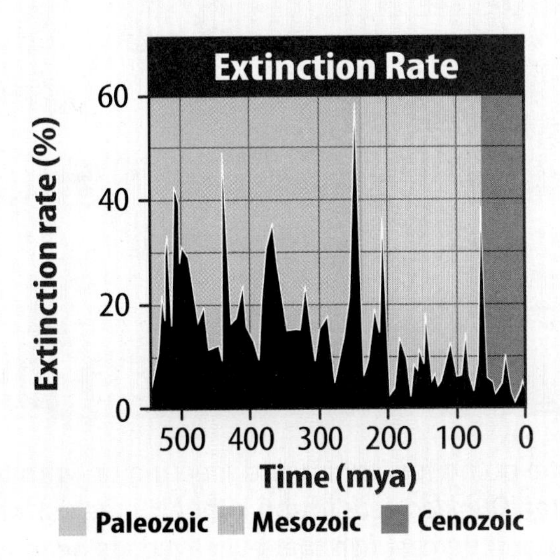

Lesson 2

The Paleozoic Era

Reading Guide

Key Concepts
ESSENTIAL QUESTIONS

- What major geologic events occurred during the Paleozoic era?
- What does fossil evidence reveal about the Paleozoic era?

Vocabulary

Paleozoic era p. 609
Mesozoic era p. 609
Cenozoic era p. 609
inland sea p. 610
coal swamp p. 612
supercontinent p. 613

g Multilingual eGlossary

Inquiry What animal was this?

Imagine going for a swim and meeting up with this Paleozoic monster. *Dunkleosteus* (duhn kuhl AHS tee us) was one of the largest and fiercest fish that ever lived. Its head was covered in bony armor 5 cm thick—even its eyes had bony armor. It had razor-sharp teethlike plates that bit with a force like that of present-day alligators.

Launch Lab

20 minutes

What can you learn about your ancestors?

Scientists use fossils and rocks to learn about Earth's history. What could you use to research your past?

1. Write as many facts as you can about one of your grandparents or other older adult family members or friends.
2. What items, such as photos, do you have that can help you?

Think About This

1. If you wanted to know about a great-great-great grandparent, what clues do you think you could find?
2. How does knowledge about past generations in your family benefit you today?
3. **Key Concept** How do you think learning about distant relatives is like studying Earth's past?

Early Paleozoic

In many families, three generations—grandparents, parents, and children—live closely together. You could call them the old generation, the middle generation, and the young generation. These generations are much like the three eras of the Phanerozoic eon. *The* **Paleozoic** (pay lee uh ZOH ihk) **era** *is the oldest era of the Phanerozoic eon. The* **Mesozoic** (mez uh ZOH ihk) **era** *is the middle era of the Phanerozoic eon. The* **Cenozoic** (sen uh ZOH ihk) **era** *is the youngest era of the Phanerozoic eon.*

Word Origin

Paleozoic
from Greek *palai,* means "ancient"; and Greek *zoe,* means "life"

As shown in **Figure 8,** the Paleozoic era lasted for more than half the Phanerozoic eon. Because it was so long, it is often divided into three parts: early, middle, and late. The Cambrian and Ordovician periods make up the Early Paleozoic.

The Age of Invertebrates

The organisms from the Cambrian explosion were invertebrates (ihn VUR tuh brayts) that lived only in the oceans. Invertebrates are animals without backbones. So many kinds of invertebrates lived in Early Paleozoic oceans that this time is often called the age of invertebrates.

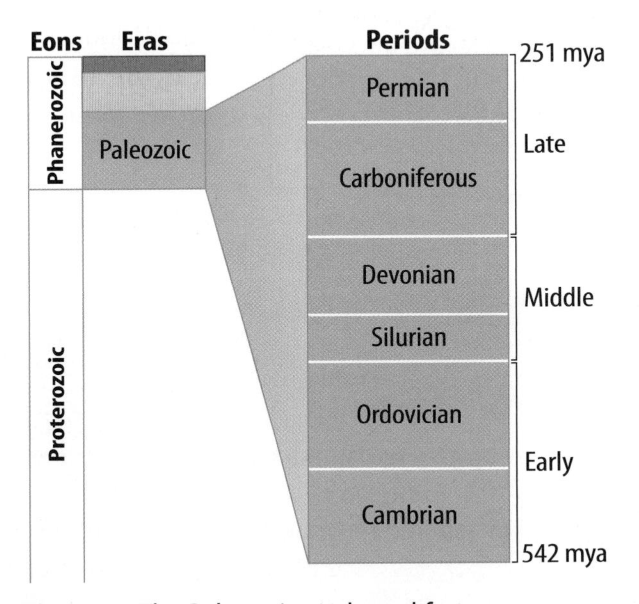

Figure 8 The Paleozoic era lasted for 291 million years. It is divided into six periods.

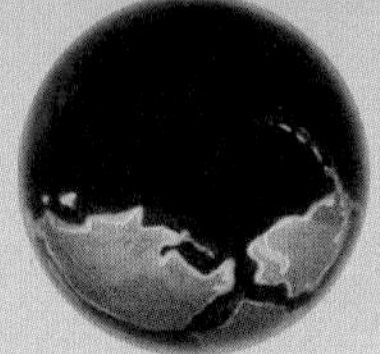

Cambrian Period
542 – 488
million years ago

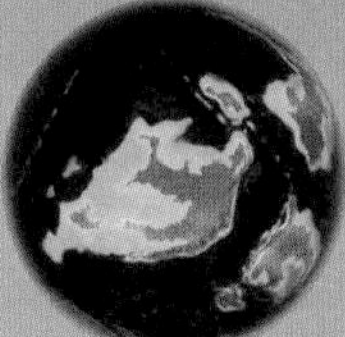

Ordovician Period
488 – 444
million years ago

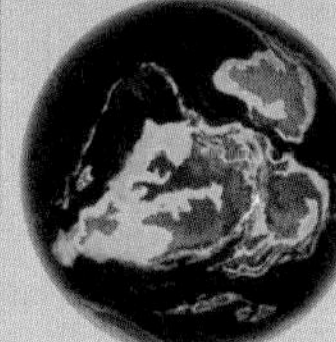

Silurian Period
444 – 416
million years ago

Figure 9 Earth's continents and life-forms changed dramatically during the Paleozoic era.

Visual Check In what period did life first appear on land?

Geology of the Early Paleozoic

If you could have visited Earth during the Early Paleozoic, it would have seemed unfamiliar to you. As shown in **Figure 9**, there was no life on land. All life was in the oceans. The shapes and locations of Earth's continents also would have been unfamiliar, as shown in **Figure 10.** Notice that the landmass that would become North America was on the equator.

Earth's climate was warm during the Early Paleozoic. Rising seas flooded the continents and formed many shallow inland seas. *An* **inland sea** *is a body of water formed when ocean water floods continents.* Most of North America was covered by an inland sea.

Reading Check How do inland seas form?

FOLDABLES®

Make a horizontal, three-tab book. Label it as shown. Use your book to record information about changes during the Paleozoic Era.

Early Paleozoic | Middle Paleozoic | Late Paleozoic

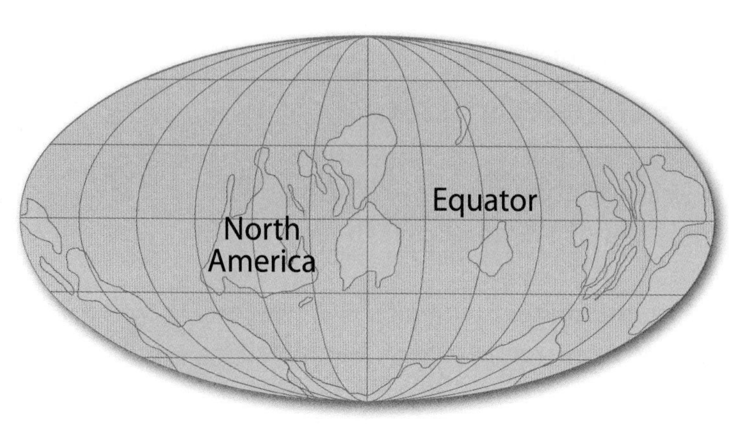

Figure 10 During the Early Paleozoic, North America straddled the equator.

Middle Paleozoic

The Early Paleozoic ended with a mass extinction event, but many invertebrates survived. New forms of life lived in huge coral reefs along the edges of the continents. Soon, animals with backbones, called vertebrates, evolved.

The Age of Fishes

Some of the earliest vertebrates were fishes. So many types of fishes lived during the Silurian (suh LOOR ee un) and Devonian (dih VOH nee un) periods that the Middle Paleozoic is often called the age of fishes. Some fishes, such as the *Dunkleosteus* pictured at the beginning of this lesson, were heavily armored. **Figure 11** also shows what a *Dunkleosteus* might have looked like. On land, cockroaches, dragonflies, and other insects evolved. Earth's first plants appeared. They were small and lived in water.

Figure 11 *Dunkleosteus* was a top Devonian predator.

Geology of the Middle Paleozoic

Middle Paleozoic rocks contain evidence of major collisions between moving continents. These collisions created mountain ranges. When several landmasses collided with the eastern coast of North America, the Appalachian (ap uh LAY chun) Mountains began to form. By the end of the Paleozoic era, the Appalachians were probably as high as the Himalayas are today.

Key Concept Check How did the Appalachian Mountains form?

Late Paleozoic

Like the Early Paleozoic, the Middle Paleozoic ended with a mass extinction event. Many marine invertebrates and some land animals disappeared.

The Age of Amphibians

In the Late Paleozoic, some fishlike organisms spent part of their lives on land. *Tiktaalik* (tihk TAH lihk) was an organism that had lungs and could breathe air. It was one of the earliest amphibians. Amphibians were so common in the Late Paleozoic that this time is known as the age of amphibians.

Ancient amphibian species adapted to land in several ways. As you read, they had lungs and could breathe air. Their skins were thick, which slowed moisture loss. Their strong limbs enabled them to move around on land. However, all amphibians, even those living today, must return to the water to mate and lay eggs.

Reptile species evolved toward the end of the Paleozoic era. Reptiles were the first animals that did not require water for reproduction. Reptile eggs have tough, leathery shells that protect them from drying out.

 **Key Concept Check** How did different species adapt to land?

Coal Swamps

During the Late Paleozoic, dense, tropical forests grew in swamps along shallow inland seas. When trees and other plants died, they sank into the swamps, such as the one illustrated in **Figure 12.** *A* **coal swamp** *is an oxygen-poor environment where, over time, plant material changes into coal.* The coal swamps of the Carboniferous (car buhn IF er us) and Permian periods eventually became major sources of coal that we use today.

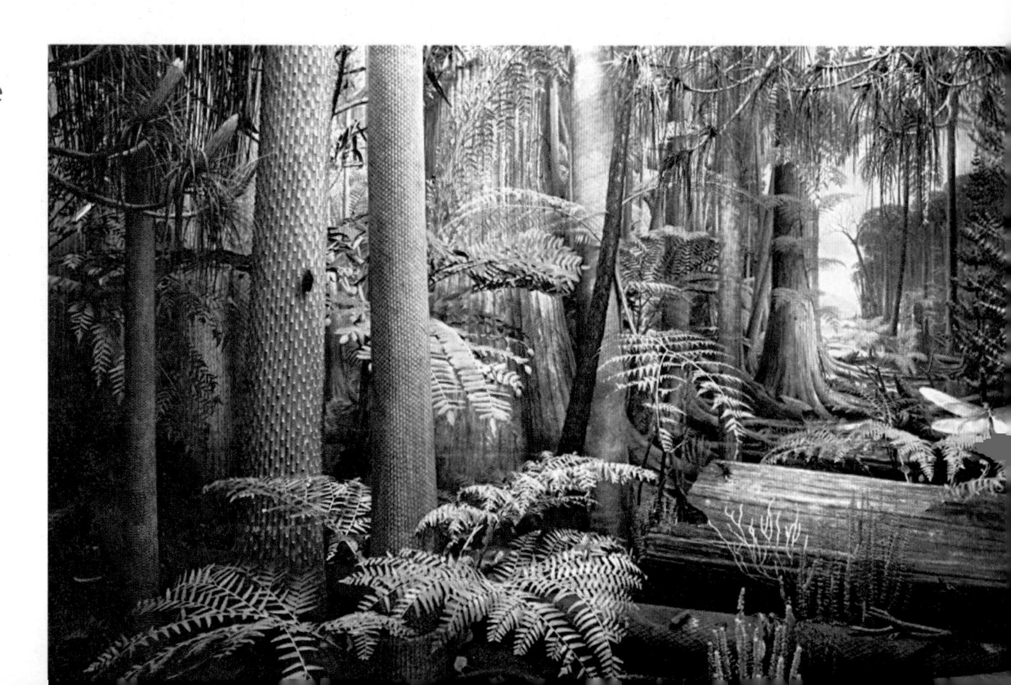

Figure 12 Plants buried in ancient coal swamps became coal.

Formation of Pangaea

Geologic evidence indicates that many continental collisions occurred during the Late Paleozoic. As continents moved closer together, new mountain ranges formed. By the end of the Paleozoic era, Earth's continents had formed a giant supercontinent—Pangaea. A **supercontinent** *is an ancient landmass which separated into present-day continents.* Pangaea formed close to Earth's equator, as shown in **Figure 13.** As Pangaea formed, coal swamps dried up and Earth's climate became cooler and drier.

The Permian Mass Extinction

The largest mass extinction in Earth's history occurred at the end of the Paleozoic era. Fossil evidence indicates that 95 percent of marine life-forms and 70 percent of all life on land became extinct. This extinction event is called the Permian mass extinction.

Key Concept Check What does fossil evidence reveal about the end of the Paleozoic era?

Scientists debate what caused this mass extinction. The formation of Pangaea likely decreased the amount of space where marine organisms could live. It would have contributed to changes in ocean currents, making the center of Pangaea drier. But Pangaea formed over many millions of years. The extinction event occurred more suddenly.

Some scientists hypothesize that a large meteorite impact caused drastic climate change. Others propose that massive volcanic eruptions changed the global climate. Both a meteorite impact and large-scale eruptions would have ejected ash and rock into the atmosphere, blocking out sunlight, reducing temperatures, and causing a collapse of food webs.

Whatever caused it, Earth had fewer species after the Permian mass extinction. Only species that could adapt to the changes survived.

Pangaea

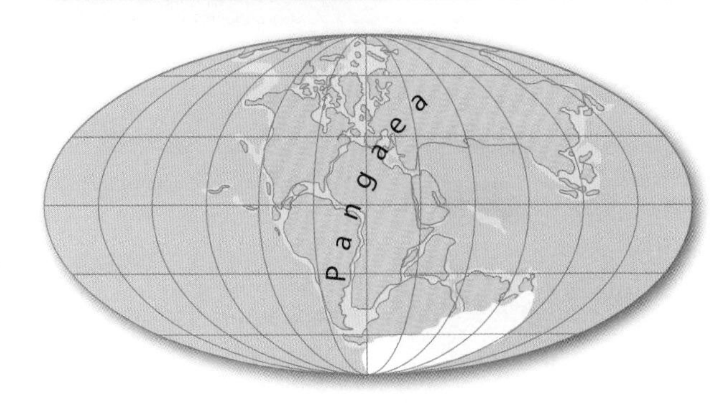

Figure 13 The supercontinent Pangaea formed at the end of the Paleozoic era.

Inquiry MiniLab

20 minutes

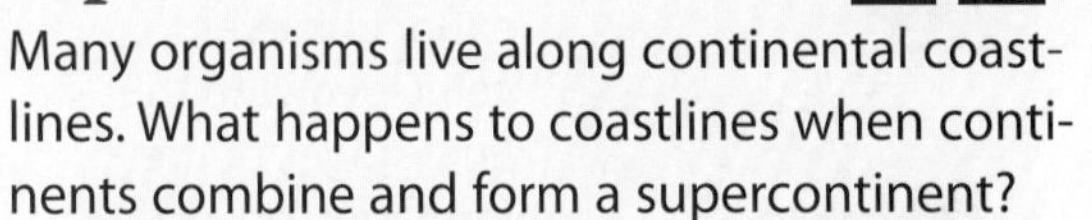

What would happen if a supercontinent formed?

Many organisms live along continental coastlines. What happens to coastlines when continents combine and form a supercontinent?

1. Read and complete a lab safety form.
2. Form a stick of **modeling clay** into a flat pancake shape. Form three pancake shapes from an identical stick of clay. Make all four shapes equal thicknesses.
3. With a **flexible tape measure,** measure the perimeter of each shape.

Analyze and Conclude

1. **Compare** Is the perimeter of the larger shape more or less than the combined perimeters of the three smaller shapes?
2. **Key Concept** How might the formation of Pangaea have affected life on Earth?

Lesson 2 Review

Visual Summary

Life slowly moved to land during the Paleozoic era as amphibians and reptiles evolved.

In the Late Paleozoic, massive coal swamps formed along inland seas.

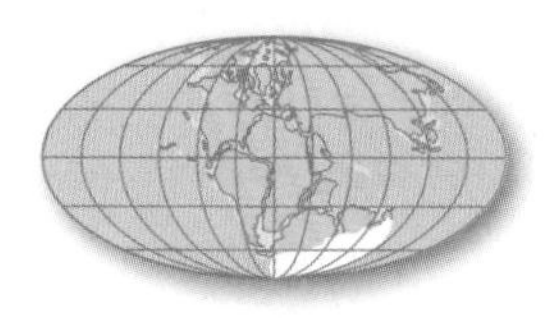

At the end of the Paleozoic era, a mass extinction event coincided with the final stages of the formation of Pangaea.

FOLDABLES®

Use your lesson Foldable to review the lesson. Save your Foldable for the project at the end of the chapter.

What do you think NOW?

You first read the statements below at the beginning of the chapter.

3. North America was once on the equator.

4. All of Earth's continents were part of a huge supercontinent 250 million years ago.

Did you change your mind about whether you agree or disagree with the statements? Rewrite any false statements to make them true.

Use Vocabulary

1. **Distinguish** between the Paleozoic era and the Mesozoic era.
2. When ocean water covers part of a continent, a(n) ________ forms.
3. **Use the term** *supercontinent* in a complete sentence.

Understand Key Concepts 

4. Which was true of North America during the Early Paleozoic?
   - **A.** It had many glaciers.
   - **B.** It was at the equator.
   - **C.** It was part of a supercontinent.
   - **D.** It was populated by reptiles.
5. **Compare** ancient amphibians and reptiles and explain how each group adapted to live on land.
6. **Draw** a cartoon that shows how the Appalachian Mountains formed.

Interpret Graphics

7. **Organize** A time line of the Paleozoic era is pictured below. Copy the time line and fill in the missing periods.

Paleozoic					
	Ordovician	Silurian	Devonian	Carboniferous	

8. **Sequence** Copy and fill in the graphic organizer below. Start with Precambrian time, then list the eras in order.

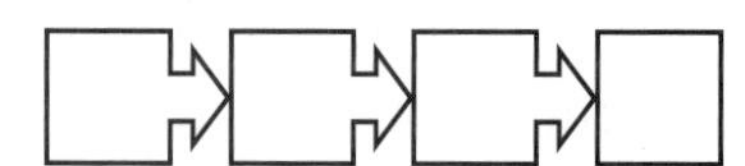

Critical Thinking

9. **Consider** What if 100 percent of organisms had become extinct at the end of the Paleozoic era?
10. **Evaluate** the possible effects of climate change on present-day organisms.

Inquiry **Skill Practice** **Interpret Data** **30 minutes**

When did coal form?

Coal is fossilized plant material. When swamp plants die, they become covered by oxygen-poor water and change to peat. Over time, high temperatures and pressure from sediments transform the peat into coal. When did the plants live that formed the coal we use today?

Learn It

A bar graph can display the same type of information as a line graph. However, instead of data points and a line that connects them, a bar graph uses rectangular bars to show how values compare. **Interpret the data** below to learn when most coal formed.

Try It

1. Carefully study the bar graph. Notice that time is plotted on the *x*-axis (as geologic periods), and that coal deposits (as tons accumulated per year) are plotted on the *y*-axis.
2. Use the graph and what you know about coal formation to answer the following questions.

Apply It

3. Which coal deposits are oldest? Which are youngest?
4. During which geologic period did most of the coal form?
5. Approximately how much coal accumulated during the Paleozoic era? The Mesozoic era?
6. Why are there no data on the graph for the Cambrian, Ordovician, and Silurian periods of geologic time?
7. **Key Concept** What does fossil evidence reveal about the Paleozoic era?

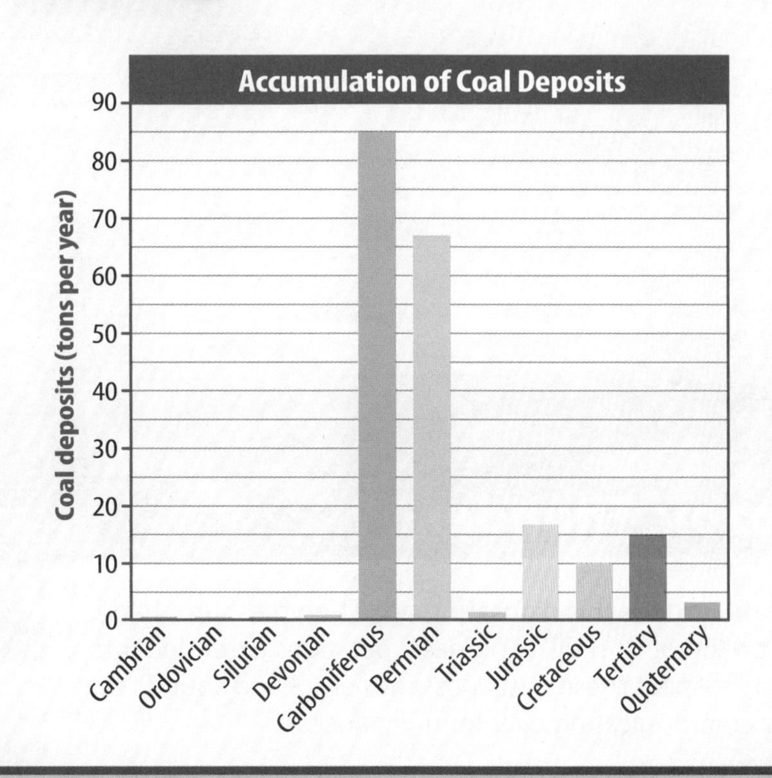

Lesson 3

Reading Guide

Key Concepts
ESSENTIAL QUESTIONS

- What major geologic events occurred during the Mesozoic era?
- What does fossil evidence reveal about the Mesozoic era?

Vocabulary

dinosaur p. 620
plesiosaur p. 621
pterosaur p. 621

Multilingual eGlossary

The Mesozoic Era

Inquiry Mesozoic Thunder?

Can you imagine the sounds this dinosaur made? *Corythosaurus* had a tall, bony crest on top of its skull. Long nasal passages extended into the crest. Scientists suspect these nasal passages amplified sounds that could be used for communicating over long distances.

Launch Lab

20 minutes

How diverse were dinosaurs?

How many different dinosaurs were there?

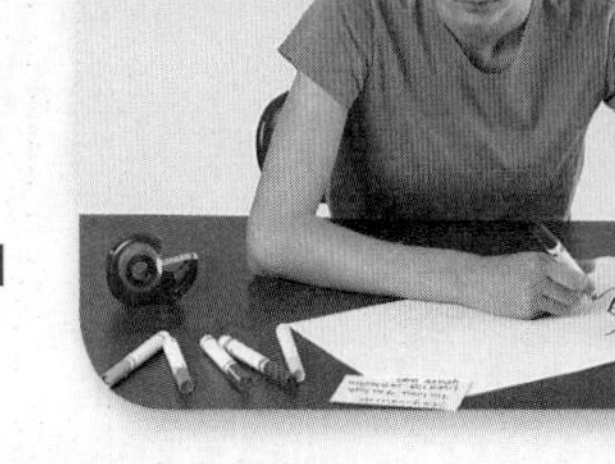

1. Read and complete a lab safety form.
2. Your teacher will give you an **index card** listing a species name of a dinosaur, the dinosaur's dimensions, and the time when it lived.
3. Draw a picture of what you imagine your dinosaur looked like. Before you begin, decide with your classmates what common scale you should use.
4. **Tape** your dinosaur drawing to the Mesozoic time line your teacher provides.

Think About This

1. What was the biggest dinosaur? The smallest? Can you see any trends in size on the time line?
2. Did all the dinosaurs live at the same time?
3. **Key Concept** Dinosaurs were numerous and diverse. Do you think any dinosaurs could swim or fly?

Geology of the Mesozoic Era

When people imagine what Earth looked like millions of years ago, they often picture a scene with dinosaurs, such as the *Corythosaurus* shown on the opposite page. Dinosaurs lived during the Mesozoic era. The Mesozoic era lasted from 251 mya to 65.5 mya. As shown in **Figure 14,** it is divided into three periods: the Triassic (tri A sihk), the Jurassic (joo RA sihk), and the Cretaceous (krih TAY shus).

Breakup of Pangaea

Recall that the supercontinent Pangaea formed at the end of the Paleozoic era. The breakup of Pangaea was the dominant geologic event of the Mesozoic era. Pangaea began to break apart in the Late Triassic. Eventually, Pangaea split into two separate landmasses—Gondwanaland (gahn DWAH nuh land) and Laurasia (la RAY shzah). Gondwanaland was the southern continent. It included the future continents of Africa, Antarctica, Australia, and South America. Laurasia, the northern continent, included the future continents of North America, Europe, and Asia.

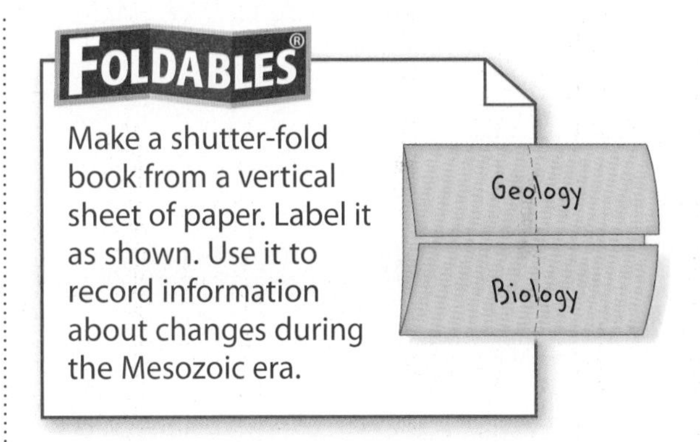

FOLDABLES®

Make a shutter-fold book from a vertical sheet of paper. Label it as shown. Use it to record information about changes during the Mesozoic era.

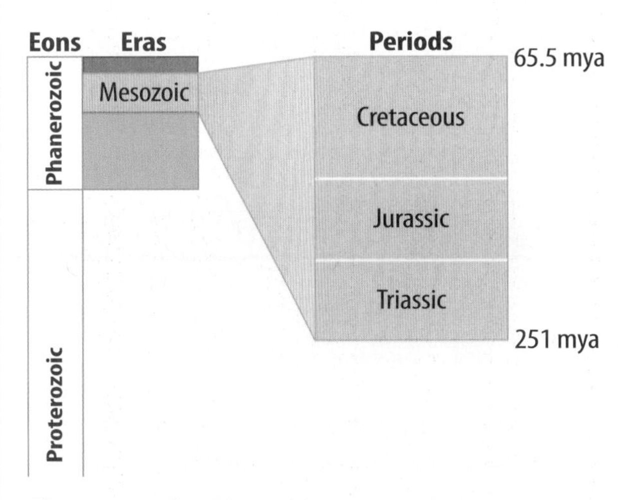

Figure 14 The Mesozoic era was the middle era of the Phanerozoic eon. It lasted for 185.5 million years.

Figure 15 Dinosaurs dominated the Mesozoic era, but many other species also lived during this time in Earth's history.

Return of Shallow Seas

The type of species represented in **Figure 15** adapted to an environment of lush tropical forests and warm ocean waters. That is because the climate of the Mesozoic era was warmer than the climate of the Paleozoic era. It was so warm that, for most of the era, there were no ice caps, even at the poles. With no glaciers, the oceans had more water. Some of this water flowed onto the continents as Pangaea split apart. This created narrow channels that grew larger as the continents moved apart. Eventually, the channels became oceans. The Atlantic Ocean began to form at this time.

Key Concept Check When did the Atlantic Ocean begin to form?

Sea level rose during most of the Mesozoic era, as shown in **Figure 16.** Toward the end of the era, sea level was so high that inland seas covered much of Earth's continents. This provided environments for the evolution of new organisms.

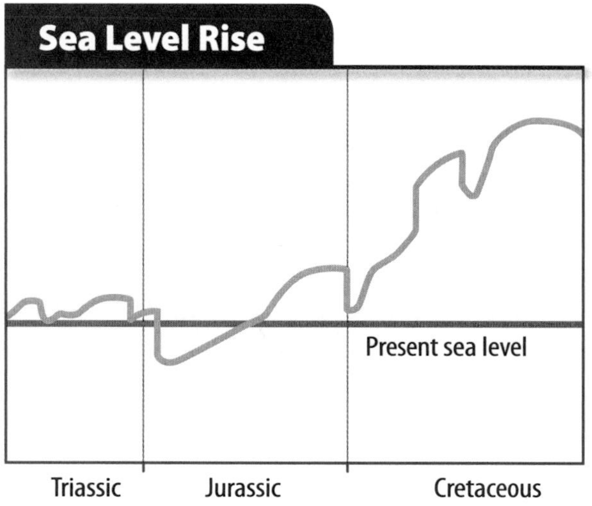

Figure 16 Earth's sea level rose during the Mesozoic era.

Visual Check In which period was sea level at its highest?

Mesozoic North America

Along North America's eastern coast and the Gulf of Mexico, sea level rose and receded over millions of years. As this happened, seawater **evaporated,** leaving massive salt deposits behind. Some of these salt deposits are sources of salt today. Other salt deposits later became traps for oil. Today, salt traps in the Gulf of Mexico are an important source of oil.

Throughout the Mesozoic era, the North American continent moved slowly and steadily westward. Its western edge collided with several small landmasses carried on an ancient oceanic plate. As this plate subducted beneath the North American continent, the crust buckled inland, slowly pushing up the Rocky Mountains, shown on the map in **Figure 17.** In the dry southwest, windblown sand formed huge dunes. In the middle of the continent, a warm inland sea formed.

Key Concept Check How did the Rocky Mountains form?

REVIEW VOCABULARY
evaporated
changed from liquid to gas

Figure 17 The Rocky Mountains began forming during the Mesozoic era. By the end of the era, an inland sea covered much of the central part of North America.

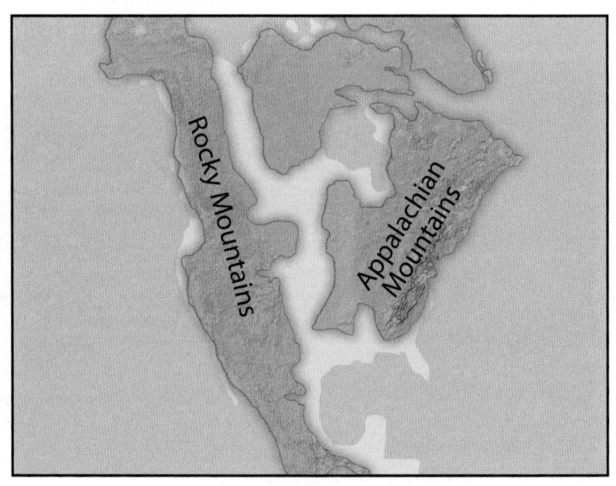

MiniLab — 20 minutes

Can you run like a reptile?

Unlike the limbs of crocodiles and other modern reptiles, dinosaur limbs were positioned directly under their bodies. What did this mean?

1. Pick a partner. One of you—the dinosaur—will run on all fours with arms held straight directly below the shoulders. The other—the crocodile—will run with arms bent and positioned out from the body.
2. Race each other, then reverse positions.

Analyze and Conclude

1. **Compare** Which could move faster—the dinosaur or the crocodile?
2. **Infer** Which posture do you think could support more weight?
3. **Key Concept** How might their posture have enabled dinosaurs to become so successful? How might it have helped them become so large?

Mesozoic Life

The species of organisms that survived the Permian mass extinction event lived in a world with few species. Vast amounts of unoccupied space were open for organisms to inhabit. New types of cone-bearing trees, such as pines and cycads, began to appear. Toward the end of the era, the first flowering plants evolved. Dominant among vertebrates living on land were the dinosaurs. Hundreds of species of many sizes existed.

Dinosaurs

Though dinosaurs have long been considered reptiles, scientists today actively debate dinosaur classification. Dinosaurs share a common ancestor with present-day reptiles, such as crocodiles. However, dinosaurs differ from present-day reptiles in their unique hip structure, as shown in **Figure 18.** **Dinosaurs** *were dominant Mesozoic land vertebrates that walked with legs positioned directly below their hips.* This meant that many walked upright. In contrast, the legs of a crocodile stick out sideways from its body. It appears to drag itself along the ground.

Scientists hypothesize that some dinosaurs are more closely related to present-day birds than they are to present-day reptiles. Dinosaur fossils with evidence of feathery exteriors have been found. For example, *Archaeopteryx* (ar kee AHP tuh rihks), a small bird the size of a pigeon, had wings and feathers but also claws and teeth. Many scientists suggest it was an ancestor to birds.

Figure 18 Fossils provide evidence that the hip structure of a dinosaur enabled it to walk upright.

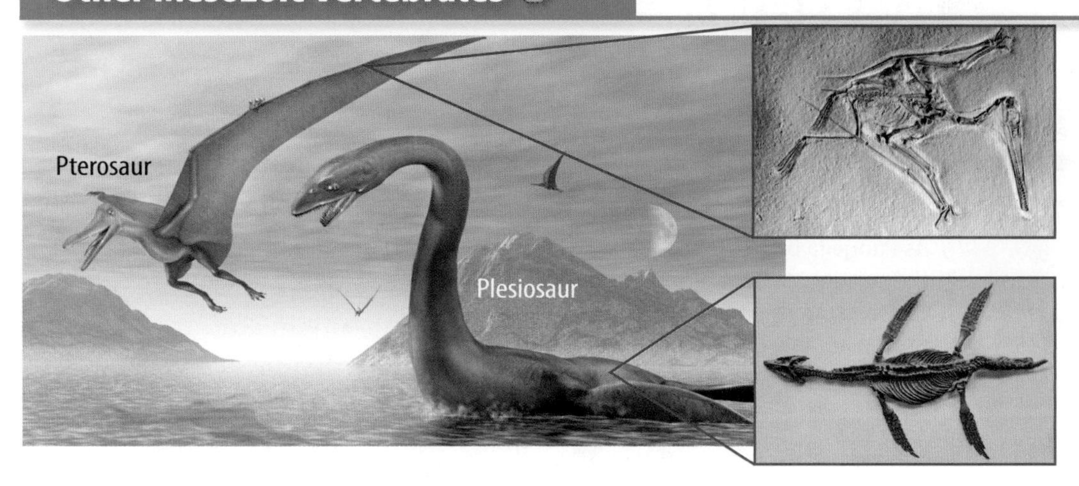

Figure 19 Not all large Mesozoic vertebrates were dinosaurs.

Visual Check How did the limbs of these reptiles compare to the limbs of dinosaurs?

Other Mesozoic Vertebrates

Dinosaurs dominated land. But, fossils indicate that other large vertebrates swam in the seas and flew in the air, as shown in Figure 19. **Plesiosaurs** (PLY zee oh sorz) *were Mesozoic marine reptiles with small heads, long necks, and flippers.* Through much of the Mesozoic, these reptiles dominated the oceans. Some were as long as 14 m.

Other Mesozoic reptiles could fly. **Pterosaurs** (TER oh sorz) *were Mesozoic flying reptiles with large, batlike wings.* One of the largest pterosaurs, the *Quetzalcoatlus* (kwetz oh koh AHT lus), had a wingspread of nearly 12 m. Though they could fly, pterosaurs were not birds. As you have read, birds are more closely related to dinosaurs.

Word Origin
pterosaur
from Greek *pteron,* means "wing"; and *sauros,* means "lizard"

 **Key Concept Check** How could you distinguish fossils of plesiosaurs and pterosaurs from fossils of dinosaurs?

Appearance of Mammals

Dinosaurs and reptiles dominated the Mesozoic era, but another kind of animal also lived during this time—mammals. Mammals evolved early in the Mesozoic and remained small in size throughout the era. Few were larger than present-day cats.

Cretaceous Extinction Event

The Mesozoic era ended 65.5 mya with a mass extinction called the Cretaceous extinction event. You read in Lesson 1 that scientists propose a large meteorite impact contributed to this extinction. This crash would have produced enough dust to block sunlight for a long time. There is evidence that volcanic eruptions also occurred at the same time. These eruptions would have added more dust to the atmosphere. Without light, plants died. Without plants, animals died. Dinosaur species and other large Mesozoic vertebrate species could not adapt to the changes. They became extinct.

Lesson 3 Review

Assessment Online Quiz

Visual Summary

As Pangaea broke up, the continents began to move into their present-day positions.

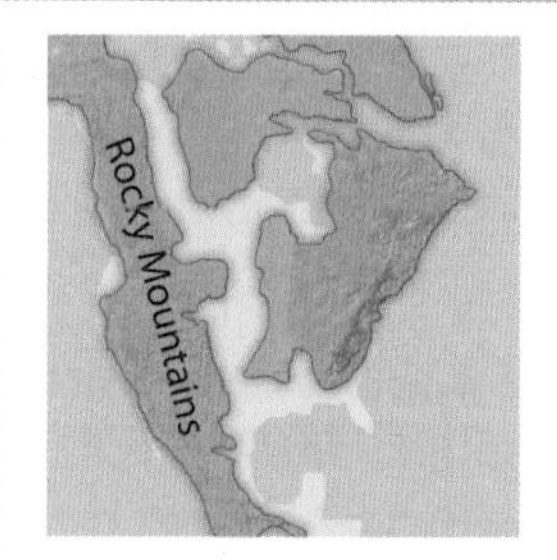

The Mesozoic climate was warm and sea level was high.

Dinosaurs were not the only large vertebrates that lived during the Mesozoic era.

FOLDABLES®

Use your lesson Foldable to review the lesson. Save your Foldable for the project at the end of the chapter.

What do you think NOW?

You first read the statements below at the beginning of the chapter.

5. All large Mesozoic vertebrates were dinosaurs.

6. Dinosaurs disappeared in a large mass extinction event.

Did you change your mind about whether you agree or disagree with the statements? Rewrite any false statements to make them true.

Use Vocabulary

1. A(n) ________ was a marine Mesozoic reptile.
2. A(n) ________ was a Mesozoic reptile that could fly.

Understand Key Concepts 

3. Which major event happened during the Mesozoic era?
   - **A.** Humans evolved.
   - **B.** Life moved onto land.
   - **C.** The Appalachian Mountains formed.
   - **D.** The Atlantic Ocean formed.
4. **Compare** the sizes of reptiles and mammals during the Mesozoic era.
5. **Explain** how the Rocky Mountains formed.

Interpret Graphics

6. **Identify** Which type of vertebrate does each skeletal figure below represent?

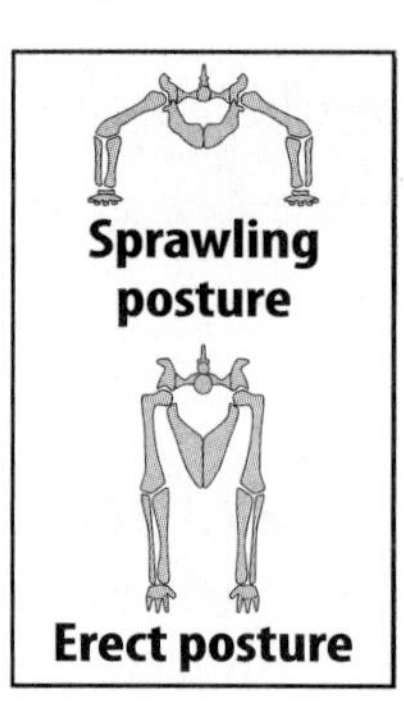

7. **Sequence** Copy and fill in the graphic organizer below to list the periods of the Mesozoic era in order.

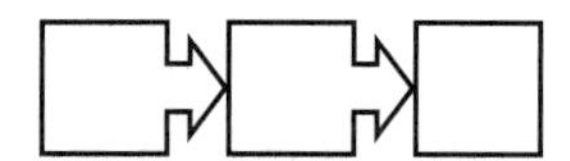

Critical Thinking

8. **Infer** how Earth might be different if there had been no extinction event at the end of the Mesozoic era.
9. **Propose** how the breakup of Pangaea might have affected evolution.

AMERICAN MUSEUM OF NATURAL HISTORY

Digging Up a Surprise

CAREERS in SCIENCE

A fossil discovery in China reveals some unexpected clues about early mammals.

The Mesozoic era, 251 to 65.5 million years ago, was the age of the dinosaurs. Many species of dinosaurs roamed Earth, from the ferocious tyrannosaurs to the giant, long-necked brachiosaurs. What other animals lived among the dinosaurs? For years, paleontologists assumed that the only mammals that lived at that time were no bigger than mice. They were no match for the dinosaurs.

Recent fossil discoveries revealed new information about these early mammals. Jin Meng is a paleontologist at the American Museum of Natural History in New York City. In northern China, Meng and other paleontologists discovered fossils of animals that probably died in volcanic eruptions 130 million years ago. Among these fossils were the remains of a mammal over 1 foot long—about the size of a small dog. A representation of the mammal, *Repenomamus robustus* (reh peh noh MA muhs • roh BUS tus), is shown to the right.

This fossil would reveal an even bigger surprise. When examined under microscopes in the lab, scientists discovered small bones in the fossil's rib cage where its stomach had been. The bones were the tiny limbs, fingers, and teeth of a young plant-eating dinosaur. The mammal's last meal had been a young dinosaur!

This was an exciting discovery. Meng and his team learned that early mammals were larger than they thought and were meat eaters, too. Those tiny bones proved to be a huge find. Paleontologists now have a new picture of how animals interacted during the age of dinosaurs.

IVPP V13605

▲ **Paleontologists studying a fossil of the mammal *Repenomamus robustus* found tiny *Psittacosaurus* bones in its stomach.**

This is a representation of a young *Psittacosaurus*—only 12 cm long. ▶

It's Your Turn

DIAGRAM With a group, research the plants and the animals that lived in the same environment as *Repenomamus*. Create a drawing showing the relationships among the organisms. Compare your drawing to those of other groups.

Lesson 4

Reading Guide

Key Concepts
ESSENTIAL QUESTIONS

- What major geologic events occurred during the Cenozoic era?
- What does fossil evidence reveal about the Cenozoic era?

Vocabulary

Holocene epoch p. 625
Pleistocene epoch p. 627
ice age p. 627
glacial groove p. 627
mega-mammal p. 628

g Multilingual eGlossary

The Cenozoic Era

Inquiry Is this animal alive?

No, this is a statue in a Los Angeles, California, pond that has been oozing tar for thousands of years. It shows how a mammoth might have become stuck in a tar pit. Mammoths lived at the same time as early humans. What do you think it was like to live alongside these animals?

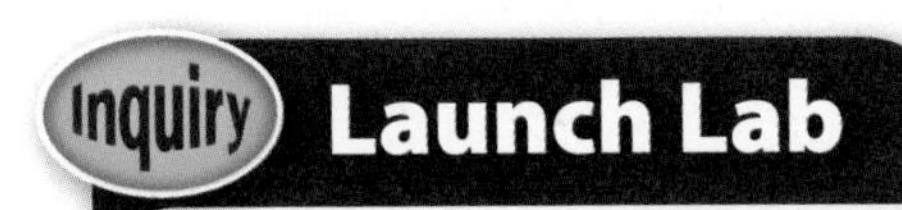

Launch Lab

10 minutes

What evidence do you have that you went to kindergarten?

Rocks and fossils provide evidence about Earth's past. The more recent the era, the more evidence exists. Is this true for you, too?

1. Make a list of items you have, such as a diploma, that could provide evidence about what you did and what you learned in kindergarten.
2. Make another list of items that could provide evidence about your school experience during the past year.

Think About This

1. Which list is longer? Why?
2. **Key Concept** How do you think the items on your lists are like evidence from the first and last eras of the Phanerozoic eon?

Geology of the Cenozoic Era

Have you ever experienced a severe storm? What did your neighborhood look like afterward? Piles of snow, rushing water, or broken trees might have made your neighborhood seem like a different place. In a similar way, the landscapes and organisms of the Paleozoic and Mesozoic eras might have been strange and unfamiliar to you. Though some unusual animals lived during the Cenozoic era, this era is more familiar. People know more about the Cenozoic era than they know about any other era because we live in the Cenozoic era. Its fossils and its rock record are better preserved.

As shown in **Figure 20,** the Cenozoic era spans the time from the end of the Cretaceous period, 65.5 mya, to present day. Geologists divide it into two periods—the Tertiary (TUR shee ayr ee) period and the Quaternary (KWAH tur nayr ee) period. These periods are further subdivided into epochs. *The most recent epoch, the* **Holocene** (HOH luh seen) **epoch,** *began 10,000 years ago.* You live in the Holocene epoch.

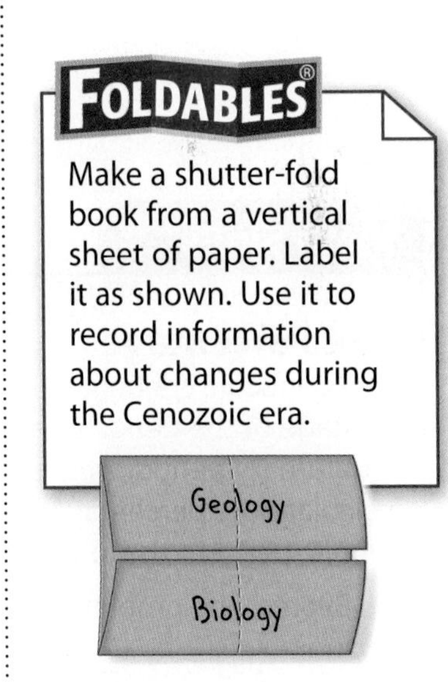

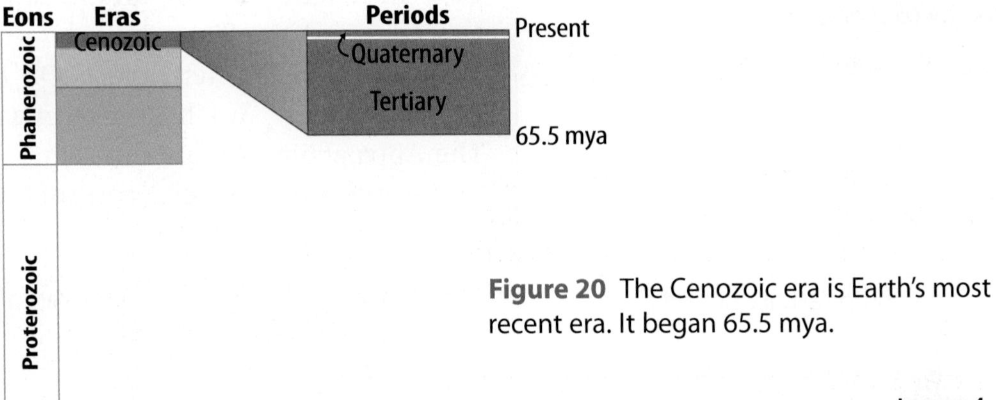

Figure 20 The Cenozoic era is Earth's most recent era. It began 65.5 mya.

Figure 21 Mammals dominated the landscapes of the Cenozoic era.

Word Origin

Cenozoic
From Greek *kainos,* means "new"; and *zoic,* means "life"

Math Skills

Review
Math Practice
Personal Tutor

Use Percentages

The Cenozoic era began 65.5 mya. What percentage of the Cenozoic era is taken up by the Quaternary period, which began 2.6 mya? To calculate the percentage of a part to the whole, perform the following steps:

a. Express the problem as a fraction.

$$\frac{2.6 \text{ mya}}{65.5 \text{ mya}}$$

b. Convert the fraction to a decimal. 2.6 mya divided by 65.5 mya = 0.040

c. Multiply by 100 and add %.
$0.040 \times 100 = 4.0\%$

Practice

What percent of the Cenozoic era is represented by the Tertiary period, which lasted from 65.5 mya to 2.6 mya? [Hint: Subtract to find the length of the Tertiary period.]

Cenozoic Mountain Building

As shown in the globes in **Figure 21,** Earth's continents continued to move apart during the Cenozoic era, and the Atlantic Ocean continued to widen. As the continents moved, some landmasses collided. Early in the Tertiary period, India crashed into Asia. This collision began to push up the Himalayas—the highest mountains on Earth today. At about the same time, Africa began to push into Europe, forming the Alps. These mountains continue to get higher today.

In North America, the western coast continued to push against the seafloor next to it, and the Rocky Mountains continued to grow in height. New mountain ranges—the Cascades and the Sierra Nevadas—began to form along the western coast. On the eastern coast, there was little tectonic activity. The Appalachian Mountains, which formed during the Paleozoic era, continue to erode today.

Reading Check Why are the Appalachian Mountains relatively small today?

Pleistocene Ice Age

Like the Mesozoic era, the early part of the Cenozoic era was warm. In the middle of the Tertiary period, the climate began to cool. By the Pliocene (PLY oh seen) epoch, ice covered the poles as well as many mountaintops. It was even colder during the next epoch—the Pleistocene (PLY stoh seen).

The **Pleistocene epoch** *was the first epoch of the Quaternary period.* During this time, glaciers advanced and retreated many times. They covered as much as 30 percent of Earth's land surface. *An* **ice age** *is a time when a large proportion of Earth's surface is covered by glaciers.* Sometimes, rocks carried by glaciers created deep gouges or grooves, as shown in **Figure 22.** **Glacial grooves** *are grooves made by rocks carried in glaciers.*

The glaciers contained huge amounts of water. This water originated in the oceans. With so much water in glaciers, sea level dropped. As sea level dropped, inland seas drained away, exposing dry land. When sea level was at its lowest, the Florida peninsula was about twice as wide as it is today.

Pleistocene Ice Age

Figure 22 Glacial grooves in Ohio are evidence that glaciers extended far into North America during the Pleistocene ice age.

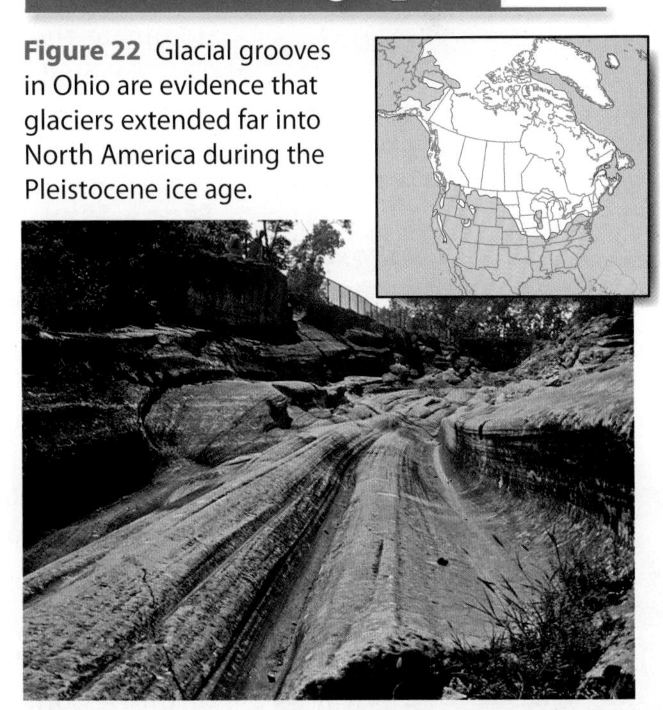

Visual Check Approximately what percentage of the United States was covered with ice?

Figure 23 These mega-mammals lived at different times during the Cenozoic era. They are all extinct today. The human is included for reference.

Cenozoic Life—The Age of Mammals

The mass extinction event at the end of the Mesozoic era meant that there was more space for each surviving species. Flowering plants, including grasses, evolved and began to dominate the land. These plants provided new food sources. This enabled the evolution of many types of animal species, including mammals. Mammals were so successful that the Cenozoic era is sometimes called the age of mammals.

Mega-Mammals

Recall that mammals were small during the Mesozoic era. Many new types of mammals appeared during the Cenozoic era. Some were very large, such as those shown in **Figure 23.** *The large mammals of the Cenozoic era are called* **mega-mammals.** Some of the largest lived during the Oligocene and Miocene periods, from 34 mya to 5 mya. Others, such as woolly mammoths, giant sloths, and saber-toothed cats, lived during the cool climate of the Pliocene and Pleistocene periods, from 5 mya to 10,000 years ago. Many fossils of these animals have been discovered. The saber-toothed cat skull in **Figure 24** was discovered in the Los Angeles tar pits pictured at the beginning of this lesson. A few mummified mammoth bodies also have been discovered preserved for thousands of years in glacial ice.

Key Concept Check How do scientists know that mega-mammals lived during the Cenozoic era?

Figure 24 The saber-toothed cat was a fierce Pleistocene predator.

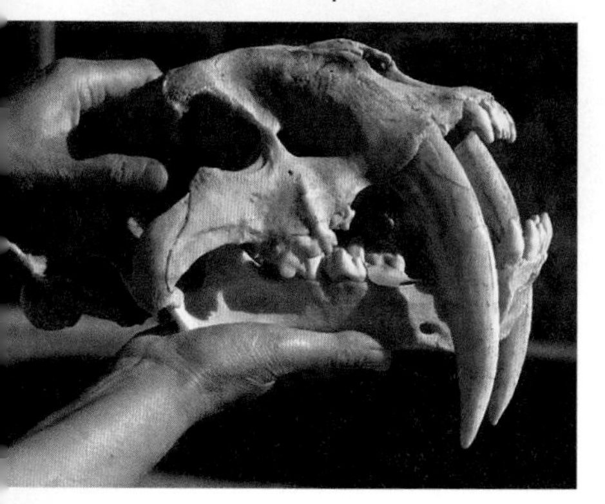

Isolated Continents and Land Bridges

The mammals depicted in **Figure 23** lived in North America, South America, Europe, and Asia. Different mammal species evolved in Australia. This is mostly because of the movement of Earth's tectonic plates. You read earlier that land bridges can connect continents that were once separated. You also read that when continents are separated, species that once lived together can become geographically isolated.

Most of the mammals that live in Australia today are marsupials (mar SOO pee ulz). These mammals, like kangaroos, carry their young in pouches. Some scientists suggest that marsupials did not evolve in Australia. Instead, they **hypothesize** that marsupial ancestors migrated to Australia from South America when South America and Australia were connected to Antarctica by land bridges, as shown in **Figure 25.** After ancestral marsupials arrived in Australia, Australia moved away from Antarctica, and water covered the land bridges between South America, Antarctica, and Australia. Over time, the ancestral marsupials evolved into the types of marsupials that live in Australia today.

ACADEMIC VOCABULARY
hypothesize
(verb) To make an assumption about something that is not positively known

 **Reading Check** What major geologic events affected the evolution of marsupials in Australia?

Land Bridges

Figure 25 At the beginning of the Cenozoic era, Australia was linked to South America via Antarctica, which was then warm. This provided a route for animal migration.

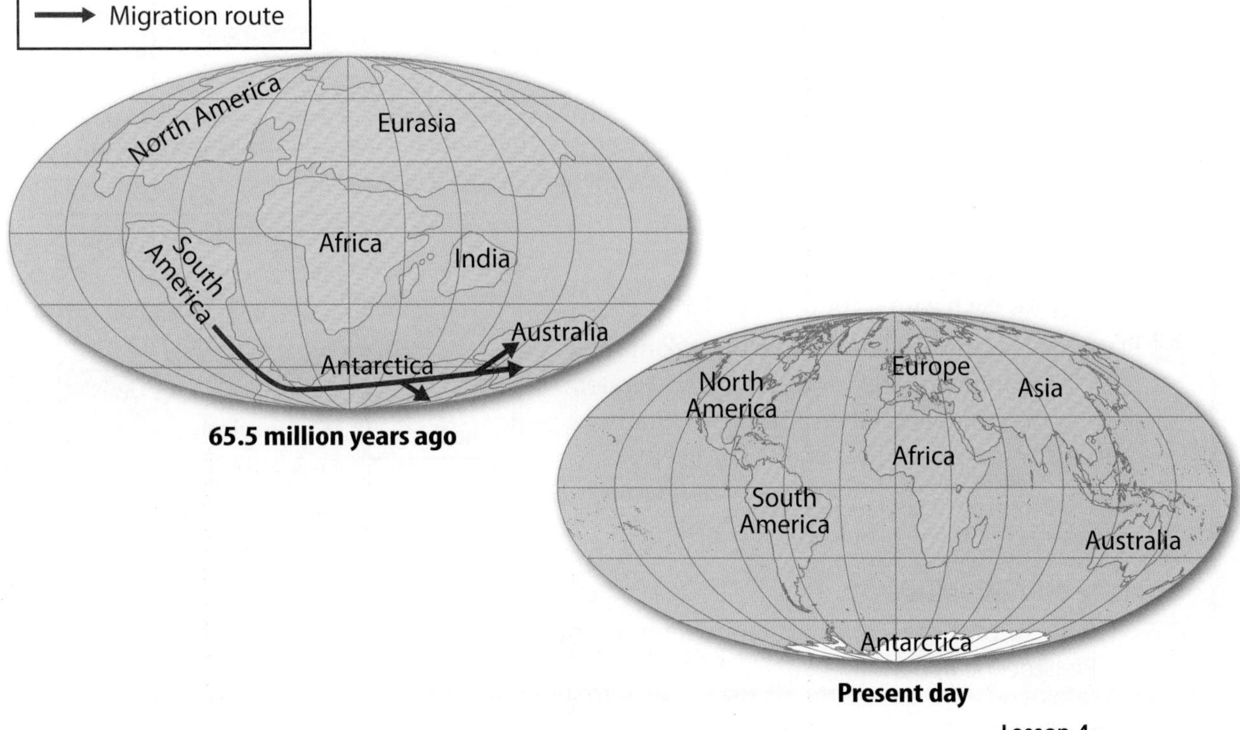

Figure 26 *Lucy* is the name scientists have given this 3.2-million-year-old human ancestor.

Rise of Humans

The oldest fossil remains of human ancestors have been found in Africa, where scientists think humans first evolved. These fossils are nearly 6 million years old. A skeleton of a 3.2-million-year-old human ancestor is shown in **Figure 26.**

Modern humans—called *Homo sapiens*—didn't evolve until the Pleistocene epoch. Early *Homo sapiens* migrated to Europe, Asia, and eventually North America. Early humans likely migrated to North America from Asia using a land bridge that connected the continents during the Pleistocene ice age. This land bridge is now covered with water.

Pleistocene Extinctions

Climate changed at the close of the Pleistocene epoch 10,000 years ago. The Holocene epoch was warmer and drier. Forests replaced grasses. The mega-mammals that lived during the Pleistocene became extinct. Some scientists suggest that mega-mammal species could not adapt fast enough to survive the environmental changes.

Key Concept Check How did climate change at the end of the Pleistocene epoch?

Future Changes

There is evidence that present-day Earth is undergoing a global-warming climate change. Many scientists suggest that humans have contributed to this change because of their use of coal, oil, and other fossil fuels over the past few centuries.

Inquiry MiniLab

20 minutes

What happened to the Bering land bridge?

Pleistocene animals and humans likely crossed into North America from Asia using the Bering land bridge. Why did this bridge disappear?

1. Read and complete a lab safety form
2. Form two pieces of **modeling clay** into continents, each with a continental shelf.
3. Place the clay models into a **watertight container** with the continental shelves touching. Add water, leaving the continental shelves exposed. Place a dozen or more **ice cubes** on the continents.
4. During your next science class, observe the container and record your observations.

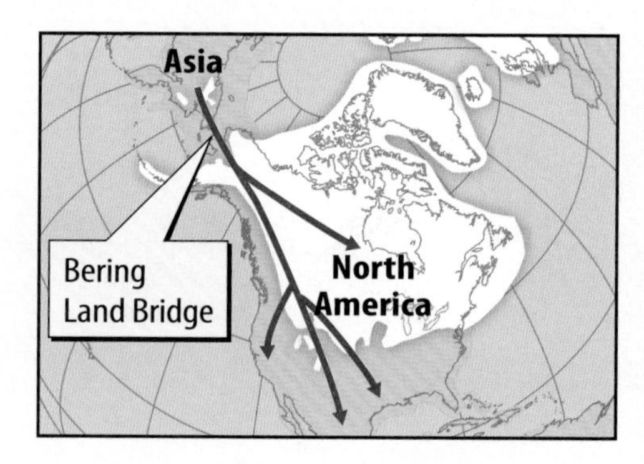

Analyze and Conclude

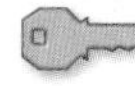

Key Concept How does your model represent what happened at the end of the Pleistocene epoch?

Lesson 4 Review

Visual Summary

The mega-mammals that lived during most of the Cenozoic era are extinct.

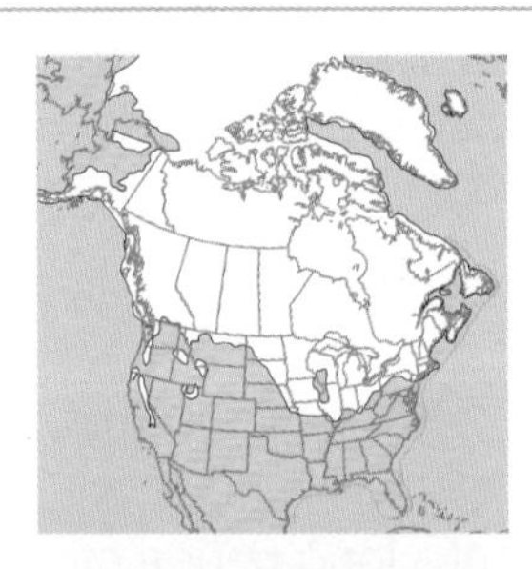

Glaciers extended well into North America during the Pleistocene ice age.

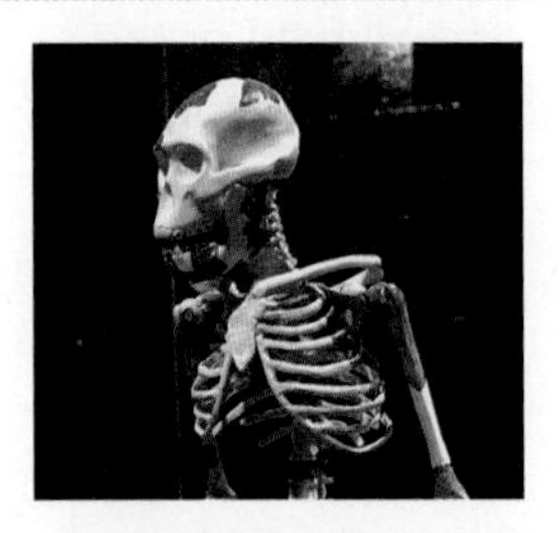

Lucy is a 3.2-million-year-old human ancestor.

FOLDABLES®

Use your lesson Foldable to review the lesson. Save your Foldable for the project at the end of the chapter.

What do you think NOW?

You first read the statements below at the beginning of the chapter.

7. Mammals evolved after dinosaurs became extinct.

8. Ice covered nearly one-third of Earth's land surface 10,000 years ago.

Did you change your mind about whether you agree or disagree with the statements? Rewrite any false statements to make them true.

Use Vocabulary

1. Gouges made by ice sheets are ________.
2. You live in the ________ epoch.

Understand Key Concepts 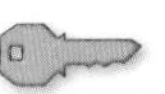

3. Which organism lived during the Cenozoic era?
   - **A.** *Brachiosaurus*
   - **B.** *Dunkleosteus*
   - **C.** saber-toothed cats
   - **D.** trilobites
4. **Classify** Which terms are associated with the Cenozoic era: *Homo sapiens,* mammoth, dinosaur, grass?

Interpret Graphics

5. **Determine** The map below shows coastlines of the southeastern U.S. at three times during the Cenozoic era. Which choice represents the coastline at the height of the Pleistocene ice age?

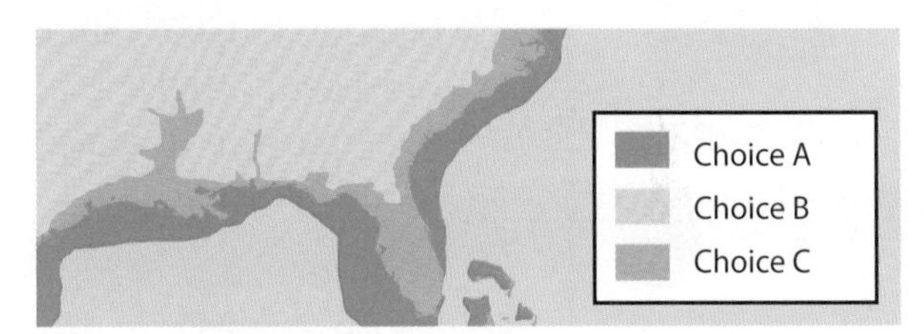

6. **Summarize** Copy and fill in the graphic organizer below to list living mammals that might be considered mega-mammals today.

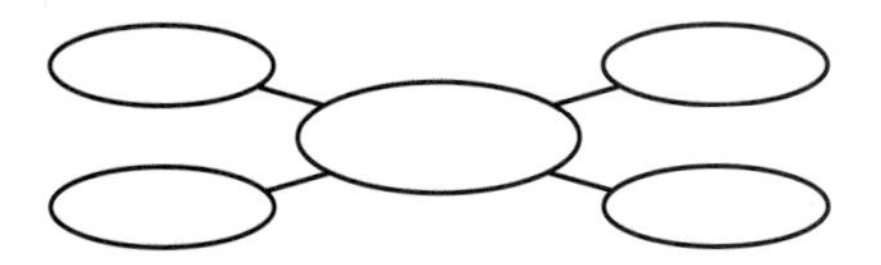

Critical Thinking

7. **Suggest** what might happen if the Australian continent crashed into Asia.

Math Skills

Math Practice

8. The Cenozoic era began 65.5 mya. The Oligocene and Miocene epochs extended from 34 mya to 5 mya. What percentage of the Cenozoic era is represented by the Oligocene and Miocene epochs?

Lab — 90 minutes

Modeling Geologic Time

Materials

meterstick

tape measure

poster board

colored markers

colored paper

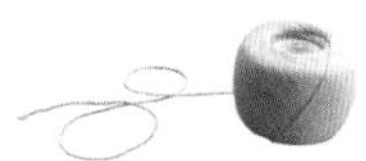

string

maps

Evidence suggests that Earth formed approximately 4.6 billion years ago. But how long is 4,600,000,000 years? It is difficult to comprehend time that extends so far into the past unless you can relate it to your own experience. In this activity, you will develop a metaphor for geologic time using a scale that is familiar to you. Then, you will create a model to share with your class.

Question

How can you model geologic time using a familiar scale?

Procedure

1. Think of something you are familiar with that can model a long period of time. For example, you might choose the length of a football field or the distance between two U.S. cities on a map—one on the east coast and one on the west coast.
2. Make a model of your metaphor using a metric scale. On your model, display the events listed in the table on the next page. Use the equation below to generate true-to-scale dates in your model.

$$\frac{\text{Known age of past event (years before present)}}{\text{Known age of Earth (years before present)}} = \frac{X \text{ time scale unit location}}{\text{Maximum distance or extent of metaphor}}$$

Example: To find where "first fish" would be placed on your model if you used a meterstick (100 cm), set up your equation as follows:

$$\frac{500{,}000{,}000 \text{ years}}{4{,}600{,}000{,}000 \text{ years}} = \frac{X \text{ (location on meterstick)}}{100 \text{ cm}}$$

3. In your Science Journal, keep a record of all the math equations you used. You can use a calculator, but show all equations.

Analyze and Conclude

4. **Calculate** What percentage of geologic time have modern humans occupied? Set up your equation as follows:

$$\frac{100{,}000}{4{,}600{,}000{,}000} \times 100 = \text{\% of time occupied by } H.\ sapiens$$

5. **Estimate** Where does the Precambrian end on your model? Estimate how much of geologic time falls within the Precambrian.
6. **Evaluate** What other milestone events in Earth's history, other than those listed in the table, could you include on your model?
7. **Appraise** the following sentence as it relates to your life: "Time is relative."
8. **The Big Idea** The Earth events on your model are based mostly on fossil evidence. How are fossils useful in understanding Earth's history? How are they useful in the development of the geologic time scale?

Communicate Your Results

Share your model with the class. Explain why you chose the model you did, and demonstrate how you calculated the scale on your model.

Imagine that you were asked to teach a class of kindergartners about Earth's time. How would you do it? What metaphor would you use? Why?

Some Important Approximate Dates in the History of Earth:

MYA	Event
4,600	Origin of Earth
3,500	Oldest evidence of life
500	First fish
375	*Tiktaalik* appears
320	First reptiles
250	Permian extinction event
220	Mammals and dinosaurs appear
155	Archaeopteryx appears
145	Atlantic Ocean forms
65	Cretaceous extinction event
6	Human ancestors appear
2	Pleistocene Ice Age begins
0.1	*Homo sapiens* appear
0.00052	Columbus lands in New World
??	Your birth date

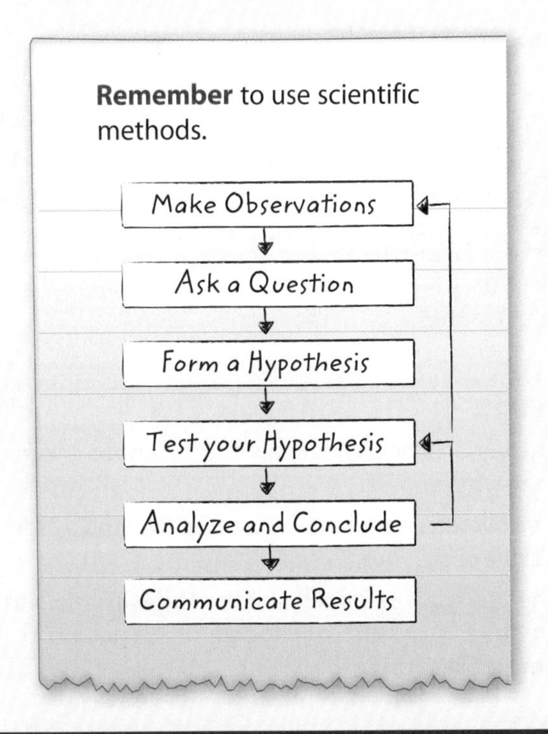

Chapter 17 Study Guide

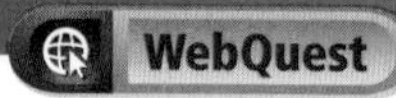

The geologic changes that have occurred during the billions of years of Earth's history have strongly affected the evolution of life.

Key Concepts Summary

Lesson 1: Geologic History and the Evolution of Life

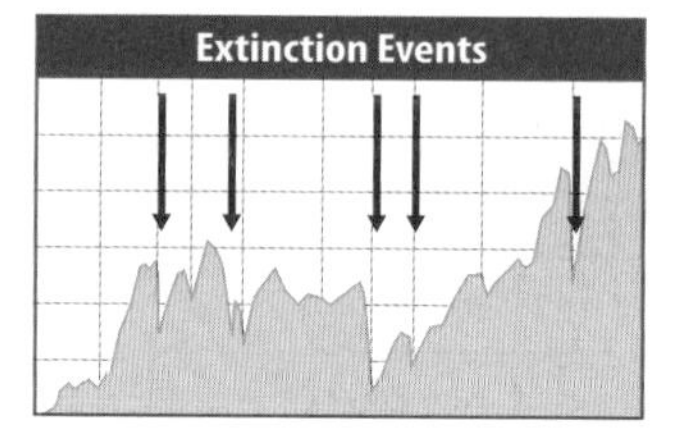

- Geologists organize Earth's history into **eons, eras, periods,** and **epochs.**
- Life evolves over time as Earth's continents move, forming **land bridges** and causing **geographic isolation.**
- **Mass extinctions** occur if many species of organisms cannot adapt to sudden environmental change.

Vocabulary

eon p. 601
era p. 601
period p. 601
epoch p. 601
mass extinction p. 603
land bridge p. 604
geographic isolation p. 604

Lesson 2: The Paleozoic Era

- Life diversified during the **Paleozoic era** as organisms moved from water to land.
- **Coal swamps** formed along **inland seas.** Later, land became drier as the **supercontinent** Pangaea formed.
- The largest mass extinction in Earth's history occurred at the end of the Permian period.

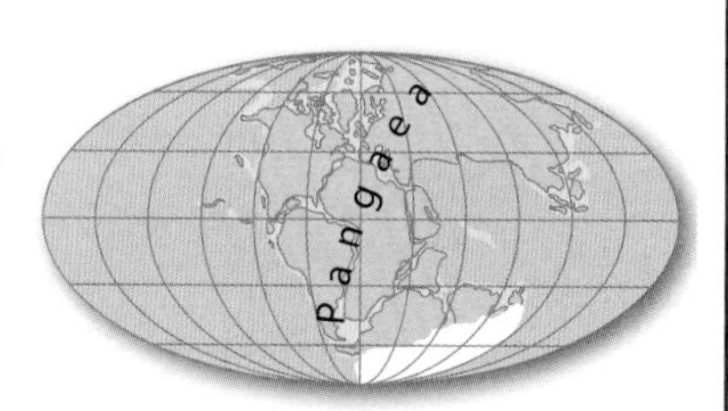

Vocabulary

Paleozoic era p. 609
Mesozoic era p. 609
Cenozoic era p. 609
inland sea p. 610
coal swamp p. 612
supercontinent p. 613

Lesson 3: The Mesozoic Era

- Sea level rose as the climate warmed.
- The Atlantic Ocean and the Rocky Mountains began to form as Pangaea broke apart.
- **Dinosaurs, plesiosaurs, pterosaurs,** and other large Mesozoic vertebrates became extinct at the end of the era.

Vocabulary

dinosaur p. 620
plesiosaur p. 621
pterosaur p. 621

Lesson 4: The Cenozoic Era

- The large, extinct mammals of the Cenozoic were **mega-mammals.**
- Ice covered nearly one-third of Earth's land at the height of the Pleistocene **ice age.**
- The **Pleistocene epoch** and the **Holocene epoch** are the two most recent epochs of the geologic time scale.

Vocabulary

Holocene epoch p. 625
Pleistocene epoch p. 627
ice age p. 627
glacial groove p. 627
mega-mammal p. 628

- Personal Tutor
- Vocabulary eGames
- Vocabulary eFlashcards

FOLDABLES® Chapter Project

Assemble your lesson Foldables as shown to make a Chapter Project. Use the project to review what you have learned in this chapter.

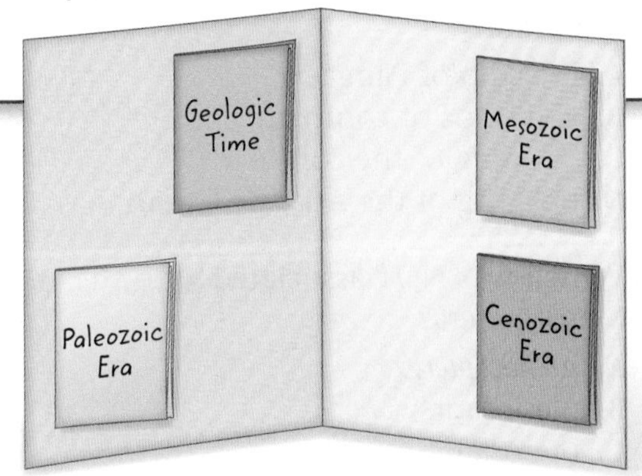

Use Vocabulary

1. The longest time unit in the geologic time scale is the ________.
2. Eras are subdivided into ________.
3. Many boundaries in the geologic time scale are marked by the occurrence of ________.
4. When glaciers melt, shallow ________ form in the interiors of continents.
5. The ________ was the first era of the Phanerozoic eon.
6. A(n) ________ can form when plants are buried in an oxygen-poor environment.
7. Marine Mesozoic reptiles included ________.
8. Modern humans evolved during the ________.

Link Vocabulary and Key Concepts

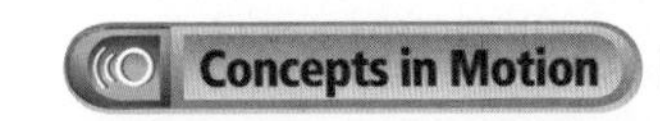

Interactive Concept Map

Copy this concept map, and then use vocabulary terms from the previous page and other terms from the chapter to complete the concept map.

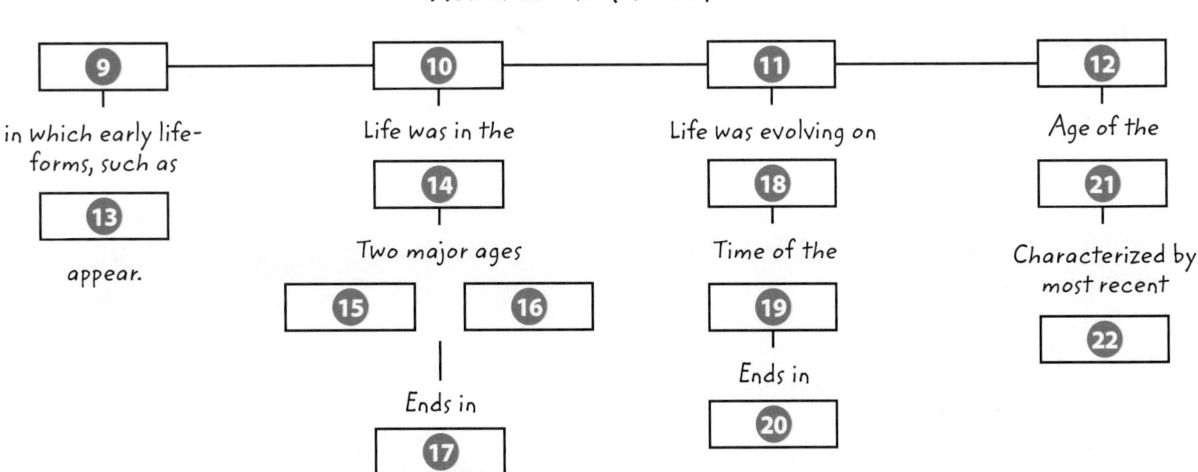

Chapter 17 Review

Understand Key Concepts 

1. The trilobite fossil below represents an organism that lived during the Cambrian period.

What distinguished this organism from organisms that lived earlier in time?

A. It had hard parts.
B. It lived on land.
C. It was a reptile.
D. It was multicellular.

2. What are the many divisions in the geologic time scale based on?

A. changes in the fossil record every billion years
B. changes in the fossil record every million years
C. gradual changes in the fossil record
D. sudden changes in the fossil record

3. Which is NOT a cause of a mass extinction event?

A. meteorite collision
B. severe hurricane
C. tectonic activity
D. volcanic activity

4. Which is the correct order of eras, from oldest to youngest?

A. Cenozoic, Mesozoic, Paleozoic
B. Mesozoic, Cenozoic, Paleozoic
C. Paleozoic, Cenozoic, Mesozoic
D. Paleozoic, Mesozoic, Cenozoic

5. Which were the first organisms to inhabit land environments?

A. amphibians
B. plants
C. reptiles
D. trilobites

6. Which event(s) produced the Appalachian Mountains?

A. breakup of Pangaea
B. collisions of continents
C. flooding of the continent
D. opening of the Atlantic Ocean

7. Which was NOT associated with the Mesozoic era?

A. *Archaeopteryx*
B. plesiosaurs
C. pterosaurs
D. *Tiktaalik*

8. Which is true for the beginning of the Cenozoic era?

A. Mammals and dinosaurs lived together.
B. Mammals first evolved.
C. Dinosaurs had killed all mammals.
D. Dinosaurs were extinct.

9. What is unrealistic about the picture on this stamp?

A. Dinosaurs were not this large.
B. Dinosaurs did not have long necks.
C. Humans did not live with dinosaurs.
D. Early humans did not use stone tools.

Assessment

Online Test Practice

Critical Thinking

10. **Hypothesize** how a major change in global climate could lead to a mass extinction.

11. **Evaluate** how the Permian-Triassic mass extinction affected the evolution of life.

12. **Predict** what Earth's climate might be like if sea level were very low.

13. **Differentiate** between amphibians and reptiles. What feature enabled reptiles—but not amphibians—to be successful on land?

14. **Hypothesize** how the bone structure of dinosaur limbs might have contributed to the success of dinosaurs during the Mesozoic era.

15. **Debate** Some scientists argue that humans have changed Earth so much that a new epoch—the Anthropocene epoch—should be added to the geologic time scale. Explain whether you think this is a good idea and, if so, when it should begin.

16. **Interpret Graphics** What is wrong with the geologic time line shown below?

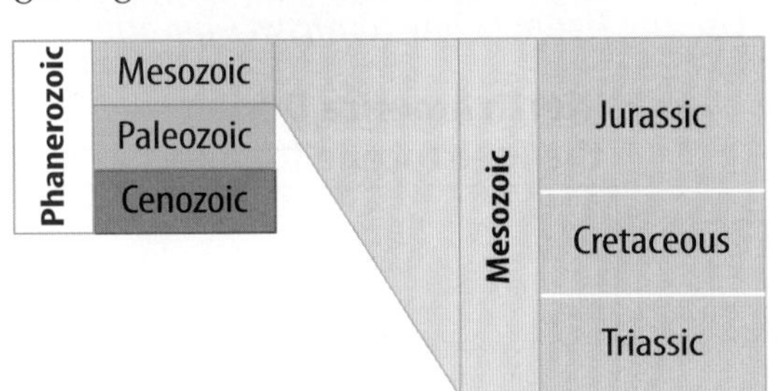

Writing in Science

17. **Decide** which period of Earth's history you would want to visit if you could travel back in time. Write a letter to a friend about your visit, describing the climate, the organisms, and the positions of Earth's continents at the time of your visit. Include a main idea, supporting details and examples, and a concluding sentence.

REVIEW THE BIG IDEA

18. What have scientists learned about Earth's past by studying rocks and fossils? How is the evolution of Earth's life-forms affected by geologic events? Provide examples.

19. The photo below shows an extinct dinosaur. What changes on Earth can cause organisms to become extinct?

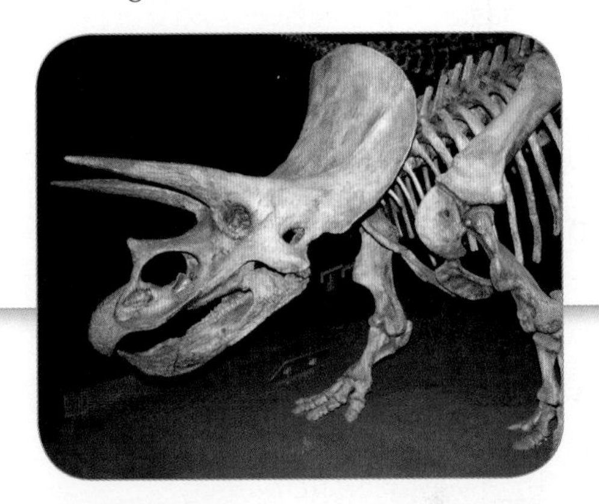

Math Skills

Math Practice

Use Percentages

Use the table to answer the questions.

Era	Period	Epoch	Time Scale
Cenozoic	Quaternary	Holocene	10,000 years ago
		Pleistocene	1.8 mya
	Tertiary	Pliocene	5.3 mya
		Miocene	23.8 mya
		Oligocene	33.7 mya
		Eocene	54.8 mya
		Paleocene	65.5 mya

20. What percentage of the Quaternary period is represented by the Holocene epoch?

21. What percentage of the Tertiary period is represented by the Pliocene epoch?

Standardized Test Practice

Record your answers on the answer sheet provided by your teacher or on a sheet of paper.

Multiple Choice

Use the figure below to answer question 1.

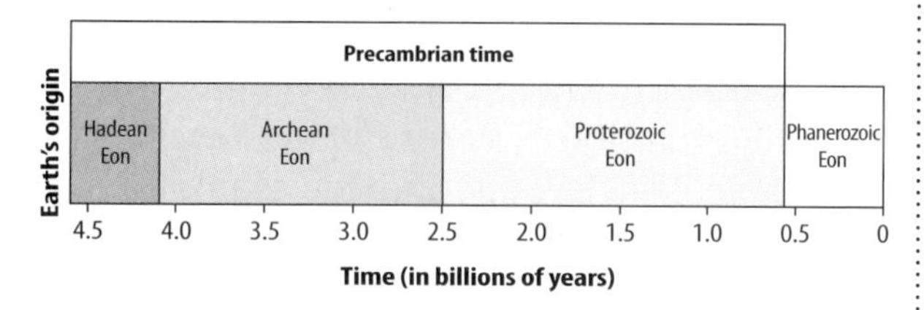

1 Approximately how long did Precambrian time last?

A 0.5 billion years
B 3.5 billion years
C 4.0 billion years
D 4.25 billion years

2 Which is the smallest unit of geologic time?

A eon
B epoch
C era
D period

3 Which is known as the age of invertebrates?

A Early Cenozoic
B Early Paleozoic
C Late Mesozoic
D Late Precambrian

4 Which made dinosaurs different from modern-day reptiles?

A head shape
B hip structure
C jaw alignment
D tail length

5 What is the approximate age of the oldest fossils of early human ancestors?

A 10,000 years
B 6 million years
C 65 million years
D 1.5 billion years

6 Which was NOT an adaptation that enabled amphibians to live on land?

A ability to breathe oxygen
B ability to lay eggs on land
C strong limbs
D thick skin

7 Which is considered a mega-mammal?

A Archaeopteryx
B plesiosaur
C Tiktaalik
D woolly mammoth

Use the figure below to answer question 8.

North America During the Pleistocene Ice Age

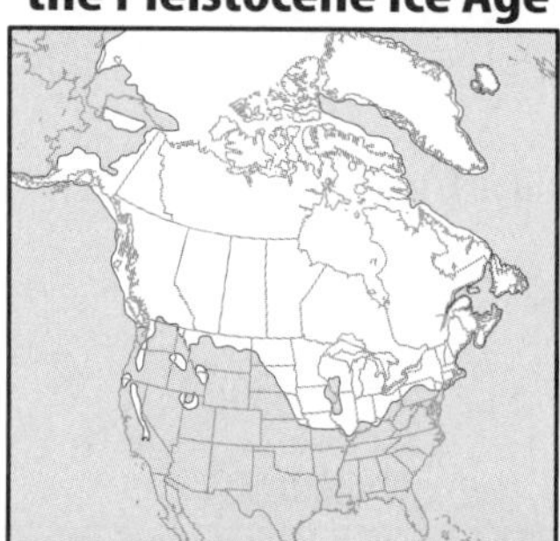

8 The figure above is a map of glacial coverage in North America. Which section of the United States would most likely have the greatest number of glacial grooves?

A the Northeast
B the Northwest
C the Southeast
D the Southwest

Use the graph below to answer question 9.

Sea Level Rise During Mesozoic

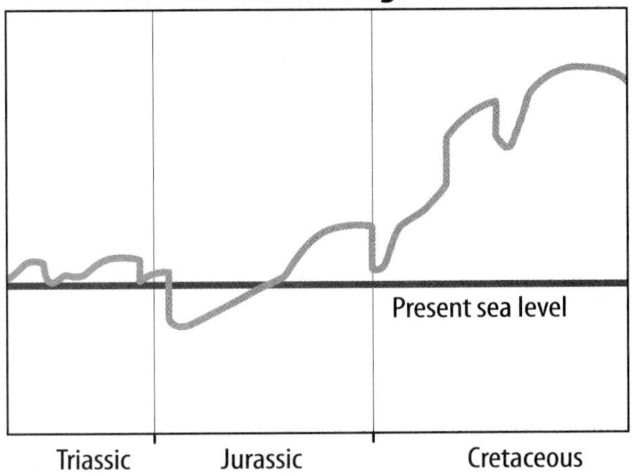

9 Based on the graph above, when might inland seas have covered much of Earth's continents?

A Early Cretaceous
B Early Jurassic
C Middle Triassic
D Late Cretaceous

10 Which did NOT occur in the Paleozoic era?

A appearance of mammals
B development of coal swamps
C evolution of invertebrates
D formation of Pangaea

11 What do geologists use to mark divisions in geologic time?

A abrupt changes in the fossil record
B frequent episodes of climate change
C movements of Earth's tectonic plates
D rates of radioactive mineral decay

Constructed Response

Use the graph below to answer questions 12 and 13.

Extinction Events

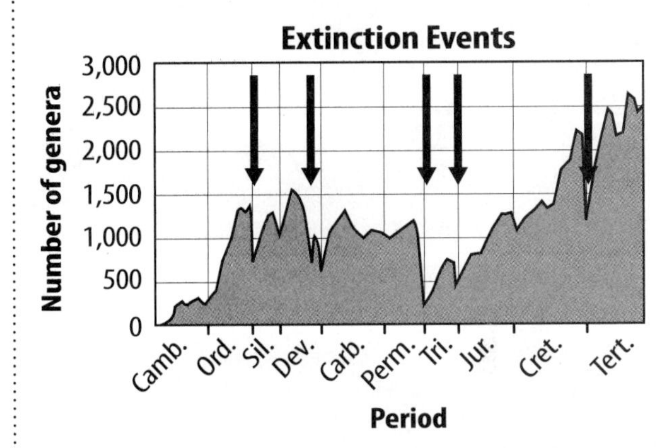

12 In the graph above, what events do the arrows mark? What happens during these events?

13 What event appears to have had the greatest impact? Explain your answer in terms of the graph.

14 What are two possible reasons why large populations of organisms die?

15 What is the relationship between the evolution of marsupials and the movement of Earth's tectonic plates?

16 Why did new and existing aquatic organisms flourish during the Mesozoic era? Use the terms *glaciers, Pangaea,* and *sea level* in your explanation.

17 What is the link between iridium and the mass extinction of dinosaurs?

NEED EXTRA HELP?																	
If You Missed Question...	1	2	3	4	5	6	7	8	9	10	11	12	13	14	15	16	17
Go to Lesson...	1	1	2	3	4	2	4	4	3	2	1	1	1	1-3	4	3	1

Unit 5

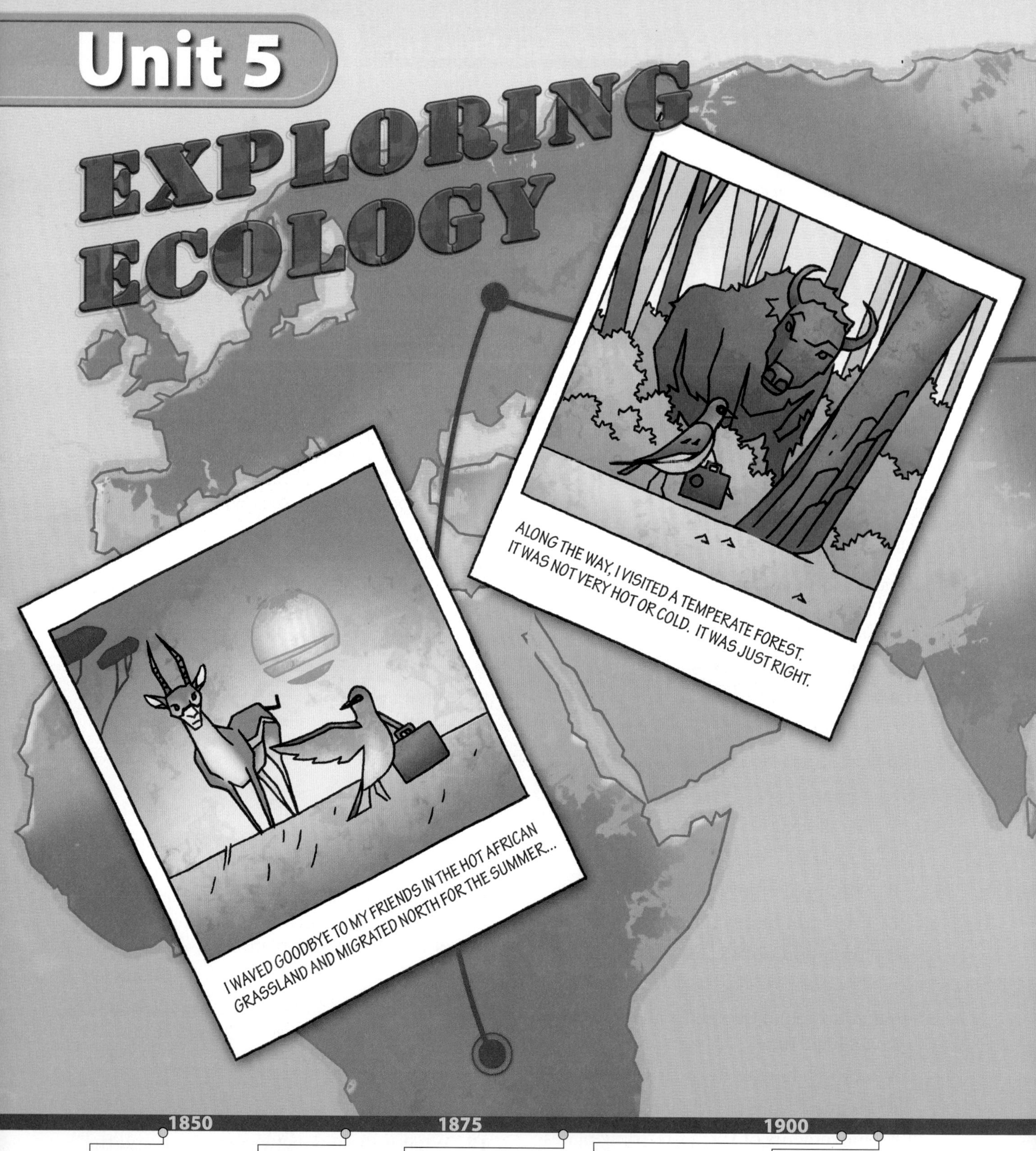

1850 | 1875 | 1900

1849
The U.S. Department of Interior is established and is responsible for the management and conservation of most federal land.

1872
The world's first national park, Yellowstone, is created.

1892
The Sierra Club is founded in San Francisco by John Muir. It goes on to be the oldest and largest grassroots environmental organization in the United States.

1915
Congress passes a bill establishing Rocky Mountain National Park in Colorado.

1920
Congress passes the Federal Water Power Act. This act creates a Federal Power Commission with authority over waterways, and the construction and use of water-power projects.

1955
The Air Pollution Control Act is the first of several United States Clean Air Acts to control air pollution on a national level.

1990
The Clean Air Act Amendments propose emissions trading and add provisions for reducing acid rain, ozone depletion, and toxic air pollution. They also establish a national permits program.

2006
The documentary *An Inconvenient Truth* is released to educate about global warming. The film's popularity raises international awareness of the cause.

Inquiry
Visit ConnectED for this unit's **STEM** activity.

Unit 5

Nature of SCIENCE

Graphs

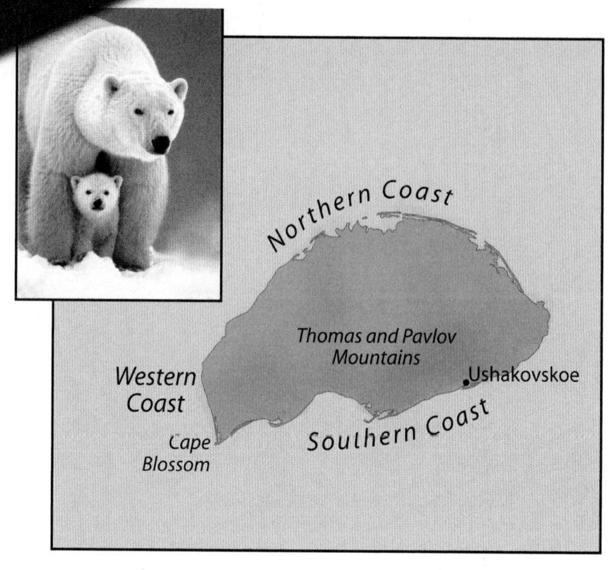

Figure 1 Scientists collect data about polar bears on Wrangel Island, Russia.

Polar bears are one of the largest land mammals. They hunt for food on ice packs that stretch across the Arctic Ocean. Recently, ice in the Arctic has not been as thick as it has been in the past. In addition, the ice does not cover as much area as it used to, making it difficult for polar bears to hunt. Scientists collect data about how these changes in the ice affect polar bear populations. One well-studied population of polar bears is on Wrangel Island, Russia, shown in **Figure 1.** Scientists collect and study data on polar bears to draw conclusions and make predictions about a possible polar bear extinction. Scientists often use graphs to better understand data. A **graph** is a type of chart that shows relationships between variables. Scientists use graphs to visually organize and summarize data. You can use different types of graphs to present different kinds of data.

Types of Graphs

Bar Graphs

The horizontal *x*-axis on a bar graph often contains categories rather than measurements. The heights of the bars show the measured quantity. For example, the *x*-axis on this bar graph contains different locations on Wrangel Island. The heights of the bars show how many bears researchers observed. The different colors show the age categories of polar bears. Where were ten adult polar bears observed?

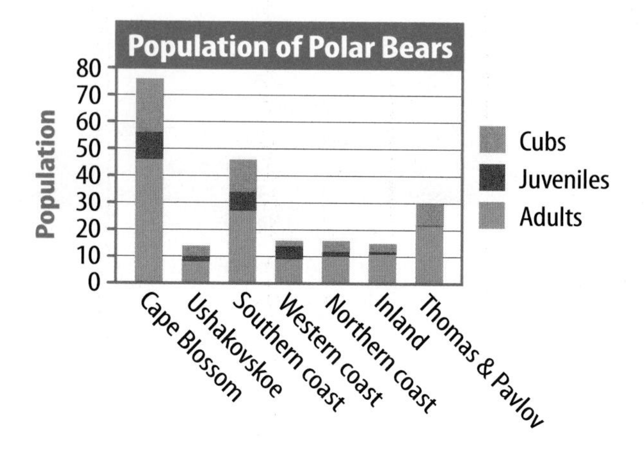

Circle Graphs

A circle graph usually illustrates the percentage of each category of data as it relates to the whole. This circle graph shows the percentage of different age categories of polar bears on Wrangel Island. Adults, shown by the blue color, make up the largest percentage of the total population. This circle graph contains similar data to the bar graph but presents it in a different way. What percentage of the total polar bear population are cubs?

Age Distribution of Polar Bears

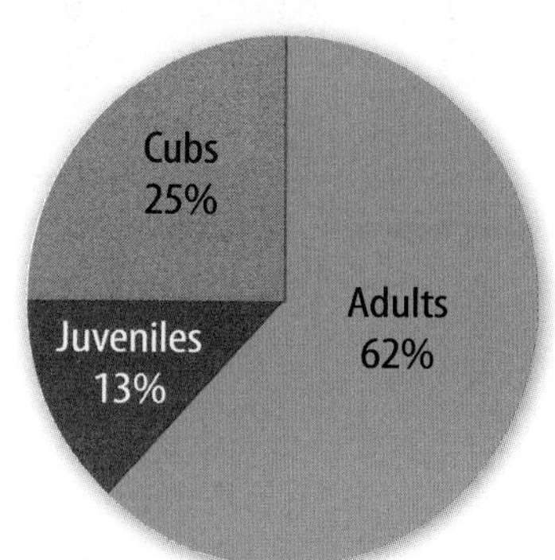

30 minutes

What can graphs tell you about polar bears?

A colleague gives you some data she collected about polar bears on Wrangel Island. She observed the condition of bears near Cape Blossom and classified the bears as starving, average, or healthy. She also recorded the age category of each bear. What can you learn by graphing these data?

	Starving	Average	Healthy
Adult	3	11	14
Juvenile	4	33	13
Cub	3	12	6

1. Make a bar graph of the number of bears in each category that are starving, in average condition, or healthy.
2. Add the numbers of starving bears. Add the total number of bears. Divide the number of starving bears by the total number of bears and multiply by 100 to calculate the percentage of starving bears. Repeat the calculations to find the percentages of average-condition and healthy bears. Make a circle graph showing the different conditions of the bears. For more information on how to make circle graphs, go to the Science Skill Handbook in the back of your book.

Analyze and Conclude

1. **Analyze** On your bar graph, indicate how you can tell which age category of bears is the healthiest.
2. **Determine** What group of bears do you think left the most walrus carcasses? Explain.

Line Graphs

A line graph helps you analyze how a change in one variable affects another variable. Scientists on Wrangel Island counted all the polar bears on the island each year for 10 years. They plotted each year of the survey on the horizontal *x*-axis and the bear population on the vertical *y*-axis. The population decreased between years 2 and 4. It increased between years 6 and 8. How did the population change during the last three years of the survey?

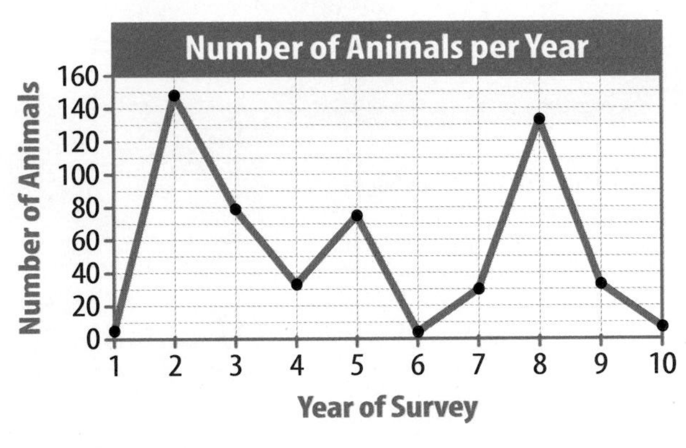

Double Line Graphs

You can use a double line graph to compare relationships of two sets of data. The blue line represents the population of polar bears. The orange line represents the number of walrus carcasses found on Wrangel Island. You can see that the blue and orange lines follow a similar pattern. This tells scientists that these two sets of data are related. Walruses are an important food source for polar bears on the island.

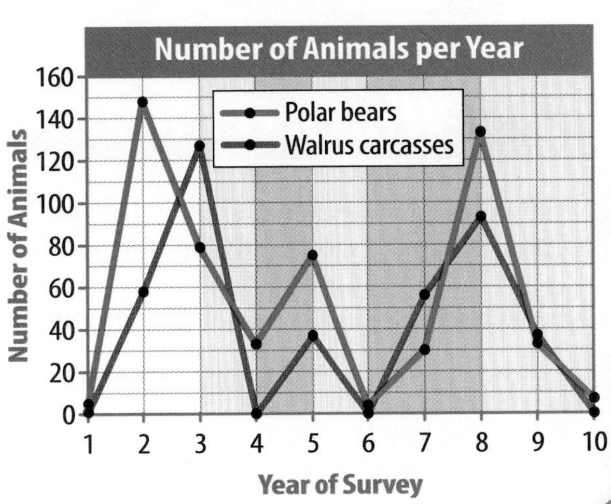

Chapter 18

Interactions Within Ecosystems

THE BIG IDEA **How do living things interact with each other and the environment?**

Who lives here?

Many different organisms live in this wetland. You might find birds, frogs, fish, plants, or even alligators here.

- What do you think each organism needs to survive?
- How do you think organisms that live here get the things they need?
- How do you think organisms interact with each other and the environment?

Get Ready to Read

What do you think?

Before you read, decide if you agree or disagree with each of these statements. As you read this chapter, see if you change your mind about any of the statements.

1. In symbiosis, two species cooperate in a way that benefits both species.
2. Overpopulation can be damaging to an ecosystem.
3. Sunlight provides the energy at the base of all food chains on Earth.
4. A detritivore is a type of carnivore.
5. Human actions can have unintended effects on the environment.
6. The only job of the U.S. Environmental Protection Agency is to enforce environmental laws.

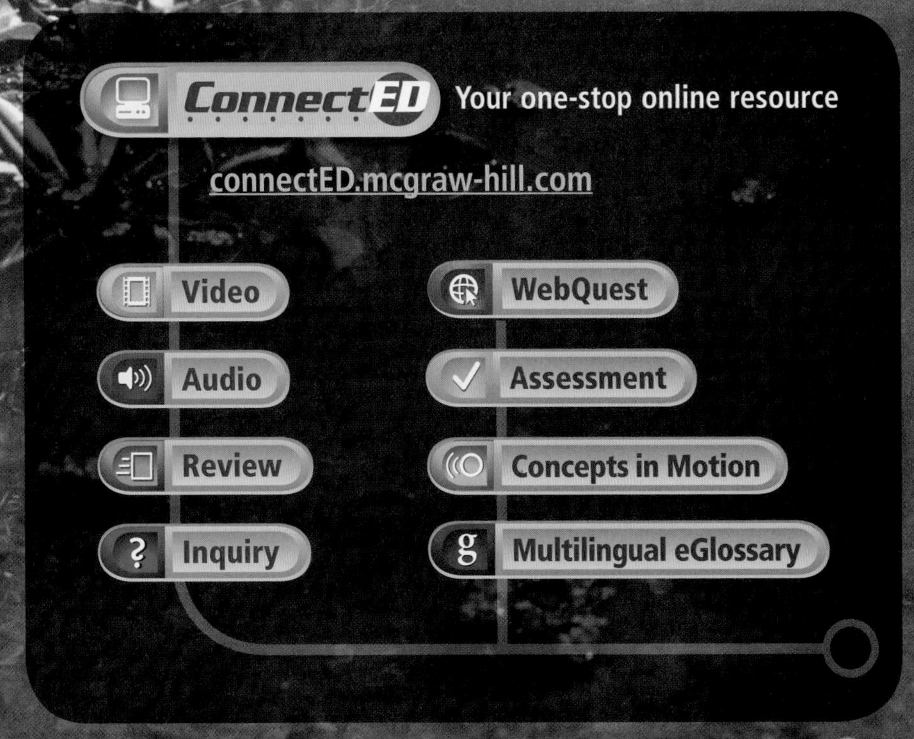

Lesson 1

Ecosystems

Reading Guide

Key Concepts
ESSENTIAL QUESTIONS

- How can you describe an ecosystem?
- In what ways do living organisms interact?
- How do population changes affect ecosystems?

Vocabulary

habitat p. 648
population p. 648
community p. 648
niche p. 649
predation p. 649
symbiosis p. 649
carrying capacity p. 651

g Multilingual eGlossary

Video

- BrainPOP®
- What's Science Got to do With It?

Inquiry What's happening here?

What do you think these bees are doing to this tree? All of the organisms in an area interact with the living and nonliving things around them. How do you think the bees and the tree in the photograph are interacting? Do you think either species is harmed or helped by the interaction?

20 minutes

Who's who?

Each member of an ecosystem plays a specific role and interacts with other factors in the environment. This helps maintain a functioning system. Nonliving factors in an ecosystem include the Sun, air, water, and soil; biotic factors include organisms.

1. Have another member of your group use **tape** to attach a **picture of an organism** to your back. Repeat for each group member.
2. Play a game of "Twenty Ecosystem Questions" with your group.
3. Take turns asking one *yes* or *no* question at a time to guess what organism you are in the ecosystem. Record the questions and answers about your organism in your Science Journal.

Think About This

1. Describe what you are in the ecosystem and what part you play.
2. In what ways does your part of the ecosystem interact with those of other members of your group?
3. **Key Concept** Draw a diagram explaining how you think the different parts of the ecosystem represented by your group interact.

Abiotic and Biotic Factors

What do you need to plant a garden? No garden can grow without sunlight, soil, water, and plants. Gardens also need compost to enrich the soil and insects or birds to pollinate flowers. Gardens are home to birds, rabbits, beetles, and many other organisms that feed on plants. A garden is an example of an ecosystem.

Ecosystems contain all the nonliving and living parts of the environment in a given area. The nonliving parts of an ecosystem are called abiotic factors. They include sunlight, water, soil, and air. The abiotic factors of an ecosystem, such as those in the wetland shown in **Figure 1,** determine what kinds of organisms can live there.

The living or once-living parts of an ecosystem are called biotic factors. These factors include living organisms, wastes produced by living organisms, and the decayed remains of dead organisms.

Key Concept Check How can you describe an ecosystem?

Figure 1 Many of these organisms—including water lilies and dragonflies—can survive only in wetland ecosystems.

Visual Check What abiotic factors can you identify in this ecosystem?

FOLDABLES®

Create a horizontal four-door book. Label it as shown. Use it to organize your notes on ecosystems.

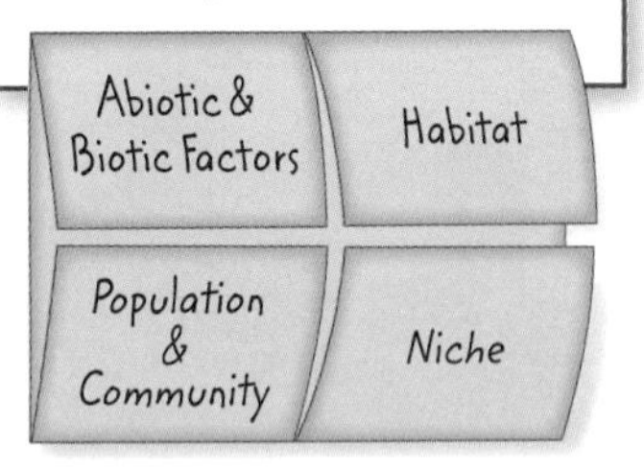

Habitats

Every species lives in a particular habitat. A **habitat** *is the place within an ecosystem that provides the biotic and abiotic factors an organism needs to survive and reproduce.* For example, a habitat for dragonflies includes plants that grow in shallow water. The plant leaves provide a place for adult dragonflies to lay eggs. The underwater stems and leaves provide shelter for young dragonflies.

Populations and Communities

Each species of dragonfly in a wetland ecosystem forms a population. A **population** *is all the organisms of the same species that live in the same area at the same time.* Species that have populations in an African savanna ecosystem include giraffes, kudus, wildebeests, zebras, and the other species shown in **Figure 2.** *All the populations living in an ecosystem at the same time form a* **community.**

WORD ORIGIN

community
from Latin *communitatem,* means "fellowship"

 **Reading Check** How are populations and communities similar? How are they different?

Figure 2 This savanna is a grassland ecosystem that provides habitats for many populations.

Interactions of Living Things

More than one population can live in the same habitat. For example, giraffes and two types of antelope—kudus and steenboks—feed on trees that grow in the African savanna. How can these three populations share the same habitat? Each species uses resources in their habitat, such as water, food, and shelter. However, each species has a different way of using the resources.

A **niche** (NICH) *is the way a species interacts with abiotic and biotic factors to obtain food, find shelter, and fulfill other needs.* As shown in **Figure 2,** giraffes feed at the tops of the trees. Kudus eat from mid-level branches, and steenboks feed on the lowest branches. Although all of these animals live in the same area, they use resources in their habitat in different ways. Giraffes, kudus, and steenboks have different niches.

Predation

A predator is an organism that hunts and kills other organisms for food. Prey is an organism caught and eaten by a predator. **Predation** *is the act of one organism, the predator, feeding on another organism, its prey.* African lions, such as the one in **Figure 2,** are predators that eat zebras and other savanna species. These animals are the lions' prey.

Symbiosis

Another type of interaction between organisms is symbiosis. **Symbiosis** *is a close, long-term relationship between two species that usually involves an exchange of food or energy.* Examples of the three types of symbiosis are shown in **Table 1.** In mutualism—one type of symbiosis—both species benefit from the relationship. In commensalism, one species benefits from the relationship. The other species is neither harmed nor benefited. In parasitism, one species (the parasite) benefits and the other (the host) is harmed.

Key Concept Check What is one way that living things interact?

Table 1 Symbiosis

Mutualism

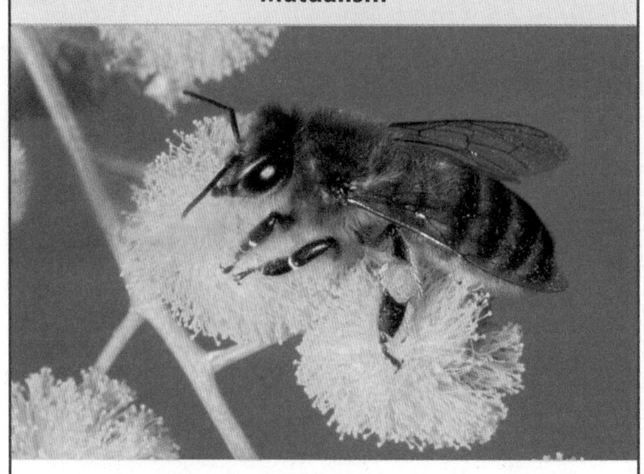

Honeybees pollinate acacia flowers as they collect nectar for their hives. Without pollinators such as these, acacias could not produce fertile seeds.

Commensalism

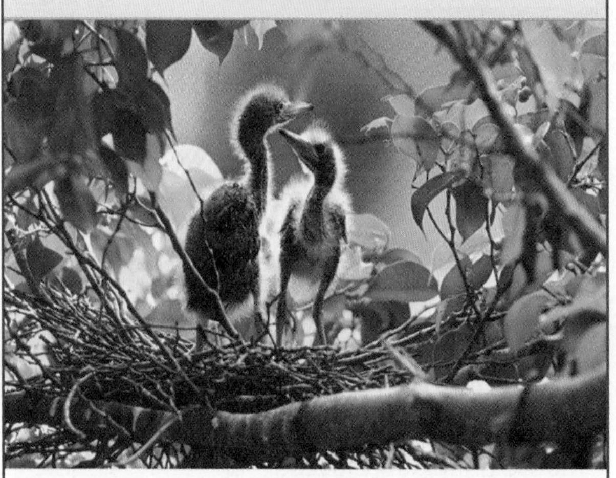

A bird nesting in a tree has a place to raise its young. It neither harms nor benefits the tree.

Parasitism

The roots of *Striga* plants grow into host plants, robbing the hosts of water and nutrients.

Figure 3 Wolves prey on moose in this ecosystem. The wolf population can keep increasing until there are no longer enough moose to support it.

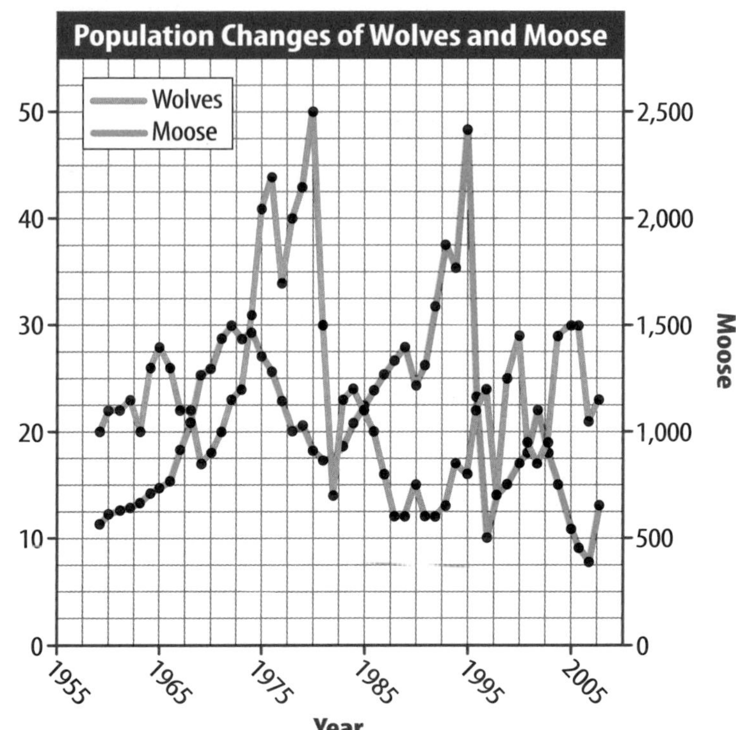

Math Skills

Use Graphs

To interpret the data on a line graph, do the following:

- Identify the information being presented.
- Read the units on the *x*- and *y*-axes. For example, the graph in **Figure 3** compares the wolf and moose populations over a 50-year period.
- Use the graph grid to locate specific information. Sometimes you have to estimate. For example, in 1980, there were 50 wolves and about 900 moose.

Practice

Based on **Figure 3,** what happens to the moose population as the wolf population increases?

- **Math Practice**
- **Personal Tutor**

Competition

Organisms that share the same habitat often compete for resources. Competition describes interactions between two or more organisms that need the same resource at the same time. For example, trees compete for sunlight, and the shade from tall trees can slow the growth of younger trees. Wolves compete with ravens for meat from the animals that wolves kill.

Population Changes

The number of individuals in a population is always changing. Populations increase when offspring are produced or when new individuals move into a community. Populations decrease when individuals die or move away.

Changes in the abiotic factors of an ecosystem affect population size. If a drought reduces plant growth, less food is available for plant eaters, which can lead to a decrease in plant-eater populations. Interactions between organisms also affect population size. As shown in **Figure 3,** predators help control the size of prey populations.

Population density is the size of a population compared to the amount of space available. A high population density means individuals live closer together. This can increase competition and make it easier for disease to be transmitted from one individual to another.

Overpopulation

In summer, an adult moose eats about 18 kg of leaves and twigs every day. As long as food is available, a moose population can continue to grow. But there is a limit to the resources an ecosystem can provide. **Carrying capacity** *is the largest number of individuals of one species that an ecosystem can support over time.* A habitat's carrying capacity depends on its abiotic and biotic factors.

What happens if a population exceeds its carrying capacity? The area becomes overpopulated. If a moose population gets too large, the moose will eat so many plants that they can damage or kill plant life. The destruction of plants destroys habitat for the moose and other species. When food runs out, individuals move away, starve, or are more likely to become sick and die.

Changes in an ecosystem can increase or decrease its carrying capacity for a particular species. Drought, flood, or the arrival of a competing species can reduce carrying capacity. On the other hand, good growing conditions or the disappearance of a competing species can increase carrying capacity.

Extinction

If all the members of a population die or move away from an area, that population becomes extinct. If all populations of a species disappear from Earth, the entire species becomes extinct.

Extinction of one population can affect other populations. What if all the individuals in a moose population died from starvation or disease? Carrying capacity for wolves in that population's area could decrease because no moose would be available as a source of food. Plant-eating animals would no longer be competing with moose for food, so their population's carrying capacity would increase.

Key Concept Check How do population changes affect ecosystems?

Inquiry MiniLab — 20 minutes

What did the rabbits do?

In 1859, a hunter brought 24 rabbits from England to Australia and released them to establish a population for sport hunting. Rabbits had no natural predators in Australia, competed with other grazers for food, and took over the burrows of other animals.

1. Examine the rabbit population map (Map A). The rabbit population in Australia increased from 24 rabbits in 1859 to 600 million rabbits in 1950. Draw a possible graph of this population change in your Science Journal.
2. In your Science Journal, describe the information that Map A provides about the extent of Australia's rabbit population today.

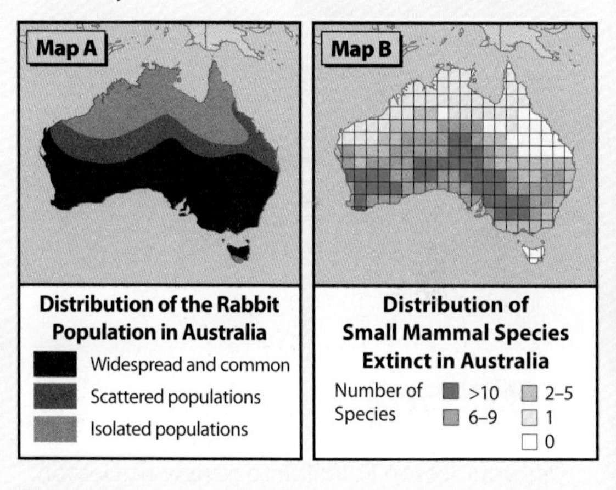

3. Discuss some reasons why North America has not experienced the overpopulation of rabbits that is seen in Australia.

Analyze and Conclude

1. **Infer** why the rabbit population in Australia experienced exponential growth.
2. **Compare** Map A and Map B (number of extinct animal species). Describe the relationship between the information given in the two maps.
3. **Key Concept** Explain how the introduction of rabbits and the rapid increase in their population affected the ecosystem in Australia.

Lesson 1 Review

Online Quiz

Visual Summary

Ecosystems are all the living and nonliving things in a given area.

Species in the same habitat have different niches.

Populations can increase and decrease.

FOLDABLES®

Use your lesson Foldable to review the lesson. Save your Foldable for the project at the end of the chapter.

What do you think NOW?

You first read the statements below at the beginning of the chapter.

1. In symbiosis, two species cooperate in a way that benefits both species.
2. Overpopulation can be damaging to an ecosystem.

Did you change your mind about whether you agree or disagree with the statements? Rewrite any false statements to make them true.

Use Vocabulary

1. **Distinguish** between a population and a community.
2. The way in which an organism uses the resources in its habitat is its ________.
3. **Define** *carrying capacity* in your own words.

Understand Key Concepts 

4. Which is an abiotic factor?
   - **A.** sunlight
   - **B.** trees
   - **C.** animal wastes
   - **D.** decayed leaves
5. **Describe** two ways in which two species might interact.
6. **Explain** how the extinction of one population can affect other populations in the same community.

Interpret Graphics

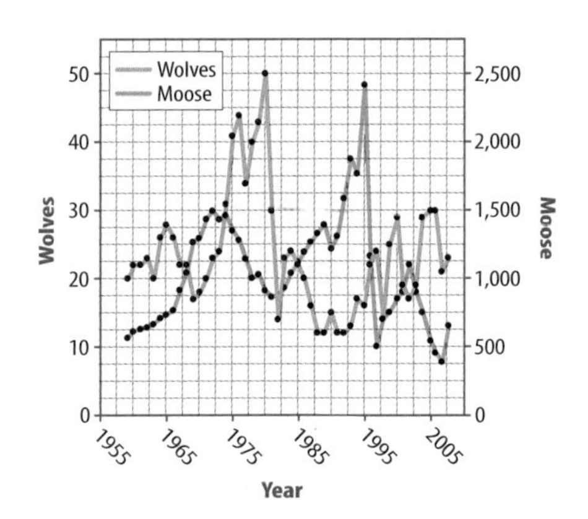

7. **Explain** the changes in the populations shown in the graph above.

Critical Thinking

8. **Hypothesize** The number of mice in a grassland community rises every summer and falls every winter. What could cause the changes in population size?

Math Skills

Math Practice

9. Based on the graph in Question 7 above, what was the maximum population size for both wolves and moose?

River Repair

AMERICAN MUSEUM OF NATURAL HISTORY

SCIENCE & SOCIETY

Restoring the Bronx River Ecosystem in New York City

Today, the Bronx River flows past highways, buildings, and factories. But 400 years ago, there were thick forests, clean water, and salt marshes filled with wildlife such as bears, beavers, and oysters. Over the centuries, as New York City grew, the Bronx River became a dumping ground for factory waste. Construction took the place of the river's surrounding forests and salt marshes. The plants and animals that had once inhabited this environment disappeared.

Today scientists, conservationists, and volunteers are working together to restore the river's ecosystem.

Bringing in Native Plants and Healthy Soil Volunteers are clearing invasive plants, such as mugwort, and planting native shrubs and grasses, such as cordgrass. These native plants help reduce erosion along the riverbank. Contaminated soil has been removed and replaced with clean soil. In some areas, grassy ditches have been filled with porous soil that absorbs runoff and prevents erosion.

▲ A biologist releases herring into the river in the hopes they will reproduce.

Reintroducing Native Fish into the Bronx River The team released alewives, a species of herring, into the river. Alewives once were an important part of the Bronx River's ecosystem, serving as food for birds, raccoons, and river otters. The reintroduction was successful, and new generations of alewives are living in the river.

Rebuilding Habitats for Shellfish A team of scientists, conservationists, and volunteers create artificial reefs using hundreds of mesh bags filled with clamshells. Young oysters attach to the bags and start to grow. Oysters help filter pollution from the water. These artificial reefs also help provide protection for animals such as crabs and snails and serve as food for shorebirds.

People and wildlife are benefiting from the restored river habitat. In 2006, a beaver was seen in the Bronx River. It was the first beaver spotted in the river in more than 200 years!

▲ Volunteers helped the Bronx River ecosystem by planting trees.

It's Your Turn

WRITE Choose a natural area in your neighborhood that is in need of restoration. Write a persuasive paragraph in which you give reasons why it should be restored.

Lesson 2

Reading Guide

Key Concepts 
ESSENTIAL QUESTIONS

- How does energy move through an ecosystem?
- How does matter move through an ecosystem?

Vocabulary

producer p. 655
consumer p. 656
detritivore p. 656
food web p. 657
energy pyramid p. 658

g Multilingual eGlossary

Energy and Matter

Inquiry Where do they live?

These organisms, including the clusters of tubeworms, the eelpout fish, and the crabs, live near a hydrothermal vent. These vents are cracks in Earth's surface in the depths of the ocean. Organisms there obtain energy from the chemicals released from these cracks. In what other ways do you think organisms obtain energy?

Inquiry Launch Lab

20 minutes

Where is the matter?

Matter can change form but cannot be destroyed as it is recycled through ecosystems. Matter exists as a solid, a liquid, or a gas as it performs different functions in an ecosystem.

1. Read and complete a lab safety form.
2. Measure the height of a **votive candle and wick** with a **ruler.** Place the candle in a **petri dish,** cover it with a **jar,** and find the mass of the setup to within 0.01 g on a **balance.** Record the results in your Science Journal.

   ⚠ Tie back any long hair or loose clothing.

3. Lift the jar and light the candle using a **match.** Extinguish the match. Quickly cover the candle with the jar, and let it burn until the flame goes out.

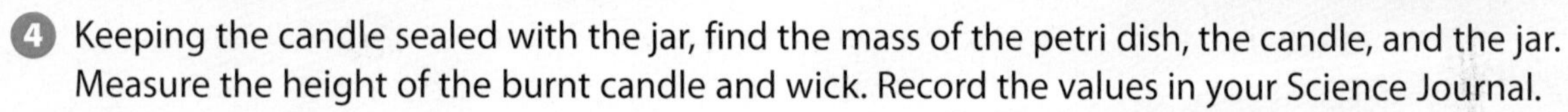

4. Keeping the candle sealed with the jar, find the mass of the petri dish, the candle, and the jar. Measure the height of the burnt candle and wick. Record the values in your Science Journal.

Think About This

1. What changes did you observe during the burning of the candle?
2. How did the mass of the candle and the other equipment after the burning compare with their mass before the burning?
3. **Key Concept** How do you think matter changed form in this lab?

Food Energy

Cars must have gasoline to keep running. Lamps must have electricity to stay lit. Organisms also need a constant supply of energy, in the form of food, to stay alive. How a species obtains energy is an important part of its niche.

Producers *are organisms that use an outside energy source, such as the Sun, and produce their own food.* Most producers—green plants, algae, and some kinds of bacteria—make energy-rich compounds through photosynthesis. Recall that photosynthesis is the chemical process that uses carbon dioxide, water, and light energy—usually from the Sun—to produce glucose (a type of sugar) and oxygen.

Some producers make energy-rich compounds through chemosynthesis, a chemical process similar to photosynthesis. Chemosynthesis uses a chemical such as hydrogen sulfide or methane, instead of light, to produce glucose. Producers that use chemosynthesis include bacteria that live in hot springs or near deep-sea thermal vents, such as the one shown on the previous page.

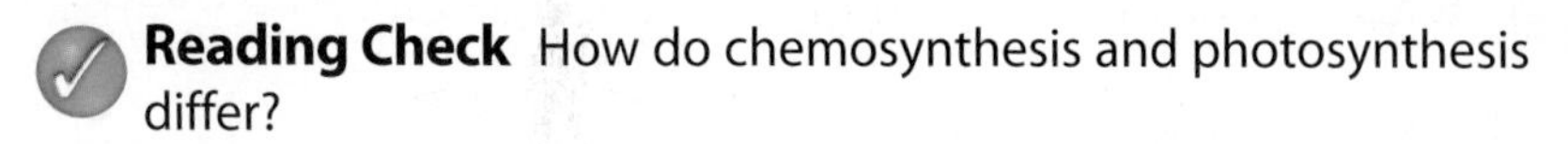

Reading Check How do chemosynthesis and photosynthesis differ?

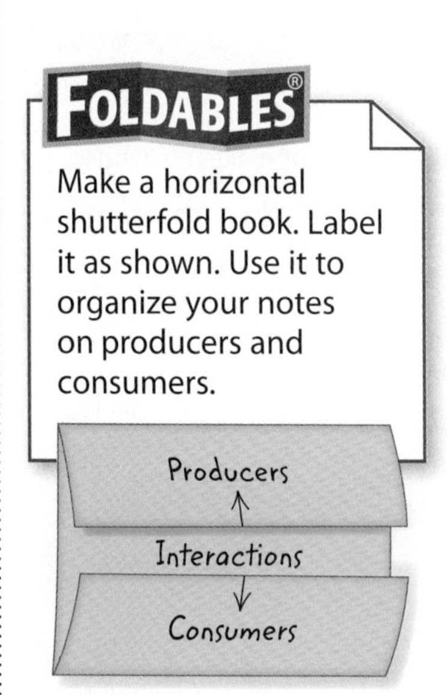

Make a horizontal shutterfold book. Label it as shown. Use it to organize your notes on producers and consumers.

Herbivore Aphids suck fluids from plants.

Omnivore Raccoons will eat almost anything, from fruit and insects to fish and garbage.

Carnivore Moles eat small animals such as earthworms.

Detritivore Termites feed on decaying plant materials.

Figure 4 Consumers fall into one of four types: herbivores, omnivores, carnivores, or detritivores.

Science Use v. Common Use

consumer

Science Use an organism that cannot make its own food

Common Use a person who uses economic goods

Consumers

Organisms that cannot make their own food are **consumers,** as shown in **Figure 4.** They obtain energy and nutrients by consuming other organisms or compounds produced by other organisms.

- Herbivores eat producers. They include butterflies, aphids, snails, mice, rabbits, fruit-eating bats, gorillas, and cows.
- Omnivores eat producers and consumers. They include corals, crickets, ants, bears, robins, raccoons, and humans.
- Carnivores eat herbivores, omnivores, and other carnivores. They include scorpions, octopuses, sharks, tuna, frogs, insect-eating bats, moles, and owls.
- **Detritivores** (duh TRI tuh vorz) *consume the bodies of dead organisms and wastes produced by living organisms.* They include termites, wood lice, and earthworms. Scavengers are detritivores that eat the bodies of animals killed by carnivores or omnivores. Examples include hyenas, jackals, and vultures. Decomposers are microscopic detritivores. They cause the decay of dead organisms or wastes produced by living organisms. Most decomposers are fungi and bacteria.

 **Reading Check** How do organisms obtain energy?

The Flow of Energy

Recall that producers use energy from the environment and make their own food. This is the first step in the flow of energy through an ecosystem. Most producers use photosynthesis and convert the energy from sunlight into chemical energy stored in food molecules. Others use chemosynthesis. Once energy from the environment is converted into food energy, it can be transferred to other organisms.

In an ecosystem, food energy is transferred from one organism to another through feeding relationships. Food chains and food webs are models that describe how energy is transferred through an ecosystem.

Food Chains

A food chain is a simple model that shows how energy moves from a producer to one or more consumers through feeding relationships. Every food chain begins with a producer because producers are the source of all food energy in an ecosystem.

Food Webs

Most ecosystems contain many food chains. *A* **food web** *is a model of energy transfer that can show how the food chains in a community are interconnected.* A food web for a rain-forest community is shown in **Figure 5.**

Key Concept Check How does energy move through an ecosystem?

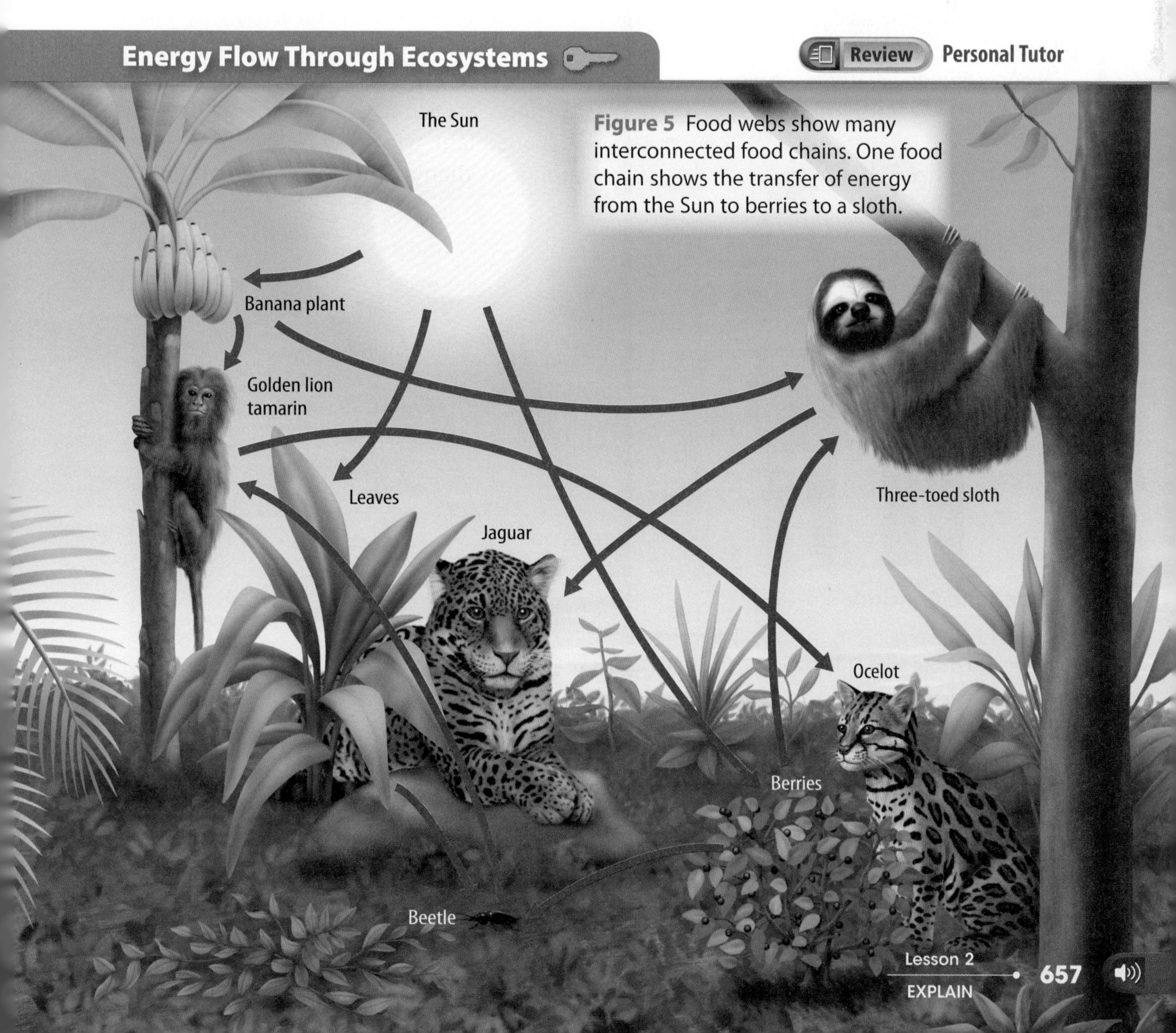

Figure 5 Food webs show many interconnected food chains. One food chain shows the transfer of energy from the Sun to berries to a sloth.

20 minutes

Who's in the web?

All ecosystems, from your backyard or school playground to a tropical rain forest or arctic area, operate under the same ecological principles. In every ecosystem, plants and animals are connected with one another in interdependent relationships.

1. Read and complete a lab safety form.
2. Display your **Organism ID** on your head, arm, or shirt. Take your assigned game position.
3. With a piece of **yarn,** link up with another organism that you need to survive.
4. Link with different-colored yarn to a plant or an animal that needs you to survive.
5. Make a yarn link with another possible connection, and react to scenarios announced by the group.

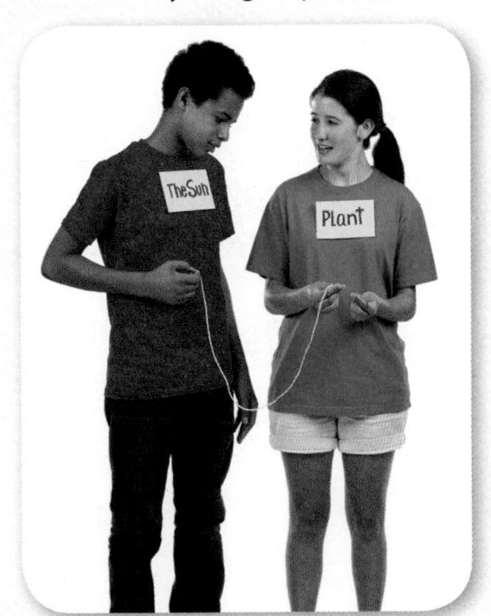

Analyze and Conclude

1. **Describe** the organisms that you linked with and those that linked to you. Explain why each link formed.
2. **Explain** how the food web reacted to one of the scenarios.
3. **Key Concept** Draw a diagram of the connections that included your organism. Using a different color, trace the path of the Sun's energy through the ecosystem.

Energy Pyramids

Most food chains have at least three links, but no more than five. Why? Think about a blackberry plant. It converts light energy from the Sun to chemical energy stored in the plant's tissues. The plant uses some of that energy to perform life processes, including growing blackberries. Some of the stored energy is lost as heat.

When a robin eats blackberries, only part of the energy stored in the plant is transferred to the bird. The robin uses some of the energy for its own life processes, and again some of the energy is lost as heat. A falcon that eats the robin receives even less energy for its life processes.

As shown in **Figure 6,** *an* **energy pyramid** *is a model that shows the amount of energy available in each link of a food chain.* The loss of energy at each level of an energy pyramid helps explain why there are always more producers than carnivores in a community.

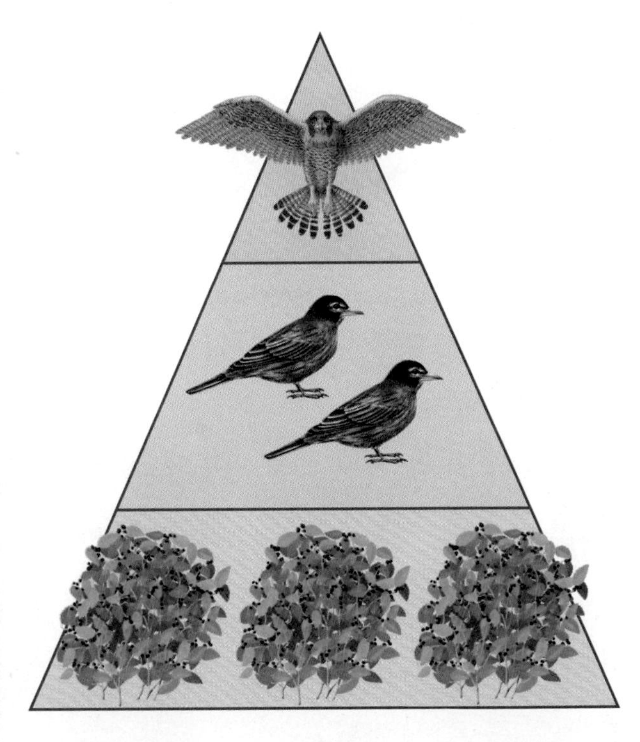

Figure 6 An energy pyramid shows that each step in a food chain contains less energy than previous steps.

Animation

Cycling Materials

Living organisms need more than a constant supply of energy. They also need matter to build cells and tissues. Virtually all the matter on Earth today has been here since the planet formed. The law of conservation of matter states that matter cannot be created or destroyed, but it can change form. Matter is recycled through ecosystems, changing form along the way. Three of the most important pathways of matter moving through an ecosystem are described by the nitrogen **cycle**, the water cycle, and the oxygen-carbon dioxide cycle.

Word Origin
cycle
from Greek *kyklos*, means "circle or wheel"

 **Key Concept Check** How does matter move through an ecosystem?

Nitrogen Cycle

Proteins are essential to all life. An important component of every protein molecule is the element nitrogen. Nitrogen gas makes up about 78 percent of Earth's atmosphere, but most organisms cannot obtain nitrogen from the air. Nitrogen-fixing bacteria live in the soil or on the roots of plants. They convert nitrogen gas into compounds that plants and other producers can absorb. The nitrogen cycle describes how nitrogen moves from the atmosphere to the soil into the bodies of living organisms and back to the atmosphere, as shown in **Figure 7.**

 **Reading Check** What is the niche of nitrogen-fixing bacteria?

Figure 7 Nitrogen changes form as it cycles through an ecosystem.

Concepts in Motion **Animation**

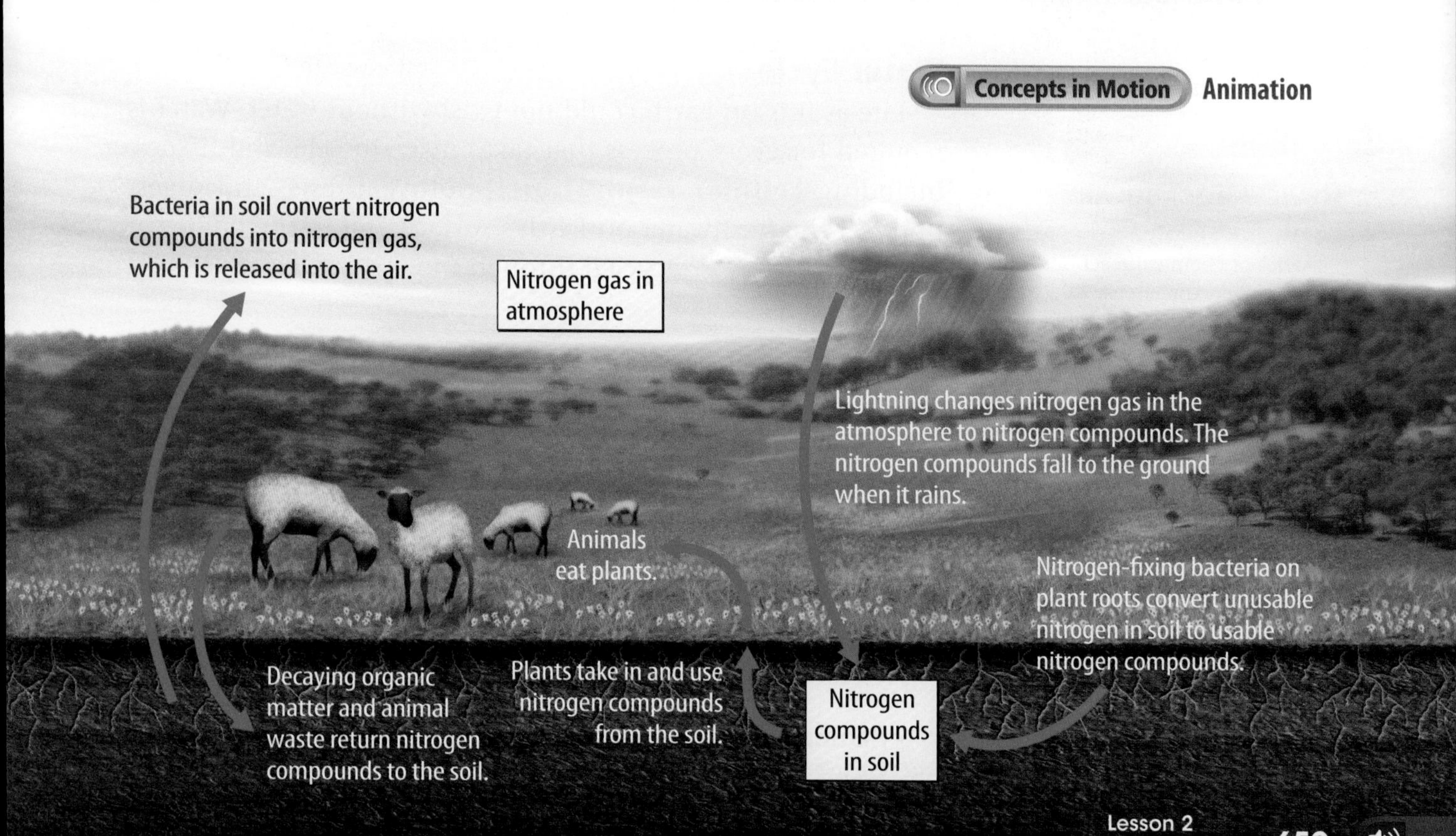

The Water Cycle

Review Personal Tutor

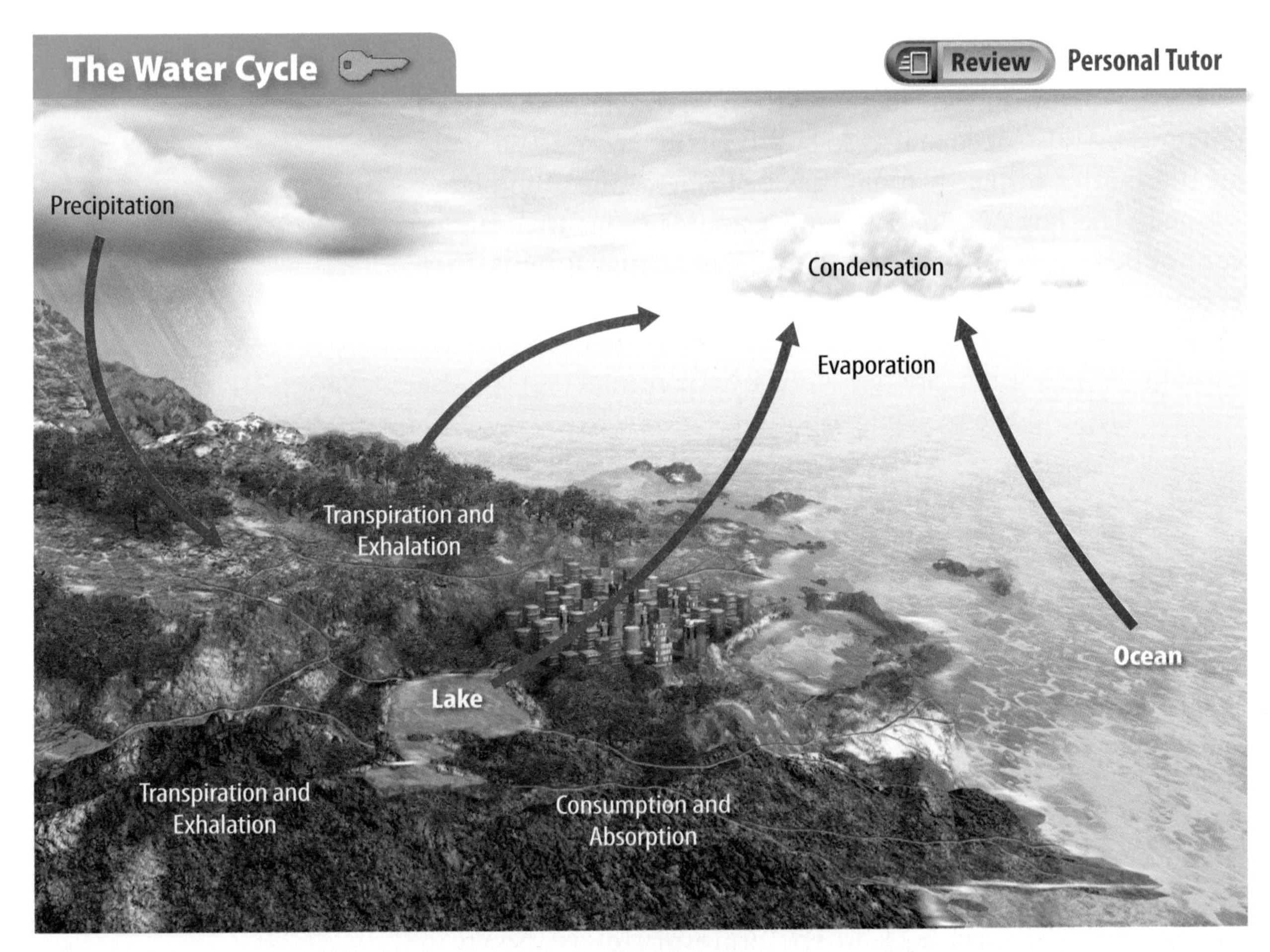

Figure 8 During the water cycle, water moves from the Earth's surface into the atmosphere and then back again.

Visual Check Which processes send water vapor into the atmosphere?

Water Cycle

Life as it is on Earth could not exist without water. Water is required for every process that takes place in cells and tissues, including **cellular respiration,** photosynthesis, and digestion. All the freshwater on Earth's surface and in the bodies of living organisms is recycled through the water cycle, as shown in **Figure 8.**

REVIEW VOCABULARY

cellular respiration
a series of chemical reactions that convert the energy in food molecules into a usable source of energy called ATP

Water evaporates from Earth's surface and rises into the atmosphere as water vapor. Water vapor also is released from the leaves of plants in the process called transpiration and from animals when they exhale. When water vapor comes into contact with cooler air, it condenses and forms clouds. Clouds condense further, and it rains or snows, returning water to Earth's surface as precipitation. Plants and other organisms absorb water from the soil. Animals consume water by drinking fluids or from foods that they eat.

Reading Check What is the role of condensation in the water cycle?

The Oxygen and Carbon Dioxide Cycles

Oxygen and carbon dioxide also cycle through Earth's ecosystems. The cells of most organisms, including those in all plants and animals, require oxygen for cellular respiration. Cells use the energy released during cellular respiration for life processes.

Cellular respiration releases carbon dioxide into the atmosphere, where it can be taken in by plant leaves and used for photosynthesis. Photosynthesis releases oxygen into the atmosphere, where it can be taken in by animals, plants, and other organisms. As shown in **Figure 9,** photosynthesis and cellular respiration are important processes in the cycling of carbon dioxide and oxygen through the living and nonliving parts of ecosystems.

Reading Check How do cellular respiration and photosynthesis compare?

Figure 9 The oxygen and carbon dioxide cycles include processes such as photosynthesis and the formation of fossil fuels.

Lesson 2 Review

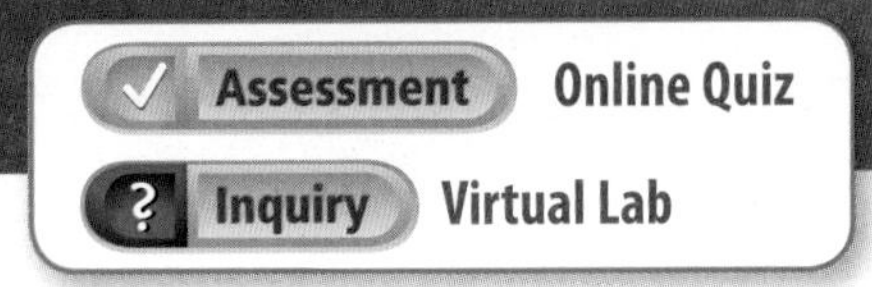

Visual Summary

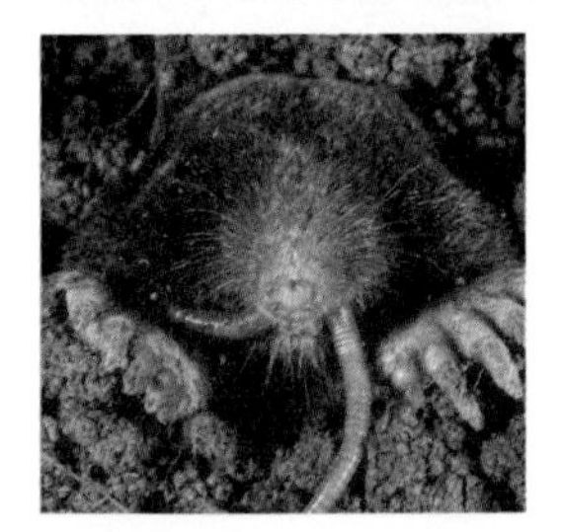

Organisms are classified as producers or consumers.

Energy is transferred from one organism to another through feeding relationships.

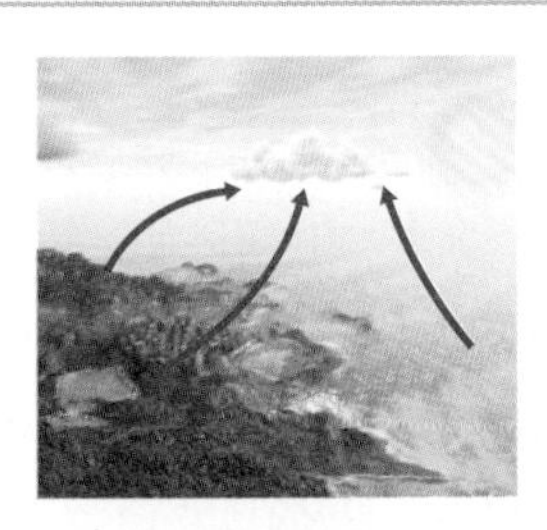

Matter can be changed into different forms and cycles through ecosystems. It cannot be destroyed.

Use your lesson Foldable to review the lesson. Save your Foldable for the project at the end of the chapter.

What do you think NOW?

You first read the statements below at the beginning of the chapter.

3. Sunlight provides the energy at the base of all food chains on Earth.

4. A detritivore is a type of carnivore.

Did you change your mind about whether you agree or disagree with the statements? Rewrite any false statements to make them true.

Use Vocabulary

1. **Define** *food web* in your own words.
2. **Compare** detritivores and scavengers.
3. **Distinguish** between a producer and a consumer.

Understand Key Concepts

4. Which is an animal that eats both producers and consumers?
   - **A.** carnivore
   - **B.** detritivore
   - **C.** herbivore
   - **D.** omnivore
5. **Compare and contrast** the niches of scavengers and decomposers. Give an example of each type of organism.
6. **Explain** how water changes form in the water cycle.

Interpret Graphics

7. **Identify** the two chemical processes occurring in the diagram below.

8. **Sequence** Draw a graphic organizer like the one below and use it to show important steps in the water cycle, beginning with evaporation and ending with consumption.

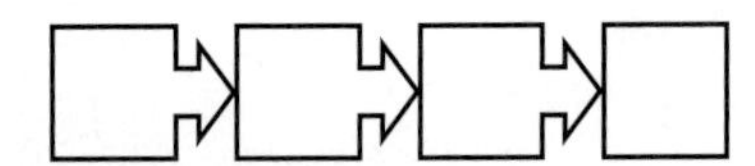

Critical Thinking

9. **Analyze** Some biologists classify scavengers as carnivores. Explain why you agree or disagree.

Inquiry **Skill Practice** **Analyze Data** 40 minutes

How much water can be conserved in greenhouses and nurseries?

Water-conservation scientists collect and study data on, among other things, how water is used in greenhouses. These studies involve observing the relationships between environmental factors and how much water is used to grow crops.

Learn It

In science, **data analysis** involves classifying, comparing, and recognizing cause and effect. Patterns that help scientists determine the meaning of the data then can be identified.

Try It

1. Copy Table 1 into your Science Journal, and calculate the percentage of applied water actually used by the plants. Fill in the table with those values.

Table 1 Greenhouse Watering Data

Day	Applied Water (m3)	Water Drainage (m3)	Solar Radiation (Jcm2)	Percent Water Used
1	72	26	2,649	
2	72	38	1,696	
3	72	42	1,459	
4	72	34	1,977	
5	72	39	1,745	
6	72	38	1,518	
7	72	43	1,665	

2. Draw a bar graph showing the percentage of applied water used at different levels of solar radiation.
3. From the information in Table 1, identify which growing method consumes the largest amount of water. Record your answer in your Science Journal.
4. From the information in Table 2, identify the year in which there were no greenhouse operations.

Apply It

5. **Analyze** your data, and graph the relationship between solar radiation and the amount of water used. Describe this relationship.
6. **Evaluate** the information in Table 3 and describe the growing method that uses water most efficiently. Explain how you reached your conclusion.

Table 3 Water Consumption in the Naivasha Basin (Million cubic meters/year)

Crops	Total Water Consumed (Mm3/year)
Greenhouse flowers	20.44
Outdoor flowers	22.37
Grass/fodder	0.86
Vegetables	22.15

7. **Compare** the data in Table 2 to determine the growing method that had the largest area increase between 1996 and 2006.
8. **Key Concept** Predict which method of cultivation most likely will be used the most in the future.

Table 2 Indoor v. Outdoor Cultivation Estimated Irrigated Area (all figures in hectare (ha))

Type of Area	Study A 1996–97	Study B 1999	Study C 2001	Study D 2004	Study E 2006
Outdoor	3,581	3,548	4,417	3,101	3,800
Greenhouse	---	1,020	614	1,191	1,600
Total	3,581	4,568	5,031	4,292	5,400

Lesson 3

Reading Guide

Key Concepts 
ESSENTIAL QUESTIONS

- In what ways do humans affect ecosystems?
- What can humans do to protect ecosystems and their resources?

Vocabulary

renewable resource p. 666
nonrenewable resource p. 666
resource depletion p. 666

Multilingual eGlossary

Video
What's Science Got to do With It?

Humans and Ecosystems

Inquiry Washing a Bird?

Organisms that live in the same area, including humans, interact with one another. Sometimes these interactions can be harmful, such as for the marine bird pictured that is recovering from an oil spill. However, some can be beneficial, such as for the workers cleaning the spill. What other ways do you think humans interact with ecosystems?

Launch Lab

20 minutes

How can you conserve resources by reusing items?

One of the simplest actions individuals can take to protect the environment is to reuse items. Many of the items that you use today can be made into something else and used again.

1. Read and complete a lab safety form.
2. Think about items you have seen that can have more than one use. Write your list in your Science Journal.
3. From the **selection of items** provided, choose one item and design a new use for it.
4. Describe or sketch the new reusable item you designed. Explain how it can be used.

Think About This

1. Choose an item from those presented by your classmates. Brainstorm another way the item could be reused.
2. **Key Concept** How does the practice of reusing items protect ecosystems?

Affecting the Environment

About 3.5 billion years ago, microscopic organisms called cyanobacteria (si an oh bak TIH ree uh) began to make a big change in Earth's environment. Cyanobacteria are unicellular organisms that probably were the first life forms capable of photosynthesis. Over a period of approximately one billion years, photosynthesis increased the concentration of oxygen in the atmosphere from almost zero to higher than it is today. Without oxygen in the atmosphere, life as it is on Earth might not have been possible.

All organisms change the environment in some way. Beavers cut down trees to build dams and form ponds, as shown in **Figure 10.** Humans also change the environment. We change ecosystems by replacing wildlife habitats with buildings, roads, farms, and mines. Our use of energy resources such as coal and natural gas can create pollutants that affect plant and animal life in the air, the water, and on land.

Reading Check How did cyanobacteria change Earth's environment?

Figure 10 All organisms affect their environment. When beavers build dams, they reduce forest habitat. At the same time, they create ponds— a new habitat for dragonflies, frogs, ducks, and other aquatic species.

Using Natural Resources

Humans use many of the same resources as other species, including renewable resources such as food, water, and oxygen. **Renewable resources** *are resources that can be replenished by natural processes at least as quickly as they are used.*

Unlike most species, humans also use nonrenewable resources, especially fossil fuels—coal, oil, and natural gas. **Nonrenewable resources** *are natural resources that are used up faster than they can be replaced by natural processes.* One example of a nonrenewable resource is fossil fuels. Earth's supply of fossil fuels is dwindling, and people are looking for ways to replace them with renewable sources.

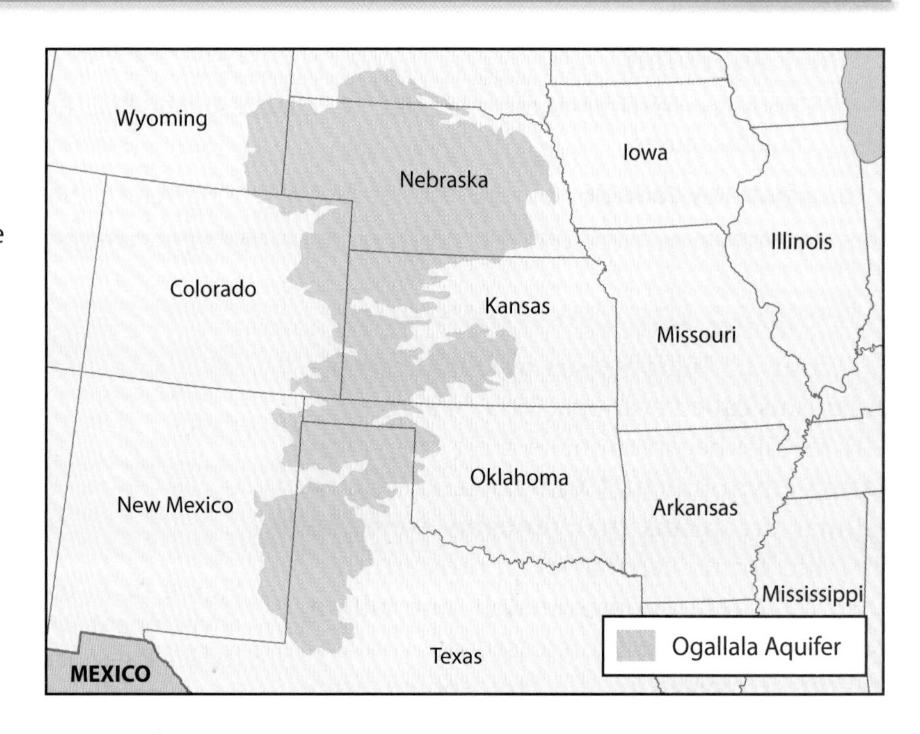

Figure 11 Today, the water in the Ogallala Aquifer is still pumped out faster than it is replaced, but the rate of depletion has slowed from an average of about 3 m per year to 1.3 cm per year.

Any resource becomes nonrenewable if it is used up faster than it can be replaced. **Resource depletion**—*the exhaustion of one or more resources in an area*—is happening in the United States and throughout the world. One example of this is the Ogallala Aquifer. The Ogallala Aquifer is a vast reservoir of underground water in the Great Plains region, as shown in **Figure 11.** The aquifer is an important source of irrigation water for U.S. farmland. From the 1940s to the 1970s, people pumped water out of the aquifer much faster than it could be replenished by natural processes. Authorities predicted that by the year 2000 there would be little water left in the aquifer. In response, people began developing and using water-saving irrigation technologies.

WORD ORIGIN

depletion
from Latin *deplere,* means "to empty"

Understanding the Effects of Pollution

When people use resources, their actions can have unintended consequences. No one could have predicted that burning fossil fuels would lead to smog formation or acid rain. It took some time before people learned that applying chemical fertilizers to farmland could lead to harmful algal blooms in nearby lakes and streams. The more people can learn about how their actions affect the environment, the better their ability to make good environmental choices in the future.

Unintended Consequences The invention of chlorofluorocarbons (klor oh flor oh KAR buhnz), commonly called CFCs, in the early 1900s is another example of unintended consequences. CFCs are chemical coolants used in refrigerators and air conditioners. In the 1970s, it was discovered that CFCs and related chemicals damage the ozone layer. The ozone layer is the layer of the atmosphere that shields Earth's surface from ultraviolet (UV) light from the Sun. High UV levels damage DNA, increase skin cancer rates, and disrupt photosynthesis.

As shown in **Figure 12,** potential further damage to the ozone layer was avoided by an international treaty called the Montreal Protocol. Signed by numerous countries in the 1980s, the treaty phased out CFC use worldwide.

Key Concept Check In what ways do humans affect ecosystems?

FOLDABLES®

Make a vertical trifold book. Label it as shown. Use it to summarize the human impact on natural resources, the human impact on pollution, and the EPA.

Human Impact on Natural Resources

Human Impact on Pollution

Environmental Protection Agency

Figure 12 This simulation shows what could have happened to Earth's ozone layer without the Montreal Protocol.

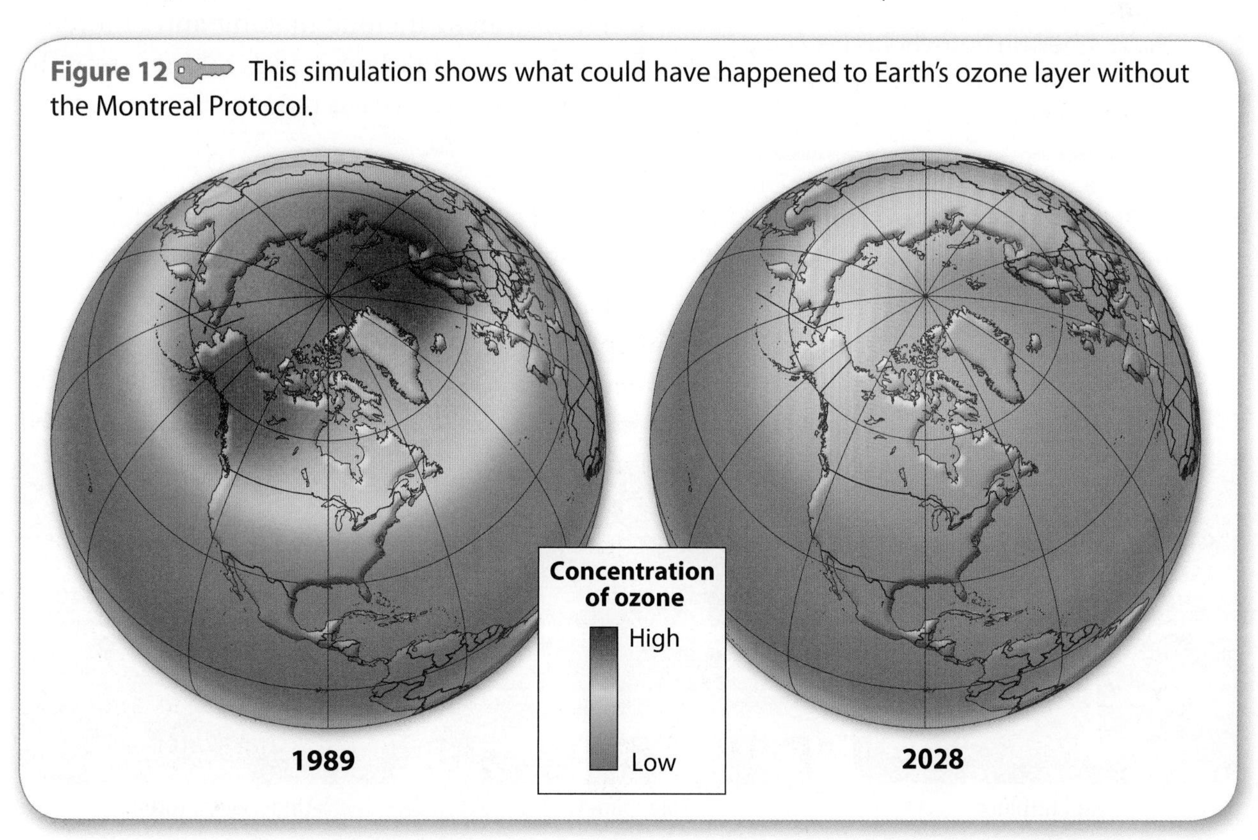

MiniLab

20 minutes

How can you calculate a carbon footprint?

A carbon footprint is an estimate of the amount of carbon dioxide that is produced through home energy use, transportation, diet, and household waste. CO_2 emissions are called greenhouse gas emissions because of the ways in which they impact the environment.

1. Study the **table** provided. This table describes home energy use, transportation, diet, and household waste of the households of two fictitious students. Note the differences between households.
2. Calculate the carbon footprint for each household by entering the data into a **carbon-footprint calculator.**

Analyze and Conclude

1. **Compare** the carbon footprints of the two households. How are they different?
2. **Analyze** How would you calculate the individual carbon footprint values for students A and B?
3. **Key Concept** Evaluate which of these 4-person households has the greatest effect on the ecosystem. Explain your answer.

Global Climate Change Most scientists agree that fossil-fuel use also increases concentrations of carbon dioxide and other greenhouse gases in the atmosphere. The amount of greenhouse gases emitted by a person, an organization, an event, or a product is called its carbon footprint. An increase in greenhouse gases is contributing to global warming—a rise in Earth's average surface temperature. Global climate change is a result of global warming and could lead to a variety of environmental consequences. It could change the kinds of crops that can be grown in various parts of the world, increase the number and severity of floods and droughts, and even raise sea levels.

Protecting the World

The better people understand how human actions affect the environment, the more people can avoid causing harm. Scientists are working to develop renewable energy resources that can reduce pollution and people's dependence on fossil fuels, as shown in **Figure 13.** Other strategies for protecting the environment include making and enforcing environmental laws, and taking steps in people's daily lives to reduce their impact on the planet.

Figure 13 Harnessing the power of the wind and the Sun to generate electricity already is reducing our consumption of fossil fuels. Underwater turbines that capture the energy of tides and ocean currents also are being tested.

Wind turbines

Solar panels

Underwater turbine

Enacting Environmental Laws

Who is responsible for preventing pollution, cleaning it up if it happens, or repairing damage to the environment? Protecting the environment can be expensive, and people don't always agree about who should pay the costs. The U.S. government passes laws to help protect the environment. Many of these laws are **enforced** by the U.S. Environmental Protection Agency (EPA). Some of these laws include the following:

- The Endangered Species Act, signed into law in 1973, lists species threatened or endangered with extinction, designates habitat to protect them, and outlaws actions that would harm them.
- The Clean Air Act, enacted in 1970, gives the EPA the power to create emissions standards for automobiles, industries, and power plants to reduce the amount of pollutants in the air. The law includes regulations designed to reduce carbon dioxide emissions.
- The Clean Water Act, first enacted in 1972 and expanded in 1977 and 1987, regulates the discharge of pollutants into waterways. This set of laws has helped make significant improvements to water quality.

The EPA also monitors environmental health, looks for ways to reduce human impacts, develops plans for cleaning up polluted areas such as one shown in **Figure 14,** and supports environmental research at universities and national laboratories.

Reading Check What is the EPA's role in keeping the environment healthy?

Figure 14 Water that collected in this abandoned mine contains toxic levels of copper. Part of the environmental cleanup involves building dams to prevent the water from flowing into nearby streams.

ACADEMIC VOCABULARY
enforce
(verb) to carry out effectively

Making a Difference

Everyone can take action to help keep the environment healthy and to make sure that future generations of life on Earth have the resources they need to survive. Collectively, these actions are known as the 5Rs.

Restore Habitats and ecosystems that have been damaged can be restored—brought back to their original state. Examples include planting trees to restore forests or removing trash to clean up streams and beaches.

Rethink Great ideas come from reconsidering, or rethinking, the way people carry out daily tasks. For example, a gym owner in Oregon invented a way for people exercising on spin bikes to generate electricity for the gym's video and sound systems. One train station in Tokyo, shown in **Figure 15,** uses tiles that convert people's footsteps into electricity.

Figure 15 The floor of this train station in Tokyo, Japan, is made of special tiles that convert the energy from people's footsteps into electricity. The tiles help power ticket gates, lights, and signs.

Reduce People can reduce waste and pollution by using fewer resources whenever possible. Turn off lights when they aren't being used. Put on a sweater instead of turning up the heat. Walk or ride your bicycle instead of riding in a vehicle.

Reuse Have broken items repaired instead of replacing them. Buy used items instead of new items. Invent new uses for things instead of throwing them in the trash.

Recycle People can conserve resources by recycling—processing things so they can be used again for another purpose. Paper, plastic, glass, metal, yard waste, and used appliances and electronics can be recycled instead of dumped in landfills.

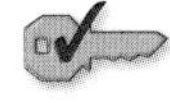
Key Concept Check How can people protect ecosystems and conserve resources?

Lesson 3 Review

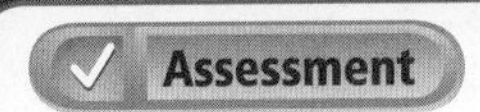

Online Quiz

Visual Summary

Organisms affect their environment in both positive and negative ways.

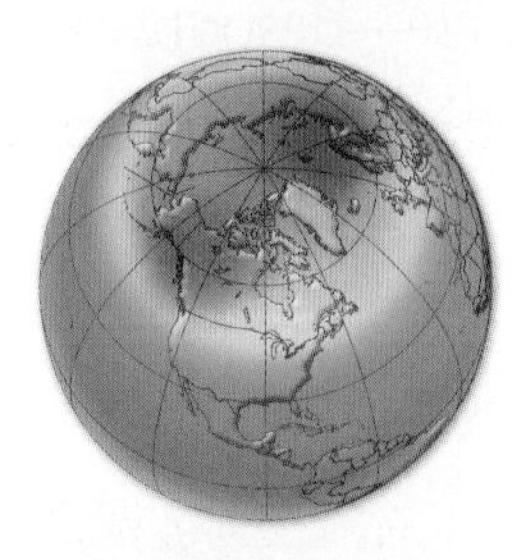

Making and enforcing environmental laws can protect the environment from further damage.

More industries and people are turning to renewable resources rather than using nonrenewable resources.

FOLDABLES®

Use your lesson Foldable to review the lesson. Save your Foldable for the project at the end of the chapter.

What do you think

You first read the statements below at the beginning of the chapter.

5. Human actions can have unintended effects on the environment.

6. The only job of the U.S. Environmental Protection Agency is to enforce environmental laws.

Did you change your mind about whether you agree or disagree with the statements? Rewrite any false statements to make them true.

Use Vocabulary

1. **Distinguish** between renewable resources and nonrenewable resources.
2. **Define** *resource depletion* in your own words.

Understand Key Concepts

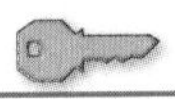

3. Which term describes changes to the atmosphere that could disrupt ecosystems and cause sea levels to rise?
   - **A.** climate change
   - **B.** renewable resources
   - **C.** resource depletion
   - **D.** water pollution
4. **List** Name three sets of environmental laws passed by the U.S. government, and explain how each one helps prevent environmental problems.

Interpret Graphics

5. **Identify** What kind of renewable energy resource does the technology shown to the right use for generating electricity?

6. **Summarize Information** Copy the table below and fill in actions that can be taken in each category to help conserve resources and protect the environment.

Restore	
Rethink	
Reduce	
Reuse	
Recycle	

Critical Thinking

7. **Summarize** Explain how an environmental law that requires better gas mileage for cars and trucks could affect global climate change.

2 class periods

Measure Your Carbon Footprint

Materials

graph paper

computer with internet access

The amount of carbon emitted by the actions you take during your daily life can be calculated and described by a value called the carbon footprint. The greater the size of your carbon footprint, the greater the negative impact your actions have on the environment. The 5Rs—restore, rethink, reduce, reuse, and recycle—describe ways in which people can reduce their carbon footprints in an effort to conserve Earth's resources.

Question

What information is used to measure a carbon footprint? How can the size of a person's carbon footprint be reduced?

Procedure

1. Read and complete a lab safety form.
2. Copy the table below into your Science Journal, and use it to record your household information.
3. Calculate your carbon emissions score using the table and the information provided. Record it in your Science Journal.
4. Compare your score with the average per capita score in the United States and the average per capita score in the world. Using these values, create a bar graph in your Science Journal to show how the three compare.

Energy Use	Household
Home energy (household type and number of bedrooms)	
Heating and cooling (level of efficiency or amount of electricity used)	
Lighting (type and number of bulbs)	
Appliances (energy star? unplug when not in use?)	
Hot water (conservation behaviors?)	
Personal vehicles (type of car, annual mileage)	
Air travel (miles per trip)	
Food and diet (heavy, average, or low use of meat in diet?)	
Recycling and waste (amount of recycling and composting)	

5. Using what you know about carbon footprints and what affects them, estimate your carbon footprint 5 years ago and predict what your carbon footprint will be in 5 years. Include any changes that you might make to decrease your carbon footprint.
6. Modify your plan to include as many 5R practices as possible. Record your revisions in your Science Journal.

Analyze and Conclude

7. **Draw** a line graph showing your estimated carbon footprint of 5 years ago, your carbon footprint today, and your predicted carbon footprint after you have used your 5R policies. What does the graph show about the changes to your carbon footprint over time?
8. **The Big Idea** A shoe company states that the carbon footprints of its boots range from 0.055–0.090 tons per pair. Describe the factors that might result in this carbon footprint and the ways that it could be reduced. What could you do to offset the amount of carbon released by the production of these boots?

Communicate Your Results

A spokesperson for a major food industry says that the company does not plan to list a product's carbon footprint on its packaging. Create a presentation that could convince the company that including this information is important.

Inquiry Extension

Research the carbon footprint of a commonly used product. The total carbon footprint includes production, shipping, storing, retailing, and use of the product. Use a graphic organizer to trace the original sources of all of the materials and processes used in the production of the product.

Lab Tips

- ☑ When calculating your carbon footprint, think about the habits of your entire household.
- ☑ Remember that many factors are involved in calculating a carbon footprint.
- ☑ Make sure your plan to reduce your carbon footprint contains all of the 5 Rs.

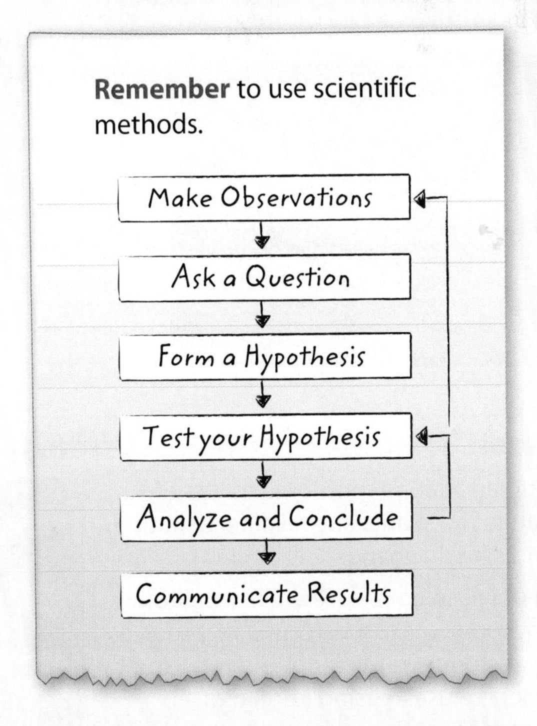

Chapter 18 Study Guide

Organisms interact with each other and the environment around them to obtain food, shelter, living space, and other resources needed for life.

Key Concepts Summary

Lesson 1: Ecosystems

- An ecosystem consists of all the living and nonliving parts of the environment in a given area and the interactions among them.
- Organisms cooperate with, compete with, or feed on one another to obtain the resources they need for survival.
- **Populations** that grow larger than an ecosystem's **carrying capacity** are overpopulated. Overpopulation can harm the ecosystem by depleting resources. Extinction—the complete disappearance of a population from a **community**—can alter the ways in which remaining populations interact.

Vocabulary

habitat p. 648
population p. 648
community p. 648
niche p. 649
predation p. 649
symbiosis p. 649
carrying capacity p. 651

Lesson 2: Energy and Matter

- Energy, usually from the Sun, moves through an ecosystem by being transferred from one organism to another.
- Matter changes form as it cycles through an ecosystem.

Vocabulary

producer p. 655
consumer p. 656
detritivore p. 656
food web p. 657
energy pyramid p. 658

Lesson 3: Humans and Ecosystems

- Human actions contribute to loss of habitat for plants and wildlife, pollution, and climate change.
- People can educate themselves about environmental issues; conserve resources by restoring, rethinking, and reducing resource use; reusing instead of replacing; and recycling.

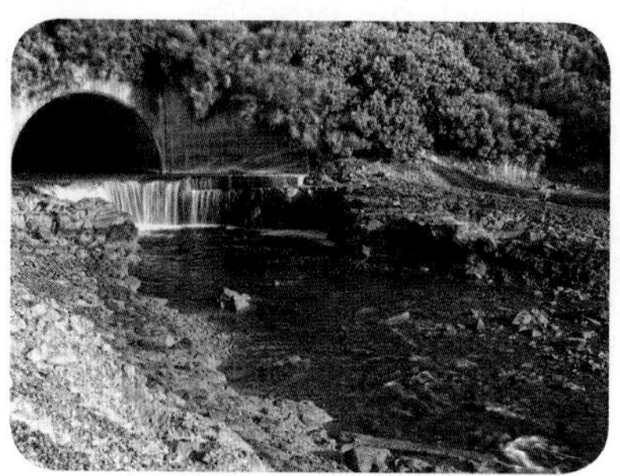

Vocabulary

renewable resource p. 666
nonrenewable resource p. 666
resource depletion p. 666

- Personal Tutor
- Vocabulary eGames
- Vocabulary eFlashcards

FOLDABLES® Chapter Project

Assemble your lesson Foldables as shown to make a Chapter Project. Use the project to review what you have learned in this chapter.

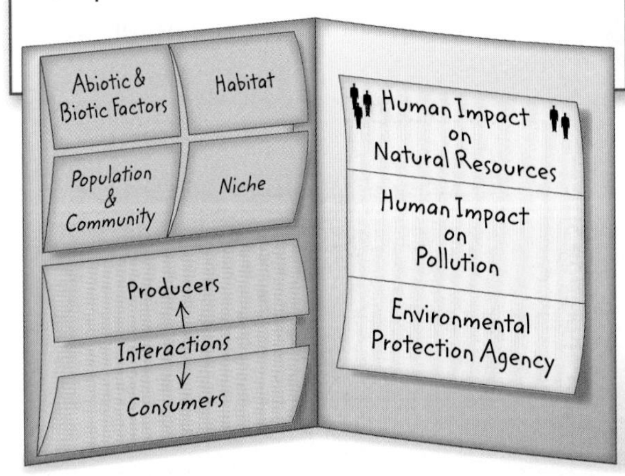

Use Vocabulary

Write the vocabulary term that best matches each phrase.

1. the portion of the environment that provides an organism with all the abiotic and biotic factors it needs for life
2. the manner in which an organism obtains food, shelter, and other needs
3. the largest number of individuals of a species that an ecosystem can support over time
4. organisms that use an outside energy source and produce their own food
5. a model of energy transfer that shows how the food chains in a community are interconnected
6. the exhaustion of one or more resources in an area

Link Vocabulary and Key Concepts

Concepts in Motion Interactive Concept Map

Copy this concept map, and then use vocabulary terms from the previous page to complete the concept map.

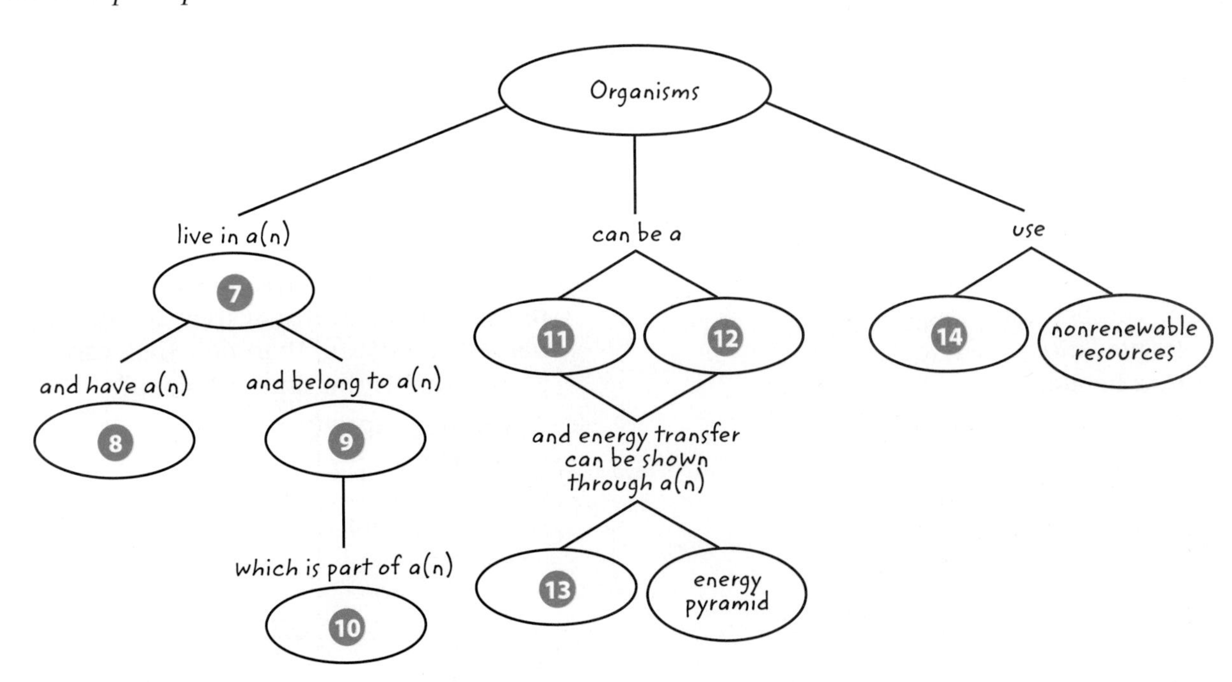

Chapter 18 Review

Understand Key Concepts

1. Which part of an ecosystem do the organisms in the figure below represent?

A. a community
B. a habitat
C. a population
D. a species

2. Which is a relationship in which two organisms cooperate and neither is harmed?

A. competition
B. mutualism
C. parasitism
D. predation

3. Earthworms, bacteria, and millipedes are examples of which type of organisms?

A. competitors
B. detritivores
C. parasites
D. predators

4. Leafcutter ants bring bits of leaves to their nest as food for fungus. The ants feed on the fungus. What does this describe?

A. competition
B. predation
C. a habitat
D. a niche

5. Which models the changes in the availability of food energy as energy moves through an ecosystem?

A. an energy pyramid
B. a food chain
C. a food web
D. a population graph

6. What does the red line in the graph below represent?

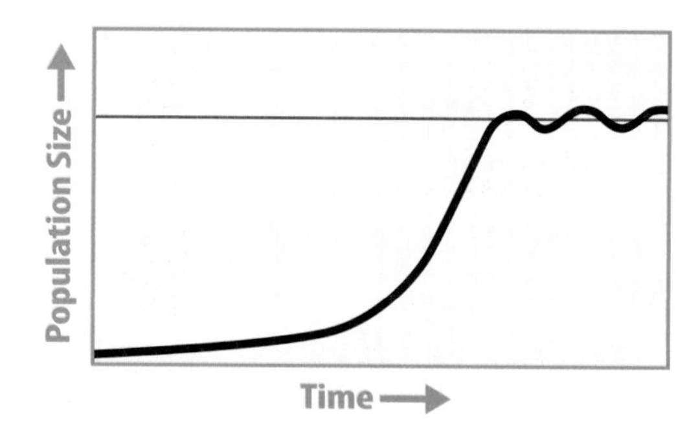

A. carrying capacity
B. ecosystem size
C. overpopulation
D. population density

7. Which model includes the action of bacteria in soil taking a gas from the atmosphere and converting it into a material that plants can absorb through their roots?

A. carbon dioxide and oxygen cycle
B. nitrogen cycle
C. phosphate cycle
D. water cycle

8. In some parts of Africa, people's need for firewood is so great that they are forced to harvest trees faster than new trees can grow. What is this an example of?

A. carrying capacity
B. recycling
C. renewable resources
D. resource depletion

Critical Thinking

9 **Analyze** the graph below. What happened to the moose and wolf populations between 2000 and 2009? Explain how the predator–prey relationship could have resulted in these changes.

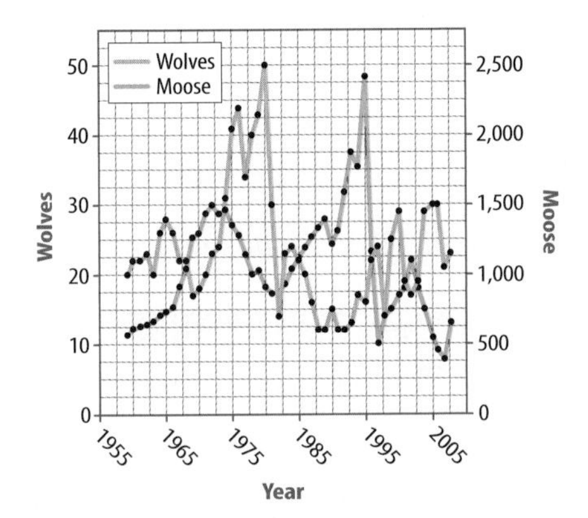

10 **Hypothesize** Why do seed packets include information about how far apart the seeds should be planted? Use the term *competition* in your answer.

11 **Compare and contrast** the movement of nitrogen and water through ecosystems.

12 **Imagine** what happens to a carbon dioxide molecule that enters the leaf of a plant. Explain how the carbon atom and the two oxygen atoms each could end up in the bodies of different carnivores.

13 **Infer** In many regions, people are removing large areas of forest to make room for farmland and homes. In what way could this action contribute to global warming?

Writing in Science

14 **Write** a five-sentence paragraph that describes how you and your friends or your family could make changes to help protect the environment and conserve resources. Base your ideas on at least one of the 5Rs—restore, rethink, reduce, reuse, recycle.

REVIEW THE BIG IDEA

15 Describe at least two ways in which you interact with abiotic factors in your environment and two ways in which you interact with biotic factors in your environment.

16 The alligator pictured below lives in a wetland environment. What are the abiotic and biotic factors that it interacts with? In what ways does it interact?

Math Skills

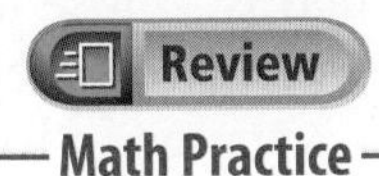

Math Practice

Use Graphs

Use the graph in the previous column to answer questions 17 and 18.

17 a. What is the greatest number of moose recorded during the counting period?

b. What was the approximate number of wolves during that same year?

18 a. During what two periods of time did the wolf population drop the fastest?

b. What happened to the moose population during those same periods?

Standardized Test Practice

Record your answers on the answer sheet provided by your teacher or on a sheet of paper.

Multiple Choice

Use the figure below to answer question 1.

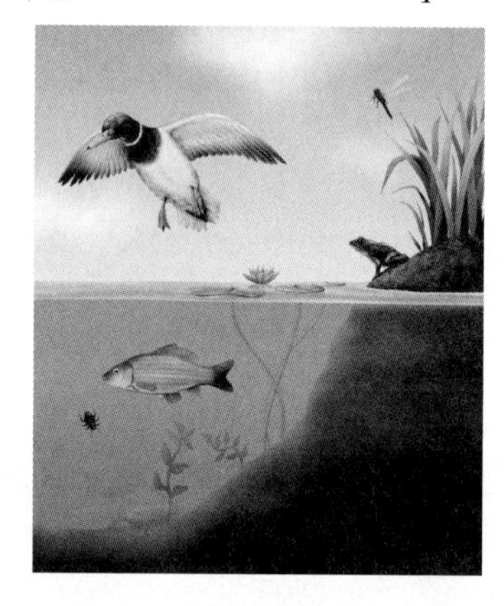

1 Which are the abiotic factors in the ecosystem shown above?

A frogs, fish, and ducks

B water, soil, and rocks

C frogs, fish, lilies, and algae

D grass, insects, lilies, and algae

2 Which is a close long-term relationship between two species that usually involves an exchange of food or energy?

A competition

B extinction

C predation

D symbiosis

3 Which is an example of predation?

A birds nesting in a tree

B giraffes feeding on tree tops

C honeybees pollinating flowers

D lions hunting zebras

4 In which category do deep–sea-vent bacteria belong if they use chemosynthesis and make food?

A consumer

B detrivore

C omnivore

D producer

Use the figure below to answer question 5.

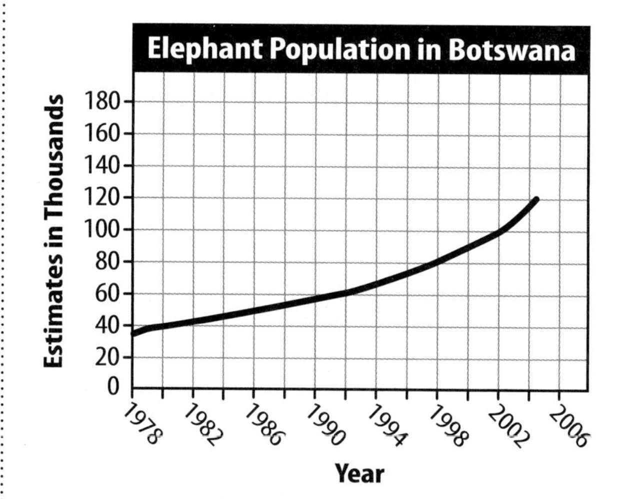

5 What does the graph above show about the elephants in Botswana?

A The population decreased.

B The population increased.

C The population became extinct.

D The population stayed the same.

6 A mole eats small animals such as larvae and worms. Which type of consumer is the mole?

A carnivore

B detrivore

C herbivore

D omnivore

Assessment
Online Standardized Test Practice

7 Which cycle of materials includes the process of photosynthesis removing CO_2 from the atmosphere and releasing oxygen?

A nitrogen cycle

B water cycle

C oxygen-carbon dioxide cycle

D plant life cycle

Use the figure below to answer question 8.

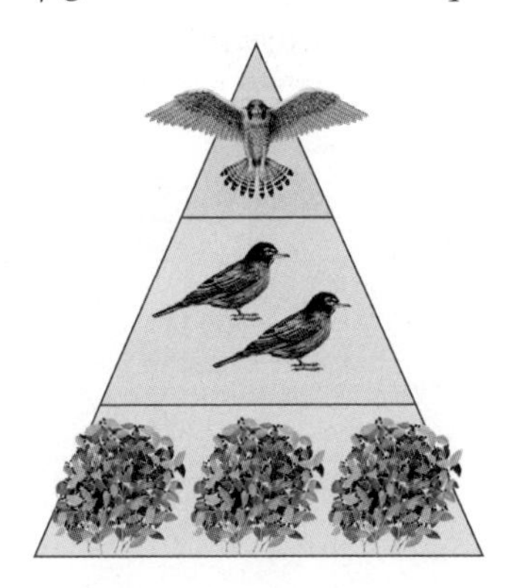

8 Which describes the flow of energy in the food chain shown in the diagram above?

A blackberries, robin, falcon

B blackberries, robin, leaves

C falcon, robin, blackberries

D the Sun, blackberries, robin

9 Because Earth's supply of fossil fuels is dwindling, people should try to replace them with which type of resources?

A carbon resources

B depleted resources

C nonrenewable resources

D renewable resources

Constructed Response

Use the figure below to answer questions 10 and 11.

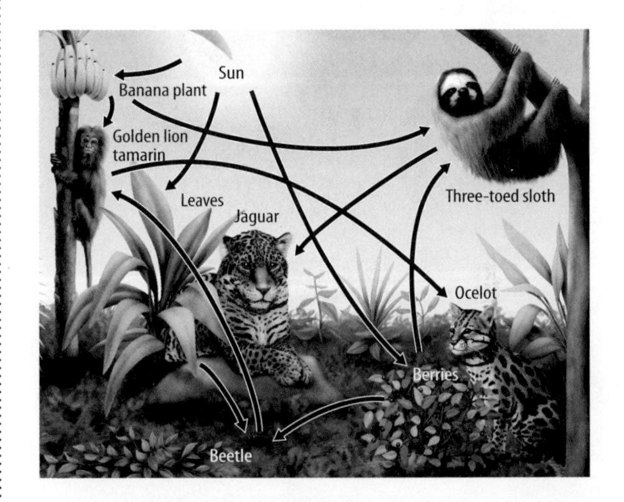

10 Select and compare two food chains from the diagram above, identifying the producers and consumers in each.

11 What might happen to the three-toed sloth population in the rainforest ecosystem shown if the jaguars were to become extinct? What other effects might this have on the ecosystem?

12 A local wetland is home to endangered frogs, many species of plants, fish, crustaceans, and nesting birds. A construction company wants to fill in part of it to put up a shopping mall. Make an argument against the project explaining the possible effects on the wetland ecosystem.

13 Compare mutualism to commensalism. Provide an example for each.

NEED EXTRA HELP?													
If You Missed Question...	1	2	3	4	5	6	7	8	9	10	11	12	13
Go to Lesson...	2	1	2	2	3	2	2	2	2	2	1, 2	1, 2, 3	1

Chapter 19

Biomes and Ecosystems

THE BIG IDEA **How do Earth's biomes and ecosystems differ?**

Inquiry Modern Art?

Although it might look like a piece of art, this structure was designed to replicate several ecosystems. When Biosphere 2 was built in the 1980s near Tucson, Arizona, it included a rain forest, a desert, a grassland, a coral reef, and a wetland. Today, it is used mostly for research and education.

- How realistic do you think Biosphere 2 is?
- Is it possible to make artificial environments as complex as those in nature?
- How do Earth's biomes and ecosystems differ?

Get Ready to Read

What do you think?

Before you read, decide if you agree or disagree with each of these statements. As you read this chapter, see if you change your mind about any of the statements.

1. Deserts can be cold.
2. There are no rain forests outside the tropics.
3. Estuaries do not protect coastal areas from erosion.
4. Animals form coral reefs.
5. An ecosystem never changes.
6. Nothing grows in the area where a volcano has erupted.

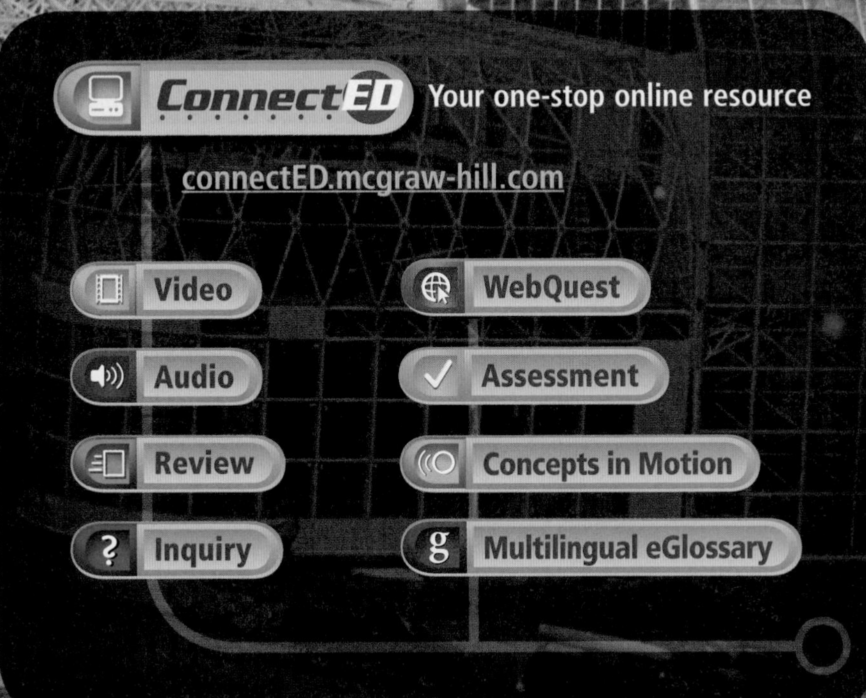

Lesson 1

Reading Guide

Key Concepts
ESSENTIAL QUESTIONS

- How do Earth's land biomes differ?
- How do humans impact land biomes?

Vocabulary

biome p. 683
desert p. 684
grassland p. 685
temperate p. 687
taiga p. 689
tundra p. 689

Multilingual eGlossary

Video BrainPOP®

Land Biomes

Inquiry Plant or Animal?

Believe it or not, this is a flower. One of the largest flowers in the world, *Rafflesia* (ruh FLEE zhuh), grows naturally in the tropical rain forests of southeast Asia. What do you think would happen if you planted a seed from this plant in a desert? Would it survive?

Inquiry Launch Lab

10 minutes

What is the climate in China?

Beijing, China, and New York, New York, are about the same distance from the equator but on opposite sides of Earth. How do temperature and rainfall compare for these two cities?

1. Locate Beijing and New York on a world map.
2. Copy the table to the right in your Science Journal. From the data and charts provided, find and record the average high and low temperatures in January and in June for each city.
3. Record the average rainfall in January and in June for each city.

High Temperature (°C)	January	June
Beijing		
New York		
Low Temperature (°C)	**January**	**June**
Beijing		
New York		
Rainfall (mm)	**January**	**June**
Beijing		
New York		

Think About This

1. What are the temperature and rainfall ranges for each city?
2. **Key Concept** How do you think the climates of these cities differ year-round?

Land Ecosystems and Biomes

When you go outside, you might notice people, grass, flowers, birds, and insects. You also are probably aware of nonliving things, such as air, sunlight, and water. The living or once-living parts of an environment are the biotic parts. The nonliving parts that the living parts need to survive are the abiotic parts. The biotic and abiotic parts of an environment together make up an ecosystem.

Earth's continents have many different ecosystems, from deserts to rain forests. Scientists classify similar ecosystems in large geographic areas as biomes. *A* **biome** *is a geographic area on Earth that contains ecosystems with similar biotic and abiotic features.* As shown in **Figure 1,** Earth has seven major land biomes. Areas classified as the same biome have similar climates and organisms.

Figure 1 Earth contains seven major biomes.

Concepts in Motion Animation

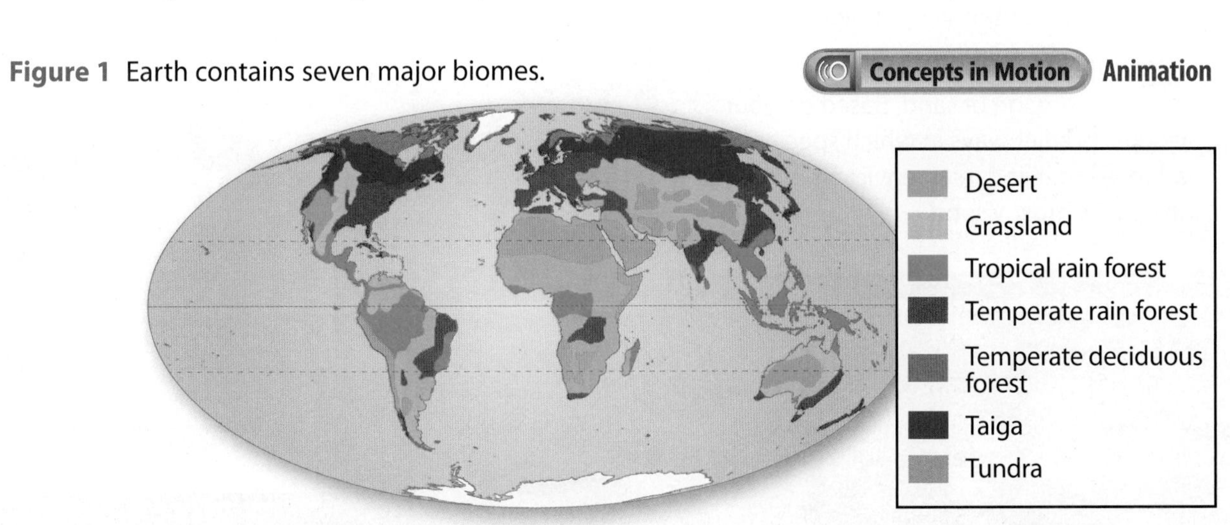

20 minutes

How hot is sand?

If you have ever walked barefoot on a sandy beach on a sunny day, you know how hot sand can be. But how hot is the sand below the surface?

1. Read and complete a lab safety form.
2. Position a **desk lamp** over a **container** of **sand** that is at least 7 cm deep.
3. Place one **thermometer** on the surface of the sand and bury the tip of another **thermometer** about 5 cm below the surface. Record the temperature on each thermometer in your Science Journal.
4. Turn on the lamp and record the temperatures again after 10 minutes.

Analyze and Conclude

1. **Describe** the temperatures of the sand at the surface and below the surface.
2. **Predict** what would happen to the temperature of the sand at night.
3. **Key Concept** Desert soil contains a high percentage of sand. Based on your results, predict ways in which species are adapted to living in an environment where the soil is mostly sand.

Desert Biome

Woodpeckers

Deserts *are biomes that receive very little rain.* They are on nearly every continent and are Earth's driest ecosystems.

- Most deserts are hot during the day and cold at night. Others, like those in Antarctica, remain cold all of the time.
- Rainwater drains away quickly because of thin, porous soil. Large patches of ground are bare.

Biodiversity

- Animals include lizards, bats, woodpeckers, and snakes. Most animals avoid activity during the hottest parts of the day.
- Plants include spiny cactus and thorny shrubs. Shallow roots absorb water quickly. Some plants have accordion-like stems that expand and store water. Small leaves or spines reduce the loss of water.

Human Impact

- Cities, farms, and recreational areas in deserts use valuable water.
- Desert plants grow slowly. When they are damaged by people or livestock, recovery takes many years.

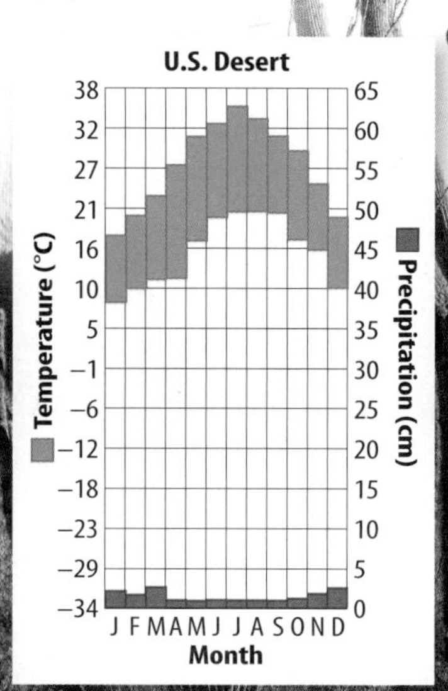

Grassland Biome

Black-footed ferret

Grassland *biomes are areas where grasses are the dominant plants.* Also called prairies, savannas, and meadows, grasslands are the world's "breadbaskets." Wheat, corn, oats, rye, barley, and other important cereal crops are grasses. They grow well in these areas.

- Grasslands have a wet and a dry season.
- Deep, fertile soil supports plant growth.
- Grass roots form a thick mass, called sod, which helps soil absorb and hold water during periods of drought.

Reading Check Why are grasslands called "breadbaskets"?

Biodiversity

- Trees grow along moist banks of streams and rivers. Wildflowers bloom during the wet season.
- In North America, large herbivores, such as bison and elk, graze here. Insects, birds, rabbits, prairie dogs, and snakes find shelter in the grasses.
- Predators in North American grasslands include hawks, ferrets, coyotes, and wolves.
- African savannas are grasslands that contain giraffes, zebras, and lions. Australian grasslands are home to kangaroos, wallabies, and wild dogs.

Human Impact

- People plow large areas of grassland to raise cereal crops. This reduces habitat for wild species.
- Because of hunting and loss of habitat, large herbivores—such as bison—are now uncommon in many grasslands.

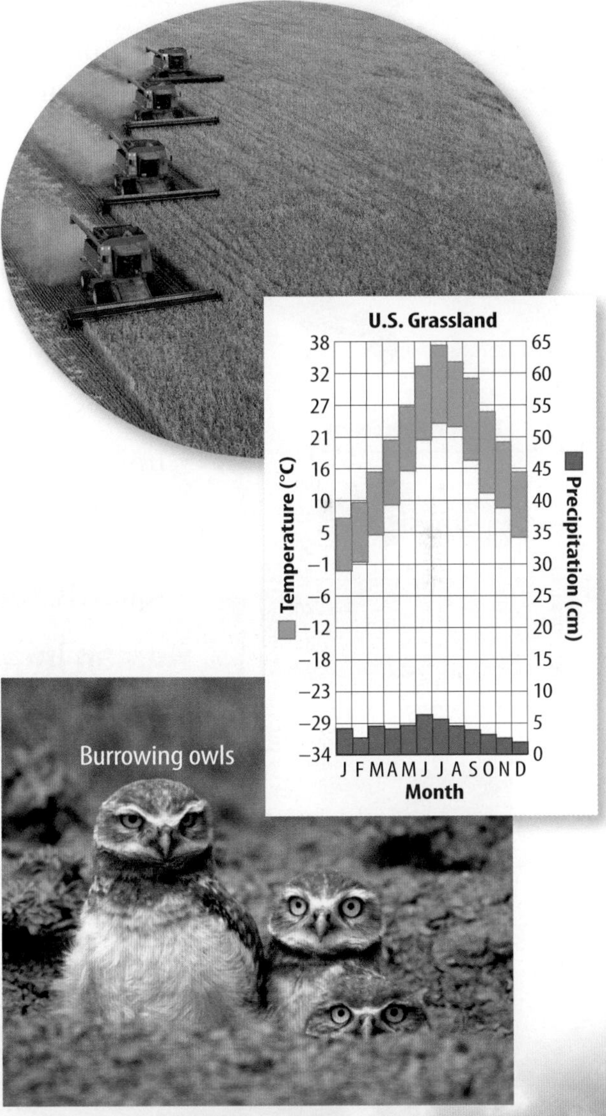

Burrowing owls

Tropical Rain Forest Biome

Toucan

The forests that grow near the equator are called tropical rain forests. These forests receive large amounts of rain and have dense growths of tall, leafy trees.

- Weather is warm and wet year-round.
- The soil is shallow and easily washed away by rain.
- Less than 1 percent of the sunlight that reaches the top of forest trees also reaches the forest floor.
- Half of Earth's species live in tropical rain forests. Most live in the canopy—the uppermost part of the forest.

Reading Check Where do most organisms live in a tropical rain forest?

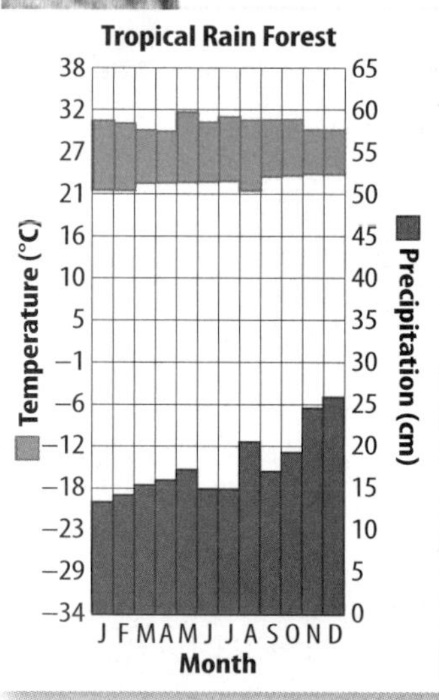

Biodiversity

- Few plants live on the dark forest floor.
- Vines climb the trunks of tall trees.
- Mosses, ferns, and orchids live on branches in the canopy.
- Insects make up the largest group of tropical animals. They include beetles, termites, ants, bees, and butterflies.
- Larger animals include parrots, toucans, snakes, frogs, flying squirrels, fruit bats, monkeys, jaguars, and ocelots.

Human Impact

- People have cleared more than half of Earth's tropical rain forests for lumber, farms, and ranches. Poor soil does not support rapid growth of new trees in cleared areas.
- Some organizations are working to encourage people to use less wood harvested from rain forests.

Temperate Rain Forest Biome

Regions of Earth between the tropics and the polar circles are **temperate** *regions.* Temperate regions have relatively mild climates with distinct seasons. Several biomes are in temperate regions, including rain forests. Temperate rain forests are moist ecosystems mostly in coastal areas. They are not as warm as tropical rain forests.

- Winters are mild and rainy.
- Summers are cool and foggy.
- Soil is rich and moist.

Biodiversity

- Forests are dominated by spruce, hemlock, cedar, fir, and redwood trees, which can grow very large and tall.
- Fungi, ferns, mosses, vines, and small flowering plants grow on the moist forest floor.
- Animals include mosquitoes, butterflies, frogs, salamanders, woodpeckers, owls, eagles, chipmunks, raccoons, deer, elk, bears, foxes, and cougars.

Human Impact

- Temperate rain forest trees are a source of lumber. Logging can destroy the habitat of forest species.
- Rich soil enables cut forests to grow back. Tree farms help provide lumber without destroying habitat.

Key Concept Check In what ways do humans affect temperate rain forests?

Elk

FOLDABLES®

Use a sheet of paper to make a horizontal two-tab book. Record what you learn about desert and temperate rain forest biomes under the tabs, and use the information to compare and contrast these biomes.

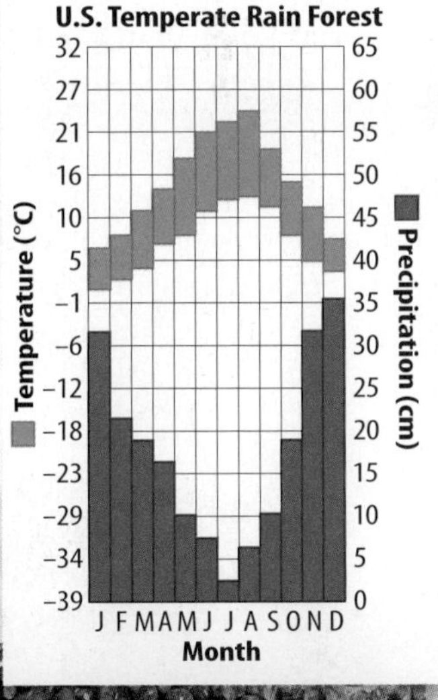

Temperate Deciduous Forest Biome

Temperate deciduous forests grow in temperate regions where winter and summer climates have more variation than those in temperate rain forests. These forests are the most common forest ecosystems in the United States. They contain mostly deciduous trees, which lose their leaves in the fall.

- Winter temperatures are often below freezing. Snow is common.
- Summers are hot and humid.
- Soil is rich in nutrients and supports a large amount of diverse plant growth.

Biodiversity

- Most plants, such as maples, oaks, birches, and other deciduous trees, stop growing during the winter and begin growing again in the spring.
- Animals include snakes, ants, butterflies, birds, raccoons, opossums, and foxes.
- Some animals, including chipmunks and bats, spend the winter in hibernation.
- Many birds and some butterflies, such as the monarch, migrate to warmer climates for the winter.

Human Impact

Over the past several hundred years, humans have cleared thousands of acres of Earth's deciduous forests for farms and cities. Today, much of the clearing has stopped and some forests have regrown.

Key Concept Check How are temperate deciduous rain forests different from temperate rain forests?

Red fox

Taiga Biome

A **taiga** (TI guh) *is a forest biome consisting mostly of cone-bearing evergreen trees.* The taiga biome exists only in the northern hemisphere. It occupies more space on Earth's continents than any other biome.

- Winters are long, cold, and snowy. Summers are short, warm, and moist.
- Soil is thin and acidic.

Biodiversity

- Evergreen trees, such as spruce, pine, and fir, are thin and shed snow easily.
- Animals include owls, mice, moose, bears, and other cold-adapted species.
- Abundant insects in summer attract many birds, which migrate south in winter.

Human Impact

- Tree harvesting reduces taiga habitat.

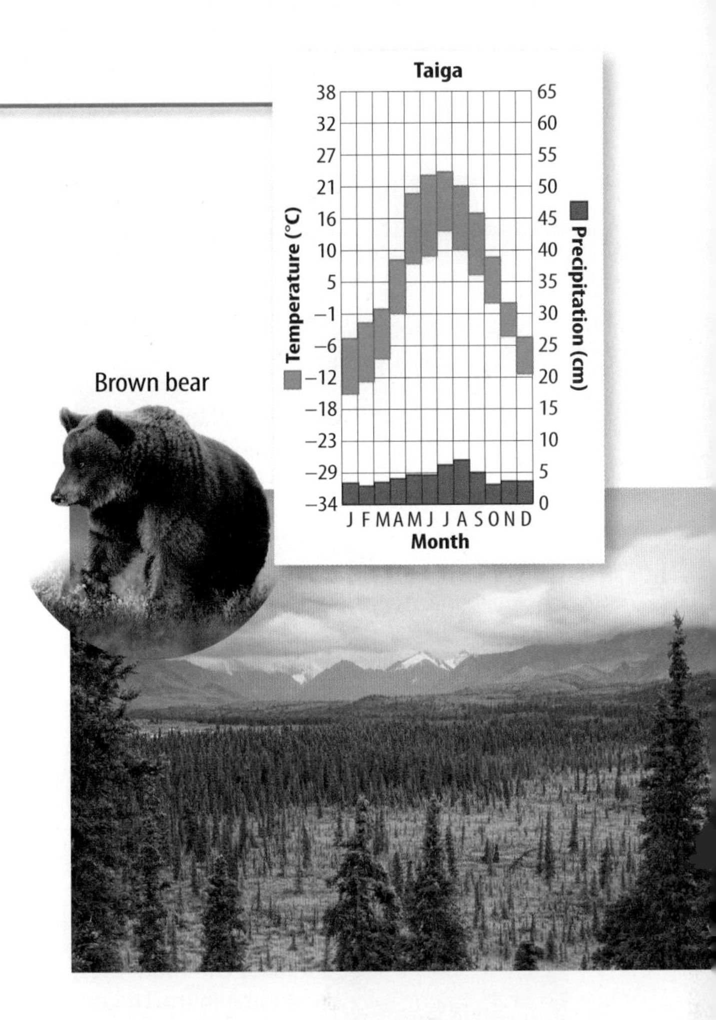

Tundra Biome

A **tundra** (TUN druh) *biome is cold, dry, and treeless.* Most tundra is south of the North Pole, but it also exists in mountainous areas at high altitudes.

- Winters are long, dark, and freezing; summers are short and cool; the growing season is only 50–60 days.
- Permafrost—a layer of permanently frozen soil—prevents deep root growth.

Biodiversity

- Plants include shallow-rooted mosses, lichens, and grasses.
- Many animals hibernate or migrate south during winter. Few animals, including lemmings, live in tundras year-round.

Human Impact

- Drilling for oil and gas can interrupt migration patterns.

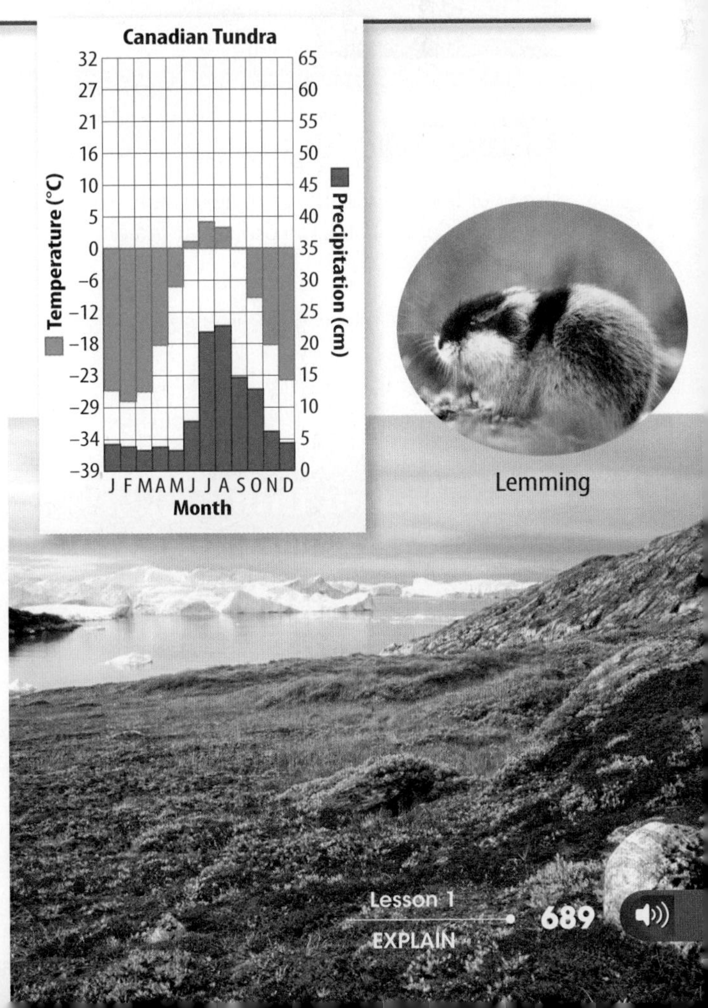

Lesson 1 Review

Assessment Online Quiz

Inquiry Virtual Lab

Visual Summary

Earth has seven major land biomes, ranging from hot, dry deserts to cold, forested taigas.

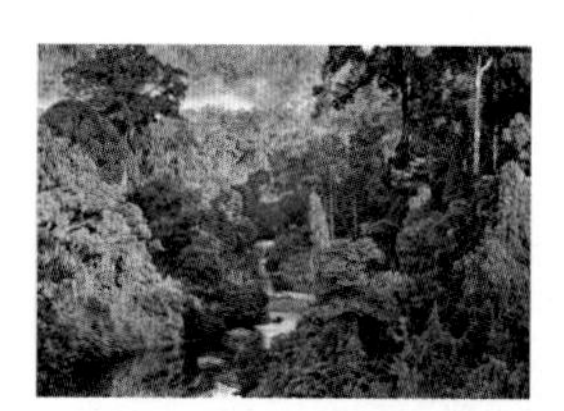

Half of Earth's species live in rain forest biomes.

Temperate deciduous forests are the most common forest biome in the United States.

FOLDABLES®

Use your lesson Foldable to review the lesson. Save your Foldable for the project at the end of the chapter.

What do you think NOW?

You first read the statements below at the beginning of the chapter.

1. Deserts can be cold.

2. There are no rain forests outside the tropics.

Did you change your mind about whether you agree or disagree with the statements? Rewrite any false statements to make them true.

Use Vocabulary

1. **Define** *biome* using your own words.
2. **Distinguish** between tropical rain forests and temperate rain forests.
3. A cold, treeless biome is a(n) ________.

Understand Key Concepts 

4. **Explain** why tundra soil cannot support the growth of trees.
5. **Give examples** of how plants and animals adapt to temperate deciduous ecosystems.

Interpret Graphics

6. **Determine** What is the average annual rainfall for the biome represented by the chart to the right?

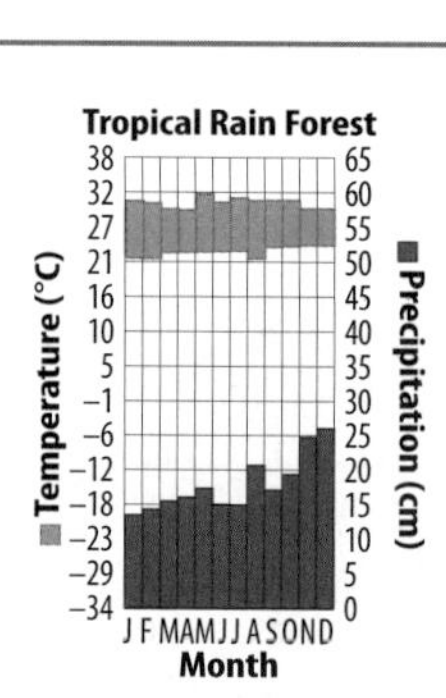

7. **Summarize Information** Copy the graphic organizer below and fill it in with animals and plants of the biome you live in.

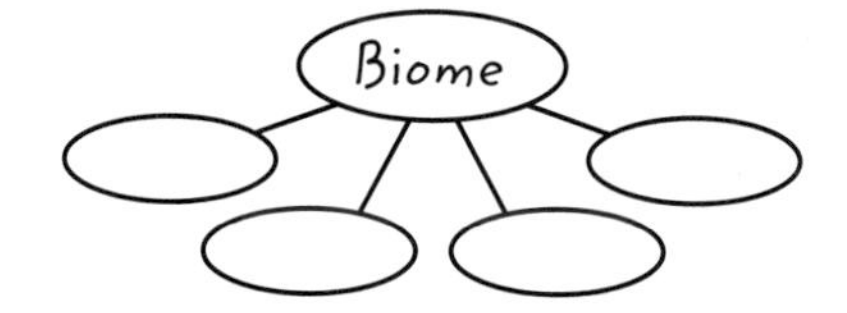

Critical Thinking

8. **Plan** an enclosed zoo exhibit for a desert ecosystem. What abiotic factors should you consider?
9. **Recommend** one or more actions people can take to reduce habitat loss in tropical and taiga forests.

Inquiry **Skill Practice** **Interpret Data** 20 minutes

Which biome is it?

Materials

biome data

You have read about the major land biomes found on Earth. Within each biome are ecosystems with similar biotic and abiotic factors. In this lab, you will **interpret data** describing a particular area on Earth to identify which biome it belongs to.

Learn It

Scientists collect and present data in a variety of forms, including graphs and tables. In this activity, you will interpret data in a graph and apply the information to the ideas you learned in the lesson.

Try It

1. Examine the temperature and precipitation data in the graph given you by your teacher.
2. Create a table from these data in your Science Journal. Calculate the average temperature and precipitation during the winter and the summer.
3. Examine the image of the biome and identify some plants and animals in the image.
4. Compare your data to the information on land biomes presented in Lesson 1. Which biome is the most similar?

Apply It

5. Which land biome did your data come from? Why did you choose this biome?
6. Are the data in your graph identical to the data in the graph of the biome in Lesson 1 to which it belongs? Why or why not?
7. Describe this biome. What do you think your biome will be like six months from now?
8. **Key Concept** How might humans affect the organisms in your biome?

Lesson 2

Aquatic Ecosystems

Reading Guide

Key Concept
ESSENTIAL QUESTIONS

- How do Earth's aquatic ecosystems differ?
- How do humans impact aquatic ecosystems?

Vocabulary

salinity p. 693
wetland p. 696
estuary p. 697
intertidal zone p. 699
coral reef p. 699

g Multilingual eGlossary

Inquiry Floating Trees?

These plants, called mangroves, are one of the few types of plants that grow in salt water. They usually live along ocean coastlines in tropical ecosystems. What other organisms do you think live near mangroves?

Launch Lab

10 minutes

What happens when rivers and oceans mix?

Freshwater and saltwater ecosystems have different characteristics. What happens in areas where freshwater rivers and streams flow into oceans?

1. Read and complete a lab safety form.
2. In a **plastic tub,** add 100 g of **salt** to 2 L of water. Stir with a **long-handled spoon** until the salt dissolves.
3. In another **container,** add 5 drops of **blue food coloring** to 1 L of water. Gently pour the colored water into one corner of the plastic tub. Observe how the color of the water changes in the tub.
4. Observe the tub again in 5 minutes.

Think About This

1. What bodies of water do the containers represent?
2. What happened to the water in the tub after 5 minutes? What do you think happens to the salt content of the water?
3.  **Key Concept** How do you think the biodiversity of rivers and oceans differ? What organisms do you think might live at the place where the two meet?

Aquatic Ecosystems

If you've ever spent time near an ocean, a river, or another body of water, you might know that water is full of life. There are four major types of water, or aquatic, ecosystems: freshwater, wetland, estuary, and ocean. Each type of ecosystem contains a unique variety of organisms. Whales, dolphins, and corals live only in ocean ecosystems. Catfish and trout live only in freshwater ecosystems. Many other organisms that do not live under water, such as birds and seals, also depend on aquatic ecosystems for food and shelter.

Important abiotic factors in aquatic ecosystems include temperature, sunlight, and dissolved oxygen gas. Aquatic species have adaptations that enable them to use the oxygen in water. The gills of a fish separate oxygen from water and move it into the fish's bloodstream. Mangrove plants, pictured on the previous page, take in oxygen through small pores in their leaves and roots.

Salinity (say LIH nuh tee) is another important abiotic factor in aquatic ecosystems. **Salinity** *is the amount of salt dissolved in water.* Water in saltwater ecosystems has high salinity compared to water in freshwater ecosystems, which contains little salt.

Math Skills

Use Proportions

Salinity is measured in parts per thousand (PPT). One PPT water contains 1 g salt and 1,000 g water. Use proportions to calculate salinity. What is the salinity of 100 g of water with 3.5 g of salt?

$$\frac{3.5 \text{ g salt}}{100 \text{ g seawater}} = \frac{x \text{ g salt}}{1{,}000 \text{ g seawater}}$$

$$100\,x = 3500$$

$$x = \frac{3500}{100} = 35 \text{ PPT}$$

Practice

A sample contains 0.1895 g of salt per 50 g of seawater. What is its salinity?

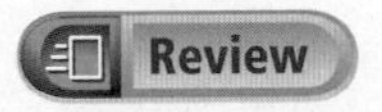

- **Math Practice**
- **Personal Tutor**

Great Blue Heron

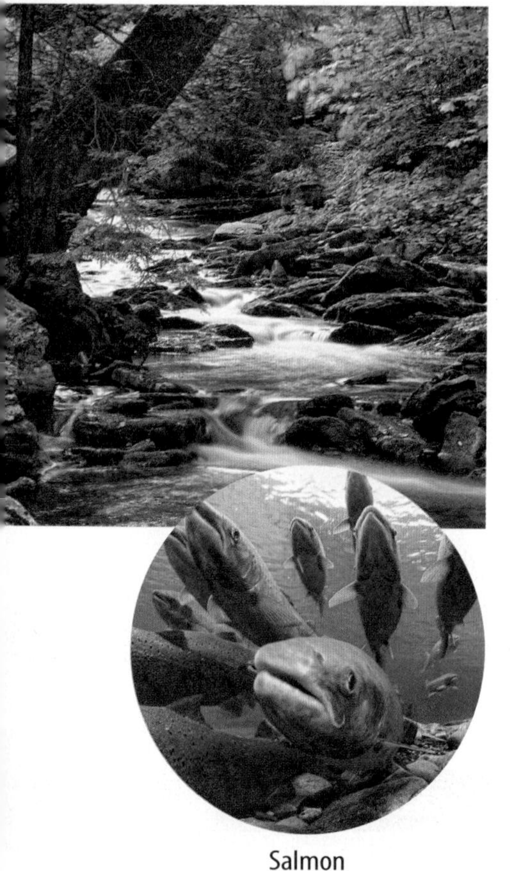

Freshwater ecosystems include streams, rivers, ponds, and lakes. Streams are usually narrow, shallow, and fast-flowing. Rivers are larger, deeper, and flow more slowly.

- Streams form from underground sources of water, such as springs or from runoff from rain and melting snow.
- Stream water is often clear. Soil particles are quickly washed downstream.
- Oxygen levels in streams are high because air mixes into the water as it splashes over rocks.
- Rivers form when streams flow together.
- Soil that washes into a river from streams or nearby land can make river water muddy. Soil also introduces nutrients, such as nitrogen, into rivers.
- Slow-moving river water has higher levels of nutrients and lower levels of dissolved oxygen than fast-moving water.

Salmon

Biodiversity

- Willows, cottonwoods, and other water-loving plants grow along streams and on riverbanks.
- Species adapted to fast-moving water include trout, salmon, crayfish, and many insects.
- Species adapted to slow-moving water include snails and catfish.

Stonefly larva

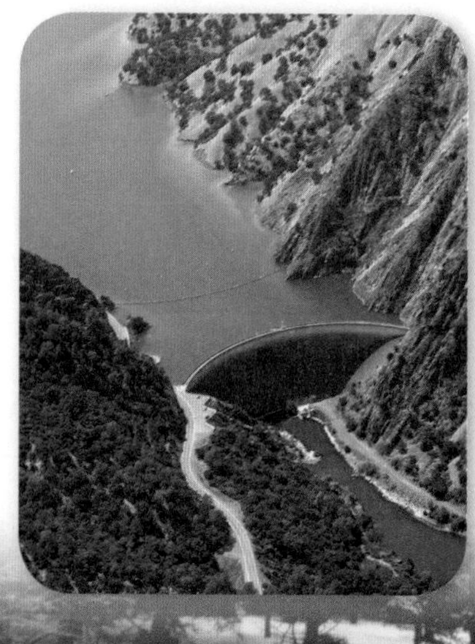

Human Impact

- People take water from streams and rivers for drinking, laundry, bathing, crop irrigation, and industrial purposes.
- Hydroelectric plants use the energy in flowing water to generate electricity. Dams stop the water's flow.
- Runoff from cities, industries, and farms is a source of pollution.

Freshwater: Ponds and Lakes

Ponds and lakes contain freshwater that is not flowing downhill. These bodies of water form in low areas on land.

- Ponds are shallow and warm.
- Sunlight reaches the bottom of most ponds.
- Pond water is often high in nutrients.
- Lakes are larger and deeper than ponds.
- Sunlight penetrates into the top few feet of lake water. Deeper water is dark and cold.

Biodiversity

- Plants surround ponds and lake shores.
- Surface water in ponds and lakes contains plants, algae, and microscopic organisms that use sunlight for photosynthesis.
- Organisms living in shallow water near shorelines include cattails, reeds, insects, crayfish, frogs, fish, and turtles.
- Fewer organisms live in the deeper, colder water of lakes where there is little sunlight.
- Lake fish include perch, trout, bass, and walleye.

Smallmouth bass

Reading Check Why do few organisms live in the deep water of lakes?

Human Impact

- Humans fill in ponds and lakes with sediment to create land for houses and other structures.
- Runoff from farms, gardens, and roads washes pollutants into ponds and lakes, disrupting food webs.

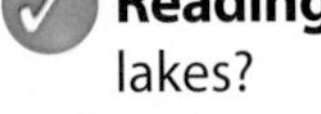

Key Concept Check How do ponds and lakes differ?

Wetlands

Common loon

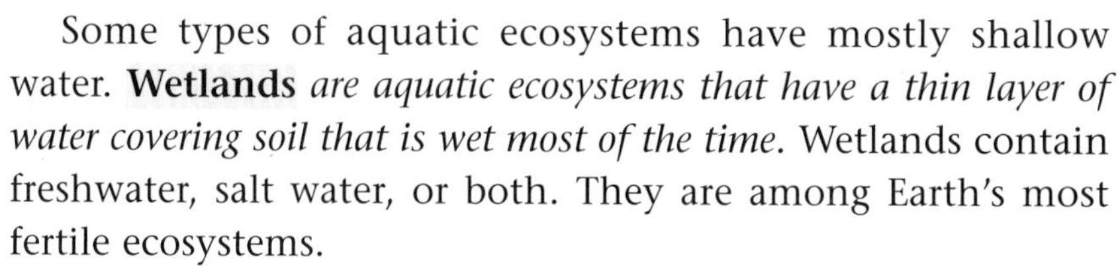

Some types of aquatic ecosystems have mostly shallow water. **Wetlands** *are aquatic ecosystems that have a thin layer of water covering soil that is wet most of the time.* Wetlands contain freshwater, salt water, or both. They are among Earth's most fertile ecosystems.

- Freshwater wetlands form at the edges of lakes and ponds and in low areas on land. Saltwater wetlands form along ocean coasts.
- Nutrient levels and biodiversity are high.
- Wetlands trap sediments and purify water. Plants and microscopic organisms filter out pollution and waste materials.

Biodiversity

- Water-tolerant plants include grasses and cattails. Few trees live in saltwater wetlands. Trees in freshwater wetlands include cottonwoods, willows, and swamp oaks.
- Insects are abundant and include flies, mosquitoes, dragonflies, and butterflies.
- More than one-third of North American bird species, including ducks, geese, herons, loons, warblers, and egrets, use wetlands for nesting and feeding.
- Other animals that depend on wetlands for food and breeding grounds include alligators, turtles, frogs, snakes, salamanders, muskrats, and beavers.

Human Impact

- In the past, many people considered wetlands as unimportant environments. Water was drained away to build homes and roads and to raise crops.
- Today, many wetlands are being preserved, and drained wetlands are being restored.

 **Key Concept Check** How do humans impact wetlands?

Estuaries

Estuaries (ES chuh wer eez) *are regions along coastlines where streams or rivers flow into a body of salt water.* Most estuaries form along coastlines, where freshwater in rivers meets salt water in oceans. Estuary ecosystems have varying degrees of salinity.

- Salinity depends on rainfall, the amount of freshwater flowing from land, and the amount of salt water pushed in by tides.
- Estuaries help protect coastal land from flooding and erosion. Like wetlands, estuaries purify water and filter out pollution.
- Nutrient levels and biodiversity are high.

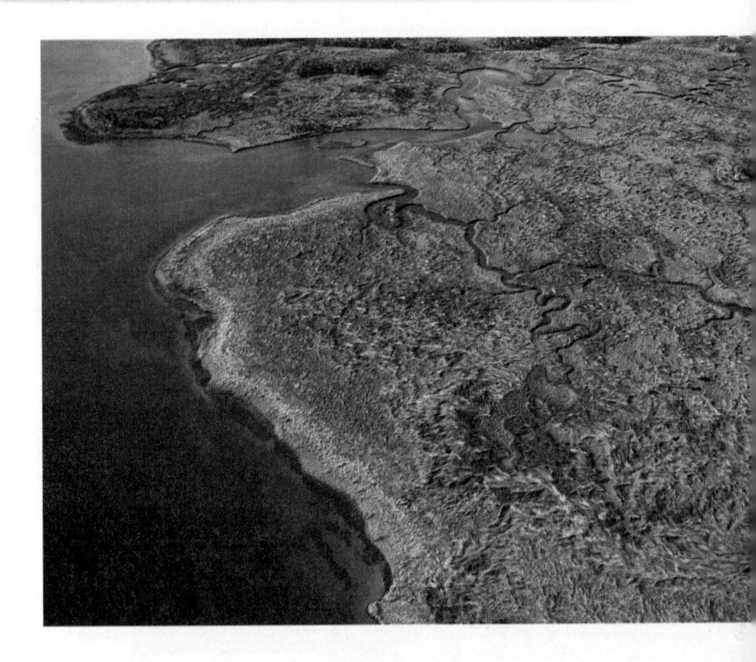

Biodiversity

- Plants that grow in salt water include mangroves, pickleweeds, and seagrasses.
- Animals include worms, snails, and many species that people use for food, including oysters, shrimp, crabs, and clams.
- Striped bass, salmon, flounder, and many other ocean fish lay their eggs in estuaries.
- Many species of birds depend on estuaries for breeding, nesting, and feeding.

Human Impact

- Large portions of estuaries have been filled with soil to make land for roads and buildings.
- Destruction of estuaries reduces habitat for estuary species and exposes the coastline to flooding and storm damage.

Word Origin

estuary
from Latin *aestuarium,* means "a tidal marsh or opening."

FOLDABLES®

Make a horizontal two-tab book and label it as shown. Use it to compare how biodiversity and human impact differ in wetlands and estuaries.

Wetlands | Estuaries

Harvest mouse

Jellyfish

Most of Earth's surface is covered by ocean water with high salinity. The oceans contain different types of ecosystems. If you took a boat trip several kilometers out to sea, you would be in the open ocean—one type of ocean ecosystem. The open ocean extends from the steep edges of continental shelves to the deepest parts of the ocean. The amount of light in the water depends on depth.

- Photosynthesis can take place only in the uppermost, or sunlit, zone. Very little sunlight reaches the twilight zone. None reaches the deepest water, known as the dark zone.
- Decaying matter and nutrients float down from the sunlit zone, through the twilight and dark zones, to the seafloor.

Biodiversity

- Microscopic algae and other producers in the sunlit zone form the base of most ocean food chains. Other organisms living in the sunlit zone are jellyfish, tuna, mackerel, and dolphins.
- Many species of fish stay in the twilight zone during the day and swim to the sunlit zone at night to feed.
- Sea cucumbers, brittle stars, and other bottom-dwelling organisms feed on decaying matter that drifts down from above.
- Many organisms in the dark zone live near cracks in the seafloor where lava erupts and new seafloor forms.

Reading Check Which organisms are at the base of most ocean food chains?

Human Impact

- Overfishing threatens many ocean fish.
- Trash discarded from ocean vessels or washed into oceans from land is a source of pollution. Animals such as seals become tangled in plastic or mistake it for food.

Fur seal

Ocean: Coastal Oceans

Sea stars

Coastal oceans include several types of ecosystems, including continental shelves and intertidal zones. *The* **intertidal zone** *is the ocean shore between the lowest low tide and the highest high tide.*

- Sunlight reaches the bottom of shallow coastal ecosystems.
- Nutrients washed in from rivers and streams contribute to high biodiversity.

Biodiversity

- The coastal ocean is home to mussels, fish, crabs, sea stars, dolphins, and whales.
- Intertidal species have adaptations for surviving exposure to air during low tides and to heavy waves during high tides.

Human Impact

- Oil spills and other pollution harm coastal organisms.

Ocean: Coral Reefs

Another ocean ecosystem with high biodiversity is the coral reef. *A* **coral reef** *is an underwater structure made from outside skeletons of tiny, soft-bodied animals called coral.*

- Most coral reefs form in shallow tropical oceans.
- Coral reefs protect coastlines from storm damage and erosion.

Biodiversity

- Coral reefs provide food and shelter for many animals, including parrotfish, groupers, angelfish, eels, shrimp, crabs, scallops, clams, worms, and snails.

Human Impact

- Pollution, overfishing, and harvesting of coral threaten coral reefs.

Inquiry MiniLab — 15 minutes

How do ocean ecosystems differ?

Ocean ecosystems include open oceans, coastal oceans, and coral reefs—each one a unique environment with distinctive organisms.

1. Read and complete a lab safety form.
2. In a **large plastic tub,** use **rocks** and **sand** to make a structure representing an open ocean, a coastal ocean, or a coral reef.

3. Fill the tub with **water.**
4. Make waves by gently moving your hand back and forth in the water.

Analyze and Conclude

1. **Observe** What happened to your structure when you made waves? How might a hurricane affect the organisms that live in the ecosystem you modeled?
2. **Key Concept** Compare your results with results of those who modeled other ecosystems. Suggest what adaptations species might have in each ecosystem.

Grouper

Lesson 2 Review

Visual Summary

Freshwater ecosystems include ponds and lakes.

Wetlands can be saltwater ecosystems or freshwater ecosystems.

Coral reefs and coastal ecosystems have high levels of biodiversity.

FOLDABLES®

Use your lesson Foldable to review the lesson. Save your Foldable for the project at the end of the chapter.

What do you think NOW?

You first read the statements below at the beginning of the chapter.

3. Estuaries do not protect coastal areas from erosion.

4. Animals form coral reefs.

Did you change your mind about whether you agree or disagree with the statements? Rewrite any false statements to make them true.

Use Vocabulary

1. **Define** the term *salinity*.
2. **Distinguish** between a wetland and an estuary.
3. An ocean ecosystem formed from the skeletons of animals is a(n) ________.

Understand Key Concepts 

4. Which ecosystem contains both salt water and freshwater?
   A. estuary
   B. lake
   C. pond
   D. stream
5. **Describe** what might happen to a coastal area if its estuary were filled in to build houses.

Interpret Graphics

6. **Describe** Copy the drawing to the right and label the light zones. Describe characteristics of each zone.

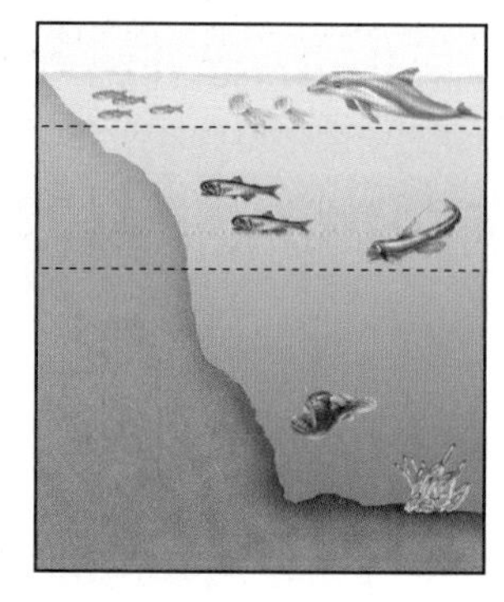

Critical Thinking

7. **Recommend** actions people might take to prevent pollutants from entering coastal ecosystems.

Math Skills

Math Practice

8. The salinity of the Baltic Sea is about 10 PPT. What weight of salt is present in 2,000 g of its seawater?

SCIENCE & SOCIETY

Saving an Underwater Wilderness

A researcher takes a water sample from a marine reserve. ▼

How do scientists help protect coral reefs?

Pollution and human activities, such as mining and tourism, have damaged many ecosystems, including coral reefs. Scientists and conservation groups are working together to help protect and restore coral reefs and areas that surround them. One way is to create marine reserves where no fishing or collection of organisms is allowed.

A team of scientists, including marine ecologists Dr. Dan Brumbaugh and Kate Holmes from the American Museum of Natural History, are investigating how well reserves are working. These scientists compare how many fish of one species live both inside and outside reserves. Their results indicate that more species of fish and greater numbers of each species live inside reserves than outside—one sign that reefs in the area are improving.

Reef ecosystems do not have to be part of a reserve in order to improve, however. Scientists can work with local governments to find ways to limit damage to reef ecosystems. One way is to prevent overfishing by limiting the number of fish caught. Other ways include eliminating the use of destructive fishing practices that can harm reefs and reducing runoff from farms and factories.

By creating marine reserves, regulating fishing practices, and reducing runoff, humans can help reefs that were once in danger become healthy again.

Kate Holmes examines a coral reef. ▶

It's Your Turn

WRITE Write a persuasive essay to a town near a marine reserve describing why coral reefs are important habitats.

AMERICAN MUSEUM OF NATURAL HISTORY

Lesson 3

Reading Guide

Key Concepts

ESSENTIAL QUESTIONS

- How do land ecosystems change over time?
- How do aquatic ecosystems change over time?

Vocabulary

ecological succession p. 703
climax community p. 703
pioneer species p. 704
eutrophication p. 706

Multilingual eGlossary

Video

- Science Video
- What's Science Got to do With It?

How Ecosystems Change

Inquiry How did this happen?

This object was once part of a mining system used to move copper and iron ore. Today, so many forest plants have grown around it that it is barely recognizable. How do you think this happened? What do you think this object will look like after 500 more years?

Launch Lab

15 minutes

How do communities change?

An ecosystem can change over time. Change usually happens so gradually that you might not notice differences from day to day.

1. Your teacher has given you **two pictures of ecosystem communities.** One is labeled *A* and the other is labeled *B*.
2. Imagine community A changed and became like community B. On a blank piece of **paper**, draw what you think community A might look like midway in its change to becoming like community B.

Think About This

1. What changes did you imagine? How long do you think it would take for community A to become like community B?
2. **Key Concept** Summarize the changes you think would happen as the community changed from A to B.

How Land Ecosystems Change

Have you ever seen weeds growing up through cracks in a concrete sidewalk? If they were not removed, the weeds would keep growing. The crack would widen, making room for more weeds. Over time, the sidewalk would break apart. Shrubs and vines would move in. Their leaves and branches would grow large enough to cover the concrete. Eventually, trees could start growing there.

This process is an example of **ecological succession**—*the process of one ecological community gradually changing into another.* Ecological succession occurs in a series of steps. These steps can usually be predicted. For example, small plants usually grow first. Larger plants, such as trees, usually grow last.

The final stage of ecological succession in a land ecosystem is a **climax community**—*a stable community that no longer goes through major ecological changes.* Climax communities differ depending on the type of biome they are in. In a tropical forest biome, a climax community would be a mature tropical forest. In a grassland biome, a climax community would be a mature grassland. Climax communities are usually stable over hundreds of years. As plants in a climax community die, new plants of the same species grow and take their places. The community will continue to contain the same kinds of plants as long as the climate remains the same.

Key Concept Check What is a climax community?

FOLDABLES®

Fold a sheet of paper into fourths. Use two sections on one side of the paper to describe and illustrate what land might look like before secondary succession and the other side to describe and illustrate the land after secondary succession is complete.

REVIEW VOCABULARY

community
all the organisms that live in one area at the same time

Science Use v. Common Use

pioneer

Science Use the first species that colonize new or undisturbed land

Common Use the first human settlers in an area

Primary Succession

What do you think happens to a lava-filled landscape when a volcanic eruption is over? As shown in **Figure 2**, volcanic lava eventually becomes new soil that supports plant growth. Ecological succession in new areas of land with little or no soil, such as on a lava flow, a sand dune, or exposed rock, is primary succession. *The first species that colonize new or undisturbed land are* **pioneer species.** The lichens and mosses in **Figure 2** are pioneer species.

Figure 2 Following a volcanic eruption, a landscape undergoes primary succession.

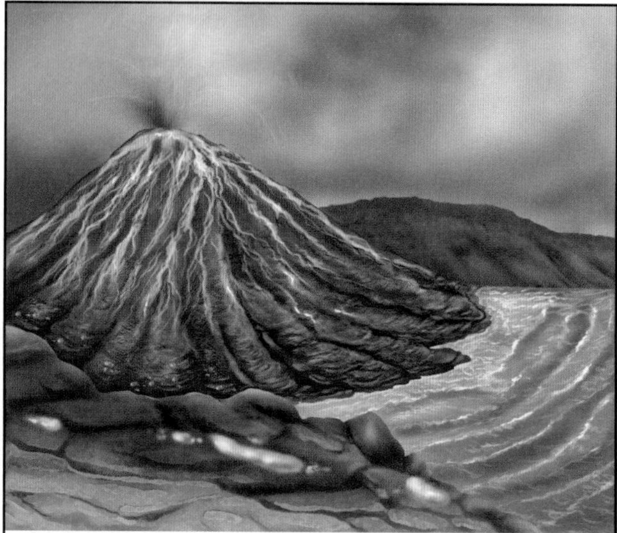

During a volcanic eruption, molten lava flows over the ground and into the water. After the eruption is over, the lava cools and hardens into bare rock.

Lichen spores carried on the wind settle on the rock. Lichens release acid that helps break down the rock and create soil. Lichens add nutrients to the soil as they die and decay.

Airborne spores from mosses and ferns settle onto the thin soil and add to the soil when they die. The soil gradually becomes thick enough to hold water. Insects and other small organisms move into the area.

After many years the soil is deep and has enough nutrients for grasses, wildflowers, shrubs, and trees. The new ecosystem provides habitats for many animals. Eventually, a climax community develops.

Secondary Succession

In areas where existing ecosystems have been disturbed or destroyed, secondary succession can occur. One example is forestland in New England that early colonists cleared hundreds of years ago. Some of the cleared land was not planted with crops. This land gradually grew back to a climax forest community of beech and maple trees, as illustrated in **Figure 3.**

Reading Check Where does secondary succession occur?

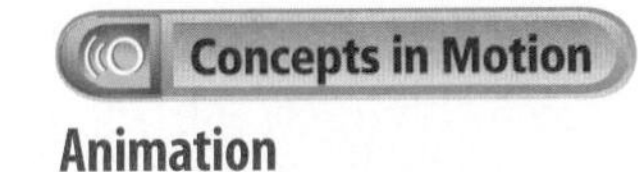

Animation

Figure 3 When disturbed land grows back, secondary succession occurs.

Settlers in New England cleared many acres of forests to create cropland. In places where people stopped planting crops, the forest began to grow back.

Seeds of grasses, wildflowers, and other plants quickly began to sprout and grow. Young shrubs and trees also started growing. These plants provided habitats for insects and other small animals, such as mice.

White pines and poplars were the first trees in the area to grow to their full height. They provided shade and protection to slower growing trees, such as beech and maple.

Eventually, a climax community of beech and maple trees developed. As older trees die, new beech and maple seedlings grow and replace them.

Aquatic Succession

Review Personal Tutor

Figure 4 The water in a pond is slowly replaced by soil. Eventually, land plants take over and the pond disappears.

How Freshwater Ecosystems Change

Like land ecosystems, freshwater ecosystems change over time in a natural, predictable process. This process is called aquatic succession.

Aquatic Succession

Aquatic succession is illustrated in **Figure 4.** Sediments carried by rainwater and streams accumulate on the bottoms of ponds, lakes, and wetlands. The decomposed remains of dead organisms add to the buildup of soil. As time passes, more and more soil accumulates. Eventually, so much soil has collected that the water disappears and the area becomes land.

Key Concept Check What happens to a pond, a lake, or a wetland over time?

Eutrophication

As decaying organisms fall to the bottom of a pond, a lake, or a wetland, they add nutrients to the water. **Eutrophication** (yoo troh fuh KAY shun) *is the process of a body of water becoming nutrient-rich.*

WORD ORIGIN
eutrophication
from Greek *eutrophos,* means "nourishing"

Eutrophication is a natural part of aquatic succession. However, humans also contribute to eutrophication. The fertilizers that farmers use on crops and the waste from farm animals can be very high in nutrients. So can other forms of pollution. When fertilizers and pollution run off into a pond or lake, nutrient concentrations increase. High nutrient levels support large populations of algae and other microscopic organisms. These organisms use most of the dissolved oxygen in the water and less oxygen is available for fish and other pond or lake organisms. As a result, many of these organisms die. Their bodies decay and add to the buildup of soil, speeding up succession.

Lesson 3 Review

Assessment Online Quiz

Visual Summary

Ecosystems change in predictable ways through ecological succession.

The final stage of ecological succession in a land ecosystem is a climax community.

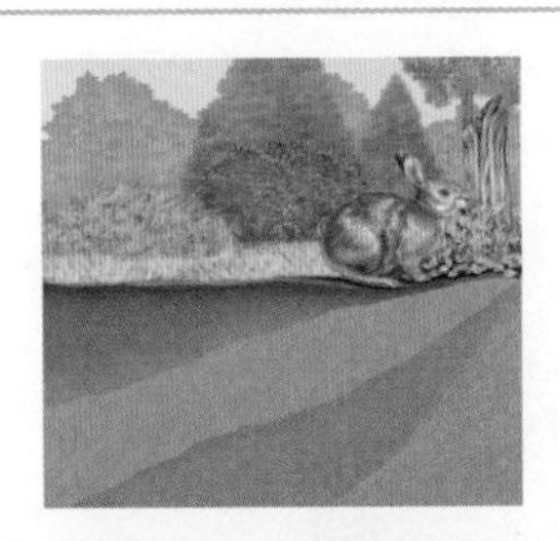

The final stage of aquatic succession is a land ecosystem.

FOLDABLES®

Use your lesson Foldable to review the lesson. Save your Foldable for the project at the end of the chapter.

What do you think NOW?

You first read the statements below at the beginning of the chapter.

5. An ecosystem never changes.

6. Nothing grows in the area where a volcano has erupted.

Did you change your mind about whether you agree or disagree with the statements? Rewrite any false statements to make them true.

Use Vocabulary

1. **Define** *pioneer species* in your own words.
2. The process of one ecological community changing into another is ________.
3. **Compare and contrast** succession and eutrophication in freshwater ecosystems.

Understand Key Concepts 

4. **Draw** a picture of what your school might look like in 500 years if it were abandoned.
5. Which process occurs after a forest fire?
   - **A.** eutrophication
   - **B.** photosynthesis
   - **C.** primary succession
   - **D.** secondary succession

Interpret Graphics

6. **Determine** What kind of succession—primary or secondary—might occur in the environment pictured to the right? Explain.

7. **Summarize Information** Copy the graphic organizer below and fill it with the types of succession an ecosystem can go through.

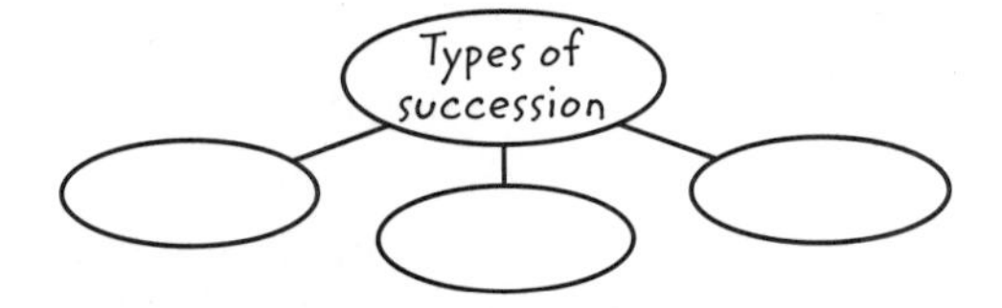

Critical Thinking

8. **Reflect** What kinds of abiotic factors might cause a grassland climax community to slowly become a forest?
9. **Recommend** actions people can take to help prevent the loss of wetland and estuary habitats.

Lab

60 minutes

A Biome for Radishes

Materials

paper towels

small jar

plastic wrap

jar lid

radish seeds

desk lamp

magnifying lens

Safety

Biomes contain plant and animal species adapted to particular climate conditions. Many organisms can live only in one type of biome. Others can survive in more than one biome. A radish is a plant grown around the world. How do you think radish seeds grow in different biomes? In this lab, you will model four different biomes and ecosystems—a temperate deciduous forest, a temperate rain forest, a desert, and a pond—and determine which biome the radishes grow best in.

Ask a Question

Which biome do radishes grow best in?

Make Observations

1. Read and complete a lab safety form.
2. Fold two pieces of paper towel lengthwise. Place the paper towels on opposite sides of the top of a small jar, as shown, with one end of each towel inside the jar and one end outside. Add water until about 10 cm of the paper towels are in the water. The area inside the jar models a pond ecosystem.
3. Place a piece of plastic wrap loosely over the end of one of the paper towels hanging over the jar's edge. Do not completely cover the paper towel. This paper towel models a temperate rain forest ecosystem. The paper towel without plastic wrap models a temperate deciduous forest.
4. Place the jar lid upside-down on the top of the jar. The area in the lid models a desert.

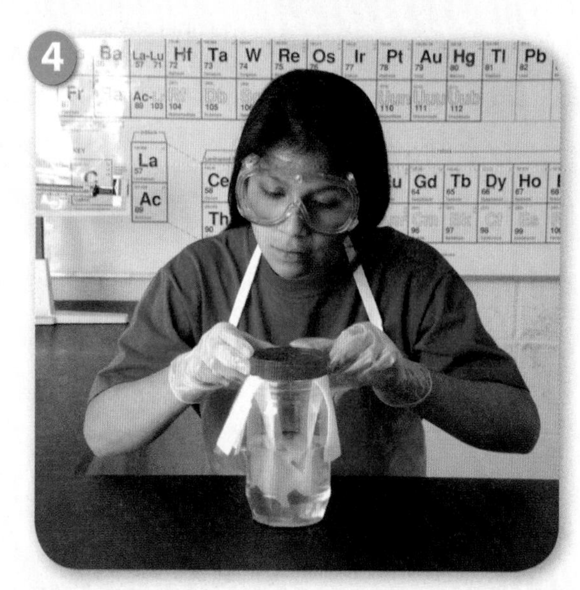

Form a Hypothesis

5. Observe the four biomes and ecosystems you have modeled. Based on your observations and your knowledge of the abiotic factors a plant requires, hypothesize which biome or ecosystem you think radish seeds will grow best in.

By permission of TOPS Learning Systems, www.topscience.org.

Test Your Hypothesis

6. Place three radish seeds in each biome: pond, temperate forest, temperate rain forest, and desert. Gently press the seeds to the paper towel until they stick.
7. Place your jar near a window or under a desk lamp that can be turned on during the day.
8. In your Science Journal, record your observations of the seeds and the paper towel.
9. After five days, use a magnifying lens to observe the seeds and the paper towels again. Record your observations.

Analyze and Conclude

10. **Compare and Contrast** How did the appearance of the seeds change after five days in each model biome?
11. **Critique** Evaluate your hypothesis. Did the seeds grow the way you expected? In which biome did the seeds grow the most?
12. **The Big Idea** In the biome with the most growth, what characteristics do you think made the seeds grow best?

Lab Tips

- ☑ Do not eat the radish seeds.
- ☑ If your seeds fall off the paper towel strips, do not replace them.

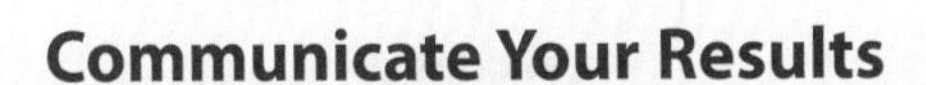

Communicate Your Results

Working in a group of three or four, create a table showing results for each biome. Present the table to the class.

Inquiry Extension

In this lab, you determined which biome produced the most growth of radish seeds. Seeds of different species might sprout in several different biomes. However, not all sprouted seeds grow to adulthood. Design a lab to test what conditions are necessary for radishes to grow to adulthood.

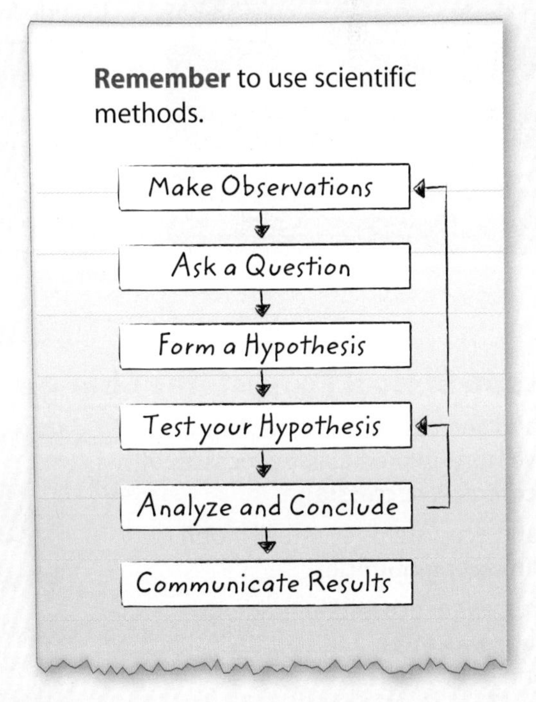

By permission of TOPS Learning Systems, www.topscience.org.

Chapter 19 Study Guide

Each of Earth's land biomes and aquatic ecosystems is characterized by distinct environments and organisms. Biomes and ecosystems change by natural processes of ecological succession and by human activities.

Key Concepts Summary

Lesson 1: Land Biomes

- Each land **biome** has a distinct climate and contains animals and plants well adapted to the environment. Biomes include **deserts**, **grasslands**, tropical rain forests, **temperate** rain forests, deciduous forests, **taigas**, and **tundras.**
- Humans affect land biomes through agriculture, construction, and other activities.

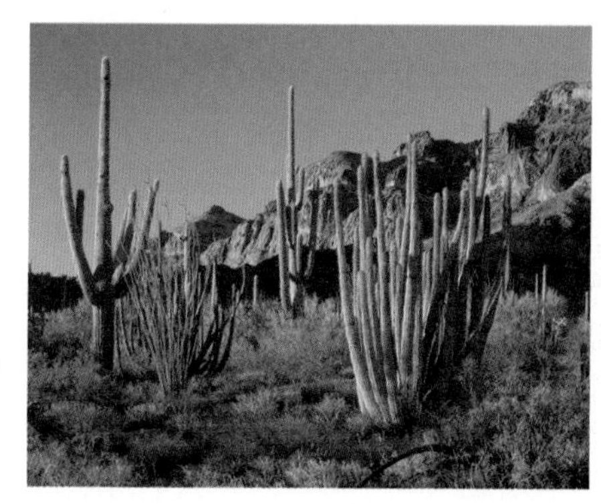

Vocabulary

biome p. 683
desert p. 684
grassland p. 685
temperate p. 687
taiga p. 689
tundra p. 689

Lesson 2: Aquatic Ecosystems

- Earth's aquatic ecosystems include freshwater and saltwater ecosystems. **Wetlands** can contain either salt water or freshwater. The **salinity** of **estuaries** varies.
- Human activities such as construction and fishing can affect aquatic ecosystems.

Vocabulary

salinity p. 693
wetland p. 696
estuary p. 697
intertidal zone p. 699
coral reef p. 699

Lesson 3: How Ecosystems Change

- Land and aquatic ecosystems change over time in predictable processes of **ecological succession.**
- Land ecosystems eventually form **climax communities.**
- Freshwater ecosystems undergo **eutrophication** and eventually become land ecosystems.

Vocabulary

ecological succession p. 703
climax community p. 703
pioneer species p. 704
eutrophication p. 706

- Personal Tutor
- Vocabulary eGames
- Vocabulary eFlashcards

FOLDABLES® Chapter Project

Assemble your lesson Foldables as shown to make a Chapter Project. Use the project to review what you have learned in this chapter.

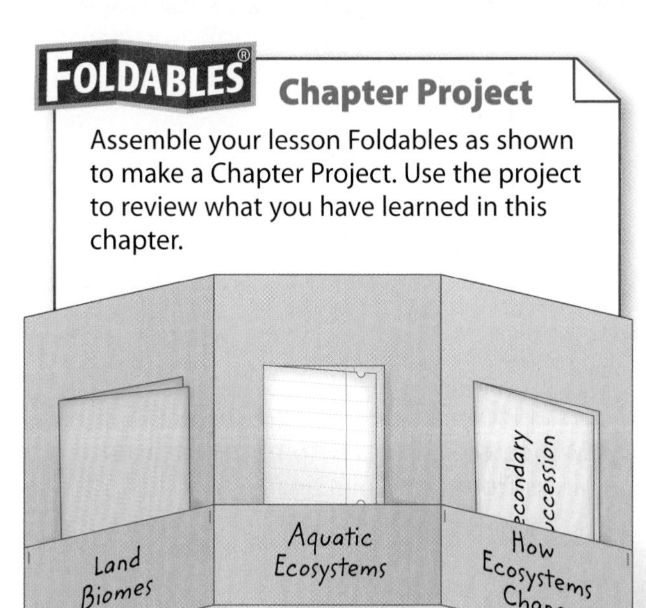

Use Vocabulary

Choose the vocabulary word that fits each description.

1. group of ecosystems with similar climate
2. area between the tropics and the polar circles
3. land biome with a layer of permafrost
4. the amount of salt dissolved in water
5. area where a river empties into an ocean
6. coastal zone between the highest high tide and the lowest low tide
7. process of one ecological community gradually changing into another
8. a stable community that no longer goes through major changes
9. the first species to grow on new or disturbed land

Link Vocabulary and Key Concepts

Concepts in Motion Interactive Concept Map

Copy this concept map, and then use vocabulary terms from the previous page and other terms from this chapter to complete the concept map.

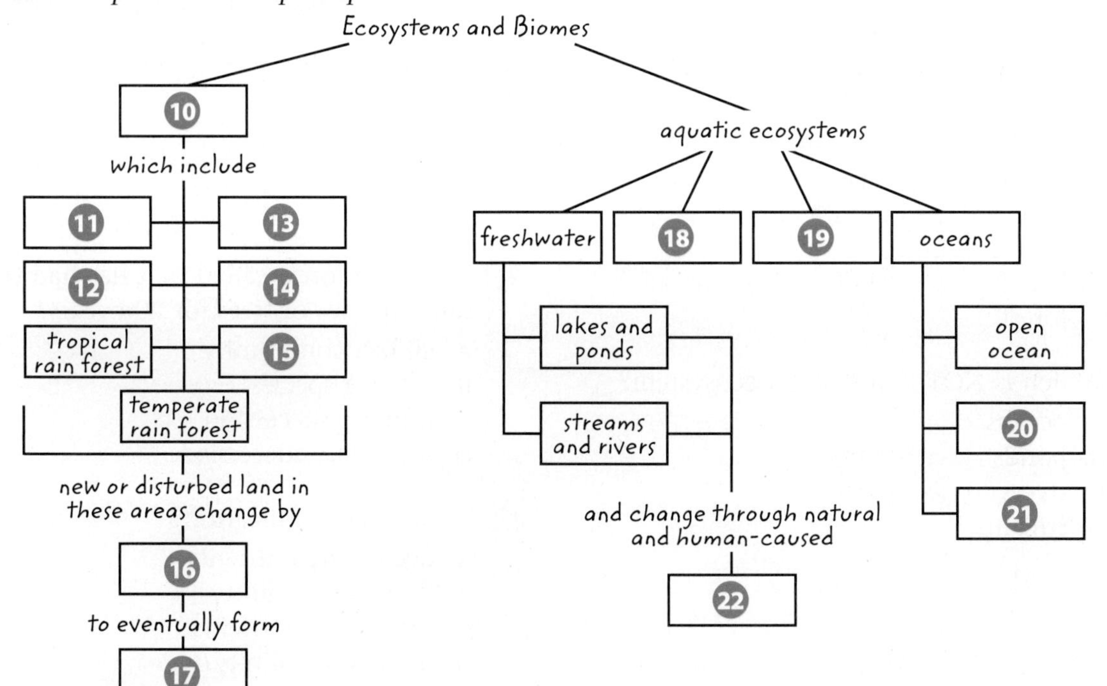

Chapter 19 Review

Understand Key Concepts 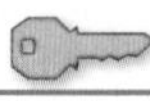

1. Where would you find plants with stems that can store large amounts of water?
   A. desert
   B. grassland
   C. taiga
   D. tundra

2. What does the pink area on the map below represent?

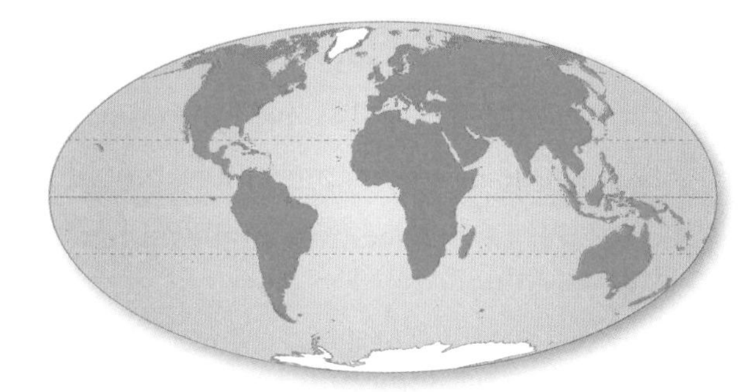

   A. taiga
   B. tundra
   C. temperate deciduous forest
   D. temperate rain forest

3. Where would you find trees that have no leaves during the winter?
   A. estuary
   B. tundra
   C. temperate deciduous forest
   D. temperate rain forest

4. Which biomes have rich, fertile soil?
   A. grassland and taiga
   B. grassland and tundra
   C. grassland and tropical rain forest
   D. grassland and temperate deciduous forest

5. Which is NOT a freshwater ecosystem?
   A. oceans
   B. ponds
   C. rivers
   D. streams

6. Where would you find species adapted to withstand strong wave action?
   A. estuaries
   B. wetlands
   C. intertidal zone
   D. twilight zone

7. Which ecosystem has flowing water?
   A. estuary
   B. lake
   C. stream
   D. wetland

8. Which ecosystems help protect coastal areas from flood damage?
   A. estuaries
   B. ponds
   C. rivers
   D. streams

9. Which organism below would be the first to grow in an area that has been buried in lava?

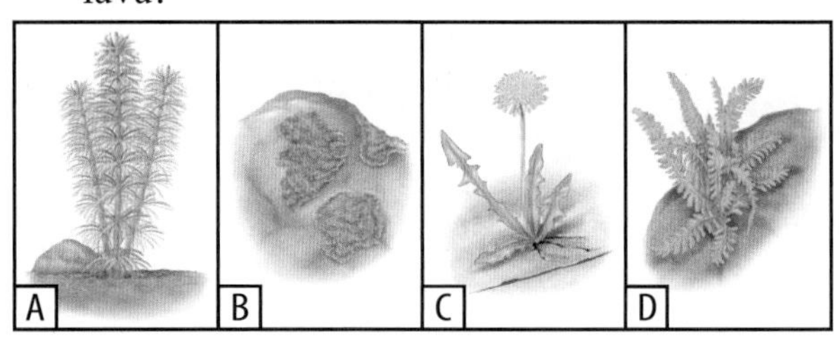

   A. A
   B. B
   C. C
   D. D

10. What is a forest called that has had the same species of trees for 200 years?
    A. climax community
    B. pioneer species
    C. primary succession
    D. secondary succession

11. What is eutrophication?
    A. decreasing nutrients
    B. decreasing salinity
    C. increasing nutrients
    D. increasing salinity

Assessment
Online Test Practice

Critical Thinking

12. **Compare** mammals that live in tundra biomes with those that live in desert biomes. What adaptations does each group have that help them survive?

13. **Analyze** You are invited to go on a trip to South America. Before you leave, you read a travel guide that says the country you will be visiting has hot summers, cold winters, and many wheat farms. What biome will you be visiting? Explain your reasoning.

14. **Contrast** How are ecosystems in the deep water of lakes and oceans different?

15. **Analyze** Which type of ocean ecosystem is likely to have the highest levels of dissolved oxygen? Why?

16. **Hypothesize** Why are the first plants that appear in primary succession small?

17. **Interpret Graphics** The following climate data were recorded for a forest ecosystem. To which biome does this ecosystem likely belong?

Climate Data	June	July	August
Average temperature (°C)	16.0	16.5	17.0
Average rainfall (cm)	3.0	2.0	2.0

Writing in Science

18. **Write** a paragraph explaining the succession process that might occur in a small pond on a cow pasture. Include a main idea, supporting details, and concluding sentence.

REVIEW THE BIG IDEA

19. Earth contains a wide variety of organisms that live in different conditions. How do Earth's biomes and ecosystems differ?

20. The photo below shows Biosphere 2, built in Arizona as an artificial Earth. Imagine that you have been asked to build a biome of your choice for Biosphere 3. What biotic and abiotic features should you consider?

Math Skills

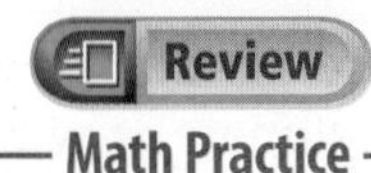

Math Practice

Use Proportions

21. At its highest salinity, the water in Utah's Great Salt Lake contained about 14.5 g of salt in 50 g of lake water. What was the salinity of the lake?

22. The seawater in Puget Sound off the coast of Oregon has a salinity of about 24 PPT. What weight of salt is there in 1,000 g of seawater?

Standardized Test Practice

Record your answers on the answer sheet provided by your teacher or on a sheet of paper.

Multiple Choice

1 Which aquatic ecosystem contains a mixture of freshwater and salt water?

A coral reef

B estuary

C pond

D river

Use the diagram below to answer question 2.

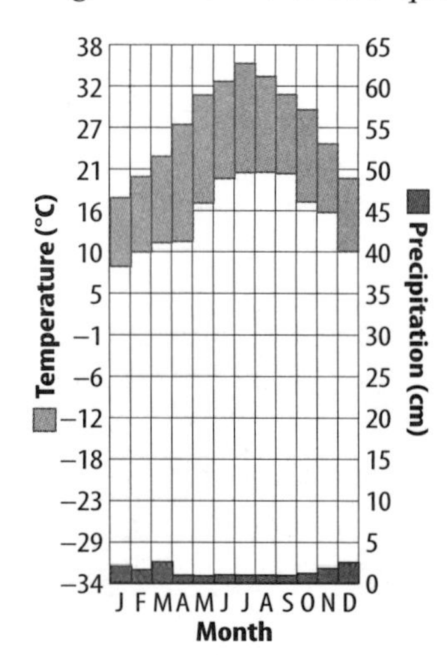

2 The diagram above most likely illustrates the climate of which biome?

A desert

B grassland

C tropical rain forest

D tundra

3 Which occurs during the first stage of ecological succession?

A eutrophication

B settlement

C development of climax community

D growth of pioneer species

4 Which biome has lost more than half its trees to logging activity?

A grassland

B taiga

C temperate deciduous forest

D tropical rain forest

Use the diagram below to answer question 5.

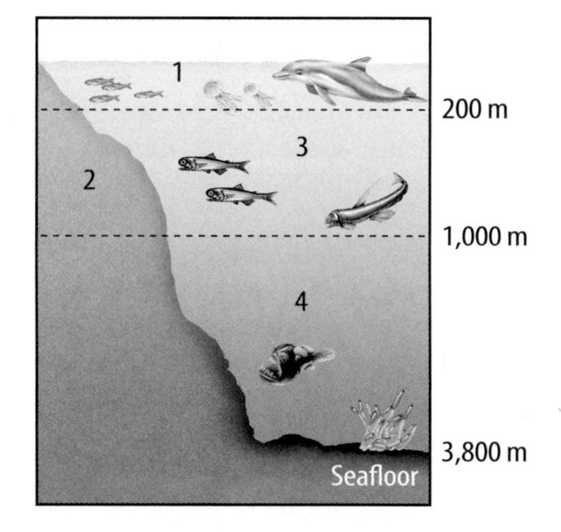

5 In the diagram above, where might you find microscopic photosynthetic organisms?

A 1

B 2

C 3

D 4

6 During aquatic succession, freshwater ponds

A become saltwater ponds.

B fill with soil.

C gain organisms.

D increase in depth.

Assessment

Online Standardized Test Practice

Use the diagram below to answer question 7.

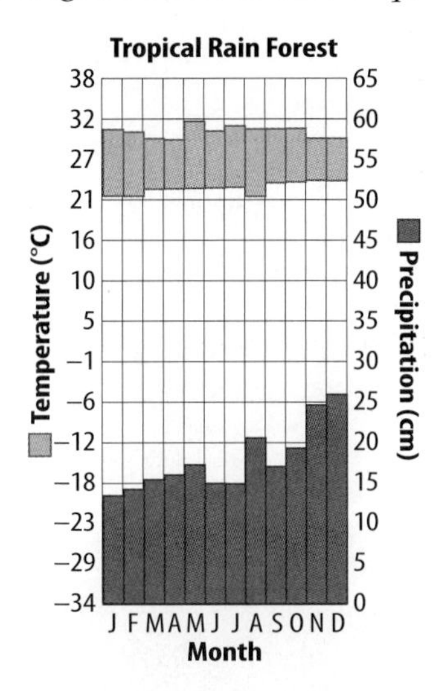

7 Based on the diagram above, which is true of the tropical rain forest biome?

A Precipitation increases as temperatures rise.

B Rainfall is greatest mid-year.

C Temperatures rise at year-end.

D Temperatures vary less than rainfall amounts.

8 Which aquatic biome typically has many varieties of nesting ducks, geese, herons, and egrets?

A coral reefs

B intertidal zones

C lakes

D wetlands

Constructed Response

Use the table below to answer questions 9 and 10.

Land Biome	Climate and Plant Life	Location
Desert		
Grassland		
Taiga		
Temperate deciduous forest		
Temperate rain forest		
Tropical rain forest		
Tundra		

9 Briefly describe the characteristics of Earth's seven land biomes. List one example of each biome, including its location.

10 How does human activity affect each land biome?

Use the table below to answer question 11.

Aquatic Ecosystem	Aquatic Animal
Coastal ocean	
Coral reefs	
Estuaries	
Lakes and ponds	
Open ocean	

11 Complete the table above with the name of an aquatic animal that lives in each of Earth's aquatic ecosystems.

NEED EXTRA HELP?											
If You Missed Question...	1	2	3	4	5	6	7	8	9	10	11
Go to Lesson...	2	1	2	3	2	2	2	2	1	1	2

Chapter 20

Environmental Impacts

THE BIG IDEA **How do human activities impact the environment?**

How many people are there?

More than 6 billion people live on Earth. Every day, people all over the world travel, eat, use water, and participate in recreational activities.

- What resources do people need and use?
- What might happen if any resources run out?
- How do human activities impact the environment?

Get Ready to Read

What do you think?

Before you read, decide if you agree or disagree with each of these statements. As you read this chapter, see if you change your mind about any of the statements.

1. Earth can support an unlimited number of people.
2. Humans can have both positive and negative impacts on the environment.
3. Deforestation does not affect soil quality.
4. Most trash is recycled.
5. Sources of water pollution are always easy to identify.
6. The proper method of disposal for used motor oil is to pour it down the drain.
7. The greenhouse effect is harmful to life on Earth.
8. Air pollution can affect human health.

Lesson 1

Reading Guide

Key Concepts
ESSENTIAL QUESTIONS

- What is the relationship between resource availability and human population growth?
- How do daily activities impact the environment?

Vocabulary
population p. 719
carrying capacity p. 720

Multilingual eGlossary

Video

- BrainPOP®
- Science Video
- What's Science Got to do With It?

People and the Environment

Inquiry What's the impact?

This satellite image shows light coming from Europe and Africa at night. You can see where large cities are located. What do you think the dark areas represent? When you turn on the lights at night, where does the energy to power the lights come from? How might this daily activity impact the environment?

Inquiry Launch Lab

20 minutes

What happens as populations increase in size?

In the year 200, the human population consisted of about a quarter of a billion people. By the year 2000, it had increased to more than 6 billion, and by 2050, it is projected to be more than 9 billion. However, the amount of space available on Earth will remain the same.

1. Read and complete a lab safety form.
2. Place 10 **dried beans** in a **100-mL beaker.**
3. At the start signal, double the number of beans in the beaker. There should now be 20 beans.

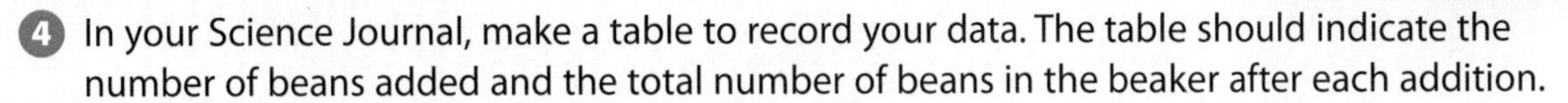

4. In your Science Journal, make a table to record your data. The table should indicate the number of beans added and the total number of beans in the beaker after each addition.

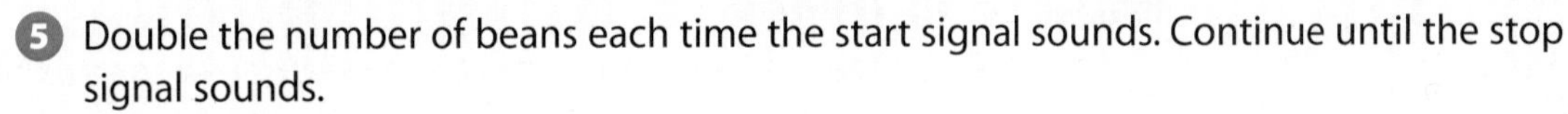

5. Double the number of beans each time the start signal sounds. Continue until the stop signal sounds.

Think About This

1. Can you add any more beans to the beaker? Why or why not?
2. How many times did you have to double the beans to fill the beaker?
3. **Key Concept** How might the growth of a population affect the availability of resources, such as space?

Population and Carrying Capacity

Have you ever seen a sign such as the one shown in **Figure 1**? The sign shows the population of a city. In this case, population means how many people live in the city. Scientists use the term *population,* too, but in a slightly different way. For scientists, *a* **population** *is all the members of a species living in a given area.* You are part of a population of humans. The other species in your area, such as birds or trees, each make up a separate population.

The Human Population

When the first American towns were settled, most had low populations. Today, some of those towns are large cities, crowded with people. In a similar way, Earth was once home to relatively few humans. Today, about 6.7 billion people live on Earth. The greatest increase in human population occurred during the last few centuries.

Figure 1 This sign shows the population of the city. Scientists use the word *population* to describe all the members of a species in an area.

Figure 2 Human population stayed fairly steady for most of history and then "exploded" in the last few hundred years.

Visual Check How does the rate of human population growth from the years 200 to 1800 compare to the rate of growth from 1800 to 2000?

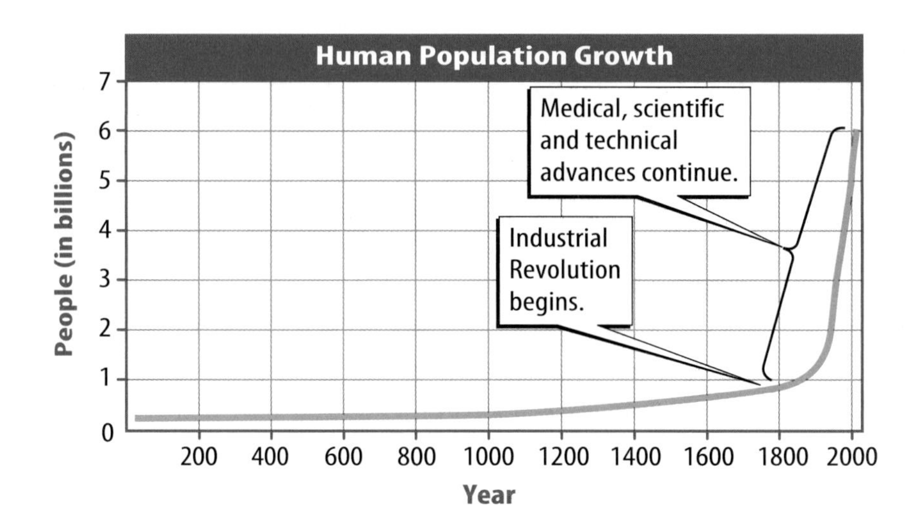

WORD ORIGIN

population
from Latin *populus,* means "people"

Population Trends

Have you ever heard the phrase ***population*** *explosion?* Population explosion describes the sudden rise in human population that has happened in recent history. The graph in **Figure 2** shows how the human population has changed. The population increased at a fairly steady rate for most of human history. In the 1800s, the population began to rise sharply.

What caused this sharp increase? Improved health care, clean water, and other technological advancements mean that more people are living longer and reproducing. In the hour or so it might take you to read this chapter, about 15,000 babies will be born worldwide.

Reading Check What factors contributed to the increase in human population?

FOLDABLES®

Use a sheet of paper to make a small vertical shutterfold. Draw the arrows on each tab and label as illustrated. Use the Foldable to discuss how human population growth is related to resources.

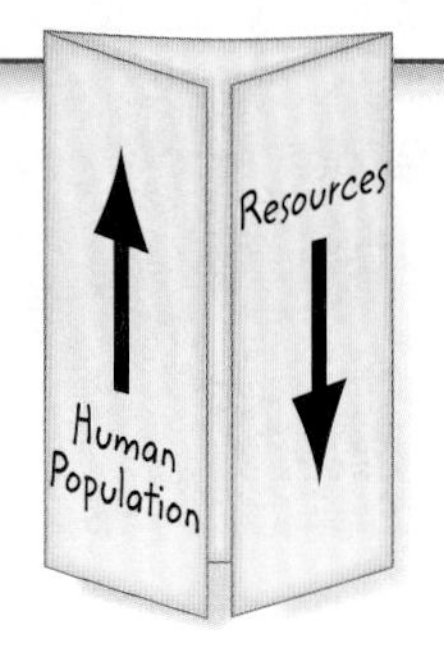

Population Limits

Every human being needs certain things, such as food, clean water, and shelter, to survive. People also need clothes, transportation, and other items. All the items used by people come from resources found on Earth. Does Earth have enough resources to support an unlimited number of humans?

Earth has limited resources. It cannot support a population of any species in a given environment beyond its carrying capacity. **Carrying capacity** *is the largest number of individuals of a given species that Earth's resources can support and maintain for a long period of time.* If the human population continues to grow beyond Earth's carrying capacity, eventually Earth will not have enough resources to support humans.

Key Concept Check What is the relationship between the availability of resources and human population growth?

Impact of Daily Actions

Each of the 6.7 billion people on Earth uses **resources** in some way. The use of these resources affects the environment. Consider the impact of one activity—a shower.

Science Use v. Common Use

resource

Science Use a natural source of supply or support

Common Use a source of information or expertise

Consuming Resources

Like many people, you might take a shower each day. The metal in the water pipes comes from resources mined from the ground. Mining can destroy habitats and pollute soil and water. Your towel might be made of cotton, a resource obtained from plants. Growing plants often involves the use of fertilizers and other chemicals that run off into water and affect its quality.

The water itself also is a resource—one that is scarce in some areas of the world. Most likely, fossil fuels are used to heat the water. Recall that fossil fuels are nonrenewable resources, which means they are used up faster than they can be replaced by natural processes. Burning fossil fuels also releases pollution into the atmosphere.

Now, think about all the activities that you do in one day, such as going to school, eating meals, or playing computer games. All of these activities use resources. Over the course of your lifetime, your potential impact on the environment is great. Multiply this impact by 6.7 billion, and you can understand why it is important to use resources wisely.

Key Concept Check What are three things you did today that impacted the environment?

Positive and Negative Impacts

As shown in **Figure 3**, not all human activities have a negative impact on the environment. In the following lessons, you will learn how human activities affect soil, water, and air quality. You will also learn things you can do to help reduce the impact of your actions on the environment.

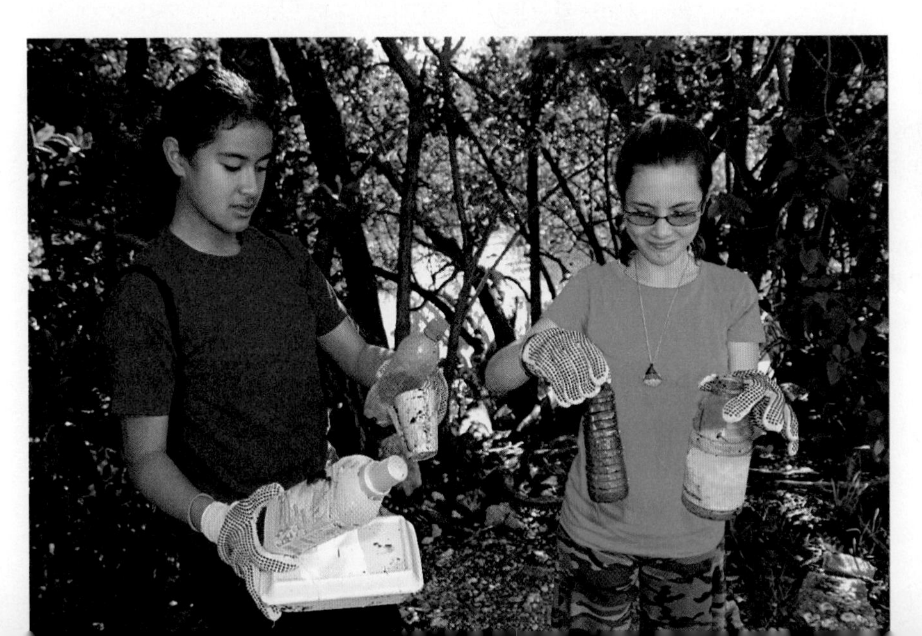

Figure 3 Cleaning up streams and picking up litter are ways people can positively impact the environment.

Lesson 1 Review

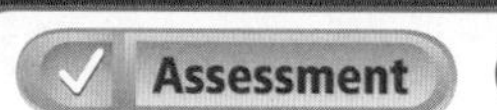

Online Quiz

Visual Summary

Human population has exploded since the 1800s. Every day billions of people use Earth's resources. The human population will eventually reach its carrying capacity.

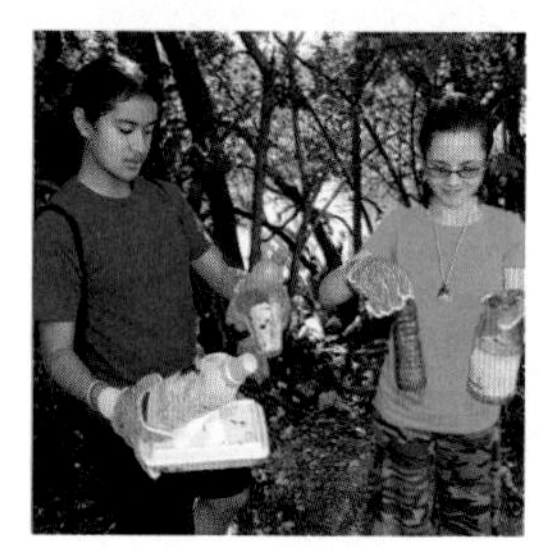

When humans use resources, they can have both negative and positive impacts on the environment. It is important for humans to use resources wisely.

FOLDABLES®

Use your lesson Foldable to review the lesson. Save your Foldable for the project at the end of the chapter.

What do you think NOW?

You first read the statements below at the beginning of the chapter.

1. Earth can support an unlimited number of people.
2. Humans can have both positive and negative impacts on the environment.

Did you change your mind about whether you agree or disagree with the statements? Rewrite any false statements to make them true.

Use Vocabulary

1. **Define** *carrying capacity* in your own words.
2. All the members of a certain species living in a given area is a(n) ________.

Understand Key Concepts 

3. Approximately how many people live on Earth?
   - **A.** 2.4 billion
   - **B.** 6.7 billion
   - **C.** 7.6 billion
   - **D.** 12.1 billion
4. **Identify** something you could do to reduce your impact on the environment.
5. **Reason** Why do carrying capacities exist for all species on Earth?

Interpret Graphics

6. **Take Notes** Copy the graphic organizer below. List two human activities and the effect of each activity on the environment.

Activity	Effect on the Environment

7. **Summarize** how human population growth has changed over time, using the graph below.

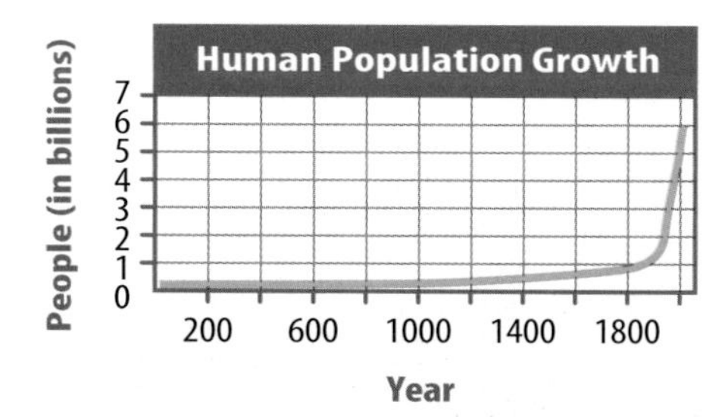

Critical Thinking

8. **Predict** What might happen if a species reaches Earth's carrying capacity?
9. **Reflect** Technological advances allow farmers to grow more crops. Do you think these advances affect the carrying capacity for humans? Explain.

What amount of Earth's resources do you use in a day?

Many of the practices we engage in today became habits long before we realized the negative effects that they have on the environment. By analyzing your daily resource use, you might identify some different practices that can help protect Earth's resources.

Learn It

In science, **data** are **collected** as accurate numbers and descriptions and organized in specific ways. The meaning of your observations can be determined by **analyzing** the data you collected.

Try It

1. With your group, design a data collection form for recording each group member's resource use for one 24-h period.
2. You should include space to collect data on water use, fossil fuel use (which may include electricity use and transportation), how much meat and dairy products you eat, how much trash you discard, and any other resources you might use in a typical 24-h period. Indicate the units in which you will record the data.
3. Share your form with the other groups, and complete a final draft using the best features from each group's design.
4. Distribute copies of the form to each group member. Record each instance and quantity of resource use during a 24-h period.
5. For each resource, calculate how much you would use in 1 year, based on your usage in the 24-h period.

Apply It

6. Consider whether a single 24-h period is representative of each of the 365 days of your year. Explain your answer.
7. How would you modify your data collection design to reflect a more realistic measure of your resource use over a year?
8. **Key Concept** Explain how two of the activities that you recorded deplete resources or pollute the soil, the water, or the air. How can you change your activities to reduce your impact or have a positive impact?

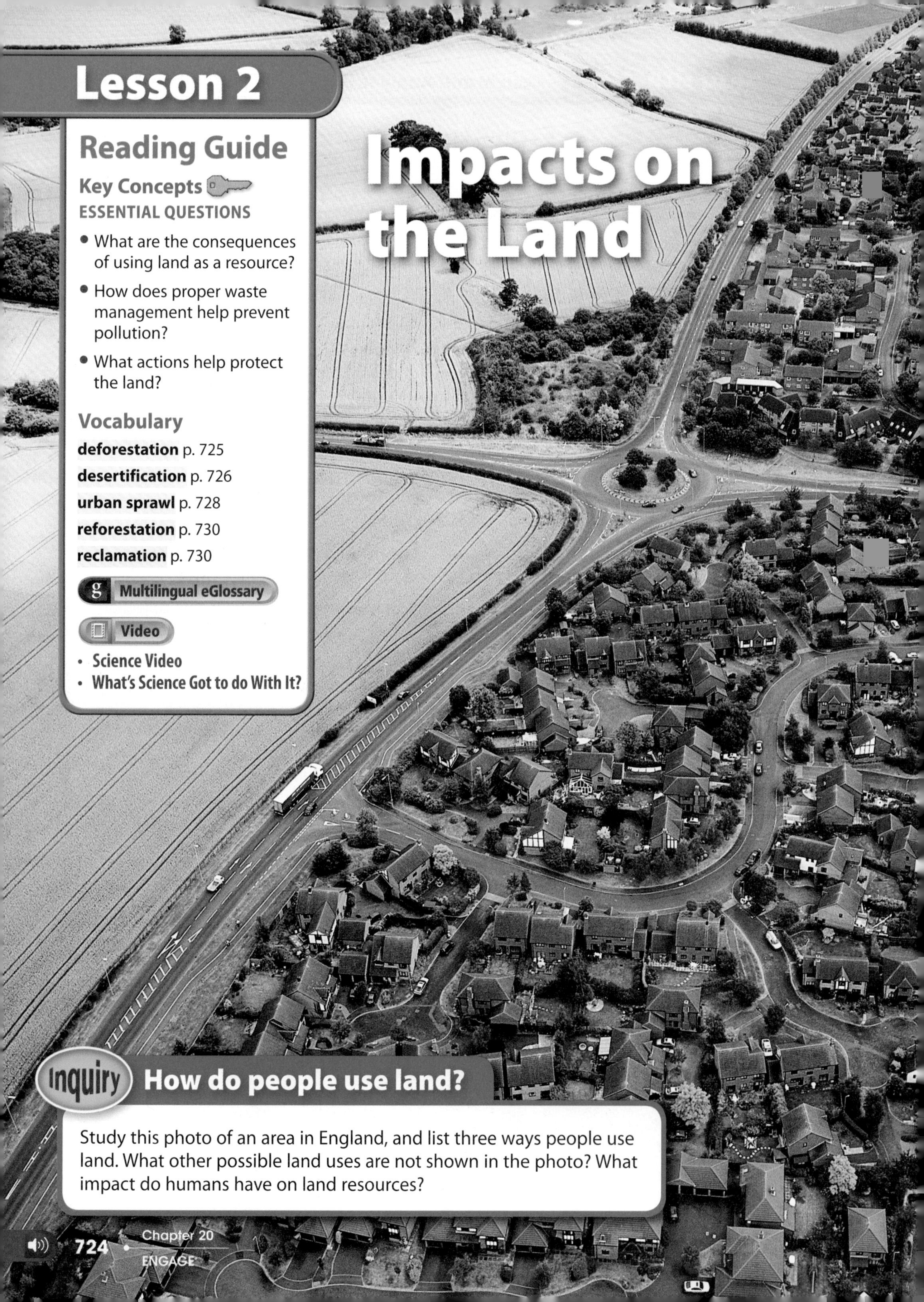

Lesson 2

Reading Guide

Key Concepts
ESSENTIAL QUESTIONS

- What are the consequences of using land as a resource?
- How does proper waste management help prevent pollution?
- What actions help protect the land?

Vocabulary

deforestation p. 725
desertification p. 726
urban sprawl p. 728
reforestation p. 730
reclamation p. 730

Multilingual eGlossary

Video

- Science Video
- What's Science Got to do With It?

Impacts on the Land

Inquiry How do people use land?

Study this photo of an area in England, and list three ways people use land. What other possible land uses are not shown in the photo? What impact do humans have on land resources?

Inquiry Launch Lab

20 minutes

How can items be reused?

As an individual, you can have an effect on the use and the protection of Earth's resources by reducing, reusing, and recycling the materials you use every day.

1. Read and complete a lab safety form.
2. Have one member of your group pull an item from the **item bag.**
3. Discuss the item with your group and take turns describing as many different ways to reuse it as possible.
4. List the different uses in your Science Journal.
5. Repeat steps 2–4.
6. Share your lists with the other groups. What uses did other groups think of that were different from your group's ideas for the same item?

Think About This

1. Describe your group's items and three different ways that you thought to reuse each item.
2. How does reusing these items help to reduce the use of Earth's resources?
3. **Key Concept** How do you think the action of reusing items helps to protect the land?

Using Land Resources

What do the metal in staples and the paper in your notebook have in common? Both come from resources found in or on land. People use land for timber production, agriculture, and mining. All of these activities impact the environment.

Forest Resources

Trees are cut down to make wood and paper products, such as your notebook. Trees are also cut for fuel and to clear land for agriculture, grazing, or building houses or highways.

Sometimes forests are cleared, as shown in **Figure 4.** **Deforestation** *is the removal of large areas of forests for human purposes.* Approximately 130,000 km^2 of tropical rain forests are cut down each year, an area equal in size to the state of Louisiana. Tropical rain forests are home to an estimated 50 percent of all the species on Earth. Deforestation destroys habitats, which can lead to species' extinction.

Deforestation also can affect soil quality. Plant roots hold soil in place. Without these natural anchors, soil erodes away. In addition, deforestation affects air quality. Recall that trees remove carbon dioxide from the air when they undergo photosynthesis. When there are fewer trees on Earth, more carbon dioxide remains in the atmosphere. You will learn more about carbon dioxide in Lesson 4.

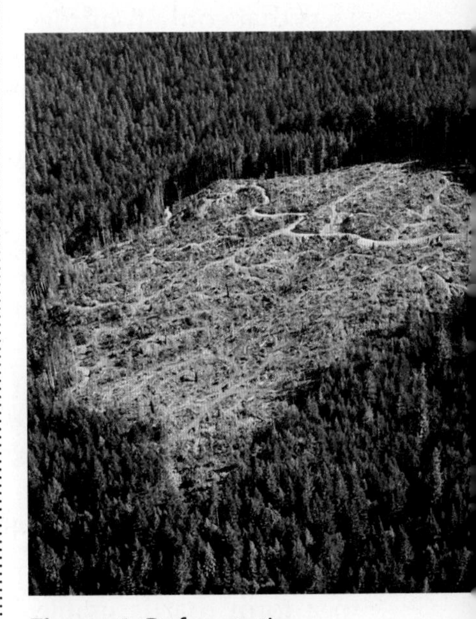

Figure 4 Deforestation occurs when forests are cleared for agriculture, grazing, or other purposes.

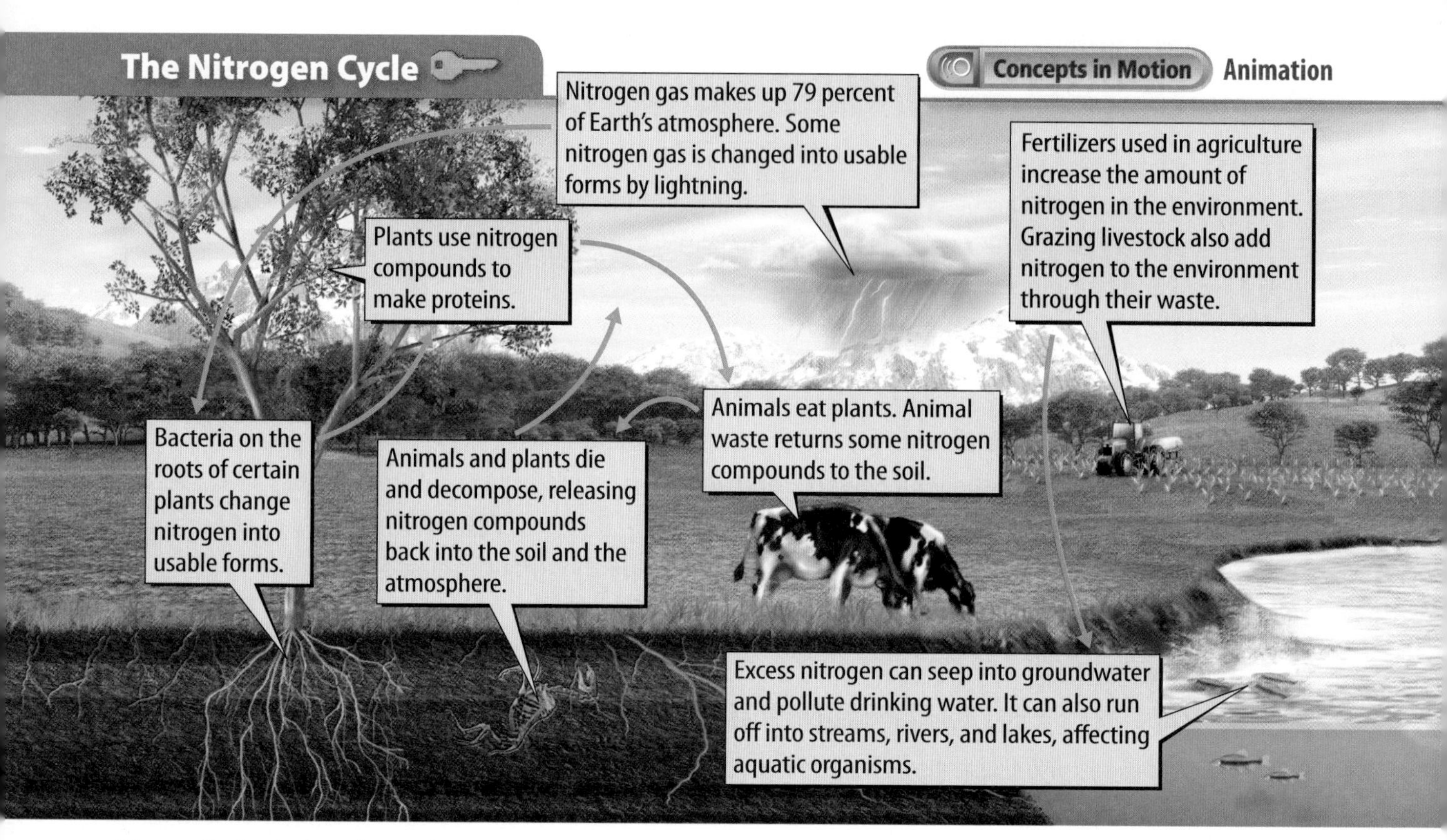

Figure 5 Agricultural practices can increase the amount of nitrogen that cycles through ecosystems.

Visual Check How does the use of fertilizers affect the environment?

Agriculture and the Nitrogen Cycle

It takes a lot of food to feed 6.7 billion people. To meet the food demands of the world's population, farmers often add fertilizers that contain nitrogen to soil to increase crop yields.

As shown in **Figure 5,** nitrogen is an element that naturally cycles through ecosystems. Living things use nitrogen to make proteins. And when these living things die and decompose or produce waste, they release nitrogen into the soil or the atmosphere.

Although nitrogen gas makes up about 79 percent of Earth's atmosphere, most living things cannot use nitrogen in its gaseous form. Nitrogen must be converted into a usable form. Bacteria that live on the roots of certain plants convert atmospheric nitrogen to a form that is usable by plants. Modern agricultural practices include adding fertilizer that contains a usable form of nitrogen to soil.

Scientists estimate that human activities such as manufacturing and applying fertilizers to crops have doubled the amount of nitrogen cycling through ecosystems. Excess nitrogen can kill plants adapted to low nitrogen levels and affect organisms that depend on those plants for food. Fertilizers can seep into groundwater supplies, polluting drinking water. They can also run off into streams and rivers, affecting aquatic organisms.

Other Effects of Agriculture

Agriculture can impact soil quality in other ways, too. Soil erosion can occur when land is overfarmed or overgrazed. High rates of soil erosion can lead to desertification. **Desertification** *is the development of desert-like conditions due to human activities and/or climate change.* A region of land that undergoes desertification is no longer useful for food production.

Reading Check What causes desertification?

Figure 6 Some resources must be mined from the ground.

Mining

Many useful rocks and minerals are removed from the ground by mining. For example, copper is removed from the surface by digging a strip mine, such as the one shown in **Figure 6.** Coal and other in-ground resources also can be removed by digging underground mines.

Mines are essential for obtaining much-needed resources. However, digging mines disturbs habitats and changes the landscape. If proper regulations are not followed, water can be polluted by **runoff** that contains heavy metals from mines.

Review Vocabulary
runoff
the portion of precipitation that moves over land and eventually reaches streams, rivers, lakes, and oceans

 **Key Concept Check** What are some consequences of using land as a resource?

Construction and Development

You have read about important resources that are found on or in land. But did you know that land itself is a resource? People use land for living space. Your home, your school, your favorite stores, and your neighborhood streets are all built on land.

Inquiry MiniLab — 20 minutes

What happens when you mine?

Coal is a fossil fuel that provides energy for many activities. People obtain coal by mining, or digging, into Earth's surface.

1. Read and complete a lab safety form.
2. Research the difference between strip-mining and underground mining.
3. Use **salt dough** and **other materials** to build a model hill that contains coal deposits. Follow the instructions provided on how to build the model.
4. Sketch the profile of the hill. Use a **ruler** to measure the dimensions of the hill.
5. Decide which mining method to use to remove the coal. Mine the coal.
6. Try to restore the hill to its original size, shape, and forest cover.

Analyze and Conclude

1. **Compare** the appearance of your restored hill to the drawing of the original hill.
2. **Key Concept** Describe two consequences of the lost forest cover and loose soil on the mined hill.

Figure 7  Urban sprawl can lead to habitat destruction as forests are cut down to make room for housing developments.

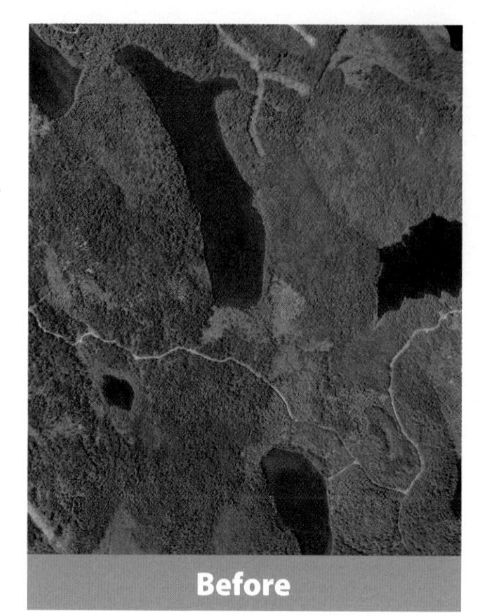

Urban Sprawl

In the 1950s, large tracts of rural land in the United States were developed as suburbs, residential areas on the outside edges of a city. When the suburbs became crowded, people moved farther out into the country. More open land was cleared for still more development. *The development of land for houses and other buildings near a city is called* **urban sprawl.** The impacts of urban sprawl include habitat destruction, shown in **Figure 7,** and loss of farmland. Increased runoff also occurs, as large areas are paved for sidewalks and streets. An increase in runoff, especially if it contains sediments or chemical pollutants, can reduce the water quality of streams, rivers, and groundwater.

Roadways

Urban sprawl occurred at the same time as another trend in the United States—increased motor vehicle use. Only a small percentage of Americans owned cars before the 1940s. By 2005, there were 240 million vehicles for 295 million people, greatly increasing the need for roadways. In 1960, the United States had about 16,000 km of interstate highways. Today, the interstate highway system includes 47,000 km of paved roadways. Like urban sprawl, roadways increase runoff and disturb habitats.

Reading Check What two trends triggered the need for more highways?

Recreation

Not all of the land used by people is paved and developed. People also use land for recreation. They hike, bike, ski, and picnic, among other activities. In urban areas, some of these activities take place in public parks. As you will learn later in this lesson, parks and other green spaces help decrease runoff.

Math Skills

Use Percentages

Between 1960 and today, interstate highways increased from a total of 16,000 km to 47,000 km. What percent increase does this represent?

1. Subtract the starting value from the final value.
   47,000 km − 16,000 km = 31,000 km
2. Divide the difference by the starting value.
   $\frac{31{,}000 \text{ km}}{16{,}000 \text{ km}} = 1.94$
3. Multiply by 100 and add a % sign.
   $1.94 \times 100 = 194\%$

Practice

In 1950, the U.S. population was about 150,000,000. By 2007, it was nearly 300,000,000. What was the percent increase?

- Math Practice
- Personal Tutor

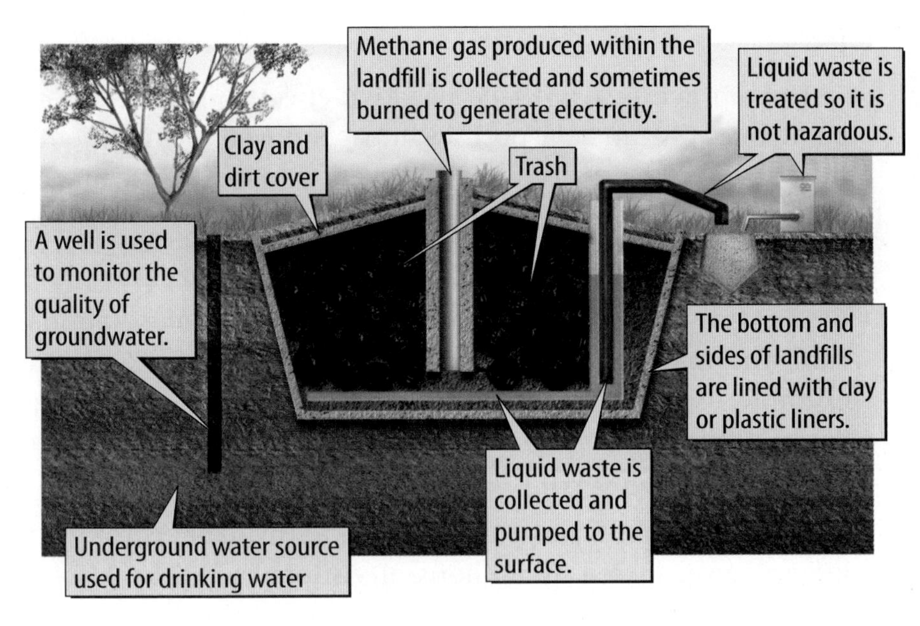

Figure 8 About 54 percent of the trash in the United States is disposed of in landfills.

Visual Check How can the methane gas produced within a landfill be used?

Waste Management

On a typical day, each person in the United States generates about 2.1 kg of trash. That adds up to about 230 million metric tons per year! Where does all that trash go?

Landfills

About 31 percent of the trash is recycled and composted. About 14 percent is burned, and the remaining 55 percent is placed in landfills, such as the one shown in **Figure 8.** Landfills are areas where trash is buried. Landfills are another way that people use land.

A landfill is carefully designed to meet government regulations. Trash is covered by soil to keep it from blowing away. Special liners help prevent pollutants from leaking into soil and groundwater supplies.

Key Concept Check What is done to prevent the trash in landfills from polluting air, soil, and water?

Hazardous Waste

Some trash cannot be placed in landfills because it contains harmful substances that can affect soil, air, and water quality. This trash is called hazardous waste. The substances in hazardous waste also can affect the health of humans and other living things.

Both industries and households generate hazardous waste. For example, hazardous waste from the medical industry includes used needles and bandages. Household hazardous waste includes used motor oil and batteries. The U.S. Environmental Protection Agency (EPA) works with state and local agencies to help people safely **dispose** of hazardous waste.

FOLDABLES®

Use a sheet of notebook paper to make a horizontal two-tab concept map. Label and draw arrows as illustrated. Use the Foldable to identify positive and negative factors that have an impact on land.

ACADEMIC VOCABULARY

dispose
(verb) to throw away

Figure 9 Yellowstone Falls are in Yellowstone National Park, which was created in 1872.

Positive Actions

Human actions can have negative effects on the environment, but they can have positive impacts as well. Governments, society, and individuals can work together to reduce the impact of human activities on land resources.

Protecting the Land

The area shown in **Figure 9** is part of Yellowstone National Park, the first national park in the world. The park was an example for the United States and other countries as they began setting aside land for preservation. State and local governments also followed this example.

Protected forests and parks are important habitats for wildlife and are enjoyed by millions of visitors each year. Mining and logging are allowed on some of these lands. However, the removal of resources must meet environmental regulations.

Reforestation and Reclamation

A forest is a complex ecosystem. With careful planning, it can be managed as a renewable resource. For example, trees can be select-cut. That means that only some trees in one area are cut down, rather than the entire forest. In addition, people can practice reforestation. **Reforestation** *involves planting trees to replace trees that have been cut or burned down.* Reforestation can keep a forest healthy or help reestablish a deforested area.

WORD ORIGIN
reclamation
from Latin *reclamare,* means "to call back"

Mined land also can be made environmentally healthy through reclamation. **Reclamation** *is the process of restoring land disturbed by mining.* The before and after photos in **Figure 10** show that the mined area has been reshaped, covered with soil, and then replanted with trees and other vegetation.

Reading Check How do reforestation and reclamation positively impact land?

Figure 10 As part of reclamation, grasses and trees were planted on this coal mine in Indiana.

Green Spaces

In urban areas, much of the land is covered with parking lots, streets, buildings, and sidewalks. Many cities use green spaces to create natural environments in urban settings. Green spaces are areas that are left undeveloped or lightly developed. They include parks within cities and forests around suburbs. Green spaces, such as the park shown in **Figure 11,** provide recreational opportunities for people and shelter for wildlife. Green spaces also reduce runoff and improve air quality as plants remove excess carbon dioxide from the air.

How can you help?

Individuals can have a big impact on land-use issues by practicing the three Rs—reusing, reducing, and recycling. Reusing is using an item for a new purpose. For example, you might have made a bird feeder from a used plastic milk jug. Reducing is using fewer resources. You can turn off the lights when you leave a room to reduce your use of electricity.

Recycling is making a new product from a used product. Plastic containers can be recycled into new plastic products. Recycled aluminum cans are used to make new aluminum cans. Paper, shown in **Figure 11,** also can be recycled.

Figure 11 shows another way people can lessen their environmental impact on the land. The student in the bottom photo is composting food scraps into a material that is added to soil to increase its fertility. Compost is a mixture of decaying organic matter, such as leaves, food scraps, and grass clippings. It is used to improve soil quality by adding nutrients to soil. Composting and reusing, reducing, and recycling all help reduce the amount of trash that ends up in landfills.

Key Concept Check What can you do to help lessen your impact on the land?

Figure 11 Green spaces, recycling, and composting are three things that can have positive impacts on land resources.

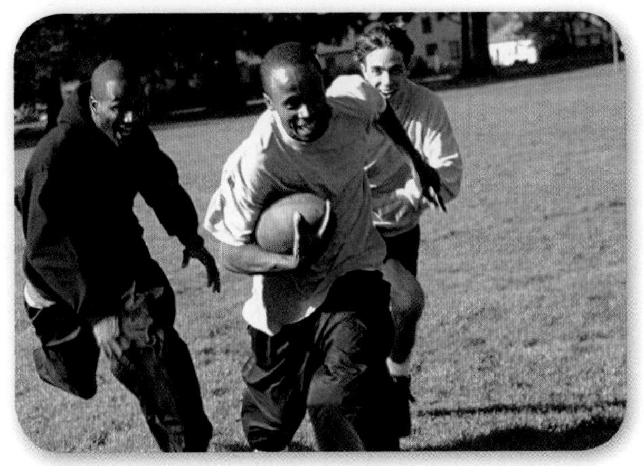

Parks provide recreational opportunities for people and habitats for wildlife, such as birds.

Using recycled paper to make new paper reduces deforestation as well as water use during paper production.

Composting speeds up the rate of decomposition for vegetable scraps, leaving a rich material that can be used as natural fertilizer.

Lesson 2 Review

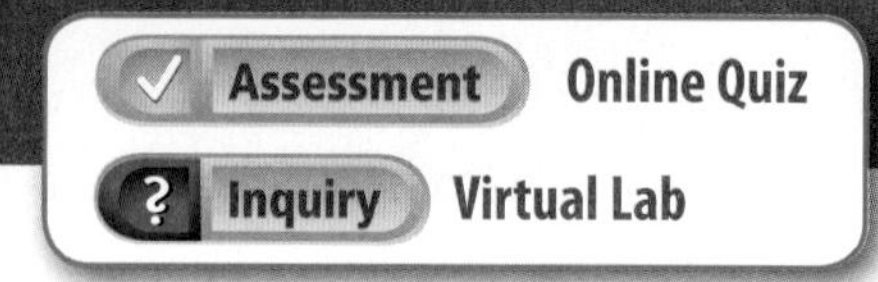

Visual Summary

Deforestation, agriculture, and mining for useful rocks and minerals all can affect land resources negatively.

People use land for living space, which can lead to urban sprawl, an increase in roadways, and the need for proper waste disposal.

Creating national parks, preserves and local green spaces, reforestation, and practicing the three Rs are all ways people can positively impact land resources.

FOLDABLES®

Use your lesson Foldable to review the lesson. Save your Foldable for the project at the end of the chapter.

What do you think NOW?

You first read the statements below at the beginning of the chapter.

3. Deforestation does not affect soil quality.

4. Most trash is recycled.

Did you change your mind about whether you agree or disagree with the statements? Rewrite any false statements to make them true.

Use Vocabulary

1. **Distinguish** between deforestation and reforestation.
2. **Use the term** *urban sprawl* in a sentence.
3. **Define** *desertification.*

Understand Key Concepts 

4. Which has a positive impact on land?
   A. composting
   B. deforestation
   C. mining
   D. urban sprawl
5. **Apply** How can the addition of fertilizers to crops affect the nitrogen cycle?
6. **Analyze** Why must waste disposal be carefully managed?

Interpret Graphics

7. **Organize** Copy and fill in the graphic organizer below. In each oval, list one way that people use land.

8. **Describe** the function of the liner in the diagram below.

Math Skills

Math Practice

9. In 1950, 35.1 million people lived in suburban areas. By 1990, the number had increased to 120 million people. What was the percent increase in suburban population?

Inquiry **Skill Practice** **Design an Experiment** 45 minutes

Materials

creative construction materials

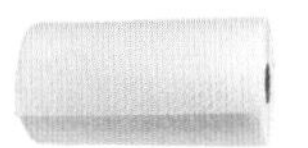

paper towels

scissors

masking tape

Safety

How will you design an environmentally safe landfill?

Your city is planning to build an environmentally safe landfill and is accepting design proposals. Your task is to develop and test a design to submit to city officials.

Learn It

When you **design an experiment,** you consider the variables you want to test and how you will measure the results.

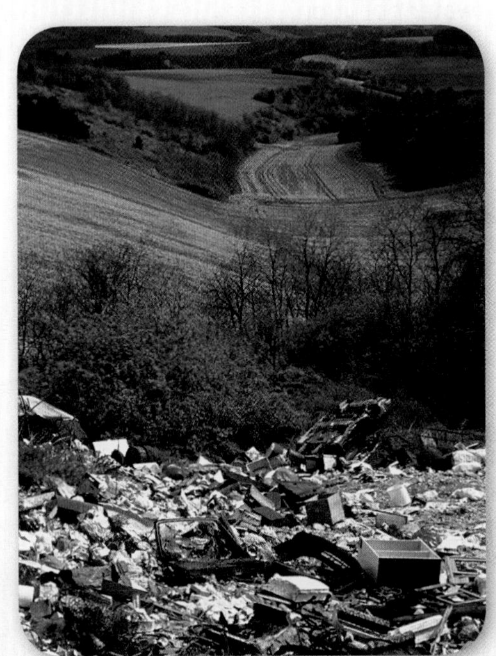

Try It

1. Read and complete a lab safety form.
2. Read the information provided about landfill requirements as set by the Environmental Protection Agency.
3. Plan and diagram your landfill design.
4. Use the materials to build your landfill model. Add waste materials.
5. Pour 350 mL of water on your landfill to simulate rain. Observe the path the water takes.
6. Collect the leachate and compare its volume with that of the other groups. Leachate is the liquid that seeps out of your landfill.
7. Compare your landfill design to that of other groups.

Apply It

8. Explain how you designed your landfill to meet requirements and function efficiently.
9. How did your landfill design compare to those of other groups? How much leachate did your group collect compared to other groups?
10. What changes would you make to the design of your landfill? What changes would you make to your procedure to test the effectiveness of your landfill?
11. **Key Concept** Explain how your landfill helped to prevent the pollution of soil and water.

Lesson 3

Reading Guide

Key Concepts
ESSENTIAL QUESTIONS

- How do humans use water as a resource?
- How can pollution affect water quality?
- What actions help prevent water pollution?

Vocabulary

point-source pollution p. 736
nonpoint-source pollution p. 737

Multilingual eGlossary

Impacts on Water

Inquiry How Much Water?

About 34 percent of all water used in the United States is used to irrigate crops. Where does all this water come from? What other ways do humans use water? What happens when water is polluted or runs out?

20 minutes

Which water filter is the most effective?

Suppose you have been hired by the Super-Clean Water Treatment Plant to test new water filters. Their old filters do not remove all of the particles from the treated water. Your job is to design an effective water filter.

1. Read and complete a lab safety form.
2. Obtain a **water sample,** a **funnel,** and two **500-mL beakers.**
3. Use **coffee filters, paper towels, cotton,** and **gravel** to make a filter in the funnel. Hold the funnel over the first beaker. Pour half of your water sample into the funnel and collect the water in the beaker. Record your results in your Science Journal.
4. Remove the filter and rinse the funnel. Based on your results, make a second, more efficient filter. Repeat step 3 using the second beaker.
5. Draw a diagram of both filtering methods in your Science Journal.

Think About This

1. Were either of your filters successful in removing the particles from the water? Why or why not?
2. What changes would you make to your filter to make it work more efficiently?
3. **Key Concept** How do water treatment plants make more water available for human use?

Water as a Resource

Most of Earth's surface is covered with water, and living things on Earth are made mostly of water. Neither the largest whale nor the smallest algae can live without this important resource. Like other organisms, humans need water to survive. Humans also use water in ways that other organisms do not. People wash cars, do laundry, and use water for recreation and transportation.

Household activities, however, make up only a small part of human water use. As shown in **Figure 12,** most water in the United States is used by power plants. The water is used to generate electricity and to cool equipment. Like the land uses you learned about earlier, the use of water as a resource also impacts the environment.

Key Concept Check How do humans use water as a resource?

Water Use

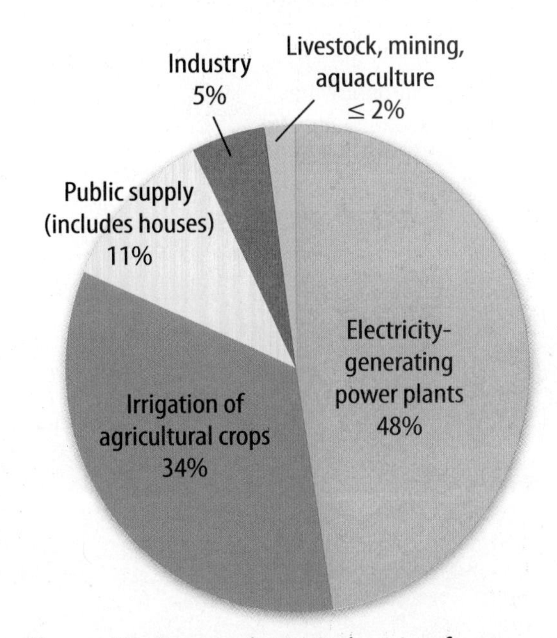

Figure 12 Power plants, industries, farms, and households all use water.

Sources of Water Pollution

Water moves from Earth's surface to the atmosphere and back again in the water cycle. Thermal energy from the Sun causes water at Earth's surface to evaporate into the atmosphere. Water vapor in the air cools as it rises, then condenses and forms clouds. Water returns to Earth's surface as precipitation. Runoff reenters oceans and rivers or it can seep into the ground. Pollution from a variety of sources can impact the quality of water as it moves through the water cycle.

Point-Source Pollution

Point-source pollution *is pollution from a single source that can be identified.* The discharge pipe in **Figure 13** that is releasing industrial waste directly into a river is an example of point-source pollution. Other examples of point-source pollution in **Figure 13** are the oil spilling from the tanker and the runoff from the mining operation.

WORD ORIGIN
pollution
from Latin *polluere,* means "to contaminate"

Sources of Water Pollution

Review Personal Tutor

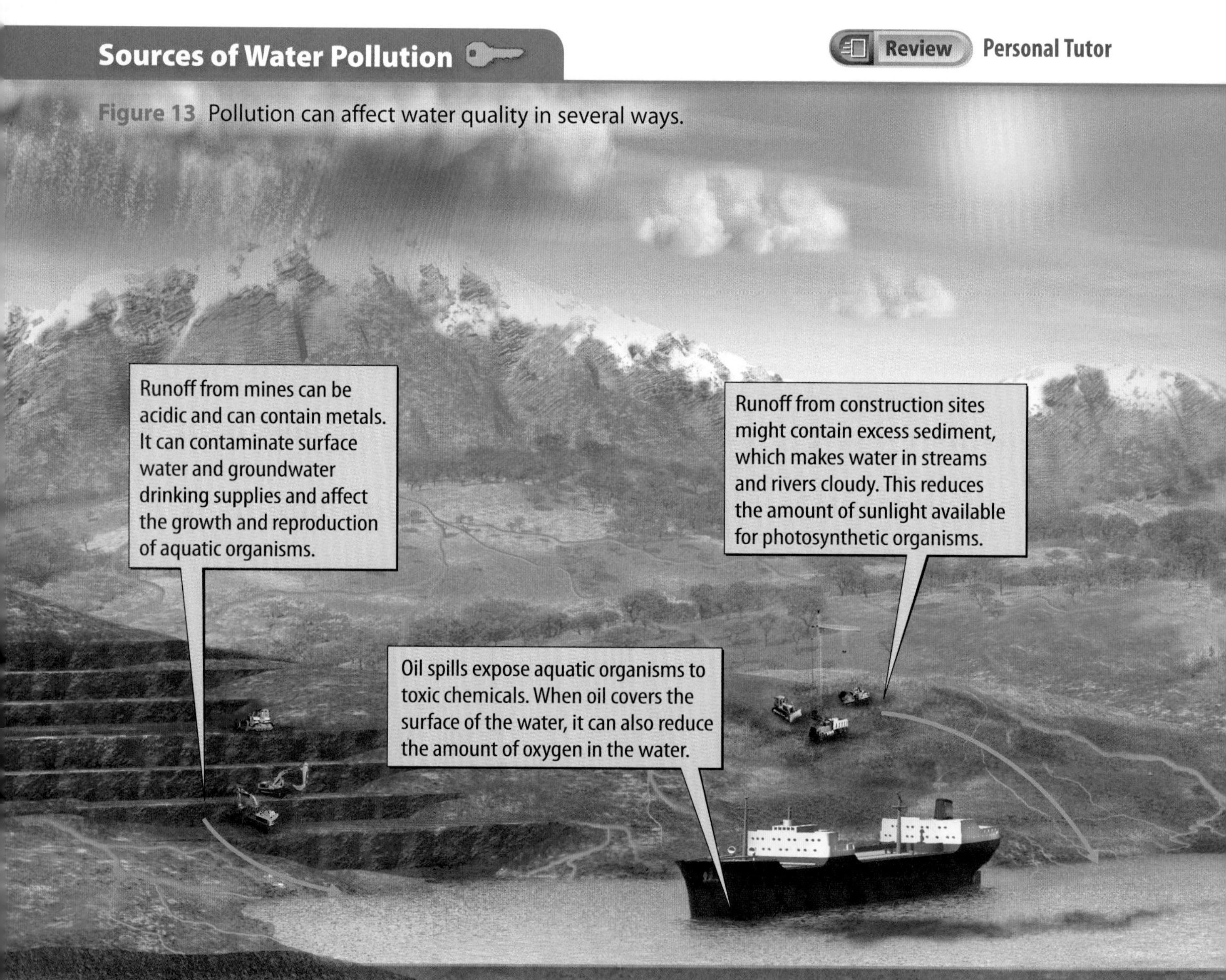

Figure 13 Pollution can affect water quality in several ways.

Nonpoint-Source Pollution

Pollution from several widespread sources that cannot be traced back to a single location is called **nonpoint-source pollution.** As precipitation runs over Earth's surface, the water picks up materials and substances from farms and urban developments, such as the ones shown in **Figure 13.** These different sources might be several kilometers apart. This makes it difficult to trace the pollution in the water back to one specific source. Runoff from farms and urban developments are examples of nonpoint-source pollution. Runoff from construction sites, which can contain excess amounts of sediment, is another example of nonpoint-source pollution.

Most of the water pollution in the United States comes from nonpoint sources. This kind of pollution is harder to pinpoint and therefore harder to control.

Key Concept Check How can pollution affect water quality?

FOLDABLES®

Make a vertical three-tab book. Draw a Venn diagram on the front. Cut the folds to form three tabs. Label as illustrated. Use the Foldable to compare and contrast sources of pollution.

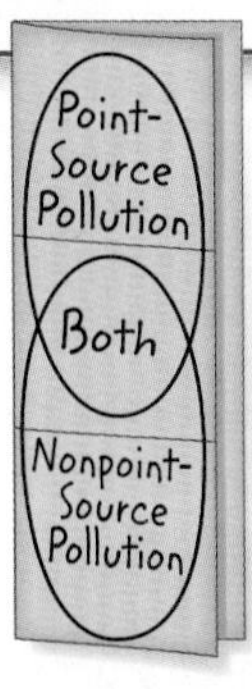

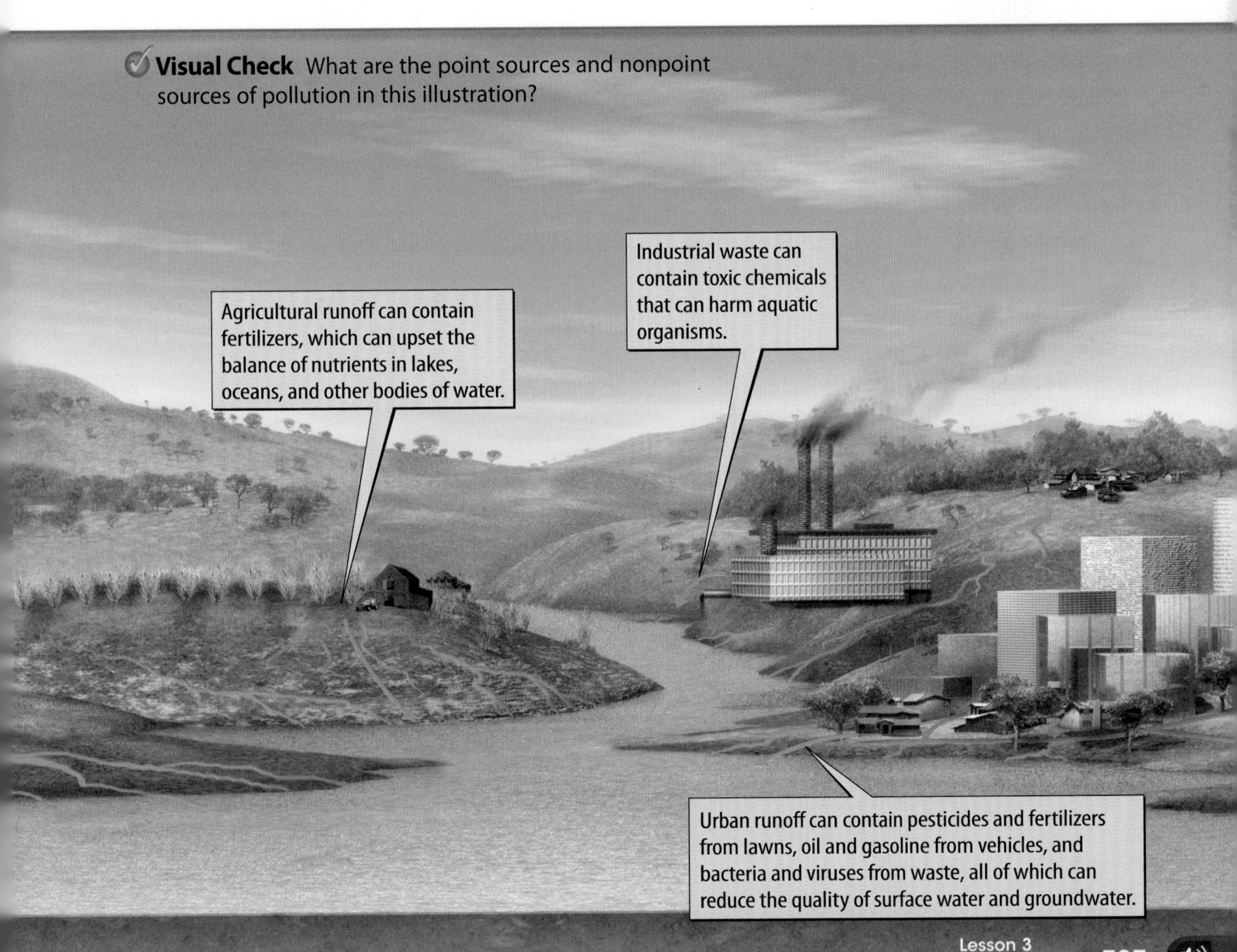

Figure 14 In 1969, burning litter and chemical pollution floating on the Cuyahoga River in northeastern Ohio inspired international efforts to clean up the Great Lakes.

Positive Actions

Once pollution enters water, it is difficult to remove. In fact, it can take decades to clean polluted groundwater! That is why most efforts to reduce water pollution focus on preventing it from entering the environment, rather than cleaning it up.

International Cooperation

In the 1960s, Lake Erie, one of the Great Lakes, was heavily polluted by runoff from fertilized fields and industrial wastes. Rivers that flowed into the lake were polluted, too. Litter soaked with chemicals floated on the surface of one of these rivers—the Cuyahoga River. As shown in **Figure 14,** the litter caught fire. The fire spurred Canada and the United States—the two countries that border the Great Lakes—into action.

The countries formed several agreements to clean up the Great Lakes. The goals of the countries are pollution prevention, as well as cleanup and research. Although, the Great Lakes still face challenges from aquatic species that are not native to the lakes and from the impact of excess sediments, pollution from toxic chemicals has decreased.

Reading Check Why is it important to focus on preventing water pollution before it happens?

Inquiry MiniLab

20 minutes

What's in well water?

The graph shows the level of nitrates in well water in Spanish Springs Valley, Nevada, over a 10-year period. Nitrate is a form of nitrogen that can contaminate groundwater when it leaches out of septic systems.

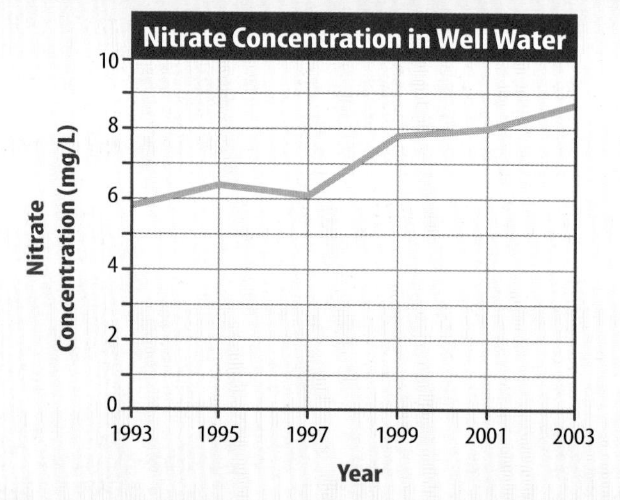

Analyze and Conclude

1. **Describe** what happened to the average level of nitrate in the well water of Spanish Springs Valley between 1993 and 2003.
2. **Analyze** Excess nitrate in drinking water can cause serious illness, especially in infants. The maximum allowable level in public drinking water is 10 mg/L. How close did the highest level of nitrate concentration come to the maximum level allowed?
3. **Key Concept** An article in the newspaper described a Spanish Springs Valley project to connect all houses to the sewer system. Predict how this will affect nitrate levels in the well water.

National Initiatives

In addition to working with other governments, the United States has laws to help maintain water quality within its borders. The Clean Water Act, for example, regulates sources of water pollution, including sewage systems. The Safe Drinking Water Act protects supplies of drinking water throughout the country.

How can you help?

Laws are effective ways to reduce water pollution. But simple actions taken by individuals can have positive impacts, too.

Reduce Use of Harmful Chemicals Many household products, such as paints and cleaners, contain harmful chemicals. People can use alternative products that do not contain toxins. For example, baking soda and white vinegar are safe, inexpensive cleaning products. In addition, people can reduce their use of artificial fertilizers on gardens and lawns. As you read earlier, compost can enrich soils without harming water quality.

Dispose of Waste Safely Sometimes using products that contain pollutants is necessary. Vehicles, for example, cannot run without motor oil. This motor oil has to be replaced regularly. People should never pour motor oil or other hazardous substances into drains, onto the ground, or directly into streams or lakes. These substances must be disposed of safely. Your local waste management agency has tips for safe disposal of hazardous waste.

Conserve Water Water pollution can be reduced simply by reducing water use. Easy ways to conserve water include taking shorter showers and turning off the water when you brush your teeth. **Figure 15** shows other ways to reduce water use.

Key Concept Check How can individuals help prevent water pollution?

Figure 15 People can help reduce water pollution by conserving water.

Visual Check How does sweeping a deck help reduce water pollution?

Keeping water in the refrigerator instead of running water from a faucet until the water is cold helps conserve water.

Sweeping leaves and branches from a deck instead of spraying them off using water from a hose helps conserve water.

Lesson 3 Review

Online Quiz

Visual Summary

Water is an important resource; all living things need water to survive. Water is used for agriculture, for electricity production, and in homes and businesses every day.

Water pollution can come from many sources, including chemicals from agriculture and industry and oil spills.

International cooperation and national laws help prevent water pollution. Individuals can help conserve water by reducing water use and disposing of wastes properly.

FOLDABLES®

Use your lesson Foldable to review the lesson. Save your Foldable for the project at the end of the chapter.

What do you think NOW?

You first read the statements below at the beginning of the chapter.

5. Sources of water pollution are always easy to identify.

6. The proper method of disposal for used motor oil is to pour it down the drain.

Did you change your mind about whether you agree or disagree with the statements? Rewrite any false statements to make them true.

Use Vocabulary

1. **Define** *nonpoint-source pollution* and *point-source pollution* in your own words.
2. **Use the term** *nonpoint-source pollution* in a sentence.

Understand Key Concepts 

3. Which uses the most water in the United States?
   A. factories
   B. farms
   C. households
   D. power plants
4. **Survey** three classmates to find out how they conserve water at home.
5. **Diagram** Make a diagram showing how runoff from lawns can impact water quality.

Interpret Graphics

6. **Sequence** Draw a graphic organizer such as the one below to illustrate the cleanup of Lake Erie, beginning with the pollution of the lake.

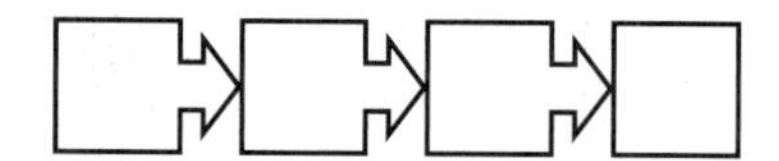

7. **Classify** the pollution source shown below as point-source or nonpoint-source. Explain your reasoning.

Critical Thinking

8. **Visualize** a map of a river that flows through several countries. Explain why international cooperation is needed to reduce water pollution.
9. **Identify** a human activity that impacts water quality negatively. Then describe a positive action that can help reduce the pollution caused by the activity.

Dead Zones

GREEN SCIENCE

What causes lifeless areas in the ocean?

For thousands of years, people have lived on coasts, making a living by shipping goods or by fishing. Today, fisheries in the Gulf of Mexico provide jobs for thousands of people and food for millions more. Although humans and other organisms depend on the ocean, human activities can harm marine ecosystems. Scientists have been tracking dead zones in the ocean for several decades. They believe that these zones are a result of human activities on land.

A large dead zone in the Gulf of Mexico forms every year when runoff from spring and summer rain in the Midwest drains into the Mississippi River. The runoff contains nitrogen and phosphorous from fertilizer, animal waste, and sewage from farms and cities. This nutrient-rich water flows into the gulf. Algae feed on excess nutrients and multiply rapidly, creating an algal bloom. The results of the algal bloom are shown below.

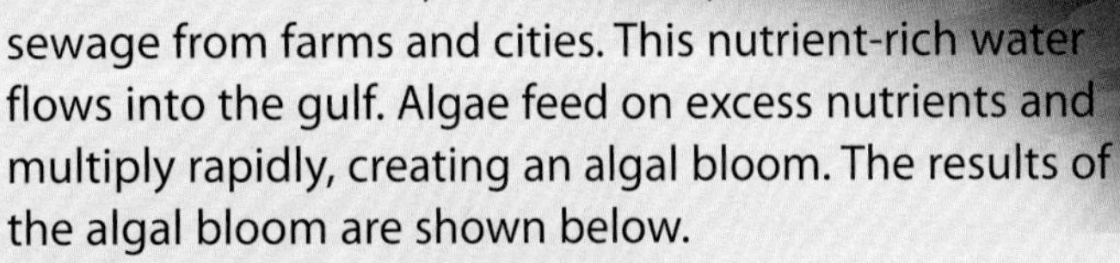

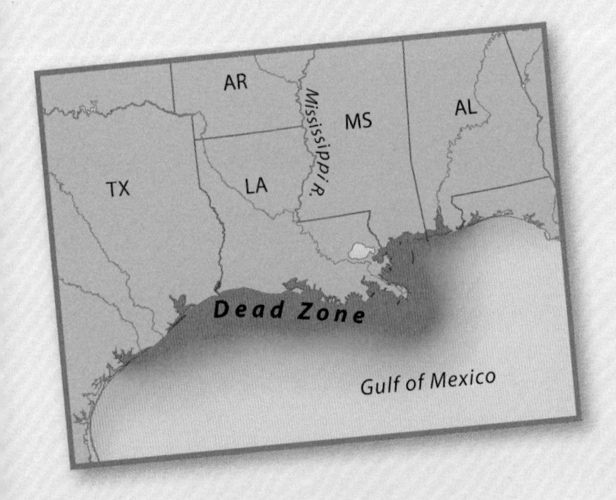

Some simple changes in human activity can help prevent dead zones. People upstream from the Gulf can decrease the use of fertilizer and apply it at times when it is less likely to be carried away by runoff. Picking up or containing animal waste can help, too. Also, people can modernize and improve septic and sewage systems. How do we know these steps would work? Using them has already restored life to dead zones in the Great Lakes!

Freshwater runoff

1 River water containing nitrogen and phosphorous flows into the Gulf of Mexico.

Dead algae
Algal bloom

2 After the algal bloom, dead algae sink to the ocean floor.

Dead fish

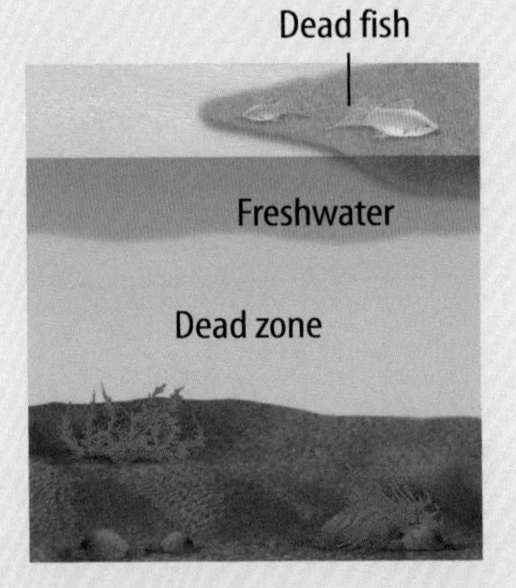

3 Decomposing algae deplete the water's oxygen, killing other organisms.

It's Your Turn

RESEARCH AND REPORT Earth's oceans contain about 150 dead zones. Choose three. Plot them on a map and write a report about what causes each dead zone.

Lesson 4

Impacts on the Atmosphere

Reading Guide

Key Concepts
ESSENTIAL QUESTIONS

- What are some types of air pollution?
- How are global warming and the carbon cycle related?
- How does air pollution affect human health?
- What actions help prevent air pollution?

Vocabulary

photochemical smog p. 743
acid precipitation p. 744
particulate matter p. 744
global warming p. 745
greenhouse effect p. 746
Air Quality Index p. 747

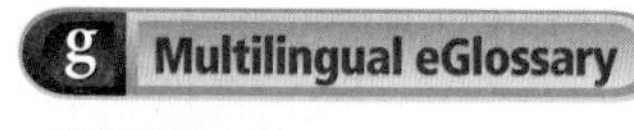

Video BrainPOP®

Inquiry Why wear a mask?

In some areas of the world, people wear masks to help protect themselves against high levels of air pollution. Where does this pollution come from? How do you think air pollution affects human health and the environment?

20 minutes

Where's the air?

In 1986, an explosion at a nuclear power plant in Chernobyl, Russia, sent radioactive pollution 6 km into the atmosphere. Within three weeks, the radioactive cloud had reached Italy, Finland, Iceland, and North America.

1. Read and complete a lab safety form.
2. With your group, move to your assigned area of the room.
3. Lay out **sheets of paper** to cover the table.
4. When the **fan** starts blowing, observe whether water droplets appear on the paper. Record your observations in your Science Journal.
5. Lay out another set of paper sheets and record your observations when the fan blows in a different direction.

Think About This

1. Did the water droplets reach your location? Why or why not?
2. How is the movement of air and particles by the fan similar to the movement of the pollution from Chernobyl? How does the movement differ?
3. **Key Concept** How do you think the health of a person in Iceland could be affected by the explosion in Chernobyl?

Importance of Clean Air

Your body, and the bodies of other animals, uses oxygen in air to produce some of the energy it needs. Many organisms can survive for only a few minutes without air. But the air you breathe must be clean or it can harm your body.

Types of Air Pollution

Human activities can produce pollution that enters the air and affects air quality. Types of air pollution include smog, acid precipitation, particulate matter, chlorofluorocarbons (CFCs), and carbon monoxide.

Smog

The brownish haze in the sky in **Figure 16** is photochemical smog. **Photochemical smog** *is caused when nitrogen and carbon compounds in the air react in sunlight.* Nitrogen and carbon compounds are released when fossil fuels are burned to provide energy for vehicles and power plants. These compounds react in sunlight and form other substances. One of these substances is ozone. Ozone high in the atmosphere helps protect living things from the Sun's ultraviolet radiation. However, ozone close to Earth's surface is a major component of smog.

Figure 16 Burning fossil fuels releases compounds that can react in sunlight and form smog.

Acid Precipitation

Another form of pollution that occurs as a result of burning fossil fuels is acid precipitation. **Acid precipitation** *is rain or snow that has a lower pH than that of normal rainwater.* The pH of normal rainwater is about 5.6. Acid precipitation forms when gases containing nitrogen and sulfur react with water, oxygen, and other chemicals in the atmosphere. Acid precipitation falls into lakes and ponds or onto the ground. It makes the water and the soil more acidic. Many living things cannot survive if the pH of water or soil becomes too low. The trees shown in **Figure 17** have been affected by acid precipitation.

Figure 17 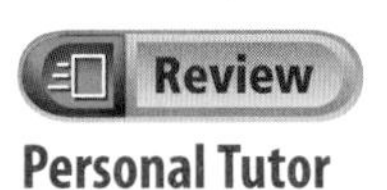 Acid precipitation can make the soil acidic and kill trees and other plant life.

Personal Tutor

Particulate Matter

The mix of both solid and liquid particles in the air is called **particulate matter.** Solid particles include smoke, dust, and dirt. These particles enter the air from natural processes, such as volcanic eruptions and forest fires. Human activities, such as burning fossil fuels at power plants and in vehicles, also release particulate matter. Inhaling particulate matter can cause coughing, difficulty breathing, and other respiratory problems.

WORD ORIGIN

particulate
from Latin *particula,* means "small part"

FOLDABLES®

Make a two-tab book. Label the tabs as illustrated. Use your Foldable to record factors that increase or decrease air pollution.

Factors That Increase Air Pollution | Factors That Decrease Air Pollution

CFCs

Ozone in the upper atmosphere absorbs harmful ultraviolet (UV) rays from the Sun. Using products that contain CFCs, such as air conditioners and refrigerators made before 1996, affects the ozone layer. CFCs react with sunlight and destroy ozone molecules. As a result, the ozone layer thins and more UV rays reach Earth's surface. Increased skin cancer rates have been linked with an increase in UV rays.

Carbon Monoxide

Carbon monoxide is a gas released from vehicles and industrial processes. Forest fires also release carbon monoxide into the air. Wood-burning and gas stoves are sources of carbon monoxide indoors. Breathing carbon monoxide reduces the amount of oxygen that reaches the body's tissues and organs.

 **Key Concept Check** What are some types of air pollution?

The Carbon Cycle

Concepts in Motion Animation

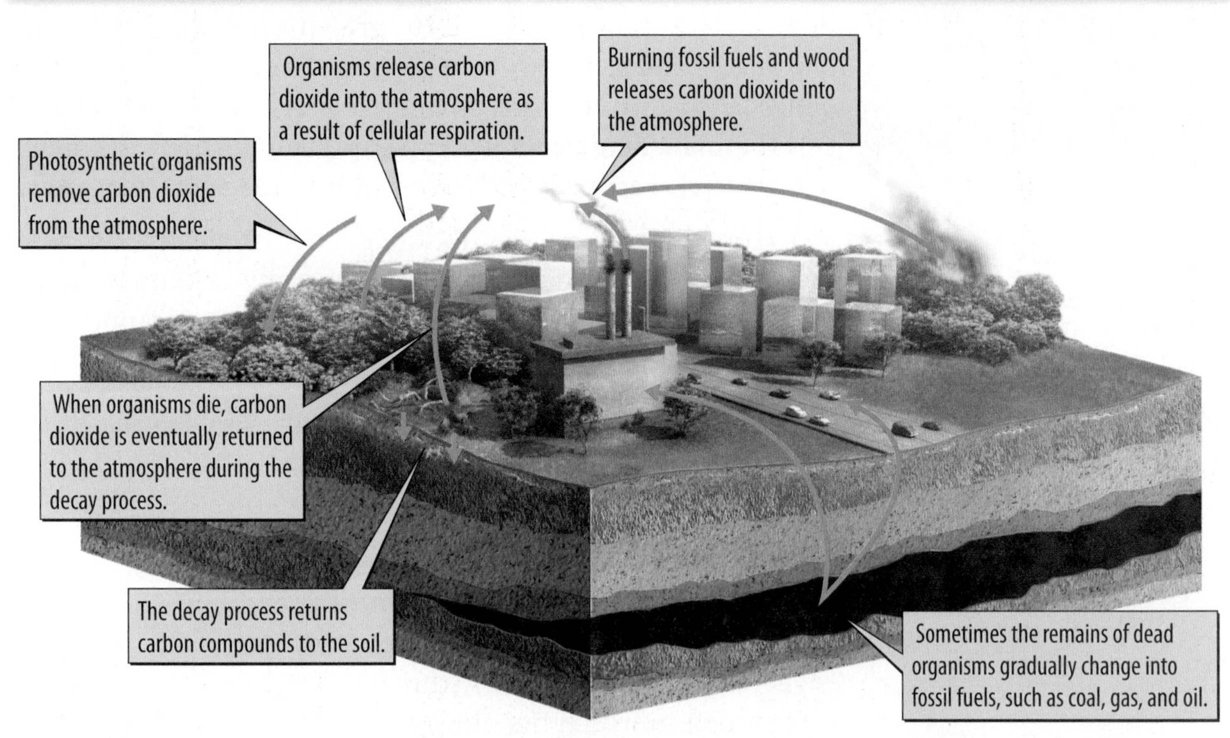

Figure 18 Some human activities can increase the amount of carbon dioxide in the atmosphere.

Visual Check Which processes add carbon to the atmosphere?

Global Warming and the Carbon Cycle

Air pollution affects natural cycles on Earth. For example, burning fossil fuels for electricity, heating, and transportation releases substances that cause acid precipitation. Burning fossil fuels also releases carbon dioxide into the atmosphere, as shown in **Figure 18.** An increased concentration of carbon dioxide in the atmosphere can lead to **global warming,** *an increase in Earth's average surface temperature.* Earth's temperature has increased about 0.7°C over the past 100 years. Scientists estimate it will rise an additional 1.8 to 4.0°C over the next 100 years. Even a small increase in Earth's average surface temperature can cause widespread problems.

Effects of Global Warming

Warmer temperatures can cause ice to melt, making sea levels rise. Higher sea levels can cause flooding along coastal areas. In addition, warmer ocean waters might lead to an increase in the intensity and frequency of storms.

Global warming also can affect the kinds of living things found in ecosystems. Some hardwood trees, for example, do not thrive in warm environments. These trees will no longer be found in some areas if temperatures continue to rise.

 **Key Concept Check** How are global warming and the carbon cycle related?

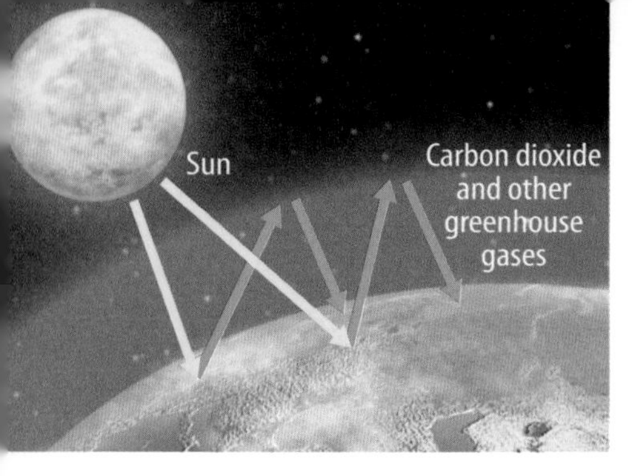

Figure 19 Greenhouse gases absorb and reradiate thermal energy from the Sun and warm Earth's surface.

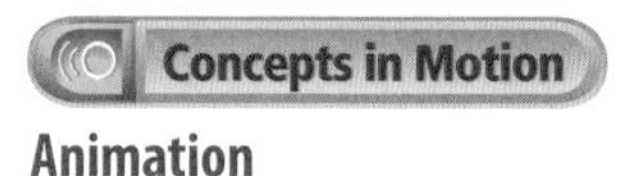

Animation

The Greenhouse Effect

Why does too much carbon dioxide in the atmosphere increase Earth's temperature? *The* **greenhouse effect** *is the natural process that occurs when certain gases in the atmosphere absorb and reradiate thermal energy from the Sun.* As shown in **Figure 19,** this thermal energy warms Earth's surface. Without the greenhouse effect, Earth would be too cold for life as it exists now.

Carbon dioxide is a greenhouse gas. Other greenhouse gases include methane and water vapor. When the amount of greenhouse gases increases, more thermal energy is trapped and Earth's surface temperature rises. Global warming occurs.

Reading Check How are the greenhouse effect and global warming related?

Health Disorders

Air pollution affects the environment and human health as well. Air pollution can cause respiratory problems, including triggering asthma attacks. Asthma is a disorder of the respiratory system in which breathing passageways narrow during an attack, making it hard for a person to breathe. **Figure 20** shows some health disorders caused by pollutants in the air.

Key Concept Check How can air pollution affect human health?

Figure 20 Air pollution can harm the environment and your health.

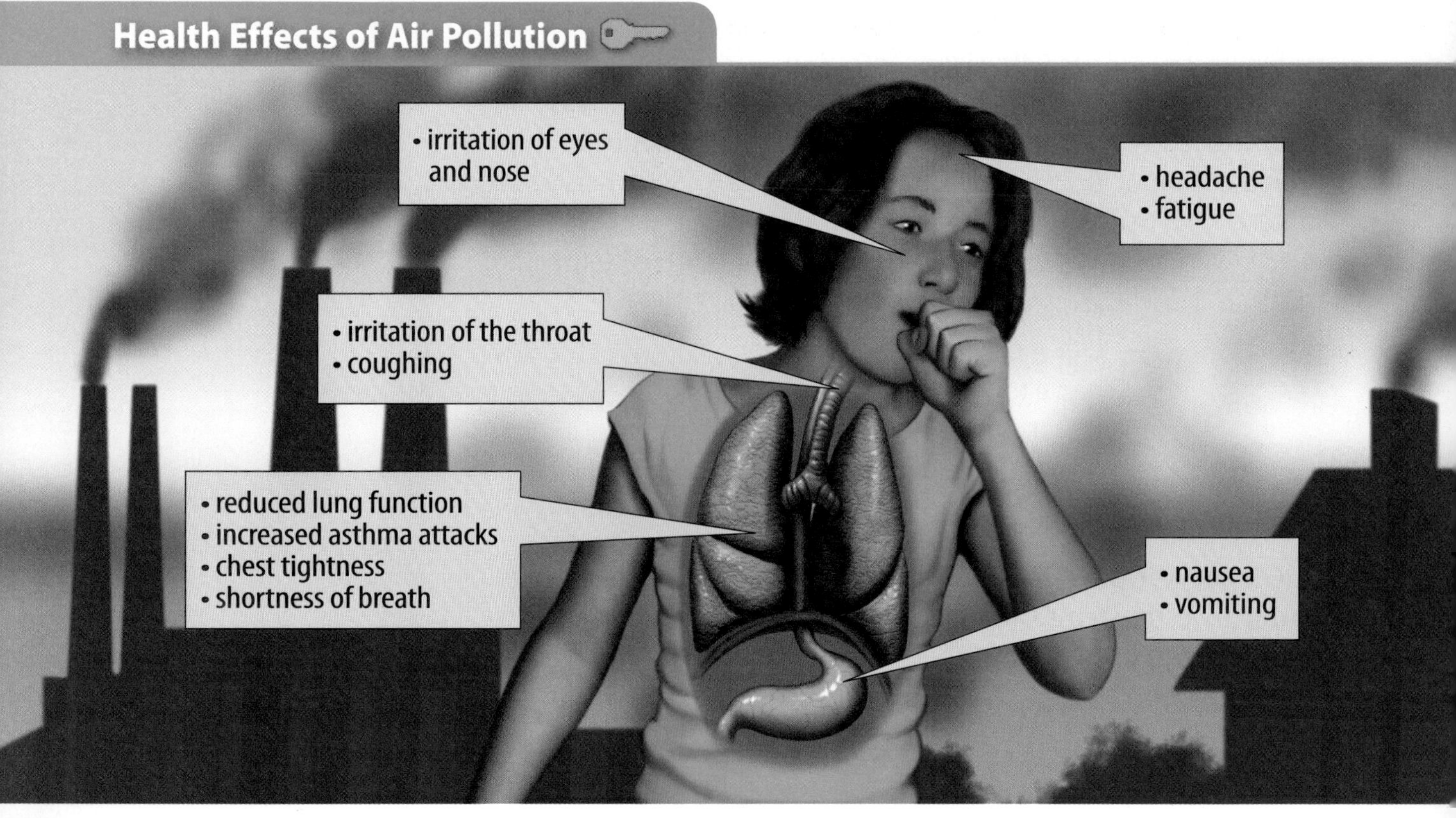

Table 1 Air Quality Index

Ozone Concentration (parts per million)	Air Quality Index Values	Air Quality Description	Preventative Actions
0.0 to 0.064	0 to 50	good	No preventative actions needed.
0.065 to 0.084	51 to 100	moderate	Highly sensitive people should limit prolonged outdoor activity.
0.085 to 0.104	101 to 150	unhealthy for sensitive groups	Sensitive people should limit prolonged outdoor activity.
0.105 to 0.124	151 to 200	unhealthy	All groups should limit prolonged outdoor activity.
0.125 to 0.404	201 to 300	very unhealthy	Sensitive people should avoid outdoor activity. All other groups should limit outdoor activity.

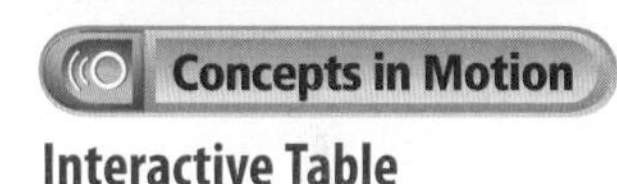

Interactive Table

Measuring Air Quality

Some pollutants, such as smoke from forest fires, are easily seen. Other pollutants, such as carbon monoxide, are invisible. How can people know when levels of air pollution are high?

The EPA works with state and local agencies to measure and report air quality. *The* **Air Quality Index** (AQI) *is a scale that ranks levels of ozone and other air pollutants.* Study the AQI for ozone in **Table 1.** It uses color codes to rank ozone levels on a scale of 0 to 300. Although ozone in the upper atmosphere blocks harmful rays from the Sun, ozone that is close to Earth's surface can cause health problems, including throat irritation, coughing, and chest pain. The EPA cautions that no one should do physical activities outside when AQI values reach 300.

Inquiry MiniLab

10 minutes

What's in the air?

Suppose your friend suffers from asthma. People with respiratory problems such as asthma are usually more sensitive to air pollution. *Sensitive* is a term used on the AQI. Use the AQI in **Table 1** to answer the questions below.

Analyze and Conclude

1. **Identify** Today's AQI value is 130. What is the concentration of ozone in the air?
2. **Decide** Is today a good day for you and your friend to go to the park to play basketball for a few hours? Why or why not?
3. **Key Concept** Predict how you and your friend might be affected by the air if you played basketball today.

Hybrid car

Solar car

Figure 21 Energy-efficient and renewable-energy vehicles help reduce air pollution.

Visual Check How does driving a solar car help reduce air pollution?

Positive Actions

Countries around the world are working together to reduce air pollution. For example, 190 countries, including the United States, have signed the Montreal Protocol to phase out the use of CFCs. Levels of CFCs have since decreased. The Kyoto Protocol aims to reduce emissions of greenhouse gases. Currently, 184 countries have accepted the agreement.

National Initiatives

In the United States, the Clean Air Act sets limits on the amount of certain pollutants that can be released into the air. Since the law was passed in 1970, amounts of carbon monoxide, ozone near Earth's surface, and acid precipitation-producing substances have decreased by more than 50 percent. Toxins from industrial factories have gone down by 90 percent.

Cleaner Energy

Using renewable energy resources such as solar power, wind power, and geothermal energy to heat homes helps reduce air pollution. Recall that renewable resources are resources that can be replaced by natural processes in a relatively short amount of time. People also can invest in more energy-efficient appliances and vehicles. The hybrid car shown in **Figure 21** uses both a battery and fossil fuels for power. It is more energy efficient and emits less pollution than vehicles that are powered by fossil fuels alone. The solar car shown in **Figure 21** uses only the Sun's energy for power.

How can you help?

Reducing energy use means that fewer pollutants are released into the air. You can turn the thermostat down in the winter and up in the summer to save energy. You can walk to the store or use public transportation. Each small step you take to conserve energy helps improve air, water, and soil quality.

Key Concept Check How can people help prevent air pollution?

Lesson 4 Review

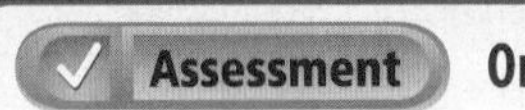

Visual Summary

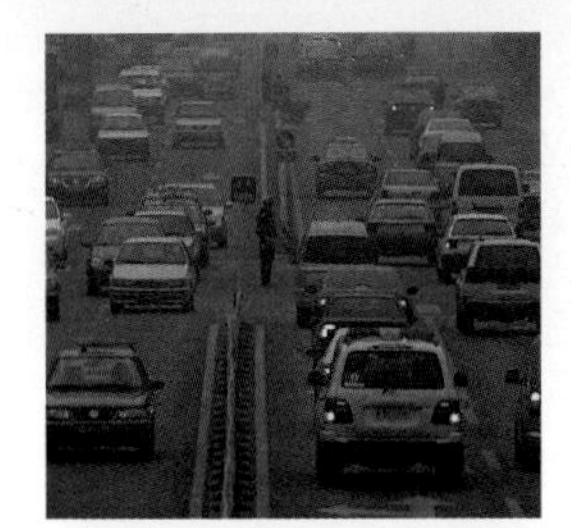

Burning fossil fuels releases nitrogen and carbon compounds and particulate matter into the air.

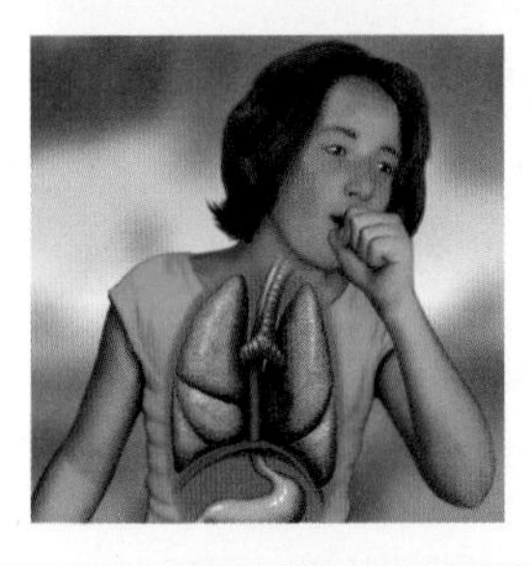

Air pollution can affect human health, causing eye, nose, and throat irritation, increased asthma, and headaches.

Certain laws and international agreements require people to reduce air pollution. Individuals can reduce air pollution by using alternative forms of energy to heat homes and power vehicles.

Use your lesson Foldable to review the lesson. Save your Foldable for the project at the end of the chapter.

What do you think NOW?

You first read the statements below at the beginning of the chapter.

7. The greenhouse effect is harmful to life on Earth.

8. Air pollution can affect human health.

Did you change your mind about whether you agree or disagree with the statements? Rewrite any false statements to make them true.

Use Vocabulary

1. **Use the term** *air quality index* in a sentence.
2. The natural heating of Earth's surface that occurs when certain gases absorb and reradiate thermal energy from the Sun is ________.
3. **Define** *global warming* in your own words.

Understand Key Concepts

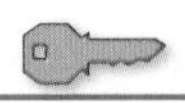

4. Which is NOT a possible heath effect of exposure to air pollution?
   A. chest tightness
   B. eye irritation
   C. increased lung function
   D. shortness of breath
5. **Relate** What happens in the carbon cycle when fossil fuels are burned for energy?
6. **Compare** the goals of the Montreal Protocol and the Kyoto Protocol.

Interpret Graphics

7. **Sequence** Copy and fill in the graphic organizer below to identify types of air pollution.

8. **Describe** air quality when the ozone concentration is 0.112 ppm using the table below.

Ozone Concentration (ppm)	Air Quality Index Values	Air Quality Description
0.105 to 0.124	151 to 200	unhealthy
0.125 to 0.404	201 to 300	very unhealthy

Critical Thinking

9. **Predict** Some carbon is stored in frozen soils in the Arctic. What might happen to Earth's climate if these soils thawed?

Inquiry Lab

3 class periods

Materials

office supplies

magazines

computer with Internet access

Safety

Design a Green City

City planners have asked the architectural firms in town to design an eco-friendly city that will be based on an environmentally responsible use of land, water, and energy. The city should include homes, businesses, schools, green spaces, industry, waste management, and transportation options.

Question

What are the most environmentally friendly materials and practices to use when designing a green city?

Procedure

1. Make a list of the things you will include in your city.
2. Research environmentally responsible structures and practices for the elements of your city. Your research may include using the library or talking with owners, employees, or patrons of businesses to identify existing environmental problems. Use the questions below to help guide your research.

- What materials can you use for building the structures?
- What building practices and designs can you use to minimize environmental impact?
- How will you address energy use by homeowners, businesses, and industry?
- How will you address water use for homes, businesses, and industry?
- How will you address energy use and pollution issues related to public transportation?
- What is the most environmentally friendly system of waste management?

3. Analyze the information you gathered in step 2. Discuss how you will use what you have learned as you design your city.
4. Design your city. Use the colored pencils and markers to draw a map of the city, including all of the elements of the city. Add captions, other graphics, and/or a key to explain any elements or actions and their intended results.
5. Copy and complete the *Required Elements and Actions* chart on the following page. For each element in your city, explain the environmental issue associated with the element and what action you took in your city to address the issue. Does your design include an action for each element?

Required Elements and Actions		
Element	**Environmental Issue**	**Action Taken**
Waste management	All waste goes into landfills.	designed a curbside recycling program

6. If needed, modify your design to include any other actions you need to take.

Analyze and Conclude

7. Describe one identified environmental issue, the action taken, and the intended outcome in your design plans.
8. Compare your design to the designs of other groups. What did they do differently?
9. **The Big Idea** Predict whether there will be any changes in the quality of your city's water resources after years of use of your design. Explain your answer.

Communicate Your Results

Suppose your classmates are members of the city planning board. Present your design to the board. Explain the structures and practices that are intended to make the city environmentally responsible.

Make a 3-D model of your city. Try to use recycled or environmentally friendly materials to represent the structures in your city.

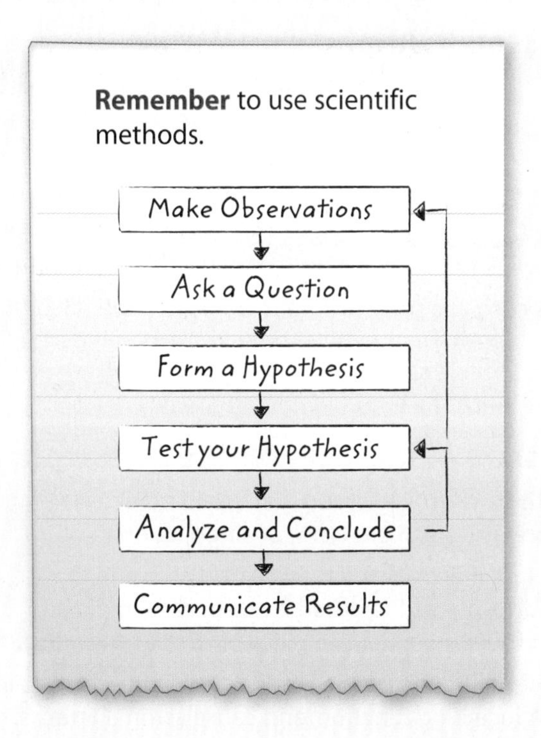

Chapter 20 Study Guide

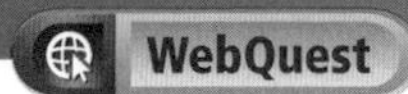

Human activities can impact the environment negatively, including deforestation, water pollution, and global warming, and positively, such as through reforestation, reclamation, and water conservation.

Key Concepts Summary

Lesson 1: People and the Environment

- Earth has limited resources and cannot support unlimited human **population** growth.
- Daily actions can deplete resources and pollute soil, water, and air.

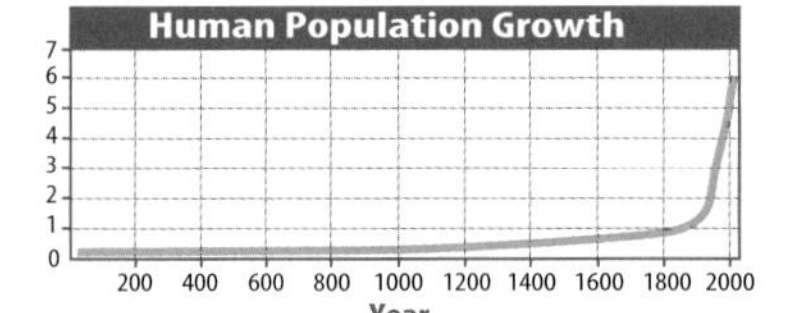

Vocabulary

population p. 719
carrying capacity p. 720

Lesson 2: Impacts on the Land

- **Deforestation, desertification,** habitat destruction, and increased rates of extinction are associated with using land as a resource.
- Landfills are constructed to prevent contamination of soil and water by pollutants from waste. Hazardous waste must be disposed of in a safe manner.
- Positive impacts on land include preservation, **reforestation,** and **reclamation.**

Vocabulary

deforestation p. 725
desertification p. 726
urban sprawl p. 728
reforestation p. 730
reclamation p. 730

Lesson 3: Impacts on Water

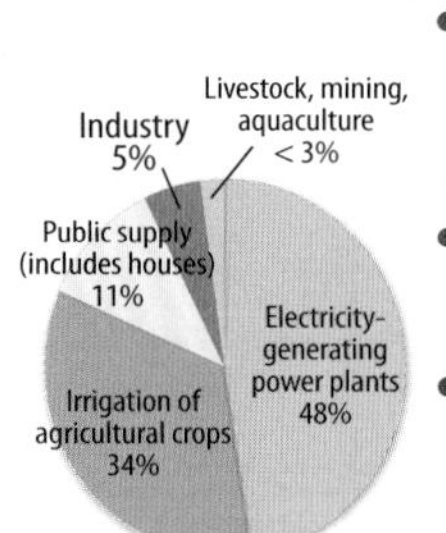

- Humans use water in electricity production, industry, and agriculture, as well as for recreation and transportation.
- **Point-source pollution** and **nonpoint-source pollution** can reduce water quality.
- International agreements and national laws help prevent water pollution. Other positive actions include disposing of waste safely and conserving water.

Vocabulary

point-source pollution p. 736
nonpoint-source pollution p. 737

Lesson 4: Impacts on the Atmosphere

- **Photochemical smog,** CFCs, and **acid precipitation** are types of air pollution.
- Human activities can add carbon dioxide to the atmosphere. Increased levels of carbon dioxide in the atmosphere can lead to **global warming.**
- Air pollutants such as ozone can irritate the respiratory system, reduce lung function, and cause asthma attacks.
- International agreements, laws, and individual actions such as conserving energy help decrease air pollution.

Vocabulary

photochemical smog p. 743
acid precipitation p. 744
particulate matter p. 744
global warming p. 745
greenhouse effect p. 746
Air Quality Index p. 747

- Personal Tutor
- Vocabulary eGames
- Vocabulary eFlashcards

FOLDABLES® Chapter Project

Assemble your lesson Foldables as shown to make a Chapter Project. Use the project to review what you have learned in this chapter.

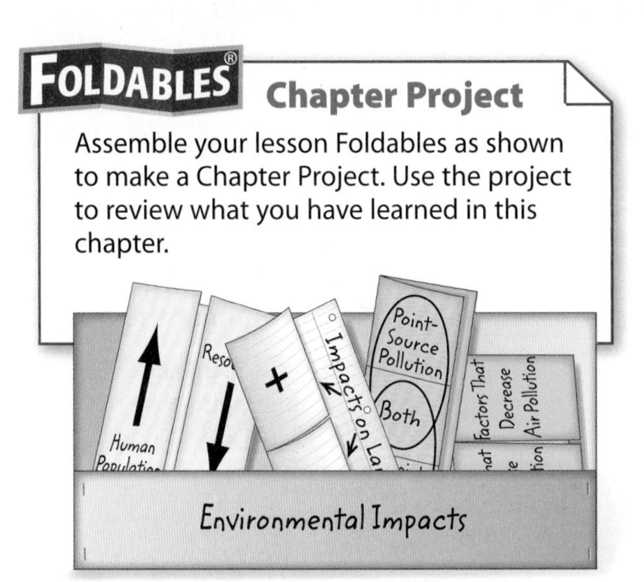

Use Vocabulary

1. Use the term *carrying capacity* in a sentence.
2. Distinguish between desertification and deforestation.
3. Planting trees to replace logged trees is called ________.
4. Distinguish between point-source and nonpoint-source pollution.
5. Define the greenhouse effect in your own words.
6. Solid and liquid particles in the air are called ________.

Link Vocabulary and Key Concepts

Copy this concept map, and then use vocabulary terms from the previous page to complete the concept map.

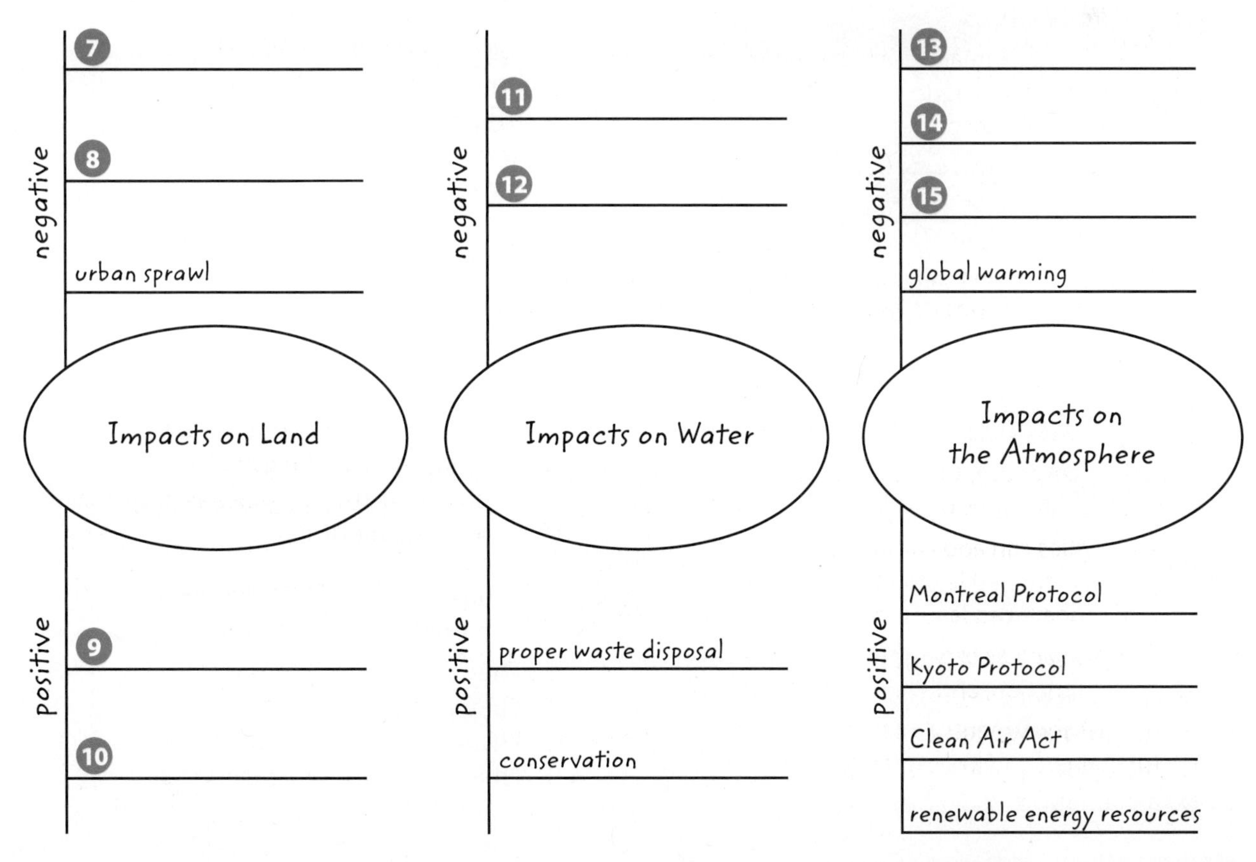

Chapter 20 Review

Understand Key Concepts 

1. Which is a population?
   A. all the animals in a zoo
   B. all the living things in a forest
   C. all the people in a park
   D. all the plants in a meadow

2. Which caused the greatest increase of the growth of the human population?
   A. higher death rates
   B. increased marriage rates
   C. medical advances
   D. widespread disease

3. What percentage of species on Earth live in tropical rain forests?
   A. 10 percent
   B. 25 percent
   C. 50 percent
   D. 75 percent

4. What process is illustrated in the diagram below?

   A. desertification
   B. recycling
   C. reforestation
   D. waste management

5. Which could harm human health?
   A. compost
   B. hazardous waste
   C. nitrogen
   D. reclamation

6. Which source of pollution would be hardest to trace and control?
   A. runoff from a city
   B. runoff from a mine
   C. an oil leak from an ocean tanker
   D. water from a factory discharge pipe

7. According to the diagram below, which is the correct ranking of water use in the United States, in order from most to least?

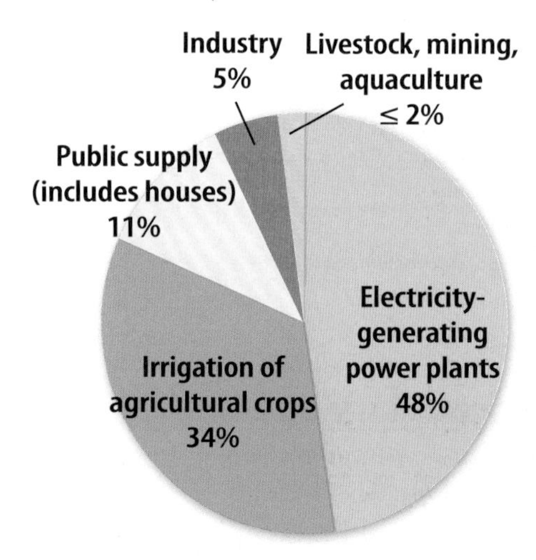

   A. industrial, public supply, irrigation, power plants
   B. irrigation, industrial, public supply, power plants
   C. power plants, irrigation, public supply, industrial
   D. public supply, power plants, industrial, irrigation

8. What is the main purpose of the Safe Drinking Water Act?
   A. to ban point-source pollution
   B. to clean up the Great Lakes
   C. to protect drinking-water supplies
   D. to regulate landfills

9. Why has the use of CFCs been phased out?
   A. They cause acid rain.
   B. They produce smog.
   C. They destroy ozone molecules.
   D. They impact the nitrogen cycle.

Assessment
Online Test Practice

Critical Thinking

10 **Decide** Rates of human population growth are higher in developing countries than in developed countries. Yet, people in developed countries use more resources than those in developing countries. Should international efforts focus on reducing population growth or reducing resource use? Explain.

11 **Relate** How does the carrying capacity for a species help regulate its population growth?

12 **Assess** your personal impact on the environment today. Include both positive and negative impacts on soil, water, and air.

13 **Infer** How does deforestation affect levels of carbon in the atmosphere?

14 **Role-Play** Suppose you are a soil expert advising a farmer on the use of fertilizers. What would you tell the farmer about the environmental impact of the fertilizers?

15 **Create** Use the data below to create a circle graph showing waste disposal methods in the United States.

Waste Disposal Methods—United States	
Method	**Percent of Waste Disposed**
Landfill	55%
Recycling/composting	31%
Incineration	14%

Writing in Science

16 **Compose** a letter to a younger student to help him or her understand air pollution. The letter should identify the different kinds of pollution and explain their causes.

REVIEW THE BIG IDEA

17 How do human activities impact the environment? Give one example each of how human activities impact land, water, and air resources.

18 What positive actions can people take to reduce or reverse negative impacts on the environment?

Math Skills

Review
Math Practice

Use Percentages

19 Between 1960 and 1990, the number of people per square mile in the United States grew from 50.7 people to 70.3 people. What was the percent change?

20 Between 1950 and 1998, the rural population in the United States decreased from 66.2 million to 53.8 million people. What was the percent change in rural population?

21 During the twentieth century, the population of the western states increased from 4.3 million people to 61.2 million people. What was the percent change during the century?

Standardized Test Practice

Record your answers on the answer sheet provided by your teacher or on a sheet of paper.

Multiple Choice

1 Which action can help restore land that has been disturbed by mining?

A deforestation

B desertification

C preservation

D reclamation

2 Which is a consequence of deforestation?

A Animal habitats are destroyed.

B Carbon in the atmosphere is reduced.

C Soil erosion is prevented.

D The rate of extinction is slowed.

Use the graph below to answer question 3.

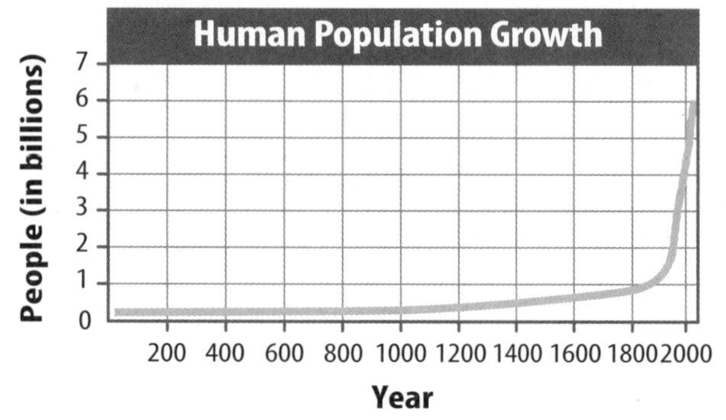

3 During which time span did the human population increase most?

A 1400–1600

B 1600–1800

C 1800–1900

D 1900–2000

4 Which accounts for the least water use in the United States?

A electricity-generating power plants

B irrigation of agricultural crops

C mines, livestock, and aquaculture

D public supply, including houses

5 Which is a point source of water pollution?

A discharge pipes

B runoff from farms

C runoff from construction sites

D runoff from urban areas

6 Which air pollutant contains ozone?

A acid precipitation

B carbon monoxide

C CFCs

D smog

Use the figure below to answer question 7.

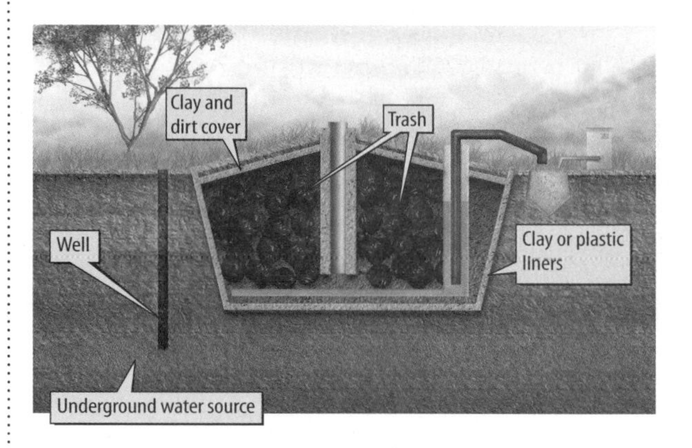

7 What is the function of the well in the figure above?

A to generate electricity

B to monitor quality of groundwater

C to prevent pollution of nearby land

D to treat hazardous water

8 Which action helps prevent water pollution?

A pouring motor oil on the ground

B putting hazardous wastes in the trash

C using fertilizers when gardening

D using vinegar when cleaning

9 What effect does ozone near Earth's surface have on the human body?

A It increases lung function.

B It increases throat irritation.

C It reduces breathing problems.

D It reduces skin cancer.

Use the figure below to answer question 10.

10 Which term describes what is shown in the figure above?

A acid precipitation

B global warming

C greenhouse effect

D urban sprawl

11 Which results in habitat destruction?

A reclamation

B reforestation

C urban sprawl

D water conservation

Constructed Response

Use the figure below to answer questions 12 and 13.

12 Which events shown in the figure remove carbon dioxide from the atmosphere?

13 Relate the carbon cycle shown in the figure to global warming and the greenhouse effect.

14 List two actions that help prevent air pollution. Then explain the pros and the cons of taking each action.

15 Explain how taking a hot shower can impact the environment.

16 Create an advertisement for a solar car. Include information about the environmental impacts of the car in your ad.

NEED EXTRA HELP?																
If You Missed Question...	1	2	3	4	5	6	7	8	9	10	11	12	13	14	15	16
Go to Lesson...	2	2	1	3	3	4	2	3	4	2	2	4	4	4	1	4

Unit 6

Heredity and Human Body Systems

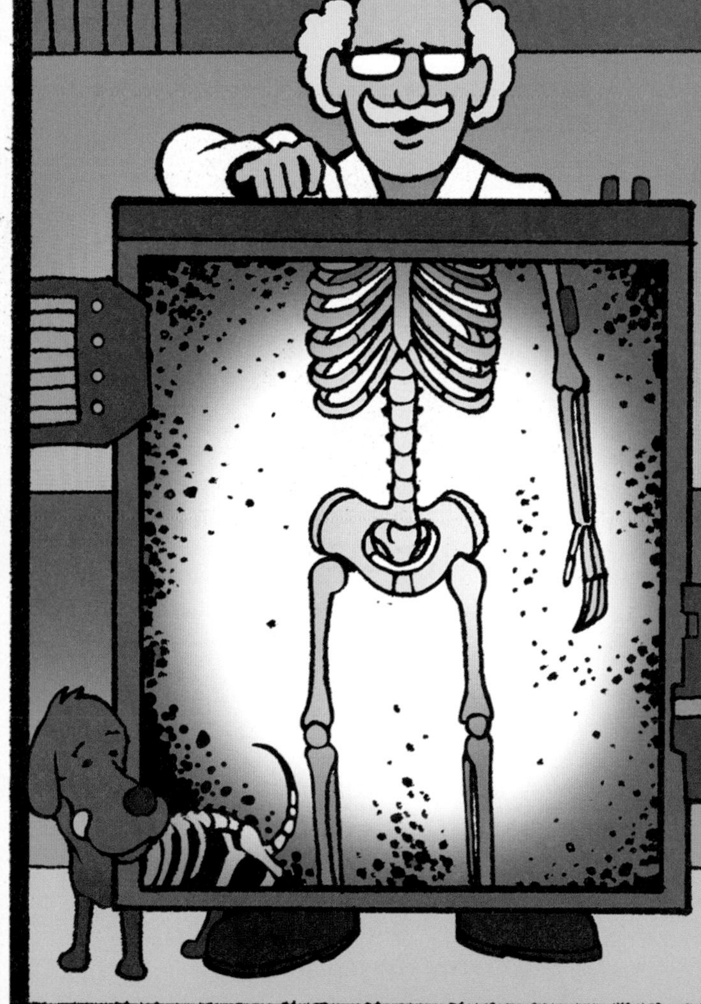

1800 1850 1900 1950

1818
The first well-documented case of a person-to-person blood transfusion is performed. British obstetrician James Blundell transfuses 4 oz. of blood from a man to his wife who had just given birth.

1823
German surgeon Christian Bünger performs the first autograft, replacing skin on a man's nose with some from his thigh.

1905
Dr. Eduard Zirm performs the first successful cornea transplant on patient Alois Glogar.

1954
The first successful organ transplant between living relatives (a kidney transplant between twins) is completed by Dr. Joseph Murray and Dr. David Hume in Boston.

1967
The first successful liver and heart transplants are completed. Dr. Thomas Starzl performs the liver transplant in Denver, Colorado. Dr. Christiaan Barnard performs the heart transplant in Cape Town, South Africa.

Turn it yet again, and the circulatory systems appears.

Amazing, no?

1960 1980 2000

1968
The Uniform Anatomical Gift Act establishes the Uniform Donor Card as a legal document that allows anyone 18 years of age or older to legally donate their organs upon death.

1984
The National Organ Transplant Act (NOTA) establishes a nationwide computer registry, authorizes financial support for organ procurement organizations, and prohibits the buying or selling of organs in the United States.

2001
Due to widespread use of advanced surgical techniques and higher success rates for surgeries, the number of living donors passes the number of deceased donors.

? Inquiry
Visit ConnectED for this unit's **STEM** activity.

Technology

Some people wonder why governments have invested so much money to explore space. Why not solve problems here on Earth using that money? What did all of that money buy?

Technology is the practical application of science to commerce or industry. Science and technology depend on each other. Once scientists understand a scientific concept, they apply the science to new technologies. Today, many technologies originally developed for space are solving problems for people worldwide. For example, sensors designed to remotely measure the temperature of distant stars led to the development of the thermometer shown in **Figure 1**. When pointed toward the ear canal, the thermometer provides an accurate temperature reading in 2 seconds. The images in **Figure 2** show other technologies developed for space that are now used on Earth.

Figure 1 A technology originally developed to measure the temperature of stars now enables a parent to easily and quickly determine whether a child has a fever.

Figure 2 Medical professionals use technologies originally developed for space to help improve the health of patients.

To monitor the health of astronauts during space walks, space suits contain tiny sensors that measure astronauts' temperature, respiration, and cardiac activity. The technology that led to the development of these sensors is now used on Earth to monitor people's health. ▶

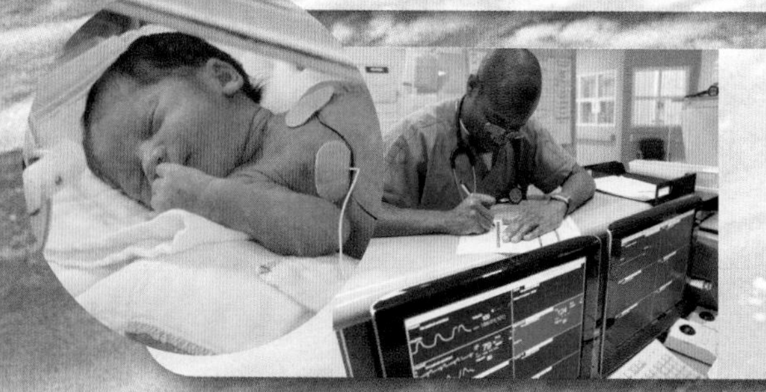

◀ Hospitals can monitor patients from a central nurses' station using electronics similar to those used in space suits. Sensors in infants' clothes can monitor a baby's breathing while the baby sleeps.

Doctors use sensors developed to study the effect of weightlessness on the muscles of astronauts to monitor repeated muscle movements that can lead to carpal tunnel syndrome in the workplace. ▶

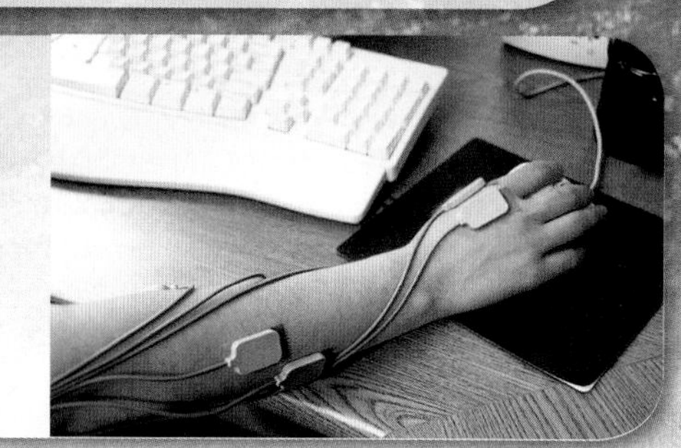

Space technology improves health worldwide.

Worldwide, people contribute to and benefit from space technologies. For example, a Canadian company produced robotic arms for the space shuttle. They collaborated with a medical school to develop instruments that enable surgeons to perform microscopic surgery on the brain. Chinese scientists developed a high-resolution X-ray imaging system for spacecraft. These X-rays are safer, faster, and more accurate than previous X-rays. Now, doctors can use the system to more accurately diagnose diseases. In addition, a Spanish company developed a navigation system for the blind based on space navigation technology.

Science and technology cannot solve all problems. But together, they help medical professionals keep people healthy and improve the quality of life for everyone on Earth.

MiniLab
25 minutes

What can you invent from space technology?

Space technology may have more uses than are currently known. Your job is to think of a new use for a joystick.

1. Work with a partner to discuss ways in which technology used in space might be used to improve health. Consider various types of illnesses or medical conditions in which a person's abilities to do everyday tasks might be helped by space technology.
2. Select one idea and develop it. Draw pictures of your device to show how it will work.

Analyze and Conclude

1. **Explain** How does your invention improve the ability of medical professionals or the lives of people with medical conditions?
2. **Key Concept** Why is your device a space technology?

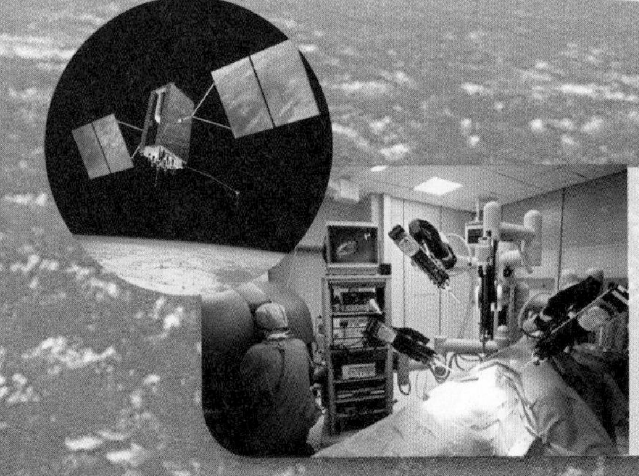

Surgeons use joysticks, similar to those used to control the Lunar Rover, to perform surgery on a patient who is thousands of miles away.

The temperature inside an astronaut's space suit can become extremely hot. So NASA developed a technology that circulates a cool fluid through tubes built into the suit.

Scientists applied this technology and developed a therapy for people with multiple sclerosis (MS). MS is a disease that slows the transfer of nerve signals from the brain. MS can affect the ability to think, speak, and control movement. Studies show that a slight decrease in body temperature can restore the transfer of some nerves signals. Therefore, scientists developed cooling suits for patients with MS based on NASA's cooling space suits.

Chapter 21

Interactions of Human Body Systems

THE BIG IDEA
How do human body systems interact and support life?

Inquiry What is going on here?

In this photo, the small vessels shown in red are part of the circulatory system. The vessels surround structures in the lungs called alveoli (al VEE uh li), shown in blue.

- Why do you think these vessels surround the alveoli?
- What do you think would happen if the vessels were not there?
- How do human body systems interact and support life?

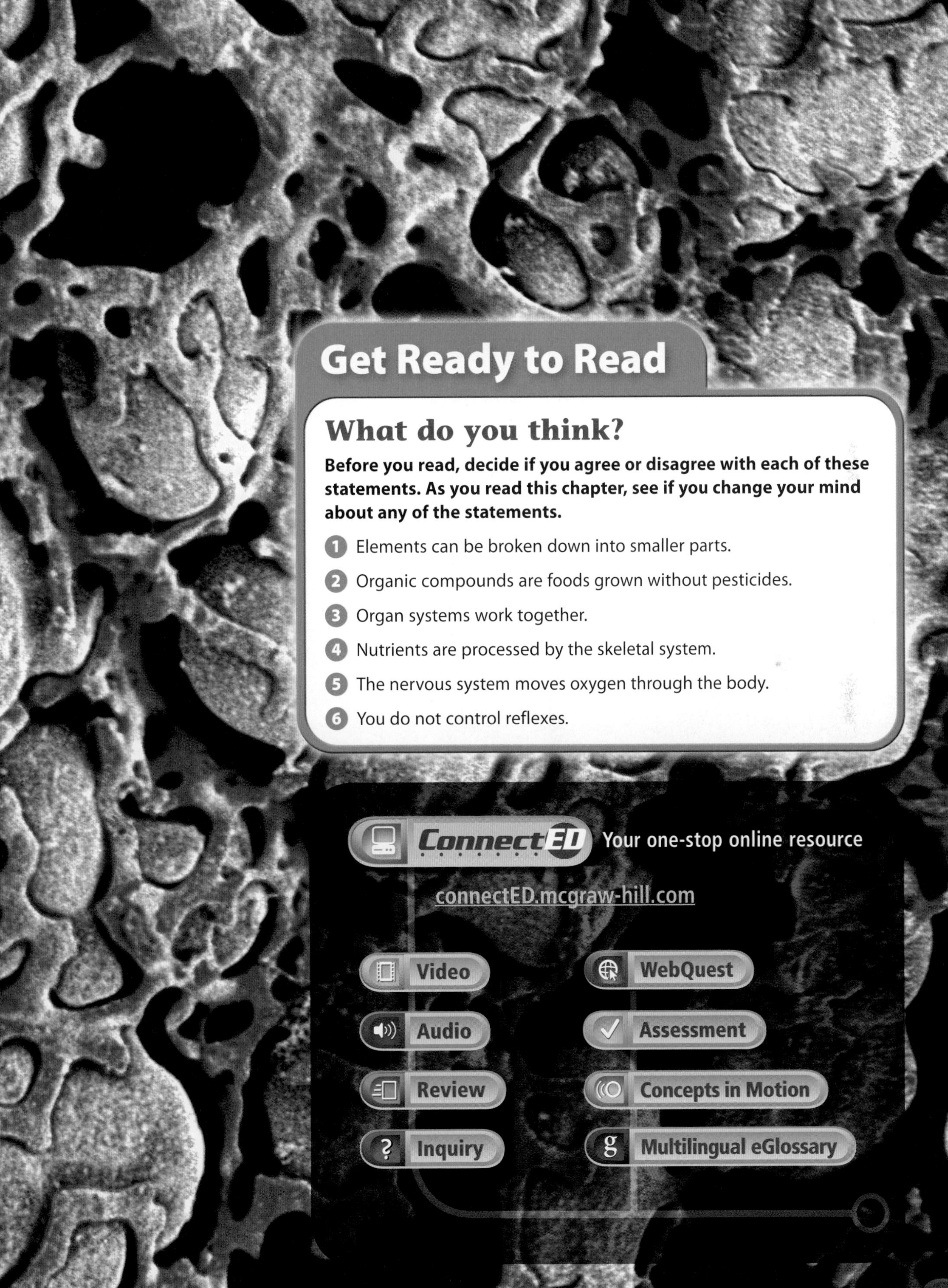

Get Ready to Read

What do you think?

Before you read, decide if you agree or disagree with each of these statements. As you read this chapter, see if you change your mind about any of the statements.

1. Elements can be broken down into smaller parts.
2. Organic compounds are foods grown without pesticides.
3. Organ systems work together.
4. Nutrients are processed by the skeletal system.
5. The nervous system moves oxygen through the body.
6. You do not control reflexes.

ConnectED Your one-stop online resource

connectED.mcgraw-hill.com

- Video
- Audio
- Review
- Inquiry
- WebQuest
- Assessment
- Concepts in Motion
- Multilingual eGlossary

Lesson 1

Reading Guide

Key Concepts 
ESSENTIAL QUESTIONS

- What are the functions of inorganic substances in the human body?
- What are the functions of organic substances in the human body?
- How does the body's organization enable it to function?

Vocabulary
macromolecule p. 768
monosaccharide p. 768
amino acid p. 768
nucleotide p. 768

g Multilingual eGlossary

The Human Body

Inquiry Strange Clothing?

Have you ever seen a multicolor scan of a human body? This is a thermal scan that shows the temperature in different areas of the body. Red, or "hot," areas might indicate additional blood flow that could affect body function.

Launch Lab

10 minutes

Will it disappear?

Over 60 percent of your body's weight is due to water. Water in our cells and body fluids serves an important role in the movement of materials throughout our bodies.

1. Read and complete a lab safety form.
2. Half fill a **beaker** with water.
3. Add a **spoonful of sugar,** and stir for 15 seconds.
4. Observe what occurs in the beaker. Record your observations in your Science Journal.
5. Empty the beaker's contents and clean the beaker according to your teacher's instructions.
6. Repeat steps 2–5 replacing the sugar with **salt, corn syrup, baking soda,** an **antacid tablet,** a **multivitamin,** and **candy bits.**

Think About This

1. What happened to each of the substances when they were stirred into the water?
2. **Key Concept** Why do you think water makes up such a large portion of your body?

Life and Chemistry

Have you ever modeled a volcanic eruption by mixing vinegar and baking soda? When baking soda—also called sodium bicarbonate ($NaHCO_3$)—combines with the acetic acid (CH_3COOH) in vinegar, a chemical reaction occurs. You might recall that a chemical reaction is the process that occurs when compounds, called reactants, form one or more new substances, called products.

During a chemical reaction, bonds are broken and new bonds are formed. When acetic acid and sodium bicarbonate react, water (H_2O), carbon dioxide (CO_2), and sodium acetate ($NaCH_3COO$) form. Bubbles form when vinegar and baking soda mix, as shown in **Figure 1.** The bubbles are caused by the CO_2 gas and the water that are released as products of this chemical reaction.

Chemical reactions are everywhere. Moldy bread and green pennies are the result of chemical reactions. As you will read in this lesson, chemical reactions also occur in your body. These chemical reactions take place in your body's cells. They are essential for human life.

Figure 1 When baking soda and acetic acid combine, water, CO_2 gas, and sodium acetate form.

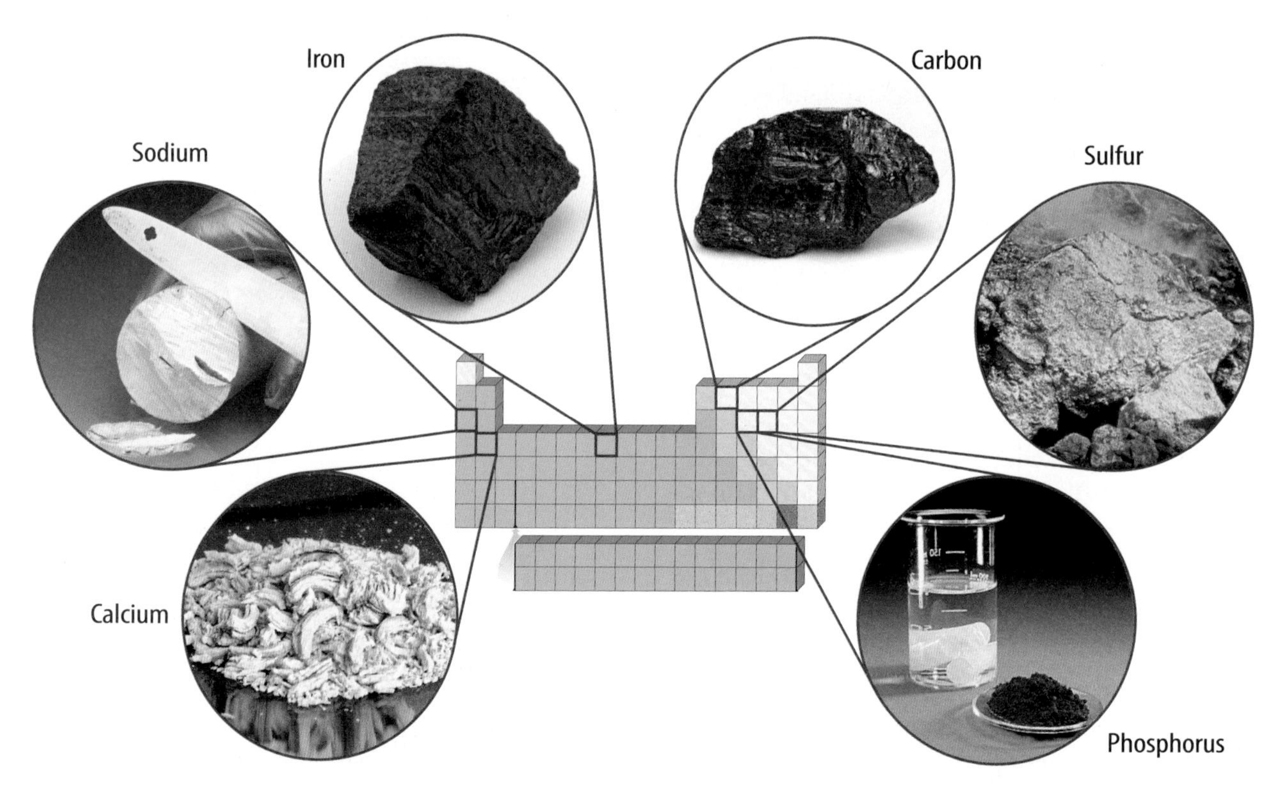

Figure 2 Elements have different textures, colors, and properties.

Elements and Compounds

Recall that elements are the basic units that make up chemicals. Elements are substances that cannot be broken down or transformed into another element during a chemical reaction. Elements have different physical properties, as shown in **Figure 2.** Although there are almost 100 elements found in nature, six elements—carbon, oxygen, hydrogen, nitrogen, calcium, and phosphorus—make up 99 percent of your body's mass.

Reading Check Which elements make up most of the body's mass?

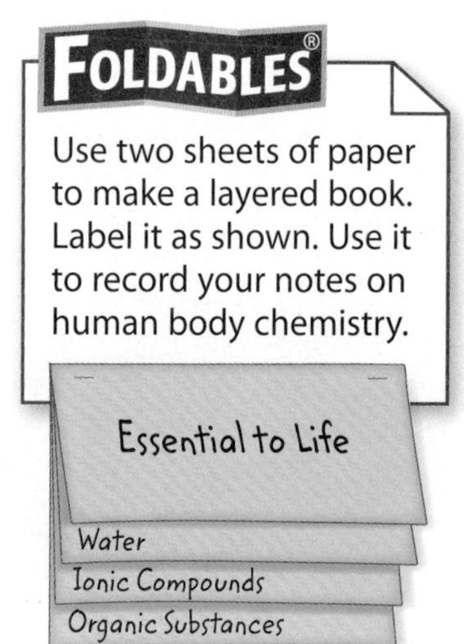

Compounds are substances made of two or more elements. Unlike elements, compounds can be broken down into simpler substances. The acetic acid and sodium bicarbonate you read about at the beginning of this lesson are compounds. Acetic acid is formed when the elements carbon, hydrogen, and oxygen combine.

Compounds can be made by binding elements together in two different ways—ionic or covalent bonds. Ionic bonds are formed when electrons travel from one element to another. One element has a positive charge, and the other has a negative charge. The opposite charges attract. Table salt, or sodium chloride (NaCl), is an example of a compound that has an ionic bond. Covalent bonds are formed when the electrons in each element are shared. Many gases in the atmosphere, such as oxygen (O_2) and nitrogen (N_2), form by covalent bonds.

Inorganic Substances

Inorganic compounds are everywhere on Earth. Inorganic compounds are substances that do not contain carbon-hydrogen bonds. Substances such as ammonia (NH_3) and NaCl are inorganic compounds. Many inorganic compounds are essential for human life. Water and oxygen gas are inorganic compounds that humans need to survive.

Ionic Compounds As you read earlier, NaCl is an inorganic compound that forms by ionic bonding. Recall that ionic bonds are formed when a positive ion is attracted to a negative ion. When a substance gives up or gains an electron, it is called an ion. In the compound NaCl, Na^+ is a positively charged ion, and Cl^- is a negatively charged ion. Many ions are important for survival. For example, calcium (Ca^{2+}) helps nerve and muscle cells function and makes up bone. These compounds rely on water to move through the body.

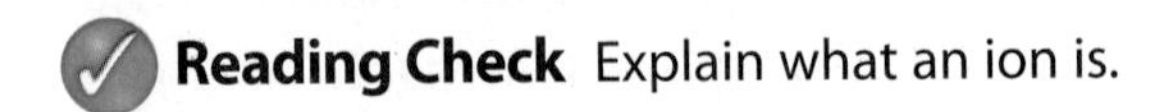

Reading Check Explain what an ion is.

Water You might know that water is called the universal solvent. What does that mean? A solvent is a substance that dissolves other substances. Ionic compounds dissolve well in water and allow ions such as Na^+, Cl^-, and Ca^{2+} to travel through the body dissolved in water.

Water is able to dissolve many substances because of its polarity. Water molecules are formed by covalent bonds that link the hydrogen (H) to the oxygen (O).

As shown in **Figure 3,** water is able to easily dissolve ionic substances because the positive ions in the compound are attracted to the oxygen end of the molecules and the negative ions are attracted to the hydrogen end of the molecules.

Key Concept Check How does water help the body obtain ionic substances?

The Structure of a Water Molecule

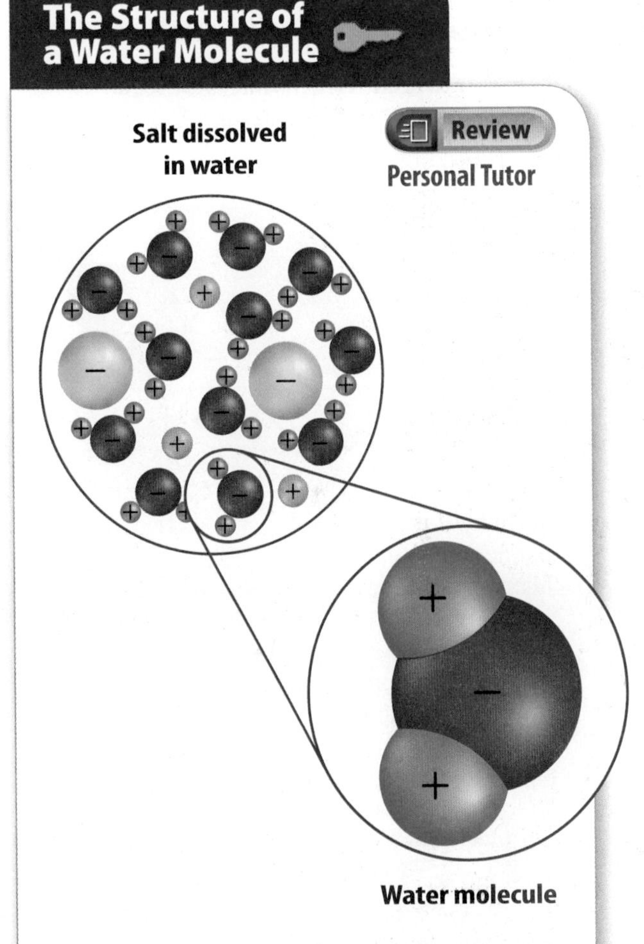

Figure 3 A water molecule has a bent shape with the large oxygen atom at one end of the molecule. This bent shape results in the oxygen end of the molecule having a negative charge and the hydrogen end having a positive charge. H^+ ions are shown in blue, O^- in red, Na^+ in grey, and Cl^- in green.

Visual Check Which ion is attracted to the oxygen end of the water molecule?

WORD ORIGIN

macromolecule
from Greek *makro-*, means "long"; and Latin *molecula*, means "mass"

Organic Substances

You might have heard the word *organic* used to describe certain fruits and vegetables. However, when used in science, the term *organic* describes compounds that contain carbon and other elements, such as hydrogen, oxygen, phosphorus, nitrogen, or sulfur, held together by covalent bonds. Organic compounds carry out many different functions. *Substances that form from joining many small molecules together are called* **macromolecules.** The four macromolecules in the body are carbohydrates, lipids, proteins, and nucleic acids, shown in **Figure 4.**

Carbohydrates Sugars, starches, and cellulose are carbohydrates. Carbohydrates are formed when *simple sugars, called* **monosaccharides** (mah nuh SA kuh ridez), are joined together. Carbohydrates are the body's major source of energy.

Lipids Triglycerides and cholesterol are lipids (LIH pihdz), and, like carbohydrates, they are made from carbon, hydrogen, and oxygen. Lipids help insulate your body and are a major part of cell membranes.

Proteins The adult human body is made up of 10–20 percent protein. Proteins form when **amino** (uh MEE noh) **acids,** *the building blocks of protein,* join together. Some proteins give cells structure, some help cells communicate, and some are enzymes.

Nucleic Acids Much as computer chips store information, nucleic acids are macromolecules that store information used by the body to perform different functions. Nucleic acids are formed when **nucleotides** (NEW klee uh tidez), *molecules made of a nitrogen base, a sugar, and a phosphate group,* join together. The body contains two types of nucleic acids, DNA and RNA.

Key Concept Check What are the functions of organic compounds in the human body?

Macromolecules 

Figure 4 Much like a train contains many boxcars linked together, macromolecules are many organic compounds joined together.

Visual Check Which macromolecule is made of nucleotide polymers?

Carbohydrate
During digestion, humans break down carbohydrates into glucose, a monosaccharide, and store it as glycogen.

Lipid
Lipids, also called fats, contain fewer oxygen atoms than carbohydrates and do not dissolve in water.

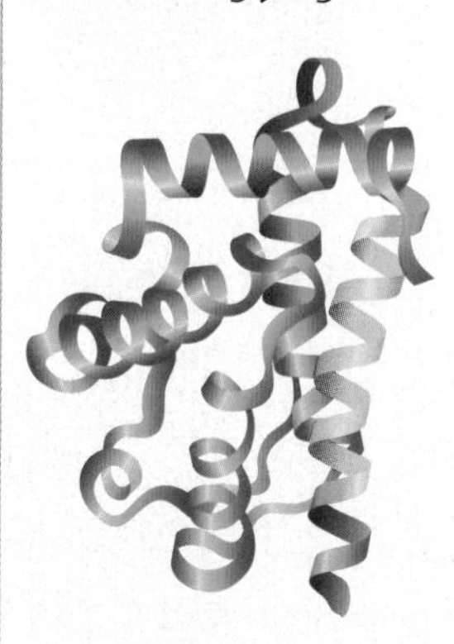

Protein
All amino acids consist of carbon, oxygen, hydrogen, and nitrogen. Some also contain sulfur.

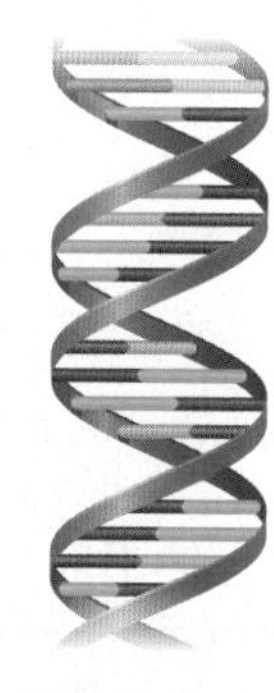

Nucleic acid
DNA is made of two strands of nucleotide polymers. RNA is made of a single strand.

The Body's Organization

In order to function and survive, macromolecules in the human body must be organized in different compartments. For example, most of the DNA is stored in the nuclei of cells. Cholesterol and other types of lipids are used to form cell membranes. Organizing macromolecules in specific locations helps cells carry out specific functions.

Recall that cells are the building blocks of all living things. Cells have different shapes depending on their function. Neurons are long and slender so they can carry information over long distances. Red blood cells are flexible disks and can move easily through blood vessels.

Tissues are made of a group of cells that work together and perform a function. Cardiac muscle cells form a tissue that helps the heart pump blood throughout the body.

An organ is a group of tissues that work together and perform a function. The liver, spleen, and lungs are all organs.

An organ system, such as the one in **Figure 5,** is a group of organs that works together and performs a specific task. Organ systems work together and help the body communicate, defend itself, process energy, transport substances, and move.

Key Concept Check How does the body's organization enable it to function?

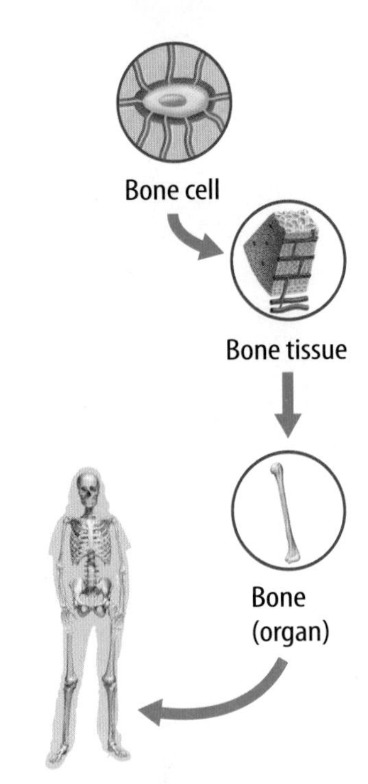

Figure 5 The human body is made of cells, tissues, organs, and organ systems that work together.

Inquiry MiniLab

20 minutes

Does teamwork stack up?

For your body to function properly, your body systems must work together. How can your team work together to complete a task? By doing this lab, you will understand how important teamwork is for your body functions.

1. Read and complete a lab safety form.
2. Obtain **a piece of string** for each group member, three **plastic cups,** and a **rubber band.**
3. Using only the materials provided, stack the cups in a pyramid. Do not touch the cups with any part of your body.

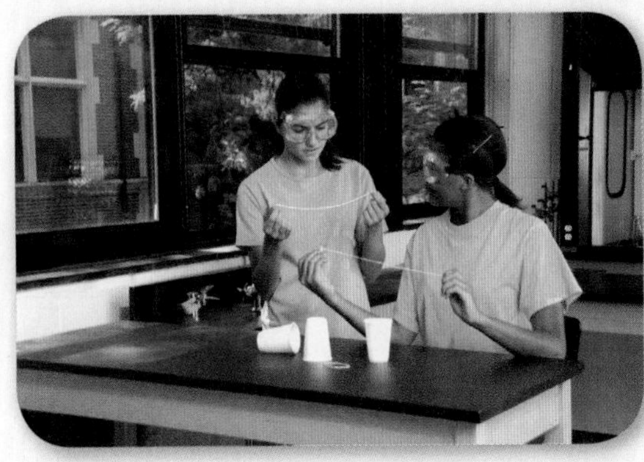

Analyze and Conclude

1. **Explain** What method did you use to form the pyramid?
2. **Evaluate** How did your team work together to form the pyramid?
3. **Key Concept** Compare how your team worked together to how systems work together in the body.

Lesson 1 Review

Visual Summary

Elements are substances that cannot be broken down or transformed into other elements during chemical reactions.

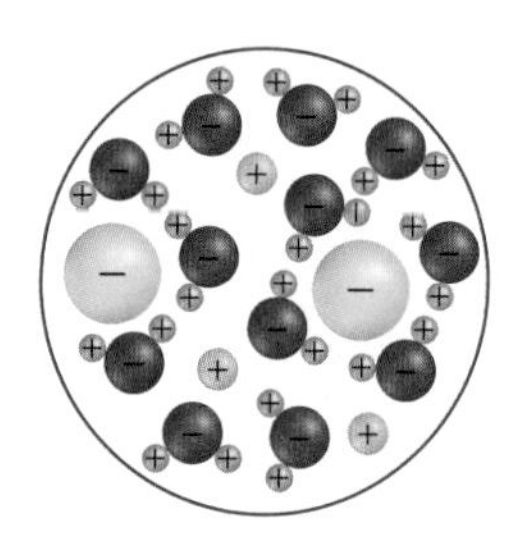

Water dissolves ionic substances easily.

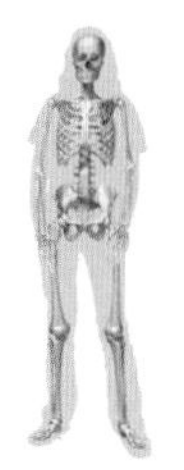

Organ systems are groups of organs that work together and perform a specific task.

FOLDABLES®

Use your lesson Foldable to review the lesson. Save your Foldable for the project at the end of the chapter.

What do you think NOW?

You first read the statements below at the beginning of the chapter.

1. Elements can be broken down into smaller parts.
2. Organic compounds are foods grown without pesticides.
3. Organ systems work together.

Did you change your mind about whether you agree or disagree with the statements? Rewrite any false statements to make them true.

Use Vocabulary

1. **Distinguish** between nucleotides and amino acids.
2. **Use the term** *monosaccharides* in a sentence.
3. **Define** *macromolecule* in your own words.

Understand Key Concepts 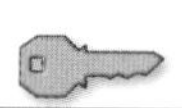

4. Which is a lipid?
   - **A.** cellulose
   - **B.** cholesterol
   - **C.** DNA
   - **D.** starch
5. **Distinguish** between proteins and nucleic acids.
6. **Draw** a picture that illustrates how water dissolves NaCl.

Interpret Graphics

7. **Summarize** Copy and fill in the graphic organizer below to show how the human body is organized.

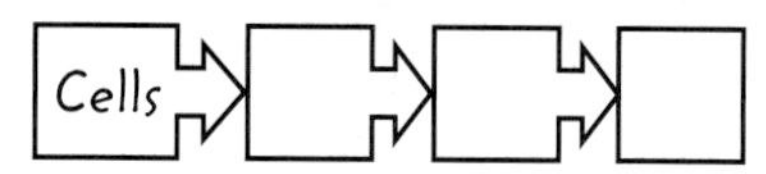

8. **Describe** the differences in the structures of the macromolecules shown below.

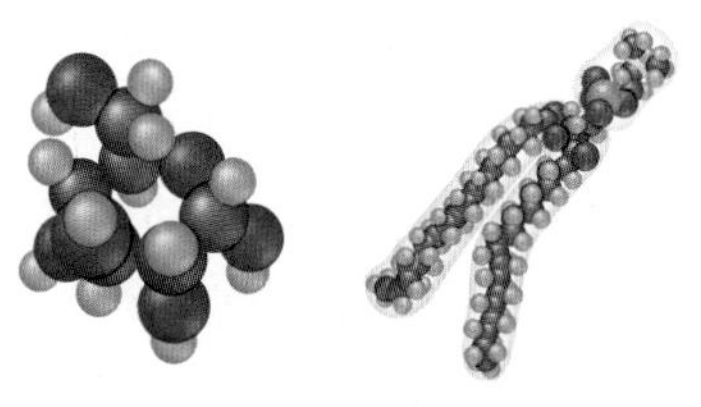

Critical Thinking

9. **Hypothesize** how the body would function if cells were not organized into tissues.
10. **Relate** the four macromolecules to their functions in a cell.

Your Remarkable, Renewable Skin

SCIENCE & SOCIETY

Keeping You Safe from Head to Toe

Lots of organs keep your body working, but only one covers you from head to toe—your skin. In many ways, your skin is like body armor. It protects you from bacteria and other germs, absorbs impacts, and blocks ultraviolet (UV) light from the Sun.

Luckily, when there's a crack in this armor, such as a cut or a burn, your skin can repair itself, or regenerate. To do this, it works closely with the circulatory system. When a wound bleeds, platelets in your blood rush to the injury. They clump together, forming a blood clot and stopping the bleeding. Then the clot dries into a hard scab that protects the area. Underneath the scab, red blood cells deliver nutrients and oxygen to help new skin cells form. As the skin regenerates, the wound closes.

This process works well for most skin injuries. But sometimes a wound is too large, and the skin can't heal on its own. When this happens, doctors can treat the area with a skin graft. In this process, the injured skin is replaced with healthy skin taken from another part of the injured person's body. However, a skin graft depends on the circulatory system to succeed. For the new skin to stay healthy, it needs nutrients and oxygen that are carried by blood. What happens if an injured person doesn't have enough healthy skin to use as a skin graft? Skin injuries can sometimes heal using artificial skin.

To make artificial skin, scientists begin with a mesh of connective tissue called collagen. Skin cells are grown on the collagen to form sheets of skin tissue. Placing the artificial skin over large wounds encourages new skin to grow and protects the wounds from infection.

Skin graft site

Skin graft

Skin Graft Process

1. Healthy skin is taken from another part of the body. For small, deep wounds, a thick layer of skin is used. Thinner layers are used for larger wounds.
2. If more skin is needed, the skin is run through a machine, called a mesher, that makes it bigger.
3. The healthy skin is attached with stitches or special staples.

Healed skin graft

It's Your Turn

RESEARCH Work with a partner to research how artificial skin can be used to help large wounds heal. Share your findings with your class.

Lesson 2

Reading Guide

Key Concepts
ESSENTIAL QUESTIONS

- How are nutrients processed in the body?
- How does the body transport and process oxygen and wastes?
- How does the body coordinate movement and respond to stimuli?
- How do feedback mechanisms help maintain homeostasis?

Vocabulary

homeostasis p. 773
negative feedback p. 780
positive feedback p. 780

Multilingual eGlossary

BrainPOP®

How Body Systems Interact

Have you ever seen a circus aerial act? When people work together, many things can get done. Systems in the human body work together, too. In this lesson, you will learn how important teamwork is for the body.

Launch Lab

15 minutes

How can you model homeostasis?

The environment around you is constantly changing; however, your body must know how to maintain a delicate balancing act to continue operating. A thermostat in a building helps maintain a constant inside temperature no matter what the outside air temperature is. Like that thermostat, our bodies maintain a constant temperature by sensing the environment and responding.

1. Read and complete a lab safety form.
2. Put one drop of **blue food coloring** into a **beaker** of water.
3. Hold one end of **clear, flexible tubing.** Have your partner hold the other end of the tubing to form a U shape. Have a third student pour the water into the tubing until it is half full.
4. Place your thumb over one end of the tubing to prevent the water from spilling out. Your partner should also place his or her thumb over the other end of the tubing.
5. Try several variations of tubing position and observe what happens. Record your observations in your Science Journal.

Think About This

1. What happened to the level of the water in the tubing as the ends were raised and lowered?
2.  **Key Concept** What methods do you think your body might use to maintain body temperature on a hot or a cold day?

Homeostasis

Did you know that your body has a system to keep its internal temperature constant, much as a thermostat helps keep temperature constant in a building? Most parts of the body function best at 37°C. When temperatures in the body fall below 37°C, you might have goose bumps, as shown in **Figure 6.** The endocrine system, which regulates body temperature, sends messages through the nervous system. The nervous system signals the muscular system to cause the body to shiver. When you shiver, muscles move. This movement generates thermal energy and helps raise body temperature. Keeping the body's temperature constant requires that the endocrine system, the nervous system, and the muscular system work together.

Your body's organ systems work together and maintain temperature, nutrient levels, oxygen, fluid levels, pH, and many other types of homeostasis (hoh mee oh STAY sus). **Homeostasis** *is the ability to maintain constant internal conditions when outside conditions change.* In this lesson, you will read how organ systems work together and maintain homeostasis.

Figure 6 Goose bumps form when tiny muscles attached to hairs on the skin contract and pull the hairs up straight.

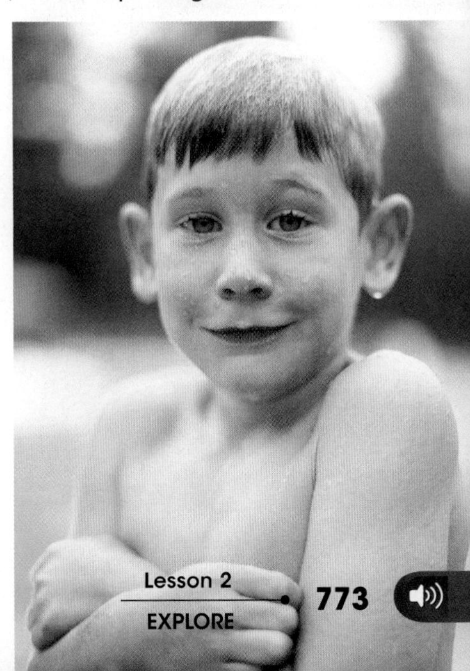

Processing Nutrients

Science Use v. Common Use

organ

Science Use a group of tissues performing a specific function

Common Use a keyboard instrument in which pipes are sounded by compressed air

Maintaining homeostasis keeps the internal environment in the body functioning properly. Many **organ** systems work together and maintain energy homeostasis.

Recall that the body gets most of its energy from carbohydrates. Lipids and proteins also provide energy. Food is broken down in the digestive system by chemical and mechanical digestion. Chemical digestion occurs when enzymes in saliva and acid in your stomach break down food. Mechanical digestion happens when you chew your food. The digestive system, the circulatory system, and the muscular system work together and process and obtain nutrients from food. The skeletal system, the endocrine system, and the lymphatic system also work with the digestive system and process those nutrients.

Muscles and Digestion

Food enters the body through the digestive system and is broken down into nutrients that can be absorbed into the body. However, the muscular system is needed to get food through the digestive system. Muscles that surround the stomach contract and move food to the small intestine. These contractions, shown in **Figure 7,** are called peristalsis (per uh STAHL sus).

Muscles help the jaw move when you chew. They work with the digestive system and help you swallow. Muscles also surround the esophagus, stomach, the small intestine, and the large intestine and help move food through the digestive system.

Reading Check Where are muscles found in the digestive system?

Figure 7 The muscular system and digestive system work together and process food.

Concepts in Motion Animation

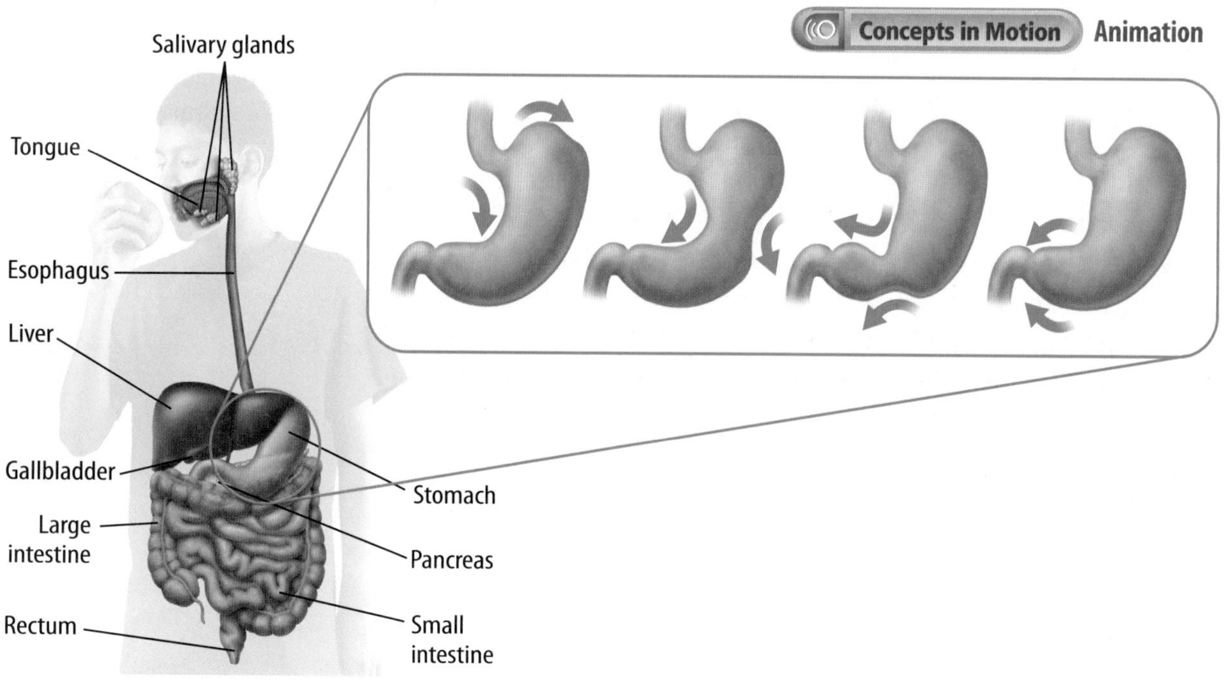

Absorbing Nutrients

Concepts in Motion Animation

Figure 8 The digestive system and the circulatory system work together and absorb nutrients and move them throughout the body.

Visual Check What does each villus contain?

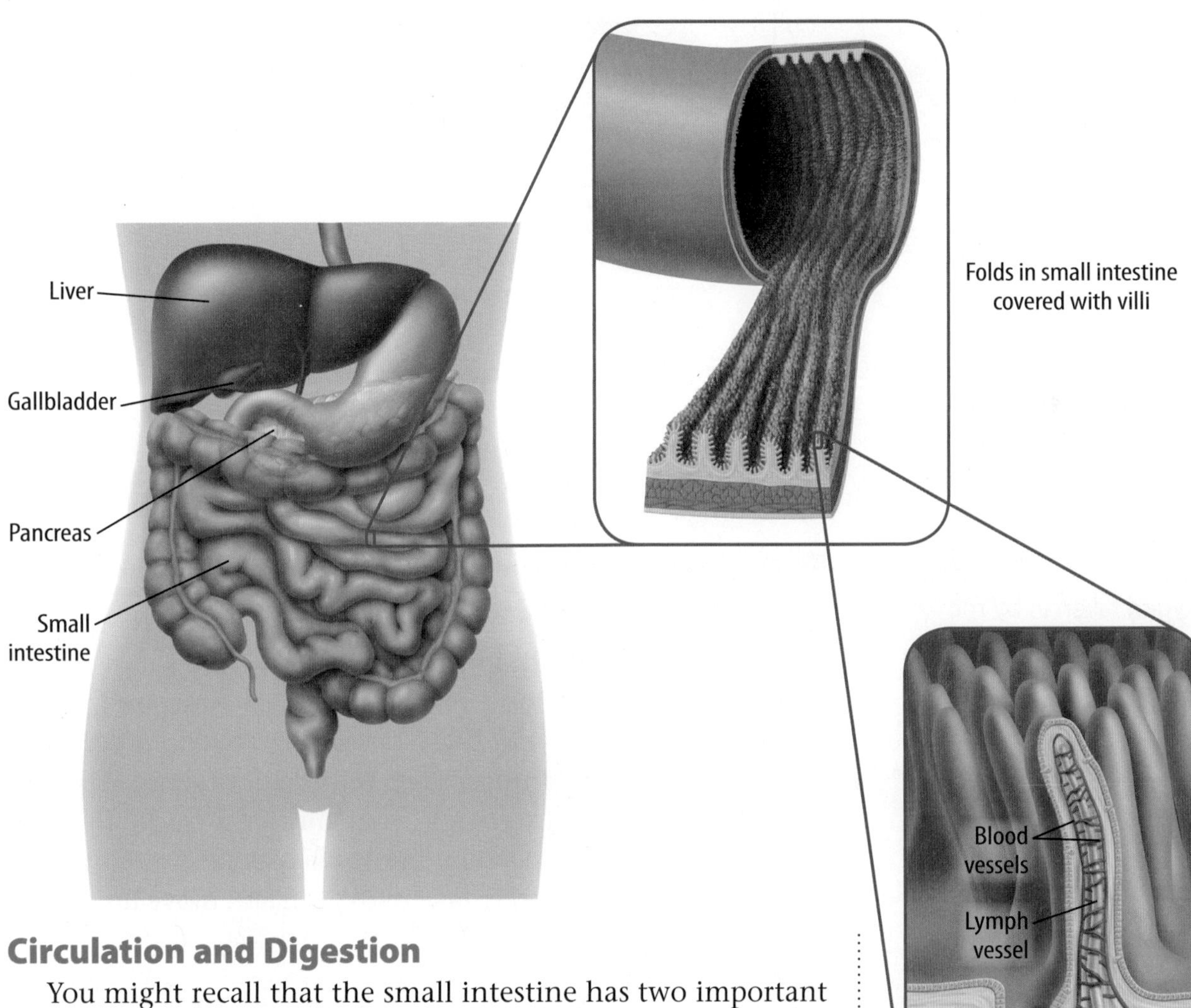

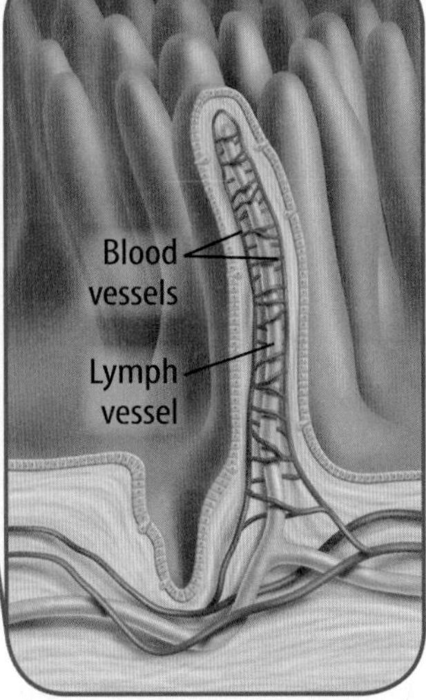

One villus

Circulation and Digestion

You might recall that the small intestine has two important jobs—breaking down food and absorbing nutrients. Two other systems—the muscular system and the circulatory system—work with the small intestine. The muscular system helps the small intestine break down food. The circulatory system works with the small intestine and gets nutrients to the rest of the body. Nutrients are absorbed by small fingerlike projections, called villi (VIH li; singular, villus), in the small intestine. The villi have blood vessels inside them, which are part of the circulatory system, as shown in **Figure 8.** Nutrients enter these blood vessels and are then transported to the rest of the body. The muscular system also surrounds the blood vessels and helps blood and nutrients move through the body.

Key Concept Check How are nutrients processed in the body?

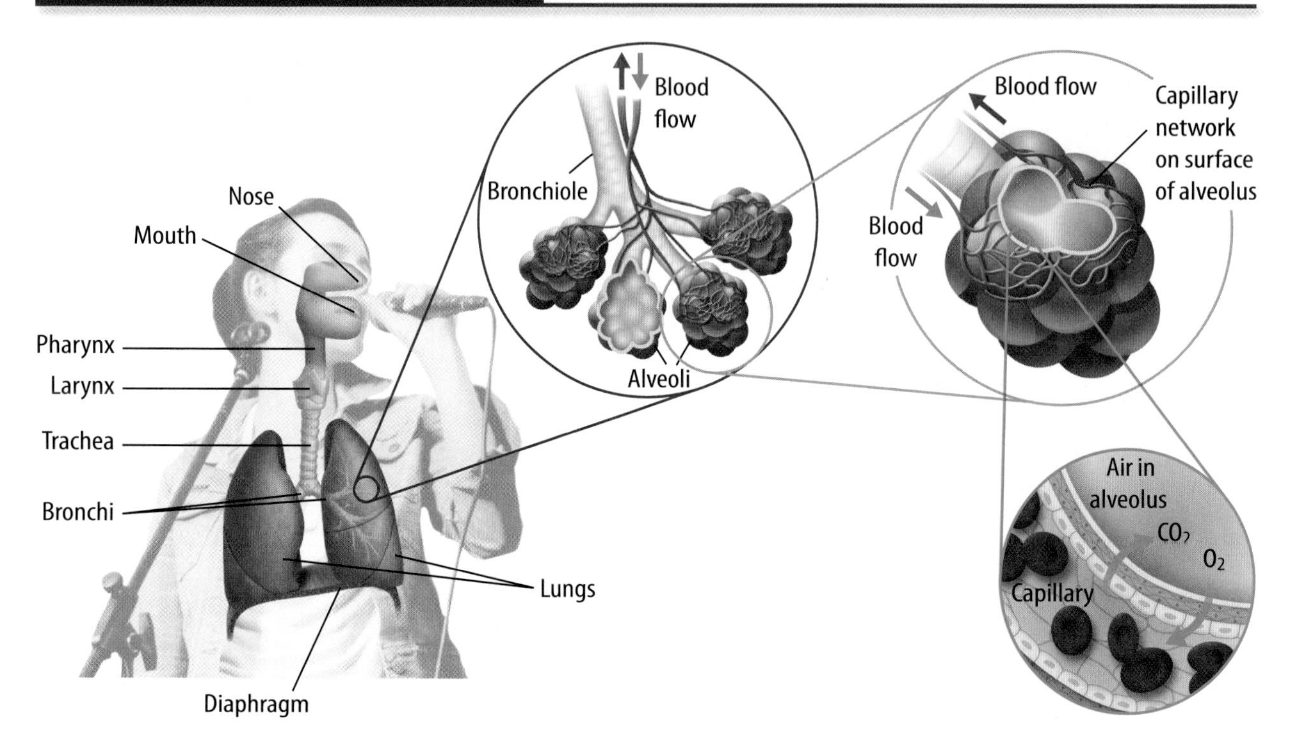

Figure 9 When you inhale, oxygen taken in by the respiratory system enters the circulatory system and is transported to all cells in the body.

Processing Oxygen and Wastes

The body cannot function without its systems working together. For example, humans require oxygen to survive. Your lungs take in oxygen and release carbon dioxide. You might recall that all cells in your body use oxygen to help process the energy in nutrients into energy that cells can use. Oxygen helps the body obtain energy from nutrients by performing **cellular respiration.** Do you know what organ systems work together and help the body take in oxygen and move it through the body?

Review Vocabulary

cellular respiration
a series of chemical reactions that convert the energy in food molecules into a usable source of energy called ATP

 **Reading Check** How does oxygen help the body obtain energy?

Oxygen Transport

Oxygen enters the body through the respiratory system, as shown in **Figure 9.** When you inhale, the respiratory system works with the circulatory system and transports oxygen to all cells in the body. The muscular system also helps the respiratory system by expanding your chest so that cells in the lungs fill up with oxygen.

Recall that the circulatory system works with the small intestine and moves nutrients into the body. The circulatory system also works with the lungs and helps oxygen travel through the body. Oxygen that is taken in by capillaries, as shown in **Figure 9,** is transported to the rest of the body through larger blood vessels.

Eliminating Wastes

The excretory system works with several other organ systems and eliminates wastes. Recall that the body processes food, oxygen, and liquids. Food and liquids are processed by the digestive system. After nutrients are absorbed during digestion, the excretory system removes solid waste products, called feces, through the rectum.

The excretory system also works with the respiratory and circulatory systems and removes carbon dioxide (CO_2) from the body. Oxygen is used in all organs of the body. The CO_2 produced by cells throughout the body enters capillaries and is transported to the lungs, where it is exhaled. These three systems work together and maintain oxygen homeostasis by making sure that CO_2 is removed.

The excretory system also maintains fluid homeostasis. Liquid waste travels through the circulatory system to the kidneys, as shown in **Figure 10,** which make urine. Liquid waste also travels to the skin where fluid is released during sweating.

Key Concept Check How does the body transport and process oxygen and wastes?

Figure 10 The kidneys remove liquids, salts, and other wastes from the body by making urine.

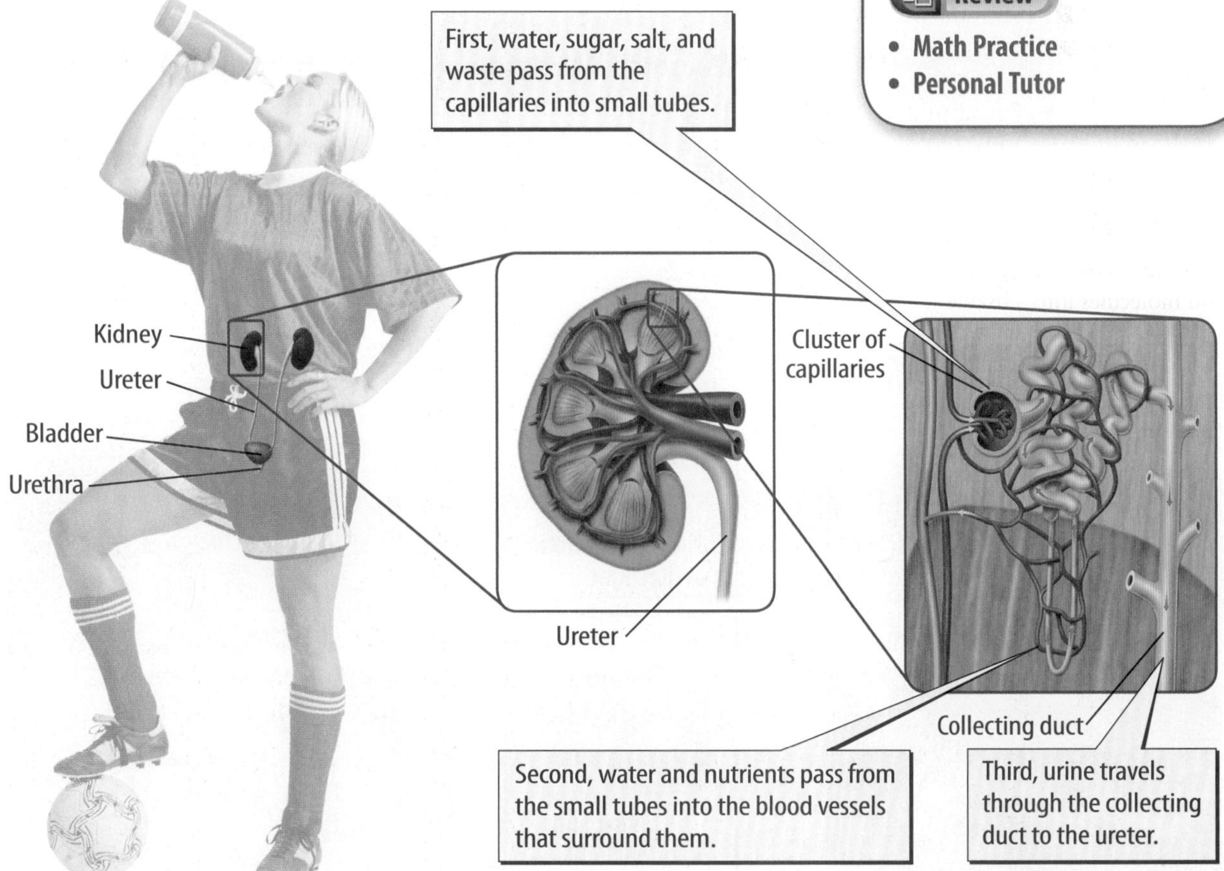

Math Skills

Use Volume

Volume is a measure of the amount of matter that a hollow object, such as the stomach or the lungs, will hold. For example, the volume of an empty stomach is about 0.08 L. After eating, a person's stomach is 1.5 L. What volume of food was consumed?

Subtract the starting volume from the final volume.

$1.5 \text{ L} - 0.08 \text{ L} = 1.42 \text{ L}$

Practice

A certain person's bladder has a volume of 550 mL. The person has the urge to urinate when the bladder contains 200 mL of urine. What volume of the bladder remains empty?

Review

- **Math Practice**
- **Personal Tutor**

Control and Coordination

Have you ever wondered how your heart beats without you thinking about it? The heart contains a group of specialized cells called pacemaker cells. These cells control the rate at which the heart beats by responding to signals from the nervous system. When exercising, the nervous system speeds up the heartbeat. When sleeping, the nervous system slows the rate at which the heart beats. The nervous system also works with other organ systems and controls the body's functions.

The nervous system uses electrical signals and helps organ systems of the body respond quickly to changes in the internal and external environments. The body also uses the endocrine system to help it respond to changes and maintain homeostasis. The nervous system coordinates rapid changes, and the endocrine system coordinates slower responses.

Key Concept Check How does the body coordinate movement and respond to stimuli?

Sensory Input

The nervous system coordinates the body's response to external stimuli. For example, when you dim lights to watch a movie, your pupils change in size, as shown in **Figure 11.** The nervous system also coordinates your response to the sight, the smell, the touch, and the taste of popcorn. The nervous system works with the respiratory and muscular systems to detect the popcorn's aroma. It coordinates muscles in the eyes to see the popcorn. The nervous system also works with the digestive system and prepares for eating the popcorn by producing saliva. It also coordinates the digestive and muscular systems so that the popcorn is broken down and moved through the body.

Figure 11 Your pupils increase in size when you enter a darkened room.

Visual Check When does the iris contract?

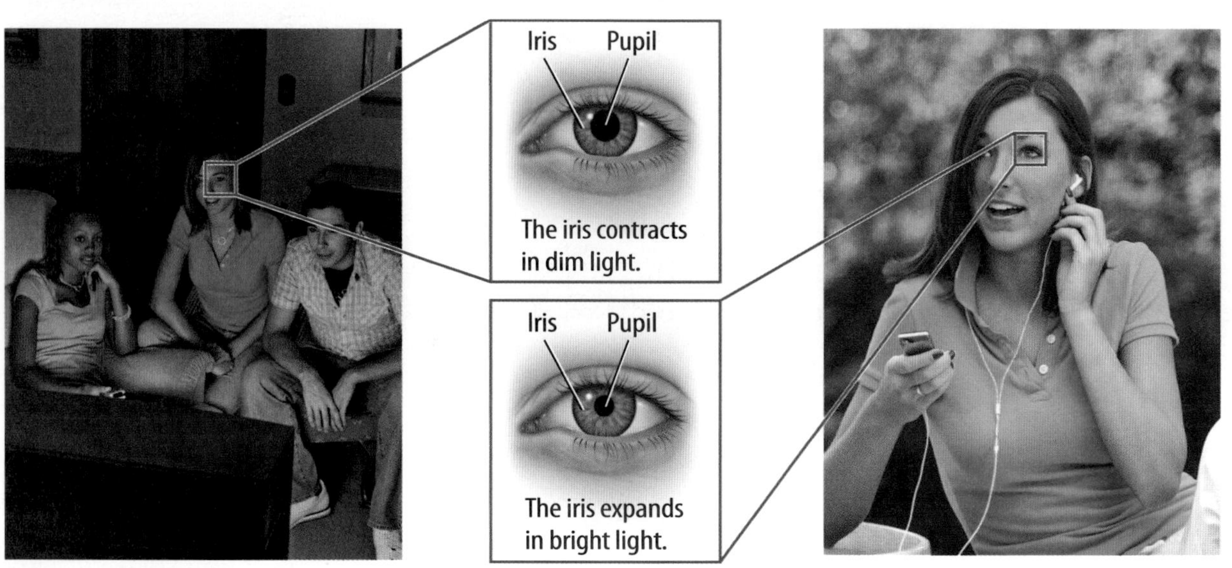

Reflexes

Recall that the nervous system uses electrical signals, coordinates responses to stimuli, and signals other organ systems. Neurons send electrical signals to the brain for processing so the nervous system can coordinate a response.

However, when a person touches a hot stove, the nervous system coordinates the response so quickly that the brain does not first process information about the hot stove. In fact, the response to touching a hot stove is so fast that thought is not required before the person removes his or her hand. This is because the nervous system has a rapid response system, called a reflex, that reacts to stimuli without sending information to the brain for processing. Reflexes allow the nervous system to coordinate a rapid response and tell the muscular system and skeletal system to move without thought.

Hormones

The nervous system coordinates the responses of other organ systems by using electrical signals. The endocrine system coordinates other organ systems by using chemical signals called hormones. Hormones are secreted from endocrine organs such as the thyroid gland, the adrenal gland, and the pancreas.

Reading Check What are hormones?

These chemical signals travel through the circulatory system to organ systems such as the digestive and muscular systems. They also control processes that maintain homeostasis. In the beginning of this lesson, you read that temperature homeostasis is maintained by producing thermal energy. The endocrine system, the nervous system, and the muscular system work together and maintain temperature homeostasis. Insulin, a hormone released from the pancreas, works with the digestive system and maintains energy homeostasis.

Inquiry MiniLab — 20 minutes

Is the hand quicker than the eye?

Your body sometimes can play tricks on you. When you look at something, the optic nerve sends a message to your brain. This experiment will explore how the body responds to a stimulus with the optic nerve and sometimes misinterprets what is seen.

1. Read and complete a lab safety form.
2. Using an **index card** folded in half, draw a bird on one side and a bird cage on the other side with a **permanent marker.** Both pictures should be drawn in the center of the card. Be sure to make the cage big enough so the bird can fit inside.
3. Slip the folded card over the pointed end of a **pencil.** The point should break through the card slightly. **Tape** both sides of the card to the pencil.
4. Twirl the pencil back and forth rapidly.
5. In your Science Journal, record what you see happening.

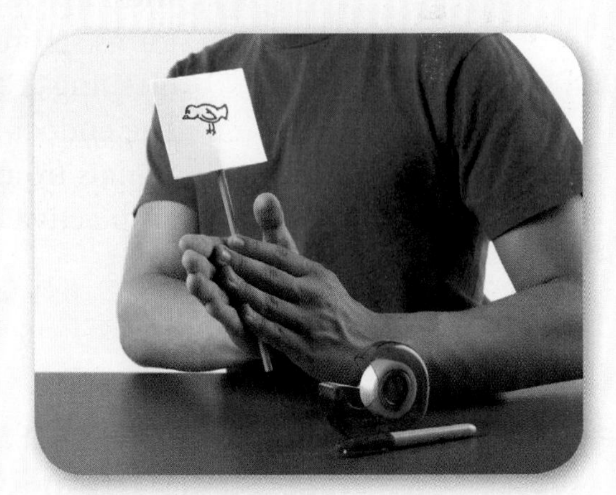

Analyze and Conclude

1. **Describe** What happened to the bird as you twirled the pencil?
2. **Draw Conclusions** What body functions worked together to see the illusion? Discuss real-world applications.
3. **Key Concept** How does the body respond to a stimulus?

Feedback Mechanisms

As you have read, homeostasis helps the body maintain a constant internal environment. The endocrine and nervous systems help **detect** changes in either the internal or the external environment and respond to those changes. Organ systems use feedback mechanisms to maintain homeostasis.

ACADEMIC VOCABULARY
detect
(verb) to identify the presence of something

Negative Feedback

Negative feedback *is a control system that helps the body maintain homeostasis by sending a signal to stop a response.* Negative feedback is used when you are hungry because the digestive system receives signals that it is time to eat. When you eat, the digestive and circulatory systems then work together and increase the amount of nutrients in the body. As the nutrients are being processed, the stomach sends signals to the brain to tell the body that you are full and to stop eating.

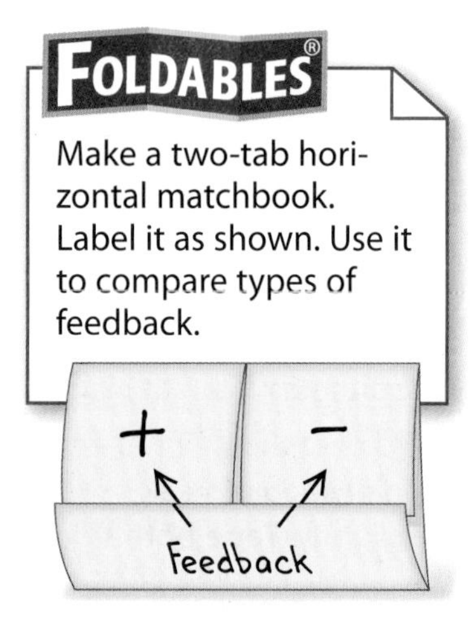

Positive Feedback

In contrast to negative feedback, **positive feedback** *is a control system that sends a signal to increase a response.* Blood clotting is an example of positive feedback. When you are bleeding, the circulatory system maintains homeostasis by controlling blood loss. Blood cells called platelets move to the site of the wound and help control bleeding by forming a clot with a protein called fibrin, as shown in **Figure 12.** As the clot forms, more platelets travel to the clot and help control the bleeding. Childbirth is also an example of positive feedback. The endocrine system signals the muscular system to contract. Signals from the muscular system tell the endocrine system to keep activating the muscular system until the baby is born.

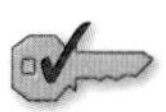

Key Concept Check How do feedback systems help maintain homeostasis?

Figure 12 The body uses positive feedback to clot blood. **Concepts in Motion** Animation

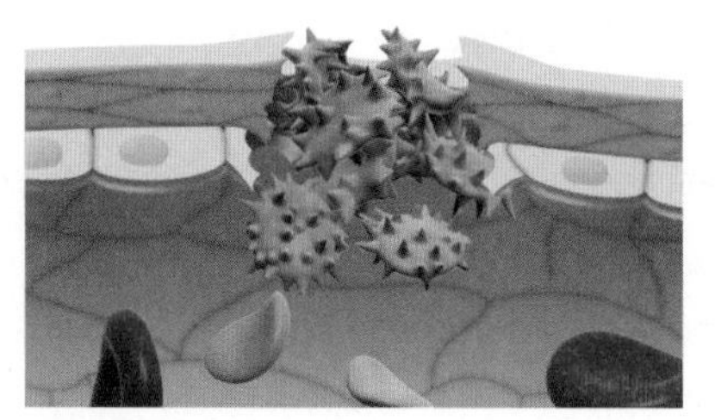

Step 1
Platelets rush to the tear and form a plug that stops the bleeding.

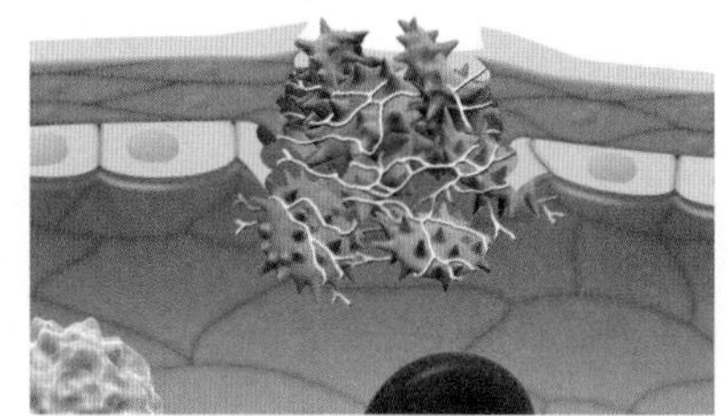

Step 2
A web of fibrin forms around the platelets and holds them in place.

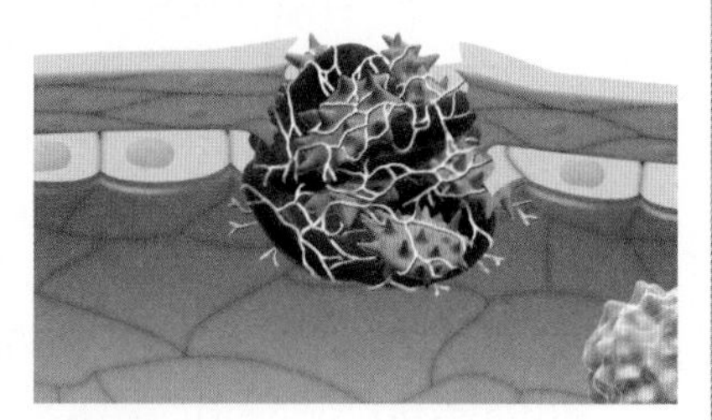

Step 3
The fibrin web catches more platelets and red blood cells, and these form a blood clot.

Lesson 2 Review

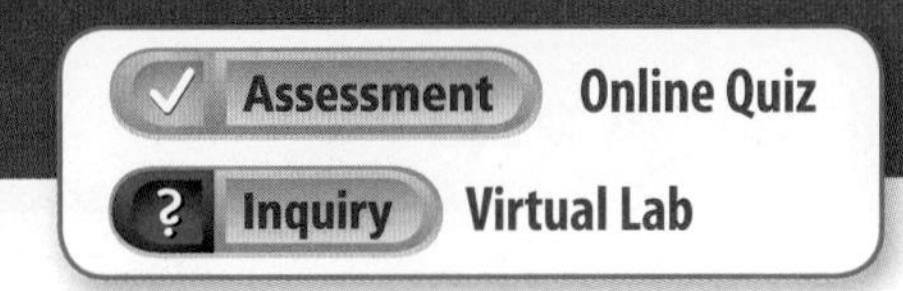

Visual Summary

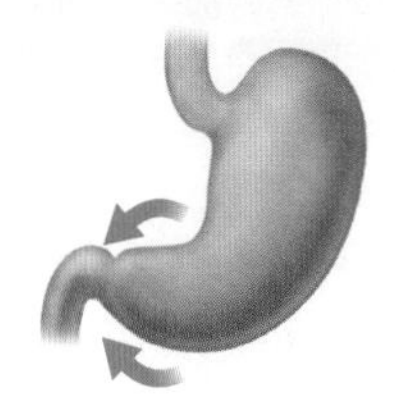

Muscles that surround the stomach help move food to the small intestine.

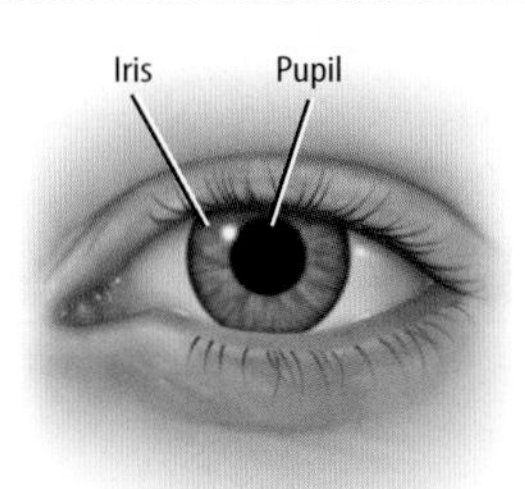

The nervous system coordinates the body's response to external stimuli.

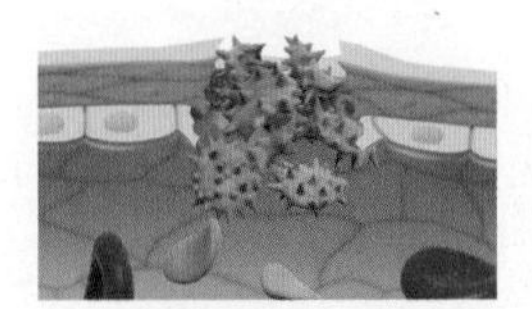

When you are bleeding, the circulatory system maintains homeostasis by controlling blood loss.

Use your lesson Foldable to review the lesson. Save your Foldable for the project at the end of the chapter.

What do you think NOW?

You first read the statements below at the beginning of the chapter.

4. Nutrients are processed by the skeletal system.

5. The nervous system moves oxygen through the body.

6. You do not control reflexes.

Did you change your mind about whether you agree or disagree with the statements? Rewrite any false statements to make them true.

Use Vocabulary

1. **Distinguish** between negative feedback and positive feedback.
2. The ability to maintain a constant environment in the body is ________.
3. **Use the term** *positive feedback* in a sentence.

Understand Key Concepts 

4. Which system sends electrical signals?
   - **A.** endocrine
   - **B.** muscular
   - **C.** nervous
   - **D.** respiratory
5. **Explain** the role of the muscular system in maintaining temperature homeostasis.
6. **Describe** how the muscular system helps substances such as nutrients and oxygen travel through the body.

Interpret Graphics

7. **Summarize** Copy the graphic organizer below. Use it to show how organ systems work together and process nutrients and remove food waste from the body.

Critical Thinking

8. **Evaluate** which body systems are working together in the figure shown below.

Math Skills

Math Practice

9. During normal breathing, the average human inhales about 0.5 L of air per breath. If a person takes 15 breaths per minute, what volume of air does the person inhale in 1 min?

Materials

stopwatch

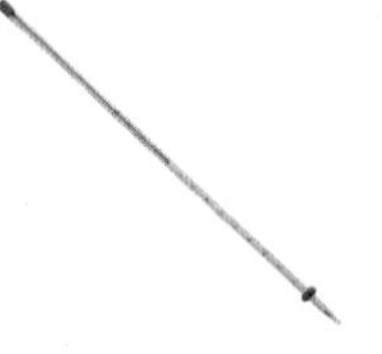

thermometer

Also needed: other materials approved by your teacher

Safety

How can a stimulus affect homeostasis?

Have you ever been on a roller coaster? Most people love the thrill and excitement of an amusement park ride. As the ride speeds up, slows down, and spins around, your heart races, you catch your breath, and goose bumps form on your arms. Because of outside stimuli, your body systems work overtime to keep homeostasis. This experiment will give you an opportunity to test the results of stimuli on yourself and your classmates.

Ask a Question

How does a stimulus affect your body?

Make Observations

1. Read and complete a lab safety form.
2. With your lab team, brainstorm a list of stimuli that affect the human body, such as a hearing a loud noise or viewing a roller coaster ride. From this list, select one stimulus that appeals to the group.
3. Create a plan to introduce the same stimulus to each member of your group. Make sure to get your teacher's approval before proceeding.

4. Copy the data table below in your Science Journal.
5. Sit quietly for 1 min. Measure your heart rate, your breathing rate, and your temperature. Record the measurements in your table. Also be sure to note the status of body systems, such as skin color, saliva production, and sweating, that might be affected by the stimulus.

Measurement	Before Stimulus	After Stimulus	10 Min After Stimulus
Heart rate			
Breathing rate			
Temperature			

Form a Hypothesis

6. Based on your knowledge of the stimulus you selected, form a hypothesis about the effect of your stimulus on heart rate, breathing rate, and body temperature.

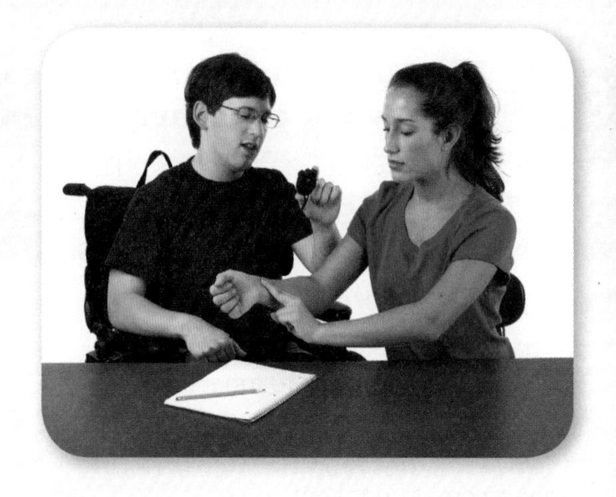

Test Your Hypothesis

7. Introduce the selected stimulus to each member of your team.
8. Immediately after the stimulus has been completed, measure your heart and breathing rates and your body temperature. Record the data. Note any changes in the other systems you were watching, and record their statuses.
9. Sit quietly for 10 min. Measure and record your heart rate, your breathing rate, and your body temperature.
10. Graph the results of all three intervals for each category—heart rate, breathing rate, and body temperature.

Lab Tips

- ☑ Heart and breathing rates are measured in occurrences per minute.
- ☑ Make sure all students are tested as uniformly as possible.

Analyze and Conclude

11. **Examine** the data from the procedure. How did your body respond to the stimulus? Was your hypothesis correct?
12. **Analyze** what happened to your heart rate, your breathing rate, and your body temperature during the experiment.
13. **The Big Idea** Explain how your results relate to homeostasis.

Communicate Your Results

Design a 30-second news flash skit to announce your results in a creative way. Have each team member take part in the presentation.

Extension

How do you think an athlete's body would respond to your stimulus? Could diet play a role? Explore how environment and lifestyle could alter the body's response to a stimulus and affect homeostasis.

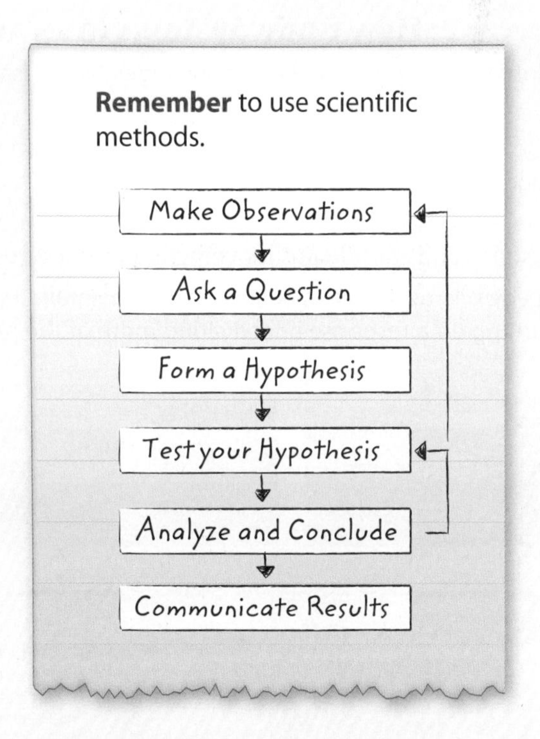

Chapter 21 Study Guide

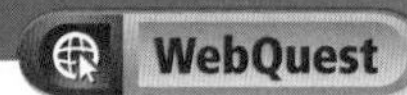

Human body systems are organized into systems that work together and maintain homeostasis.

Key Concepts Summary

Lesson 1: The Human Body

- Inorganic compounds are required for human body systems to function. Water is a universal solvent and essential for life.
- Organic compounds contain carbon and are essential to support life. **Macromolecules** are organic compounds that make up all cells.
- The body's organization plan helps groups of organs that perform the same function work together.

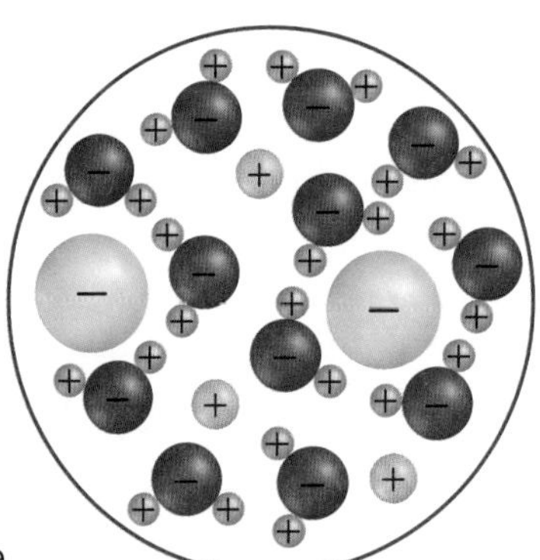

Vocabulary

macromolecule p. 768
monosaccharide p. 768
amino acid p. 768
nucleotide p. 768

Lesson 2: How Body Systems Interact

- Nutrients are processed when the digestive, muscular, and circulatory systems work together.
- The respiratory, circulatory, muscular, and digestive systems work together and process oxygen and remove wastes from the body.
- The nervous system communicates with the muscular and skeletal systems and coordinates movement and responds to stimuli.
- Feedback mechanisms help the body maintain **homeostasis** by turning off a response or activating more of the response.

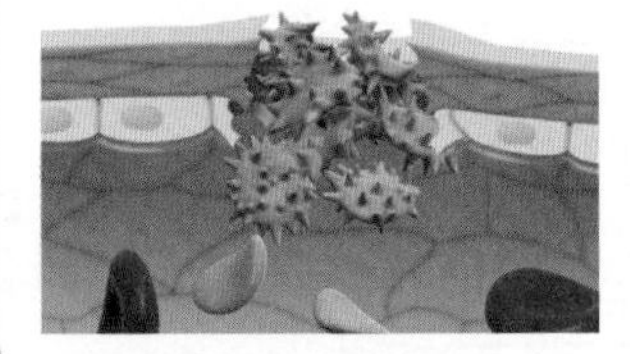

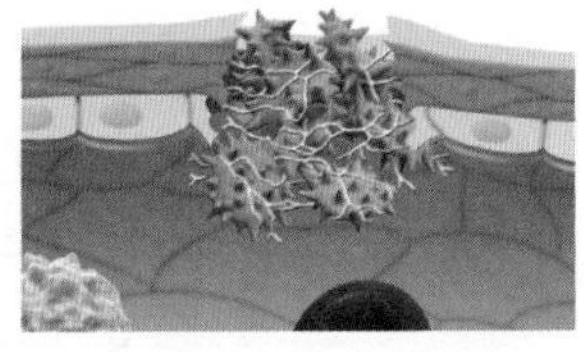

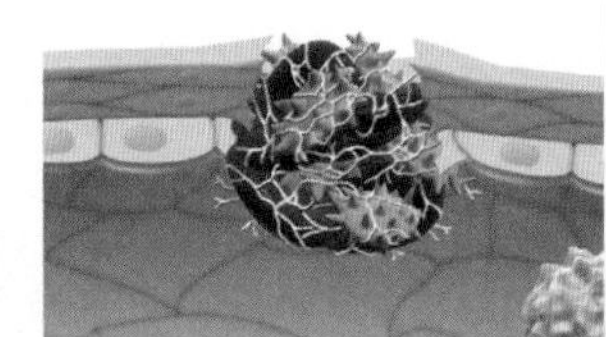

Vocabulary

homeostasis p. 773
negative feedback p. 780
positive feedback p. 780

Study Guide

- Personal Tutor
- Vocabulary eGames
- Vocabulary eFlashcards

FOLDABLES® Chapter Project

Assemble your lesson Foldables as shown to make a Chapter Project. Use the project to review what you have learned in this chapter.

Essential to Life
Water
Feedback
Interactions of Human Body Systems

Use Vocabulary

1. Lipids, proteins, carbohydrates, and nucleic acids are all ________.
2. DNA and RNA are made of ________.
3. Write the definition of *amino acids* in your own words.
4. Use the term *negative feedback* in a sentence.
5. Use the term *homeostasis* in a sentence.
6. When a response causes more of a response, ________ is occurring.

Link Vocabulary and Key Concepts

Copy this concept map, and then use vocabulary terms from the previous page and other terms from the chapter to complete the concept map.

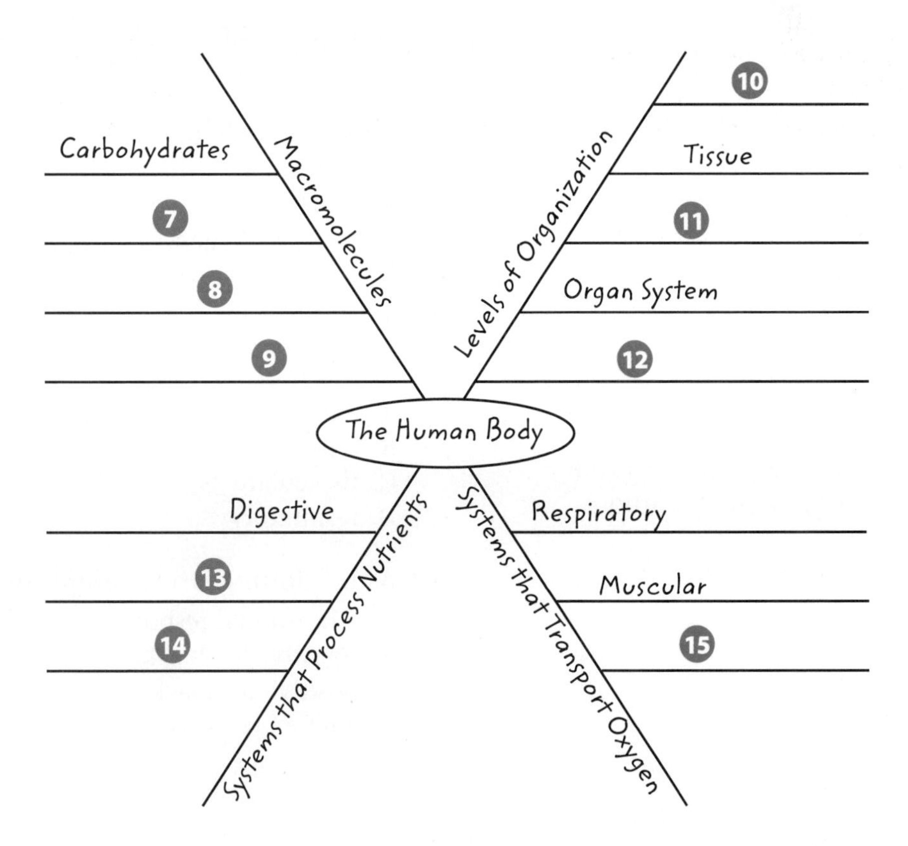

Chapter 21 Review

Understand Key Concepts

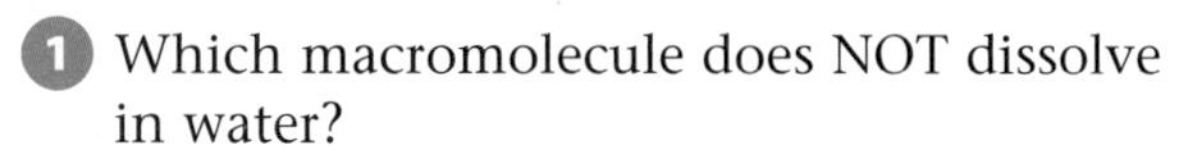

1. Which macromolecule does NOT dissolve in water?
   A. carbohydrate
   B. lipid
   C. nucleic acid
   D. protein

2. Which CANNOT be broken down into smaller parts?
   A. compounds
   B. elements
   C. macromolecules
   D. water

3. What is the structure shown below?

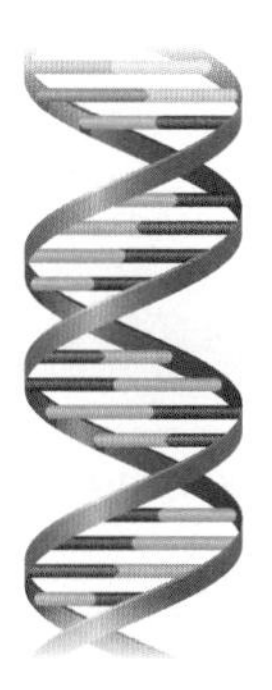

   A. amino acid
   B. DNA
   C. monosaccharide
   D. triglyceride

4. What is NaCl?
   A. an element
   B. a lipid
   C. an inorganic compound
   D. an organic compound

5. Which organ system works with the excretory system to remove carbon dioxide?
   A. circulatory
   B. digestive
   C. endocrine
   D. nervous

6. What is formed in the figure below?

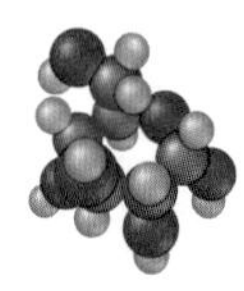

   A. carbohydrate
   B. lipid
   C. nucleic acid
   D. protein

7. What is the smallest living unit of the human body?
   A. cell
   B. organ
   C. system
   D. tissue

8. What are chemical signals produced by the endocrine system called?
   A. hormones
   B. reflexes
   C. ionic compounds
   D. nucleic acids

9. The muscular system and the nervous system work together and
   A. coordinate movement.
   B. digest food.
   C. process oxygen.
   D. transport blood.

10. Which organ system is NOT used to obtain nutrients from food?
   A. circulatory
   B. digestive
   C. muscular
   D. respiratory

11. Blood clotting is an example of
   A. a hormonal response.
   B. negative feedback.
   C. positive feedback.
   D. a reflex response.

Critical Thinking

12 **Summarize** how the body's organization helps it function.

13 **Compare** the compositions and functions of carbohydrates and lipids.

14 **Relate** the structure of the molecule shown below to its ability to dissolve ionic compounds.

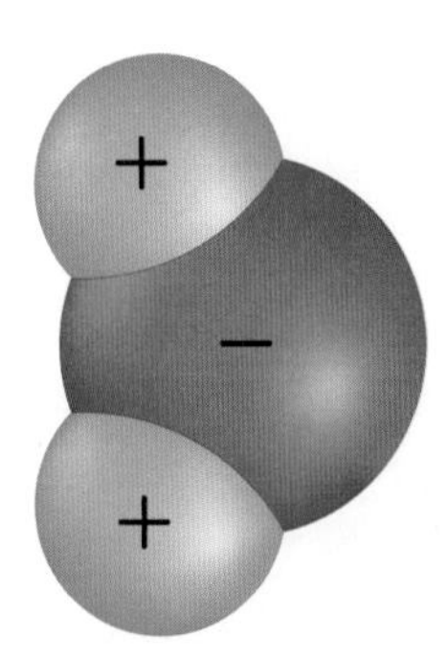

15 **Summarize** the role of carbon in enabling the body to function.

16 **Relate** the two functions of the small intestine to obtaining nutrients from food.

17 **Evaluate** the effect of anhydrosis, the inability to sweat, on temperature homeostasis.

18 **Compare** the roles of the circulatory system in obtaining nutrients and wastes.

19 **Relate** homeostasis to negative feedback.

Writing in Science

20 **Write** a five-sentence paragraph that describes how organ systems interact to carry out functions in the body. Be sure your paragraph includes an example of how organ systems interact to carry out a specific function.

REVIEW THE BIG IDEA

21 **Assess** how the nervous system interacts with other organ systems and coordinates responses to the external environment.

22 How do human body systems in the lungs interact and support life?

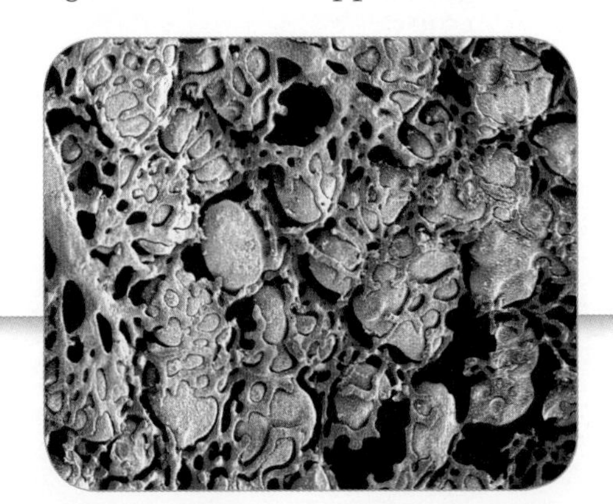

Math Skills

Math Practice

Use Volume

23 The average human stomach ranges from a volume of 0.08 L empty to 4 L full. What is the total volume of food and gastric secretions required to fill the stomach?

24 A person consumes 2 L of fluid per day. If the person urinates a total of 1,400 mL in 24 h, what volume of liquid is lost in other ways or retained in the body? (Hint: 1 L = 1,000 mL)

25 Normal feces are about 75% solid and 25% water. If the volume of a person's feces is 0.6 L, what is the volume of water in the feces? (Hint: What is 25% of 0.6 L?)

Dr. Richard Kessel & Dr. Randy Kardon/Tissues & Organs/Visuals Unlimited.

Standardized Test Practice

Record your answers on the answer sheet provided by your teacher or on a sheet of paper.

Multiple Choice

1 Which ion is found in bone and plays a role in nerve and muscle cell function?

A calcium (Ca^{2+})

B chloride (Cl^{-})

C fluoride (F^{-})

D sodium (Na^{+})

2 Which is NOT a function of proteins?

A to dissolve ionic substances

B to give cells structure

C to help cells communicate

D to serve as enzymes

Use the figure below to answer question 3.

3 The macromolecule shown in the figure belongs to which class of compounds?

A carbohydrates

B lipids

C nucleotides

D proteins

4 Which organs help maintain fluid homeostasis?

A heart and kidneys

B kidneys and skin

C lungs and heart

D skin and lungs

5 The digestive system and the circulatory system work together and absorb nutrients in which structure?

A alveolus

B kidney

C thyroid

D villus

Use the figure below to answer question 6.

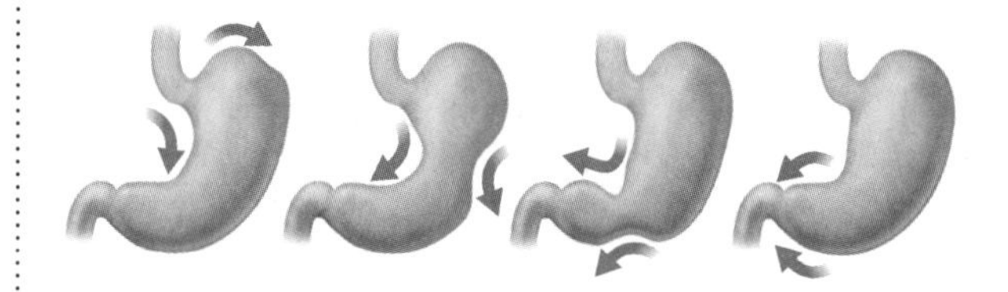

6 The figure above shows a step in moving food through the body. Which body systems are involved in this step?

A digestive and respiratory

B muscular and circulatory

C muscular and digestive

D skeletal and respiratory

7 Which systems work together and transport oxygen and carbon dioxide through the body?

A circulatory and respiratory

B digestive and excretory

C endocrine and nervous

D muscle and skeletal

Assessment
Online Standardized Test Practice

Use the figure below to answer question 8.

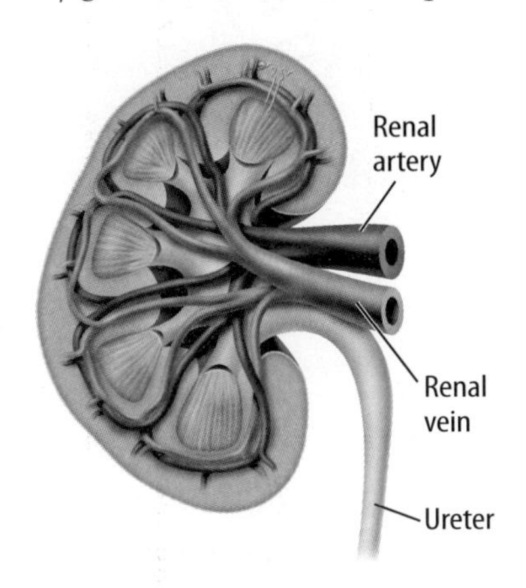

8 Which body systems are linked by the role of the organ shown in the figure?

A circulatory and digestive
B digestive and respiratory
C excretory and circulatory
D skeletal and excretory

9 Which body systems help the body respond to stimuli?

A circulatory and respiratory
B digestive and excretory
C endocrine and nervous
D immune and skeletal

10 Which is a major role of water in the body?

A It forms cell membranes.
B It joins together amino acids.
C It stores information in the body.
D It transports ions through the body.

Constructed Response

Use the figure below to answer questions 11 and 12.

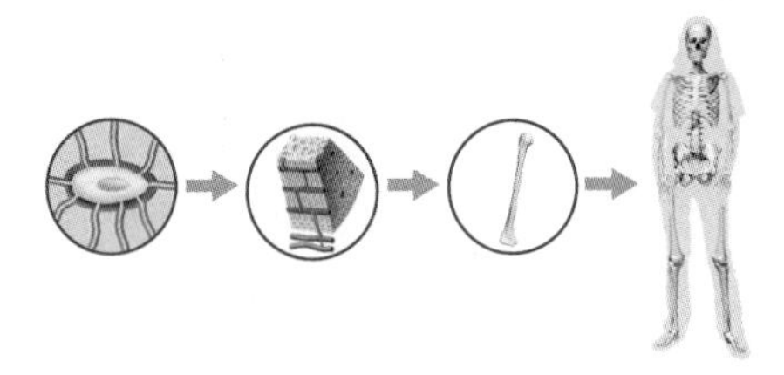

11 Use the figure above to explain how the human body is organized into four levels. Identify and define each level in your response.

12 Based on the figure, describe the four levels of organization in the circulatory system. Give an example of structures at each level.

13 Explain how the body responds to increases and decreases in temperature. Are these responses negative or positive feedback?

14 Polio is a disease of the nervous system that paralyzes muscles. In the mid-twentieth century many children contracted polio. Some died because they could not breathe. Use your knowledge of the interactions between body systems to explain how polio could affect breathing.

NEED EXTRA HELP?														
If You Missed Question...	1	2	3	4	5	6	7	8	9	10	11	12	13	14
Go to Lesson...	1	1	1	2	2	2	2	2	2	1	1	1	2	2

Chapter 22

Heredity and How Traits Change

THE BIG IDEA **How do species adapt to new environments over time?**

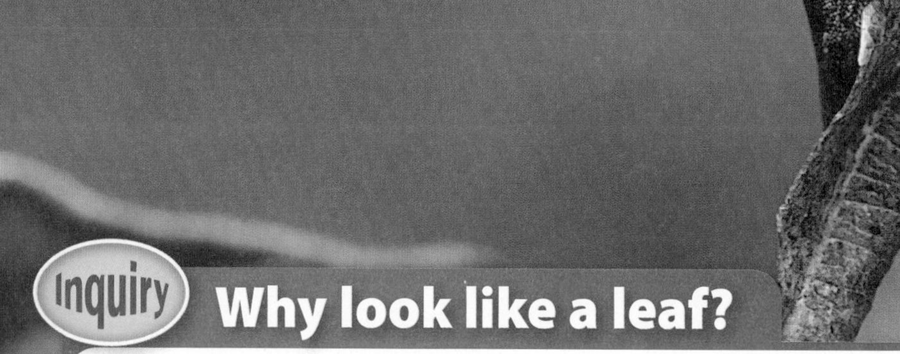

Inquiry Why look like a leaf?

This gecko looks like the dead leaves of this plant. It also blends in well with dead leaves on the forest floor.

- What is the advantage of this gecko looking like a plant leaf?
- What environmental factors might explain how the gecko looks?
- How do species adapt to new environments over time?

Get Ready to Read

What do you think?

Before you read, decide if you agree or disagree with each of these statements. As you read this chapter, see if you change your mind about any of the statements.

1. Genes are on chromosomes.
2. Only dominant genes are passed on to offspring.
3. Modern-day genetics disproved Gregor Mendel's ideas about inheritance.
4. Mutations can cause disease in an individual.
5. A population that lacks variation among its individuals might not be able to adapt to a changing environment.
6. Extinction occurs when the last individual of a species dies.

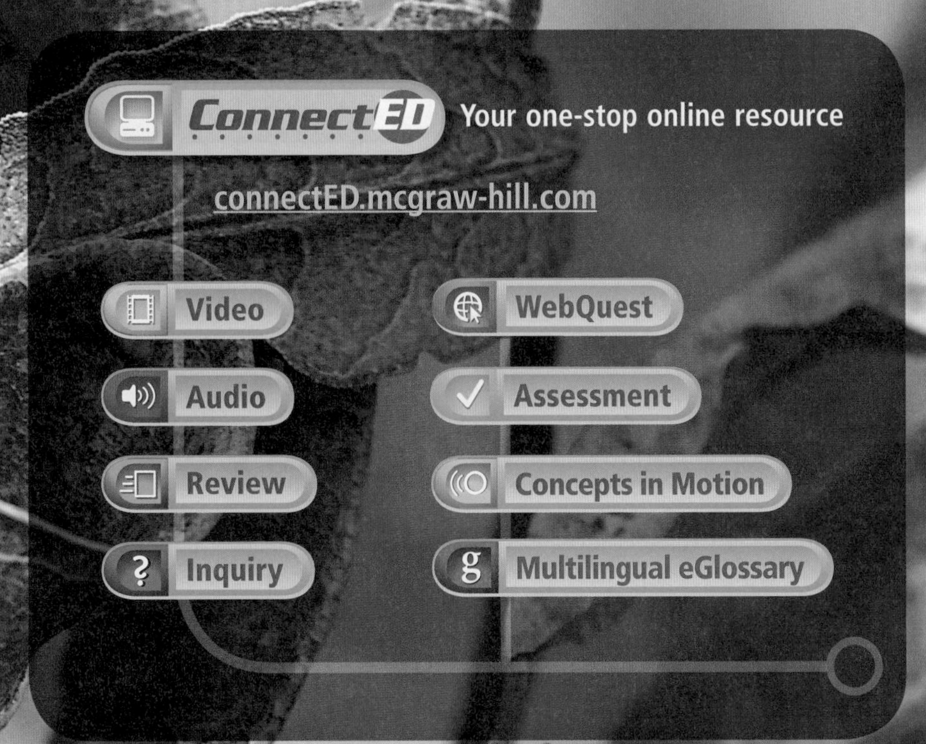

Lesson 1

Reading Guide

Key Concepts

ESSENTIAL QUESTIONS

- How are traits inherited?
- Why do scientists study genetics?
- What did Gregor Mendel investigate and discover about heredity?

Vocabulary

heredity p. 793
genetics p. 793
selective breeding p. 796
dominant trait p. 797
recessive trait p. 797
genotype p. 798
phenotype p. 798
heterozygous p. 798
homozygous p. 798

Multilingual eGlossary

Video BrainPOP®

How are traits inherited?

Inquiry Like Mother, Like Daughter?

How similar or different do you think these sandhill crane nestlings are from their mother? How similar or different are they from each other? Why? Eventually, they will make the bugling and rattling sounds of sandhill cranes. Are they born knowing how to make these sounds, or do they have to learn by listening to them?

Launch Lab

10 minutes

How are traits expressed?

Some of the traits you inherited from your parents are expressed in one of two forms. You can determine which form of a trait you express through observation.

1. Study the pictures in the table to the right. Note the different forms in which some traits can be expressed.
2. Have a classmate help you identify which form of each trait you express. Record your results in the class data sheet.
3. Calculate the percentage of each form of the traits using data from the entire class.

Trait Expression		
	Dimples	No dimples
Trait 1		
	Cleft chin	Smooth chin
Trait 2		

Think About This

1. Which traits do you express? What were the percentages for your class for each trait?
2. Based on the percentages, which form of each trait do you think might be expressed more often?
3. **Key Concept** How do you think you inherited these traits?

From Parent to Offspring

Although you and your classmates are all the same species, *Homo sapiens,* you have different eye colors, hair colors, and heights. Why are you all so different? How do you inherit traits from your parents? Do all organisms inherit traits through the same processes? **Heredity** (huh REH duh tee) *is the passing of traits from parents to offspring.* Scientists studying heredity have been answering questions such as these for centuries.

What is genetics?

Genetics (juh NE tihks) *is the study of how traits pass from parents to offspring.* All organisms have genes. Genes determine everything from an organism's shape to its life functions. Genes can even control how an organism behaves.

For most organisms, genes are sections of DNA that contain information about a specific trait of that organism. This information can vary. A gene with different information for a trait is called an allele. For example, dimples on some of your classmates' faces result from alleles on a pair of chromosomes. **Figure 1** illustrates the relationship between alleles on chromosomes and the traits they express. Recall that chromosomes are coiled strands of DNA.

Concepts in Motion
Animation

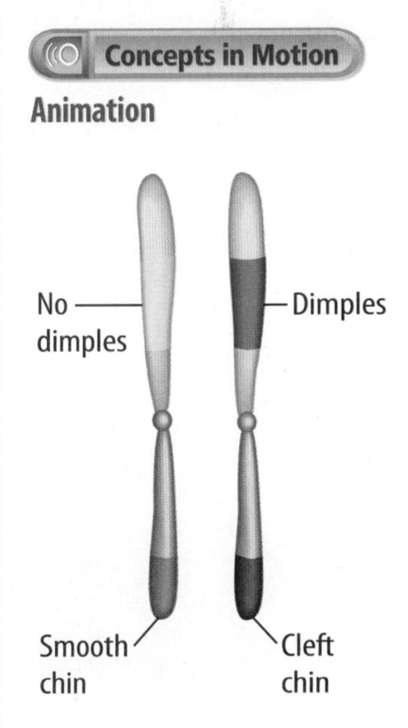

Figure 1 Each chromosome pair has genes for the same traits. A gene's alleles are in the same location on each chromosome of a pair.

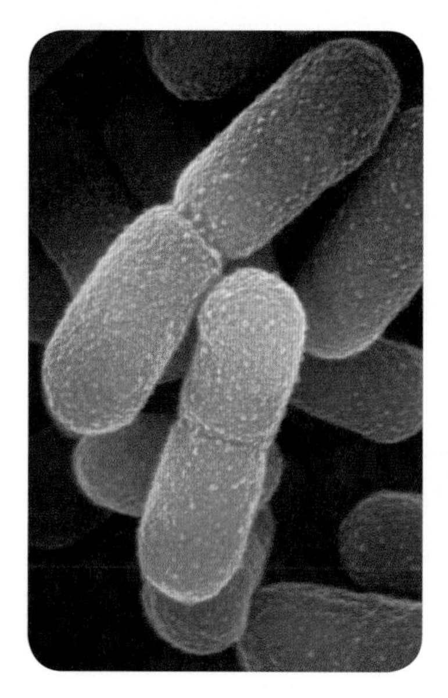

Bacteria, such as these *E. coli,* can reproduce asexually.

Some plants, such as this spider plant, can reproduce asexually.

In sexual reproduction, an offspring's genes come from a sperm and an egg.

Figure 2  Asexual and sexual reproduction both pass traits to offspring, but in different ways and with different results.

REVIEW VOCABULARY

trait
a distinguishing characteristic of an organism

How are traits inherited?

An organism passes its **traits** to its offspring in one of two ways—through asexual reproduction or through sexual reproduction. Unicellular organisms, such as the bacteria shown in **Figure 2,** and some multicellular organisms, such as the spider plant in **Figure 2,** can reproduce asexually. In asexual reproduction, one organism makes a copy of its genes and itself. In sexual reproduction, offspring, such as the golden orioles in **Figure 2,** receive half of their genes from an egg cell and the other half from a sperm cell.

An individual organism expresses the traits in the genes it inherited. For example, if your eyes are blue, it is because you inherited the genes for blue eye color from your parents. In fact, the specific combination of the genes you inherited from your parents for all of your traits is unique. Unless you are an identical twin, triplet, or quadruplet, there is no other person with the same combination of all of your genes!

Inherited traits differ from traits that an individual acquires, or learns, during its lifetime. For example, a bird's size primarily is inherited from its parents. However, the song it sings primarily is learned. The juvenile golden orioles in **Figure 2** will learn to sing by listening to their parents' songs. Obedience in domesticated animals is another example of an acquired trait. You can teach a dog to sit, but its puppies will not be born already knowing that they should sit when you tell them.

Key Concept Check How are traits inherited?

Why do scientists study genetics?

Scientists began studying genetics to understand how traits are inherited. They soon learned that genes control how an organism develops. They also learned that sometimes genes play a role in the development of disease. In addition, scientists use genetics to find out more about how species are related.

Development and Disease By studying genetics, scientists have learned that genes control how organisms develop. For example, genes control limb development, body segmentation, and the formation of organs, such as eyes and ears. Studying the genetics of development also can help scientists understand more about disease in humans. For example, scientists have learned how problems with genes in fish can result in diseases in those fish. Scientists can use what they have learned and apply it the study of human genetics.

Common Ancestors Studying genetics also can help scientists determine how organisms are related. **Figure 3** shows how scientists placed a gene that controls eye development in mice into a fruit fly. The fly developed normal eyes! In another experiment, scientists placed the gene that controls eye development in fruit flies into a frog. Like the fly that received the mouse gene, the frog developed normal eyes. Because the genes from these organisms are similar enough to produce normal eyes when the genes are exchanged, scientists suspect that these species share a common, ancient ancestor.

Key Concept Check Why do scientists study genetics?

FOLDABLES®

Make a horizontal two-tab book with a tab top. Label it as shown. Use it to organize information about the study of genetics.

Development and Disease | Common Ancestors

Why Scientists Study Genetics

Figure 3 When genes that control eye development were transferred between invertebrates and vertebrates, each organism still developed normal eyes.

Visual Check Explain the difference between Experiment A and Experiment B.

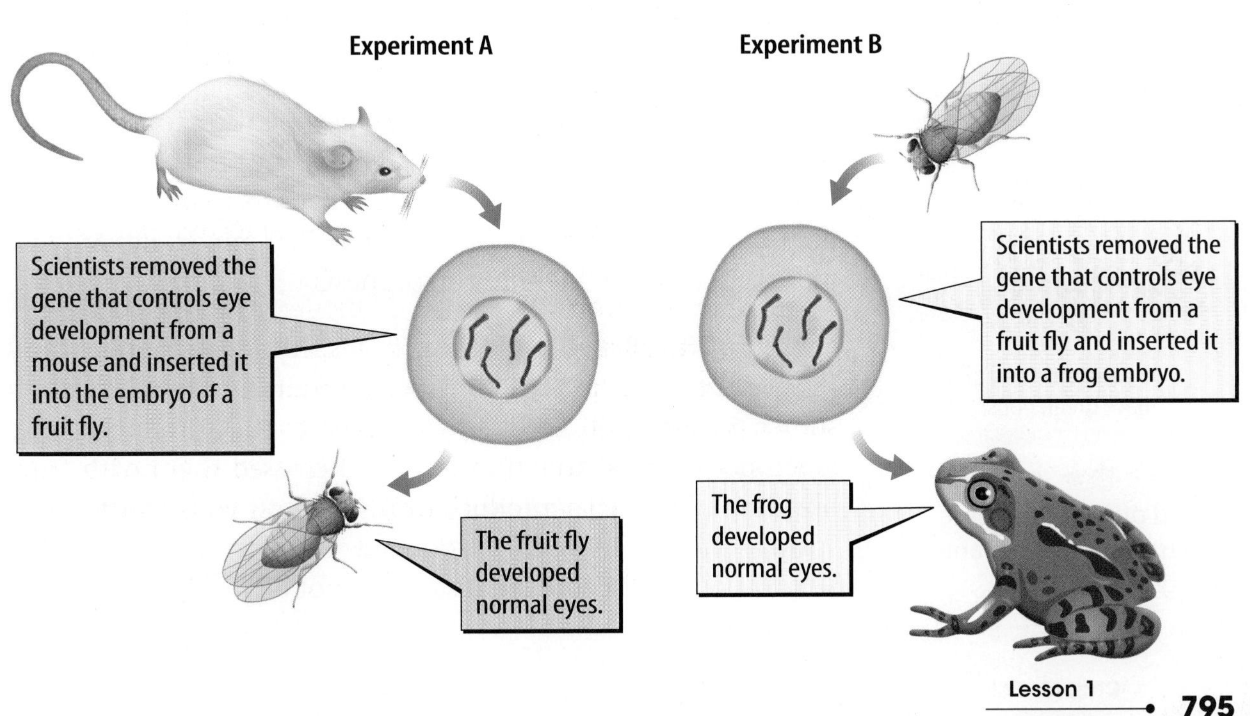

▲ **Figure 4** In selective breeding, plants or animals with desired traits are bred to produce offspring with those traits.

Heredity—the History and the Basics

For thousands of years, humans have been slowly improving their crops, such as corn and apples, as well as their farm animals through a method called selective breeding. **Selective breeding** *is the selection and breeding of organisms for desired traits.* Suppose you are a farmer who owns the hens shown in **Figure 4.** The average number of eggs per year produced by each hen is shown below its nest. If you wanted to breed hens that produced more eggs, which hen would you breed with the neighbor's rooster? Why?

Throughout history, people have successfully used selective breeding to produce trees that grow larger fruit or cows that produce more milk. However, people did not always get the results they expected. It was not until an Austrian friar named Gregor Mendel began experimenting with pea plants that people understood more about how selective breeding works.

Reading Check What is selective breeding?

Mendel's Experiments

In 1856, Gregor Mendel began experimenting to answer the question of how traits are inherited. At the time, most scientists thought that traits blended from parents to offspring, similar to the way two paint colors blend when mixed. But Mendel did not accept the blending hypothesis.

Crossing True-Breeding Plants To test his ideas, Mendel carefully selected pea plants with specific traits and bred them. As shown in **Figure 5,** Mendel chose plants that produced only green pods, called true-breeding, and crossed them with true-breeding plants that produced only yellow pods. All the offspring, called hybrids, produced only green pods. The yellow-pod trait seemed to disappear, not blend with the green-pod trait.

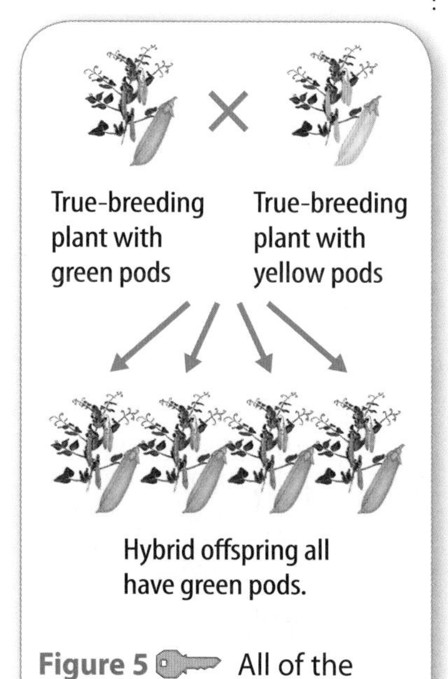

Figure 5 All of the hybrid offspring of these two true-breeding plants had green pods.

Crossing Hybrids When Mendel crossed two hybrid plants with green pods, the cross resulted in offspring with green pods and offspring with yellow pods, as shown in **Figure 6.** These offspring were in a ratio of about 3:1, green to yellow. Mendel tested thousands of pea plants. He tracked traits such as seed shape and flower color. The crosses between hybrids for each trait produced a similar 3:1 ratio. Mendel proposed several ideas to explain his results.

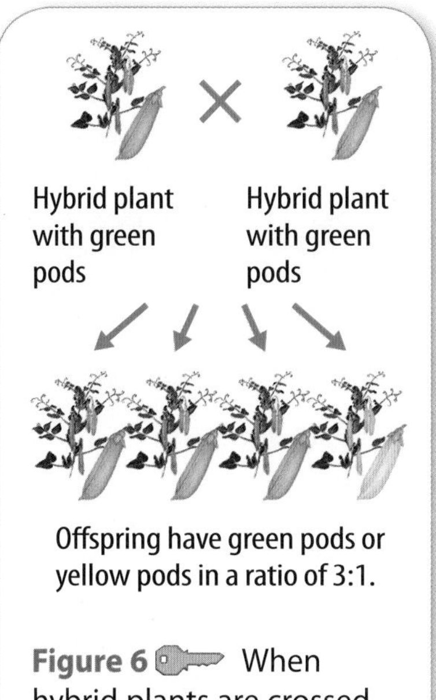

Figure 6 When hybrid plants are crossed, offspring produced are in a 3:1 ratio.

Dominant and Recessive Alleles

Mendel proposed that instead of blending, some traits of organisms are dominant, while others are recessive. *A* **dominant trait** *is a genetic factor that blocks another genetic factor. A* **recessive trait** *is a genetic factor that is blocked by the presence of a dominant factor.* When an individual has one dominant allele and one recessive allele for a trait, the dominant trait is expressed. This explains why the offspring of the true-breeding green-pod plants and the true-breeding yellow-pod plants all produce green pods. Green pods are a dominant trait, and yellow pods are a recessive trait.

Inquiry MiniLab

How can you model Mendel's principles?

Recall that during sex-cell formation, different combinations of alleles can occur randomly.

1. Obtain two **allele cards.** Determine if your parent plant is green or yellow. Join the group that has the same color plant as you do.
2. Pair up with one student from the other group. Randomly choose one of your allele cards, and have your partner do the same. Write the allele combination of these cards in your Science Journal. Repeat this process to create the genetic identity of a second offspring.
3. Assume the identity of one of the first-generation offspring. Take the appropriate allele cards to represent your identity. Declare if you are a green or a yellow plant. Record your results on a class data sheet.
4. Pair up with another person in your class. Choose a card at random, and have your partner do the same. Write down the allele combination of the second-generation offspring you created. Repeat three more times. Then record your data on the class data sheet.

Analyze and Conclude

1. **Identify** What color was your parent plant? What allele combination gave the plant its color?
2. **Analyze** What were the totals for yellow and green plants in the first-generation offspring? What allele combinations made up these plants?
3. **Key Concept** Determine the ratio of yellow to green plants in the second-generation offspring. How does this compare with the ratio that Mendel discovered with his crosses?

Mendel's Principles of Inheritance

Earlier in this lesson, you read that an allele is one form of a gene. Mendel did not know about genes or alleles. Mendel did suspect, however, that a physical factor was responsible for the traits of his pea plants. Mendel was right. The factors Mendel proposed now are called genes. Which alleles are present on a pair of chromosomes determines whether an individual has the dominant or the recessive trait. *The alleles of all the genes on an organism's chromosomes make up the organism's* **genotype** (JEE nuh tipe). *How the traits appear, or are expressed, is the organism's* **phenotype** (FEE nuh tipe).

WORD ORIGIN

phenotype
from Greek *phainein,* means "to show"; and *typos,* means "type"

Mendel's hybrid pea plants had genotypes of one allele for green pods and one allele for yellow pods. The phenotypes of these plants were green pods. *When an organism's genotype has two different alleles for a trait, it is called* **heterozygous** (he tuh roh ZI gus). The hybrid plants were heterozygous for pod color. *When an organism's genotype has two identical alleles for a trait, it is called* **homozygous** (hoh muh ZI gus). Mendel's true-breeding plants were homozygous for pod color. Homozygous and heterozygous genotypes also affect the phenotypes of other organisms, such as the gerbils shown in **Figure 7.**

Next, you will read about the rediscovery of Mendel's work. Scientists have confirmed and built upon Mendel's ideas as they learned more about genetics and heredity.

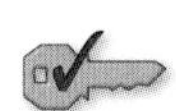

Key Concept Check What did Mendel investigate and discover about heredity?

Figure 7 Agouti is a dominant trait *(A)* and produces a phenotype of hairs with alternating bands of color and no color. The recessive trait *(a)* produces hairs of all one color.

Visual Check What is the genotype of the agouti gerbil?

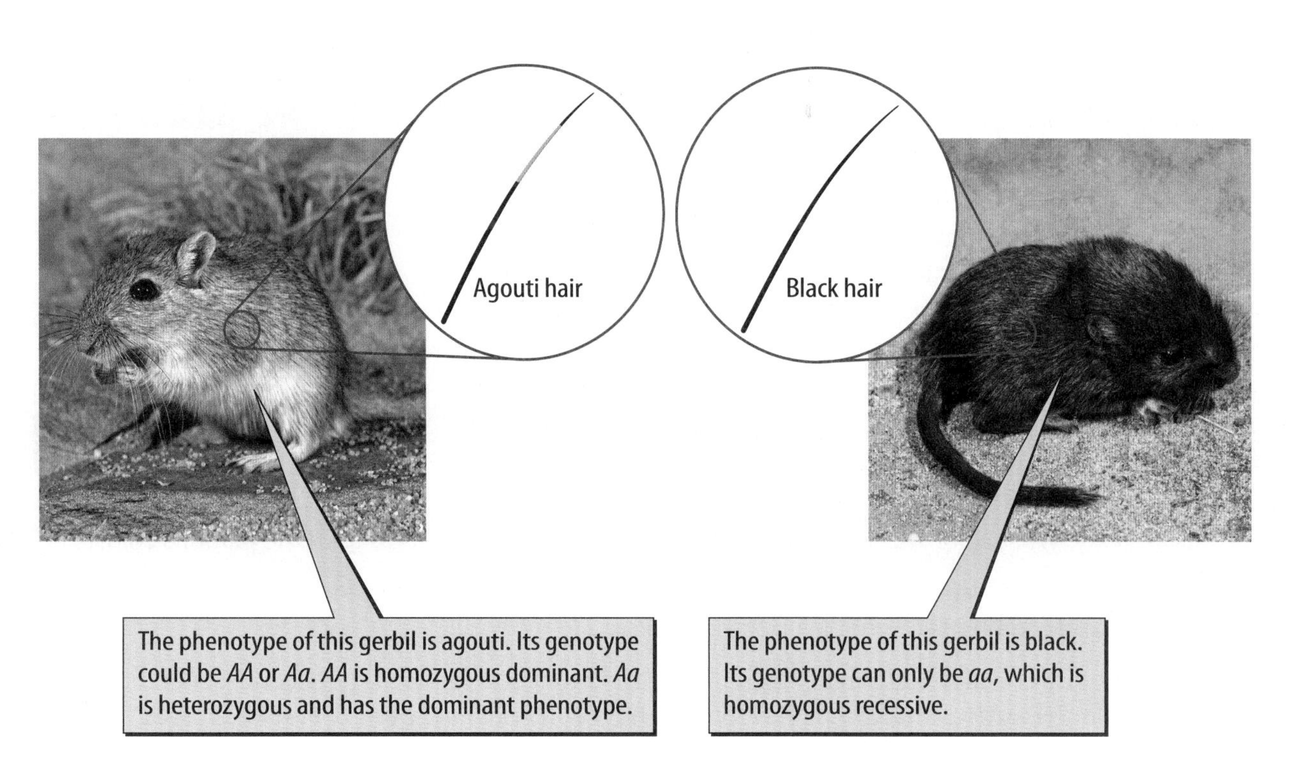

Lesson 1 Review

Assessment Online Quiz
Inquiry Virtual Lab

Visual Summary

Traits are passed from parents to offspring during asexual or sexual reproduction.

Through selective breeding, Mendel showed that some traits are dominant and some traits are recessive.

An organism's genotype can be homozygous or heterozygous.

FOLDABLES®

Use your lesson Foldable to review the lesson. Save your Foldable for the project at the end of the chapter.

What do you think 

You first read the statements below at the beginning of the chapter.

1. Genes are on chromosomes.
2. Only dominant genes are passed on to offspring.

Did you change your mind about whether you agree or disagree with the statements? Rewrite any false statements to make them true.

Use Vocabulary

1. **Distinguish** between genotype and phenotype.
2. **Define** *homozygous* and *heterozygous.*
3. **Use the term** *selective breeding* in a complete sentence.

Understand Key Concepts 

4. A genetic factor that blocks another genetic factor is called a ________ trait.
   A. dominant
   B. heterozygous
   C. homozygous
   D. recessive
5. **Summarize** how studying genetics helps scientists learn more about how species are related.
6. **Describe** how traits are inherited from parents by offspring.

Interpret Graphics

7. **Predict** Each of the organisms below is carrying a recessive allele for white feathers. Predict the ratio of black-feathered to white-feathered chicks in their offspring. Explain your prediction.

8. **Identify** Copy and fill in the table below to differentiate between true-breeding plants and hybrids.

True-breeding	
Hybrid	

Critical Thinking

9. **Design an experiment** to test whether curled wings are dominant over straight wings in fruit flies.

AMERICAN MUSEUM OF NATURAL HISTORY

When Asthma Attacks

Is asthma caused by genes, the environment, or both?

More than 6 million children under the age of 18 in the United States have asthma. During an asthma attack, airways tighten up and become filled with mucus. This makes it hard to breathe. If you or someone in your family has asthma, you might know the signs of an asthma attack—difficulty breathing, wheezing, coughing, and tightness of the chest. While the symptoms can come and go, the disease is always there.

Asthma is caused by a combination of genes and environmental factors. Children are more likely to have asthma if a parent or close relative has it. Allergies and some respiratory infections can cause asthma. If asthma runs in a person's family, exposure to pollen, dust mites, animal hair, cockroaches, and cigarette smoke can trigger the development of asthma.

Currently, there is no cure for asthma, but medications can help people prevent or treat asthma attacks. Scientists hope to learn more about the genetic and the environmental causes of asthma so they can improve medicines and someday find a cure.

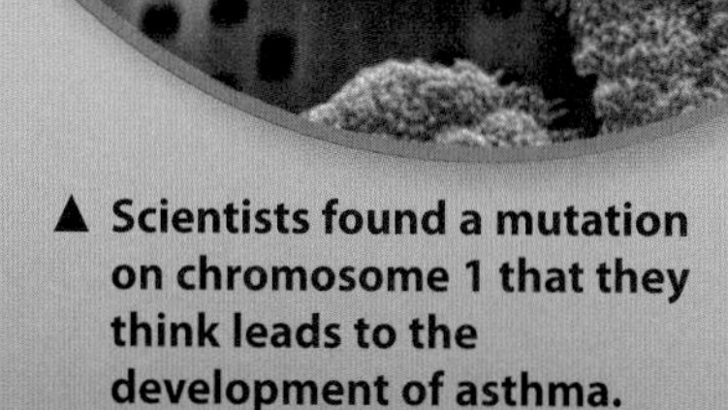

▲ **Scientists found a mutation on chromosome 1 that they think leads to the development of asthma.**

To learn more about the genetic causes of asthma, scientists studied a group of people called the Hutterites. The Hutterites live in small, isolated communities in rural parts of the north-central United States and central Canada. The Hutterites all share the same diet and lifestyle, but only some individuals have asthma. Researchers learned that Hutterites with asthma have a tiny mutation on a particular gene. Researchers suspect this mutation makes Hutterites more likely to have or develop asthma. This discovery might help doctors predict and prevent asthma in Hutterites and people worldwide.

Scientists studied the genes of the Hutterites to learn more about the causes of asthma. ▶

INTERVIEW How does your school help reduce the presence of asthma-attack triggers in the environment? To find out, make a list of questions and interview a school official. Then, list ways to make your school a more healthful environment.

Lesson 2

Reading Guide

Key Concepts
ESSENTIAL QUESTIONS

- How can you use tools to predict genetic outcomes?
- What are the other patterns of inheritance?
- What role can mutations play in the inheritance of disease?

Vocabulary

monohybrid cross p. 803
Punnett square p. 803
incomplete dominance p. 804
codominance p. 804
multiple alleles p. 804
sex-linked trait p. 805
polygenic inheritance p. 805
pedigree p. 806
mutation p. 806
genetic engineering p. 807

Multilingual eGlossary

Genetics After Mendel

Inquiry White Eyes?

About 100 years ago, a mutant white-eyed fruit fly just like this one was suddenly born into a population of normal fruit flies. Since fruit flies usually have red eyes, the white-eyed fly surprised scientists. Would Gregor Mendel have been able to explain the appearance of this white-eyed fly?

Inquiry Launch Lab

20 minutes

Can you model probable outcomes?

The study of heredity includes studying probability. Probability is the likelihood that a specific outcome will occur and usually is measured in terms of percentages.

1. Read and complete a lab safety form.
2. Obtain a **Group ID card.**
3. If you are in group A, obtain a **coin** and flip it 100 times. For each flip, record whether the coin landed heads-up or tails-up in your Science Journal.
4. If you are in group B, use your imagination to flip a coin 100 times. For each flip, record whether the coin landed heads-up or tails-up.
5. Hand in your data to your teacher. Your teacher will use the data to try to determine which group actually flipped a coin.

Think About This

1. How did your teacher determine which data were real and which were not real?
2.  **Key Concept** How are data used to help determine probability? How do you think the sample size used in an experiment affects the reliability of data?

Rediscovering Mendel's Work

Today, Gregor Mendel is considered the father of genetics. Soon after he published his results on pea-plant genetics, however, his work mostly was forgotten. Scientists continued to support the idea that traits blended from parents to offspring. They did not know about genes and chromosomes.

Mendel's work was rediscovered in 1900. At this time, genetics was a rapidly growing field of science. Scientists had discovered chromosomes and could see them inside cells. They also thought that the cell nucleus contained genes. Scientists soon realized that genes were on chromosomes in the nucleus. They confirmed that genes were Mendel's dominant and recessive factors. The next step was to learn more about how to predict patterns of inheritance.

Reading Check Explain why the rediscovery of Mendel's pea-plant experiments was important to genetics.

Predicting Genetic Outcomes

Think about flipping a coin into the air. The chance that the coin will land heads-up is one-half, or 50 percent. The chance that the coin will land tails-up is also one-half, or 50 percent. The chance of a coin landing heads-up two times in a row is $\frac{1}{2} \times \frac{1}{2}$, which equals one-quarter, or 25 percent.

Math Skills

Use Probability

Probability is a ratio that compares the number of ways a certain outcome occurs to the number of possible outcomes. If you have a regular six-sided die, what is the probability of rolling an even number?

$$\frac{\text{number of sides with even numbers}}{\text{total number of sides}} = \frac{3}{6}$$

Reduce to lowest terms.

$$\frac{3}{6} = \frac{1}{2}$$

Practice

With the same six-sided die, what is the probability of rolling either a 3 or a 5?

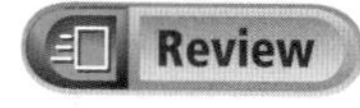

- Math Practice
- Personal Tutor

Probability

If you flipped a coin ten times in a row, what is the chance that half of your flips will be heads? Using probability, you might predict a heads-to-tails ratio of 5:5. However, while the 5:5 ratio is the most probable outcome, any outcome is possible. Probabilities are predictions; they do not guarantee outcomes. Your coin flips could result in ten heads in a row.

Punnett Squares and Predicting Genetic Outcomes

With enough data, Mendel was able to predict the outcome of a monohybrid cross. *A cross between two individuals that are hybrids for one trait is a* **monohybrid cross.** Mendel predicted a 3:1 ratio of the dominant phenotype to the recessive phenotype. When Mendel crossed many sets of heterozygous plants that produced green pods, 428 of the offspring produced green pods, and 152 produced yellow pods. The results were close to the 3:1 ratio he predicted. Mendel knew that the probability of getting green pods was three-quarters, or 75 percent.

A **Punnett square** *shows the probability of all possible genotypes and phenotypes of offspring.* **Figure 8** shows a Punnett square for a monohybrid cross between pea plants with green pods. The Punnett square predicts that 75 percent of the offspring will express the dominant phenotype of green pods.

Key Concept Check How does a Punnett square help scientists predict genetic outcomes?

Personal Tutor

	G	g
G	GG	Gg
g	Gg	gg

Phenotypes—3 green, 1 yellow
Genotypes—1 *GG*, 2 *Gg*, 1 *gg*

Figure 8 The Punnett square shows the predicted outcome of the cross between two heterozygous green-pod pea plants.

Visual Check What percentage of the offspring will be heterozygous?

Inquiry MiniLab

10 minutes

How can you predict outcomes using a Punnett square?

The Punnett square, right, shows a cross between two plants and their alleles for flower color. The dominant allele *R* indicates purple flowers, and the recessive allele *r* indicates white flowers.

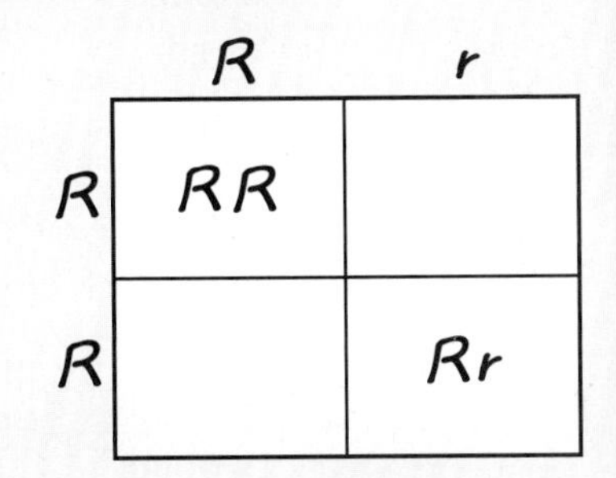

1. Copy and complete the Punnett square in your Science Journal.
2. Create a second Punnett square using the two offspring you filled in from the first cross.

Analyze and Conclude

1. **Describe** What color were the flowers of the parent plants, crossed in the first Punnett square? What was the ratio of offspring with purple flowers to offspring with white flowers?
2. **Analyze** What was the ratio of purple flowers to white flowers in the offspring of the second Punnett square you completed?
3. **Key Concept** Evaluate what the results of the Punnett square mean in terms of what you could expect in real life.

Incomplete Dominance

Parent plant produces round radishes.

Parent plant produces oblong radishes.

Offspring produce oval radishes.

Figure 9 Incomplete dominance produces offspring that are a combination of the parental traits.

Other Patterns of Inheritance

Not all traits are dominant or recessive. Some traits are expressed by two alleles at the same time, resulting in either incomplete dominance or codominance. Other traits are controlled by many alleles.

Incomplete Dominance

When an offspring's phenotype is a combination of its parents' phenotypes, it is called **incomplete dominance.** Neither allele is dominant. Instead, both alleles are expressed, producing a phenotype that looks like a combination, or blend, of the parental traits. The oval radishes in **Figure 9** resulted from a cross between a radish plant that produces round radishes and one that produces oblong radishes.

Codominance and Multiple Alleles

When both alleles can be independently observed in a phenotype, it is called **codominance.** Some human blood types show codominance. If a person receives a type A allele from one parent and a type B allele from the other parent, he or she will have type AB blood.

Human blood type is also an example of **multiple alleles,** *or a gene that has more than two alleles.* There are three different alleles for the ABO blood type in humans—I^A, I^B, and i. The I^A and I^B alleles are codominant to each other, but both are dominant to the i allele. Even though there are multiple alleles, a person only can inherit two of these alleles—one from each parent, as shown in **Table 1.**

Table 1 The blood-group system in humans has three alleles. It is an example of multiple alleles as well as codominant alleles.

Table 1 Human ABO Blood Types

Phenotype	Possible Genotypes
Type A	I^AI^A or I^Ai
Type B	I^BI^B or I^Bi
Type O	ii
Type AB	I^AI^B

Sex-Linked Traits

Sex chromosomes determine an organism's gender, or sex. Females have two X chromosomes, and males have an X and a Y chromosome. *When the allele for a trait is on an X or Y chromosome, it is called a* **sex-linked trait.** Recall the male fruit fly with white eyes shown at the beginning of this lesson. Normal fruit-fly eyes are red. If the white-eyed male breeds with normal females, all the male and female offspring have red eyes. When these red-eyed females breed with normal males, every female offspring has red eyes. But half of the males have white eyes. Why?

In fruit flies, the allele for eye color is on only the X chromosome, not on the Y chromosome, as shown in **Figure 10.** Females with white eyes have two white-eye alleles. But males with white eyes have only one white-eye allele.

FOLDABLES®

Make a vertical four-tab book. Label it as shown. Use it to organize your notes on inheritance patterns.

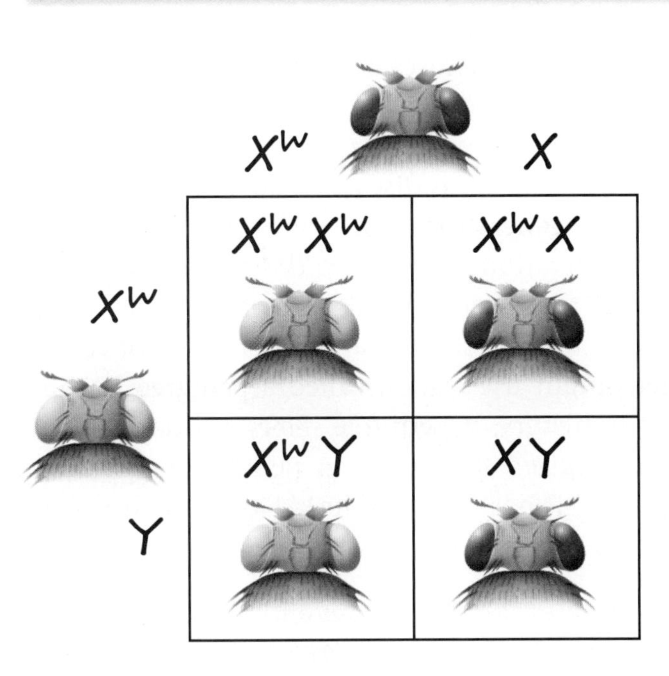

Figure 10 In this Punnett square, the superscripted *w* on the X chromosome represents the sex-linked white-eye allele.

Visual Check What is the genotype of a white-eyed female fly?

Polygenic Inheritance

Think about the various heights of all your classmates. Some traits, such as human height, are controlled by many genes, and the genes express a range of outcomes. **Polygenic inheritance** *occurs when multiple genes determine the phenotype of a trait.* Other examples of polygenic inheritance include the number of petals on a daisy and the length of flowers on tomato plants.

Key Concept Check What are the other patterns of inheritance?

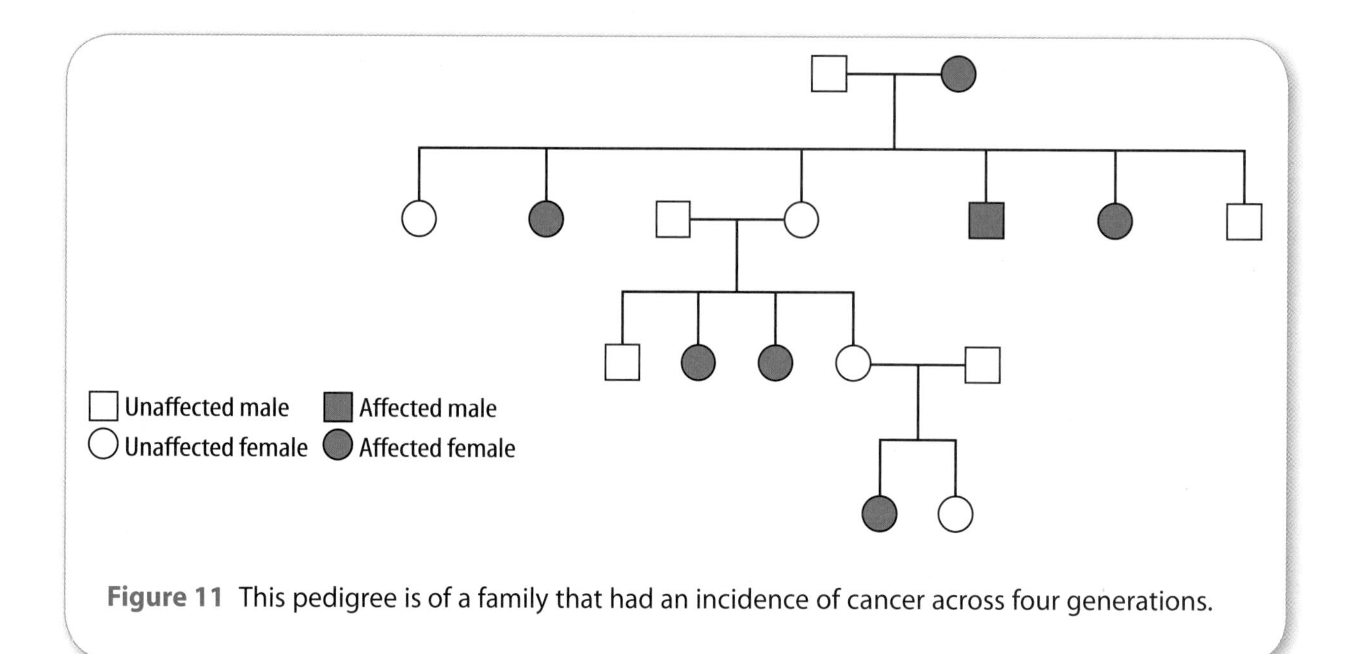

Figure 11 This pedigree is of a family that had an incidence of cancer across four generations.

Inheritance of Disease

Understanding human diseases and how they are inherited is one area of research in genetics. For centuries, scientists have known that many diseases tend to be more common in some families than in others. These diseases also tend to produce patterns across many generations. For example, a disease might skip every other generation or show up only in males. Today, scientists use tools to understand disease patterns. A doctor might ask a patient if he or she has a family history of cancer. The doctor then might use a tool, called a pedigree, to help analyze the family history. *A* **pedigree** *shows genetic traits that were inherited by members of a family.* A pedigree helps determine whether the trait has a genetic link. **Figure 11** shows the pedigree for a family in which cancer was common in each generation.

Reading Check Explain how a pedigree is used.

Mutations

WORD ORIGIN
mutation
from Latin *mutare,* means "to change"

Diseases are often the result of a mutation in one or more genes of an individual. *A* **mutation** *is any permanent change in the sequence of DNA in a gene or a chromosome of a cell.* Mutations often, but not always, result in a change in the appearance or the function of an organism. Mutations can be dominant or recessive. If mutations occur in reproductive cells, they can be passed from parent to offspring. Cancer, diabetes, and birth defects all result from mutations in genes. You will read more about the role of mutations in populations in Lesson 3.

Key Concept Check What role can mutations play in the inheritance of disease?

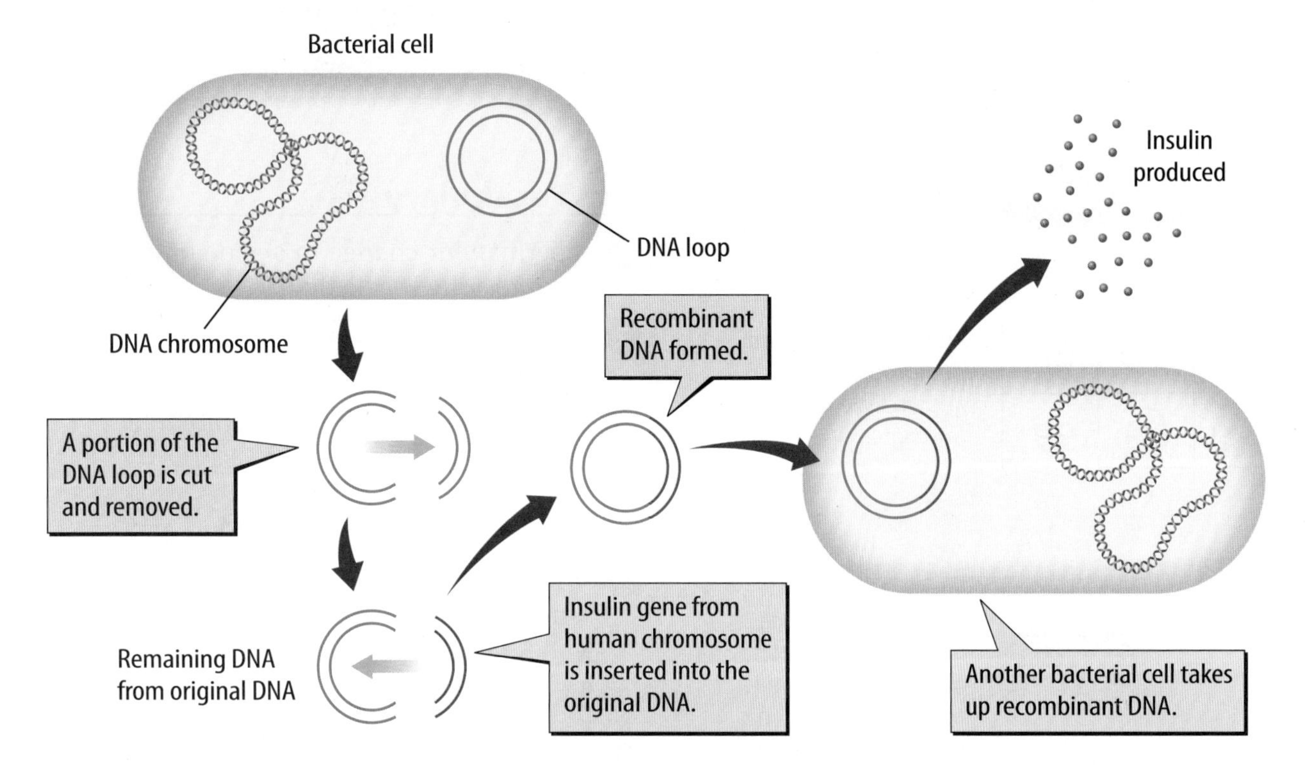

▲ **Figure 12** Bacteria have been genetically engineered to produce human insulin.

Genetic Engineering

Scientists today are using what they have learned about genetics to help people. For example, some people with diabetes cannot produce the protein insulin. Insulin helps control sugar levels in blood. Scientists discovered the insulin gene in human cells. They inserted the gene into the DNA of bacteria. The bacteria now make human insulin, as shown in **Figure 12.** This is an example of genetic engineering. *In* **genetic engineering**, *the genetic material of an organism is modified by inserting DNA from another organism.*

Genetic engineering today is used for many purposes. For example, in the 1960s, a unique gene was discovered in a species of jellyfish. The gene, called GFP, makes a protein that glows green. Scientists can use GFP to learn how plants respond to changes in their environments. A plant, for example, can be made to glow when it is under stress. Certain parts in animal embryos also can be made to glow when that part is beginning to grow. **Figure 13** shows mosquito larvae into which the GFP gene has been inserted. Because this experiment was successful, scientists hope to use genetic engineering to prevent mosquitoes from hosting the organism that causes malaria. This disease kills millions of people each year and is transmitted by mosquito bites.

Reading Check What are some different uses of genetic engineering?

Figure 13 The gene that makes a jellyfish glow has been inserted into the genetic code of mosquito larvae. ▼

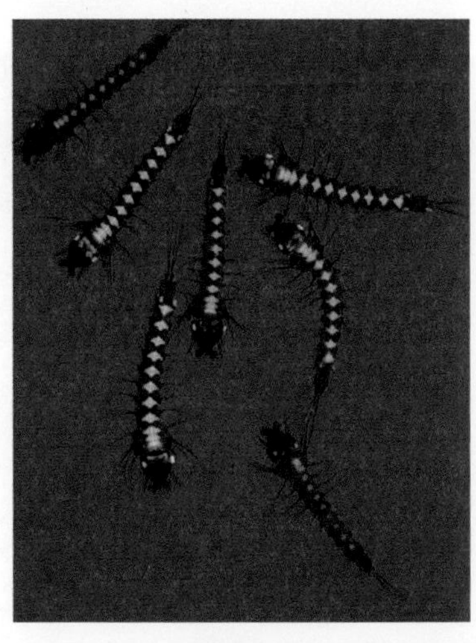

Lesson 2 Review

Assessment | Online Quiz

Visual Summary

	G	g
G	GG	Gg
g	Gg	gg

Scientists use Punnett squares and pedigrees to predict and analyze genetic outcomes.

There are many patterns of inheritance, including incomplete dominance, codominance, and polygenic inheritance.

Scientists use genetic engineering to help treat diseases and learn more about how organisms develop.

FOLDABLES®

Use your lesson Foldable to review the lesson. Save your Foldable for the project at the end of the chapter.

What do you think NOW?

You first read the statements below at the beginning of the chapter.

3. Modern-day genetics disproved Gregor Mendel's ideas about inheritance.

4. Mutations can cause disease in an individual.

Did you change your mind about whether you agree or disagree with the statements? Rewrite any false statements to make them true.

Use Vocabulary

1 A permanent change in a gene or a chromosome is called a(n) ________.

2 **Define** *incomplete dominance* and *codominance* in your own words.

3 **Use the term** *polygenic inheritance* in a complete sentence.

Understand Key Concepts 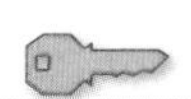

4 Which pattern of inheritance produces a range of outcomes in a trait?

A. codominance
B. sex-linked
C. incomplete dominance
D. polygenic inheritance

5 **Explain** how a doctor might use a pedigree to understand whether a person's disease has a genetic link.

6 **Describe** how a Punnett square helps scientists understand the results of some genetic crosses.

Interpret Graphics

7 **Summarize** Copy and fill in the graphic organizer shown below to summarize the negative effects mutations can have on individuals.

Mutations

Think Critically

8 **Predict** the outcome of a cross between the two fruit flies shown below.

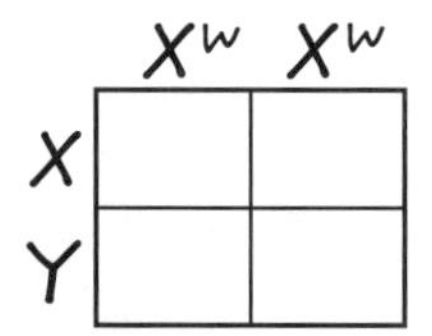

Math Skills

Math Practice

9 A glass jar contains 6 red marbles, 5 green marbles, and 9 blue marbles. What is the probability of drawing a green marble?

Inquiry Skill Practice Analyze Data

20 minutes

What can you learn by analyzing a pedigree?

Although most squirrels have gray or reddish fur and dark-colored eyes, some have white fur and pink eyes. This condition—albinism—occurs when a squirrel inherits genes that cannot produce the pigment that gives color to hair and eyes. The pedigree below shows fur color for three generations of squirrels.

Learn It

Scientists **analyze data** to learn about cause and effect and to answer questions.

Try It

1. Study the pedigree below. The top two symbols represent the parent squirrels. Examine the key to see what the symbols mean. What is the fur color of each parent?
2. The second row represents the four offspring of the parents, or the first generation. Lines connect the symbols that represent the offspring to the parents' symbols. Describe the fur color of the offspring.
3. Two offspring in the first generation reproduced with other squirrels that were not related to them. The symbols for these squirrels are connected to the squirrel they reproduced with.
4. The offspring of the squirrels from the first generation are represented by symbols in the third row. How many squirrels were produced in the second generation? How many of them have gray fur? White fur?
5. Based on your observations of the pedigree, do you think the gene for albinism is dominant or recessive? Explain your answer.

Apply It

6. Suppose a girl bought a female zebra finch. She bred her finch with a friend's male zebra finch. Her bird produced one male and two female offspring. As the offspring grew, one of the female birds looked different from its siblings. It did not have black stripes near its eyes or a black bar under its neck as the other birds did. She took it to a pet store and learned that the female was exhibiting a recessive trait called penguin. Construct a pedigree that illustrates the parents and the first generation of the five birds in this scenario. Include a key.
7. **Key Concept** Describe the possible genotypes and phenotypes of the parents of the girl's zebra finch.

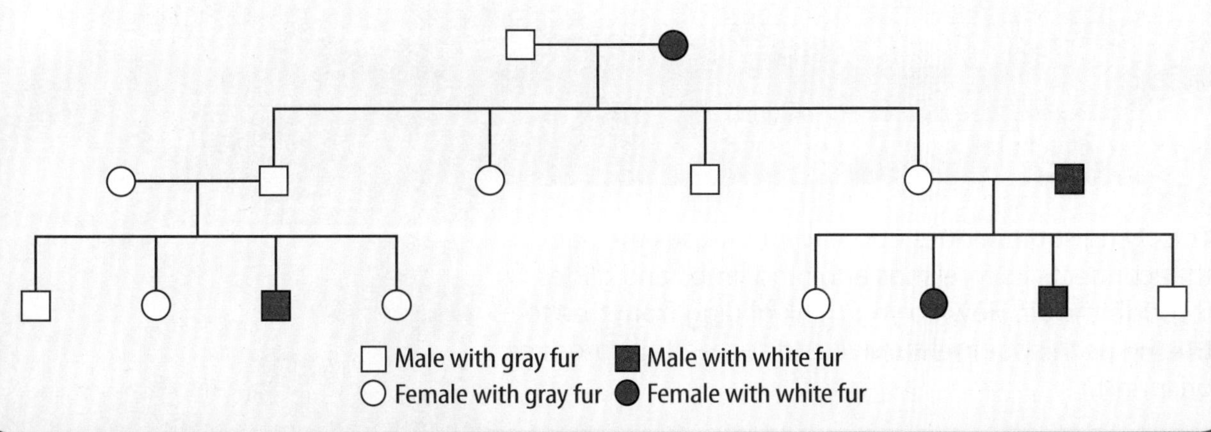

Lesson 3

Adaptation and Evolution

Reading Guide

Key Concepts
ESSENTIAL QUESTIONS

- How does natural selection occur?
- What is an adaptation?
- Why do traits change over time?

Vocabulary

variation p. 811
natural selection p. 812
adaptation p. 813
evolution p. 815
extinction p. 816
conservation biology p. 816

Multilingual eGlossary

Video

- Science Video
- What's Science Got to do With It?

Inquiry Fight or Flight?

This flying squirrel does not really fly. It uses the flap that connects its forelimbs and hind limbs and glides through the air. How do you think gliding from tree to tree helps this squirrel survive and successfully produce offspring?

Launch Lab

10 minutes

How does variation help survival?

Mutations cause differences, or variations, in traits. How can variations help or hinder an organism's survival?

1. Read and complete a lab safety form.
2. Examine various **cell phones** or **pictures of cell phones.** Note the different features that each cell phone offers. For example, some might have a full keyboard, some might have a camera, and some might have a hands-free option.
3. Make a data table in your Science Journal, and record your observations about the characteristics of each phone.

Think About This

1. What are some of the variations among the phones?
2. If each variation represents a mutation in the population of phones, how might each mutation have a positive or a negative impact on the population?
3. **Key Concept** How do you think the development of new characteristics might help a population of organisms survive?

Mutations, Variation, and Natural Selection

Recall that mutations can lead to changes in traits. Therefore, mutations can produce differences among individuals. *Slight differences in inherited traits among individuals in a population are called* **variations.** For example, birds in a population might have variations in feather color or nest-building skills.

In 1976, scientists measured several traits in a population of medium ground finches on one of the Galápagos Islands. They discovered that the birds had variations in beak size. Most had smaller beaks, but all of the birds of this species, such as the one in **Figure 14,** preferred to eat small, soft seeds.

The next year, it never rained on the island. None of the plants reproduced, so no new seeds formed. After the finches ate all of the small, soft seeds, many could no longer survive. The few seeds left were relatively large and hard, so the birds that survived were those that could crack and eat these seeds. These tended to be the finches with relatively large beaks.

In 1978, the scientists measured the beaks of the surviving birds' offspring. They compared the average beak size of birds hatched in 1978 to that of birds hatched in 1976. In just two years, the average beak size of birds in the population had increased. How did this happen?

Figure 14 Medium ground finches eat seeds and live on the Galápagos Islands.

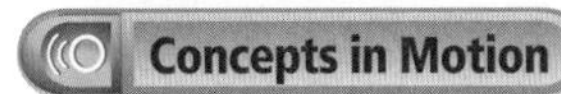

Concepts in Motion

Animation

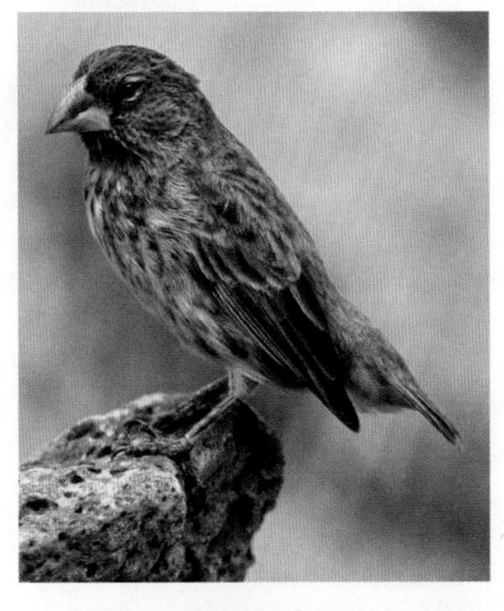

Natural Selection

The process by which individuals with variations that help them survive in their environment live longer, compete better, and reproduce more than those individuals without these variations is called **natural selection.** For the finches, there was variation among individual birds for the trait of beak size. Some of the birds had small beaks, and some of them had larger beaks. When a change in the environment—a drought—occurred, the birds with larger beaks were better able to survive than birds with smaller beaks. The surviving individuals then passed on the favorable trait to their offspring. Over the two-year span of natural selection and reproduction, the average beak size of the birds in the population increased. The birds with larger beaks were naturally selected by environmental conditions and survived. An example of natural selection in plants is shown in **Figure 15.**

Key Concept Check How does natural selection occur?

Figure 15 In this example, due to natural selection, the average height of the sunflowers in a population decreases.

Visual Check What happened when individual sunflowers competed for limited resources?

Natural Selection

Review Personal Tutor

❶ Variation Individuals in a population differ from one another. In this population, some sunflowers are taller than others.

❷ Inheritance Traits are inherited from parents. Tall sunflowers produce tall sunflowers. Short sunflowers produce short sunflowers.

❸ Competition Due to limited resources, not all offspring will survive. Individuals with a trait that better suits the environment are more likely to survive and reproduce. In this environment, short sunflowers are more successful.

❹ Natural Selection Over time, the average height of the sunflower population is short if the short sunflowers continue to reproduce successfully.

Adaptations

The traits of surviving individuals, such as larger beaks or shorter sunflowers, become more common as the survivors reproduce and pass the genes for their traits to their offspring. *An* **adaptation** *is an inherited trait that increases an organism's chance of surviving and reproducing in a particular environment.* Adaptations can be structural, functional, or behavioral.

Structural Adaptations

The flap of skin on the flying squirrel shown at the beginning of the lesson enables it to glide distances of up to 45 m. The flap, which may help the squirrel escape from a predator, is an example of a structural adaptation. Structural adaptations involve physical characteristics, such as color or shape. Another example of a structural adaptation occurs in many desert plants, such as the cactus shown in **Figure 16.** These plants have leaves that are reduced in size. The adaptation of smaller leaves helps reduce water loss in a dry environment.

Figure 16 The spines on this cactus are leaves that are reduced in size. Decreased leaf size helps reduce water loss from the plant.

Inquiry MiniLab

15 minutes

How can you observe change over time?

The data collected by the scientists who studied the finches on the Galápagos Islands is shown in the graphs on the right. Recall that the scientists measured the beak size of the adult finches in 1976 then, in 1978, measured the beaks of all the birds that survived the drought. Examine the graphs, and then answer the following questions.

Finch Beak Size, Galápagos Islands

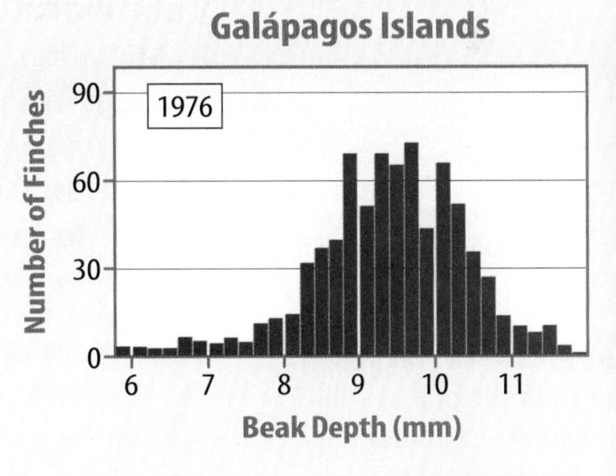

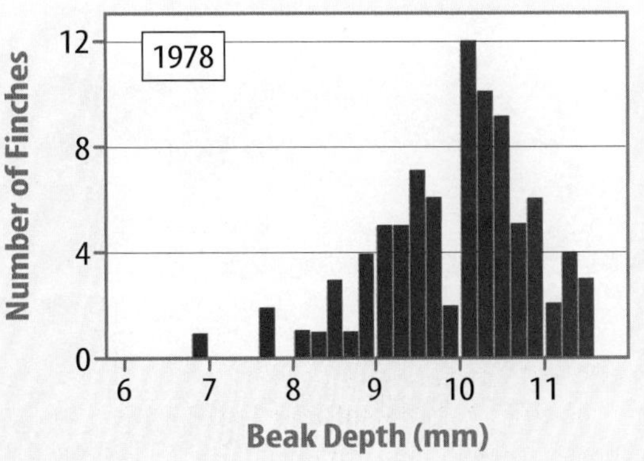

Analyze and Conclude

1. **Hypothesize** Write a hypothesis that explains why some birds might have survived the drought of 1977, while most birds did not.
2. **Analyze** What change occurred in the average beak size of the finch population after the drought of 1977?
3. **Infer** How might differences in beak size have affected which birds survived the drought? Do the data in the graphs support your hypothesis? If not, revise your hypothesis to reflect this information.
4. **Key Concept** Infer how the change in the environment might have affected the beak size of the finch population for the years following 1978.

Figure 17 The alpine snowbell is adapted to a short growing season, blooming in the spring while still surrounded by snow.

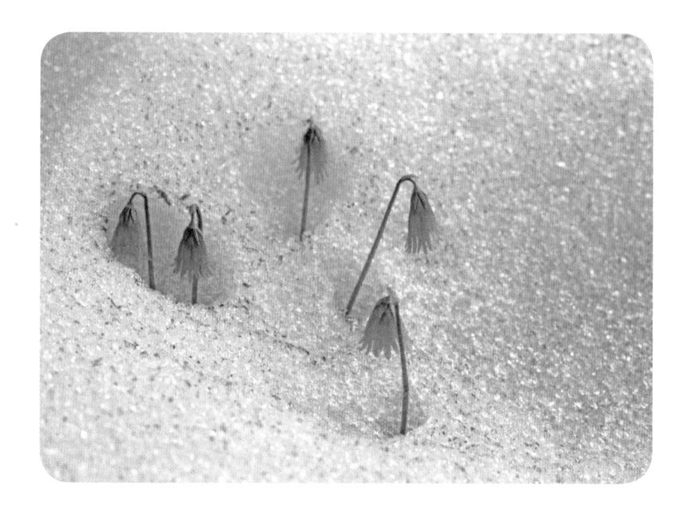

Functional Adaptations

Functional adaptations involve internal systems that affect an organism's physiology or biochemistry. For example, the alpine snowbell shown in **Figure 17** has adapted to survive with a short growing season. In the high altitude of the mountains, light and temperature conditions for flowering are optimal for only a short period during the summer. The alpine snowbell produces flower buds at the end of the previous season. The buds remain dormant over the winter. In the spring, increased light triggers the plant to bloom even if it is still surrounded by snow. These adaptations enable the survival of this species.

Behavioral Adaptations

Migration, and other behavioral adaptations, involve the ways an organism behaves or acts. The caribou in **Figure 18** are migrating south for the winter. Other animals, such as birds, whales, and butterflies, also migrate. Animal species that migrate to find adequate food and suitable temperatures survive and reproduce more successfully.

Key Concept Check Describe three types of adaptations.

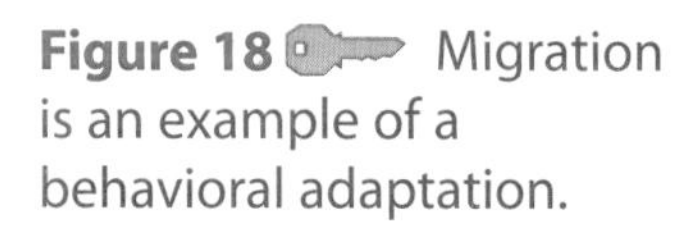

Figure 18 Migration is an example of a behavioral adaptation.

Evolution of Populations—Why Traits Change

Once an inherited trait has become more frequent in a population, the population has adapted and evolved. **Evolution** *is change over time.* Evolution by natural selection is a way that populations change over time. When populations evolve, species can look and behave differently than their ancestors. This happens because the frequency of genetic traits changes over time. As the environment changes, different inherited traits might enable survival, and the population can evolve again.

Key Concept Check Why do traits change over time?

A Modern Example of Change Over Time

You already might know that bacteria can cause infections in your body, such as strep throat or pneumonia. Sometimes a doctor might prescribe an antibiotic to help you fight an infection. Antibiotics, first used in the 1940s, are drugs that kill bacteria. Although, in many cases, antibiotics effectively kill bacteria, variation exists within a population of bacteria. As shown in **Figure 19,** some bacteria in a population already might have a mutation that enables them to **survive** when exposed to an antibiotic. When the surviving bacteria reproduce, that trait passes to their offspring. Soon, most individuals in the population survive when exposed to the antibiotic. Bacteria that survive when exposed to an antibiotic are called antibiotic-resistant. Antibiotic-resistant bacteria have caused deadly infections and are of great concern to scientists.

FOLDABLES®

Make a vertical three-tab book with a tab-top. Label it as shown. Use it to organize your notes on the different types of adaptations.

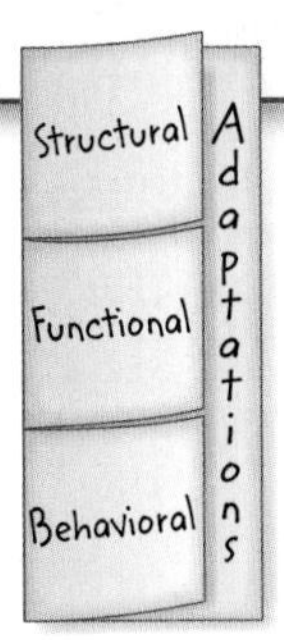

Academic Vocabulary

survive
(verb) to remain alive

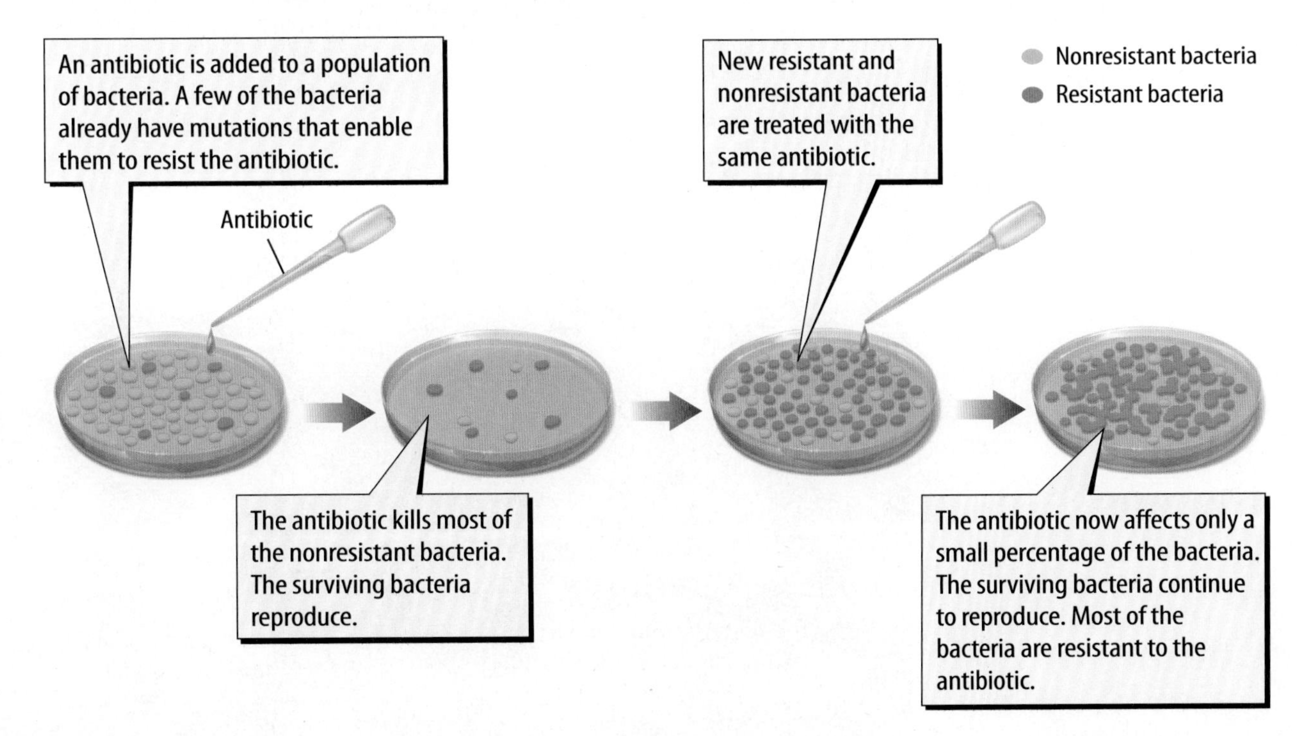

Figure 19 A population of bacteria can evolve antibiotic resistance.

Extinction and Conservation Biology

You already have read that for evolution by natural selection to occur, variation within a population is needed. A population with diversity can survive changes in its environment and persist through time. What happens when a population lacks variation among its individuals, and the environment changes? The population might lose its ability to reproduce successfully and fail to survive. *When the last individual of a species dies, the species has undergone* **extinction.**

WORD ORIGIN

extinction
from Latin *extinctus,* means "wipe out"

Today, many species are threatened with extinction. A species' habitat might have been altered or destroyed. Some species have been hunted to extinction. For others, new species **introduced** into many habitats make it difficult for some native species to survive and reproduce.

Some species can be saved with a relatively new field of science. **Conservation biology** *is a branch of biology that studies why many species are in trouble and what can be done to save them.* Sometimes scientists' knowledge of genetics helps species that are in danger of extinction. For example, by 1995, the population of Florida panthers, such as the one shown in **Figure 20,** was between 20 and 30 individuals. The population had lost much of its natural variation and was struggling to survive. Scientists' understanding of genetics and heredity saved the population from extinction. Scientists introduced into the Florida population several female panthers from a population in Texas. This was done to increase genetic diversity in the Florida population. By 2003, the Florida panther population had increased to 80 individuals, and the effort was considered a success.

SCIENCE USE V. COMMON USE

introduce
Science Use to bring a substance or organism into a habitat or a population

Common Use to make someone known to others

Figure 20 An understanding of genetics and heredity has helped restore the Florida panther population.

Lesson 3 Review

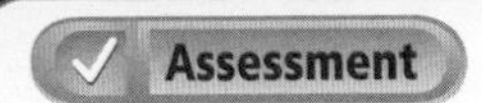

Visual Summary

Natural selection occurs when individuals with traits that better suit the environment survive longer and reproduce more successfully than individuals without the traits.

An adaptation is an inherited trait that increases an organism's chance of surviving and reproducing in an environment.

Conservation biologists work to save species from extinction.

FOLDABLES®

Use your lesson Foldable to review the lesson. Save your Foldable for the project at the end of the chapter.

What do you think NOW?

You first read the statements below at the beginning of the chapter.

5. A population that lacks variation among its individuals might not be able to adapt to a changing environment.

6. Extinction occurs when the last individual of a species dies.

Did you change your mind about whether you agree or disagree with the statements? Rewrite any false statements to make them true.

Use Vocabulary

1. **Define** *variation* in your own words.
2. **Use the term** *adaptation* in a complete sentence.
3. **Define** *evolution* and *extinction.*

Understand Key Concepts 

4. A population of bacteria becoming resistant after being exposed to antibiotics is an example of
   - **A.** evolution.
   - **B.** extinction.
   - **C.** conservation biology.
   - **D.** selective breeding.
5. **Explain** how mutations can benefit an organism.
6. **Summarize** why traits change over time.

Interpret Graphics

7. **Identify** The markings on the gecko below make it look like a leaf. Identify and explain what type of adaptation this is.

8. **Summarize** Copy and fill in the graphic organizer below to summarize the relationships among natural selection, mutation, variation, and adaptation.

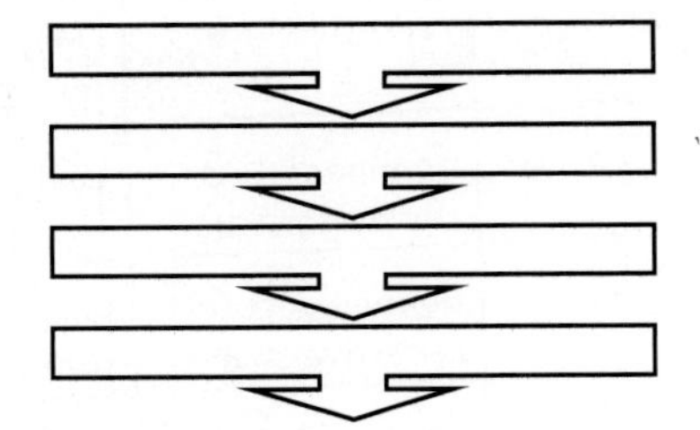

Critical Thinking

9. **Assess** the importance of genetics to the field of conservation biology.

1 class period

What's in a face?

Materials

coin

Safety

The traits that make us each look different are caused by thousands of combinations of different pairs of alleles. Each gene in a pair comes from one parent, and each gene is selected randomly.

Ask a Question

How can you model the allele contribution of each parent to determine the genotype and the phenotype of an organism's traits?

Make Observations

1. Read and complete a lab safety form.
2. Obtain your group number. If you are in group 1, you will make a female face. If you are in group 2, you will make a male face.
3. Copy the table below in your Science Journal. The table lists the traits you will use to make your face. For the first trait, flip a coin to determine the allele contributed by the mother. For all traits, heads will represent the dominant allele, and tails will represent the recessive allele. Record your results.
4. Flip the coin a second time to determine the allele contributed by the father. Record your results. Record the genotype for this trait.
5. Repeat steps 3 and 4 for each of the remaining traits.
6. Using the genotypes, determine and record the phenotype your face will express for each trait. Draw your face in your Science Journal.

Data Table—Facial Traits

Trait	Possible Alleles	Possible Phenotypes	Allele from Mother	Allele from Father	Genotype	Phenotype
Face shape	dominant = R recessive = r	round: RR or Rr squarish: rr				
Chin shape	dominant = Q recessive = q	very prominent: QQ or Qq less prominent: qq				
Earlobes	dominant = E recessive = e	hang free: EE or Ee attached: ee				
Hair texture	dominant = A recessive = a	curly: AA wavy: Aa straight: aa				
Eyebrows	dominant = B recessive = b	bushy: BB or Bb fine: bb				
Eyelashes	dominant = G recessive = g	long: GG or Gg short: gg				
Lips	dominant = H recessive = h	thick: HH or Hh thin: hh				
Nose	dominant = N recessive = n	large: NN medium: Nn small: nn				

7. Find a partner who has created a face of the opposite gender of yours. Based on the genotypes of each face, determine the possible genotypes of a second generation. Use Punnett squares to track the different genotypes.
8. For each trait, determine and record the most probable phenotype. If there is an equal chance of either genotype, flip a coin to decide which phenotype you will use.

Form a Hypothesis

9. Form a hypothesis to explain the relationship between the genotypes of the first generation and the phenotype of the second generation. Predict the phenotype of your second-generation face.

Test Your Hypothesis

10. Flip the coin to determine which allele the mother will contribute to the second-generation face for the first trait. Heads represents the first allele in the genotype. Tails represents the second allele in the genotype. Record your results.
11. Repeat step 10 to determine the allele contributed by the father. Record the genotype for this trait.
12. Repeat steps 10 and 11 for the remaining traits.
13. Using the genotypes, determine and record the phenotype your face will express for each trait. Draw your face in your Science Journal.
14. Compare the phenotype of the second-generation face to your hypothesis.

Analyze and Conclude

15. **Analyze** Did your results match your hypothesis? Why or why not?
16. **Critique** Were there some genetic combinations that were more likely for the offspring than others? Were there some combinations that were not possible? Why?
17. **The Big Idea** How are genes and traits passed on from parents to offspring?

Communicate Your Results

With your partner, make a poster that shows the cross between the two faces and the second-generation face that resulted. Include information about how you determined the genotypes and the phenotypes for each trait.

Choose one of the facial characteristics and do several genetic crosses to create many offspring. Then draw a pedigree that illustrates the presence of the trait over the two generations.

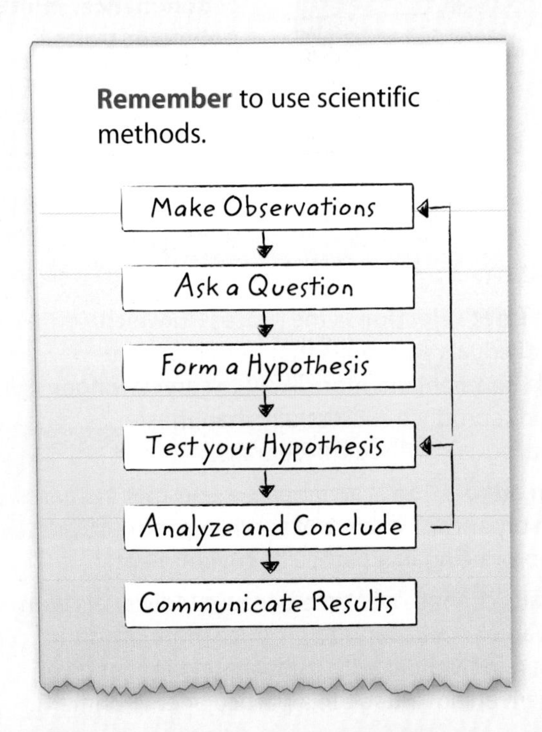

Chapter 22 Study Guide

Mutations can change the traits in individuals that are passed from parents to offspring, producing variation in a population that can lead to greater success and survival in new environments.

Key Concepts Summary

Lesson 1: How are traits inherited?

- Traits are inherited as genes pass from parent to offspring.
- Scientists study **genetics** to learn more about the development of organisms as well as the development of disease. Scientists also can learn more about how organisms are related by studying genetics.
- Mendel studied how traits are passed from parents to offspring. He discovered that traits can be **dominant** or **recessive** and that the outcome of crosses between parents can be predicted.

Vocabulary

heredity p. 793
genetics p. 793
selective breeding p. 796
dominant trait p. 797
recessive trait p. 797
genotype p. 798
phenotype p. 798
heterozygous p. 798
homozygous p. 798

Lesson 2: Genetics After Mendel

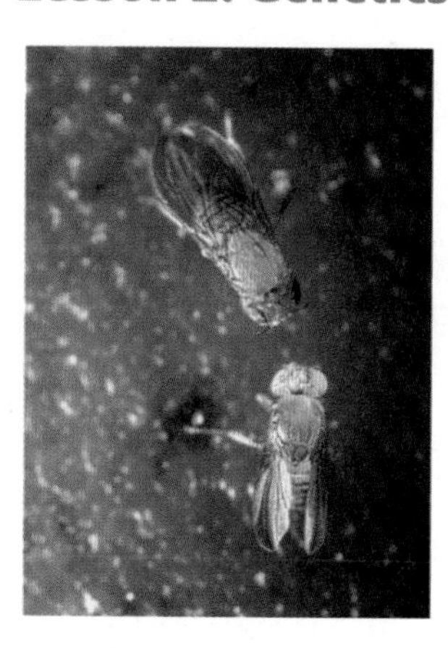

- Tools, such as **Punnett squares,** can be used to predict the probability of certain allele combinations given the genotypes of the parents of a genetic cross.
- Traits can be inherited in ways other than dominant or recessive. Traits can be inherited as **incomplete dominance, codominance, multiple alleles,** and **polygenic traits.**
- **Mutations** are changes in genes and can cause disease. As genes are passed from parent to offspring, so is the disease.

monohybrid cross p. 803
Punnett square p. 803
incomplete dominance p. 804
codominance p. 804
multiple alleles p. 804
sex-linked trait p. 805
polygenic inheritance p. 805
pedigree p. 806
mutation p. 806
genetic engineering p. 807

Lesson 3: Adaptation and Evolution

- **Natural selection** is the process in which individuals with traits that better suit the environment are more likely to survive longer and reproduce successfully than those individuals without these traits.
- An **adaptation** is an inherited trait that increases an organism's chance of surviving and reproducing in a particular environment.
- Traits change over time because individuals with adaptive traits tend to produce more offspring than individuals with traits that no longer give them an advantage in a particular environment.

variation p. 811
natural selection p. 812
adaptation p. 813
evolution p. 815
extinction p. 816
conservation biology p. 816

- Personal Tutor
- Vocabulary eGames
- Vocabulary eFlashcards

FOLDABLES® Chapter Project

Assemble your lesson Foldables as shown to make a Chapter Project. Use the project to review what you have learned in this chapter.

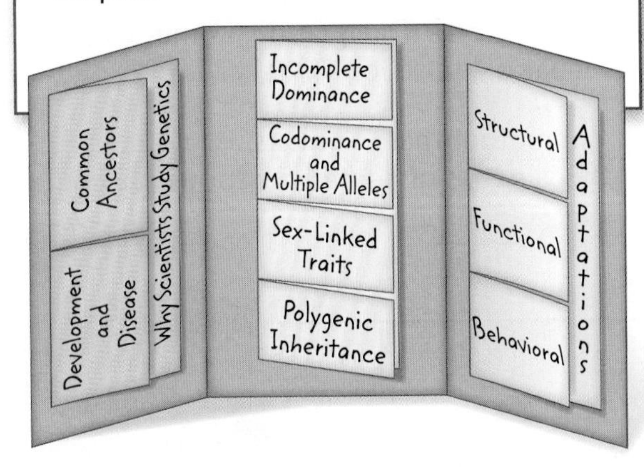

Use Vocabulary

1. The study of how traits are passed from parents to offspring is ________.
2. Individual organisms that have desirable traits are bred to produce offspring in ________.
3. Explain a monohybrid cross.
4. When alleles produce a phenotype that is a combination of the parents' phenotypes, it is called ________.
5. Change over time is ________.
6. Define *conservation biology.*

Link Vocabulary and Key Concepts

Concepts in Motion Interactive Concept Map

Copy this concept map, and then use vocabulary terms from the previous page to complete the concept map.

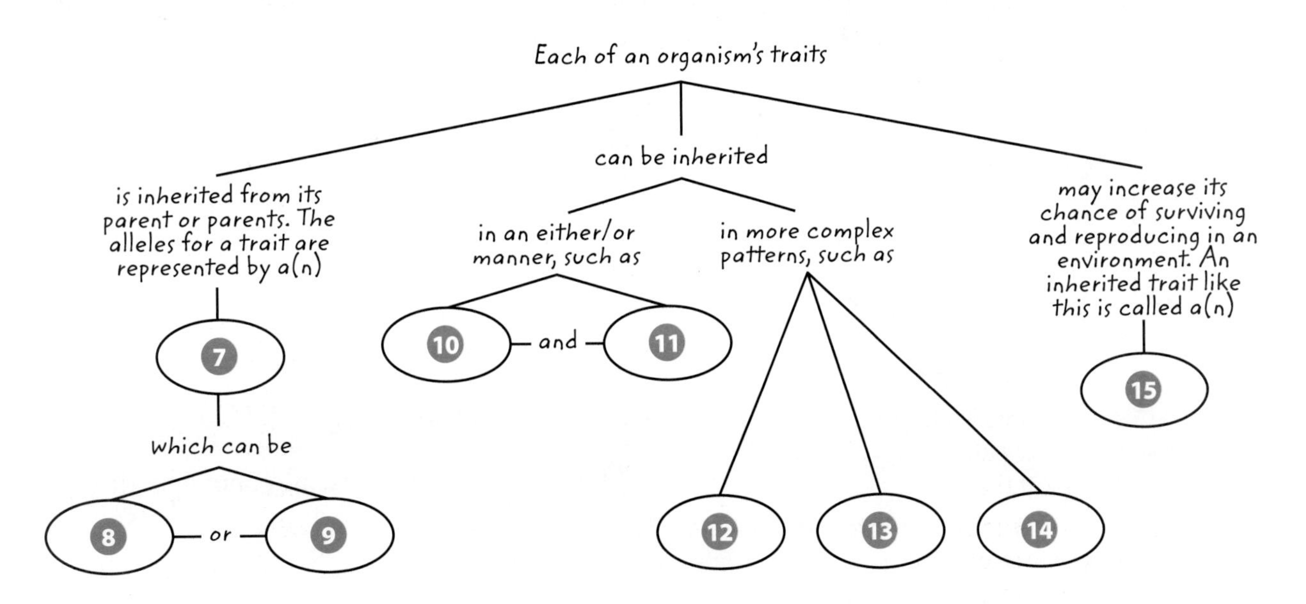

Chapter 22 Review

Understand Key Concepts 

1. Bird migration in the spring is an example of
   A. a behavioral adaptation.
   B. a heterozygous cross.
   C. a homozygous cross.
   D. a structural adaptation.

2. Which is NOT a reason why scientists study genetics?
   A. to discover how organisms are related
   B. to find cures for disease in humans
   C. to conserve species that are in danger of extinction
   D. to learn more about how to conserve ecosystems

3. Which best describes codominance?
   A. Both alleles are expressed for a trait.
   B. More than two alleles control a trait.
   C. The allele for a trait is on the X or the Y chromosome.
   D. Two different alleles produce an intermediate form.

4. In the Punnett square below, which genotype correctly fits in the blank cell?

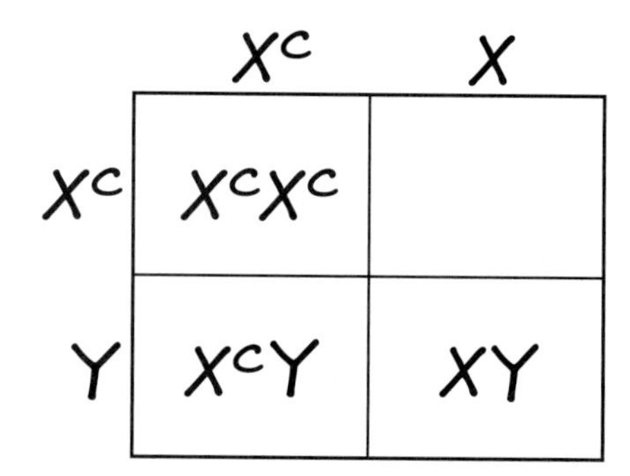

   A. XX
   B. X^cX
   C. XY
   D. X^cY

5. Which list of the processes of evolution is in the correct order?
   A. adaptation → mutation → variation → natural selection
   B. mutation → natural selection → variation → adaptation
   C. mutation → variation → natural selection → adaptation
   D. variation → adaptation → mutation → natural selection

6. Which tool that scientists use to analyze the genetics of a family is shown below?

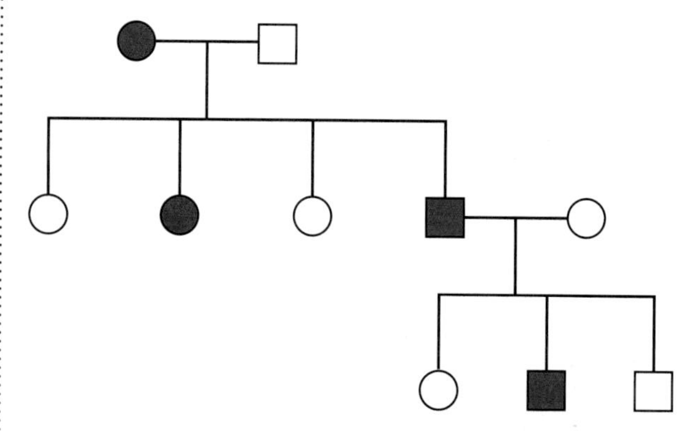

   A. a pedigree
   B. a phenotype
   C. genetic engineering
   D. a Punnett square

7. A farmer crosses an apple tree that produces large apples with an apple tree that produces apples with a popular taste. Which process does this scenario describe?
   A. adaptation
   B. mutation
   C. natural selection
   D. selective breeding

8. In which process are microbes genetically changed to produce animal proteins?
   A. adaptation
   B. variation
   C. genetic engineering
   D. selective breeding

✓ Assessment

Online Test Practice

Critical Thinking

9. **Evaluate** the importance of genetics to human health.

10. **Compare** genotype and phenotype.

11. **Summarize** what Gregor Mendel discovered about heredity.

12. **Justify** why a range of beak sizes might exist in a population of birds.

13. **Infer** How can two parents with Type A blood produce a child with Type O blood?

14. **Compare** the three types of adaptations—structural, functional, and behavioral.

15. **Assess** the role of natural selection in evolution.

16. **Justify** why knowledge of genetics is important in conserving endangered species.

17. **Justify** how variation in a population affects adaptation to new environments.

18. **Infer** In the population of bugs shown below, suppose the red bugs are seen and eaten by birds with higher frequency than the brown bugs. Infer what changes will occur in the population over time.

Writing in Science

19. **Write** a paragraph explaining the difference between selective breeding and natural selection to a third-grade student who is having difficulty understanding the terms.

REVIEW THE BIG IDEA

20. How do heredity and evolution explain why and how species adapt to new environments over time?

21. What is the advantage for this gecko to look like a plant leaf? What behavioral adaptation, along with its appearance, might also help to it survive? Explain your answer.

Math Skills

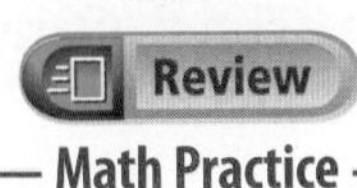

Review

Math Practice

Use Probability

22. There are five vowels in the alphabet. What is the probability of drawing a vowel from 26 tiles each containing a different single letter of the alphabet?

23. A number from 1–11 is chosen randomly. What is the probability of choosing an even number?

24. What is the probability of picking a red chip from a bag that contains six red chips, four blue chips, two yellow chips, and one white chip?

Standardized Test Practice

Record your answers on the answer sheet provided by your teacher or on a sheet of paper.

Multiple Choice

1 Mendel found that some pea plants only produced offspring with green pods. Other plants only produced offspring with yellow pods. He called these true-breeding plants. Which statement describes all true-breeding plants?

A They are heterozygous.

B They are homozygous.

C They are genetically engineered.

D They have dominant alleles.

2 What happened when Mendel crossed a true-breeding pea plant with green pods with a true-breeding plant with yellow pods?

A All of the offspring had green pods.

B Half of the offspring had green pods.

C Only one of the offspring had green pods.

D Three-quarters of the offspring had green pods.

Use the figure below to answer question 3.

	G	G
g	Gg	Gg
g		Gg

3 Which genotype belongs in the empty box in the Punnett square?

A GG

B gg

C Gg

D GGgg

4 Which term describes an inheritance pattern in which many different genes determine a trait's phenotype?

A codominance

B incomplete dominance

C multiple alleles

D polygenic inheritance

5 How are mutations and variation related?

A Mutations and variation are not related.

B Mutations are caused by variation.

C Mutations can cause variation.

D Mutations prevent variation.

Use the figure below to answer question 6.

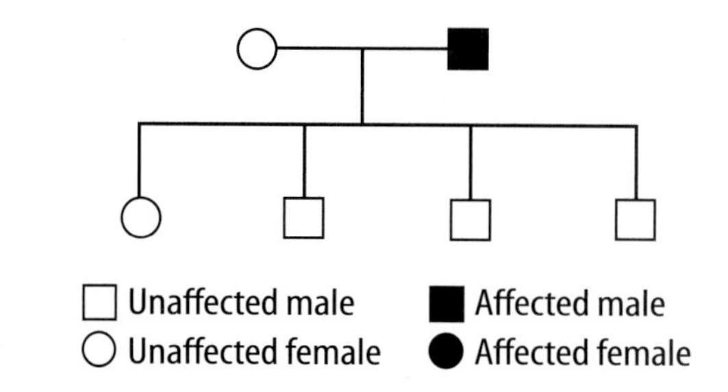

6 The pedigree chart above shows the occurrence of a disease in a family. Which statement summarizes the occurrence of the disease on the chart?

A All males have the disease.

B All of the children have the disease.

C One son has the disease.

D Only the father has the disease.

7 Which is a mutation?

A a factor that causes all human disease

B any gene that is harmful

C a permanent change in a gene

D a piece of recombinant DNA

8 Which describes how sexual reproduction and asexual reproduction are similar?

A Both are ways to acquire traits during an organism's lifetime.

B Both are processes that require two different parents.

C Both are ways that genes are passed from parents to offspring.

D Both result in offspring that are heterozygous for all traits.

9 Which is a behavioral adaptation?

A a puppy learning to roll over

B a skunk's ability to release a foul odor

C caribou migrating south for the winter

D the bright coloration on a snake's back

Use the figure below to answer question 10.

	X^w	X
X		
Y		

10 Predict the outcome of this cross between two fruit flies. The *w* represents a sex-linked allele for white eyes. What percentage of male offspring will have white eyes?

A 0%

B 25%

C 50%

D 100%

Constructed Response

Use the figure below to answer question 11.

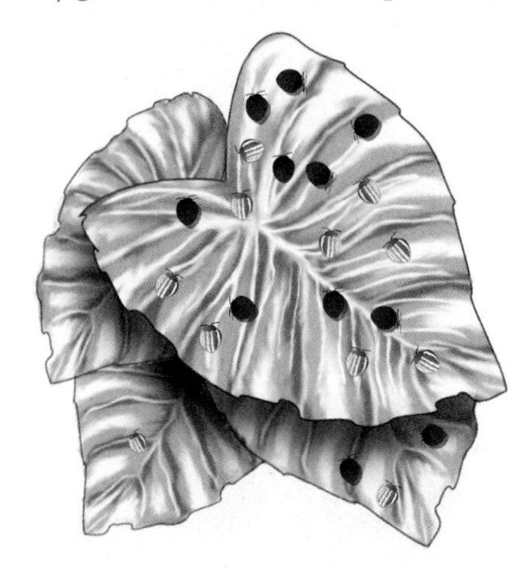

11 The beetles shown are members of the same species. Beetles with striped wings blend with the leaves and are not eaten by predators as frequently as the solid-colored beetles. Predict how the population will change over time due to natural selection. Explain your answer.

12 "Natural selection cannot occur without variation in a population." Explain the meaning of this statement.

13 Antibiotics are substances that kill bacteria. Some bacteria have evolved resistance to certain antibiotics. Explain how this resistance developed.

NEED EXTRA HELP?													
If You Missed Question...	1	2	3	4	5	6	7	8	9	10	11	12	13
Go to Lesson...	1	1	2	2	3	2	2	1	3	2	1	3	3

Student Resources

For Students and Parents/Guardians

These resources are designed to help you achieve success in science. You will find useful information on laboratory safety, math skills, and science skills. In addition, science reference materials are found in the Reference Handbook. You'll find the information you need to learn and sharpen your skills in these resources.

Table of Contents

Scientific Methods

Scientists use an orderly approach called the scientific method to solve problems. This includes organizing and recording data so others can understand them. Scientists use many variations in this method when they solve problems.

Identify a Question

The first step in a scientific investigation or experiment is to identify a question to be answered or a problem to be solved. For example, you might ask which gasoline is the most efficient.

Gather and Organize Information

After you have identified your question, begin gathering and organizing information. There are many ways to gather information, such as researching in a library, interviewing those knowledgeable about the subject, and testing and working in the laboratory and field. Fieldwork is investigations and observations done outside of a laboratory.

Researching Information Before moving in a new direction, it is important to gather the information that already is known about the subject. Start by asking yourself questions to determine exactly what you need to know. Then you will look for the information in various reference sources, like the student is doing in **Figure 1.** Some sources may include textbooks, encyclopedias, government documents, professional journals, science magazines, and the Internet. Always list the sources of your information.

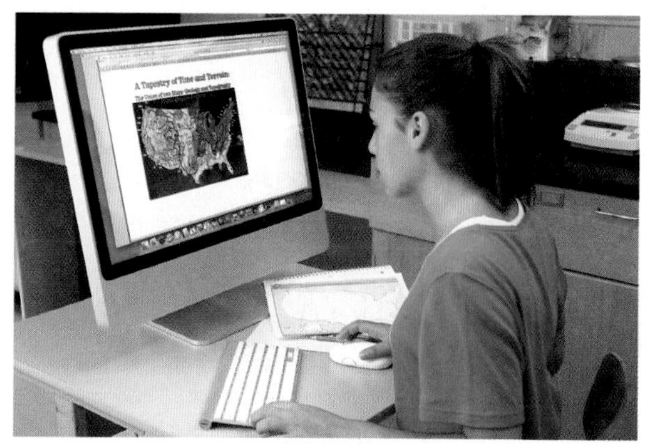

Figure 1 The Internet can be a valuable research tool.

Evaluate Sources of Information Not all sources of information are reliable. You should evaluate all of your sources of information, and use only those you know to be dependable. For example, if you are researching ways to make homes more energy efficient, a site written by the U.S. Department of Energy would be more reliable than a site written by a company that is trying to sell a new type of weatherproofing material. Also, remember that research always is changing. Consult the most current resources available to you. For example, a 1985 resource about saving energy would not reflect the most recent findings.

Sometimes scientists use data that they did not collect themselves, or conclusions drawn by other researchers. This data must be evaluated carefully. Ask questions about how the data were obtained, if the investigation was carried out properly, and if it has been duplicated exactly with the same results. Would you reach the same conclusion from the data? Only when you have confidence in the data can you believe it is true and feel comfortable using it.

Interpret Scientific Illustrations As you research a topic in science, you will see drawings, diagrams, and photographs to help you understand what you read. Some illustrations are included to help you understand an idea that you can't see easily by yourself, like the tiny particles in an atom in **Figure 2.** A drawing helps many people to remember details more easily and provides examples that clarify difficult concepts or give additional information about the topic you are studying. Most illustrations have labels or a caption to identify or to provide more information.

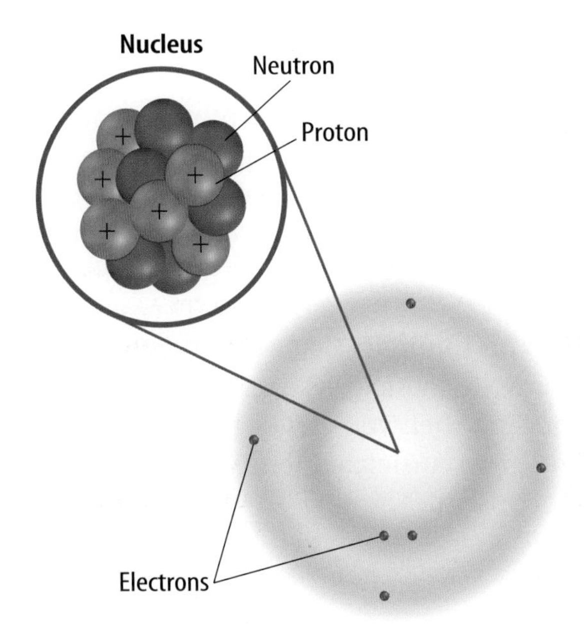

Figure 2 This drawing shows an atom of carbon with its six protons, six neutrons, and six electrons.

Concept Maps One way to organize data is to draw a diagram that shows relationships among ideas (or concepts). A concept map can help make the meanings of ideas and terms more clear, and help you understand and remember what you are studying. Concept maps are useful for breaking large concepts down into smaller parts, making learning easier.

Network Tree A type of concept map that not only shows a relationship, but how the concepts are related is a network tree, shown in **Figure 3.** In a network tree, the words are written in the ovals, while the description of the type of relationship is written across the connecting lines.

When constructing a network tree, write down the topic and all major topics on separate pieces of paper or notecards. Then arrange them in order from general to specific. Branch the related concepts from the major concept and describe the relationship on the connecting line. Continue to more specific concepts until finished.

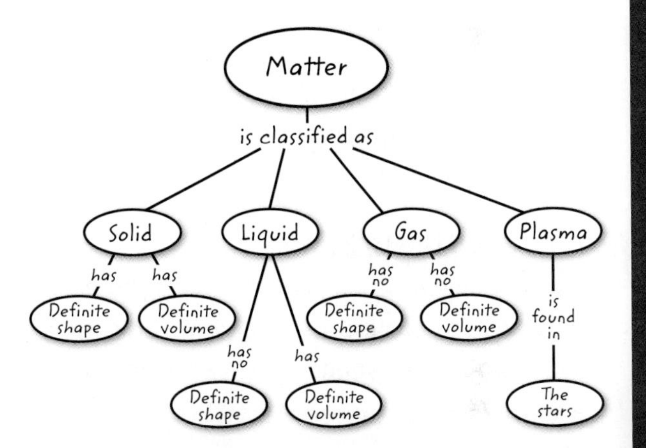

Figure 3 A network tree shows how concepts or objects are related.

Events Chain Another type of concept map is an events chain. Sometimes called a flow chart, it models the order or sequence of items. An events chain can be used to describe a sequence of events, the steps in a procedure, or the stages of a process.

When making an events chain, first find the one event that starts the chain. This event is called the initiating event. Then, find the next event and continue until the outcome is reached, as shown in **Figure 4** on the next page.

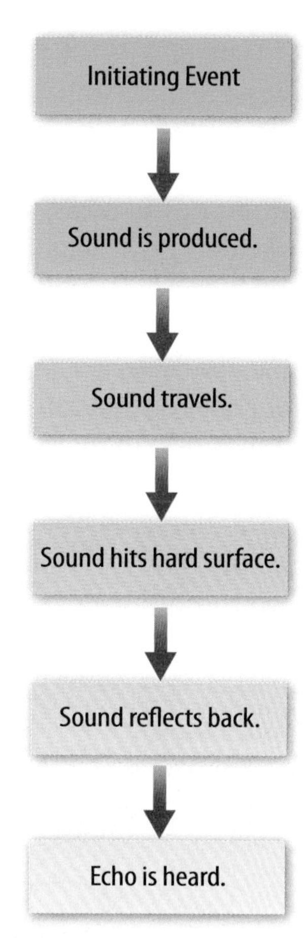

Figure 4 Events-chain concept maps show the order of steps in a process or event. This concept map shows how a sound makes an echo.

Cycle Map A specific type of events chain is a cycle map. It is used when the series of events do not produce a final outcome, but instead relate back to the beginning event, such as in **Figure 5.** Therefore, the cycle repeats itself.

To make a cycle map, first decide what event is the beginning event. This is also called the initiating event. Then list the next events in the order that they occur, with the last event relating back to the initiating event. Words can be written between the events that describe what happens from one event to the next. The number of events in a cycle map can vary, but usually contain three or more events.

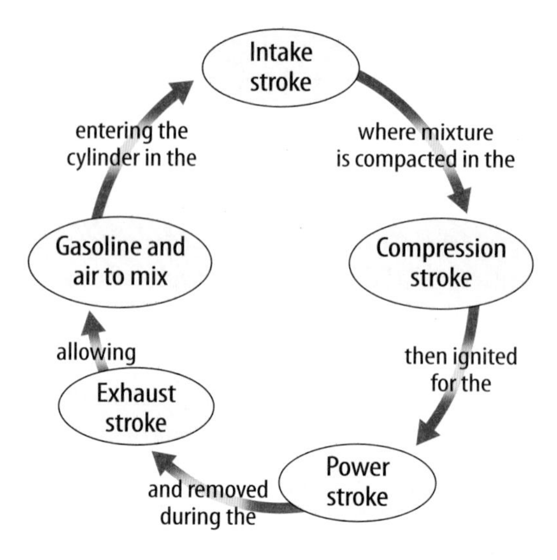

Figure 5 A cycle map shows events that occur in a cycle.

Spider Map A type of concept map that you can use for brainstorming is the spider map. When you have a central idea, you might find that you have a jumble of ideas that relate to it but are not necessarily clearly related to each other. The spider map on sound in **Figure 6** shows that if you write these ideas outside the main concept, then you can begin to separate and group unrelated terms so they become more useful.

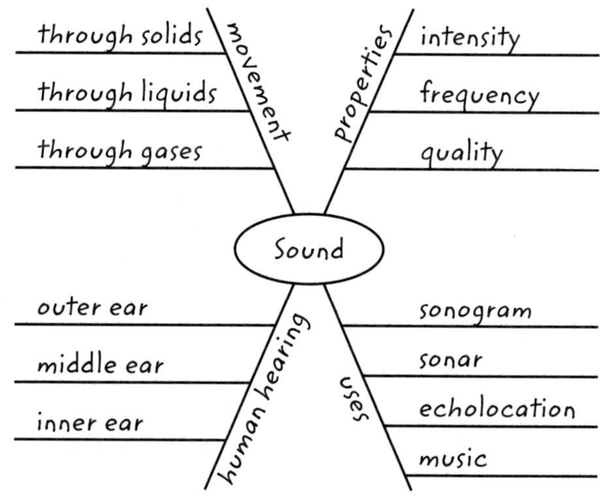

Figure 6 A spider map allows you to list ideas that relate to a central topic but not necessarily to one another.

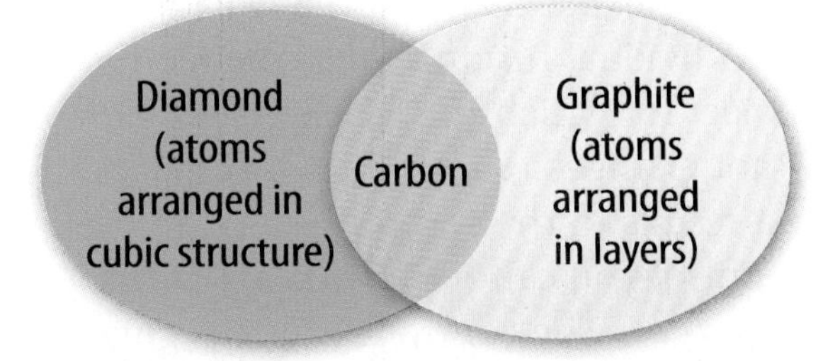

Figure 7 This Venn diagram compares and contrasts two substances made from carbon.

Venn Diagram To illustrate how two subjects compare and contrast you can use a Venn diagram. You can see the characteristics that the subjects have in common and those that they do not, shown in **Figure 7.**

To create a Venn diagram, draw two overlapping ovals that are big enough to write in. List the characteristics unique to one subject in one oval, and the characteristics of the other subject in the other oval. The characteristics in common are listed in the overlapping section.

Make and Use Tables One way to organize information so it is easier to understand is to use a table. Tables can contain numbers, words, or both.

To make a table, list the items to be compared in the first column and the characteristics to be compared in the first row. The title should clearly indicate the content of the table, and the column or row heads should be clear. Notice that in **Table 1** the units are included.

Table 1 Recyclables Collected During Week

Day of Week	Paper (kg)	Aluminum (kg)	Glass (kg)
Monday	5.0	4.0	12.0
Wednesday	4.0	1.0	10.0
Friday	2.5	2.0	10.0

Make a Model One way to help you better understand the parts of a structure, the way a process works, or to show things too large or small for viewing is to make a model. For example, an atomic model made of a plastic-ball nucleus and chenille stem electron shells can help you visualize how the parts of an atom relate to each other. Other types of models can be devised on a computer or represented by equations.

Form a Hypothesis

A possible explanation based on previous knowledge and observations is called a hypothesis. After researching gasoline types and recalling previous experiences in your family's car you form a hypothesis—our car runs more efficiently because we use premium gasoline. To be valid, a hypothesis has to be something you can test by using an investigation.

Predict When you apply a hypothesis to a specific situation, you predict something about that situation. A prediction makes a statement in advance, based on prior observation, experience, or scientific reasoning. People use predictions to make everyday decisions. Scientists test predictions by performing investigations. Based on previous observations and experiences, you might form a prediction that cars are more efficient with premium gasoline. The prediction can be tested in an investigation.

Design an Experiment A scientist needs to make many decisions before beginning an investigation. Some of these include: how to carry out the investigation, what steps to follow, how to record the data, and how the investigation will answer the question. It also is important to address any safety concerns.

Science Skill Handbook
Math Skill Handbook
Foldables Handbook
Reference Handbook
Glossary/Glosario
Index

Test the Hypothesis

Now that you have formed your hypothesis, you need to test it. Using an investigation, you will make observations and collect data, or information. This data might either support or not support your hypothesis. Scientists collect and organize data as numbers and descriptions.

Follow a Procedure In order to know what materials to use, as well as how and in what order to use them, you must follow a procedure. **Figure 8** shows a procedure you might follow to test your hypothesis.

Procedure

Step 1	Use regular gasoline for two weeks.
Step 2	Record the number of kilometers between fill-ups and the amount of gasoline used.
Step 3	Switch to premium gasoline for two weeks.
Step 4	Record the number of kilometers between fill-ups and the amount of gasoline used.

Figure 8 A procedure tells you what to do step-by-step.

Identify and Manipulate Variables and Controls In any experiment, it is important to keep everything the same except for the item you are testing. The one factor you change is called the independent variable. The change that results is the dependent variable. Make sure you have only one independent variable, to assure yourself of the cause of the changes you observe in the dependent variable. For example, in your gasoline experiment the type of fuel is the independent variable. The dependent variable is the efficiency.

Many experiments also have a control—an individual instance or experimental subject for which the independent variable is not changed. You can then compare the test results to the control results. To design a control you can have two cars of the same type. The control car uses regular gasoline for four weeks. After you are done with the test, you can compare the experimental results to the control results.

Collect Data

Whether you are carrying out an investigation or a short observational experiment, you will collect data, as shown in **Figure 9.** Scientists collect data as numbers and descriptions and organize them in specific ways.

Observe Scientists observe items and events, then record what they see. When they use only words to describe an observation, it is called qualitative data. Scientists' observations also can describe how much there is of something. These observations use numbers, as well as words, in the description and are called quantitative data. For example, if a sample of the element gold is described as being "shiny and very dense" the data are qualitative. Quantitative data on this sample of gold might include "a mass of 30 g and a density of 19.3 g/cm^3."

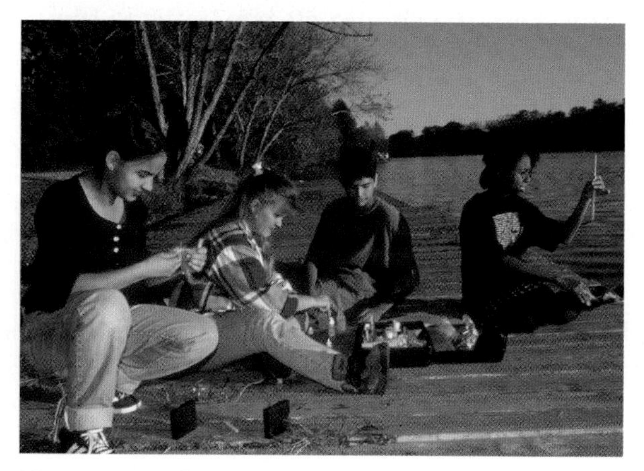

Figure 9 Collecting data is one way to gather information directly.

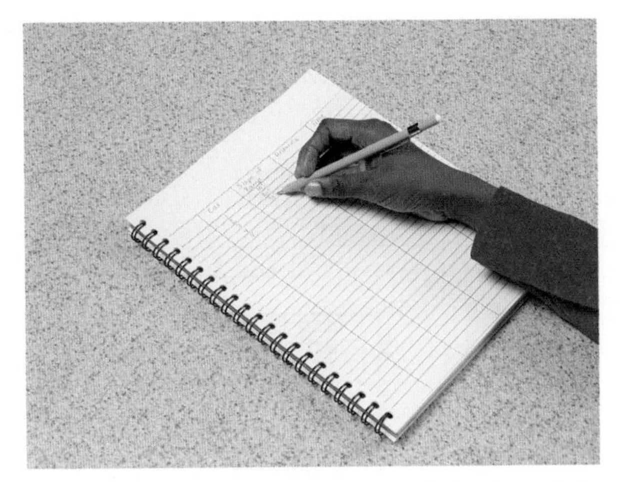

Figure 10 Record data neatly and clearly so it is easy to understand.

When you make observations you should examine the entire object or situation first, and then look carefully for details. It is important to record observations accurately and completely. Always record your notes immediately as you make them, so you do not miss details or make a mistake when recording results from memory. Never put unidentified observations on scraps of paper. Instead they should be recorded in a notebook, like the one in **Figure 10.** Write your data neatly so you can easily read it later. At each point in the experiment, record your observations and label them. That way, you will not have to determine what the figures mean when you look at your notes later. Set up any tables that you will need to use ahead of time, so you can record any observations right away. Remember to avoid bias when collecting data by not including personal thoughts when you record observations. Record only what you observe.

Estimate Scientific work also involves estimating. To estimate is to make a judgment about the size or the number of something without measuring or counting. This is important when the number or size of an object or population is too large or too difficult to accurately count or measure.

Sample Scientists may use a sample or a portion of the total number as a type of estimation. To sample is to take a small, representative portion of the objects or organisms of a population for research. By making careful observations or manipulating variables within that portion of the group, information is discovered and conclusions are drawn that might apply to the whole population. A poorly chosen sample can be unrepresentative of the whole. If you were trying to determine the rainfall in an area, it would not be best to take a rainfall sample from under a tree.

Measure You use measurements every day. Scientists also take measurements when collecting data. When taking measurements, it is important to know how to use measuring tools properly. Accuracy also is important.

Length To measure length, the distance between two points, scientists use meters. Smaller measurements might be measured in centimeters or millimeters.

Length is measured using a metric ruler or meterstick. When using a metric ruler, line up the 0-cm mark with the end of the object being measured and read the number of the unit where the object ends. Look at the metric ruler shown in **Figure 11.** The centimeter lines are the long, numbered lines, and the shorter lines are millimeter lines. In this instance, the length would be 4.50 cm.

Figure 11 This metric ruler has centimeter and millimeter divisions.

Science Skill Handbook
Math Skill Handbook
Foldables Handbook
Reference Handbook
Glossary/Glosario
Index

Mass The SI unit for mass is the kilogram (kg). Scientists can measure mass using units formed by adding metric prefixes to the unit gram (g), such as milligram (mg). To measure mass, you might use a triple-beam balance similar to the one shown in **Figure 12.** The balance has a pan on one side and a set of beams on the other side. Each beam has a rider that slides on the beam.

When using a triple-beam balance, place an object on the pan. Slide the largest rider along its beam until the pointer drops below zero. Then move it back one notch. Repeat the process for each rider proceeding from the larger to smaller until the pointer swings an equal distance above and below the zero point. Sum the masses on each beam to find the mass of the object. Move all riders back to zero when finished.

Instead of putting materials directly on the balance, scientists often take a tare of a container. A tare is the mass of a container into which objects or substances are placed for measuring their masses. To find the mass of objects or substances, find the mass of a clean container. Remove the container from the pan, and place the object or substances in the container. Find the mass of the container with the materials in it. Subtract the mass of the empty container from the mass of the filled container to find the mass of the materials you are using.

Figure 12 A triple-beam balance is used to determine the mass of an object.

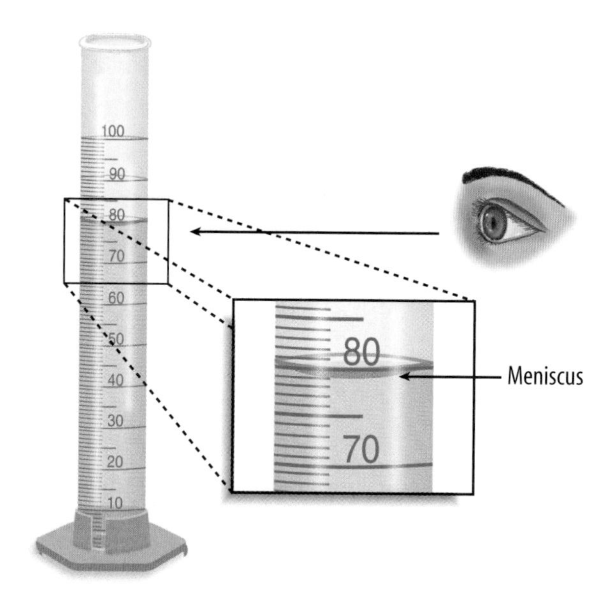

Figure 13 Graduated cylinders measure liquid volume.

Liquid Volume To measure liquids, the unit used is the liter. When a smaller unit is needed, scientists might use a milliliter. Because a milliliter takes up the volume of a cube measuring 1 cm on each side it also can be called a cubic centimeter ($cm^3 = cm \times cm \times cm$).

You can use beakers and graduated cylinders to measure liquid volume. A graduated cylinder, shown in **Figure 13,** is marked from bottom to top in milliliters. In lab, you might use a 10-mL graduated cylinder or a 100-mL graduated cylinder. When measuring liquids, notice that the liquid has a curved surface. Look at the surface at eye level, and measure the bottom of the curve. This is called the meniscus. The graduated cylinder in **Figure 13** contains 79.0 mL, or 79.0 cm^3, of a liquid.

Temperature Scientists often measure temperature using the Celsius scale. Pure water has a freezing point of 0°C and boiling point of 100°C. The unit of measurement is degrees Celsius. Two other scales often used are the Fahrenheit and Kelvin scales.

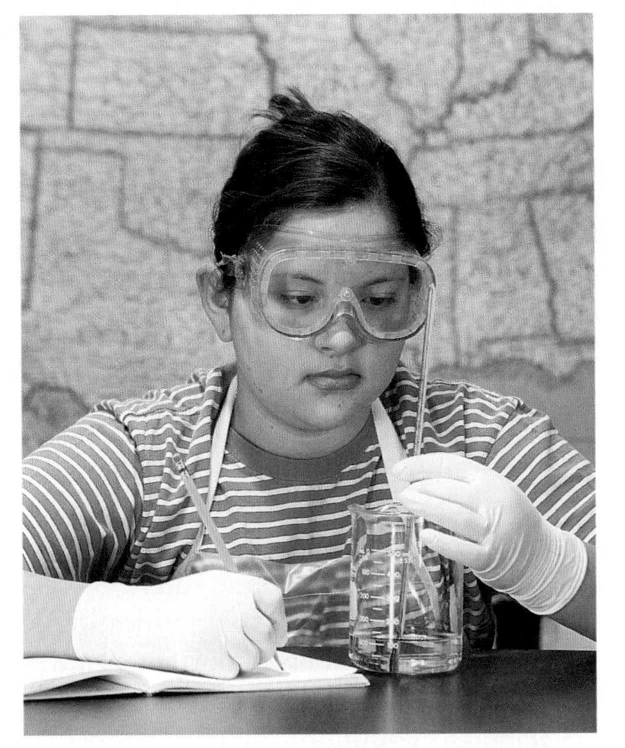

Figure 14 A thermometer measures the temperature of an object.

Scientists use a thermometer to measure temperature. Most thermometers in a laboratory are glass tubes with a bulb at the bottom end containing a liquid such as colored alcohol. The liquid rises or falls with a change in temperature. To read a glass thermometer like the thermometer in **Figure 14,** rotate it slowly until a red line appears. Read the temperature where the red line ends.

Form Operational Definitions An operational definition defines an object by how it functions, works, or behaves. For example, when you are playing hide and seek and a tree is home base, you have created an operational definition for a tree.

Objects can have more than one operational definition. For example, a ruler can be defined as a tool that measures the length of an object (how it is used). It can also be a tool with a series of marks used as a standard when measuring (how it works).

Analyze the Data

To determine the meaning of your observations and investigation results, you will need to look for patterns in the data. Then you must think critically to determine what the data mean. Scientists use several approaches when they analyze the data they have collected and recorded. Each approach is useful for identifying specific patterns.

Interpret Data The word *interpret* means "to explain the meaning of something." When analyzing data from an experiment, try to find out what the data show. Identify the control group and the test group to see whether changes in the independent variable have had an effect. Look for differences in the dependent variable between the control and test groups.

Classify Sorting objects or events into groups based on common features is called classifying. When classifying, first observe the objects or events to be classified. Then select one feature that is shared by some members in the group, but not by all. Place those members that share that feature in a subgroup. You can classify members into smaller and smaller subgroups based on characteristics. Remember that when you classify, you are grouping objects or events for a purpose. Keep your purpose in mind as you select the features to form groups and subgroups.

Compare and Contrast Observations can be analyzed by noting the similarities and differences between two or more objects or events that you observe. When you look at objects or events to see how they are similar, you are comparing them. Contrasting is looking for differences in objects or events.

Recognize Cause and Effect A cause is a reason for an action or condition. The effect is that action or condition. When two events happen together, it is not necessarily true that one event caused the other. Scientists must design a controlled investigation to recognize the exact cause and effect.

Draw Conclusions

When scientists have analyzed the data they collected, they proceed to draw conclusions about the data. These conclusions are sometimes stated in words similar to the hypothesis that you formed earlier. They may confirm a hypothesis, or lead you to a new hypothesis.

Infer Scientists often make inferences based on their observations. An inference is an attempt to explain observations or to indicate a cause. An inference is not a fact, but a logical conclusion that needs further investigation. For example, you may infer that a fire has caused smoke. Until you investigate, however, you do not know for sure.

Apply When you draw a conclusion, you must apply those conclusions to determine whether the data supports the hypothesis. If your data do not support your hypothesis, it does not mean that the hypothesis is wrong. It means only that the result of the investigation did not support the hypothesis. Maybe the experiment needs to be redesigned, or some of the initial observations on which the hypothesis was based were incomplete or biased. Perhaps more observation or research is needed to refine your hypothesis. A successful investigation does not always come out the way you originally predicted.

Avoid Bias Sometimes a scientific investigation involves making judgments. When you make a judgment, you form an opinion. It is important to be honest and not to allow any expectations of results to bias your judgments. This is important throughout the entire investigation, from researching to collecting data to drawing conclusions.

Communicate

The communication of ideas is an important part of the work of scientists. A discovery that is not reported will not advance the scientific community's understanding or knowledge. Communication among scientists also is important as a way of improving their investigations.

Scientists communicate in many ways, from writing articles in journals and magazines that explain their investigations and experiments, to announcing important discoveries on television and radio. Scientists also share ideas with colleagues on the Internet or present them as lectures, like the student is doing in **Figure 15.**

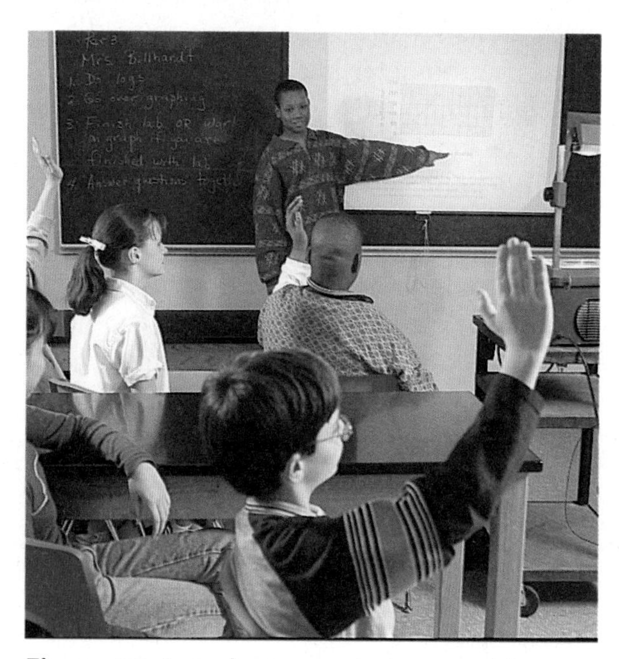

Figure 15 A student communicates to his peers about his investigation.

These safety symbols are used in laboratory and field investigations in this book to indicate possible hazards. Learn the meaning of each symbol and refer to this page often. *Remember to wash your hands thoroughly after completing lab procedures.*

PROTECTIVE EQUIPMENT

Do not begin any lab without the proper protection equipment.

GOGGLES 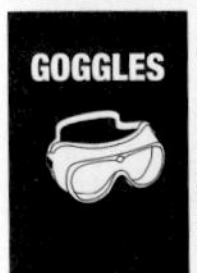	Proper eye protection must be worn when performing or observing science activities that involve items or conditions as listed below.
APRON 	Wear an approved apron when using substances that could stain, wet, or destroy cloth.
SOAP 	Wash hands with soap and water before removing goggles and after all lab activities.
GLOVES 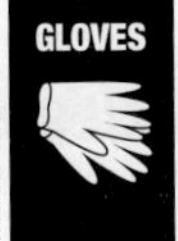	Wear gloves when working with biological materials, chemicals, animals, or materials that can stain or irritate hands.

LABORATORY HAZARDS

Symbols	Potential Hazards	Precaution	Response
DISPOSAL	contamination of classroom or environment due to improper disposal of materials such as chemicals and live specimens	• DO NOT dispose of hazardous materials in the sink or trash can. • Dispose of wastes as directed by your teacher.	• If hazardous materials are disposed of improperly, notify your teacher immediately.
EXTREME TEMPERATURE	skin burns due to extremely hot or cold materials such as hot glass, liquids, or metals; liquid nitrogen; dry ice	• Use proper protective equipment, such as hot mitts and/or tongs, when handling objects with extreme temperatures.	• If injury occurs, notify your teacher immediately.
SHARP OBJECTS	punctures or cuts from sharp objects such as razor blades, pins, scalpels, and broken glass	• Handle glassware carefully to avoid breakage. • Walk with sharp objects pointed downward, away from you and others.	• If broken glass or injury occurs, notify your teacher immediately.
ELECTRICAL	electric shock or skin burn due to improper grounding, short circuits, liquid spills, or exposed wires	• Check condition of wires and apparatus for fraying or uninsulated wires, and broken or cracked equipment. • Use only GFCI-protected outlets	• DO NOT attempt to fix electrical problems. Notify your teacher immediately.
CHEMICAL	skin irritation or burns, breathing difficulty, and/or poisoning due to touching, swallowing, or inhalation of chemicals such as acids, bases, bleach, metal compounds, iodine, poinsettias, pollen, ammonia, acetone, nail polish remover, heated chemicals, mothballs, and any other chemicals labeled or known to be dangerous	• Wear proper protective equipment such as goggles, apron, and gloves when using chemicals. • Ensure proper room ventilation or use a fume hood when using materials that produce fumes. • NEVER smell fumes directly. • NEVER taste or eat any material in the laboratory.	• If contact occurs, immediately flush affected area with water and notify your teacher. • If a spill occurs, leave the area immediately and notify your teacher.
FLAMMABLE	unexpected fire due to liquids or gases that ignite easily such as rubbing alcohol	• Avoid open flames, sparks, or heat when flammable liquids are present.	• If a fire occurs, leave the area immediately and notify your teacher.
OPEN FLAME	burns or fire due to open flame from matches, Bunsen burners, or burning materials	• Tie back loose hair and clothing. • Keep flame away from all materials. • Follow teacher instructions when lighting and extinguishing flames. • Use proper protection, such as hot mitts or tongs, when handling hot objects.	• If a fire occurs, leave the area immediately and notify your teacher.
ANIMAL SAFETY	injury to or from laboratory animals	• Wear proper protective equipment such as gloves, apron, and goggles when working with animals. • Wash hands after handling animals.	• If injury occurs, notify your teacher immediately.
BIOLOGICAL	infection or adverse reaction due to contact with organisms such as bacteria, fungi, and biological materials such as blood, animal or plant materials	• Wear proper protective equipment such as gloves, goggles, and apron when working with biological materials. • Avoid skin contact with an organism or any part of the organism. • Wash hands after handling organisms.	• If contact occurs, wash the affected area and notify your teacher immediately.
FUME	breathing difficulties from inhalation of fumes from substances such as ammonia, acetone, nail polish remover, heated chemicals, and mothballs	• Wear goggles, apron, and gloves. • Ensure proper room ventilation or use a fume hood when using substances that produce fumes. • NEVER smell fumes directly.	• If a spill occurs, leave area and notify your teacher immediately.
IRRITANT	irritation of skin, mucous membranes, or respiratory tract due to materials such as acids, bases, bleach, pollen, mothballs, steel wool, and potassium permanganate	• Wear goggles, apron, and gloves. • Wear a dust mask to protect against fine particles.	• If skin contact occurs, immediately flush the affected area with water and notify your teacher.
RADIOACTIVE	excessive exposure from alpha, beta, and gamma particles	• Remove gloves and wash hands with soap and water before removing remainder of protective equipment.	• If cracks or holes are found in the container, notify your teacher immediately.

Safety in the Science Laboratory

Introduction to Science Safety

The science laboratory is a safe place to work if you follow standard safety procedures. Being responsible for your own safety helps to make the entire laboratory a safer place for everyone. When performing any lab, read and apply the caution statements and safety symbol listed at the beginning of the lab.

General Safety Rules

1. Complete the *Lab Safety Form* or other safety contract BEFORE starting any science lab.
2. Study the procedure. Ask your teacher any questions. Be sure you understand safety symbols shown on the page.
3. Notify your teacher about allergies or other health conditions that can affect your participation in a lab.
4. Learn and follow use and safety procedures for your equipment. If unsure, ask your teacher.
5. Never eat, drink, chew gum, apply cosmetics, or do any personal grooming in the lab. Never use lab glassware as food or drink containers. Keep your hands away from your face and mouth.
6. Know the location and proper use of the safety shower, eye wash, fire blanket, and fire alarm.

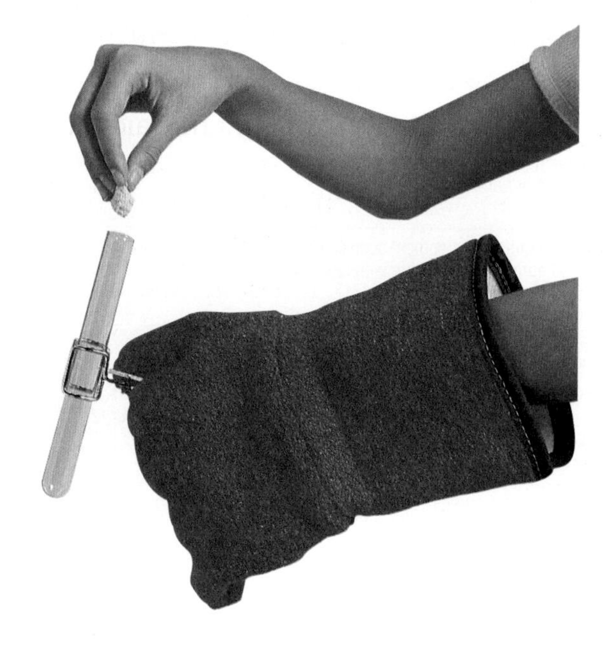

Prevent Accidents

1. Use the safety equipment provided to you. Goggles and a safety apron should be worn during investigations.
2. Do NOT use hair spray, mousse, or other flammable hair products. Tie back long hair and tie down loose clothing.
3. Do NOT wear sandals or other open-toed shoes in the lab.
4. Remove jewelry on hands and wrists. Loose jewelry, such as chains and long necklaces, should be removed to prevent them from getting caught in equipment.
5. Do not taste any substances or draw any material into a tube with your mouth.
6. Proper behavior is expected in the lab. Practical jokes and fooling around can lead to accidents and injury.
7. Keep your work area uncluttered.

Laboratory Work

1. Collect and carry all equipment and materials to your work area before beginning a lab.
2. Remain in your own work area unless given permission by your teacher to leave it.

3. Always slant test tubes away from yourself and others when heating them, adding substances to them, or rinsing them.
4. If instructed to smell a substance in a container, hold the container a short distance away and fan vapors toward your nose.
5. Do NOT substitute other chemicals/substances for those in the materials list unless instructed to do so by your teacher.
6. Do NOT take any materials or chemicals outside of the laboratory.
7. Stay out of storage areas unless instructed to be there and supervised by your teacher.

Laboratory Cleanup

1. Turn off all burners, water, and gas, and disconnect all electrical devices.
2. Clean all pieces of equipment and return all materials to their proper places.
3. Dispose of chemicals and other materials as directed by your teacher. Place broken glass and solid substances in the proper containers. Never discard materials in the sink.
4. Clean your work area.
5. Wash your hands with soap and water thoroughly BEFORE removing your goggles.

Emergencies

1. Report any fire, electrical shock, glassware breakage, spill, or injury, no matter how small, to your teacher immediately. Follow his or her instructions.
2. If your clothing should catch fire, STOP, DROP, and ROLL. If possible, smother it with the fire blanket or get under a safety shower. NEVER RUN.
3. If a fire should occur, turn off all gas and leave the room according to established procedures.
4. In most instances, your teacher will clean up spills. Do NOT attempt to clean up spills unless you are given permission and instructions to do so.
5. If chemicals come into contact with your eyes or skin, notify your teacher immediately. Use the eyewash, or flush your skin or eyes with large quantities of water.
6. The fire extinguisher and first-aid kit should only be used by your teacher unless it is an extreme emergency and you have been given permission.
7. If someone is injured or becomes ill, only a professional medical provider or someone certified in first aid should perform first-aid procedures.

Math Review

Use Fractions

A fraction compares a part to a whole. In the fraction $\frac{2}{3}$, the 2 represents the part and is the numerator. The 3 represents the whole and is the denominator.

Reduce Fractions To reduce a fraction, you must find the largest factor that is common to both the numerator and the denominator, the greatest common factor (GCF). Divide both numbers by the GCF. The fraction has then been reduced, or it is in its simplest form.

Example

Twelve of the 20 chemicals in the science lab are in powder form. What fraction of the chemicals used in the lab are in powder form?

Step 1 Write the fraction.

$$\frac{\text{part}}{\text{whole}} = \frac{12}{20}$$

Step 2 To find the GCF of the numerator and denominator, list all of the factors of each number.

Factors of 12: 1, 2, 3, 4, 6, 12
(the numbers that divide evenly into 12)

Factors of 20: 1, 2, 4, 5, 10, 20
(the numbers that divide evenly into 20)

Step 3 List the common factors.

1, 2, 4

Step 4 Choose the greatest factor in the list. The GCF of 12 and 20 is 4.

Step 5 Divide the numerator and denominator by the GCF.

$$\frac{12 \div 4}{20 \div 4} = \frac{3}{5}$$

In the lab, $\frac{3}{5}$ of the chemicals are in powder form.

Practice Problem At an amusement park, 66 of 90 rides have a height restriction. What fraction of the rides, in its simplest form, has a height restriction?

Add and Subtract Fractions with Like Denominators To add or subtract fractions with the same denominator, add or subtract the numerators and write the sum or difference over the denominator. After finding the sum or difference, find the simplest form for your fraction.

Example 1

In the forest outside your house, $\frac{1}{8}$ of the animals are rabbits, $\frac{3}{8}$ are squirrels, and the remainder are birds and insects. How many are mammals?

Step 1 Add the numerators.

$$\frac{1}{8} + \frac{3}{8} = \frac{(1 + 3)}{8} = \frac{4}{8}$$

Step 2 Find the GCF.

$$\frac{4}{8} \text{ (GCF, 4)}$$

Step 3 Divide the numerator and denominator by the GCF.

$$\frac{4 \div 4}{8 \div 4} = \frac{1}{2}$$

$\frac{1}{2}$ of the animals are mammals.

Example 2

If $\frac{7}{16}$ of the Earth is covered by freshwater, and $\frac{1}{16}$ of that is in glaciers, how much freshwater is not frozen?

Step 1 Subtract the numerators.

$$\frac{7}{16} - \frac{1}{16} = \frac{(7 - 1)}{16} = \frac{6}{16}$$

Step 2 Find the GCF.

$$\frac{6}{16} \text{ (GCF, 2)}$$

Step 3 Divide the numerator and denominator by the GCF.

$$\frac{6 \div 2}{16 \div 2} = \frac{3}{8}$$

$\frac{3}{8}$ of the freshwater is not frozen.

Practice Problem A bicycle rider is riding at a rate of 15 km/h for $\frac{4}{9}$ of his ride, 10 km/h for $\frac{2}{9}$ of his ride, and 8 km/h for the remainder of the ride. How much of his ride is he riding at a rate greater than 8 km/h?

Add and Subtract Fractions with Unlike Denominators To add or subtract fractions with unlike denominators, first find the least common denominator (LCD). This is the smallest number that is a common multiple of both denominators. Rename each fraction with the LCD, and then add or subtract. Find the simplest form if necessary.

Example 1

A chemist makes a paste that is $\frac{1}{2}$ table salt (NaCl), $\frac{1}{3}$ sugar ($C_6H_{12}O_6$), and the remainder is water (H_2O). How much of the paste is a solid?

Step 1 Find the LCD of the fractions.

$\frac{1}{2} + \frac{1}{3}$ (LCD, 6)

Step 2 Rename each numerator and each denominator with the LCD.

Step 3 Add the numerators.

$\frac{3}{6} + \frac{2}{6} = \frac{(3 + 2)}{6} = \frac{5}{6}$

$\frac{5}{6}$ of the paste is a solid.

Example 2

The average precipitation in Grand Junction, CO, is $\frac{7}{10}$ inch in November, and $\frac{3}{5}$ inch in December. What is the total average precipitation?

Step 1 Find the LCD of the fractions.

$\frac{7}{10} + \frac{3}{5}$ (LCD, 10)

Step 2 Rename each numerator and each denominator with the LCD.

Step 3 Add the numerators.

$\frac{7}{10} + \frac{6}{10} = \frac{(7 + 6)}{10} = \frac{13}{10}$

$\frac{13}{10}$ inches total precipitation, or $1\frac{3}{10}$ inches.

Practice Problem On an electric bill, about $\frac{1}{8}$ of the energy is from solar energy and about $\frac{1}{10}$ is from wind power. How much of the total bill is from solar energy and wind power combined?

Example 3

In your body, $\frac{7}{10}$ of your muscle contractions are involuntary (cardiac and smooth muscle tissue). Smooth muscle makes $\frac{3}{15}$ of your muscle contractions. How many of your muscle contractions are made by cardiac muscle?

Step 1 Find the LCD of the fractions.

$\frac{7}{10} - \frac{3}{15}$ (LCD, 30)

Step 2 Rename each numerator and each denominator with the LCD.

$\frac{7 \times 3}{10 \times 3} = \frac{21}{30}$

$\frac{3 \times 2}{15 \times 2} = \frac{6}{30}$

Step 3 Subtract the numerators.

$\frac{21}{30} - \frac{6}{30} = \frac{(21 - 6)}{30} = \frac{15}{30}$

Step 4 Find the GCF.

$\frac{15}{30}$ (GCF, 15)

$\frac{1}{2}$

$\frac{1}{2}$ of all muscle contractions are cardiac muscle.

Example 4

Tony wants to make cookies that call for $\frac{3}{4}$ of a cup of flour, but he only has $\frac{1}{3}$ of a cup. How much more flour does he need?

Step 1 Find the LCD of the fractions.

$\frac{3}{4} - \frac{1}{3}$ (LCD, 12)

Step 2 Rename each numerator and each denominator with the LCD.

$\frac{3 \times 3}{4 \times 3} = \frac{9}{12}$

$\frac{1 \times 4}{3 \times 4} = \frac{4}{12}$

Step 3 Subtract the numerators.

$\frac{9}{12} - \frac{4}{12} = \frac{(9 - 4)}{12} = \frac{5}{12}$

$\frac{5}{12}$ of a cup of flour

Practice Problem Using the information provided to you in Example 3 above, determine how many muscle contractions are voluntary (skeletal muscle).

Multiply Fractions To multiply with fractions, multiply the numerators and multiply the denominators. Find the simplest form if necessary.

Example

Multiply $\frac{3}{5}$ by $\frac{1}{3}$.

Step 1 Multiply the numerators and denominators.

$\frac{3}{5} \times \frac{1}{3} = \frac{(3 \times 1)}{(5 \times 3)}\ \frac{3}{15}$

Step 2 Find the GCF.

$\frac{3}{15}$ (GCF, 3)

Step 3 Divide the numerator and denominator by the GCF.

$\frac{3 \div 3}{15 \div 3} = \frac{1}{5}$

$\frac{3}{5}$ multiplied by $\frac{1}{3}$ is $\frac{1}{5}$.

Practice Problem Multiply $\frac{3}{14}$ by $\frac{5}{16}$.

Find a Reciprocal Two numbers whose product is 1 are called multiplicative inverses, or reciprocals.

Example

Find the reciprocal of $\frac{3}{8}$.

Step 1 Inverse the fraction by putting the denominator on top and the numerator on the bottom.

$\frac{8}{3}$

The reciprocal of $\frac{3}{8}$ is $\frac{8}{3}$.

Practice Problem Find the reciprocal of $\frac{4}{9}$.

Divide Fractions To divide one fraction by another fraction, multiply the dividend by the reciprocal of the divisor. Find the simplest form if necessary.

Example 1

Divide $\frac{1}{9}$ by $\frac{1}{3}$.

Step 1 Find the reciprocal of the divisor.

The reciprocal of $\frac{1}{3}$ is $\frac{3}{1}$.

Step 2 Multiply the dividend by the reciprocal of the divisor.

$\frac{\frac{1}{9}}{\frac{1}{3}} = \frac{1}{9} \times \frac{3}{1} = \frac{(1 \times 3)}{(9 \times 1)} = \frac{3}{9}$

Step 3 Find the GCF.

$\frac{3}{9}$ (GCF, 3)

Step 4 Divide the numerator and denominator by the GCF.

$\frac{3 \div 3}{9 \div 3} = \frac{1}{3}$

$\frac{1}{9}$ divided by $\frac{1}{3}$ is $\frac{1}{3}$.

Example 2

Divide $\frac{3}{5}$ by $\frac{1}{4}$.

Step 1 Find the reciprocal of the divisor.

The reciprocal of $\frac{1}{4}$ is $\frac{4}{1}$.

Step 2 Multiply the dividend by the reciprocal of the divisor.

$\frac{\frac{3}{5}}{\frac{1}{4}} = \frac{3}{5} \times \frac{4}{1} = \frac{(3 \times 4)}{(5 \times 1)} = \frac{12}{5}$

$\frac{3}{5}$ divided by $\frac{1}{4}$ is $\frac{12}{5}$ or $2\frac{2}{5}$.

Practice Problem Divide $\frac{3}{11}$ by $\frac{7}{10}$.

Use Ratios

When you compare two numbers by division, you are using a ratio. Ratios can be written 3 to 5, 3:5, or $\frac{3}{5}$. Ratios, like fractions, also can be written in simplest form.

Ratios can represent one type of probability, called odds. This is a ratio that compares the number of ways a certain outcome occurs to the number of possible outcomes. For example, if you flip a coin 100 times, what are the odds that it will come up heads? There are two possible outcomes, heads or tails, so the odds of coming up heads are 50:100. Another way to say this is that 50 out of 100 times the coin will come up heads. In its simplest form, the ratio is 1:2.

Example 1

A chemical solution contains 40 g of salt and 64 g of baking soda. What is the ratio of salt to baking soda as a fraction in simplest form?

Step 1 Write the ratio as a fraction.

$$\frac{\text{salt}}{\text{baking soda}} = \frac{40}{64}$$

Step 2 Express the fraction in simplest form. The GCF of 40 and 64 is 8.

$$\frac{40}{64} = \frac{40 \div 8}{64 \div 8} = \frac{5}{8}$$

The ratio of salt to baking soda in the sample is 5:8.

Example 2

Sean rolls a 6-sided die 6 times. What are the odds that the side with a 3 will show?

Step 1 Write the ratio as a fraction.

$$\frac{\text{number of sides with a 3}}{\text{number of possible sides}} = \frac{1}{6}$$

Step 2 Multiply by the number of attempts.

$\frac{1}{6} \times 6$ attempts $= \frac{6}{6}$ attempts $= 1$ attempt

1 attempt out of 6 will show a 3.

Practice Problem Two metal rods measure 100 cm and 144 cm in length. What is the ratio of their lengths in simplest form?

Use Decimals

A fraction with a denominator that is a power of ten can be written as a decimal. For example, 0.27 means $\frac{27}{100}$. The decimal point separates the ones place from the tenths place.

Any fraction can be written as a decimal using division. For example, the fraction $\frac{5}{8}$ can be written as a decimal by dividing 5 by 8. Written as a decimal, it is 0.625.

Add or Subtract Decimals When adding and subtracting decimals, line up the decimal points before carrying out the operation.

Example 1

Find the sum of 47.68 and 7.80.

Step 1 Line up the decimal places when you write the numbers.

$$\begin{array}{r} 47.68 \\ +\ 7.80 \\ \hline \end{array}$$

Step 2 Add the decimals.

$$\begin{array}{r} \scriptstyle{1\ 1} \\ 47.68 \\ +\ 7.80 \\ \hline 55.48 \end{array}$$

The sum of 47.68 and 7.80 is 55.48.

Example 2

Find the difference of 42.17 and 15.85.

Step 1 Line up the decimal places when you write the number.

$$\begin{array}{r} 42.17 \\ -15.85 \\ \hline \end{array}$$

Step 2 Subtract the decimals.

$$\begin{array}{r} \scriptstyle{3\ 11\ 1} \\ 42.17 \\ -15.85 \\ \hline 26.32 \end{array}$$

The difference of 42.17 and 15.85 is 26.32.

Practice Problem Find the sum of 1.245 and 3.842.

Science Skill Handbook | Math Skill Handbook | Foldables Handbook | Reference Handbook | Glossary/Glosario | Index

Multiply Decimals To multiply decimals, multiply the numbers like numbers without decimal points. Count the decimal places in each factor. The product will have the same number of decimal places as the sum of the decimal places in the factors.

Example

Multiply 2.4 by 5.9.

Step 1 Multiply the factors like two whole numbers.

$24 \times 59 = 1416$

Step 2 Find the sum of the number of decimal places in the factors. Each factor has one decimal place, for a sum of two decimal places.

Step 3 The product will have two decimal places.

14.16

The product of 2.4 and 5.9 is 14.16.

Practice Problem Multiply 4.6 by 2.2.

Divide Decimals When dividing decimals, change the divisor to a whole number. To do this, multiply both the divisor and the dividend by the same power of ten. Then place the decimal point in the quotient directly above the decimal point in the dividend. Then divide as you do with whole numbers.

Example

Divide 8.84 by 3.4.

Step 1 Multiply both factors by 10.

$3.4 \times 10 = 34, 8.84 \times 10 = 88.4$

Step 2 Divide 88.4 by 34.

```
    2.6
34)88.4
  -68
   204
  -204
     0
```

8.84 divided by 3.4 is 2.6.

Practice Problem Divide 75.6 by 3.6.

Use Proportions

An equation that shows that two ratios are equivalent is a proportion. The ratios $\frac{2}{4}$ and $\frac{5}{10}$ are equivalent, so they can be written as $\frac{2}{4} = \frac{5}{10}$. This equation is a proportion.

When two ratios form a proportion, the cross products are equal. To find the cross products in the proportion $\frac{2}{4} = \frac{5}{10}$, multiply the 2 and the 10, and the 4 and the 5. Therefore $2 \times 10 = 4 \times 5$, or $20 = 20$.

Because you know that both ratios are equal, you can use cross products to find a missing term in a proportion. This is known as solving the proportion.

Example

The heights of a tree and a pole are proportional to the lengths of their shadows. The tree casts a shadow of 24 m when a 6-m pole casts a shadow of 4 m. What is the height of the tree?

Step 1 Write a proportion.

$$\frac{\text{height of tree}}{\text{height of pole}} = \frac{\text{length of tree's shadow}}{\text{length of pole's shadow}}$$

Step 2 Substitute the known values into the proportion. Let h represent the unknown value, the height of the tree.

$$\frac{h}{6} \times \frac{24}{4}$$

Step 3 Find the cross products.

$h \times 4 = 6 \times 24$

Step 4 Simplify the equation.

$4h \times 144$

Step 5 Divide each side by 4.

$$\frac{4h}{4} \times \frac{144}{4}$$

$h = 36$

The height of the tree is 36 m.

Practice Problem The ratios of the weights of two objects on the Moon and on Earth are in proportion. A rock weighing 3 N on the Moon weighs 18 N on Earth. How much would a rock that weighs 5 N on the Moon weigh on Earth?

Science Skill Handbook
Math Skill Handbook
Foldables Handbook
Reference Handbook
Glossary/Glosario
Index

Use Percentages

The word *percent* means "out of one hundred." It is a ratio that compares a number to 100. Suppose you read that 77 percent of Earth's surface is covered by water. That is the same as reading that the fraction of Earth's surface covered by water is $\frac{77}{100}$. To express a fraction as a percent, first find the equivalent decimal for the fraction. Then, multiply the decimal by 100 and add the percent symbol.

Example 1

Express $\frac{13}{20}$ as a percent.

Step 1 Find the equivalent decimal for the fraction.

$$\begin{array}{r} 0.65 \\ 20\overline{)13.00} \\ \underline{12\ 0} \\ 1\ 00 \\ \underline{1\ 00} \\ 0 \end{array}$$

Step 2 Rewrite the fraction $\frac{13}{20}$ as 0.65.

Step 3 Multiply 0.65 by 100 and add the % symbol.

$0.65 \times 100 = 65 = 65\%$

So, $\frac{13}{20} = 65\%$.

This also can be solved as a proportion.

Example 2

Express $\frac{13}{20}$ as a percent.

Step 1 Write a proportion.

$\frac{13}{20} = \frac{x}{100}$

Step 2 Find the cross products.

$1300 = 20x$

Step 3 Divide each side by 20.

$\frac{1300}{20} = \frac{20x}{20}$

$65\% = x$

Practice Problem In one year, 73 of 365 days were rainy in one city. What percent of the days in that city were rainy?

Solve One-Step Equations

A statement that two expressions are equal is an equation. For example, $A = B$ is an equation that states that A is equal to B.

An equation is solved when a variable is replaced with a value that makes both sides of the equation equal. To make both sides equal the inverse operation is used. Addition and subtraction are inverses, and multiplication and division are inverses.

Example 1

Solve the equation $x - 10 = 35$.

Step 1 Find the solution by adding 10 to each side of the equation.

$$x - 10 = 35$$
$$x - 10 + 10 = 35 - 10$$
$$x = 45$$

Step 2 Check the solution.

$$x - 10 = 35$$
$$45 - 10 = 35$$
$$35 = 35$$

Both sides of the equation are equal, so $x = 45$.

Example 2

In the formula $a = bc$, find the value of c if $a = 20$ and $b = 2$.

Step 1 Rearrange the formula so the unknown value is by itself on one side of the equation by dividing both sides by b.

$$a = bc$$
$$\frac{a}{b} = \frac{bc}{b}$$
$$\frac{a}{b} = c$$

Step 2 Replace the variables a and b with the values that are given.

$$\frac{a}{b} = c$$
$$\frac{20}{2} = c$$
$$10 = c$$

Step 3 Check the solution.

$$a = bc$$
$$20 = 2 \times 10$$
$$20 = 20$$

Both sides of the equation are equal, so $c = 10$ is the solution when $a = 20$ and $b = 2$.

Practice Problem In the formula $h = gd$, find the value of d if $g = 12.3$ and $h = 17.4$.

Use Statistics

The branch of mathematics that deals with collecting, analyzing, and presenting data is statistics. In statistics, there are three common ways to summarize data with a single number—the mean, the median, and the mode.

The **mean** of a set of data is the arithmetic average. It is found by adding the numbers in the data set and dividing by the number of items in the set.

The **median** is the middle number in a set of data when the data are arranged in numerical order. If there were an even number of data points, the median would be the mean of the two middle numbers.

The **mode** of a set of data is the number or item that appears most often.

Another number that often is used to describe a set of data is the range**.** The **range** is the difference between the largest number and the smallest number in a set of data.

Example

The speeds (in m/s) for a race car during five different time trials are 39, 37, 44, 36, and 44.

To find the mean:

Step 1 Find the sum of the numbers.

$39 + 37 + 44 + 36 + 44 = 200$

Step 2 Divide the sum by the number of items, which is 5.

$200 \div 5 = 40$

The mean is 40 m/s.

To find the median:

Step 1 Arrange the measures from least to greatest.

36, 37, 39, 44, 44

Step 2 Determine the middle measure.

36, 37, <u>39</u>, 44, 44

The median is 39 m/s.

To find the mode:

Step 1 Group the numbers that are the same together.

44, 44, 36, 37, 39

Step 2 Determine the number that occurs most in the set.

<u>44, 44,</u> 36, 37, 39

The mode is 44 m/s.

To find the range:

Step 1 Arrange the measures from greatest to least.

44, 44, 39, 37, 36

Step 2 Determine the greatest and least measures in the set.

<u>44,</u> 44, 39, 37, <u>36</u>

Step 3 Find the difference between the greatest and least measures.

$44 - 36 = 8$

The range is 8 m/s.

Practice Problem Find the mean, median, mode, and range for the data set 8, 4, 12, 8, 11, 14, 16.

A **frequency table** shows how many times each piece of data occurs, usually in a survey. **Table 1** below shows the results of a student survey on favorite color.

Table 1 Student Color Choice

Color	Tally	Frequency
red	IIII	4
blue	~~IIII~~	5
black	II	2
green	III	3
purple	~~IIII~~ II	7
yellow	~~IIII~~ I	6

Based on the frequency table data, which color is the favorite?

Use Geometry

The branch of mathematics that deals with the measurement, properties, and relationships of points, lines, angles, surfaces, and solids is called geometry.

Perimeter The **perimeter** (P) is the distance around a geometric figure. To find the perimeter of a rectangle, add the length and width and multiply that sum by two, or $2(l + w)$. To find perimeters of irregular figures, add the length of the sides.

Example 1

Find the perimeter of a rectangle that is 3 m long and 5 m wide.

Step 1 You know that the perimeter is 2 times the sum of the width and length.

$P = 2(3 \text{ m} + 5 \text{ m})$

Step 2 Find the sum of the width and length.

$P = 2(8 \text{ m})$

Step 3 Multiply by 2.

$P = 16 \text{ m}$

The perimeter is 16 m.

Example 2

Find the perimeter of a shape with sides measuring 2 cm, 5 cm, 6 cm, 3 cm.

Step 1 You know that the perimeter is the sum of all the sides.

$P = 2 + 5 + 6 + 3$

Step 2 Find the sum of the sides.

$P = 2 + 5 + 6 + 3$

$P = 16$

The perimeter is 16 cm.

Practice Problem Find the perimeter of a rectangle with a length of 18 m and a width of 7 m.

Practice Problem Find the perimeter of a triangle measuring 1.6 cm by 2.4 cm by 2.4 cm.

Area of a Rectangle The **area** (A) is the number of square units needed to cover a surface. To find the area of a rectangle, multiply the length times the width, or $l \times w$. When finding area, the units also are multiplied. Area is given in square units.

Example

Find the area of a rectangle with a length of 1 cm and a width of 10 cm.

Step 1 You know that the area is the length multiplied by the width.

$A = (1 \text{ cm} \times 10 \text{ cm})$

Step 2 Multiply the length by the width. Also multiply the units.

$A = 10 \text{ cm}^2$

The area is 10 cm^2.

Practice Problem Find the area of a square whose sides measure 4 m.

Area of a Triangle To find the area of a triangle, use the formula:

$A = \frac{1}{2}(\text{base} \times \text{height})$

The base of a triangle can be any of its sides. The height is the perpendicular distance from a base to the opposite endpoint, or vertex.

Example

Find the area of a triangle with a base of 18 m and a height of 7 m.

Step 1 You know that the area is $\frac{1}{2}$ the base times the height.

$A = \frac{1}{2}(18 \text{ m} \times 7 \text{ m})$

Step 2 Multiply $\frac{1}{2}$ by the product of 18×7. Multiply the units.

$A = \frac{1}{2}(126 \text{ m}^2)$

$A = 63 \text{ m}^2$

The area is 63 m^2.

Practice Problem Find the area of a triangle with a base of 27 cm and a height of 17 cm.

Circumference of a Circle The **diameter** (d) of a circle is the distance across the circle through its center, and the **radius** (r) is the distance from the center to any point on the circle. The radius is half of the diameter. The distance around the circle is called the **circumference** (C). The formula for finding the circumference is:

$C = 2\pi r$ or $C = \pi d$

The circumference divided by the diameter is always equal to 3.1415926... This nonterminating and nonrepeating number is represented by the Greek letter π (pi). An approximation often used for π is 3.14.

Example 1

Find the circumference of a circle with a radius of 3 m.

Step 1 You know the formula for the circumference is 2 times the radius times π.

$C = 2\pi(3)$

Step 2 Multiply 2 times the radius.

$C = 6\pi$

Step 3 Multiply by π.

$C \approx 19$ m

The circumference is about 19 m.

Example 2

Find the circumference of a circle with a diameter of 24.0 cm.

Step 1 You know the formula for the circumference is the diameter times π.

$C = \pi(24.0)$

Step 2 Multiply the diameter by π.

$C \approx 75.4$ cm

The circumference is about 75.4 cm.

Practice Problem Find the circumference of a circle with a radius of 19 cm.

Area of a Circle The formula for the area of a circle is: $A = \pi r^2$

Example 1

Find the area of a circle with a radius of 4.0 cm.

Step 1 $A = \pi(4.0)^2$

Step 2 Find the square of the radius.

$A = 16\pi$

Step 3 Multiply the square of the radius by π.

$A \approx 50 \text{ cm}^2$

The area of the circle is about 50 cm^2.

Example 2

Find the area of a circle with a radius of 225 m.

Step 1 $A = \pi(225)^2$

Step 2 Find the square of the radius.

$A = 50625\pi$

Step 3 Multiply the square of the radius by π.

$A \approx 159043.1$

The area of the circle is about 159043.1 m^2.

Example 3

Find the area of a circle whose diameter is 20.0 mm.

Step 1 Remember that the radius is half of the diameter.

$A = \pi\left(\frac{20.0}{2}\right)^2$

Step 2 Find the radius.

$A = \pi(10.0)^2$

Step 3 Find the square of the radius.

$A = 100\pi$

Step 4 Multiply the square of the radius by π.

$A \approx 314 \text{ mm}^2$

The area of the circle is about 314 mm^2.

Practice Problem Find the area of a circle with a radius of 16 m.

Volume The measure of space occupied by a solid is the **volume** (V). To find the volume of a rectangular solid multiply the length times width times height, or $V = l \times w \times h$. It is measured in cubic units, such as cubic centimeters (cm^3).

Example

Find the volume of a rectangular solid with a length of 2.0 m, a width of 4.0 m, and a height of 3.0 m.

Step 1 You know the formula for volume is the length times the width times the height.

$V = 2.0 \text{ m} \times 4.0 \text{ m} \times 3.0 \text{ m}$

Step 2 Multiply the length times the width times the height.

$V = 24 \text{ m}^3$

The volume is 24 m^3.

Practice Problem Find the volume of a rectangular solid that is 8 m long, 4 m wide, and 4 m high.

To find the volume of other solids, multiply the area of the base times the height.

Example 1

Find the volume of a solid that has a triangular base with a length of 8.0 m and a height of 7.0 m. The height of the entire solid is 15.0 m.

Step 1 You know that the base is a triangle, and the area of a triangle is $\frac{1}{2}$ the base times the height, and the volume is the area of the base times the height.

$V = \left[\frac{1}{2}(b \times h)\right] \times 15$

Step 2 Find the area of the base.

$V = \left[\frac{1}{2}(8 \times 7)\right] \times 15$

$V = \left(\frac{1}{2} \times 56\right) \times 15$

Step 3 Multiply the area of the base by the height of the solid.

$V = 28 \times 15$

$V = 420 \text{ m}^3$

The volume is 420 m^3.

Example 2

Find the volume of a cylinder that has a base with a radius of 12.0 cm, and a height of 21.0 cm.

Step 1 You know that the base is a circle, and the area of a circle is the square of the radius times π, and the volume is the area of the base times the height.

$V = (\pi r^2) \times 21$

$V = (\pi 12^2) \times 21$

Step 2 Find the area of the base.

$V = 144\pi \times 21$

$V = 452 \times 21$

Step 3 Multiply the area of the base by the height of the solid.

$V \approx 9{,}500 \text{ cm}^3$

The volume is about 9,500 cm^3.

Example 3

Find the volume of a cylinder that has a diameter of 15 mm and a height of 4.8 mm.

Step 1 You know that the base is a circle with an area equal to the square of the radius times π. The radius is one-half the diameter. The volume is the area of the base times the height.

$V = (\pi r^2) \times 4.8$

$V = \left[\pi\left(\frac{1}{2} \times 15\right)^2\right] \times 4.8$

$V = (\pi 7.5^2) \times 4.8$

Step 2 Find the area of the base.

$V = 56.25\pi \times 4.8$

$V \approx 176.71 \times 4.8$

Step 3 Multiply the area of the base by the height of the solid.

$V \approx 848.2$

The volume is about 848.2 mm^3.

Practice Problem Find the volume of a cylinder with a diameter of 7 cm in the base and a height of 16 cm.

Science Applications

Measure in SI

The metric system of measurement was developed in 1795. A modern form of the metric system, called the International System (SI), was adopted in 1960 and provides the standard measurements that all scientists around the world can understand.

The SI system is convenient because unit sizes vary by powers of 10. Prefixes are used to name units. Look at **Table 2** for some common SI prefixes and their meanings.

Table 2 Common SI Prefixes

Prefix	Symbol	Meaning	
kilo–	k	1,000	thousandth
hecto–	h	100	hundred
deka–	da	10	ten
deci–	d	0.1	tenth
centi–	c	0.01	hundreth
milli–	m	0.001	thousandth

Example

How many grams equal one kilogram?

Step 1 Find the prefix *kilo–* in **Table 2.**

Step 2 Using **Table 2,** determine the meaning of *kilo–*. According to the table, it means 1,000. When the prefix *kilo–* is added to a unit, it means that there are 1,000 of the units in a "kilounit."

Step 3 Apply the prefix to the units in the question. The units in the question are grams. There are 1,000 grams in a kilogram.

Practice Problem Is a milligram larger or smaller than a gram? How many of the smaller units equal one larger unit? What fraction of the larger unit does one smaller unit represent?

Dimensional Analysis

Convert SI Units In science, quantities such as length, mass, and time sometimes are measured using different units. A process called dimensional analysis can be used to change one unit of measure to another. This process involves multiplying your starting quantity and units by one or more conversion factors. A conversion factor is a ratio equal to one and can be made from any two equal quantities with different units. If 1,000 mL equal 1 L then two ratios can be made.

$$\frac{1{,}000 \text{ mL}}{1 \text{ L}} = \frac{1 \text{ L}}{1{,}000 \text{ mL}} = 1$$

One can convert between units in the SI system by using the equivalents in **Table 2** to make conversion factors.

Example

How many cm are in 4 m?

Step 1 Write conversion factors for the units given. From **Table 2,** you know that 100 cm = 1 m. The conversion factors are

$$\frac{100 \text{ cm}}{1 \text{ m}} \textit{ and } \frac{1 \text{ m}}{100 \text{ cm}}$$

Step 2 Decide which conversion factor to use. Select the factor that has the units you are converting from (m) in the denominator and the units you are converting to (cm) in the numerator.

$$\frac{100 \text{ cm}}{1 \text{ m}}$$

Step 3 Multiply the starting quantity and units by the conversion factor. Cancel the starting units with the units in the denominator. There are 400 cm in 4 m.

$$4 \cancel{\text{m}} = \frac{100 \text{ cm}}{1 \cancel{\text{m}}} = 400 \text{ cm}$$

Practice Problem How many milligrams are in one kilogram? (Hint: You will need to use two conversion factors from **Table 2.**)

Science Skill Handbook
Math Skill Handbook
Foldables Handbook
Reference Handbook
Glossary/Glosario
Index

Table 3 Unit System Equivalents	
Type of Measurement	**Equivalent**
Length	1 in = 2.54 cm 1 yd = 0.91 m 1 mi = 1.61 km
Mass and weight*	1 oz = 28.35 g 1 lb = 0.45 kg 1 ton (short) = 0.91 tonnes (metric tons) 1 lb = 4.45 N
Volume	1 in^3 = 16.39 cm^3 1 qt = 0.95 L 1 gal = 3.78 L
Area	1 in^2 = 6.45 cm^2 1 yd^2 = 0.83 m^2 1 mi^2 = 2.59 km^2 1 acre = 0.40 hectares
Temperature	$°C = \frac{(°F - 32)}{1.8}$ K = °C + 273

*Weight is measured in standard Earth gravity.

Convert Between Unit Systems **Table 3** gives a list of equivalents that can be used to convert between English and SI units.

Example

If a meterstick has a length of 100 cm, how long is the meterstick in inches?

Step 1 Write the conversion factors for the units given. From **Table 3,** 1 in = 2.54 cm.

$$\frac{1 \text{ in}}{2.54 \text{ cm}} \textit{ and } \frac{2.54 \text{ cm}}{1 \text{ in}}$$

Step 2 Determine which conversion factor to use. You are converting from cm to in. Use the conversion factor with cm on the bottom.

$$\frac{1 \text{ in}}{2.54 \text{ cm}}$$

Step 3 Multiply the starting quantity and units by the conversion factor. Cancel the starting units with the units in the denominator. Round your answer to the nearest tenth.

$$100 \cancel{\text{cm}} \times \frac{1 \text{ in}}{2.54 \cancel{\text{cm}}} = 39.37 \text{ in}$$

The meterstick is about 39.4 in long.

Practice Problem 1 A book has a mass of 5 lb. What is the mass of the book in kg?

Practice Problem 2 Use the equivalent for in and cm (1 in = 2.54 cm) to show how 1 $in^3 \approx$ 16.39 cm^3.

Precision and Significant Digits

When you make a measurement, the value you record depends on the precision of the measuring instrument. This precision is represented by the number of significant digits recorded in the measurement. When counting the number of significant digits, all digits are counted except zeros at the end of a number with no decimal point such as 2,050, and zeros at the beginning of a decimal such as 0.03020. When adding or subtracting numbers with different precision, round the answer to the smallest number of decimal places of any number in the sum or difference. When multiplying or dividing, the answer is rounded to the smallest number of significant digits of any number being multiplied or divided.

Example

The lengths 5.28 and 5.2 are measured in meters. Find the sum of these lengths and record your answer using the correct number of significant digits.

Step 1 Find the sum.

5.28 m	2 digits after the decimal
+ 5.2 m	1 digit after the decimal
10.48 m	

Step 2 Round to one digit after the decimal because the least number of digits after the decimal of the numbers being added is 1.

The sum is 10.5 m.

Practice Problem 1 How many significant digits are in the measurement 7,071,301 m? How many significant digits are in the measurement 0.003010 g?

Practice Problem 2 Multiply 5.28 and 5.2 using the rule for multiplying and dividing. Record the answer using the correct number of significant digits.

Scientific Notation

Many times numbers used in science are very small or very large. Because these numbers are difficult to work with scientists use scientific notation. To write numbers in scientific notation, move the decimal point until only one non-zero digit remains on the left. Then count the number of places you moved the decimal point and use that number as a power of ten. For example, the average distance from the Sun to Mars is 227,800,000,000 m. In scientific notation, this distance is 2.278×10^{11} m. Because you moved the decimal point to the left, the number is a positive power of ten.

The mass of an electron is about 0.000 000 000 000 000 000 000 000 000 000 911 kg. Expressed in scientific notation, this mass is 9.11×10^{-31} kg. Because the decimal point was moved to the right, the number is a negative power of ten.

Example

Earth is 149,600,000 km from the Sun. Express this in scientific notation.

Step 1 Move the decimal point until one non-zero digit remains on the left.

1.496 000 00

Step 2 Count the number of decimal places you have moved. In this case, eight.

Step 2 Show that number as a power of ten, 10^8.

Earth is 1.496×10^8 km from the Sun.

Practice Problem 1 How many significant digits are in 149,600,000 km? How many significant digits are in 1.496×10^8 km?

Practice Problem 2 Parts used in a high performance car must be measured to 7×10^{-6} m. Express this number as a decimal.

Practice Problem 3 A CD is spinning at 539 revolutions per minute. Express this number in scientific notation.

SCIENCE SKILL HANDBOOK
MATH SKILL HANDBOOK
FOLDABLES HANDBOOK
REFERENCE HANDBOOK
GLOSSARY/GLOSARIO
INDEX

Make and Use Graphs

Data in tables can be displayed in a graph—a visual representation of data. Common graph types include line graphs, bar graphs, and circle graphs.

Line Graph A line graph shows a relationship between two variables that change continuously. The independent variable is changed and is plotted on the x-axis. The dependent variable is observed, and is plotted on the *y*-axis.

Example

Draw a line graph of the data below from a cyclist in a long-distance race.

Table 4 Bicycle Race Data

Time (h)	Distance (km)
0	0
1	8
2	16
3	24
4	32
5	40

Step 1 Determine the *x*-axis and *y*-axis variables. Time varies independently of distance and is plotted on the *x*-axis. Distance is dependent on time and is plotted on the *y*-axis.

Step 2 Determine the scale of each axis. The *x*-axis data ranges from 0 to 5. The *y*-axis data ranges from 0 to 50.

Step 3 Using graph paper, draw and label the axes. Include units in the labels.

Step 4 Draw a point at the intersection of the time value on the *x*-axis and corresponding distance value on the *y*-axis. Connect the points and label the graph with a title, as shown in **Figure 8.**

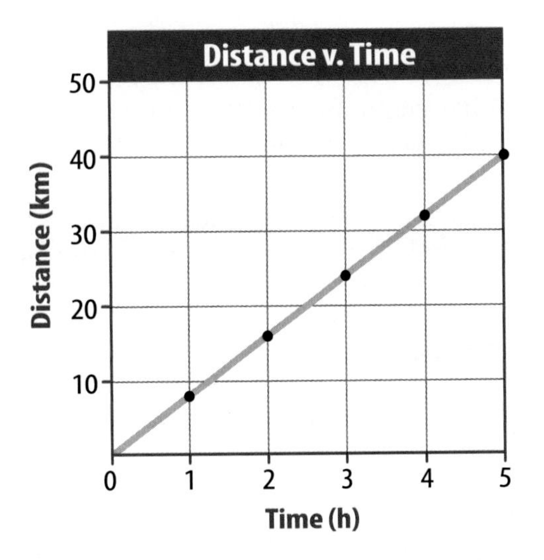

Figure 8 This line graph shows the relationship between distance and time during a bicycle ride.

Practice Problem A puppy's shoulder height is measured during the first year of her life. The following measurements were collected: (3 mo, 52 cm), (6 mo, 72 cm), (9 mo, 83 cm), (12 mo, 86 cm). Graph this data.

Find a Slope The slope of a straight line is the ratio of the vertical change, rise, to the horizontal change, run.

$$\text{Slope} = \frac{\text{vertical change (rise)}}{\text{horizontal change (run)}} = \frac{\text{change in } y}{\text{change in } x}$$

Example

Find the slope of the graph in **Figure 8.**

Step 1 You know that the slope is the change in *y* divided by the change in *x*.

$$\text{Slope} = \frac{\text{change in } y}{\text{change in } x}$$

Step 2 Determine the data points you will be using. For a straight line, choose the two sets of points that are the farthest apart.

$$\text{Slope} = \frac{(40 - 0) \text{ km}}{(5 - 0) \text{ h}}$$

Step 3 Find the change in *y* and *x*.

$$\text{Slope} = \frac{40 \text{ km}}{5 \text{ h}}$$

Step 4 Divide the change in *y* by the change in *x*.

$$\text{Slope} = \frac{8 \text{ km}}{\text{h}}$$

The slope of the graph is 8 km/h.

Bar Graph To compare data that does not change continuously you might choose a bar graph. A bar graph uses bars to show the relationships between variables. The *x*-axis variable is divided into parts. The parts can be numbers such as years, or a category such as a type of animal. The *y*-axis is a number and increases continuously along the axis.

Example

A recycling center collects 4.0 kg of aluminum on Monday, 1.0 kg on Wednesday, and 2.0 kg on Friday. Create a bar graph of this data.

Step 1 Select the *x*-axis and *y*-axis variables. The measured numbers (the masses of aluminum) should be placed on the *y*-axis. The variable divided into parts (collection days) is placed on the *x*-axis.

Step 2 Create a graph grid like you would for a line graph. Include labels and units.

Step 3 For each measured number, draw a vertical bar above the *x*-axis value up to the *y*-axis value. For the first data point, draw a vertical bar above Monday up to 4.0 kg.

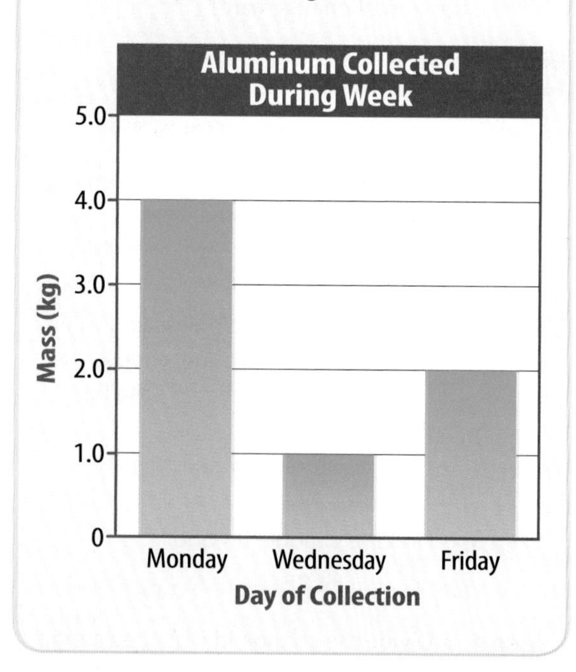

Practice Problem Draw a bar graph of the gases in air: 78% nitrogen, 21% oxygen, 1% other gases.

Circle Graph To display data as parts of a whole, you might use a circle graph. A circle graph is a circle divided into sections that represent the relative size of each piece of data. The entire circle represents 100%, half represents 50%, and so on.

Example

Air is made up of 78% nitrogen, 21% oxygen, and 1% other gases. Display the composition of air in a circle graph.

Step 1 Multiply each percent by 360° and divide by 100 to find the angle of each section in the circle.

$$78\% \times \frac{360°}{100} = 280.8°$$

$$21\% \times \frac{360°}{100} = 75.6°$$

$$1\% \times \frac{360°}{100} = 3.6°$$

Step 2 Use a compass to draw a circle and to mark the center of the circle. Draw a straight line from the center to the edge of the circle.

Step 3 Use a protractor and the angles you calculated to divide the circle into parts. Place the center of the protractor over the center of the circle and line the base of the protractor over the straight line.

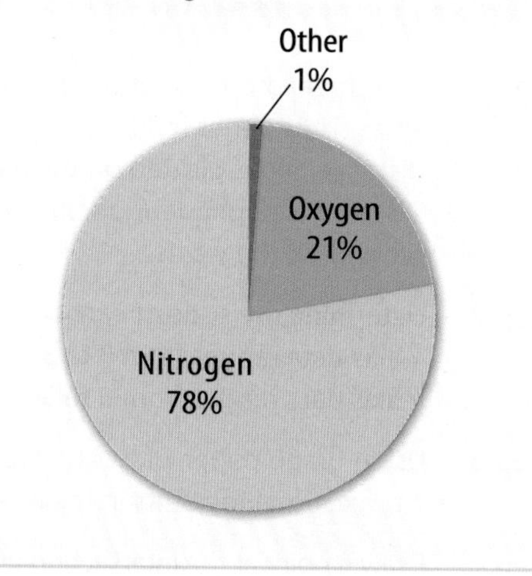

Practice Problem Draw a circle graph to represent the amount of aluminum collected during the week shown in the bar graph to the left.

Student Study Guides & Instructions

By Dinah Zike

1. You will find suggestions for Study Guides, also known as Foldables or books, in each chapter lesson and as a final project. Look at the end of the chapter to determine the project format and glue the Foldables in place as you progress through the chapter lessons.

2. Creating the Foldables or books is simple and easy to do by using copy paper, art paper, and internet printouts. Photocopies of maps, diagrams, or your own illustrations may also be used for some of the Foldables. Notebook paper is the most common source of material for study guides and 83% of all Foldables are created from it. When folded to make books, notebook paper Foldables easily fit into 11″ × 17″ or 12″ × 18″ chapter projects with space left over. Foldables made using photocopy paper are slightly larger and they fit into Projects, but snugly. Use the least amount of glue, tape, and staples needed to assemble the Foldables.

3. Seven of the Foldables can be made using either small or large paper. When 11″ × 17″ or 12″ × 18″ paper is used, these become projects for housing smaller Foldables. Project format boxes are located within the instructions to remind you of this option.

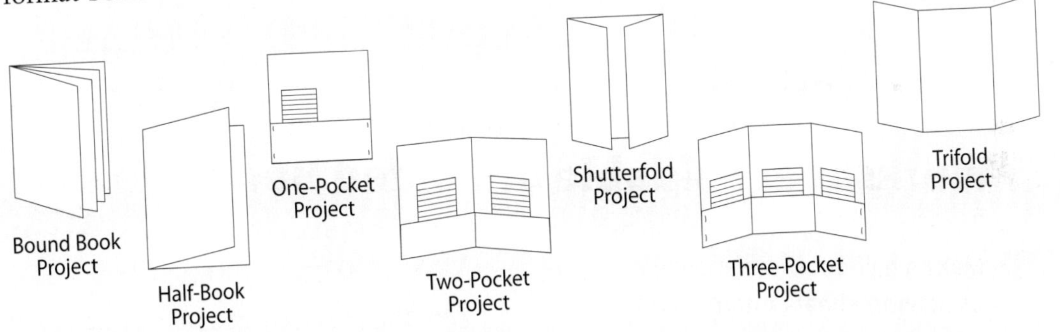

4. Use one-gallon self-locking plastic bags to store your projects. Place strips of two-inch clear tape along the left, long side of the bag and punch holes through the taped edge. Cut the bottom corners off the bag so it will not hold air. Store this Project Portfolio inside a three-hole binder. To store a large collection of project bags, use a giant laundry-soap box. Holes can be punched in some of the Foldable Projects so they can be stored in a three-hole binder without using a plastic bag. Punch holes in the pocket books before gluing or stapling the pocket.

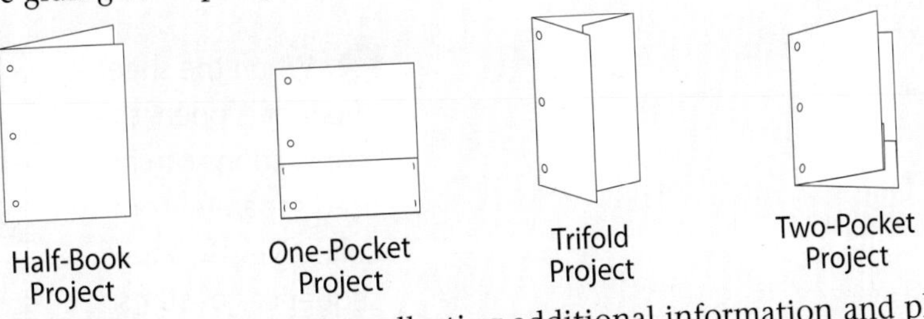

5. Maximize the use of the projects by collecting additional information and placing it on the back of the project and other unused spaces of the large Foldables.

Half-Book Foldable® By Dinah Zike

Step 1 Fold a sheet of notebook or copy paper in half.

Label the exterior tab and use the inside space to write information.

PROJECT FORMAT

Use 11″ × 17″ or 12″ × 18″ paper on the horizontal axis to make a large project book.

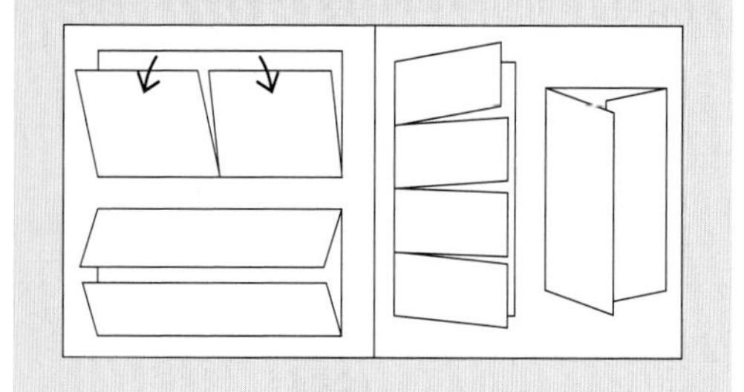

Variations

Paper can be folded horizontally, like a *hamburger* or vertically, like a *hot dog.*

A 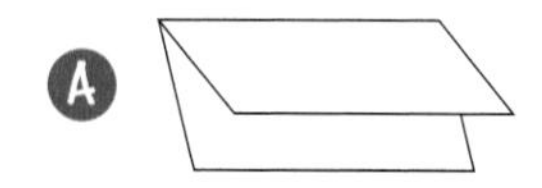

B 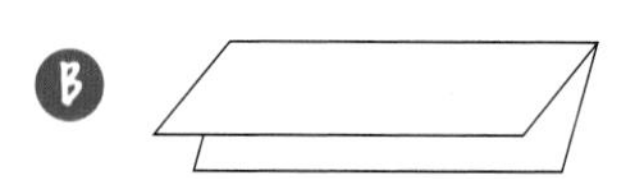

C Half-books can be folded so that one side is ½ inch longer than the other side. A title or question can be written on the extended tab.

Worksheet Foldable or Folded Book® By Dinah Zike

Step 1 Make a half-book (see above) using work sheets, internet print-outs, diagrams, or maps.

Step 2 Fold it in half again.

Variations

A This folded sheet as a small book with two pages can be used for comparing and contrasting, cause and effect, or other skills.

B When the sheet of paper is open, the four sections can be used separately or used collectively to show sequences or steps.

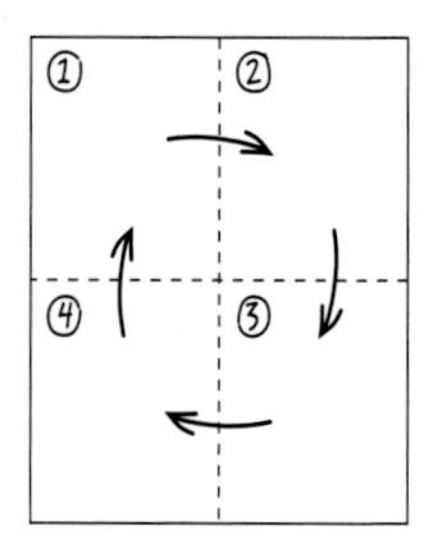

Two-Tab and Concept-Map Foldable® By Dinah Zike

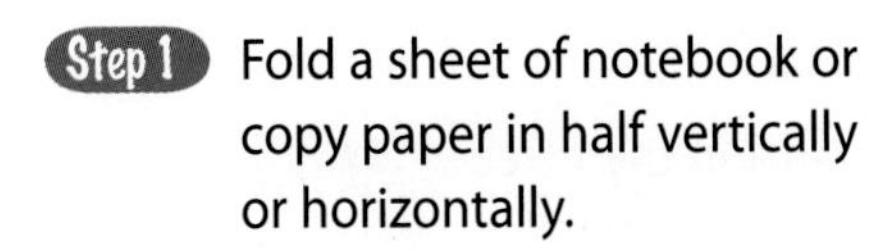

Step 1 Fold a sheet of notebook or copy paper in half vertically or horizontally.

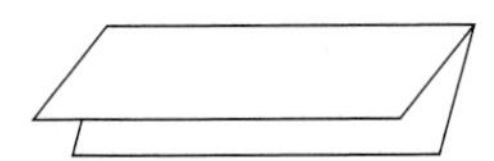

Step 2 Fold it in half again, as shown.

Step 3 Unfold once and cut along the fold line or valley of the top flap to make two flaps.

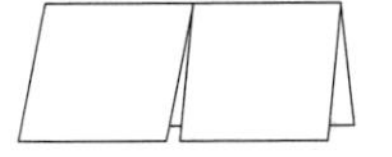

Variations

A Concept maps can be made by leaving a ½ inch tab at the top when folding the paper in half. Use arrows and labels to relate topics to the primary concept.

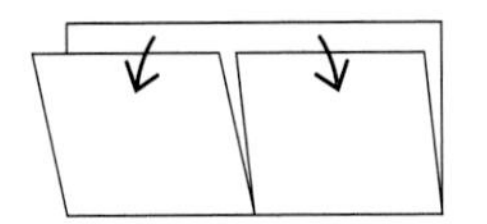

B Use two sheets of paper to make multiple page tab books. Glue or staple books together at the top fold.

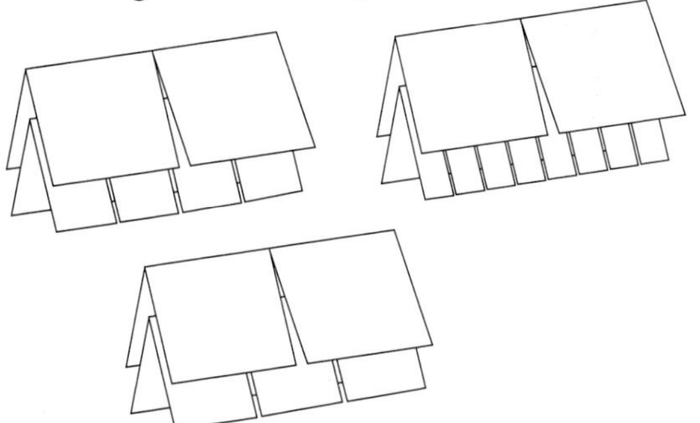

Three-Quarter Foldable® By Dinah Zike

Step 1 Make a two-tab book (see above) and cut the left tab off at the top of the fold line.

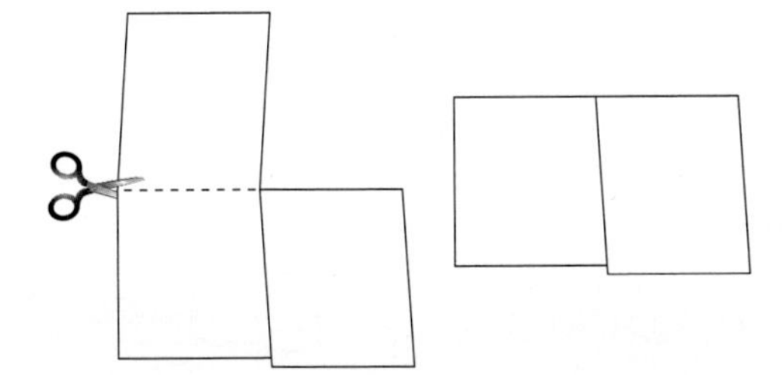

Variations

A Use this book to draw a diagram or a map on the exposed left tab. Write questions about the illustration on the top right tab and provide complete answers on the space under the tab.

B Compose a self-test using multiple choice answers for your questions. Include the correct answer with three wrong responses. The correct answers can be written on the back of the book or upside down on the bottom of the inside page.

Three-Tab Foldable® By Dinah Zike

Step 1 Fold a sheet of paper in half horizontally.

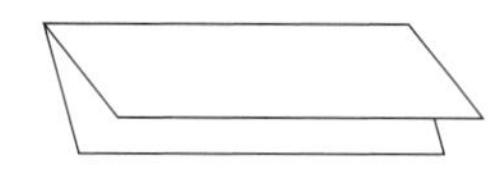

Step 2 Fold into thirds.

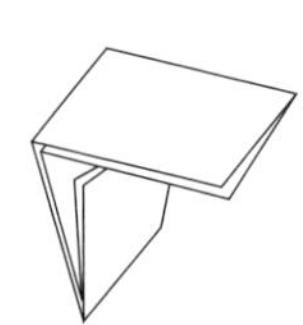

Step 3 Unfold and cut along the folds of the top flap to make three sections.

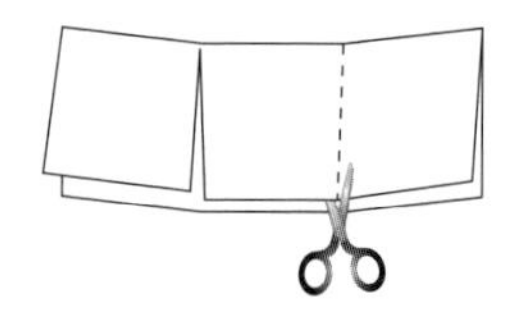

Variations

A Before cutting the three tabs draw a Venn diagram across the front of the book.

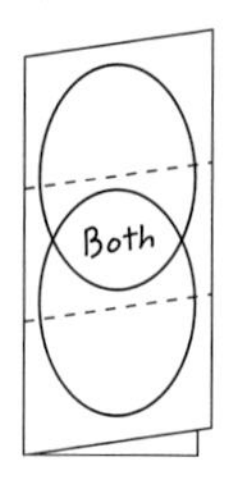

B Make a space to use for titles or concept maps by leaving a ½ inch tab at the top when folding the paper in half.

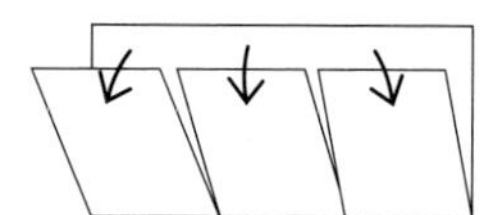

Four-Tab Foldable® By Dinah Zike

Step 1 Fold a sheet of paper in half horizontally.

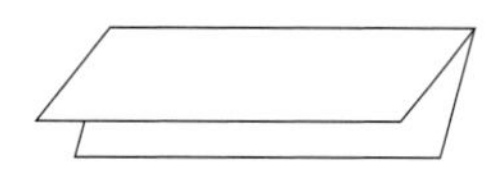

Step 2 Fold in half and then fold each half as shown below.

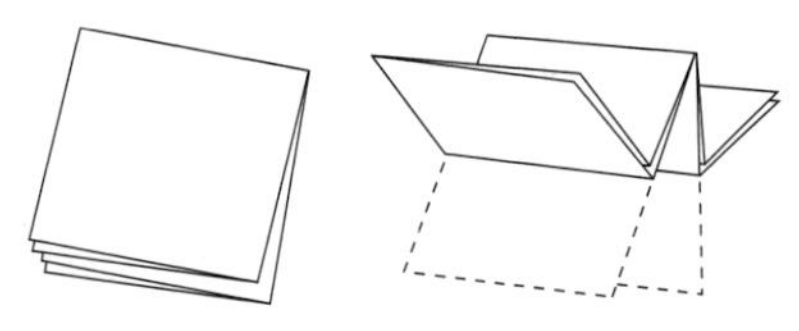

Step 3 Unfold and cut along the fold lines of the top flap to make four tabs.

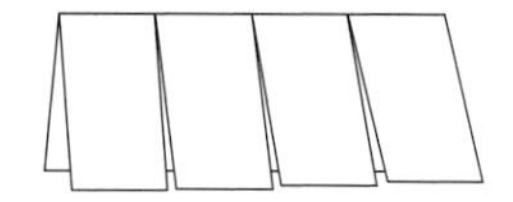

Variations

A Make a space to use for titles or concept maps by leaving a ½ inch tab at the top when folding the paper in half.

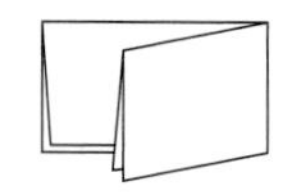

B Use the book on the vertical axis, with or without an extended tab.

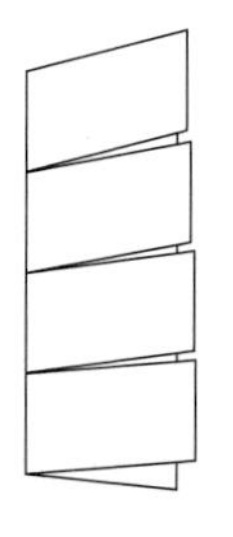

Folding Fifths for a Foldable® By Dinah Zike

Step 1 Fold a sheet of paper in half horizontally.

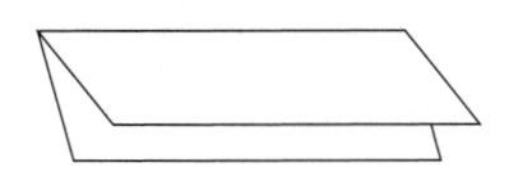

Step 2 Fold again so one-third of the paper is exposed and two-thirds are covered.

Step 3 Fold the two-thirds section in half.

Step 4 Fold the one-third section, a single thickness, backward to make a fold line.

Variations

A Unfold and cut along the fold lines to make five tabs.

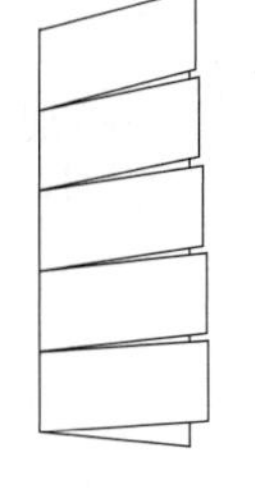

B Make a five-tab book with a ½ inch tab at the top (see two-tab instructions).

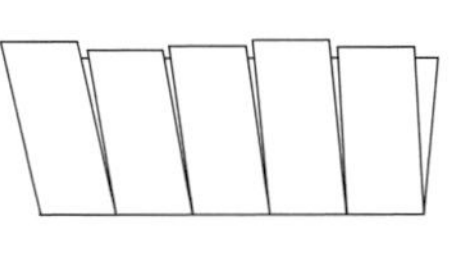

C Use 11″ × 17″ or 12″ × 18″ paper and fold into fifths for a five-column and/or row table or chart.

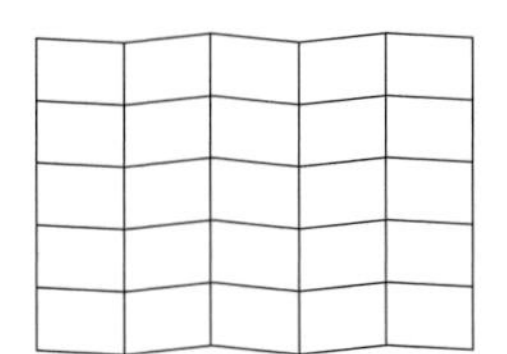

Folded Table or Chart, and Trifold Foldable® By Dinah Zike

Step 1 Fold a sheet of paper in the required number of vertical columns for the table or chart.

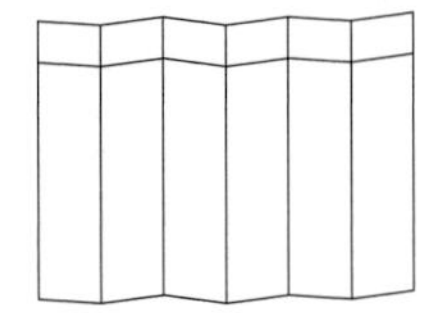

Step 2 Fold the horizontal rows needed for the table or chart.

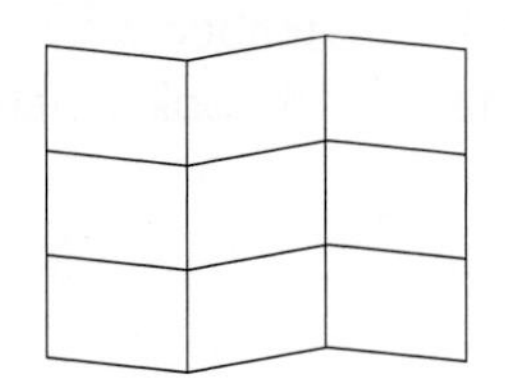

PROJECT FORMAT
Use 11″ × 17″ or 12″ × 18″ paper and fold it to make a large trifold project book or larger tables and charts.

Variations

A Make a trifold by folding the paper into thirds vertically or horizontally.

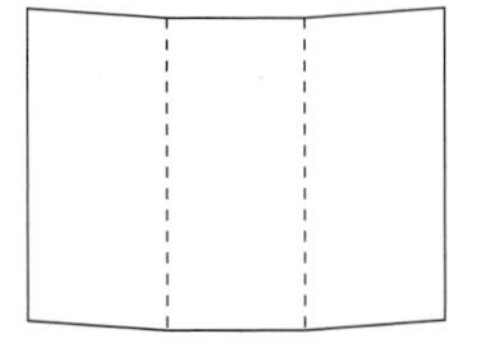

B Make a trifold book. Unfold it and draw a Venn diagram on the inside.

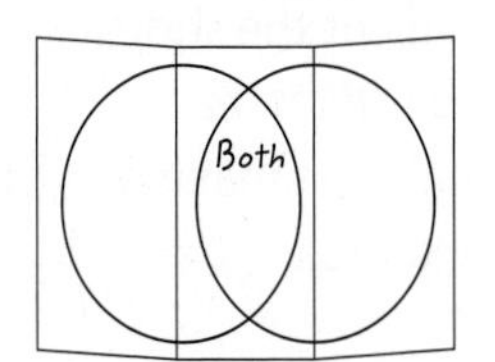

Two or Three-Pockets Foldable® By Dinah Zike

Step 1 Fold up the long side of a horizontal sheet of paper about 5 cm.

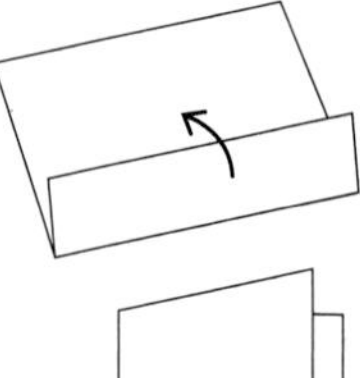

Step 2 Fold the paper in half.

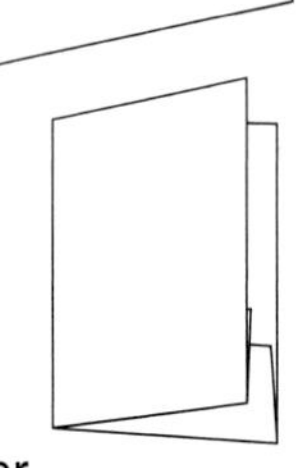

Step 3 Open the paper and glue or staple the outer edges to make two compartments.

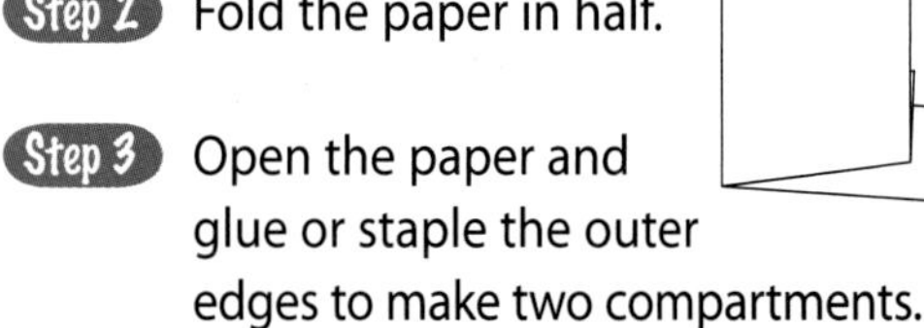

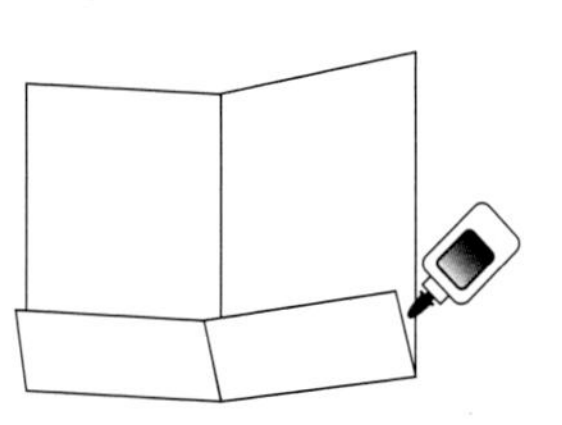

Variations

A Make a multi-page booklet by gluing several pocket books together.

B Make a three-pocket book by using a trifold (see previous instructions).

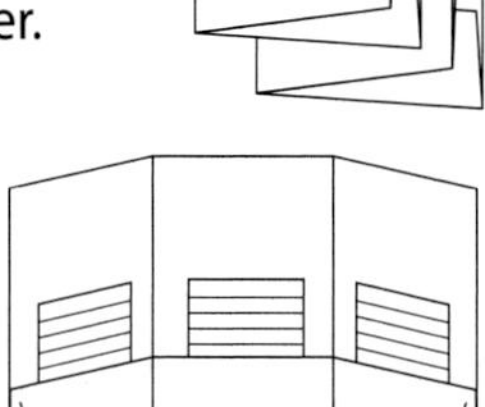

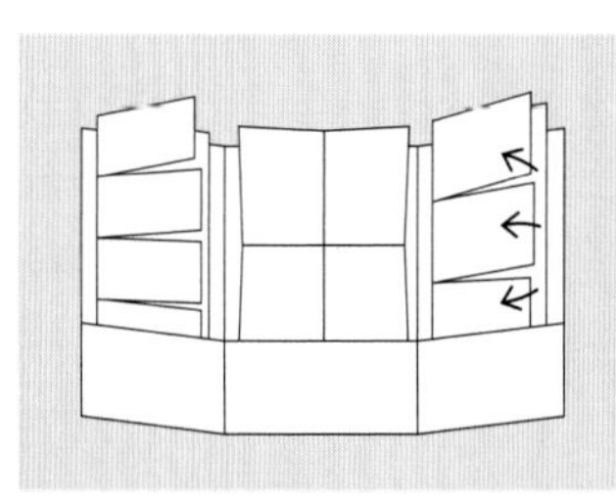

PROJECT FORMAT
Use 11″ × 17″ or 12″ × 18″ paper and fold it horizontally to make a large multi-pocket project.

Matchbook Foldable® By Dinah Zike

Step 1 Fold a sheet of paper almost in half and make the back edge about 1–2 cm longer than the front edge.

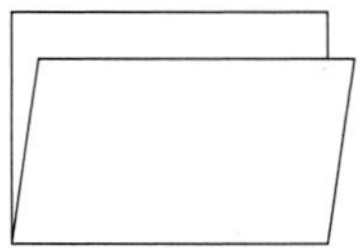

Step 2 Find the midpoint of the shorter flap.

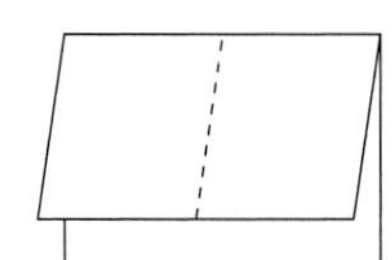

Step 3 Open the paper and cut the short side along the midpoint making two tabs.

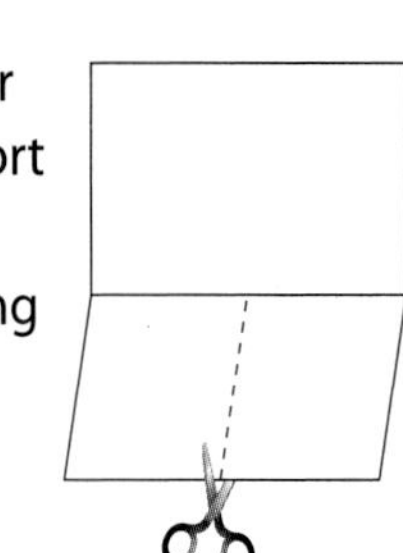

Step 4 Close the book and fold the tab over the short side.

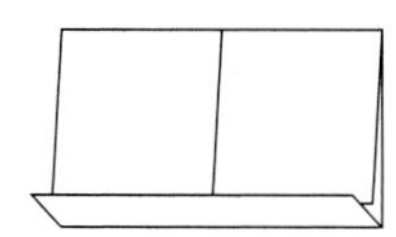

Variations

A Make a single-tab matchbook by skipping Steps 2 and 3.

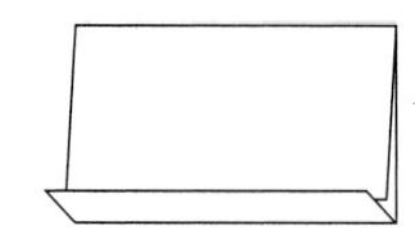

B Make two smaller matchbooks by cutting the single-tab matchbook in half.

Shutterfold Foldable® By Dinah Zike

Step 1 Begin as if you were folding a vertical sheet of paper in half, but instead of creasing the paper, pinch it to show the midpoint.

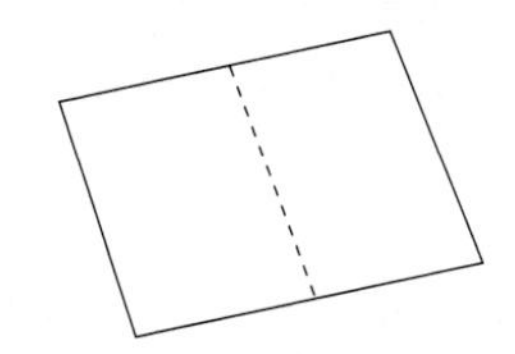

PROJECT FORMAT
Use 11″ × 17″ or 12″ × 18″ paper and fold it to make a large shutterfold project.

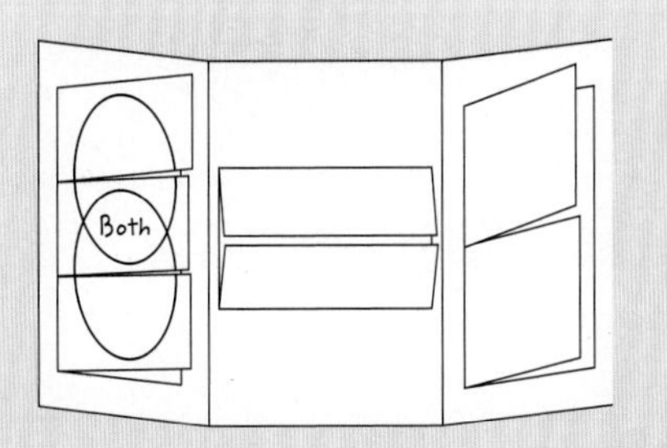

Step 2 Fold the top and bottom to the middle and crease the folds.

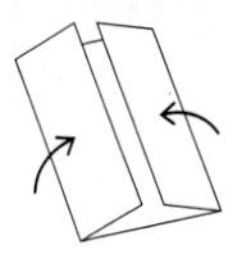

Variations

A Use the shutterfold on the horizontal axis.

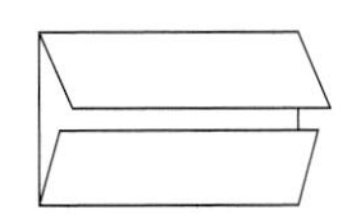

B Create a center tab by leaving .5–2 cm between the flaps in Step 2.

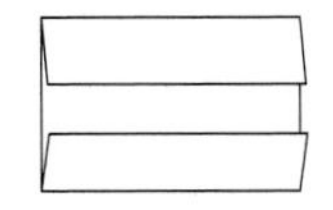

Four-Door Foldable® By Dinah Zike

Step 1 Make a shutterfold (see above).

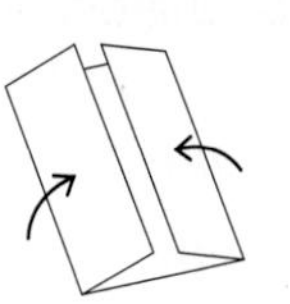

Step 2 Fold the sheet of paper in half.

Step 3 Open the last fold and cut along the inside fold lines to make four tabs.

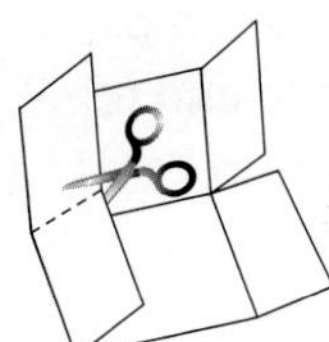

Variations

A Use the four-door book on the opposite axis.

B Create a center tab by leaving .5–2 cm between the flaps in Step 1.

Bound Book Foldable® By Dinah Zike

Step 1 Fold three sheets of paper in half. Place the papers in a stack, leaving about .5 cm between each top fold. Mark all three sheets about 3 cm from the outer edges.

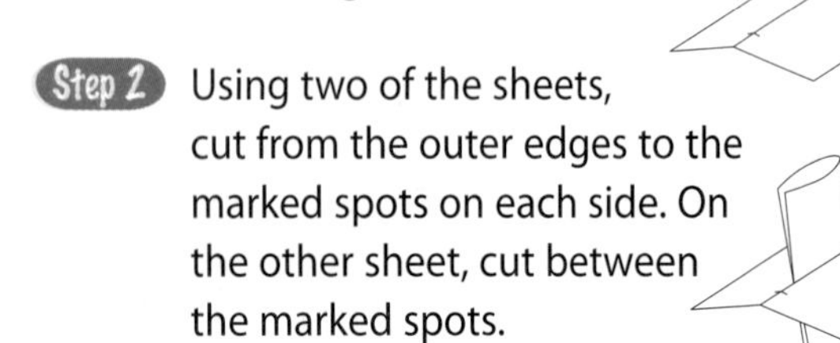

Step 2 Using two of the sheets, cut from the outer edges to the marked spots on each side. On the other sheet, cut between the marked spots.

Step 3 Take the two sheets from Step 1 and slide them through the cut in the third sheet to make a 12-page book.

Step 4 Fold the bound pages in half to form a book.

Variation

A Use two sheets of paper to make an eight-page book, or increase the number of pages by using more than three sheets.

PROJECT FORMAT
Use two or more sheets of 11″ × 17″ or 12″ × 18″ paper and fold it to make a large bound book project.

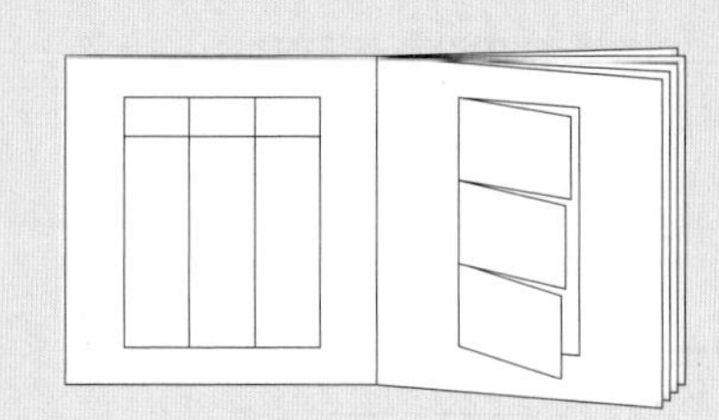

Accordian Foldable® By Dinah Zike

Step 1 Fold the selected paper in half vertically, like a *hamburger*.

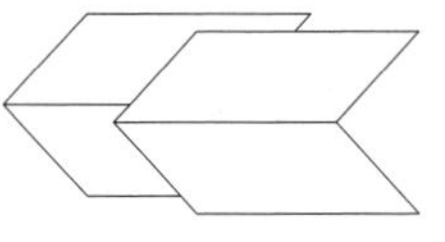

Step 2 Cut each sheet of folded paper in half along the fold lines.

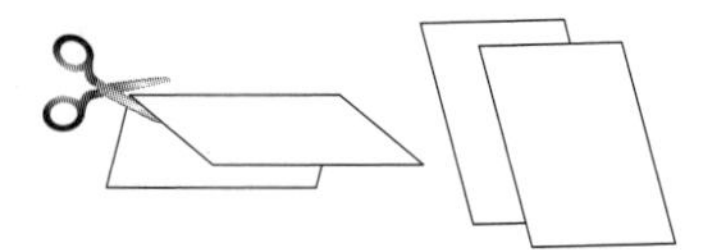

Step 3 Fold each half-sheet almost in half, leaving a 2 cm tab at the top.

Step 4 Fold the top tab over the short side, then fold it in the opposite direction.

Variations

A Glue the straight edge of one paper inside the tab of another sheet. Leave a tab at the end of the book to add more pages.

B Tape the straight edge of one paper to the tab of another sheet, or just tape the straight edges of nonfolded paper end to end to make an accordian.

C Use whole sheets of paper to make a large accordian.

Layered Foldable® By Dinah Zike

Step 1 Stack two sheets of paper about 1–2 cm apart. Keep the right and left edges even.

Step 2 Fold up the bottom edges to form four tabs. Crease the fold to hold the tabs in place.

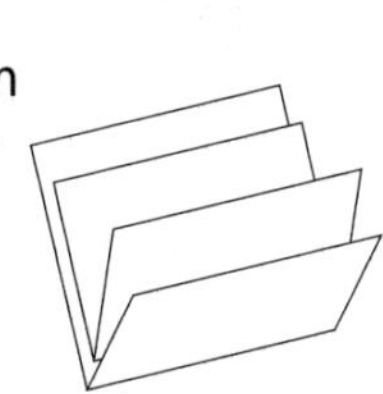

Step 3 Staple along the folded edge, or open and glue the papers together at the fold line.

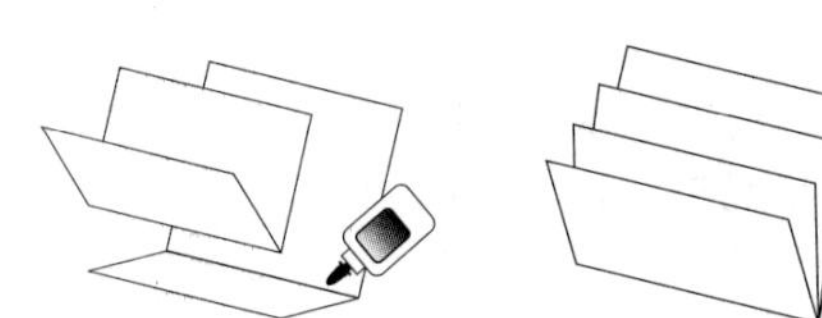

Variations

A Rotate the book so the fold is at the top or to the side.

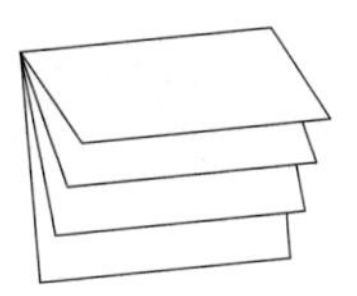

B Extend the book by using more than two sheets of paper.

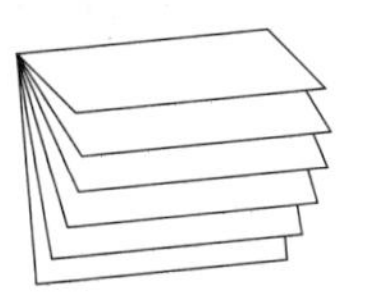

Envelope Foldable® By Dinah Zike

Step 1 Fold a sheet of paper into a *taco*. Cut off the tab at the top.

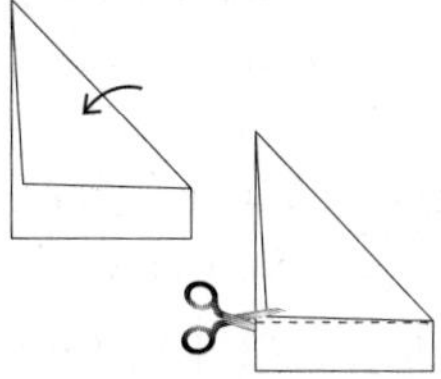

Step 2 Open the *taco* and fold it the opposite way making another *taco* and an X-fold pattern on the sheet of paper.

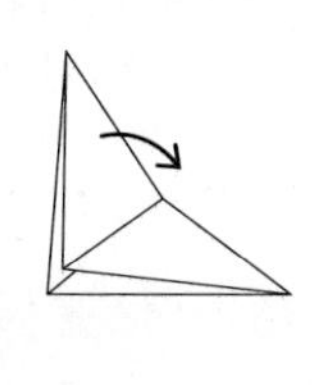

Step 3 Cut a map, illustration, or diagram to fit the inside of the envelope.

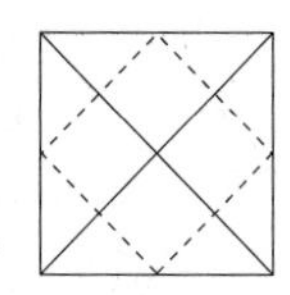

Step 4 Use the outside tabs for labels and inside tabs for writing information.

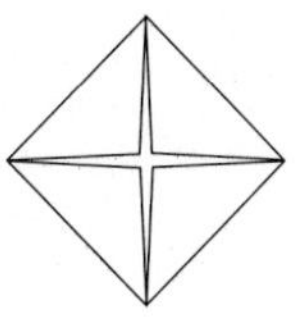

Variations

A Use 11″ × 17″ or 12″ × 18″ paper to make a large envelope.

B Cut off the points of the four tabs to make a window in the middle of the book.

Sentence Strip Foldable® By Dinah Zike

Step 1 Fold two sheets of paper in half vertically, like a *hamburger*.

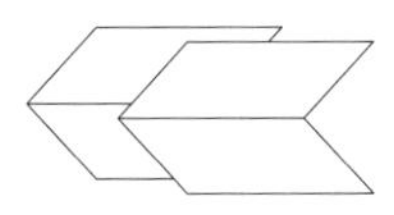

Step 2 Unfold and cut along fold lines making four half sheets.

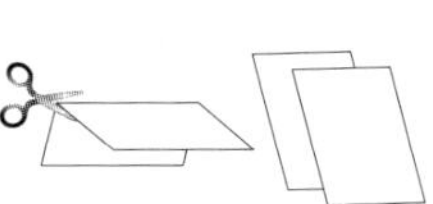

Step 3 Fold each half sheet in half horizontally, like a *hot dog*.

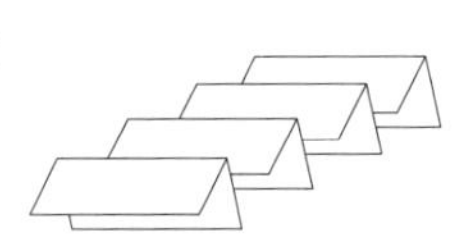

Step 4 Stack folded horizontal sheets evenly and staple together on the left side.

Step 5 Open the top flap of the first sentence strip and make a cut about 2 cm from the stapled edge to the fold line. This forms a flap that can be raisied and lowered. Repeat this step for each sentence strip.

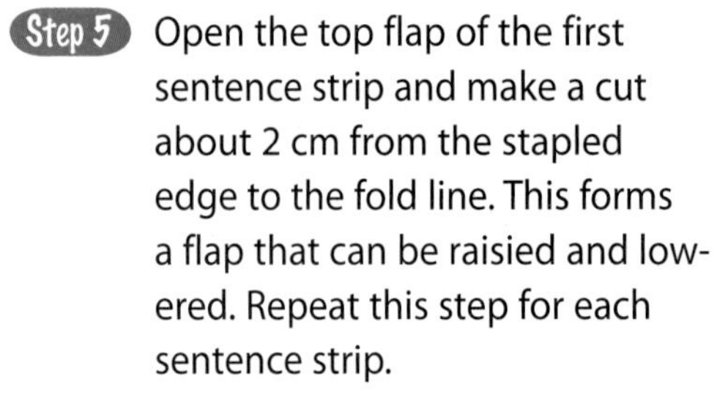

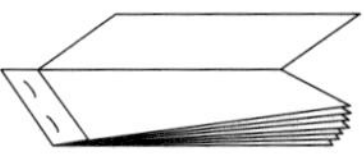

Variations

A Expand this book by using more than two sheets of paper.

B Use whole sheets of paper to make large books.

Pyramid Foldable® By Dinah Zike

Step 1 Fold a sheet of paper into a *taco*. Crease the fold line, but do not cut it off.

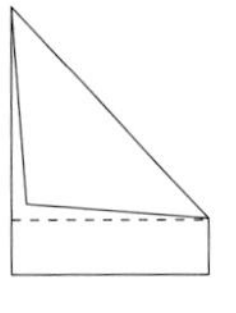

Step 2 Open the folded sheet and refold it like a *taco* in the opposite direction to create an X-fold pattern.

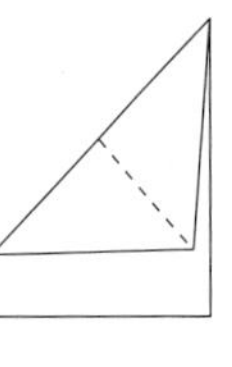

Step 3 Cut one fold line as shown, stopping at the center of the X-fold to make a flap.

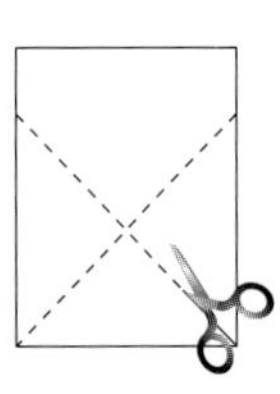

Step 4 Outline the fold lines of the X-fold. Label the three front sections and use the inside spaces for notes. Use the tab for the title.

Step 5 Glue the tab into a project book or notebook. Use the space under the pyramid for other information.

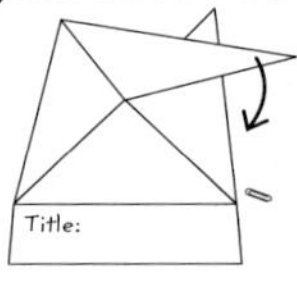

Step 6 To display the pyramid, fold the flap under and secure with a paper clip, if needed.

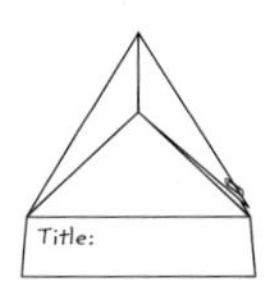

Single-Pocket or One-Pocket Foldable® By Dinah Zike

Step 1 Using a large piece of paper on a vertical axis, fold the bottom edge of the paper upwards, about 5 cm.

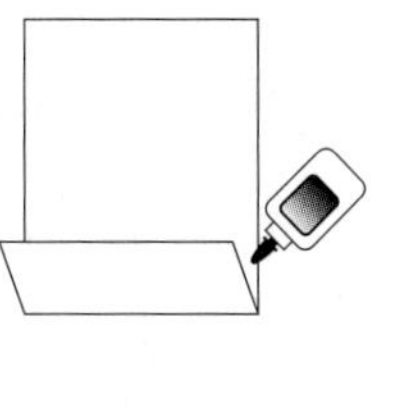

Step 2 Glue or staple the outer edges to make a large pocket.

PROJECT FORMAT
Use 11″ × 17″ or 12″ × 18″ paper and fold it vertically or horizontally to make a large pocket project.

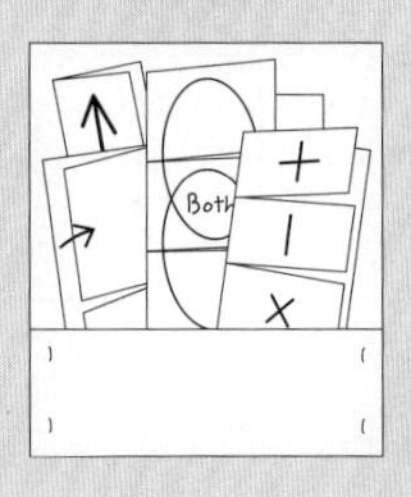

Variations

A Make the one-pocket project using the paper on the horizontal axis.

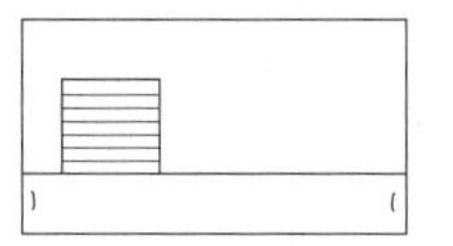

B To store materials securely inside, fold the top of the paper almost to the center, leaving about 2–4 cm between the paper edges. Slip the Foldables through the opening and under the top and bottom pockets.

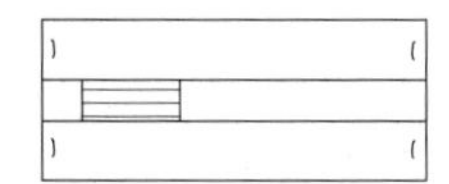

Multi-Tab Foldable® By Dinah Zike

Step 1 Fold a sheet of notebook paper in half like a *hot dog*.

Step 2 Open the paper and on one side cut every third line. This makes ten tabs on wide ruled notebook paper and twelve tabs on college ruled.

Step 3 Label the tabs on the front side and use the inside space for definitions or other information.

Variation

A Make a tab for a title by folding the paper so the holes remain uncovered. This allows the notebook Foldable to be stored in a three-hole binder.

Science Skill Handbook
Math Skill Handbook
Foldables Handbook
Reference Handbook
Glossary/Glosario
Index

PERIODIC TABLE OF THE ELEMENTS

Element — Hydrogen
Atomic number — 1
Symbol — H
Atomic mass — 1.01
State of matter

Gas
Liquid
Solid
Synthetic

A column in the periodic table is called a **group.**

A row in the periodic table is called a **period.**

Period	1	2	3	4	5	6	7	8	9
1	Hydrogen 1 **H** 1.01								
2	Lithium 3 **Li** 6.94	Beryllium 4 **Be** 9.01							
3	Sodium 11 **Na** 22.99	Magnesium 12 **Mg** 24.31							
4	Potassium 19 **K** 39.10	Calcium 20 **Ca** 40.08	Scandium 21 **Sc** 44.96	Titanium 22 **Ti** 47.87	Vanadium 23 **V** 50.94	Chromium 24 **Cr** 52.00	Manganese 25 **Mn** 54.94	Iron 26 **Fe** 55.85	Cobalt 27 **Co** 58.93
5	Rubidium 37 **Rb** 85.47	Strontium 38 **Sr** 87.62	Yttrium 39 **Y** 88.91	Zirconium 40 **Zr** 91.22	Niobium 41 **Nb** 92.91	Molybdenum 42 **Mo** 95.96	Technetium 43 **Tc** (98)	Ruthenium 44 **Ru** 101.07	Rhodium 45 **Rh** 102.91
6	Cesium 55 **Cs** 132.91	Barium 56 **Ba** 137.33	Lanthanum 57 **La** 138.91	Hafnium 72 **Hf** 178.49	Tantalum 73 **Ta** 180.95	Tungsten 74 **W** 183.84	Rhenium 75 **Re** 186.21	Osmium 76 **Os** 190.23	Iridium 77 **Ir** 192.22
7	Francium 87 **Fr** (223)	Radium 88 **Ra** (226)	Actinium 89 **Ac** (227)	Rutherfordium 104 **Rf** (267)	Dubnium 105 **Db** (268)	Seaborgium 106 **Sg** (271)	Bohrium 107 **Bh** (272)	Hassium 108 **Hs** (270)	Meitnerium 109 **Mt** (276)

The number in parentheses is the mass number of the longest lived isotope for that element.

Series						
Lanthanide series	Cerium 58 **Ce** 140.12	Praseodymium 59 **Pr** 140.91	Neodymium 60 **Nd** 144.24	Promethium 61 **Pm** (145)	Samarium 62 **Sm** 150.36	Europium 63 **Eu** 151.96
Actinide series	Thorium 90 **Th** 232.04	Protactinium 91 **Pa** 231.04	Uranium 92 **U** 238.03	Neptunium 93 **Np** (237)	Plutonium 94 **Pu** (244)	Americium 95 **Am** (243)

Metal

Metalloid

Nonmetal

Recently discovered

10	11	12	13	14	15	16	17	18
								Helium 2 He 4.00
			Boron 5 B 10.81	Carbon 6 C 12.01	Nitrogen 7 N 14.01	Oxygen 8 O 16.00	Fluorine 9 F 19.00	Neon 10 Ne 20.18
			Aluminum 13 Al 26.98	Silicon 14 Si 28.09	Phosphorus 15 P 30.97	Sulfur 16 S 32.07	Chlorine 17 Cl 35.45	Argon 18 Ar 39.95
Nickel 28 Ni 58.69	Copper 29 Cu 63.55	Zinc 30 Zn 65.38	Gallium 31 Ga 69.72	Germanium 32 Ge 72.64	Arsenic 33 As 74.92	Selenium 34 Se 78.96	Bromine 35 Br 79.90	Krypton 36 Kr 83.80
Palladium 46 Pd 106.42	Silver 47 Ag 107.87	Cadmium 48 Cd 112.41	Indium 49 In 114.82	Tin 50 Sn 118.71	Antimony 51 Sb 121.76	Tellurium 52 Te 127.60	Iodine 53 I 126.90	Xenon 54 Xe 131.29
Platinum 78 Pt 195.08	Gold 79 Au 196.97	Mercury 80 Hg 200.59	Thallium 81 Tl 204.38	Lead 82 Pb 207.20	Bismuth 83 Bi 208.98	Polonium 84 Po (209)	Astatine 85 At (210)	Radon 86 Rn (222)
Darmstadtium 110 Ds (281)	Roentgenium 111 Rg (280)	Copernicium 112 Cn (285)	Ununtrium * 113 Uut (284)	Ununquadium * 114 Uuq (289)	Ununpentium * 115 Uup (288)	Ununhexium * 116 Uuh (293)		Ununoctium * 118 Uuo (294)

* The names and symbols for elements 113-116 and 118 are temporary. Final names will be selected when the elements' discoveries are verified.

Gadolinium 64 Gd 157.25	Terbium 65 Tb 158.93	Dysprosium 66 Dy 162.50	Holmium 67 Ho 164.93	Erbium 68 Er 167.26	Thulium 69 Tm 168.93	Ytterbium 70 Yb 173.05	Lutetium 71 Lu 174.97
Curium 96 Cm (247)	Berkelium 97 Bk (247)	Californium 98 Cf (251)	Einsteinium 99 Es (252)	Fermium 100 Fm (257)	Mendelevium 101 Md (258)	Nobelium 102 No (259)	Lawrencium 103 Lr (262)

Science Skill Handbook
Math Skill Handbook
Foldables Handbook
Reference Handbook
Glossary/Glosario
Index

Topographic Map Symbols

Topographic Map Symbols	
Primary highway, hard surface	Index contour
Secondary highway, hard surface	Supplementary contour
Light-duty road, hard or improved surface	Intermediate contour
Unimproved road	Depression contours
Railroad: single track	
Railroad: multiple track	Boundaries: national
Railroads in juxtaposition	State
	County, parish, municipal
Buildings	Civil township, precinct, town, barrio
cem Schools, church, and cemetery	Incorporated city, village, town, hamlet
Buildings (barn, warehouse, etc.)	Reservation, national or state
Wells other than water (labeled as to type)	Small park, cemetery, airport, etc.
Tanks: oil, water, etc. (labeled only if water)	Land grant
Located or landmark object; windmill	Township or range line, U.S. land survey
Open pit, mine, or quarry; prospect	
	Township or range line, approximate location
Marsh (swamp)	
Wooded marsh	Perennial streams
Woods or brushwood	Elevated aqueduct
Vineyard	Water well and spring
Land subject to controlled inundation	Small rapids
Submerged marsh	Large rapids
Mangrove	Intermittent lake
Orchard	Intermittent stream
Scrub	Aqueduct tunnel
Urban area	Glacier
	Small falls
x7369 Spot elevation	Large falls
670 Water elevation	Dry lake bed

Rocks

Rocks		
Rock Type	**Rock Name**	**Characteristics**
Igneous (intrusive)	Granite	Large mineral grains of quartz, feldspar, hornblende, and mica. Usually light in color.
	Diorite	Large mineral grains of feldspar, hornblende, and mica. Less quartz than granite. Intermediate in color.
	Gabbro	Large mineral grains of feldspar, augite, and olivine. No quartz. Dark in color.
Igneous (extrusive)	Rhyolite	Small mineral grains of quartz, feldspar, hornblende, and mica, or no visible grains. Light in color.
	Andesite	Small mineral grains of feldspar, hornblende, and mica or no visible grains. Intermediate in color.
	Basalt	Small mineral grains of feldspar, augite, and possibly olivine or no visible grains. No quartz. Dark in color.
	Obsidian	Glassy texture. No visible grains. Volcanic glass. Fracture looks like broken glass.
	Pumice	Frothy texture. Floats in water. Usually light in color.
Sedimentary (detrital)	Conglomerate	Coarse grained. Gravel or pebble-size grains.
	Sandstone	Sand-sized grains 1/16 to 2 mm.
	Siltstone	Grains are smaller than sand but larger than clay.
	Shale	Smallest grains. Often dark in color. Usually platy.
Sedimentary (chemical or organic)	Limestone	Major mineral is calcite. Usually forms in oceans and lakes. Often contains fossils.
	Coal	Forms in swampy areas. Compacted layers of organic material, mainly plant remains.
Sedimentary (chemical)	Rock Salt	Commonly forms by the evaporation of seawater.
Metamorphic (foliated)	Gneiss	Banding due to alternate layers of different minerals, of different colors. Parent rock often is granite.
	Schist	Parallel arrangement of sheetlike minerals, mainly micas. Forms from different parent rocks.
	Phyllite	Shiny or silky appearance. May look wrinkled. Common parent rocks are shale and slate.
	Slate	Harder, denser, and shinier than shale. Common parent rock is shale.
Metamorphic (nonfoliated)	Marble	Calcite or dolomite. Common parent rock is limestone.
	Soapstone	Mainly of talc. Soft with greasy feel.
	Quartzite	Hard with interlocking quartz crystals. Common parent rock is sandstone.

Minerals

Minerals					
Mineral (formula)	**Color**	**Streak**	**Hardness Pattern**	**Breakage Properties**	**Uses and Other**
Graphite (C)	black to gray	black to gray	1–1.5	basal cleavage (scales)	pencil lead, lubricants for locks, rods to control some small nuclear reactions, battery poles
Galena (PbS)	gray	gray to black	2.5	cubic cleavage perfect	source of lead, used for pipes, shields for X rays, fishing equipment sinkers
Hematite (Fe_2O_3)	black or reddish-brown	reddish-brown	5.5–6.5	irregular fracture	source of iron; converted to pig iron, made into steel
Magnetite (Fe_3O_4)	black	black	6	conchoidal fracture	source of iron, attracts a magnet
Pyrite (FeS_2)	light, brassy, yellow	greenish-black	6–6.5	uneven fracture	fool's gold
Talc ($Mg_3 Si_4O_{10}(OH)_2$)	white, greenish	white	1	cleavage in one direction	used for talcum powder, sculptures, paper, and tabletops
Gypsum ($CaSO_4 \cdot 2H_2O$)	colorless, gray, white, brown	white	2	basal cleavage	used in plaster of paris and dry wall for building construction
Sphalerite (ZnS)	brown, reddish-brown, greenish	light to dark brown	3.5–4	cleavage in six directions	main ore of zinc; used in paints, dyes, and medicine
Muscovite ($KAl_3Si_3O_{10}(OH)_2$)	white, light gray, yellow, rose, green	colorless	2–2.5	basal cleavage	occurs in large, flexible plates; used as an insulator in electrical equipment, lubricant
Biotite ($K(Mg,Fe)_3(AlSi_3O_{10})(OH)_2$)	black to dark brown	colorless	2.5–3	basal cleavage	occurs in large, flexible plates
Halite (NaCl)	colorless, red, white, blue	colorless	2.5	cubic cleavage	salt; soluble in water; a preservative

Minerals

Minerals					
Mineral (formula)	**Color**	**Streak**	**Hardness**	**Breakage Pattern**	**Uses and Other Properties**
Calcite ($CaCO_3$)	colorless, white, pale blue	colorless, white	3	cleavage in three directions	fizzes when HCl is added; used in cements and other building materials
Dolomite ($CaMg(CO_3)_2$)	colorless, white, pink, green, gray, black	white	3.5–4	cleavage in three directions	concrete and cement; used as an ornamental building stone
Fluorite (CaF_2)	colorless, white, blue, green, red, yellow, purple	colorless	4	cleavage in four directions	used in the manufacture of optical equipment; glows under ultraviolet light
Hornblende ($(CaNa)_{2-3}(Mg,Al,Fe)_5\text{-}(Al,Si)_2Si_6O_{22}(OH)_2$)	green to black	gray to white	5–6	cleavage in two directions	will transmit light on thin edges; 6-sided cross section
Feldspar ($(KAlSi_3O_8)$ $(NaAlSi_3O_8)$, $(CaAl_2Si_2O_8)$)	colorless, white to gray, green	colorless	6	two cleavage planes meet at 90° angle	used in the manufacture of ceramics
Augite ($((Ca,Na)(Mg,Fe,Al)(Al,Si)_2O_6)$)	black	colorless	6	cleavage in two directions	square or 8-sided cross section
Olivine ($((Mg,Fe)_2SiO_4)$)	olive, green	none	6.5–7	conchoidal fracture	gemstones, refractory sand
Quartz (SiO_2)	colorless, various colors	none	7	conchoidal fracture	used in glass manufacture, electronic equipment, radios, computers, watches, gemstones

Weather Map Symbols

Sample Station Model

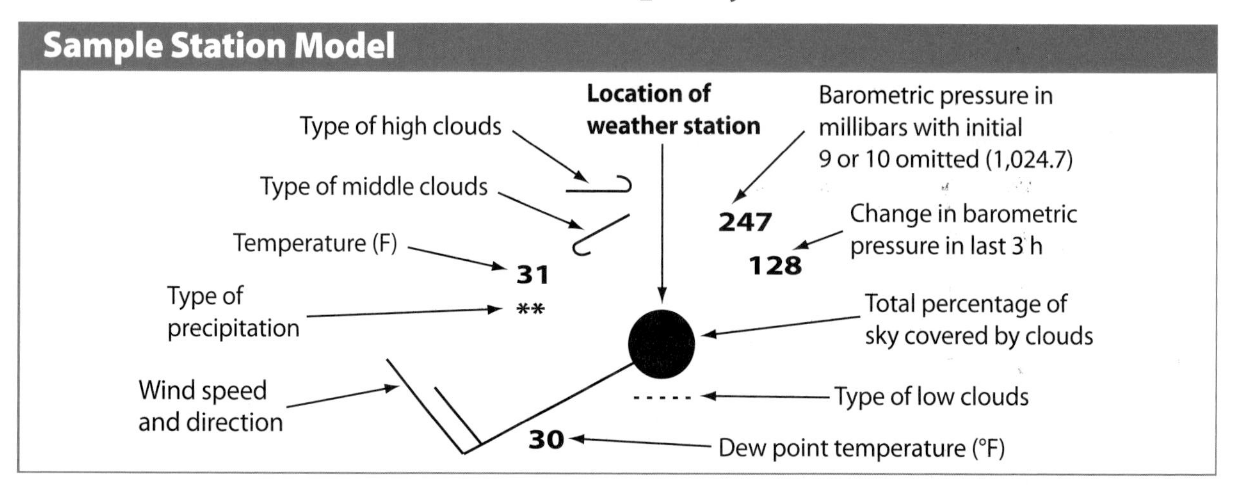

Sample Plotted Report at Each Station

Precipitation	Wind Speed and Direction	Sky Coverage	Some Types of High Clouds
Fog	0 calm	No cover	Scattered cirrus
Snow	1–2 knots	1/10 or less	Dense cirrus in patches
Rain	3–7 knots	2/10 to 3/10	Veil of cirrus covering entire sky
Thunderstorm	8–12 knots	4/10	Cirrus not covering entire sky
Drizzle	13–17 knots	–	
Showers	18–22 knots	6/10	
	23–27 knots	7/10	
	48–52 knots	Overcast with openings	
	1 knot = 1.852 km/h	Completely overcast	

Some Types of Middle Clouds	Some Types of Low Clouds	Fronts and Pressure Systems	
Thin altostratus layer	Cumulus of fair weather	(H) or High (L) or Low	Center of high- or low-pressure system
Thick altostratus layer	Stratocumulus		Cold front
Thin altostratus in patches	Fractocumulus of bad weather		Warm front
Thin altostratus in bands	Stratus of fair weather		Occluded front
			Stationary front

Use and Care of a Microscope

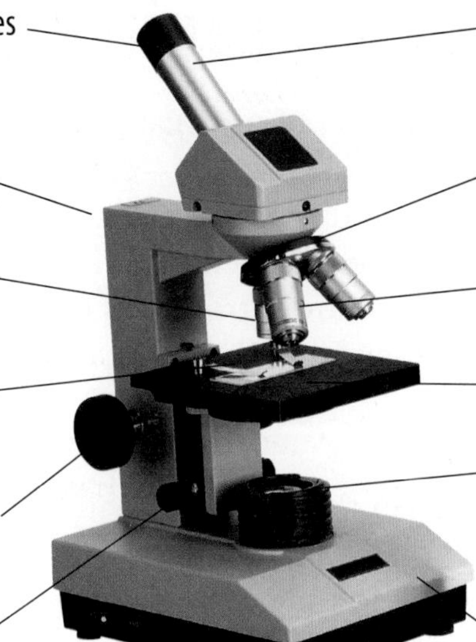

Caring for a Microscope

1. Always carry the microscope holding the arm with one hand and supporting the base with the other hand.
2. Don't touch the lenses with your fingers.
3. The coarse adjustment knob is used only when looking through the lowest-power objective lens. The fine adjustment knob is used when the high-power objective is in place.
4. Cover the microscope when you store it.

Using a Microscope

1. Place the microscope on a flat surface that is clear of objects. The arm should be toward you.
2. Look through the eyepiece. Adjust the diaphragm so light comes through the opening in the stage.
3. Place a slide on the stage so the specimen is in the field of view. Hold it firmly in place by using the stage clips.
4. Always focus with the coarse adjustment and the low-power objective lens first. After the object is in focus on low power, turn the nosepiece until the high-power objective is in place. Use ONLY the fine adjustment to focus with the high-power objective lens.

Making a Wet-Mount Slide

1. Carefully place the item you want to look at in the center of a clean, glass slide. Make sure the sample is thin enough for light to pass through.
2. Use a dropper to place one or two drops of water on the sample.
3. Hold a clean coverslip by the edges and place it at one edge of the water. Slowly lower the coverslip onto the water until it lies flat.
4. If you have too much water or a lot of air bubbles, touch the edge of a paper towel to the edge of the coverslip to draw off extra water and draw out unwanted air.

Diversity of Life: Classification of Living Organisms

A six-kingdom system of classification of organisms is used today. Two kingdoms—Kingdom Archaebacteria and Kingdom Eubacteria—contain organisms that do not have a nucleus and that lack membrane-bound structures in the cytoplasm of their cells. The members of the other four kingdoms have a cell or cells that contain a nucleus and structures in the cytoplasm, some of which are surrounded by membranes. These kingdoms are Kingdom Protista, Kingdom Fungi, Kingdom Plantae, and Kingdom Animalia.

Kingdom Archaebacteria

one-celled; some absorb food from their surroundings; some are photosynthetic; some are chemosynthetic; many are found in extremely harsh environments including salt ponds, hot springs, swamps, and deep-sea hydrothermal vents

Kingdom Eubacteria

one-celled; most absorb food from their surroundings; some are photosynthetic; some are chemosynthetic; many are parasites; many are round, spiral, or rod-shaped; some form colonies

Kingdom Protista

Phylum Euglenophyta one-celled; photosynthetic or take in food; most have one flagellum; euglenoids

Kingdom Eubacteria
Bacillus anthracis

Phylum Chlorophyta
Desmids

Phylum Bacillariophyta one-celled; photosynthetic; have unique double shells made of silica; diatoms

Phylum Dinoflagellata one-celled; photosynthetic; contain red pigments; have two flagella; dinoflagellates

Phylum Chlorophyta one-celled, many-celled, or colonies; photosynthetic; contain chlorophyll; live on land, in freshwater, or salt water; green algae

Phylum Rhodophyta most are many-celled; photosynthetic; contain red pigments; most live in deep, saltwater environments; red algae

Phylum Phaeophyta most are many-celled; photosynthetic; contain brown pigments; most live in saltwater environments; brown algae

Phylum Rhizopoda one-celled; take in food; are free-living or parasitic; move by means of pseudopods; amoebas

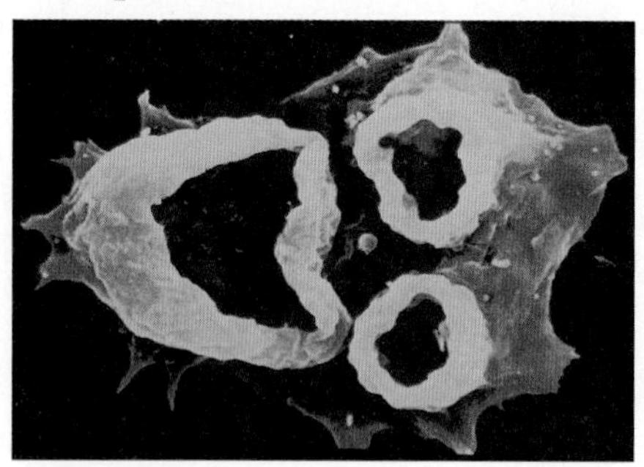

Amoeba

Phylum Zoomastigina one-celled; take in food; free-living or parasitic; have one or more flagella; zoomastigotes

Phylum Ciliophora one-celled; take in food; have large numbers of cilia; ciliates

Phylum Sporozoa one-celled; take in food; have no means of movement; are parasites in animals; sporozoans

Phylum Myxomycota
Slime mold

Phylum Oomycota
Phytophthora infestans

Phyla Myxomycota and Acrasiomycota one- or many-celled; absorb food; change form during life cycle; cellular and plasmodial slime molds

Phylum Oomycota many-celled; are either parasites or decomposers; live in freshwater or salt water; water molds, rusts and downy mildews

Kingdom Fungi

Phylum Zygomycota many-celled; absorb food; spores are produced in sporangia; zygote fungi; bread mold

Phylum Ascomycota one- and many-celled; absorb food; spores produced in asci; sac fungi; yeast

Phylum Basidiomycota many-celled; absorb food; spores produced in basidia; club fungi; mushrooms

Phylum Deuteromycota members with unknown reproductive structures; imperfect fungi; *Penicillium*

Phylum Mycophycota organisms formed by symbiotic relationship between an ascomycote or a basidiomycote and green alga or cyanobacterium; lichens

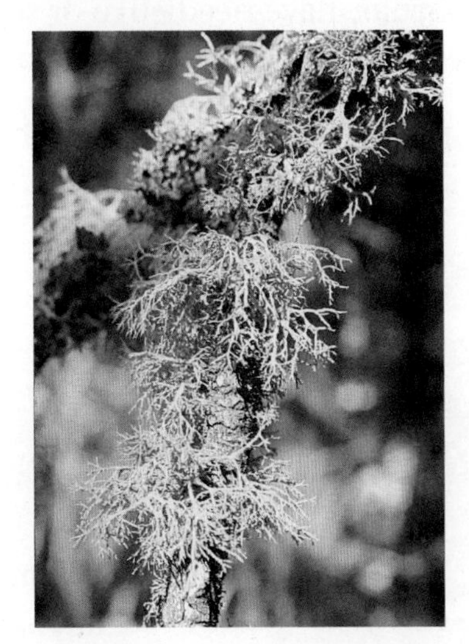

Lichens

Kingdom Plantae

Divisions Bryophyta (mosses), **Anthocerophyta** (hornworts), **Hepaticophyta** (liverworts), **Psilophyta** (whisk ferns) many-celled non-vascular plants; reproduce by spores produced in capsules; green; grow in moist, land environments

Division Lycophyta many-celled vascular plants; spores are produced in conelike structures; live on land; are photosynthetic; club mosses

Division Arthrophyta vascular plants; ribbed and jointed stems; scalelike leaves; spores produced in conelike structures; horsetails

Division Pterophyta vascular plants; leaves called fronds; spores produced in clusters of sporangia called sori; live on land or in water; ferns

Division Ginkgophyta deciduous trees; only one living species; have fan-shaped leaves with branching veins and fleshy cones with seeds; ginkgoes

Division Cycadophyta palmlike plants; have large, featherlike leaves; produces seeds in cones; cycads

Division Coniferophyta deciduous or evergreen; trees or shrubs; have needlelike or scalelike leaves; seeds produced in cones; conifers

Division Anthophyta
Tomato plant

Division Gnetophyta shrubs or woody vines; seeds are produced in cones; division contains only three genera; gnetum

Division Anthophyta dominant group of plants; flowering plants; have fruits with seeds

Kingdom Animalia

Phylum Porifera aquatic organisms that lack true tissues and organs; are asymmetrical and sessile; sponges

Phylum Cnidaria radially symmetrical organisms; have a digestive cavity with one opening; most have tentacles armed with stinging cells; live in aquatic environments singly or in colonies; includes jellyfish, corals, hydra, and sea anemones

Phylum Platyhelminthes bilaterally symmetrical worms; have flattened bodies; digestive system has one opening; parasitic and free-living species; flatworms

Division Bryophyta
Liverwort

Phylum Platyhelminthes
Flatworm

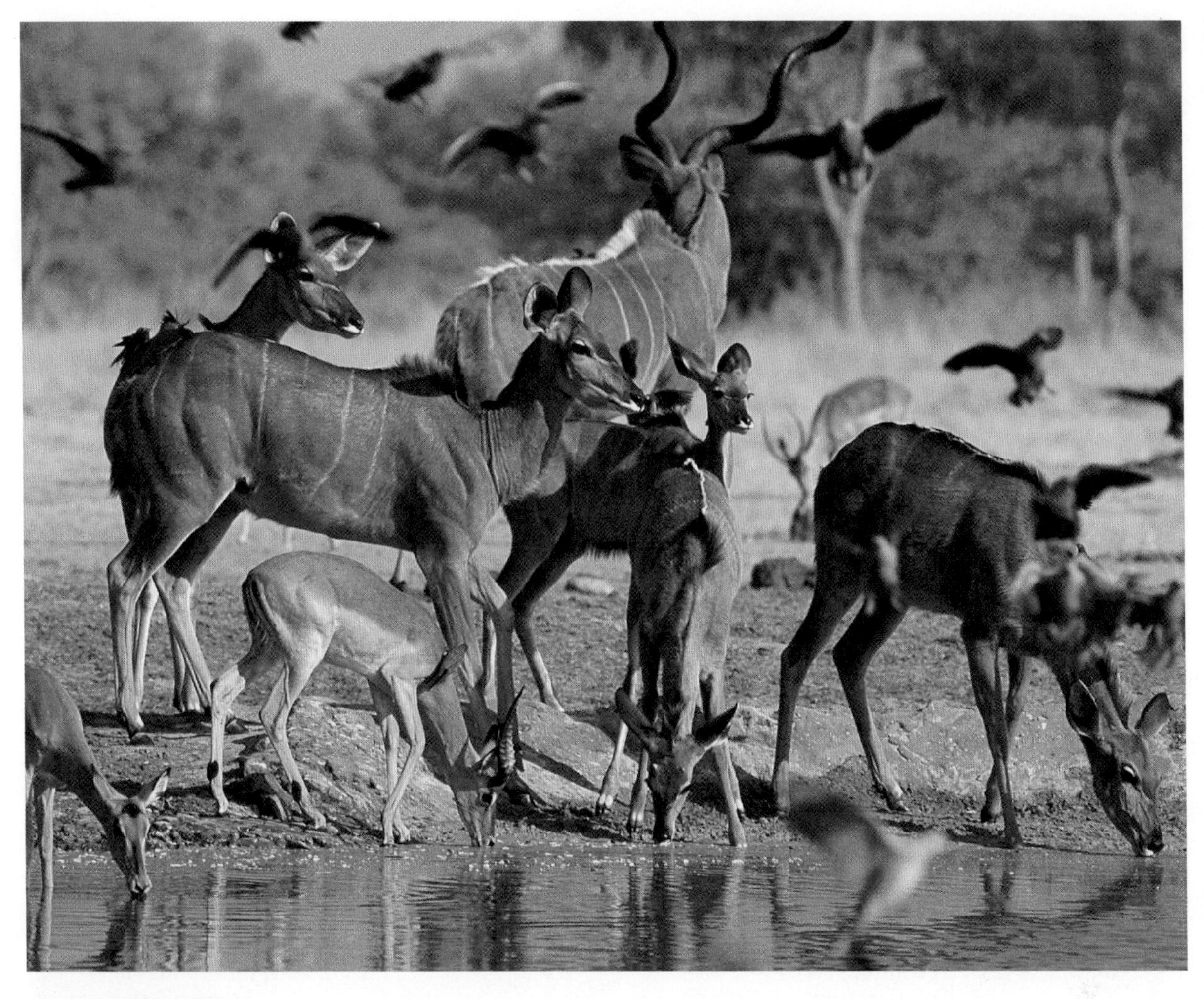

Phylum Chordata

Phylum Nematoda round, bilaterally symmetrical body; have digestive system with two openings; free-living forms and parasitic forms; roundworms

Phylum Mollusca soft-bodied animals, many with a hard shell and soft foot or footlike appendage; a mantle covers the soft body; aquatic and terrestrial species; includes clams, snails, squid, and octopuses

Phylum Annelida bilaterally symmetrical worms; have round, segmented bodies; terrestrial and aquatic species; includes earthworms, leeches, and marine polychaetes

Phylum Arthropoda largest animal group; have hard exoskeletons, segmented bodies, and pairs of jointed appendages; land and aquatic species; includes insects, crustaceans, and spiders

Phylum Echinodermata marine organisms; have spiny or leathery skin and a water-vascular system with tube feet; are radially symmetrical; includes sea stars, sand dollars, and sea urchins

Phylum Chordata organisms with internal skeletons and specialized body systems; most have paired appendages; all at some time have a notochord, nerve cord, gill slits, and a post-anal tail; include fish, amphibians, reptiles, birds, and mammals

Glossary/Glosario

Multilingual eGlossary

A science multilingual glossary is available on the ConnectEd. The glossary includes the following languages:

Arabic	Hmong	Tagalog
Bengali	Korean	Urdu
Chinese	Portuguese	Vietnamese
English	Russian	
Haitian Creole	Spanish	

Cómo usar el glosario en español:

1. Busca el término en inglés que desees encontrar.
2. El término en español, junto con la definición, se encuentran en la columna de la derecha.

Pronunciation Key

Use the following key to help you sound out words in the glossary:

a	back (BAK)	**ew**	food (FEWD)
ay	day (DAY)	**yoo**	pure (PYOOR)
ah	father (FAH thur)	**yew**	few (FYEW)
ow	flower (FLOW ur)	**uh**	comma (CAH muh)
ar	car (CAR)	**u (+ con)**	rub (RUB)
e	less (LES)	**sh**	shelf (SHELF)
ee	leaf (LEEF)	**ch**	nature (NAY chur)
ih	trip (TRIHP)	**g**	gift (GIHFT)
i (i + con + e)	idea (i DEE uh)	**j**	gem (JEM)
oh	go (GOH)	**ing**	sing (SING)
aw	soft (SAWFT)	**zh**	vision (VIH zhun)
or	orbit (OR buht)	**k**	cake (KAYK)
oy	coin (COYN)	**s**	seed, cent (SEED, SENT)
oo	foot (FOOT)	**z**	zone, raise (ZOHN, RAYZ)

English — A — Español

absolute age: the numerical age, in years, of a rock or object. (p. 583)

acceleration: a measure of the change in velocity during a period of time. (p. 27)

acid: a substance that produces a hydronium ion (H_3O^+) when dissolved in water. (p. 354)

acid precipitation: precipitation that has a lower pH than that of normal rainwater (pH 5.6). (p. 744)

activation energy: the minimum amount of energy needed to start a chemical reaction. (p. 320)

adaptation: an inherited trait that increases an organism's chance of surviving and reproducing in a particular environment. (p. 813)

edad absoluta: edad numérica, en años, de una roca o de un objeto. (pág. 583)

aceleración: medida del cambio de velocidad durante un período de tiempo. (pág. 27)

ácido: sustancia que produce ión hidronio (H_3O^+) cuando se disuelve en agua. (pág. 354)

precipitación ácida: precipitación que tiene un pH más bajo que el del agua de la lluvia normal (pH 5.6). (pág. 744)

energía de activación: cantidad mínima de energía necesaria para iniciar una reacción química. (pág. 320)

adaptación: rasgo heredado que aumenta la oportunidad de un organismo de sobrevivir y reproducirse en un medioambiente. (pág. 813)

Air Quality Index (AQI): a scale that ranks levels of ozone and other air pollutants. (p. 747)

Índice de calidad del aire (ICA): escala que clasifica los niveles de ozono y de otros contaminantes del aire. (pág. 747)

amino (uh MEE noh) acid: the building block of protein. (p. 768)

aminoácido: bloque constructor de proteína. (pág. 768)

asteroid: a small, rocky object that orbits the Sun. (p. 377)

asteroide: objeto pequeño y rocoso que orbita el Sol. (pág. 377)

astronomical unit: the average distance from Earth to the Sun—about 150 million km. (p. 378)

unidad astronómica: distancia media entre la Tierra y el Sol, aproximadamente 150 millones de km. (pág. 378)

atom: the smallest piece of an element that still represents that element. (p. 237)

átomo: parte más pequeña de un elemento que mantiene la identidad de dicho elemento. (pág. 237)

atomic number: the number of protons in an atom of an element. (p. 249)

número atómico: número de protones en el átomo de un elemento. (pág. 249)

average atomic mass: the average mass of the element's isotopes, weighted according to the abundance of each isotope. (p. 251)

masa atómica promedio: masa atómica promedio de los isótopos de un elemento, ponderado según la abundancia de cada isótopo. (pág. 251)

average speed: the total distance traveled divided by the total time taken to travel that distance. (p. 19)

rapidez promedio: distancia total recorrida dividida por el tiempo usado para recorrerla. (pág. 19)

B

balanced forces: forces acting on an object that combine and form a net force of zero. (p. 56)

fuerzas en equilibrio: fuerzas que actúan sobre un objeto, se combinan y forman una fuerza neta de cero. (pág. 56)

base: a substance that produces hydroxide ions (OH–) when dissolved in water. (p. 354)

base: sustancia que produce iones hidróxido (OH–) cuando se disuelve en agua. (pág. 354)

biome: a geographic area on Earth that contains ecosystems with similar biotic and abiotic features. (p. 683)

bioma: área geográfica en la Tierra que contiene ecosistemas con características bióticas y abióticas similares. (pág. 683)

Boyle's Law: the law that pressure of a gas increases if the volume decreases and pressure of a gas decreases if the volume increases, when temperature is constant. (p. 220)

Ley de Boyle: ley que afirma que la presión de un gas aumenta si el volumen disminuye y que la presión de un gas disminuye si el volumen aumenta, cuando la temperatura es constante. (pág. 220)

C

carbon film: the fossilized carbon outline of an organism or part of an organism. (p. 568)

película de carbono: contorno de carbono fosilizado de un organismo o parte de un organismo. (pág. 568)

carrying capacity: the largest number of individuals of one species that an ecosystem can support over time. (pp. 651, 720)

capacidad de carga: número mayor de individuos de una especie que un medioambiente puede mantener. (pág. 651, 720)

Science Skill Handbook
Math Skill Handbook
Foldables Handbook
Reference Handbook
Glossary/Glosario
Index

cast: a fossil copy of an organism made when a mold of the organism is filled with sediment or mineral deposits. (p. 579)

catalyst: a substance that increases reaction rate by lowering the activation energy of a reaction. (p. 322)

catastrophism: the idea that conditions and organisms on Earth change in quick, violent events. (p. 565)

Cenozoic era: the youngest era of the Phanerozoic eon. (p. 609)

centripetal (sen TRIH puh tuhl) force: in circular motion, a force that acts perpendicular to the direction of motion, toward the center of the curve. (p. 66)

Charles's Law: the law that the volume of a gas increases with increasing temperature, if the pressure is constant. (p. 221)

chemical bond: a force that holds two or more atoms together. (p. 268)

chemical energy: energy that is stored in and released from the bonds between atoms. (p. 90)

chemical equation: a description of a reaction using element symbols and chemical formulas. (p. 304)

chemical formula: a group of chemical symbols and numbers that represent the elements and the number of atoms of each element that make up a compound. (p. 280)

chemical reaction: a process in which atoms of one or more substances rearrange to form one or more new substances. (p. 301)

cinder cone: a small, steep-sided volcano that erupts gas-rich, basaltic lava. (p. 548)

circular motion: any motion in which an object is moving along a curved path. (p. 66)

cleavage: the breaking of a mineral along a smooth, flat surface. (p. 465)

climax community: a stable community that no longer goes through major ecological changes. (p. 703)

contramolde: copia fósil de un organismo producida cuando un molde del organismo se llena con depósitos de sedimento o mineral. (pág. 579)

catalizador: sustancia que aumenta la velocidad de reacción al disminuir la energía de activación de una reacción. (pág. 322)

catatrofismo: idea de que las condiciones y los organismos en la Tierra cambian mediante eventos rápidos y violentos. (pág. 565)

era Cenozoica: era más joven del eón Fanerozoico. (pág. 609)

fuerza centrípeta: en movimiento circular, la fuerza que actúa de manera perpendicular a la dirección del movimiento, hacia el centro de la curva. (pág. 66)

Ley de Charles: ley que afirma que el volumen de un gas aumenta cuando la temperatura aumenta, si la presión es constante. (pág. 221)

enlace químico: fuerza que mantiene unidos dos o más átomos. (pág. 268)

energía química: energía almacenada en y liberada por los enlaces entre los átomos. (pág. 90)

ecuación química: descripción de una reacción con símbolos de los elementos y fórmulas químicas. (pág. 304)

fórmula química: grupo de símbolos químicos y números que representan los elementos y el número de átomos de cada elemento que forman un compuesto. (pág. 280)

reacción química: proceso en el cual átomos de una o más sustancias se acomodan para formar una o más sustancias nuevas. (pág. 301)

cono de ceniza: volcán pequeño de lados empinados que expulsa lava rica en gas basáltico. (pág. 548)

movimiento circular: cualquier movimiento en el cual un objeto se mueve a lo largo de una trayectoria curva. (pág. 66)

exfoliación: rompimiento de un mineral en láminas o superficies planas. (pág. 465)

comunidad clímax: comunidad estable que ya no sufrirá mayores cambios ecológicos. (pág. 703)

coal swamp: an oxygen-poor environment where, over a period of time, decaying plant material changes into coal. (p. 612)

codominance: an inheritance pattern in which both alleles can be observed in a phenotype. (p. 804)

coefficient: a number placed in front of an element symbol or chemical formula in an equation. (p. 308)

combustion: a chemical reaction in which a substance combines with oxygen and releases energy. (p. 314)

comet: a small, rocky and icy object that orbits the Sun. (p. 377)

community: all the populations living in an ecosystem at the same time. (p. 648)

complex machine: two or more simple machines working together. (p. 107)

composite volcano: a large, steep-sided volcano that results from explosive eruptions of andesitic and rhyolitic lavas along convergent plate boundaries. (p. 548)

concentration: the amount of a particular solute in a given amount of solution. (p. 346)

condensation: the change of state from a gas to a liquid. (p. 212)

conduction: the transfer of thermal energy due to collisions between particles. (p. 174)

conservation biology: a branch of biology that studies why many species are in trouble and what can be done to save them. (p. 816)

constants: the factors in an experiment that remain the same. (p. NOS 21)

constant speed: the rate of change of position in which the same distance is traveled each second. (p. 18)

consumer: an organism that cannot make its own food and gets energy by eating other organisms. (p. 656)

contact force: a push or a pull on one object by another object that is touching it. (p. 45)

pantano de carbón: medioambiente pobre en oxígeno donde, al paso de un período de tiempo, el material en descomposición de plantas, se transforma en carbón. (pág. 612)

condominante: patrón heredado en el cual los dos alelos se observan en un fenotipo. (pág. 804)

coeficiente: número colocado en frente del símbolo de un elemento o de una fórmula química en una ecuación. (pág. 308)

combustión: reacción química en la cual una sustancia se combina con oxígeno y libera energía. (pág. 314)

cometa: objeto pequeño, rocoso y helado que orbita el Sol. (pág. 377)

comunidad: todas las poblaciones que viven en un ecosistema al mismo tiempo. (pág. 648)

máquina compleja: dos o más máquinas simples que trabajan juntas. (pág. 107)

volcán compuesto: volcán grande de lados empinados producido por erupciones explosivas de lavas andesíticas y riolíticas a lo largo de límites convergentes. (pág. 548)

concentración: cantidad de cierto soluto en una cantidad dada de solución. (pág. 346)

condensación: cambio de estado gaseoso a líquido. (pág. 212)

conducción: transferencia de energía térmica debido a colisiones entre partículas. (pág. 174)

biología conservacionista: rama de la biología que estudia por qué muchas especies están en problemas y qué se puede hacer para salvarlas. (pág. 816)

constantes: factores que no cambian en un experimento. (pág. NOS 21)

velocidad constante: velocidad a la que se cambia de posición, en la cual se recorre la misma distancia por segundo. (pág. 18)

consumidor: organismo que no elabora su propio alimento y obtiene energía comiendo otros organismos. (pág. 656)

fuerza de contacto: empuje o arrastre ejercido sobre un objeto por otro que lo está tocando. (pág. 45)

continental drift: Wegener's hypothesis which suggested that the continents are in constant motion on Earth's surface. (p. 495)

deriva continental: hipótesis de Wegener que sugirió que los continentes están en constante movimiento en la superficie de la Tierra. (pág. 495)

control group: the part of a controlled experiment that contains the same factors as the experimental group, but the independent variable is not changed. (p. NOS 21)

grupo de control: parte de un experimento controlado que contiene los mismos factores que el grupo experimental, pero la variable independiente no se cambia. (pág. NOS 21)

convection: the transfer of thermal energy by the movement of particles from one part of a material to another. (p. 178); the circulation of particles within a material caused by differences in thermal energy and density. (p. 516)

convección: transferencia de energía térmica por el movimiento de partículas de una parte de la materia a otra. (pág. 178); circulación de partículas en el interior de un material causada por diferencias en la energía térmica y la densidad. (pág. 516)

convection current: the movement of fluids in a cycle because of convection. (p. 179)

corriente de convección: movimiento de fluidos en un ciclo debido a la convección. (pág. 179)

convergent plate boundary: the boundary between two plates that move toward each other. (p. 513)

límite convergente de placas: límite entre dos placas que se acercan una hacia la otra. (pág. 513)

coral reef: an underwater structure made from outside skeletons of tiny, soft-bodied animals called coral. (p. 699)

arrecife de coral: estructura bajo el agua formada por exoesqueletos de animales diminutos y de cuerpo blando. (pág. 699)

cornea (KOR nee uh): a convex lens made of transparent tissue located on the outside of the eye. (p. 144)

córnea: lente convexo hecho de tejido transparente, ubicado en la parte externa del ojo. (pág. 144)

correlation: a method used by geologists to fill in the missing gaps in an area's rock record by matching rocks and fossils from separate locations. (p. 578)

correlación: método utilizado por los geólogos para completar vacios en un área de registro de rocas, comparando rocas y fósiles de lugares distanciados. (pág. 578)

covalent bond: a chemical bond formed when two atoms share one or more pairs of valence electrons. (p. 277)

enlace covalente: enlace químico formado cuando dos átomos comparten uno o más pares de electrones de valencia. (pág. 277)

critical thinking: comparing what you already know information you are given in order to decide whether you agree with it. (p. NOS 10)

pensamiento crítico: comparación que se hace cuando algo acerca de información nueva, y se decide si se está o no de acuerdo con ella. (pág. NOS 10)

crystallization: the process by which atoms form a solid with an orderly, repeating pattern. (p. 463)

cristalización: proceso por el cual los átomos forman un sólido con un patrón ordenado y repetitivo. (pág. 463)

crystal structure: the orderly, repeating pattern of atoms in a crystal. (p. 462)

estructura del cristal: patrón repetitivo y ordenado de los átomos en un cristal. (pág. 462)

D

decomposition: a type of chemical reaction in which one compound breaks down and forms two or more substances. (p. 313)

descomposición: tipo de reacción química en la que un compuesto se descompone y forma dos o más sustancias. (pág. 313)

Science Skill Handbook
Math Skill Handbook
Foldables Handbook
Reference Handbook
Glossary/Glosario
Index

deforestation: the removal of large areas of forests for human purposes. (p. 725)

dependent variable: the factor a scientist observes or measures during an experiment. (p. NOS 21)

deposition: the process of changing directly from a gas to a solid. (p. 212); the laying down or settling of eroded material. (p. 481)

description: a spoken or written summary of an observation. (p. NOS 21)

desert: a biome that receives very little rain. (p. 684)

desertification: the development of desertlike conditions due to human activities and/or climate change. (p. 726)

detritivore (duh TRI tuh vor): an organism that consumes the bodies of dead organisms and wastes produced by living organisms. (p. 656)

dinosaur: dominant Mesozoic land vertebrates that walked with their legs positioned directly below their hips. (p. 620)

displacement: the difference between the initial, or starting, position and the final position of an object that has moved. (p. 13)

divergent plate boundary: the boundary between two plates that move away from each other. (p. 513)

dominant trait: a genetic factor that blocks another genetic factor. (p. 797)

double-replacement reaction: a type of chemical reaction in which the negative ions in two compounds switch places, forming two new compounds. (p. 314)

deforestación: eliminación de grandes áreas de bosques con propósitos humanos. (pág. 725)

variable dependiente: factor que el científico observa o mide durante un experimento. (pág. NOS 21)

deposición: proceso de cambiar directamente de gas a sólido. (pág. 212); establecimiento o asentamiento de material erosionado. (pág. 481)

descripción: resumen oral o escrito de una observación. (pág. NOS 21)

desierto: bioma que recibe muy poca lluvia. (pág. 684)

desertificación: desarrollo de condiciones parecidas a las del desierto debido a actividades humanas y/o al cambio en el clima. (pág. 726)

detritívoro: organismo que consume los cuerpos de organismos muertos y los residuos producidos por organismos vivos. (pág. 656)

dinosaurio: vertebrados dominantes de la tierra del Mesozoico que caminaban con las extremidades ubicadas justo debajo de las caderas. (pág. 620)

desplazamiento: diferencia entre la posición inicial, o salida, y la final de un objeto que se ha movido. (pág. 13)

límite divergente de placas: límite entre dos placas que se alejan una de la otra. (pág. 513)

rasgo dominante: factor genético que bloquea otro factor genético. (pág. 797)

reacción de sustitución doble: tipo de reacción química en la que los iones negativos de dos compuestos intercambian lugares, para formar dos compuestos nuevos. (pág. 314)

E

earthquake: vibrations caused by the rupture and sudden movement of rocks along a break or a crack in Earth's crust. (p. 531)

echo: a reflected sound wave. (p. 129)

ecological succession: the process of one ecological community gradually changing into another. (p. 703)

efficiency: the ratio of output work to input work. (p. 109)

terremoto: vibraciones causadas por la ruptura y el movimiento repentino de las rocas en una fractura o grieta en la corteza de la Tierra. (pág. 531)

eco: onda sonora reflejada. (pág. 129)

sucesión ecológica: proceso en el que una comunidad ecológica cambia gradualmente en otra. (pág. 703)

eficiencia: relación entre energía invertida y energía útil. (pág. 109)

electric energy: energy carried by an electric current. (p. 88)

electron: a negatively charged particle that occupies the space in an atom outside the nucleus. (p. 239)

electron cloud: the region surrounding an atom's nucleus where one or more electrons are most likely to be found. (p. 244)

electron dot diagram: a model that represents valence electrons in an atom as dots around the element's chemical symbol. (p. 271)

endothermic reaction: a chemical reaction that absorbs thermal energy. (p. 319)

energy: the ability to cause change. (p. 87)

energy pyramid: a model that shows the amount of energy available in each link of a food chain. (p. 658)

energy transformation: the conversion of one form of energy to another. (p. 97)

enzyme: a catalyst that speeds up chemical reactions in living cells. (p. 322)

eon: the longest unit of geologic time (p. 601)

epicenter: the location on Earth's surface directly above an earthquake's focus. (p. 534)

epoch: a division of geologic time smaller than a period. (p. 601)

era: a large division of geologic time that is smaller than an eon. (p. 601)

estuary (ES chuh wer ee): a coastal area where freshwater from rivers and streams mixes with salt water from seas or oceans. (p. 697)

eutrophication (yoo troh fuh KAY shun): the process of a body of water becoming nutrient-rich. (p. 706)

evaporation: the process of a liquid changing to a gas at the surface of the liquid. (p. 212)

evolution: change over time. (p. 815)

exothermic reaction: a chemical reaction that releases thermal energy. (p. 319)

experimental group: the part of the controlled experiment used to study relationships among variables. (p. NOS 21)

energía eléctrica: energía transportada por una corriente eléctrica. (pág. 88)

electrón: partícula cargada negativamente que ocupa el espacio por fuera del núcleo de un átomo. (pág. 239)

nube de electrones: región que rodea el núcleo de un átomo en donde es más probable encontrar uno o más electrones. (pág. 244)

diagrama de puntos de Lewis: modelo que representa electrones de valencia en un átomo a manera de puntos alrededor del símbolo químico del elemento. (pág. 271)

reacción endotérmica: reacción química que absorbe energía térmica. (pág. 319)

energía: capacidad de ocasionar cambio. (pág. 87)

pirámide energética: modelo que muestra la cantidad de energía disponible en cada enlace de una cadena alimentaria. (pág. 658)

transformación de energía: conversión de una forma de energía a otra. (pág. 97)

enzima: catalizador que acelera reacciones químicas en las células vivas. (pág. 322)

eón: unidad más larga del tiempo geológico. (pág. 601)

epicentro: lugar en la superficie de la Tierra justo encima del foco de un terremoto. (pág. 534)

época: división del tiempo geológico más pequeña que un período. (pág. 601)

era: división grande del tiempo geológico que es más pequeña que un eón. (pág. 601)

estuario: zona costera donde el agua dulce de los ríos y arroyos se mezcla con el agua salada de los mares y los océanos. (pág. 697)

eutrofización: proceso por el cual un cuerpo de agua se vuelve rico en nutrientes. (pág. 706)

evaporación: proceso por el cual un líquido cambia a gas en la superficie de dicho líquido. (pág. 212)

evolución: cambio con el paso del tiempo. (pág. 815)

reacción exotérmica: reacción química que libera energía térmica. (pág. 319)

grupo experimental: parte del experimento controlado que se usa para estudiar las relaciones entre las variables. (pág. NOS 21)

explanation: an interpretation of observations. (p. NOS 12)

extinction (ihk STINGK shun): event that occurs when the last individual of a species dies. (p. 816)

extrusive rock: igneous rock that forms when volcanic material erupts, cools, and crystallizes on Earth's surface. (p. 480)

explicación: interpretación de las observaciones. (pág. NOS 12)

extinción: evento que ocurre cuando el último individuo de una especie muere. (pág. 816)

roca extrusiva: roca ígnea que se forma cuando el material volcánico sale, se enfría y se cristaliza en la superficie de la Tierra. (pág. 480)

F

fault: a crack or a fracture in Earth's lithosphere along which movement occurs. (p. 533)

focus: a location inside Earth where rocks first move along a fault and from which seismic waves originate. (p. 534)

foliation [foh lee AY shun]: rock texture that forms when uneven pressures cause flat minerals to line up, giving the rock a layered appearance. (p. 475)

food web: a model of energy transfer that can show how the food chains in a community are interconnected. (p. 657)

force: a push or a pull on an object. (p. 45)

force pair: the forces two objects apply to each other. (p. 71)

fossil: the preserved remains or evidence of past living organisms. (p. 565)

fracture: the breaking of a mineral along a rough or irregular surface. (p. 465)

friction: a contact force that resists the sliding motion of two surfaces that are touching. (p. 49)

falla: grieta o fractura en la litosfera de la Tierra en la cual ocurre el movimiento. (pág. 533)

foco: lugar en el interior de la Tierra donde se originan las ondas sísmicas, las cuales son producidas por el movimiento de las rocas a lo largo de un falla. (pág. 534)

foliación: textura de la roca que se forma cuando presiones disparejas causan que los minerales planos se alineen, dándole a la roca una apariencia de capas. (pág. 475)

red alimentaria: modelo de transferencia de energía que explica cómo las cadenas alimentarias están interconectadas en una comunidad. (pág. 657)

fuerza: empuje o arrastre ejercido sobre un objeto. (pág. 45)

par de fuerzas: fuerzas que dos objetos se aplican entre sí. (pág. 71)

fósil: restos conservados o evidencia de organismos vivos del pasado. (pág. 565)

fractura: rompimiento de un mineral en una superficie desigual o irregular. (pág. 465)

fricción: fuerza que resiste el movimiento de dos superficies que están en contacto. (pág. 49)

G

Galilean moons: the four largest of Jupiter's 63 moons; discovered by Galileo. (p. 393)

gas: matter that has no definite volume and no definite shape. (p. 204)

genetic engineering: the modification of genetic material of an organism by inserting DNA from another organism. (p. 807)

lunas de Galileo: las cuatro lunas más grandes de las 63 lunas de Júpiter; descubiertas por Galileo. (pág. 393)

gas: materia que no tiene volumen ni forma definidos. (pág. 204)

ingeniería genética: modificación del material genético de un organismo insertándole ADN de otro organismo. (pág. 807)

genetics (juh NEH tihks): the study of how traits are passed from parents to offspring. (p. 793)

genotype (JEE nuh tipe): the alleles of all the genes on an organism's chromosomes; controls an organism's phenotype. (p. 798)

geographic isolation: the separation of a population of organisms from the rest of its species due to some physical barrier such as a mountain range or an ocean. (p. 604)

glacial grooves: grooves in solid rock formations made by rocks that are carried by glaciers. (p. 627)

global warming: an increase in the average temperature of Earth's surface. (p. 745)

grain: an individual particle in a rock. (p. 471)

grassland: a biome where grasses are the dominant plants. (p. 685)

gravity: an attractive force that exists between all objects that have mass. (p. 47)

greenhouse effect: the natural process that occurs when certain gases in the atmosphere absorb and reradiate thermal energy from the Sun. (pp. 385, 746)

genética: estudio de cómo los rasgos pasan de los padres a los hijos. (pág. 793)

genotipo: de los alelos de todos los genes en los cromosomas de un organismo, los controles de fenotipo de un organismo. (pág. 798)

aislamiento geográfico: separación de una población de organismos del resto de su especie debido a alguna barrera física, tal como una cordillera o un océano. (pág. 604)

surcos glaciales: surcos en las formaciones de roca sólida producidos por las rocas transportadas por los glaciares. (pág. 627)

calentamiento global: incremento en la temperatura promedio de la superficie de la Tierra. (pág. 745)

grano: partícula individual de una roca. (pág. 471)

pradera: bioma donde los pastos son las plantas dominantes. (pág. 685)

gravedad: fuerza de atracción que existe entre todos los objetos que tienen masa. (pág. 47)

efecto invernadero: proceso natural que ocurre cuando ciertos gases en la atmósfera absorben y vuelven a irradiar la energía térmica del Sol. (pág. 385, 746)

H

habitat: the place within an ecosystem where an organism lives; provides the biotic and abiotic factors an organism needs to survive and reproduce. (p. 648)

half-life: the time required for half of the amount of a radioactive parent element to decay into a stable daughter element. (p. 585)

heat: the movement of thermal energy from a region of higher temperature to a region of lower temperature. (p. 169)

heat engine: a machine that converts thermal energy into mechanical energy. (p. 186)

heating appliance: a device that converts electric energy into thermal energy. (p. 183)

heredity (huh REH duh tee): the passing of traits from parents to offspring. (p. 793)

heterogeneous mixture: a mixture in which substances are not evenly mixed. (p. 337)

hábitat: lugar en un ecosistema donde vive un organismo; proporciona los factores bióticos y abióticos que un organismo necesita para vivir y reproducirse. (pág. 648)

vida media: tiempo requerido para que la mitad de cierta cantidad de un elemento radiactivo se desintegre en otro elemento estable. (pág. 585)

calor: movimiento de energía térmica de una región de alta temperatura a una región de baja temperatura. (pág. 169)

motor térmico: máquina que convierte energía térmica en energía mecánica. (pág. 186)

calentador: aparato que convierte energía eléctrica en energía térmica. (pág. 183)

herencia: paso de rasgos de los padres a los hijos. (pág. 793)

mezcla heterogénea: mezcla en la cual las sustancias no están mezcladas de manera uniforme. (pág. 337)

heterozygous (he tuh roh ZI gus): a genotype in which the two alleles of a gene are different. (p. 798)

heterocigoto: genotipo en el cual los dos alelos de un gen son diferentes. (pág.798)

Holocene epoch: the current epoch of geologic time which began 10,000 years ago. (p. 625)

Holoceno: época actual del tiempo geológico que comenzó hace 10.000 años. (pág. 625)

homeostasis (hoh mee oh STAY sus): an organism's ability to maintain steady internal conditions when outside conditions change. (p. 773)

homeostasis: capacidad de un organismo de mantener las condiciones internas estables cuando las condiciones externas cambian. (pág. 773)

homogeneous mixture: a mixture in which two or more substances are evenly mixed but not bonded together. (p. 337)

mezcla homogénea: mezcla en la cual dos o más sustancias están mezcladas de manera uniforme, pero no están unidas químicamente. (pág. 337)

homozygous (hoh muh ZI gus): a genotype in which the two alleles of a gene are the same. (p. 798)

homocigoto: genotipo en el cual los dos alelos de un gen son iguales. (pág. 798)

hot spot: a location where volcanoes form far from plate boundaries. (p. 546)

punto caliente: lugar lejos de los límites de las placas donde se forman volcanes. (pág. 546)

hydronium ion (H_3O^+): a positively charged ion formed when an acid dissolves in water. (p. 354)

ión hidronio (H_3O^+): ión cargado positivamente que se forma cuando un ácido se disuelve en agua. (pág. 354)

hypothesis: a possible explanation for an observation that can be tested by scientific investigations. (p. NOS 6)

hipótesis: explicación posible para una observación que puede ponerse a prueba en investigaciones científicas. (pág. NOS 6)

I

ice age: a period of time when a large portion of Earth's surface is covered by glaciers. (p. 627)

era del hielo: período de tiempo cuando los glaciares cubren una gran porción de la superficie de la Tierra. (pág. 627)

impact crater: a round depression formed on the surface of a planet, moon, or other space object by the impact of a meteorite. (p. 402)

cráter de impacto: depresión redonda formada en la superficie de un planeta, luna u otro objeto espacial debido al impacto de un meteorito. (pág. 402)

inclined plane: a simple machine that consists of a ramp, or a flat, sloped surface. (p. 106)

plano inclinado: máquina simple que consiste en una rampa, o superficie plana inclinada. (pág. 106)

inclusion: a piece of an older rock that becomes a part of a new rock. (p. 577)

inclusión: pedazo de una roca antigua que se convierte en parte de una roca nueva. (pág. 577)

incomplete dominance: an inheritance pattern in which an offspring's phenotype is a combination of the parents' phenotypes. (p. 804)

dominancia incompleta: patrón hereditario en el cual el fenotipo del hijo es una mezcla de los fenotipos de los padres. (pág. 804)

independent variable: the factor that is changed by the investigator to observe how it affects a dependent variable. (p. NOS 21)

variable independiente: factor que el investigador cambia para observar cómo afecta la variable dependiente. (pág. NOS 21)

index fossil: a fossil representative of a species that existed on Earth for a short length of time, was abundant, and inhabited many locations. (p. 579)

fósil índice: fósil representativo de una especie que existió en la Tierra por un período de tiempo corto, ésta era abundante y habitaba en varios lugares. (pág. 579)

indicator: a compound that changes color at different pH values when it reacts with acidic or basic solutions. (p. 358)

indicador: compuesto que cambia de color a diferentes valores de pH cuando reacciona con soluciones ácidas o básicas. (pág. 358)

inertia (ihn UR shuh): the tendency of an object to resist a change in its motion. (p. 58)

inercia: tendencia de un objeto a resistirse al cambio en su movimiento. (pág. 58)

inference: a logical explanation of an observation that is drawn from prior knowledge or experience. (p. NOS 6)

inferencia: explicación lógica de una observación que se obtiene a partir de conocimiento previo o experiencia. (pág. NOS 6)

inhibitor: a substance that slows, or even stops, a chemical reaction. (p. 322)

inhibidor: sustancia que disminuye, o incluso detiene, una reacción química. (pág. 322)

inland sea: a body of water formed when ocean water floods continents. (p. 610)

mar interior: cuerpo de agua formado cuando el agua del océano inunda los continentes. (pág. 610)

instantaneous speed: an object's speed at a specific instant in time. (p. 18)

velocidad instantánea: velocidad de un objeto en un instante específico en el tiempo. (pág. 18)

International System of Units (SI): the internationally accepted system of measurement. (p. NOS 12)

Sistema Internacional de Unidades (SI): sistema de medidas aceptado internacionalmente. (pág. NOS 12)

intertidal zone: the ocean shore between the lowest low tide and the highest high tide. (p. 699)

zona intermareal: playa en medio de la marea baja más baja y la marea alta más alta. (pág. 699)

intrusive rock: igneous rock that forms as magma cools underground. (p. 480)

roca intrusiva: roca ígnea que se forma cuando el magma se enfría bajo el suelo. (pág. 480)

ion (I ahn): an atom that is no longer neutral because it has lost or gained valence electrons. (pp. 254, 284)

ión: átomo que no es neutro porque ha ganado o perdido electrones de valencia. (pág. 254, 284)

ionic bond: the attraction between positively and negatively charged ions in an ionic compound. (p. 286)

enlace iónico: atracción entre iones cargados positiva y negativamente en un compuesto iónico. (pág. 286)

iris: the colored part of the eye. (p. 145)

iris: parte coloreada del ojo. (pág. 145)

isotopes: atoms of the same element that have different numbers of neutrons. (pp. 250, 584)

isótopos: átomos del mismo elemento que tienen números diferentes de neutrones. (pág. 250, 584)

K

kinetic (kuh NEH tik) energy: energy due to motion. (pp. 88, 208)

energía cinética: energía debida al movimiento. (pág. 88, 208)

kinetic molecular theory: an explanation of how particles in matter behave. (p. 218)

teoría cinética molecular: explicación de cómo se comportan las partículas en la materia. (pág. 218)

L

land bridge: a landform that connects two continents that were previously separated. (p. 604)

lava: magma that erupts onto Earth's surface. (pp. 472, 546)

law of conservation of energy: law that states that energy can be transformed from one form to another, but it cannot be created or destroyed. (p. 97)

law of conservation of mass: law that states that the total mass of the reactants before a chemical reaction is the same as the total mass of the products after the chemical reaction. (p. 306)

law of conservation of momentum: a principle stating that the total momentum of a group of objects stays the same unless outside forces act on the objects. (p. 74)

lens: a transparent object with at least one curved side that causes light to change direction. (p. 143)

lever: a simple machine that consists of a bar that pivots, or rotates, around a fixed point. (p. 106)

light ray: represents a narrow beam of light that travels in a straight line. (p. 135)

light source: something that emits light. (p. 135)

liquid: matter with a definite volume but no definite shape. (p. 202)

lithification: the process through which sediment turns into rock. (p. 473)

lithosphere (LIH thuh sfihr): the rigid outermost layer of Earth that includes the uppermost mantle and crust. (p. 512)

luster: the way a mineral reflects or absorbs light at its surface. (p. 465)

puente terrestre: accidente geográfico que conecta dos continentes que anteriormente estaban separados. (pág. 604)

lava: magma que llega a la superficie de la Tierra. (pág. 472, 546)

ley de la conservación de la energía: ley que plantea que la energía puede transformarse de una forma a otra, pero no puede crearse ni destruirse. (pág. 97)

ley de la conservación de la masa: ley que plantea que la masa total de los reactivos antes de una reacción química es la misma que la masa total de los productos después de la reacción química. (pág. 306)

ley de la conservación del momentum: principio que establece que el momentum total de un grupo de objetos permanece constante a menos que fuerzas externas actúen sobre los objetos. (pág. 74)

lente: objeto transparente que tiene, al menos, un lado curvo que hace que la luz cambie de dirección. (pág. 143)

palanca: máquina simple que consiste en una barra que gira, o rota, alrededor de un punto fijo. (pág. 106)

rayo de luz: haz de luz angosto que viaja en línea recta. (pág. 135)

fuente lumínica: algo que emite luz. (pág. 135)

líquido: materia con volumen definido y forma indefinida. (pág. 202)

litificación: proceso mediante el cual el sedimento se vuelve roca. (pág. 473)

litosfera: capa rígida más externa de la Tierra formada por el manto superior y la corteza. (pág. 512)

brillo: forma en que un mineral refleja o absorbe la luz en su superficie. (pág. 465)

M

macromolecule: substance that forms from joining many small molecules together. (p. 768)

magma: molten rock stored beneath Earth's surface. (pp. 472, 545)

macromolécula: sustancia que se forma al unir muchas moléculas pequeñas. (pág. 768)

magma: roca derretida almacenada debajo de la superficie de la Tierra. (pág. 472, 545)

Science Skill Handbook
Math Skill Handbook
Foldables Handbook
Reference Handbook
Glossary/Glosario
Index

magnetic reversal: an event that causes a magnetic field to reverse direction. (p. 506)

mass: the amount of matter in an object. (p. 47)

mass extinction: the extinction of many species on Earth within a short period of time. (p. 603)

mass number: the sum of the number of protons and neutrons in an atom. (p. 250)

mechanical energy: sum of the potential energy and the kinetic energy in a system. (p. 91)

mega-mammal: large mammal of the Cenozoic era. (p. 628)

Mesozoic era: the middle era of the Phanerozoic eon. (p. 609)

metallic bond: a bond formed when many metal atoms share their pooled valence electrons. (p. 287)

meteor: a meteoroid that has entered Earth's atmosphere and produces a streak of light. (p. 402)

meteorite: a meteoroid that strikes a planet or a moon. (p. 402)

meteoroid: a small rocky particle that moves through space. (p. 402)

mid-ocean ridge: long, narrow mountain range on the ocean floor; formed by magma at divergent plate boundaries. (p. 503)

mineral: a solid that is naturally occurring, inorganic, and has a crystal structure and definite chemical composition. (p. 461)

mirror: any reflecting surface that forms an image by regular reflection. (p. 142)

mixture: matter that can vary in composition. (p. 336)

mold: the impression of an organism in a rock. (p. 569)

molecule (MAH lih kyewl): two or more atoms that are held together by covalent bonds and act as a unit. (p. 278)

momentum: a measure of how hard it is to stop a moving object. (p. 73)

monohybrid cross: a cross between two individuals that are hybrids for one trait. (p. 803)

inversión magnética: evento que causa que un campo magnético invierta su dirección. (pág. 506)

masa: cantidad de materia en un objeto. (pág. 47)

extinción en masa: extinción de muchas especies en la Tierra dentro de un período de tiempo corto. (pág. 603)

número de masa: suma del número de protones y neutrones de un átomo. (pág. 250)

energía mecánica: suma de la energía potencial y de la energía cinética en un sistema. (pág. 91)

mega mamífero: mamífero enorme de la era Cenozoica. (pág. 628)

era Mesozoica: era media del eón Fanerozoico. (pág. 609)

enlace metálico: enlace formado cuando muchos átomos metálicos comparten su banco de electrones de valencia. (pág. 287)

meteoro: meteorito que ha entrado a la atmósfera de la Tierra y produce un haz de luz. (pág. 402)

meteorito: meteoroide que impacta un planeta o una luna. (pág. 402)

meteoroide: partícula rocosa pequeña que se mueve por el espacio. (pág. 402)

dorsal oceánica: cordillera larga y angosta en el lecho del océano, formada por magma en los límites de las placas divergentes. (pág. 503)

mineral: sólido inorgánico que se encuentra en la naturaleza, tiene una estructura cristalina y una composición química definida. (pág. 461)

espejo: cualquier superficie reflectora que forma una imagen por reflexión común. (pág. 142)

mezcla: materia cuya composición puede variar. (pág. 336)

molde: impresión de un organismo en una roca. (pág. 569)

molécula: dos o más átomos que están unidos mediante enlaces covalentes y actúan como una unidad. (pág. 278)

momentum: medida de qué tan difícil es detener un objeto en movimiento. (pág. 73)

cruce monohíbrido: cruce entre dos individuos que son híbridos para un rasgo. (pág. 803)

monosaccharide (mah nuh SA kuh ride): a simple sugar. (p. 768)

monosacárido: azúcar simple. (pág. 768)

motion: the process of changing position. (p. 13)

movimiento: proceso de cambiar de posición. (pág. 13)

multiple alleles: a gene that has more than two alleles. (p. 804)

alelos múltiples: gen que tiene más de dos alelos. (pág. 804)

mutation (myew TAY shun): a permanent change in the sequence of DNA, or the nucleotides, in a gene or a chromosome. (p. 806)

mutación: cambio permanente en la secuencia de ADN, de los nucleótidos, en un gen o en un cromosoma. (pág. 806)

N

natural selection: the process by which organisms with variations that help them survive in their environment live longer, compete better, and reproduce more than those that do not have the variations. (p. 812)

selección natural: proceso por el cual los organismos con variaciones que las ayudan a sobrevivir en sus medioambientes viven más, compiten mejor y se reproducen más que aquellas que no tienen esas variaciones. (pág. 812)

negative feedback: a control system where the effect of a hormone inhibits further release of the hormone; sends a signal to stop a response. (p. 780)

retroalimentación negativa: sistema de control donde el efecto de una hormona inhibe la liberación de cantidades adicionales de la misma hormona; envía una señal para detener la respuesta. (pág. 780)

net force: the combination of all the forces acting on an object. (p. 55)

fuerza neta: combinación de todas las fuerzas que actúan sobre un objeto. (pág. 55)

neutron: a neutral particle in the nucleus of an atom. (p. 243)

neutrón: partícula neutra en el núcleo de un átomo. (pág. 243)

Newton's first law of motion: law that states that if the net force acting on an object is zero, the motion of the object does not change. (p. 57)

primera ley del movimiento de Newton: ley que establece que si la fuerza neta ejercida sobre un objeto es cero, el movimiento de dicho objeto no cambia. (pág. 57)

Newton's second law of motion: law that states that the acceleration of an object is equal to the net force exerted on the object divided by the object's mass. (p. 65)

segunda ley del movimiento de Newton: ley que establece que la aceleración de un objeto es igual a la fuerza neta que actúa sobre él divida por su masa. (pág. 65)

Newton's third law of motion: law that states that for every action there is an equal and opposite reaction. (p. 71)

tercera ley del movimiento de Newton: ley que establece que para cada acción hay una reacción igual en dirección opuesta. (pág. 71)

niche (NICH): the way a species interacts with abiotic and biotic factors to obtain food, find shelter, and fulfill other needs. (p. 649)

nicho: forma como una especie interactúa con los factores abióticos y bióticos para obtener alimento, encontrar refugio y satisfacer otras necesidades. (pág. 649)

noncontact force: a force that one object applies to another object without touching it. (p. 46)

fuerza de no contacto: fuerza que un objeto puede aplicar sobre otro sin tocarlo. (pág. 46)

nonpoint-source pollution: pollution from several widespread sources that cannot be traced back to a single location. (p. 737)

contaminación de fuente no puntual: contaminación de varias fuentes apartadas que no se pueden rastrear hasta una sola ubicación. (pág. 737)

Science Skill Handbook
Math Skill Handbook
Foldables Handbook
Reference Handbook
Glossary/Glosario
Index

nonrenewable resource: a natural resource that is used up faster than it can be replaced by natural processes. (p. 666)

normal polarity: when magnetized objects, such as compass needles, orient themselves to point north. (p. 506)

nuclear decay: a process that occurs when an unstable atomic nucleus changes into another more stable nucleus by emitting radiation. (p. 253)

nuclear energy: energy stored in and released from the nucleus of an atom. (p. 90)

nucleotide (NEW klee uh tide): a molecule made of a nitrogen base, a sugar, and a phosphate group. (p. 768)

nucleus: the region in the center of an atom where most of an atom's mass and positive charge are concentrated. (p. 242)

recurso no renovable: recurso natural que se usa más rápidamente de lo que se puede reemplazar por procesos naturales. (pág. 666)

polaridad normal: ocurre cuando los objetos magnetizados, tales como las agujas de la brújula, se orientan a sí mismas para apuntar al norte. (pág. 506)

desintegración nuclear: proceso que ocurre cuando un núcleo atómico inestable cambia a otro núcleo atómico más estable mediante emisión de radiación. (pág. 253)

energía nuclear: energía almacenada en y liberada por el núcleo de un átomo. (pág. 90)

nucelótido: molécula constituida por una base nitrogenada, azúcar y un grupo fosfato. (pág. 768)

núcleo: región en el centro de un átomo donde se concentra la mayor cantidad de masa y las cargas positivas. (pág. 242)

O

observation: the act of using one or more of your senses to gather information and take note of what occurs. (p. NOS 6)

opaque: a material through which light does not pass. (p. 136)

ore: a deposit of minerals that is large enough to be mined for a profit. (p. 467)

observación: acción de usar uno o más sentidos para reunir información y tomar notar de lo que ocurre. (pág. NOS 6)

opaco: material por el que no pasa la luz. (pág. 136)

mena: depósito de minerales suficientemente grandes como para ser explotados con un beneficio. (pág. 467)

P

paleontologist: scientist who studies fossils. (p. 570)

Paleozoic era: the oldest era of the Phanerozoic eon. (p. 609)

Pangaea (pan JEE uh): name given to a supercontinent that began to break apart approximately 200 million years ago. (p. 495)

particulate matter: the mix of both solid and liquid particles in the air. (p. 744)

pedigree: a model that shows genetic traits that were inherited by members of a family. (p. 806)

percent error: the expression of error as a percentage of the accepted value. (p. NOS 15)

paleontólogo: científico que estudia los fósiles. (pág. 570)

era Paleozoica: era más antigua del eón Fanerozoico. (pág. 609)

Pangea: nombre dado a un supercontinente que empezó a separarse hace aproximadamente 200 millones de años. (pág. 495)

partículas en suspensión: mezcla de partículas sólidas y líquidas en el aire. (pág. 744)

pedigrí: modelo que muestra los rasgos genéticos heredados por los miembros de una familia. (pág. 806)

error porcentual: expresión del error como porcentaje del valor aceptado. (pág. NOS 15)

period: a unit of geologic time smaller than an era. (p. 601)
period of revolution: the time it takes an object to travel once around the Sun. (p. 378)
period of rotation: the time it takes an object to complete one rotation. (p. 378)
pH: an inverse measure of the concentration of hydronium ions (H_3O^+) in a solution. (p. 356)
phenotype (FEE nuh tipe): how a trait appears or is expressed. (p. 798)
photochemical smog: air pollution that forms from the interaction between chemicals in the air and sunlight. (p. 743)
pioneer species: the first species that colonizes new or undisturbed land. (p. 704)
pitch: the perception of how high or low a sound is; related to the frequency of a sound wave. (p. 127)
plate tectonics: theory that Earth's surface is broken into large, rigid pieces that move with respect to each other. (p. 511)
Pleistocene epoch: the first epoch of the Quaternary period. (p. 627)
plesiosaur: Mesozoic marine reptile with a small head, long neck, and flippers. (p. 621)
point-source pollution: pollution from a single source that can be identified. (p. 736)
polar molecule: a molecule with a slight negative charge in one area and a slight positive charge in another area. (pp. 279, 344)
polygenic inheritance: an inheritance pattern in which multiple genes determine the phenotype of a trait. (p. 805)
population: all the organisms of the same species that live in the same area at the same time. (pp. 648, 719)
position: an object's distance and direction from a reference point. (p. 9)
positive feedback: a control system in which the effect of a hormone causes more of the hormone to be released; sends a signal to increase a response. (p. 780)

período: unidad del tiempo geológico más pequeña que una era. (pág. 601)
período de revolución: tiempo que gasta un objeto en dar una vuelta alrededor del Sol. (pág. 378)
período de rotación: tiempo que gasta un objeto para completar una rotación. (pág. 378)
pH: medida inversa de la concentración de iones hidronio (H_3O^+) en una solución. (pág. 356)
fenotipo: forma como aparece o se expresa un rasgo. (pág. 798)
smog fotoquímico: polución del aire que se forma de la interacción entre los químicos en el aire y la luz solar. (pág. 743)
especie pionera: primera especie que coloniza tierra nueva o tierra virgen. (pág. 704)
tono: percepción de qué tan alto o bajo es el sonido; relacionado con la frecuencia de la onda sonora. (pág. 127)
tectónica de placas: teoría que afirma que la superficie de la Tierra está divida en piezas enormes y rígidas que se mueven una con respecto a la otra. (pág. 511)
época del Pleistoceno: primera época del período Cuaternario. (pág. 627)
plesiosaurio: reptil marino del Mesozoico de cabeza pequeña, cuello largo y aletas. (pág. 621)
contaminación de fuente puntual: contaminación de una sola fuente que se puede identificar. (pág. 736)
molécula polar: molécula con carga ligeramente negativa en una parte y ligeramente positiva en otra. (pág. 279, 344)
herencia poligénica: patrón de herencia en el cual genes múltiples determinan el fenotipo de un rasgo. (pág. 805)
población: todos los organismos de la misma especie que viven en la misma área al mismo tiempo. (pág. 648, 719)
posición: distancia y dirección de un objeto según un punto de referencia. (pág. 9)
retroalimentación positiva: sistema de control en el cual el efecto de una hormona causa más liberación de la hormona; envía una señal para aumentar la respuesta. (pág. 780)

potential (puh TEN chul) energy: stored energy due to the interactions between objects or particles. (p. 89)

energía potencial: energía almacenada debido a las interacciones entre objetos o partículas. (pág. 89)

predation: the act of one organism, the predator, feeding on another organism, its prey. (p. 649)

depredación: acción en la cual un organismo, el depredador, come a otro organismo, la presa. (pág. 649)

prediction: a statement of what will happen next in a sequence of events. (p. NOS 6)

predicción: afirmación de lo que ocurrirá después en una secuencia de eventos. (pág. NOS 6)

pressure: the amount of force per unit area applied to an object's surface. (p. 219)

presión: cantidad de fuerza por unidad de área aplicada a la superficie de un objeto. (pág. 219)

primary wave (also P-wave): a type of seismic wave which causes particles in the ground to move in a push-pull motion similar to a coiled spring. (p. 535)

onda primaria (también, onda P): tipo de onda sísmica que causa un movimiento de atracción y repulsión en las partículas del suelo, similar a un resorte. (pág. 535)

producer: an organism that uses an outside energy source, such as the Sun, and produces its own food. (p. 655)

productor: organismo que usa una fuente de energía externa, como el Sol, para elaborar su propio alimento. (pág. 655)

product: a substance produced by a chemical reaction. (p. 305)

producto: sustancia producida por una reacción química. (pág. 305)

proton: positively charged particle in the nucleus of an atom. (p. 242)

protón: partícula cargada positivamente en el núcleo de un átomo. (pág. 242)

pterosaur: Mesozoic flying reptile with large, batlike wings. (p. 621)

pterosaurio: reptil volador del Mesozoico de alas grandes parecidas a las del murciélago. (pág. 621)

pulley: a simple machine that consists of a grooved wheel with a rope or cable wrapped around it. (p. 107)

polea: máquina simple que consiste en una rueda acanalada rodeada por una cuerda o cable. (pág. 107)

Punnett square: a model that is used to show the probability of all possible genotypes and phenotypes of offspring. (p. 803)

cuadro de Punnett: modelo usado para mostrar la probabilidad de todos los genotipos y fenotipos posibles de la cría. (pág. 803)

pupil: an opening into the interior of the eye at the center of the iris. (p. 145)

pupila: abertura en el interior del ojo y en el centro del iris. (pág. 145)

Q

qualitative data: the use of words to describe what is observed in an experiment. (p. NOS 21)

datos cualitativos: uso de palabras para describir lo que se observa en un experimento. (pág. NOS 21)

quantitative data: the use of numbers to describe what is observed in an experiment. (p. NOS 21)

datos cuantitativos: uso de números para describir lo que se observa en un experimento. (pág. NOS 21)

R

radiant energy: energy carried by an electromagnetic wave. (p. 93)

energía radiante: energía que transporta una onda electromagnética. (pág. 93)

radiation: the transfer of thermal energy by electromagnetic waves. (p. 173)

radioactive: any element that spontaneously emits radiation. (p. 252)

radioactive decay: the process by which an unstable element naturally changes into another element that is stable. (p. 584)

reactant: a starting substance in a chemical reaction. (p. 305)

recessive trait: a genetic factor that is blocked by the presence of a dominant factor. (p. 797)

reclamation: a process in which mined land must be recovered with soil and replanted with vegetation. (p. 730)

reference point: the starting point you use to describe the motion or the position of an object. (p. 9)

reforestation: process of planting trees to replace trees that have been cut or burned down. (p. 730)

refrigerator: a device that uses electric energy to pump thermal energy from a cooler location to a warmer location. (p. 184)

relative age: the age of rocks and geologic features compared with other nearby rocks and features. (p. 575)

renewable resource: a natural resource that can be replenished by natural processes at least as quickly as it is used. (p. 666)

resource depletion: the exhaustion of one or more resources in an area. (p. 666)

retina (RET nuh): an area at the back of the eye that includes special light-sensitive cells—rod cells and cone cells. (p. 146)

reversed polarity: when magnetized objects reverse direction and orient themselves to point south. (p. 506)

ridge push: the process that results when magma rises at a mid-ocean ridge and pushes oceanic plates in two different directions away from the ridge. (p. 517)

radiación: transferencia de energía térmica por ondas electromagnéticas. (pág. 173)

radiactivo: cualquier elemento que emite radiación de manera espontánea. (pág. 252)

desintegración radioactiva: proceso poer el cual un elemento inestable cambia naturalmente en otro elemento que es estable. (pág. 584)

reactivo: sustancia inicial en una reacción química. (pág. 305)

rasgo recesivo: factor genético boqueado por la presencia de un factor dominante. (pág. 797)

recuperación: proceso por el cual las tierras explotadas se deben recubrir con suelo y se deben replantar con vegetación. (pág. 730)

punto de referencia: punto que se escoge para describir la ubicación, o posición, de un objeto. (pág. 9)

reforestación: proceso de siembra de árboles para reemplazar los árboles que se han cortado o quemado. (pág. 730)

refrigerador: aparato que usa energía eléctrica para bombear energía térmica desde un lugar más frío hacia uno más caliente. (pág. 184)

edad relativa: edad de las rocas y de las características geológicas comparada con otras rocas cercanas y sus características. (pág. 575)

recurso renovable: recurso natural que se reabastece por procesos naturales al menos tan rápidamente como se usa. (pág. 666)

agotamiento de un recurso: agotamiento de uno o más recursos en un área. (pág. 666)

retina: área en la parte posterior del ojo que incluye especiales sensibles a la luz—bastones y conos. (pág. 146)

polaridad inversa: ocurre cuando los objetos magnetizados invierten la dirección y se orientan a sí mismos para apuntar al sur. (pág. 506)

empuje de dorsal: proceso que resulta cuando el magma se levanta en la dorsal oceánica y empuja las placas oceánicas en dos direcciones diferentes, lejos de la dorsal. (pág. 517)

rock: a naturally occurring solid composed of minerals, rock fragments, and sometimes other materials such as organic matter. (p. 471)

rock cycle: the series of processes that change one type of rock into another type of rock. (p. 479)

roca: sólido de origen natural compuesto de minerales, acumulación de fragmentos y algunas veces de otros materiales como materia orgánica. (pág. 471)

ciclo geológico: series de procesos que cambian un tipo de roca en otro tipo de roca. (pág. 479)

S

salinity (say LIH nuh tee): a measure of the mass of dissolved salts in a mass of water. (p. 693)

saturated solution: a solution that contains the maximum amount of solute the solution can hold at a given temperature and pressure. (p. 348)

science: the investigation and exploration of natural events and of the new information that results from those investigations. (p. NOS 4)

scientific law: a rule that describes a pattern in nature. (p. NOS 9)

scientific notation: a method of writing or displaying very small or very large values in a short form. (p. NOS 15)

scientific theory: an explanation of observations or events that is based on knowledge gained from many observations and investigations. (p. NOS 9)

screw: a simple machine that consists of an inclined plane wrapped around a cylinder. (p. 106)

seafloor spreading: the process by which new oceanic crust forms along a mid-ocean ridge and older oceanic crust moves away from the ridge. (p. 504)

secondary wave (also S-wave): a type of seismic wave that causes particles to move at right angles relative to the direction the wave travels. (p. 535)

sediment: rock material that forms when rocks are broken down into smaller pieces or dissolved in water as rocks erode. (p. 473)

salinidad: medida de la masa de sales disueltas en una masa de agua. (pág. 693)

solución saturada: solución que contiene la cantidad máxima de soluto que la solución puede sostener a cierta temperatura y presión. (pág. 348)

ciencia: investigación y exploración de eventos naturales y la información nueva que resulta de dichas investigaciones. (pág. NOS 4)

ley científica: regla que describe un patrón en la naturaleza. (pág. NOS 9)

notación científica: método para escribir o expresar números muy pequeños o muy grandes en una forma corta. (pág. NOS 15)

teoría científica: explicación de las observaciones y los eventos basada en conocimiento obtenido en muchas observaciones e investigaciones. (pág. NOS 9)

tornillo: máquina simple que consiste en un plano inclinado incrustado alrededor de un cilindro. (pág. 106)

expansión del lecho marino: proceso mediante el cual se forma corteza oceánica nueva en la dorsal oceánica, y la corteza oceánica vieja se aleja de la dorsal. (pág. 504)

onda secundaria (también, onda S): tipo de onda sísmica que causa que las partículas se muevan en ángulos rectos respecto a la dirección en que la onda viaja. (pág. 535)

sedimento: material rocoso que se forma cuando las rocas se rompen en piezas pequeñas o cuando se disuelven en agua al erosionarse. (pág. 473)

seismic energy: the energy transferred by waves moving through the ground. (p. 92)

seismic wave: energy that travels as vibrations on and in Earth. (p. 534)

seismogram: a graphical illustration of seismic waves. (p. 537)

seismologist (size MAH luh just): scientist that studies earthquakes. (p. 536)

seismometer (size MAH muh ter): an instrument that measures and records ground motion and can be used to determine the distance seismic waves travel. (p. 537)

selective breeding: the selection and breeding of organisms for desired traits. (p. 796)

sex-linked trait: a trait with the allele on an X or Y chromosome. (p. 805)

shield volcano: a large volcano with gentle slopes of basaltic lavas, common along divergent plate boundaries and oceanic hotspots. (p. 548)

simple machine: a machine that does work using one movement. (p. 105)

single-replacement reaction: a type of chemical reaction in which one element replaces another element in a compound. (p. 314)

slab pull: the process that results when a dense oceanic plate sinks beneath a more buoyant plate along a subduction zone, pulling the rest of the plate that trails behind it. (p. 517)

solid: matter that has a definite shape and a definite volume. (p. 201)

solubility (sahl yuh BIH luh tee): the maximum amount of solute that can dissolve in a given amount of solvent at a given temperature and pressure. (p. 348)

solute: any substance in a solution other than the solvent. (p. 343)

solution: another name for a homogeneous mixture. (p. 337)

solvent: the substance that exists in the greatest quantity in a solution. (p. 343)

sound energy: energy carried by sound waves. (p. 92)

sound wave: a longitudinal wave that can travel only through matter. (p. 123)

energía sísmica: energía transferida por ondas que se mueven a través del suelo. (pág. 92)

onda sísmica: energía que viaja en forma de vibraciones por encima y dentro de la Tierra. (pág. 534)

sismograma: ilustración gráfica de las ondas sísmicas. (pág. 537)

sismólogo: científico que estudia los terremotos. (pág. 536)

sismómetro: instrumento que mide y registra el movimiento del suelo y que determina la distancia de las ondas sísmicas. (pág. 537)

cría selectiva: proceso de cría de organismos para características deseadas. (pág. 796)

rasgo ligado al sexo: rasgo con el alelo en un cromosoma X o Y. (pág. 805)

volcán escudo: volcán grande con ligeras pendientes de lavas basálticas, común a lo largo de los límites de placas divergentes y puntos calientes oceánicos. (pág. 548)

máquina simple: máquina que hace trabajo con un movimiento. (pág. 105)

reacción de sustitución sencilla: tipo de reacción química en la que un elemento reemplaza a otro en un compuesto. (pág. 314)

convergencia de placas: proceso que resulta cuando una placa oceánica densa se hunde debajo de una placa flotante en una zona de subducción, arrastrando el resto de la placa detrás suyo. (pág. 517)

sólido: materia con forma y volumen definidos. (pág. 201)

solubilidad: cantidad máxima de soluto que puede disolverse en una cantidad dada de solvente a temperatura y presión dadas. (pág. 348)

soluto: cualquier sustancia en una solución diferente del solvente. (pág. 343)

solución: otro nombre para una mezcla homogénea. (pág. 337)

solvente: sustancia que existe en mayor cantidad en una solución. (pág. 343)

energía sonora: energía que transportan las ondas sonoras. (pág. 92)

onda sonora: onda longitudinal que sólo viaja a través de la materia. (pág. 123)

specific heat: the amount of thermal energy it takes to increase the temperature of 1 kg of a material by 1°C. (p. 175)

calor específico: cantidad de energía térmica necesaria para aumentar la temperatura de 1 Kg de un material en 1°C. (pág. 175)

speed: the distance an object moves divided by the time it takes to move that distance. (p. 17)

rapidez: distancia que un objeto recorre dividida por el tiempo que éste tarda en recorrer dicha distancia. (pág. 17)

streak: the color of a mineral's powder. (p. 465)

raya: color del polvo de un mineral. (pág. 465)

subduction: the process that occurs when one tectonic plate moves under another tectonic plate. (p. 513)

subducción: proceso que ocurre cuando una placa tectónica se mueve debajo de otra placa tectónica. (pág. 513)

sublimation: the process of changing directly from a solid to a gas. (p. 212)

sublimación: proceso de cambiar directamente de sólido a gas. (pág. 212)

substance: matter with a composition that is always the same. (p. 336)

sustancia: materia cuya composición es siempre la misma. (pág. 336)

supercontinent: an ancient landmass which separated into present-day continents. (p. 613)

supercontinente: antigua masa de tierra que se dividió en los continentes actuales. (pág. 613)

superposition: the principle that in undisturbed rock layers, the oldest rocks are on the bottom. (p. 576)

superposición: principio que establece que en las capas de rocas inalteradas, la rocas más viejas se encuentran en la parte inferior. (pág. 576)

surface tension: the uneven forces acting on the particles on the surface of a liquid. (p. 203)

tensión superficial: fuerzas desiguales que actúan sobre las partículas en la superficie de un líquido. (pág. 203)

surface wave: a type of seismic wave that causes particles in the ground to move up and down in a rolling motion. (p. 535)

onda superficial: tipo de onda sísmica que causa un movimiento de rodamiento hacia arriba y hacia debajo de las partícula en el suelo. (pág. 535)

symbiosis (sihm bee OH sus): a close, long-term relationship between two species that usually involves an exchange of food or energy. (p. 649)

simbiosis: relación estrecha a largo plazo entre dos especies que generalmente involucra intercambio de alimento o energía. (pág. 649)

synthesis (SIHN thuh sus): a type of chemical reaction in which two or more substances combine and form one compound. (p. 313)

síntesis: tipo de reacción química en el que dos o más sustancias se combinan y forman un compuesto. (pág. 313)

T

taiga (TI guh): a forest biome consisting mostly of cone-bearing evergreen trees. (p. 689)

taiga: bioma de bosque constituido en su mayoría por coníferas perennes. (pág. 689)

technology: the practical use of scientific knowledge, especially for industrial or commercial use. (p. NOS 9)

tecnología: uso práctico del conocimiento científico, especialmente para empleo industrial o comercial. (pág. NOS 9)

temperate: the term describing any region of Earth between the tropics and the polar circles. (p. 687)

temperatura: término que describe cualquier región de la Tierra entre los trópicos y los círculos polares. (pág. 687)

temperature: the measure of the average kinetic energy of the particles in a material. (pp. 167, 208)

temperatura: medida de la energía cinética promedio de las partículas de un material. (pág. 167, 208)

terrestrial planets: Mercury, Venus, Earth, and Mars—the planets closest to the Sun that are made of rock and metallic materials and have solid outer layers. (p. 383)

planetas terrestres: Mercurio, Venus, Tierra, y Marte—los planetas que están más cercanos al Sol y que están compuestos por roca, materiales metálicos, y tienen capas externas sólidas. (pág. 383)

texture: a rock's grain size and the way the grains fit together. (p. 472)

textura: tamaño del grano de una roca y la forma como los granos encajan. (pág. 472)

thermal conductor: a material through which thermal energy flows quickly. (p. 174)

conductor térmico: material mediante el cual la energía térmica se mueve con rapidez. (pág. 174)

thermal contraction: a decrease in a material's volume when the temperature is decreased. (p. 176)

contracción térmica: disminución del volumen de un material cuando disminuye la temperatura. (pág. 176)

thermal energy: the sum of the kinetic energy and the potential energy of the particles that make up an object. (pp. 91, 166, 209)

energía térmica: suma de la energía cinética y potencial de las partículas que componen un objeto. (pág. 91, 166, 209)

thermal expansion: an increase in a material's volume when the temperature is increased. (p. 176)

expansión térmica: aumento en el volumen de un material cuando aumenta la temperatura. (pág. 176)

thermal insulator: a material through which thermal energy flows slowly. (p. 174)

aislante térmico: material en el cual la energía térmica se mueve con lentitud. (pág. 174)

thermostat: a device that regulates the temperature of a system. (p. 184)

termostato: aparato que regula la temperatura de un sistema. (pág. 184)

trace fossil: the preserved evidence of the activity of an organism. (p. 569)

fósil traza: evidencia conservada de la actividad de un organismo. (pág. 569)

transform plate boundary: the boundary between two plates that slide past each other. (p. 513)

límite de placas transcurrente: límite entre dos placas que se deslizan una con respecto a la otra. (pág. 513)

translucent: a material that allows most of the light that strikes it to pass through, but through which objects appear blurry. (p. 136)

translúcido: material que permite el paso de la mayor cantidad de luz que lo toca, pero a través del cual los objetos se ven borrosos. (pág. 136)

transparent: a material that allows almost all of the light striking it to pass through, and through which objects can be seen clearly. (p. 136)

transparente: material que permite el paso de la mayor cantidad de luz que lo toca, y a través del cual los objetos pueden verse con nitidez. (pág. 136)

tundra (TUN druh): a biome that is cold, dry, and treeless. (p. 689)

tundra: bioma frío, seco y sin árboles. (pág. 689)

U

unbalanced forces: forces acting on an object that combine and form a net force that is not zero. (p. 56)

fuerzas no balanceadas: fuerzas que actúan sobre un objeto, se combinan y forman una fuerza neta diferente de cero. (pág. 56)

unconformity: a surface where rock has eroded away, producing a break, or gap, in the rock record. (p. 578)

discontinuidad: superficie donde la roca se ha erosionado, produciendo un vacío en el registro geológico sedimentario. (pág. 578)

uniformitarianism: a principle stating that geologic processes that occur today are similar to those that occurred in the past. (p. 566)

uniformimsmo: principio que establece que los procesos geológicos que ocurren actualmente son similares a aquellos que ocurrieron en el pasado. (pág. 566)

unsaturated solution: a solution that can still dissolve more solute at a given temperature and pressure. (p. 348)

solución insaturada: solución que aún puede disolver más soluto a cierta temperatura y presión. (pág. 348)

uplift: the process that moves large bodies of Earth materials to higher elevations. (p. 480)

levantamiento: proceso por el cual se mueven grandes cuerpos de materiales de la Tierra hacia elevaciones mayores. (pág. 480)

urban sprawl: the development of land for houses and other buildings near a city. (p. 728)

expansión urbana: urbanización de tierra para viviendas y otras construcciones cerca de la ciudad. (pág. 728)

valence electron: the outermost electron of an atom that participates in chemical bonding. (p. 270)

electrón de valencia: electrón más externo de un átomo que participa en el enlace químico. (pág. 270)

vapor: the gas state of a substance that is normally a solid or a liquid at room temperature. (p. 204)

vapor: estado gaseoso de una sustancia que normalmente es sólida o líquida a temperatura ambiente. (pág. 204)

vaporization: the change in state from a liquid to a gas. (p. 211)

vaporización: cambio de estado líquido a gaseoso. (pág. 211)

variable: any factor that can have more than one value. (p. NOS 21)

variable: cualquier factor que tenga más de un valor. (pág. NOS 21)

variation (ver ee AY shun): a slight difference in an inherited trait among individual members of a species. (p. 811)

variación: ligera diferencia en un rasgo hereditario entre los miembros individuales de una especie. (pág. 811)

velocity: the speed and the direction of a moving object. (p. 23)

velocidad: rapidez y dirección de un objeto en movimiento. (pág. 23)

viscosity (vihs KAW sih tee): a measurement of a liquid's resistance to flow. (pp. 202, 549)

viscosidad: medida de la resistencia de un líquido a fluir. (pág. 202, 549)

volcanic ash: tiny particles of pulverized volcanic rock and glass. (p. 549)

ceniza volcánica: partículas diminutas de roca y vidrio volcánicos pulverizados. (pág. 549)

volcano: a vent in Earth's crust through which molten rock flows. (p. 545)

volcán: abertura en la corteza terrestre por donde fluye la roca derretida. (pág. 545)

wedge: a simple machine that consists of an inclined plane with one or two sloping sides; it is used to split or separate an object. (p. 106)

cuña: máquina simple que consiste en un plano inclinado con uno o dos lados inclinados; se usa para partir o separar un objeto. (pág. 106)

weight: the gravitational force exerted on an object. (p. 48)

peso: fuerza gravitacional ejercida sobre un objeto. (pág. 48)

wetland: an aquatic ecosystem that has a thin layer of water covering soil that is wet most of the time. (p. 696)

humedal: ecosistema acuático que tiene una capa delgada de suelo cubierto de agua que permanece húmedo la mayor parte del tiempo. (pág. 696)

wheel and axle: a simple machine that consists of an axle attached to the center of a larger wheel, so that the shaft and wheel rotate together. (p. 106)

rueda y eje: máquina simple que consiste en un eje insertado en el centro de una rueda grande, de manera que el eje y la rueda rotan juntos. (pág. 106)

work: the amount of energy used as a force moves an object over a distance. (p. 99)

trabajo: cantidad de energía usada como fuerza que mueve un objeto a cierta distancia. (pág. 99)

Italic numbers = illustration/photo **Bold numbers = vocabulary term**
lab = indicates entry is used in a lab on this page

A

B

D

E

F

I

Science Skill Handbook
Math Skill Handbook
Foldables Handbook
Reference Handbook
Glossary/Glosario
Index

P

R

Science Skill Handbook
Math Skill Handbook
Foldables Handbook
Reference Handbook
Glossary/Glosario
Index

S

V

W

X

Credits

Art Acknowledgments: The McGraw-Hill Companies, Argosy, Epigraphics, MCA+, Articulate Graphics, John E Kaufmann Graphic Arts, Mapping Specialists

Photo Credits

Cover Russell Burden/Photolibrary; **iv** Ransom Studios; **ix** (b)Fancy Photography/Veer; **NOS 02–NOS 03** Photo by John Kaplan, NASA; **NOS 04** (inset)SMC Images/Getty Images, (bkgd)Popperfoto/Getty Images; **NOS 05** (t)Maria Stenzel/Getty, (c)Stephen Alvarez/Getty Images, (b)Science Source/Photo Researchers, Inc.; **NOS 06** Martyn Chillmaid/photolibrary.com; **NOS 07** NASA; **NOS 09** (t)Andy Sacks/Getty Images, (c)NASA, H. Ford (JHU), G. Illingworth (UCSC/LO), M.Clampin (STScl), G. Hartig (STScl), the ACS Science Team, and ESA, (b)Brand X Pictures/PunchStock; **NOS 10** StockShot/Alamy; **NOS 11** Michael Newman/PhotoEdit; **NOS 12** Tim Wright/CORBIS; **NOS 13–NOS 14** Hutchings Photography/Digital Light Source; **NOS 16** (tl)Matt Meadows, (tr)ASP/YPP/age fotostock, (bl)Blair Seitz/photolibrary.com, (br)The McGraw-Hill Companies; **NOS 17** (t)The McGraw-Hill Companies, (c)photostock1/Alamy, (b)Blend Images/Alamy; **NOS 18** (tl)The McGraw-Hill Companies; **NOS 19** The McGraw-Hill Companies, (5)Carlos Casariego/Getty Images; **NOS 20** MANDEL NGAN/AFP/Getty Images; **NOS 21** Brian Stevenson/Getty Images; **NOS 22** (t)Hisham Ibrahim/Getty Images, (bl)Bordner Aerials, (other)AP Photo/NTSB; **NOS 23** AP Photo/The Minnesota Daily, Stacy Bengs; **NOS 24** LARRY DOWNING/Reuters/Landov; **NOS 25** Plus Pix/age fotostock; **NOS 26** ASSOCIATED PRESS; **NOS 28** The McGraw-Hill Companies; **NOS 29** Hutchings Photography/Digital Light Source; **NOS 31** (l)Andy Sacks/Getty Images, (r)Photo by John Kaplan, NASA; **1** ACE STOCK LIMITED/Alamy; **4** (t)Wildscape/Alamy, (b)Andy Crawford/BBC Visual Effects - modelmaker/Dorling Kindersley; **5** (t)Miriam Maslo/Science Photo Library/CORBIS, (b)The Gravity Group, LLC; **6–7** Andrew Holt/Getty Images; **8** Camille Moirenc/Getty Images; **9** Hutchings Photography/Digital Light Source; **12** Aerial Photos of New Jersey; **14** Camille Moirenc/Getty Images; **15** David J. Green - technology/Alamy; **16** Heinrich van den Berg/Getty Images; **17** (t to b)Hutchings Photography/Digital Light Source, (2)Department of Defense, (3)Car Culture/CORBIS, (4)Andersen Ross/Jupiter Images; **20** Hutchings Photography/Digital Light Source; **22** Michael Dunning/Photo Researchers; **23–24** Hutchings Photography/Digital Light Source; **25** Hutchings Photography/Digital Light Source, (5)Aaron Haupt; **26** Leo Dennis Productions/Brand X/CORBIS; **30** (t)Hutchings Photography/Digital Light Source, (b)Robert Holmes/CORBIS; **31** (t)Comstock/PunchStock, (b)Eyecon Images/Alamy Images; **32** (t)NASA, (b)Agence Zoom/Getty Images; **33** Leo Dennis Productions/Brand X/CORBIS; **34** Hutchings Photography/Digital Light Source; **36** (t)Camille Moirenc/Getty Images, (c)Steve Bloom Images/Alamy, (b)Agence Zoom/Getty Images; **39** Andrew Holt/Getty Images; **42–43** Jason Horowitz/zefa/CORBIS; **44** NASA; **45** (t)Hutchings Photography/Digital Light Source, (b)Terje Rakke/Getty Images; **46** (t)Hutchings Photography/Digital Light Source, (c)Daniel Smith/Getty Images, (b)Ryuhei Shindo/CORBIS/Jupiter Images; **48** NASA; **49** (tl)(tr)Horizons Companies, (b)Hutchings Photography/Digital Light Source; **50** Hutchings Photography/Digital Light Source; **51** (t)Terje Rakke/Getty Images, (b)Hutchings Photography/Digital Light Source; **52** (t)NASA/Goddard Space Flight Center Scientific Visualization Studio, (c)VisionsofAmerica/Joe Sohm/Getty Images, (b)Ursula Gahwiler/photolibrary.com, (bkgd)StockTrek/Getty Images; **53** Bryan Mullennix/Getty Images; **54** (t)Hutchings Photography/Digital Light Source, (b)Marvin E. Newman/Getty Images; **55–56** (l)Hutchings Photography/Digital Light Source; **57** (t)Tim Keatley/Alamy, (b)Hutchings Photography/Digital Light Source; **58** AP Photo/Insurance Institute for Highway Safety; **59** (t)Hutchings Photography/Digital Light Source, (c)Tim Keatley/Alamy, (b)AP Photo/Insurance Institute for Highway Safety; **60** (t)Macmillan/McGraw-Hill, (b)Hutchings Photography/Digital Light Source; **61** Michael Steele/Getty Images; **62** Hutchings Photography/Digital Light Source; **63** (l)Tim Garcha/CORBIS, (r)Myrleen Ferguson Cate/PhotoEdit; **64** Hutchings Photography/Digital Light Source; **67** Tim Garcha/CORBIS; **68** Hutchings Photography/Digital Light Source; **69** Patrik Giardino/CORBIS; **70** Hutchings Photography/Digital Light Source; **71** (t)LIN HUI/Xinhua/Landov, (b)Duncan Soar/Alamy; **72** (tl)David Madison/Getty Images, (r)Design Pics Inc./Alamy, (bl)U.S. Air Force photo by Carleton Bailie; **74** (tl)Richard Megna, Fundamental Photographs, NYC, (r)Hutchings Photography/Digital Light Source, (bl)Richard Megna, Fundamental Photographs, NYC; **75** (t)Hutchings Photography/Digital Light Source, (c)David Madison/Getty Images, (b)Richard Megna, Fundamental Photographs, NYC; **76** (inset) C Squared Studios/Getty Images, (other)The McGraw-Hill Companies; **78** (t to b)Hutchings Photography/Digital Light Source, (2)Marvin E. Newman/Getty Images, (3)Myrleen Ferguson Cate/PhotoEdit, (4)LIN HUI/Xinhua/Landov; **81** Jason Horowitz/zefa/CORBIS; **84–85** Jan & Rhys Hanna/Getty Images; **86** Giles Barnard/photolibrary.com; **87** (t)Hutchings Photography/Digital Light Source, (b)NASA; **88** Glen Allison/Getty Images; **89** Bettmann/CORBIS; **90** (l)Photodisc/Getty Images, (cl)San Juan County, New Mexico, USA, (cr)RIA Novosti/Photo Researchers, Inc., (r)euroluftbildde euroluftbildde/photolibrary.com; **91** (t)Jim West/Alamy, (b)ARCTIC IMAGES/Alamy; **92** (t)Carol Farneti Foster/OSF/photolibrary.com, (b)Roger Ressmeyer/CORBIS; **94** (t)NASA, (c)Glen Allison/Getty Images; **95** (tr)Dieter Möbus/age fotostock, (bkgd)Getty Images; **96** REUTERS/Kimimasa Mayama/Landov; **97** (t)Hutchings Photography/Digital Light Source, (b)AP Photo/Douglas C. Pizac; **98** (t)Shenval/Alamy Images, (b)RIA Novosti/Photo Researchers, Inc., (b)Tony Hertz/age fotostock; **101** Glowimages/Getty Images; **102** (tl)AP Photo/Douglas C. Pizac, (r)CMB/age fotostock, (bl)REUTERS/Kimimasa Mayama/Landov; **103** The McGraw-Hill Companies; **104** Joe McBride/Getty Images; **105** (t)Hutchings Photography/Digital Light Source, (b)Lew Robertson/Getty Images; **106** (l)Drive Images/Alamy, (c)Marc Volk/photolibrary.com, (r)VStock LLC/age fotostock; **107** (tl)Medioimages/Photodisc/Getty Images, (tc)Polka Dot Images/photolibrary.com, (tr)Dorling Kindersley, (b)Brand X Pictures; **108** Hutchings Photography/Digital Light Source; **111** (tl)Lew Robertson/Getty Images, (cl)Drive Images/Alamy, (r)Eric Fowke/PhotoEdit, (bl)Brand X Pictures; **112** The McGraw-Hill Companies; **113** Hutchings Photography/Digital Light Source; **114** (t)Carol Farneti Foster/OSF/photolibrary.com, (c)Glowimages/Getty Images, (b)VStock LLC/age fotostock; **116** CMB/age fotostock; **117** (l)Thinkstock/Getty Images, (r)Jan & Rhys Hanna/Getty Images; **120–121** Art on File/CORBIS; **122** (t)Hutchings Photography/Digital Light Source, (b)Lauren Zeid/eStock Photo, (c)Stephen Dalton/Minden Pictures; **123** BABU/Reuters/Landov; **125** Hutchings Photography/Digital Light Source; **126** SPL/Photo Researchers, Inc.; **127** CNRI/Photo Researchers, Inc.; **128** (l)Getty Images/SW Productions, (cl)Reed Kaestner/CORBIS, (cr)Creatas/Jupiter Images, (r)U.S. Air Force photo/Senior Airman Julianne Showalter; **129** (t)NURC-UConn and SBNMS/NOAA, (b)Owen Franken/CORBIS; **130** CNRI/Photo Researchers, Inc.; **131** UHB Trust/Getty Images; **132** Mark Thompson/Getty Images for Reebok; **133** Hutchings Photography/Digital Light Source; **134** (tl)Johnny Greig Cut-Outs/Alamy, (tr)Gregor Schuster/Getty Images, (cl)TOM MARESCHAL/Alamy, (bl)Jeff Curtes/CORBIS, (br)Untitled X-Ray/Nick Veasey/Getty Images; **135** (t)Kelly Redinger/Design Pics/CORBIS, (b)Blend Images/Alamy; **136** (t)Michael & Patricia Fogden/Minden Pictures/Getty Images, (c)Jasper James/Getty Images, (b)Brian Atkinson/Alamy; **137** (t)Hutchings Photography/Digital Light Source, (b)Clive Streeter/Dorling Kindersley; **138** (t)Blend Images/Alamy, (c)Untitled X-Ray/Nick Veasey/Getty Images, (b)Michael & Patricia Fogden/Minden Pictures/Getty Images; **139** Hutchings Photography/Digital Light Source; **140** altrendo nature/Getty Images; **141** Hutchings Photography/Digital Light Source; **142** (t)(c)Richard Megna, Fundamental Photographs, NYC, (b)AP Photo/Eric Gay; **143** (l to r, t to b)Don Farrall,

Credits

(2)Derrick Alderman/Alamy, (3)Dennis Hallinan/Alamy, (4)Don Farrall/Getty Images, (5)JEROME WEXLER/Photo Researchers, Inc.; **146** (t)Eye of Science/Photo Researchers, Inc., (b)Steve Allen/Brand X/CORBIS; **147** (t)Image Plan/CORBIS, (b)Acme Food Arts/Getty Images; **148** (t)David. Parker/Photo Researchers, Inc., (other)Hutchings Photography/Digital Light Source; **149** (t)Richard Megna, Fundamental Photographs, NYC, (b)Dennis Hallinan/Alamy; **150–151** Hutchings Photography/Digital Light Source; **152** (t)BABU/Reuters/Landov, (c)Clive Streeter/Dorling Kindersley, (b)JEROME WEXLER/Photo Researchers, Inc.; **155** (l)AP Photo/Eric Gay, (r)Art on File/CORBIS; **162–163** Tyrone Turner/National Geographic Stock; **164** Philip Scalia/Alamy; **165** (t)Hutchings Photography/Digital Light Source, (b)Jamie Sabau/Getty Images; **167** Johner/Getty Images; **169** (t)PARIS PIERCE/Alamy, (c)University Library, Leipzig, Germany/Archives Charmet/The Bridgeman Art Library, (b)British Library, London, UK/© British Library Board. All Rights Reserved/The Bridgeman Art Library; **171** Hutchings Photography/Digital Light Source; **172** Alaska Stock LLC/Alamy; **173** (t)Hutchings Photography/Digital Light Source, (b)The McGraw-HIll Companies; **174** Anthony-Masterson/Getty Images; **176** (t)(c)Matt Meadows, (b)The McGraw-Hill Companies; **177** (t)Royalty-Free/CORBIS, (b)Hutchings Photography/Digital Light Source; **178 180** Matt Meadows; **181** (tl)Rachael Bowes/Alamy, (tr)CHRISTOPHER WEDDLE/UPI/Landov, (b)Paul Glendell/Alamy, (bkgd)David Papazian/Beateworks/CORBIS; **182** Kevin Foy/Alamy; **184** (inset)Thomas Northcut/Getty Images, (bkgd)steven langerman/Alamy; **185** Hutchings Photography/Digital Light Source; **187** (l)Steven Langerman/Alamy, (r)Thomas Northcut/Getty Images; **188** The McGraw-Hill Companies; **189** Hutchings Photography/Digital Light Source; **190** (t)Summer Jones/Alamy, (b) Royalty-Free/CORBIS; **192** (l)Anthony-Masterson/Getty Images, (r)Thomas Northcut/Getty Images; **193** (l)Tom Uhlman/Alamy, (r)Tyrone Turner/National Geographic Stock; **196–197** Gregor M. Schmid/CORBIS; **198** Atlantide Phototravel/CORBIS; **199** Hutchings Photography/Digital Light Source; **201** (t)Royalty Free/CORBIS, (c)Steve Hamblin/Alamy, (b)The McGraw-Hill Companies; **202** (t)Vito Palmisano/Getty Images, (b)creativ collection/age fotostock; **203** (l)Mauritius/SuperStock, (r)Hutchings Photography/Digital Light Source; **204** Alberto Coto/Getty Images; **206** (t)Finley - StockFood Munich, (b)The McGraw-Hill Companies, (bkgd)Michael Rosenfeld/Getty Images; **207** Alaska Stock/age fotostock; **208 211** Hutchings Photography/Digital Light Source; **212** (l)Charles D. Winters/Photo Researchers, Inc., (r)Jean du Boisberranger/Getty Images; **214** Hutchings Photography/Digital Light Source; **215** (t)Alaska Stock/age fotostock, (c)Hutchings Photography/Digital Light Source, (bl)Jean du Boisberranger/Getty Images, (br)Charles D. Winters/Photo Researchers, Inc.; **216** Hutchings Photography/Digital Light Source; **217** Check Six/Getty Images; **218 221 224 225** Hutchings Photography/Digital Light Source; **229** Gregor M. Schmid/CORBIS; **232** (inset)CERN PHOTO/Frédéric Pitchal/Sygma/CORBIS; **232–233** (bkgd)JAMES BRITTAIN/Photolibrary; **234** (inset)Drs. Ali Yazdani & Daniel J. Hornbaker/Photo Researchers, Inc, (bkgd)Alessandro Della Bella/Keystone/CORBIS; **235** (l to r, t to b)CORBIS, (t)Hutchings Photography/Digital Light Source, (2)Creatas/PunchStock, (3)CORBIS, (4)DAJ/Getty Images; **236** (l)Scala/Art Resource, NY, (r)The Royal Institution, London, UK/The Bridgeman Art Library; **237** (t)Horizons Companies, (b)Drs. Ali Yazdani & Daniel J. Hornbaker/Photo Researchers, Inc.; **242** Hutchings Photography/Digital Light Source; **245** (t)Horizons Companies, (b)Drs. Ali Yazdani & Daniel J. Hornbaker/Photo Researchers, Inc.; **246** Royalty-Free/CORBIS; **247** (inset)(bkgd)Derrick Alderman/Alamy; **248** Hutchings Photography/Digital Light Source; **251** The McGraw-Hill Companies; **252** (t)SPL/Photo Researchers, Inc., (b)Time Life Pictures/Mansell/Time Life Pictures/Getty Images, (inset)figure 13 John Cancalosi/age fotostock; **256** The McGraw-Hill Companies; **257** Hutchings Photography/Digital Light Source; **261** JAMES BRITTAIN/Photolibrary; **264–265** altrendo images/Getty Images; **266** Douglas Fisher/Alamy; **267 272** Hutchings Photograpy/Digital Light Source; **274** (t)Popperfoto/Getty Images, (c)Underwood & Underwood/CORBIS, (bl)John Meyer, (br)Ilene MacDonald/Alamy; **275** Gazimal/Getty Images; **276 280** Hutchings Photography/Digital Light Source; **282** (l)Macmillan/McGraw-Hill, (r)Hutchings Photography/Digital Light Source; **283** Brent Winebrenner/Photolibrary.com; **284** Hutchings Photography/Digital Light Source; **288** (t)Photodisc/Getty Images, (c)C Squared Studios/Getty Images, (b)Jennifer Martine/Jupiter Images; **290** The McGraw-Hill Companies; **291** Hutchings Photography/Digital Light Source; **295** altrendo images/Getty Images; **298–299** (bkgd)Anton Luhr/Photolibrary; **300** Darwin Dale/Photo Researchers, Inc.C5454; **301** Hutchings Photography/Digital Light Source; **302** (tl)CORBIS, (tr)Charles D. Winters/Photo Researchers, Inc., (cl)London Scientific Films/photolibrary.com, (cr)sciencephotos/Alamy, (bl)Brand X Pictures, (br)Dante Fenolio/Photo Researchers, Inc.; **305–306** Hutchings Photography/Digital Light Source; **310** The McGraw-Hill Companies; **311** sciencephotos/Alamy; **312** AP Photo/Greg Campbell; **313** (t)Charles D. Winters/Photo Researchers, Inc., (b)The McGraw-Hill Companies, Inc./Jacques Cornell photographer; **314** (t)The McGraw-Hill Companies, Inc./Stephen Frisch, photographer, (c)sciencephotos/Alamy, (b)Park Dale/Alamy; **315** The McGraw-Hill Companies, Inc./Stephen Frisch, photographer; **316** Joel Sartore/National Geographic/Getty Images; **317** Brand X Pictures/PunchStock; **321** (tl)McGraw-Hill Companies, (tc)Brand X Pictures/PunchStock, (tr)Tetra Images/Getty Images, (bl)The McGraw-Hill Companies, (bc)CORBIS, (br)Alexis Grattier/Getty Images; **323** Hutchings Photography/Digital Light Source; **324** The McGraw-Hill Companies; **325** Hutchings Photography/Digital Light Source; **326** Park Dale/Alamy; **329** Anton Luhr/Photolibrary; **332–333** Keith Kapple/SuperStock; **334** Visual&Written SL/Alamy; **335** Hutchings Photography/Digital Light Source; **336** (tl)PHOTOTAKE Inc./Alamy, (tr)CORBIS, (cl)Reuters/Landov, (cr)Flirt/SuperStock, (bl)Don Farrall/Getty Images, (br)Royalty-Free/CORBIS; **337** (tl)liquidlibrary/PictureQuest, (tcl)Mark Steinmetz, (tcr)Michael Maes/Getty Images, (tr)C Squared Studios/Getty Images, (bl)Dennis Kunkel Microscopy, Inc., (br)Mark Steinmetz; **338** Hutchings Photography/Digital Light Source; **340** (tl)CORBIS, (tr)C Squared Studios/Getty Images, (b)Hutchings Photography/Digital Light Source; **341** (br)Stockdisc/PunchStock, Getty Images/Photodisc; **342** Chris Cheadle/Getty Images; **343** Hutchings Photography/Digital Light Source; **344** (t)Ingram Publishing/Alamy, (c)Michael Maes/Getty Images, (b)SuperStock/age fotostock; **345** The McGraw-Hill Companies; **346–347** Hutchings Photography/Digital Light Source; **349** (t to b)Stock Connection Distribution/Alamy, (other)Hutchings Photography/Digital Light Source; **350** Hutchings Photography/Digital Light Source; **351** (other)Hutchings Photography/Digital Light Source, (4)Macmillan/McGraw-Hill; **352** Fletcher & Baylis/Photo Researchers, Inc.; **353** (t)Hutchings Photography/Digital Light Source, (b)Horizons Companies; **355** (cw from top)Charles D. Winters/SPL/Photo Researchers, (2)Steve Mason/Getty Images, (3)Yasuhide Fumoto/Getty Images, (4)Phil Degginger/Alamy, (5)Westend61/SuperStock, (6)F. Schussler/PhotoLink/Getty Images, (7)Richard Megna, Fundamental Photographs, NYC; **356** (t)CORBIS, (bl)Busse Yankushev/age fotostock, (br)Spencer Jones/Getty Images; **357** (tl)liquidlibrary/PictureQuest, (tr)CORBIS, (bl)The McGraw-Hill Companies, Inc./Jacques Cornell photographer, (br)Robert Manella for MMH; **358** L. S. Stepanowicz/Visuals Unlimited; **360** (t to b)Hutchings Photography/Digital Light Source, (4)(5)Macmillan/McGraw-Hill; **361** Hutchings Photography/Digital Light Source; **362** The McGraw-Hill Companies; **365** Keith Kapple/SuperStock; **370** Jason Reed/Photodisc/Getty Images; **371** Edwin Stranner/Photolibrary; **372–373** NASA/JPL/Space Science Institute; **374** UVimages/amanaimages/Corbis; **375** (t)Hutchings Photography/Digital Light Source, (b)Diego Barucco/Alamy; **376** NASA/JPL; **379** Hutchings Photography/Digital Light Source; **380** UVimages/amanaimages/Corbis; **381** (t)Ambient Images Inc./Alamy, (cl)American Museum of Natural History, (bkgd)NASA and H. Richer (University of British Columbia); **382** ESA/DLR/FU Berlin (G. Neukum); **383** (t)Hutchings Photography/Digital

Credits

Light Source, (bl)NASA/Johns Hopkins University Applied Physics Laboratory/Carnegie Institution of Washington, (bcl)NASA, (bcr)NASA Goddard Space Flight Center, (br)NASA/JPL/Malin Space Science Systems; **384** (l)(r)NASA/Johns Hopkins University Applied Physics Laboratory/Carnegie Institution of Washington; **385** NASA; **386** (tl)NASA, (c)NASA Goddard Space Flight Center, (r)Image Ideas/PictureQuest, (bl)Comstock/JupiterImages; **387** (tl)NASA/JPL, (tc)NASA/JPL/Malin Space Science Systems, (tr)ESA/DLR/FU Berlin (G. Neukum), (b)NASA/JPL/University of Arizona; **388** (tl)NASA/JPL/Malin Space Science Systems, (cl)Comstock/Jupiter Images, (bl)(bcr)NASA, (bcl)NASA/Johns Hopkins University Applied Physics Laboratory/Carnegie Institution of Washington, (br)NASA/JPL/Malin Space Science Systems; **391** (l)NASA/JPL/USGS, (cl)NASA and The Hubble Heritage Team (STScI/AURA)Acknowledgment: R.G. French (Wellesley College), J. Cuzzi (NASA/Ames), L. Do, (c)NASA Goddard Space Flight Center, (cr)(r)NASA/JPL; **392** (l)NASA/JPL/USGS, (r)NASA/JPL; **393** NASA and The Hubble Heritage Team (STScI/AURA) Acknowledgment: R.G. French (Wellesley College), J. Cuzzi (NASA/Ames), L. Dones **394** (bl)NASA/ESA and Erich Karkoschka, University of Arizona, (other)NASA/JPL/Space Science Institute; **395** NASA/JPL; **396** (t)NASA/JPL, (c)NASA/JPL/USGS, (b)NASA/ESA and Erich Karkoschka, University of Arizona; **397** Frederick M. Brown/Getty Images; **398** Gordon Garradd/SPL/Photo Researchers, Inc.; **399** (t)Hutchings Photography/Digital Light Source; **400** (l)Dr. R. Albrecht, ESA/ESO Space Telescope European Coordinating Facility; NASA, (tr)NASA, ESA, and J. Parker (Southwest Research Institute), (br)NASA, ESA, and M. Brown (California Institute of Technology); **401** (l to r, t to b)NASA/JPL/JHUAPL, (2)NASA/Goddard Space Flight Center Scientific Visualization Studio, (3)NASA/JPL/USGS, (4)Ben Zellner (Georgia Southern University), Peter Thomas (Cornell University), NASA/ESA, (5)Roger Ressmeyer, (6)NASA/JPL-Caltech; **402** (t)Jonathan Blair/CORBIS, (b)Hutchings Photography/Digital Light Source; **403** (tl)NASA/JPL/USGS, (cl)Gordon Garradd/SPL/Photo Researchers, Inc., (bl)Jonathan Blair/CORBIS, (br)Roger Ressmeyer; **404** (t to b)Hutchings Photography/Digital Light Source, (4)Macmillan/McGraw-Hill; **405** Hutchings Photography/Digital Light Source; **406** (t to b)UVimages/amanaimages/Corbis, (2)NASA Goddard Space Flight Center, (3)NASA/JPL/USGS, (4)Roger Ressmeyer; **408** NASA Goddard Space Flight Center; **409** NASA/JPL/Space Science Institute; **412–413** NASA, ESA, and S. Beckwith (STScI) and the HUDF Team; **414** Stephen & Donna O'Meara/Photo Researchers, Inc.; **415** (t)Hutchings Photography/Digital Light Source, (b)Joseph Baylor Roberts/Getty Images; **417** (tl)NRAO/AUI/NSF/SCIENCE PHOTO LIBRARY, (tr)NASA/CXC/MIT/H. Marshall et al, (cl)NASA/JPL-Caltech/E. Churchwell (University of Wisconsin), (cr)NASA/JPL-Caltech/Univ. of Virginia, (b)Hutchings Photography/Digital Light Source; **421** (t to b)Macmillan/McGraw-Hill, (2)Brand X Pictures/PunchStock, (3)Aaron Haupt, (4)Hutchings Photography/Digital Light Source; **422** Science Source/Photo Researchers, Inc; **423** Digital Vision/PunchStock; **424** Hutchings Photography/Digital Light Source; **425** (t to b)Jerry Lodriguss/Photo Researchers, Inc., (2)Naval Research Laboratory, (3)SOHO Consortium, ESA, NASA, (4)Arctic-Images/Getty Images; **426** (t)(b)Photo courtesy of NASA/CORBIS; **428** (t)Jerry Lodriguss/Photo Researchers, Inc., (b)Photo courtesy of NASA/CORBIS; **429** STEREO Stereoscopic Observations Constraining the Initiation of Polar Coronal JetsS. Patsourakos, E. Pariat, A. Vourlidas, S. K. Antiochos, J. P. Wuesler/NASA; **430** NASA; **431 434** Hutchings Photography/Digital Light Source; **435** (t)X-ray: NASA/CXC/SAO; Optical: NASA/STScI, (b)NASA, The Hubble Heritage Team (STScI/AURA), Y.-H. Chu (UIUC), S. Kulkarni (Caltech) and R. Rothschild (UCSD); **436** (l)NASA, The Hubble Heritage Team (STScI/AURA), Y.-H. Chu (UIUC), S. Kulkarni (Caltech) and R. Rothschild (UCSD), (r)NASA; **437** Aaron Haupt; **438** NASA/Hubble Heritage Team; **439** NASA/Alamy; **440** (t)CORBIS, (c)Robert Gendler/NASA, (b)NASA, ESA, and The Hubble Heritage Team (STScI/AURA); **441** STS-82 Crew/STScI/NASA; **445** NASA/Alamy; **446** (t to b)Hutchings Photography/Digital Light Source, (2)(3)(5)(6)The McGraw-Hill Companies; **447** (t)Hutchings Photography/Digital Light Source, (b)NASA/JPL-Caltech, (inset)Brand X Pictures/PunchStock, (inset)NASA/JPL-Caltech/S. Willner (Harvard-Smithsonian Center for Astrophysics); **448** X-ray: NASA/CXC/SAO; Optical: NASA/STScI; **451** NASA, ESA, and S. Beckwith (STScI) and the HUDF Team; **456** (cw from top)Robert Harding/photolibrary.com, (2)Christine Strover/Alamy, (3)Photolibrary/age fotostock, (4)Neal & Molly Jansen/age fotostock, (5)maurizio grimaldi/age fotostock, (6)Royalty-Free/CORBIS; **457** (tl)Jeff Vinnick/Getty Images, (tr)Jeff Greenberg/age fotostock, (c)Tom Grill/age fotostock, (b)Neale Clark/Robert Harding World Imagery/Getty Images; **458–459** Ric Ergenbright/CORBIS; **460** Evan Collis/CORBIS; **461** (t)Hutchings Photography/Digital Light Source, (bl)Andrew Silver/U.S. Geological Survey, (br)Mark Steinmetz; **462** (l)Ryan McVay/Getty Images, (r)Jeffrey Hamilton/Getty Images; **463** (t)PROF. H. HASHIMOTO, OSAKA UNIVERSITY/SCIENCE PHOTO LIBRARY, (c)Mark A. Schneider/Photo Researchers, Inc., (b)Hutchings Photography/Digital Light Source; **464** (l to r, t to b)Smithsonian Institution/CORBIS, (2)(3)Andrew Silver/U.S. Geological Survey, (4)Doug Sherman/Geofile, (5)(6)(7)José Manuel Sanchis Calvete/CORBIS, (8)Harry Taylor/Getty Images, (9)Mark Schneider/Getty Images, (10)Dr. Parvinder Sethi; **465** (t)(c)Mark Steinmetz, (b)Phil Degginger/Alamy; **466** (tl)Dorling Kindersley/Getty Images, (cl)Daniel Sambraus/Photo Researchers, Inc., (bc)Mark Steinmetz, (br)Visuals Unlimited/CORBIS; **467** (l)CORBIS, (tr)(cr)Getty Images, (b)Luis Veiga/Getty Images; **468** (t)Dr. Parvinder Sethi, (c)Mark Steinmetz, (b)Getty Images; **469** (inset)Mark Steinmetz, (bkgd)CARSTEN PETER/SPELEORESEARCH & FILMS/National Geographic Stock; **470** Peter Essick/National Geographic Stock; **471** (t)Hutchings Photography/Digital Light Source, (c)(b)Mark Steinmetz; **472** (l)The McGraw-Hill Companies Inc./Ken Cavanagh Photographer, (r)Mark Schneider/Getty Images; **473** Mark Steinmetz; **474** (l to r, t to b)Macmillan/McGraw-Hill, (2)Brent Turner/BLT Productions, (5)Hutchings Photography/Digital Light Source; **475** (t)T.Lehne/Lotuseaters/Alamy, (c)R. Strange/PhotoLink/Getty Images, (b)David Williams/Alamy; **476** (t)Brent Turner/BLT Productions, (cl)(bl)Mark Steinmetz, (bc)(br) Macmillan/McGraw-Hill; **477** (r)Hutchings Photography/Digital Light Source, (9)Harry Taylor/Getty Images, (10)Hutchings Photography/Digital Light Source, (11)The McGraw-Hill Companies; **478** Michael T. Sedam/CORBIS; **479** (t)Hutchings Photography/Digital Light Source, (b)Masterfile; **481** Hutchings Photography/Digital Light Source; **482** (t to b)Mark Steinmetz, (2)studiomode/Alamy, (3)Joel Arem/Photo Researchers, Inc., (4)Visuals Unlimited/CORBIS, (5)Mark Steinmetz; **483** (t)Michael T. Sedam/CORBIS, (b)Masterfile; **484** (t to b)(2)The McGraw-Hill Companies, (3)(5)Hutchings Photography/Digital Light Source; **485** Hutchings Photography/Digital Light Source; **486** (l)Mark Steinmetz, (t)Mark Schneider/Getty Images; **488** (l)Daniel Sambraus/Photo Researchers, Inc., (r)CORBIS, Stephen Reynolds; **489** Ric Ergenbright/CORBIS; **492–493** Arctic-Images/Getty Images; **494** Oddur Sigurdsson/Visuals Unlimited, Inc.; **495** Hutchings Photography/Digital Light Source; **497** Walter Geiersperger/CORBIS; **498** Tim Fitzharris/Minden Pictures; **499** Hutchings Photography/Digital Light Source; **501** (l)Peter Johnson/CORBIS, (r)Clare Flemming; **502** Science Source/Photo Researchers; **503** Hutchings Photography/Digital Light Source; **504 508** Image courtesy of Submarine Ring of Fire 2002 Exploration, NOAA-OE.; **509** (t to b)(2)(4)Hutchings Photography/Digital Light Source, (r)Dr. Peter Sloss, formerly of NGDC/NOAA/NGDC, (3)Macmillan/McGraw-Hill; **510** NASA; **511** Hutchings Photography/Digital Light Source; **514** (t to b)Dr. Ken MacDonald/Photo Researchers, Inc., (2)Lloyd Cluff/CORBIS, (3)Jim Richardson/CORBIS, (4)Tony Waltham/Getty Images; **516 519** Richard Megna/Fundamental Photographs; **520** (2)(4)The McGraw-Hill Companies, (other)Hutchings Photography/Digital Light Source; **528–529** Alberto Garcia/CORBIS; **530** Roger Ressmeyer/CORBIS; **531** (t)Hutchings Photography/Digital Light Source, (b)William S Helsel/Getty Images; **533** Tom Bean/CORBIS; **538** Hutchings Photography/Digital Light Source; **539** (t)Grant Smith/CORBIS, (c)Roger Ressmeyer/CORBIS, (b)Photodisc/Alamy; **542** (t)U.S. Geological Survey, (b)Hutchings

Credits

Photography/Digital Light Source; **543** U.S. Geological Survey; **544** Douglas Peebles/CORBIS; **545** Hutchings Photography/Digital Light Source; **546** (inset)National Oceanic and Atmospheric Administration (NOAA); **547** CORBIS Premium RF/Alamy; **548** (tl)Bernd Mellmann/Alamy, (tr)Robert Glusic/Getty Images, (bl)Michael S. Yamashita/CORBIS, (br)National Geographic/Getty Images; **549** (t)Digital Vision/Getty Images, (b)David Weintraub/Photo Researchers, Inc.; **550** (t)Tony Lilley/Alamy, (b)Hutchings Photography/Digital Light Source; **551** (t)Art Wolfe/Getty Images, (b)Image courtesy of Alaska Volcano Observatory/U.S. Geological Survey; **552** (t)Courtesy Chris G. Newhall/U.S. Geological Survey, (b)Science Source/Photo Researchers; **553** (t)Digital Vision/Getty Images, (c)Bernd Mellmann/Alamy, (bl)Robert Glusic/Getty Images, (br)Courtesy Chris G. Newhall/U.S. Geological Survey; **554** (t to b)Macmillan/McGraw-Hill, (2)Macmillan/McGraw-Hill, (3)Hutchings Photography/Digital Light Source, (4)Karl Weatherly/CORBIS; **556** Robert Glusic/Getty Images; **558** David Weintraub/Photo Researchers, Inc.; **559** Alberto Garcia/CORBIS; **562–563** CORBIS/SuperStock; **564** Howard Grey/Getty Images; **565** Hutchings Photography/Digital Light Source; **566** (t)Walt Anderson/Visuals Unlimited, Inc., (b)Hutchings Photography/Digital Light Source; **567** Wim van Egmond/Visuals Unlimited, Inc.; **568** (t)Staffan Widstrand/CORBIS, (c)David Lyons/Alamy, (b)Scientifica/Getty Images; **569** (t)Dick Roberts/Visuals Unlimited, (c)Dick Roberts/Visuals Unlimited, (b)age fotostock/SuperStock; **570** (tl)John Cancalosi/Alamy, (tr)David Troy/Alamy, (b)Chase Studio Inc.; **571** (t)Jim Zuckerman/CORBIS, (c)JTB Photo Communications, Inc./Alamy, (b)Jonathan Blair/CORBIS; **572** (t)Walt Anderson/Visuals Unlimited, Inc., (c)Scientifica/Getty Images, (b)Chase Studio Inc.; **573** (t)(cl)American Museum of Natural History, (b)Ted Mead/Photolibrary/Getty Images; **574** Tom Bean; **575** (t)Hutchings Photography/Digital Light Source, (b)Hemis/CORBIS; **577** Hutchings Photography/Digital Light Source; **578** (t)Ashley Cooper/Alamy, (c)Stephen Reynolds, (b)Marli Miller/Visuals Unlimited, Inc.; **580** Ashley Cooper/Alamy; **581** (t to b)Hutchings Photography/Digital Light Source, (2)(4)Macmillan/McGraw-Hill; **582** Richard T. Nowitz/Photo Researchers, Inc.; **583** (t)Hutchings Photography/Digital Light Source, (b)Stockbyte/PunchStock; **586** Hutchings Photography/Digital Light Source; **590** (l to r, t to b)DK Limited/CORBIS, (2)Andrew Ward/Life File/Getty Images, (3)John Cancalosi/Alamy, (4)(5)Kjell B. Sandved/Visuals Unlimited, (6)Andrew Ward/Life File/Getty Images, (7)(9)(13)(14)DK Limited/CORBIS, (8)(10)The Natural History Museum/Alamy, (11)Andrew Ward/Life File/Getty Images, (12)Kjell B. Sandved/Visuals Unlimited, (15)Tom Bean/CORBIS; **591** (cw from top)John Cancalosi/Alamy, (2)Andrew Ward/Life File/Getty Images, (3)Kjell B. Sandved/Visuals Unlimited, (4)(6)DK Limited/CORBIS, (5)Tom Bean/CORBIS, (7)The Natural History Museum/Alamy; **592** (t)David Lyons/Alamy, (b)Hemis/CORBIS; **595** CORBIS/SuperStock; **598–599** Kevin Schafer/CORBIS; **600** Francois Gohier/Photo Researchers, Inc.; **601** Hutchings Photography/Digital Light Source; **602** (tl)Andy Crawford/Getty Images, (tr)(bl)DK Limited/CORBIS; **603** Jonathan Blair/CORBIS; **604** (t)Hutchings Photography/Digital Light Source, (bl)Robert Clay/Alamy, (bc)Panoramic Images/Getty Images, (br)Tom Bean/CORBIS; **605** 2007 Photograph of Ediacara Biota diorama at Smithsonian Institute/Joshua Sherurcij; **606** (t)Andy Crawford/Getty Images, (b)Panoramic Images/Getty Images; **607** (tl)Andrew Ward/Life File/Getty Images, (tc)DK Limited/CORBIS, (tr)Tom Bean/CORBIS, (bl)John Cancalosi/Alamy; **608** Mark Steinmetz; **609** Hutchings Photography/Digital Light Source; **611** (b)DEA PICTURE LIBRARY/Getty Images; **612** The Field Museum, GEO85637c; **613** Hutchings Photography/Digital Light Source; **614** The Field Museum, GEO85637c; **615** Photodisc/Getty Images; **616** DEA PICTURE LIBRARY/Getty Images; **617 620** Hutchings Photography/Digital Light Source; **621** (t)Naturfoto Honal/CORBIS, (b)Nigel Reed QEDimages/Alamy; **623** (tl)Jin Meng, (tr)American Museum of Natural History, (bl)AMNH/Denis Finnin, (br)American Museum of Natural History; **624** Nik Wheeler/CORBIS; **625** Hutchings Photography/Digital Light Source; **627** Mark Steinmetz; **628** Sinclair Stammers/Photo Researchers, Inc.; **630** (t)Ariadne Van Zandbergen/Lonely Planet Images, Inc., (b)Getty Images; **631** Ariadne Van Zandbergen/Lonely Planet Images, Inc.; **632** (t to b)Hutchings Photography/Digital Light Source, (2)(4)Macmillan/McGraw-Hill, (3)(5)(6)(7)Hutchings Photography/Digital Light Source; **636** Tom Bean/CORBIS; **637** Kevin Schafer/CORBIS; **642** (t)Wayne R Bilenduke/Getty Images; **644–645** Christian Arnal/Photolibrary; **646** Ines Roberts/Flowerphotos/Photoshot; **647** Hutchings Photography/Digital Light Source; **649** (t)DWSPL/Visuals Unlimited, (c)MedioImages/SuperStock; **649** (b)Biosphoto/Nicolet Gilles/Peter Arnold, Inc.; **650** Tom Ulrich/Visuals Unlimited; **652** Tom Ulrich/Visuals Unlimited; **653** (t)Joyce Dopkeen/The New York Times/Redux, (b)A. Kudryavtsev, (bkgd)Diane Shapiro/Peter Arnold, Inc.; **654** EMORY KRISTOF/National Geographic Stock; **655** Hutchings Photography/Digital Light Source; **656** (tl)WILDLIFE/Peter Arnold, Inc., (tr)Jeffrey Lepore/Photo Researchers, (bl)Bryan Mullennix/Getty Images, (br)Biosphoto/Cavignaux Régis/Peter Arnold, Inc; **658** Hutchings Photography/Digital Light Source; **662** Biosphoto/Cavignaux Régis/Peter Arnold, Inc.; **664** Bradley Kanaris/Getty Images; **665** (t)Hutchings Photography/Digital Light Source, (b)John Foster/Masterfile; **668** (l)John Lee/Getty Images, (c)Raymond Forbes/Masterfile, (r)Andrew Vaughan/AP Images; **669** Inga Spence/Visuals Unlimited; **670** Kyodo/AP Images; **671** (t)John Foster/Masterfile, (c)John Lee/Getty Images, (b)Andrew Vaughan/AP Images; **672** (t)Aaron Haupt, (b)The McGraw-Hill Companies; **673** Hutchings Photography/Digital Light Source; **674** Inga Spence/Visuals Unlimited; **677** Christian Arnal/Photolibrary; **680–681** Roger Ressmeyer/CORBIS; **682** Frans Lanting/CORBIS; **684** (t)Tom Vezo/Minden Pictures/Getty Images, (c)Hutchings Photography/Digital Light Source, (b)David Muench; **685** (t)Laura Romin & Larry Dalton/Alamy, (tc)Jim Brandenburg/Minden Pictures, (b)Chuck Haney/DanitaDelimont.com, (bc)Comstock/PunchStock; **686** (tl)age fotostock/Photolibrary, (tr)Gavriel Jecan/Getty Images, (bl)Jacques Jangoux/Mira.com, (br)Frans Lanting/CORBIS; **687** (t)M DeFreitas/Getty Images, (bl)Doug Sherman/Geofile, (br)Comstock Images/Alamy; **688** (t)Tom Till, (bl)Bruce Lichtenberger/Peter Arnold, Inc., (br)Russ Munn/CORBIS; **689** (t)age fotostock/SuperStock, (tc)Marvin Dembinsky Photo Associates/Alamy, (b)Andre Gallant/Getty Images, (bc)age fotostock/SuperStock; **690** (t)Frans Lanting/CORBIS, (b)Tom Till; **691** Andre Gallant/Getty Images; **692** Woodfall/Photoshot; **693** (tr)Hutchings Photography/Digital Light Source; **694** (tl)Comstock/PunchStock, (tr)Arthur Morris/Visuals Unlimited/Getty Images, (cl)Paul Nicklen/NGS/Getty Images, (cr)Stephen Dalton/Minden Pictures, (b)Medioimages/PunchStock, (bcl)Dr. Marli Mill/Visuals Unlimited/Getty Images; **695** (t)Jean-Paul Ferrero/Minden Picture, (tc)Lynn & Donna Rogers/Peter Arnold, Inc., (b)Image Ideas/PictureQuest, (bc)Photograph by Tim McCabe, courtesy USDA Natural Resources Conservation Service; **696** (t)Michael S. Quinton/National Geographic/Getty Images, (c)James L. Amos/Peter Arnold, Inc., (b)Steve Bly/Getty Images; **697** (t)Tom & Therisa Stack/Tom Stack & Associates, (bl)B. Moose Peterson, (br)David Noton Photography/Alamy; **698** (t)Camille Lusardi/Photolibrary, (c)Gregory Ochocki/Photo Researchers, (b)Doug Allan/Getty Images; **699** (t)Gavriel Jecan/Photolibrary, (b)Comstock Images/PictureQuest, (br)Hutchings Photography/Digital Light Source; **700** (t)Jean-Paul Ferrero/Minden Picture, (c)Steve Bly/Getty Images, (b)Comstock Images/PictureQuest; **701** (t)K. Holmes/American Museum of Natural History, (c)K. Frey/American Museum of Natural History, (b)Stephen Frink/Getty Images, (bkgd)Wayne Levin/Getty Images; **702** Paul Bradforth/Alamy; **708** (c)Hutchings Photography/Digital Light Source; **708** (t to b)(5)(7)Hutchings Photography/Digital Light Source, (2)(4)Macmillan/McGraw-Hill; **709** (tr)Hutchings Photography/Digital Light Source; **710** (t)David Muench, (c)Tom & Therisa Stack/Tom Stack & Associates; **713** Roger Ressmeyer/CORBIS; **714** Macmillan/McGraw-Hill; **716–717** Grant Faint/Getty Images; **718** Data courtesy Marc Imhoff of NASA GSFC and Christopher Elvidge of NOAA NGDC. Image by Craig Mayhew and Robert Simmon, NASA **719** (t)Hutchings

Credits

Photography/Digital Light Source, (b)Douglas McLaughlin; **721** Jeff Greenberg/Alamy; **722** (t)Douglas McLaughlin, (b)Jeff Greenberg/Alamy; **723** Hutchings Photography/Digital Light Source; **724** Jason Hawkes/CORBIS; **725** (t)Hutchings Photography/Digital Light Source, (b)Steven Weinberg/Getty Images; **727** (t)Beth Wald/Getty Images, (b)Hutchings Photography/Digital Light Source; **728** (l)(r)Aerial Photography Copyright Province of Nova Scotia; **730** (t)Karl Weatherly/Getty Images, (bl)(br)Indiana DNR; **731** (t)Doug Menuez/Getty Images, (c)Lester Lefkowitz/CORBIS, (b)Wave Royalty Free/Alamy; **732** (t)Steven Weinberg/Getty Images, (c)Aerial Photography Copyright Province of Nova Scotia, (b)Jeff Greenberg/Alamy; **733** (t to b)The McGraw-Hill Companies, (r)Alain Choisnet/Getty Images, (2)(4)Hutchings Photography/Digital Light Source, (3)Macmillan/McGraw-Hill; **734** Noah Clayton/Getty Images; **735** Hutchings Photography/Digital Light Source; **738** AP Images; **739** (l)Adrian Weinbrecht/PhotoLibrary, (r)David Young-Wolff/PhotoEdit; **740** (t)Noah Clayton/Getty Images, (b)David Young-Wolff/PhotoEdit; **741** Bo Zaunders/CORBIS; **742** Fritz Hoffmann/CORBIS; **743** (t)Hutchings Photography/Digital Light Source, (b)Getty Images; **744** SIMON FRASER/SCIENCE PHOTO LIBRARY; **747** Nancy Ney/Getty Images; **748** (l)Toru Hanai/Reuters/CORBIS, (r, b)AP Images; **749** (t)Getty Images; **750** (t)(b)Macmillan/McGraw-Hill, (3)(7)Hutchings Photography/Digital Light Source, (8)The McGraw-Hill Companies; **751** Mark Scheuern/Alamy; **752** (t)Doug Menuez/Getty Images, (b)AP Images; **755** Grant Faint/Getty Images; **760** (t to b)Eric Audras/Getty Images, (2)NASA Spinoff, (3)CORBIS/age fotostock, (4)David Fischer/Getty Images, (5)NASA Spinoff; **760–761** (bkgd)StockTrek/Getty Images; **761** (t)NOAA, (c)Patrick Landmann/Photo Researchers, Inc., (bl)courtesy of Polar Products, Inc., (br)NASA; **762–763** Dr. Richard Kessel & Dr. Randy Kardon/Tissues & Organs/Visuals Unlimited; **763** Scientifica/Visuals Unlimited/Getty Images; **764** (b)Mark Steinmetz; **765** (t)Hutchings Photography/Digital Light Source; **766** (tl)Wally Eberhart/Visuals Unlimited/Getty Images, (tr)Mark Steinmetz, (cl)Charles D. Winters/Photo Researchers, (cr)USGS, (bl)Jill Braaten/The McGraw-Hill Companies, Inc., (br)Stephen Frisch/The McGraw-Hill Companies; **769** Hutchings Photography/Digital Light Source; **770** USGS; **771** (inset)Matt Rainey/Star Ledger/CORBIS, (bkgd)Micro Discovery/CORBIS; **772** WireImage/Getty Images; **773** (t)Hutchings Photography/Digital Light Source, (b)Kevin R. Morris/CORBIS; **774** Hutchings Photography/Digital Light Source; **776** Holger Winkler/zefa/CORBIS; **777** Photodisc/Getty Images; **778** (l)Jim and Mary Whitmer/Digital Light Source, (r)Jim Whitmer/Digitial Light Source; **779** Hutchings Photography/Digital Light Source; **782** (t)(c)Hutchings Photography/Digital Light Source, (b)Purestock/Getty Images; **783** Hutchings Photography/Digital Light Source; **787** Dr. Richard Kessel & Dr. Randy Kardon/Tissues & Organs/Visuals Unlimited; **790–791** Thomas Marent/Minden Pictures; **791** Arthur Morris/CORBIS; **792** (tr)Spencer Grant/PhotoEdit, (bl)UpperCut Images/Alamy; **793** (tl)ImageShop/CORBIS, (br)Image Source/Getty Images; **794** (l)G. Wanner/Getty Images, (c)Nigel Cattlin/Alamy, (r)KLAUS NIGGE/National Geographic Stock; **797** Hutchings Photography/Digital Light Source; **798** (l)Juniors Bildarchiv/Photolibrary, (r)Joe Blossom/NHPA/Photoshot; **799** (t)KLAUS NIGGE/National Geographic Stock, (b)Joe Blossom/NHPA/Photoshot; **800** (inset)Andrew Syred/Photo Researchers, (bkgd)Curt Maas/Getty Images; **801** Dr. Jeremy Burgess/Photo Researchers; **802** Hutchings Photography/Digital Light Source; **807 808** Sinclair Stammers/Photo Researchers; **809** John Daniels/Bruce Coleman, Inc./Photoshot; **810** (inset)Joe McDonald/CORBIS, (bkgd)Whit Preston/Getty Images; **811** (t)Mark Steinmetz; **812** (b)age fotostock/SuperStock; **813** Roine Magnusson/Getty Images; **814** (t)WILDLIFE/Peter Arnold, Inc., (b)Tom Walker/Photolibrary; **816** JH Pete Carmichael/Getty Images; **817** (t)Roine Magnusson/Getty Images, (c)JH Pete Carmichael/Getty Images, (b)Thomas Marent/Minden Pictures; **818** Macmillan/McGraw-Hill; **819** Hutchings Photography/Digital Light Source; **820** (t)Dr. Jeremy Burgess/Photo Researchers, (b)age fotostock/SuperStock; **823** Thomas Marent/Minden Pictures; **SR-00–SR-01** Gallo Images - Neil Overy/Getty Images; **SR-2** Hutchings Photography/Digital Light Source; **SR-6** Michell D. Bridwell/PhotoEdit; **SR-7** (t)The McGraw-Hill Companies; **SR-7** (b)Dominic Oldershaw; **SR-8** StudiOhio; **SR-9** Timothy Fuller; **SR-10** Aaron Haupt; **SR-47** Matt Meadows; **SR-48** (c)NIBSC/Photo Researchers, Inc., (r)Science VU/Drs. D.T. John & T.B. Cole/Visuals Unlimited, Inc., Stephen Durr; **SR-49** (t)Mark Steinmetz, (r)Andrew Syred/Science Photo Library/Photo Researchers, (br)Rich Brommer; **SR-50** (l)Lynn Keddie/Photolibrary, (tr)G.R. Roberts, David Fleetham/Visuals Unlimited/Getty Images; **SR-51** Gallo Images/CORBIS.

PERIODIC TABLE OF THE ELEMENTS

Element — Hydrogen
Atomic number — 1
Symbol — H
Atomic mass — 1.01
State of matter

Gas
Liquid
Solid
Synthetic

A column in the periodic table is called a **group.**

A row in the periodic table is called a **period.**

Period	1	2	3	4	5	6	7	8	9
1	Hydrogen 1 **H** 1.01								
2	Lithium 3 **Li** 6.94	Beryllium 4 **Be** 9.01							
3	Sodium 11 **Na** 22.99	Magnesium 12 **Mg** 24.31							
4	Potassium 19 **K** 39.10	Calcium 20 **Ca** 40.08	Scandium 21 **Sc** 44.96	Titanium 22 **Ti** 47.87	Vanadium 23 **V** 50.94	Chromium 24 **Cr** 52.00	Manganese 25 **Mn** 54.94	Iron 26 **Fe** 55.85	Cobalt 27 **Co** 58.93
5	Rubidium 37 **Rb** 85.47	Strontium 38 **Sr** 87.62	Yttrium 39 **Y** 88.91	Zirconium 40 **Zr** 91.22	Niobium 41 **Nb** 92.91	Molybdenum 42 **Mo** 95.96	Technetium 43 **Tc** (98)	Ruthenium 44 **Ru** 101.07	Rhodium 45 **Rh** 102.91
6	Cesium 55 **Cs** 132.91	Barium 56 **Ba** 137.33	Lanthanum 57 **La** 138.91	Hafnium 72 **Hf** 178.49	Tantalum 73 **Ta** 180.95	Tungsten 74 **W** 183.84	Rhenium 75 **Re** 186.21	Osmium 76 **Os** 190.23	Iridium 77 **Ir** 192.22
7	Francium 87 **Fr** (223)	Radium 88 **Ra** (226)	Actinium 89 **Ac** (227)	Rutherfordium 104 **Rf** (267)	Dubnium 105 **Db** (268)	Seaborgium 106 **Sg** (271)	Bohrium 107 **Bh** (272)	Hassium 108 **Hs** (270)	Meitnerium 109 **Mt** (276)

The number in parentheses is the mass number of the longest lived isotope for that element.

Lanthanide series	Cerium 58 **Ce** 140.12	Praseodymium 59 **Pr** 140.91	Neodymium 60 **Nd** 144.24	Promethium 61 **Pm** (145)	Samarium 62 **Sm** 150.36	Europium 63 **Eu** 151.96
Actinide series	Thorium 90 **Th** 232.04	Protactinium 91 **Pa** 231.04	Uranium 92 **U** 238.03	Neptunium 93 **Np** (237)	Plutonium 94 **Pu** (244)	Americium 95 **Am** (243)